AF867779

Quantum Electrodynamics

Atoms, Lasers and Gravity

Quantum Electrodynamics

Atoms, Lasers and Gravity

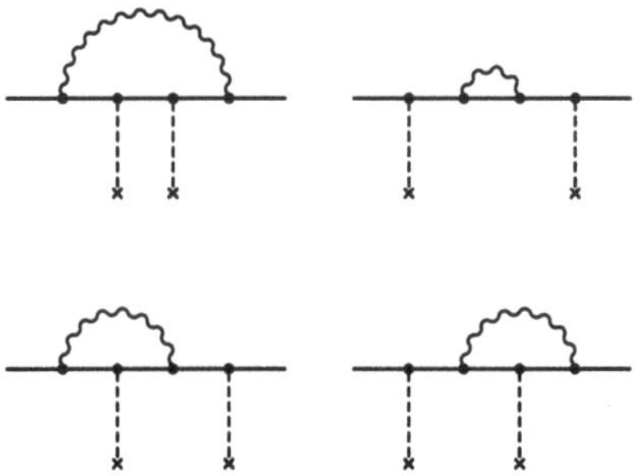

Ulrich D Jentschura
Missouri University of Science and Technology, USA

Gregory S Adkins
Franklin & Marshall College, USA

World Scientific

NEW JERSEY • LONDON • SINGAPORE • BEIJING • SHANGHAI • HONG KONG • TAIPEI • CHENNAI • TOKYO

Published by

World Scientific Publishing Co. Pte. Ltd.
5 Toh Tuck Link, Singapore 596224
USA office: 27 Warren Street, Suite 401-402, Hackensack, NJ 07601
UK office: 57 Shelton Street, Covent Garden, London WC2H 9HE

British Library Cataloguing-in-Publication Data
A catalogue record for this book is available from the British Library.

QUANTUM ELECTRODYNAMICS: ATOMS, LASERS AND GRAVITY

Copyright © 2022 by World Scientific Publishing Co. Pte. Ltd.

All rights reserved. This book, or parts thereof, may not be reproduced in any form or by any means, electronic or mechanical, including photocopying, recording or any information storage and retrieval system now known or to be invented, without written permission from the publisher.

For photocopying of material in this volume, please pay a copying fee through the Copyright Clearance Center, Inc., 222 Rosewood Drive, Danvers, MA 01923, USA. In this case permission to photocopy is not required from the publisher.

ISBN 978-981-125-225-9 (hardcover)
ISBN 978-981-125-226-6 (ebook for institutions)
ISBN 978-981-125-227-3 (ebook for individuals)

For any available supplementary material, please visit
https://www.worldscientific.com/worldscibooks/10.1142/12722#t=suppl

Desk Editor: Nur Syarfeena Binte Mohd Fauzi

Dedicated to All Those Who Seek Beauty in Nature

Preface

This book is about quantum electrodynamics (QED). In the strict sense of the word, QED is the theory of the *quantized* electromagnetic field, as opposed to classical electromagnetism. The characteristic element of quantization is the existence of so-called *loop* corrections which are due to the quantum fluctuations of the electromagnetic field, and of the fermion fields. Quantum fluctuations of the electromagnetic field lead to, e.g., back-reaction of the radiation emitted by a moving electron onto its own trajectory (its energy level). This process is known as the self-energy of the electron; the back-reaction "feedback *loop*" can also be understood on the basis of a fully quantum theory of the electromagnetic field. Quantum fluctuations of the fermion fields lead to the emergence of short-lived, so-called virtual electron-positron (fermion) pairs [1, 2]; these pairs can be emitted virtually by a photon. Their back-reaction onto the photon propagator (or, *loop* correction to the photon propagator) leads to a correction to how light travels in vacuum or in a medium. As such, QED is an extension of classical electrodynamics into the quantum domain.

However, if we were to restrict the subject of this book to calculations that actually require field quantization, then we would miss some of the most interesting and application-oriented research areas connected with electromagnetism and atoms. Specifically, a large number of very interesting calculations can be done without quantizing the fermion (electron-positron) field, relying only on the quantization of the electromagnetic field. Again, many basic facts can even be inferred on the basis of photon exchange processes (electromagnetic interaction processes) that do not require full quantization of the electromagnetic field. The situation is a little subtle.

Let us start by enumerating just the main application areas of QED. *(i)* QED describes the theory needed in order to infer *fundamental constants*, and to constrain their conceivable variation with space and time. The spectra of bound atomic systems are explained, and lepton moments (notably, the electron and muon anomalous magnetic moments) can be calculated. *(ii)* Atomic physics and QED theory are important tools in the description of *physical mechanisms* of rather cross-disciplinary importance: An example is cosmological hydrogen recombination, which describes the absorption of free electrons by protons in the Universe, to form hydrogen atoms. An incomplete list of other processes which require a QED description is as follows: For example, we have resonant, radiationless recombination processes (dielectronic recombination), Casimir–Polder long-range interactions among atoms and molecules, and the atom-surface long-range interaction. The modification of the lifetime of atomic states and of the electromagnetic self-energy shift in cavities also profit from the QED formalism. Finally, we have radiation and particle creation processes in strong light fields, in regimes where "matter perturbs light" instead of the opposite. *(iii)* Last, but certainly not least, QED is important for a number of *practically important applications*, both within fundamental physics but also, in technologically oriented areas. These include synchrotron radiation, Compton backscattering and the concomitant generation of high-energy photons, and Bhabha scattering (electron-positron scattering) which has been used for the normalization of cross sections at electron-positron colliders such as LEP (CERN). Finally, the Casimir–Polder interaction of atoms in the vicinity of extended objects (plates) has technological relevance in the nanoworld and needs to be described by QED.

This book is thus written with the following three main notions in mind. *(i)* The book should be readable in today's relatively fast-paced academic environment, where different obligations and continuous interruptions plague both students and researchers on different levels in their daily life. Consequently, the representation is as self-contained as possible in all sections in order to ensure ease of reading. *(ii)* The book is meant to provide information on different levels. The book is written with both the advanced student as well as the interested senior physicist in mind. Students of field theory may enhance their knowledge by studying applications of field theory in atomic physics; senior atomic physicists may widen their horizon, becoming familiar with quantum-field theoretical techniques, and finally, researchers in bound-state quantum electrodynamics may find a compendium of useful formulas to consider and as reference material. We have tried to write the book so that it is intermediate between a textbook and a monograph, providing both information accompanying, say, a graduate course on quantum electrodynamics, but also, containing information on the status of current research in atomic physics and quantum electrodynamics of bound states. Reference questions ("exercises") are included for each chapter; these are mostly designed to act as checks for the reader, testing the grasp of a particular section or chapter.

The contents of the book can be divided into two parts, the first of which could be entitled as *Quantum Mechanics with Field Quantization* (Chaps. 2–6). This part of the book is meant to provide a useful complement to a course on quantum mechanics, illustrating the main concepts by a number of interesting applications while expanding on the concepts taught in quantum mechanics by quantizing the electromagnetic field. Most applications relevant to quantum chemistry and to technology can be described using a hybrid approach with a quantization of the electromagnetic field, but without quantization of the fermion field. The formalism of nonrelativistic

QED (NRQED) is used at a rudimentary level. We need to couple the bound electron to the quantized electromagnetic field and perform a perturbative expansion in electron-photon interactions. The quantized formalism will be contrasted with time-independent perturbation theory based on interactions with a classical field. The full propagator formalism will not even be necessary at this stage; the quantization of the electromagnetic field will be fully sufficient. In particular, we shall derive the low-energy part of the Lamb shift and include the high-energy part based on formal perturbative operators which will be assumed valid, with a precise derivation given, at a later stage. Also, we will include tiny variations of the Coulomb potential on a phenomenological basis, with a special emphasis on concrete calculations and theoretical predictions. This part will accompany a course on quantum mechanics and will also be interesting for students who wish to broaden the knowledge acquired in quantum mechanics with a perspective on possible applications. Some knowledge of relativistic quantum mechanics, i.e., of the Dirac equation, will also be useful, but not strictly required.

The second part is composed of Chaps. 7–19 and could be entitled as the *Quantum Field Theory of Bound States*. The second part of the book could provide for a useful complement to a graduate course on field theory, illustrating the somewhat abstract concepts on a number of experimentally verifiable calculations concerning atomic physics phenomena. The experimental verification using high-precision spectroscopy can be accomplished in a laboratory of moderate dimensions; the multi-billion dollar investment necessary to construct an accelerator is not required. Still, the theoretical description requires all the tools provided by quantum field theory. Measured on this scale, quantum electrodynamics actually is a very gratifying subject: One can be in contact with experimentalists who are able to measure the effects in laboratories close by, and one is able to talk to each in a very friendly and fruitious atmosphere. In that sense, QED in its purest form represents an interesting subfield of physics. Finally, the modification necessary for the description of laser-modified processes (laser-dressed Furry picture) will also be treated.

The second part also contains an introduction to the (ordinary) Furry picture, in which one incorporates the Coulomb interaction into the unperturbed Hamiltonian (see Chap. 15). We explain how to apply the vertex calculations from field theory in order to describe bound-state effects. The emphasis is the interplay of dimensional, photon mass, and photon energy regularizations. Also, the matching of the S-matrix element against the effective Hamiltonian will be explained for the $\alpha(Z\alpha)^5$ terms, which correspond to the two-Coulomb–Vertex scattering diagrams involving high-energy photons. An introduction to the Bethe–Salpeter equation (BSEQ) and Nonrelativistic Quantum Electrodynamics (NRQED, an effective theory for bound states) also forms part of this book. The second part may accompany a course on quantum field theory, and it may also be interesting for the phenomenologically interested student who wishes to learn how to apply the formalism to experiments carried out in the laboratory.

As a monograph with a textbook component, this book tries to bridge areas that have been covered in a number of dedicated books elsewhere in the literature. Topics related to atomic physics and quantum electrodynamics have been discussed in the literature for a number of decades. This introduction would thus be incomplete without a brief survey of existing books in related areas. A necessarily incomplete list of books can be given as follows: V. B. Berestetskii, E. M. Lifshitz, and L. P. Pitaevskii [3] have written a classic treatise which was published as volume 4 of the textbook series on physics by L. D. Landau and E. M. Lifshitz (completed after the death of L. D. Landau). A very comprehensive introduction to quantum electrodynamics has been given by A. I. Akhiezer and V. B. Berestetskii [4]. Formal aspects of the problems have been stressed in the book of I. Białynicki-Birula and Z. Białynicka-Birula [5] (as usual in the Polish language, the name of the second author, who is the wife of the first author, is being spelled a little differently). The "ł" is pronounced like the French h-aspiré and the notation is reminiscent to the quantum unit of action, or reduced Planck constant, $\hbar$. Furthermore, we should mention the monograph of H. A. Bethe and E. E. Salpeter [6], which stresses basic processes of atomic physics, the monograph of W. Heitler [7], which contains a comprehensive introduction to radiation theory, and the book of H. A. Bethe and R. Jackiw [8], which contains a good and easily readable overview of the basic structural and dynamical properties of atoms. Regarding the field-theoretical foundations of quantum electrodynamics, we would like to mention the comprehensive works of J. Schwinger [9], S. S. Schweber [10], and a brief supplemental monograph by R. P. Feynman [11] which contains a number of detailed calculations, as well as the monograph by C. Itzykson and J. B. Zuber [2], which summarizes essentially everything known in field theory up to 1980, in a very concise but comprehensive form. No other book rivals the textbook of W. Greiner and J. Reinhardt [12] in terms of its excellent presentation of scattering processes (Compton, Bhabha, and Møller scattering). The didactic value of the book of Greiner and Reinhardt should not be underestimated. The book of J. M. Jauch and F. Rohrlich [13] gives an elegant introduction into basic concepts of QED as well.

The intricacies of angular momentum algebra are treated in the monographs by A. R. Edmonds [14], D. M. Brink, and G. R. Satchler [15], D. A. Varshalovich, A. N. Moskalev and V. K. Khersonskii [16]. Foundations of atomic physics, with a special emphasis on the quantum mechanical formulation and basic principles, are discussed in the monographs of H. A. Bethe and E. E. Salpeter [6], mentioned previously, M. E. Rose [17], E. Merzbacher [18], B. H. Bransden and C. J. Joachain [19], and M. Weissbluth [20]. The monographs by W. R. Johnson [21] and by I. P. Grant [22] stress the computational aspects of relativistic many-body atomic physics, whereas the work of I. Lindgren and J. Morrison [23] emphasizes the angular momentum algebra needed for the many-body problem. L. Labzowsky [24] gives an overview of the relativistic many-body formalism. Two very interesting books by J. J. Sakurai [25, 26] on the foundations of quantum mechanics should also be mentioned. On the more specialized side, the monograph by M. I. Eides, H. Grotch and V. A. Shelyuto [27] contains a recent comprehensive status report on light atoms. The concepts of field quantization, in the nonrelativistic domain, are discussed in the book of P. W. Milonni [28]. Finally, W. Dittrich and H. Gies [29] have given an account of the modifications of the Maxwell equations in nontrivial background fields.

We tried to mention the names of a few researchers who were actively involved in advancements of bound-state QED over the last few decades. The purpose of the book, though, is not to provide an all-inclusive review of all results obtained so far, as the status is evolving and would otherwise constitute a "moving target". On the occasion, we apologize to all colleagues whose works may have been overlooked in the discussion. This book project was supported by the National Science Foundation under grants PHY–1707489, PHY–2011762, PHY–14030973, PHY–1710856 and PHY–2110294. Insightful conversations with Professors K. Pachucki, P. J. Mohr, V. A. Yerokhin, C. M. Adhikari, G. W. F. Drake, Y. Itin, I. Nándori and Professor T. W. Hänsch are gratefully acknowledged. The authors also express gratitude to Dr. B. J. Wundt for elucidating conversations, and they acknowledge Dr. N. C. James, B. Robinson, D. McNamara, J. Padron, and A. Wetzel for proofreading the manuscript. The Hungarian Academy Institute for Nuclear Physics in Debrecen (ATOMKI) is gratefully acknowledged for hospitality during a Sabbatical stay in the Academic year 2018–2019, when an important part of the book was completed.

Rolla, Missouri, and Lancaster, Pennsylvania, May 2021

U. D. Jentschura
Ulrich D. Jentschura
Missouri University of Science and Technology
Department of Physics and LAMOR
Rolla, Missouri 65409, USA
ulj@mst.edu

G. S. Adkins
Gregory S. Adkins
Franklin & Marshall College
Department of Physics and Astronomy
Lancaster, Pennsylvania 17603, USA
gadkins@fandm.edu

Contents

Chapter 1

Introduction

1.1 Accurate Numbers

Quantum electrodynamics (QED) is a theory known for the occurrence of long and accurate numbers. At present, QED still remains in a somewhat unique position within the standard model of fundamental interactions, in the sense that it is required to describe both high- and low-energy phenomena. So far, it is the only quantum field theory that has enabled us to infer fundamental constants to better than 8 digits of accuracy, and in the case of the Rydberg constant, 12 decimals (see Refs. [30, 31]).

Among other notable achievements, there are three paradigmatic QED effects related to bound-state physics which have been measured to impressive accuracy. They can certainly be called cornerstones for the current belief that QED can describe both light and heavy bound systems (atoms and ions) better than any other theory. Furthermore, these effects have been used to determine various fundamental constants such as the Rydberg constant and the electron mass. The first of these are *(i)* measurements of atomic transition frequencies which include contributions from QED effects such as the Lamb shift. These transitions entail a change in the principal or angular momentum quantum numbers of the atomic electrons. The second paradigmatic effect is related to *(ii)* hyperfine-structure transitions which correspond to a spin-flip of the atomic nucleus. These transitions also receive corrections from QED. Finally, *(iii)* measurements of free-electron and bound-electron g factors have contributed in determining the fine-structure constant and the electron mass.

A book on QED would be incomplete without listing some of the impressive experimental results, some of which may have come of age but are far from outdated. Indeed, they have an almost historic character and have advanced physics quite considerably. We should note that the experimental difficulties connected with precise measurements are often underestimated by theorists; the analysis of the systematic effects relevant to a particular apparatus can be full of traps. Likewise, the conceptual and practical difficulties in obtaining accurate theoretical predictions are often underestimated by experimentalists.

In our necessarily incomplete list of QED-related results, we start with the hyperfine structure. The 1S hydrogen hyperfine structure has been measured in Ref. [32] with the result

$$\Delta E_{1S} = 1\,420\,405\,751.7667(10)\,\mathrm{Hz}\,, \tag{1.1}$$

whose uncertainty was, however, later amplified by a factor of three (see Ref. [33]). The value of

$$\Delta E_{1S} = 1\,420\,405\,751.768(1)\,\mathrm{Hz} \tag{1.2}$$

has been obtained in Ref. [34] as a conservative average of various experimental investigations of comparable accuracy. For the $2S$ hydrogen hyperfine structure, the most recent experimental result comes from Ref. [35],

$$\Delta E_{2S} = 177\,556\,860(15)\,\mathrm{Hz}\,. \tag{1.3}$$

The latter is in a $1.4\,\sigma$ disagreement with theory [35–37].

For the hyperfine splitting in muonium, a bound system of a positively charged muon and an electron ($\mu^+ e^-$), the experimental result is (see Ref. [38])

$$\Delta E_{1S}^{\mu\, e^-} = 4\,463\,302\,765(53)\,\mathrm{Hz}\,. \tag{1.4}$$

Current theory benefits from the availability of nonperturbative self-energy calculations which include the hyperfine effects [39, 40]. The most recent theoretical result is [see Eq. (221) of Ref. [30]]

$$\Delta E_{1S}^{\mu\, e^-} = 4\,463\,302\,891(272)\,\mathrm{Hz}\,. \tag{1.5}$$

For positronium, the experimental result for the $1S$ hyperfine splitting interval reads [41–43]

$$\Delta E_{1S}^{e^- e^+} = 203\,389.10(74)\,\mathrm{MHz}\,. \tag{1.6}$$

The extreme accuracy of these results implies a certain challenge to theory which provides the result [44–47]

$$\Delta E_{1S}^{e^- e^+} = 203\,391.69(16)\,\mathrm{MHz}\,, \tag{1.7}$$

which differs from the experimental result by about three standard deviations. A recent experiment [48]

$$\Delta E_{1S}^{e^- e^+} = 203\,394.2(20)\,\mathrm{MHz} \tag{1.8}$$

is in better agreement with theory but less precise.

We continue with gyromagnetic ratios, or g factors. According to Dirac theory, the g factor of the electron exactly is $g = 2$. However, taking into account the electron anomaly, which is due to QED effects, we can parameterize the g factor of the electron as

$$g = 2\,(1 + \kappa)\,, \tag{1.9}$$

where κ is known as the electron anomaly. Throughout this book, we will use Eq. (1.9) for the g factor of a lepton, i.e., the symbol κ is understood to denote the g factor anomaly of the lepton unless otherwise stated or implied by the context. The equality of κ for electron and positron constitutes a test of charge-parity-time-inversion (CPT) symmetry. In an experimental *tour-de-force*, the following results have been obtained [49],

$$\kappa = 1\,159\,652\,188.4(4.3) \times 10^{-12} \qquad \text{(electron)}\,, \tag{1.10a}$$

$$\kappa = 1\,159\,652\,187.9(4.3) \times 10^{-12} \qquad \text{(positron)}\,. \tag{1.10b}$$

Recently, these results have been improved [50–52] in a series of equally impressive experiments. The electron g factor is connected with Sommerfeld's fine-structure constant α. The two leading terms (in the fine-structure constant α) of the theoretical expression relating κ to the quantum electrodynamic expression for the light fermion anomalous magnetic moment (of electron and positron) are given by [53,54]

$$\kappa = \frac{\alpha}{2\pi} - 0.328\,478\,444 \left(\frac{\alpha}{\pi}\right)^2 + \mathcal{O}(\alpha^3)\,. \tag{1.11}$$

Recent theoretical efforts have culminated in analytic four-loop calculations [55], as well as numerical five-loop evaluations (see Refs. [56–58] and Table 3 of Ref. [59]). The relation (1.11) constitutes the main input datum for the determination of the fine-structure constant α, which according to Refs. [60, 61] reads as

$$\alpha = \frac{e^2}{4\pi\epsilon_0 \hbar c} = \frac{1}{137.035\,999\,084(21)}\,, \tag{1.12}$$

where e is the electron charge (the elementary charge is $|e| = -e$).

Some of the most accurately measured transitions in simple atomic systems concern atomic hydrogen and deuterium, i.e., one-electron atoms. Indeed, the hyperfine centroid of the $1S$–$2S$ transition in hydrogen has been measured as [62–64]

$$\Delta E_{1S-2S} = 2\,466\,061\,413\,187\,035(10)\,\text{Hz}\,. \tag{1.13}$$

The accuracy is 4.2 parts in 10^{15}. It is experimentally more difficult to measure transitions to higher excited states. For the $1S$–$3S$ transition, a recent experimental value from the Paris group reads as [65]

$$\begin{aligned} \Delta E_{1S-3S}^{F=1\to F=1} &= 2\,922\,742\,936.729(13)\,\text{MHz}\,, \\ \Delta E_{1S-3S} &= 2\,922\,743\,278.678(13)\,\text{MHz}\,. \end{aligned} \tag{1.14}$$

The second entry refers to the hyperfine centroid and is obtained from the first by a subtraction of hyperfine effects. The most recent measurement [66] at the Max–Planck Institute for Quantum Optics gives the result

$$\begin{aligned} \Delta E_{1S-3S}^{F=1\to F=1} &= 2\,922\,742\,936.711(17)\,\text{MHz}\,, \\ \Delta E_{1S-3S} &= 2\,922\,743\,278.659(17)\,\text{MHz}\,, \end{aligned} \tag{1.15}$$

where the latter is the hyperfine centroid.

Theoretical predictions for other hydrogen transitions that have not yet been measured can be obtained from the web site [67]. For the hyperfine centroids of the $1S$–$3S$ and $2S$–$3S$ transitions, which are of particular spectroscopic interest, the theoretical predictions are

$$E^{\rm th}_{1S-3S} = 2\,922\,743\,278.671\,6(14)\,{\rm MHz}\,, \tag{1.16}$$

$$E^{\rm th}_{1S-3S} = 456\,681\,865.484\,5(14)\,{\rm MHz}\,. \tag{1.17}$$

The theoretical predictions are the result of decade-long theoretical efforts. A comparison of observed hydrogen and deuterium transition frequencies to theoretical expressions can be used to determine the Rydberg constant R_∞. The most recent value of this constant [60, 61] reads as

$$R_\infty = \frac{\alpha^2 m_e c^2}{2hc} = 10\,973\,731.568\,160(21)\,{\rm m}^{-1}\,. \tag{1.18}$$

The Rydberg constant is related to transition frequencies in a fictitious hydrogen atom with an infinitely heavy point nucleus and with relativistic effects "switched off". The Rydberg constant is a wave number; it indicates the number of wavelengths (cycles) per meter. It can be converted to an angular wavenumber k (the modulus of the wave vector $\vec{k}$) via multiplication by a factor 2π. The angular wave number has units of radians per meter. The Rydberg constant can be converted to a frequency using the formula

$$R_\infty\, c = 3.289\,841\,960\,355(19) \times 10^{15}\,{\rm Hz}\,. \tag{1.19}$$

Any book on atomic physics would be incomplete without also indicating the numerical value of the Bohr radius, which reads as [60, 61]

$$a_0 = \frac{\hbar}{\alpha m_e c} = 0.529\,177\,210\,903(80) \times 10^{-10}\,{\rm m}\,. \tag{1.20}$$

As described in Ref. [68], experimental input data from more than one transition are necessary in order to infer fundamental constants (e.g., the Rydberg constant) from measurements. The reason is that there is no one-to-one mapping of a transition to a particular physical constant but rather, adopting a quotation often ascribed to OPPENHEIMER,[1] "understanding hydrogen is understanding all of physics." Indeed, a "biography of hydrogen" in terms of its discovery potential for all of physics has been given in Ref. [69].

1.2 Fundamental Constants

QED theory and experiment have often been used in order to constrain possible extensions of the standard model. Tiny conceivable variations of the fundamental forces from their commonly accepted nature would imply small variations in observable phenomena. Because QED effects have been measured very precisely, they can be used to constrain such models.

[1] Julius Robert Oppenheimer (1904–1967).

One might ask about the foundations of the hypothesis that fundamental constants need to be "constant" through space and time; this hypothesis can ultimately be traced to the Einstein equivalence principle (EEP), which is based on three foundations. Namely, the Einstein equivalence principle affirms *(i)* the validity of the weak equivalence principle, *(ii)* local position invariance (LPI) and *(iii)* local Lorentz invariance (LLI). The weak equivalence principle states that free-fall at a given point in space-time is the same for all physical systems. NEWTON[2] stated that the property of a physical object called "mass" ("inertial mass") is proportional to the "weight" (which enters the gravitational force law), conjecturing that the two are proportional. LPI implies that the outcome of any local non-gravitational experiment (e.g., a frequency measurement) is independent of when and where in the universe it is performed, and LLI states that the outcome of any local non-gravitational experiment is independent of the velocity of the free-falling reference frame in which it is being performed.

A conceivable violation of any of these three axioms would induce a position-dependent or time-dependent shift of electromagnetically realized standards of units within the SI system. This could potentially induce a dependence on the location of a laboratory or on the time of measurement. Recently, it has been claimed [70] that a gravitational modification of the vacuum polarization tensor in strong gravitational fields (deep gravitational potentials) could lead to a modification of the speed of light near gravitational centers such as stars or black holes, which would violate the EEP. The mechanism proposed in Ref. [70] involves a gravitational interaction of the virtual electrons and positrons within the quantum electrodynamic "loop" that corrects the photon propagator (i.e., the gravitational correction to "vacuum polarization"). More colloquially speaking, the quantum fluctuations of the fermion field are in turn modified by the influence of the gravitational field onto the fermionic quantum fluctuations. However, in Ref. [71], the claim made in Ref. [70] could be refuted. A relativistically covariant evaluation of the vacuum polarization integral, corrected for gravitational effects, reveals that the speed of light is not altered locally in deep gravitational fields.

Two caveats need to be mentioned: *(i)* The vacuum polarization tensor is modified in a nontrivial way for so-called off-shell photons, which are themselves virtual, as they mediate the Coulomb interaction. *(ii)* Globally, the speed of light may be altered in the vicinity of strong gravitational centers such as the Sun. Indeed, the Shapiro time delay describes a delay induced by space-time curvature for photons as they propagate through strong gravitational fields. This delay, however, does not affect the outcome of a locally performed experiment, while it has been confirmed with the Cassini spacecraft in superior conjunction on its way to Saturn [72], using Doppler tracking with both X-band (7175 MHz) as well as Ka-band (34 316 MHz) radar. The rather subtle clarification of the effect proposed in Ref. [70] and refuted in Ref. [71] reveals how easy it is to be mistaken in effects whose theoretical

[2]Sir Isaac Newton PRS (1642–1726).

description requires a combination of quantum electrodynamics and general relativity. Very recently, tiny violations (limitations) of Einstein's equivalence principle, based on quantum effects, have been discussed in Ref. [73].

An even more intriguing conjecture concerns the possible time variation of fundamental constants. In Ref. [74], BEKENSTEIN[3] gave a thorough analysis of the consequences of promoting the fine-structure constant to a scalar field. While the time variation of fundamental constants would entail a violation of the EEP, a number of theoretical models actually allow, or even predict, such variations. One example is given by so-called KALUZA[4]–KLEIN[5] theories, where the fine-structure "constant" emerges after an integration over a compactified, extra dimension, and is related to the vacuum expectation value of a scalar quantum field ϕ [see Eq. (37) of Ref. [75]]. String theories predict interrelations among the fundamental constants [see Eq. (42) of Ref. [75]], e.g., they predict that the fine-structure constant be proportional to the square root of the gravitational constant. Finally, the Higgs mechanism (see Chap. 12 of Ref. [2]) relates fundamental constants such as the masses of elementary particles to the vacuum expectation value of a background field known as the Higgs field, which could in principle change over space and time. In Ref. [74], it has been observed that a part of a charged particle's mass is the electromagnetic self mass, which also depends on the fine-structure constant. In all of these theories, the fine-structure constant is far from being a fundamental, time-independent constant. An almost historic (from today's point of view), but very conspicuous, theoretical basis for the variation of fundamental constants is given by Dirac's large number hypothesis. Let us consider the ratio of the age T of the Universe and the time it takes light to travel a distance equal to the classical electron radius r_e. One finds that the ratio is of the same order-of-magnitude as the ratio of the electric and gravitational interactions of electrons and protons,

$$r_e = \frac{\alpha \hbar}{m_e c}, \qquad \frac{T}{(r_e/c)} \sim 10^{40} \sim \frac{e^2}{4\pi\epsilon_0 G m_e m_p}. \tag{1.21}$$

Dirac conjectured [76] that the equality holds exactly, implying that the gravitational interaction constant G might be inversely proportional to the age T of the Universe.

The task of spectroscopy and QED theory is to relate these conjectures to model-independent, laboratory-based bounds on the temporal variation of constants. A rather illustrative application of QED theory and measurement, constraining various theoretical models, has been given in Ref. [63]. Input data from more than one measurement and more than one transition are needed in order to constrain the variation of fundamental constants. The hydrogen $1S$–$2S$ frequency measurement from Eq. (1.13) has been repeated in 1999 and 2003 [62, 63]. Another important input datum is given by the $^{199}\mathrm{Hg}^{+}$ $5d^{10}6s\ ^2S_{1/2}(F = 0) \Leftrightarrow 5d^9 6s^2\ ^2D_{5/2}(F' = 2)$

[3] Jacob David Bekenstein (1947–2015).
[4] Theodor Franz Eduard Kaluza (1885–1954).
[5] Oskar Benjamin Klein (1894–1977).

electric quadrupole ("clock") transition frequency $\nu_{\rm Hg}$, which was measured at the National Institute of Standards and Technology (NIST) between July 2000 and December 2002 (see Ref. [77]).

Using theoretical considerations [63], one may show that the two measurements actually test different combinations of the nuclear magnetic moment of cesium $\mu_{\rm Cs}$ and of the fine-structure constant α. Specifically, in the NIST measurement [77], one investigates the temporal change in the frequency ratio $\nu_{\rm Cs}/\nu_{\rm Hg}$, which is related to the drift of $\mu_{\rm Cs}$ and α as follows,

$$\frac{\partial}{\partial t} \ln \frac{\nu_{\rm Cs}}{\nu_{\rm Hg}} = \frac{\partial}{\partial t} \ln \left(\frac{\mu_{\rm Cs}}{\mu_B} \alpha^{2.0+0.8+3.2} \right) . \tag{1.22}$$

Here, the quantity μ_B is an in principle arbitrary normalization which makes the argument of the logarithm dimensionless (the notation is of course inspired by the Bohr magneton). The basis for this assumption lies in the fact that a frequency "measurement" precisely corresponds to a comparison of two frequencies, that of the measured transition and that of the cesium reference standard. The latter in turn is proportional to the Bohr magneton that defines the frequency standard. The $1S$–$2S$ experiment [63] tests the drift of

$$\frac{\partial}{\partial t} \ln \frac{\nu_{\rm Cs}}{\nu_{\rm H}} = \frac{\partial}{\partial t} \ln \left(\frac{\mu_{\rm Cs}}{\mu_B} \alpha^{2.0+0.8} \right) , \tag{1.23}$$

where the subscript H stands for hydrogen in contrast to Hg which denotes mercury. One may solve Eqs. (1.22) and (1.23) for the temporal variations of $\mu_{\rm Cs}$ and α. The results are (see Ref. [63])

$$\frac{\partial}{\partial t} \ln \alpha = (-0.9 \pm 2.9) \times 10^{-15} \ {\rm yr}^{-1} , \tag{1.24}$$

and

$$\frac{\partial}{\partial t} \ln \frac{\mu_{\rm Cs}}{\mu_B} = (0.6 \pm 1.3) \times 10^{-14} \ {\rm yr}^{-1} . \tag{1.25}$$

The drifts are model-independent and consistent with zero. If more than two transitions are used as input data, linear regression must be used in order to solve for the drifts in a model-independent calculation.

One may summarize that while some astrophysical observations are claimed to suggest a slight violation of the EEP in the form of time-evolving fundamental constants, no such effects have been detected in laboratory experiments. Recent laboratory experiments yield bounds on the temporal variation of the fine-structure constant (logarithmic time derivative $\dot{\alpha}/\alpha$) smaller than 10^{-16} per year [78–80].

1.3 Overview of Chapter Contents

In order to guide the reader in the study of this book, we briefly outline the contents of each chapter. Many phenomena which influence our daily lives are caused by the interaction of the quantized electromagnetic field with atoms. The description of atomic (quantum) processes requires a "new" language and a unit system tailored to the physical entities to be described; this is discussed in Chap. 2.

One can describe a large number of important phenomena, starting from a nonrelativistic theory of the atom and a quantized theory of the electromagnetic field. These include the dynamic ("ac") Stark effect and spontaneous atomic decay (see Chap. 3). The approach requires us to become familiar with time-dependent perturbation theory, which is a very useful concept beyond the realm of application discussed here. In many cases, the frequencies of the interacting light are so low that a nonrelativistic treatment on the level of the dipole approximation is fully sufficient.

In second-order perturbation theory in the quantized interaction, one can already enter the field of bound-state quantum electrodynamics, on the basis of the calculation of the so-called BETHE[6] logarithm (see Chap. 4). In order to proceed with the calculation, we shall obtain the SCHRÖDINGER[7]–COULOMB[8] Green function. We will do this by matching homogeneous solutions of the time-independent radial Schrödinger–Coulomb wave equation, at the cusp $r_1 = r_2$ of equal radial arguments, using ideas inspired by the treatment of similar problems in classical electrodynamics [81].

Interatomic interactions and interactions of an atom with a dielectric surface require the use of the quantized–field formalism, as well as second-order and fourth-order perturbation theory. This approach is described in Chap. 5.

The predominant symmetry that governs the physics of one-electron, and many-electron, atoms is the rotational symmetry of the Coulomb potential. It has been observed that scalars and vectors, and tensors of second rank transform separately under rotations; this has led to the introduction of so-called Wigner matrices which describe the individual transformation of tensor components under rotations. Also, it has been observed that the tensorial addition of vectors leads to the formation of higher-order tensors; this naturally implies the introduction of vector-addition coefficients, which are otherwise known as CLEBSCH[9]–GORDAN[10] coefficients. The Racah–Wigner algebra, which describes the formalism, is summarized in Chap. 6.

Regarding a fully quantum, and relativistic, field theory of atoms, we start with the free DIRAC[11] equation in Chap. 7. This equation has originally been obtained as a linearization of the KLEIN–GORDON[12] equation, using matrix algebra which implements a noncommutative multiplication operation. However, the Dirac equation soon turned out to be "more intelligent than its inventor" (this quote has been coined by Dirac himself). Indeed, the Dirac equation was the first equation to describe an "inner" degree of freedom of a particle, in this case, the spin degree of

[6]Hans Albrecht Bethe (1906–2005).
[7]Erwin Rudolf Josef Alexander Schrödinger (1887–1961).
[8]Charles-Augustin de Coulomb (1736–1806).
[9]Rudolf Friedrich Alfred Clebsch (1833–1872).
[10]Paul Albert Gordan (1837–1912).
[11]Paul Adrien Maurice Dirac OM FRS (1902–1984).
[12]Walter Gordon (1893–1939).

freedom [82,83]. We recall that the electron spin was discovered by GOUDSMIT[13] and UHLENBECK[14] in 1925, in two rather experimentally inspired theoretical papers [84, 85]. A brief, but informative account of the history of the concept of "spin" has been recently given in Ref. [86]. The inclusion of the spin degree of freedom leads to more complicated angular algebra, because the spin needs to be added to the orbital angular momentum. Also, in Chap. 7, we take a look at the quantized Dirac field, and calculate the propagator for a free Dirac particle.

The case of an interacting Dirac particle is discussed in Chap. 8. Three situations are of particular significance. First, we treat the Dirac–Coulomb problem, which describes a particle bound to the Coulomb field of a nucleus. The wave function solutions and the propagator are discussed. Second, we consider the Dirac–Volkov problem, which describes a relativistic electron under the influence of a strong laser field which is described as a classical background field. Here, again, the wave function solutions and the propagator are treated. The third problem considered in Chap. 8 is the Dirac equation in curved space-time, for which we present a (hopefully) easily digestible derivation.

Up to this point in the current book, we still have not covered, to a full extent, the quantum formalism for the electromagnetic and fermion fields, in fully relativistic form. In Chap. 9, we start this program with the quantized electromagnetic field and discuss field commutators and field propagators. The subject matter goes beyond the contents of Chap. 2, where we only discuss the field operators in the Schrödinger and interaction pictures, but not their time-ordered products, which lead to the propagators. The photon propagator is treated in both Coulomb as well as Lorenz gauges. In Coulomb gauge, the instantaneous Coulomb interaction has to be added (in non-quantized form) to the quantized interaction mediated by the spatial degrees of the freedom of the vector potential operator. Finally, we also consider the generation of photons by a classical, external field configuration, where the photon field is quantized, but the current distribution is kept as a classical entity.

In Chap. 10, we catch up on the basic quantum electrodynamic processes, profiting from our exposure to the quantized electromagnetic field (in Chaps. 2 and 9), as well as the quantized fermion field (in Chap. 7). We consider tree-level diagrams involving (free) electron propagators, and loop diagrams (vertex correction and vacuum polarization). The reader is exposed, among other aspects, to the ideas of dimensional regularization and on-shell renormalization. Finally, the forest formula is introduced, which allows one to renormalized nested divergences. The considerations described in Chap. 10 are primarily focused on particle-physics techniques, and therefore are not specific to bound states. They do not require the rather involved angular integrals over spin-angular, and related functions which naturally occur in nonperturbative bound-state calculations. The latter take the Coulomb interaction into account to all orders.

[13]Samuel Goudsmit (1902–1978).

[14]George Eugene Uhlenbeck (1900–1988).

In Chap. 11, we engage in the so-called "Foldy–Wouthuysen program" of identifying the nonrelativistic expansion (the "leading terms" and the "relativistic corrections") relevant to the description of bound states. In particular, the Foldy–Wouthuysen transformation allows us to separate the particle from the antiparticle degrees of freedom. The approach is based on the observation that the average speed of an electron in a Bohr orbit in a light atom is of the order of $v \sim Z\alpha c$, where Z is the nuclear charge number, and α is the fine-structure constant. For not too large Z, one thus has the relation $v \ll c$. The binding parameter $Z\alpha$ describes the strength of the coupling of the bound electron to the nucleus. It is thus possible to expand scattering amplitudes, transition matrix elements, and energy shifts in the powers of the dimensionless parameter $Z\alpha$. In Chap. 11, we begin the related considerations by an expansion about the nonrelativistic limit of the Dirac equation via the Foldy–Wouthuysen transformation for a single particle. In the treatment of the Lamb shift, which also is carried out in Chap. 11, we augment the discussion by considering the effective Dirac equation, corrected by radiative effects previously discussed in Chap. 10 (form factors of the electron). A serious problem arises: The expansion in $Z\alpha$ eventually turns out to be nonanalytic, i.e., it gives rise to powers of $Z\alpha$ and logarithms $\ln(Z\alpha)$. We have to approach the formalism at the heart of the $Z\alpha$-expansion carefully.

In Chap. 12, we continue the "Foldy–Wouthuysen program" and use the results obtained in Chap. 11, in particular, for the relativistic corrections to the transition current, in order to calculate the two-particle BREIT[15] Hamiltonian. This Hamiltonian can be used for many-particle systems and describes the interaction of two charged particles in the non-retardation limit, i.e., in the limit of a static interaction (where the frequency of the exchanged photon vanishes). Corrections of relative order $(Z\alpha)^2$ to the leading Coulomb interaction can be described by the Breit Hamiltonian. A further issue also is given attention in Chap. 12: Namely, when external fields interact with atomic systems, it is necessary to separate the internal degrees of freedom from the external ones, and to identify the generalized correction terms. This poses additional difficulties for many-particle systems, described in Chap. 12.

In Chap. 13, the theory of fully correlated basis sets and helium is described. It is well known that the eigenfunctions of many-electron systems cannot be written in closed analytic form. One may try to reduce the problem to the calculation of self-consistent orbitals, which describe the probability density associated with individual electron orbits. However, in this case, one factorizes the dependence on the coordinates of the two electrons and loses the possibility to describe the full correlation among the electrons. If the quantum Hamiltonian cannot be diagonalized in closed analytic form, then one must resort to approximation methods. A fully correlated basis of trial functions, which contain the interelectron distance $r_{12} = |\vec{r}_1 - \vec{r}_2|$, leads to a satisfactory answer. In choosing the trial functions ("basis functions"),

[15] Gregory Breit, transcribed from Russian: Grigory Alfredovich Breit-Shneider (1899–1981).

one must exercise extreme caution. In general, one has to trade suitability in regard to the modeling of the electron correlation against analytic tractability of the basis integrals. The approach has meanwhile been applied with very high accuracy and can be generalized to many-electron systems. It is described in Chap. 13.

In Chap. 14, we shall make a certain detour and consider relativistic bound systems with high nuclear charge numbers Z, with an emphasis on the many-electron case. The binding parameter $Z\alpha$ is treated to all orders in the "non-perturbative" bound-state calculations. Also, at high nuclear charge numbers, the preferred method for the treatment of the electron correlation changes from the so-called LS coupling to jj coupling. The electron correlation term (electron-electron repulsion), which couples the orbital angular momenta, is of order $Z\alpha^2 mc^2$, where m is the electron mass and c is the speed of light. The spin-orbit coupling term for each electron is of order $(Z\alpha)^4 mc^2$, while the spin-spin coupling among electrons is of order $\alpha\,(Z\alpha)^3 mc^2$. The electron correlation term couples the orbital angular momenta of the electrons together, and eventually leads to LS coupling, while the spin-orbit term couples the orbital angular momentum ℓ of each electron to the spin s, as summarized in the total angular momentum j. The transition occurs at $Z \approx (1/\alpha)^{2/3} = 26.58$, that is, between iron (Fe, $Z = 26$) and cobalt (Co, $Z = 27$) in the periodic table of elements. While a comprehensive discussion of all intricacies of the many-electron problem at high nuclear charge would go beyond the scope of this book, we include a brief survey of existing codes for the treatment of many-electron systems in Chap. 14. Furthermore, we focus on the angular integrals which are useful for the treatment of photon exchange with tightly bound electrons, where the relativistic corrections can no longer be ignored.

In order to go beyond the Breit approximation, one has to include retardation, i.e., one cannot ignore the frequency of the exchanged photon any longer. This is described in Chap. 15, where we carry out example calculations of higher-order effective operators. This includes those related to the $\alpha\,(Z\alpha)^5$ correction to the Lamb shift. We also investigate the recoil correction (SALPETER[16] correction) beyond the Breit terms. This approach allows us to better identify the terms which contribute to a particular energy shift in a given order in the $Z\alpha$-expansion. A discussion of the intricacies of nonperturbative (in the binding parameter $Z\alpha$) bound-state calculations, for quantum electrodynamic loop corrections, also is included in Chap. 15. This includes a brief glance at the Furry picture.

Any book on quantum electrodynamics would be incomplete without an overview of the Bethe–Salpeter equation, which determines the bound-state energies of two-particle systems and allows one, in principle, to take all quantum electrodynamic loop and recoil corrections into account to all orders. Furthermore, as we see in Chap. 16, the Bethe–Salpeter equation actually can be characterized as the generalization of the Lippmann–Schwinger equation for one-particle states, to two constituent particles. Essentially, the self-energy operator in the Lippmann–

[16] Edwin Ernest Salpeter (1924–2008).

Schwinger equation is replaced by a two-particle interaction kernel. The theory is based on a sixteen-component, two-particle wave function. The difficulties encountered in a systematic analysis of the Bethe–Salpeter equation are tremendous, and we illustrate its use on the basis of an example calculation, namely, a reevaluation of the Salpeter correction, where we put special emphasis on the need to sum certain classes of diagrams in the low-energy domain, in order to obtain a consistent result. Furthermore, in comparison to our treatment in Chap. 15, we investigate the gauge dependence of the correction, as well as the use of dimensional regularization in its evaluation. A discussion of the leading recoil correction to hyperfine splitting in muonium completes Chap. 16. The latter correction could be characterized as the nuclear-spin-dependent part of the recoil correction.

The development of nonrelativistic quantum electrodynamics (NRQED) has been inspired by modern techniques developed within effective field theory. In Chap. 17, we give a corresponding overview. Roughly speaking, one uses the concept that upon a Foldy–Wouthuysen transformation, one can actually separate particle from antiparticle terms in the theory, and promotes this idea to a field theory of its own. That is, one initially pretends that one does not know about the existence of fully relativistic quantum electrodynamics at all. Instead, one tries to write down a theory, i.e., a Lagrangian, in which one uses nonrelativistic fermionic field operators (in particular, one uses different field operators for electrons and positrons). Additional operators are used for the description of the electromagnetic field. The latter are, preferably, written down in Coulomb gauge. Because the theory is inherently nonrelativistic, one also obtains nonrelativistic expressions for the loop corrections. An example is the electron self-energy, for which we have examined the nonrelativistic setting in Chap. 4. The question then is how the high-energy part of the Lamb shift, discussed in Chap. 11, could be incorporated into the framework of NRQED. The answer is based on the observation that the exchange of high-energy photons promotes the virtual electron to a high-energy state, i.e., into a kinematic region which is detached from the applicability of a purely nonrelativistic theory. This implies that it is possible to formulate the effect of the high-energy photons in terms of effective operators, to be added to the NRQED Lagrangian. These operators are not required to lead to renormalizable interactions; the theory is only required to reproduce the result of the full QED theory up to a given order in the α-expansion. Even four-fermion vertices [see Fig. 17.4(k)] are allowed, as we illustrate in a treatment of the Salpeter correction based on NRQED, which supplements the treatments presented in Chaps. 15 and 16. The evaluation of the effective operators by a matching calculation, and the general applicability of these operators for any desired process, are cornerstones of the effective-field-theoretical approach.

In Chap. 18, we follow our field-theoretical ambitions and derive the effective Lagrangian of quantum electrodynamics that extends the Maxwell Lagrangian to higher orders in the electromagnetic fields. The derivation augments and contrasts a number of concepts introduced earlier. In contrast to the treatment of photon emission by a given classical current distribution (see Chap. 9), we now consider classical electromagnetic fields, but quantize the fermion field. In the sense of Chap. 17, the so-called "QED effective Lagrangian" derived by HEISENBERG[17] and EULER[18] (see Refs. [87, 88]), and SCHWINGER[19] (see Ref. [89]) qualifies itself as an effective operator derived by integrating out the high-energy fermionic degrees of freedom. Very much in line with the concept of NRQED, the fourth-order terms (in the electric and magnetic fields) lead to four-photon (four-boson) vertices which would otherwise constitute a nonrenormalizable interaction. The general applicability of the QED effective action is derived from the fact that one may insert any combination of either stationary or oscillating fields, and make physical predictions. E.g., one derives the correction to the light propagation speed in a strong electromagnetic background field due to virtual excitations of the fermion fields. This constitutes a paradigmatic application of the QED effective action to oscillatory fields. A different application of the QED effective action, this time to a static Coulomb field, complements the discussion: It is given by the derivation of the leading term (for large distances) of the three-Coulomb-vertex correction to the electrostatic interaction of the bound electron and the atomic nucleus, otherwise known as the WICHMANN[20]–KROLL[21] correction [90, 91].

In Chap. 19, we discuss the renormalization group and its relation to quantum electrodynamics. From the discussion of vacuum polarization, we know that the effective strength of the quantum electrodynamic coupling constant grows at small distances, or large momentum scales. The coupling "runs" with the distance, or momentum, scale at which phenomena are observed. The decisive consideration leading to the renormalization group in quantum electrodynamics is as follows. In the limit of small distances, the rate of change of the coupling constant (the fine-structure constant) with the momentum scale is independent of the mass of the electron. Instead, it depends only on the value of the coupling at the distance scale in question. A specific functional relation fulfilled by the "running coupling" leads to the introduction of the so-called QED β function which describes the running of the charge with the momentum scale. This leads to the renormalization group (RG) of GELL-MANN[22] and LOW.[23] The RG program is then applied to a number of phenomena of physical interest such as the RG flow of the QED effective action, and the choice of an "optimal" value of the coupling constant for the description of a specific process [so-called BRODSKY[24]–LEPAGE[25]–MACKENZIE[26] (BLM) scale

---

[17] Werner Karl Heisenberg (1901–1976).
[18] Hans Heinrich Euler (1909–1941).
[19] Julian Seymour Schwinger (1918–1994).
[20] Eyvind Hugo Wichmann (1928–2019).
[21] Norman Myles Kroll (1922–2004).
[22] Murray Gell-Mann (1929–2019).
[23] Francis Eugene Low (1921–2007).
[24] Stanley Jerome Brodsky (b. 1940).
[25] Gerard Peter Lepage (b. 1952).
[26] Paul Blanchard Mackenzie (b. 1950).

setting [92]]. Finally, we investigate possible corollaries from RG transformations, relevant to the high-energy limit of QED.

1.4 Miscellaneous Remarks

One normally does not include remarks on the notation used in a book in the Preface. However, in QED calculations, due to the elevated level of complexity of calculations, one needs to use many symbols, some of which are "overloaded" with different meanings, depending on the context. In order to alleviate at least some conceivable ambiguities, we have used a slightly different font in some cases. For example, we use μ for the reduced mass of a two-body bound system, and μ for the renormalization scale (see Chap. 10). Despite our efforts at disambiguation, μ also appears as the subscript denoting the magnetic projection in the spin-angular function $\chi_{\kappa\,\mu}(\hat{r})$. This should not lead to confusion due to the different context in which the symbol appears. Throughout the treatise, κ is the Dirac angular quantum number, defined in Chap. 6, while we denote the anomalous contribution κ to the electron g factor as $g = 2(1 + \kappa)$. We also note that the notation ϵ is used for the overlapping parameter that separates the low-energy from the high-energy scale in Lamb shift calculations (see Sec. 4.5.2), while we use ϵ for the infinitesimal imaginary part in propagators. Finally, the notation $d = 4 - 2\varepsilon$ is used for the non-integer space-time dimension in dimensional regularization (see Chap. 10).

1.5 Further Thoughts

Here are some suggestions for further thought.

(1) **Covariances.** Study some of the recent CODATA papers [30, 60, 61, 93]. They use the method of covariances for the adjustment of fundamental physical constants and combine the results of a number of high-precision experiments in order to determine, in some cases, just one single physical constant. What is a covariance? How does it enter the adjustment of the fundamental constants?

(2) **Fine-Structure Constant.** Read up on different determination of the fine-structure constant α, e.g., using photon recoil (see Refs. [94–96]). In particular, convince yourself that the recoil energy E_R of the atom upon absorption/emission of a photon of wave vector k by an atom of mass M is given as

$$E_R = \frac{(\hbar k)^2}{2M}, \tag{1.26}$$

so that a photon recoil measurement eventually leads to a determination of $\hbar/M$. The fine-structure constant α can eventually be determined as

$$\alpha = \left[2\frac{R_\infty}{c} \times \frac{u}{m_e} \times \frac{M}{u} \times \frac{\hbar}{M}\right]^{1/2} \tag{1.27}$$

where u is the atomic mass unit and R_∞ is the Rydberg constant.

Investigate the following questions:

- How is the Rydberg constant being determined? Study the paper [97] as well as Refs. [30, 60, 93] in order to elucidate the connection of high-precision QED calculations of hydrogen and deuterium energy levels, and the determination of the Rydberg constant, as well as the proton and deuteron radii.
- The primary "other" method for determining the fine-structure constant is based on free-electron g factor measurements [49, 52]. From the bound-electron g factor, using the value of α determined from the free-electron g factor, one infers u/m_e (see Refs. [60, 98]). So, the value of α determined from the free-electron g factor also influences u/m_e, and in consequence, the determination of α from photon recoil [see Eq. (1.27)] is not completely independent of the determination of α from the free-electron g factor. Do an approximate calculation to determine if the uncertainty in α, from the current free-electron g factor measurements, could influence the determination of α from the recoil measurement, in an appreciable fashion.

Chapter 2

From Unit Systems for the Microworld to Field Quantization

2.1 Overview

On the journey toward the microworld described by quantum electrodynamics, we shall meet matter waves in the Hamilton–Jacobi formalism in Sec. 2.2. This episode illustrates the transition from classical mechanics to quantum mechanics. The transition to unit systems pertinent to the microworld is illustrated in Sec. 2.3. We emphasize the distinctions between the natural and the atomic unit systems. Field quantization is discussed in Sec. 2.4, with an emphasis on the electromagnetic field, in the Coulomb gauge. At the end of the chapter, we have the main ingredients at our disposal, in regard to the description of the interaction of atoms with the quantized radiation field.

2.2 Atoms and Field Quantization

2.2.1 *Matter Waves and Hamilton–Jacobi Formalism*

Atoms are described by quantum mechanics. The Schrödinger equation describes the propagation (in space and time) of matter waves which correspond to the wave functions of the electrons. Let us consider the transition from the classical theory of an electron orbiting a proton (two oppositely charged objects attracting each other via the Coulomb force) to the quantum mechanical description. *A priori*, one might think that this transition is equivalent to the transition from ray optics, where light is considered to be a classical "ray", to wave optics, where light is described by electromagnetic waves. If that analogy were perfect, then quantum mechanics would still be deterministic: Matter waves, which describe the propagation of matter, would correspond to the electromagnetic waves whose propagation is described by the Maxwell equations; matter wave propagation would be deterministic in much the same way as the propagation of electromagnetic waves is deterministic. If we knew the electron's wave function at time t_0, then we would precisely know what its wave function will be in the future, no matter what interactions the electron went through on its course.

What is so fundamentally new about quantum mechanics is the postulate of the collapse of the wave function upon measurement. This postulate states that after a measurement of a quantum system, the system is in an eigenstate of the quantum mechanical operator that describes the measured physical quantity. This postulate, which incorporates the role of the observer in the outcome of the measurement, implies that after the measurement, the evolution of the wave function of the electron, starts "from scratch" and there is no way to predict the future time development of the system before the measurement was taken. Indeed, all we can say about a quantum mechanical measurement is to give a probability for its outcome; the theory is fundamentally nondeterministic.

We investigate the Schrödinger Hamiltonian describing an electron,

$$H = \frac{\vec{p}^2}{2m} + V(\vec{r}) = -\frac{\hbar^2}{2m}\vec{\nabla}^2 + V(\vec{r}), \tag{2.1}$$

where $\vec{p} = -\mathrm{i}\hbar\vec{\nabla}$ is the momentum operator and $V(\vec{r})$ is the potential. This equation can be easily motivated: The paradigmatic example of a wave describing a particle is a plane wave of the form $\exp(\mathrm{i}\vec{k}\cdot\vec{r})$, where $\vec{k}$ is the wave vector and $\vec{r}$ is the spatial coordinate. It is an eigenfunction of the operator $\vec{p}$ with eigenvalue $\hbar\vec{k}$, which according to DE BROGLIE,[1] describes the relation of the wavelength of the matter wave to the momentum. Furthermore, it is clear that the potential $V(\vec{r})$ should enter the Hamiltonian.

It is rather surprising to realize that in comparison to what would be expected of the "wave fronts" (planes of constant classical action) derived from Hamiltonian–Jacobi theory, the Schrödinger equation adds a dissipative term to the propagation of a quantum matter wave. This can be seen as follows. We start from the classical action S_0 for a classical particle,

$$L = T - V, \qquad T = \frac{\vec{p}^2}{2m} = \frac{1}{2}m\dot{\vec{r}}^2, \qquad S_0 = \int_0^t \mathrm{d}t'\, L(\vec{r}(t'), \dot{\vec{r}}(t')), \tag{2.2}$$

where $\dot{\vec{r}}(t') = \mathrm{d}\vec{r}(t')/\mathrm{d}t'$. Because S_0 is obtained by integrating the trajectory of the particle from time zero to time t, S_0 actually is a function of t and of the endpoint of the trajectory, $\vec{r}(t)$, and so $S_0 = S_0(t, \vec{r}(t))$. We find

$$\frac{\mathrm{d}S_0}{\mathrm{d}t} = L, \qquad \frac{\mathrm{d}S_0}{\mathrm{d}t} = \frac{\partial S_0}{\partial t} + \vec{\nabla}S_0 \cdot \frac{\partial \vec{r}}{\partial t} = \frac{\partial S_0}{\partial t} + \vec{p}\cdot\dot{\vec{r}}, \tag{2.3}$$

where the Euler–Lagrange equation

$$\frac{\partial L}{\partial \vec{r}(t)} = \frac{\mathrm{d}}{\mathrm{d}t}\frac{\partial L}{\partial \dot{\vec{r}}(t)} \tag{2.4}$$

has been used. Using $\dot{\vec{r}} = \vec{p}/m$, this can be rewritten as

$$-\frac{\partial S_0}{\partial t} = (\vec{p}\cdot\dot{\vec{r}} - L) = H = \frac{\vec{p}^2}{2m} + V = \frac{(\vec{\nabla}S_0)^2}{2m} + V(\vec{r}). \tag{2.5}$$

The gradient vector $\vec{p} = \vec{\nabla}S_0$ is necessarily perpendicular to the surfaces of constant classical action S_0. Indeed, the whole point of the Hamilton–Jacobi formalism is

[1] Louis-Victor-Pierre Ramond, 7ème duc de Broglie (1892–1987).

that the classically allowed trajectories are perpendicular to the surfaces of constant action [99]. Likewise, wave fronts of a traveling wave are characterized by a constant phase and move perpendicular to the surfaces where the phase is constant. This suggests the *ansatz*

$$\psi_0(t,\vec{r}) = \exp\left(\frac{\mathrm{i}}{\hbar}\, S_0(t,\vec{r})\right) \tag{2.6}$$

for the matter wave function. This equation for the wave function $\psi_0(t,\vec{r})$ would hold if the dynamics of the matter wave were purely governed by the laws of classical mechanics. Let us now formulate the true quantum mechanical wave function $\psi(t,\vec{r})$ in terms of an exponential,

$$\psi(t,\vec{r}) = \exp\left(\frac{\mathrm{i}}{\hbar}\, S(t,\vec{r})\right), \tag{2.7}$$

where we drop the subscript zero on both ψ and S. Inserting this *ansatz* into the Schrödinger equation $\mathrm{i}\,\partial_t\psi = H\,\psi$, one gets

$$-\frac{\partial S}{\partial t} = \frac{(\vec{\nabla} S)^2}{2m} + V(\vec{r}) - \frac{\mathrm{i}\,\hbar}{2m}\,\vec{\nabla}^2\, S\,. \tag{2.8}$$

This has to be compared with Eq. (2.5),

$$-\frac{\partial S_0}{\partial t} = \frac{(\vec{\nabla} S_0)^2}{2m} + V(\vec{r})\,. \tag{2.9}$$

The third term on the right-hand side of Eq. (2.8) characterizes a diffusive process in the context of the Brownian motion of the quantum particle [100] and describes the spreading of quantum mechanical wave packets. It may be interpreted as a quantum mechanical correction to the classical equations of motion for matter waves. The correction is of order $\hbar$. Indeed, the expansion in powers of $\hbar$ is known as the semiclassical or Wentzel–Kramers–Brioullin (WKB) expansion [101].

2.2.2 *Field Quantization as Second Quantization*

We have just discussed the transition from classical mechanics to wave mechanics. This is known as first quantization because a fundamentally classical theory (without waves) has been quantized and formulated as a quantum theory (which describes matter waves). Classical electrodynamics describes electromagnetic waves and one might thus consider it to be "first-quantized" in the sense that any wave theory contains aspects otherwise reserved for *wave* mechanics. The next natural question to ask concerns the nature of second quantization and the quantization of a theory that already describes waves. One may argue that, for the electromagnetic field, first quantization is unnecessary because the field itself is already a spatially distributed quantity which is not necessarily localized at one and only one place, as it would be for a point particle in classical mechanics. In second quantization, one identifies fundamental excitations of the electromagnetic field as "photons" which can be created and annihilated. The wave functions describing the fundamental excitations are multiplied with creation and annihilation operators.

The same procedure can also be applied to the matter waves which describe electrons. This eventually leads to the "second-quantized" description of the fermion field. Here, we shall first be concerned with the quantized electromagnetic field and describe the atom on the level of first quantization before we finally go on to discuss the quantized fermion field in Chap. 7. The reason is as follows. Typically, one has to quantize a theory when the fundamental energy scale or de Broglie wavelength of a physical process is commensurate with the fundamental excitations of the quantum theory. For the quantized theory of the fermion field, this energy scale is the electron mass scale $m_e c^2 \approx 0.511\,\mathrm{MeV}$, which corresponds to a wavelength of the order of the electron Compton wavelength

$$\lambda_e = 2\pi \lambdabar_e = \frac{h}{m_e c} = 2.426 \times 10^{-12}\,\mathrm{m}. \tag{2.10}$$

The Bohr radius of a typical atom is $a_0 = 0.529 \times 10^{-10}\,\mathrm{m} = 0.529\,\text{Å}$, where we recall the unit symbol Å for the Angstrom. The Bohr radius is larger than the electron Compton wavelength λ_e by a factor $(2\pi\alpha)^{-1} = 21.81$, where we recall the value given in Eq. (1.12) for the fine-structure constant. The physics relevant to the quantized fermion field happens at length scales which are a lot smaller than the typical dimension of an atom. In a first approximation, we do not need to quantize the fermion field if we want to describe the interaction of the atom with an external electromagnetic field. In higher orders of approximation, though, one cannot avoid describing the physics that happens at the scale of the electron Compton wavelength; this leads to vacuum polarization and other effects.

The fundamental excitations of the electromagnetic field (the photons) can have an arbitrarily small energy, or, expressed differently, an arbitrarily long wavelength. One might think that a quantum description of any electromagnetic interaction is indispensable on all energy scales. This is, however, not the case. The decisive parameter here is the strength of the electromagnetic field. Photons are bosons; they enjoy occupying one and the same quantum state collectively, and in great numbers. One photon more (or less) occupying an already well-populated state (e.g., a laser mode) will not make a large difference. A strong laser field can be described as a classical electromagnetic wave up to an excellent approximation.

One has to quantize the electromagnetic field if the number of excitations (the number of photons) is small. In the case of the CASIMIR[2]–POLDER[3] interaction between macroscopic plates, the geometry of the apparatus imposes boundary conditions that change the shape of all of the photon wave functions collectively. The modification of the zero-point energy of the photon modes then leads to a measurable effect even if no real, and no virtual, photons are present (see Chap. 6 of Ref. [81]). The self-energy of a bound electron involves at least one virtual photon. The correct result for this effect [81] can only be obtained if one quantizes the electromagnetic field. A third example combines both mentioned criteria: An

[2]Hendrik Brugt Gerhard Casimir (1909–2000).
[3]Dirk Polder (1919–2001).

atom tries to emit a single photon, making a quantum jump, and is either hindered or assisted by the specific geometry of the apparatus, e.g., a perfectly conducting cavity. In this case, the quantum description is required for two reasons: *(i)* because a description of the spontaneous emission of a photon by the atom would be impossible without taking the vacuum fluctuations of the electromagnetic field into account, and *(ii)* because the modification of the electromagnetic field operators (modification of the photon modes) by the boundary conditions could not be described without quantization.

In the formalism to be developed in the following, we shall introduce the fundamental annihilation and creation operators $a_\lambda(\vec{k})$ and $a^+_\lambda(\vec{k})$, which act as particle creation and annihilation operators for photons of polarization λ and wave vector $\vec{k}$. They are multiplying the photon wave functions, which are proportional to $\exp[-\mathrm{i}\,(\omega\,t - \vec{k}\cdot\vec{r})]$ and represent the wave nature of light. The specific form of the photon field operator is a mathematical manifestation of the wave-particle duality.

2.3 Unit Systems Scaled to the Microworld

2.3.1 *Unit Systems and Observation Scales*

The crossroads of atomic physics and field theory share a number of intricacies and peculiarities. Among these, we notice that three different systems of physical units are currently in use in electrodynamics and in bound-state calculations. All three of these have a *raison d'être*; they are specially suited for a particular regime of applications. The *SI mksA unit system* (commonly referred to as the SI or the SI mksA system) is the basic unit system accepted worldwide for the description of physical processes and recommended by CODATA (Committee on Data for Science and Technology). Here, SI stands for *Système International*, while mksA denotes the units of meter, kilogram, second, and Ampere. The distinctive advantage of this unit system is that all physical quantities can immediately be related to physical units as they are read off from ordinary gauged instruments. There is no need for any unit conversion, or for supplementing any prefactors at the end of a calculation. The disadvantage is that a number of formulas become more lengthy than necessary, because of the appearance of additional prefactors.

The *natural unit system*, derived from the Heaviside–Lorentz system, is defined so that the numerical values of the reduced Planck constant, denoted as $\hbar = h/(2\pi)$, the speed of light c and the vacuum permittivity ϵ_0 attain values of unity, i.e., $\hbar = c = \epsilon_0 = 1$. One might ask how this is possible because the Coulomb law, $F = q_1\, q_2/(4\pi\epsilon_0 r^2)$, relates the force F between two charges q_1 and q_2 and their mutual distance r. Moreover, q_1 and q_2 are measured in units of charge (Coulomb in the SI mksA system), whereas r is measured in meters. So, in order for us to be able to set $\epsilon_0 = 1$, the units of charge, length and force must be related. This seems counterintuitive since charge is fundamentally different from length and force — but is it? The rationale behind setting $\epsilon_0 = 1$ is that the units of length, force, and charge can be related if we understand that force is generated by a configuration of two charges at a distance r from each other — as expressed by the Coulomb law. Likewise, we can measure length in terms of the distance traveled by time in a second. If we define our unit of length to be equal to the distance traveled by time in a second, then the speed of light c in our new unit system is one unit of length, divided by one unit of time, and the numerical value of c (the *reduced quantity*) multiplying this ratio is just one. We now remember that a relation like $F = q_1 q_2/(4\pi\epsilon_0 r^2)$ allows us to calculate F using a computer that knows nothing about physical units. This is because if q_1 and q_2 are measured in Coulombs, then ϵ_0 furnishes the conversion factor so that F comes out in units of kilogram times meter per second squared. If our base units are chosen so that the numerical value of ϵ_0 becomes unity, then we might just as well pretend that we think like a computer, and calculate using the numerical values (the reduced quantities) corresponding to the physical quantities only, dropping the units altogether. Thus, we set $c = 1$, as well as $\hbar = \epsilon_0 = 1$. Near the end of a calculation, it is the quantity calculated that tells us how to resupply the missing factors of $\hbar$, c, and ϵ_0 that we have neglected so far.

In the natural unit system, one has $e^2 = 4\pi\alpha$. Because a factor e^2 is supplied in every order of QED scattering theory (loop calculations), the natural unit system is well adapted to high-energy processes. In this system, for bound-state theory, the basic unit of energy is $mc^2 = m$, where m is the electron mass. Using the same unit system for bound-state theory as for high-energy QED scattering theory, one obtains uniformity with respect to the prefactors and the general formulas for Feynman diagram calculations. The natural unit system also is very well adapted for identifying the terms in bound-state theory that contribute at a particular order in the fine-structure constant α to a particular atomic property such as the energy of a particular atomic level. Specifically, high-energy virtual photons have an energy $\sim m$, the Schrödinger energy and the interelectron interaction are of order $\alpha^2 m$, and relativistic corrections are of order $\alpha^4 m$. Genuine QED energy corrections originating from loop diagrams start at order $\alpha^5 m$ in electronic bound systems. In general, if one enters QED bound-state theory from the particle physics side, then the natural unit system appears to be the natural choice for calculations.

The *atomic unit system* is different; it originates from the days when the properties of atoms were being explored by physicists, and the most basic unit of length was naturally identified as the Bohr radius of an atom. In atomic units, all physical quantities (energy, length, velocity) are expressed so that they assume values of order unity for the motion of the electrons in typical atomic systems. For example, the basic unit of energy is known as the Hartree[4] energy or simply the "Hartree", being equal to $E_h = \alpha^2 mc^2$. The ionization potential of hydrogen (the negative of the ground-state binding energy) is equal to one half in atomic units (measured in units of Hartrees). The atomic unit system is very useful for doing calculations

[4]Douglas Rayner Hartree FRS (1897–1956).

in fundamentally nonrelativistic atomic systems. A particular example would consist in a nonrelativistic calculation of the energy levels of helium or lithium, i.e., a nonrelativistic calculation of a few-body system. Here, the unit system of choice is the atomic unit system; one is interested in atomic level energies and ionization potentials scaled to the systems under investigation. As humans, we have chosen the meter (or the foot) as our basic unit of length because it is adapted to the dimensions of the human body, which first and foremost defines the length scale on which we are operating. A human being measures in between 1.5 and 2.0 meters or in between 5 and 7 feet. The numerical values (the reduced quantities) multiplying the base unit of length are in between $10^{-1} \ldots 10^{1}$. Thus, it is natural to choose atomic units so that the typical ionization potential of an atom, or binding energies, are obtained as quantities of the order of one Hartree. In order to figure yet a supplementary analogy, let us mention that in currency trading, the unit of a transaction is the million. Buying "two dollars", for a currency trader, in reality, means to buy two million dollars. Simply, if the scales involved in a transaction (or, physically, in an interacting system) are different, then it is customary to choose a system of units adapted to the problem at hand.

In contrast to the atomic unit system, the base unit of the natural unit system is the reduced Compton wavelength of the electron, which is smaller than the Bohr radius by the fine-structure constant $\alpha \approx 1/137.036$. One of the difficulties of the bound-state formalism is to accommodate effects on both of these length scales and to describe them in a such a way that they can be consistently added at the end of a calculation. This matching will be one of the recurrent themes treated in this book.

2.3.2 *Natural Unit System*

As we enter the quantum world, the fundamental length and energy scales change from mascroscopic to microscopic dimensions. Therefore, it is a good idea to introduce an appropriate system of physical units which is adapted to the physical dimensions. The natural unit system is very well adapted to the quantum world in general, while the atomic unit system leads to additional simplifications for computationally intensive problems where the only relevant degrees of freedom are atomic and all quantities can be scaled to the atomic length and mass scales.

We start with the natural unit system. In many works on quantum field theory, it is simply stated that "natural units are used in which $\hbar = c = \epsilon_0 = 1$". This small digression may serve as a further explanation for the "procedure of going into natural units". Somewhat non-standard symbols shall be used. It may be surprising to observe that there exists no commonly accepted, standard choice for the natural unit of mass. However, it is natural to choose this scale as $\hat{m}$, where $\hat{m}$ is the mass of the lightest stable particle, i.e., the mass of the electron [60],

$$\hat{m} = m = m_e = 9.109\,383\,56(11) \times 10^{-31}\ \mathrm{kg}. \tag{2.11}$$

The natural unit of length, $\hat{\ell}$, is equal to the reduced Compton wavelength of the electron, $ƛ_C$,

$$\hat{\ell} = ƛ_C = \frac{\hbar}{m_e\, c} = 3.861\,592\,6764(18) \times 10^{-13}\ \text{m}\,, \tag{2.12}$$

where c is the speed of light in vacuum. Note that the *reduced* Compton wavelength of the electron is just equal to the Compton wavelength divided by a factor 2π and its designation has nothing to do with the distinction of physical and reduced quantities. The natural unit of time, $\hat{t}$, is the reduced Compton wavelength of the electron, $ƛ_C$, divided by a factor of c,

$$\hat{t} = \frac{\hbar}{m_e\, c^2}. \tag{2.13}$$

Here, we will use the paradigm that within a given unit system, products and ratios of fundamental units yield derived fundamental units. The natural unit of velocity, $\hat{v}$, is obtained by dividing the natural unit of length by the natural unit of time,

$$\hat{v} = \hat{\ell} \cdot \hat{t}^{-1} = c = 299\,792\,458\ \text{m} \cdot \text{s}^{-1}. \tag{2.14}$$

Therefore, the speed of light in vacuum has the numerical value of one in natural units,

$$c = 1\ \hat{v} = 1\ \hat{\ell} \cdot \hat{t}^{-1}\,. \tag{2.15}$$

The SI mksA unit of action, which is $\text{kg}\cdot\text{m}^2\cdot\text{s}^{-1}$, is obtained by multiplying suitable powers of the SI mksA units of mass, length, and time. The natural unit of action, $\hat{S}$, is obtained as

$$\hat{S} = \hat{m} \cdot \hat{\ell}^2 \cdot \hat{t}^{-1} = \hbar. \tag{2.16}$$

Therefore, the reduced Planck constant $\hbar$ (the Planck constant h divided by 2π) has a numerical value of one in natural units,

$$\hbar = 1\ \hat{S} = 1\ \hat{m} \cdot \hat{\ell}^2 \cdot \hat{t}^{-1} \equiv 1\,. \tag{2.17}$$

The "reduced quantity" is defined as the "numerical value before the unit symbol". Within a given unit system, we can eventually drop the unit symbol and identify a physical quantity with its reduced quantity. In natural units, the speed of light is $c = 1\,\hat{\ell}/\hat{t} \equiv 1$. This consideration justifies the identity after the "≡" sign in Eq. (2.17). The equivalence amounts to a rescaling of all physical quantities to their natural scales. Instead of working with the physical quantities, one is working with *reduced* quantities which are real numbers defined as the quotients of the physical quantities and the natural unit in which they are measured [102, 103]. Pedantically, it would be more correct to denote the reduced quantities with an asterisk [102, 103], i.e., to write $\hbar^* = 1$ for the result of the operation of dividing the reduced Planck constant $\hbar$ by the quantity $1\ \hat{m} \cdot \hat{\ell}^2 \cdot \hat{t}^{-1}$, which describes its natural unit. However, most authors denote physical and reduced quantities with the same symbol, as we have done in Eq. (2.17).

Alternatively, in assigning the value $\epsilon_0 = 1$ to the vacuum permittivity, we identify the unit of charge with the square root of the unit of force times the unit of distance, according to the formula $F = q_1 q_2 (4\pi\epsilon_0 r^2)^{-1}$ for the Coulomb law. The fact that one can do this is a consequence of the so-called "Buckingham π theorem" (see Ref. [104]), which bears the name of BUCKINGHAM.[5] This theorem relates the number of equations one has in order to describe a given set of physical processes to the number of dimensionless physical constants that one can derive from the description of the process. It determines the number of dimensionful quantities one can eliminate in a unit system. So, in natural units, the unit of charge $\hat{Q}$ reads as

$$\hat{Q} = \sqrt{\epsilon_0 \hbar c} = \frac{e}{\sqrt{4\pi \alpha}} . \tag{2.18}$$

If we identify physical and reduced quantities, we can write $\hat{Q} = 1$. As one can easily verify, in the natural unit system the vacuum permittivity is obtained by multiplying powers of base units:

$$\epsilon_0 = 1 \, \hat{Q}^2 \cdot \hat{t}^2 \cdot \hat{m}^{-1} \cdot \hat{\ell}^{-3} . \tag{2.19}$$

In view of Eqs. (2.15), (2.17), and (2.19), the constants $\hbar$, c, and ϵ_0 all have the numerical value of one in natural units. However, one can now *rescale* all physical quantities to their natural scales. That is to say, instead of working with the actual physical quantities, one is working with *reduced* quantities [102, 103] which are real numbers defined as the quotients of the actual quantities and the natural unit in which they are measured. It is generally recommended to denote the reduced quantities with an asterisk, i.e., to write λ^* for a wavelength divided by the electron Compton wavelength. We recall once more that it is customary, within the natural unit system, to denote actual and reduced quantities with the same symbol.

After the introduction of natural units, the rescaling to natural scales and the identification of physical and reduced quantities, one indeed obtains

$$\hbar = c = \epsilon_0 = 1. \tag{2.20}$$

After this identification, Eq. (2.18) also implies that $e^2 = 4\pi\alpha$. One can easily check that this relation is fulfilled irrespective of the mass scale or mass unit $\hat{m}$. Alternatively, we can say that if the natural unit system is defined to be a unit system in which $\hbar = c = \epsilon_0 = 1$, then we are free to choose a mass scale m, or, keep a mass scale m in all formulas.

The *Système International* with its fundamental units of meter (m), kilogram (kg), second (s), and Ampere (A) is most accurately called the SI mksA system. The recent acceptance of a particular value for the Planck constant in the SI mksA unit system has led to a redefinition of the kilogram as a standard mass and to a redefinition of the SI as a quantum SI unit system [105–109]. The basic idea is as follows: One defines the second as a unit of time; it is fixed by setting the numerical value of the ground state hyperfine splitting of cesium (isotope ^{133}Cs) to

[5] Edgar Buckingham (1867–1940).

be equal to $\nu_{\rm Cs} = 9\,192\,631\,770\,{\rm s}^{-1}$ (exactly). The meter is derived from the unit of time by setting the numerical value of the speed of light to be exactly equal to $c = 299\,792\,458\,{\rm m/s}$. In the quantum SI unit system, one compares mechanical and electrical work using a Watt balance, i.e., one "weighs the Watt" (see Ref. [110]). The "weighing of the Watt" leads to a accurate determination of the Planck constant if one measures the voltage using a Programmable Josephson Voltage System (PJVS). Indeed, the Josephson constant $K_J = 2e/h = 483\,597.8525(30) \times 10^9\,{\rm Hz/V}$ is very well known [60]. A PJVS generates a voltage $\mathcal{V} = n_V\,\nu_0/K_J$, where n_V is the number of Josephson junctions and ν_0 is the microwave frequency at which the system is being irradiated. In summary, using a Watt balance and knowing the local value of the acceleration due to gravity to high accuracy, one may convert a test mass into an accurately measured voltage $\mathcal{V}$, which, in turn, based on the Josephson effect, is converted into an accurate value of the Planck constant h. Conversely, one may fix the numerical value of the Planck constant equal to exactly

$$h = 6.626\,070\,15 \times 10^{-34}\,{\rm kg\,m^2/s}, \tag{2.21}$$

which was adopted on 16 November 2018 by the International Bureau of Weights and Measures (BIPM). Having a defined value of h, one can map the definition of the kilogram onto a sum of photon frequencies. Namely, one may define that a kilogram is equal to a mass, $m_{\rm kg}$, whose energy equivalent $m_{\rm kg}\,c^2$ is equal to the energy sum $h\sum_i \nu_i$ of a collection of photons whose sum frequency is

$$\sum_i \nu_i = [10^{34}/6.626\,070\,15]\,{\rm Hz}\,. \tag{2.22}$$

The advantage of defining the (reduced) Planck constant $\hbar$ in the present context is the following. Observe that the natural units $\hat{m}$, $\hat{\ell}$, $\hat{t}$, and $\hat{Q}$ involve one mass scale and are otherwise related by the constants $\hbar$, c, and ϵ_0. For different problems, it is often convenient to choose slightly modified natural units with a mass or, equivalently, an energy scale different from the mass of the electron. If all three constants $\hbar$, c and ϵ_0 are defined in the SI mksA unit system, it implies that when a mass scale is chosen for a generalized natural unit system, all other natural units follow from the mass scale via exact factors.

A very important role is played by the Coulomb potential. The name "Coulomb potential" is usually associated with the potential energy of an electron in the electrostatic field of a nucleus with charge number Z. In natural units, it is given by the simple expression

$$V(\vec{r}) = V(r) = e\left(\frac{-Ze}{4\pi\,\epsilon_0\,r}\right) = \frac{-Z\,e^2}{4\pi\,\epsilon_0\,r} = \frac{-4\pi\,Z\alpha}{4\pi\,r} = -\frac{Z\alpha}{r}\,, \tag{2.23}$$

where we have used Eq. (1.12) expressed in natural units [see Eq. (2.20)]. Here and in the following, we denote the electron charge by $e = -|e|$, so that the charge of the nucleus is $-Z\,e$.

2.3.3 *Atomic Unit System*

The atomic system is defined through the relations

$$\hbar = m_e = |e| = 4\pi\,\epsilon_0 = 1\,, \tag{2.24}$$

where we recall that m_e denotes the electron mass. As a consequence,

$$\alpha = \frac{e^2}{4\pi\epsilon_0\hbar c} = \frac{1}{(4\pi\epsilon_0)c} = \frac{1}{c} \qquad \text{(atomic units)}\,, \tag{2.25}$$

and the Bohr radius is

$$a_0 = \frac{\hbar}{\alpha\, m_e\, c} = \frac{1}{c^{-1}\times 1\times c} = 1 \qquad \text{(atomic units)}\,. \tag{2.26}$$

The atomic unit of energy E_h is the Hartree,

$$E_h = \alpha^2\, m_e\, c^2. \tag{2.27}$$

In SI mksA units, we have [31]

$$E_h = 2R_\infty\, h\, c = 27.211\,386\,245\,988(53)\,\text{eV}\,. \tag{2.28}$$

Let us illustrate various unit conversions by way of example. The energy of a bound state in a one-electron atom, including the first relativistic corrections but excluding the reduced-mass effects, can be obtained from the Dirac equation. Without going into details of the derivation, we simply state the result for a quantum level of principal quantum number n and total electron angular momentum j,

$$E_{nj} = m\,c^2 - \frac{(Z\alpha)^2\, mc^2}{2n^2} - \frac{(Z\alpha)^4\, mc^2}{n^3}\left(\frac{1}{2j+1} - \frac{3}{8\,n}\right) \qquad \text{(SI mksA units)}\,, \tag{2.29}$$

where we neglect terms of order $\alpha^6\, m\,c^2$. (Here and in the following, m is the mass of the bound particle, which usually is equal to the electron mass.) The first term on the right-hand side qualifies itself as the electron rest mass converted into an energy. The second term is the Schrödinger energy, and the third term is the first relativistic correction. In natural units with $\hbar = c = \epsilon_0 = 1$, yet keeping the mass scale arbitrary, we have from Eq. (2.29),

$$E_{nj} = m - \frac{(Z\alpha)^2\, m}{2n^2} - \frac{(Z\alpha)^4\, m}{n^3}\left(\frac{1}{2j+1} - \frac{3}{8\,n}\right) \qquad \text{(natural units)}\,. \tag{2.30}$$

The leading term takes a simple form and reminds us of the mass-energy equivalence that follows from relativity theory. In atomic units, we have

$$E_{nj} = \frac{1}{\alpha^2} - \frac{Z^2}{2n^2} - \frac{Z^4\,\alpha^2}{n^3}\left(\frac{1}{2j+1} - \frac{3}{8\,n}\right) \qquad \text{(atomic units)}\,. \tag{2.31}$$

This is obtained by simply dividing the corresponding formula by a factor $\alpha^2 m$.

Let us illustrate the unit conversion once more, but this time considering a part of the $(Z\alpha)^4 mc^2$ correction which is due to the *zitterbewegung* term. In SI mksA units, the corresponding Hamiltonian reads

$$\delta H_Z = \frac{\pi\,\hbar^3}{2m^2 c}\, Z\alpha\,\delta^{(3)}(\vec{r}) \qquad \text{(SI mksA units)}\,, \tag{2.32}$$

where $\delta^{(3)}(\vec{r})$ denotes the Dirac-δ function. The probability density at the origin in a hydrogen eigenstate with quantum numbers n, $\ell = L$, and j (spectroscopic notation nL_j, example: $nS_{1/2}$ for $\ell = L = 0$) is

$$\langle nL_j|\delta^{(3)}(\vec{r})|nL_j\rangle = \frac{1}{\pi\, n^3}\left(\frac{Z\alpha mc}{\hbar}\right)^3 \delta_{\ell 0} \qquad \text{(SI mksA units)}. \tag{2.33}$$

The energy shift is obtained as

$$\begin{aligned}\Delta E_Z &= \langle nL_j|\delta H_Z|nL_j\rangle = \frac{\pi\,\hbar^3}{2m^2c} Z\alpha\,\langle nL_j|\delta^{(3)}(\vec{r})|nL_j\rangle \\ &= \frac{\pi\,\hbar^3}{2m^2c}\frac{Z\alpha}{\pi\,n^3}\left(\frac{Z\alpha mc}{\hbar}\right)^3 = \frac{(Z\,\alpha)^4\,m\,c^2}{2\,n^3}\,\delta_{\ell 0} \qquad \text{(SI mksA units)}.\end{aligned} \tag{2.34}$$

In atomic units, the expression for the *zitterbewegung* term converts as follows,

$$\begin{aligned}\delta H_Z &= \frac{\pi\,\hbar^3}{2m^2c} Z\alpha \left(\frac{\alpha mc}{\hbar}\right)^3 \delta^{(3)}\left(\frac{\alpha mc\vec{r}}{\hbar}\right) \\ &= \frac{\pi\,Z\,\alpha^4\,mc^2}{2}\,\delta^{(3)}(\vec{r}/a_0) = \frac{\pi\,Z\,\alpha^2}{2}\,\delta^{(3)}(\vec{r}) \qquad \text{(atomic units)}.\end{aligned} \tag{2.35}$$

The probability density at the origin is

$$\langle nL_j\left|\delta^{(3)}(\vec{r})\right|nL_j\rangle = \frac{Z^3}{\pi\,n^3}\,\delta_{\ell 0} \qquad \text{(atomic units)}. \tag{2.36}$$

The energy shift reads as follows,

$$\Delta E_Z = \langle nL_j\left|\delta H_Z\right|nL_j\rangle = \frac{\pi\,Z\,\alpha^2}{2}\langle nL_j|\delta^{(3)}(\vec{r})|nL_j\rangle = \frac{Z^4\,\alpha^2}{2\,n^3}\,\delta_{\ell 0} \qquad \text{(atomic units)}. \tag{2.37}$$

The quantity ΔE_Z obtained is the reduced quantity; one has to multiply it by the atomic energy scale $\alpha^2 mc^2$ in order to obtain the actual energy which reproduces Eq. (2.34). In natural units, the *zitterbewegung* Hamiltonian is

$$\begin{aligned}\delta H_Z &= \frac{\pi\,\hbar^3}{2m^2c} Z\alpha \left(\frac{mc}{\hbar}\right)^3 \delta^{(3)}\left(\frac{mc\vec{r}}{\hbar}\right) \\ &= \frac{\pi\,Z\alpha\,m}{2}\,\delta^{(3)}\left(\frac{\vec{r}}{\lambda\!\!\!^{-}_e}\right) = \frac{\pi\,Z\alpha\,m}{2}\,\delta^{(3)}(\vec{r}) \qquad \text{(natural units)}.\end{aligned} \tag{2.38}$$

The square of the wave function at the origin reads as

$$\langle nL_j\left|\delta^{(3)}(\vec{r})\right|nL_j\rangle = \frac{(Z\alpha)^3}{\pi\,n^3}\,\delta_{\ell 0} \qquad \text{(natural units)}. \tag{2.39}$$

For the energy shift, we have

$$\Delta E_Z = \langle nL_j\left|\delta H_Z\right|nL_j\rangle = \frac{\pi\,Z\alpha\,m}{2}\,\frac{(Z\alpha)^3}{\pi\,n^3} = \frac{(Z\alpha)^4 m}{2n^3}\,\delta_{\ell 0}\,. \tag{2.40}$$

Again, one may identify this result as the reduced quantity and multiply by c^2 to reproduce Eq. (2.34).

In the remainder of this book, we shall use the SI mksA, the natural, and the atomic system interchangeably. Indeed, all formulations have their advantages and disadvantages. For instance, if we strive to calculate the helium ground-state energy, then the only practical way to set up the calculation is with a computer program where all quantities are scaled to the dimensions of the atom in question, i.e., in atomic units. If we strive to keep a compact notation in writing down a field operator for an electric or magnetic field, in order to carry out a complex calculation while keeping intermediate expressions compact, then the natural unit system is most appropriate. The same is true if we attempt to systematically order the terms of a quantum electrodynamic perturbation series in powers of the fine-structure constant; this is most practical in natural units. Finally, if we strive to carry out a model calculation for a process while making sure that we do not miss any important terms in intermediate steps, then it sometimes pays off to fully retain SI mksA units in all intermediate steps of the calculation. Intermediate attempts like the Gaussian system, where one sets $\epsilon_0 = 1/(4\pi)$ but keeps most other physical quantities and dimensions unchanged, are not used in the following, and they seem to be disfavored by the scientific community in recent years. Typically, in the Gaussian system, formulas cannot be reduced too much in complexity, but one still loses the tracking of the SI mksA units and incurs several possible sources of confusion in formulas. Examples include the tracking of a factor 4π to an inverse power of ϵ_0 or an integration over a solid angle.

2.4 Field Quantization for the Electromagnetic Field

2.4.1 *Quantization of the Free Electromagnetic Field*

We have already mentioned that the process of field quantization can be interpreted in a natural way as "second quantization" (see Sec. 2.2.2). The quantization of the electromagnetic field is quite problematic from a theoretical point of view because it is a massless spin-1 field. Normally, the quantization of a theory proceeds via the correspondence principle: One writes down the Lagrangian, identifies the coordinates, determines the conjugate variables (canonical momenta), and postulates the canonical dispersion relations. However, in a massless spin-1 theory like the Maxwell theory of electrodynamics, as we shall discuss in more detail below, the canonical momentum corresponding to the scalar potential vanishes whereas the conjugate variable of the vector potential is the electric field. One can remedy the problem in various ways, but here, we shall pursue a very direct way which uses a few particular properties of the Coulomb gauge.

We recall from classical electrodynamics the Maxwell equations, which read as follows (in natural units)

$$\vec{\nabla} \cdot \vec{E}(t, \vec{r}) = \rho(\vec{r}, t), \tag{2.41a}$$

$$\vec{\nabla} \cdot \vec{B}(t, \vec{r}) = 0, \tag{2.41b}$$

$$\vec{\nabla} \times \vec{E}(t, \vec{r}) = -\frac{\partial}{\partial t}\vec{B}(t, \vec{r}), \tag{2.41c}$$

$$\vec{\nabla} \times \vec{B}(t, \vec{r}) = \frac{\partial}{\partial t}\vec{E}(t, \vec{r}) + \vec{J}(t, \vec{r}). \tag{2.41d}$$

Here, in contrast to classical electrodynamics, we shall order the space-time arguments of the electric and magnetic fields as $(t, \vec{r})$, as inspired by the four-vector notation $x^\mu = (t, \vec{r})$. The electric field $\vec{E}$ and the magnetic induction field $\vec{B}$ are related to the scalar potential Φ and the vector potential $\vec{A}$,

$$\vec{E}(t, \vec{r}) = -\vec{\nabla}\Phi(t, \vec{r}) - \frac{\partial}{\partial t}\vec{A}(t, \vec{r}), \tag{2.42a}$$

$$\vec{B}(t, \vec{r}) = \vec{\nabla} \times \vec{A}(t, \vec{r}). \tag{2.42b}$$

A vector field $\vec{A}$ can be uniquely decomposed into a transverse component $\vec{A}_\perp$ and a longitudinal component $\vec{A}_\parallel$, with $\vec{\nabla} \cdot \vec{A}_\perp = 0$ and $\vec{\nabla} \times \vec{A}_\parallel = \vec{0}$. In Coulomb gauge, the longitudinal component of the vector potential vanishes,

$$\vec{\nabla} \cdot \vec{A}(t, \vec{r}) = 0, \qquad \vec{A}(t, \vec{r}) = \vec{A}_\perp(t, \vec{r}), \qquad \vec{A}_\parallel(t, \vec{r}) = 0, \tag{2.43}$$

and the potentials are coupled to the sources as follows (see Chap. 1 of Ref. [81]),

$$\vec{\nabla}^2 \Phi(t, \vec{r}) = -\rho(t, \vec{r}), \tag{2.44a}$$

$$\left(\frac{\partial^2}{\partial t^2} - \vec{\nabla}^2\right) \vec{A}_\perp(t, \vec{r}) = \vec{J}_\perp(t, \vec{r}), \tag{2.44b}$$

$$\vec{\nabla}\frac{\partial}{\partial t}\Phi(t, \vec{r}) = \vec{J}_\parallel(t, \vec{r}). \tag{2.44c}$$

In Coulomb gauge, the longitudinal part of the electric field is given as the gradient of the scalar potential, as a consequence of Eqs. (2.42a) and (2.43),

$$\vec{E}_\parallel(t, \vec{r}) = -\vec{\nabla}\Phi(t, \vec{r}) - \frac{\partial}{\partial t}\vec{A}_\parallel(t, \vec{r}) = -\vec{\nabla}\Phi(t, \vec{r}), \tag{2.45}$$

while the transverse part is

$$\vec{E}_\perp(t, \vec{r}) = -\frac{\partial}{\partial t}\vec{A}_\perp(t, \vec{r}). \tag{2.46}$$

In the following, we shall perform a quantization of the electromagnetic field alluding, exclusively, to the transverse components of the vector potential, $\vec{A} = \vec{A}_\perp$. The scalar potential Φ will not even take part in the quantization process. The justification for this approach in a more general context will be given later in Chap. 9. We also recall from Chap. 8 of Ref. [81] the Lagrangian density of the free electromagnetic field,

$$\mathcal{L} = -\frac{1}{4} F^{\mu\nu} F_{\mu\nu} = -\frac{1}{4} (\partial_\mu A_\nu - \partial_\nu A_\mu)(\partial^\mu A^\nu - \partial^\nu A^\mu) = \frac{1}{2} \left(\vec{E}^2 - \vec{B}^2\right), \tag{2.47}$$

where $F^{\mu\nu}$ is the electromagnetic field strength tensor. The quantization procedure in Coulomb gauge, which is especially suited for nonrelativistic problems, is somewhat noncovariant. According to Eq. (2.44a), in the Coulomb gauge, the scalar potential would couple to the charge distribution instantaneously, i.e., in the form of an action-at-a-distance law, were it not for the additional consistency condition (2.44c). The latter forces the (otherwise problematic) longitudinal component of the electric field to be given by a retarded integral (see Sec. 4.2.4 of Ref. [81]).

From the Maxwell Lagrangian (2.47), the canonical momenta corresponding to the spatial components $\vec{A}_\perp$ of the four-vector potential are easily found as follows,

$$\Pi^0 = \frac{\partial \mathcal{L}}{\partial(\partial_0 A^0)} = 0\,, \qquad \Pi^k = \frac{\partial \mathcal{L}}{\partial(\partial_0 A_{\perp,k})} = -\partial_0 A_\perp^k = E_\perp^k\,. \tag{2.48}$$

Here, by convention, $k = 1, 2, 3$ is a spatial index; we use the (flat-)space metric in the convention

$$g_{\mu\nu} = \mathrm{diag}(1, -1, -1, -1)\,, \qquad \mu, \nu = 0, 1, 2, 3\,. \tag{2.49}$$

The conjugate variable of the vector potential is obtained as the electric field; it is extremely useful to observe that the conjugate variable of $A_{\perp,k}$ is $E_\perp^k$, which implies that the conjugate variable of $A_\perp^k$ is $-E_\perp^k$.

The Hamiltonian density is obtained from the Lagrangian density by a Legendre transformation,

$$\mathcal{H} = \mathcal{H}(t, \vec{r}) = \Pi_k \left(+ \frac{\partial A_\perp^k}{\partial t} \right) - \mathcal{L} = \vec{E}_\perp^2 - \mathcal{L} = \frac{1}{2} \left(\vec{E}_\perp^2 + \vec{B}^2 \right) - \frac{1}{2} \vec{E}_\parallel^2\,. \tag{2.50}$$

Finally, the Hamiltonian is

$$H = \int \mathrm{d}^3 r\, \mathcal{H}(t, \vec{r}) = \frac{1}{2} \int \mathrm{d}^3 r \left(\vec{E}_\perp^2 + \vec{B}^2 - \vec{E}_\parallel^2 \right). \tag{2.51}$$

It may appear as disturbing to see the term $\vec{E}_\parallel^2$ have a negative sign; this phenomenon finds a natural explanation once the term $\rho\,\Phi$ is added to the Hamiltonian [see Eq. (9.48) and the following discussion in Chap. 9]. We now use the correspondence principle and postulate some fundamental equal-time commutator relations for the quantized theory. Essentially, the paradigm is to reduce the quantization problem, for equal times, to ordinary canonical quantization as known from quantum mechanics, where we would postulate the commutations relation $[p, x] = -\mathrm{i}$ for the position operator x and the conjugate momentum operator $p = -\mathrm{i}\partial/\partial x$. The commutators of the vector potential and of the electric field with themselves vanish,

$$\left[A_\perp^i(t, \vec{r}), A_\perp^j(t, \vec{r}\,') \right] = 0\,, \qquad \left[\Pi^i(t, \vec{r}), \Pi^j(t, \vec{r}\,') \right] = 0\,. \tag{2.52}$$

In analogy to quantum mechanical coordinates and momenta, one might now assume the following commutation relations for the vector potential and its conjugate momenta,

$$\left[\Pi^i(t, \vec{r}), A_{\perp,j}(t, \vec{r}\,') \right] = -\left[\Pi^i(t, \vec{r}), A_\perp^j(t, \vec{r}\,') \right] \overset{?!}{=} -\mathrm{i}\,\delta^{ij}\,\delta^{(3)}(\vec{r} - \vec{r}\,')\,. \tag{2.53}$$

Once again, the sign is motivated by the fact that the conjugate variable of the covariant component $A_{\perp,j}$ is the contravariant field component Π^j. However, because we have imposed additional constraints on the vector potential in the form $\vec{A} = \vec{A}_\perp$ [see Eq. (2.43)], Eq. (2.53) is not compatible with the Coulomb gauge condition $\vec{\nabla} \cdot \vec{A}(t, \vec{r}) = 0$. This can be seen by taking the divergence of both sides of Eq. (2.53). The compatibility with the Coulomb gauge condition is restored if we replace the Dirac-δ function on the right-hand side of Eq. (2.53) by the transverse Dirac-δ. It reads as $\delta^{\perp,ij}(\vec{r} - \vec{r}')$, where

$$\delta^{\perp,ij}(\vec{r} - \vec{r}') = \int \frac{d^3k}{(2\pi)^3} \left(\delta^{ij} - \frac{k^i k^j}{\vec{k}^2} \right) e^{i\vec{k}\cdot(\vec{r}-\vec{r}')} . \tag{2.54}$$

In coordinate space, it is hard to give the transverse Dirac-δ in a simpler form than

$$\delta^{\perp,ij}(\vec{r} - \vec{r}') = \left(\delta^{ij} - \frac{\nabla^i \nabla^j}{\vec{\nabla}^2} \right) \delta^{(3)}(\vec{r} - \vec{r}') . \tag{2.55}$$

The transverse Dirac-δ is a tensor of rank two whose divergence with respect to both spatial indices vanishes,

$$\nabla^i \delta^{\perp,ij}(\vec{r} - \vec{r}') = \nabla^j \delta^{\perp,ij}(\vec{r} - \vec{r}') = 0 , \qquad \nabla^i \equiv \frac{\partial}{\partial x^i} . \tag{2.56}$$

Our notation implies that

$$\vec{\nabla} \cdot \vec{V} = \nabla^j V^j = \frac{\partial}{\partial x} V^x + \frac{\partial}{\partial y} V^y + \frac{\partial}{\partial z} V^z , \tag{2.57}$$

for any vector $\vec{V}$. We shall consistently use superscripts for the "special and spatial" vector components ∇^i of the Cartesian three-vector $\vec{\nabla}$. Our slightly nonstandard notation is motivated by our desire to keep the components of the vector $\vec{\nabla}$ (with an upper, contravariant index, see Chap. 6 of Ref. [81]) identical to those we would expect for a Cartesian vector. There is a further certain subtlety in the notation, which we would like to comment on. In this treatise, the spatial components x^i ($i = 1, 2, 3$) of the position operator $\vec{r}$ are denoted as

$$\vec{r} = x\, \hat{e}_x + y\, \hat{e}_y + z\, \hat{e}_z = x^1\, \hat{e}_x + x^2\, \hat{e}_y + x^3\, \hat{e}_z = \sum_{i=1}^{3} x^i\, \hat{e}_i . \tag{2.58}$$

The superscript (rather than subscript) of the Cartesian component stresses the contravariant (rather than covariant) nature of the component (see also Chap. 8). The notation is in part motivated by the desire to avoid confusion between the mth power $r^m = |\vec{r}|^m$ of the radial distance and the mth component x^m of the position operator. In Chap. 12, though, we shall slightly change this convention, in view of the complex nature of composite systems, where it becomes necessary to further distinguish position vectors measured in regard to different points of origin. For the time being, the notation (2.58) appears to be a convenient choice.

Employing the correspondence principle and the transverse Dirac-δ function, and taking into account that $\vec{A}$ is transverse, the modified postulate for the commutator reads

$$\left[\Pi^i(t, \vec{r}), A_{\perp,j}(t, \vec{r}')\right] = -\left[\Pi^i(t, \vec{r}), A_\perp^j(t, \vec{r}')\right] = -i\, \delta^{\perp,ij}(\vec{r} - \vec{r}') . \tag{2.59}$$

In terms of the electric field operator and the vector potential, the commutation relation reads

$$\left[\Pi^i(t,\vec{r}),\, A^j_\perp(t,\vec{r}\,')\right] = \left[E^i_\perp(t,\vec{r}),\, A^j_\perp(t,\vec{r}\,')\right] = \mathrm{i}\,\delta^{\perp,ij}(\vec{r}-\vec{r}\,')\,. \tag{2.60}$$

We decompose the field operator into fundamental modes as follows,

$$\vec{A}_\perp(t,\vec{r}) = \sum_{\lambda=1}^{2}\int \frac{\mathrm{d}^3k}{\sqrt{2\omega\,(2\pi)^3}}\,\hat{\epsilon}_\lambda(\vec{k})\left[a_\lambda(\vec{k})\,\mathrm{e}^{-\mathrm{i}\,(\omega t-\vec{k}\cdot\vec{r})} + a^+_\lambda(\vec{k})\,\mathrm{e}^{\mathrm{i}\,(\omega t-\vec{k}\cdot\vec{r})}\right], \tag{2.61a}$$

$$\vec{A}_\perp(x) = \sum_{\lambda=1}^{2}\int \frac{\mathrm{d}^3k}{\sqrt{2\omega\,(2\pi)^3}}\,\hat{\epsilon}_\lambda(\vec{k})\left[a_\lambda(\vec{k})\,\mathrm{e}^{-\mathrm{i}\,k\cdot x} + a^+_\lambda(\vec{k})\,\mathrm{e}^{\mathrm{i}\,k\cdot x}\right], \tag{2.61b}$$

$$\{k^{\mu=0,1,2,3}\} = (\omega,\vec{k}) = (|\vec{k}|,\vec{k})\,. \tag{2.61c}$$

Here, $\lambda = 1,2$ denotes the photon polarization and $k\cdot x = k^\mu x_\mu = \omega\,t - \vec{k}\cdot\vec{r}$. The decomposition is in fact somewhat arbitrary with respect to prefactors and integration measures because a change in the integration measure can always be reabsorbed in a redefinition of the fundamental photon creation and annihilation operators $a^+_\lambda(\vec{k})$ and $a_\lambda(\vec{k})$. Later, we will discuss alternative representations. The decomposition of the electromagnetic field operator given in Eq. (2.61) is especially suited for non-relativistic calculations. In d^3k integrations like the one performed in Eq. (2.61b), it is important to note that the zeroth component, $k^0 = \omega$, of the four-vector k^μ needs to be defined in terms of the spatial components; otherwise, the result of the integral could not be expressed solely in terms of t and $\vec{r}$. For particles of mass m, the component k^0 is determined as the positive square root

$$k^0 \equiv \omega \equiv \sqrt{\vec{k}^2 + m^2}\,, \tag{2.62}$$

where m is the particle mass ($m = 0$ for photons). In the literature on field theory, one finds both integrals of the type $\int \mathrm{d}^3k f(k)$, as well as $\int \mathrm{d}^4k f(k)$, where $k = (k^\mu)$ is a four-vector. In the first case, the component k^0 is determined by the general condition (2.62), i.e., always positive. In the latter case, the k^0 component is not determined by any additional condition but simply integrated over. Both Eqs. (2.61a) and (2.61b) are examples of the foremost interpretation.

For a given wave vector $\vec{k}$ of the photon, the two polarization vectors $\hat{\epsilon}_\lambda(\vec{k})$ with $\lambda = 1,2$ fulfill the relations

$$\hat{\epsilon}_\lambda(\vec{k})\cdot\hat{\epsilon}_{\lambda'}(\vec{k}) = \delta_{\lambda\lambda'}\,, \qquad \vec{k}\cdot\hat{\epsilon}_\lambda(\vec{k}) = 0\,, \qquad \sum_{\lambda=1}^{2}\epsilon^i_\lambda(\vec{k})\,\epsilon^j_\lambda(\vec{k}) = \delta^{ij} - \frac{k^i\,k^j}{\vec{k}^2}\,. \tag{2.63}$$

Using spherical coordinates (r, θ and φ), we can express the (originally Cartesian) components of $\vec{k}$ as follows,

$$\vec{k} = |\vec{k}|\,(\hat{e}_x\,\sin\theta\,\cos\varphi + \hat{e}_y\,\sin\theta\,\sin\varphi + \hat{e}_z\,\cos\theta)\,. \tag{2.64}$$

The two polarization vectors may be chosen as

$$\hat{e}_1(\vec{k}) = \hat{e}_x\,\cos\theta\,\cos\varphi + \hat{e}_y\,\cos\theta\,\sin\varphi - \hat{e}_z\,\sin\theta\,, \tag{2.65a}$$

$$\hat{e}_2(\vec{k}) = -\,\hat{e}_x\,\sin\varphi + \hat{e}_y\,\cos\varphi\,, \tag{2.65b}$$

$$\vec{k}/|\vec{k}| = \hat{e}_1(\vec{k})\times\hat{e}_2(\vec{k})\,. \tag{2.65c}$$

The wave equation $\square \vec{A} = 0$, with the "quabla" operator $\square = \partial_t^2 - \vec{\nabla}^2$, corresponds to the equation $k^2 = k^\mu k_\mu = \omega^2 - \vec{k}^2 = 0$ in Fourier space (where we anticipate the Lorentz-covariant notation and the Einstein summation convention). Inverting (2.61), we can express the fundamental creation and annihilation operators as follows,

$$
\begin{aligned}
a_\lambda(\vec{k}) &= \int \mathrm{d}^3 r \, \frac{\mathrm{e}^{\mathrm{i}(\omega t - \vec{k}\cdot\vec{r})}}{\sqrt{2\omega(2\pi)^3}} \, \hat{\epsilon}_\lambda(\vec{k}) \cdot \left(\omega \vec{A}_\perp(t,\vec{r}) + \mathrm{i} \frac{\partial}{\partial t} \vec{A}_\perp(t,\vec{r}) \right) \\
&= \int \mathrm{d}^3 r \, \frac{\exp(\mathrm{i}\, k \cdot x)}{\sqrt{2\omega(2\pi)^3}} \, \hat{\epsilon}_\lambda(\vec{k}) \cdot \left(\omega \vec{A}_\perp(x) + \mathrm{i} \frac{\partial}{\partial t} \vec{A}_\perp(x) \right) ,
\end{aligned} \tag{2.66a}
$$

and

$$
\begin{aligned}
a^+_\lambda(\vec{k}) &= \int \mathrm{d}^3 r \, \frac{\mathrm{e}^{-\mathrm{i}(\omega t - \vec{k}\cdot\vec{r})}}{\sqrt{2\omega(2\pi)^3}} \, \hat{\epsilon}_\lambda(\vec{k}) \cdot \left(\omega \vec{A}_\perp(t,\vec{r}) - \mathrm{i} \frac{\partial}{\partial t} \vec{A}_\perp(t,\vec{r}) \right) \\
&= \int \mathrm{d}^3 r \, \frac{\exp(-\mathrm{i}\, k \cdot x)}{\sqrt{2\omega(2\pi)^3}} \, \hat{\epsilon}_\lambda(\vec{k}) \cdot \left(\omega \vec{A}_\perp(x) - \mathrm{i} \frac{\partial}{\partial t} \vec{A}_\perp(x) \right) .
\end{aligned} \tag{2.66b}
$$

Here, $x = (t, \vec{r})$ is the space-time coordinate. The derivation of Eqs. (2.66a) and (2.66b) is simplified if one observes that the operator $\mathrm{i}\partial_t$, when acting on the positive-frequency (annihilation) part of $\vec{A}_\perp(t,\vec{r})$, "pulls down" a factor ω from the exponential, whereas it "pulls down" a factor $-\omega$ from the negative-frequency part of $\vec{A}_\perp(t,\vec{r})$. Thus, the positive- and negative-frequency parts can be determined by forming appropriate linear combinations.

The definition of the operators $a_\lambda(\vec{k})$ and $a^+_\lambda(\vec{k})$ suggests that they should be time-independent. It is instructive to verify this fact for $a_\lambda(\vec{k})$ by an explicit calculation, because the representation on the right-hand side of Eq. (2.66a) could otherwise suggest a possible time dependence. From Eq. (2.66a), one infers that

$$
\begin{aligned}
\frac{\mathrm{d}}{\mathrm{d}t} a_\lambda(\vec{k}) &= \mathrm{i} \int \mathrm{d}^3 r \, \frac{\mathrm{e}^{\mathrm{i}(\omega t - \vec{k}\cdot\vec{r})}}{\sqrt{2\omega(2\pi)^3}} \, \hat{\epsilon}_\lambda(\vec{k}) \cdot \left(\omega^2 \vec{A}_\perp(t,\vec{r}) + \frac{\partial^2}{\partial t^2} \vec{A}_\perp(t,\vec{r}) \right) \\
&= \mathrm{i} \int \mathrm{d}^3 r \, \frac{\mathrm{e}^{\mathrm{i}(\omega t - \vec{k}\cdot\vec{r})}}{\sqrt{2\omega(2\pi)^3}} \, \hat{\epsilon}_\lambda(\vec{k}) \cdot \left(\omega^2 \vec{A}_\perp(t,\vec{r}) + \vec{\nabla}^2 \vec{A}_\perp(t,\vec{r}) \right) \\
&= \mathrm{i} \int \mathrm{d}^3 r \, \hat{\epsilon}_\lambda(\vec{k}) \cdot \left(\omega^2 \vec{A}_\perp(t,\vec{r}) + \vec{A}_\perp(t,\vec{r}) \overleftarrow{\nabla}^2 \right) \frac{\mathrm{e}^{\mathrm{i}(\omega t - \vec{k}\cdot\vec{r})}}{\sqrt{2\omega(2\pi)^3}} = 0 .
\end{aligned} \tag{2.67}
$$

In the transition from the first to the second line, we have used the wave equation for the free vector potential operator in the form $(\partial_t^2 - \vec{\nabla}^2)\, \vec{A}_\perp(t,\vec{r}) = 0$. In going from the second to the third line, we have integrated by parts (twice). Finally, we can replace $\vec{\nabla}^2 \to -\vec{k}^2 = -\omega^2$, verifying that the time derivative of $a_\lambda(\vec{k})$ vanishes.

We can now calculate the commutator $[a_\lambda(\vec{k}), a^+_{\lambda'}(\vec{k}')]$. In writing down the integrands, we exploit the time-independence and choose $t = t'$. Furthermore, we take notice of the fundamental commutators (2.52),

$$
\begin{aligned}
[a_\lambda(\vec{k}), a^+_{\lambda'}(\vec{k}')] &= \int \mathrm{d}^3r\,\mathrm{d}^3r'\,\frac{\mathrm{e}^{\mathrm{i}\,(\vec{k}'\cdot\vec{r}'-\vec{k}\cdot\vec{r})}}{2\,(2\pi)^3\,\sqrt{\omega\,\omega'}}\,\epsilon^i_\lambda(\vec{k})\,\epsilon^j_{\lambda'}(\vec{k}') \\
&\quad\times\left[\omega\,A^i_\perp(t,\vec{r}) + \mathrm{i}\,\frac{\partial}{\partial t}A^i_\perp(t,\vec{r})\,,\,\omega'\,A^j_\perp(t,\vec{r}\,') - \mathrm{i}\,\frac{\partial}{\partial t}A^j_\perp(t,\vec{r}\,')\right] \\
&= \int \mathrm{d}^3r\,\mathrm{d}^3r'\,\frac{\mathrm{e}^{\mathrm{i}\,(\vec{k}'\cdot\vec{r}'-\vec{k}\cdot\vec{r})}}{2\,(2\pi)^3\,\sqrt{\omega\,\omega'}}\,\epsilon^i_\lambda(\vec{k})\,\epsilon^j_{\lambda'}(\vec{k}') \\
&\quad\times(-\mathrm{i})\,\left\{\omega\left[E^j_\perp(t,\vec{r}\,'), A^i_\perp(t,\vec{r})\right] + \omega'\left[E^i_\perp(t,\vec{r}), A^j_\perp(t,\vec{r}\,')\right]\right\}\,. \qquad (2.68)
\end{aligned}
$$

Using the field commutators from Eq. (2.60), we can reformulate this as

$$
\begin{aligned}
[a_\lambda(\vec{k}), a^+_{\lambda'}(\vec{k}')] &= \int \mathrm{d}^3r\,\mathrm{d}^3r'\,\frac{\mathrm{e}^{\mathrm{i}\,(\vec{k}'\cdot\vec{r}'-\vec{k}\cdot\vec{r})}}{2\,(2\pi)^3\,\sqrt{\omega\,\omega'}}\,\epsilon^i_\lambda(\vec{k})\,\epsilon^j_{\lambda'}(\vec{k}')\;(\omega+\omega')\;\delta^{\perp,ij}(\vec{r}-\vec{r}\,') \\
&= \int \mathrm{d}^3r\,\mathrm{d}^3r'\,\frac{\mathrm{e}^{\mathrm{i}\,(\vec{k}'\cdot\vec{r}'-\vec{k}\cdot\vec{r})}}{2\,(2\pi)^3\,\sqrt{\omega\,\omega'}}\,\epsilon^i_\lambda(\vec{k})\,\epsilon^j_{\lambda'}(\vec{k}')\;(\omega+\omega') \\
&\quad\times\left(\delta^{ij}\,\delta^{(3)}(\vec{r}-\vec{r}\,') - \frac{\nabla^i\,\nabla^j}{\vec{\nabla}^2}\delta^{(3)}(\vec{r}-\vec{r}\,')\right) \\
&= \int \frac{\mathrm{d}^3r}{(2\pi)^3}\,\mathrm{e}^{\mathrm{i}\,(\vec{k}'-\vec{k})\cdot\vec{r}}\,\left(\vec{\epsilon}_\lambda(\vec{k})\cdot\vec{\epsilon}_{\lambda'}(\vec{k}')\right)\,\frac{\omega+\omega'}{2\sqrt{\omega\,\omega'}} = \delta^{(3)}(\vec{k}-\vec{k}')\,\delta_{\lambda\lambda'}\,.
\end{aligned}
\tag{2.69}
$$

We observe that $(\omega+\omega')\,/(2\sqrt{\omega\,\omega'}) \to 1$ for $\omega \to \omega'$ (or $\vec{k} = \vec{k}'$). In the transformations made in Eq. (2.69), we have integrated by parts with respect to both $\vec{r}$ and $\vec{r}\,'$; the orthogonality $\vec{k}\cdot\vec{\epsilon}_\lambda(\vec{k})$ also has been used. The final result for the commutator of the fundamental creation and annihilation operators is

$$
[a_\lambda(\vec{k}), a^+_{\lambda'}(\vec{k}')] = \delta^{(3)}(\vec{k}-\vec{k}')\,\delta_{\lambda\lambda'}\,. \tag{2.70}
$$

We note that its precise form relies on our choice of weight factors under the integral sign in Eq. (2.61). The functional form (2.70) reminds us of the analogy of the electromagnetic oscillators with harmonic oscillator modes of the vacuum.

Let us summarize once more the results of the quantization program in Coulomb gauge. We only take into account the two transverse polarization states of the electromagnetic field, not the longitudinal and scalar photons. The scalar potential will not take part in the quantization because we are unable to find a corresponding canonical momentum. This raises the question of how to deal with the Coulomb interaction in the quantized formalism. The answer is as follows: From the quantized formulas in the Coulomb gauge, we obtain the photon propagator (in the Coulomb gauge) after the addition of the instantaneous Coulomb interaction (see Chap. 6 of Ref. [111] and Sec. 9.3 in Chap. 9 here). The propagator looks different from the one obtained in, say, the Lorenz gauge. However, one can show that the difference does not play a role in the calculation of fundamental processes. Indeed, the photon propagator obtained in Coulomb gauge, as a sum of the instantaneous and retarded interactions, is related by a gauge transformation to the photon propagator obtained in a covariant gauge (e.g., the Lorenz gauge). The difference in the calculation of fundamental processes vanishes after current conservation has been taken into account (see Chap. 9).

Yet another issue, which will be dealt with in greater detail in Chap. 9, concerns the fact that the fully quantized electromagnetic field in Lorenz gauge actually contains two unphysical virtual degrees of freedom (scalar and longitudinal photons). These must later be excluded from the physical spectrum via the GUPTA[6]–BLEULER[7] condition. They are still present in the propagator obtained in Lorenz gauge. Still, we can use both propagators, obtained in Coulomb and in Lorenz gauge, and calculate physical processes. The advantage of the Coulomb gauge is that for low-energy processes, we have two photon polarizations less to sum over, whereas the advantage of the Lorenz gauge is that it yields a manifestly covariant form of the propagator. Also, in Coulomb gauge, it is not necessary to introduce a Gupta–Bleuler condition: The two degrees of freedom taking part in the quantization are the physical ones; the scalar potential is not quantized and the scalar and longitudinal photons never appear.

2.4.2 *Field Operators and Quantized Hamiltonian*

In classical mechanics, momentum is just a number that can be measured to essentially arbitrary accuracy. In quantum mechanics, momentum is a (Hermitian) operator. In the same light, an electric or magnetic field is a number in classical field theory whereas in quantum field theory, it is given by a Hermitian operator. It is instructive to write down the operators which actually correspond to the electric and magnetic fields; these are Hermitian. The expectation value of these operators in a given state of the photon field is a real quantity while the operators themselves may contain manifestly complex terms. Before discussing the electric and magnetic field operators, we first recall the vector potential operator (2.61)

$$\vec{A}_\perp(t,\vec{r}) = \sum_{\lambda=1}^{2} \int \frac{\mathrm{d}^3k}{\sqrt{2\omega\,(2\pi)^3}}\, \hat{\epsilon}_\lambda(\vec{k})\, \left(a_\lambda(\vec{k})\,\mathrm{e}^{-\mathrm{i}\,k\cdot x} + a^+_\lambda(\vec{k})\,\mathrm{e}^{\mathrm{i}\,k\cdot x}\right). \tag{2.71}$$

In Coulomb gauge, the electric field operator $\vec{E}_\perp = \vec{E}^+_\perp = -\partial\vec{A}_\perp/\partial t$ is simply obtained as the time derivative of the vector potential operator and reads as follows,

$$\vec{E}_\perp(t,\vec{r}) = \sum_{\lambda=1}^{2} \int \frac{\mathrm{d}^3k}{\sqrt{2\omega\,(2\pi)^3}}\, \hat{\epsilon}_\lambda(\vec{k})\,\mathrm{i}\,\omega\, \left(a_\lambda(\vec{k})\,\mathrm{e}^{-\mathrm{i}\,k\cdot x} - a^+_\lambda(\vec{k})\,\mathrm{e}^{\mathrm{i}\,k\cdot x}\right). \tag{2.72}$$

---

[6]Suraj North Gupta (b. 1924).
[7]Konrad Bleuler (1912–1992).

The magnetic field operator is obtained as the curl $\vec{B} = \vec{B}^{+} = \vec{\nabla} \times \vec{A}_{\perp}$ of the vector potential,

$$\vec{B}(t,\vec{r}) = \sum_{\lambda=1}^{2} \int \frac{\mathrm{d}^3 k}{\sqrt{2\omega\,(2\pi)^3}}\, \mathrm{i}\left(\vec{k} \times \hat{\epsilon}_\lambda(\vec{k})\right)\left(a_\lambda(\vec{k})\,\mathrm{e}^{-\mathrm{i}\,k\cdot x} - a^{+}_\lambda(\vec{k})\,\mathrm{e}^{\mathrm{i}\,k\cdot x}\right). \tag{2.73}$$

Both $\vec{E}(t,\vec{r})$ and $\vec{B}(t,\vec{r})$ are given by Hermitian operators and contain positive-frequency terms (proportional to $\mathrm{e}^{-\mathrm{i}k\cdot x}$) as well as negative-frequency terms (proportional to $\mathrm{e}^{\mathrm{i}k\cdot x}$). The positive-frequency as well as negative-frequency parts alone could not act as field operators because these would not correspond to Hermitian operators.

Starting from the postulated fundamental commutator relation given in Eq. (2.60), and using the inversion formulas (2.66), we had derived, in Eq. (2.70), the result $[a_\lambda(\vec{k}), a^{+}_{\lambda'}(\vec{k}')] = \delta^{(3)}(\vec{k} - \vec{k}')\,\delta_{\lambda\lambda'}$. It is instructive to verify once more that this result is consistent with the fundamental commutator of the electric field of the vector potential operator at equal times $t = t'$,

$$\begin{aligned}
\left[E^i_\perp(t,\vec{r}), A^j_\perp(t,\vec{r}')\right] &= \sum_{\lambda=1}^{2} \int \frac{\mathrm{d}^3 k}{\sqrt{2\omega\,(2\pi)^3}} \sum_{\lambda'=1}^{2} \int \frac{\mathrm{d}^3 k'}{\sqrt{2\omega'\,(2\pi)^3}}\, \epsilon^i_\lambda(\vec{k})\,\epsilon^j_{\lambda'}(\vec{k}') \\
&\quad \times \Big\{\mathrm{i}\omega\left[a_\lambda(\vec{k}), a^{+}_{\lambda'}(\vec{k}')\right]\mathrm{e}^{-\mathrm{i}(\omega-\omega')t}\mathrm{e}^{\mathrm{i}\vec{k}\cdot\vec{r}-\mathrm{i}\,\vec{k}'\cdot\vec{r}'} \\
&\quad - \mathrm{i}\omega\left[a^{+}_\lambda(\vec{k}), a_{\lambda'}(\vec{k}')\right]\mathrm{e}^{\mathrm{i}(\omega-\omega')t}\mathrm{e}^{-\mathrm{i}\vec{k}\cdot\vec{r}+\mathrm{i}\vec{k}'\cdot\vec{r}'}\Big\} \\
&= \frac{\mathrm{i}}{2} \sum_{\lambda=1}^{2} \int \frac{\mathrm{d}^3 k}{(2\pi)^3}\, \epsilon^i_\lambda(\vec{k})\,\epsilon^j_\lambda(\vec{k})\left(\mathrm{e}^{\mathrm{i}\,\vec{k}\cdot(\vec{r}-\vec{r}')} + \mathrm{e}^{-\mathrm{i}\,\vec{k}\cdot(\vec{r}-\vec{r}')}\right) \\
&= \mathrm{i}\,\delta^{\perp,ij}(\vec{r} - \vec{r}')\,.
\end{aligned} \tag{2.74}$$

In the derivation, we have used Eq. (2.54). By virtue of the Heisenberg uncertainty relation, the nonvanishing fundamental commutators of the vector potential and its conjugate momentum imply that the electric field and the vector potential cannot be measured simultaneously with arbitrary accuracy. This statement, though, does not really matter because the gauge-dependent vector potential does not represent a physically measurable quantity. Therefore, it is of prime interest to investigate whether electric and magnetic fields can be measured simultaneously in the quantum world. Indeed, the commutator of the electric and magnetic fields reads as

$$\begin{aligned}
\left[E^i_\perp(t,\vec{r}),\, B^j(t,\vec{r}')\right] &= \sum_{\lambda=1}^{2} \int \frac{\mathrm{d}^3 k}{\sqrt{2\omega\,(2\pi)^3}} \sum_{\lambda'=1}^{2} \int \frac{\mathrm{d}^3 k'}{\sqrt{2\omega'\,(2\pi)^3}}\, \epsilon^i_\lambda(\vec{k})\left(\vec{k}' \times \hat{\epsilon}_{\lambda'}(\vec{k}')\right)^j \\
&\quad \times \left\{\mathrm{i}^2\,\omega\left[a_\lambda(\vec{k}),\, a^{+}_{\lambda'}(\vec{k}')\right]\mathrm{e}^{-\mathrm{i}\,(\omega-\omega')t}\,\mathrm{e}^{\mathrm{i}\,\vec{k}\cdot\vec{r}-\mathrm{i}\,\vec{k}'\cdot\vec{r}'} + \text{h.c.}\right\},
\end{aligned} \tag{2.75}$$

where "h.c." stands for the Hermitian conjugate. It can be simplified to read

$$
\begin{aligned}
\left[E^i_\perp(t,\vec{r}), B^j(t,\vec{r}\,')\right] &= \frac{1}{2}\sum_{\lambda=1}^{2}\int \frac{\mathrm{d}^3k}{(2\pi)^3}\epsilon^i_\lambda(\vec{k})\left(\vec{k}\times\hat{\epsilon}_\lambda(\vec{k})\right)^j \mathrm{e}^{\mathrm{i}\,\vec{k}\cdot(\vec{r}-\vec{r}\,')} + \text{h.c.} \\
&= \frac{1}{2}\sum_{\lambda=1}^{2}\int \frac{\mathrm{d}^3k}{(2\pi)^3}\,\epsilon^i_\lambda(\vec{k})\,\epsilon^{j\ell m}\,k^\ell\,\epsilon^m_\lambda(\vec{k})\,\mathrm{e}^{\mathrm{i}\,\vec{k}\cdot(\vec{r}-\vec{r}\,')} + \text{h.c.} \\
&= \frac{1}{2}\int \frac{\mathrm{d}^3k}{(2\pi)^3}\,\epsilon^{j\ell m}\,k^\ell\left(\delta^{im} - \frac{k^i\,k^m}{\vec{k}^2}\right)\mathrm{e}^{\mathrm{i}\,\vec{k}\cdot(\vec{r}-\vec{r}\,')} + \text{h.c.} \\
&= -\mathrm{i}\,\epsilon^{ij\ell}\,\nabla^\ell\delta^{(3)}(\vec{r}-\vec{r}\,') = -\mathrm{i}\,\epsilon^{ij\ell}\,\frac{\partial}{\partial x^\ell}\delta^{(3)}(\vec{r}-\vec{r}\,')\,.
\end{aligned}
\tag{2.76}
$$

This is manifestly different from zero. The presence of the Levi-Cività[8] tensor ϵ^{ijk} implies that, e.g., the x component of the electric field and the y component of the magnetic field cannot be measured simultaneously to arbitrary precision at the same time t.

Let us now shift our attention to the Hamiltonian operator H_F of the free electromagnetic field. It is time-independent in the absence of source terms. We evaluate it at time $t = 0$ and take into account that the electric and magnetic field operators represent Hermitian operators ($\vec{E}^+_\perp = \vec{E}$, $\vec{B}^+_\perp = \vec{B}$),

$$
\begin{aligned}
H_F =& \frac{1}{2}\int \mathrm{d}^3r\,\left(\vec{E}^2_\perp(t=0,\vec{r}) + \vec{B}^2(t=0,\vec{r})\right) \qquad (2.77)\\
=& -\frac{1}{2}\sum_{\lambda=1}^{2}\int\frac{\mathrm{d}^3k}{\sqrt{2\omega\,(2\pi)^3}}\sum_{\lambda'=1}^{2}\int\frac{\mathrm{d}^3k'}{\sqrt{2\omega'\,(2\pi)^3}} \\
&\times\left\{\omega\,\omega'\,\hat{\epsilon}_\lambda(\vec{k})\cdot\hat{\epsilon}_{\lambda'}(\vec{k}') + \left(\vec{k}\times\hat{\epsilon}_\lambda(\vec{k})\right)\cdot\left(\vec{k}'\times\hat{\epsilon}_{\lambda'}(\vec{k}')\right)\right\} \\
&\times\int\mathrm{d}^3r\,\left(a_\lambda(\vec{k})\,\mathrm{e}^{\mathrm{i}\,\vec{k}\cdot\vec{r}} - a^+_\lambda(\vec{k})\,\mathrm{e}^{-\mathrm{i}\,\vec{k}\cdot\vec{r}}\right)\left(a_{\lambda'}(\vec{k}')\,\mathrm{e}^{\mathrm{i}\,\vec{k}'\cdot\vec{r}} - a^+_{\lambda'}(\vec{k}')\,\mathrm{e}^{-\mathrm{i}\,\vec{k}'\cdot\vec{r}}\right).
\end{aligned}
$$

The integral over d^3r leads to

$$
\begin{aligned}
&\int\mathrm{d}^3r\,\left(a_\lambda(\vec{k})\,\mathrm{e}^{\mathrm{i}\,\vec{k}\cdot\vec{r}} - a^+_\lambda(\vec{k})\,\mathrm{e}^{-\mathrm{i}\,\vec{k}\cdot\vec{r}}\right)\left(a_{\lambda'}(\vec{k}')\,\mathrm{e}^{\mathrm{i}\,\vec{k}'\cdot\vec{r}} - a^+_{\lambda'}(\vec{k}')\,\mathrm{e}^{-\mathrm{i}\,\vec{k}'\cdot\vec{r}}\right) \\
&\quad= (2\pi)^3\left[\left(a_\lambda(\vec{k})\,a_{\lambda'}(-\vec{k}) + a^+_\lambda(\vec{k})\,a^+_{\lambda'}(-\vec{k})\right)\delta^{(3)}(\vec{k}+\vec{k}')\right. \\
&\qquad\left.-\left(a^+_\lambda(\vec{k})\,a_{\lambda'}(\vec{k}) + a_\lambda(\vec{k})\,a^+_{\lambda'}(\vec{k})\right)\delta^{(3)}(\vec{k}-\vec{k}')\right].
\end{aligned}
\tag{2.78}
$$

The Dirac-δ's lead to the following vector structures in Eq. (2.77),

$$
\left(\vec{k}\times\hat{\epsilon}_\lambda(\vec{k})\right)\cdot\left(\vec{k}\times\hat{\epsilon}_{\lambda'}(\vec{k})\right) = \vec{k}^2\,\delta_{\lambda\lambda'}\,, \tag{2.79a}
$$

$$
\left(\vec{k}\times\hat{\epsilon}_\lambda(\vec{k})\right)\cdot\left(-\vec{k}\times\hat{\epsilon}_{\lambda'}(-\vec{k})\right) = -\vec{k}^2\,\hat{\epsilon}_\lambda(\vec{k})\cdot\hat{\epsilon}_{\lambda'}(-\vec{k})\,. \tag{2.79b}
$$

We now insert Eqs. (2.78), (2.79a), and (2.79b) into Eq. (2.77); this gives rise to a term in the integrand proportional to $\omega^2 - \vec{k}^2 = 0$, which vanishes. The surviving

[8] Tullio Levi-Cività (1873–1941).

term is the Hamilton operator of the free electromagnetic field,

$$\begin{aligned} H_F &= \frac{1}{2} \sum_{\lambda=1}^{2} \sum_{\lambda'=1}^{2} \int \frac{\mathrm{d}^3 k}{2\omega} \left(\omega^2 + \vec{k}^2\right) \delta_{\lambda\lambda'} \left(a_\lambda^+(\vec{k})\, a_{\lambda'}(\vec{k}) + a_\lambda(\vec{k})\, a_{\lambda'}^+(\vec{k})\right) \\ &= \frac{1}{2} \sum_{\lambda=1}^{2} \int \mathrm{d}^3 k\, \omega \left(a_\lambda^+(\vec{k})\, a_\lambda(\vec{k}) + a_\lambda(\vec{k})\, a_{\lambda'}^+(\vec{k})\right) \\ &= \sum_{\lambda=1}^{2} \int \mathrm{d}^3 k\, \omega \left(a_\lambda^+(\vec{k})\, a_\lambda(\vec{k}) + \frac{1}{2} \delta^{(3)}(\vec{0})\right). \end{aligned} \tag{2.80}$$

The term $\delta^{(3)}(\vec{0})$ is the consequence of the relation $[a_\lambda(\vec{k}), a_\lambda^+(\vec{k})] = \delta^{(3)}(\vec{k} - \vec{k})$. For $\vec{k} = \vec{k}'$, the Dirac-δ is evaluated at the origin and the expression $\delta^{(3)}(\vec{0})$ is obtained. *A priori*, the expression $\delta^{(3)}(\vec{0})$ is undefined and has to be given a physical interpretation. Let us consider the transition from a continuous photon spectrum to a discrete spectrum with the latter being normalized to a quantization volume V. In a finite normalization volume, we evaluate a discrete sum over the available photon wave vectors $\vec{k}$; this sum needs to be matched against the integral $\mathrm{d}^3 k$ over the wave vectors k in the volume V. This integral, in turn, gives rise to a Dirac-δ term $\delta^{(3)}(\vec{0})$ in the continuum limit. Let us indicate the matching relation [112],

$$\sum_{\vec{k}} \to \frac{V}{(2\pi)^3} \int \mathrm{d}^3 k \to \delta^{(3)}(\vec{0}) \int \mathrm{d}^3 k\,, \tag{2.81}$$

and discuss its justification. The summation over $\vec{k}$ on the left is to be understood as a multi-index vector $\vec{n}$ composed of integers, which are related to the components of $\vec{k}$ by $\vec{k} = 2\pi\, \vec{n}/L$. The first replacement "$\to$" simply follows from counting the available modes satisfying periodic boundary conditions in a cubic volume $V = L^3$, which results in one mode per fundamental volume $(2\pi/L)$ in k-space. This is discussed at various places in the literature; one particular source is Eq. (1.1.39) of Ref. [113]. The second replacement ("$\to$") in Eq. (2.81) can be justified as follows: The Dirac-δ function in the wave-vector space is

$$\delta^{(3)}(\vec{k}) = \int_{\mathbb{R}^3} \frac{\mathrm{d}^3 r}{(2\pi)^3} \exp(\mathrm{i}\,\vec{k} \cdot \vec{r})\,. \tag{2.82}$$

Here, the factor $(2\pi)^{-3}$ is due to the definition of the three-dimensional Dirac-δ function. If we restrict the $\mathrm{d}^3 r$-integration to a finite volume V, then the Dirac-δ of vanishing argument receives a natural interpretation as

$$\delta^{(3)}(\vec{k} = \vec{0}) \approx \int_V \frac{\mathrm{d}^3 r}{(2\pi)^3} \exp(\mathrm{i}\,\vec{0} \cdot \vec{r}) = \frac{V}{(2\pi)^3}\,. \tag{2.83}$$

Applying Eq. (2.83) backwards, the term proportional to $\delta^{(3)}(\vec{0})$ in Eq. (2.80) can be identified as a sum over the zero-point energies $\frac{1}{2}\omega$ of the "oscillator" modes,

$$H_0 = \frac{1}{2} \sum_{\vec{k}} \sum_{\lambda=1}^{2} \omega = \frac{1}{2} \sum_{\vec{k}} \sum_{\lambda=1}^{2} |\vec{k}|\,. \tag{2.84}$$

The zero-point Hamiltonian operator is instrumental in the treatment of the Casimir interaction between plates (see Chap. 8 of Ref. [81]). The zero-point energy is relevant for configurations with nontrivial boundary conditions. One cannot always simply shift the reference point of the energy scale and ignore this term. By contrast, in situations with an undistorted vacuum structure, it is indeed permissible to leave out the zero-point energy from the Hamiltonian of the electromagnetic field.

In the absence of these subtle effects caused by external conditions, one can indeed subtract the vacuum-expectation value of the Hamiltonian in Eq. (2.80) (which is proportional to the Dirac-δ of vanishing argument) and rewrite the Hamiltonian as the normal-ordered operator

$$:H_F: = H_F - \langle 0|H_F|0\rangle = \sum_{\lambda=1}^{2} \int \mathrm{d}^3k\, \omega\, a_\lambda^+(\vec{k})\, a_\lambda(\vec{k})\,. \tag{2.85}$$

The vacuum expectation value $\langle 0|H_F|0\rangle$ of H_F has been explicitly subtracted. The operator corresponding to the field momentum (Poynting vector operator) is given by

$$\vec{S} = \int \mathrm{d}^3r\, (\vec{E}\times\vec{B}) = \sum_{\lambda=1}^{2} \int \mathrm{d}^3k\, \vec{k}\, a_\lambda^+(\vec{k})\, a_\lambda(\vec{k})\,. \tag{2.86}$$

The norm of the Fock-space vector $|1_{\vec{k}\lambda}\rangle = a_\lambda^+(\vec{k})|0\rangle$, which describes a state of the electromagnetic field with one photon in the mode $\vec{k}\lambda$, is finally calculated as follows,

$$\|\, |1_{\vec{k}\lambda}\rangle\, \|^2 = \langle 0|a_\lambda(\vec{k})\, a_\lambda^+(\vec{k})|0\rangle = \delta^{(3)}(\vec{k}-\vec{k})\, \langle 0|0\rangle = \delta^{(3)}(\vec{0})\,. \tag{2.87}$$

A normalized state vector is therefore $|1_{\vec{k}\lambda}\rangle/\sqrt{\delta^{(3)}(\vec{0})}$, which is of course ill-defined, but the singular Dirac-δ term again finds a natural explanation within a calculation that uses a finite normalization volume, as outlined in Eq. (2.83). The energy expectation value of a one-photon state $|1_{\vec{k}\lambda}\rangle$, calculated with the free Hamiltonian (2.85), reads as follows,

$$\begin{aligned}\langle 1_{\vec{k}\lambda}|\, :H_F:\, |1_{\vec{k}\lambda}\rangle &= \sum_{\lambda'=1}^{2} \int \mathrm{d}^3k'\, \omega\, \langle 0|\left[a_\lambda(\vec{k}),\, a_{\lambda'}^+(\vec{k}')\right]\left[a_{\lambda'}(\vec{k}'),\, a_\lambda^+(\vec{k})\right]|0\rangle \\ &= \omega\, \delta^{(3)}(\vec{0}) = \omega\, \|\, |1_{\vec{k}\lambda}\rangle\, \|^2\,. \end{aligned} \tag{2.88}$$

Again, the somewhat intriguing formally infinite norm of one-photon states disappears as we employ a finite normalization volume V.

2.4.3 *Discretized Formulation of the Field Operators*

We shall now discuss the transition to the discretized representation with an explicit normalization volume V for the electromagnetic field modes. In Eq. (2.81), we already have encountered terms like $\delta^{(3)}(\vec{0})$, which can only be interpreted consistently if one transforms the expressions to a finite normalization volume. We will therefore treat alternative, discretized representations of the field operators in the current chapter and start with the vector potential operator, whose form is given by

$$\vec{A}_\perp(t,\vec{r}) = \sum_{\vec{k}\lambda} \frac{1}{\sqrt{V}} \frac{1}{\sqrt{2\omega}} \hat{\epsilon}_\lambda(\vec{k}) \left(\mathrm{a}_\lambda(\vec{k})\,\mathrm{e}^{-\mathrm{i}\,k\cdot x} + \mathrm{a}^+_\lambda(\vec{k})\,\mathrm{e}^{\mathrm{i}\,k\cdot x}\right), \tag{2.89}$$

where again $\omega = k^0 = |\vec{k}|$. The normalization of the discretized annihilation and creation operators, $\mathrm{a}_\lambda(\vec{k})$ and $\mathrm{a}^+_\lambda(\vec{k})$ remains to be discussed. We observe the different font used for the discrete-space and continuum-space creation and annihilation operators. The summation over $\vec{k}$ is to be understood as a multi-index governing the summations over the allowed quantum numbers n_1, n_2, and n_3, which are related to the components of $\vec{k}$ as in $\vec{k} = 2\pi\,\vec{n}/L$. Applying the transformation given in Eq. (2.81), $\int \frac{\mathrm{d}^3k}{(2\pi)^3} \to \frac{1}{V}\sum_{\vec{k}}$, the vector potential operator transforms into

$$\vec{A}_\perp(t,\vec{r}) \to \sqrt{\frac{V}{(2\pi)^3}} \sum_\lambda \int \frac{\mathrm{d}^3k}{\sqrt{(2\pi)^3}} \frac{1}{\sqrt{2\omega}} \hat{\epsilon}_\lambda(\vec{k}) \left(\mathrm{a}_\lambda(\vec{k})\,\mathrm{e}^{-\mathrm{i}\,k\cdot x} + \mathrm{a}^+_\lambda(\vec{k})\,\mathrm{e}^{\mathrm{i}\,k\cdot x}\right). \tag{2.90}$$

The normalization volume V is naturally absorbed into the discretized annihilation and creation operators $\mathrm{a}_\lambda(\vec{k})$ and $\mathrm{a}^+_\lambda(\vec{k})$. We recall the commutation relation of the fundamental creation and annihilation operators in continuum space (2.70),

$$[a_\lambda(\vec{k}), a^+_{\lambda'}(\vec{k}')] = \delta^{(3)}(\vec{k}-\vec{k}')\,\delta_{\lambda\lambda'}\,. \tag{2.91}$$

The transition to the discretized representation preserves physical units if we identify the peak of the Dirac-δ function with a large normalization volume, inspired by Eq. (2.83),

$$\delta^{(3)}(\vec{k}-\vec{k}') \to \sqrt{\frac{V}{(2\pi)^3}}\,\delta_{\vec{k}\vec{k}'} \qquad a_\lambda(\vec{k}) \to \sqrt{\frac{V}{(2\pi)^3}}\,\mathrm{a}_\lambda(\vec{k})\,. \tag{2.92}$$

In this case, the discrete-space commutators read as follows,

$$[\mathrm{a}_\lambda(\vec{k}),\, \mathrm{a}^+_{\lambda'}(\vec{k}')] = \delta_{\vec{k}\vec{k}'}\,\delta_{\lambda\lambda'}\,. \tag{2.93}$$

Again, we observe the font in the expression $\mathrm{a}_\lambda(\vec{k})$ which is different from the one used in the continuum-space operator $a_\lambda(\vec{k})$. We now reassure ourselves of the consistency of the commutator relation (2.91) with the transition (2.92). On one hand, we have the following relation, using Eq. (2.91),

$$\int \mathrm{d}^3k\,[a_\lambda(\vec{k}), a^+_\lambda(\vec{k}')] = \int \mathrm{d}^3k\,\delta(\vec{k}-\vec{k}') = 1\,, \tag{2.94}$$

while on the other hand, the transition to discrete space is mediated by Eq. (2.92),

$$\int \mathrm{d}^3k\,[a_\lambda(\vec{k}), a^+_\lambda(\vec{k}')] \to \left(\sum_{\vec{k}} \frac{(2\pi)^3}{V}\right)\left(\sqrt{\frac{V}{(2\pi)^3}}\right)^2 [\mathrm{a}_\lambda(\vec{k}), \mathrm{a}^+_\lambda(\vec{k}')] = \sum_{\vec{k}} \delta_{\vec{k}\vec{k}'} = 1\,. \tag{2.95}$$

The discrete-space version of the normal-ordered field Hamiltonian (2.85) reads as

$$:H_F: \; = \sum_{\vec{k}\lambda} \omega\, \mathrm{a}^+_\lambda(\vec{k})\,\mathrm{a}_\lambda(\vec{k})\,. \tag{2.96}$$

The discrete-space one-photon state $|1^{(d)}_{\vec{k}\lambda}\rangle = \mathrm{a}^+_\lambda(\vec{k})|0\rangle$, in contrast to Eq. (2.87), is normalized to unity, as an elementary calculation involving Eq. (2.92) shows. Its expectation value on the discrete Hamiltonian (2.96) is

$$\begin{aligned}\langle 1^{(d)}_{\vec{k}\lambda}|H_F|1^{(d)}_{\vec{k}\lambda}\rangle &= \sum_{\vec{k}'\lambda'} \omega\, \langle 0|[\mathrm{a}_{\lambda'}(\vec{k}'), \mathrm{a}^+_\lambda(\vec{k})]\,[\mathrm{a}_\lambda(\vec{k}), \mathrm{a}^+_{\lambda'}(\vec{k}')]|0\rangle \\ &= \sum_{\vec{k}'\lambda'} \omega\, \delta^2_{\vec{k}\vec{k}'}\, \delta^2_{\lambda\lambda'} = \omega\,. \end{aligned} \tag{2.97}$$

The discretized electric-field operator is obtained by differentiation of the vector potential and reads as,

$$\vec{E}_\perp(t,\vec{r}) = \sum_{\vec{k}\lambda} \frac{\mathrm{i}}{\sqrt{V}} \sqrt{\frac{\omega}{2}}\, \hat{\epsilon}_\lambda(\vec{k}) \left(\mathrm{a}_\lambda(\vec{k})\, \mathrm{e}^{-\mathrm{i}\,k\cdot x} - \mathrm{a}^+_\lambda(\vec{k})\, \mathrm{e}^{\mathrm{i}\,k\cdot x}\right), \tag{2.98}$$

whereas the magnetic field operator $\vec{B}(t,\vec{r}) = \vec{\nabla} \times \vec{A}_\perp(t,\vec{r})$ can be given as follows,

$$\vec{B}(t,\vec{r}) = \sum_{\vec{k}\lambda} \frac{\mathrm{i}}{\sqrt{V}} \frac{1}{\sqrt{2\omega}} (\vec{k} \times \hat{\epsilon}_\lambda(\vec{k})) \left(\mathrm{a}_\lambda(\vec{k})\, \mathrm{e}^{-\mathrm{i}\,k\cdot x} - \mathrm{a}^+_\lambda(\vec{k})\, \mathrm{e}^{\mathrm{i}\,k\cdot x}\right). \tag{2.99}$$

The discrete-space version of the commutator of the electric field and the vector potential is a generalization of Eq. (2.75),

$$\left[E^i_\perp(t,\vec{r}), A^j_\perp(t,\vec{r}\,')\right] = \sum_{\vec{k}\lambda} \frac{\mathrm{i}}{V} \left(\delta^{ij} - \frac{k^i\, k^j}{\vec{k}^2}\right) \mathrm{e}^{\mathrm{i}\,\vec{k}\cdot(\vec{r}-\vec{r}\,')}\,. \tag{2.100}$$

For $V \to \infty$, the latter expression approaches the transverse Dirac-δ function.

Finally, we should remember that the discrete as well as the continuum representations of the field operators have important applications. For the derivation of the Lamb shift, the continuum representation is preferable because we have to sum over a continuum of modes. For the ac Stark shift, to be discussed in Sec. 3.3.2, we need the discretized representation because only the laser mode is populated by a large number of photons. The normalization to a finite volume V helps in matching the result with the intensity of the incoming laser radiation. For the treatment of the laser-dressed Lamb shift [114], one needs both forms because both a laser field and the interaction with the continuum of modes of the radiation field need to be described simultaneously.

2.5 Interaction Picture and Phase Conventions

2.5.1 *Field Operators in the Schrödinger and Interaction Pictures*

The last part of the current chapter will be focused on two subtle points, which are normally not treated in detail in quantum field theory books, but which are very useful to know. One of these is connected with the transition from the Schrödinger picture to the interaction picture, and another one with certain normalizations and global phases of the field operators.

It is known that quantum mechanics can be formulated in the Schrödinger and in the interaction picture. Likewise, it is possible to formulate second-quantized operators in both pictures. The operators correspond to the quantum mechanical observables, and the field Hamiltonian corresponds to the quantum mechanical Hamilton operator. In general, a Hamiltonian generates time translations, whereas the momentum operator generates spatial translations. Let us consider the expression $\exp(\mathrm{i}Ht)\,A\,\exp(-\mathrm{i}Ht)$, where A is an operator and H is the Hamiltonian. If A commutes with H, then the expectation value of A in a particular quantum state is time-independent. If a complete set of commuting observables (including the Hamiltonian) is found, then we can identify completely specified eigenvectors.

In the Heisenberg picture of quantum mechanics, the states are time-independent, whereas the operators evolve with time. By contrast, in the Schrödinger picture, the states evolve, whereas the operators are time-independent. The interaction picture interpolates between these viewpoints: The problem is reformulated so that the quantum states evolve only with the interaction Hamiltonian, as opposed to the complete Hamiltonian. It is useful to observe that the entire canonical formalism of quantum field theory (scattering theory, see Ref. [2]), actually uses the interaction picture without stating this fact explicitly. However, a complementary formulation of field operators in the Schrödinger picture is sometimes useful, especially when treating nonrelativistic problems.

Let the time evolution of a system be governed by an unperturbed Hamiltonian H_0 which, by assumption, carries no explicit time dependence. The operator A, by assumption, carries no explicit time dependence either. Under the action of H_0, a matrix element of the operator A acquires a time dependence according to

$$[\langle a|A|b\rangle]\,(t) = \langle a|\exp(\mathrm{i}\,H_0 t)\,A\,\exp(-\mathrm{i}\,H_0 t)|b\rangle\,. \tag{2.101}$$

The time evolution is a unitary transformation. In the Schrödinger picture, one attributes the time evolution $\exp(-\mathrm{i}\,H_0 t)$ to the wave function. So, in the Schrödinger picture, the unperturbed time evolution of the state vector is given by

$$\mathrm{i}\frac{\mathrm{d}}{\mathrm{d}t}|b(t)\rangle = H_0|b(t)\rangle\,, \qquad |b(t)\rangle = \mathrm{e}^{-\mathrm{i}\,H_0\,t}|b\rangle\,. \tag{2.102}$$

In the interaction picture, in contrast to the wave function, the operator A acquires a time dependence according to

$$\mathrm{i}\frac{\mathrm{d}}{\mathrm{d}t}A(t) = [A(t), H_0]\,, \qquad A(t) = \exp(\mathrm{i}\,H_0 t)\,A\,\exp(-\mathrm{i}\,H_0 t)\,. \tag{2.103}$$

If A has an additional explicit time dependence, then one has to add a partial-derivative term on the right-hand side, $\mathrm{i}\,\mathrm{d}A(t)/\mathrm{d}t = [A(t), H_0] + \mathrm{i}\,\partial_t A(t)$, but we shall not dwell on this further. Let us now add a further contribution (perturbation) H_1 to the Hamiltonian,

$$H = H_0 + H_1\,. \tag{2.104}$$

Again, we assume that H_1 carries no explicit time dependence. In many practical applications, H_1 in itself has no time-dependence, but it is acquired via a time-dependent transformation (action of H_0 on H_1). In the Schrödinger picture, the equation of motion for the state vector $|\psi(t)\rangle$ is

$$\mathrm{i}\frac{\mathrm{d}}{\mathrm{d}t}|\psi(t)\rangle = (H_0 + H_1)\,|\psi(t)\rangle\,. \tag{2.105}$$

The transition to the interaction picture is done by defining the wave function $|\psi_I(t)\rangle$ according to

$$|\psi_I(t)\rangle = \exp(\mathrm{i}\,H_0 t)\,|\psi(t)\rangle\,. \tag{2.106}$$

If H_1 vanishes, then $|\psi_I(t)\rangle$ is time-independent. Inserting the *ansatz* (2.106) into (2.105), we obtain the equation of motion for $|\psi_I(t)\rangle$,

$$\mathrm{i}\,\frac{\mathrm{d}}{\mathrm{d}t}|\psi_I(t)\rangle = \exp(\mathrm{i}\,H_0 t)\,H_1\,\exp(-\mathrm{i}\,H_0 t)|\psi_I(t)\rangle = H_1(t)\,|\psi_I(t)\rangle\,. \tag{2.107}$$

Thus, in the interaction picture, the interaction Hamiltonian is manifestly time-dependent even if H_1 initially carries no time-dependence,

$$H_I(t) = \exp(\mathrm{i}\,H_0\,t)\,H_1\,\exp(-\mathrm{i}\,H_0\,t)\,. \tag{2.108}$$

Denoting by T the time ordering, we observe that the following form of the time-ordered time-evolution operator solves Eq. (2.107),

$$|\psi_I(t)\rangle = \mathrm{T}\left[\exp\left(-\mathrm{i}\int_0^t \mathrm{d}t'\,H_I(t')\right)\right]|\psi_I(0)\rangle\,. \tag{2.109}$$

The interaction picture is intermediate between the Schrödinger and Heisenberg pictures; in the latter case, the entire time evolution due to both H_0 and H_1 is ascribed to the operator, and the state vectors are time-independent. We should illustrate these general statements by way of example. For a continuum of field modes, we recall the field Hamiltonian,

$$:H_F: = \int \mathrm{d}^3k\,\omega\,a^+_\lambda(\vec{k})\,a_\lambda(\vec{k})\,, \qquad [a_\lambda(\vec{k}), a^+_{\lambda'}(\vec{k}')] = \delta^{(3)}(\vec{k}-\vec{k}')\,\delta_{\lambda\lambda'}\,. \tag{2.110}$$

We identify the time-independent fundamental creation and annihilation operators with the time-dependent ones at time $t = 0$, i.e., $a_\lambda(\vec{k}) = a_\lambda(\vec{k}, t = 0)$ and $a^+_\lambda(\vec{k}) = a^+_\lambda(\vec{k}, t = 0)$. We then start from the Heisenberg equation of motion, $\mathrm{i}(\mathrm{d}A(t)/\mathrm{d}t) = [A(t), H_0]$, and calculate the time dependence of $a_\lambda(\vec{k}, t)$ induced by the free normal-ordered Hamiltonian, with the initial condition $a_\lambda(\vec{k}, t = 0) = a_\lambda(\vec{k})$,

$$\mathrm{i}\,\frac{\mathrm{d}}{\mathrm{d}t}a_\lambda(\vec{k},t)\bigg|_{t=0} = \left[a_\lambda(\vec{k}, t=0),\,\sum_{\lambda'=1}^{2}\int \mathrm{d}^3k'\,\omega'\,a^+_{\lambda'}(\vec{k}')\,a_{\lambda'}(\vec{k}')\right] = \omega\,a_\lambda(\vec{k})\,, \tag{2.111}$$

where $\omega' = |\vec{k}'|$ and we recall that $\omega = |\vec{k}|$. The equation of motion is immediately solved by

$$a_\lambda(\vec{k},t) = a_\lambda(\vec{k})\,\mathrm{e}^{-\mathrm{i}\,\omega\,t}\,, \qquad a^+_\lambda(\vec{k},t) = a_\lambda(\vec{k})\,\mathrm{e}^{\mathrm{i}\,\omega\,t}\,. \tag{2.112}$$

The exponential factor (dependence on time) of the fundamental creation and annihilation operators is thus identified as the time dependence of the operators in the interaction picture, due to the action of the free field Hamiltonian.

This consideration allows us to identify the "usual" field operators for the electromagnetic vector potentials given in Eqs. (2.71) and (2.89) as interaction-picture field operators. Conversely, the transformation from the interaction picture to the Schrödinger picture can be accomplished very easily, by setting the time equal to zero.

2.5.2 *Integration Measure and Phase Conventions*

A second subtle point concerns the normalizations and phases of field operators, which are not uniformly assigned in the literature. We shall demonstrate that the difference in the integration measures can be absorbed in different conventions for the field commutators. The commutators of the fundamental field operators we have used so far,

$$[a_\lambda(\vec{k}), a^+_{\lambda'}(\vec{k}')] = \delta^{(3)}(\vec{k} - \vec{k}')\, \delta_{\lambda\lambda'}\,, \qquad [\mathrm{a}_\lambda(\vec{k}), \mathrm{a}^+_{\lambda'}(\vec{k}')] = \delta_{\vec{k}\,\vec{k}'}\, \delta_{\lambda\lambda'}\,, \tag{2.113}$$

display a particularly simple structure with regard to the prefactors. This structure is connected with the integration or summation "measure" used in the field operators (2.71) and (2.89),

$$\int \frac{\mathrm{d}^3 k}{\sqrt{2\omega\,(2\pi)^3}}\,, \qquad \sum_{\vec{k}\lambda} \frac{1}{\sqrt{V}}\, \frac{1}{\sqrt{2\omega}}\,, \tag{2.114}$$

in the continuum and discrete representations, respectively. Our conventions coincide with those adopted in Ref. [115].

Otherwise, one may point out that the integration measure

$$\frac{\mathrm{d}^3 k}{(2\pi)^3\,(2\,\omega)} = \int \frac{\mathrm{d}^4 k}{(2\pi)^3}\, \delta(k^2)\, \Theta(k^0)\,, \qquad k^2 = (k^0)^2 - \vec{k}^{\,2} = \omega^2 - \vec{k}^{\,2}\,, \tag{2.115}$$

has the advantage of being Lorentz-invariant, in contrast to the integration measure in Eq. (2.114). We also recall that by convention, in three-dimensional integrations one sets $\omega = |\vec{k}|$, while $\omega = k^0$ is being integrated over in the four-dimensional integral in Eq. (2.115). We may now transform the fundamental creation and annihilation operators as follows [see Eq. (3.110) of Ref. [2]],

$$a_\lambda(\vec{k}) = \frac{1}{\sqrt{(2\pi)^3}\,\sqrt{2\omega}}\, a_{\vec{k}\lambda}\,, \qquad a^+_\lambda(\vec{k}) = \frac{1}{\sqrt{(2\pi)^3}\,\sqrt{2\omega}}\, a^+_{\vec{k}\lambda}\,. \tag{2.116}$$

Then,

$$[a_{\vec{k}\lambda}, a^+_{\vec{k}'\lambda'}] = (2\,\omega)\,(2\pi)^3\,\delta^{(3)}(\vec{k} - \vec{k}')\,\delta_{\lambda\lambda'}\,. \tag{2.117}$$

The vector potential operator for the transverse modes reads as

$$\vec{A}_\perp(t, \vec{r}) = \sum_\lambda \int \frac{\mathrm{d}^3 k}{(2\pi)^3\,(2\,\omega)}\, \hat{\epsilon}_\lambda(\vec{k})\, \left(a_{\vec{k}\lambda}\, \mathrm{e}^{-\mathrm{i}\,k\cdot x} + a^+_{\vec{k}\lambda}\, \mathrm{e}^{\mathrm{i}\,k\cdot x} \right)\,. \tag{2.118}$$

The operators for the electric and magnetic field follow suit, incorporating the Lorentz-invariant integration measure (2.115). In fully relativistic calculations, it is therefore preferable to use the conventions of Eq. (2.117). In nonrelativistic calculations, the simple structure of the prefactors in Eq. (2.113) is very appealing. It corresponds to the "harmonic-oscillator style" commutator of the discrete-space fundamental commutators given in the rightmost equation in (2.113).

Another change in the conventions, which is sometimes useful, concerns the phase of the field operators. In Ref. [113], the authors start formulating the quantized electromagnetic theory from the electric field operator. Naturally, one would like to have unit prefactors (no imaginary unit) in front of the fundamental creation and annihilation operators, when considering the "most fundamental" field. This can be achieved by transforming the phases of the fundamental operators according to

$$a_{\vec{k}\lambda} \to -\mathrm{i}\,\mathcal{A}_{\vec{k}\lambda}\,, \quad a^{+}_{\vec{k}\lambda} \to \mathrm{i}\,\mathcal{A}^{+}_{\vec{k}\lambda}\,. \tag{2.119}$$

The fundamental commutator is left invariant, i.e., $[\mathcal{A}_{\vec{k}\lambda}, \mathcal{A}^{+}_{\vec{k}'\lambda'}] = \delta_{\vec{k}\,\vec{k}'}\,\delta_{\lambda\lambda'}$. The electric field operator then has a certain similarity to the position operator for a harmonic oscillator,

$$\vec{E}_{\perp}(t,\vec{r}) = \sum_{\vec{k}\lambda} \frac{1}{\sqrt{V}} \sqrt{\frac{\omega}{2}}\, \hat{\epsilon}_{\lambda}(\vec{k}) \left(\mathcal{A}_{\vec{k}\lambda}\, \mathrm{e}^{-\mathrm{i}\,k\cdot x} + \mathcal{A}^{+}_{\vec{k}\lambda}\, \mathrm{e}^{\mathrm{i}\,k\cdot x} \right). \tag{2.120}$$

The vector potential operator now involves a number of imaginary units,

$$\vec{A}_{\perp}(t,\vec{r}) = \sum_{\vec{k}\lambda} \frac{1}{\sqrt{V}} \frac{1}{\sqrt{2\omega}}\, \hat{\epsilon}_{\lambda}(\vec{k}) \left(-\mathrm{i}\,\mathcal{A}_{\vec{k}\lambda}\, \mathrm{e}^{-\mathrm{i}\,k\cdot x} + \mathrm{i}\,\mathcal{A}^{+}_{\vec{k}\lambda}\, \mathrm{e}^{\mathrm{i}\,k\cdot x} \right). \tag{2.121}$$

The operator for the magnetic field then reads as

$$\vec{B}(t,\vec{r}) = \sum_{\vec{k}\lambda} \frac{1}{\sqrt{V}} \frac{1}{\sqrt{2\omega}} \left(\vec{k} \times \hat{\epsilon}_{\lambda}(\vec{k}) \right) \left(\mathcal{A}_{\vec{k}\lambda}\, \mathrm{e}^{-\mathrm{i}\,k\cdot x} + \mathcal{A}^{+}_{\vec{k}\lambda}\, \mathrm{e}^{\mathrm{i}\,k\cdot x} \right). \tag{2.122}$$

The phase conventions have no consequences for the calculation of physical processes but severely alter the course of intermediate steps in calculations.

2.6 Further Thoughts

Here are some suggestions for further thought.

(1) **Field Hamiltonian.** Verify that the field Hamiltonian given in Eq. (2.77) is time-independent, either by calculating its vanishing time derivative, or by first restoring the time dependence of the field operators and verifying that it cancels.

(2) **Discretized Field Operator.** Repeat all the derivations in Sec. 2.4.2, ranging from Eq. (2.71) to Eq. (2.88), with the discretized version of the field operators given in Sec. 2.4.3.

(3) **Field Commutator.** Based on Eq. (2.60), which we recall for convenience,

$$\left[E^{i}_{\perp}(t,\vec{r}),\, A^{j}_{\perp}(t,\vec{r}\,') \right] = \mathrm{i}\,\delta^{\perp,ij}(\vec{r}-\vec{r}\,')\,, \tag{2.123}$$

show that the commutator of the electric and magnetic fields is

$$\left[E^{i}_{\perp}(t,\vec{r}),\, B^{j}(t,\vec{r}\,') \right] = -\mathrm{i}\,\epsilon^{jk\ell}\,\nabla^{k}\,\delta^{\perp,i\ell}(\vec{r}-\vec{r}\,')\,, \tag{2.124}$$

i.e., verify Eq. (2.76). Can one measure the electric and magnetic field at a given point in space simultaneously, and if not, which components of the electric and magnetic fields fail to commute?

(4) **Field Commutator.** Complete the missing intermediate steps in Eq. (2.100), i.e., in the derivation of the field commutator $\left[E_\perp^i(t,\vec{r}),\, A_\perp^j(t,\vec{r}\,')\right]$ in the discrete representation.

(5) **Norm of One-Photon State.** Use Eq. (2.93) for the commutator of the discrete-space photon creation and annihilation operators,

$$[\mathrm{a}_\lambda(\vec{k}),\, \mathrm{a}^+_{\lambda'}(\vec{k}')] = \delta_{\vec{k}\vec{k}'}\,\delta_{\lambda\lambda'}. \tag{2.125}$$

In order to show that the one-photon state in the discrete representation is normalized to unity,

$$\| |1_{\vec{k}\lambda}\rangle \|^2 = \langle 0|\mathrm{a}_\lambda(\vec{k})\,\mathrm{a}^+_\lambda(\vec{k})|0\rangle = 1\,. \tag{2.126}$$

Convince yourself that the problematic infinite norm in the continuum representation, given in Eq. (2.87),

$$\| |1_{\vec{k}\lambda}\rangle \|^2 = \langle 0|a_\lambda(\vec{k})\,a^+_\lambda(\vec{k})|0\rangle = \delta^{(3)}(\vec{0})\,, \tag{2.127}$$

can be avoided for a finite normalization volume V.

(6) **Missing Steps.** Complete the missing intermediate steps in Eq. (2.111), using the relation $[AB,C] = A\,[B,C] + [A,C]B$. Then, repeat the derivation using the discrete-space version of the field operators and Hamiltonian.

(7) **Missing SI Units.** Restore the missing prefactors of $\hbar$, ϵ_0 and c in the continuum and discrete-space versions of the electromagnetic field operators. Your answer should look like

$$\vec{A}(\vec{r}) = \sum_{\lambda=1}^{2}\int \frac{\mathrm{d}^3k}{\sqrt{(2\pi)^3}}\sqrt{\frac{\hbar}{2\epsilon_0\,\omega}}\,\hat{\epsilon}_\lambda(\vec{k})\left(a_\lambda(\vec{k})\,\mathrm{e}^{\mathrm{i}\,\vec{k}\cdot\vec{r}} + a^+_\lambda(\vec{k})\,\mathrm{e}^{-\mathrm{i}\,\vec{k}\cdot\vec{r}}\right), \tag{2.128a}$$

$$\vec{E}(\vec{r}) = \sum_{\lambda=1}^{2}\int \frac{\mathrm{d}^3k}{\sqrt{(2\pi)^3}}\sqrt{\frac{\hbar\,\omega}{2\epsilon_0}}\,\hat{\epsilon}_\lambda(\vec{k})\left(\mathrm{i}\,a_\lambda(\vec{k})\,\mathrm{e}^{\mathrm{i}\,\vec{k}\cdot\vec{r}} - \mathrm{i}\,a^+_\lambda(\vec{k})\,\mathrm{e}^{-\mathrm{i}\,\vec{k}\cdot\vec{r}}\right), \tag{2.128b}$$

$$\vec{B}(\vec{r}) = \sum_{\lambda=1}^{2}\frac{\mathrm{d}^3k}{\sqrt{(2\pi)^3}}\sqrt{\frac{\hbar}{2\epsilon_0\,\omega}}\left(\vec{k}\times\hat{\epsilon}_\lambda(\vec{k})\right)\left(\mathrm{i}\,a_\lambda(\vec{k})\,\mathrm{e}^{\mathrm{i}\,\vec{k}\cdot\vec{r}} - \mathrm{i}\,a^+_\lambda(\vec{k})\,\mathrm{e}^{-\mathrm{i}\,\vec{k}\cdot\vec{r}}\right), \tag{2.128c}$$

where $[a_\lambda(\vec{k}), a^+_{\lambda'}(\vec{k}')] = \delta^{(3)}(\vec{k}-\vec{k}')\,\delta_{\lambda\lambda'}$. The discrete representation in the Schrödinger picture should read as

$$\vec{A}(\vec{r}) = \sum_{\vec{k}\lambda}\sqrt{\frac{\hbar}{2\,\epsilon_0\,\omega\,V}}\,\hat{\epsilon}_\lambda(\vec{k})\left(\mathrm{a}_\lambda(\vec{k})\,\mathrm{e}^{\mathrm{i}\,\vec{k}\cdot\vec{r}} + \mathrm{a}^+_\lambda(\vec{k})\,\mathrm{e}^{-\mathrm{i}\,\vec{k}\cdot\vec{r}}\right), \tag{2.129a}$$

$$\vec{E}(\vec{r}) = \sum_{\vec{k}\lambda}\sqrt{\frac{\hbar\,\omega}{2\,\epsilon_0\,V}}\,\hat{\epsilon}_\lambda(\vec{k})\left(\mathrm{i}\,\mathrm{a}_\lambda(\vec{k})\,\mathrm{e}^{\mathrm{i}\,\vec{k}\cdot\vec{r}} - \mathrm{i}\,\mathrm{a}^+_\lambda(\vec{k})\,\mathrm{e}^{-\mathrm{i}\,\vec{k}\cdot\vec{r}}\right), \tag{2.129b}$$

$$\vec{B}(\vec{r}) = \sum_{\vec{k}\lambda}\sqrt{\frac{\hbar}{2\,\epsilon_0\,\omega\,V}}\left(\vec{k}\times\hat{\epsilon}_\lambda(\vec{k})\right)\left(\mathrm{i}\,\mathrm{a}_\lambda(\vec{k})\,\mathrm{e}^{\mathrm{i}\,\vec{k}\cdot\vec{r}} - \mathrm{i}\,\mathrm{a}^+_\lambda(\vec{k})\,\mathrm{e}^{-\mathrm{i}\,\vec{k}\cdot\vec{r}}\right), \tag{2.129c}$$

where $[\mathrm{a}_\lambda(\vec{k}), \mathrm{a}^+_{\lambda'}(\vec{k}')] = \delta_{\vec{k}\,\vec{k}'}\,\delta_{\lambda\lambda'}$.

(8) **Dimensional Analysis of QED.** Consider the natural unit system introduced in Sec. 2.3.2 and consider the QED Lagrangian in D spatial and one time dimension. Show that charge is dimensionless in natural units only if $D = 3$. (This consideration has important consequences for dimensional regularization introduced in Chap. 10.)

Chapter 3

Time-Ordered Perturbations

3.1 Overview

A canonical quantum mechanics course covers a number of problems such as the one-dimensional harmonic oscillator, or first- and second-order perturbation theory. For the current endeavor, we need to recall in more detail a few special chapters, related to atomic physics. This will be necessary in order to tackle a few more problems which are of prime relevance later for all those atomic physics processes, which cannot be described without an understanding of field quantization.

The purpose of the current chapter is to first give a coherent description of three aspects which are not normally treated in depth in a canonical quantum mechanics course. Specifically, in Sec. 3.2, we discuss time-ordered perturbation theory, and the derivation of FERMI's[1] Golden Rule. We proceed to a derivation of decay rates of excited atomic states due to the interaction with the quantized radiation field. In Sec. 3.3, the dynamic (ac) Stark shift is calculated using a Schrödinger-picture operator for the electromagnetic vector potential. The zero-frequency (static) limit of the dynamic Stark shift (in this limit, called the dc Stark shift) is discussed in Sec. 3.4. Surprisingly, the energy perturbation due to the dc Stark shift allows for an analysis up to very high orders in perturbation theory. Perhaps even more surprisingly, the perturbation series turns out to be a divergent series in large order. As we shall see in Sec. 3.4, this aspect has important consequences and can even be generalized to field theories such as quantum electrodynamics (QED).

3.2 Time-Ordered Perturbations and Fermi's Golden Rule

3.2.1 *Derivation of Fermi's Golden Rule*

Fermi's golden rule is a nontrivial result that can be derived from time-ordered perturbation theory, and it is rather interesting to consider its derivation in detail. We temporarily revert to SI mksA units; the number of factors of $\hbar$ and c to be supplemented is not excessive. We start from the time-dependent Schrödinger

[1]Enrico Fermi (1901–1954).

equation

$$H\,\psi(t,\vec{r}) = \mathrm{i}\hbar\frac{\partial}{\partial t}\psi(t,\vec{r})\,. \tag{3.1}$$

The Hamiltonian is composed of an unperturbed part and a perturbation,

$$H = H_0 + H_1\,, \tag{3.2}$$

where H_1 is the interaction Hamiltonian in the Schrödinger picture (see Sec. 2.5.1). The operator H_1 is used as defined in Eq. (2.104), not as outlined in Eq. (2.108), where it automatically acquires an explicit time dependence. As we shall see in Sec. 3.2.3, the use of Schrödinger-picture field operators in the description of the interaction with the electromagnetic field makes the calculation of decay rates of atomic states easier. But first, let us dwell on a more general setting; we suppose that a complete set of eigenstates is available,

$$H_0\,\psi_m(\vec{r}) = E_m\,\psi_m(\vec{r})\,, \qquad \langle\psi_m|\psi_n\rangle = \delta_{mn}\,. \tag{3.3}$$

The unperturbed energy eigenvalues are denoted as E_m. The states acquire a time-dependence according to

$$\psi(t,\vec{r}) = \sum_m a_m(t)\,\psi_m(\vec{r})\,\mathrm{e}^{-\mathrm{i}\,E_m\,t/\hbar}\,. \tag{3.4}$$

The time-dependence $\exp(-\mathrm{i}\,E_m\,t/\hbar)$ is due only to the action of the free Hamiltonian. Any additional time-dependence due to the interaction is described by the expansion coefficients $a_m(t)$. One must project onto a complete set of states and take into account Eq. (3.3), establishing that the state evolution is described by the equation

$$\mathrm{i}\hbar\,\frac{\mathrm{d}a_k(t)}{\mathrm{d}t} = \sum_m \langle\psi_k|H_1|\psi_m\rangle\,a_m(t)\,\mathrm{e}^{\mathrm{i}\,(E_k-E_m)\,t/\hbar}\,. \tag{3.5}$$

Time-dependent perturbation theory is based on an expansion of the time-dependence of the $a_m(t)$ in powers of H_1. From this definition, it is immediately clear that for a small perturbation H_1, a better approximation $a_k^{(p+1)}(t)$ to $a_k(t)$ of order $p+1$ is obtained by integrating the approximation $a_k^{(p)}(t)$ of order p,

$$\mathrm{i}\hbar\,\frac{\mathrm{d}a_k^{(p+1)}(t)}{\mathrm{d}t} = \sum_m \langle\psi_k|H_1|\psi_m\rangle\,a_m^{(p)}(t)\,\mathrm{e}^{\mathrm{i}\,(E_k-E_m)\,t/\hbar}\,. \tag{3.6}$$

This is because each time we integrate the right-hand side, one more power of H_1 appears in the result for $a_k(t)$. The approximation $a_m^{(p)}(t)$ is of order g^p if H_1 is proportional to a coupling parameter g. The zeroth-order approximation is obtained by letting $H_1 \to 0$ on the right-hand side of Eq. (3.6),

$$\frac{\mathrm{d}a_m^{(0)}(t)}{\mathrm{d}t} = 0\,, \qquad a_m^{(0)}(t) = \delta_{im}\,, \tag{3.7}$$

which implies that we assume the system to be initially prepared in the state $|\psi_i\rangle$ at time $t = 0$.

The first-order approximation for the final-state amplitude (denoted by the subscript f) is given as

$$\begin{aligned} \mathrm{i}\hbar\, a_f^{(1)}(t) &= \sum_m \int_0^t \mathrm{d}t' \, \langle\psi_f|H_1|\psi_m\rangle \, a_m^{(0)}(t')\, \mathrm{e}^{\mathrm{i}\,(E_f - E_m)\,t'/\hbar} \\ &= \int_0^t \mathrm{d}t' \, \langle\psi_f|H_1|\psi_i\rangle \, \mathrm{e}^{\mathrm{i}\,(E_f - E_i)\,t'/\hbar} \,. \end{aligned} \tag{3.8}$$

Integrating Eq. (3.8) to first order, one obtains

$$\mathrm{i}\hbar\, a_f^{(1)}(t) = 2\, \langle\psi_f|H_1|\psi_i\rangle \exp\left(\frac{\mathrm{i}\,(E_f - E_i)\,t}{2\hbar}\right) \frac{\sin\left[(E_f - E_i)\,t/(2\hbar)\right]}{(E_f - E_i)/\hbar} \,. \tag{3.9}$$

The time-dependent probability $P_f(t) = \left|a_f^{(1)}(t)\right|^2$ for the transition from state $|\psi_i\rangle$ to state $|\psi_f\rangle$ then reads as

$$P_f(t) = \left|a_f^{(1)}(t)\right|^2 = 4\, \frac{|\langle\psi_f|H_1|\psi_i\rangle|^2}{\hbar^2} \, \frac{\sin^2\left[(E_f - E_i)\,t/(2\hbar)\right]}{\left[(E_f - E_i)/\hbar\right]^2} \,. \tag{3.10}$$

Expanding for $t \to 0$, we would tentatively obtain that $P_f(t)$ goes quadratically for small t. However, that picture would be misleading.

While the literature is abundant with discussions of Fermi's golden rule, the details regarding the elimination of the quadratic time-dependence are often obscured. We should first remember that most practically relevant quantum mechanical transitions actually involve a discrete state and a continuum; Fermi's golden rule first and foremost applies to transitions from discrete states coupled to a continuum. Let us suppose that a bound electron gets ionized upon absorption of an energetic photon. The continuum electron in the final state lies in the positive-energy continuum of the Schrödinger spectrum. Let us also suppose that a bound electron, initially in an excited bound state, makes a transition to the ground state, emitting a photon. While the electron undergoes a transition from one bound state to the other, one might think that neither the initial nor the final state are in a continuum. However, that is not the case. The final state is in a continuum, albeit not in the continuum of the Schrödinger Hamiltonian of the atom, but of the field Hamiltonian of the electromagnetic field. If the final state is in a continuum, then there are quantum states which are energetically displaced only infinitesimally from the actual final state of the process. In the case of a bound-bound transition, the infinitesimally displaced states would be those with near-resonant photons. Furthermore, while not obvious, energy conservation actually holds during a quantum transition from a bound state to a continuum, if the energies of all involved particles are considered. For an atomic deexcitation process, the energy of the initial excited state is equal to the energy of the emitted photon plus the energy of the final state; the latter can be the ground state. Another example can be indicated as follows: During an ionization process, the energy of the initial electron plus the energy of the absorbed photon is equal to the energy of the final continuum electron.

The essence of Fermi's golden rule is as follows: We first observe that the expression $\sin^2(\omega t)/(\omega t)^2$ with $\omega = (E_f - E_i)/\hbar$ is highly peaked near $E_f \approx E_i$. Our assumption of the availability of a continuum of states around the final state forces us to collect the contribution from neighboring states. Energy conservation demands that $E_f \approx E_i$, because E_f is the energy of the final state. The energy E_f is calculated taking into account all of its constituent particles. Let us therefore assume that we know the density of states $\rho(E_f) = \mathrm{d}n/\mathrm{d}E_f$ in an infinitesimal interval $\mathrm{d}E_f$ around the final state,

$$\mathrm{d}n = \rho(E_f)\,\mathrm{d}E_f\,, \tag{3.11}$$

where $\mathrm{d}E_f$ measures the interval in the continuum part of the spectrum. Then, the transition rate is obtained as the integral over the available final states,

$$W_{fi} = \frac{1}{t}\int \mathrm{d}n\, P_f(t)\,. \tag{3.12}$$

This is the essence of the golden rule: For large t, the transition probability $P_f(t)$ is highly peaked near the particular final state $|\psi_f\rangle$ that fulfills the energy conservation condition. So, we can in fact sum or integrate over all final states $\mathrm{d}n$ in the immediate vicinity of the energy-conserving final state, and this integral can be extended to infinite integration limits in the energy domain because as $t \to \infty$, we are in fact picking up a contribution only from a very limited energy region. Furthermore, in this limit, the energy conservation condition has to be fulfilled exactly, and it is thus possible to pull the factor $\langle\psi_f|H_1|\psi_i\rangle$ out of the integral in Eq. (3.12), provided the energy conservation condition is fulfilled. The resulting integral

$$\int_{-\infty}^{\infty} \mathrm{d}\omega\, \frac{\sin^2(\omega t)}{\omega^2} = \pi\, t \tag{3.13}$$

is proportional to t, not t^2. Alternatively, one can match the expression to a Dirac-δ function,

$$\frac{\sin^2(\omega t)}{\omega^2} \to \pi\, t\, \delta(\omega)\,, \qquad t \to \infty\,. \tag{3.14}$$

Fermi's golden rule is finally obtained by summing over all the available final states, realizing that the contributing states are those with $E_f \approx E_i$ because of the peaked structure of $\sin^2(\omega t)/\omega^2$ near $\omega = 0$,

$$\begin{aligned}
W_{fi} &= \frac{1}{t}\int \mathrm{d}n\, P_f(t) = \frac{1}{t}\int \mathrm{d}E_f\, \rho(E_f)\, P_f(t) \\
&= \frac{4}{t}\int \mathrm{d}E_f\, \rho(E_f)\, \frac{|\langle\psi_f|H_1|\psi_i\rangle|^2}{\hbar^2}\, \frac{\sin^2[(E_f - E_i)\,t/(2\hbar)]}{[(E_f - E_i)/\hbar]^2} \\
&= \frac{4}{\hbar\, t}\left(|\langle\psi_f|H_1|\psi_i\rangle|^2\, \rho(E_f)\Big|_{E_f = E_i}\right) \int_{-\infty}^{\infty} \mathrm{d}\omega\, \frac{\sin^2(\omega t/2)}{\omega^2} \\
&= \frac{2\pi}{\hbar}\left(|\langle\psi_f|H_1|\psi_i\rangle|^2\, \rho(E_f)\Big|_{E_f = E_i}\right).
\end{aligned} \tag{3.15}$$

We have extracted the value of the integrand in front of the peak as a prefactor. It is easy to convince oneself that the result, Fermi's golden rule,

$$W_{fi} = \frac{2\pi}{\hbar} \left(|\langle \psi_f | H_1 | \psi_i \rangle|^2 \, \rho(E_f) \Big|_{E_f = E_i} \right), \tag{3.16}$$

has the right physical dimension: Namely, $\rho(E_f)$ has dimension of inverse energy, because it counts the available states per unit energy interval, per Eq. (3.11). Then, the square of the energy (the dimension of $| \langle \psi_f | H_1 | \psi_i \rangle |^2$) is divided by the product of $\hbar$ and energy (contained in ρ), and $\hbar$ has dimension of the product of energy and time. So, W_{fi} has dimension of inverse time, which is correct for a transition rate.

3.2.2 *Fermi's Golden Rule and Nuclear Beta Decay*

As an example, let us consider nuclear beta decay. This example will provide us with a straightforward generalization of Fermi's golden rule, namely, a situation with two particles in the final state, where available final states are degenerate because the transition energy may be distributed among the two particles. In beta decay, a nucleus with charge number Z and neutron number N undergoes a transition $(Z, N) \to (Z+1, N-1)$ with a neutron becoming a proton, emitting an electron and an electron antineutrino,

$$n \to p + e^- + \bar{\nu}_e \,. \tag{3.17}$$

Beta decay is mediated by weak interactions, whose energy scale is a lot higher than that of a typical nuclear transition. We shall therefore assume that the Hamiltonian matrix element in question is energy-independent and given as

$$\langle \psi_f | H_1 | \psi_i \rangle = \frac{G_F \mathcal{M}}{V}, \tag{3.18}$$

where G_F is the Fermi coupling constant, $\mathcal{M}$ is an invariant matrix element, and V is a (large) normalization volume. The matrix element $\mathcal{M}$ describes the transition of the nucleus, i.e., the transition of neutron to proton. The recoil momentum and recoil energy of the heavy nucleus will be neglected in the final state of the process. The two emitted particles are the electron and the electron antineutrino. Let the momentum of the electron be $\vec{p}_e$, and the momentum of the antineutrino be $\vec{p}_\nu$. Energy conservation implies that the initial excitation energy E_0 of the nucleus is equal to the energy $E_e \approx c|\vec{p}_e| = c p_e$ of the outgoing electron, plus the energy $E_\nu \approx c p_\nu$ of the outgoing antineutrino. Both electron and antineutrino masses will be neglected in the following. Energy conservation implies that

$$E_0 = E_e + c p_\nu \approx c p_e + c p_\nu \,. \tag{3.19}$$

The recoil energy of the decaying nucleus will be neglected in the following. The density of two-particle final states in the normalization volume V can be written as follows,

$$\mathrm{d}^6 n = \frac{V \, \mathrm{d}^3 p_e}{(2\pi\hbar)^3} \frac{V \, \mathrm{d}^3 p_\nu}{(2\pi\hbar)^3}, \tag{3.20}$$

based on the wisdom that every final quantum state occupies a phase-space volume $\hbar$ [see also Eq. (2.81)]. Because of rotational symmetry, we can write after the integration over the solid angles Ω_e and Ω_ν,

$$\int_{\Omega_e} \mathrm{d}^3 p_e = 4\pi p_e^2 \,\mathrm{d}p_e\,, \qquad \int_{\Omega_\nu} \mathrm{d}^3 p_\nu = \frac{4\pi}{c^3}\,(E_0 - E_e)^2\,\mathrm{d}E_\nu\,. \tag{3.21}$$

In the normalization volume V, we have

$$\begin{aligned} \mathrm{d}^2 n &= \int_{\Omega_e} \frac{V\,\mathrm{d}^3 p_e}{(2\pi\hbar)^3} \int_{\Omega_\nu} \frac{V\,\mathrm{d}^3 p_\nu}{(2\pi\hbar)^3} = \frac{V\,4\pi p_e^2\,\mathrm{d}p_e}{(2\pi\hbar)^3}\,\frac{V\,4\pi (E_0 - E_e)^2\,\mathrm{d}E_\nu}{c^3\,(2\pi\hbar)^3} \\ &= \frac{V^2}{4\pi^4\,\hbar^6\,c^3}\,p_e^2\,(E_0 - E_e)^2\,\mathrm{d}p_e\,\mathrm{d}E_\nu = \mathrm{d}\rho(E_e)\,\mathrm{d}E_\nu\,, \end{aligned} \tag{3.22}$$

where the density of electron states is given by

$$\mathrm{d}\rho(E_e) = \frac{\mathrm{d}\rho(E_e)}{\mathrm{d}p_e}\,\mathrm{d}p_e = \frac{V^2}{4\pi^4\,\hbar^6\,c^3}\,p_e^2\,(E_0 - E_e)^2\,\mathrm{d}p_e\,. \tag{3.23}$$

The density $\mathrm{d}^2 n$ is double-differential; the final electron momentum still needs to be integrated over. An increase in the energy of the electron is compensated by a decrease in the energy of the emitted antineutrino, and we have written the double-differential density of states so that the dependence on the electron energy is singled out and can be integrated over separately. In the sense of Eq. (3.11), we have for the single-differential density of states,

$$\mathrm{d}n = \mathrm{d}E_\nu \int_0^{E_0/c} \mathrm{d}p_e\,\frac{\mathrm{d}\rho(E_e)}{\mathrm{d}p_e}\,, \tag{3.24}$$

where E_0/c is the highest allowed electron momentum. Now, we use Fermi's golden rule (3.16) in the form

$$W_{fi} = \frac{2\pi}{\hbar}\left(\left|\langle\psi_f|H_1|\psi_i\rangle\right|^2 \left.\frac{\mathrm{d}n}{\mathrm{d}E_\nu}\right|_{E_0 = cp_e + cp_\nu} \right). \tag{3.25}$$

The volume V^2 cancels against the V^{-2} factor from the matrix element of H_1 given in Eq. (3.18), and we obtain

$$\begin{aligned} W_{fi} &= \frac{2\pi}{\hbar} \int_0^{E_0/c} \mathrm{d}p_e \left(\frac{G_F\,|\mathcal{M}|}{V}\right)^2 V^2\,\frac{1}{4\pi^4\,\hbar^6\,c^3}\,p_e^2\,(E_0 - E_e)^2 \\ &= \int_0^{E_0/c} \mathrm{d}p_e\,\frac{G_F^2}{2\pi^3\,\hbar^7\,c^3}\,|\mathcal{M}|^2\,p_e^2\,(E_0 - E_e)^2\,. \end{aligned} \tag{3.26}$$

The integral over the available momenta evaluates to

$$\int_0^{E_0/c} p_e^2\,(E_0 - E_e)^2\,\mathrm{d}p_e = \frac{E_0^5}{30\,c^3}\,, \tag{3.27}$$

where $E_e = cp_e$. The total decay rate, within our approximations, is thus,

$$W_{fi} = \int \mathrm{d}W_{fi} = \frac{1}{\tau} = \frac{G_F^2\,|\mathcal{M}|^2\,E_0^5}{60\,\pi^3\,(\hbar c)^6\,\hbar}\,, \tag{3.28}$$

where τ is the lifetime of the nucleus against beta decay. The functional dependence on E_0 has been confirmed in observations of beta decay lifetimes over many orders of magnitude and is commonly referred to as "SARGENT's[2] Law".

[2] Bernice Weldon Sargent (1904–1967).

3.2.3 *Fermi's Golden Rule and Atomic Decay Rates*

One of the most paradigmatic applications of Fermi's golden rule within atomic physics consists in the derivation of the decay rate of an excited atomic state against one-photon (electric-dipole) decay to a state of lower energy. We start from the electric field operator given in Eq. (2.129b) in the Schrödinger picture,

$$\vec{E}(\vec{r}) = \sum_{\vec{k}\lambda} \sqrt{\frac{\hbar\,\omega_{\vec{k}}}{2\epsilon_0\,V}}\,\hat{\epsilon}_\lambda(\vec{k})\,\left(\mathrm{i}\,\mathbb{a}_\lambda(\vec{k})\,\mathrm{e}^{\mathrm{i}\,\vec{k}\cdot\vec{r}} - \mathrm{i}\,\mathbb{a}^+_\lambda(\vec{k})\,\mathrm{e}^{-\mathrm{i}\,\vec{k}\cdot\vec{r}}\right). \tag{3.29}$$

(We again recall that the field operators $\mathbb{a}_\lambda(\vec{k})$, in contrast to the operators $a_\lambda(\vec{k})$, pertain to the discrete representation.) The interaction Hamiltonian that describes the dipole coupling of the electron to the electric field is

$$\begin{aligned} H_1 &= -\,e\,\vec{r}\cdot\vec{E}(\vec{r}) \approx -e\,\vec{r}\cdot\vec{E}(\vec{0}) \\ &= \sum_{\vec{k}\lambda} \sqrt{\frac{\hbar\,\omega_{\vec{k}}}{2\,\epsilon_0\,V}}\,\left(-e\,\vec{r}\cdot\hat{\epsilon}_\lambda(\vec{k})\right)\left(\mathrm{i}\,\mathbb{a}_\lambda(\vec{k}) - \mathrm{i}\,\mathbb{a}^+_\lambda(\vec{k})\right), \end{aligned} \tag{3.30}$$

where we apply the dipole approximation $\vec{k}\cdot\vec{r} \ll 1$. For typical atomic transitions, the wavelength of the emitted (visible) light is in the range of some 1000 Å, whereas the atomic dimensions are in the range of a Bohr radius (on the order of 1 Å). The wavelength emitted by the atom is much longer than the dimension of the atom itself, which justifies the approximation $\vec{E}(\vec{r}) \approx \vec{E}(\vec{0})$, where we assume the atom to be located at the origin.

The most interesting part of the calculation concerns the proper identification of the quantum states involved in the process. Let $|\psi_i\rangle = |\phi_i\rangle \otimes |0\rangle$ be the initial excited atomic state, where $|\phi_i\rangle$ is the initial atomic state, and $|0\rangle$ is vacuum state of the electromagnetic field (no photons present). The final state is $|\psi_f\rangle = |\phi_f\rangle \otimes |1_{\vec{k}\lambda}\rangle$, where $|\psi_f\rangle$ is the final (energetically lower) state of the process, and $|1_{\vec{k}\lambda}\rangle = \mathbb{a}^+_{\vec{k}\lambda}|0\rangle$ is a one-photon state of wave vector $\vec{k}$ and polarization λ. The unperturbed Hamiltonian is the sum of the atomic Hamiltonian and the unperturbed electromagnetic field Hamiltonian,

$$H_0 = \sum_m E^{(0)}_m\,|\phi_m\rangle\,\langle\phi_m| + \sum_{\vec{k}\lambda} \hbar\omega\,\mathbb{a}^+_\lambda(\vec{k})\,\mathbb{a}_\lambda(\vec{k}), \tag{3.31}$$

where the unperturbed energies of the atomic states are denoted as $E^{(0)}_m$. Because the energy of the product state (atom ⊗ field) before and after the interaction has to be the same, the photon has to be emitted exactly on resonance, $E_i = E^{(0)}_i = E_f = E^{(0)}_f + \hbar\,\omega$. Here, E_i and E_f are the energies of the initial and final product states of the atom and field, and $E^{(0)}_i$ and $E^{(0)}_f$ are the energies of the atomic states involved in the process. As already stressed in Sec. 3.2.1, energy conservation $E_i = E_f$ holds provided we count the energies of all involved particles in the initial and final states, and that includes the emitted photons.

For the interaction, the photon creation operator in the electric-field operator (2.129b) leads to the nonvanishing matrix element

$$\left\langle \phi_f, 1_{\vec{k}\lambda} \left| H_1 \right| \phi_i, 0 \right\rangle = \mathrm{i}\, e \sqrt{\frac{\hbar\,\omega}{2\,\epsilon_0\,V}} \left\langle \phi_f \left| \vec{r} \cdot \hat{\epsilon}_\lambda(\vec{k}) \right| \phi_i \right\rangle . \tag{3.32}$$

The density of available final states for the photon is

$$\mathrm{d}^3 n = \frac{V\,\mathrm{d}^3 p}{(2\pi\hbar)^3} = \frac{V\,\mathrm{d}^3 k}{(2\pi)^3} = \frac{V\,k^2\,\mathrm{d}k\,\mathrm{d}^2\Omega}{(2\pi)^3} = \frac{V\,k^2}{(2\pi)^3\,\hbar c}\,\mathrm{d}^2\Omega\,\mathrm{d}E_k\,, \tag{3.33a}$$

$$\mathrm{d}^2\rho(f) = \frac{V\,k^2}{(2\pi)^3\,\hbar c}\,\mathrm{d}^2\Omega\,, \tag{3.33b}$$

where $\mathrm{d}^2\rho(f)$ is still differential in the solid angle into which the photon is emitted. The photon momentum is $\vec{p} = \hbar\,\vec{k}$ and the photon energy is $E_k = \hbar\omega = \hbar\,c\,k$ where $k = |\vec{k}|$. The differential transition rate (decay rate) is

$$\begin{aligned} \mathrm{d}^2 W_{fi} &= \frac{2\pi}{\hbar} \left| \left\langle \phi_f, 1_{\vec{k}\lambda} \left| H_1 \right| \phi_i, 0 \right\rangle \right|^2 \mathrm{d}^2\rho(f) \\ &= \frac{2\pi}{\hbar}\, e^2\, \frac{\hbar\,\omega}{2\,\epsilon_0\,V} \left| \left\langle \phi_f \left| \vec{r} \cdot \hat{\epsilon}_\lambda(\vec{k}) \right| \phi_i \right\rangle \right|^2 \frac{V\,k^2}{(2\pi)^3\,\hbar c}\,\mathrm{d}^2\Omega\,. \end{aligned} \tag{3.34}$$

This expression is the differential rate for emission into the solid angle $\mathrm{d}^2\Omega$ with definite photon polarization λ. In this derivation, we meticulously denote the degree of the differentials; the solid angle is thus denoted as $\mathrm{d}^2\Omega = \sin\theta\,\mathrm{d}\theta\,\mathrm{d}\varphi$, where θ is the polar, and φ is the azimuth angle. Note that we cannot simply integrate over $\mathrm{d}\Omega \to 4\pi$ because the photon polarization vectors $\hat{\epsilon}_{\vec{k}\lambda}$ depend on the emission angle. In order to obtain the total rate, we have to sum over the polarizations of the emitted photon and integrate over the solid angle. Remembering in each step that $\omega = ck = c\left|\vec{k}\right|$, we obtain

$$\begin{aligned} W_{fi} &= \frac{2\pi}{\hbar} \sum_\lambda \int \mathrm{d}^2\Omega\; e^2\, \frac{\hbar\,\omega}{2\,\epsilon_0\,V} \left| \left\langle \phi_f \left| \vec{r} \cdot \hat{\epsilon}_\lambda(\vec{k}) \right| \phi_i \right\rangle \right|^2 \frac{V\,k^2}{(2\pi)^3\,\hbar c} \\ &= \frac{e^2}{4\pi\,\epsilon_0\,\hbar c}\, \frac{\omega}{2} \sum_\lambda \int \frac{\mathrm{d}^2\Omega}{4\pi} \left| \left\langle \phi_f \left| \vec{r} \cdot \hat{\epsilon}_\lambda(\vec{k}) \right| \phi_i \right\rangle \right|^2 \frac{(2\pi)\,(4\pi)^2\,k^2}{8\,\pi^3} \\ &= 2\alpha\,\omega\,k^2 \sum_\lambda \int \frac{\mathrm{d}^2\Omega}{4\pi} \left| \left\langle \phi_f \left| \vec{r} \cdot \hat{\epsilon}_\lambda(\vec{k}) \right| \phi_i \right\rangle \right|^2 , \end{aligned} \tag{3.35}$$

where $\alpha = e^2/(4\pi\epsilon_0\hbar c)$ is the fine-structure constant. The angular integral over the polarization vectors is finally carried out with the help of Eq. (2.63),

$$\begin{aligned} W_{fi} &= 2\alpha\, \frac{\omega^3}{c^2} \sum_{ij} \int \frac{\mathrm{d}^2\Omega}{4\pi} \left\langle \phi_f \left| r^j \right| \phi_i \right\rangle \left\langle \phi_i \left| r^\ell \right| \phi_f \right\rangle \left(\delta^{j\ell} - \frac{k^j\,k^\ell}{\vec{k}^2} \right) \\ &= 2\alpha\, \frac{\omega^3}{c^2} \left\langle \phi_f \left| r^j \right| \phi_i \right\rangle \left\langle \phi_i \left| r^j \right| \phi_f \right\rangle \left(1 - \frac{1}{3} \right) . \end{aligned} \tag{3.36}$$

The result

$$W_{fi} = \frac{4\alpha}{3}\,\frac{\omega^3}{c^2} \left| \left\langle \phi_f \left| \vec{r} \right| \phi_i \right\rangle \right|^2 , \qquad \omega = \frac{E_i - E_f}{\hbar}\,, \tag{3.37}$$

is of fundamental importance for the description of atomic processes. In Eq. (3.37), we assume an average over the magnetic quantum numbers of the initial state, and a sum over the magnetic quantum numbers of the final state. One should still be aware that if any degeneracy persists in the final state (e.g., magnetic quantum numbers), then a sum over the degenerate final states needs to be performed in order to obtain the total decay rate. A potentially necessary averaging over initial-state configurations also is implied in Eq. (3.37).

3.3 Dynamic Stark Effect

3.3.1 *Way 1: Time-Dependent Perturbation Theory*

The dynamic Stark effect (sometimes called ac Stark shift, where "ac" stands for "alternating current") is an energy shift of an atomic reference state, caused by far off-resonant electromagnetic driving. Better stated, it is an energy shift of the atomic level, averaged over a period of the irradiating electromagnetic field, which is present if the frequency of the radiation cannot excite a transition to an excited state because it is far off resonance. The ac Stark shift, therefore, constitutes a perturbation and can be treated using time-dependent perturbation theory, but we have to go to second order (in contrast to Fermi's golden rule, where a first-order treatment was fully sufficient). Let us figure an analogy, in order to better understand the physical essence of the effect: If one drives a swing far off resonance, the child sitting on the swing will never get very high up, but will feel an average "quiver" energy due to an oscillatory motion of the swing. The position of the child, somewhat uncomfortably, will oscillate, and due to the average displacement from the equilibrium position of the swing, the child will acquire an average potential energy. By analogy, in a situation where an atom subject to far off-resonance electromagnetic radiation is unable to excite transitions to excited states, it feels a moderate "quiver" energy equal to the ac Stark shift.

The most "classical" way of calculating this energy shift is given by time-dependent perturbation theory, with a classical, oscillating, electric laser field. Thus, in this section, we rederive the classical expressions for the ac Stark shift, using a classical description of the laser field, inspired by Chap. 5 of Ref. [26]. The Hamiltonian is

$$H = H_0 + H_1(t)\,, \qquad H_0 = \sum_j E_j |\phi_j\rangle\,\langle\phi_j|\,,$$
$$H_1(t) = H_d \cos(\omega_{\rm L}\, t)\,, \qquad H_d = -e\, z\, E_{\rm L}\,, \tag{3.38}$$

where H_d is the dipole interaction Hamiltonian for the laser field, polarized in the z direction. The atomic eigenstates have energies E_j, and the monochromatic laser light has an angular frequency $\omega_{\rm L}$. We expect the wavelength of the laser radiation to be large compared to the spatial extent of the atomic wave function, so that the dipole approximation implicit in formulating H_d is justified (see also the pertinent discussion in Sec. 3.2.3).

In contrast to our discussion of Fermi's golden rule, we now use time-dependent perturbation theory in the interaction picture. Furthermore, because an energy shift is calculated, we have to employ adiabatic damping in the infinite past and in the infinite future. Similar considerations can be employed in the derivation of quantum electrodynamic energy shifts of atomic levels (for a review see, e.g., Chap. 2 of Ref. [116]). Thus, the interaction Hamiltonian is represented as

$$H_{\mathrm{I}}(\epsilon, t) = \mathrm{e}^{-\epsilon |t|} \left(\mathrm{e}^{\mathrm{i}\, H_0\, t/\hbar}\, H_1(t)\, \mathrm{e}^{-\mathrm{i}\, H_0\, t/\hbar} \right), \tag{3.39}$$

where H_0 is the unperturbed Hamiltonian of the atom. This interaction Hamiltonian is different from the one introduced in Eq. (2.108), in the sense that $H_1(t)$ has an explicit time dependence (laser field oscillation). It also contains exponentials which transform it into the interaction picture, and an exponential prefactor which leads to adiabatic damping in the distant past ($t \to -\infty$) and the distant future ($t \to \infty$), based on an infinitesimal positive damping parameter ϵ. In contrast to the derivation of Fermi's golden rule, we do not consider a decay process, which starts at a definite time with a prepared state, but we consider the energy perturbation of a stationary state. For this reason, we need to introduce some adiabatic damping, so that the reference state can be used as an adiabatic eigenstate of the full Hamiltonian in the infinite future and infinite past, where the interaction is not present.

Formally, the equation of motion of the wave function $|\phi_{\mathrm{I}}(t)\rangle$ in the interaction picture,

$$\mathrm{i}\,\hbar \frac{\partial}{\partial t} |\phi_{\mathrm{I}}(t)\rangle = H_{\mathrm{I}}(\epsilon, t)\, |\phi_{\mathrm{I}}(t)\rangle, \tag{3.40}$$

is solved by the time-ordered exponential

$$U_{\mathrm{I}}(\epsilon, t) = \mathrm{T}\,\exp\left(-\frac{\mathrm{i}}{\hbar} \int_{-\infty}^{t} \mathrm{d}t'\, H_{\mathrm{I}}(\epsilon, t) \right), \tag{3.41}$$

where T denotes the time ordering, and $U_{\mathrm{I}}(\epsilon, t)$ is a unitary time-translation operator. Expanding up to second order in the interaction, we calculate $U_{\mathrm{I}}(t)$ as follows,

$$\begin{aligned} U_{\mathrm{I}}(\epsilon, t) &\approx 1 - \frac{\mathrm{i}}{\hbar} \int_{-\infty}^{t} \mathrm{d}t'\, H_{\mathrm{I}}(\epsilon, t') + \frac{(-\mathrm{i})^2}{2!\,\hbar^2} \int_{-\infty}^{t} \mathrm{d}t' \int_{-\infty}^{t} \mathrm{d}t''\, \mathrm{T}\left(H_{\mathrm{I}}(\epsilon, t')\, H_{\mathrm{I}}(\epsilon, t'') \right) \\ &= 1 - \frac{\mathrm{i}}{\hbar} \int_{-\infty}^{t} \mathrm{d}t'\, H_{\mathrm{I}}(\epsilon, t') - \frac{1}{\hbar^2} \int_{-\infty}^{t} \mathrm{d}t' \int_{-\infty}^{t'} \mathrm{d}t''\, H_{\mathrm{I}}(\epsilon, t')\, H_{\mathrm{I}}(\epsilon, t''). \end{aligned} \tag{3.42}$$

We are now interested in the time evolution of the initial atomic reference state, which we denote as $|\phi_n\rangle$, with $H_0\,|\phi_n\rangle = E_n\,|\phi_n\rangle$. The time-dependent wave function in the interaction picture is

$$|\phi_{\mathrm{I}}(t)\rangle = U_{\mathrm{I}}(\epsilon, t)\;|\phi_{\mathrm{I}}(t = -\infty)\rangle = \sum_m c_m(t)\,|\phi_m\rangle. \tag{3.43}$$

The initial boundary condition is $c_m(-\infty) = \delta_{mn}$ for the reference state $|\phi_n\rangle$, with all other $c_m(-\infty)$ vanishing in the infinite past. In the interaction picture, $|\phi_n\rangle$ is stationary, because it is an eigenstate of the unperturbed Hamiltonian. We are interested in the projection

$$c_n(t) = \langle\phi_n|\phi_{\rm I}(t)\rangle = \langle\phi_n|U_{\rm I}(\epsilon,t)|\phi_n\rangle\,. \tag{3.44}$$

We now insert $U_{\rm I}(\epsilon,t)$ from Eq. (3.42) and calculate the matrix element

$$\begin{aligned} c_n(t) = \langle\phi_n|U_{\rm I}(\epsilon,t)|\phi_n\rangle &= 1 - \frac{1}{\hbar^2}\int_{-\infty}^{t} {\rm d}t' \int_{-\infty}^{t'} {\rm d}t''\,\langle\phi_n|H_{\rm I}(\epsilon,t')\,H_{\rm I}(\epsilon,t'')|\phi_n\rangle \\ &= 1 - \frac{1}{\hbar^2}\sum_m \int_{-\infty}^{t} {\rm d}t' \int_{-\infty}^{t'} {\rm d}t''\,\langle\phi_n|H_{\rm I}(\epsilon,t')|\phi_m\rangle\,\langle\phi_m|\,H_{\rm I}(\epsilon,t'')|\phi_n\rangle\,, \end{aligned} \tag{3.45}$$

where m is a multi-index which counts all bound and continuum states of the unperturbed atom. *Throughout this book, we will understand the summation symbol $\sum$ in a generalized sense, to include an integral over a continuous spectrum when appropriate.* The first-order perturbation vanishes because the reference state $|\phi_n\rangle$ cannot be coupled to itself in a dipole transition. Based on the result

$$\langle\phi_n|H_{\rm I}(\epsilon,t)|\phi_m\rangle = {\rm e}^{-\epsilon|t|}\,{\rm e}^{{\rm i}\,(E_n-E_m)t/\hbar}\,\cos(\omega_{\rm L}\,t)\,\langle\phi_n|H_d|\phi_m\rangle\,, \tag{3.46}$$

the time integrals can be carried out without much difficulty. The final result for the matrix element $c_n(t)$ in second-order perturbation theory is

$$c_n(t) = 1 - \frac{\exp(2\,\epsilon\,t)}{2\,\hbar\,\epsilon}\left(\frac{\rm i}{4}\sum_{m,\pm}\frac{\langle\phi_n|H_d|\phi_m\rangle\,\langle\phi_m|H_d|\phi_n\rangle}{E_n - E_m \pm \hbar\,\omega_{\rm L} + {\rm i}\hbar\epsilon}\right)\,, \tag{3.47}$$

where we ignore terms that vanish in the limit $\epsilon \to 0$. The easiest way to derive Eq. (3.47) is to assume that $t < 0$, which implies that $t' < 0$, so that we do not have to split the integration intervals into domains with negative and positive integration variables. Furthermore, we note that differentiation with respect to t of the numerator $\exp(2\,\epsilon\,t)$ brings down another factor ϵ. The $\pm$ summation over the signs $(\pm)$ in Eq. (3.47) covers two terms which differ in the sign of $\omega_{\rm L}$ in the denominator. The logarithmic derivative reads as

$$\lim_{\epsilon\to 0}\left(\frac{\partial}{\partial t}\ln c_n(t)\right) = \lim_{\epsilon\to 0}\left\{-\frac{\rm i}{4\hbar}\sum_{m,\pm}\frac{\langle\phi_n|H_d|\phi_m\rangle\,\langle\phi_m|H_d|\phi_n\rangle}{E_n - E_m \pm \hbar\,\omega_{\rm L} + {\rm i}\hbar\epsilon}\right\} = -{\rm i}\,\frac{\Delta E_n}{\hbar}\,, \tag{3.48}$$

if we interpret $c_n(t)$ as $c_n(t) = \exp(-{\rm i}\,\Delta E_n\,t/\hbar)$. From now on, we shall assume that the limit $\epsilon \to 0$ is taken at the end of a calculation, if an infinitesimal parameter ϵ is introduced into an expression. A remark is in order. We had calculated our matrix element $c_n(t) = \langle\phi_n|U_{\rm I}(\epsilon,t)|\phi_n\rangle$. Had we projected onto a different final state, as in the matrix element $\langle\phi_m|U_{\rm I}(\epsilon,t)|\phi_n\rangle$ with $E_m \neq E_n$, then we would not have obtained the infinitesimal ϵ term in the denominator, as in Eq. (3.47). Then, the logarithmic derivative calculated below in Eq. (3.48) would be of order ϵ, and the result would vanish in the limit $\epsilon \to 0$. This is a manifestation of a general principle that in adiabatic time-dependent perturbation theory, the final and initial states should be energetically degenerate in order to obtain nonvanishing final results for transition rates and energy shifts. We recall that energy conservation only holds if all emitted and absorbed quanta are counted in, so the degeneracy effectively amounts to an energy conservation condition. We have already encountered this principle in our derivation of Fermi's golden rule.

The dynamic Stark shift of the reference state $|\phi_n\rangle$ is thus obtained as

$$\Delta E_n = \frac{1}{4} \sum_{m,\pm} \frac{\langle \phi_n | H_d | \phi_m \rangle \langle \phi_m | H_d | \phi_n \rangle}{E_n - E_m \pm \hbar\omega_{\rm L}} . \tag{3.49}$$

In order to gain additional insight into the time evolution, we observe that the process is calculated in the interaction picture. Our *ansatz* for the wave function in the interaction picture has been given in Eq. (3.43), and we single out the component of the reference state $|n\rangle$,

$$|\psi_{\rm I}(t)\rangle = U_{\rm I}(\epsilon, t)\, |\psi_{\rm I}(0)\rangle = \sum_m c_m(t)\, |\phi_m\rangle \approx c_n(t)\, |\phi_n\rangle . \tag{3.50}$$

This translates to a Schrödinger-picture time evolution

$$|\psi_{\rm I}(t)\rangle = {\rm e}^{-{\rm i}\, H_0\, t/\hbar}\, |\psi_{\rm I}(t)\rangle \approx {\rm e}^{-{\rm i}\,(E_n + \Delta E_n)\, t}\, |\phi_n\rangle . \tag{3.51}$$

This justifies, *a posteriori*, the identification $c_n(t) = \exp(-{\rm i}\,\Delta E_n\, t)$ given in Eq. (3.48).

Complementing the sum over the spectrum of virtual states in operator form in Eq. (3.49), we can express the dynamic Stark shift of the reference state $|\phi_n\rangle$ as

$$\Delta E_n = \frac{1}{4} \sum_{\pm} \left\langle \phi_n \left| H_d \frac{1}{E_n - H_0 \pm \hbar\omega_{\rm L}} H_d \right| \phi_n \right\rangle , \tag{3.52}$$

where H_d is time-independent [see Eq. (3.38)]. For a spherically symmetric reference state $|\phi_n\rangle$, we define the dynamic dipole polarizability $\alpha(\omega_{\rm L})$ as

$$\begin{aligned} \alpha(\omega_{\rm L}) &= e^2 \sum_{\pm} \left\langle \phi_n \left| z \frac{1}{H_0 - E_n \pm \hbar\omega_{\rm L}} z \right| \phi_n \right\rangle \\ &= \frac{e^2}{3} \sum_{\pm} \left\langle \phi_n \left| \vec{r} \frac{1}{H_0 - E_n \pm \hbar\omega_{\rm L}} \vec{r} \right| \phi_n \right\rangle , \end{aligned} \tag{3.53}$$

which is not to be confused with the fine-structure constant α. The scalar product of the two coordinate vectors $\vec{r}$ is implied (for a complex atom, one has to sum over the coordinates $\vec{r}_a$ of all electrons labeled by the index a). The polarizability is normalized in such a way that it has the physical dimension of a dipole moment when multiplied by an electric field. This dynamically induced dipole is the dipole moment dynamically induced by the exciting, oscillatory electric field. In SI mksA units, the unit of polarizability is $\rm C^2\, m^2/J$.

The ac Stark shift of the reference state is finally expressed as

$$\Delta E_n = -\frac{E_{\rm L}^2}{4}\, \alpha(\omega_{\rm L}) = -\frac{I_{\rm L}}{2\, \epsilon_0\, c}\, \alpha(\omega_{\rm L}) , \tag{3.54}$$

where the laser field intensity of a plane electromagnetic wave in SI mksA units reads as

$$I_{\rm L} = \frac{1}{2}\,\epsilon_0\,c\,E_{\rm L}^2\,. \tag{3.55}$$

The real part of ΔE_n represents the shift of the energy eigenvalue, induced by the perturbation. If the frequency of the laser field is too low to excite a transition to an energetically upper state, then the imaginary part of the energy shift vanishes. However, if the energy $E_n + \hbar\omega_{\rm L}$ is located within a continuum of states in the spectrum of H_0, then the ac Stark shift develops an imaginary part. The imaginary part changes the amplitude modulus of $|n\rangle$ with time, and represents a "width" of the state $|\phi_n\rangle$ acquired in the laser field, due to ionization into the continuum,

$$E_n = {\rm Re}\,E_n - \frac{\rm i}{2}\,\Gamma_n\,. \tag{3.56}$$

This leads to the following evolution of the state vector with time, with an exponential factor $\exp(-\frac{\rm i}{\hbar}\,E_n\,t)$,

$$|\phi_n(t)\rangle = \exp\left(-\tfrac{1}{2\hbar}\,\Gamma_n\,t\right)\,\exp\left[-\frac{\rm i}{\hbar}\,(E_n + {\rm Re}\,\Delta E_n)\,t\right]\,|\phi_n(0)\rangle\,. \tag{3.57}$$

Formally, we can establish a time-dependent norm (a nonvanishing decay rate) of the state $|\phi_n\rangle$ in the laser field, because of depletion into the continuum by ionization, according to the formula $\||\phi_n(t)\rangle|^2 = \exp\left(-\Gamma_n\,t\right)$. It is instructive to match this result with the transition rate, calculated using Fermi's Golden Rule in first-order time-dependent perturbation theory. The result for the ionization cross section $\sigma_{\rm ioni}$ reads [6, 117]

$$\sigma_{\rm ioni} = \frac{\omega_{\rm L}}{I_{\rm L}}\,\Gamma_n = \frac{\omega_{\rm L}}{\epsilon_0\,c}\,{\rm Im}\,\alpha(\omega_{\rm L})\,. \tag{3.58}$$

This result has the right physical dimension: The angular frequency has units of radians per second, whereas Γ_n is (the imaginary part of) an energy. The intensity has units of energy per time per area; the ratio given in Eq. (3.58) therefore is a cross sectional area, i.e., it has a physical dimension of length squared. It is highly instructive to rewrite Eq. (3.58) as follows,

$$\gamma_n = \frac{\Gamma_n}{\hbar} = \sigma_{\rm ioni}\,\frac{I_{\rm L}}{\hbar\omega_L}\,. \tag{3.59}$$

Here, γ_n is the rate of transition from the bound to the continuum state, while $I_{\rm L}/(\hbar\omega)$ is the number of photons per cross sectional area per time. Multiplied by the ionization cross section $\sigma_{\rm ioni}$, the latter quantity gives the transition rate. For typical cases, ionization cross sections are of the order of the square of the Bohr radius, and they scale with $1/Z^4$ for hydrogenlike ions (of nuclear charge number Z), because these ions become smaller with increasing nuclear charge.

3.3.2 *Way 2: Quantized Fields*

It is highly instructive to repeat the above derivation using time-independent perturbation theory in the second-quantized formalism. To be specific, we adopt the electromagnetic field in the second-quantized formalism, whereas the atom is first-quantized. In Eqs. (2.85) and (2.96), we had encountered the Hamiltonian of the electromagnetic field as a sum over the available modes. For a single, highly populated laser mode, the Hamiltonian $H_{\rm L}$ of the laser mode simply counts the laser photons on the basis of the number operator $a^+_{\rm L}\,a_{\rm L}$. The angular laser frequency is denoted as $\omega_{\rm L}$, and $a_{\rm L} \equiv a_{\lambda_{\rm L}}(\vec{k}_{\rm L})$ is the annihilation operator of a laser mode photon (in the discretized representation (2.96), with $[a_{\rm L}, a^+_{\rm L}] = 1$). The Hamiltonian for the atom is kept in symbolic form, as in Eq. (3.38), and the unperturbed Hamiltonian of the atom and laser field (hence, with two zero indices) is given as

$$H_{00} = \sum_m E_m \,|\phi_m\rangle\,\langle\phi_m| + \hbar\,\omega_{\rm L}\, a^+_{\rm L}\, a_{\rm L}\,, \tag{3.60}$$

where the sum over m covers the entire bound and continuous spectra of the atom. We reserve the notation H_0 for the unperturbed Hamiltonian of the atom alone, in agreement with Eq. (3.38). Furthermore, in agreement with the discussion in Sec. 3.3.1, we assume that the polarization vector $\hat{\epsilon}_{\rm L} = \hat{\rm e}_z$ of the laser field is aligned along the z axis.

It is necessary to restrict the operator of the electric field in the discrete representation, given in Eq. (2.129b), to the laser mode,

$$\vec{E}_{\rm L}(\vec{r}) = {\rm i}\sqrt{\frac{\hbar\omega_{\rm L}}{2\,\epsilon_0\,V}}\,\hat{\epsilon}_{\rm L}\,\left(a_{\rm L}\,{\rm e}^{{\rm i}\,\vec{k}\cdot\vec{r}} - a^+_{\rm L}\,{\rm e}^{-{\rm i}\,\vec{k}\cdot\vec{r}}\right)\,. \tag{3.61}$$

In the dipole approximation, we have

$$\vec{E}_{\rm L}(\vec{r}) \approx \vec{E}_{\rm L}(\vec{r} = 0) = {\rm i}\sqrt{\frac{\hbar\omega_{\rm L}}{2\,\epsilon_0\,V}}\,\hat{\epsilon}_{\rm L}\,\left(a_{\rm L} - a^+_{\rm L}\right)\,. \tag{3.62}$$

On the level of second quantization, the electromagnetic field operator can thus be formulated as a time-independent quantity, whereas on the level of classical field theory, one cannot transform the time evolution of the electromagnetic fields away. The number states in the Fock space are the states $|n_{\rm L}\rangle$ with $a^+_{\rm L}a_{\rm L}|n_{\rm L}\rangle = n_{\rm L}|n_{\rm L}\rangle$. They fulfill the following relations,

$$a_{\rm L}|n_{\rm L}\rangle = \sqrt{n_{\rm L}}\,|n_{\rm L} - 1\rangle\,, \qquad a^+_{\rm L}|n_{\rm L}\rangle = \sqrt{n_{\rm L}+1}\,|n_{\rm L}+1\rangle\,. \tag{3.63}$$

The quantized electric-dipole interaction of the electron enters the interaction Hamiltonian, which reads as follows,

$$H_1 = -e\,\vec{r}\cdot\vec{E}_{\rm L} = -{\rm i}\,e\,z\sqrt{\frac{\hbar\,\omega_{\rm L}}{2\,\epsilon_0\,V}}\,\left(a_{\rm L} - a^+_{\rm L}\right)\,. \tag{3.64}$$

In second quantization, the Hamiltonian for the coupled system consisting of atom and radiation field reads

$$H = H_{00} + H_1\,, \tag{3.65}$$

where H_{00} is given in Eq. (3.60). Eigenstates of H_{00} are product states with the atom in state $|\phi_n\rangle$ and $n_{\rm L}$ photons in the laser mode. We may choose $|\phi_n, n_{\rm L}\rangle$ as the reference state, where the subscript n in ϕ_n and $n_{\rm L}$ have totally different physical meanings. The Hamiltonian multiplies the state by the (unperturbed) energy eigenvalue,

$$H_{00}\,|\phi_n, n_{\rm L}\rangle = (E_n + \hbar\, n_{\rm L}\, \omega_{\rm L})\,|\phi_n, n_{\rm L}\rangle\,. \tag{3.66}$$

The first-order perturbation vanishes, $\langle \phi_n, n_{\rm L}\,|H_1|\,\phi_n, n_{\rm L}\rangle = 0$. For second-order energy shifts, we need the reduced Green function which excludes the reference state, i.e., the product state of the atom being in the reference state and the photon field being in the state with $n_{\rm L}$ laser photons. The spectral decomposition of the reduced Green function reads as

$$\left(\frac{1}{z - H_{00}}\right)' = \underbrace{\sum_{m_{\rm L}}\sum_{m}}_{|\phi_m, m_{\rm L}\rangle \neq |\phi_n, n_{\rm L}\rangle} \frac{|\phi_m, m_{\rm L}\rangle\,\langle\phi_m, m_{\rm L}|}{z - E_m - m_{\rm L}\,\hbar\,\omega_{\rm L}}\,. \tag{3.67}$$

Here, we exclude only the reference state $|\phi_n, n_{\rm L}\rangle$ from the sum over virtual, intermediate states. For the state $|\phi_n, n_{\rm L}\rangle$, both the atomic part as well as the field part of the product state are equal to the reference state. The second-order energy shift is

$$\begin{aligned}
\Delta E_n &= \left\langle \phi_n, n_{\rm L} \left| H_1 \left(\frac{1}{E_n + n_{\rm L}\,\hbar\,\omega_{\rm L} - H_{00}}\right)' H_1 \right| \phi_n, n_{\rm L} \right\rangle \\
&= \frac{e^2\,\hbar\,\omega_{\rm L}}{2\,V} \left(\left\langle \phi_n, n_{\rm L} \left| z\, a_{\rm L} \frac{1}{E_n + n_{\rm L}\,\hbar\,\omega_{\rm L} - H_{00}} z\, a^+_{\rm L} \right| \phi_n, n_{\rm L} \right\rangle \right. \\
&\quad \left. + \left\langle \phi_n, n_{\rm L} \left| z\, a^+_{\rm L} \frac{1}{E_n + n_{\rm L}\,\hbar\,\omega_{\rm L} - H_{00}} z\, a_{\rm L} \right| \phi_n, n_{\rm L} \right\rangle \right).
\end{aligned} \tag{3.68}$$

Because $a_{\rm L}$ annihilates, while $a^+_{\rm L}$ generates, exactly one photon, the second-order energy shift involves a very restricted number of intermediate photon states, with one photon more, or one photon less, than the reference state,

$$\begin{aligned}
\Delta E_n &= \frac{e^2\,\hbar\,\omega_{\rm L}}{2\,\epsilon_0\,V} \sum_m \left\{ \frac{\langle \phi_n, n_{\rm L}\,|\,z\,a_{\rm L}|\,\phi_m, n_{\rm L}+1\rangle\,\langle \phi_m, n_{\rm L}+1\,|\,z\,a^+_{\rm L}|\,\phi_n, n_{\rm L}\rangle}{E_n + n_{\rm L}\,\hbar\,\omega_{\rm L} - (E_m + (n_{\rm L}+1)\,\hbar\,\omega_{\rm L})} \right. \\
&\quad \left. + \frac{\langle \phi_n, n_{\rm L}\,|\,z\,a^+_{\rm L}|\,\phi_m, n_{\rm L}-1\rangle\,\langle \phi_m, n_{\rm L}-1\,|\,z\,a_{\rm L}|\,\phi_n, n_{\rm L}\rangle}{E_n + n_{\rm L}\,\hbar\,\omega_{\rm L} - (E_m + (n_{\rm L}-1)\,\hbar\,\omega_{\rm L})} \right\} \\
&= \frac{e^2\,\hbar\,\omega_{\rm L}}{2\,\epsilon_0\,V} \left((n_{\rm L}+1) \left\langle \phi_n \left| z \frac{1}{E_n - \hbar\,\omega_{\rm L} - H_0} z \right| \phi_n \right\rangle \right. \\
&\quad \left. + n_{\rm L} \left\langle \phi_n \left| z \frac{1}{E_n + \hbar\,\omega_{\rm L} - H_0} z \right| \phi_n \right\rangle \right).
\end{aligned} \tag{3.69}$$

Here, H_0 refers to the atomic part of the unperturbed Hamiltonian. For large photon number $n_{\rm L}$ and large normalization volume V, with a finite ratio $n_{\rm L}/V$, we can replace $(n_{\rm L}+1)/V \to n_{\rm L}/V$ and obtain

$$\Delta E_n = \frac{n_{\rm L}\,e^2\,\omega_{\rm L}}{2\epsilon_0\,V} \left\langle \phi_n \left| z \frac{1}{E_n - \hbar\omega_{\rm L} - H_0} z + z \frac{1}{E_n + \hbar\omega_{\rm L} - H_0} z \right| \phi_n \right\rangle. \tag{3.70}$$

With the dynamic polarizability $\alpha(\omega_{\rm L})$ as defined in Eq. (3.53), we can write

$$\Delta E_n = -\frac{n_{\rm L}\,\hbar\,\omega_{\rm L}}{2\,\epsilon_0\,V}\,\alpha(\omega_{\rm L})\,. \tag{3.71}$$

It remains to clarify the role of the laser intensity in this result. The energy density w of the laser field and the intensity $I_{\rm L}$ are related by

$$w = \frac{n_{\rm L}\,\hbar\,\omega_{\rm L}}{V}\,, \qquad I_{\rm L} = w\,c = \frac{n_{\rm L}\,\hbar\,\omega_{\rm L}\,c}{V}\,. \tag{3.72}$$

For ΔE_n, we finally obtain the same expression as in Eq. (3.54),

$$\Delta E_n = -\frac{I_{\rm L}}{2\,\epsilon_0\,c}\,\alpha(\omega_{\rm L})\,. \tag{3.73}$$

We thus conclude that the classical-field and the quantized-field approach give consistent results in the limit of a strong, macroscopic laser field.

3.3.3 *Way 3: Gell-Mann–Low–Sucher Theorem*

It is quite surprising that there actually is a third way to derive the energy shift of an atom in an oscillatory laser field. The third way is almost completely independent of the two derivations given above and exclusively relies on a rather abstract theorem, known as the Gell-Mann–Low–Sucher theorem, which is given in Eq. (2) of Ref. [118]. This theorem relates the auto-correlative matrix element $\langle\phi_n|U_{\rm I}(\epsilon,0)|\phi_n\rangle$ for a reference state $|\phi_n\rangle$ to the energy shift experienced by the state under a perturbation (3.39),

$$H_{\rm I}(\epsilon,t) = {\rm e}^{-\epsilon|t|}\,H_{\rm I}(t) = {\rm e}^{-\epsilon|t|}\left({\rm e}^{{\rm i}\,H_0\,t/\hbar}\,H_1(t)\,{\rm e}^{-{\rm i}\,H_0\,t/\hbar}\right)\,, \tag{3.74}$$

where $H_{\rm I}(t)$ is the interaction Hamiltonian in the interaction picture. From the matrix element $\langle\phi_n|U_{\rm I}(\epsilon,t=0)|\phi_n\rangle$ alone, which describes the time evolution from the infinite past, where the interaction is adiabatically switched off, to the time $t=0$, one can derive the energy shift. Let us carry out the calculation for the ac Stark shift. We first recall the result given in Eq. (3.47), which reads

$$c_n(t) = \langle\phi_n|U_{\rm I}(\epsilon,t)|\phi_n\rangle = 1 - \frac{\exp(2\epsilon t)}{2\hbar\epsilon}\left(\frac{\rm i}{4}\sum_{m,\pm}\frac{\langle\phi_n|H_d|\phi_m\rangle\langle\phi_m|H_d|\phi_n\rangle}{E_n - E_m \pm \hbar\omega_{\rm L} + {\rm i}\epsilon}\right)\,. \tag{3.75}$$

The matrix element of the time evolution operator, at time zero, through second-order time-dependent perturbation theory, therefore reads as

$$\langle\phi_n|U_{\rm I}(\epsilon,0)|\phi_n\rangle = 1 - \frac{1}{2\hbar\epsilon}\left(\frac{\rm i}{4}\sum_{m,\pm}\frac{\langle\phi_n|H_d|\phi_m\rangle\langle\phi_m|H_d|\phi_n\rangle}{E_n - E_m \pm \hbar\omega_{\rm L} + {\rm i}\epsilon}\right)\,. \tag{3.76}$$

One first augments the interaction Hamiltonian $H_{\rm I}(\epsilon,t)$ defined in Eq. (3.39) by an artificial coupling parameter g, which serves to characterize the coupling strength and orders the perturbation series,

$$H_{\rm I}(\epsilon,t) \to g\,H_{\rm I}(\epsilon,t)\,. \tag{3.77}$$

Then, according to the Gell-Mann–Low–Sucher theorem, we have

$$\begin{aligned}
\Delta E_n &= \lim_{\epsilon\to 0}\lim_{g\to 1}\left(\mathrm{i}\,\hbar\epsilon\, g\,\frac{\partial}{\partial g}\,\ln\langle\phi|U_{\mathrm{I}}(\epsilon,0)|\phi\rangle\right)\\
&= \lim_{\epsilon\to 0}\lim_{g\to 1}\left(\mathrm{i}\,\hbar\epsilon\, g\,\frac{\partial}{\partial g}\,\ln\left(1-\frac{g^2}{2\hbar\epsilon}\,\frac{\mathrm{i}}{4}\sum_{m,\pm}\frac{\langle\phi_n|H_d|\phi_m\rangle\langle\phi_m|H_d|\phi_n\rangle}{E_n-E_m\pm\hbar\omega_{\mathrm{L}}+\mathrm{i}\epsilon}\right)\right)\\
&\approx \lim_{\epsilon\to 0}\lim_{g\to 1}\frac{g}{8}\,\frac{\partial}{\partial g}\,g^2\sum_{m,\pm}\frac{\langle\phi_n|H_d|\phi_m\rangle\langle\phi_m|H_d|\phi_n\rangle}{E_n-E_m\pm\hbar\omega_{\mathrm{L}}+\mathrm{i}\epsilon}\\
&= \frac{1}{4}\sum_{m,\pm}\frac{\langle\phi_n|H_d|\phi_m\rangle\langle\phi_m|H_d|\phi_n\rangle}{E_n-E_m\pm\hbar\omega_{\mathrm{L}}+\mathrm{i}\epsilon}\,.
\end{aligned}\tag{3.78}$$

The latter result for ΔE_n is in full agreement with the intermediate result given in Eq. (3.49) in Sec. 3.3.1. The next steps toward Eqs. (3.54) and (3.73) are completed as outlined in Sec. 3.3.1, demonstrating that the Gell-Mann–Low–Sucher theorem is in agreement with the alternative derivations presented in Secs. 3.3.1 and 3.3.2. In the transformation step denoted by "≈" in Eq. (3.78), an expansion of the logarithm in powers of g is implied, which is equivalent to a second-order expansion in the time-dependent perturbation H_{I}. Higher-order terms in g would correspond to fourth-order interactions of the atom with the laser field, and would be proportional to so-called hyperpolarizabilities.

3.4 Static Stark Effect and Level Width

3.4.1 *Stark Shift and Large-Order Perturbation Theory*

It is quite intriguing to observe that the above second-order perturbation theory treatment of the ac Stark effect has an interesting static limit that corresponds to a time-independent electric field. In this case, perturbation theory can be calculated to very high order, and a number of interesting phenomena can be observed, which will be described in the following. These observations are in fact connected with a set of deep questions regarding the asymptotic character of perturbation series in high orders. They are discussed here using the static Stark effect as a particularly simple, conceptually evident example, where a number of phenomena find an immediate physical interpretation.

Let us therefore consider the static limit of the above treatment, i.e., the case $\omega_{\mathrm{L}} \to 0$. With $H_d = -e\,z\,E_{\mathrm{L}}$, the energy shift simplifies to

$$\Delta E_n = \lim_{\omega_{\mathrm{L}}\to 0}\frac{1}{4}\sum_{m,\pm}\frac{\langle\phi_n|H_d|\phi_m\rangle\,\langle\phi_m|H_d|n\rangle}{E_n-E_m\pm\hbar\omega_{\mathrm{L}}+\mathrm{i}\,\epsilon} = \frac{e^2\,E_{\mathrm{L}}^2}{2}\sum_{m\neq n}\frac{\langle\phi_n|z|\phi_m\rangle\,\langle\phi_m|z|\phi_n\rangle}{E_n-E_m}\,.\tag{3.79}$$

The symmetric limit $\pm\omega_{\mathrm{L}} \to 0$ eliminates the reference state $|\phi_n\rangle$ from the sum over intermediate states $|\phi_j\rangle$ and leads to an overall factor 2. However, in order to calculate the static limit properly, we have to remember that this result has been obtained for an oscillatory electric field of the form $E_{\mathrm{L}}\,\cos(\omega_{\mathrm{L}}\,t)$. In this result, the term $E_{\rm L}^2/2$ represents the square of the electric field, averaged over a laser period. For a static field $\vec{E} = E_z\,\hat{e}_z$, with an interaction Hamiltonian

$$H_1 = -e\,\vec{E}\cdot\vec{r} = -e\,E_z\,z\,, \tag{3.80}$$

the second-order energy shift therefore reads as

$$\Delta E_n = e^2\,E_z^2 \sum_{m\neq n} \frac{\langle\phi_n|z|\phi_m\rangle\,\langle\phi_j|z|\phi_n\rangle}{E_n - E_m} = -e^2\,E_z^2\left\langle\phi_n\left|z\left(\frac{1}{H_0 - E_0}\right)'z\right|\phi_n\right\rangle. \tag{3.81}$$

Here, $H_0 = \sum_m E_m|\phi_m\rangle\langle\phi_m|$ is the unperturbed atomic Hamiltonian, and the prime denotes the (atomic) reduced Green function. In general, when an atom or ion is exposed to a static electric field, the energy levels are displaced from their field-free positions (Stark effect). For a nondegenerate reference state, the expectation value of the electric dipole operator in a particular quantum eigenstate (e.g., the ground state) of an atom vanishes due to parity symmetry, and the leading effect is of order E_z^2. For the ground state $|\phi_{1S}\rangle$ of the hydrogen atom, the second-order shift reads, in operator notation,

$$\Delta E_0 = -e^2\,E_z^2\left\langle\phi_{1S}\left|z\left(\frac{1}{H_0 - E_0}\right)'z\right|\phi_{1S}\right\rangle = -\frac{9}{4}\,E_h\,G^2 < 0\,, \tag{3.82}$$

where the numerical result refers to hydrogen. Here, E_h is the Hartree energy, as defined in Eq. (2.27), while G is a coupling parameter,

$$G = E_z/\hat{E}_{\rm a.u.}\,. \tag{3.83}$$

The atomic unit $\hat{E}_{\rm a.u.}$ of the electric field strength is equal to the electric field felt by a bound electron at a distance of one Bohr radius from a proton (see Sec. 2.3.3). (In atomic units, we therefore have $\hat{E}_{\rm a.u.} = 1$.) In SI mksA units, the atomic field strength is given by

$$\hat{E}_{\rm a.u.} = \left|-\frac{\partial}{\partial r}\left(-\frac{e}{4\pi\epsilon_0\,r}\right)\right|_{r=a_0}\right| = \frac{e}{4\pi\epsilon_0\,a_0^2} = \frac{\alpha^3\,m_e^2\,c^3}{e\hbar}\,. \tag{3.84}$$

Here, m_e is the electron mass (we calculate in the non-recoil limit, with an infinitely heavy atomic nucleus). Numerically, we have $\hat{E}_{\rm a.u.} = 5.1422\times10^{11}$V/m, illustrating that the electric fields, which hold the atoms together, are typically much larger than macroscopically relevant electric fields, but smaller than the electric fields needed to "make the vacuum spark", i.e., to generate electron-positron pairs from the vacuum. This latter effect will be discussed in greater detail in Chap. 18. The energy shift in Eq. (3.82) is negative, because all virtual states entering H_0 have a higher energy than the reference state.

An interesting phenomenon associated with the static Stark effect has given rise to some discussion. Namely, the perturbation series that describes the static Stark energy shift is nonalternating (the terms have the same sign), and it is factorially divergent in large orders in the coupling parameter G. For the ground state of hydrogen, which we denote by the subscript $|0\rangle$, the first perturbative terms read as follows,

$$E_0(G) = E_h \left(-\frac{1}{2} - \frac{9}{4} G^2 - \frac{3\,555}{64} G^4 - \frac{2\,512\,779}{512} G^6 - \frac{13\,012\,777\,803}{16\,384} G^8 + \cdots \right). \tag{3.85}$$

Here, E_0 is the shifted ground-state energy of the hydrogen atom as a function of coupling parameter G. By inspection, one can infer that the magnitude of the perturbative coefficients grows rather drastically with the order of perturbation theory. One can formally write the perturbation series as [119–122]

$$E_0(G) \sim -E_h \sum_{K=0,2,4,\ldots}^{\infty} a_K\, G^K\,, \qquad a_K \sim \frac{3\,K!}{\pi} \left(\frac{3}{2}\right)^K, \qquad K \to \infty\,. \tag{3.86}$$

The series in Eq. (3.85) is divergent, no matter how small the electric field strength is. The ratio of two successive terms in the perturbation series is

$$\frac{a_{K+2}\, G^{K+2}}{a_K\, G^K} \sim \frac{9}{4} K\,(K+1)\, G^2 \sim \frac{9}{4} K^2\, G^2 \tag{3.87}$$

for large K. This implies that the perturbative terms grow with the order K when $K \geq 2\,(3\,G)^{-1}$ no matter how small G is. The minimal term in the perturbation series is reached at order $K \sim 2/(3\,G)$, and it is at this order that the perturbation series can be "optimally truncated".

The physics behind the divergent perturbation series can be analyzed as follows: First, it is necessary to remark that the atom will *not* experience an arbitrarily large energy shift in a small electric field. Rather, the rigid interpretation of a perturbation series as a convergent rock-solid physical prediction should be reconsidered. A perturbation series, in general, is an asymptotic series which does not converge, but whose first terms still give us a good indication of the physically relevant answer, provided the coupling parameter G is not too large, i.e., for the case $G \ll 1$. In this case, it is possible to obtain a good prediction by considering only the first few terms of the asymptotic series until the minimum term is reached.

For the Stark effect, the nonalternating divergent perturbation series given in Eqs. (3.85) and (3.86) has a further physical interpretation. This additional point can be understood if we realize that non-alternating divergent series typically correspond to Borel-type integrals like those given in Eq. (3.93), which are evaluated "on the cut" in the complex plane, i.e., in a parameter region where a slight displacement of the coupling parameter leads to the emergence of a nonvanishing imaginary part for the resonance energy. In the current case, the physical interpretation for the emerging imaginary part is evident: The bound electron is tied to the nucleus by the Coulomb potential. However, in a uniform, infinitely extended electric field, it may eventually tunnel through the barrier into a region where the total electrostatic potential (sum of Coulomb field and external static field) is lower than the binding energy of the electron. Quantum mechanically, the tunneling probability is nonvanishing even for a small external electric field (small coupling parameter G), and it gives rise to a decay width $\Gamma_0(G)$ which represents the imaginary part of the perturbed ground-state energy $E_0(G)$,

$$E_0(G) = \mathrm{Re}\,E_0(G) - \frac{\mathrm{i}}{2}\,\Gamma_0(G)\,, \tag{3.88}$$

in full analogy to Eq. (3.56). The decay width $\Gamma_0(G)$ can be approximated, for small coupling parameter G, as follows (see Refs. [119–122]),

$$\Gamma_0(G) \sim \frac{4\,E_h}{G}\,\exp\left(-\frac{2}{3\,G}\right)\,, \qquad G \to 0\,. \tag{3.89}$$

The leading term in the large-order asymptotics of the perturbation series (3.86) and the leading term in the decay width (3.89) are related by a dispersion relation. For the Stark effect, the dispersion relation is not completely trivial; it is given in Eq. (1) of Ref. [122]. One has to take into account cuts of the resonance energy $E_0(G)$ for both positive as well as negative coupling parameter G. A detailed analysis of the implications of the complex structure of the resonance energies can be found in Ref. [123]. Under certain conditions, it is possible to calculate the decay width using Borel–Padé summation [124], using nontrivial complex contours in the evaluation of the Borel integral [125, 126]. The resonance energy is obtained as a complex quantity even if the perturbative coefficients are purely real. In these cases, it is possible to overcome the limit set by the smallest term in the asymptotic series. The numerical values for the real and imaginary parts of the resonance energies of the Schrödinger–Stark problem, obtained using resummation, can be verified against directly obtained solutions of complex-scaled radial differential equations (see Refs. [127, 128]).

One may ask if the divergent perturbation series (3.85) can be mapped onto a mathematical model problem. To this end, let us consider the integral

$$\mathcal{A}(g) = -\frac{1}{g}\,\exp\left(\frac{1}{g}\right)\,\mathrm{Ei}\left(-\frac{1}{g}\right) = \frac{1}{g}\int_0^\infty \frac{1}{1+u}\,\exp\left(-\frac{u}{g}\right)\,\mathrm{d}u\,. \tag{3.90}$$

Here, the so-called exponential integral Ei is defined by a principal-value prescription, denoted as (P.V.),

$$\mathrm{Ei}(x) \;=\; -(\mathrm{P.V.})\int_{-\infty}^{x} t^{-1}\,\mathrm{e}^{t}\,\mathrm{d}t\,. \tag{3.91}$$

Upon expansion in powers of g, we obtain the perturbation series,

$$\mathcal{A}(g) \sim \sum_{K=0}^{\infty} (-1)^K\,K!\,g^K\,, \qquad g \to 0\,, \tag{3.92}$$

which is divergent no matter how small g is. Note that the "∼" sign means that the series on the right is an asymptotic, alternating series.

An even more interesting example is the integral

$$\mathcal{N}(g) = \frac{1}{g}\int_C \frac{1}{1-u}\,\exp\left(-\frac{u}{g}\right)\,\mathrm{d}u\,, \tag{3.93}$$

for $g > 0$. Here, C is a contour which starts at $u = 0$, ends at $u = +\infty$, and encircles the pole at $u = 1$ as specified below. Upon expansion in powers of g, $\mathcal{N}(g)$ gives rise to a nonalternating series,

$$\mathcal{N}(g) = \sum_{K=0}^{\infty} K!\, g^K\,, \qquad g \to 0\,. \tag{3.94}$$

From the integral representation in Eq. (3.93), it is clear that there is some sort of ambiguity for real, positive g in the final evaluation of $\mathcal{N}(g)$, depending on how the pole at $u = 1$ is encircled. Indeed, depending on how the pole term is encircled in Eq. (3.93), we have the result

$$\mathcal{N}(g > 0) = \begin{cases} \dfrac{1}{g} \exp\left(-\dfrac{1}{g}\right) \mathrm{Ei}(g^{-1}) & \text{(principal value)} \\ \dfrac{1}{g} \exp\left(-\dfrac{1}{g}\right) \mathrm{Ei}(g^{-1}) + \dfrac{\mathrm{i}\pi}{g} \exp\left(-\dfrac{1}{g}\right) & \text{(upper contour)} \\ \dfrac{1}{g} \exp\left(-\dfrac{1}{g}\right) \mathrm{Ei}(g^{-1}) - \dfrac{\mathrm{i}\pi}{g} \exp\left(-\dfrac{1}{g}\right) & \text{(lower contour)} \end{cases}. \tag{3.95}$$

An interesting observation can be made: Although the original perturbation series (3.94) is purely real for real coupling g, the result of the "summation" of the series can be manifestly complex. The "summation" in this case consists in the evaluation of the integral (3.93). We recall that the series (3.94) was generated upon expansion of (3.93) in powers of g. A numerical evaluation of (3.93) along a contour infinitesimally above or below the pole at $u = 1$ therefore corresponds to an evaluation of the very expression from which the perturbation series (3.94) originated by a perturbative expansion. This is the meaning of the idea of resummation as first imagined by EULER[3] who wrote in a letter to GOLDBACH[4] in 1745: *"Summa cuiusque seriei est valor expressionis illius finitae, ex cuius evolutione illa series oritur."* This is translated as follows: The sum of any (possibly divergent) series is equal to the numerical value of that particular finite expression, which gave rise to the (possibly divergent) series upon asymptotic expansion in the first place. There is some ambiguity in identifying the latter finite expression: Namely, as illustrated in Eq. (3.95), one may add or subtract a term whose perturbation series vanishes, i.e., a non-analytic term for $g \to 0$. Indeed, the Taylor expansion of $\exp(-1/g)$ vanishes to all orders in g.

Still, it is intriguing to observe that the imaginary parts of the expressions on the right-hand side of Eq. (3.95) come in discrete steps: Either one has no imaginary part at all, or one has a positive, or a negative, nonanalytic expression. For a hydrogen atom in an electrostatic field, the imaginary part of the energy shift has a physical interpretation: It describes the probability for the bound electron to escape to infinity, in any electric field, by tunneling through the potential barrier. This imaginary part of the energy shift should be negative, therefore describing a decaying state [see also Eq. (3.56)]. It is very intriguing to ask whether or not, given the

[3]Leonhard Euler (1707–1783).
[4]Christian Goldbach (1690–1764).

purely real character of the perturbation series (3.94), and of the series (3.85) which describes a hydrogen atom in an electrostatic field, one may in effect recalculate the decay width of the state against tunneling, i.e., the imaginary part of the energy shift. We have already mentioned the related investigations in Refs. [123, 125, 126]. For our mathematical model problem (3.94), the negative imaginary part of the final answer is obtained when using the "lower" contour, i.e., the contour which encircles the pole in the mathematically positive direction. In the power series (3.94), all coefficients are positive. By contrast, the coefficients are negative in the series (3.85). So, the "upper" contour should be used, in order to obtain a negative imaginary part of the energy, as it was employed in Refs. [123, 125]. According to Table I of Ref. [123], by a resummation of the perturbation series (3.85), one obtains a numerical result for the manifestly complex ground-state resonance energy E_0 of

$$\frac{E_0(G = 0.04)}{E_h} = -0.503\,771\,591\,013\,654\,2(5) - \frac{\mathrm{i}}{2}\,3.892\,699\,990(1) \times 10^{-6}\,. \quad (3.96)$$

The numerical accuracy, which is quoted in parentheses, deserves a comment. Namely, in the resummation methods described in Refs. [123, 125], one first has to perform an analytic continuation of the so-called Borel transform of the perturbation series (3.85) beyond its original circle of convergence, in order to be able to evaluate a numerical approximation to the integral representation (3.93) corresponding to the shifted energy (3.85) in the electrostatic field. This analytic continuation can be accomplished by Padé approximants, conformal mappings in the Borel plane, and combinations thereof. So-called order-dependent mappings represent a viable alternative [129–131]. A full optimization of these methods for the (dc) Stark effect, to go beyond the numerical accuracy of reference data such as the result given in Eq. (3.96), remains an open problem.

3.4.2 *Quantum Electrodynamics and Large-Order Perturbations*

Considerations regarding the asymptotic character of a perturbation series have important consequences for QED. As argued in Ref. [132], the vacuum of quantum electrodynamics becomes unstable against pair creation if one reverses the sign of the fine-structure constant α. This is because $e^2 = 4\pi\alpha$. So, in a hypothetical world in which $e^2 = -4\pi\alpha$, opposite charges would repel each other, and the vacuum would become unstable against the creation of electron-positron pairs. These would move away from each other, thereby accumulating positrons at one end of the Universe, and electrons at the other. Dyson [132] argued that this consideration conclusively shows that the QED perturbation series (which describe energy shifts, corrections to the photon propagator, or vacuum-to-vacuum transition amplitudes) must constitute a divergent series, because its radius of convergence about the origin must vanish. Otherwise, there would be no room for the decay processes in the case $\alpha < 0$.

From today's point of view, this semi-quantitative consideration of Dyson is quite self-evident. This is because one has obtained a much more intuitive understanding of the mechanism that generates the divergence. E.g., Zinn-Justin and coworkers [129, 130] have investigated the perturbation series describing so-called renormalization-group function of various scalar theories, or theories describing self-interacting fields with an $O(N)$ symmetry group, with the notion of determining critical exponents, inspired by the numerical evaluations of higher-order Feynman diagrams performed by other groups [133]. In $O(N)$-symmetric field theories with self-interaction terms, the divergence of the perturbation series follows quite naturally from the path integral formulation. Let us consider a one-dimensional analogy of a path integral, namely, an integral of the type

$$J(g) = \int_{-\infty}^{\infty} \mathrm{d}x \exp\left(-\frac{1}{2} x^2 - g\, x^4\right), \tag{3.97}$$

where the factor $\exp\left(-\frac{1}{2} x^2\right)$ models the Gaussian measure of the full field theory and the term $\exp\left(-g\, x^4\right)$ describes the quartic self-interaction. Upon expansion in g, formula (3.97) generates a perturbation series whose coefficients entail integrals that evaluate to Euler Γ functions. The resulting perturbation series in g,

$$J(g) \sim \sum_{K=0}^{\infty} C_K\, g^K, \tag{3.98}$$

is divergent. In general, the structure of the coefficients C_K of order K in a divergent perturbation series in scalar field theories with self-interactions is known [130] to be of the form

$$C_K \sim A^K K^b \Gamma(K)\, g^K, \qquad K \to \infty, \tag{3.99}$$

with appropriate parameters A and b and coupling parameter g. For $O(N)$ symmetric ϕ^4 field theories, all renormalization-group functions (β functions and the anomalous field dimensions) have been shown to exhibit a divergent character for large orders, and adapted summation methods ("order-dependent mappings") have been applied to the resummation of the series [129, 130].

Indeed, the analysis of large orders in perturbation theory now is a well established field of research, and there is no doubt about its divergent character, for wide classes of quantum mechanical perturbations, and field theories. The analysis of quantum mechanical operators [134–137] has revealed that any perturbation series in quantum mechanics must be divergent unless the perturbation is Kato-bounded with regard to the terms in the unperturbed Hamiltonian. Paradigmatic examples for the divergence of perturbation theory (almost in the status of a "toy model") are anharmonic oscillators. For even anharmonic oscillators, described by Hamiltonians of the form

$$H = -\tfrac{1}{2}\,\partial_x^2 + \tfrac{1}{2}x^2 + g\, x^N, \qquad N = 4, 6, 8, \dots, \tag{3.100}$$

the large-order asymptotics of the perturbative coefficients, for arbitrary even N and an arbitrary bound-state energy level, are given by the famous Bender–Wu formulas [138]. For the level of the Hamiltonian (3.100) with quantum number n, we have

$$E_n^{(N)}(g) \sim \sum_K E_{n,K}^{(N)} g^K \,, \tag{3.101a}$$

$$E_{n,K}^{(N)} \sim \frac{(-1)^{K+1}\,(N-2)}{\pi^{3/2}\,n!\,2^{K+1-n}} \Gamma\left(\frac{N-2}{2}K + n + \tfrac{1}{2}\right) \times \left[B\left(\frac{N}{N-2}, \frac{N}{N-2}\right)\right]^{-\frac{N-2}{2}K-n-\frac{1}{2}} \,, \tag{3.101b}$$

for $K \to \infty$. Here, $B(x,y) = \Gamma(x)\Gamma(y)/\Gamma(x+y)$ is the Euler Beta Function. The formulas have recently been generalized to odd anharmonic oscillators with a Hamiltonian

$$H = -\tfrac{1}{2}\,\partial_x^2 + \tfrac{1}{2}x^2 + \sqrt{g}\,x^M \,, \qquad M = 3, 5, 7, \ldots \,. \tag{3.102}$$

In this case, the result for the nth resonance energy of an odd anharmonic oscillator of degree M, denoted as $\epsilon_n^{(M)}(g)$, reads as

$$\epsilon_n^{(M)}(g) \sim \sum_K \epsilon_{n,K}^{(M)} g^K \,, \tag{3.103}$$

$$\epsilon_{n,K}^{(M)} \sim \frac{2-M}{\pi^{3/2}\,n!\,2^{2K+1-n}} \Gamma\left((M-2)K + n + \tfrac{1}{2}\right) \times \left[B\left(\frac{M}{M-2}, \frac{M}{M-2}\right)\right]^{-(M-2)K-n-\frac{1}{2}} \,. \tag{3.104}$$

This result is derived in Refs. [139, 140]. A deeper analysis of the problem has meanwhile led to the conclusion that so-called resurgent trans-series, which involve powers of the structures $\exp(-1/g)$, $\ln(-g)$, and g itself, adequately describe the resonance energies of rather wide classes of anharmonic oscillators [141, 142]. Furthermore, the utility of these resurgent transseries for field-theoretical problems has been investigated [143], and discussed at international meetings (see Ref. [144]). This includes the nonlinear σ model, as a model problem for quantum chromodynamics which, nevertheless, exceeds the complexity of a pure "toy" model (see Ref. [145]).

For QED, the situation is complicated because one has both bosons (photons) as well as fermions (leptons) in the theory. The fermions are massive, so one can integrate them out, in order to obtain an estimate of the resulting, nonlinear interaction of the photons. The integration of the fermions leads to a so-called fermionic determinant. This has been analyzed in Refs. [146–150]. The result for the photon-photon interaction can be inserted in various perturbation series, in order to estimate the large-order growth of vacuum-polarization insertions, or loop diagrams which contribute to the anomalous magnetic moment of the electron, and other effects. In general, the structure of the perturbative coefficients C_K of the series describing the QED effects is conjectured to be proportional to (see Chap. 39 of Ref. [130]),

$$C_K \propto (-1)^K A^{-K}\,\Gamma(K/2) \,, \qquad A = 4.886 \,, \qquad K \to \infty \,, \tag{3.105}$$

which is different from the estimate (3.99) in the sense that the Γ function is evaluated at argument $K/2$, not K. The perturbation coefficients in QED do not grow as drastically with the order of perturbation theory as in scalar and $O(N)$ symmetric field theories. The conceivable nonperturbative contributions are thus smaller than otherwise expected. One can then infer from Eq. (3.105) that the nonperturbative contributions in QED are proportional to

$$\exp\left(-\frac{A^2}{\alpha^2}\right) = 2.55 \times 10^{-194\,698}\,, \tag{3.106}$$

relative to the leading term. So, the original argument supporting the divergent character of the perturbation series, originally brought forward by Dyson [132], is correct, but the limitations to the predictive power of QED are not practically relevant. The situation is different for the analysis of critical exponents in field theory, where the zero of the so-called β function is obtained at a critical coupling $g = g^*$ of order unity [129, 130]. The static Stark effect, which initiated this discussion in the context of the current treatise, remains a transparent example case for the realization of such a divergence, with a particularly transparent interpretation, namely, the tunneling autoionization width of a level in the external, static, electric, field.

3.5 Further Thoughts

Here are some suggestions for further thought.

(1) **Time Evolution.** Using Eq. (3.4), derive Eq. (3.5).

(2) **Density of States.** Reconsider Eq. (3.20). Show that the equation is consistent with the density $\mathrm{d}n = L\,\mathrm{d}p/\hbar$, which pertains to a one-dimensional quantum system. Here, L is the overall length dimension of the one-dimensional system.

(3) **Coefficients and Expansions.** Derive the time-dependent expression Eq. (3.47) based on Eqs. (3.45) and (3.46).

(4) **Laser Intensity.** Verify Eq. (3.55) by calculating the contribution of the electric-field component of the electromagnetic wave, adding the magnetic-field component (same magnitude) and averaging over a laser period.

(5) **Ionization Cross-Section.** Derive Eq. (3.58) on the basis of Fermi's Golden Rule, as introduced in Sec. 3.2. Start from the overlap integral of a bound and a continuum wave function. Then, study the evaluation of the overlap integrals for hydrogen, on the basis of Chap. 71 of Ref. [6] and Chap. 4 which follows here.

(6) **Ionization Cross-Section.** Verify Eq. (3.58) by matching the imaginary part of the ac Stark shift with a beam of incoming photons.

(7) **Large-Order Perturbation Theory.** Do a literature search on Eq. (3.85) and try to find ways to obtain as many terms in the perturbation series in powers of G as possible.

(8) **Numerical Calculations.** It is a good exercise to evaluate the integral $\mathcal{A}(g)$ numerically and accurately, for a few numerical example cases such as, for

example, to pick a certain value, the case $g = 0.07$. For comparison, one may sum the first few terms of the asymptotic expansion for the same value of g, and compare. One will find that the first terms give an excellent approximation of the numerical value of the entire series even if the series eventually diverges. Also, one should investigate the integral (3.93), substituting $u \to \exp(\mathrm{i}\theta)\, u$ with $\theta > 0$ (for the "upper" contour) and $u \to \exp(-\mathrm{i}\theta)\, u$ with $\theta > 0$ (for the "lower" contour). The numerical result, obtained using Gauss–Laguerre quadrature, should be quite independent of the angle $0 < \theta < \pi/2$ because the integrand is suppressed for $u \to \infty$. Automatic integration routines in modern computer algebra systems [151] may help in the endeavour.

Chapter 4

Bound-Electron Self–Energy and Bethe Logarithm

4.1 Overview

At this point, we have prepared ourselves for the analysis of a number of physical processes which involve atoms and the quantized electromagnetic field. In the current chapter, concrete applications of the quantized field-theoretical formalism will be discussed, with particular emphasis on one-electron (hydrogenlike) ions. It will turn out to be advantageous that a number of quantities of interest, which can only be calculated numerically for many-electron ions, are amenable to an analytic treatment in the case of a hydrogenlike ion.

Two examples are the bound-state and continuum wave functions of the eigenstates of the hydrogenlike ion, and the Green function, which can be calculated either as a double-sum over associated Laguerre polynomials (so-called Sturmian decomposition), or in terms of Whittaker functions. The final expressions of the Green functions, which we derive, may be used for calculations; yet, these expressions provide little insight without a deeper understanding of how these results can actually be derived. Analytic expressions for the Green functions will be developed based on the "concatenation approach", which is based on the matching of solutions of the homogeneous radial differential equation at the cusp (see Chaps. 2 and 4 of Ref. [81]). A further, very elegant way of deriving the Schrödinger–Coulomb Green function, based on a momentum-space approach, profits from the $SO(4)$ symmetry of the hydrogen atom, and leads to a very elegant splitting of the hydrogen Green function into a free term, a one-potential term, and a many-potential term. These can be combined toward the end of the calculation into a compact integral representation.

We apply the results to the calculation of the dynamic polarizability of the hydrogen atom, and use the result to calculate the Bethe logarithm of the hydrogen atom, which is a sum over the virtual excitations of the bound system, mediated by interactions of the bound electron with its own radiation field. This calculation is presented in detail, and reference values of Bethe logarithms of unprecedented accuracy are given, as a benchmark for future studies in this and related areas.

It should be noted that atomic physics calculations probably belong to the most sophisticated calculations carried out by mankind so far. The analytic techniques combine mathematical insight, physical inspiration and a great deal of calculational dedication. Nevertheless, a deeper look into these calculations is of utmost interest to the phenomenologically interested theoretical physicist because it combines two established areas of theory: atomic physics of bound states and field theory. Training in the solution of bound-state differential equations will find applications later, as we discuss solutions of the Dirac equation in Chaps. 7 and 8. The calculation of bound-state wave functions, and Green functions, involves finding solutions to bound-state differential equations with specific boundary conditions, and matching them appropriately at the cusp, i.e., at the point where the radial arguments are equal.

4.2 Schrödinger–Coulomb Hamiltonian and Wave Functions

4.2.1 *Spectrum of the Hydrogen Atom and SO(4) Symmetry*

When discussing the hydrogen atom, it is often stated (see, e.g., Ref. [152]) that one-electron ions possess an $SO(4)$ symmetry. However, this statement is sometimes used as a tongue-in-the-cheek remark, to imply that whoever is not familiar with the $SO(4)$ symmetry fails to understand what the hydrogen atom is all about. So, it is justified to ask in which sense the $SO(4)$ symmetry needs to be understood. One might guess, at first, that spherically symmetric bound states of the hydrogen atom exhibit $SO(3)$ (rotational) symmetry, and that the $SO(4)$ symmetry is nothing but a typographical error, mixing up the dimensionality of the space on which the Schrödinger Hamiltonian is acting. This is not the case; yet, it makes the problem only more intriguing.

Our treatment below is inspired by Chap. 30 of Ref. [152]. We switch to natural units with $\hbar = c = \epsilon_0 = 1$. After the transformation to center-of-mass coordinates, the Schrödinger–Coulomb Hamiltonian of the hydrogen atom can be written as

$$H_S = \frac{\vec{p}^{\,2}}{2\mu} - \frac{Z\alpha}{r}, \qquad \vec{p} = -\mathrm{i}\,\vec{\nabla}\,, \tag{4.1}$$

where $\mu = m_1\, m_2/(m_1 + m_2)$ is the reduced mass, Z is the nuclear charge number, and α is the fine-structure constant. The masses of the two constituent particles (electron and nucleus) are denoted as m_1 and m_2, and $\vec{r}$ is the distance between the two constituent particles ("relative coordinate"). The first decisive observation is that, in addition to the angular momentum vector

$$\vec{L} = \vec{r} \times \vec{p} = -\mathrm{i}\vec{r} \times \vec{\nabla}\,, \tag{4.2}$$

the Runge–Lenz operator also is conserved, i.e., it commutes with H_S. Indeed, the vector-valued Runge–Lenz operator $\vec{M}$ is given as

$$\vec{M} = \frac{1}{2\mu}\left(\vec{p} \times \vec{L} - \vec{L} \times \vec{p}\right) - \frac{Z\alpha}{r}\,\vec{r}\,, \tag{4.3}$$

and it commutes with the Hamiltonian, $[H_S, \vec{M}] = 0$. The Runge–Lenz operator is perpendicular to the angular momentum. In operator notation, we have

$$\vec{M} \cdot \vec{L} = \vec{L} \cdot \vec{M} = 0\,. \tag{4.4}$$

The two statements above are not redundant; $\vec{L}$ does not commute with $\vec{M}$. The square of the Runge–Lenz operator is

$$\vec{M}^2 = \frac{2H_S}{\mu}\left(\vec{L}^2 + 1\right) + (Z\alpha)^2\,. \tag{4.5}$$

Some important commutator relations are

$$\left[L^i, L^j\right] = \mathrm{i}\,\epsilon^{ijk}\,L^k\,, \qquad \left[M^i, L^j\right] = \mathrm{i}\,\epsilon^{ijk}\,M^k\,, \tag{4.6a}$$

$$\left[M^i, M^j\right] = \mathrm{i}\left(-\frac{2H_S}{\mu}\right)\epsilon^{ijk}\,L^k\,. \tag{4.6b}$$

The algebra involved in verifying Eq. (4.6) is rather lengthy and left as an exercise to the reader.

It is helpful to notice that in verifying Eq. (4.6b), the Schrödinger Hamiltonian (4.1) appears on the right-hand side of the equation explicitly, i.e., one has to keep track of all momentum operators in the intermediate stages of the calculation. We now restrict the considerations to a bound-state subspace with (possibly degenerate) energy $\langle H_S \rangle = E_n < 0$, and define

$$\vec{M}' = \sqrt{-\frac{\mu}{2\,E_n}}\,\vec{M} \approx \sqrt{-\frac{\mu}{2\,H_S}}\,\vec{M}\,. \tag{4.7}$$

In natural units, the operator $\vec{M}'$ has the same physical dimension as $\vec{M}$, because of mass-energy equivalence. Then we get the following algebraic relations,

$$\left[L^i, L^j\right] = \mathrm{i}\,\epsilon^{ijk}\,L^k\,, \qquad \left[M'^i, L^j\right] = \mathrm{i}\,\epsilon^{ijk}\,M'^k\,, \qquad \left[M'^i, M'^j\right] = \mathrm{i}\,\epsilon^{ijk}\,L^k\,. \tag{4.8}$$

These relations can be mapped onto those that are fulfilled by the six generators of rotations in four-dimensional space. In three-dimensional space, we can rotate about all three axes, but in four-dimensional space, there are three more rotations possible, which correspond to rotations of the three "spatial" axes about the fourth axis. The $SO(4)$ generators can thus be ordered into a six-dimensional vector structure and identified as follows,

$$\tilde{L} = \begin{pmatrix} L^x \\ L^y \\ L^z \\ M'^x \\ M'^y \\ M'^z \end{pmatrix} \triangleq \begin{pmatrix} \tilde{L}^{23} \\ \tilde{L}^{31} \\ \tilde{L}^{12} \\ \tilde{L}^{14} \\ \tilde{L}^{24} \\ \tilde{L}^{34} \end{pmatrix} = \begin{pmatrix} X_2\,P_3 - X_3\,P_2 \\ X_3\,P_1 - X_1\,P_3 \\ X_1\,P_2 - X_2\,P_1 \\ X_1\,P_4 - X_4\,P_1 \\ X_2\,P_4 - X_4\,P_2 \\ X_3\,P_4 - X_4\,P_3 \end{pmatrix}. \tag{4.9}$$

They fulfill the same commutation relations as the components $\tilde{L}^{ij} = X^i\,P^j - X^j\,P^i$ of angular momentum in four-dimensional space, but with a Euclidean metric. We have $[P^i, X^j] = -\mathrm{i}\,\delta^{ij}$, and the commutator relations are

$$[\tilde{L}^{ij}, \tilde{L}^{k\ell}] = \mathrm{i}\left(\delta^{ik}\,\tilde{L}^{j\ell} + \delta^{j\ell}\,\tilde{L}^{ik} - \delta^{jk}\,\tilde{L}^{i\ell} - \delta^{i\ell}\,\tilde{L}^{jk}\right). \tag{4.10}$$

In our discussion of the angular momentum in special relativity (see Chap. 8 of Ref. [81]), we had encountered the relation $[p^\mu, x^\nu] = \mathrm{i}\, g^{\mu\nu}$ for the components p^μ of the four-momentum and space-time coordinates x^ν, and thus

$$[L^{\mu\nu}, L^{\kappa\lambda}] = \mathrm{i}\left(g^{\mu\lambda} L^{\nu\kappa} + g^{\nu\kappa} L^{\mu\lambda} - g^{\mu\kappa} L^{\nu\lambda} - g^{\nu\lambda} L^{\mu\kappa}\right). \tag{4.11}$$

Here, $L^{\mu\nu} = x^\mu p^\nu - x^\nu p^\mu$ is the generalized angular momentum in four-dimensional space-time. We recall that the metric of space-time is $g^{\rho\sigma} = \mathrm{diag}(1,-1,-1,-1)$ which is the metric of $SO(3,1)$. The identification of Eqs. (4.10) and (4.11) proceeds via the replacement $g^{\mu\lambda} \to -\delta_{ij}$ and an appropriate change in the indices. Furthermore, we can transform the generators as

$$\vec{I} = \frac{1}{2}\left(\vec{L} + \vec{M}'\right), \qquad \vec{K} = \frac{1}{2}\left(\vec{L} - \vec{M}'\right), \tag{4.12}$$

with $[I_i, K_j] = 0$. The algebra is thus separated into two separate $SO(3)$ algebras,

$$[I_i, I_j] = \mathrm{i}\,\epsilon_{ijk}\, I_k, \qquad [K_i, K_j] = \mathrm{i}\,\epsilon_{ijk}\, K_k, \qquad [I_i, K_j] = 0. \tag{4.13}$$

We recall that the angular momentum generates rotations, and fulfills the algebraic relations of the Lie group that generates length-preserving rotations of three-dimensional space, which corresponds to the $SO(3)$ group. Then, by virtue of the angular momentum algebra, the eigenvalues of $\vec{I}$ and $\vec{K}$ are

$$\vec{I}^{\,2} = i\,(i+1), \qquad \vec{K}^{\,2} = k\,(k+1). \tag{4.14}$$

Here, one should remember that the spinor representations of the rotation group allow i and k to be half integer, while in the usual multipole representations (see Chap. 2 of Ref. [81]), i and k assume non-negative integer values. So, the possible values are $i, k = 0, \frac{1}{2}, 1, \frac{2}{3}, 2, \ldots$. Because the Runge–Lentz vector is perpendicular to the angular momentum [see Eq. (4.4)], we have

$$\vec{I}^{\,2} = \frac{1}{4}\left(\vec{L}^2 + \vec{M}'^2\right), \qquad \vec{K}^{\,2} = \frac{1}{4}\left(\vec{L}^2 + \vec{M}'^2\right). \tag{4.15}$$

The following two operators, commonly referred to as Casimir operators, thus commute with all components of $\vec{I}$ and $\vec{K}$,

$$C_1 = \vec{I}^2 + \vec{K}^2 = i\,(i+1) + k\,(k+1) = \frac{1}{2}\left(\vec{L}^2 + \vec{M}'^2\right), \tag{4.16a}$$

$$C_2 = \vec{I}^2 - \vec{K}^2 = i\,(i+1) - k\,(k+1) = 0. \tag{4.16b}$$

The latter relation follows by direct inspection of Eq. (4.15) and implies that we are only dealing with the subspace of $SO(4)$ for which $i = k$. We now project Eq. (4.5) onto the eigenspace of the hydrogen atom with $\langle H_S \rangle = E_n$,

$$\vec{M}^2 = \frac{2E_n}{\mu}\left(\vec{L}^2 + 1\right) + (Z\alpha)^2. \tag{4.17}$$

Upon multiplication by $[-\mu/(2E_n)]$, we obtain the corresponding relation of $\vec{M}'$,

$$\vec{M}'^2 = -\vec{L}^{\,2} - 1 - \frac{\mu}{2E_n}(Z\alpha)^2. \tag{4.18}$$

We now combine Eqs. (4.14), (4.15) and (4.18) to obtain the following relation in the space with $\langle H_S \rangle = E_n$,

$$\begin{aligned} 2k(k+1) = 2\vec{K}^2 &= \frac{1}{2}\left(\vec{L}^2 + \vec{M}'^2\right) \\ &= \frac{1}{2}\left(\vec{L}^2 - \vec{L}^2 - 1 - \frac{\mu}{2E_n}(Z\alpha)^2\right) = \frac{1}{2}\left(-1 - \frac{\mu}{2E_n}(Z\alpha)^2\right). \end{aligned} \tag{4.19}$$

Solving for E_n, we find

$$E_n = -\frac{\mu\,(Z\alpha)^2}{2(2k+1)^2} = -\frac{(Z\alpha)^2\,\mu}{2n^2}, \qquad n \equiv (2k+1). \tag{4.20}$$

We have obtained the Schrödinger spectrum on the basis of purely algebraic manipulations, because $k \geq 0$ may be integer or half integer.

4.2.2 *Differential Equations and the Hydrogen Atom*

We have just determined the Schrödinger spectrum but not the wave functions of the hydrogen atom. We shall now find the hydrogen wave functions, using natural units with $\hbar = c = \epsilon_0 = 1$. The three-dimensional Schrödinger equation for a two-body system of reduced mass μ in a scalar, time-independent potential $V(\vec{r})$ is given by

$$H_S\,\psi_{E\ell m}(\vec{r}) = \left(-\frac{\vec{\nabla}^2}{2\mu} + V(\vec{r})\right)\psi_{E\ell m}(\vec{r}) = E\,\psi_{E\ell m}(\vec{r}), \tag{4.21}$$

where the subscript of the wave function ψ denotes the energy E, the angular momentum ℓ, and the magnetic quantum number m. For an electron of charge e in the Coulomb potential of a nucleus of charge $-Ze$, the potential is spherically symmetric and given by

$$V(\vec{r}) = -\frac{Ze^2}{4\pi\epsilon_0 r}, \tag{4.22}$$

where $e^2 = 4\pi\alpha$. In spherical coordinates, the Laplace operator is given by

$$\vec{\nabla}^2 = \frac{\partial^2}{\partial r^2} + \frac{2}{r}\frac{\partial}{\partial r} - \frac{\vec{L}^2}{r^2}, \qquad \vec{L} = -\mathrm{i}\,\vec{r} \times \vec{\nabla}, \tag{4.23}$$

where $\vec{L}$ is the angular momentum operator. The full equation in spherical coordinates then reads as

$$\left[-\frac{1}{2\mu}\left(\frac{\partial^2}{\partial r^2} + \frac{2}{r}\frac{\partial}{\partial r}\right) + \frac{\vec{L}^2}{2\,\mu\,r^2} - \frac{Z\alpha}{r}\right]\psi_{E\ell m}(r,\theta,\varphi) = E\,\psi_{E\ell m}(r,\theta,\varphi). \tag{4.24}$$

It is solved by the separation *ansatz*

$$\psi_{E\ell m}(r,\theta,\varphi) = \mathcal{R}_{E\ell}(r)\,Y_{\ell m}(\theta,\varphi), \tag{4.25}$$

where $Y_{\ell m}(\theta,\varphi)$ is the spherical harmonic defined by

$$Y_{\ell m}(\theta,\varphi) = \sqrt{\frac{2\ell+1}{4\pi}\frac{(\ell-m)!}{(\ell+m)!}}\,P_\ell^m(\cos(\theta))\,\mathrm{e}^{\mathrm{i}m\varphi}, \tag{4.26}$$

with the associated Legendre polynomials $P_\ell^m(\cos(\theta))$. For an introduction to these quantities, see Chap. 2 of Ref. [81]. The spherical harmonic has the properties

$$\vec{L}^2 \, Y_{\ell m}(\theta,\varphi) = \ell(\ell+1)\, Y_{\ell m}(\theta,\varphi)\,, \qquad L_z \, Y_{\ell m}(\theta,\varphi) = m \, Y_{\ell m}(\theta,\varphi)\,. \tag{4.27}$$

After the separation, we have to solve the radial equation,

$$\left(-\frac{1}{2\mu}\left(\frac{\mathrm{d}^2}{\mathrm{d}r^2} + \frac{2}{r}\frac{\mathrm{d}}{\mathrm{d}r}\right) + \frac{\ell(\ell+1)}{2\,\mu\, r^2} - \frac{Z\alpha}{r} \right) \mathcal{R}_{E\ell}(r) = E\,\mathcal{R}_{E\ell}(r)\,. \tag{4.28}$$

One uses the substitution $u_{E\ell}(r) = r\mathcal{R}_{E\ell}(r)$ and arrives at the differential equation

$$\left(-\frac{1}{2\mu}\frac{\mathrm{d}^2}{\mathrm{d}r^2} + \frac{\ell(\ell+1)}{2\mu\, r^2} - \frac{Z\alpha}{r} - E \right) u_{E\ell}(r) = 0\,. \tag{4.29}$$

It is helpful to rescale the differential equation, and to express it in terms of dimensionless quantities,

$$\rho = (-8\,\mu\, E)^{1/2}\, r\,, \qquad \nu = Z\alpha\left(-\frac{\mu}{2E}\right)^{1/2}, \qquad \widetilde{u}_{E\ell}(\rho) = u_{E\ell}\,(r)\,. \tag{4.30}$$

Here, ν is a generalized energy variable, which can be real or imaginary. It generalizes the concept of the principal quantum number,

$$E = -\frac{(Z\alpha)^2\,\mu}{2\,\nu^2}\,. \tag{4.31}$$

For bound states with negative energy E, the generalized quantum number $\nu = n$ is real. For continuum states with $E > 0$, ν is purely imaginary. We choose the branch cut of the square root in such a way as to give ρ a positive imaginary part. Let us assume that the positive energy E obtains a small negative imaginary part (in the sense of a nonvanishing decay width Γ, so that $E \to \mathrm{Re}\,E - \mathrm{i}\Gamma/2$). Then, the argument $-8\,\mu\,E$ of the square root in the expression $\rho = (-8\,\mu\,E)^{1/2}\, r$ obtains a small positive imaginary part, while the real part of the expression $-8\,\mu\,E$ is negative. With the standard choice of the branch cut of the complex square root, the variable ρ becomes purely imaginary, with a positive imaginary part, while ν becomes negative imaginary [see also Eq. (4.70)]. The dimensionless equation then reads as

$$\left(\frac{\mathrm{d}^2}{\mathrm{d}\rho^2} - \frac{\ell(\ell+1)}{\rho^2} + \frac{\nu}{\rho} - \frac{1}{4} \right) \widetilde{u}_{E\ell}(\rho) = 0\,. \tag{4.32}$$

We can trivially rewrite the equation as

$$\left(\frac{\mathrm{d}^2}{\mathrm{d}\rho^2} - \frac{1}{4} + \frac{\nu}{\rho} + \frac{\frac{1}{4} - (\ell + \frac{1}{2})^2}{\rho^2} \right) \widetilde{u}_{E\ell}(\rho) = 0\,, \tag{4.33}$$

and match it with the definition of Whittaker's differential equation [153],

$$\frac{\mathrm{d}^2 w}{\mathrm{d}\rho^2} + \left(-\frac{1}{4} + \frac{\kappa}{\rho} + \frac{\frac{1}{4} - \mu^2}{\rho^2} \right) w = 0\,, \tag{4.34}$$

where $\kappa = \nu$ and $\mu = \ell + \frac{1}{2}$. The linearly independent solutions of this differential equation are called the Whittaker M and W functions, where

$$w = M_{\kappa,\mu}(\rho) = \mathrm{e}^{-\frac{1}{2}\rho}\,\rho^{\mu+1/2}\,{}_1F_1(\mu + \tfrac{1}{2} - \kappa, 2\mu + 1, \rho)\,, \tag{4.35a}$$

$$w = W_{\kappa,\mu}(\rho) = \mathrm{e}^{-\frac{1}{2}\rho}\,\rho^{\mu+1/2}\,U(\mu + \tfrac{1}{2} - \kappa, 2\mu + 1, \rho)\,. \tag{4.35b}$$

Here, the confluent hypergeometric function is denoted as ${}_1F_1$, and we also encounter the hypergeometric U function [153]. A priori, the equation (4.33) defines a valid differential equation for any values of ℓ, ν, and ρ. However, we shall see below that a special case is encountered for $\nu = n$ where n is a positive integer.

We first discuss the M function,

$$\begin{aligned}\widetilde{u}_{E\ell}(\rho) &= M_{\nu,\ell+\frac{1}{2}}(\rho) = \mathrm{e}^{-\frac{1}{2}\rho}\,\rho^{\ell+1}\,{}_1F_1(\ell + 1 - \nu, 2\ell + 2, \rho)\\ &= \mathrm{e}^{-\frac{1}{2}\rho}\,\rho^{\ell+1}\,\frac{\Gamma(2\ell+2)}{\Gamma(\ell+1-\nu)\Gamma(\ell+1+\nu)}\int_0^1 \mathrm{e}^{u\rho}\,u^{\ell-\nu}\,(1-u)^{\ell+\nu}\,\mathrm{d}u\,.\end{aligned} \tag{4.36}$$

Let us dwell a little on the definition of the so-called confluent (Kummer's) hypergeometric function. In order to understand this function, we first remember that the exponential function is given by a convergent series as

$$\sum_{n=0}^{\infty} \frac{z^n}{n!} = \exp(z)\,. \tag{4.37}$$

When we decorate the terms in the infinite sum over n by appropriate prefactors, which contain Gamma functions, we get

$${}_1F_1(a, c; z) = \sum_{n=0}^{\infty} \frac{\Gamma(a+n)}{\Gamma(a)}\,\frac{\Gamma(c)}{\Gamma(c+n)}\,\frac{z^n}{n!}\,, \tag{4.38}$$

where the definition of the Gamma function for $z > 0$ is [154, 155]

$$\Gamma(z) = \int_0^{\infty} t^{z-1}\exp(-t)\,\mathrm{d}t\,, \qquad \Gamma(n) = (n-1)!\,. \tag{4.39}$$

In particular, we have as a special case of the confluent hypergeometric function,

$${}_1F_1(1, 1; z) = \sum_{n=0}^{\infty} \frac{\Gamma(1+n)}{\Gamma(1)}\,\frac{\Gamma(1)}{\Gamma(1+n)}\,\frac{z^n}{n!} = \sum_{n=0}^{\infty} \frac{z^n}{n!} = \exp(z)\,. \tag{4.40}$$

From the functional form of $M_{\nu,\ell+1/2}(\rho)$ in Eq. (4.36), we can immediately conclude that $M_{\nu,\ell+1/2}(\rho)$ is regular at $\rho = 0$ because of the fundamental properties of the confluent hypergeometric ${}_1F_1$ function. In many cases, it is useful to introduce the Pochhammer symbol

$$(a)_n = \frac{\Gamma(a+n)}{\Gamma(a)}\,, \tag{4.41}$$

while emphasizing that a thorough discussion of properties of hypergeometric, and other special functions, can be found in Refs. [154, 156, 157].

Let us consider the case of a so-called terminating hypergeometric function, which occurs for a nonpositive integer first argument $a = -m \leq 0$. We then have the following situation: When $n \leq m$, we have a Γ function of negative argument in both the numerator as well as the denominator of the expression $\Gamma(a+n)/\Gamma(a) = \Gamma(-m+n)/\Gamma(-m)$, and the whole term therefore is finite. In order to see this, one displaces the arguments of the Gamma function according to $\Gamma(-m+n+\epsilon)/\Gamma(-m+\epsilon)$ and invokes the recursion relation $\Gamma(x) = (x-1)\,\Gamma(x-1)$ until the Gamma function in the numerator and denominator cancel. On the other hand, when $n > m$, the expression $\Gamma(a+n)/\Gamma(a) = \Gamma(-m+n)/\Gamma(-m) = 0$, because the denominator is infinite, whereas the numerator is finite. The series thus terminates at the summation index $n = m$.

Furthermore, a remark is in order. One may write the general solution of Eq. (4.33) either in terms of $M_{\nu,\ell+1/2}(\rho)$ and $W_{\nu,\ell+1/2}(\rho)$, or in terms of $M_{\nu,\ell+1/2}(\rho)$ and $M_{\nu,-\ell-1/2}(\rho)$. One might thus think that the second required solution to Eq. (4.33) is given by $M_{\nu,-\ell-1/2}(\rho)$. However, one runs into trouble because upon substituting $\ell \to -\ell - 1$ in Eq. (4.36), the argument of the Gamma function transforms as $\Gamma(2\ell+2) \to \Gamma(-2\ell)$. Here, the quantity $\Gamma(-2\ell)$ becomes ill-defined for integer ℓ. This is problematic because the physically interesting case corresponds to an integer angular momentum quantum number ℓ.

The complementary solution to Eq. (4.36) is found by replacing in Eq. (4.36) the hypergeometric ${}_1F_1$ function by the hypergeometric U function,

$$\begin{aligned} \widetilde{u}_{E\ell}(\rho) &= W_{\nu,\ell+\frac{1}{2}}(\rho) = \mathrm{e}^{-\frac{1}{2}\rho}\,\rho^{\ell+1}\,U(\ell+1-\nu, 2\ell+2, \rho) \\ &= \frac{\mathrm{e}^{-\frac{1}{2}\rho}\rho^{\nu}}{\Gamma(\ell+1-\nu)} \int_0^\infty e^{-t}\,t^{\ell-\nu}\left(1+\frac{t}{\rho}\right)^{\ell+\nu} \mathrm{d}t\,. \end{aligned} \tag{4.42}$$

For the integral representation to be valid, we must exclude the case of $\ell + 1 - \nu$ being a negative integer. The function $\widetilde{u}_{E\ell}(\rho) = W_{\nu,\ell+\frac{1}{2}}(\rho)$ has a cut along the negative ρ axis. A very useful relation connects the Whittaker W function with the hypergeometric ${}_2F_0$ function, namely [153]

$$W_{\nu,\ell+\frac{1}{2}}(\rho) = \mathrm{e}^{-\frac{1}{2}\rho}\rho^{\nu}\,{}_2F_0\left(\ell+1-\nu, -\ell-\nu, -\rho^{-1}\right). \tag{4.43}$$

The W function is regular at infinity due to the asymptotic expansion [153]

$$W_{\nu,\ell+\frac{1}{2}}(\rho) \sim \mathrm{e}^{-\frac{1}{2}\rho}\rho^{\nu} \sum_{n=0}^{\infty} \frac{(\ell+1-\nu)_\nu\,(-\ell-\nu)_\nu}{n!\,(-\rho)^n}, \qquad \rho \to \infty\,. \tag{4.44}$$

The solution $M_{\nu,\ell+1/2}(\rho)$ is regular at the origin, whereas $W_{\nu,\ell+1/2}(\rho)$ is not. The situation is inverted at $\rho = \infty$.

In order to fix ideas, let us summarize the unnormalized eigenstates of the hydrogen atom that we have found so far. Specifically, we have encountered solutions of the differential equation (4.29) for arbitrary real energy parameter E. The solutions of the eigenvalue problem $H_S\,\psi_{E\ell m}(\vec{r}) = E\,\psi_{E\ell m}(\vec{r})$ read as follows [see Eq. (4.30)],

$$u_{E\ell}(r) = M_{\nu,\ell+\frac{1}{2}}(\rho) = M_{\nu,\ell+\frac{1}{2}}\left((-8\,\mu\,E)^{1/2}\,r\right), \tag{4.45a}$$

$$u_{E\ell}(r) = W_{\nu,\ell+\frac{1}{2}}(\rho) = W_{\nu,\ell+\frac{1}{2}}\left((-8\,\mu\,E)^{1/2}\,r\right), \qquad \nu = Z\alpha\left(-\frac{\mu}{2E}\right)^{1/2}, \tag{4.45b}$$

$$\mathcal{R}_{E\ell}(r) = \frac{u_{E\ell}(r)}{r}, \qquad \psi_{E\ell m}(\vec{r}) = \mathcal{R}_{E\,\ell}(r)\,Y_{\ell m}(\theta,\varphi). \tag{4.45c}$$

The Schrödinger–Coulomb Hamiltonian H_S is given in Eq. (4.1). Because we would like to describe bound and continuum wave functions, the remaining task is to find appropriate normalizations for the wave functions. For bound states, we must choose the solution regular at the origin, with the exception of a few cases where the two solutions actually coincide. Up to now, we have solved the radial Schrödinger equation for all possible values of the energy. This is not sufficient, though, for a particular value of the energy to be an eigenvalue. At the bound-state eigenvalues, the solution must the regular at both the origin as well as at infinity, and for a continuum eigenstate, the solution must be normalized to a Dirac-δ in the energy variable. The remaining task in finding the eigenstates therefore implies the calculation of appropriate normalization factors.

4.2.3 *Schrödinger–Coulomb Bound States*

In Eq. (4.31), we have defined the generalized energy variable ν. For $\nu = n$, where n is a positive integer, we have

$$E_n = -\frac{(Z\alpha)^2\,\mu}{2\,n^2}, \tag{4.46}$$

which is a bound-state energy of the Schrödinger Hamiltonian. In this case, the Whittaker M function given in Eq. (4.36) assumes the following form,

$$M_{n,\ell+\frac{1}{2}}(\rho) = \mathrm{e}^{-\frac{1}{2}\rho}\,\rho^{\ell+1}\;{}_1F_1(\ell+1-n, 2\ell+2; \rho). \tag{4.47}$$

For $n \geq \ell+1$, the hypergeometric series is terminating, the state is normalizable, and we can write the M function as follows,

$$M_{n,\ell+\frac{1}{2}}(\rho) = e^{-\frac{1}{2}\rho}\rho^{\ell+1}\sum_{j=0}^{n-\ell-1}\frac{\Gamma(\ell+1-n+j)}{\Gamma(\ell+1-n)}\,\frac{\Gamma(2\ell+2)}{\Gamma(2\ell+2+j)}\,\frac{\rho^j}{j!}. \tag{4.48}$$

This function is finite at both the origin $\rho = 0$ as well as for $\rho \to \infty$. With the recursion relation of the Γ function, one may simplify the ratio $\Gamma(\ell+1-n+j)/\Gamma(\ell+1-n)$, which leads to the expression

$$M_{n,\ell+\frac{1}{2}}(\rho) = e^{-\frac{1}{2}\rho}\rho^{\ell+1}\sum_{j=0}^{n-\ell-1}(-1)^j\,\frac{(n-\ell-1)!}{(n-\ell-1-j)!}\,\frac{(2\ell+1)!}{(2\ell+1+j)!}\,\frac{\rho^j}{j!}. \tag{4.49}$$

This is (almost) the canonical form of the radial bound-state wave function. We now define the associated Laguerre polynomials,

$$L_\nu^m(\rho) = \sum_{p=0}^{\nu}(-1)^p\,\frac{(\nu+m)!}{(\nu-p)!(m+p)!}\,\frac{\rho^p}{p!} = \frac{(m+\nu)!}{m!\,\nu!}\;{}_1F_1(-\nu, m+1, \rho). \tag{4.50}$$

For illustration, it is useful to indicate the first few polynomials explicitly,

$$L_0^0(\rho) = 1\,, \qquad L_1^0(\rho) = 1 - \rho\,, \qquad L_1^1(\rho) = 2 - \rho\,,$$
$$L_2^0(\rho) = 1 - 2\rho + \frac{\rho^2}{2}\,, \qquad L_2^1(\rho) = 3 - 3\rho + \frac{\rho^2}{2}\,, \qquad L_2^2(\rho) = 6 - 4\rho + \frac{\rho^2}{2}\,. \tag{4.51}$$

In terms of the associated Laguerre polynomials, we can finally write Eq. (4.49) as follows,

$$M_{n,\ell+1/2}(\rho) = \mathrm{e}^{-\frac{1}{2}\rho}\,\rho^{\ell+1}\,\frac{(n-\ell-1)!\,(2\ell+1)!}{(n+\ell)!}\,L_{n-\ell-1}^{2\ell+1}(\rho)\,, \tag{4.52}$$

provided n and ℓ are positive integers with $n \geq \ell + 1$. The solution $\widetilde{u}_{n\ell}(\rho) = M_{n,\ell+1/2}(\rho)$ fulfills the boundary conditions for a bound state in quantum mechanics, namely according to Ref. [158],

$$\lim_{\rho\to 0}\widetilde{u}_{n\ell}(\rho) = \lim_{\rho\to\infty}\widetilde{u}_{n\ell}(\rho) = 0\,, \qquad \widetilde{u}_{n\ell}(\rho) = \rho\,\mathcal{R}_{\ell}(\rho)\,, \qquad \widetilde{u}_{n\ell}(\rho) \in L^2(\mathbb{R})\,. \tag{4.53}$$

According to Eq. (4.30), we have the following identification for the ρ variable, for a bound state of quantum number n,

$$\rho = (-8\,\mu\,E_n)^{1/2}\,r = \frac{2Z\alpha\mu r}{n} = \frac{2Zr}{a_0\,n}\,, \qquad a_0 = \frac{1}{\alpha\mu}\,, \tag{4.54}$$

where we use the convenient notation a_0 for the Bohr radius of the bound system with reduced mass μ. The radial Whittaker function becomes

$$M_{n,\ell+1/2}\left(\frac{2Zr}{a_0 n}\right) = \exp\left(-\frac{Zr}{a_0 n}\right)\left(\frac{2Zr}{a_0 n}\right)^{\ell+1}\frac{(n-\ell-1)!\,(2\ell+1)!}{(n+\ell)!}\,L_{n-\ell-1}^{2\ell+1}\left(\frac{2Zr}{a_0 n}\right)\,. \tag{4.55}$$

At the bound-state energy, the solutions regular at the origin, and regular at infinity, actually have to coincide up to a constant prefactor, because the eigenstate is normalizable, and the M and W function span the space of solutions. Indeed, we have for $\nu = n$,

$$M_{n,\ell+1/2}\left(\frac{2Zr}{a_0 n}\right) = (-1)^{n+\ell+1}\,\frac{(2\ell+1)!}{(n+\ell)!}\,W_{n,\ell+1/2}\left(\frac{2Zr}{a_0\,n}\right)\,. \tag{4.56}$$

We remember that $M_{n,\ell+1/2}(\rho)$ was the solution for $u_{n\ell}(\rho)$, so in order to obtain the radial wave function $\mathcal{R}_{n\ell}(r)$ of the bound state, we have to calculate $\mathcal{R}_{n\ell}(r) = u_{n\ell}(r)/r$, within the separation *ansatz* given in Eq. (4.25). Furthermore, we apply a normalization prefactor

$$\mathcal{N} = \frac{\sqrt{(n+\ell)!}}{n\,(2\ell+1)!\,\sqrt{(n-\ell-1)!}}\left(\frac{Z}{a_0}\right)^{1/2}\,. \tag{4.57}$$

The normalized radial wave function

$$R_{n\ell}(r) = \mathcal{N}\,\mathcal{R}_{E\ell}(r) = \frac{\mathcal{N}}{r}\,u_{E\ell}(r) = \frac{\mathcal{N}}{r}\,M_{n,\ell+1/2}\left(\frac{2Zr}{a_0 n}\right) \tag{4.58}$$

can be written as

$$R_{n\ell}(r) = \frac{\sqrt{(n-\ell-1)!}}{\sqrt{(n+\ell)!}} \frac{2^{\ell+1}}{n^2} \left(\frac{Z}{a_0}\right)^{3/2} \left(\frac{Zr}{a_0\, n}\right)^{\ell} \exp\left(-\frac{Zr}{a_0 n}\right) L_{n-\ell-1}^{2\ell+1}\left(\frac{2Zr}{a_0 n}\right). \quad (4.59)$$

Combining the radial with the angular part, we find for the discrete bound-state eigenfunctions of the Schrödinger–Coulomb Hamiltonian,

$$\psi_{n\ell m}(r,\theta,\varphi) = R_{n\ell}(r)\, Y_{\ell m}(\theta,\varphi)\,, \quad (4.60a)$$

$$H_S\, \psi_{n\ell m}(r,\theta,\varphi) = E_n\, \psi_{n\ell m}(r,\theta,\varphi) = \left(-\frac{(Z\alpha)^2\,\mu}{2n^2}\right) \psi_{n\ell m}(r,\theta,\varphi)\,, \quad (4.60b)$$

where H_S is given in Eq. (4.1). The normalization prefactor $\mathcal{N}$ ensures that the wave functions $\psi_{n\ell m}(r,\theta,\varphi)$ form an orthonormal set of basis functions,

$$\int \mathrm{d}^3 r\, \psi^*_{n\ell m}(r,\theta,\varphi)\, \psi_{n'\ell' m'}(r,\theta,\varphi) = \delta_{nn'}\, \delta_{\ell\ell'}\, \delta_{mm'}\,. \quad (4.61)$$

Further useful properties of the associated Laguerre polynomials are summarized in Ref. [159].

4.2.4 *Schrödinger–Coulomb Virial Theorem*

The virial theorem connects the expectation values of the kinetic operator and the potential in an eigenstate of the Schrödinger–Coulomb Hamiltonian. It is useful to afford a small digression on this topic as the virial immediately leads to analytic formulas for the expectation values of the kinetic and potential energy operators. We recall the explicit form of the Schrödinger–Coulomb Hamiltonian,

$$H_S = \frac{\vec{p}^{\,2}}{2m} + V\,, \qquad V = -\frac{Z\alpha}{r}\,, \qquad \vec{p} = \mathrm{i}\, m\, [H_S, \vec{r}\,]\,. \quad (4.62)$$

Let us write the expectation value of an operator A in an eigenstate of the Schrödinger–Coulomb Hamiltonian as $\langle A \rangle \equiv \langle \phi | A | \phi \rangle$. For any operator A, in view of the eigenvalue property $(H_S - E_S)|\phi\rangle = 0$, we have

$$[H_S - E_S, A]\, |\phi\rangle = (H_S - E_S) A |\phi\rangle\,. \quad (4.63)$$

The expectation value of the kinetic operator in a Schrödinger–Coulomb energy eigenstate is given as

$$\begin{aligned} \left\langle \frac{\vec{p}^{\,2}}{2m} \right\rangle &= \frac{1}{2} \langle \mathrm{i}\, [H_S - E_S, \vec{r}\,] \cdot \vec{p} \rangle = -\frac{1}{2} \langle \mathrm{i}\, \vec{r} \cdot (H_S - E_S)\, \vec{p} \rangle \quad (4.64) \\ &= -\frac{1}{2} \langle \mathrm{i}\, \vec{r} \cdot [H_S - E_S, \vec{p}\,] \rangle = \frac{1}{2} \langle \mathrm{i}\, \vec{r} \cdot [\vec{p}, H_S - E_S] \rangle = \frac{1}{2} \left\langle \vec{r} \cdot \vec{\nabla}(V) \right\rangle . \end{aligned}$$

Here, we use the shorthand notation $\langle A \rangle$ for the expectation value $\langle \phi | A | \phi \rangle$ of an operator A in the quantum state $|\phi\rangle$. If V is the Coulomb potential, then

$$\frac{1}{2} \left\langle \vec{r} \cdot \vec{\nabla}(V) \right\rangle = \frac{1}{2} \left\langle \vec{r} \cdot \vec{\nabla} \left(-\frac{Z\alpha}{r}\right) \right\rangle = \frac{1}{2} \left\langle \vec{r} \cdot \frac{Z\alpha \vec{r}}{r^3} \right\rangle = \frac{1}{2} \left\langle \frac{Z\alpha}{r} \right\rangle = -\frac{1}{2} \langle V \rangle\,, \quad (4.65)$$

and the virial theorem follows as

$$\left\langle \frac{\vec{p}^{\,2}}{2m} \right\rangle = -\frac{1}{2} \langle V \rangle \,. \tag{4.66}$$

Furthermore, for a bound-state $|\phi\rangle$ with

$$\langle \phi | H_S | \phi \rangle = \left\langle \phi \left| \frac{\vec{p}^{\,2}}{2m} + V \right| \phi \right\rangle = -\frac{(Z\alpha)^2 \mu}{2n^2} \,, \tag{4.67}$$

one has

$$\langle \phi | \vec{p}^{\,2} | \phi \rangle = \frac{(Z\alpha)^2 \mu^2}{n^2} \,, \qquad \langle \phi | V | \phi \rangle = -\frac{(Z\alpha)^2 \mu}{n^2} \,, \tag{4.68}$$

as a consequence of the virial theorem.

4.2.5 *Schrödinger–Coulomb Continuum States*

The continuum wave functions cover the energy spectrum from $E = 0$ to $E = \infty$, and it is customary to parameterize the energy according to the modulus $k = |\vec{k}|$ of the wave vector, writing

$$E = \frac{k^2}{2\mu} \,, \qquad E > 0 \,. \tag{4.69}$$

We recall the essential formulas for the radial part of the wave function regular at the origin, and convert these for continuum wave functions,

$$\rho = (-8\,\mu\,E)^{1/2}\, r = 2\mathrm{i}\, k\, r \,, \tag{4.70a}$$

$$\nu = Z\alpha \left(-\frac{\mu}{2E} \right)^{1/2} = -\mathrm{i}\, \frac{Z\alpha\mu}{k} \,. \tag{4.70b}$$

According to Eqs. (4.30) and (4.36), the radial (Whittaker) wave function is a ${}_1F_1$ hypergeometric function

$$\widetilde{u}_{E\ell}(\rho) = u_{E\ell}(r) = M_{n,\ell+\frac{1}{2}}(\rho) = \mathrm{e}^{-\frac{1}{2}\rho}\, \rho^{\ell+1}\, {}_1F_1(\ell + 1 - \nu, 2\ell + 2, \rho) \,, \tag{4.70c}$$

which evaluates as follows, for a continuum state,

$$u_{E\ell}(r) = (2\,\mathrm{i}\,k\,r)^{\ell+1}\, \mathrm{e}^{-\mathrm{i}\,k r}\, {}_1F_1\left(\ell + 1 + \mathrm{i}\, \frac{Z\alpha\mu}{k}, 2\ell + 2, 2\mathrm{i}kr \right) . \tag{4.70d}$$

It is extremely important to get the analytic continuation into the complex plane right for all quantities mentioned above; it helps to assign an infinitesimal imaginary part to the energy E and to lay the branch cut of the square root along the negative real axis. The exponential factor $\exp(-\mathrm{i}\,kr)$ is oscillatory for the continuum functions, and this behavior is typical for the continuum wave functions, which are normalized to a Dirac-δ. We write the continuum Schrödinger wave function $\psi_{k\ell m}(\vec{r})$ as follows,

$$\psi_{k\ell m}(\vec{r}) = R_{k\ell}(r)\, Y_{\ell m}(\theta, \varphi) \,, \qquad R_{k\ell}(r) = \frac{\mathcal{N}'}{r}\, u_{E\ell}(r) \,, \qquad E = \frac{k^2}{2\mu} \,. \tag{4.71}$$

These functions should now be normalized to a Dirac-δ in the momentum scale, and the normalization factor $\mathcal{N}'$ needs to be determined. Solutions for manifestly different wave vectors k and k' are orthogonal to each other, and the result for the overlap integral in the limit $k \to k'$ can therefore only be proportional to the Dirac-δ, which in turn implies that the result should be obtained by integration over a wide integration region. The only option for this is to integrate over the region of large radial variable r. We thus need to calculate the behavior of the confluent hypergeometric function for large argument. It is given by

$$ {}_1F_1(a,c,z) \sim \Gamma(c) \left(\frac{\mathrm{e}^z \, z^{a-c}}{\Gamma(a)} + \frac{(-z)^{-a}}{\Gamma(c-a)} \right), \qquad |z| \to \infty . \tag{4.72} $$

Substituting this asymptotic form into Eq. (4.70), one can obtain the following result after a somewhat tedious calculation [both terms in Eq. (4.72) contribute],

$$ \begin{aligned} R^*_{k'\ell}(r)\, R_{k\ell}(r) \sim {} & \frac{|\mathcal{N}'|^2}{r^2} \, \mathrm{e}^{\mathrm{i}\,(k'-k)\,r} \exp\left(-\frac{\pi Z\alpha\mu}{k}\right) \frac{\Gamma^2(2\ell+2)}{\left|\Gamma\left(\ell+1+\mathrm{i}\,\frac{Z\alpha\mu}{k}\right)\right|^2} \\ & + \frac{|\mathcal{N}'|^2}{r^2} \, \mathrm{e}^{-\mathrm{i}\,(k'-k)\,r} \exp\left(-\frac{\pi Z\alpha\mu}{k}\right) \frac{\Gamma^2(2\ell+2)}{\left|\Gamma\left(\ell+1+\mathrm{i}\,\frac{Z\alpha\mu}{k}\right)\right|^2} , \qquad r \to \infty . \end{aligned} \tag{4.73} $$

On the right-hand side, we may ignore terms proportional to $\mathrm{e}^{\mathrm{i}\,(k'+k)\,r}$ and $\mathrm{e}^{-\mathrm{i}\,(k'+k)\,r}$ which do not contribute to the Dirac-δ. The normalization integral becomes

$$ \int_0^\infty \mathrm{d}r\, r^2 \, R^*_{k'\ell}(r)\, R_{k\ell}(r) = |\mathcal{N}'|^2 \, \exp\left(-\frac{\pi Z\alpha\mu}{k}\right) \frac{\Gamma^2(2\ell+2)}{\left|\Gamma\left(\ell+1+\mathrm{i}\,\frac{Z\alpha\mu}{k}\right)\right|^2} \, 2\pi\,\delta(k'-k) . \tag{4.74} $$

The normalization constant may thus be chosen as follows,

$$ \mathcal{N}' = \mathrm{i}^{-(\ell+1)} \exp\left(\frac{\pi Z\alpha\mu}{2k}\right) \frac{\left|\Gamma\left(\ell+1+\mathrm{i}\,\frac{Z\alpha\mu}{k}\right)\right|}{\Gamma(2\ell+2)} . \tag{4.75} $$

This leads to the following form for the radial part of the continuum wave functions,

$$ \begin{aligned} R_{k\ell}(r) = {} & \frac{2k\,(2\,k\,r)^\ell}{(2\ell+1)!} \exp\left(\frac{\pi Z\alpha\mu}{2k}\right) \left|\Gamma\left(\ell+1+\mathrm{i}\,\frac{Z\alpha\mu}{k}\right)\right| \\ & \times \mathrm{e}^{-\mathrm{i}\,kr} \; {}_1F_1\left(\ell+1+\mathrm{i}\,\frac{Z\alpha\mu}{k}, 2\ell+2, 2\mathrm{i}kr\right) . \end{aligned} \tag{4.76a} $$

The eigenvalue properties and normalization conditions are as follows,

$$ \psi_{k\ell m}(\vec{r}) = R_{k\ell}(r)\, Y_{\ell m}(\theta,\varphi) , \qquad H_S\, \psi_{k\ell m}(\vec{r}) = \frac{k^2}{2\mu}\, \psi_{k\ell m}(\vec{r}) , \tag{4.76b} $$

$$ \int \mathrm{d}^3 r\, \psi^*_{k\ell m}(\vec{r})\, \psi_{k'\ell' m'}(\vec{r}) = 2\pi\delta(k'-k)\, \delta_{\ell\ell'}\, \delta_{mm'} . \tag{4.76c} $$

In order to renormalize to the energy scale in the Eq. (4.76a), we observe that

$$ \delta\,(E'-E) = \delta\left(\frac{k'^2}{2\mu} - \frac{k^2}{2\mu}\right) = \frac{\mu}{k}\,\delta(k'-k) , \qquad \frac{\partial}{\partial k}\left(\frac{k^2}{2\mu}\right) = \frac{k}{\mu} . \tag{4.77} $$

Therefore, the continuum wave functions, normalized to the energy scale, are obtained from those normalized to the $2\pi\,\delta(k-k')$ scale via a multiplicative factor $\sqrt{\mu/(2\pi k)}$ and read as

$$R_{E\ell}(r) = \sqrt{\frac{2k\,\mu}{\pi}}\,\frac{(2\,k\,r)^\ell}{(2\ell+1)!}\,\exp\left(\frac{\pi Z\alpha\mu}{2k}\right)\left|\Gamma\left(\ell+1+\mathrm{i}\,\frac{Z\alpha\mu}{k}\right)\right|$$
$$\times\,\mathrm{e}^{-\mathrm{i}\,kr}\;{}_1F_1\left(\ell+1+\mathrm{i}\,\frac{Z\alpha\mu}{k},2\ell+2,2\mathrm{i}kr\right),\qquad k=\sqrt{2\mu\,E}\,. \tag{4.78a}$$

The eigenvalue properties and normalization conditions change into

$$\psi_{E\ell m}(\vec{r}) = R_{E\ell}(r)\,Y_{\ell m}(\theta,\varphi)\,,\qquad H_S\,\psi_{E\ell m}(\vec{r}) = E\,\psi_{k\ell m}(\vec{r})\,, \tag{4.78b}$$

$$\int \mathrm{d}^3r\,\psi^*_{E\ell m}(\vec{r})\,\psi_{E'\ell' m'}(\vec{r}) = \delta(E'-E)\,\delta_{\ell\ell'}\,\delta_{mm'}\,. \tag{4.78c}$$

A few remarks regarding the momentum-normalized, and energy-normalized, solutions are in order. For momentum normalization, our formulas are in agreement with the ones given in Eqs. (36.18) and (36.22) of Chap. 36 of Ref. [160]. The normalization integral is given in Eq. (33.4) of Chap. 33 of Ref. [160]. Our results also are in agreement with the results given in Eqs. (4.2), (4.8) and (4.20) of Ref. [6]. However, one has to take notice of the fact that the wave functions in Ref. [6] are normalized to the $\delta(k'-k)$-scale (without the 2π prefactor). Of course, wave functions normalized to the scale $\delta(k'-k)$ can be obtained from Eq. (4.76a) via multiplication by a factor $1/\sqrt{2\pi}$. This amounts to a replacement $2\to\sqrt{2/\pi}$ in the prefactor in Eq. (4.76a).

A remark is in order concerning Chap. 4 of Ref. [6]. Namely, one has to read Chap. 4 of Ref. [6] carefully. The formulas given in Eqs. (4.22) and (4.23) of Ref. [6], in contrast to the normalization factor c_k given in Eq. (4.20) of Ref. [6], are normalized to the energy scale, not to the wave number scale. This is implicitly stated in Ref. [6] but may give rise to confusion, including confusion in rather recent publications (see, e.g., the Erratum published for Ref. [161]). One may observe that in Refs. [6,160], formulas for the continuum wave functions and normalization factors are scattered over a range of pages.

Natural units are used in Eq. (4.76a). In atomic units, one would use the relation $k = \alpha\,m_e\,k_{\text{a.u.}} \approx \alpha\,\mu\,k_{\text{a.u.}}$, where m_e is the electron mass and $k_{\text{a.u.}}$ is the momentum expressed in atomic units. This is because $\alpha m_e c$ is the unit of momentum in atomic units. In natural units, we have $c=1$. Furthermore, in atomic units, we have $\mu \approx m_e = 1$, because the atomic mass scale is the electron mass scale, which is approximately equal to the reduced mass μ. The natural unit system used in Eq. (4.76a) appears to be a good compromise between compact notation on one hand, and the retained possibility of keeping track of physical mass and length scales on the other hand, for the calculation of ionization cross sections. Our k in the prefactor in Eq. (4.76a) is dimensionful; all other factors in the wave function are dimensionless. Therefore, the physical dimension of $R_{k\ell}$ equals that of a wave number.

For the calculation of transition rates and cross sections, with the help of Fermi's golden rule as discussed in Chap. 3, the energy-normalized wave functions have to be used. This is immediately evident if one takes into account that the final result for Fermi's golden rule, given in Eq. (3.16), assumes that the transition is into a single, well-defined final state, which lies within an "allowed" energy band, over which one integrates in Eq. (3.15). In order for this integration to be well-defined, one has to discretize the energy continuum of the final states, say, in equal (small) intervals ΔE, with one state occupying an energy interval $(E - \Delta E/2, E + \Delta E/2)$, in much the same way that one keeps a finite, but large normalization volume V in the derivation of Fermi's golden rule. If one picks a representative energy E^* within the interval $(E - \Delta E/2, E + \Delta E/2)$ and postulates that exactly one state should be found within the allowed interval ΔE, then the uniqueness of the final state in Eq. (3.16) is ensured. In the limit $\Delta E \to 0$, this condition is exactly fulfilled by wave functions normalized to a Dirac-δ in the energy variable.

4.2.6 *Continuum States for a Repulsive Potential*

For a repulsive Coulomb potential, the Hamiltonian (4.1) changes to

$$H_S = \frac{\vec{p}^{\,2}}{2\mu} + \frac{Z\alpha}{r}, \qquad \vec{p} = -\mathrm{i}\vec{\nabla}. \tag{4.79}$$

This Hamiltonian is obtained from Eq. (4.1) by the formal replacement $r \to -r$ in the potential term. A decomposition of the Laplacian operator in spherical coordinates reveals that the kinetic term in (4.79) is invariant under the substitution $r \to -r$. Therefore, the eigenstates of the Hamiltonian (4.79) are obtained by the corresponding replacement $r \to -r$ in the Whittaker functions. As a consequence, there are no bound states for a repulsive Coulomb interaction. The exponential suppression of the bound-state wave function for large r is replaced by exponential growth under the replacement $r \to -r$; the solutions are not normalizable. By contrast, continuum wave functions proportional to $\exp(-\mathrm{i}\,k\,r)$ remain oscillatory under the replacement $r \to -r$. A careful analysis shows that the argument of the exponential prefactor in Eq. (4.76a) also changes sign as one normalizes the continuum radial wave functions.

When normalized to the energy scale, the continuum eigenstates of the repulsive potential can be written as

$$\begin{aligned} \psi_{E\ell m}(\vec{r}) &= R_{E\ell}(r)\, Y_{\ell m}(\theta,\varphi)\,, \\ R_{E\ell}(r) &= \sqrt{\frac{2k\,\mu}{\pi}}\, \frac{(2\,k\,r)^\ell}{(2\ell+1)!} \exp\left(-\frac{\pi Z\alpha\mu}{2k}\right) \left|\Gamma\left(\ell+1+\mathrm{i}\,\frac{Z\alpha\mu}{k}\right)\right| \\ &\quad \times \mathrm{e}^{\mathrm{i}\,kr}\, {}_1F_1\left(\ell+1+\mathrm{i}\,\frac{Z\alpha\mu}{k}, 2\ell+2, -2\mathrm{i}kr\right), \qquad k = \sqrt{2\mu E}\,. \end{aligned} \tag{4.80a}$$

The eigenvalue properties are given as follows,

$$\psi_{E\ell m}(\vec{r}) = R_{E\ell}(r)\, Y_{\ell m}(\theta,\varphi)\,, \qquad H_S\, \psi_{E\ell m}(\vec{r}) = E\, \psi_{E\ell m}(\vec{r})\,, \tag{4.80b}$$

$$\int \mathrm{d}^3 r\, \psi^{\star}_{E\ell m}(\vec{r})\, \psi_{E'\ell' m'}(\vec{r}) = \delta(E'-E)\, \delta_{\ell\ell'}\, \delta_{mm'}\,. \tag{4.80c}$$

The latter equation, again, implies that the wave functions are normalized to the energy scale.

4.3 Schrödinger–Coulomb Green Function in Coordinate Space

4.3.1 *Green Function and Radial Equation*

The Schrödinger–Coulomb Green function $G(\vec{r}_1, \vec{r}_2, E)$ is the Green function of the Hamiltonian (4.1), which we recall for convenience,

$$H_S = \frac{\vec{p}^{\,2}}{2\mu} - \frac{Z\alpha}{r}\,, \qquad \vec{p} = -\mathrm{i}\vec{\nabla}\,, \tag{4.81}$$

which fulfills the defining differential equation

$$(H_S - E)\, G(\vec{r}_1, \vec{r}_2, E) = \delta^{(3)}(\vec{r}_1 - \vec{r}_2)\,, \tag{4.82}$$

where H_S acts on the variable $\vec{r}_1$. We shall use techniques familiar from classical electrodynamics, namely, an algorithm which we would like to refer to as the concatenation approach for the calculation of the Green function (see Chaps. 2 and 4 of Ref. [81]), after the separation into radial and angular variables. For the Schrödinger–Coulomb Hamiltonian, the angular structure of the Hamiltonian (4.81) helps in reducing Eq. (4.82) to a radial equation. The quantity E in Eq. (4.82) is not necessarily real, and in some problems of practical interest, it is helpful to promote the energy variable E to a general complex variable z. One sometimes sets $z = E_n - \omega$ where E_n is a bound-state energy, and ω is the (complex) energy of a virtual photon.

We use the generalized principal quantum number ν in order to parameterize the energy argument E [see also Eq. (4.30)],

$$\nu^2 = -\frac{Z^2\alpha^2\mu}{2E}\,. \tag{4.83}$$

Our separation *ansatz* for the Green function is familiar from Chaps. 2 and 4 of Ref. [81],

$$G(\vec{r}_1, \vec{r}_2, \nu) = \sum_{\ell=0}^{\infty} \sum_{m=-\ell}^{\ell} f_{\ell m}(r_1, r_2, \theta_2, \varphi_2)\, Y_{\ell m}(\theta_1, \varphi_1)\,. \tag{4.84}$$

Inserting the *ansatz* into the defining equation (4.82), we find

$$\left[-\frac{1}{2\mu}\left(\frac{\partial^2}{\partial r_1^2} + \frac{2}{r_1}\frac{\partial}{\partial r_1}\right) + \frac{\vec{L}_1^2}{2\mu r_1^2} - \frac{Z\alpha}{r_1} + \frac{Z^2\alpha^2\mu}{2\nu^2}\right] \sum_{\ell=0}^{\infty} \sum_{m=-\ell}^{\ell} f_{\ell m}(r_1, r_2, \theta_2, \varphi_2)\, Y_{\ell m}(\theta_1, \varphi_1)$$
$$= \frac{1}{r_1 r_2}\, \delta(r_1 - r_2)\, \delta(\cos\theta_1 - \cos\theta_2)\, \delta(\varphi_1 - \varphi_2)\,, \tag{4.85}$$

where we have used $(r_1 r_2)^{-1}$ as a symmetric prefactor. One multiplies the equation by $-2\mu/\hbar^2$ and obtains

$$\left(\frac{\partial^2}{\partial r_1^2} + \frac{2}{r_1}\frac{\partial}{\partial r_1} - \frac{\vec{L}_1^2}{r_1^2} + \frac{2Z\alpha\mu}{r_1} - \frac{(Z\alpha\mu)^2}{\nu^2}\right)\sum_{\ell=0}^{\infty}\sum_{m=-\ell}^{\ell} f_{\ell m}(r_1,r_2,\theta_2,\varphi_2)\,Y_{\ell m}(\theta_1,\varphi_1)$$
$$= -\frac{2\mu}{r_1 r_2}\,\delta(r_1 - r_2)\,\delta(\cos\theta_1 - \cos\theta_2)\,\delta(\varphi_1 - \varphi_2)\,. \tag{4.86}$$

We multiply both sides by $Y^*_{\ell' m'}(\theta_1,\varphi_1)$ and integrate over the solid angle of $\vec{r}_1$,

$$\int \mathrm{d}\Omega_1 = \int_{-1}^{1} \mathrm{d}(\cos\theta_1)\int_0^{2\pi} \mathrm{d}\varphi_1\,. \tag{4.87}$$

The orthonormality of the spherical harmonics dictates that

$$\sum_{\ell=0}^{\infty}\sum_{m=-\ell}^{\ell}\left[\left(\frac{\partial^2}{\partial r_1^2} + \frac{2}{r_1}\frac{\partial}{\partial r_1} - \frac{\ell(\ell+1)}{r_1^2} + \frac{2Z\alpha\mu}{r_1} - \frac{(Z\alpha\mu)^2}{\nu^2}\right) f_{\ell m}(r_1,r_2,\theta_2,\varphi_2)\right]\delta_{\ell\ell'}\delta_{mm'}$$
$$= \left(\frac{\partial^2}{\partial r_1^2} + \frac{2}{r_1}\frac{\partial}{\partial r_1} - \frac{\ell(\ell+1)}{r_1^2} + \frac{2Z\alpha\mu}{r_1} - \frac{(Z\alpha\mu)^2}{\nu^2}\right) f_{\ell' m'}(r_1,r_2,\theta_2,\varphi_2)$$
$$= -\frac{2\mu}{r_1 r_2}\,\delta(r_1 - r_2)\,Y^*_{\ell' m'}(\theta_2,\varphi_2)\,. \tag{4.88}$$

As expected, we find that the angular dependence of $f_{\ell' m'}(r_1,r_2,\theta_2,\varphi_2)$ is given by $Y^*_{\ell' m'}(\theta_2,\varphi_2)$. One now redefines $\ell' \to \ell$ and $m' \to m$ and separates $f_{\ell m}$ into a radial and angular part, writing

$$f_{\ell m}(r_1,r_2,\theta_2,\varphi_2) = g_\ell(r_1,r_2,\nu)\,Y^*_{\ell m}(\theta_2,\varphi_2)\,, \tag{4.89}$$

which amounts to the following representation of the Green function,

$$G(\vec{r}_1,\vec{r}_2,\nu) = \sum_{\ell=0}^{\infty}\sum_{m=-\ell}^{\ell} g_\ell(r_1,r_2,\nu)\,Y_{\ell m}(\theta_1,\varphi_1)\,Y^*_{\ell m}(\theta_2,\varphi_2)\,. \tag{4.90}$$

We thus have to find a purely radial function $g_\ell(r_1,r_2,\nu)$ satisfying the equation

$$\left(\frac{\partial^2}{\partial r_1^2} + \frac{2}{r_1}\frac{\partial}{\partial r_1} - \frac{\ell(\ell+1)}{r_1^2} + \frac{2Z\alpha\mu}{r_1} - \frac{(Z\alpha\mu)^2}{\nu^2}\right) g_\ell(r_1,r_2,\nu) = -\frac{2\mu}{r_1 r_2}\,\delta(r_1 - r_2)\,. \tag{4.91}$$

In analogy to our analysis of bound-state wave functions, where we wrote the radial part as $\mathcal{R}_{E\ell}(r) = u_{E\ell}(r)/r$, we now define the rescaled radial function $\mathcal{G}$,

$$\mathcal{G}_\ell(r_1,r_2,\nu) = r_1\, r_2\, g_\ell(r_1,r_2,\nu)\,, \tag{4.92}$$

which fulfills

$$\frac{1}{r_1 r_2}\left(\frac{\partial^2}{\partial r_1^2} - \frac{\ell(\ell+1)}{r_1^2} + \frac{2Z\alpha\mu}{r_1} - \frac{(Z\alpha\mu)^2}{\nu^2}\right)\mathcal{G}_\ell(r_1,r_2,\nu) = -\frac{2\mu}{r_1 r_2}\,\delta(r_1 - r_2)\,. \tag{4.93}$$

Our experience with the Schrödinger–Coulomb Hamiltonian suggests a rescaling to dimensionless quantities,

$$\rho_1 = \frac{2Z\alpha\,\mu}{\nu} r_1\,, \qquad \rho_2 = \frac{2Z\alpha\,\mu}{\nu} r_2\,, \tag{4.94}$$

to find the dimensionless equation

$$\left(\frac{\partial^2}{\partial\rho_1^2} - \frac{1}{4} + \frac{\nu}{\rho_1} + \frac{\frac{1}{4} - \left(\ell + \frac{1}{2}\right)^2}{\rho_1^2}\right) \mathcal{G}_\ell(\rho_1, \rho_2, \nu) = -\frac{\nu}{Z\alpha}\,\delta(\rho_1 - \rho_2)\,. \tag{4.95}$$

We will solve this equation in two ways, *(i)* in the same way as for the Poisson equation (Chap. 2 of Ref. [81]), by an appropriate construction using Whittaker functions, and *(ii)* using a separation *ansatz* based on the associated Laguerre polynomials.

4.3.2 *Solution using Whittaker Functions*

We recall from Sec. 4.2.2 that the general eigenfunction solution of the Schrödinger–Coulomb Hamiltonian is given by the linear combination of Whittaker functions M and W, where the M function is regular at the origin while the W function is not. We thus have two solutions to the homogeneous equation for $\rho_1 < \rho_2$ and $\rho_1 > \rho_2$. Inspired by our treatment of Green functions in classical electrodynamics (Chaps. 2 and 4 of Ref. [81]), we concatenate the radial Green function $\mathcal{G}_\ell(\rho_1, \rho_2, \nu)$ at $\rho_1 = \rho_2$ by integrating Eq. (4.95) from $\rho_1 = \rho_2 - \epsilon$ to $\rho_1 = \rho_2 + \epsilon$.

A comparison of Eqs. (4.33) and (4.95) reveals that the general solution to the homogeneous equation corresponding to Eq. (4.95) (i.e., the equation obtained by replacing the Dirac-δ function by zero) reads as follows,

$$\mathcal{G}_\ell(\rho_1, \rho_2, \nu) = \alpha_\ell(\rho_2)\, M_{\nu,\ell+\frac{1}{2}}(\rho_1) + \beta_\ell(\rho_2)\, W_{\nu,\ell+\frac{1}{2}}(\rho_1)\,. \tag{4.96}$$

For $\rho_1 < \rho_2$, we assume that $\mathcal{G}_\ell(\rho_1, \rho_2, \nu) = \alpha_\ell(\rho_2)\, M_{\nu,\ell+1/2}(\rho_1)$ because the solution needs to be regular at the origin. For $\rho_1 > \rho_2$ on the other hand, we have $\mathcal{G}_\ell(\rho_1, \rho_2, \nu) = \beta_\ell(\rho_2)\, W_{\nu,\ell+1/2}(\rho_1)$ because the solution needs to be regular at infinity. Continuity at $\rho_1 = \rho_2$ requires that (we assume that a is a constant)

$$\mathcal{G}_\ell(\rho_1, \rho_2, \nu) = a\, M_{\nu,\ell+\frac{1}{2}}(\rho_<)\, W_{\nu,\ell+\frac{1}{2}}(\rho_>)\,, \quad \rho_< = \min(\rho_1, \rho_2)\,, \quad \rho_> = \max(\rho_1, \rho_2)\,. \tag{4.97}$$

We now integrate Eq. (4.95) from $\rho_1 = \rho_2 - \epsilon$ to $\rho_1 = \rho_2 + \epsilon$ to obtain

$$\int_{\rho_2-\epsilon}^{\rho_2+\epsilon} \mathrm{d}\rho_1 \left(\frac{\partial^2}{\partial\rho_1^2} - \frac{1}{4} + \frac{\nu}{\rho_1} + \frac{\frac{1}{4} - \left(\ell + \frac{1}{2}\right)^2}{\rho_1^2}\right) \mathcal{G}_\ell(\rho_1, \rho_2, \nu)$$
$$= -\frac{\nu}{Z\alpha} \int_{\rho_2-\epsilon}^{\rho_2+\epsilon} \mathrm{d}\rho_1\, \delta(\rho_1 - \rho_2)\,. \tag{4.98}$$

The terms without derivative operators on the left-hand side drop out because $\mathcal{G}_\ell(\rho_1, \rho_2, \nu)$ is assumed to be continuous at $\rho_1 = \rho_2$,

$$\int_{\rho_2-\epsilon}^{\rho_2+\epsilon} \mathrm{d}\rho_1 \left[\frac{\partial^2}{\partial\rho_1^2} \mathcal{G}_\ell(\rho_1, \rho_2, \nu)\right] = -\frac{\nu}{Z\alpha}\,. \tag{4.99}$$

An integration immediately leads to

$$\lim_{\epsilon\to 0}\left[\left(\frac{\partial \mathcal{G}_\ell(\rho_1, \rho_2, \nu)}{\partial\rho_1}\right)_{\rho_1=\rho_2+\epsilon} - \left(\frac{\partial \mathcal{G}_\ell(\rho_1, \rho_2, \nu)}{\partial\rho_1}\right)_{\rho_1=\rho_2-\epsilon}\right] = -\frac{\nu}{Z\alpha}\,. \tag{4.100}$$

We now use the specific forms of $\mathcal{G}_\ell(\rho_1, \rho_2, \nu)$ for the two terms in Eq. (4.97), observing that in each case, only one of the Whittaker functions depends on ρ_1. Specifically, $\rho_1 = \rho_2 + \epsilon$ implies that $\rho_1 = \rho_>$, while $\rho_1 = \rho_2 - \epsilon$ implies that $\rho_2 = \rho_>$. We can thus pull out the respective other Whittaker function to write

$$a\left[M_{\nu,\ell+\frac{1}{2}}(\rho_1)\frac{\partial}{\partial\rho_1}W_{\nu,\ell+\frac{1}{2}}(\rho_1) - W_{\nu,\ell+\frac{1}{2}}(\rho_1)\frac{\partial}{\partial\rho_1}M_{\nu,\ell+\frac{1}{2}}(\rho_1)\right] = -\frac{\nu}{Z\alpha}. \tag{4.101}$$

The Wronskian of the two solutions (M and W functions) is independent of ρ and can therefore be calculated for any ρ. We take notice of the following asymptotic forms,

$$M_{\nu,\ell+\frac{1}{2}}(\rho) \to \rho^{\ell+1}, \qquad W_{\nu,\ell+\frac{1}{2}}(\rho) \to \frac{\Gamma(2\ell+1)}{\Gamma(1+\ell-\nu)\,\rho^\ell}, \qquad \rho \to 0. \tag{4.102}$$

The Wronskian thus evaluates to

$$M_{\nu,\ell+\frac{1}{2}}(\rho)\left(\frac{\partial}{\partial\rho}W_{\nu,\ell+\frac{1}{2}}(\rho)\right) - W_{\nu,\ell+\frac{1}{2}}(\rho)\left(\frac{\partial}{\partial\rho}M_{\nu,\ell+\frac{1}{2}}(\rho)\right) = -\frac{\Gamma(2\ell+2)}{\Gamma(1+\ell-\nu)}. \tag{4.103}$$

Solving Eq. (4.101) for a, we find

$$a = \frac{\nu\,\Gamma(1+\ell-\nu)}{Z\alpha\,(2\ell+1)!}. \tag{4.104}$$

We immediately find the $\mathcal{G}_\ell$ function as

$$\mathcal{G}_\ell(\rho_1, \rho_2, \nu) = \frac{\nu\,\Gamma(1+\ell-\nu)}{Z\alpha\,(2\ell+1)!}\, M_{\nu,\ell+\frac{1}{2}}(\rho_<)\, W_{\nu,\ell+\frac{1}{2}}(\rho_>), \tag{4.105}$$

and g_ℓ is obtained by dividing $\mathcal{G}_\ell$ by $r_1 r_2$,

$$g_\ell(\rho_1, \rho_2, \nu) = \frac{\Gamma(1+\ell-\nu)}{(2\ell+1)!}\,\frac{\nu}{Z\alpha\, r_1\, r_2}\, M_{\nu,\ell+\frac{1}{2}}(\rho_<)\, W_{\nu,\ell+\frac{1}{2}}(\rho_>). \tag{4.106}$$

Furthermore,

$$G(\vec{r}_1, \vec{r}_2, \nu) = \sum_{\ell=0}^{\infty}\sum_{m=-\ell}^{\ell} g_\ell(r_1, r_2, \nu)\, Y_{\ell m}(\theta_1, \varphi_1)\, Y^*_{\ell m}(\theta_2, \varphi_2), \tag{4.107}$$

so that

$$g_\ell(r_1, r_2, \nu) = \frac{\Gamma(1+\ell-\nu)}{(2\ell+1)!}\,\frac{\nu}{Z\alpha\, r_1\, r_2}\, M_{\nu,\ell+\frac{1}{2}}\left(\frac{2Z\alpha\mu}{\nu} r_<\right) W_{\nu,\ell+\frac{1}{2}}\left(\frac{2Z\alpha\mu}{\nu} r_>\right). \tag{4.108}$$

A final word on the integration ranges for $r_<$ and $r_>$ is in order. Within classical electrodynamics (Chap. 5 of Ref. [81]), we can generally associate $r_<$ with the point at which radiation is generated, whereas $r_>$ is the point at which radiation is observed. Within typical atomic physics calculations, one cannot avoid integrations over the entire range of $r \in (0, \infty)$, and the integration over the full range of r is suitable for numerical evaluation in many cases.

4.3.3 *Solution using Laguerre Polynomials*

For analytic calculations, a representation in terms of so-called Sturmian functions is sometimes preferred over the representation (4.108). The so-called Sturmians are special forms of Laguerre polynomials, and the Sturmian representation reminds us of an eigenfunction representation (see Sec. 2.5 of Ref. [81]), because it is formulated in terms of the variables r_1 and r_2, not $r_<$ and $r_>$, and therefore facilitates integrations over r_1 and r_2 separately. We incur a further explicit sum over the Sturmians, though, in exchange for discarding the need to use the variables $r_<$ and $r_>$, *in lieu* of the more convenient r_1 and r_2.

We use the *ansatz*

$$\mathcal{G}_\ell(\rho_1, \rho_2, \nu) = \sum_{k=0}^{\infty} R_{k\ell}(\rho_1)\, f_{k\ell}(\rho_2, \nu)\,. \tag{4.109}$$

We recall Eq. (4.95),

$$\left(\frac{\partial^2}{\partial \rho_1^2} - \frac{1}{4} + \frac{\nu}{\rho_1} + \frac{\frac{1}{4} - \left(\ell + \frac{1}{2}\right)^2}{\rho_1^2}\right) \mathcal{G}_\ell(\rho_1, \rho_2, \nu) = -\frac{\nu}{Z\alpha}\, \delta(\rho_1 - \rho_2)\,. \tag{4.110}$$

This leads to

$$\sum_{k=0}^{\infty} f_{k\ell}(\rho_2, \nu) \left(\frac{\partial^2}{\partial \rho_1^2} - \frac{1}{4} + \frac{\nu}{\rho_1} + \frac{\frac{1}{4} - \left(\ell + \frac{1}{2}\right)^2}{\rho_1^2}\right) R_{k\ell}(\rho_1) = -\frac{\nu}{Z\alpha}\, \delta(\rho_1 - \rho_2)\,. \tag{4.111}$$

We know that $R_{k\ell}(\rho_1) = M_{\nu,\ell+1/2}(\rho_1)$ and $R_{k\ell}(\rho_1) = W_{\nu,\ell+1/2}(\rho)$ are solutions of the homogeneous equation. Let us pretend, for a moment, that $\nu = n$ is an integer. Then, the solution to the homogeneous equation is [see Eq. (4.52)]

$$M_{n,\ell+\frac{1}{2}}(\rho) = \mathrm{e}^{-\frac{1}{2}\rho}\, \rho^{\ell+1}\, \frac{(n-\ell-1)!\,(2\ell+1)!}{(n+\ell)!}\, L_{n-\ell-1}^{2\ell+1}(\rho)\,. \tag{4.112}$$

Identifying $k = n - \ell - 1$, we enter into Eq. (4.109) with the *ansatz*

$$R_{k\ell}(\rho) = \frac{k!\,(2\ell+1)!}{(k+2\ell+1)!}\, \rho^{\ell+1}\, L_k^{2\ell+1}(\rho)\, \mathrm{e}^{-\rho/2}\,, \tag{4.113}$$

which, for $\nu = n$, would just be the radial eigenfunction. For non-integer ν, however, a residual term remains which can be used in order to fulfill the inhomogeneous equation.

Performing elementary differentiations, we obtain the following result,

$$\begin{aligned} &\left(\frac{\partial^2}{\partial \rho_1^2} - \frac{1}{4} + \frac{\nu}{\rho_1} + \frac{\frac{1}{4} - (\ell + \frac{1}{2})^2}{\rho_1^2}\right) R_{k\ell}(\rho_1) \\ &= \frac{k!\,(2\ell+1)!}{(k+2\ell+1)!} \mathrm{e}^{-\rho_1/2} \rho_1^{\ell} \left(\rho_1 \frac{\partial^2}{\partial \rho_1^2} + (2\ell + 2 - \rho_1)\frac{\partial}{\partial \rho_1} + (\nu - \ell - 1)\right) L_k^{2\ell+1}(\rho_1)\,. \end{aligned} \tag{4.114}$$

The last part of Eq. (4.114) is close to the defining differential equation of the Laguerre polynomials [159]. We can easily rearrange the differential equation in Eq. (4.114) to the following form:

$$\begin{aligned}&\left(\rho_1\frac{\partial^2}{\partial\rho_1^2}+(2\ell+2-\rho_1)\frac{\partial}{\partial\rho_1}+(\nu-\ell-1)\right)L_k^{2\ell+1}(\rho_1)\\&=\left(\rho_1\frac{\partial^2}{\partial\rho_1^2}+(2\ell+2-\rho_1)\frac{\partial}{\partial\rho_1}+k\right)L_k^{2\ell+1}(\rho_1)+(\nu-\ell-1-k)\,L_k^{2\ell+1}(\rho_1)\\&=(\nu-\ell-1-k)\,L_k^{2\ell+1}(\rho_1)\,.\end{aligned}\tag{4.115}$$

We have used the differential equation solved by the associated Laguerre polynomials $L_k^{\xi}(\rho_1)$, which reads as follows [see Eq. (22.6.15) of Ref. [159]]

$$\left(\rho_1\frac{\partial^2}{\partial\rho_1^2}+(\xi+1-\rho_1)\frac{\partial}{\partial\rho_1}+k\right)L_k^{\xi}(\rho_1)=0\,,\tag{4.116}$$

where in our case $\xi=2\ell+1$. Combining Eqs. (4.114) and (4.115), we find

$$\left(\frac{\partial^2}{\partial\rho_1^2}-\frac{\ell(\ell+1)}{\rho_1^2}+\frac{\nu}{\rho_1}-\frac{1}{4}\right)R_{k\ell}(\rho_1)=\frac{k!\,(2\ell+1)!}{(k+2\ell+1)!}\mathrm{e}^{-\rho_1/2}\rho_1^{\ell}\,(\nu-\ell-1-k)\,L_k^{2\ell+1}(\rho_1)\,.\tag{4.117}$$

Plugging the result back into Eq. (4.111) yields

$$\sum_{k=0}^{\infty}f_{k\ell}(\rho_2,\nu)\frac{k!\,(2\ell+1)!}{(k+2\ell+1)!}e^{-\rho_1/2}\rho_1^{\ell}\,(\nu-\ell-1-k)\,L_k^{2\ell+1}(\rho_1)=-\frac{\nu}{Z\alpha}\,\delta(\rho_1-\rho_2)\,.\tag{4.118}$$

We multiply both sides with $e^{-\rho_1/2}\rho_1^{\ell+1}\,L_{k'}^{2\ell+1}(\rho_1)$ and integrate over the remaining radial part of the volume integral $\int\mathrm{d}\rho_1\rho_1^2$, leading to the expression

$$\begin{aligned}&\sum_{k=0}^{\infty}(\nu-\ell-1-k)\,\frac{k!\,(2\ell+1)!}{(k+2\ell+1)!}f_{k\ell}(\rho_2,\nu)\int_0^{\infty}\mathrm{d}\rho_1\,e^{-\rho_1}\,\rho_1^{2\ell+1}L_k^{2\ell+1}(\rho_1)L_{k'}^{2\ell+1}(\rho_1)\\&=-\frac{\nu}{Z\alpha}\int_0^{\infty}\mathrm{d}\rho_1\,\rho_1^{\ell+1}\,\mathrm{e}^{-\rho_1/2}L_{k'}^{2\ell+1}(\rho_1)\,\delta(\rho_1-\rho_2)\,.\end{aligned}\tag{4.119}$$

The orthogonality relation of the associated Laguerre polynomials is given by [see Eq. (22.2.12) of Ref. [159]]

$$\int_0^{\infty}\mathrm{d}\rho_1\,\mathrm{e}^{-\rho_1}\,\rho_1^{2\ell+1}L_k^{2\ell+1}(\rho_1)L_{k'}^{2\ell+1}(\rho_1)=\frac{(k+2\ell+1)!}{k!}\delta_{kk'}\,.\tag{4.120}$$

Consequently, if we carry out the integration, we obtain

$$\begin{aligned}&\sum_{k=0}^{\infty}(\nu-\ell-1-k)\,\frac{k!\,(2\ell+1)!}{(k+2\ell+1)!}f_{k\ell}(\rho_2,\nu)\frac{(k+2\ell+1)!}{k!}\delta_{kk'}\\&=-\frac{\nu}{Z\alpha}\,\rho_2^{\ell+1}\,\mathrm{e}^{-\rho_2/2}L_{k'}^{2\ell+1}(\rho_2)\,.\end{aligned}\tag{4.121}$$

Collapsing the sum using the Kronecker-δ and redefining $k'\to k$, one finds that

$$(\nu-\ell-1-k)\;(2\ell+1)!\,f_{k\ell}(\rho_2,\nu)=-\frac{\nu}{Z\alpha}\,\rho_2^{\ell+1}\,\mathrm{e}^{-\rho_2/2}L_k^{2\ell+1}(\rho_2)\,,\tag{4.122}$$

which immediately gives a result for $f_{k\ell}(\rho_2,\nu)$. We plug the result for $f_{k\ell}(\rho_2,\nu)$ back into Eq. (4.109), and use Eq. (4.113),

$$\begin{aligned}\mathcal{G}_\ell(\rho_1,\rho_2,\nu) &= \frac{\nu}{Z\alpha}\sum_{k=0}^{\infty} R_{k\ell}(\rho_1)\,\rho_2^{\ell+1}\,\frac{\mathrm{e}^{-\rho_2/2}L_k^{2\ell+1}(\rho_2)}{(2\ell+1)!\,(k+\ell+1-\nu)}\\ &= \frac{\nu}{Z\alpha}\sum_{k=0}^{\infty}\mathrm{e}^{-\frac{1}{2}(\rho_1+\rho_2)}(\rho_1\rho_2)^{\ell+1}\frac{k!\,L_k^{2\ell+1}(\rho_1)\,L_k^{2\ell+1}(\rho_2)}{(k+2\ell+1)!\,(k+\ell+1-\nu)}\,. \end{aligned} \tag{4.123}$$

Recalling the definitions from Eq. (4.94), one arrives at

$$\begin{aligned}\mathcal{G}_\ell(\rho_1,\rho_2,\nu) &= 2\mu\left(\frac{2Z}{a_0\nu}\right)^{2\ell+1}\exp\left(-Z\frac{r_1+r_2}{a_0\nu}\right)(r_1\,r_2)^{\ell+1}\\ &\quad\times\sum_{k=0}^{\infty}\frac{k!\,L_k^{2\ell+1}\left(\frac{2Zr_1}{a_0\nu}\right)L_k^{2\ell+1}\left(\frac{2Zr_2}{a_0\nu}\right)}{(k+2\ell+1)!\,(k+\ell+1-\nu)}\,. \end{aligned} \tag{4.124}$$

Using Eq. (4.92), we obtain the radial Green function $g_\ell(r_1,r_2,\nu)$ by dividing $\mathcal{G}_\ell$ by the product $r_1\,r_2$,

$$g_\ell(r_1,r_2,\nu) = 2\mu\left(\frac{2Z}{a_0\nu}\right)^{2\ell+1}\mathrm{e}^{-\frac{Z(r_1+r_2)}{a_0\nu}}(r_1\,r_2)^{\ell}\sum_{k=0}^{\infty}\frac{k!\,L_k^{2\ell+1}\left(\frac{2Zr_1}{a_0\nu}\right)L_k^{2\ell+1}\left(\frac{2Zr_2}{a_0\nu}\right)}{(k+2\ell+1)!\,(k+\ell+1-\nu)}\,, \tag{4.125}$$

where $a_0 = 1/(\alpha\mu)$ is the Bohr radius including the reduced-mass dependence. Our result is in agreement with the one derived in Ref. [158] by a Laplace transform of the differential equation for the radial Green function. For convenience, we recall that the total Green function is given as

$$G(\vec{r}_1,\vec{r}_2,\nu) = \sum_{\ell=0}^{\infty}\sum_{m=-\ell}^{\ell} g_\ell(r_1,r_2,\nu)\,Y_{\ell m}(\theta_1,\varphi_1)\,Y^*_{\ell m}(\theta_2,\varphi_2)\,. \tag{4.126}$$

The representation (4.125) separates the two variables r_1 and r_2 and is especially suited for analytic calculations, because the radial integrations can be carried out independently, before the sum over k is taken.

4.3.4 *Green Function and Dynamic Polarizability*

The Green function as derived in Eq. (4.125) is to be applied to the evaluation of matrix elements of the Schrödinger–Coulomb propagator. Two important matrix elements are as follows,

$$P(\phi_n,\omega) = \frac{1}{3\mu}\left\langle\phi_n\left|p^i\frac{1}{H_S-E_n+\omega}p^i\right|\phi_n\right\rangle, \tag{4.127a}$$

$$Q(\phi_n,\omega) = \frac{(Z\alpha)^4\,\mu^3}{3}\left\langle\phi_n\left|x^i\frac{1}{H_S-E_n+\omega}x^i\right|\phi_n\right\rangle, \tag{4.127b}$$

where the sum over $i = 1,2,3$ is implicit in view of the summation convention. We use superscripts in order to denote the Cartesian components, and we have to include prefactors of μ in order to render the matrix elements dimensionless. In

Eqs. (4.127a) and (4.127b), ϕ_n is the bound reference state with energy $E_n < 0$. We assume that the subscript n in ϕ_n is a multi-index that specifies the principal quantum number and the orbital angular momentum quantum number ℓ. The matrix elements P and Q are independent of the magnetic projection quantum number m. They are related to the polarizability $\alpha(\omega)$ defined in Eq. (3.53).

While implicit formulas for these matrix elements were derived around 1970 in Ref. [162], it was not until 1993 that convenient expressions were derived for the matrix elements of the propagator, in Ref. [163]. The calculation has to be well organized, and one has to be familiar with a number of rather peculiar identities fulfilled by hypergeometric functions (so-called contiguous relations). We shall describe the evaluation in some detail, for the reference state being the $1S$ ground state of the hydrogen atom, and first concentrate on the matrix element

$$P(1S,\omega) = \frac{1}{3\mu}\left\langle 1S \left| p^i \frac{1}{H_S - E_n + \omega} p^i \right| 1S \right\rangle = \frac{1}{3\mu}\left\langle 1S \left| p^i \frac{1}{H_S - z} p^i \right| 1S \right\rangle , \quad z = E_n - \omega , \tag{4.128}$$

where the Schrödinger–Coulomb Hamiltonian H_S [see also Eq. (4.1)] and the $1S$ wave function are given as follows,

$$H_S = \frac{\vec{p}^{\,2}}{2\mu} - \frac{Z\alpha}{r} , \qquad \psi_{1S}(\vec{r}) = R_{1S}(r)\, Y_{00}(\theta,\varphi) , \qquad R_{1S}(r) = 2\left(\frac{Z}{a_0}\right)^{3/2} \mathrm{e}^{-Zr/a_0} . \tag{4.129}$$

We had previously parameterized the energy argument z of the Green function in terms of the generalized principal quantum number ν as follows [see Eq. (4.83)],

$$z = E_n - \omega = -\frac{(Z\alpha)^2\mu}{2\nu^2} = E_n \frac{n^2}{\nu^2} , \qquad t = \sqrt{\frac{E_n}{z}} = \frac{\nu}{n} . \tag{4.130}$$

For the ground state, we have $\nu = t$ because $n = 1$. The t variable has certain advantages. Namely, if $z = E_n - \omega$, then $\omega = 0$ corresponds to $t = 1$ and $\omega = \infty$ corresponds to $t = 0$. Any conceivable integral over ω, which we may have to do in the following, therefore is mapped onto an integral over t in the range $t \in (0,1)$. In terms of the photon energy ω, we write

$$t = \sqrt{\frac{E_n}{E_n - \omega}} = \left(1 + \frac{2n^2\omega}{(Z\alpha)^2\,\mu}\right)^{-1/2} , \qquad \omega = \frac{(Z\alpha)^2\,\mu}{2n^2}\,\frac{1-t^2}{t^2} . \tag{4.131}$$

An integral over a range of ω values can be transformed into a t integral as follows,

$$\int_{\omega_0}^{\omega_1} \mathrm{d}\omega\, \omega\, f(\omega) = \frac{(Z\alpha)^4\,\mu^2}{2n^4} \int_{t_1}^{t_0} \mathrm{d}t\, \frac{1-t^2}{t^5}\, \tilde{f}(t) , \tag{4.132}$$

where $\tilde{f}(t) \equiv f(\omega(t))$ is the original function, expressed in terms of t. All of these identities will become useful in the following. We now return to the evaluation of the matrix element of the propagator and use the coordinate-space representation of the Schrödinger–Coulomb Green function,

$$G(\vec{r}_1,\vec{r}_2,\nu) = \sum_{\ell=0}^{\infty} \sum_{m=-\ell}^{\ell} g_\ell(r_1,r_2,\nu)\, Y_{\ell m}(\theta_1,\varphi_1)\, Y^*_{\ell m}(\theta_2,\varphi_2) . \tag{4.133}$$

For the radial Green function, we use the Sturmian decomposition given in Eq. (4.125),

$$g_\ell(r_1, r_2, \nu) = 2\mu \left(\frac{2Z}{a_0 \nu}\right)^{2\ell+1} \mathrm{e}^{-Z\frac{r_1+r_2}{a_0\nu}} (r_1 r_2)^\ell \sum_{k=0}^{\infty} \frac{k!\, L_k^{2\ell+1}\left(\frac{2Zr_1}{a_0\nu}\right) L_k^{2\ell+1}\left(\frac{2Zr_2}{a_0\nu}\right)}{(k+2\ell+1)!\,(k+\ell+1-\nu)}. \tag{4.134}$$

The ($\ell = 1$)-component reads as follows,

$$g_{\ell=1}(r_1, r_2, \nu) = 16\mu \left(\frac{Z}{a_0 \nu}\right)^3 \exp\left(-Z\frac{r_1+r_2}{a_0\nu}\right) r_1 r_2 \sum_{k=0}^{\infty} \frac{k!\, L_k^3\left(\frac{2Zr_1}{a_0\nu}\right) L_k^3\left(\frac{2Zr_2}{a_0\nu}\right)}{(k+3)!\,(2+k-\nu)}. \tag{4.135}$$

After the angular integration, which is easy to perform using the explicit representation of the spherical harmonics, we reduce $P(1S, \omega)$ to the following radial integral,

$$P(1S,\omega) = \frac{1}{3\mu} \int_0^\infty \mathrm{d}r_1\, r_1^2 \int_0^\infty \mathrm{d}r_2\, r_2^2 \left(\frac{\partial}{\partial r_1} R_{1S}(r_1)\right) g_{\ell=1}(r_1, r_2, t) \left(\frac{\partial}{\partial r_2} R_{1S}(r_2)\right). \tag{4.136}$$

Here and elsewhere, partial derivatives are indicated in order to denote the differentiation with respect to the radial variable only (strictly speaking, the wave function also depends on other parameters like the Bohr radius). We use the rescaling of variables as in Eq. (4.94),

$$\rho_1 = \frac{2Z\alpha\,\mu}{t} r_1, \qquad \rho_2 = \frac{2Z\alpha\,\mu}{t} r_2, \tag{4.137}$$

where we take into account that $\nu = t$ for the ground state, to bring the integral into the form

$$P(1S,t) = \sum_{k=0}^{\infty} \frac{t^5\, k!}{12\,(k+3)!\,(2+k-t)} \times \int_0^\infty \mathrm{d}\rho_1\, \rho_1^2 \int_0^\infty \mathrm{d}\rho_2\, \rho_2^2\, \mathrm{e}^{-\frac{1}{2}(1+t)\rho_1 - \frac{1}{2}(1+t)\rho_2}\, L_k^3(\rho_1)\, L_k^3(\rho_2), \tag{4.138}$$

changing the argument of the function from ω to t, which parameterizes the energy according to Eq. (4.131). The integrals over ρ_1 and ρ_2 may be evaluated using the formula (10.12.33) of Ref. [156],

$$\int_0^\infty \mathrm{d}\rho\, \mathrm{e}^{-s\rho} \rho^{\gamma-1} L_k^m(\rho) = \frac{\Gamma(\gamma)\,\Gamma(k+m+1)}{k!\,\Gamma(m+1)} s^{-\gamma}\, {}_2F_1\left(-k, \gamma, 1+m, \frac{1}{s}\right). \tag{4.139}$$

The hypergeometric function resulting from the integration is terminating in view of the negative first argument $-k < 0$. In typical cases of practical interest, after the use of contiguous relations, the hypergeometrics can be expressed in closed analytic form, using formulas such as

$${}_2F_1(-k, a, a, z) = (1-z)^k, \quad {}_2F_1(-k, a+1, a, z) = (1-z)^k \frac{(z-1)\,a + k\,z}{a\,(z-1)}. \tag{4.140}$$

The matrix element $P(1S,t)$ can thus be written as a rather compact sum,

$$P(1S,t) = \frac{64\,t^5}{3\,(1+t)^8} \sum_{k=0}^{\infty} \frac{(k+1)\,(k+2)\,(k+3)}{(2+k-t)} (-\zeta)^{2k}\,, \qquad \zeta = \frac{1-t}{1+t}. \tag{4.141}$$

The summation over k can be carried out using various formulas. We here give a selection of useful sums, which are relevant to matrix elements which involve the Schrödinger–Coulomb propagator, and refer to standard encyclopediae for larger collections of formulas [164–169]. Three relatively easy formulas are as follows,

$$\sum_{k=0}^{\infty} k\,s^k = \frac{s}{(1-s)^2}\,, \qquad \sum_{k=0}^{\infty} k^2\,s^k = \frac{s+s^2}{(1-s)^3}\,, \qquad \sum_{k=0}^{\infty} k^3\,s^k = \frac{s+4\,s^2+s^3}{(1-s)^4}. \tag{4.142}$$

A slightly more involved formula is

$$\sum_{k=0}^{\infty} \frac{s^k}{a+k} = \Phi(s,1,a) = \frac{{}_2F_1\,(1,a,1+a,s)}{a}. \tag{4.143}$$

Here, Φ is the Lerch transcendent, whose general defining relation is

$$\Phi(s,\gamma,a) = \sum_{k=0}^{\infty} \frac{s^k}{(a+k)^\gamma}. \tag{4.144}$$

A formula valid in a larger domain of the complex plane is

$$\Phi(s,\gamma,a) = \frac{1}{1-s} \sum_{k=0}^{\infty} \left(\frac{-s}{1-s}\right)^k \sum_{j=0}^{k} \frac{(-1)^j}{(a+j)^\gamma} \binom{k}{j}. \tag{4.145}$$

The Lerch Φ transcendent can be thought of as a generalization of the Riemann zeta function, because $\Phi(1,s,1) = \zeta(s)$. Sometimes, more complex mathematical formulae have to be used, like the following one, found in Ref. [154],

$$\sum_{k=0}^{\infty} \frac{\Gamma(k+\lambda)}{k!}\, s^k\, {}_2F_1(-k,b,c,z) = \Gamma(\lambda)\,(1-s)^{-\lambda}\, {}_2F_1\Big(\lambda,b,c,-\frac{s\,z}{1-s}\Big). \tag{4.146}$$

For our expression given in Eq. (4.141), the summation over k leads to the following result,

$$P(1S,t) = \frac{2t^2(1+5t+7t^2+11t^3)}{3\,(1+t)^5} - \frac{64\,(1-t)\,t^6}{3\,(2-t)\,(1+t)^7}\, {}_2F_1\left(1,2-t,3-t,\left(\frac{1-t}{1+t}\right)^2\right). \tag{4.147}$$

This expression already is quite compact, but it is possible to bring it into some more standard form, using the so-called contiguous relations for hypergeometric functions. These are given in Sec. 2.8 of Ref. [154]. Two of them, conveniently written to allow for a lowering of the second and third arguments, read as follows,

$${}_2F_1(a,b,c,z) = \frac{1}{z}\,\frac{c-1}{(c-a-1)}\left[{}_2F_1(a,b-1,c-1,z) + (z-1)\,{}_2F_1(a,b,c-1,z)\right], \tag{4.148a}$$

$${}_2F_1(a,b,c,z) = \frac{1}{b-1}\,\left[(c-1)\,{}_2F_1(a,b-1,c-1,z) + (b-c)\,{}_2F_1(a,b-1,c,z)\right]. \tag{4.148b}$$

Judicious repeated application of the contiguous relations, lowering the second and third arguments of the hypergeometric functions in turn, then leads to the result

$$P(1S,t) = \frac{2t^2(1-t-5t^2-11t^3)}{3\,(1+t)^2\,(1-t)^3} + \frac{64\,t^5}{3\,(1-t^2)^3}\,{}_2F_1\left(1,-t,1-t,\left(\frac{1-t}{1+t}\right)^2\right). \quad (4.149)$$

For an excited reference state with principal quantum number $n \geq 2$, the hypergeometric function in the result takes a slightly more general form,

$$ {}_2F_1\left(1,-n\,t,1-n\,t,\left(\frac{1-t}{1+t}\right)^2\right), \quad (4.150)$$

as exemplified in the results

$$P(2S,t) = \frac{2t^2\left(1-2\,t-8\,t^2-46\,t^3+151\,t^4\right)}{3\,(1-t)^4\,(1+t)^2} + \frac{256\,t^5\left(1-4\,t^2\right)}{3\,(1-t)^4\,(1+t)^4}\,{}_2F_1\left(1,-2t,1-2t,\left(\frac{1-t}{1+t}\right)^2\right), \quad (4.151)$$

and

$$P(2P,t) = \frac{2t^2\left(3-6\,t-3\,t^2+12\,t^3+29\,t^4+122\,t^5-413\,t^6\right)}{9\,(1-t)^5\,(1+t)^3} + \frac{256\,t^7\left(-3+11\,t^2\right)}{9\,(1-t)^5\,(1+t)^5}\,{}_2F_1\left(1,-2t,1-2t,\left(\frac{1-t}{1+t}\right)^2\right). \quad (4.152)$$

In the results (4.149), (4.151), and (4.152), t parameterizes the photon energy ω according to Eq. (4.131); it is important that t in this case carries a dependence on the principal quantum number n. As we are using natural units with $\hbar = c = \epsilon_0 = 1$, it may be helpful to convert Eq. (4.131) to SI mksA units,

$$t = \left(1 + \frac{2n^2\hbar\omega}{(Z\alpha)^2\,\mu\,c^2}\right)^{-1/2}. \quad (4.153)$$

The result (4.152) for the dynamic polarizability, expressed in terms of the energy variable, has been derived in Ref. [170]. We recall the conventions implied by Eq. (4.130). For completeness, we also indicate explicit results for Q matrix elements, for the states with principal quantum numbers $n = 1$ and $n = 2$. For the ground state, we have

$$Q(1S,t) = \frac{2t^2(3-3t-12t^2+12t^3+19t^4-19t^5-26t^6-38t^7)}{3\,(1-t)^5\,(1+t)^4} + \frac{256\,t^9}{3\,(1+t)^5\,(1-t)^5}\,{}_2F_1\left(1,-t,1-t,\left(\frac{1-t}{1+t}\right)^2\right). \quad (4.154)$$

The static limit of the polarizability is proportional to the limit as $t \to 1$ of the matrix element (4.154).

An expansion of the hypergeometric ${}_2F_1$ function in powers of its argument $[(1-t)/(1+t)]^2$, amounts to an expansion for $t \to 1$, or $\omega \to 0$. With the resulting expression, one easily obtains the well-known result $\lim_{\omega\to 0} Q(1S,\omega) = 9/4$. The $2S$ state has the following matrix element,

$$Q(2S,t) = \frac{16t^2\,(21 - 42\,t - 48\,t^2 + 138\,t^3 + 14\,t^4 - 166\,t^5 - 16\,t^6 - 314\,t^7 + 1181\,t^8)}{3\,(1-t)^6\,(1+t)^4} + \frac{16384\,t^9\,(1-4t^2)}{3\,(1-t)^6\,(1+t)^6}\,{}_2F_1\left(1,-2t,1-2t,\left(\frac{1-t}{1+t}\right)^2\right). \tag{4.155}$$

This result is divergent in the limit $\omega \to 0$, because of the virtual $2P$ state, which is energetically degenerate with respect to the $2S$ reference state. This behavior has been studied in detail in Ref. [171] and, in the context of black-body energy shifts, in Ref. [172]. For $t \to 1$, one has

$$Q(2S,t) = \frac{36}{1-t} + 60 + \mathcal{O}(t-1) = \frac{9(Z\alpha)^2\mu}{\omega} + 60 + \mathcal{O}(\omega)\,. \tag{4.156}$$

Finally, for the $2P$ reference state, we have a more complex result, which reads

$$\begin{aligned} Q(2P,t) = {} & \frac{16t^2}{9\,(1-t)^7\,(1+t)^5}\,\bigl(45 - 90t - 153t^2 + 396t^3 + 150t^4 - 696t^5 \\ & + 66t^6 + 564t^7 + 61t^8 + 850t^9 - 3241t^{10}\bigr) \\ & - \frac{16384\,t^{11}\,(3-11\,t^2)}{9\,(1+t)^7\,(1-t)^7}\,{}_2F_1\left(1,-2t,1-2t,\left(\frac{1-t}{1+t}\right)^2\right). \end{aligned} \tag{4.157}$$

For $t \to 1$, one has

$$Q(2P,t) = \frac{12}{1-t} + 70 + \mathcal{O}(t-1) = \frac{3(Z\alpha)^2\mu}{\omega} + 88 + \mathcal{O}(\omega)\,. \tag{4.158}$$

The results is Eqs. (4.152) and (4.157) require the use of the $(\ell = 0)$- and $(\ell = 2)$-components in the Schrödinger–Coulomb propagator (4.134).

Let us conclude the discussion of dipole matrix elements of the Schrödinger–Coulomb Hamiltonian, by establishing the connection to the dynamic polarizability defined in Eq. (3.53), which we recall as

$$\alpha(\phi_n,\omega) = \frac{e^2}{3}\left(\left\langle\phi_n\left|x^i\,\frac{1}{H_S - E_n + \omega}\,x^i\right|\phi_n\right\rangle + \left\langle\phi_n\left|x^i\,\frac{1}{H_S - E_n - \omega}\,x^i\right|\phi_n\right\rangle\right), \tag{4.159}$$

where the two terms with the different sign of ω are explicitly written out. The matrix element involving the denominator $H_S - E_n + \omega$ is proportional to $Q(\phi_n, t)$, whereas the one with the denominator $H_0 - E_n - \omega$ is proportional to $Q(\phi_n, s)$ with

$$s = \left(1 - \frac{2n^2\hbar\omega}{(Z\alpha)^2\,\mu c^2}\right)^{-1/2}. \tag{4.160}$$

If the energy $\hbar\omega$ exceeds the bound-state energy $(Z\alpha)^2\,\mu c^2/(2n^2)$, the argument s acquires an imaginary part which eventually gives rise to the imaginary part of the dipole polarizability [which is proportional to the one-photon ionization cross-section, see Eq. (3.58)]. The results for the Q matrix elements in Eqs. (4.154) and (4.155) are dimensionless. In SI mksA units, the polarizability has a dimension of the square of the dipole moment, divided by the energy. Applying this scaling consistently, we have in SI mksA units

$$\begin{aligned}\alpha(\phi_n,\omega) &= e^2\left(\frac{a_0}{Z}\right)^2 \frac{1}{(Z\alpha)^2\mu c^2}\,(Q(\phi_n,t)+Q(\phi_n,s))\\ &= \frac{e^2\,\hbar^2}{(Z\alpha)^4\mu^3c^4}\,(Q(\phi_n,t)+Q(\phi_n,s))\,. \end{aligned} \tag{4.161}$$

The static limits for $1S$, $2S$, and $2P$ states are

$$\alpha(1S,\omega\to 0) = \frac{9}{2}\,\frac{e^2\,\hbar^2}{(Z\alpha)^4\mu^3c^4}\,, \tag{4.162a}$$

$$\alpha(2S,\omega\to 0) = 120\,\frac{e^2\,\hbar^2}{(Z\alpha)^4\mu^3c^4}\,, \tag{4.162b}$$

$$\alpha(2P,\omega\to 0) = 176\frac{e^2\,\hbar^2}{(Z\alpha)^4\mu^3c^4}\,. \tag{4.162c}$$

For the $2S$ and $2P$ states, the divergence of the Q matrix elements as $\omega\to 0$ cancels out in the final result because the limit is approached symmetrically in the two terms involving the two signs $\pm\omega$ in Eq. (4.159). From the atomic physics point of view, we now have all necessary analytic techniques at our disposal, which are required in order to calculate the Bethe logarithm, which is a quantity obtained as an integral over matrix elements of the P type, i.e., over (a part of) the dynamic polarizability of the atom. However, before we can finally proceed to the calculation of this quantity, and be able to interpret the result in a suitable context, we still need to discuss two further ideas: the renormalization and the overlapping parameter.

4.4 Schrödinger–Coulomb Green Function in Momentum Space

4.4.1 *Free Green Function in Momentum Space*

We have just finished our discussion of the Schrödinger–Coulomb Green function in coordinate space, and evaluated a number of phenomenologically interesting matrix elements. For a number of calculations, though, which include the quadrupole radiation correction to the Lamb shift [163,173], a different formulation is useful, namely, in momentum space. In addition, the derivation of the Schrödinger–Coulomb Green function in momentum space offers the opportunity for an ideal illustration of the algebraic beauty of the $SO(4)$ symmetry of the hydrogen atom, based on ideas originally formulated in Ref. [174]. However, a close inspection of Ref. [174] reveals that a number of details of the calculation are being "swept under the rug", and the connection of the Schrödinger–Coulomb wave functions in momentum space to the Green function is not very clearly displayed.

We shall attempt to fill the missing pieces in the derivation presented in Ref. [174] and start with the clarification of certain normalization factors relevant to the formulation of the Green function in momentum space. First, one recalls that the Green function of the free Schrödinger equation fulfills

$$(H_0 - E)\, G(\vec{r}, \vec{r}\,') = \left(-\frac{\vec{\nabla}^2}{2m} - E\right) G(\vec{r}, \vec{r}\,') = \delta^{(3)}(\vec{r} - \vec{r}\,')\,, \tag{4.163}$$

or

$$\left(\vec{\nabla}^2 + 2mE\right) G(\vec{r}, \vec{r}\,') = -2m\delta^{(3)}(\vec{r} - \vec{r}\,')\,. \tag{4.164}$$

So, we have, with reference to the well-known Green function of the Helmholtz equation,

$$G(\vec{r}, \vec{r}\,') = \frac{(2m)\, \mathrm{e}^{-\sqrt{2mE}\,|\vec{r}-\vec{r}\,'|}}{4\pi|\vec{r} - \vec{r}\,'|}\,. \tag{4.165}$$

In calculating the Fourier transform, the second argument gets transformed with the complex conjugate $\exp(+\mathrm{i}\vec{p}\cdot\vec{r})$ of the Fourier factor $\exp(-\mathrm{i}\vec{p}\cdot\vec{r})$. This is because we associate $G(\vec{r}, \vec{r}\,')$, in an eigenfunction decomposition, with

$$G(\vec{r}, \vec{r}\,') = \sum_n \frac{\psi_n(\vec{r})\, \psi_n^*(\vec{r}\,')}{\lambda_n - E}\,, \qquad -\frac{\vec{\nabla}^2}{2m}\, \psi_n(\vec{r}) = \lambda_n\, \psi_n(\vec{r})\,, \tag{4.166}$$

and the completeness relation for the wave functions is

$$\sum_n \psi_n(\vec{r})\, \psi_n^*(\vec{r}\,') = \delta^{(3)}(\vec{r} - \vec{r}\,')\,, \tag{4.167}$$

where n is a multi-index that enumerates the eigenstates in a particular basis, say, the plane-wave basis or the spherical basis. The latter involves the spherical Bessel functions in the radial variable (see Chap. 2.5 of Ref. [81]). In Eq. (4.166), the sum over states may include a continuum.

Let us investigate the Fourier transform of Eq. (4.163), with the proviso that the second argument is transformed according to the complex conjugate:

$$\begin{aligned} &\int \mathrm{d}^3 r \int \mathrm{d}^3 r' \left(-\frac{\vec{\nabla}^2}{2m} - E\right) G(\vec{r} - \vec{r}\,') \mathrm{e}^{-\mathrm{i}\vec{p}\cdot\vec{r} + \mathrm{i}\vec{p}\,'\cdot\vec{r}\,'} \\ &= \int \mathrm{d}^3 r \int \mathrm{d}^3 r' \delta^{(3)}(\vec{r} - \vec{r}\,') \mathrm{e}^{-\mathrm{i}\vec{p}\cdot\vec{r} + \mathrm{i}\vec{p}\,'\cdot\vec{r}\,'}\,. \end{aligned} \tag{4.168}$$

A double partial integration of the term on the left-hand side leads to

$$\left(\frac{\vec{p}^{\,2}}{2m} - E\right) G(\vec{p}, \vec{p}\,') = (2\pi)^3\, \delta^{(3)}(\vec{p} - \vec{p}\,')\,, \tag{4.169}$$

where

$$G(\vec{p}, \vec{p}\,') = \int \mathrm{d}^3 r \int \mathrm{d}^3 r' \mathrm{e}^{-\mathrm{i}\vec{p}\cdot\vec{r} + \mathrm{i}\vec{p}\,'\cdot\vec{r}\,'}\, G(\vec{r}, \vec{r}\,')\,. \tag{4.170}$$

Within the conventions usually adopted in the physics literature for the Fourier amplitudes, one thus incurs a factor of $(2\pi)^3$ in momentum space, in the defining equation of the Green function (4.169).

4.4.2 Toward the $SO(4)$ Symmetry

The defining equation of the Schrödinger–Coulomb Green function is

$$\left(\frac{\vec{p}^{\,2}}{2m} + V - E\right) G(\vec{r},\vec{r}\,') = \delta^{(3)}(\vec{r} - \vec{r}\,') \,. \tag{4.171}$$

The Coulomb potential, in momentum space, is

$$V(\vec{p} - \vec{p}\,') = -\frac{4\pi Z\alpha}{(\vec{p} - \vec{p}\,')^2} \,. \tag{4.172}$$

The defining equation for the Green function thus becomes, in momentum space,

$$\left(\frac{\vec{p}^{\,2}}{2m} - E\right) G(\vec{p},\vec{p}\,') - \int \frac{d^3p''}{(2\pi)^3} \frac{4\pi Z\alpha}{(\vec{p} - \vec{p}\,'')^2} G(\vec{p}\,'',\vec{p}\,') = (2\pi)^3 \, \delta^{(3)}(\vec{p} - \vec{p}\,') \,. \tag{4.173}$$

One may go into (Euclidean) momentum space, setting

$$E = -\frac{X^2}{2m} \,, \tag{4.174}$$

where $X = p_0$ has the interpretation of the zeroth component of the Euclidean four-momentum. For a free particle, the parameter X has the interpretation of being equal to

$$X = p_0 = \sqrt{-2mE} = \mathrm{i}\,|\vec{p}_{\rm free}| \,, \qquad E = \frac{\vec{p}_{\rm free}^{\,2}}{2m} \,. \tag{4.175}$$

The Schrödinger–Coulomb Green function describes an energy-conserving system, and hence one needs only one energy argument, $X = X'$. One can map the momentum space onto the unit sphere, via the conformal transformation

$$\xi_0 = \frac{X^2 - \vec{p}^{\,2}}{X^2 + \vec{p}^{\,2}} \,, \qquad \vec{\xi} = \frac{2X}{X^2 + \vec{p}^{\,2}} \, \vec{p} \,, \qquad \xi_0^2 + \vec{\xi}^{\,2} = 1 \,. \tag{4.176}$$

It is useful to derive the equation

$$d^3\Omega = \frac{d^3\xi}{|\xi_0|} \tag{4.177}$$

for the area element on the three-dimensional unit sphere, embedded in four-dimensional space. Here, one should remember that the three-dimensional components of the four-dimensional vector $(\xi_0, \vec{\xi})$ may have varying magnitude and we only have the restriction $\xi_0^2 + \vec{\xi}^{\,2} = 1$.

The derivation of Eq. (4.177) is less trivial than one might think. One first remembers that the infinitesimal element of a two-dimensional surface $\vec{x} = \vec{x}(t_1, t_2)$, parameterized by the variables t_1 and t_2, reads as:

$$d^2S = \left| \begin{pmatrix} \frac{\partial x}{\partial t_1} \\ \frac{\partial y}{\partial t_1} \\ \frac{\partial z}{\partial t_1} \end{pmatrix} \times \begin{pmatrix} \frac{\partial x}{\partial t_2} \\ \frac{\partial y}{\partial t_2} \\ \frac{\partial z}{\partial t_2} \end{pmatrix} \right| dt_1 \, dt_2 = \left| \det \begin{pmatrix} \hat{e}_x & \hat{e}_y & \hat{e}_z \\ \frac{\partial x}{\partial t_1} & \frac{\partial y}{\partial t_1} & \frac{\partial z}{\partial t_1} \\ \frac{\partial x}{\partial t_2} & \frac{\partial y}{\partial t_2} & \frac{\partial z}{\partial t_2} \end{pmatrix} \right| dt_1 \, dt_2 \,. \tag{4.178}$$

The appropriate generalization to the three-dimensional "surface" of a manifold embedded into four-dimensional space is, with $\xi_0 = \xi_0(t_1, t_2, t_3)$, $\xi_x = \xi_x(t_1, t_2, t_3)$, $\xi_y = \xi_y(t_1, t_2, t_3)$, $\xi_z = \xi_z(t_1, t_2, t_3)$, where ξ_0 is the fourth coordinate,

$$\mathrm{d}^3\Omega = \left| \det \begin{pmatrix} \hat{\mathrm{e}}_0 & \hat{\mathrm{e}}_x & \hat{\mathrm{e}}_y & \hat{\mathrm{e}}_z \\ \frac{\partial \xi_0}{\partial t_1} & \frac{\partial \xi_x}{\partial t_1} & \frac{\partial \xi_y}{\partial t_1} & \frac{\partial \xi_z}{\partial t_1} \\ \frac{\partial \xi_0}{\partial t_2} & \frac{\partial \xi_x}{\partial t_2} & \frac{\partial \xi_y}{\partial t_2} & \frac{\partial \xi_z}{\partial t_2} \\ \frac{\partial \xi_0}{\partial t_3} & \frac{\partial \xi_x}{\partial t_3} & \frac{\partial \xi_y}{\partial t_3} & \frac{\partial \xi_z}{\partial t_3} \end{pmatrix} \right| \mathrm{d}t_1 \, \mathrm{d}t_2 \, \mathrm{d}t_3 \,. \tag{4.179}$$

First, one calculates the four-dimensional vector described by the determinant, and then calculates its vector modulus. The three-dimensional unit sphere, embedded in four-dimensional space, can be interpreted as a three-dimensional manifold, parameterized by the coordinates $\xi_x = t_1$, $\xi_y = t_2$, and $\xi_z = t_3$, while $\xi_0 = \sqrt{1 - \xi_x^2 - \xi_y^2 - \xi_z^2}$. One finds that

$$\mathrm{d}^3\Omega = \frac{\mathrm{d}\xi_x \, \mathrm{d}\xi_y \, \mathrm{d}\xi_z}{\sqrt{1 - \xi_x^2 - \xi_y^2 - \xi_z^2}} = \frac{\mathrm{d}^3\xi}{|\xi_0|} \,. \tag{4.180}$$

One should now transform $\mathrm{d}^3\xi$ into d^3p. To this end, one interprets the second member of the relation (4.176), keeping X constant, in the sense that

$$\mathrm{d}^3\xi = \left| \det \begin{pmatrix} \frac{\partial \xi_x}{\partial p_x} & \frac{\partial \xi_y}{\partial p_x} & \frac{\partial \xi_z}{\partial p_x} \\ \frac{\partial \xi_x}{\partial p_y} & \frac{\partial \xi_y}{\partial p_y} & \frac{\partial \xi_z}{\partial p_y} \\ \frac{\partial \xi_x}{\partial p_z} & \frac{\partial \xi_y}{\partial p_z} & \frac{\partial \xi_z}{\partial p_z} \end{pmatrix} \right| \mathrm{d}p_x \, \mathrm{d}p_y \, \mathrm{d}p_z = \frac{8X^3 \, (X^2 - \vec{p}^{\,2})}{(X^2 + \vec{p}^{\,2})^4} \, \mathrm{d}^3p \,. \tag{4.181}$$

We remember that according to Eq. (4.176), we have $\xi_0 = (X^2 - \vec{p}^{\,2})/(X^2 + \vec{p}^{\,2})$, and so one obtains

$$\mathrm{d}^3\Omega = \frac{\mathrm{d}^3\xi}{|\xi_0|} = \left(\frac{2X}{X^2 + \vec{p}^{\,2}} \right)^3 \mathrm{d}^3p \,, \tag{4.182}$$

and, as a trivial consequence,

$$\delta^{(3)}(\vec{\Omega} - \vec{\Omega}') = \left(\frac{X^2 + \vec{p}^{\,2}}{2X} \right)^3 \delta^{(3)}(\vec{p} - \vec{p}\,') \,. \tag{4.183}$$

By an elementary calculation, one can show that

$$(\xi - \xi')^2 = (\xi_0 - \xi_0')^2 + (\vec{\xi} - \vec{\xi}\,')^2 = \frac{4\,X^2}{(X^2 + \vec{p}^{\,2})\,(X^2 + \vec{p}\,'^2)} \, (\vec{p} - \vec{p}\,')^2 \,. \tag{4.184}$$

One observes that every point $\vec{\Omega}$ on the three-dimensional unit sphere can be mapped uniquely onto a vector $\vec{p}$. Hence, suppressing in the notation the dependence on X or E, one may define the generalized Green function $\Gamma(\vec{\Omega}, \vec{\Omega}')$ as follows,

$$\Gamma(\vec{\Omega}, \vec{\Omega}') = -\frac{1}{16mX^3} \, (X^2 + \vec{p}^{\,2})^2 \, G(\vec{p}, \vec{p}\,') \, (X^2 + \vec{p}\,'^2)^2 \,. \tag{4.185}$$

It is advantageous to define $D(\xi - \xi') = 1/[4\pi^2 (\xi - \xi')^2]$, which according to Eq. (4.184) can be simplified as follows for $|\xi| = |\xi'| = 1$ (in the sense of a modulus of a four-dimensional unit vector),

$$D(\xi - \xi') = \frac{1}{4\pi^2} \frac{1}{(\xi - \xi')^2} = \frac{1}{4\pi^2} \frac{1}{(\xi_0 - \xi_0')^2 + (\vec{\xi} - \vec{\xi}')^2}$$
$$= \frac{(X^2 + \vec{p}^{\,2})\,(X^2 + \vec{p}^{\,\prime 2})}{16\,\pi^2\,X^2} \frac{1}{(\vec{p} - \vec{p}^{\,\prime})^2} \,. \tag{4.186}$$

Let us reformulate the expression $H = \int \mathrm{d}^3\Omega''\, D(\xi - \xi'')\Gamma(\vec{\Omega}'', \vec{\Omega}')$ as

$$H = \int \mathrm{d}^3 p'' \left(\frac{2X}{X^2 + \vec{p}^{\,\prime\prime 2}}\right)^3 \left\{\frac{(X^2 + \vec{p}^{\,2})\,(X^2 + \vec{p}^{\,\prime\prime 2})}{16\,\pi^2\,X^2} \frac{1}{(\vec{p} - \vec{p}^{\,\prime\prime})^2}\right\}$$
$$\times \left\{-\frac{1}{16 m X^3} (X^2 + \vec{p}^{\,\prime\prime 2})^2\, G(\vec{p}^{\,\prime\prime}, \vec{p}^{\,\prime})\,(X^2 + \vec{p}^{\,\prime 2})^2\right\}$$
$$= \int \mathrm{d}^3 p''\, 8X^3 \left\{\frac{(X^2 + \vec{p}^{\,2})(X^2 + \vec{p}^{\,\prime 2})^2}{16\,\pi^2\,X^2} \frac{1}{(\vec{p} - \vec{p}^{\,\prime\prime})^2}\right\} \left\{-\frac{1}{16 m X^3}\, G(\vec{p}^{\,\prime\prime}, \vec{p}^{\,\prime})\right\}$$
$$= -\frac{1}{4\,m\,X^2} \int \frac{\mathrm{d}^3 p''}{8\pi^3}\, \pi \left\{(X^2 + \vec{p}^{\,2})\,(X^2 + \vec{p}^{\,\prime 2})^2 \frac{1}{(\vec{p} - \vec{p}^{\,\prime\prime})^2}\right\} G(\vec{p}^{\,\prime\prime}, \vec{p}^{\,\prime})$$
$$= -\frac{1}{16\,m\,X^2} (X^2 + \vec{p}^{\,2})\,(X^2 + \vec{p}^{\,\prime 2})^2 \int \frac{\mathrm{d}^3 p''}{(2\pi)^3} \frac{4\pi}{(\vec{p} - \vec{p}^{\,\prime\prime})^2}\, G(\vec{p}^{\,\prime\prime}, \vec{p}^{\,\prime})\,. \tag{4.187}$$

Therefore,

$$\int \mathrm{d}^3\Omega''\, D(\xi - \xi'')\, \Gamma(\vec{\Omega}'', \vec{\Omega}')$$
$$= -\frac{X}{2\,m} \left(\frac{X^2 + \vec{p}^{\,2}}{2X}\right)^3 \left(\frac{X^2 + \vec{p}^{\,\prime 2}}{X^2 + \vec{p}^{\,2}}\right)^2 \int \frac{\mathrm{d}^3 p''}{(2\pi)^3} \frac{4\pi}{(\vec{p} - \vec{p}^{\,\prime\prime})^2}\, G(\vec{p}^{\,\prime\prime}, \vec{p}^{\,\prime})\,. \tag{4.188}$$

We define the energy variable [see Eq. (4.83)],

$$\nu = \frac{Z\alpha m}{X} = \frac{Z\alpha m}{\sqrt{-2mE}}\,, \tag{4.189}$$

which is some sort of generalized principal quantum number, as it attains an integer value at a bound-state energy $E = -(Z\alpha)^2 m/(2n^2)$,

$$\nu = \frac{Z\alpha m}{\sqrt{-2m\,[-(Z\alpha)^2 m/(2n^2)]}} = \frac{Z\alpha m}{\sqrt{(Z\alpha)^2 m^2/n^2}} = n\,. \tag{4.190}$$

It is useful to investigate the combination

$$\Gamma(\vec{\Omega},\vec{\Omega}') - 2\nu \int \mathrm{d}^3\Omega''\, D(\xi-\xi'')\Gamma(\vec{\Omega}'',\vec{\Omega}') = -\frac{(X^2+\vec{p}^{\,2})^2\, G(\vec{p},\vec{p}')\,(X^2+\vec{p}^{\,\prime 2})^2}{16mX^3}$$
$$-2\frac{Z\alpha m}{X}\left\{-\frac{X}{2m}\left(\frac{X^2+\vec{p}^{\,2}}{2X}\right)^3\left(\frac{X^2+\vec{p}^{\,\prime 2}}{X^2+\vec{p}^{\,2}}\right)^2 \int \frac{\mathrm{d}^3p''}{(2\pi)^3}\,\frac{4\pi}{(\vec{p}-\vec{p}'')^2}\, G(\vec{p}'',\vec{p}')\right\}$$
$$=\left(\frac{X^2+\vec{p}^{\,\prime 2}}{X^2+\vec{p}^{\,2}}\right)^2\left(\frac{X^2+\vec{p}^{\,2}}{2X}\right)^3\left[-\frac{X^2+\vec{p}^{\,2}}{2m}\, G(\vec{p},\vec{p}') + \int \frac{\mathrm{d}^3p''}{8\pi^3}\,\frac{4\pi Z\alpha}{(\vec{p}-\vec{p}'')^2}\, G(\vec{p}'',\vec{p}')\right]$$
$$=\left(\frac{X^2+\vec{p}^{\,\prime 2}}{X^2+\vec{p}^{\,2}}\right)^2\left(\frac{X^2+\vec{p}^{\,2}}{2X}\right)^3\left[-(2\pi)^3\,\delta^{(3)}(\vec{p}-\vec{p}')\right] = -(2\pi)^3\,\delta^{(3)}(\vec{\Omega}-\vec{\Omega}')\,, \tag{4.191}$$

where we have used Eq. (4.183). So, we have derived the compact formula

$$\Gamma(\vec{\Omega},\vec{\Omega}') - 2\nu \int \mathrm{d}^3\Omega''\, D(\xi-\xi'')\,\Gamma(\vec{\Omega}'',\vec{\Omega}') = -(2\pi)^3\,\delta^{(3)}(\vec{\Omega}-\vec{\Omega}')\,. \tag{4.192}$$

The Fourier transformation of $D(\xi-\xi')$ is

$$D(k) = \int \mathrm{d}^4\xi\, \mathrm{e}^{-\mathrm{i}k\cdot(\xi-\xi')}\,\frac{1}{4\pi^2}\,\frac{1}{(\xi-\xi')^2} = \frac{1}{k^2}\,, \tag{4.193}$$

where $k^2 = \sum_{\mu=1}^{4} k_\mu\, k_\mu$ is the Euclidean vector square, and the four-vectors ξ and ξ' are, a priori, not restricted to be unit vectors. Let us suppose now that we had a decomposition of the form

$$D(\xi-\xi') = \sum_{n=0}^{\infty}\sum_{\ell=0}^{n}\sum_{m=-\ell}^{\ell}\frac{1}{2(n+1)}\,\frac{\rho_<^n}{\rho_>^{n+2}}\, Y_{n\ell m}(\chi,\theta,\varphi)\, Y^*_{n\ell m}(\chi',\theta',\varphi')\,, \tag{4.194}$$

where the $Y_{n\ell m}(\chi,\theta,\varphi)$ are four-dimensional spherical harmonics. (See Eq. (4.229) for a proof.) These depend on the angular variables χ, θ and φ which refer to the angular components of the ξ vector,

$$\xi_x = |\xi|\,\cos\varphi\,\sin\theta\,\sin\chi\,, \qquad \xi_y = |\xi|\,\sin\varphi\,\sin\theta\,\sin\chi\,, \tag{4.195a}$$
$$\xi_z = |\xi|\,\cos\theta\,\sin\chi\,, \qquad \xi_0 = |\xi|\,\cos\chi\,. \tag{4.195b}$$

For completeness, we note that

$$\mathrm{d}^4\xi = \xi^3\,\mathrm{d}\xi\,\mathrm{d}^3\Omega\,, \qquad \mathrm{d}^3\Omega = \sin^2\chi\,\sin\theta\,\mathrm{d}\chi\,\mathrm{d}\theta\,\mathrm{d}\varphi\,. \tag{4.196}$$

For $|\xi| = 1$, the angular variables alone suffice to define the unit vector. In Eq. (4.194), one has $\rho = |\xi|$ and $\rho' = |\xi'|$. So, for two points on the unit sphere, we would set $\rho = \rho' = 1$. We assume that the four-dimensional spherical harmonics $Y_{n\ell m}$ have the properties

$$\int \mathrm{d}^3\Omega\, Y^*_{n\ell m}(\chi,\theta,\varphi)\, Y_{n'\ell'm'}(\chi,\theta,\varphi) = \delta_{nn'}\,\delta_{\ell\ell'}\,\delta_{mm'}\,, \tag{4.197}$$

$$\sum_{n\ell m} Y_{n\ell m}(\chi,\theta,\varphi)\, Y^*_{n\ell m}(\chi',\theta',\varphi') = \delta^{(3)}(\vec{\Omega}-\vec{\Omega}')\,. \tag{4.198}$$

Let us guess that the Green function Γ can be expressed as

$$\Gamma(\vec{\Omega},\vec{\Omega}') = -(2\pi)^3 \sum_{n\ell m} \frac{Y_{n\ell m}(\chi,\theta,\varphi)\, Y^*_{n\ell m}(\chi',\theta',\varphi')}{1-\nu/(n+1)}\,, \tag{4.199}$$

and verify this guess with the help of Eq. (4.192). Using Eq. (4.194) for $\rho = |\xi| = \rho' = |\xi'| = 1$, one can derive for the expression $\Xi = \Gamma(\vec{\Omega},\vec{\Omega}') - 2\nu \int \mathrm{d}^3\Omega''\, D(\xi-\xi'')\, \Gamma(\vec{\Omega}'',\vec{\Omega}')$ the representation

$$\begin{aligned}
\Xi &= -\,(2\pi)^3 \sum_{n\ell m} \frac{Y_{n\ell m}(\chi,\theta,\varphi)\, Y^*_{n\ell m}(\chi',\theta',\varphi')}{1-\nu/(n+1)} \\
&\quad + (2\pi)^3\, 2\nu \sum_{n\ell m} \sum_{n'\ell' m'} \int \mathrm{d}^3\Omega''\, \frac{1}{2(n+1)}\, Y_{n\ell m}(\chi,\theta,\varphi)\, Y^*_{n\ell m}(\chi'',\theta'',\varphi'') \\
&\quad \times \frac{Y_{n'\ell' m'}(\chi'',\theta'',\varphi'')\, Y^*_{n\ell m}(\chi',\theta',\varphi')}{1-\nu/(n+1)} \\
&= -\,(2\pi)^3 \sum_{n\ell m} \left(\frac{1}{1-\nu/(n+1)} - \frac{\nu}{n+1}\, \frac{1}{1-\nu/(n+1)} \right) \\
&\quad \times Y_{n\ell m}(\chi,\theta,\varphi)\, Y^*_{n\ell m}(\chi',\theta',\varphi') = -(2\pi)^3\, \delta^{(3)}(\vec{\Omega}-\vec{\Omega}')\,.
\end{aligned} \tag{4.200}$$

We have thus shown Eq. (4.199) as a remarkable consequence of the $SO(4)$ symmetry of the hydrogen atom. The bound-state spectrum is obtained in view of the poles at

$$\nu = n + 1 = 1, 2, 3, \dots\,. \tag{4.201}$$

This consideration suggests that $n+1$ takes the role of the principal quantum number within the hydrogen spectrum; hence, the four-dimensional spherical harmonic with $n = 0$ is associated with the (innermost) K shell of hydrogen, while $n = 1$ is associated with the (next in order) L shell, and so on. We will see, in a moment, that $Y_{n\ell m}$ is proportional to the Schrödinger–Coulomb momentum-space eigenfunction of principal quantum number $n+1$. Around the pole, one may expand, using

$$E = -\frac{(Z\alpha)^2 m}{2(n+1)^2} + \epsilon\,, \qquad X = \frac{Z\alpha m}{n+1} - \frac{(n+1)}{Z\alpha}\,\epsilon + \mathcal{O}(\epsilon^2)\,, \tag{4.202}$$

with $n = 0, 1, 2, 3, \dots$, and obtain, on one hand,

$$\begin{aligned}
\Gamma(\vec{\Omega},\vec{\Omega}') &\approx (2\pi)^3 \frac{X^2}{m\,\epsilon} Y_{n\ell m}(\chi,\theta,\varphi)\, Y^*_{n\ell m}(\chi',\theta',\varphi') \\
&= \frac{1}{m\,\epsilon} \left[(2\pi)^{3/2}\, X\, Y_{n\ell m}(\chi,\theta,\varphi)\right] \left[(2\pi)^{3/2}\, X\, Y^*_{n\ell m}(\chi',\theta',\varphi')\right].
\end{aligned} \tag{4.203}$$

On the other hand, we have from Eq. (4.185),

$$\begin{aligned}
\Gamma(\vec{\Omega},\vec{\Omega}') &= -\,\frac{1}{16 m X^3}\, (X^2 + \vec{p}^{\,2})^2\, G(\vec{p},\vec{p}')\, (X^2 + \vec{p}^{\,\prime 2})^2 \\
&= -\,\frac{1}{16 m X^3} (X^2 + \vec{p}^{\,2})^2\, \frac{\psi_{(n+1)\ell m}(\vec{p})\, \psi_{(n+1)\ell m}(\vec{p}')}{(-\epsilon)}\, (X^2 + \vec{p}^{\,\prime 2})^2 \\
&= \frac{1}{m\,\epsilon} \left[\frac{(X^2 + \vec{p}^{\,2})^2}{4 X^{3/2}}\, \psi_{(n+1)\ell m}(\vec{p}) \right] \left[\frac{(X^2 + \vec{p}^{\,\prime 2})^2}{4 X^{3/2}}\, \psi_{(n+1)\ell m}(\vec{p}') \right].
\end{aligned} \tag{4.204}$$

A comparison of terms in square brackets in Eqs. (4.203) and (4.204) yields the wave functions (we replace $n \to n-1$)

$$\psi_{n\ell m}(\vec{p}) = (2\pi)^{3/2} \frac{4X^{5/2}}{(X^2+\vec{p}^2)^2} Y_{(n-1)\ell m}(\chi,\theta,\varphi)\,. \tag{4.205}$$

The four-dimensional spherical harmonics are normalized according to Eq. (4.197), whereas the wave functions in momentum space are normalized according to

$$\int \frac{\mathrm{d}^3 p}{(2\pi)^3} \left|\psi_{n\ell m}(\vec{p})\right|^2 = 1\,. \tag{4.206}$$

The integration measure explains the prefactor $(2\pi)^{3/2}$ in Eq. (4.205).

There is something to be said about the comparison to Eq. (4.182), which we recall for convenience,

$$\mathrm{d}^3 p = \left(\frac{X^2+\vec{p}^2}{2X}\right)^3 \mathrm{d}^3\Omega\,. \tag{4.207}$$

Namely, in view of the orthogonality properties (4.197), one may define the functions

$$\chi_{n\ell m}(\vec{p}) = (2\pi)^{3/2} \left(\frac{2X}{X^2+\vec{p}^2}\right)^{3/2} Y_{n\ell m}(\chi,\theta,\varphi)\,, \tag{4.208a}$$

$$\xi_0 = \cos\chi = \frac{X^2-\vec{p}^2}{X^2+\vec{p}^2}\,, \qquad \vec{\xi} = \frac{2X}{X^2+\vec{p}^2}\,\vec{p}\,, \tag{4.208b}$$

$$\theta = \theta_{\vec{\xi}} = \theta_{\vec{p}}\,, \qquad \varphi = \varphi_{\vec{\xi}} = \varphi_{\vec{p}}\,. \tag{4.208c}$$

For amusement, one notes the identity

$$\begin{aligned} \mathrm{d}^3\Omega &= \sin^2\chi\,\sin\theta\,\mathrm{d}\chi\,\mathrm{d}\theta\,\mathrm{d}\varphi = (1-\cos^2\chi)\,\frac{\partial\chi}{\partial|\vec{p}|}\mathrm{d}|\vec{p}|\,\sin\theta\,\mathrm{d}\theta\,\mathrm{d}\varphi \\ &= \left(\frac{2X}{X^2+\vec{p}^2}\right)^3 \vec{p}^{\,2}\mathrm{d}|\vec{p}|\,\sin\theta\,\mathrm{d}\theta\,\mathrm{d}\varphi = \left(\frac{2X}{X^2+\vec{p}^2}\right)^3 \mathrm{d}^3 p\,, \end{aligned} \tag{4.209}$$

verifying Eq. (4.182) once more. The functions defined in Eq. (4.208) have the property

$$\int \frac{\mathrm{d}^3 p}{(2\pi)^3}\, \chi^*_{n\ell m}(\vec{p})\,\chi_{n'\ell'm'}(\vec{p}) = \int \mathrm{d}^3\Omega\, Y^*_{n\ell m}(\vec{p})\,Y_{n'\ell'm'}(\vec{p}) = \delta_{nn'}\,\delta_{\ell\ell'}\,\delta_{mm'}\,. \tag{4.210}$$

The prefactor in Eq. (4.205) is manifestly different from the one in Eq. (4.208); still, both $\psi_{n\ell m}(\vec{p})$ and $\chi_{n\ell m}(\vec{p})$ are normalized to unity when integrated over $\mathrm{d}^3 p/(2\pi)^3$.

4.4.3 *Four-Dimensional Spherical Harmonics*

We now need to define the four-dimensional spherical harmonics used in Eq. (4.194), and it will turn out that the Gegenbauer polynomials $C_n^{(1)}(x)$ are instrumental in our investigations. Indeed, the Gegenbauer polynomials are generated by the relation

$$\frac{1}{1-2xt+t^2} = \sum_{n=0}^{\infty} C_n^{(1)}(x)\,t^n\,. \tag{4.211}$$

Here, we shall elect to define associated Gegenbauer polynomials in a somewhat non-standard way, which is designed to highlight the connection of the associated Gegenbauer polynomials with the associated Legendre polynomials. The associated Legendre polynomials are usually denoted as P_n^ℓ. One may thus define the Gegenbauer polynomials as Q,

$$Q_n(x) = C_n^{(1)}(x). \tag{4.212}$$

They have the property

$$\frac{1}{\sin^2\chi}\frac{\partial}{\partial\chi}\sin^2\chi\frac{\partial}{\partial\chi}Q_n(\cos\chi) = -n\,(n+2)\,Q_n(\cos\chi). \tag{4.213}$$

In full analogy with the associated Legendre polynomials, one may define the associated Gegenbauer-type polynomials as

$$Q_n^\ell(x) = (-1)^\ell\left(1-x^2\right)^{\ell/2}\frac{\partial^\ell}{\partial x^\ell}Q_n(x), \qquad \ell \le n. \tag{4.214}$$

For $\ell > n$, the associated Gegenbauer polynomials vanish because $Q_n(x)$ is of order n. The associated Gegenbauer polynomials have the property

$$\left(\frac{1}{\sin^2\chi}\frac{\partial}{\partial\chi}\sin^2\chi\frac{\partial}{\partial\chi} - \frac{\ell\,(\ell+1)}{\sin^2\chi}\right)Q_n^\ell(\cos\chi) = -n\,(n+2)\,Q_n^\ell(\cos\chi). \tag{4.215}$$

For the normalization, one finds

$$\int_0^\pi \mathrm{d}\chi\,\sin^2\chi\,[Q_n(\cos\chi)]^2 = \frac{\pi}{2}, \tag{4.216a}$$

$$\int_0^\pi \mathrm{d}\chi\,\sin^2\chi\,\left[Q_n^\ell(\cos\chi)\right]^2 = \frac{\pi}{2}\,\frac{(n+\ell+1)!}{(n+1)\,(n-\ell)!}. \tag{4.216b}$$

One may thus define the four-dimensional spherical harmonics as

$$Y_{n\ell m}(\chi,\theta,\varphi) = \sqrt{\frac{2}{\pi}}\sqrt{\frac{(n+1)\,(n-\ell)!}{(n+\ell+1)!}}\,Q_n^\ell(\cos\chi)\,Y_{\ell m}(\theta,\varphi), \tag{4.217}$$

where the $Y_{\ell m}$'s are the usual spherical harmonics, and verify that they fulfill Eq. (4.197),

$$\int_0^{2\pi}\mathrm{d}\varphi\int_0^\pi\mathrm{d}\theta\int_0^\pi\mathrm{d}\chi\,\sin^2\chi\,\sin\theta\,Y_{n\ell m}(\chi,\theta,\varphi)\,Y^*_{n'\ell'm'}(\chi,\theta,\varphi) = \delta_{nn'}\delta_{\ell\ell'}\delta_{mm'}. \tag{4.218}$$

The addition theorem is

$$\sum_{n\ell m} Y_{n\ell m}(\chi,\theta,\varphi)\,Y^*_{n\ell m}(\chi',\theta',\varphi') = \frac{n+1}{2\pi^2}\,Q_n(\hat{x}\cdot\hat{x}'). \tag{4.219}$$

Here, $\hat{x}$ is a four-dimensional unit vector, associated with the angles χ, θ and φ. Analogously, $\hat{x}'$ is a four-dimensional unit vector, associated with the angles χ', θ' and φ'. The four-dimensional Laplacian is

$$\vec{\nabla}^2 = \frac{\partial^2}{\partial\rho^2} + \frac{3}{\rho}\frac{\partial}{\partial\rho} + \frac{1}{\rho^2}\left(\frac{1}{\sin^2\chi}\frac{\partial}{\partial\chi}\sin^2\chi\frac{\partial}{\partial\chi} - \frac{\vec{L}^2}{\sin^2\chi}\right). \tag{4.220}$$

Let us return to the explicit form of the D function, given in Eq. (4.193),

$$D(\xi - \xi') = \frac{1}{4\pi^2} \frac{1}{(\xi - \xi')^2}, \qquad D(k) = \int \mathrm{d}^4\xi \, \mathrm{e}^{-\mathrm{i}k\cdot(\xi-\xi')} \frac{1}{4\pi^2} \frac{1}{(\xi - \xi')^2} = \frac{1}{k^2}. \tag{4.221}$$

The Fourier transform dictates that

$$\nabla^2 D(\xi - \xi') = -\delta^{(4)}(\xi - \xi'). \tag{4.222}$$

Here, ∇^2 is the four-dimensional Laplacian. Let us also determine A, in coordinate space, from an *ansatz* borrowed from Eq. (4.194),

$$D(\xi - \xi') = \sum_{n=0}^{\infty} \sum_{\ell=0}^{n} \sum_{m=-\ell}^{\ell} A \frac{\rho_<^n}{\rho_>^{n+2}} Y_{n\ell m}(\chi, \theta, \varphi) \, Y^*_{n\ell m}(\chi', \theta', \varphi'). \tag{4.223}$$

Of course,

$$\delta^{(4)}(\xi - \xi') = \frac{1}{\rho^3 \sin^2\chi \sin\theta} \delta(\rho - \rho') \, \delta(\chi - \chi') \, \delta(\theta - \theta') \, \delta(\varphi - \varphi'). \tag{4.224}$$

We already know that

$$\sum_{n\ell m} Y_{n\ell m}(\chi, \theta, \varphi) \, Y^*_{n\ell m}(\chi', \theta', \varphi') = \frac{1}{\sin^2\chi \sin\theta} \delta(\chi - \chi') \, \delta(\theta - \theta') \, \delta(\varphi - \varphi'). \tag{4.225}$$

So, we need to solve the radial equation

$$A \left(\frac{\partial^2}{\partial\rho^2} + \frac{3}{\rho} \frac{\partial}{\partial\rho} - n(n+2) \right) \frac{\rho_<^n}{\rho_>^{n+2}} = -\frac{1}{\rho^3} \delta(\rho - \rho'). \tag{4.226}$$

Integrating this equation from $\rho = \rho' - \epsilon$ to $\rho = \rho' + \epsilon$ using techniques discussed in Chap. 2 of Ref. [81], we are led to

$$A \left(\left. \frac{\partial}{\partial\rho} \frac{\rho_<^n}{\rho_>^{n+2}} \right|_{\rho=\rho'+\epsilon} - \left. \frac{\partial}{\partial\rho} \frac{\rho_<^n}{\rho_>^{n+2}} \right|_{\rho=\rho'-\epsilon} \right) = -\frac{1}{\rho'^3}, \tag{4.227}$$

and so

$$A \left. \left(\frac{\partial}{\partial\rho} \frac{\rho'^n}{\rho^{n+2}} - \frac{\partial}{\partial\rho} \frac{\rho^n}{\rho'^{n+2}} \right) \right|_{\rho=\rho'} = -\frac{1}{\rho'^3}, \qquad A = \frac{1}{2(n+1)}. \tag{4.228}$$

We thus confirm Eq. (4.194),

$$\begin{aligned} D(\xi - \xi') &= \frac{1}{4\pi^2} \frac{1}{(\xi - \xi')^2} \\ &= \sum_{n=0}^{\infty} \sum_{\ell=0}^{n} \sum_{m=-\ell}^{\ell} \frac{1}{2(n+1)} \frac{\rho_<^n}{\rho_>^{n+2}} Y_{n\ell m}(\chi, \theta, \varphi) \, Y^*_{n\ell m}(\chi', \theta', \varphi'). \end{aligned} \tag{4.229}$$

4.4.4 Wave Functions in Momentum Space

According to Eq. (4.205), one has the wave functions

$$\psi_{n\ell m}(\vec{p}) = (2\pi)^{3/2} \frac{4X^{5/2}}{(X^2+\vec{p}^{\,2})^2} Y_{(n-1)\,\ell m}(\chi,\theta,\varphi)\,, \tag{4.230}$$

where we recall from Eq. (4.217),

$$Y_{n\ell m}(\chi,\theta,\varphi) = \sqrt{\frac{2}{\pi}} \sqrt{\frac{(n+1)\,(n-\ell)!}{(n+\ell+1)!}}\, Q_n^\ell(\cos\chi)\, Y_{\ell m}(\theta,\varphi)\,. \tag{4.231}$$

We also recall that according to Eq. (4.176),

$$\xi_0 = \cos\chi = \frac{X^2 - \vec{p}^{\,2}}{X^2+\vec{p}^{\,2}}\,, \qquad \vec{\xi} = \frac{2X}{X^2+\vec{p}^{\,2}}\,\vec{p}\,, \qquad \hat{\xi} = \hat{p}\,. \tag{4.232}$$

Then,

$$\begin{aligned}\psi_{n\ell m}(\vec{p}) &= (2\pi)^{3/2} \frac{4X^{5/2}}{(X^2+\vec{p}^{\,2})^2} Y_{(n-1)\,\ell m}(\chi,\theta,\varphi)\\ &= (2\pi)^{3/2} \frac{4X^{5/2}}{(X^2+\vec{p}^{\,2})^2} \sqrt{\frac{2}{\pi}} \sqrt{\frac{n\,(n-1-\ell)!}{(n+\ell)!}}\, Q_{n-1}^\ell(\cos\chi)\, Y_{\ell m}(\theta,\varphi)\\ &= 16\,\pi \sqrt{\frac{n\,(n-1-\ell)!}{(n+\ell)!}} \frac{X^{5/2}}{(X^2+\vec{p}^{\,2})^2} Q_{n-1}^\ell\left(\frac{X^2-\vec{p}^{\,2}}{X^2+\vec{p}^{\,2}}\right) Y_{\ell m}(\theta,\varphi)\,, \end{aligned} \tag{4.233}$$

where $\theta = \theta_{\vec{p}}$ and $\varphi = \varphi_{\vec{p}}$. A comparison of Eq. (4.174) with the bound-state energies,

$$E = -\frac{X^2}{2m} = -\frac{(Z\alpha m)^2}{2mn^2}\,, \qquad X = \frac{Z\alpha m}{n}\,, \tag{4.234}$$

allows us to bring the bound-state wave function into the form

$$\psi_{n\ell m}(\vec{p}) = \frac{16\,\pi\, a_0^{3/2}\, n^2}{Z^{3/2}} \sqrt{\frac{(n-1-\ell)!}{(n+\ell)!}} \frac{1}{\left(1+\frac{n^2 a_0^2 \vec{p}^{\,2}}{Z^2}\right)^2} Q_{n-1}^\ell\left(\frac{1-\frac{n^2 a_0^2 \vec{p}^{\,2}}{Z^2}}{1+\frac{n^2 a_0^2 \vec{p}^{\,2}}{Z^2}}\right) Y_{\ell m}(\theta,\varphi)\,, \tag{4.235}$$

where $a_0 = 1/(\alpha m)$ is the Bohr radius. The identity

$$Q_{n-1}^\ell(x) = (-1)^{n-1}\, 2^\ell\, \ell!\, (1-x^2)^{\ell/2}\, C_{n-\ell-1}^{\ell+1}(-x)\,, \tag{4.236}$$

where the $C_{n-\ell-1}^{\ell+1}(x)$ are the Gegenbauer polynomials in the conventional definition (see Ref. [159]), can be used in order to bring the wave function, in momentum space, into the familiar form [see page 39 of Ref. [6]],

$$\begin{aligned}\psi_{n\ell m}(\vec{p}) &= 2^{2\ell+4}\,\pi\,\ell! \sqrt{\frac{(n-1-\ell)!}{(n+\ell)!}} \left(1+\frac{n^2 a_0^2 \vec{p}^{\,2}}{Z^2}\right)^{-(\ell+2)}\\ &\quad\times n^{\ell+2}\,(a_0/Z)^{\ell+3/2}\,|\vec{p}|^\ell\, C_{n-\ell-1}^{(\ell+1)}\left(\frac{\frac{n^2 a_0^2 \vec{p}^{\,2}}{Z^2}-1}{\frac{n^2 a_0^2 \vec{p}^{\,2}}{Z^2}+1}\right) Y_{\ell m}(\theta,\varphi)\,. \end{aligned} \tag{4.237}$$

Factors of Z are suppressed in Ref. [6] because of the atomic unit system used. We absorb the overall prefactor $(-1)^{n-1}$ incurred in the conversion from Eq. (4.235) to (4.237) into the global phase of the wave function.

4.4.5 *Integral Representation*

The formula in Eq. (4.199) leads to a convenient representation of the wave functions in momentum space. One may obtain a convenient integral representation of the Green function by exploring the connection of the four-dimensional multipole decomposition (4.194) with the explicit representation (4.199) of the Green function. However, in order to do so, it is necessary to use an additional, not very obvious trick and specialize (4.194) to the cases $\xi = r_1$, and $\xi' = r_2$, as follows,

$$\begin{aligned} r_1 &= \rho\,\xi\,, & |\xi| &= 1\,, & |r_1| &= \rho\,, & \hat{r}_1 &= \hat{\xi}\,, \\ r_2 &= \xi'\,, & |\xi'| &= 1\,, & |r_2| &= 1\,, & \hat{r}_2 &= \hat{\xi}'\,, \end{aligned} \tag{4.238}$$

with $0 < \rho < 1$, so that $|r_1| = \rho = r_<$, and $|r_2| = 1 = r_>$. Then,

$$\begin{aligned} |r_1 - r_2|^2 &= |r_1|^2 + |r_2|^2 - 2\rho\,(\xi\cdot\xi') = 1 + \rho^2 - 2\rho\,(\xi\cdot\xi') \\ &= (1+\rho^2 - 2\rho) + 2\rho - 2\rho\,(\xi\cdot\xi') \\ &= (1-\rho)^2 + \rho(\xi^2 + \xi'^2) - 2\rho\,(\xi\cdot\xi') = (1-\rho)^2 + \rho(\xi-\xi')^2\,. \end{aligned} \tag{4.239}$$

The decomposition of Eq. (4.194) becomes

$$\begin{aligned} D(r_1 - r_2) &= \frac{1}{4\pi^2}\,\frac{1}{(1-\rho)^2 + \rho(\xi-\xi')^2} \\ &= \sum_{n=0}^{\infty}\sum_{\ell=0}^{n}\sum_{m=-\ell}^{\ell} \frac{\rho^n}{2(n+1)}\,Y_{n\ell m}(\chi,\theta,\varphi)\,Y^*_{n\ell m}(\chi',\theta',\varphi')\,. \end{aligned} \tag{4.240}$$

Incidentally, one can verify the multiplicity of the hydrogen spectrum in this way. One simply sets $\xi = \xi'$, but keeps ρ variable. To this end, one observes that the four-dimensional total solid angle is

$$\int \mathrm{d}\Omega = \int_0^{2\pi}\mathrm{d}\varphi \int_0^{\pi}\mathrm{d}\theta \int_0^{\pi}\mathrm{d}\chi\,\sin^2\chi\,\sin\theta = 2\pi^2\,. \tag{4.241}$$

Now, setting $\xi = \xi'$ in Eq. (4.240), and integrating over $\mathrm{d}\Omega$, one finds that

$$\frac{2\pi^2}{4\pi^2}\,\frac{1}{(1-\rho)^2} = \sum_{n=0}^{\infty}\frac{\rho^n}{2(n+1)}\sum_{\ell=0}^{n}\sum_{m=-\ell}^{\ell} 1\,. \tag{4.242}$$

In view of the result

$$\sum_{\ell=0}^{n}\sum_{m=-\ell}^{\ell} 1 = (n+1)^2\,, \qquad \frac{1}{(1-\rho)^2} = \sum_{n=0}^{\infty}\frac{\rho^n}{(n+1)}\,(n+1)^2 = \sum_{n=0}^{\infty}\rho^n\,(n+1)\,, \tag{4.243}$$

one confirms the multiplicity of the eigenvalue for the principal quantum number $n{+}1$, as $(n{+}1)^2$, where n is a nonnegative integer. Let us trivially rewrite Eq. (4.240) in the form

$$\frac{1}{2\pi^2}\,\frac{1}{(1-\rho)^2 + \rho(\xi-\xi')^2} = \sum_{n=0}^{\infty}\sum_{\ell=0}^{n}\sum_{m=-\ell}^{\ell}\frac{\rho^n}{n+1}\,Y_{n\ell m}(\chi,\theta,\varphi)\,Y^*_{n\ell m}(\chi',\theta',\varphi')\,. \tag{4.244}$$

It is *a priori* not obvious how the latter expression could be brought into correspondence with Eq. (4.199), in view of the presence of the factor $1/[1-\nu/(n+1)]$ in the latter case. However, one may use the algebraic identity

$$\frac{1}{1-\nu/(n+1)} = 1 + \frac{\nu}{n+1} + \nu^2 \frac{1}{(n+1)(n+1-\nu)}$$
$$= 1 + \frac{\nu}{n+1} + \frac{\nu^2}{(n+1)} \int_0^1 \mathrm{d}\rho\, \rho^{n-\nu}, \tag{4.245}$$

to reformulate the expression $\mathcal{S} = -(2\pi)^{-3}\,\Gamma(\vec{\Omega},\vec{\Omega}')$ as follows, starting from Eq. (4.199),

$$\mathcal{S} = \sum_{n\ell m} \frac{Y_{n\ell m}(\chi,\theta,\varphi)\, Y^*_{n\ell m}(\chi',\theta',\varphi')}{1-\nu/(n+1)}$$
$$= \sum_{n\ell m} Y_{n\ell m}(\chi,\theta,\varphi)\, Y^*_{n\ell m}(\chi',\theta',\varphi') \left(1 + \frac{\nu}{n+1} + \frac{\nu^2}{n+1} \int_0^1 \mathrm{d}\rho\, \rho^{n-\nu}\right)$$
$$= \sum_{n\ell m} Y_{n\ell m}(\chi,\theta,\varphi)\, Y^*_{n\ell m}(\chi',\theta',\varphi') + \nu \sum_{n\ell m} \frac{Y_{n\ell m}(\chi,\theta,\varphi)\, Y^*_{n\ell m}(\chi',\theta',\varphi')}{n+1}$$
$$+ \nu^2 \int_0^1 \mathrm{d}\rho\, \rho^{-\nu} \sum_{n\ell m} \frac{\rho^n}{n+1} Y_{n\ell m}(\chi,\theta,\varphi)\, Y^*_{n\ell m}(\chi',\theta',\varphi')$$
$$= \delta^{(3)}(\vec{\Omega}-\vec{\Omega}') + \frac{\nu}{2\pi^2} \frac{1}{(\xi-\xi')^2} + \frac{\nu^2}{2\pi^2} \int_0^1 \mathrm{d}\rho\, \rho^{-\nu} \frac{1}{(1-\rho)^2 + \rho(\xi-\xi')^2}. \tag{4.246}$$

With one partial integration, one can show that

$$-\frac{1}{(2\pi)^3}\Gamma(\vec{\Omega},\vec{\Omega}') = \delta^{(3)}(\vec{\Omega}-\vec{\Omega}') + \frac{\nu}{2\pi^2} \int_0^1 \mathrm{d}\rho\, \rho^{-\nu} \frac{\partial}{\partial\rho} \frac{\rho}{(1-\rho)^2 + \rho(\xi-\xi')^2}. \tag{4.247}$$

In view of the identity

$$\frac{\partial}{\partial\rho} \frac{\rho}{(1-\rho)^2 + \rho(\xi-\xi')^2} = \frac{1-\rho^2}{[(1-\rho)^2 + \rho(\xi-\xi')^2]^2} \tag{4.248}$$

and the relation

$$\delta^{(3)}(\vec{\Omega}-\vec{\Omega}') = \lim_{\rho\to 1} \frac{1}{2\pi^2} \frac{1-\rho^2}{[(1-\rho)^2 + \rho(\xi-\xi')^2]^2} \tag{4.249}$$

one can even write the very compact formula

$$-\frac{1}{(2\pi)^3}\Gamma(\vec{\Omega},\vec{\Omega}') = \frac{1}{2\pi^2} \int_0^1 \mathrm{d}\rho\, \rho^{-\nu} \frac{\partial}{\partial\rho} \frac{\rho(1-\rho^2)}{[(1-\rho)^2 + \rho(\xi-\xi')^2]^2}, \tag{4.250}$$

because the latter expression is equivalent to

$$-\frac{1}{(2\pi)^3}\Gamma(\vec{\Omega},\vec{\Omega}') = \frac{1}{2\pi^2} \int_0^1 \mathrm{d}\rho \frac{\partial}{\partial\rho} \frac{\rho^{1-\nu}(1-\rho^2)}{[(1-\rho)^2 + \rho(\xi-\xi')^2]^2}$$
$$- \frac{1}{2\pi^2} \int_0^1 \mathrm{d}\rho\, (-\nu) \rho^{-\nu-1} \frac{\rho(1-\rho^2)}{[(1-\rho)^2 + \rho(\xi-\xi')^2]^2}$$
$$= \frac{1}{2\pi^2} \lim_{\rho\to 1} \frac{\rho^{1-\nu}(1-\rho^2)}{[(1-\rho)^2 + \rho(\xi-\xi')^2]^2}$$
$$+ \frac{\nu}{2\pi^2} \int_0^1 \mathrm{d}\rho\, \rho^{-\nu} \frac{\rho(1-\rho^2)}{[(1-\rho)^2 + \rho(\xi-\xi')^2]^2}$$
$$= \delta^{(3)}(\vec{\Omega}-\vec{\Omega}') + \frac{\nu}{2\pi^2} \int_0^1 \mathrm{d}\rho\, \rho^{-\nu} \frac{\partial}{\partial\rho} \frac{\rho}{(1-\rho)^2 + \rho(\xi-\xi')^2}. \tag{4.251}$$

Using Eq. (4.250) and recalling Eqs. (4.184) and (4.185),

$$G(\vec{p},\vec{p}\,') = -16mX^3 \frac{1}{(X^2+\vec{p}^{\,2})^2} \Gamma(\vec{\Omega},\vec{\Omega}') \frac{1}{(X^2+\vec{p}\,'^2)^2}, \tag{4.252a}$$

$$(\xi-\xi')^2 = \frac{4X^2(\vec{p}-\vec{p}\,')^2}{(X^2+\vec{p}^{\,2})(X^2+\vec{p}\,'^2)}. \tag{4.252b}$$

One may write $\mathcal{G} = -(2\pi)^{-3} G(\vec{p},\vec{p}\,')$ as

$$\begin{aligned}
\mathcal{G} &= -\frac{16mX^3}{2\pi^2} \frac{1}{(X^2+\vec{p}^{\,2})^2} \int_0^1 \mathrm{d}\rho\,\rho^{-\nu} \frac{\partial}{\partial\rho} \frac{\rho(1-\rho^2)}{[(1-\rho)^2+\rho(\xi-\xi')^2]^2} \frac{1}{(X^2+\vec{p}\,'^2)^2} \\
&= -\frac{16mX^3}{2\pi^2} \int_0^1 \mathrm{d}\rho\,\rho^{-\nu} \frac{\partial}{\partial\rho} \frac{\rho(1-\rho^2)}{[(1-\rho)^2(X^2+\vec{p}^{\,2})(X^2+\vec{p}\,'^2)+4\rho X^2(\vec{p}-\vec{p}\,')^2]^2} \\
&= -\frac{mX^3}{2\pi^2} \frac{\mathrm{i}e^{\mathrm{i}\pi\nu}}{2\sin(\pi\nu)} \int_C \mathrm{d}\rho\rho^{-\nu} \frac{\partial}{\partial\rho} \frac{(1-\rho^2)/\rho}{\left[X^2(\vec{p}-\vec{p}\,')^2 + \frac{(1-\rho)^2}{4\rho}(X^2+\vec{p}^{\,2})(X^2+\vec{p}\,'^2)\right]^2}.
\end{aligned} \tag{4.253}$$

The latter formula demands an explanation; we have replaced the simple integral over $\mathrm{d}\rho$ over the interval $\rho \in (0,1)$ by a contour integral. This is done in order to lift the restriction $\nu < 1$. The precise definition of the contour C is illustrated in the following. With the standard definition of the logarithm, the function

$$\rho^{-\nu} = \exp[-\nu \ln(\rho)] \tag{4.254}$$

has a branch cut along the negative ρ axis. So, with the standard definition of the branch cut of the logarithm, one has

$$\ln(-|\rho|+\mathrm{i}\epsilon) = \ln(|\rho|) + \mathrm{i}\,\pi, \qquad \ln(-|\rho|-\mathrm{i}\epsilon) = \ln(|\rho|) - \mathrm{i}\,\pi. \tag{4.255}$$

Otherwise, with the branch cut of the logarithm along the positive real axis, one has

$$\ln(|\rho|+\mathrm{i}\epsilon) = \ln(|\rho|), \qquad \ln(|\rho|-\mathrm{i}\epsilon) = \ln(|\rho|) + 2\pi\mathrm{i}, \tag{4.256}$$

because the imaginary part has to vary continuously as the phase of the complex argument is being ramped up from zero to 2π. Let us define an integration contour C that starts at $\rho = 1+\mathrm{i}\epsilon$, follows infinitesimally above the real axis, encircles the origin in the mathematically positive sense, and finishes at $\rho = 1-\mathrm{i}\epsilon$. One can replace the contour C equivalently by an expression which is minus the contribution from the "upper" contour from $\rho = 0$ to $\rho = 1+\mathrm{i}\epsilon$, plus the contribution from the contour from $\rho = 0$ to $\rho = 1-\mathrm{i}\epsilon$. Then, for any test function $F(\rho)$, with the contour C as just defined, one has

$$\begin{aligned}
\frac{\mathrm{i}e^{\mathrm{i}\pi\nu}}{2\sin(\pi\nu)} \int_C \mathrm{d}\rho\,\rho^{-\nu} F(\rho) &= \frac{\mathrm{i}e^{\mathrm{i}\pi\nu}}{2[\exp(\mathrm{i}\pi\nu)-\exp(-\mathrm{i}\pi\nu)]/(2\mathrm{i})} \\
&\quad \times \left(-\int_0^1 \mathrm{d}\rho\,\rho^{-\nu} + \int_0^1 \mathrm{d}\rho\,\rho^{-\nu} e^{-2\pi\mathrm{i}\nu}\right) F(\rho) \\
&= \frac{1}{1-\exp(-2\mathrm{i}\pi\nu)} \left(1-e^{-2\pi\mathrm{i}\nu}\right) \int_0^1 \mathrm{d}\rho\,\rho^{-\nu} F(\rho) \\
&= \int_0^1 \mathrm{d}\rho\,\rho^{-\nu} F(\rho).
\end{aligned} \tag{4.257}$$

We have derived the operator identity

$$\frac{\mathrm{i}e^{\mathrm{i}\pi\nu}}{2\sin(\pi\nu)} \int_C \mathrm{d}\rho\, \rho^{-\nu}\, F(\rho) = \int_0^1 \mathrm{d}\rho\, \rho^{-\nu}\, F(\rho)\,. \tag{4.258}$$

Our investigations have culminated in the compact integral representation

$$G(\vec{p},\vec{p}\,') = 4\pi m X^3 \left(\frac{\mathrm{i}e^{\mathrm{i}\pi\nu}}{2\sin(\pi\nu)}\right) \tag{4.259}$$

$$\times \int_C \mathrm{d}\rho\, \rho^{-\nu}\, \frac{\partial}{\partial\rho} \frac{(1-\rho^2)/\rho}{\left[X^2\,(\vec{p}-\vec{p}\,')^2 + \frac{(1-\rho)^2}{4\rho}\,(X^2+\vec{p}^{\,2})\,(X^2+\vec{p}\,'^2)\right]^2}\,.$$

We also recall that the energy argument of the Green function is parameterized as

$$\nu = \frac{Z\alpha m}{X} = \frac{Z\alpha m}{\sqrt{-2mE}}\,. \tag{4.260}$$

For Lamb shift calculations, one typically replaces

$$E \to E - \omega\,, \tag{4.261}$$

where E is the reference state energy, and therefore [see also Eq. (4.130)]

$$\nu = \frac{Z\alpha m}{\sqrt{-2m(E-\omega)}} = \frac{Z\alpha m}{\sqrt{-2m\left(-\frac{(Z\alpha)^2 m}{2n^2} - \omega\right)}} = n\,\frac{1}{\sqrt{1+\frac{2n^2\omega}{(Z\alpha)^2 m}}} = n\,t\,. \tag{4.262}$$

The integral representation (4.259) has been used with good effect in Lamb shift treatments [163, 173].

4.5 Modern Ideas and Bound-State Self–Energy

4.5.1 *Essence of Renormalization*

The renormalization program of quantum electrodynamics has been outlined at various places in the literature, e.g., in Sec. 8.4 of Ref. [81], based on the Casimir effect (attractive force between two perfectly conducting plates). The calculation involves the zero-point energy of the electromagnetic field and its modification in the presence of nontrivial boundary conditions. We remember the general paradigmatic equation [see Eq. (8.144) of Ref. [81]]

$$\text{Physical\_Observable} = \text{Bare\_Observable}\Big|_{\text{reg.}} - \text{Counter\_Term}\Big|_{\text{reg.}}$$

$$= \lim_{\text{regulator}\to 0} \int \left(\text{Bare\_Observable}\Big|_{\text{reg.-int.}} - \text{Counter\_Term}\Big|_{\text{reg.-int.}}\right)\,, \tag{4.263}$$

where we recall that both the bare observable as well as the counter term need to be regularized. Here, we understand that "reg." denotes the regularized physical quantity, while "reg-int." denotes the regularized integrand. Integrals are typically involved in the calculation.

In our derivation of the Bethe logarithm, which is an energy shift to be discussed in detail below, we have to regularize a logarithmic divergence. On the level of classical field theory, we may actually encounter a divergence of this type in the calculation of an electrostatic potential, in quite a different setting, and we would like to illustrate the concept of renormalization once more, in a toy-model calculation [see also Sec. 8.5 of Ref. [81]].

We consider an infinitely long, charged linear structure (long rod) that extends along the x axis, and evaluate the electrostatic potential at a point $\vec{r} = (0, 0, z)$, elevated from the x axis by a distance z. Although the resulting integrals are divergent (at face value), one can do this calculation using the standard formalism for the evaluation of an electrostatic potential, but only if one subtracts a counterterm. We start from the familiar equation

$$\Phi(\vec{r}) = \frac{1}{4\pi\epsilon_0} \int \mathrm{d}^3 r' \, \frac{\rho(\vec{r}')}{|\vec{r} - \vec{r}'|} \tag{4.264}$$

and use a charge density ρ of a long charged rod along the x-axis,

$$\rho(\vec{r}') = \lambda \, \delta(y') \, \delta(z') \,. \tag{4.265}$$

The potential is to be evaluated at the point $\vec{r} = (0, 0, z)$,

$$\Phi(z) = \frac{\lambda}{4\pi\epsilon_0} \int_{-\infty}^{\infty} \frac{\mathrm{d}x'}{\sqrt{z^2 + (x')^2}} = \frac{\lambda}{2\pi\epsilon_0} \int_0^{\infty} \frac{\mathrm{d}x'}{\sqrt{z^2 + (x')^2}} \,. \tag{4.266}$$

This expression is formally infinite, but the divergence at the upper limit of the integration domain is only logarithmic. First, we regularize this integral by introducing an upper cutoff Λ for the x'-integration,

$$\Phi(z)\Big|_{\mathrm{reg.}} = \frac{\lambda}{2\pi\epsilon_0} \int_0^{\Lambda} \frac{\mathrm{d}x'}{\sqrt{z^2 + (x')^2}} = \frac{\lambda}{2\pi\epsilon_0} \left(\ln(2) - \ln(z) - \ln(\Lambda) + \mathcal{O}(\Lambda^{-1}) \right) . \tag{4.267}$$

Because the integral diverges, we have to subtract a counterterm. Fixing the potential to be zero at $z = z_0$, we can devise the following regularized counterterm,

$$C\Big|_{\mathrm{reg.}} = \frac{\lambda}{2\pi\epsilon_0} \int_0^{\Lambda} \frac{\mathrm{d}x'}{\sqrt{z_0 + (x')^2}} = \frac{\lambda}{2\pi\epsilon_0} \left(\ln(2) - \ln(z_0) - \ln(\Lambda) + \mathcal{O}(\Lambda^{-1}) \right) , \tag{4.268}$$

which is infinite in the limit $\Lambda \to \infty$ and independent of z. The renormalized expression reads

$$\Phi(z) = \lim_{\Lambda \to \infty} \left(\Phi(z)\Big|_{\mathrm{reg.}} - C\Big|_{\mathrm{reg.}} \right) = -\frac{\lambda}{2\pi\epsilon_0} \ln\left(\frac{z}{z_0} \right) , \tag{4.269}$$

where the precise value of z_0 is physically irrelevant as it vanishes in calculating any potential differences; the term with z_0 can otherwise be thought of as fixing a zero of the potential. This limit $\Lambda \to \infty$ has been taken after evaluating the difference $\Phi(z)\big|_{\mathrm{reg.}} - C\big|_{\mathrm{reg.}}$. Our model example illustrates that after regularization, one has to evaluate both the original integral as well as the counterterm as a function of the regulator, while keeping the regulator Λ finite up to the last step in the calculation.

4.5.2 *Overlapping Parameter*

The Bethe logarithm term, to be discussed below, is only part of the bound-state self-energy of an electron. The calculation of the self-energy necessitates a global renormalization and requires the introduction of a scale-separation parameter, in order to distinguish the self-interactions due to high-energy, and low-energy photons. The scale-separation parameter is alternatively called an "overlapping parameter". Indeed, the introduction of such a parameter is nothing unnatural, but, as we shall see below, can be necessary even on a very basic level, namely, in the evaluation of innocent-looking integrals such as

$$F(Z\alpha) = \int_0^\infty \mathrm{d}\omega\, \mathrm{e}^{-\omega} \sqrt{(Z\alpha)^2 + \omega^2}\,. \tag{4.270}$$

Here, $Z\alpha$ is a coupling parameter which we might as well have called g. However, in quantum electrodynamic bound-state calculations, the expansion parameter is $Z\alpha$, where Z is the nuclear charge number and α is the fine-structure constant. For any value of $Z\alpha$, the integral (4.270) can easily be evaluated by numerical quadrature. One surprisingly encounters problems when trying to evaluate the asymptotic expansion of the integral as $Z\alpha \to 0$. If we naively try to expand the integrand in powers of $Z\alpha$, in order to obtain an expansion of $F(Z\alpha)$, then we run into a problem,

$$\sqrt{\omega^2 + (Z\alpha)^2} = \omega + \frac{(Z\alpha)^2}{2\,\omega} - \frac{(Z\alpha)^4}{8\,\omega^3} + \mathcal{O}((Z\alpha)^6)\,, \tag{4.271}$$

because the expansion generates inverse powers of ω, and these lead to divergences at the lower limit of integration in Eq. (4.270).

If we think of ω as a dimensionless photon energy variable, then the physics of the problem is completely different, depending on the energy of the photon ω, which necessitates a different treatment depending on whether ω is commensurate with the electron-mass scale $\omega \sim 1$ or commensurate with the atomic binding-energy scale $\omega \sim (Z\alpha)^2$ (we set the electron mass scale equal to unity in our model problem). Equally surprisingly, the problem generated by the inverse powers of ω in Eq. (4.271) can be solved if one introduces a so-called overlapping parameter, or scale-separation parameter ϵ which fulfills the inequality $(Z\alpha)^2 \ll \epsilon \ll 1$. We cut off the integral (4.270) at a lower limit $\omega = \epsilon$, and expand first in $Z\alpha$, then in ϵ,

$$\begin{aligned} H(Z\alpha, \epsilon) &= \int_\epsilon^\infty \mathrm{d}\omega\, \mathrm{e}^{-\omega} \sqrt{(Z\alpha)^2 + \omega^2} \\ &= 1 + (Z\alpha)^2 \left(-\frac{\gamma_E}{2} - \frac{\ln(\epsilon)}{2} \right) \\ &\quad + (Z\alpha)^4 \left(-\frac{3}{32} + \frac{\gamma_E}{16} - \frac{1}{16\,\epsilon^2} + \frac{1}{8\,\epsilon} + \frac{\ln(\epsilon)}{16} \right) + \mathcal{O}(Z\alpha)^5\,, \end{aligned} \tag{4.272}$$

where we neglect all terms of order $\mathcal{O}(Z\alpha)^5$ and all terms of order ϵ and higher. The Euler–Mascheroni constant is $\gamma_E = 0.57721\,56649\ldots$. Only the divergent terms in ϵ are kept, including the logarithmic ones. Because $\omega > \epsilon$ in Eq. (4.272), we can intuitively identify $H(Z\alpha, \epsilon)$ as the high-energy part of the photon integral (4.270).

The original integral (4.270) is independent of ϵ, so, there must be a compensating term which eliminates the dependence on ϵ. It can only come from the low-energy region $\omega < \epsilon$, i.e., from the integral

$$L(Z\alpha, \epsilon) = \int_0^\epsilon \mathrm{d}\omega\, \mathrm{e}^{-\omega} \sqrt{(Z\alpha)^2 + \omega^2}\,. \tag{4.273}$$

In this region, we cannot apply the expansion (4.271), but ω is "genuinely" small in this region, and so we can expand the exponential as follows,

$$\mathrm{e}^{-\omega} = 1 - \omega + \frac{\omega^2}{2!} + \cdots, \tag{4.274}$$

while keeping the square root $\sqrt{(Z\alpha)^2 + \omega^2}$ in unexpanded form. We should point out that each and every term in the ω-expansion of the exponential actually gives rise to terms of arbitrarily high order in the $Z\alpha$-expansion. However, if we are interested in the expansion of $L(Z\alpha, \epsilon)$ in powers of $Z\alpha$, then we are still justified in applying the expansion, because higher-order terms in the ω-expansion of the exponential give rise to higher orders in the $Z\alpha$-expansion (for the low-energy part). The low-energy part finally evaluates to

$$\begin{aligned} L(Z\alpha, \epsilon) &= (Z\alpha)^2 \left(\frac{1}{4} + \frac{1}{2} \ln\left(\frac{2\epsilon}{Z\alpha} \right) \right) + \frac{(Z\alpha)^3}{3} \\ &+ (Z\alpha)^4 \left(+ \frac{1}{64} + \frac{1}{16\,\epsilon^2} - \frac{1}{8\,\epsilon} - \frac{1}{16} \ln\left(\frac{2\epsilon}{Z\alpha} \right) \right) + \mathcal{O}(Z\alpha)^5 \,. \end{aligned} \tag{4.275}$$

Again, we neglect all terms of order ϵ and higher, and we group the terms in powers of $Z\alpha$. This is achieved by first letting $Z\alpha \to 0$, then $\epsilon \to 0$. The sum of $H(Z\alpha, \epsilon)$ and $L(Z\alpha, \epsilon)$ is finite and free of any dependence on ϵ,

$$\begin{aligned} F(Z\alpha) &= H(Z\alpha, \epsilon) + L(Z\alpha, \epsilon) \\ &= 1 + (Z\alpha)^2 \left(\frac{1}{4} - \frac{\gamma_E}{2} + \frac{1}{2} \ln\left(\frac{2}{Z\alpha} \right) \right) + \frac{(Z\alpha)^3}{3} \\ &+ (Z\alpha)^4 \left(-\frac{5}{64} + \frac{\gamma_E}{16} - \frac{1}{16} \ln\left(\frac{2}{Z\alpha} \right) \right) + \mathcal{O}(Z\alpha)^5 \,. \end{aligned} \tag{4.276}$$

The result is correct up to terms of order $\mathcal{O}(Z\alpha)^5$. One may still have doubts about the necessity of including logarithmic terms in the semi-analytic expansion. However, for $Z\alpha = 1/1000$, a numerical evaluation of the integral (4.270), in agreement with the first few terms in powers of $Z\alpha$, evaluates to 1.00000 37621 76214.

Indeed, the overlapping parameter is an utmost versatile instrument for handling logarithmic terms in physical calculations where a simple power series is not sufficient to represent a semi-analytic expansion. In Lamb shift calculations, it is customary to label the coefficients in the semi-analytic expansion systematically, indicating both the power of $Z\alpha$ as well as the power of the logarithm. Formally, we can write the result for $F(Z\alpha)$ as

$$F(Z\alpha) = \sum_{m=0}^{\infty} \sum_{n=0}^{1} A_{mn}\, (Z\alpha)^m \ln^n\left(\frac{2}{Z\alpha} \right). \tag{4.277}$$

By comparison of coefficients, we read off the following results from Eq. (4.276),

$$A_{00} = 1\,, \qquad A_{20} = \frac{1}{4} - \frac{\gamma_E}{2}\,, \qquad A_{21} = \frac{1}{2}\,, \qquad A_{30} = \frac{1}{3}\,, \tag{4.278}$$

as well as $A_{41} = -\frac{1}{16}$. For $Z\alpha \to 0$, the numerical contribution of the logarithmic terms is enhanced. Let us try to analyze the ϵ-dependence somewhat more deeply, by writing

$$H(Z\alpha,\epsilon) = \sum_m (Z\alpha)^m \, H_m(Z\alpha,\epsilon)\,, \qquad L(Z\alpha,\epsilon) = \sum_m (Z\alpha)^m \, L_m(Z\alpha,\epsilon)\,, \tag{4.279}$$

for the high- and low-energy parts, respectively. Here, the coefficient functions $H_m(Z\alpha,\epsilon)$ and $L_m(Z\alpha,\epsilon)$ typically contain logarithmic terms of the form $\ln(\epsilon)$ as well as powers of ϵ. Because the original function $F(Z\alpha)$ is independent of ϵ, the dependence on ϵ separately cancels for each order in $Z\alpha$ expansion, i.e., we have for any $m = 0, 1, 2, \ldots$,

$$\sum_{n=0}^{1} A_{mn} \ln^n(Z\alpha) = H_m(Z\alpha,\epsilon) + L_m(Z\alpha,\epsilon)\,, \tag{4.280}$$

independent of the numerical value of ϵ (m is not repeated and not summed over). We have obtained our functions H_m and L_m by first expanding in $Z\alpha \to 0$, then in $\epsilon \to 0$. This procedure is justified provided we assume that $Z\alpha \ll \epsilon \ll 1$. Under this assumption, we are free to choose the value of ϵ. We first choose a value of ϵ larger than $Z\alpha$, so that the expansion first in $Z\alpha$, then in ϵ is indicated. We should also stress that we are free to choose the degree to which the functions H_m and L_m should be expanded in ϵ, provided all logarithmic and constant terms are kept. Namely, in the sum $H_m(Z\alpha,\epsilon) + L_m(Z\alpha,\epsilon)$, the dependence on ϵ necessarily has to cancel at the end of the calculation. We can choose to expand the functions $H_m(Z\alpha,\epsilon)$ and $L_m(Z\alpha,\epsilon)$ only up to the constant term in powers of ϵ for $\epsilon \to 0$. All the dependence on ϵ, both in negative as well as positive powers of ϵ, has to cancel between the low- and high-energy parts. Furthermore, the positive powers of ϵ vanish for both L and H, separately, in the limit $\epsilon \to 0$, and, therefore, do not have to be kept. Keeping these terms in the calculation for illustrative purposes, it is an instructive exercise to show the additional cancelation of the positive powers of ϵ for the first few terms as given in Eqs. (4.272) and (4.273).

This scale separation parameter ϵ therefore acts as an "overlapping parameter" for the two energy scales present in the self-energy of the bound electron. The method is necessary for the separation of the two different energy scales for the virtual photons: the nonrelativistic domain, in which the virtual photon assumes values of the order of the atomic binding energy, and the relativistic domain, in which the virtual photon assumes values of the order of the electron rest mass. The analogy to self-energy calculations is approximately as follows: The quantity $\sqrt{\omega^2 + (Z\alpha)^2}$ in the numerator is approximately equivalent to the electron propagator, where $Z\alpha$ is the coupling strength to the nucleus. We can directly expand in $Z\alpha$ for the high-energy part, but we have to keep the entire bound-state "propagator" expression $\sqrt{\omega^2 + (Z\alpha)^2}$ with all powers of the Coulomb field strength for the low-energy part.

4.6 Bound-State Self–Energy: Low-Energy Part

4.6.1 *Length Gauge*

The Bethe logarithm arises naturally from the low-energy part of the self-energy of a bound electron. In its evaluation, we shall need the general formalism of the quantized electromagnetic field derived in Chap. 3, the concept of second-order perturbation theory in both atomic and field variables, as derived in Sec. 3.3.2, the radial Schrödinger–Coulomb Green function as derived in Sec. 4.3.3, and the concept of renormalization explained in Sec. 4.5.1 as well as the concept of an overlapping parameter developed in Sec. 4.5.2. Indeed, in the first derivation of the Lamb shift [175], an upper cutoff $\omega \sim \epsilon$ was used for the energy of the virtual photon, in order to cut off a logarithmic ultraviolet divergence of the bound-electron self-energy; later, the parameter ϵ was identified as a scale-separation parameter that separates the scale of the binding energy from that of the electron mass in much the same way as in our toy model in Sec. 4.5.1. The Bethe logarithm was thus identified as the low-energy part of the energy shift.

We shall evaluate the starting expression for the evaluation of the Bethe logarithm in two ways, namely, in the so-called length gauge, where the atom-field interaction Hamiltonian is

$$H_1 = -e\,\vec{r}\cdot\vec{E}\,, \tag{4.281}$$

and in the so-called velocity gauge, where one uses a $\vec{p}\cdot\vec{A}$ coupling. Here, e is the charge of the orbiting particle (electron), $\vec{r}$ is the position coordinate of the electron, and $\vec{E}$ is the electric-dipole field operator.

There is an important subtlety with regard to the length gauge in field-theoretical calculations. Namely, the self-energy of a bound electron is calculated with regard to the interaction of the bound electron with the radiation field; the interaction of the compound two-body system with the radiation field would have to be formulated differently. The Schrödinger–Coulomb Hamiltonian (4.1) is expressed in terms of the relative coordinate of electron and proton, not in terms of the electron coordinate. In a system with an infinitely heavy nucleus, where the reduced mass μ is equal to the electron mass m, the electron position operator $\vec{r}$ is equal to the relative coordinate. We thus assume that

$$\mu = m \qquad \text{(length gauge)} \tag{4.282}$$

for the rest of the current section on the length gauge. The interpretation of the renormalization of the self-energy relies on the comparison of the bound-electron self-energy with the free self-energy of a free electron, and is facilitated by the assumption (4.282). In our treatment of the length-gauge self-energy, we shall therefore assume that the nucleus is infinitely heavy. The correct reduced-mass dependence is easier to evaluate in the velocity gauge, while the length gauge has the advantage that is formulated in terms of the physically observable electric field strength (operator), rather than the gauge-dependent vector potential.

In the length gauge, the interaction Hamiltonian is the "classic" dipole interaction (4.281). The task is to evaluate the energy shift of an atom-field state $|\phi_n, 0\rangle$ with the atom in state $|\phi_n\rangle$ (and no photons) in second order of stationary perturbation theory. The electric-dipole field operator, in the discrete representation, is given by Eq. (2.129b) and recalled here for convenience, in the dipole approximation and in units with $\hbar = c = \epsilon_0 = 1$,

$$\vec{E} \approx \vec{E}(\vec{r} = 0) = \sum_{\vec{k}\lambda} \sqrt{\frac{\omega_{\vec{k}}}{2V}} \hat{\epsilon}_\lambda(\vec{k}) \left(\mathrm{i}\, \mathrm{a}_\lambda(\vec{k}) - \mathrm{i}\, \mathrm{a}_\lambda^+(\vec{k}) \right) . \tag{4.283}$$

The sum extends over all modes of the vacuum, in contrast to Eq. (3.64) where only the laser mode is treated. The atom-field interaction thus reads

$$H_1 = -e\vec{r} \cdot \vec{E} = -e \sum_{\vec{k}\lambda} \sqrt{\frac{\omega_{\mathrm{L}}}{2V}} \left(\hat{\epsilon}_\lambda(\vec{k}) \cdot \vec{r} \right) \left(\mathrm{i}\, \mathrm{a}_\lambda(\vec{k}) - \mathrm{i}\, \mathrm{a}_\lambda^+(\vec{k}) \right) . \tag{4.284}$$

The unperturbed atom+field system is described by the Hamiltonian (3.31), which we recall for convenience,

$$H_0 = \sum_m E_m |\phi_m\rangle \langle \phi_m| + \sum_{\vec{k}\lambda} \omega_{\vec{k}}\, \mathrm{a}_\lambda^+(\vec{k})\, \mathrm{a}_\lambda(\vec{k}) . \tag{4.285}$$

The sum over m contains the bound and the continuous spectrum (m is a multi-index that contains all necessary quantum numbers to specify the state). We here assume that the energies E_m and eigenstates $|\phi_m\rangle$ correspond to those of the Schrödinger–Coulomb problem.

As in Sec. 3.3.2, we assume that the unperturbed state of the system contains the atom in the state $|\phi_n\rangle$ and the electromagnetic field in the vacuum state $|0\rangle$, without any photons present. The unpertured energy eigenvalue with respect to H_0 of this state simply is E_n, where E_n is the energy of the atomic bound state. Virtual states with the atom in the state $|\phi_m\rangle$ and one photon in the radiation field are denoted as $|\phi_m, 1_{\vec{k}\lambda}\rangle$ and have the energy

$$H_0 |\phi_m, 1_{\vec{k}\lambda}\rangle = (E_m + \omega_{\vec{k}}) |\phi_m, 1_{\vec{k}\lambda}\rangle . \tag{4.286}$$

We denote the wave vector and the polarization state of the single virtual photon, in the usual sense, by $\vec{k}$ and λ, respectively, and use the shorthand notation ω in the following derivation because there will only be one photon involved. The second-order perturbation (operators are used in the Schrödinger picture) can be written as

$$\Delta E = \left\langle \phi_n, 0 \left| H_1 \frac{1}{E_n - H_0} H_1 \right| \phi_n, 0 \right\rangle . \tag{4.287}$$

It involves the coupling of unperturbed states $|\phi_n, 0\rangle$ to virtual states $|\phi_j, 1_{\vec{k}\lambda}\rangle$. The calculation, which we give in compact but complete form, proceeds as follows,

$$
\begin{aligned}
\Delta E &= \sum_a \sum_{\vec{k}\lambda} \frac{\left\langle \phi_n, 0 \left| H_{\rm I} \right| \phi_a, 1_\lambda(\vec{k}) \right\rangle \left\langle \phi_a, 1_\lambda(\vec{k}) \left| H_{\rm I} \right| \phi_n, 0 \right\rangle}{E_n - E_a - \omega_{\vec{k}}} \\
&= \sum_a \sum_{\vec{k}\lambda} \frac{1}{E_n - E_a - \omega_{\vec{k}}} \left\langle \phi_n, 0 \left| (-\mathrm{i} e) \sum_{\vec{k}'\lambda'} \sqrt{\frac{\omega_{\vec{k}'}}{2V}} \hat{\epsilon}_{\lambda'}(\vec{k}') \cdot \vec{r}\, a_{\lambda'}(\vec{k}') \right| \phi_a, 1_\lambda(\vec{k}) \right\rangle \\
&\quad \times \left\langle \phi_a, 1_\lambda(\vec{k}) \left| \mathrm{i} e \sum_{\vec{k}''\lambda''} \sqrt{\frac{\omega_{\vec{k}''}}{2V}} \hat{\epsilon}_{\lambda''}(\vec{k}'') \cdot \vec{r}\, a^+_{\lambda''}(\vec{k}'') \right| \phi_n, 0 \right\rangle \\
&= \sum_a \sum_{\vec{k}\lambda} \frac{e^2 \omega_{\vec{k}}}{2V} \frac{\left\langle \phi_n \left| \hat{\epsilon}_\lambda(\vec{k}) \cdot \vec{r} \right| \phi_a \right\rangle \left\langle \phi_a \left| \hat{\epsilon}_\lambda(\vec{k}) \cdot \vec{r} \right| \phi_n \right\rangle}{E_n - E_a - \omega_{\vec{k}}} \\
&= \sum_a \sum_{\vec{k}} \frac{e^2 \omega_{\vec{k}}}{2V} \left(\delta^{ij} - \frac{k^i k^j}{\vec{k}^2} \right) \frac{\left\langle \phi_n \left| x^i \right| \phi_a \right\rangle \left\langle \phi_a \left| x^j \right| \phi_n \right\rangle}{E_n - E_a - \omega_{\vec{k}}} .
\end{aligned}
\tag{4.288}
$$

Using the matching relation (2.81) for the summation over continuum modes, we now obtain

$$
\begin{aligned}
\Delta E &= e^2 \sum_a \int \frac{\mathrm{d}^3 k}{(2\pi)^3} \frac{k}{2} \left(\delta^{ij} - \frac{k^i k^j}{\vec{k}^2} \right) \frac{\left\langle \phi_n \left| x^i \right| \phi_a \right\rangle \left\langle \phi_a \left| x^j \right| \phi_n \right\rangle}{E_n - E_a - k} \\
&= e^2 \sum_a \int \frac{\mathrm{d}k\, k^2}{2\pi^2} \int \frac{\mathrm{d}\Omega_{\vec{k}}}{4\pi} \frac{k}{2} \left(\delta^{ij} - \frac{k^i k^j}{\vec{k}^2} \right) \frac{\left\langle \phi_n \left| x^i \right| \phi_a \right\rangle \left\langle \phi_a \left| x^j \right| \phi_n \right\rangle}{E_n - E_a - k} \\
&= \frac{2\alpha}{3\pi} \int \mathrm{d}k\, k^3 \left\langle \phi_n \left| x^i \frac{1}{E_n - H_S - k} x^i \right| \phi_n \right\rangle \\
&= \frac{2\alpha}{3\pi} \int \mathrm{d}k\, k \left(\frac{3}{2m} - k \left\langle \phi_n \left| x^i x^i \right| \phi_n \right\rangle + \left\langle \phi_n \left| \frac{p^i}{m} \frac{1}{E_n - H_S - k} \frac{p^i}{m} \right| \phi_n \right\rangle \right) .
\end{aligned}
\tag{4.289}
$$

A few explanatory remarks are now in order. In the derivation (4.289), the Einstein summation convention is used for the spatial indices i and j with $i, j = 1, \ldots, 3$. In our unit system ($\hbar = c = \epsilon_0 = 1$), we have $e^2 = 4\pi\alpha$, where α is the fine-structure constant. The infinitesimal solid angle is $\mathrm{d}\Omega_{\vec{k}}$. Also, k is the magnitude of $\vec{k}$, and $\omega_{\vec{k}}$ is the frequency of the vacuum mode with wave vector $\vec{k}$ so we have $\omega_{\vec{k}} = c\,k = k$. The sum over the photon polarizations is carried out using Eq. (2.63). In Eq. (4.289), the integration limits for k have not been specified, but it is clear that, without a renormalization, the photon modes available range from $k = 0$ to $k = \infty$. Let us anticipate the introduction of an upper cutoff $k_{\rm max} = \epsilon$. The last line of Eq. (4.289) shows that the self-energy is strongly divergent for large ϵ. For the first two terms in round brackets in the last line of Eq. (4.289), this is obvious, and for the third term, one observes that

$$
\left\langle \phi_n \left| \frac{p^i}{m} \frac{1}{E_n - H_S - k} \frac{p^i}{m} \right| \phi_n \right\rangle \to -\frac{1}{k} \left\langle \phi_n \left| \frac{\vec{p}^{\,2}}{m^2} \right| \phi_n \right\rangle , \qquad k \to \infty . \tag{4.290}
$$

The term $E_n - H_S$ can be neglected in comparison to k for $k \to \infty$ even if $E_n - H_S$ is operator-valued. In intermediate steps of the derivation (4.289), the following two results are useful,

$$\left\langle \phi_n, 0 \left| (-\mathrm{i}\, e) \sum_{\vec{k}'\lambda'} \sqrt{\frac{\omega_{\vec{k}'}}{2V}} \left(\hat{\epsilon}_{\lambda'}(\vec{k}') \cdot \vec{r}\right) \mathrm{a}_{\lambda'}(\vec{k}') \right| \phi_a, 1_\lambda(\vec{k}) \right\rangle$$

$$= -\mathrm{i}\, e \sqrt{\frac{\omega_{\vec{k}}}{2V}} \left\langle \phi_n \left| \hat{\epsilon}_\lambda(\vec{k}) \cdot \vec{r} \right| \phi_a \right\rangle, \tag{4.291a}$$

$$\left\langle \phi_n, 1_\lambda(\vec{k}) \left| \mathrm{i}\, e \sum_{\vec{k}''\lambda''} \sqrt{\frac{\omega_{\vec{k}''}}{2V}} \left(\hat{\epsilon}_{\lambda''}(\vec{k}'') \cdot \vec{r}\right) \mathrm{a}^+_{\lambda''}(\vec{k}'') \right| \phi_a, 0 \right\rangle$$

$$= \mathrm{i}\, e \sqrt{\frac{\omega_{\vec{k}}}{2V}} \left\langle \phi_n \left| \hat{\epsilon}_\lambda(\vec{k}) \cdot \vec{r} \right| \phi_a \right\rangle. \tag{4.291b}$$

In going from the last-but-one to the last line of Eq. (4.289), the following identity has been used,

$$\left\langle \phi_n \left| \frac{p^i}{m} \frac{1}{E - H_S - k} \frac{p^i}{m} \right| \phi_n \right\rangle = -\,\mathrm{i} \left\langle \phi_n \left| \frac{p^i x^i}{m} \right| \phi_n \right\rangle$$

$$+ k \left\langle \phi_n \left| \vec{r}^{\,2} \right| \phi_n \right\rangle + k^2 \left\langle \phi_n \left| x^i \frac{1}{E - H_S - k} x^i \right| \phi_n \right\rangle. \tag{4.292}$$

The identity can be shown by applying the operator relations

$$\frac{\vec{p}}{m} = \mathrm{i}\,[H_S, \vec{r}], \qquad H_S = \frac{\vec{p}^{\,2}}{2m} - \frac{Z\alpha}{r}, \qquad \mu = m, \tag{4.293}$$

repeatedly to the left-hand side of (4.292). In reformulating the first term on the right-hand side of Eq. (4.292), use is made of the identities

$$\mathrm{i} \left\langle \phi_n \left| \frac{p^i x^i}{m} \right| \phi_n \right\rangle = \frac{3}{2m}, \qquad \langle \phi_n | \{p^i, x^i\} | \phi_n \rangle = 0. \tag{4.294}$$

This latter identity can be proven by observing that

$$\mathrm{i}\, m\, (E_n - E_a) \langle \phi_n | x^i\, x^j | \phi_a \rangle = \mathrm{i}\, m\, \langle \phi_n | [H_S, x^i\, x^j] | \phi_a \rangle = \langle \phi_n | x^i\, p^j + x^j\, p^i | \phi_a \rangle. \tag{4.295}$$

For $|\phi_n\rangle = |\phi_a\rangle$, the left-hand side of (4.295) trivially vanishes.

Now we turn to the evaluation of the mass counter term. The physically observable self-energy actually is a residual effect, equal to the self-energy of a bound electron, minus the self-energy of a free electron. However, the reasoning is a little subtle in the nonrelativistic domain. The first guess would be to consider the free self-energy operator, where the Schrödinger–Coulomb Hamiltonian is replaced by the free Schrödinger Hamiltonian in the propagator denominator. One would then calculate the expectation value of this free self-energy operator, sandwiched for an electron in a bound state. However, it is better to proceed in two steps. Namely, one first calculates the full free self-energy operator, which is equivalent to calculating all its matrix elements in a basis of eigenstates of the free Hamiltonian. In the second step, one projects this operator onto a basis of eigenstates of the free Hamiltonian. Let us therefore calculate with the eigenstates of the free Schrödinger equation,

$$H_0 = \frac{\vec{p}^{\,2}}{2m}, \quad \frac{\vec{p}}{m} = \mathrm{i}\,[H_0, \vec{r}], \quad \psi_{\vec{q}}(\vec{r}) = \mathrm{e}^{\mathrm{i}\vec{q}\cdot\vec{r}}, \quad H_0\, \psi_{\vec{q}}(\vec{r}) = E_{\vec{q}}\, \psi_{\vec{q}}(\vec{r}) = \frac{\vec{q}^{\,2}}{2m}\, \psi_{\vec{q}}(\vec{r}), \tag{4.296}$$

normalized so that $\langle\psi_{\vec{q}}|\psi_{\vec{q}'}\rangle = (2\pi)^3\,\delta^{(3)}(\vec{q}-\vec{q}')$. We should recall that p^i here is the momentum operator, which acts as follows on a momentum-space wave function,

$$\vec{p}\,\psi_{\vec{q}}(\vec{r}) = -\mathrm{i}\vec{\nabla}\mathrm{e}^{\mathrm{i}\vec{q}\cdot\vec{r}} = \vec{q}\,\mathrm{e}^{\mathrm{i}\vec{q}\cdot\vec{r}}. \tag{4.297}$$

The self-energy operator in the length gauge is not diagonal in the momentum eigenstate basis because the position operator x^i does not commute with the free Schrödinger Hamiltonian H_0. Let us denote an off-diagonal element of the second-order perturbation $\langle\psi_{\vec{q}}|H_1[1/(E_{\vec{q}'} - H_0 - k)]H_1|\psi_{\vec{q}'}\rangle$ as $\langle\psi_{\vec{q}}|\delta m|\psi_{\vec{q}'}\rangle$, reinterpreting the free self-energy operator as a mass-counter-term operator δm. We can write this matrix element as

$$\begin{aligned}\langle\psi_{\vec{q}}|\delta m|\psi_{\vec{q}'}\rangle =& \frac{2\alpha}{3\pi}\int \mathrm{d}k\,k^3 \left\langle\psi_{\vec{q}}\left|x^i\,\frac{1}{E_{\vec{q}'}-H_0-k}\,x^i\right|\psi_{\vec{q}'}\right\rangle \\ =& \frac{2\alpha}{3\pi}\int \mathrm{d}k\,k\left\{\mathrm{i}\left\langle\psi_{\vec{q}}\left|\frac{p^i x^i}{m}\right|\psi_{\vec{q}'}\right\rangle - k\,\langle\psi_{\vec{q}}|x^i\,x^i|\psi_{\vec{q}'}\rangle\right. \\ &+\left\langle\psi_{\vec{q}}\left|\frac{p^i}{m}\,\frac{1}{E_{\vec{q}'}-H_0+k}\,\frac{p^i}{m}\right|\psi_{\vec{q}'}\right\rangle \\ &\left.-k\left(\left\langle\psi'_{\vec{q}}\left|x^i\,\frac{1}{E_{\vec{q}'}-H_0-k}\,x^i\,H_0\right|\psi_{\vec{q}'}\right\rangle - \text{h.c.}\right)\right\}. \end{aligned} \tag{4.298}$$

The integrand has been transformed using identities akin to those used in deriving Eq. (4.292). Because the free Hamiltonian H_0 commutes with the momentum operator, we have

$$\left\langle\psi_{\vec{q}}\left|\frac{p^i}{m}\,\frac{1}{E_{\vec{q}'}-H_0-k}\,\frac{p^i}{m}\right|\psi_{\vec{q}'}\right\rangle = -\frac{1}{k}\left\langle\psi_{\vec{q}}\left|\frac{\vec{q}\cdot\vec{q}'}{m^2}\right|\psi_{\vec{q}'}\right\rangle = -\frac{1}{k}\left\langle\psi_{\vec{q}}\left|\frac{\vec{q}^{\,2}}{m^2}\right|\psi_{\vec{q}'}\right\rangle. \tag{4.299}$$

The first three terms in curly brackets in Eq. (4.298) are diagonal in the momentum basis, i.e., proportional to $\delta^{(3)}(\vec{q}-\vec{q}')$. Now, the expectation value of the mass counter term, in a bound state $|\phi_n\rangle$, can be easily calculated by inserting, twice, a complete basis of momentum eigenstates of the free Hamiltonian,

$$\begin{aligned}\langle\phi_n|\delta m|\phi_n\rangle =& \frac{2\alpha}{3\pi}\int \mathrm{d}k\,k\left\langle\phi_n\left|\mathrm{i}\frac{p^i x^i}{m} - k\,\vec{r}^{\,2} - \frac{1}{k}\frac{\vec{p}^{\,2}}{m^2}\right|\phi_n\right\rangle \\ &-\sum_{\vec{q}}\sum_{\vec{q}'}\frac{2\alpha}{3\pi}\int \mathrm{d}k\,k\,\langle\phi_n|\psi_{\vec{q}}\rangle\left\langle\psi_{\vec{q}}\left|x^i\,\frac{1}{E-H_0-k}\,x^i\,H_0\right|\psi_{\vec{q}'}\right\rangle\langle\psi_{\vec{q}'}|\phi_n\rangle \\ &+\sum_{\vec{q}}\sum_{\vec{q}'}\frac{2\alpha}{3\pi}\int \mathrm{d}k\,k\,\langle\phi_n|\psi_{\vec{q}}\rangle\left\langle\psi_{\vec{q}}\left|H_0\,x^i\,\frac{1}{E-H_0-k}\,x^i\right|\psi_{\vec{q}'}\right\rangle\langle\psi_{\vec{q}'}|\phi_n\rangle \\ =& \frac{2\alpha}{3\pi}\int \mathrm{d}k\,k\left(\frac{3}{2m} - k\,\langle\phi_n|\vec{r}^{\,2}|\phi_n\rangle - \frac{1}{k}\left\langle\phi_n\left|\frac{\vec{p}^{\,2}}{m^2}\right|\phi_n\right\rangle\right). \end{aligned} \tag{4.300}$$

Here, the closure relation for the spectrum of the free Hamiltonian,

$$\sum_{\vec{q}}\langle\vec{r}|\psi_{\vec{q}}\rangle\,\langle\psi_{\vec{q}}|\vec{r}'\rangle = \int\frac{\mathrm{d}^3q}{(2\pi)^3}\,\mathrm{e}^{\mathrm{i}\vec{q}\cdot(\vec{r}'-\vec{r})} = \delta^{(3)}(\vec{r}-\vec{r}'), \tag{4.301}$$

has been used repeatedly. The renormalized low-energy part $\Delta E_{\rm LEP}$ of the self-energy can be obtained by forming the difference $\Delta E - \langle\phi_n|\delta m|\phi_n\rangle$, where ΔE is given in Eq. (4.289) and $\langle\phi_n|\delta m|\phi_n\rangle$ is given in Eq. (4.300),

$$\Delta E_{\rm LEP} = \Delta E - \langle\phi_n|\delta m|\phi_n\rangle$$
$$= \frac{2\alpha}{3\pi}\int_0^\epsilon \mathrm{d}k\, k \left\langle \phi_n \left| \frac{p^i}{m}\left(\frac{1}{E_n - H_S - k} + \frac{1}{k}\right)\frac{p^i}{m}\right|\phi_n\right\rangle, \qquad \mu = m\,. \tag{4.302}$$

This length-gauge energy shift is valid in the non-recoil limit and still divergent at the upper limit of integration ϵ, but only logarithmically. The integral over k diverges at infinity and this constitutes an ultraviolet divergence. However, this divergence is of a different nature as compared to the ultraviolet divergences in quantum field theory. Rather, it is an intermediate divergence which is later canceled against the infrared divergence or "infrared catastrophe" of the vertex correction to the forward scattering amplitude for an electron in a Coulomb field, which can be expressed as a the so-called Dirac F_1 form factor correction. The cancelation of the ϵ parameter between the low- and high-energy parts has been discussed for a toy model in Sec. 4.5.2, and we shall see later that the calculation described here models the low-energy part in our toy model of Sec. 4.5.2 rather well.

4.6.2 *Velocity Gauge*

In Sec. 3.3.2, we had formulated the dipole interaction of the atom with the quantized electromagnetic field in terms of the dipole interaction with the electric field $(-e\vec{r}\cdot\vec{E})$, which is referred to as the length gauge. However, it is also possible to formulate the interaction in the velocity gauge, with the interaction Hamiltonian

$$\mathcal{H}_1 = \frac{(\vec{p} - e\,\vec{A})^2}{2m} - \frac{\vec{p}^{\,2}}{2m} = -\frac{e}{2m}\left\{\vec{A},\vec{p}\right\} + \frac{e^2}{2m}\vec{A}^2\,, \tag{4.303}$$

where $\{\cdot,\cdot\}$ is the anticommutator. The term quadratic in $\vec{A}$,

$$\mathcal{H}_S = \frac{e^2}{2m}\vec{A}^2\,, \tag{4.304}$$

is commonly referred to as the seagull term, because in a diagrammatic language, it corresponds to a graph with two photons emerging from the same interaction vertex.

In the velocity gauge, the coupling of the bound electron to the radiation field is exactly given by Eq. (4.303), with $\vec{p}$ being the relative momentum, i.e., the conjugate variable of the relative distance $\vec{r}$. There actually is a second term with the electron mass m replaced by the mass M of the nucleus, which leads to the self-energy of the nucleus. In the absence of the vector potential, the two kinetic-energy terms, for the electron and the nucleus, consistently add up to

$$\frac{\vec{p}^{\,2}}{2m} + \frac{\vec{p}^{\,2}}{2M} = \frac{\vec{p}^{\,2}}{2\mu}\,, \tag{4.305}$$

where μ is the reduced mass. Here we ignore the kinetic-energy term of the nucleus. Indeed, Eq. (4.303) exclusively describes the interaction of the bound electron with the radiation field, not that of the nucleus or the combined two-body system.

Incidentally, it is helpful to note that Power and Zienau [176] have shown that the two interaction Hamiltonians in length and velocity gauge are related by a gauge transformation, which, strictly speaking, also involves a gauge transformation of the wave function. For the moment, we shall only briefly mention questions regarding the general gauge invariance of quantum processes with regard to the transition from velocity to length gauge (work on related questions continues up to this day, see Ref. [177]). One decisive observation is that in length gauge, the interaction is formulated in terms of the gauge-independent electric field (which is a physical observable, or, in the quantized formalism, an operator corresponding to a physical observable), while in the velocity gauge, we have the gauge-dependent vector potential. In the presence of a nonvanishing vector potential, the kinetic momentum of the particle is no longer described by the canonical momentum $\vec{p}$, but by the canonical momentum $\vec{p} - e\vec{A}$. The latter is the conjugate variable of the position operator in the presence of a nonvanishing vector potential. According to pertinent remarks given by Lamb on page 268 of Ref. [178], the wave function $\phi_n(\vec{r})$ of a bound state therefore preserves its probabilistic interpretation only in the length gauge, and it is this gauge which should be used for off-resonance excitation processes [179]. Otherwise, one also has to gauge-transform the wave function (which would introduce more calculational difficulties, because the gauge-transformed wave functions differ from the Schrödinger wave functions discussed in Secs. 4.2.3, 4.2.5 and 4.2.6). Recently, limitations on the gauge invariance due to changed physical interpretations have been discussed in Refs. [177, 180].

Let us briefly examine the connection of gauge transformations, minimal coupling, and canonical momenta. Under a gauge transformation $\vec{A} \to \vec{A} + \vec{\nabla}\Lambda$, the canonical momentum transforms according to

$$\vec{\pi} = \vec{p} - e\vec{A} \to \vec{\pi}' = \vec{p} - e\vec{A} - e\vec{\nabla}\Lambda\,. \tag{4.306}$$

The gauge transformation leaves the canonical commutation relations invariant, i.e.,

$$[\pi^i, \pi^j] = [\pi'^i, \pi'^j] = -\mathrm{i}\left(\frac{\partial}{\partial x^i} A^j(x) - \frac{\partial}{\partial x^j} A^i(x)\right), \tag{4.307}$$

because partial derivatives interchange, $\partial_i\,\partial_j\Lambda = \partial_j\,\partial_i\Lambda$. Also, a quick calculation shows that

$$[x'^i, \pi'^j] = \left[x^i, \pi^j - e\frac{\partial}{\partial x^j}\Lambda\right] = [x^i, \pi^j] = \mathrm{i}\,\delta^{ij}\,. \tag{4.308}$$

We conclude that canonical commutation relations are gauge invariant, but not the conjugate momenta which receive a modification according to (4.306). The conjugate momentum is not a physical observable. This could otherwise suggest that the length gauge is far preferable over the velocity gauge.

For energy perturbations, however, both gauges can be used while ignoring the gauge transform of the wave function. This may seem quite surprising. The reason is that all quantum electrodynamic energy shifts, which involve interactions with quantized fields, can actually be formulated in terms of adiabatically damped interactions that vanish in the infinite future and the infinite past, like those described in the classical-field derivation of the dynamic Stark shift given in Sec. 3.3.1. Because the state $\phi_n(\vec{r})$ (without any photons in the radiation field) is an eigenstate of the unperturbed Hamiltonian (without interaction), and because the interaction is adiabatically switched off for the in- and out-states (past and future), both gauges are permissible. A deeper analysis of related problems will be performed in Sec. 11.2.2.

The Power–Zienau transformation which connects both gauges is discussed in Sec. 12.6.1. For the time being, we contend ourselves with a brief, but instructive derivation of the gauge-independence of the bound-state self-energy. The interaction Hamiltonian in Eq. (4.303) is a sum of the $\vec{A} \cdot \vec{p}$ coupling and the seagull term which is proportional to $\vec{A}^2$. In the discrete representation, the vector-potential operator (2.129a) in the dipole approximation is given by

$$\vec{A}(\vec{r}) \approx \vec{A}(\vec{r} = \vec{0}) = \sum_{\vec{k}\lambda} \frac{1}{\sqrt{2\omega_{\vec{k}} V}} \hat{\epsilon}_\lambda(\vec{k}) \left(\mathrm{a}_\lambda(\vec{k}) + \mathrm{a}_\lambda^+(\vec{k}) \right) . \tag{4.309}$$

In the dipole approximation, it is independent of position and commutes with the momentum operator, i.e., $\{\vec{A}, \vec{p}\} = 2\vec{A} \cdot \vec{p}$. We start from the same unperturbed state $|\phi_n, 0\rangle$ as in Sec. 4.6.2 and calculate all perturbative corrections to the energy in the order $e^2 = 4\pi\alpha$ due to the dipole interaction Hamiltonian given in Eq. (4.303). Again, the subscript n in ϕ_n is a multi-index containing all relevant quantum numbers of the reference state. The first-order perturbation due to the seagull term

$$\left\langle \phi_n, 0 \left| \frac{e^2}{2m} \vec{A}^2 \right| \phi_n, 0 \right\rangle \tag{4.310}$$

shifts all atomic levels by the same amount and can be discarded. With the unperturbed atom+field Hamiltonian given in Eq. (3.31), the second-order perturbation due to the $\vec{A} \cdot \vec{p}$ coupling is easily evaluated in analogy to Eq. (4.289),

$$\begin{aligned}
\Delta E &= \sum_a \sum_{\vec{k}\lambda} \frac{\left\langle \phi_n, 0 \left| \left(-\frac{e}{m} \vec{A} \cdot \vec{p} \right) \right| \phi_a, 1_{\vec{k}\lambda} \right\rangle \left\langle \phi_a, 1_{\vec{k}\lambda} \left| \left(-\frac{e}{m} \vec{A} \cdot \vec{p} \right) \right| \phi_n, 0 \right\rangle}{E_n - E_a - \omega_{\vec{k}}} \\
&= \frac{e^2}{m^2} \sum_a \sum_{\vec{k}\lambda} \frac{1}{2\omega_{\vec{k}} V} \frac{\langle \phi_n | \hat{\epsilon}_\lambda(\vec{k}) \cdot \vec{p} | \phi_a \rangle \langle \phi_a | \hat{\epsilon}_\lambda(\vec{k}) \cdot \vec{p} | \phi_n \rangle}{E_n - E_a + \omega_{\vec{k}}} \\
&\to \frac{e^2}{m^2} \sum_a \int \frac{\mathrm{d}^3 k}{(2\pi)^3} \frac{1}{2k} \left(\delta^{ij} - \frac{k^i k^j}{\vec{k}^2} \right) \frac{\langle \phi_n | p^i | \phi_a \rangle \langle \phi_a | p^j | \phi_n \rangle}{E_n - E_a - k} \\
&= \frac{2\alpha}{3\pi} \int \mathrm{d}k \, k \left\langle \phi_n \left| \frac{p^i}{m} \frac{1}{E_n - H_S - k} \frac{p^i}{m} \right| \phi_n \right\rangle .
\end{aligned} \tag{4.311}$$

The mass counter term is easy to calculate in the velocity gauge. We take a general matrix element of the self-energy, for a free electron, sandwiched between normalized eigenstates $|\psi_{\vec{q}}\rangle$ and $|\psi_{\vec{q}'}\rangle$ of free Hamiltonian $H_0 = \vec{p}^2/(2\mu)$. Because the free

Hamiltonian commutes with the momentum operator, we have in the same conventions as in Eq. (4.296),

$$\langle \psi_{\vec{q}} | \delta m | \psi_{\vec{q}'} \rangle = \frac{2\alpha}{3\pi} \int \mathrm{d}k\, k \left\langle \psi_{\vec{q}} \left| \frac{p^i}{m} \frac{1}{E_{\vec{q}} - H_0 - k} \frac{p^i}{m} \right| \psi_{\vec{q}'} \right\rangle$$
$$= -\frac{2\alpha}{3\pi} \int \mathrm{d}k \left\langle \psi_{\vec{q}} \left| \frac{\vec{q}^{\,2}}{m^2} \right| \psi_{\vec{q}'} \right\rangle . \tag{4.312}$$

The mass counter term for a free electron is diagonal in the momentum basis. In analogy to Eq. (4.300), we now easily calculate the expectation value of the mass counter term in a bound state $|\phi_n\rangle$, with the result

$$\langle \phi_n | \delta m | \phi_n \rangle = -\frac{2\alpha}{3\pi} \int \mathrm{d}k \left\langle \phi_n \left| \frac{\vec{p}^{\,2}}{m^2} \right| \phi_n \right\rangle . \tag{4.313}$$

The renormalized bound-state self-energy is obtained as the difference of the velocity-gauge shift (4.311) and the mass counter term (4.313),

$$\Delta E_{\mathrm{LEP}} = \Delta E - \langle \phi_n | \delta m | \phi_n \rangle$$
$$= \frac{2\alpha}{3\pi} \int_0^{\epsilon} \mathrm{d}k\, k \left\langle \phi_n \left| \frac{p^i}{m} \left(\frac{1}{E_n - H_S - k} + \frac{1}{k} \right) \frac{p^i}{m} \right| \phi_n \right\rangle , \tag{4.314}$$

in full agreement with Eq. (4.289), and with the same Schrödinger wave function (without gauge transformation). However, the velocity-gauge result is valid even if $\mu \neq m$ and clarifies the reduced-mass dependence of the bound-electron self-energy in the two-body system. In retrospect, the nonrelativistic renormalization turns out to be a little more problematic in the length gauge as opposed to the velocity gauge, because it involves fewer terms in the velocity gauge. However, the length gauge has the advantage that we do not have to worry about the seagull term in the Hamiltonian. Gauge invariance is fully recovered after renormalization.

4.6.3 *Calculation of the Bethe Logarithm*

After renormalization, as we have shown, length and velocity gauges yield identical results for the low-energy part (LEP) of the renormalized self-energy ΔE_{LEP}, as given in Eqs. (4.289) and (4.311). There are two ways to evaluate the expression given in Eqs. (4.289) and (4.311). The first way is merely symbolic and relies on an analytic evaluation of the photon energy integral, and the identification of the logarithmically divergent terms. This calculation exhibits in a particularly clear manner the origin of the name "Bethe *logarithm*". The second way, which is discussed here, relies on an analytic subtraction of the logarithmically divergent terms, and on a formulation of the finite remainder term, which contains the Bethe logarithm, in terms of the dynamic polarizability matrix elements calculated in Sec. 4.3.4.

We start from Eq. (4.314), use the expression for the Schrödinger–Coulomb Hamiltonian given in Eq. (4.1), and keep H_S in operator form until the end of the calculation,

$$\begin{aligned}
\Delta E_{\rm LEP} &= -\frac{2\,\alpha}{3\,\pi}\int_0^{\epsilon} {\rm d}k\,k\left\langle\phi_n\left|\frac{p^i}{m}\left(\frac{1}{H_S-E_n+k}-\frac{1}{k}\right)\frac{p^i}{m}\right|\phi_n\right\rangle\\
&= -\frac{2\,\alpha}{3\,\pi}\int_0^{\epsilon} {\rm d}k\left\langle\phi_n\left|\frac{p^i}{m}\left(\frac{(k+H_S-E_n)-(H_S-E_n)}{H_S-E_n+k}-1\right)\frac{p^i}{m}\right|\phi_n\right\rangle\\
&= \frac{2\,\alpha}{3\,\pi}\left(\frac{\mu}{m}\right)^2\int_0^{\epsilon}{\rm d}k\left\langle\phi_n\left|\frac{p^i}{\mu}\,\frac{H_S-E_n}{H_S-E_n+k}\,\frac{p^i}{\mu}\right|\phi_n\right\rangle. \qquad (4.315)
\end{aligned}$$

We now take the principal value of this integral. Note that the imaginary part of the energy shift otherwise incurred (for an excited reference state) finds a natural interpretation in terms of the radiative decay rate of the reference state against dipole decay (see Refs. [181–183], and also Sec. 3.2.3). Here, we are only interested in the real part of the energy shift and find

$$\begin{aligned}
\Delta E_{\rm LEP} &= \frac{2\,\alpha}{3\,\pi}\left(\frac{\mu}{m}\right)^2\left\langle\phi_n\left|\frac{p^i}{\mu}\,(H_S-E_n)\,\ln\left(\frac{\epsilon}{|H_S-E_n|}\right)\frac{p^i}{\mu}\right|\phi_n\right\rangle\\
&= \frac{2\,\alpha}{3\,\pi}\left(\frac{\mu}{m}\right)^2\left(\left\langle\phi_n\left|\frac{p^i}{\mu}\,(H_S-E_n)\,\frac{p^i}{\mu}\right|\phi_n\right\rangle\ln\left(\frac{2\,\epsilon}{(Z\alpha)^2\,\mu}\right)\right.\\
&\quad\left.-\left\langle\phi_n\left|\frac{p^i}{\mu}\,(H_S-E_n)\,\ln\left(\frac{2\,|H_S-E_n|}{(Z\alpha)^2\,\mu}\right)\frac{p^i}{\mu}\right|\phi_n\right\rangle\right)\\
&= \frac{\alpha}{3\,\pi}\left(\frac{\mu}{m}\right)^2\left\langle\phi_n\left|\left[\frac{p^i}{\mu},\left[(H_S-E_n),\frac{p^i}{\mu}\right]\right]\right|\phi_n\right\rangle\ln\left(\frac{2\,\epsilon}{(Z\alpha)^2\,\mu}\right)\\
&\quad-\frac{2\,\alpha}{3\,\pi}\left(\frac{\mu}{m}\right)^2\left\langle\phi_n\left|\frac{p^i}{\mu}\,(H_S-E_n)\,\ln\left(\frac{2\,|H_S-E_n|}{(Z\alpha)^2\,\mu}\right)\frac{p^i}{\mu}\right|\phi_n\right\rangle\\
&= \frac{4\,\alpha}{3\,\pi}\,\frac{(Z\alpha)^4\,m}{n^3}\left(\frac{\mu}{m}\right)^3\left[\ln\left(\frac{2\epsilon}{(Z\alpha)^2\,\mu}\right)\delta_{\ell 0}-\ln k_0(n\ell)\right]. \qquad (4.316)
\end{aligned}$$

In the first step, we use the fact that ϵ must be large against $|H_S - E_n|$, and ignore terms of order $\mathcal{O}(\alpha\,(Z\alpha)^6\,\mu)$. The Bethe logarithm $\ln k_0(n\ell)$ is identified as

$$\ln k_0(n\ell) = \frac{n^3}{2(Z\alpha)^4}\left\langle\phi_n\left|\frac{p^i}{\mu}\,\frac{H_{\rm S}-E_n}{\mu}\,\ln\left(\frac{2|H_{\rm S}-E_n|}{(Z\alpha)^2\,\mu}\right)\frac{p^i}{\mu}\right|\phi_n\right\rangle, \qquad (4.317)$$

where the sum over $i = 1, 2, 3$ is implied by the Einstein summation convention. The Bethe logarithm depends on the principal quantum number n, and on the orbital angular momentum ℓ of the reference state $|\phi_n\rangle$, but the reduced mass μ scales out of the problem. Written as a sum over intermediate, virtual states, we have

$$\ln k_0(n\ell) = \frac{n^3}{2(Z\alpha)^4}\sum_a\left|\left\langle\phi_n\left|\frac{\vec p}{\mu}\right|\phi_a\right\rangle\right|^2\frac{E_a-E_n}{\mu}\,\ln\left(\frac{2\,|E_a-E_n|}{(Z\alpha)^2\,\mu}\right). \qquad (4.318)$$

The absolute value is very important in the argument of the logarithm. For excited reference states, the argument of the logarithm $E_a - E_n$ might otherwise become negative, which would imply a nonvanishing imaginary part of the Bethe logarithm. While the imaginary part actually has a physical interpretation, as discussed in

Sec. 4.7.1 below, we stress that the Bethe logarithm is defined in terms of the principal-value integral given in Eq. (4.316), which implies that it is purely real. The argument of the logarithm in Eq. (4.318) contains energy differences $|E_a - E_n|$ of atomic states scaled to the reference quantity

$$Z^2 R_\infty \approx \frac{(Z\alpha)^2 \mu}{2}, \tag{4.319}$$

where we generally assume that the reduced mass μ of the system is close to the electron mass m. The formulas (4.317) and (4.318) are valid for all atomic states, including S states.

Several steps in the derivation (4.316) are in need of further explanation. We have already explained the principal-value prescription used in the step labeled (P.V.). The right-hand side of the same transformation step also contains a hint that higher-order terms $\mathcal{O}(\alpha(Z\alpha)^6)$ have been neglected in the evaluation. These terms are convergent as $\epsilon \to \infty$, for example terms proportional to $\langle\phi_n|[(Z\alpha)^2/r^4]|\phi_n\rangle/\epsilon$, and will be discussed below in Sec. 4.7.3. A further remark is in order. Namely, the general commutator relation

$$ABA = \frac{1}{2}\left([A,[B,A]] + A^2 B + BA^2\right) = \frac{1}{2}\left([A,[B,A]] + \{A^2, B\}\right) \tag{4.320}$$

implies that

$$\left\langle \phi_n \left| \frac{p^i}{\mu} (H_S - E_n) \frac{p^i}{\mu} \right| \phi_n \right\rangle = \frac{1}{2} \left\langle \phi_n \left| \left[\frac{p^i}{\mu}, \left[(H_S - E_n), \frac{p^i}{\mu} \right] \right] \right| \phi_n \right\rangle, \tag{4.321}$$

while, furthermore, one has

$$\left[\frac{p^i}{\mu}, \left[(H_S - E_n), \frac{p^i}{\mu} \right] \right] = -\frac{1}{\mu^2} \vec{p}^{\,2}(V) = \frac{1}{\mu^2} \vec{\nabla}^2(V) = \frac{1}{\mu^2} 4\pi Z\alpha\, \delta^{(3)}(\vec{r}). \tag{4.322}$$

Only S states have a nonvanishing probability density at the origin,

$$\langle\phi_n|\delta^{(3)}(\vec{r})|\phi_n\rangle = \frac{(Z\alpha)^3 \mu^3}{\pi n^3} \delta_{\ell 0}. \tag{4.323}$$

Here, as always, the subscript n in ϕ_n needs to be understood as a multi-index denoting the nonrelativistic quantum numbers n, ℓ and m, where n is the principal quantum number, ℓ is the orbital angular momentum quantum number, and m is the magnetic projection quantum number.

A further, very important remark is in order. The low-energy part of the bound-electron self-energy, after renormalization, is of order $\alpha\,(Z\alpha)^4\,\mu$, as evident from Eq. (4.316). This order-of-magnitude result could have been obtained from the first line of Eq. (4.316) independently. Bound systems follow rather strict hierarchies in terms of the momentum and energy scales associated to specific operators. Specifically, bound-state energies are of order $(Z\alpha)^2\mu$, as evident from the Schrödinger–Coulomb spectrum. The same order-of-magnitude, $(Z\alpha)^2\mu$, can be associated with the Schrödinger–Coulomb Hamiltonian as an operator, because it contains a sum over virtual bound and continuum states, and the inverse $1/((Z\alpha)^2\mu)$ should thus be associated with the Green function. The same order-of-magnitude applies to the Coulomb potential alone. A bound-state momentum is of order $Z\alpha\mu$, in accordance with the nonrelativistic identification $E \sim p^2/\mu$, where p is a characteristic momentum scale. The exchanged photon energy k in the low-energy part also is of order $(Z\alpha)^2\mu$. We thus have

$$p^i \sim Z\alpha\mu\,, \qquad r^i \sim \frac{1}{Z\alpha\,\mu}\,, \qquad H_S \sim E_n \sim k \sim (Z\alpha)^2\,\mu\,, \qquad \mu \approx m\,,$$

$$\Delta E_{\mathrm{LEP}} = -\frac{2\alpha}{3\pi}\int_0^{\epsilon} \mathrm{d}k\,k \left\langle \phi_n \left| \frac{p^i}{m} \left(\frac{1}{H_S - E_n + k} - \frac{1}{k} \right) \frac{p^i}{m} \right| \phi_n \right\rangle$$

$$\sim \alpha\,((Z\alpha)^2\,m)\,((Z\alpha)^2\,m)\,\frac{Z\alpha m}{m}\,\frac{1}{(Z\alpha)^2\,m}\,\frac{Z\alpha m}{m} = \alpha\,(Z\alpha)^4\,m\,. \tag{4.324}$$

The $Z\alpha$-scaling sometimes has to be adopted to a more general method of (integration) regions (see Chaps. 16 and 17). In the current context, it provides, in a very straightforward fashion, for a useful basis in the study of bound systems. The scaling of the atomic coordinate r^i is given for reference; it is useful analyzing the $Z\alpha$-scaling of intermediate expressions in Eq. (4.289). One can easily convince oneself that the Bethe logarithm (4.317) is independent of the values of the reduced mass μ of the bound system.

In accordance with our toy model calculation discussed at length in Sec. 4.5.2, we recall that the result obtained in Eq. (4.316) has to be interpreted as the low-energy part of the bound-electron self-energy, which is obtained after first expanding in $Z\alpha$, then in ϵ. This procedure implies, first of all, that $\epsilon \gg Z\alpha$, so that, in the integrals over the photon energy ω, one integrates over the range $\omega \in (0, \infty)$ (unless divergences are encountered), because ϵ is much *larger* than the atomic energy scale. However, based on the general paradigm discussed in Sec. 4.5.2, one later neglects all terms proportional to ϵ and higher in the final result for the low-energy part,

$$\Delta E_{\mathrm{LEP}} = \frac{4\alpha}{3\pi}\,\frac{(Z\alpha)^4\,m}{n^3}\left(\frac{\mu}{m}\right)^3 \left[\ln\left(\frac{2\epsilon}{(Z\alpha)^2\,\mu} \right) \delta_{\ell 0} - \ln k_0(n\ell) \right]. \tag{4.325}$$

This is based on the paradigm, also discussed in Sec. 4.5.2, which states that the ϵ-dependent terms have to cancel in each order of the $Z\alpha$-expansion separately, whether they be divergent or not in the limit $\epsilon \to 0$. The physically relevant regime for the ϵ parameter in this case is the regime where

$$(Z\alpha)^2\,\mu \ll \epsilon \ll \mu\,, \tag{4.326}$$

i.e., ϵ is large on the scale of the atomic binding energy, which is why it is permissible to extend all convergent integrals over ω to infinity, but small on the scale of the electron mass, so that ϵ acts as a cut-off parameter for the high-energy contributions that involve infrared divergences which are naturally cut off at the atomic binding energy scale. In Sec. 4.5.2, this subtle interplay is illustrated using a compact, analytic example calculation.

Table 4.1 Bethe Logarithms for all excited states in hydrogenlike systems with $\ell \leq 4$ and $n \leq 12$. For $\ell = 2, 3, 4$, the given numerical results represent the quantity $\ln k_0(n\ell) \times 10^2$, so the Bethe logarithm for $n = 12$ and $\ell = 2$ reads as $\ln k_0(n = 12, \ell = 2) = -0.009\,342\,953\,99$.

Bethe Logarithms $\ln k_0(n\ell)$ for $n \leq 12$ and $\ell \leq 4$

	$\ell = 0$	$\ell = 1$	$\ell = 2$	$\ell = 3$	$\ell = 4$
	$\ln k_0 \times 10^{-1}$	$\ln k_0 \times 10^{1}$	$\ln k_0 \times 10^{2}$	$\ln k_0 \times 10^{2}$	$\ln k_0 \times 10^{2}$
$n = 1$	0.298 412 856	—	—	—	—
$n = 2$	0.281 176 989	-0.300 167 086	—	—	—
$n = 3$	0.276 766 361	-0.381 902 294	-0.523 214 814	—	—
$n = 4$	0.274 981 184	-0.419 548 946	-0.674 093 888	-0.173 366 148	—
$n = 5$	0.274 082 373	-0.440 346 956	-0.760 075 126	-0.220 216 838	-0.077 209 890
$n = 6$	0.273 566 421	-0.453 121 977	-0.814 720 396	-0.250 217 976	-0.096 279 743
$n = 7$	0.273 242 913	-0.461 551 773	-0.851 922 329	-0.270 909 573	-0.109 447 274
$n = 8$	0.273 026 726	-0.467 413 520	-0.878 504 298	-0.285 911 456	-0.119 043 204
$n = 9$	0.272 875 117	-0.471 657 000	-0.898 203 229	-0.297 190 149	-0.126 309 451
$n = 10$	0.272 764 694	-0.474 828 934	-0.913 227 225	-0.305 909 428	-0.131 971 806
$n = 11$	0.272 681 778	-0.477 262 681	-0.924 957 082	-0.312 802 113	-0.136 484 485
$n = 12$	0.272 617 934	-0.479 171 116	-0.934 295 399	-0.318 351 910	-0.140 146 873

4.6.4 *Numerical Values of Bethe Logarithms*

For the numerical evaluation of the Bethe logarithm, which is important for the comparison of theoretical predictions to experimental results, a representation in terms of transition elements and energy differences to intermediate, virtual states is not of great help. One may in fact use such a representation to good effect if one considers a discretized representation of the Schrödinger–Coulomb propagator on a carefully adjusted lattice of points, which takes into account the importance of the region near the nucleus and uses a highly concentrated array of lattice points in this area [184, 185]. Another representation relies on explicit results for transition matrix elements for bound-bound, and bound-continuum transitions [186, 187]. Indeed, explicit matrix elements of discrete-discrete transitions have been given rather early in the history of quantum mechanics [188], and discrete-continuum transitions have also been analyzed [161, 189]. The results given in the literature are not free of typographical errors: In Eq. (6) of Ref. [161], the argument of the inverse cotangent (arccot) function in the exponential should be replaced according to $n'n/n \to n'/n$. Certain normalization problems of the continuum wave functions used in Ref. [161] motivate a very careful discussion, as already mentioned in the discussion following Eq. (4.76a) in Sec. 4.2.5 above. For selected low-lying states coupled to the continuum, formulas have been given as early as 1930 (see Ref. [190]), but one should be aware of multiplicative correction factors given below Eq. (11) of Ref. [191].

The numerical results obtained using either lattice calculations or explicit transition matrix elements are somewhat "hanging in the air" because they are obtained as "pure numbers", without any control over the numerics (or analytic limits) in intermediate steps of the calculation. For low-lying states of hydrogenlike systems, a transparent method in evaluating the Bethe logarithm relies on the analytic results for the dynamic polarizability matrix elements given in Sec. 4.3.4.

Let us recall from Sec. 4.3.4 the definition of the P matrix elements,

$$P(\phi_n, \omega) = \frac{1}{3\mu} \left\langle \phi_n \left| p^i \frac{1}{H_S - z} p^i \right| \phi_n \right\rangle, \qquad z = E_n - \omega. \tag{4.327}$$

The definition of the t variable according to Eq. (4.131),

$$t = \sqrt{\frac{E_n}{E_n - \omega}} = \left(1 + \frac{2n^2 \omega}{(Z\alpha)^2 \mu}\right)^{-1/2}, \qquad \omega = \frac{(Z\alpha)^2 \mu}{2n^2} \frac{1 - t^2}{t^2}, \tag{4.328}$$

implies that ultraviolet divergences from the integration region $\omega \sim \epsilon \gg (Z\alpha)^2 \mu$, i.e., $\omega \to \infty$, are mapped onto the integration region $t \to 0$. We also recall the analytic result for the dynamic polarizability of the ground state, given originally in Eq. (4.149),

$$P(1S, t) = \frac{2t^2(1 - t - 5t^2 - 11t^3)}{3\,(1+t)^2\,(1-t)^3} + \frac{64\,t^5}{3\,(1-t^2)^3}\, {}_2F_1\left(1, -t, 1-t, \left(\frac{1-t}{1+t}\right)^2\right). \tag{4.329}$$

The integration region $\omega \in (0, \infty)$ is mapped onto the integration region $t \in (0, 1)$. Following the calculation reported in Eq. (4.316) in Sec. 4.6.3, the ultraviolet divergences can easily be calculated analytically and subtracted. Inspired by Eq. (4.325), we define a scaled function $F_{\rm LEP} = F_{\rm LEP}(\phi_n, \epsilon, Z\alpha)$ as follows,

$$\Delta E_{\rm LEP} = \frac{\alpha}{\pi} \frac{(Z\alpha)^4 m}{n^3} \left(\frac{\mu}{m}\right)^3 F_{\rm LEP}, \tag{4.330}$$

where $F_{\rm LEP}$ is given by

$$F_{\rm LEP} = \int_0^1 \frac{\mathrm{d}t}{t^3} \left(\frac{t^2 - 1}{n\,t^2} P(\phi_n, t) + \frac{2}{3n} - \frac{8}{3} t^2 \,\delta_{\ell,0} \right) + \frac{4}{3} \delta_{\ell 0} \ln\left(\frac{2\epsilon\, n^2}{(Z\alpha)^2 \mu}\right). \tag{4.331}$$

A comparison of Eqs. (4.330), (4.317) and (4.325) reveals that

$$F_{\rm LEP} = \frac{4}{3} \delta_{\ell 0} \ln\left(\frac{2\epsilon}{(Z\alpha)^2 \mu}\right) - \frac{4}{3} \ln k_0(n\ell). \tag{4.332}$$

Combining (4.331) and (4.332), we immediately read off the integral representation

$$\ln k_0(n\ell) = -\frac{3}{4}\,(\mathrm{P.V.}) \int_0^1 \mathrm{d}t\,\frac{1}{t^3}\left\{\frac{t^2-1}{n\,t^2}\,P(\phi_n,t) + \frac{2}{3\,n} - \frac{8}{3}\,t^2\,\delta_{\ell 0}\right\} - 2\,\ln(n)\,\delta_{\ell 0}\,. \tag{4.333}$$

The remaining t integral can be done semi-analytically, using the series representation of the hypergeometric function in Eq. (4.329) as a guide, and the remaining terms can be expressed in terms of a generalized zeta function, which is known as the Lerch Φ transcendent. We should briefly discuss these functions. The Lerch Φ transcendent is defined as

$$\Phi(z,s,a) = \sum_{n=0}^{\infty} \frac{z^n}{(n+a)^s}\,. \tag{4.334}$$

A formula valid in a larger domain of the complex plane is

$$\Phi(z,s,a) = \frac{1}{1-z}\,\sum_{n=0}^{\infty}\left(\frac{-z}{1-z}\right)^n \sum_{k=0}^{n}\frac{(-1)^k}{(a+k)^s}\binom{n}{k}. \tag{4.335}$$

The Lerch Φ transcendent can be thought of as a generalization of the zeta function, because $\Phi(1,s,1) = \zeta(s)$. The connection of the Lerch transcendent to the hypergeometric function is

$$\Phi(z,1,a) = \frac{{}_2F_1(1,a,1+a,z)}{a}\,. \tag{4.336}$$

After the t integration, hydrogenic Bethe logarithms of $1S$, $2S$ and $2P$ states can be written as rather compact sums over Lerch Φ functions, and evaluated to essentially arbitrary accuracy. We have for the ground state,

$$\begin{aligned}\ln k_0(1S) &= 10\,\ln(2) - 2\zeta(2) - 1 + \sum_{k=2}^{\infty}\frac{16\,k}{(k-1)^2\,(k+1)^2}\,\Phi\left(\frac{1+k}{1-k},1,2\,k\right)\\ &= 2.98412\,85557\,65497\,61075\,97770\,90013\,79796\,99751\,80566\,17002\\ &\qquad 00048\,15926\,13924\,06576\,62306\,75532\,86860\,62013\,30404\,72249\,.\end{aligned} \tag{4.337}$$

Here ζ denotes the Riemann zeta function. The logarithmic sum, for the $2S$ state, is given by

$$\begin{aligned}\ln k_0(2S) &= -\frac{545}{36} + \frac{22}{3}\,\ln(2) - 14\,\zeta(2) + 24\,\zeta(3) \\ &\quad + \sum_{k=3}^{\infty}\frac{1024\,k\,(k-1)\,(k+1)}{(k-2)^3\,(k+1)^3}\,\Phi\left(\frac{2+k}{2-k},1,2\,k\right)\\ &= 2.81176\,98931\,20563\,51521\,97427\,85941\,63611\,28935\,51470\,29732\\ &\qquad 41909\,18696\,96453\,24020\,20118\,89106\,87017\,48612\,02831\,24031\,,\end{aligned} \tag{4.338}$$

whereas for the $2P$ state, it reads

$$\begin{aligned}\ln k_0(2P) &= -\frac{3437}{2916} + \frac{3280}{2187}\ln(2) - \frac{14}{3}\zeta(2) + \frac{136}{9}\zeta(3) - \frac{64}{3}\zeta(4) \\ &\quad + \sum_{k=3}^{\infty} \frac{256\,k^3\,(11\,k^2-12)}{3\,(k-2)^4\,(k+2)^4}\,\Phi\left(\frac{2+k}{2-k}, 1, 2\,k\right) \\ &= -0.03001\,67086\,30212\,90244\,36757\,10951\,14406\,39409\,33044\,23103 \\ &\qquad 04668\,98525\,32719\,44796\,89622\,57183\,26244\,10312\,70799\,73828\,.\end{aligned} \tag{4.339}$$

While these formulas do not provide direct, analytic results for Bethe logarithms, they still illustrate that one can think of the Bethe logarithm as a sum of logarithms, zeta functions, and generalized zeta functions.

For other high-precision evaluations of Bethe logarithms, see Refs. [192, 193]. While the 100-figure numerical accuracy of the results (4.337), (4.338) and (4.339) far exceeds the requirements of overall theoretical predictions for the Lamb shift of the atomic states under investigation, the reference values given here may serve as a guide for conceivable future test calculations of accurate Green functions of the hydrogen atom, obtained using alternative lattice and other methods. An extensive list of 9-figure numerical results for all bound states with quantum numbers $n \leq 200$ has been given in Refs. [186, 187]. A convenient list for $n \leq 12$ and $\ell \leq 4$ is given in Table 4.1.

4.6.5 *Outline of the High-Energy Part*

The low-energy part of the self-energy has been evaluated in Eq. (4.325), and the Bethe logarithm has been given in Eq. (4.317). Furthermore, we have discussed the numerical evaluation of the Bethe logarithm in Sec. 4.6.4. We shall now briefly discuss the high-energy part of the self-energy, which is spin-dependent and requires fully relativistic loop calculations. We formulate the high-energy part in terms of effective potentials, which are evaluated on so-called SCHRÖDINGER[1]–PAULI[2] wave functions, i.e., wave functions whose radial part is equal to the wave functions discussed in Sec. 4.2.2, but whose angular part allows for the presence of an internal degree of freedom of the electron, which is the spin. The Schrödinger–Pauli wave functions are defined in Eq. (6.99); they combine a nonrelativistic radial part with a spin-angular component which describes the coupling of the orbital angular momentum ℓ and the spin angular momentum $s = 1/2$, to result in the total angular momentum j. We can only give the essential steps here; some parts have to be relegated to a more thorough discussion of the radiatively corrected Dirac equation (see Chap. 10).

The first high-energy potential we consider is

$$\Delta V_1 = \alpha\,(Z\alpha)\left[\frac{4}{3}\ln\left(\frac{m}{2\,\epsilon}\right) + \frac{11}{18}\right]\frac{\delta^{(3)}(\vec{r})}{m^2}\,, \tag{4.340}$$

[1]Erwin Rudolf Josef Alexander Schrödinger (1887–1961).
[2]Wolfgang Ernst Pauli (1900–1958).

which is proportional to the slope of the Dirac form factor of the bound electron at zero momentum transfer. Hence, it is manifestly the electron mass m that enters the denominator, not the reduced mass μ. The corresponding energy shift is

$$\Delta E_1 = \langle \phi_n | \Delta V_1 | \phi_n \rangle = \frac{\alpha}{\pi} \frac{(Z\alpha)^4 m}{n^3} \left[\frac{4}{3} \ln\left(\frac{m}{2\epsilon}\right) + \frac{11}{18} \right] \left(\frac{\mu}{m}\right)^3 \delta_{\ell 0} \,. \tag{4.341}$$

The correction ΔE_1 is spin independent and nonvanishing only for S states with $\ell = 0$.

The second potential comes from the anomalous magnetic moment of the electron. Relativistically, it is obtained as $\Delta V_2 = F_2(0)\, \mathrm{i}\, e \vec{\gamma} \cdot \vec{E}/(2m)$, where $F_2(0) = \alpha/(2\pi)$ is the anomalous magnetic moment (Schwinger factor). The non-relativistic limit of the second perturbative, "radiative" potential is

$$\Delta V_2 = \frac{\alpha}{\pi} \frac{1}{2} \frac{\pi (Z\alpha)\, \delta^{(3)}(\vec{r})}{m^2} + \frac{\alpha}{\pi} \frac{(Z\alpha)\, \vec{\sigma} \cdot \vec{L}}{4 m \mu r^3} \,, \tag{4.342}$$

where $\vec{L}$ is the orbital angular momentum, and $\vec{\sigma}$ is the vector of Pauli spin matrices [see Eq. (8.66) of Ref. [81]], which we recall for convenience,

$$\sigma^x = \sigma^1 = \begin{pmatrix} 0 & 1 \\ 1 & 0 \end{pmatrix}, \qquad \sigma^y = \sigma^2 = \begin{pmatrix} 0 & -\mathrm{i} \\ \mathrm{i} & 0 \end{pmatrix}, \qquad \sigma^z = \sigma^3 = \begin{pmatrix} 1 & 0 \\ 0 & -1 \end{pmatrix}. \tag{4.343}$$

The reduced-mass dependence of the second term (spin-orbit term) follows from the two-body Hamiltonian of the system (Breit Hamiltonian) and is a nontrivial result that takes into account the manifestly two-body nature of the bound system. In writing Eq. (4.342), we have explicitly displayed the expression

$$\Delta V_{\mathrm{std}} = \frac{\pi (Z\alpha)\, \delta^{(3)}(\vec{r})}{m^2} \,, \qquad \langle \phi_n | \Delta V_{\mathrm{std}} | \phi_n \rangle = \frac{(Z\alpha)^4 m}{n^3} \left(\frac{\mu}{m}\right)^3 \delta_{\ell 0} \,, \tag{4.344}$$

which is a "standard" Dirac-δ potential that is sometimes useful in atomic physics calculations. We note that the evaluation of the spin-dependent term given in Eq. (4.342) actually entails the use of the Schrödinger–Pauli wave function discussed in Sec. 6.4.1 below.

The second term in Eq. (4.342) necessitates the inclusion of the electron spin into the formalism, which can easily be accomplished by the use of the Schrödinger–Pauli wave function [see Eq. (6.99) below]. After the tracing of the spin degrees of freedom, what remains is the evaluation of a matrix element of r^{-3} on a hydrogenic bound state, and we are now interested in its numerical value for a general reference state with quantum numbers n, ℓ and m. Modern computer algebra systems enable us to map all momentum operators and integrations onto a well-defined subset of integrals, once the explicit form of a reference-state wave function is known. The explicit form of the reference-state wave function can thus be calculated from the formulas given in Sec. 4.2.3, or from the general formulas given below.

Alternatively, the following recursion relation, due to Kramers and Pasternack [194, 195], can be used in order to establish relations among matrix elements of different powers of the radial variable r. It reads as

$$\frac{k+1}{n^2}\left\langle\phi_n\left|r^k\right|\phi_n\right\rangle-\frac{2k+1}{Z\alpha\mu}\left\langle\phi_n\left|r^{k-1}\right|\phi_n\right\rangle$$
$$+\frac{k}{4(Z\alpha\mu)^2}\left[(2\ell+1)^2-k^2\right]\left\langle\phi_n\left|r^{k-2}\right|\phi_n\right\rangle=0\,, \tag{4.345}$$

being valid for both positive as well as negative k. Closer inspection reveals that the result for the matrix element of $1/r$ can be derived by setting $k=0$ in Eq. (4.345), whereas the result for $1/r^3$ can be derived by setting $k=-2$ in Eq. (4.345), and having a result for $1/r^2$ handy. Furthermore, this relation can be used to derive matrix elements of higher inverse powers of r. One obtains the following results [see Eq. (3.24) ff. of Ref. [6]]

$$\left\langle\phi_n\left|\frac{1}{r}\right|\phi_n\right\rangle=Z\alpha\mu\,\frac{1}{n^2}\,, \tag{4.346a}$$

$$\left\langle\phi_n\left|\frac{1}{r^2}\right|\phi_n\right\rangle=(Z\alpha\mu)^2\,\frac{2}{n^3\,(2\,\ell+1)}\,, \tag{4.346b}$$

$$\left\langle\phi_n\left|\frac{1}{r^3}\right|\phi_n\right\rangle=(Z\alpha\mu)^3\,\frac{2}{n^3\,\ell\,(\ell+1)\,(2\ell+1)}\,, \tag{4.346c}$$

$$\left\langle\phi_n\left|\frac{1}{r^4}\right|\phi_n\right\rangle=(Z\alpha\mu)^4\,\frac{4\,[3\,n^2-\ell\,(\ell+1)]}{n^5\,\ell\,(\ell+1)\,(2\ell-1)\,(2\ell+1)\,(2\ell+3)}\,. \tag{4.346d}$$

Higher inverse power of the radial variable lead to matrix elements divergent on P states,

$$\left\langle\phi_n\left|\frac{1}{r^5}\right|\phi_n\right\rangle=(Z\alpha\mu)^5\,\frac{4\,[5\,n^2-3\ell\,(\ell+1)+1]}{n^5\,(\ell-1)\,\ell\,(\ell+1)\,(\ell+2)\,(2\ell-1)\,(2\ell+1)\,(2\ell+3)}\,, \tag{4.346e}$$

$$\left\langle\phi_n\left|\frac{1}{r^6}\right|\phi_n\right\rangle=(Z\alpha\mu)^6\,\frac{64\,[35\,n^4-5n^2(6\ell\,(\ell+1)-5)+3(\ell-1)\ell(\ell+1)(\ell+2)]}{n^7\,\prod_{j=-3}^{5}(2\ell+j)}\,. \tag{4.346f}$$

Matrix elements of $1/r^3$ and $1/r^4$ are already divergent for S states with $\ell=0$. This is reflected in the denominators of the formulas. The expectation value of ΔV_2 for a reference bound state $|\phi_n\rangle$ is

$$\begin{aligned}\Delta E_2&=\langle\phi_n\,|\Delta V_2|\,\phi_n\rangle\\&=\frac{\alpha}{\pi}\,\frac{(Z\alpha)^4\,m}{n^3}\left(\frac{\mu}{m}\right)^3\frac{\delta_{\ell 0}}{2}\\&\quad+\frac{\alpha}{\pi}\,\frac{(Z\alpha)^4\,m}{n^3}\,\frac{j(j+1)-\ell(\ell+1)-s(s+1)}{2\ell\,(\ell+1)\,(2\ell+1)}\left(\frac{\mu}{m}\right)^2(1-\delta_{\ell 0})\\&=\frac{\alpha}{\pi}\,\frac{(Z\alpha)^4\,m}{n^3}\left[\left(\frac{\mu}{m}\right)^3\frac{\delta_{\ell 0}}{2}+\left(-\frac{1}{2\kappa\,(2\,\ell+1)}\right)\left(\frac{\mu}{m}\right)^2(1-\delta_{\ell 0})\right].\end{aligned} \tag{4.347}$$

Here, the Dirac quantum number κ is obtained from the orbital angular momentum ℓ and the total momentum j of the bound electron as

$$\kappa=(-1)^{j+\ell+1/2}\left(j+\frac{1}{2}\right). \tag{4.348}$$

E.g., for S states, we have $\kappa = -1$, and $\ell = 0$, so that the term $-1/[2\kappa(2\ell+1)]$ in Eq. (4.347) evaluates to $1/2$ for S states and nominally becomes equal to the first term, in the non-recoil limit $\mu \to m$. The high-energy part $\Delta E_{\rm HEP} = \Delta E_1 + \Delta E_2$ is obtained as

$$\Delta E_{\rm HEP} = \frac{\alpha}{\pi}\frac{(Z\alpha)^4 m}{n^3}\left\{\left[\frac{4}{3}\ln\left(\frac{m}{2\epsilon}\right)+\frac{10}{9}\right]\left(\frac{\mu}{m}\right)^3 \delta_{\ell 0} - \frac{1-\delta_{\ell 0}}{2\kappa(2\ell+1)}\left(\frac{\mu}{m}\right)^2\right\}, \tag{4.349}$$

which includes the full reduced-mass dependence. In the limit $\mu \to m$, we can add the low-energy part (4.325) to obtain the full self-energy shift $\Delta E_{\rm SE} = \Delta E_{\rm HEP} + \Delta E_{\rm LEP}$ as

$$\Delta E_{\rm SE} \overset{\mu=m}{=} \frac{\alpha}{\pi}\frac{(Z\alpha)^4 m}{n^3}\left[\left(\frac{4}{3}\ln\left[(Z\alpha)^{-2}\right]+\frac{11}{18}\right)\delta_{\ell 0} - \frac{1}{2\kappa(2\ell+1)} - \frac{4}{3}\ln k_0(n\ell)\right], \tag{4.350}$$

which is a nice general formula for $\mu = m$, but it does not include the full reduced-mass dependence of the result.

If we wish to include the reduced-mass dependence, then it is easier to separate the cases of S versus non-S states. For S states, the high-energy part evaluates to

$$\Delta E_{\rm HEP}(nS) = \frac{\alpha}{\pi}\frac{(Z\alpha)^4 m}{n^3}\left(\frac{\mu}{m}\right)^3\left(\frac{4}{3}\ln\left(\frac{m}{2\epsilon}\right)+\frac{10}{9}\right). \tag{4.351}$$

Adding the low-energy part (4.325), the auxiliary overlapping parameter ϵ cancels, and we obtain the following total self-energy shift for S states,

$$\Delta E(n{\rm S}) = \frac{\alpha}{\pi}\frac{(Z\alpha)^4 m}{n^3}\left(\frac{\mu}{m}\right)^3\left[\left(\frac{4}{3}\ln\left[\frac{m}{\mu}\frac{1}{(Z\alpha)^2}\right]+\frac{10}{9}\right)-\frac{4}{3}\ln k_0(nS)\right], \tag{4.352}$$

which includes the full reduced-mass dependence, including the argument of the logarithm. By contrast, for non-S states with $\ell \neq 0$, the first term in square brackets in Eq. (4.349) vanishes. Adding the low-energy part (4.316), we thus obtain the following total self-energy shift for non-S states,

$$\Delta E(n\ell_j)|_{\ell\neq 0} = \frac{\alpha}{\pi}\frac{(Z\alpha)^4 m}{n^3}\left(\frac{\mu}{m}\right)^3\left(-\frac{1}{2\kappa(2\ell+1)}\frac{m}{\mu} - \frac{4}{3}\ln k_0(n\ell)\right), \tag{4.353}$$

which depends on the total angular momentum j of the bound electron i.e., it is fine-structure dependent. We have denoted the reference state in spectroscopic notation as $n\ell_j$, i.e., for a $2P_{3/2}$ state we would have $n = 2$, $\ell = 1$ and $j = 3/2$. The reduced-mass dependence of the anomalous-magnetic-moment correction given in Eq. (4.353) finds an explanation in Eq. (12.118). The calculation of the bound self-energy for a general state, for arbitrary masses of orbiting particle and nucleus, is a nontrivial achievement. The results (4.352) and (4.353) can directly be applied to other bound systems such as muonic hydrogen, under an appropriate replacement of m for the muon mass and μ for the reduced mass of muonic hydrogen. Our only additional assumption concerns the spin-1/2 character of the orbiting particle.

4.7 Applications of the Developed Formalism

4.7.1 *Electric Dipole Decay and Imaginary Part of Self–Energy*

The calculation of the (real part of the) self-energy shift of a bound atomic state, with final results reported in Eqs. (4.350), (4.352) and (4.353), represents the most significant result of the current chapter, together with the numerical results for Bethe logarithms listed in Table 4.1 and Eq. (4.333) (for a general integral representation of the Bethe logarithm), as well as the numerical result for $n = 1$ and $n = 2$ states in Eqs. (4.337) and (4.338). However, the potential applications of the developed formalism are much more general, and we shall indicate several of them in the following.

First, it is actually quite interesting to note that there is a connection of the imaginary part of the self-energy shift to the decay rate of an atomic state. We recall that in going from Eq. (4.315) to (4.316), we have deliberately ignored the imaginary part of the energy shift, focusing only on the real part, which we extracted using a principal-value prescription. In general, an energy shift can be written as [see also Eq. (3.56)]

$$\Delta E = \mathrm{Re}\,\Delta E - \frac{\mathrm{i}}{2}\Gamma\,, \qquad \Gamma = -2\mathrm{Im}\,\Delta E\,. \tag{4.354}$$

In specifying the imaginary part of the self-energy shift (4.315), we should first find a valid prescription concerning the poles along the contour used for the integration over k. One might ask which energy in the propagator denominator should get an imaginary part: the reference state energy E_n or the virtual-state energy described by the Hamiltonian H_S? The answer is that in our formalism, the reference state is thought of as a perfect, asymptotic state, and the interaction is switched off in the infinite past and future. So, the virtual states, not the reference state, obtain an imaginary part (infinitesimal damping term), and we have

$$\begin{aligned}\Delta E_{\mathrm{LEP}} &= \frac{2\alpha}{3\pi}\left(\frac{\mu}{m}\right)^2 \int \mathrm{d}k\, k \left\langle \phi_n \left| \frac{p^i}{\mu} \frac{1}{E_n - (H_S - \mathrm{i}\epsilon) - k} \frac{p^i}{\mu} \right| \phi_n \right\rangle + \dots \\ &= \frac{2\alpha}{3\pi}\left(\frac{\mu}{m}\right)^2 \sum_m \int \mathrm{d}k\, k \left| \left\langle \phi_n \left| \frac{\vec{p}}{\mu} \right| \phi_m \right\rangle \right|^2 \frac{1}{E_n - E_m - k + \mathrm{i}\epsilon}\,. \end{aligned} \tag{4.355}$$

The omitted terms in the first line correspond to the renormalization counter-term, which we can ignore in calculating the imaginary part of the self-energy shift, because it does not contain any virtual-state energies in the denominator and cannot lead to an imaginary energy shift. Because we are eventually interested only in the imaginary part of the energy, we can use the following formula, which follows from a theorem proven by SOCHOCKI[3] and PLEMELJ[4] (for a derivation, see page 66 of

[3] Julian Sochocki (1842–1927).
[4] Josip Plemelj (1873–1967).

Ref. [26])

$$\frac{1}{E_n - (E_m - \mathrm{i}\,\epsilon) - k} = \frac{1}{E_n - E_m - k + \mathrm{i}\,\epsilon} = (\mathrm{P.V.})\frac{1}{E_n - E_m - k} - \mathrm{i}\,\pi\,\delta(E_n - E_m - k)\,, \tag{4.356}$$

which essentially describes how the poles along the k-integration contour are to be encircled. If $E_n - E_m > 0$, i.e., if the virtual state has a lower energy than the reference state, then the Dirac-δ "fires" and a nonvanishing imaginary part is found. Extracting the imaginary part, we have

$$\begin{aligned} \mathrm{Im}\,\Delta E_{\mathrm{LEP}} &= -\frac{2\alpha}{3}\left(\frac{\mu}{m}\right)^2 \sum_m \int_0^\infty \mathrm{d}k\,k\,\left|\left\langle \phi_n \left| \frac{\vec{p}}{\mu} \right| \phi_m \right\rangle\right|^2 \pi\,\delta(E_n - E_m - k) \\ &= -\frac{2\alpha}{3}\left(\frac{\mu}{m}\right)^2 \sum_{E_m < E_n} (E_n - E_m)\left|\left\langle \phi_n \left| \frac{\vec{p}}{\mu} \right| \phi_m \right\rangle\right|^2 \\ &= -\frac{2\alpha}{3}\left(\frac{\mu}{m}\right)^2 \sum_{E_m < E_n} (E_n - E_m)^3 \left|\langle \phi_n | \vec{r} | \phi_m \rangle\right|^2 . \end{aligned} \tag{4.357}$$

We have used the formula $\vec{p} = \mathrm{i}\,\mu\,[H_S, \vec{r}]$ in going to the last line of the equation. We can extract from Eq. (4.357) the following formula for the partial decay width Γ for the decay from $\phi_n \to \phi_m$ as

$$\Gamma(\phi_n \to \phi_m) = \frac{4\alpha}{3}\left(\frac{\mu}{m}\right)^2 (E_n - E_m)^3 \left|\langle \phi_m | \vec{r} | \phi_n \rangle\right|^2 . \tag{4.358}$$

This result is in full agreement with Eq. (3.37), obtained using Fermi's golden rule. Still, the result (4.358) is useful because it clarifies the reduced-mass dependence of the decay rate. We recall that in deriving Eq. (4.358), we have used natural units with $\hbar = c = \epsilon_0 = 1$.

For a $2P$ reference state, the $2P \to 1S$ transition is the only possible (dipole) decay channel, and the result for hydrogen ($Z = 1$, $\mu \approx m$) reads as

$$\Gamma(2P \to 1S) = \frac{2^8}{3^8}\,\alpha\,(Z\alpha)^4\,m\,, \tag{4.359}$$

corresponding to a natural line width of about $\Gamma/h = 99.76\,\mathrm{MHz}$, measured in cycles per second (not radians per second). Expressed in radians per second, the width is $\Gamma(2P \to 1S)/\hbar = 2\pi \times 99.76\,\mathrm{MHz} \approx 2\pi \times 100\,\mathrm{MHz} = 2\pi \times 10^8\,\mathrm{rad/s}$. It is a very instructive exercise to verify this result by an analytic calculation, using the wave functions given in Sec. 4.2.2. The generalization to other transitions is straightforward, but requires us to remember that the magnetic quantum numbers of the final state need to be summed over, whereas the initial state may be assumed to have a definite magnetic projection.

The concept of the decay rate as the imaginary part of the self-energy has originally been developed in Ref. [181]. The calculation can be generalized to the fourth-order ("two-loop") self-energy, whose imaginary part gives the two-photon decay width, or, to the coherent two-photon correction to the one-photon decay width (the latter is relevant if cascade one-photon transitions are intertwined with the genuine two-photon correction to the decay width). This is described in Refs. [182, 183, 196, 197].

4.7.2 *Magnetic Interactions and Decay Rates*

As a complement to the rather well-known derivation of the imaginary part of the bound-electron self-energy (discussed in the preceding Sec. 4.7.1), we shall now consider a generalization which is easy enough to derive in a concise calculation but sophisticated enough to lead to a nontrivial result. Our calculation concerns a conceptually rather interesting effect, which is as yet unobserved in hydrogen (but see Ref. [198] for boronlike argon). Namely, we examine the magnetic one-photon $2P_{3/2}$–$2P_{1/2}$ decay rate. The motivation is as follows. If one adds the spin to the nonrelativistic Schrödinger–Coulomb problem, and relativistic effects, then the $2P_{3/2}$ and $2P_{1/2}$ levels, which differ only in the total angular momentum quantum number (but carry orbital angular momentum $\ell = 1$), are energetically displaced from each other by the amount

$$\Delta E_{\rm FS} = E(2P_{3/2}) - E(2P_{1/2}) \approx \frac{(Z\alpha)^4 m}{32}, \tag{4.360}$$

which is known as the fine-structure interval $\Delta E_{\rm FS}$. In a five-electron (boron-like) atomic system, provided the $1S$ and $2S$ states are all occupied, one can excite the single $2p_{1/2}$ electron into the $2p_{3/2}$ state. (Throughout this book, the state of a single electron in a multi-electron system is denoted by a lowercase letter, whereas the total quantum number is denoted by an uppercase letter. This means that if we are calculating for a single electron system, we are using uppercase letters.) Ignoring the interelectron interaction, the decay rate for the $2p_{3/2}$ electron to make a transition to the $2p_{1/2}$ state is then equal to the partial decay rate of the single-electron $2P_{3/2}$ state to the $2P_{1/2}$ state, calculated using the approach to be outlined below which is based on calculating the imaginary part of the self-energy, appropriately modified to include magnetic interactions.

We start from the following expression for the (unrenormalized) electric-dipole contribution to the self-energy,

$$\begin{aligned}\Delta E &= \frac{\alpha}{\pi}\int \mathrm{d}k\,k \int \frac{\mathrm{d}\Omega}{4\pi}\left(\delta^{ij} - \frac{k^i k^j}{\vec{k}^2}\right)\left\langle \phi_n \left| \frac{p^i}{m}\,\frac{1}{E - H_S - k}\,\frac{p^j}{m}\right| \phi_n \right\rangle \\ &= \frac{2\alpha}{3\pi}\int \mathrm{d}k\,k \left\langle \phi_n \left| \frac{p^i}{m}\,\frac{1}{E - H_S - k}\,\frac{p^i}{m}\right| \phi_n \right\rangle .\end{aligned} \tag{4.361}$$

As usual, the mass renormalization is irrelevant for the imaginary part. The first line in Eq. (4.361) highlights the fact that the angular integration needs to be considered separately and in fact, turns out to be different for the magnetic interactions; so we have to start the calculation from a term that includes the transverse photon propagator.

We can easily infer the contribution to the self-energy by the coupling of the electron's spin to the magnetic field. The p^i operators in Eq. (4.361) are coupled to the vector potential, to give the interaction Hamiltonian $-e\,\vec{A}\cdot\vec{p}/m$ (in the dipole approximation, where $\vec{A}$ and $\vec{p}$ commute). The interaction of the electron spin with the magnetic field is given by [see also Eqs. (11.54) and (11.83) below]

$$\begin{aligned}H_{\rm MAG} &= -g\,\mu_e\,\frac{\sigma}{2}\cdot\vec{B} = -2\left(\frac{e}{2m}\right)\left(\frac{\vec{\sigma}}{2}\right)\cdot\vec{B}\\ &= -\left(\frac{e}{2m}\right)\vec{\sigma}\cdot\vec{B} = -\left(\frac{e}{2m}\right)\vec{\sigma}\cdot(\vec{\nabla}\times\vec{A}) = -\left(\frac{e}{2m}\right)(\vec{\sigma}\times\vec{\nabla})\cdot\vec{A}\,. \end{aligned}\tag{4.362}$$

Here, $g \approx 2$ is the spin g factor, $\mu_e = e/(2m)$ is a quantity whose magnitude is equal to the Bohr magneton, $\vec{\sigma}/2$ is the spin operator where $\vec{\sigma}$ is the vector of Pauli matrices, and $\vec{B} = \vec{\nabla}\times\vec{A}$ is the magnetic field operator, which is equal to the curl of the quantized vector field operator $\vec{A}$. In momentum space, one replaces $\vec{\nabla} \to \mathrm{i}\,\vec{k}$. Therefore, the spin-coupling Hamiltonian (4.362) is obtained from the dipole Hamiltonian $-e\,\vec{A}\cdot\vec{p}/m$ by replacing the momentum operator in the interaction Hamiltonian as follows,

$$\frac{p^i}{m} \to \frac{\mathrm{i}}{2m}\left(\vec{\sigma}\times\vec{k}\right)^i \tag{4.363}$$

for photon annihilation, and

$$\frac{p^i}{m} \to \frac{\mathrm{i}}{2m}\left(\vec{\sigma}\times(-\vec{k})\right)^i \tag{4.364}$$

for photon creation (the self-energy integral involves both processes). The magnetic contribution to the self-energy thus reads

$$\begin{aligned}\Delta E_{\rm MAG} &= \frac{\alpha}{\pi}\int \mathrm{d}k k \int \frac{\mathrm{d}\Omega}{4\pi}\,\delta^{\mathrm{T},ij}\left\langle\phi_n\left|\frac{(\vec{\sigma}\times\vec{k})^i}{2m}\,\frac{1}{E-H_S-k}\,\frac{(\vec{\sigma}\times\vec{k})^j}{2m}\right|\phi_n\right\rangle\\ &= \frac{\alpha}{\pi}\int \mathrm{d}k\, k \int \frac{\mathrm{d}\Omega}{4\pi}\left\langle\phi_n\left|\frac{(\vec{\sigma}\times\vec{k})^i}{2m}\,\frac{1}{E-H_S-k}\,\frac{(\vec{\sigma}\times\vec{k})^i}{2m}\right|\phi_n\right\rangle\end{aligned}\tag{4.365}$$

where $\delta^{\mathrm{T},ij} = \delta^{ij} - k^i\,k^j/\vec{k}^2$. Here, in contrast to the dipole contribution, the term proportional to $k^i\,k^j$ in the transverse delta function vanishes after multiplication with $(\vec{\sigma}\times\vec{k})^i\,(\vec{\sigma}\times\vec{k})^j$. The angular integral over the photon momentum can be done easily and leads to the expression

$$\int\frac{\mathrm{d}\Omega}{4\pi}\,(\vec{\sigma}_1\times\vec{k})^i\,(\vec{\sigma}_2\times\vec{k})^i = \int\frac{\mathrm{d}\Omega}{4\pi}\,\left(\vec{\sigma}_1\cdot\vec{\sigma}_2\,\vec{k}\cdot\vec{k} - \vec{\sigma}_1\cdot\vec{k}\,\vec{\sigma}_2\cdot\vec{k}\right) = \frac{2}{3}\,\vec{\sigma}_1\cdot\vec{\sigma}_2\,\vec{k}^{\,2}\,, \tag{4.366}$$

where we temporarily distinguish the spin matrices by indices 1 and 2. So, the magnetic part of the self-energy becomes

$$\Delta E_{\rm MAG} = \frac{2\alpha}{3\pi}\int \mathrm{d}k\,k^3\left\langle\phi_n\left|\frac{\vec{\sigma}}{2m}\,\frac{1}{E-H_S-k}\,\frac{\vec{\sigma}}{2m}\right|\phi_n\right\rangle. \tag{4.367}$$

Based on the formalism developed in Sec. 4.7.1, we can thus easily infer the magnetic decay rate as

$$\begin{aligned}\Gamma(2P_{3/2}\to 2P_{1/2}) &= \frac{4\alpha}{3}\left[E(2P_{3/2}) - E(2P_{1/2})\right]^3\\ &\times\sum_{i=1}^{3}\sum_{m=\pm 1/2}\left|\left\langle 2P_{1/2}, j_z = m\left|\frac{\sigma^i}{2m}\right|2P_{3/2}\right\rangle\right|^2 = \frac{\alpha\,(Z\alpha)^{12}\,m}{2^{13}\times 3^2}.\end{aligned}\tag{4.368}$$

We have summed over the magnetic projection quantum numbers j_z of the final state. The result is of course independent of the magnetic projection of the initial state. It evaluates to a width of $\Gamma(2P_{3/2} \to 2P_{1/2}) \approx Z^{12} \times 2.788 \times 10^{-13}$ Hz, which seems very tiny. Let us assume, though, that we are working, say, with boronlike argon, with a nominal charge number of 18 for the nucleus. The effective charge number for the outermost P electron in boronlike argon is $Z \approx 14$. The magnetic width for $\Gamma(2P_{3/2} \to 2P_{1/2})$ then evaluates to 15.81 Hz, which is already nonnegligible. It means that the outer $2p_{3/2}$ electron returns to the $2p_{1/2}$ state in less than a second, due to magnetic decay. The width, expressed in radians per second, is equal to the width in Hertz, multiplied by a factor 2π. The inverse of that quantity is the lifetime τ of the level. Interestingly, our nonrelativistic calculation, which give $\tau = 10.1$ ms, already is in quite good agreement with the experimental result of $\tau = 9.573(4)(5)$ ms(stat)(sys) (see Ref. [198]), even if electron correlation and relativistic effects are not included in our treatment.

4.7.3 *Higher-Order Terms*

In our derivation of the low-energy part of the self-energy, given in Eq. (4.316), we have ignored, quite explicitly, any terms of order $\alpha(Z\alpha)^5\mu$ and higher. However, given today's experimental accuracy in high-precision spectroscopy, it is actually imperative to calculate self-energy effects at least up to order $\alpha(Z\alpha)^6\mu$, i.e., two orders of $Z\alpha$ higher than given by Eq. (4.316). The nonrelativistic dipole contribution calculated in Eq. (4.316) generates only a few higher-order terms which are important. Still, we recall the derivation in Eq. (4.316) and start from the third line,

$$\begin{aligned}
\Delta E'_{\mathrm{LEP}} &= \frac{2\,\alpha}{3\,\pi} \int_0^{\epsilon} \mathrm{d}k \left\langle \phi_n \left| \frac{p^i}{m} \frac{H_S - E_n}{H_S - E_n + k} \frac{p^i}{m} \right| \phi_n \right\rangle \\
&= \frac{2\,\alpha}{3\,\pi} \left\langle \phi_n \left| \frac{p^i}{m} (H_S - E_n) \ln\left(\frac{|\epsilon + H_S - E_n|}{|H_S - E_n|} \right) \frac{p^i}{m} \right| \phi_n \right\rangle \\
&= \frac{2\alpha}{3\pi} \left\langle \phi_n \left| \frac{p^i}{m} (H_S - E_n) \ln\left(2\,\frac{\epsilon + H_S - E_n}{(Z\alpha)^2\mu} \right) \frac{p^i}{m} \right| \phi_n \right\rangle \\
&\quad - \frac{2\alpha}{3\pi} \left\langle \phi_n \left| \frac{p^i}{m} (H_S - E_n) \ln\left(2\,\frac{|H_S - E_n|}{(Z\alpha)^2\mu} \right) \frac{p^i}{m} \right| \phi_n \right\rangle .
\end{aligned} \tag{4.369}$$

Here, we replace $|\epsilon + H_S - E_n| \to \epsilon + H_S - E_n$ which is valid because we implicitly assume that ϵ is positive and much larger than the atomic binding energy scale. With ΔE_{LEP} of order $\alpha\,(Z\alpha)^4\mu$ being precisely equal to the term given in Eq. (4.325), we now have

$$\begin{aligned}
\Delta E'_{\rm LEP} &= \frac{2\alpha}{3\pi} \left\langle \phi_n \left| \frac{p^i}{m} (H_S - E_n) \ln\left(1 + \frac{H_S - E_n}{\epsilon}\right) \frac{p^i}{m} \right| \phi_n \right\rangle + \Delta E_{\rm LEP} \\
&= \frac{2\alpha}{3\pi} \left(\frac{\mu}{m}\right)^2 \left\langle \phi_n \left| \frac{p^i}{\mu} \frac{(H_S - E_n)^2}{\epsilon} \frac{p^i}{\mu} \right| \phi_n \right\rangle + \Delta E_{\rm LEP} \\
&= \frac{2\alpha}{3\pi\,\epsilon} \left\langle \phi_n \left| \left[\frac{p^i}{m}, H_S - E_n\right] \left[H_S - E_n, \frac{p^i}{m}\right] \right| \phi_n \right\rangle + \Delta E_{\rm LEP} \\
&= \frac{2\alpha}{3\pi\, m^2\, \epsilon} \left\langle \phi_n \left| \frac{(Z\alpha)^2}{r^4} \right| \phi_n \right\rangle + \Delta E_{\rm LEP}\,.
\end{aligned}$$

We implicitly assume that all matrix elements converge, and ignore terms of order $\alpha\,(Z\alpha)^8\, m$ and higher. Based on the $Z\alpha$-scaling given in Eq. (4.324), we can easily find the following order-of-magnitude estimate,

$$\left\langle \phi_n \left| \frac{(Z\alpha)^2}{r^4} \right| \phi_n \right\rangle \sim (Z\alpha)^6\, \mu^4\,, \tag{4.370a}$$

$$\frac{\alpha}{m^2\,\epsilon} \left\langle \phi_n \left| \frac{(Z\alpha)^2}{r^4} \right| \phi_n \right\rangle \sim \left(\alpha\,(Z\alpha)^6\, m\right) \left(\frac{\mu}{m}\right)^4 \frac{m}{\epsilon}\,. \tag{4.370b}$$

This energy shift is manifestly of order $\alpha\,(Z\alpha)^6\, m$. With the help of Eq. (4.346), we have

$$\left\langle \phi_n \left| \frac{(Z\alpha)^2}{r^4} \right| \phi_n \right\rangle = (Z\alpha)^6\, \mu^4\, \frac{4\,[3\,n^2 - \ell\,(\ell+1)]}{n^5\, \ell\,(\ell+1)\,(2\ell-1)\,(2\ell+1)\,(2\ell+3)}\,. \tag{4.371}$$

For a general hydrogenic bound state, the correction term therefore reads as

$$\begin{aligned}
\Delta E_{\rm CORR} &= \Delta E'_{\rm LEP} - \Delta E_{\rm LEP} \qquad (4.372)\\
&= \frac{2\alpha}{3\pi\, m^2\, \epsilon}\, (Z\alpha)^6\, \mu^4\, \frac{4\,[3\,n^2 - \ell\,(\ell+1)]}{n^5\, \ell\,(\ell+1)\,(2\ell-1)\,(2\ell+1)\,(2\ell+3)} \\
&= \frac{\alpha}{\pi} \frac{(Z\alpha)^6\, m}{n^3} \left(\frac{\mu}{m}\right)^4 \left(\frac{m}{\epsilon}\right) \frac{8\,[3\,n^2 - \ell\,(\ell+1)]}{3n^2\, \ell\,(\ell+1)\,(2\ell-1)\,(2\ell+1)\,(2\ell+3)}\,.
\end{aligned}$$

For a reference state with nonvanishing angular momentum $\ell \neq 0$, this matrix element is convergent and well-defined. The $1/r^4$ operator from the expansion of the nonrelativistic dipole part is canceled by a two-vertex scattering amplitude in the effective approach to the high-energy part [199].

For a reference S state, the isolated ℓ term in the denominator of (4.372) means that the matrix element is divergent. Indeed, for S states, if one expands the argument of the logarithm naively, then even the leading term of $\ln[1 + (H - E)/\epsilon]$ already generates a divergence. In this case, the $Z\alpha$-expansion changes, and the emergence of a divergent matrix element in the order $\alpha(Z\alpha)^6 m$ implies that a lower-order term has been unlawfully neglected in the calculation. We shall not dwell on this problem any further here but remark that the corresponding term in the order $\alpha(Z\alpha)^5 m$ can be found based on an explicit integration, using the t variable defined in Eq. (4.131) and using the result (4.151) for the dynamic polarizability $P(2S, t)$. This is detailed in Ref. [163].

4.8 Further Thoughts

Here are some suggestions for further thought.

(1) **Center-of-Mass System.** Consider Eq. (4.1), where $\vec{p}$ denotes the momentum operator for the relative coordinate. For a general two-particle system (see also Chap. 16), define

$$\vec{p}_r = \xi_2 \, \vec{p}_1 - \xi_1 \vec{p}_2 \,, \qquad \xi_1 = \frac{m_1}{m_1 + m_2} \,, \qquad \xi_2 = \frac{m_2}{m_1 + m_2} \,. \tag{4.373}$$

Define the total momentum

$$\vec{P} = \vec{p}_1 + \vec{p}_2 \,. \tag{4.374}$$

Show, in anticipation of Secs. 13.2.2 and 16.2.1, that

$$\frac{\vec{p}_1^{\,2}}{2m_1} + \frac{\vec{p}_2^{\,2}}{2m_2} = \frac{\vec{p}_r^{\,2}}{2\mu} + \frac{\vec{P}^2}{2M} \,, \tag{4.375}$$

where

$$\mu = \frac{m_1 \, m_2}{m_1 + m_2} \,, \qquad M = m_1 + m_2 \,. \tag{4.376}$$

In the center-of-mass system, the momenta add up to zero,

$$\vec{0} = \vec{p}_1 + \vec{p}_2 \,. \tag{4.377}$$

Show that, in the center-of-mass system, one has

$$\vec{p} = \vec{p}_1 = -\vec{p}_2 = \vec{p}_r \,, \qquad \vec{P} = \vec{0} \,, \tag{4.378}$$

where $\vec{p}_r$ is defined in Eq. (4.373).

(2) **Hydrogen Ground State.** Attempt to determine the hydrogen ground state wave function from a simple guess,

$$\phi(\vec{r}) = A \, \mathrm{e}^{-\beta \, r} \,. \tag{4.379}$$

Start from Eq. (4.28), for $\ell = 0$,

$$\left[-\frac{1}{2\mu} \left(\frac{\mathrm{d}^2}{\mathrm{d}r^2} + \frac{2}{r} \frac{\mathrm{d}}{\mathrm{d}r} \right) - \frac{Z\alpha}{r} - E \right] \phi(\vec{r}) = 0 \,, \tag{4.380}$$

keeping E as a free parameter. By comparing coefficients multiplying different powers of r, obtain the equations

$$-E - \frac{\beta^2}{2\mu} = 0 \,, \qquad -Z\alpha + \beta = 0 \,, \tag{4.381}$$

which are solved by $\beta = Z\alpha$, and $E = -(Z\alpha)^2 \mu/2$. Finally, determine A by a normalization condition and verify the consistency with the general formulas (4.59) and (4.60) for $n = 1$ and $\ell = 0$.

(3) **Commutators.** Attempt to show that

$$\langle \vec{r} \cdot \vec{p} \rangle = \frac{\mathrm{i}}{2} \tag{4.382}$$

for any eigenstate of the Schrödinger–Coulomb Hamiltonian given in Eq. (4.1). *Hint:* Start from the commutator $[H_S - E, \vec{r}^2]$ and consider its expectation value in a Schrödinger–Coulomb eigenstate.

(4) **Angular Momentum Algebra.** Verify Eqs. (4.5) and (4.6).

(5) **Hydrogen and Degeneracies.** Try to obtain the degeneracies of the hydrogen spectrum on the basis of algebraic manipulations, based on the considerations presented in Sec. 4.2. *Hint:* In view of Eq. (4.16b), you have two rotation groups with $i = k$. Try to count the number of eigenstates, which are given by the magnetic projections for given angular momentum $i = k$, with $\vec{I}^2 = i\,(i+1)$ $\vec{K}^2 = k\,(k+1)$. Counting the projections, attempt to verify that the eigenvalues must be $[n^2 = (2k+1)^2]$-fold degenerate.

(6) **Radial Wave Function.** Complete the missing steps in the derivation of Eq. (4.73).

(7) **Normalization of Continuum States.** Calculate the energy-normalized solutions of the free Schrödinger equation for a hydrogen atom (continuum states).

(8) **Free Green Function.** Consider the Green function of the free Schrödinger equation, which satisfies

$$(H_0 - E)\; G(\vec{r}_1, \vec{r}_2, E) = \delta^{(3)}(\vec{r}_1 - \vec{r}_2)\,, \qquad H_0 = -\frac{\vec{\nabla}^2}{2m}\,. \tag{4.383}$$

This equation can immediately be reformulated as

$$\left(\vec{\nabla}^2 + 2\,m\,E\right)\, G(\vec{r}_1, \vec{r}_2, E) = -2m\,\delta^{(3)}(\vec{r}_1 - \vec{r}_2)\,. \tag{4.384}$$

Show the equivalence of Eq. (4.383) to the Helmholtz Green function [see Eq. (5.19) of Ref. [81]], with the identifications $k^2 = 2mE$ and the replacement $2m \to 1/\epsilon_0$ in the prefactor on the right-hand side. With reference to Eq. (5.20) of Ref. [81], show that the causality requirement implies that in the exponential $\exp(\mathrm{i}\,k\,|\vec{r} - \vec{r}\,'|)$, one has to choose $\mathrm{Im}(k) = \mathrm{Im}(\sqrt{2mE}) \geq 0$. Convince yourself that this is equivalent to choosing the branch cut of the square root along the positive real axis.

(9) **Self-Energy: An Inspection.** Consider once more Eq. (4.315). Verify the consistency of the imaginary part of the bound-electron self-energy with the result (3.37). Then, clarify the relation of the (one-particle) self-energy to the two-particle effect which is the basis of the low-energy part of the relativistic recoil correction, as discussed in Secs. 15.5.2, 16.3.6, and 17.3.2.

(10) **Derivation of the Decay Width.** Show that the omitted terms in Eq. (4.355) do not contribute to the imaginary part of the self-energy operator. Then, use Eq. (4.356) to derive Eq. (4.357).

(11) **Green Function for a Constant Magnetic Field.** Convince yourself that you can write the Green function of the Schrödinger equation, using the so-called Schwinger representation, as

$$G(\vec{r}_1, \vec{r}_2, E) = \left\langle \vec{r}_1 \left| \frac{1}{E - H} \right| \vec{r}_2 \right\rangle = -\sum_n \phi_n(\vec{r}_1)\, \phi_n^*(\vec{r}_2) \int_0^\infty \mathrm{e}^{-s(E_n - E)} \tag{4.385}$$

where n is a multi-index that sums over the energy eigenstates $\phi(\vec{r})$. Consider the Hamiltonian for an electron in a constant magnetic field, directed along the z axis,

$$H = \frac{1}{2m} \left[(p_x - eB\,y)^2 + p_y^2 + p_z^2 \right], \tag{4.386}$$

where the p_x, p_y and p_z are the Cartesian components of the momentum operator. Construct the eigenstates according to the *ansatz*

$$\psi(x, y, z) = \mathrm{e}^{\mathrm{i}(p_x\, x + p_z\, z)}\, \chi(y), \tag{4.387}$$

where now, p_x and p_z are the eigenvalues of the momentum operator, i.e., numbers, and we have set $\hbar = 1$. Show that the eigenvalue equation in the y-direction is

$$\left[\frac{p_y^2}{2m} + \frac{1}{2} m\, \omega^2\, (y - y_0)^2 \right] \chi(y) = \left(E - \frac{p_z^2}{2m} \right) \chi(y) \tag{4.388}$$

where $\omega = eB/m$ is the classical cyclotron frequency. Furthermore, show that the energy eigenvalues are

$$E_{n,p_x,p_z} = \left(n + \frac{1}{2} \right) \hbar\omega + \frac{p_z^2}{2m}, \tag{4.389}$$

with the corresponding eigenfunctions

$$\psi_{n,p_x,p_z}(x, y, z) = \mathrm{e}^{\mathrm{i}(p_x\, x + p_y\, y)}\, \chi_n(y), \tag{4.390a}$$

$$\chi_n(y) = \frac{1}{\sqrt{2^n\, n!}} \left(\frac{m\omega}{\pi\hbar} \right)^{1/4} \exp\left(-\frac{m\omega^2 y^2}{2\hbar} \right) H_n\left(\sqrt{\frac{m\omega}{\hbar}}\, (y - y_0) \right). \tag{4.390b}$$

Here, $y_0 = p_x/(eB)$, n is a discrete quantum number, while p_x and p_z are continuous quantum numbers. Of course, the H_n denote the well-known Hermite polynomials (see Chap. 22 of Ref. [159]). Note that p_x does not figure in the energy eigenvalue, but still needs to be included if we are to sum over all available eigenstates. Carry out the sum over the discrete quantum numbers using Mehler's formula

$$\sum_{n=0}^{\infty} \frac{H_n(y_1)\, H_n(y_2)}{n!} \left(\tfrac{1}{2}\, w \right)^n = (1 - w^2)^{1/2} \exp\left[\frac{2 y_1\, y_2\, w - (y_1^2 + y_2^2)\, w^2}{1 - w^2} \right], \tag{4.391}$$

calculate the Green function using Eq. (4.385), and simplify the result as much as possible.

Chapter 5

Interatomic and Atom-Surface Interactions

5.1 Overview

We wish to explore some more applications of the formalism developed in Chap. 3, which concerns the operators of the quantized electromagnetic field. Using this formalism, it is possible to describe a number of physical phenomena and interactions that involve atoms, including interatomic interactions. Typically, a physical description of these effects necessitates the use of higher orders in perturbation theory. We proceed as follows: In Sec. 5.2, we develop the general formulas for time-independent, higher-order perturbation theory, which can be applied to interactions involving quantized field operators. The derivation of the third Friar moment correction (see Ref. [200]) is included as a nontrivial application of the second-order formalism (see Sec. 5.3). Fourth-order perturbation theory leads to an understanding of van der Waals and Casimir–Polder interactions (see Sec. 5.4) which involve the exchange of virtual photons among neutral atoms. As a complement, we discuss atom-surface interactions in Sec. 5.5. This chapter illustrates both the use of second-order perturbation theory as well as the use of higher-order perturbation theory, using quantized-field operators, in nontrivial and practically important applications.

5.2 Wigner–Brioullin Perturbation Theory

5.2.1 *Spectral Representation of the Green Function*

In preceding chapters, we have already used the spectral decomposition of the Green function in terms of eigenfunctions of the underlying quantum operator. Let us now try to put things on a more secure footing and discuss the spectral decomposition of an unperturbed "atomic" Hamiltonian H_0, which involves the eigenstates of the atom,

$$H_0 = \sum_m E_m^{(0)} |\psi_m^{(0)}\rangle \langle\psi_m^{(0)}| , \tag{5.1}$$

where the superscript (0) indicates the unperturbed character of the state. We view the Hamiltonian as acting on an infinite-dimensional Hilbert space. The unperturbed energies $E_m^{(0)}$ of the unperturbed eigenstates $|\phi_m^{(0)}\rangle$ enter the spectral decomposition (5.1) of the Hamiltonian, and the summation over m is meant to include an integration over the continuous part of the spectrum.

We start from the trivial operator identity

$$(z - H_0)\,\frac{\mathbb{1}}{z - H_0} = \mathbb{1}\,, \tag{5.2}$$

where we denote the unit operator by $\mathbb{1}$, and z is a complex number. Let us project (5.2) onto a basis of unperturbed eigenstates,

$$\sum_m \langle \phi_k^{(0)}|\,(z - H_0)\,|\phi_m^{(0)}\rangle \left\langle \phi_m^{(0)} \left| \frac{1}{z - H_0} \right| \phi_n^{(0)} \right\rangle = \delta_{kn}\,. \tag{5.3}$$

We now evaluate the first matrix element using the relation $H_0\,|\phi_m^{(0)}\rangle = E_m^{(0)}\,|\phi_m^{(0)}\rangle$ and obtain

$$\sum_m (z - E_m^{(0)})\,\delta_{km}\,G_{mn} = \delta_{kn}\,, \qquad G_{mn} \equiv \left\langle \phi_m^{(0)} \left| \frac{1}{z - H_0} \right| \phi_n^{(0)} \right\rangle. \tag{5.4}$$

If Eq. (5.4) is to be fulfilled for any k and n, then we have to have

$$G_{mn} = \frac{\delta_{mn}}{z - E_m^{(0)}}\,, \tag{5.5}$$

and thus the spectral decomposition of the Green function is found as

$$\frac{1}{z - H_0} = \sum_{m,n} |\phi_m^{(0)}\rangle\,G_{mn}\,\langle \phi_n^{(0)}| = \sum_m \frac{|\phi_m^{(0)}\rangle\,\langle \phi_m^{(0)}|}{z - E_m^{(0)}}\,, \tag{5.6}$$

where the right-hand side is referred to as the spectral decomposition of the Green function $1/(z - H_0)$. Of special importance is the reduced Green function, which for a reference state $|\phi_n^{(0)}\rangle$ reads as

$$\left(\frac{1}{E_n^{(0)} - H_0}\right)' = \sum_{m \neq n} \frac{|\phi_m^{(0)}\rangle\,\langle \phi_m^{(0)}|}{E_n^{(0)} - E_m^{(0)}}\,, \tag{5.7}$$

where the reference state is excluded in the spectral decomposition. We note that if H_0 contains both an atomic as well as a field contribution, then the combined atom+field reference state needs to be excluded (see Chap. 4).

5.2.2 *First-Order and Second-Order Perturbation Theory*

For many applications, second-order perturbation theory is perfectly sufficient to describe the essential physics of a particular process, and we thus re-familiarize ourselves with second-order perturbation theory, while illustrating the difficulties that await us as we try to generalize the treatment to higher orders. Let us now suppose that the atom is subject to a time-independent perturbation, with a full Hamiltonian

$$H = H_0 + g\,H_1\,. \tag{5.8}$$

In zeroth-order, we have $H_0|\phi_n^{(0)}\rangle = E_n^{(0)}|\phi_n^{(0)}\rangle$ by assumption. We attempt to solve the eigenvalues equation $H|\phi_n\rangle = E_n|\phi_n\rangle$ using the following, self-evident *ansatz*,

$$(H_0 + g\,H_{\rm I})\,|\phi_n^{(0)} + g\,\phi_n^{(1)} + g^2\,\phi_n^{(2)}\rangle = \left(E_n^{(0)} + g\,E_n^{(1)} + g^2\,E_n^{(2)}\right)|\phi_n^{(0)} + g\,\phi_n^{(1)} + g^2\,\phi_n^{(2)}\rangle\,, \tag{5.9}$$

where we group the terms according to their order in perturbation theory, and keep terms up to second order. The extraction of terms of first order in g leads to the equation,

$$g\,H_{\rm I}\,|\phi_n^{(0)}\rangle + g\,H_0\,|\phi_n^{(1)}\rangle = g\,E_n^{(1)}\,|\phi_n^{(0)}\rangle + g\,E_n^{(0)}\,|\phi_n^{(1)}\rangle\,. \tag{5.10}$$

We now project onto the state $\langle\phi_n^{(0)}|$, assuming that $\langle\phi_n^{(0)}|\phi_n^{(1)}\rangle = 0$ in view of the conservation of the norm of the eigenstate, and obtain

$$E_n^{(1)} = \langle\phi_n^{(0)}|\,H_1\,|\phi_n^{(0)}\rangle\,. \tag{5.11}$$

We now project (5.10) onto $\langle\phi_m^{(0)}|$ with $m \neq n$,

$$g\,\langle\phi_m^{(0)}|H_1\,|\phi_n^{(0)}\rangle + g\,E_m^{(0)}\,\langle\phi_m^{(0)}|\phi_n^{(1)}\rangle = g\,E_n^{(0)}\,\langle\phi_m^{(0)}|\phi_n^{(1)}\rangle\,. \tag{5.12}$$

One cancels g and obtains

$$\langle\phi_m^{(0)}|\phi_n^{(1)}\rangle = \frac{1}{E_n^{(0)} - E_m^{(0)}}\,\langle\phi_m^{(0)}|H_1\,|\phi_n^{(0)}\rangle\,. \tag{5.13}$$

Now one uses the completeness relation for the spectrum, to obtain for $|\phi_n^{(1)}\rangle = \sum_{m\neq n}|\phi_m^{(0)}\rangle\,\langle\phi_m^{(0)}|\phi_n^{(1)}\rangle$ the result

$$|\phi_n^{(1)}\rangle = \sum_{m\neq n}\frac{1}{E_n^{(0)} - E_m^{(0)}}\,|\phi_m^{(0)}\rangle\,\langle\phi_m^{(0)}|H_1\,|\phi_n^{(0)}\rangle = \left(\frac{1}{E_n^{(0)} - H_0}\right)'\,H_1\,|\phi_n^{(0)}\rangle\,. \tag{5.14}$$

We now project (5.9) onto $\langle\phi_n^{(0)}|$ (assuming orthogonality the unperturbed wave function with respect to all perturbations) and extract the term of order g^2, to obtain

$$g^2\,\langle\phi_n^{(0)}|H_1|\phi_n^{(1)}\rangle = g^2\langle\phi_n^{(0)}|E_n^{(2)}|\phi_n^{(0)}\rangle = g^2\,E_n^{(2)}\,. \tag{5.15}$$

Finally, we have for the second-order perturbation

$$E_n^{(2)} = \langle\phi_n^{(0)}|H_1|\phi_n^{(1)}\rangle = \left\langle\phi_n^{(0)}\left|H_1\left(\frac{1}{E_n^{(0)} - H_0}\right)'H_1\right|\phi_n^{(0)}\right\rangle\,, \tag{5.16}$$

a result which we have already used in Sec. 3.3.2 and which is standard material in most quantum mechanics textbooks. However, based on the above considerations, it is already clear that it would be hard to generalize this treatment to arbitrary order, without tremendous effect. Indeed, the generalization is facilitated by the Wigner–Brioullin formula which is discussed in the following.

5.2.3 Higher-Order Perturbation Theory (Wigner–Brioullin)

The generalization of the above treatment to higher orders requires some rather sophisticated calculational tricks due to WIGNER[1] and BRIOULLIN,[2] which will be described in the following. Again, we consider an unperturbed Hamiltonian H_0 and a perturbation H_1, with unperturbed eigenstates $|\phi_n^{(0)}\rangle$ of H_0,

$$H_0|\phi_n^{(0)}\rangle = E_n^{(0)}|\phi_n^{(0)}\rangle . \tag{5.17}$$

The resolvent operator is

$$R(E) = \frac{1}{H_0 + H_1 - E} . \tag{5.18}$$

It fulfills the following relation, which holds to all orders in perturbation theory,

$$R(E) = \frac{1}{H_0 - E} - \frac{1}{H_0 - E} H_1 R(E) , \tag{5.19}$$

as can be verified immediately by multiplying both sides with $(H_0 - E)$. One might now assume that the expansion

$$R(E) = \frac{1}{H_0 - E} - \frac{1}{H_0 - E} H_1 \frac{1}{H_0 - E} + \frac{1}{H_0 - E} H_1 \frac{1}{H_0 - E} H_1 \frac{1}{H_0 - E} - \cdots \tag{5.20}$$

eventually leads to formulas useful within the context of high-order perturbation theory. However, quite surprisingly, this route of calculation leads to nowhere. Rather, we choose a different route and derive independent relations where, figuratively speaking, nonperturbative factors of $R(E)$ are manifestly kept on the right-hand side of the equation.

Let us define the diagonal element of the resolvent operator of the reference state as follows,

$$D_n(E) \equiv \langle\phi_n^{(0)}|R(E)|\phi_n^{(0)}\rangle . \tag{5.21}$$

In view of (5.19), this matrix element fulfills the relation

$$\begin{aligned} D_n(E) = {} & \frac{1}{E_n^{(0)} - E} - \frac{1}{E_n^{(0)} - E} \langle\phi_n^{(0)}|H_1|\phi_n^{(0)}\rangle \, D_n(E) \\ & - \frac{1}{E_n^{(0)} - E} \sum_{m \neq n} \langle\phi_n^{(0)}|H_1|\phi_m^{(0)}\rangle \, \langle\phi_m^{(0)}|R(E)|\phi_n^{(0)}\rangle . \end{aligned} \tag{5.22}$$

The (trivial) separation of the summation into the reference state term, and further terms, will become useful in further manipulations. Similarly, we can write an arbitrary off-diagonal matrix element of the resolvent operator as

$$\begin{aligned} \langle\phi_m^{(0)}|R(E)|\phi_n^{(0)}\rangle = {} & - \frac{1}{E_m^{(0)} - E} \langle\phi_m^{(0)}|H_1|\phi_n^{(0)}\rangle \, D_n(E) \\ & - \frac{1}{E_m^{(0)} - E} \sum_{r \neq n} \langle\phi_m^{(0)}|H_1|\phi_r^{(0)}\rangle \, \langle\phi_r^{(0)}|R(E)|\phi_n^{(0)}\rangle , \qquad m \neq n . \end{aligned} \tag{5.23}$$

[1]Eugene Paul Wigner (1902–1995).
[2]Léon Nicolas Brioullin (1889–1969).

We observe that the matrix element $\langle \phi_r^{(0)} | R(E) | \phi_n^{(0)} \rangle$ is of the same structure as the matrix element on the left-hand side, $\langle \phi_m^{(0)} | R(E) | \phi_n^{(0)} \rangle$, and we can therefore iterate Eq. (5.23) to write

$$\begin{aligned}\langle \phi_m^{(0)} | R(E) | \phi_n^{(0)} \rangle = &-\frac{1}{E_m^{(0)} - E} \langle \phi_m^{(0)} | H_1 | \phi_n^{(0)} \rangle D_n(E) - \frac{1}{E_m^{(0)} - E} \sum_{r \neq n} \langle \phi_m^{(0)} | H_1 | \phi_r^{(0)} \rangle \\ &\times \left(-\frac{1}{E_r^{(0)} - E} \langle \phi_r^{(0)} | H_1 | \phi_n^{(0)} \rangle D_n(E) \right. \\ &\left. - \frac{1}{E_r^{(0)} - E} \sum_{s \neq n} \langle \phi_r^{(0)} | H_1 | \phi_s^{(0)} \rangle \langle \phi_s^{(0)} | R(E) | \phi_n^{(0)} \rangle \right). \end{aligned} \tag{5.24}$$

Using the completeness of the spectrum, this equation can be written as

$$\begin{aligned}\langle \phi_m^{(0)} | R(E) | \phi_n^{(0)} \rangle = &\frac{1}{E_m^{(0)} - E} \left\langle \phi_m^{(0)} \left| -H_1 + \left(H_1 \frac{1}{H_0 - E} H_1 \right)' \right| \phi_n^{(0)} \right\rangle D_n(E) \\ &+ \frac{1}{E_m^{(0)} - E} \sum_{s \neq n} \left\langle \phi_m^{(0)} \left| \left(H_1 \frac{1}{H_0 - E} H_1 \right)' \right| \phi_s^{(0)} \right\rangle \langle \phi_s^{(0)} | R(E) | \phi_n^{(0)} \rangle . \end{aligned} \tag{5.25}$$

The prime in the matrix element denotes an operator from whose spectral decomposition we exclude the reference state with quantum number n. We can now again recursively apply Eq. (5.23) to the rightmost matrix element and obtain an iterated form, which has the desired structure,

$$\begin{aligned}\langle \phi_m^{(0)} | R(E) | \phi_n^{(0)} \rangle = &\frac{1}{E_m^{(0)} - E} \left\langle \phi_m^{(0)} \left| \left(-H_1 + H_1 \frac{1}{H_0 - E} H_1 \right. \right. \right. \\ &\left. \left. \left. - H_1 \frac{1}{H_0 - E} H_1 \frac{1}{H_0 - E} H_1 + \cdots \right)' \right| \phi_n^{(0)} \right\rangle D_n(E). \end{aligned} \tag{5.26}$$

The relation (5.22) enables us to write the matrix element $D_n(E)$ in terms of $\langle \phi_m^{(0)} | R(E) | \phi_n^{(0)} \rangle$, but then we can use Eq. (5.26) to substitute for $\langle \phi_m^{(0)} | R(E) | \phi_n^{(0)} \rangle$, which brings us back to $D_n(E)$. The resulting equation for $D_n(E)$ is

$$\begin{aligned} D_n(E) = &\frac{1}{E_n^{(0)} - E} + \frac{1}{E_n^{(0)} - E} \left\langle \phi_n^{(0)} \left| \left(-H_1 + H_1 \frac{1}{H_0 - E} H_1 \right. \right. \right. \\ &\left. \left. \left. - H_1 \frac{1}{H_0 - E} H_1 \frac{1}{H_0 - E} H_1 + \cdots \right)' \right| \phi_n^{(0)} \right\rangle D_n(E), \end{aligned}$$

so that

$$D_n(E) = \frac{1}{E_n^{(0)} - E} + \frac{1}{E_n^{(0)} - E} F_n(E) D_n(E), \tag{5.27a}$$

$$D_n(E) = \frac{1}{E_n^{(0)} - F_n(E) - E}, \tag{5.27b}$$

where

$$F_n(E) = \left\langle \phi_n^{(0)} \left| \left(-H_1 + H_1 \frac{1}{H_0 - E} H_1 - H_1 \frac{1}{H_0 - E} H_1 \frac{1}{H_0 - E} H_1 + \ldots \right)' \right| \phi_n^{(0)} \right\rangle . \tag{5.28}$$

The condition $E = E_n^{(0)} - F_n(E)$ for a pole of the resolvent becomes

$$E = E_n^{(0)} + \left\langle \phi_n^{(0)} \middle| H_1 + H_1 \left(\frac{1}{E - H_0} \right)' H_1 \right.$$
$$\left. + H_1 \left(\frac{1}{E - H_0} \right)' H_1 \left(\frac{1}{E - H_0} \right)' H_1 + \cdots \middle| \phi_n^{(0)} \right\rangle . \tag{5.29}$$

Observe that we have reversed the sign of the propagator denominators by writing all terms with a $1/(E - H_0)$ propagator instead to $1/(H_0 - E)$. There is a fine, but important difference between the argument E in $1/(E - H_0)$ as compared to the zero-order energy eigenvalue $E_n^{(0)}$. This distinction means that the argument E can only be determined iteratively, by evaluating lower orders of perturbation theory, and then using the result in the expressions relevant for the higher orders. One generally refers to Eq. (5.29) as the Wigner–Brioullin formula.

The various orders of perturbation theory can now be read off from Eq. (5.29), starting from the replacement $H_1 \to g\, H_1$, which introduces a coupling parameter g, and from the formal expansion,

$$E = E_n^{(0)} + g\, E_n^{(1)} + g^2\, E_n^{(2)} + g^3\, E_n^{(3)} + \cdots , \tag{5.30}$$

equating powers of g. We reproduce in first and second order, respectively,

$$E_n^{(1)} = \left\langle \phi_n^{(0)} \left| H_1 \right| \phi_n^{(0)} \right\rangle ,$$
$$E_n^{(2)} = \left\langle \phi_n^{(0)} \middle| H_1 \left(\frac{1}{E_n^{(0)} - H_0} \right)' H_1 \middle| \phi_n^{(0)} \right\rangle . \tag{5.31}$$

In third order, there is a derivative term obtained by expanding the exact energy E in the propagator denominator of the second-order matrix element, in powers of g. It reads

$$g^2 \left\langle \phi_n^{(0)} \middle| H_1 \frac{1}{E_n^{(0)} + g\, E_n^{(1)} - H_0} H_1 \middle| \phi_n^{(0)} \right\rangle \Bigg|_{\mathcal{O}(g^3)} = g^3 E_n^{(1)} \frac{\partial}{\partial E_n^{(0)}} E_n^{(2)} . \tag{5.32}$$

Here, the extraction of the term of order g^3 is denoted by the appropriate symbol. The final result in third order reads

$$E_n^{(3)} = \left\langle \phi_n^{(0)} \middle| H_1 \left(\frac{1}{E_n^{(0)} - H_0} \right)' H_1 \left(\frac{1}{E_n^{(0)} - H_0} \right)' H_1 \middle| \phi_n^{(0)} \right\rangle$$
$$- \left\langle \phi_n^{(0)} \left| H_1 \right| \phi_n^{(0)} \right\rangle \left\langle \phi_n^{(0)} \middle| H_1 \left[\left(\frac{1}{E_n^{(0)} - H_0} \right)' \right]^2 H_1 \middle| \phi_n^{(0)} \right\rangle . \tag{5.33}$$

In fourth order, many more terms come in; they read

$$
\begin{aligned}
E_n^{(4)} &= \left\langle \phi_n^{(0)} \middle| H_1 \left(\frac{1}{E_n^{(0)} - H_0} \right)' H_1 \left(\frac{1}{E_n^{(0)} - H_0} \right)' H_1 \left(\frac{1}{E_n^{(0)} - H_0} \right)' H_1 \middle| \phi_n^{(0)} \right\rangle \\
&- \left\langle \phi_n^{(0)} \middle| H_1 \left[\left(\frac{1}{E_n^{(0)} - H_0} \right)' \right]^2 H_1 \middle| \phi_n^{(0)} \right\rangle \left\langle \phi_n^{(0)} \middle| H_1 \left(\frac{1}{E_n^{(0)} - H_0} \right)' H_1 \middle| \phi_n^{(0)} \right\rangle \\
&+ \left\langle \phi_n^{(0)} \middle| H_1 \left[\left(\frac{1}{E_n^{(0)} - H_0} \right)' \right]^3 H_1 \middle| \phi_n^{(0)} \right\rangle \left(\langle \phi_n^{(0)} | H_1 | \phi_n^{(0)} \rangle \right)^2 \\
&- 2 \left\langle \phi_n^{(0)} \middle| H_1 \left(\frac{1}{E_n^{(0)} - H_0} \right)' H_1 \left[\left(\frac{1}{E_n^{(0)} - H_0} \right)' \right]^2 H_1 \middle| \phi_n^{(0)} \right\rangle \langle \phi_n^{(0)} | H_1 | \phi_n^{(0)} \rangle \,.
\end{aligned}
\tag{5.34}
$$

In higher orders, one has to include derivative terms stemming from lower orders of perturbation theory, by expansion of the energy parameter E in the propagator denominators.

Two remarks are in order. *(i)* Suppose all perturbations vanish up to the order g^k, i.e., we have $E_n^{(j)} = 0$ for all $j < k$. In this case, the leading-order effect is encountered in the kth order of perturbation theory and reads

$$
E_n^{(k)} = \left\langle \phi_n^{(0)} \middle| \underbrace{H_1 \left(\frac{1}{E_n^{(0)} - H_0} \right)' \cdots H_1 \left(\frac{1}{E_n^{(0)} - H_0} \right)'}_{k-1 \text{ times}} H_1 \middle| \phi_n^{(0)} \right\rangle . \tag{5.35}
$$

(ii) The first order of perturbation theory in which derivative terms play a role is the third, where in the propagator denominator we have to use the first-order perturbation theory expression $E_n^{(1)}$. In general, for the kth total order of perturbation theory, we need to insert the energy up to order $k-2$ into the propagator denominators (two powers of H_1 "flank" the propagator and are obtained "for free"). This is computationally feasible in our calculational scheme. We iteratively advance to higher orders. By the time we are evaluating order k, we have already treated order $k-1$, and of course $k-1 > k-2$, so that the result to order $k-2$ is available once we arrive at order k.

5.3 Application to the Finite-Size Effect

5.3.1 *Master Integrals*

The evaluation of the finite-size correction, including two-photon exchange, is a nontrivial exercise, and we shall devote a brief discussion to some important momentum-space integrals which occur in its evaluation. During this discussion, we shall foreshadow ideas discussed in Secs. 10.3.4 and 16.3.2. In fact, we shall first discuss, from a somewhat abstract point of view, the master integrals necessary for the calculation, and then apply them to the calculation of the finite-size correction.

First, we mention the following Fourier transformations,

$$\int \mathrm{d}^3r\, \mathrm{e}^{\mathrm{i}\vec{q}\cdot\vec{r}}\, r^3 = \frac{96\pi}{\vec{q}^{\,6}}\,, \qquad \int \mathrm{d}^3r\, \mathrm{e}^{\mathrm{i}\vec{q}\cdot\vec{r}}\, r = -\frac{8\pi}{\vec{q}^{\,4}}\,, \qquad \int \mathrm{d}^3r\, \mathrm{e}^{\mathrm{i}\vec{q}\cdot\vec{r}}\, \frac{1}{r} = \frac{4\pi}{\vec{q}^{\,2}}\,. \tag{5.36}$$

The proof is left as an exercise to the reader. Of course, we imply and use the corresponding formulas under Fourier backtransformation. Our first master integral is

$$\begin{aligned} J(\vec{s},\vec{u}) &= \int \mathrm{d}^3r\, \frac{|\vec{r}-\vec{s}|}{|\vec{r}-\vec{u}|} = \int \mathrm{d}^3r \int \frac{\mathrm{d}^3q}{(2\pi)^3} \left\{-\frac{8\pi}{\vec{q}^{\,4}}\, \mathrm{e}^{\mathrm{i}\vec{q}\cdot(\vec{r}-\vec{s})}\right\} \int \frac{\mathrm{d}^3k}{(2\pi)^3} \left\{\frac{4\pi}{\vec{k}^{\,2}}\, \mathrm{e}^{\mathrm{i}\vec{k}\cdot(\vec{r}-\vec{u})}\right\} \\ &= \int \frac{\mathrm{d}^3q}{(2\pi)^3} \int \frac{\mathrm{d}^3k}{(2\pi)^3} \left\{-\frac{32\pi^2}{\vec{q}^{\,4}\,\vec{k}^{\,2}}\right\} \mathrm{e}^{-\mathrm{i}\vec{q}\cdot\vec{s}}\, \mathrm{e}^{\mathrm{i}\vec{k}\cdot\vec{u}} \underbrace{\int \mathrm{d}^3r\, \mathrm{e}^{\mathrm{i}(\vec{k}+\vec{q})\cdot\vec{r}}}_{=(2\pi)^3\,\delta^{(3)}(\vec{k}+\vec{q})} \\ &= \frac{\pi}{3} \int \frac{\mathrm{d}^3q}{(2\pi)^3} \left\{-\frac{96\pi}{\vec{q}^{\,6}}\right\} \mathrm{e}^{\mathrm{i}\vec{q}\cdot(\vec{u}-\vec{s})} = -\frac{\pi}{3}\, |\vec{u}-\vec{s}|^3\,. \end{aligned} \tag{5.37}$$

This result, in particular, means that we need to assign a zero value to the integral $J(\vec{s}=\vec{0},\vec{u}=\vec{0})$,

$$J(\vec{0},\vec{0}) = \int \mathrm{d}^3r\, \frac{|\vec{r}-\vec{0}|}{|\vec{r}-\vec{0}|} = \int \mathrm{d}^3r\, 1 = 0\,. \tag{5.38}$$

This result is consistent with a vanishing value for any scaleless integral in dimensional regularization. A particularly clear exposition of this fact, together with further explanatory remarks, is given in Chap. 8 of Ref. [201] [to be specific, we refer to the text following Eq. (8.52) of Ref. [201]]. Special cases of the integral J area

$$J(\vec{s},\vec{0}) = \int \mathrm{d}^3r\, \frac{|\vec{r}-\vec{s}|}{|\vec{r}|} = -\frac{\pi}{3}\, |\vec{s}|^3\,, \qquad J(\vec{0},\vec{u}) = \int \mathrm{d}^3r\, \frac{r}{|\vec{r}-\vec{u}|} = -\frac{\pi}{3}\, |\vec{u}|^3\,. \tag{5.39}$$

Our second master integral is

$$\begin{aligned} K(\vec{s},\vec{u}) &= \int \mathrm{d}^3r\, \frac{1}{|\vec{r}-\vec{s}|\,|\vec{r}-\vec{u}|} = \int \mathrm{d}^3r \int \frac{\mathrm{d}^3q}{(2\pi)^3}\, \frac{4\pi}{\vec{q}^{\,2}}\, \mathrm{e}^{\mathrm{i}\vec{q}\cdot(\vec{r}-\vec{s})} \int \frac{\mathrm{d}^3k}{(2\pi)^3}\, \frac{4\pi}{\vec{k}^{\,2}}\, \mathrm{e}^{\mathrm{i}\vec{k}\cdot(\vec{r}-\vec{u})} \\ &= \int \frac{\mathrm{d}^3q}{(2\pi)^3} \int \frac{\mathrm{d}^3k}{(2\pi)^3}\, \frac{16\pi^2}{\vec{q}^{\,2}\,\vec{k}^{\,2}} \int \mathrm{d}^3r\, \mathrm{e}^{\mathrm{i}(\vec{k}+\vec{q})\cdot\vec{r}}\, \mathrm{e}^{-\mathrm{i}\vec{q}\cdot\vec{s}-\mathrm{i}\vec{k}\cdot\vec{u}} \\ &= \int \frac{\mathrm{d}^3q}{(2\pi)^3}\, \frac{16\pi^2}{\vec{q}^{\,4}}\, \mathrm{e}^{\mathrm{i}\vec{q}\cdot(\vec{u}-\vec{s})} = -2\pi \int \frac{\mathrm{d}^3q}{(2\pi)^3} \left(-\frac{8\pi}{\vec{q}^{\,4}}\right) \mathrm{e}^{\mathrm{i}\vec{q}\cdot(\vec{u}-\vec{s})} = -2\pi\, |\vec{u}-\vec{s}|\,. \end{aligned} \tag{5.40}$$

The third integral, which is evaluated in the limit $\lambda \to 0$, reads as follows,

$$L(\lambda,\vec{u}) = \int \mathrm{d}^3r\, \frac{\mathrm{e}^{-\lambda r}}{|\vec{r}-\vec{u}|} = \frac{4\pi}{\lambda^2} - \frac{2\pi}{3}\, \vec{u}^{\,2} + \frac{\pi}{3}\, \lambda\, |\vec{u}|^3 + \mathcal{O}(\lambda^2)\,. \tag{5.41}$$

The details of the proof are left as an exercise to the reader. Finally, we observe that, for hard momenta, we can approximate the Schrödinger–Coulomb Green function $[1/(E-H_S)']$ in momentum space by the free Green function $-2\mu/\vec{k}^{\,2}$, where μ is the reduced mass of the two-body bound system. After Fourier transformation into coordinate space, one has

$$G_0(\vec{r},\vec{s}) = -\frac{2\mu}{4\pi|\vec{r}-\vec{s}|}\,. \tag{5.42}$$

With these results, we are now in the position to evaluate the two-Coulomb-photon exchange contribution to the finite-size effect.

5.3.2 First-Order and Second-Order Finite-Size Effect

Let us now assume a charge distribution of the nucleus of the functional form $f(\vec{u})$, which is normalized to unity. The perturbation δV of the Coulomb potential due to the finite-size effect can be expressed as the difference of the exact electrostatic potential due to the extended nuclear charge distribution, and the Coulomb potential,

$$\delta V(\vec{r}) = -Z\alpha \int \mathrm{d}^3 u\, f(\vec{u}) \left(\frac{1}{|\vec{r}-\vec{u}|} - \frac{1}{r}\right), \qquad \int \mathrm{d}^3 u\, f(\vec{u}) = 1\,. \tag{5.43}$$

The evaluation of the first-order energy shift involves the master integral L, and we carry out the calculation in great detail, assuming that $\phi(\vec{r})$ is an S state with vanishing orbital angular momentum,

$$\begin{aligned}
\Delta^{(1)}E = \langle\phi|\delta V|\phi\rangle &= \int \mathrm{d}^3 r\, |\phi(\vec{r})|^2 \left\{-Z\alpha \int \mathrm{d}^3 u\, f(\vec{u}) \left(\frac{1}{|\vec{r}-\vec{u}|} - \frac{1}{r}\right)\right\} \\
&\approx -Z\alpha\, |\phi(\vec{0})|^2 \int \mathrm{d}^3 u\, f(\vec{u}) \int \mathrm{d}^3 r\, \mathrm{e}^{-2Z\alpha\mu r} \left(\frac{1}{|\vec{r}-\vec{u}|} - \frac{1}{r}\right) \\
&= -Z\alpha\, |\phi(\vec{0})|^2 \int \mathrm{d}^3 u\, f(\vec{u}) \left\{L(2Z\alpha\,\mu, \vec{u}) - L(2Z\alpha\,\mu, \vec{0})\right\} \\
&= \frac{2}{3}\frac{(Z\alpha)^4\mu^3}{\pi\, n^3}\,\delta_{\ell 0} \int \mathrm{d}^3 u\, f(\vec{u})\, \vec{u}^{\,2} - \frac{2}{3}\frac{(Z\alpha)^5\mu^4}{n^3}\,\delta_{\ell 0} \int \mathrm{d}^3 u\, f(\vec{u})\, |\vec{u}|^3\,.
\end{aligned} \tag{5.44}$$

We have used the fact that

$$\left|\phi(\vec{r}=\vec{0})\right|^2 = \frac{(Z\alpha)^3\mu^3}{\pi\, n^3}\,\delta_{\ell 0}\,, \tag{5.45}$$

and the universal behavior of the slope of the hydrogenic S-state wave functions near the origin. The evaluation of the correction $\delta\phi(\vec{r})$ to the wave function involves the master integral K,

$$\begin{aligned}
\delta\phi(\vec{r}) &= \int \mathrm{d}^3 s\, G_0(\vec{r},\vec{s})\, \delta V(\vec{s})\, \phi(\vec{s}) \\
&= \frac{2\mu Z\alpha}{4\pi} \int \mathrm{d}^3 u\, f(\vec{u}) \int \mathrm{d}^3 s \left(\frac{1}{|\vec{r}-\vec{s}|\,|\vec{s}-\vec{u}|} - \frac{1}{s\,|\vec{r}-\vec{s}|}\right) \phi(\vec{0}) \\
&= \frac{2\mu Z\alpha}{4\pi} \int \mathrm{d}^3 u\, f(\vec{u}) \left[K(\vec{r},\vec{u}) - K(\vec{r},\vec{0})\right] \phi(\vec{0}) \\
&= Z\alpha\mu \int \mathrm{d}^3 u\, f(\vec{u})\, (-|\vec{r}-\vec{u}| + r)\, \phi(\vec{0})\,.
\end{aligned} \tag{5.46}$$

The second-order energy shift, due to the finite-size correction to two-Coulomb-photon exchange, involves the master integral J given in Eq. (5.37), is given as

$$\begin{aligned}
\Delta^{(2)} E &= \langle \phi | \delta V | \delta\phi \rangle = \int \mathrm{d}^3 r \, \phi^*(\vec{r}) \, \delta V(\vec{r}) \, \delta\phi(\vec{r}) \\
&= (Z\alpha)^2 \mu |\phi(\vec{0})|^2 \int \mathrm{d}^3 s \int \mathrm{d}^3 u \int \mathrm{d}^3 r f(\vec{s}) \left(\frac{|\vec{r}-\vec{u}|}{|\vec{r}-\vec{s}|} - \frac{|\vec{r}-\vec{u}|}{r} - \frac{r}{|\vec{r}-\vec{s}|} + 1 \right) f(\vec{u}) \\
&= (Z\alpha)^2 \mu |\phi(\vec{0})|^2 \int \mathrm{d}^3 s \int \mathrm{d}^3 u f(\vec{s}) \left[J(\vec{s},\vec{u}) - J(\vec{u},\vec{0}) - J(\vec{0},\vec{s}) + J(\vec{0},\vec{0}) \right] f(\vec{u}) \\
&= \frac{2}{3} \frac{(Z\alpha)^5 \mu^4}{n^3} \delta_{\ell 0} \int \mathrm{d}^3 u \, f(\vec{u}) \, |\vec{u}|^3 \\
&\quad - \frac{1}{3} \frac{(Z\alpha)^5 \mu^4}{n^3} \delta_{\ell 0} \int \mathrm{d}^3 s \int \mathrm{d}^3 u \, f(\vec{s}) \, |\vec{u}-\vec{s}|^3 \, f(\vec{u}) \,.
\end{aligned} \tag{5.47}$$

The addition of the first-order and second-order energy shifts [see Eqs. (5.44) and (5.47)] leads to the cancelation of the term proportional to $\int \mathrm{d}^3 u \, f(\vec{u}) \, |\vec{u}|^3$ and yields the result

$$\begin{aligned}
\Delta^{(1)} E + \Delta^{(2)} E &= \frac{2}{3} \frac{(Z\alpha)^4 \mu^3}{\pi n^3} \delta_{\ell 0} \int \mathrm{d}^3 u f(\vec{u}) \vec{u}^{\,2} - \frac{1}{3} \frac{(Z\alpha)^5 \mu^4}{n^3} \delta_{\ell 0} \int \mathrm{d}^3 \chi |\vec{\chi}|^3 f^{(2)}(\vec{\chi}) \\
&= \frac{2}{3} \frac{(Z\alpha)^4 \mu^3}{\pi n^3} \left\langle r^2 \right\rangle - \frac{1}{3} \frac{(Z\alpha)^5 \mu^4}{n^3} \left\langle r^3 \right\rangle_{(2)} \,.
\end{aligned} \tag{5.48}$$

We have used the following definitions of the convoluted charge density $f^{(2)}(\vec{\chi})$ and the Friar moment $\left\langle r^3 \right\rangle_{(2)}$,

$$f^{(2)}(\vec{\chi}) = \int \mathrm{d}^3 u \, f(\vec{u}) \, f(\vec{\chi}-\vec{u}) \,, \qquad \left\langle r^3 \right\rangle_{(2)} = \int \mathrm{d}^3 \chi |\vec{\chi}|^3 f^{(2)}(\vec{\chi}) \,. \tag{5.49}$$

We have assumed that $f(\vec{r}) = f(-\vec{r})$. The third moment $\langle r^3 \rangle_{(2)}$ has been obtained, in the context of the finite-size correction, in Eqs. (21c) and (66) of Ref. [200], and is therefore commonly referred to as the "Friar moment". It is to be distinguished from the Zemach moment, which involves a combination of the electric and magnetic charge distributions [202].

The leading-order effect in $\Delta^{(1)} E$, proportional to $\int \mathrm{d}^3 u \, f(\vec{u}) \, \vec{u}^{\,2}$, is mainly responsible for the well-known "proton size discrepancy" reported in Refs. [203, 204] for muonic hydrogen is described by formula (5.44). The root-mean-square radius r_p of the proton (nucleus) is defined as

$$r_p = \sqrt{\langle r^2 \rangle} \,, \qquad \langle r^2 \rangle = \int \mathrm{d}^3 u \, f(\vec{u}) \, \vec{u}^{\,2} \,, \tag{5.50}$$

where f is the charge distribution inside the proton.

We have given the correction to the Coulomb potential due to the finite size of the nucleus exclusively in terms of the second moment of the electric charge distribution. In reality, the analysis is a bit more complicated (see Ref. [205]), and our form factor f qualifies itself as the electric G_E Sachs form factor of the nucleus. In momentum space, the third Friar moment can be expressed in terms of the electric G_E Sachs form factor as

$$\langle r^3 \rangle_{(2)} = \frac{48}{\pi} \int_0^\infty \frac{\mathrm{d}q}{q^4} \left[G_E^2(q^2) - 1 + \frac{1}{3} q^2 \left\langle r^2 \right\rangle \right] , \tag{5.51}$$

and its evaluation based on scattering data depends on a careful analysis of both the region of small momentum transfer, where the subtraction terms lead to numerical cancelations, as well as on the region of large momentum transfer, where the slowly convergent large-q tail needs to be evaluated [206, 207].

5.4 Interatomic Interactions

5.4.1 *Origin of van der Waals and Casimir–Polder Interactions*

Intuitively, the origin of the van der Waals interaction between atoms can be understood as follows: Any atom is in itself polarized, in as much as the momentary position of the electron, when measured, necessarily differs from the position of the atomic nucleus. In a practical situation, the distance of the electrons from their respective nucleus will be much smaller than the interatomic separation, and a corresponding Taylor expansion of the electrostatic interactions then leads to the familiar $1/R^6$ van der Waals interaction, where R is the interatomic distance. We shall derive this interaction in the following.

It is often stated that the van der Waals interaction, when augmented by retardation, leads to the Casimir–Polder interaction, which is the long-distance limit of the van der Waals interaction and follows a $1/R^7$ law. We shall see that this statement is in need of some qualification; indeed, the retardation expansion is an expansion in powers of $1/c$ and may be applied to the expression $\exp(-\omega R/c)$, where R is the interatomic distance and ω is an angular frequency of a typical virtual atomic transition. For large R, one cannot expand the exponential, and the nontrivial frequency dependence (dependence on ω) is often referred to as the "retardation" effect.

From the point of view of quantized field operators, the van der Waals and Casimir–Polder interactions correspond to the exchange of two virtual photons. In Refs. [208, 209] and in Chap. 85 of Ref. [3], a derivation of the van der Waals and Casimir–Polder interactions is presented which is based on the matching of a two-photon scattering amplitude with an effective interaction potential. The derivation is based on the general idea that the Feynman propagator of the electromagnetic field incorporates both an advanced as well as a retarded interaction. In the language of so-called time-ordered perturbation theory, these two contributions would correspond to different time orderings of emission and absorption of the virtual photon. Here, we shall follow a different approach and consider, in fourth-order perturbation theory, the energy shift (interaction potential) corresponding to the exchange of two virtual photons between atomic systems, using Schrödinger-picture field operators. This approach outlines, in a very intuitive manner, an application of the formalism developed in Sec. 3.3.2 in higher orders of perturbation theory.

Paradoxically, the result obtained using time-independent field operators (Schrödinger picture) is commonly referred to as a result obtained using time-ordered perturbation theory. In order to understand this apparent contradiction, we have to go back to our derivation of the dynamic Stark shift using the three ways of derivation described in Sec. 3.3. The net result obtained with quantized Schrödinger-picture operators, calculated in Sec. 3.3.2, was seen to be equal to the net result obtained with time-ordered, second-order perturbation theory using a classical field (Sec. 3.3.1). We found the use of time-independent Schrödinger-picture operators to be much more convenient than time-ordered second-order perturbation theory. Indeed, the result to be obtained below, based on quantized, time-independent, fourth-order perturbation theory, could otherwise be obtained alternatively using fourth-order time-dependent perturbation theory, with adiabatically damped field operators (for the distant past and distant future, as in the Gell-Mann–Low–Sucher theorem described in Sec. 3.3.3). We shall not describe the alternative derivation here in any further detail, except for mentioning that the correspondence of the time-independent, Schrödinger-picture perturbation theory, to be outlined below, to adiabatically damped fourth-order time-ordered perturbation theory, justifies the name "time-ordered perturbation theory" for the treatment outlined below, even if all field operators are manifestly time-independent. We note that Feynman diagrams in relativistic QED are understood to include all possible time orderings of the perturbatively treated interactions.

5.4.2 *Calculation of Interatomic Interactions*

Despite the enormous practical significance of interatomic interactions, the literature on interatomic interactions is rather scarce when detailed information is sought on manifestly non-asymptotic regimes, with an intermediate separation of the two atoms which neither corresponds to the van der Waals, nor to the Casimir–Polder limit. We hope to present a transparent and streamlined derivation in the following, which avoids the intricacies of the Feynman propagator formalism for the time being and illustrates the usefulness of plain, fourth-order perturbation theory involving quantized interactions with the photon field.

For simplicity, we consider the interatomic interaction potential between two (neutral) hydrogenlike atoms, even if the results to be derived below are actually more general. The unperturbed Hamiltonian reads as follows,

$$H_0 = \frac{\vec{p}_1^{\,2}}{2m_1} + V_A(\vec{r}_1) + \frac{\vec{p}_2^{\,2}}{2m_2} + V_B(\vec{r}_2) + H_F\,, \tag{5.52a}$$

$$V_A(\vec{r}_1) = -\frac{e^2}{4\pi\epsilon_0\,|\vec{r}_1 - \vec{R}_A|}\,, \qquad V_B(\vec{r}_2) = -\frac{e^2}{4\pi\epsilon_0\,|\vec{r}_2 - \vec{R}_B|}\,. \tag{5.52b}$$

In the derivation, we keep all factors of $\hbar$, ϵ_0 and c, because the end results will involve rather large integer powers of R, up to R^{-7}, and it is difficult to reinstate SI mksA units in the final result unless care is taken in keeping a full physical unit system in all intermediate steps.

We neglect reduced-mass corrections and write the Hamiltonian as the sum of the Schrödinger–Coulomb Hamiltonians of the two electrons whose masses we denote as

m_1 and m_2 (for definiteness), plus the electromagnetic field Hamiltonian H_F. If the nuclei have charge numbers other than unity, then the two-photon exchange energy calculated below is dominated by either the electrostatic interaction among the ions, or the ion-atom interaction. The coordinates of the two separated electrons 1 and 2 are denoted as $\vec{r}_1$ and $\vec{r}_2$, respectively, and the nuclei are at $\vec{R}_A$ and $\vec{R}_B$. If the electrons are close to their respective nuclei, then

$$|\vec{r}_1 - \vec{R}_A| \ll |\vec{r}_1 - \vec{R}_B|\,, \qquad |\vec{r}_2 - \vec{R}_B| \ll |\vec{r}_2 - \vec{R}_A|\,, \tag{5.53}$$

which justifies the approximation (5.52). The Hamiltonians governing the atoms effectively reduce to Schrödinger–Coulomb Hamiltonians near the nuclei of the two atoms. Because of the large separation of the two atoms, one can ignore so-called "exchange terms" due to the symmetrization of the wave function of the two electrons, and treat them as distinguishable particles. We shall therefore choose a simple product *ansatz* for the two-particle wave functions.

We can therefore write H_0 as the sum of two Schrödinger–Coulomb Hamiltonians H_A and H_B, and a field Hamiltonian H_F,

$$H_0 = H_A + H_B + H_F\,, \qquad H_F = \sum_{\lambda=1}^{2} \int \mathrm{d}^3k\, \hbar c k\, a^+_\lambda(\vec{k})\, a_\lambda(\vec{k})\,, \tag{5.54}$$

$$H_A = \frac{\vec{p}_1^{\,2}}{2m_1} - \frac{e^2}{4\pi\epsilon_0 |\vec{r}_1 - \vec{R}_A|}\,, \qquad H_B = \frac{\vec{p}_2^{\,2}}{2m_2} - \frac{e^2}{4\pi\epsilon_0 |\vec{r}_2 - \vec{R}_B|}\,. \tag{5.55}$$

We use the field Hamiltonian in the normal-ordered form given in Eq. (2.110), and we assume that $\omega_{\vec{k}} = ck$ with $k \equiv |\vec{k}|$, and we recall that the fundamental creation and annihilation operators fulfill the relations $[a_\lambda(\vec{k}), a^+_{\lambda'}(\vec{k}')] = \delta^{(3)}(\vec{k} - \vec{k}')\,\delta_{\lambda\lambda'}$ given in Eq. (2.70). We here use the continuum representation of the field operators (see Secs. 4.6.1 and 4.6.2).

The fundamental reference state is given as

$$|\phi_0\rangle = |1S, 1S, 0\rangle = |1S, A\rangle \otimes |1S, B\rangle \otimes |0\rangle\,, \tag{5.56}$$

which is a product state with both atoms A and B in the $1S$ ground state and no photons in the radiation field. This notation is to be understood as a simple product state (no symmetrization of the fermionic part of the wave function, because of the large separation of the two atoms). One electron is centered about $\vec{R}_A$, and the other centered about $\vec{R}_B$, the photon field is in the vacuum state, and so the coordinate-space representation of $|\phi\rangle$ reads as

$$\phi_0(\vec{r}_1, \vec{r}_2) = \phi_{1S}(\vec{r}_1 - \vec{R}_A)\,\phi_{1S}(\vec{r}_2 - \vec{R}_B)\,|0\rangle\,, \qquad H_0\,|\phi_0\rangle = (E_{1S,A} + E_{1S,B})\,|\phi_0\rangle\,, \tag{5.57}$$

where $|0\rangle$ is the fundamental state (vacuum) of the radiation field, and the electronic ground-state energies are $E_{1S,A} = -\alpha^2 m_1 c^2/2$, and $E_{1S,B} = -\alpha^2 m_2 c^2/2$.

From now on, we redefine r_1 to be the distance of electron 1 to nucleus A, and r_2 to be the distance of electron 2 to nucleus B. In the dipole approximation, we can write the interaction Hamiltonian as

$$H_1 = -e\vec{r}_1 \cdot \vec{E}(\vec{r}_1) - e\vec{r}_2 \cdot \vec{E}(\vec{r}_2) \approx -e\vec{r}_1 \cdot \vec{E}(\vec{R}_A) - e\vec{r}_2 \cdot \vec{E}(\vec{R}_B)\,. \tag{5.58}$$

This means that we neglect the phase accumulated by the virtual photons as they pass through the atoms (their wavelength is much longer than a Bohr radius), while keeping the phase accumulated along the travel paths in between the two atoms (the wavelength is assumed to be small against the separation $\vec{R}_A - \vec{R}_B$ of the atomic nuclei). We can finally write the electric field operator as

$$\vec{E}(\vec{R}_A) = \sum_{\lambda=1}^{2} \int \frac{\mathrm{d}^3k}{\sqrt{(2\pi)^3}} \sqrt{\frac{\hbar c k}{2\,\epsilon_0}}\, \hat{\epsilon}_\lambda(\vec{k}) \times \left[\mathrm{i}\, a_\lambda(\vec{k}) \exp\left(\mathrm{i}\vec{k}\cdot\vec{R}_A\right) - \mathrm{i}\, a^+_\lambda(\vec{k}) \exp\left(-\mathrm{i}\vec{k}\cdot\vec{R}_A\right)\right], \tag{5.59a}$$

$$\vec{E}(\vec{R}_B) = \sum_{\lambda=1}^{2} \int \frac{\mathrm{d}^3k}{\sqrt{(2\pi)^3}} \sqrt{\frac{\hbar c k}{2\,\epsilon_0}}\, \hat{\epsilon}_\lambda(\vec{k}) \times \left[\mathrm{i}\, a_\lambda(\vec{k}) \exp\left(\mathrm{i}\vec{k}\cdot\vec{R}_B\right) - \mathrm{i}\, a^+_\lambda(\vec{k}) \exp\left(-\mathrm{i}\vec{k}\cdot\vec{R}_B\right)\right]. \tag{5.59b}$$

This is valid for $\vec{k}\cdot\vec{r}_1 \approx \vec{k}\cdot\vec{R}_A$ and $\vec{k}\cdot\vec{r}_2 \approx \vec{k}\cdot\vec{R}_B$, i.e., for long-wavelength exchanged photons ($i = 1, 2$). The problem now is to determine at which order in perturbation theory the atom-atom interaction comes into play. We use the reference state $|\phi_0\rangle$ given in Eq. (5.56) and calculate energy perturbations. "After" the exchange of virtual photons, the combined system of the two electrons plus the radiation field should be in the same state as "before" the exchanges, with both atoms in the ground state, and no photons in the radiation field.

We shall now systematically investigate at which order of perturbation theory we encounter the first nonvanishing term for the interaction energy. At first order, we immediately have $\Delta E^{(1)} = \langle\phi_0|H_1|\phi_0\rangle = 0$, because under the exchange of only one photon, the atoms cannot return to their ground state (at least in the dipole approximation). For the exchange of photons between atoms, the phase factor $\exp[\mathrm{i}\vec{k}\cdot(\vec{R}_A - \vec{R}_B)]$ from the electric-field operator in Eq. (5.59) leads to a nontrivial dependence on the interatomic distance. However, if the photon is emitted and absorbed (created and annihilated) by the same atom, then the net phase factor is $\exp[\mathrm{i}\vec{k}\cdot(\vec{R}_i - \vec{R}_i)] = 1$ with $i = A, B$, and one obtains the same expressions as those encountered in Sec. 4.6.1 for the self-energy of a single atom. The second-order energy shift simply is a sum of two one-electron self-energies

$$\Delta E^{(2)} = \left\langle \phi_0 \left| H_1 \left(\frac{1}{E_0 - H_0}\right)' H_1 \right| \phi_0 \right\rangle = \Delta E_{\mathrm{LEP},1} + \Delta E_{\mathrm{LEP},2}\,, \tag{5.60}$$

where ΔE_{LEP} is the low-energy part of the self-energy as discussed in Sec. 4.6.1, for atoms 1 and 2. If the emission and absorption occur at the positions of different atoms, then the dipole character of the transition forbids each of the atoms to return to their ground state (the expectation value $\langle\phi_n|\vec{r}|\phi_n\rangle = \vec{0}$ if ϕ_n is a reference S state). The second-order term therefore is negligible in the context of the discussion of the Casimir–Polder interaction.

Third-order perturbation theory vanishes because of an odd number of exchanged photons, which also prohibits the return to the ground state because of the dipole character of the transition. So, we have $\Delta E^{(3)} = 0$. The first nonvanishing term for the interaction energy of the two atoms therefore is a fourth-order effect. We have derived a general expression for the fourth-order energy shift in Eq. (5.34). The mixed terms, and derivative terms listed in Eq. (5.34) also do not contribute to the atom-atom interaction, because they effectively contain lower-order terms in perturbation theory, for which we showed that the effect on the exchange energy is zero. What remains is the "plain" fourth-order exchange term [corresponding to the first term on the right-hand side of Eq. (5.34)], which reads as

$$\Delta E^{(4)} = \left\langle \phi_0 \left| H_1 \left(\frac{1}{E_0 - H_0} \right)' H_1 \left(\frac{1}{E_0 - H_0} \right)' H_1 \left(\frac{1}{E_0 - H_0} \right)' H_1 \right| \phi_0 \right\rangle . \qquad (5.61)$$

Even this fourth-order term still contains contributions which are not relevant to the current discussion of the interatomic interaction energy and must be subtracted. Namely, if one inserts the interaction Hamiltonian (5.58) into Eq. (5.61), then there will be two kinds of terms generated: those where the emission and absorption of the virtual photon happens at two different atoms, leading to a nontrivial phase factor $\exp[i\vec{k} \cdot (\vec{R}_A - \vec{R}_B)]$, and those where the emission and absorption is at the same atom, and the phase factor is unity. The latter kind leads to two energy shifts which effectively are part of the so-called two-photon self-energies of the two atoms. [In the case of the two-photon self-energy, one has to consider *all* terms on the right-hand side of Eq. (5.34); our expression (5.61) isolates the "plain" fourth-order term.] In the following, we only consider the terms of the first kind, i.e., those that include the nontrivial phase factor $\exp[i\vec{k} \cdot (\vec{R}_A - \vec{R}_B)]$. This procedure may seem unclear at first but it becomes absolutely transparent once we engage in the actual calculation.

According to Eq. (5.61), we have four interactions to consider. Starting from the rightmost state $|\phi_0\rangle$, one may use the rightmost two H_1's to emit photons. The sequence of emissions may be from atom A, then from B, or A after A, or B after A, or B after B. The atoms involved in the emission determine those that will have to absorb photons, and for each sequence of emissions we have two possible sequences of absorptions. However, one may also emit-absorb-emit-absorb two photons. This gives another $2 \times 2 = 4$ terms, as represented diagrammatically in diagrams numbered (3), (4), (9) and (10) in Fig. 5.1. All 12 contributions can be represented by the diagrams in Fig. 5.1.

Let us consider diagram (1) in Fig. 5.1 and attempt to isolate the terms that lead to the interaction energy. We have four factors, whose interpretation "in reverse chronological order" is as follows, *(i)* absorption of k_1 at $\vec{R}_A$, *(ii)* absorption of k_2 at $\vec{R}_A$, *(iii)* emission of k_1 at $\vec{R}_B$, and *(iv)* emission of k_2 at $\vec{R}_B$. With the electric-field operator given in Eq. (5.59) and the atom-field interaction given in Eq. (5.58), two imaginary units eventually cancel against the two factors $(-i)$. Including all

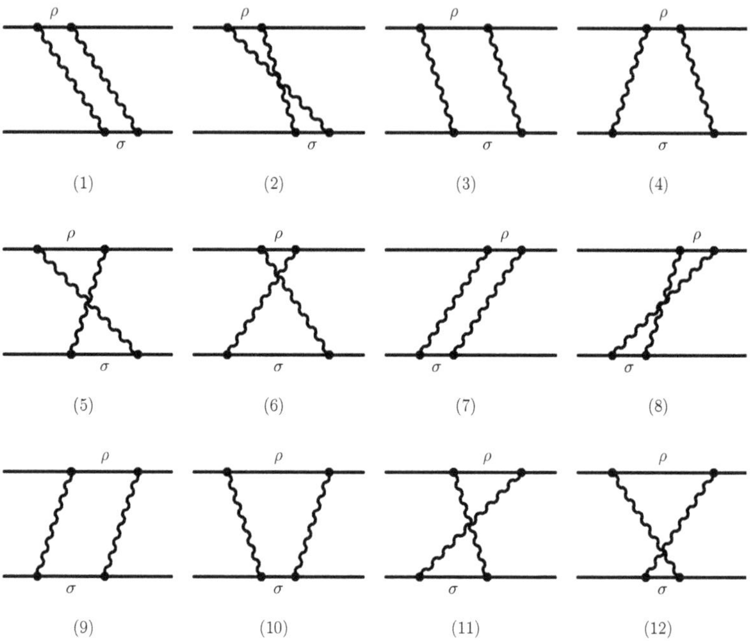

Fig. 5.1 Diagrams representing the Casimir–Polder interaction. Observe that the diagrams $7, \ldots, 12$ can be obtained from the diagrams $1, \ldots, 6$ by simply exchanging the two particles, $\rho \leftrightarrow \sigma$. This is equivalent to a reflection in a horizontal axis located half way between the electron lines. Diagrams (1), (2), (7) and (8) are dominant in the long-range limit, because the diagrams involve a virtual state with both atoms in the ground state, and two soft photons. By contrast, diagrams (3), (4), (9) and (10) are dominant in the short-range limit, because they involve a virtual state with both atoms being excited, but no (hard) virtual photons present.

prefactors, the energy shift corresponding to diagram (1) in Fig. 5.1 reads as

$$
\begin{aligned}
\Delta E_1 &= e^4 \int \frac{d^3 k_1}{(2\pi)^3} \int \frac{d^3 k_2}{(2\pi)^3} \sum_{\lambda_1,\lambda_2} \sum_{\rho,\sigma} \left(i\sqrt{\frac{\hbar c k_1}{2\,\epsilon_0}} \right) e^{i\,\vec{k}_1\cdot\vec{R}_A} \langle \phi_{1S,A} | \hat{\epsilon}_{\lambda_1}(\vec{k}_1) \cdot \vec{r}_1 | \rho \rangle \\
&\times \left(i\sqrt{\frac{\hbar c k_2}{2\,\epsilon_0}} \right) e^{i\vec{k}_2\cdot\vec{R}_A} \langle \rho | \hat{\epsilon}_{\lambda_2}(\vec{k}_2) \cdot \vec{r}_1 | \phi_{1S,A} \rangle \left(-i\sqrt{\frac{\hbar c k_1}{2\,\epsilon_0}} \right) e^{-i\vec{k}_1\cdot\vec{R}_B} \langle \phi_{1S,B} | \hat{\epsilon}_{\lambda_1}(\vec{k}_1) \cdot \vec{r}_2 | \sigma \rangle \\
&\times \left(-i\sqrt{\frac{\hbar c k_2}{2\,\epsilon_0}} \right) \frac{e^{-i\,\vec{k}_2\cdot\vec{R}_B} \langle \sigma | \hat{\epsilon}_{\lambda_2}(\vec{k}_2) \cdot \vec{r}_2 | \phi_{1S,B} \rangle}{(E_{1S,A} - E_\rho - \hbar c k_1)\,(-\hbar c k_1 - \hbar c k_2)\,(E_{1S,B} - E_\sigma - \hbar c k_2)} \,. \qquad (5.62)
\end{aligned}
$$

The virtual state ρ is associated with the upper electron line (atom A), whereas the virtual state σ is associated with the lower electron line (atom B). The two emissions are in the "σ" line (atom B), where atom A ("ρ" line) absorbs the photons. After summing over the photon polarization vectors, the expression (5.62) can be rewritten more compactly as ($\vec{R} \equiv \vec{R}_A - \vec{R}_B$)

$$\Delta E_1 = e^4 \int \frac{d^3k_1}{(2\pi)^3} \int \frac{d^3k_2}{(2\pi)^3} \frac{\hbar c k_1}{2\,\epsilon_0} \frac{\hbar c k_2}{2\,\epsilon_0} \left(\delta^{ik} - \frac{k_1^i\,k_1^k}{k_1^2}\right) \left(\delta^{j\ell} - \frac{k_2^j\,k_2^\ell}{k_2^2}\right) e^{i\,(\vec{k}_1+\vec{k}_2)\cdot\vec{R}}$$
$$\times \sum_{\rho,\sigma} \frac{\langle 1S, A|x_1^i|\rho\rangle\,\langle\rho|x_1^j|1S, A\rangle\,\langle 1S, B|x_2^k|\sigma\rangle\,\langle\sigma|x_2^\ell|1S, B\rangle}{(E_{1S,A} - E_\rho - \hbar c k_1)\,(-\hbar c k_1 - \hbar c k_2)\,(E_{1S,B} - E_\sigma - \hbar c k_2)}\,. \tag{5.63}$$

The conventions are that $\vec{k}_2$ is the momentum of the photon to the right, which is emitted "first" if one considers the diagrams from right to left, while $\vec{k}_1$ is the momentum of the photon to the left, which is the "second" emitted photon. The spatial superscripts i, j, k, and ℓ are summed over, following the Einstein summation convention. The energy shift corresponding to diagram (2) in Fig. 5.1 reads

$$\Delta E_2 = e^4 \int \frac{d^3k_1}{(2\pi)^3} \int \frac{d^3k_2}{(2\pi)^3} \frac{\hbar c k_1}{2\,\epsilon_0} \frac{\hbar c k_2}{2\,\epsilon_0} \left(\delta^{ik} - \frac{k_1^i\,k_1^k}{k_1^2}\right) \left(\delta^{j\ell} - \frac{k_2^j\,k_2^\ell}{k_2^2}\right) e^{i\,(\vec{k}_1+\vec{k}_2)\cdot\vec{R}}$$
$$\times \sum_{\rho,\sigma} \frac{\langle 1S, A|x_1^i|\rho\rangle\,\langle\rho|x_1^j|1S, A\rangle\,\langle 1S, B|x_2^k|\sigma\rangle\,\langle\sigma|x_2^\ell|1S, B\rangle}{(E_{1S,A} - E_\rho - \hbar c k_2)\,(-\hbar c k_1 - \hbar c k_2)\,(E_{1S,B} - E_\sigma - \hbar c k_2)}\,. \tag{5.64}$$

This expression is not identical to the one in Eq. (5.63); the difference can be found in the first term of the propagator denominator ($k_1 \to k_2$). In writing the expression (5.64), we use the convention that the leftmost photon absorption on the "A" line of the diagram carries an index i, the second, an index j. The energy shift corresponding to diagram (4) in Fig. 5.1 involves an absorption and an emission in both the ρ and the σ lines each, which leads to a factor $\exp[i\,(-\vec{k}_1 + \vec{k}_2)\cdot\vec{R}]$ instead of $\exp[i\,(\vec{k}_1 + \vec{k}_2)\cdot\vec{R}]$,

$$\Delta E_4 = e^4 \int \frac{d^3k_1}{(2\pi)^3} \int \frac{d^3k_2}{(2\pi)^3} \frac{\hbar c k_1}{2\,\epsilon_0} \frac{\hbar c k_2}{2\,\epsilon_0} \left(\delta^{ik} - \frac{k_1^i\,k_1^k}{k_1^2}\right) \left(\delta^{j\ell} - \frac{k_2^j\,k_2^\ell}{k_2^2}\right) e^{i\,(-\vec{k}_1+\vec{k}_2)\cdot\vec{R}}$$
$$\times \sum_{\rho,\sigma} \frac{\langle 1S, A|x_1^i|\rho\rangle\,\langle\rho|x_1^j|1S, A\rangle\,\langle 1S, B|x_2^k|\sigma\rangle\,\langle\sigma|x_2^\ell|1S, B\rangle}{(E_{1S,A} - E_\sigma - \hbar c k_1)\,(E_{1S,A} - E_\sigma + E_{1S,B} - E_\rho)\,(E_{1S,B} - E_\sigma - \hbar c k_2)}\,. \tag{5.65}$$

However, the substitution $\vec{k}_1 \to -\vec{k}_1$ under the integral sign brings this expression into a "standard form" involving the exponential $\exp[i\,(\vec{k}_1 + \vec{k}_2)\cdot\vec{R}]$. Similar operations thus permit a unified representation for all energy shifts due to diagrams (1)–(12). The complete energy shift due to two-photon exchange is the sum of twelve contributions from the diagrams in Fig. 5.1, $\Delta E^{(4)} = \sum_{i=1}^{12} \Delta E_i$,

$$\Delta E^{(4)} = \int \frac{d^3k_1}{(2\pi)^3} \int \frac{d^3k_2}{(2\pi)^3} \frac{\hbar c k_1}{2\,\epsilon_0} \frac{\hbar c k_2}{2\,\epsilon_0} \left(\delta^{ik} - \frac{k_1^i\,k_1^k}{k_1^2}\right) \left(\delta^{j\ell} - \frac{k_2^j\,k_2^\ell}{k_2^2}\right) e^{i\,(\vec{k}_1+\vec{k}_2)\cdot\vec{R}}$$
$$\times \sum_{\rho,\sigma} \langle 1S, A|x_1^i|\rho\rangle\langle\rho|x_1^j|1S, A\rangle\langle 1S, B|x_2^k|\sigma\rangle\langle\sigma|x_2^\ell|1S, B\rangle \sum_{i=1}^{12} f_i^{-1}(k_1, k_2)\,. \tag{5.66}$$

The functions f_i correspond to the diagrams in Fig. 5.1. They carry three factors each. These factors correspond to the first, second and third "section" of the respective diagrams, as separated by the four interaction vertices, as seen from left to right. For the first six diagrams, they read

$$f_1(k_1,k_2) = (E_{1S,A} - E_\rho - \hbar c k_1)(-\hbar c k_1 - \hbar c k_2)(E_{1S,B} - E_\sigma - \hbar c k_2)\,, \tag{5.67a}$$

$$f_2(k_1,k_2) = (E_{1S,A} - E_\rho - \hbar c k_2)(-\hbar c k_1 - \hbar c k_2)(E_{1S,B} - E_\sigma - \hbar c k_2)\,, \tag{5.67b}$$

$$f_3(k_1,k_2) = (E_{1S,A} - E_\rho - \hbar c k_1)(E_{1S,A} - E_\sigma + E_{1S,B} - E_\rho)(E_{1S,B} - E_\sigma - \hbar c k_2)\,, \tag{5.67c}$$

$$f_4(k_1,k_2) = (E_{1S,B} - E_\sigma - \hbar c k_1)(E_{1S,A} - E_\rho + E_{1S,B} - E_\sigma)(E_{1S,B} - E_\sigma - \hbar c k_2)\,, \tag{5.67d}$$

$$f_5(k_1,k_2) = (E_{1S,A} - E_\rho - \hbar c k_2)(E_{1S,A} - E_\rho + E_{1S,B} - E_\sigma - \hbar c k_1 - \hbar c k_2) \times (E_{1S,B} - E_\sigma - \hbar c k_2)\,, \tag{5.67e}$$

$$f_6(k_1,k_2) = (E_{1S,B} - E_\sigma - \hbar c k_1)(E_{1S,A} - E_\rho + E_{1S,B} - E_\sigma - \hbar c k_1 - \hbar c k_2) \times (E_{1S,B} - E_\sigma - \hbar c k_2)\,. \tag{5.67f}$$

Diagrams (7) to (12) in Fig. 5.1 correspond to diagrams (1) to (6) under the exchange of the two electron lines. So, the six relations for the diagrams (7) to (12) can be summarized as $f_{6+d}(k_1,k_2) = f_d(k_1,k_2,\rho \leftrightarrow \sigma, A \leftrightarrow B)$ for the diagram numbers $d = 1,\dots,6$. Still, for reference, we give them explicitly,

$$f_7(k_1,k_2) = (E_{1S,B} - E_\sigma - \hbar c k_1)(-\hbar c k_1 - \hbar c k_2)(E_{1S,A} - E_\rho - \hbar c k_2)\,, \tag{5.67g}$$

$$f_8(k_1,k_2) = (E_{1S,B} - E_\sigma - \hbar c k_2)(-\hbar c k_1 - \hbar c k_2)(E_{1S,A} - E_\rho - \hbar c k_2)\,, \tag{5.67h}$$

$$f_9(k_1,k_2) = (E_{1S,B} - E_\sigma - \hbar c k_1)(E_{1S,A} - E_\rho + E_{1S,B} - E_\sigma)(E_{1S,A} - E_\rho - \hbar c k_2)\,, \tag{5.67i}$$

$$f_{10}(k_1,k_2) = (E_{1S,A} - E_\rho - \hbar c k_1)(E_{1S,A} - E_\rho + E_{1S,B} - E_\sigma)(E_{1S,A} - E_\rho - \hbar c k_2)\,, \tag{5.67j}$$

$$f_{11}(k_1,k_2) = (E_{1S,B} - E_\sigma - \hbar c k_2)(E_{1S,A} - E_\rho + E_{1S,B} - E_\sigma - \hbar c k_1 - \hbar c k_2) \times (E_{1S,A} - E_\rho - \hbar c k_2)\,, \tag{5.67k}$$

$$f_{12}(k_1,k_2) = (E_{1S,A} - E_\rho - \hbar c k_1)(E_{1S,A} - E_\rho + E_{1S,B} - E_\sigma - \hbar c k_1 - \hbar c k_2) \times (E_{1S,A} - E_\rho - \hbar c k_2)\,. \tag{5.67l}$$

It is an excellent exercise to derive these expressions one by one, for verification and general understanding of the quantized operators.

5.4.3 *Limit of Large Separation of the Two Atoms*

In principle, Eq. (5.66) together with the denominators listed in Eq. (5.67) provides us with a formula for the interaction energy of two neutral atoms which is valid at all interatomic distance scales. However, additional insight into the physics can be gained if we isolate those terms that give the leading contributions in the limit of large, and small separations of the atoms, in a sense to be specified below.

A very large separation of the two atoms implies that the momenta of the exchanged photons k_1 and k_2 must be small in the sense of the following approximations,

$$\hbar c k_1 \ll |E_{1S,B} - E_\sigma|\,, \quad \hbar c k_1 \ll |E_{1S,A} - E_\rho|\,, \tag{5.68a}$$

$$\hbar c k_2 \ll |E_{1S,B} - E_\sigma|\,, \quad \hbar c k_2 \ll |E_{1S,A} - E_\rho|\,. \tag{5.68b}$$

The relevant wavelengths over which the two atoms "communicate" will be of the same order-of-magnitude as their separation $R = |\vec{R}| = |\vec{R}_A - \vec{R}_B|$. Thus, the exchanged photons have wave numbers in the range $k_i \sim 2\pi/R \sim 1/R$, and the condition (5.68) effectively means that

$$\frac{\hbar\, c}{R} \ll |E_{1S,A} - E_\rho|\,, \qquad R \gg \frac{\hbar c}{|E_{1S,A} - E_\rho|}\,, \tag{5.69}$$

and equivalently for atom B. The considerations below are valid for distances larger than a typical atomic transition wavelength, which is of the order of a few thousand Bohr radii.

For the denominators (5.67), the approximation (5.68) means that the dominant terms are those that contain a plain expression $(\hbar c k_1 + \hbar c k_2)$, without any further energy differences to virtual-state excitations. A quick inspection shows that these are the terms corresponding to the diagrams (1), (2), (7) and (8) in Fig. 5.1. In the rest of the propagator denominators, we can then approximate, furthermore, as follows,

$$E_{1S,A} - E_\rho - \hbar c k_i \approx E_{1S,A} - E_\rho, \qquad E_{1S,B} - E_\sigma - \hbar c k_i \approx E_{1S,B} - E_\sigma\,. \tag{5.70}$$

All four terms ΔE_d with $d = 1, 2, 7, 8$ become equivalent in this limit. The result for the fourth-order energy shift is

$$\Delta E^{(4)} \approx -4e^4 \int \frac{d^3k_1}{(2\pi)^3} \int \frac{d^3k_2}{(2\pi)^3} \frac{\hbar c k_1}{2\,\epsilon_0} \frac{\hbar c k_2}{2\,\epsilon_0} \left(\delta^{ik} - \frac{k_1^i\, k_1^k}{k_1^2}\right)\left(\delta^{j\ell} - \frac{k_2^j\, k_2^\ell}{k_2^2}\right)$$
$$\times \frac{e^{i\,(\vec{k}_1+\vec{k}_2)\cdot\vec{R}}}{\hbar c k_1 + \hbar c k_2} \left(\sum_\rho \frac{\langle 1S,A|x_1^i|\rho\rangle\,\langle\rho|x_1^j|1S,A\rangle}{E_{1S,A} - E_\rho}\right)\left(\sum_\sigma \frac{\langle 1S,B|x_2^k|\sigma\rangle\,\langle\sigma|x_2^\ell|1S,B\rangle}{E_{1S,B} - E_\sigma}\right). \tag{5.71}$$

Now we use the following property, which is valid for any S state due to radial symmetry,

$$\langle 1S|x_1^i|\rho\rangle\,\langle\rho|x^j|1S\rangle = \frac{\delta^{ij}}{3} \sum_{k=1}^{3} \langle 1S|x^k|\rho\rangle\,\langle\rho|x^k|1S\rangle\,, \tag{5.72}$$

for both atoms A and B. We also employ the Einstein summation convention to obtain the following result,

$$\left(\delta^{i\ell} - \frac{k_1^i\, k_1^\ell}{k_1^2}\right)\left(\delta^{i\ell} - \frac{k_2^i\, k_2^\ell}{k_2^2}\right) = 1 + \frac{(\vec{k}_1 \cdot \vec{k}_2)^2}{k_1^2\, k_2^2}\,. \tag{5.73}$$

With these simplifications, the long-distance limit of $\Delta E^{(4)}$ becomes

$$\Delta E^{(4)} \approx -e^4 \frac{\hbar c}{9\,\epsilon_0^2} \int \frac{d^3k_1}{(2\pi)^3} \int \frac{d^3k_2}{(2\pi)^3} \frac{k_1\, k_2\, e^{i\,(\vec{k}_1+\vec{k}_2)\cdot\vec{R}}}{k_1 + k_2} \left(1 + \frac{(\vec{k}_1 \cdot \vec{k}_2)^2}{k_1^2\, k_2^2}\right)$$
$$\times \left(\sum_\rho \frac{\langle 1S,A|x_1^i|\rho\rangle\,\langle\rho|x_1^i|1S,A\rangle}{E_{1S,A} - E_\rho}\right)\left(\sum_\sigma \frac{\langle 1S,B|x_2^k|\sigma\rangle\,\langle\sigma|x_2^k|1S,B\rangle}{E_{1S,B} - E_\sigma}\right). \tag{5.74}$$

We now define the static ground-state dipole polarizabilities $\alpha(A) \equiv \alpha(1S, A, \omega = 0)$ and $\alpha(B) \equiv \alpha(1S, B, \omega = 0)$ of atoms A and B as follows,

$$\alpha(1S, A) = \alpha(A) = \frac{2e^2}{3} \sum_\rho \frac{\langle 1S, A|x_1^i|\rho\rangle \langle\rho|x_1^i|1S, A\rangle}{E_\rho - E_{1S,A}}, \tag{5.75a}$$

$$\alpha(1S, B) = \alpha(B) = \frac{2e^2}{3} \sum_\sigma \frac{\langle 1S, B|x_2^i|\sigma\rangle \langle\sigma|x_2^i|1S, B\rangle}{E_\sigma - E_{1S,B}}, \tag{5.75b}$$

where the Cartesian component x^i is summed over. The only contributing virtual states labeled as ρ are P states, so there is no need to explicitly exclude the $1S$ reference state. We now need to evaluate

$$\Delta E^{(4)} \approx -\frac{\hbar c}{4\epsilon_0^2} \alpha(A)\,\alpha(B) \int \frac{\mathrm{d}^3 k_1}{(2\pi)^3} \int \frac{\mathrm{d}^3 k_2}{(2\pi)^3} \frac{k_1\, k_2\, \mathrm{e}^{\mathrm{i}\,(\vec{k}_1+\vec{k}_2)\cdot\vec{R}}}{k_1 + k_2} \left(1 + \frac{(\vec{k}_1 \cdot \vec{k}_2)^2}{k_1^2\, k_2^2}\right). \tag{5.76}$$

It is helpful to isolate the following sixfold integral,

$$I = \int \frac{\mathrm{d}^3 k_1}{(2\pi)^3} \int \frac{\mathrm{d}^3 k_2}{(2\pi)^3} \frac{k_1\, k_2\, \mathrm{e}^{\mathrm{i}\,(\vec{k}_1+\vec{k}_2)\cdot\vec{R}}}{k_1 + k_2} \left(1 + \frac{(\vec{k}_1 \cdot \vec{k}_2)^2}{k_1^2\, k_2^2}\right). \tag{5.77}$$

We now assume $\vec{R}$ to be aligned along the z axis without loss of generality, and express the $\vec{k}_i$ with $i = 1, 2$ in spherical coordinates as follows,

$$\vec{k}_1 = k_1 \begin{pmatrix} \sqrt{1-u_1^2}\cos\varphi_1 \\ \sqrt{1-u_1^2}\sin\varphi_1 \\ u_1 \end{pmatrix}, \quad \vec{k}_2 = k_2 \begin{pmatrix} \sqrt{1-u_2^2}\cos\varphi_2 \\ \sqrt{1-u_2^2}\sin\varphi_2 \\ u_2 \end{pmatrix}. \tag{5.78}$$

The variables $u_1 = \cos\theta_1$ and $u_2 = \cos\theta_2$ have to be integrated over the interval $[-1, 1]$. We use the parameterization

$$\frac{1}{k_1 + k_2} = R \int_0^\infty \mathrm{d}\eta\, \exp[-(k_1 + k_2)\, R\,\eta], \tag{5.79}$$

and perform a Wick–style complex rotation of the k_1 and k_2 integrations,

$$\mathrm{e}^{\mathrm{i}\,\vec{k}_1\cdot\vec{R}} \to \mathrm{e}^{\mathrm{i}k_1\, R\, u_1} = \mathrm{e}^{-k_1 R\, u_1}, \quad \mathrm{e}^{\mathrm{i}\,(\vec{k}_2\cdot\vec{R})} \to \mathrm{e}^{\mathrm{i}k_1\, R\, u_2} = \mathrm{e}^{-k_2 R\, u_2}. \tag{5.80}$$

Here, $u_1 = \cos\theta_1$ and $u_2 = \cos\theta_2$, where θ_1 and θ_2 are the direction cosines of the vector $\vec{k}_1$ and $\vec{k}_2$, with θ_1 and θ_2 being the polar angles. The integral I can finally be written as

$$\begin{aligned} I = & \int_0^\infty \mathrm{d}\eta \int_{-1}^1 \mathrm{d}u_1 \int_{-1}^1 \mathrm{d}u_2 \int_0^\infty \mathrm{d}k_1 \int_0^\infty \mathrm{d}k_2 \int_0^{2\pi} \mathrm{d}\varphi_1 \int_0^{2\pi} \mathrm{d}\varphi_2\, \frac{k_1^3\, k_2^3}{(2\pi)^6} \\ & \times \mathrm{i}\, R\, \mathrm{e}^{-R(k_1(u_1+\eta)+k_2(u_2+\eta))} \\ & \times \left[1 + \left(u_1 u_2 + \sqrt{1-u_1^2}\sqrt{1-u_2^2}\,(\cos\varphi_1\cos\varphi_2 + \sin\varphi_1\sin\varphi_2)\right)^2\right] \\ = & \frac{4\mathrm{i}}{\pi^4 R^7} \int_0^\infty \mathrm{d}\eta\, \frac{3\eta^4 + 2\eta^2 + 3}{(\eta^2 - 1)^6} = -\frac{4}{\pi^4 R^7} \int_0^\infty \mathrm{d}\xi\, \frac{3\xi^4 - 2\xi^2 + 3}{(\xi^2 + 1)^6} = -\frac{23}{16\pi^3 R^7}. \end{aligned} \tag{5.81}$$

The integrations are now easiest if performed according to the sequence as indicated in the above equation, i.e. from right to left ($d\varphi_2 \to d\varphi_1 \to \cdots \to d\eta$). In the evaluation of the final integral over η, we have again used a complex rotation, according to $d\eta \to i\,d\xi$. Using Eq. (5.76), the long-range interaction between two atoms is thus found to be

$$\Delta E^{(4)} = -\frac{23\,\hbar c}{64\,\pi^3\,\epsilon_0^2}\,\frac{\alpha(A)\,\alpha(B)}{R^7} = -\frac{23}{4\pi R^7}\,\frac{\hbar c}{(4\pi\epsilon_0)^2}\,\alpha(A)\,\alpha(B)\,, \tag{5.82}$$

$$R \gg \frac{\hbar c}{|E_{1S,A} - E_\rho|} \sim \frac{a_0}{\alpha}\,, \qquad R \gg \frac{\hbar c}{|E_{1S,B} - E_\sigma|} \sim \frac{a_0}{\alpha}\,, \tag{5.83}$$

where full SI mksA units have been kept. The order-of-magnitude estimates indicated by the "~" sign are valid for typical atomic transition frequencies, with $|E_{1S,A} - E_\rho| \sim |E_{1S,B} - E_\sigma| \sim \alpha^2 m_e c^2$ (as always, m_e denotes the electron mass). The interaction is seen to be attractive rather than repulsive.

5.4.4 *Nonretarded (van der Waals) Interatomic Interaction*

In contrast to the condition (5.69), we now consider the case where R is small compared to an average atomic transition wavelength, but large compared to the atomic radius,

$$a_0 = \frac{\hbar}{\alpha m c} \ll R \ll \frac{\hbar c}{|E_{1S,A} - E_\rho|}, \frac{\hbar c}{|E_{1S,B} - E_\sigma|} \sim \frac{a_0}{\alpha}\,. \tag{5.84}$$

The first condition ensures that there is no significant overlap of the wave functions of the two atoms, while the second condition ensures that we are still operating in the nonretardation limit, which is equivalent to the so-called van der Waals range of interatomic distances. In this case, the condition (5.68) is reversed,

$$\hbar c k_1 \gg |E_{1S,B} - E_\sigma|\,, \quad \hbar c k_1 \gg |E_{1S,A} - E_\rho|\,, \tag{5.85}$$

$$\hbar c k_2 \gg |E_{1S,B} - E_\sigma|\,, \quad \hbar c k_2 \gg |E_{1S,A} - E_\rho|\,. \tag{5.86}$$

The relevant diagrams from Fig. 5.1 contributing in this limit are (3), (4), (9) and (10), because they have the lowest number of occurrences of k_1 and k_2 in the denominators. In the relevant propagator denominators in Eq. (5.67), we may use the approximations

$$E_{1S,A} - E_\rho - \hbar c k_1 \approx -\hbar c k_1 \tag{5.87}$$

and analogously for $A \to B$, $k_1 \to k_2$, and $\rho \to \sigma$. Again, the four contributing diagrams yield equal contributions in the distance range under consideration, and we have

$$\Delta E^{(4)} \approx 4e^4 \int \frac{d^3k_1}{(2\pi)^3} \int \frac{d^3k_2}{(2\pi)^3} \frac{\hbar c k_1}{2\,\epsilon_0} \frac{\hbar c k_2}{2\,\epsilon_0} \left(\delta^{ik} - \frac{k_1^i k_1^k}{k_1^2}\right)\left(\delta^{j\ell} - \frac{k_2^j k_2^\ell}{k_2^2}\right) \frac{e^{i(\vec{k}_1 + \vec{k}_2)\cdot\vec{R}}}{(\hbar c k_1)(\hbar c k_2)}$$
$$\times \sum_{\rho,\sigma} \frac{\langle 1S, A|x_1^i|\rho\rangle\,\langle\rho|x_1^j|1S, A\rangle\,\langle 1S, B|x_2^k|\sigma\rangle\,\langle\sigma|x_2^\ell|1S, B\rangle}{E_{1S,A} - E_\rho + E_{1S,B} - E_\sigma}\,. \tag{5.88}$$

The d^3k integrations can now be performed easily, using the formula

$$\int \frac{d^3k}{(2\pi)^3}\left(\delta^{ij} - \frac{k^i\,k^j}{k^2}\right) e^{i\vec{k}\cdot\vec{R}} = \frac{1}{4\pi R^3}\left(\delta^{ij} - 3\frac{R^i\,R^j}{R^2}\right) + \frac{2}{3}\,\delta^{ij}\delta^{(3)}(\vec{R}). \tag{5.89}$$

The Dirac-δ term is irrelevant for the current discussion, as the long-range interatomic interaction is being calculated for a finite, nonvanishing separation $|\vec{R}| > a_0$, where a_0 is the Bohr radius. Thus,

$$\Delta E^{(4)} \approx -\frac{e^4}{16\pi^2\epsilon_0^2 R^6}\left(\delta^{ik} - 3\frac{R^i\,R^k}{R^2}\right)\left(\delta^{j\ell} - 3\frac{R^j\,R^\ell}{R^2}\right)\frac{\delta^{ij}}{3}\frac{\delta^{k\ell}}{3} \\ \times \sum_\rho\sum_\sigma \frac{\langle 1S,A|x_1^m|\rho\rangle\,\langle\rho|x_1^m|1S,A\rangle\,\langle 1S,B|x_2^n|\sigma\rangle\,\langle\sigma|x_2^n|1S,B\rangle}{E_\rho - E_{1S,A} + E_\sigma - E_{1S,B}}, \tag{5.90}$$

where the property (5.72) has been used, the superscripts m and n are summed over, and the sign of the terms in the propagator has been reversed. Although initially not obvious, it is advantageous to write the propagator denominator as an integral,

$$\frac{1}{E_\rho - E_{1S,A} + E_\sigma - E_{1S,B}} = \frac{1}{\pi}\int_{-\infty}^{\infty} du\,\frac{(E_\rho - E_{1S,A})\,(E_\sigma - E_{1S,B})}{[(E_\rho - E_{1S,A})^2 + u^2]\,[(E_\sigma - E_{1S,B})^2 + u^2]}, \tag{5.91}$$

and we finally obtain

$$\Delta E^{(4)} \approx -\frac{e^4}{24\pi^3\epsilon_0^2 R^6}\int_{-\infty}^{\infty} du\,\sum_{\rho,\sigma}(E_\rho - E_{1S,A})\,(E_\sigma - E_{1S,B}) \\ \times \frac{\langle 1S,A|x_1^m|\rho\rangle\,\langle\rho|x_1^m|1S,A\rangle\,\langle 1S,B|x_2^n|\sigma\rangle\,\langle\sigma|x_2^n|1S,B\rangle}{[(E_\rho - E_{1S,A})^2 + u^2]\,[(E_\sigma - E_{1S,B})^2 + u^2]}. \tag{5.92}$$

Using the substitution $u \to \hbar\omega$, this expression can be rewritten as

$$\Delta E^{(4)} = -\frac{2\hbar}{3\pi R^6}\left(\frac{1}{4\pi\epsilon_0}\right)^2\int_{-\infty}^{\infty} d\omega\left\{e^2\left\langle 1S,A\left|x_1^i\,\frac{H_A - E_{1S,A}}{(H_A - E_{1S,A})^2 + (\hbar\omega)^2}\,x_1^i\right|1S,A\right\rangle\right\} \\ \times\left\{e^2\left\langle 1S,B\left|x_2^j\,\frac{H_B - E_{1S,B}}{(H_B - E_{1S,B})^2 + (\hbar\omega)^2}\,x_2^j\right|1S,B\right\rangle\right\}, \tag{5.93}$$

where the atomic Hamiltonians H_A and H_B have been defined in Eq. (5.55). Using the decomposition

$$\frac{H_A - E_{1S,A}}{(H_A - E_{1S,A})^2 + (\hbar\omega)^2} = \frac{1}{2}\left(\frac{1}{H_A - E_{1S,A} + i\hbar\omega} + \frac{1}{H_A - E_{1S,A} - i\hbar\omega}\right), \tag{5.94}$$

and an analogous decomposition for atom B, one may identify the integrand as a sum over dynamic polarizabilities of complex argument. Since we are dealing with ground-state atoms, we can write $\alpha(A,\omega) \equiv \alpha(1S,A,\omega)$, suppressing the reference state in the notation, which results in

$$\alpha(A, i\omega) = \frac{e^2}{3}\sum_\pm\left\langle 1S,A\left|x_1^i\,\frac{1}{H_A - E_{1S,A} \pm i\hbar\omega}\,x_1^i\right|1S,A\right\rangle \\ = \frac{2e^2}{3}\left\langle 1S,A\left|x_1^i\,\frac{H_A - E_{1S_A}}{(H - E_{1S,A})^2 + (\hbar\omega)^2}\,x_1^i\right|1S,A\right\rangle. \tag{5.95}$$

When applied to Eq. (5.93), the result reads

$$\begin{aligned}\Delta E^{(4)} &= -\frac{3\hbar}{2\pi R^6}\left(\frac{1}{4\pi\epsilon_0}\right)^2 \int_{-\infty}^{\infty} \mathrm{d}\omega\, \alpha(A,\mathrm{i}\omega)\,\alpha(B,\mathrm{i}\omega) \\ &= -\frac{3\hbar}{\pi R^6}\left(\frac{1}{4\pi\epsilon_0}\right)^2 \int_{0}^{\infty} \mathrm{d}\omega\, \alpha(A,\mathrm{i}\omega)\,\alpha(B,\mathrm{i}\omega), \\ & a_0 \ll R \ll \frac{\hbar c}{|E_{1S,A} - E_\rho|}, \frac{\hbar c}{|E_{1S,B} - E_\sigma|} \sim \frac{a_0}{\alpha}. \end{aligned} \tag{5.96}$$

We observe the change in the integration domain in the first and second lines of Eq. (5.96) and find the well-known $1/R^6$ van der Waals interaction to be expressible in terms of dynamic polarizabilities of the two atoms involved in the interaction. The formula is valid only in a distance range bounded from below by the Bohr radius, and from above by a typical atomic transition wavelength.

5.4.5 *Interpolating Formula*

Summing all contributions to $\Delta E^{(4)}$ given in Eqs. (5.66) and (5.67), one would hope to obtain a formula which interpolates between the limiting cases (5.82) and (5.96), still under the assumption of non-overlapping wave functions ($a_0 \ll R$). The integrand necessarily involves the sum $\sum_{i=1}^{12}[f_i(k_1,k_2)]^{-1}$. In the calculation, one has the advantage that one integrates over both d^3k_1 and d^3k_2, over the entire space $\mathbb{R}^3$. The integrand involves a dependence on both k_1 and k_2, and it is possible to reorder the expressions in terms of symmetric or asymmetric combinations of the two photon momenta. In the following, we consider a calculational scheme which is based on an asymmetric combination of the two momenta, in order to facilitate the k_2-integration. The basic rationale is that after the reordering and replacement, one should generate a factor

$$\frac{k_2^3}{k_1+k_2} - \frac{k_2^3}{k_1-k_2} = \frac{k_2^3}{k_1+k_2} + \frac{(-k_2)^3}{k_1+(-k_2)}, \tag{5.97}$$

which eventually allows one to carry out the k_2 integration by principal value, after extending the k_2 integration interval from $(0,\infty)$ to $(-\infty,\infty)$, due to the symmetry of the integrand.

After some algebra, one can show the identity

$$\begin{aligned}&\frac{1}{f_1(k_1,k_2)} + \frac{1}{f_3(k_1,k_2)} + \frac{1}{f_4(k_1,k_2)} + \frac{1}{f_7(k_1,k_2)} + \frac{1}{f_9(k_1,k_2)} + \frac{1}{f_{10}(k_1,k_2)} \\ &= T_1 + T_2, \end{aligned} \tag{5.98}$$

where

$$T_1 = -\frac{1}{E_{\rho A}+E_{\sigma B}}\left(\frac{1}{E_{\rho A}+\hbar c k_1} + \frac{1}{E_{\sigma B}+\hbar c k_1}\right)\frac{1}{\hbar c}\left(\frac{1}{k_1+k_2} - \frac{1}{k_1-k_2}\right), \tag{5.99}$$

$$T_2 = -\frac{1}{E_{\rho A}+E_{\sigma B}}\left(\frac{1}{E_{\rho A}+\hbar c k_2} + \frac{1}{E_{\sigma B}+\hbar c k_2}\right)\frac{1}{\hbar c}\left(\frac{1}{k_1+k_2} + \frac{1}{k_1-k_2}\right), \tag{5.100}$$

where we have introduced the notation

$$E_{\rho A} = E_\rho - E_{1S,A}\,, \qquad E_{\sigma B} = E_\sigma - E_{1S,B}\,. \tag{5.101}$$

The second term T_2 is equal to the first term T_1, upon the interchange of the photon momenta k_1 and k_2. Because one integrates over d^3k_1 and d^3k_2, the two terms are equivalent, and one can replace

$$\frac{1}{f_1(k_1,k_2)} + \frac{1}{f_3(k_1,k_2)} + \frac{1}{f_4(k_1,k_2)} + \frac{1}{f_7(k_1,k_2)} + \frac{1}{f_9(k_1,k_2)} + \frac{1}{f_{10}(k_1,k_2)} \to 2\,T_1 \tag{5.102}$$

for the purpose of the later k_1 and k_2 integrations. One can also show the identities

$$\frac{1}{f_2(k_1,k_2)} + \frac{1}{f_5(k_1,k_2)} + \frac{1}{f_6(k_1,k_2)} = G_1 + G_2\,, \tag{5.103a}$$

$$G_1 = -\frac{1}{E_{\rho A} + \hbar c k_2}\,\frac{1}{E_{\sigma B} + \hbar c k_2}\,\frac{1}{\hbar c}\left(\frac{1}{k_1+k_2} + \frac{1}{k_1-k_2}\right), \tag{5.103b}$$

$$G_2 = \frac{1}{E_{\rho A} + \hbar c k_2}\,\frac{1}{E_{\sigma B} + \hbar c k_1}\,\frac{1}{\hbar c}\,\frac{1}{k_1-k_2}\,. \tag{5.103c}$$

Another combination is

$$\frac{1}{f_8(k_1,k_2)} + \frac{1}{f_{11}(k_1,k_2)} + \frac{1}{f_{12}(k_1,k_2)} = H_1 + H_2\,, \tag{5.104a}$$

$$H_1 = -\frac{1}{E_{\rho A} + \hbar c k_2}\,\frac{1}{E_{\sigma B} + \hbar c k_2}\,\frac{1}{\hbar c}\left(\frac{1}{k_1+k_2} + \frac{1}{k_1-k_2}\right), \tag{5.104b}$$

$$H_2 = \frac{1}{E_{\rho A} + \hbar c k_1}\,\frac{1}{E_{\sigma B} + \hbar c k_2}\,\frac{1}{\hbar c}\,\frac{1}{k_1-k_2}\,. \tag{5.104c}$$

If we exchange in H_2 the variables k_1 and k_2, then we end up with $-G_2$. Hence, we can ignore the sum of G_2 and H_2 for later integrations.

In the sum $G_1 + H_1$, we can exchange k_1 and k_2 and replace

$$G_1 + H_1 \to T_3 \equiv -\frac{2}{E_{\rho A} + \hbar c k_1}\,\frac{1}{E_{\sigma B} + \hbar c k_1}\,\frac{1}{\hbar c}\left(\frac{1}{k_1+k_2} - \frac{1}{k_1-k_2}\right). \tag{5.105}$$

We can then replace the entire sum of relevant terms by

$$\begin{aligned}\sum_{i=1}^{12} \frac{1}{f_i(k_1,k_2)} \to 2T_1 + T_3 &= -\frac{2}{E_{\rho A} + E_{\sigma B}}\left(\frac{1}{E_{\rho A} + \hbar c k_1} + \frac{1}{E_{\sigma B} + \hbar c k_1}\right) \\ &\quad \times \frac{1}{\hbar c}\left(\frac{1}{k_1+k_2} - \frac{1}{k_1-k_2}\right) - \frac{1}{E_{\rho A} + \hbar c k_1}\,\frac{1}{E_{\sigma B} + \hbar c k_1}\,\frac{1}{\hbar c}\left(\frac{1}{k_1+k_2} - \frac{1}{k_1-k_2}\right) \\ &= -\frac{4(E_{\rho A} + E_{\sigma B} + \hbar c k_1)}{(E_{\rho A} + E_{\sigma B})\,(E_{\rho A} + \hbar c k_1)\,(E_{\sigma B} + \hbar c k_1)}\,\frac{1}{\hbar c}\left(\frac{1}{k_1+k_2} - \frac{1}{k_1-k_2}\right).\end{aligned} \tag{5.106}$$

This is the most economic way of writing the integrand, in terms of a k_1-dependent (pre-)factor, and the term $k_2^3/(k_1+k_2) - k_2^3/(k_1-k_2)$, which is generated upon multiplication by $d^3k_2\, k_2$. This combination generates the factor given in Eq. (5.97), which eventually allows us to symmetrize the integrand for the k_2-integration.

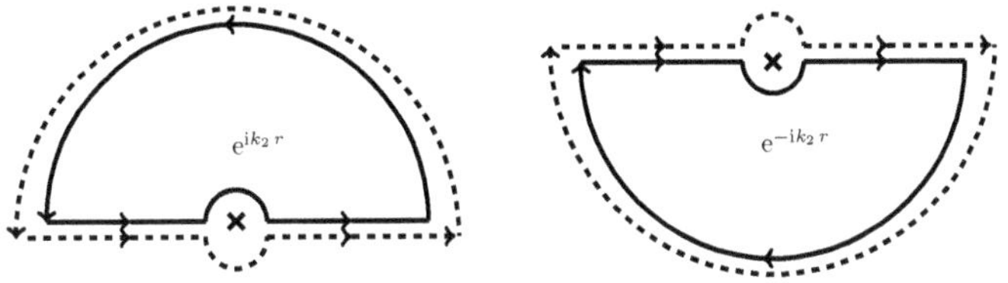

Fig. 5.2 In the evaluation of the integral (5.108), one defines the variable $x = k_2 r$ and writes the trigonometric factors in the integrand in terms of the expressions $\exp(\mathrm{i}k_2 r)$ and $\exp(-\mathrm{i}k_2 r)$. One then closes the integration contour either above or below the real axis, such as to ensure the convergence of the integral. The principal value of the integral is obtained as the arithmetic mean of the integrals infinitesimally above, and below, the real axis, i.e., as the arithmetic mean of the dashed and solid contours. By inspection of the figure, one realizes that this principal-value calculation involves taking the half pole contribution at $k_1 + k_2 = 0$.

Another important intermediate result is the following,

$$\int \frac{\mathrm{d}\Omega_k}{4\pi}\left(\delta^{ij} - \frac{k^i k^j}{\vec{k}^2}\right)\mathrm{e}^{\mathrm{i}\vec{k}\cdot\vec{R}} = \left(\delta^{ij} - \hat{R}^i \hat{R}^j\right)\frac{\sin(kR)}{kR} + \left(\delta^{ij} - 3\hat{R}^i \hat{R}^j\right)\left(\frac{\cos(kR)}{(kR)^2} - \frac{\sin(kR)}{(kR)^3}\right)$$
$$\equiv P^{ij}(k, \vec{R}). \tag{5.107}$$

In the integration, one may treat the factor $1/\vec{k}^2$ as a constant factor, then perform the $\mathrm{d}\Omega_k$ integration, and represent the k^i operators by differentiations under the integral sign, $k^i \to -\mathrm{i}\,\partial/\partial R^i$. We also need the principal-value (P.V.) integral

$$(\text{P.V.}) \int_{-\infty}^{\infty} \mathrm{d}k_2 \, \frac{k_2^3}{k_1 + k_2} \, P^{ij}(k_2 \vec{R}) = \pi \, k_1^3 \left\{ \left(\delta^{ij} - \hat{R}^i \hat{R}^j\right) \frac{\cos(k_1 R)}{k_1 R} \right.$$
$$\left. - \left(\delta^{ij} - 3\hat{R}^i \hat{R}^j\right)\left(\frac{\sin(k_1 R)}{(k_1 R)^2} + \frac{\cos(k_1 R)}{(k_1 R)^3}\right)\right\} \equiv \pi \, k_1^3 \, Q^{ij}(k_1, \vec{R}) \,. \tag{5.108}$$

In the evaluation of this integral, the principal value (P.V.) is to be understood as the arithmetic mean of results obtained by encircling the pole at $k_2 = -k_1$ either infinitesimally above or infinitesimally below the real axis (see Fig. 5.2).

The derivation starts from Eq. (5.66), which we recall for convenience,

$$\Delta E^{(4)} = \int \frac{\mathrm{d}^3 k_1}{(2\pi)^3} \int \frac{\mathrm{d}^3 k_2}{(2\pi)^3} \frac{\hbar c k_1}{2\,\epsilon_0} \frac{\hbar c k_2}{2\,\epsilon_0} \left(\delta^{ik} - \frac{k_1^i k_1^k}{k_1^2}\right)\left(\delta^{j\ell} - \frac{k_2^j k_2^\ell}{k_2^2}\right) \mathrm{e}^{\mathrm{i}(\vec{k}_1 + \vec{k}_2)\cdot\vec{R}}$$
$$\times \sum_{\rho,\sigma} \langle 1S, A|x_1^i|\rho\rangle\langle\rho|x_1^j|1S, A\rangle\langle 1S, B|x_2^k|\sigma\rangle\langle\sigma|x_2^\ell|1S, B\rangle \sum_{i=1}^{12} f_i^{-1}(k_1, k_2) \,. \tag{5.109}$$

After ordering prefactors, one can write this as

$$\Delta E^{(4)} = -\frac{\hbar c}{(4\pi\epsilon_0)^2}\frac{4e^4}{9\pi^2}\int_0^\infty dk_1\, k_1^3 \int_0^\infty dk_2\, k_2^3 \times \left(\int \frac{d\Omega_{k_1}}{4\pi}\left(\delta^{ik} - \frac{k_1^i\, k_1^k}{k_1^2}\right) e^{i\vec{k}_1\cdot\vec{R}}\right)\left(\int \frac{d\Omega_{k_2}}{4\pi}\left(\delta^{j\ell} - \frac{k_2^j\, k_2^\ell}{k_2^2}\right) e^{i\vec{k}_2\cdot\vec{R}}\right) \times \delta^{ij}\,\delta^{k\ell} \sum_{\rho,\sigma}\langle 1S, A|x_1^m|\rho\rangle\langle\rho|x_1^m|1S, A\rangle\langle 1S, B|x_2^n|\sigma\rangle\langle\sigma|x_2^n|1S, B\rangle \times \frac{E_{\rho A} + E_{\sigma B} + \hbar c k_1}{(E_{\rho A} + E_{\sigma B})\,(E_{\rho A} + \hbar c k_1)\,(E_{\sigma B} + \hbar c k_1)}\left(\frac{1}{k_1 + k_2} - \frac{1}{k_1 - k_2}\right). \tag{5.110}$$

Using the intermediate results (5.107) and (5.108), we can express the fourth-order energy shift as

$$\Delta E^{(4)} = -\frac{\hbar c}{(4\pi\epsilon_0)^2}\frac{4e^4}{9\pi}\int_0^\infty dk_1\, k_1^6\, P_{ik}(k_1, \vec{R})\, Q_{j\ell}(k_1, \vec{R})\,\delta^{ij}\,\delta^{k\ell} \times \sum_{\rho,\sigma}\frac{\langle 1S, A|x_1^m|\rho\rangle\langle\rho|x_1^m|1S, A\rangle\langle 1S, B|x_2^n|\sigma\rangle\langle\sigma|x_2^n|1S, B\rangle(E_{\rho A} + E_{\sigma B} + \hbar c k_1)}{(E_{\rho A} + E_{\sigma B})\,(E_{\rho A} + \hbar c k_1)\,(E_{\sigma B} + \hbar c k_1)}. \tag{5.111}$$

Finally,

$$\Delta E^{(4)} = -\frac{\hbar c}{(4\pi\epsilon_0)^2}\frac{4e^4}{9\pi\, R^6}\int_0^\infty dk_1 \times \left(2k_1 R\left[(k_1 R)^2 - 3\right]\cos(2k_1 R) + \left[(k_1 R)^4 - 5(k_1 R)^2 + 3\right]\sin(2k_1 R)\right) \times \sum_{\rho,\sigma}\frac{\langle 1S, A|x_1^m|\rho\rangle\langle\rho|x_1^m|1S, A\rangle\langle 1S, B|x_2^n|\sigma\rangle\langle\sigma|x_2^n|1S, B\rangle(E_{\rho A} + E_{\sigma B} + \hbar c k_1)}{(E_{\rho A} + E_{\sigma B})\,(E_{\rho A} + \hbar c k_1)\,(E_{\sigma B} + \hbar c k_1)}. \tag{5.112}$$

One now writes the trigonometric factors in terms of exponentials,

$$\cos(2k_1 R) = \frac{e^{2ik_1 R} + e^{-2ik_1 R}}{2}, \qquad \sin(2k_1 R) = \frac{e^{2ik_1 R} - e^{-2ik_1 R}}{2i}, \tag{5.113}$$

to obtain the result

$$2k_1 R\left[(k_1 R)^2 - 3\right]\cos(2k_1 R) + \left[(k_1 R)^4 - 5(k_1 R)^2 + 3\right]\sin(2k_1 R) = \frac{1}{2}\, e^{2ik_1 R}\left(-3i - 6(k_1 R) + 5i(k_1 R)^2 + 2(k_1 R)^3 - i(k_1 R)^4\right) + \text{c.c.}, \tag{5.114}$$

where c.c. denotes the complex conjugate. A Wick rotation changes the integration contour into

$$\int_0^\infty dk\, F(k) \to i\int_0^\infty \frac{d\omega}{c}\, F\left(\frac{i\omega}{c}\right), \tag{5.115}$$

where for F we take the first term on the right-hand side of in Eq. (5.114). Finally, after the addition of the complex conjugate, one obtains

$$\Delta E^{(4)} = -\frac{\hbar c}{(4\pi\epsilon_0)^2}\int_0^\infty d\omega\, \frac{\omega^4\, e^{-2\omega R/c}}{\pi\, R^2}\left(1 + \frac{2c}{\omega R} + \frac{5c^2}{(\omega R)^2} + \frac{6c^3}{(\omega R)^3} + \frac{3c^4}{(\omega R)^4}\right) \times \left[\frac{2e^2}{3}\sum_\rho \frac{E_{\rho A}}{E_{\rho A}^2 + \omega^2}\,\langle 1S, A|x_1^m|\rho\rangle\langle\rho|x_1^m|1S, A\rangle\right] \times \left[\frac{2e^2}{3}\sum_\sigma \frac{E_{\sigma B}}{E_{\sigma B}^2 + \omega^2}\langle 1S, B|x_2^n|\sigma\rangle\langle\sigma|x_2^n|1S, B\rangle\right]. \tag{5.116}$$

Using Eq. (5.95), one may write the complete result, valid for any distance between the atoms, as follows,

$$\Delta E^{(4)} = -\frac{\hbar}{\pi\, c^4\,(4\pi\epsilon_0)^2}\int_0^\infty \mathrm{d}\omega\, \alpha(A,\mathrm{i}\omega)\,\alpha(B,\mathrm{i}\omega) \times \frac{\omega^4\,\mathrm{e}^{-2\omega R/c}}{R^2}\left(1+\frac{2\,c}{\omega R}+\frac{5c^2}{(\omega R)^2}+\frac{6c^3}{(\omega R)^3}+\frac{3c^4}{(\omega R)^4}\right). \tag{5.117}$$

We now proceed to the study of the two limiting cases already discussed in Secs. 5.4.3 and 5.4.4. For very large R, where the condition (5.69) is valid, the exponential $\mathrm{e}^{-2\omega R/c}$ in Eq. (5.117) leads to a rapid damping of the contribution of finite ω, which overcompensates the large value of the speed of light in the denominator. All terms in the retardation expansion for $c \to \infty$ are thus relevant. To a very good approximation, we may therefore approximate $\alpha(A,\mathrm{i}\omega) \to \alpha(A,\omega=0) \equiv \alpha(A)$ and analogously for atom B. Then,

$$\int_0^\infty \mathrm{d}\omega\,\frac{\omega^4\,\mathrm{e}^{-2\omega R/c}}{R^2}\left(1+\frac{2c}{\omega R}+\frac{5c^2}{(\omega R)^2}+\frac{6c^3}{(\omega R)^3}+\frac{3c^4}{(\omega R)^4}\right) = \frac{23\,c^5}{4R^7}\,. \tag{5.118}$$

Thus, we may verify that for large R,

$$\Delta E^{(4)} \approx -\frac{23}{4\pi R^7}\,\frac{\hbar c}{(4\pi\epsilon_0)^2}\,\alpha(A)\,\alpha(B)\,, \qquad R \gg \frac{a_0}{\alpha}\,, \tag{5.119}$$

in exact agreement with Eq. (5.82). For intermediate distances, governed by the condition (5.84), one may ignore all but the last term in brackets in the integrand in Eq. (5.117),

$$\frac{\omega^4\,\mathrm{e}^{-2\omega R/c}}{R^2}\left(1+\frac{2c}{\omega R}+\frac{5c^2}{(\omega R)^2}+\frac{6c^3}{(\omega R)^3}+\frac{3c^4}{(\omega R)^4}\right) \to \frac{3c^4}{R^6}\,, \tag{5.120}$$

which is precisely equal to the limit of the entire expression as $c \to \infty$ (the "non-retardation limit"). Notice that the approximation $\mathrm{e}^{-2\omega R/c} \approx 1$ uniformly holds over the entire range of ω values relevant to the integration, provided R is sufficiently small. In the short-range limit, one thus obtains

$$\Delta E^{(4)} \approx -\frac{3\hbar}{\pi R^6}\,\frac{1}{(4\pi\epsilon_0)^2}\int_0^\infty \mathrm{d}\omega\,\alpha(A,\mathrm{i}\omega)\,\alpha(B,\mathrm{i}\omega)\,, \qquad a_0 \ll R \ll \frac{a_0}{\alpha}\,. \tag{5.121}$$

We take careful note of the integration limits and recover Eq. (5.93) in the short-range limit. The result (5.117) interpolates between the two limiting forms. The $1/R^7$ form is valid for large R. The long-range $1/R^7$ involves all the terms in the retardation expansion, i.e., the expansion in $\omega R/c$. One often says that the $1/R^6$ van der Waals result is changed to the $1/R^7$ Casimir–Polder interaction "due to retardation". This is slightly misleading, because it could otherwise suggest that a single "retardation correction term" is responsible for the change in the functional form, whereas from our discussion above, it is clear that the difference is due to the complete failure of the retardation expansion in the long-range limit, i.e., we cannot expand the expression $\exp(-2\omega R/c)$ in terms of its argument for very large R even if c has a large numerical value in SI mksA units.

Our consideration offers a smooth path toward an understanding of the concept of retardation. Namely, when the phase of the atomic oscillation during a virtual transition changes appreciably on the time scale it takes light to travel the interatomic separation distance R, then it is intuitively clear that some form of "retarded" form of the van der Waals interaction would have to be used. The criterion for the validity of the nonretardation approximation thus is

$$\frac{R}{c} \ll \frac{\hbar}{E_h} = \frac{a_0}{\alpha c}, \tag{5.122}$$

or, more precisely,

$$a_0 = \frac{\hbar}{\alpha m c} \ll R \ll \frac{\hbar}{\alpha^2 m c} = \frac{a_0}{\alpha}, \tag{5.123}$$

if we take into account that substantial overlap of the electronic wavefunctions is to be avoided. This picture is consistent with our calculations.

One important remark should follow our discussion: The entire treatment of the van der Waals and Casimir–Polder interaction energies has been limited to the case of ground-state interactions. For excited states, one decisive feature is the placement of the poles in the propagator denominators of the polarizability, in Eq. (5.67). A careful analysis, carried out very recently by a number of authors [208–211], reveals that in this case, the $1/R^7$ long-range asymptotics actually is changed, toward a $1/R^2$ term, though, with a numerically small coefficient. The functional form of the short-range $1/R^6$ term is not affected for excited states. Here, we were able to carry out the Wick rotation $\omega \to \mathrm{i}\omega$ in Eq. (5.115) without worrying about the infinitesimal displacement from the real axis, because both atoms are assumed to be in their respective ground states. This treatment remains approximately valid for the interaction of metastable hydrogen $2S$ atoms with ground-state atoms, because there are no lower states accessible from the $2S$ state via dipole transitions. However, one still needs to take the quasi-degenerate $2P$ states into account very carefully [212]. For highly excited states, the energetically lower states lead to the characteristic pole terms which give rise to a $1/R^2$ oscillatory tail [208–211].

5.5 Atom-Surface Interactions

5.5.1 *Perfectly Conducting Wall, Long Range*

We have already seen that the asymptotic behavior of the interaction potential for two polarizable atoms changes from $1/R^6$ for small distances to $1/R^7$ for large distances, due to retardation. This result has originally been derived in Ref. [213]. We now discuss a somewhat related effect, which is the interaction potential of an atom close to an idealized, perfectly conducting surface, or to a realistic, dielectric surface. Here, the surface takes over the role of the "other" atom. It is not charged, but the change in the boundary conditions in the vicinity of the surface provide for enough of a disturbance of the "vacuum" structure in the vicinity of the surface to induce a nontrivial interaction potential. The surface is canonically placed in the

xy plane, and the atom is supposed to be displaced from the surface by a distance z. We shall find that the behavior of the interaction potential changes from $1/z^3$ for small distances to $1/z^4$ for larger distances, where "large" and "small" distances are understood as compared to a characteristic wavelength of radiation used for inducing a virtual transition of the atom. In general, the Casimir–Polder interactions are important for micro- and nanostructure devices, where the dimensions of the involved objects become comparable to the atomic transition wavelengths of typical virtual atomic transitions.

The derivation of the Casimir–Polder interaction for arbitrary distances, as reported in Ref. [213], was very important for the understanding of QED interactions of atoms and surfaces. Nevertheless, an ideal conductor is a poor approximation for most realistic materials. For a material with an arbitrary (frequency-dependent) dielectric constant, the result has been derived in Ref. [214]. For a number of recent theoretical investigations on multipole corrections to the Casimir–Polder atom-surface interaction, we refer to the literature [215–223] and to complementary experiments [224–228].

5.5.2 *Dipole Interaction*

We assume a perfectly conducting wall in the xy plane ($z = 0$). In order to calculate the interaction energy, we assume that the surface located at $z = 0$ is made out of perfectly conducting material for all relevant incident light frequencies. The task then is to determine the allowed modes of the electromagnetic field, and calculate the second-order shift due to the virtual excitations of the modes, while subtracting the contribution of those modes that would otherwise contribute if there were no surface present (renormalization). We shall use a trick in doing the calculation: Namely, we shall assume that there is a perfectly conducting surface at $z = 0$, but we shall also assume perfectly conducting surfaces at $z = L_z$, and at $x = 0$, $x = L$, and $y = 0$, $y = L$. The atom is placed at $x = \frac{1}{2}L$, $y = \frac{1}{2}L$, and at a specific z coordinate, so that the artificially introduced boundary conditions at $x = 0$ and $x = L$, at $y = 0$ and $y = L$, and the one at $z = L_z$ move away to infinity as we let $L, L_z \to \infty$ at the end of the calculation. So, the "virtual" boundary conditions can have no effect on the final result for the interaction energy, but they help in writing down the mode functions that are essential in calculating the interaction energy in the first place. The details of the calculation will be described in the following.

We work in Cartesian coordinates, with mode functions (see Sec. 7.3.2 of Ref. [81])

$$\vec{A}_{\vec{k},\lambda}(\vec{r},t) = A_{\vec{k},\lambda,x}(\vec{r},t)\,\hat{\mathrm{e}}_x + A_{\vec{k},\lambda,y}(\vec{r},t)\,\hat{\mathrm{e}}_y + A_{\vec{k},\lambda,z}(\vec{r},t)\,\hat{\mathrm{e}}_z\,. \tag{5.124}$$

The vector potential of the fundamental modes appropriate to the interior of a rectangular parallelepiped of side lengths $L_x = L_y = L$ and L_z, with all surfaces being perfectly conducting, reads as follows (Coulomb gauge),

$$A_{\vec{k},\lambda,x} = \mathcal{A}_0 \sqrt{\frac{8}{V}}\, \epsilon_{\vec{k},\lambda,x} \cos(k_x x)\, \sin(k_y y)\, \sin(k_z z)\, \mathrm{e}^{-\mathrm{i}\,\omega\, t}\,, \tag{5.125a}$$

$$A_{\vec{k},\lambda,y} = \mathcal{A}_0 \sqrt{\frac{8}{V}}\, \epsilon_{\vec{k},\lambda,y} \sin(k_x x)\, \cos(k_y y)\, \sin(k_z z)\, \mathrm{e}^{-\mathrm{i}\,\omega\, t}\,, \tag{5.125b}$$

$$A_{\vec{k},\lambda,z} = \mathcal{A}_0 \sqrt{\frac{8}{V}}\, \epsilon_{\vec{k},\lambda,z} \sin(k_x x)\, \sin(k_y y)\, \cos(k_z z)\, \mathrm{e}^{-\mathrm{i}\,\omega\, t}\,, \tag{5.125c}$$

where $V = L^2\, L_z$, and $\mathcal{A}_0$ is a global amplitude. The energy of the electromagnetic mode is $\omega = \omega_{\vec{k}} = c\,k$. The polarization vector for the mode with wave vector $\vec{k}$ and polarization λ is $\hat{\epsilon}_{\vec{k},\lambda}$, with Cartesian components $\epsilon_{\vec{k},\lambda,x}$, $\epsilon_{\vec{k},\lambda,y}$, and $\epsilon_{\vec{k},\lambda,z}$ $(\lambda = 1, 2)$. The components of the wave vectors for the discrete modes are given as follows,

$$k_x = \frac{\ell\,\pi}{L}\,, \qquad k_y = \frac{m\,\pi}{L}\,, \qquad k_z = \frac{n\,\pi}{L_z}\,, \tag{5.126}$$

where ℓ, m , and n are integer. For the vector potential given in Eq. (5.125) not to vanish, two of these need to be nonzero. Furthermore, for $L_z \ll L$, the fundamental mode has quantum numbers $n = 0$, and $\ell = m = 1$, and therefore is z-polarized (only one polarization is available). For all modes other than the fundamental, one has two polarizations available. In the limit of a large quantization volume, the contribution of the fundamental mode to the sum over virtual excitations in Eq. (5.140) vanishes, and we can safely sum over both available polarizations (see also Sec. 7.3.2 of Ref. [81]).

The transversality (Coulomb gauge) condition for the vector potential is verified as follows,

$$\vec{\nabla} \cdot \vec{A}_{\vec{k},\lambda}(\vec{r}) = -\mathcal{A}_0 \sqrt{\frac{8}{V}} \left(\vec{k} \cdot \vec{\epsilon}_{\vec{k},\lambda}\right) \sin(k_x x)\, \sin(k_y y)\, \sin(k_z z) = 0\,, \tag{5.127}$$

which is fulfilled if

$$\vec{k} \cdot \vec{\epsilon}_{\vec{k},\lambda} = 0\,, \tag{5.128}$$

which we shall assume in the following discussion. Alternatively, we can write the vector $\vec{k}$ as follows,

$$k_x = k\, \sin\theta\, \cos\varphi\,, \qquad k_y = k\, \sin\theta\, \sin\varphi\,, \qquad k_z = k\, \cos\theta\,, \tag{5.129a}$$

$$\vec{k} = k_x\, \hat{\mathrm{e}}_x + k_y\, \hat{\mathrm{e}}_y + k_z\, \hat{\mathrm{e}}_z\,. \tag{5.129b}$$

Here, k_x, k_y and k_z must be matched against the quantum numbers ℓ, m, and n, as given in Eq. (5.126). The angles θ and φ in Eq. (5.129a) of course belong to $\vec{k}$, not to the coordinate vector $\vec{r}$. The two polarization vectors (corresponding to $\lambda = 1, 2$) for TE (transverse electric) and TM (transverse magnetic) modes can be written as follows,

$$\hat{\epsilon}_{\vec{k},\mathrm{TE}} = \sin\varphi\, \hat{\mathrm{e}}_x - \cos\varphi\, \hat{\mathrm{e}}_y\,, \tag{5.130a}$$

$$\hat{\epsilon}_{\vec{k},\mathrm{TM}} = -\cos\theta\, \cos\varphi\, \hat{\mathrm{e}}_x - \cos\theta\, \sin\varphi\, \hat{\mathrm{e}}_y + \sin\theta\, \hat{\mathrm{e}}_z\,, \tag{5.130b}$$

$$\vec{k} \cdot \hat{\epsilon}_{\vec{k},\mathrm{TE}} = \vec{k} \cdot \hat{\epsilon}_{\vec{k},\mathrm{TM}} = 0 \qquad \hat{\epsilon}_{\vec{k},\mathrm{TM}} \times \hat{\epsilon}_{\vec{k},\mathrm{TE}} = \hat{k} = \vec{k}/|\vec{k}|\,. \tag{5.130c}$$

Summing over the polarizations $\lambda = 1, 2$ where 1 stands for TE and 2 for TM, one obtains

$$\sum_\lambda \left(\hat{\epsilon}_{\vec{k},\lambda,x}\right)^2 = \frac{k_y^2 + k_z^2}{k^2}, \quad \sum_\lambda \left(\hat{\epsilon}_{\vec{k},\lambda,y}\right)^2 = \frac{k_x^2 + k_z^2}{k^2}, \quad \sum_\lambda \left(\hat{\epsilon}_{\vec{k},\lambda,z}\right)^2 = \frac{k_x^2 + k_y^2}{k^2}. \tag{5.131}$$

The mode functions in Eq. (5.125) are defined so that there is a cosine function if the Cartesian component of the wave vector in the argument of the trigonometric function is the same Cartesian component as the vector potential component itself. The presence of the sine functions implies that the tangential component of the vector potential (the component lying inside the boundary planes) vanishes for all three boundaries,

$$\begin{aligned} A_{\vec{k},\lambda,x}(x,y,z=0) &= A_{\vec{k},\lambda,x}(x,y,z=L_z) \\ &= A_{\vec{k},\lambda,x}(x,y=0,z) = A_{\vec{k},\lambda,x}(x,y=L,z) = 0\,, \end{aligned} \tag{5.132a}$$

$$\begin{aligned} A_{\vec{k},\lambda,y}(x,y,z=0) &= A_{\vec{k},\lambda,y}(x,y,z=L_z) \\ &= A_{\vec{k},\lambda,y}(x=0,y,z) = A_{\vec{k},\lambda,y}(x=L,y,z) = 0\,, \end{aligned} \tag{5.132b}$$

$$\begin{aligned} A_{\vec{k},\lambda,z}(x=0,y,z) &= A_{\vec{k},\lambda,z}(x=L,y,z) \\ &= A_{\vec{k},\lambda,z}(x,y=0,z) = A_{\vec{k},\lambda,z}(x,y=L,z) = 0\,. \end{aligned} \tag{5.132c}$$

The electric field is given by differentiation of the mode functions with respect to the time, where we assume a factor $\exp(-\mathrm{i}\omega t)$ for the time dependence [see Eq. (5.125)], so that $\vec{E} = \mathrm{i}\,\omega\,\vec{A}$. So, $\vec{E}$ is parallel to $\vec{A}$. As we can assume a perfectly conducting plate, the tangential component of the electric field has to vanish on the boundaries.

The transversal (normal) component of the electric field does not need to vanish at the boundaries, and indeed we have, e.g., for the planes with $z = 0$ and $z = L_z$,

$$\begin{aligned} E_{\vec{k},\lambda,z}(x,y,z=0) &= (-1)^n E_{\vec{k},\lambda,z}(x,y,z=L_z) \\ &= \mathrm{i}\omega\,\mathcal{A}_0 \sqrt{\frac{8}{V}}\,\epsilon_{\vec{k},\lambda,z}(\vec{k})\,\sin(k_x\,x)\,\sin(k_y\,y)\,. \end{aligned} \tag{5.133}$$

An explicit calculation of the curl of the vector potentials given in Eq. (5.125) also shows that the normal component of the magnetic field (the component in the direction of the normal to the boundary surface) vanishes on the boundaries, as it should by virtue of Faraday's law.

Let us now assume that one of the indices ℓ, m, n is zero, which implies that one Cartesian component of the wave vector $\vec{k}$ and two Cartesian components of the vector potential given in Eq. (5.125) vanish, because $\vec{A}_{\vec{k},\lambda}(\vec{r},t)$ is proportional to $\vec{\epsilon}_{\vec{k},\lambda}$. This implies that two Cartesian components of the polarization vector $\vec{\epsilon}_{\vec{k},\lambda}$ also have to vanish, and indeed, $\vec{\epsilon}_{\vec{k},\lambda}$ has to point into the same direction as the vanishing component of the vector $\vec{k}$. As an example, if $n = 0$, then $k_z = 0$, and in view of $\vec{A}_{\vec{k},\lambda,x} = \vec{A}_{\vec{k},\lambda,y} = 0$, the polarization vector $\vec{\epsilon}_{\vec{k},\lambda}$ has to point into the z direction in order to have any nonvanishing vector potential at all. Then, $A_{\vec{k},\lambda,z} = \sqrt{\frac{8}{V}}\,\vec{\epsilon}_{\vec{k},\lambda,z}\,\sin(k_x\,x)\,\sin(k_y\,y)\,\exp(-\mathrm{i}\omega t)$, so that $\hat{\epsilon}_{\vec{k},\lambda} = \hat{\mathrm{e}}_z$, and we have only one polarization, namely, z polarization. Also, we may consider, for $k_z = 0$, the curl of the vector potential, i.e., the magnetic field. It has nonvanishing components in the x and y directions. Therefore, at the "lower" and "upper" boundary planes ($z = 0$ and $z = L_z$), we have $\hat{n} \cdot \vec{B} = \hat{\mathrm{e}}_z \cdot \vec{B} = 0$, where $\hat{n} = \hat{\mathrm{e}}_z$ is the normal to the boundary surface. This is necessary in order to show consistency with Faraday's law.

For the normalization, we can use the well-known result that $\int_L \mathrm{d}x \, \cos^2(x) = L/2$ which holds whenever $\cos^2(x)$ is integrated over a whole period. Then, it is easy to show that

$$\int_V \mathrm{d}^3 r \, \cos^2(k_x x) \, \sin^2(k_y y) \, \sin^2(k_z z) = \frac{V}{8}, \tag{5.134}$$

and the same result holds for the integrals with sine functions interchanged with cosines. So,

$$\int_V \mathrm{d}^3 r \, \left| \vec{A}_{\vec{k},\lambda}(\vec{r}) \right|^2 = |\mathcal{A}_0|^2 . \tag{5.135}$$

Here and in the following, we identify $\vec{A}_{\vec{k},\lambda}(\vec{r}) \equiv \vec{A}_{\vec{k},\lambda}(\vec{r}, t = 0)$.

We must now analyze the behavior of the vector potential operator in the quantized formulation. In the presence of the boundary conditions, we can write a generalization of Eq. (2.129a), for the discretized representation of the vector potential operator in the vacuum, with normalization volume V,

$$\vec{A}(\vec{r}) = \sum_{\vec{k}\lambda} \left[\mathrm{a}_\lambda(\vec{k}) \, \vec{A}_{\vec{k},\lambda}(\vec{r}) + \mathrm{a}_\lambda^+(\vec{k}) \, \vec{A}^*_{\vec{k},\lambda}(\vec{r}) \right] . \tag{5.136}$$

In analogy with Eq. (2.93), we have

$$[\mathrm{a}_\lambda(\vec{k}), \, \mathrm{a}^+_{\lambda'}(\vec{k}')] = \delta_{\vec{k}\vec{k}'} \, \delta_{\lambda\lambda'} . \tag{5.137}$$

In vacuum, the "mode functions" are given as follows [see Eq. (2.129a)],

$$\vec{A}_{\vec{k},\lambda}(\vec{r}) = \left(\frac{\hbar}{2\epsilon_0 \, \omega \, V} \right)^{1/2} \hat{\epsilon}_{\vec{k},\lambda} \, \mathrm{e}^{\mathrm{i}\vec{k}\cdot\vec{r}}, \quad \int_V \mathrm{d}^3 r \, \left| \vec{A}_{\vec{k},\lambda}(\vec{r}) \right|^2 = \frac{\hbar}{2\epsilon_0 \omega}, \quad \mathcal{A}_0 = \sqrt{\frac{\hbar}{2 \, \epsilon_0 \, \omega}} . \tag{5.138}$$

The formula for $\mathcal{A}_0$ follows from a comparison to Eq. (5.135). We now consider an atom located at $x = \frac{1}{2} L$, $y = \frac{1}{2} L$, i.e.,

$$\vec{r} = \vec{r}_z = \frac{L}{2} \, \hat{\mathrm{e}}_x + \frac{L}{2} \, \hat{\mathrm{e}}_y + z \, \hat{\mathrm{e}}_z , \tag{5.139}$$

while the atom's z coordinate denotes the distance from the perfectly conducting plate located at $z = 0$. We use the result $\vec{E}_{\vec{k}\lambda} = -\partial \vec{A}_{\vec{k}\lambda}/\partial t = \mathrm{i}\omega \vec{A}_{\vec{k}\lambda}$. According to Eq. (8.4) of Ref. [28], one can use the following *ansatz* for the energy shift $\mathcal{E}_{\ell=1}(z)$ due to the dipole interaction with the vacuum modes (the subscript refers to the dipole interaction),

$$\begin{aligned}\mathcal{E}_1(z) &= -\frac{1}{2}\sum_{\vec{k}\lambda}\alpha(\omega)\left|\vec{E}_{\vec{k}\lambda}(\vec{r}_z)\right|^2 = -\frac{1}{2}\sum_{\vec{k}\lambda}\alpha(\omega)\,\omega^2\left(\frac{\hbar}{2\epsilon_0\omega}\right)\left|\frac{\vec{A}_{\vec{k}\lambda}(\vec{r}_z)}{\mathcal{A}_0}\right|^2\\ &= -\frac{4}{V}\sum_{\vec{k}\lambda}\alpha(\omega)\left(\frac{\hbar\omega}{2\epsilon_0}\right)\left[\left(\hat{\epsilon}_{\vec{k},\lambda,x}\right)^2\cos^2\left(\frac{k_xL}{2}\right)\sin^2\left(\frac{k_yL}{2}\right)\sin^2(k_zz)\right.\\ &\quad+\left(\hat{\epsilon}_{\vec{k},\lambda,y}\right)^2\sin^2\left(\frac{k_xL}{2}\right)\cos^2\left(\frac{k_yL}{2}\right)\sin^2(k_zz)\\ &\quad\left.+\left(\hat{\epsilon}_{\vec{k},\lambda,z}\right)^2\sin^2\left(\frac{k_xL}{2}\right)\sin^2\left(\frac{k_yL}{2}\right)\cos^2(k_zz)\right].\end{aligned} \tag{5.140}$$

Here, the fundamental modes $\vec{A}_{\vec{k}\lambda}$ are given in Eq. (5.125). We write the dipole polarizability [see Eq. (3.53)] as $\alpha(\omega)$, where we assume the reference state to be the atomic ground state. The justification of the *ansatz* given in Eq. (5.140) is not as easy as one might assume. Namely, we recall from Secs. 4.6.1 and 4.6.2 that the calculation of energy shifts involving a reference state with the atom in the ground state, and the photon field in the vacuum state, implies that the intermediate (virtual) state has to be a one-photon state. In turn, this implies that we have only one propagator denominator, as is evident from Eq. (4.302), not two, as we would otherwise need for the polarizability [see Eqs. (3.68)–(3.70)]. Indeed, the *ansatz* given in Eq. (5.140) can be justified based on a Wick rotation, as explained in Eqs. (7.116)–(7.120) of Ref. [28].

In the limit $L \to \infty$, the energy shift has to be independent of the precise location of the atom in the x and the y directions, and we can thus average over x and y, which is equivalent to an average over L. Thus, we replace every square of a trigonometric function whose argument depends only on x and y, by $\frac{1}{2}$, and obtain the result

$$\mathcal{E}_1(z) = -\sum_{\vec{k}\lambda}\frac{\hbar\alpha(\omega)\,\omega}{2\epsilon_0V}\left[\left\{\left(\hat{\epsilon}_{\vec{k},\lambda,x}\right)^2+\left(\hat{\epsilon}_{\vec{k},\lambda,y}\right)^2\right\}\sin^2(k_zz)+\left(\hat{\epsilon}_{\vec{k},\lambda,z}\right)^2\cos^2(k_zz)\right]. \tag{5.141}$$

We recall that $\omega = ck$, so that it is impossible to pull the ω's out of the sum. We can renormalize this expression by considering the limit $z \to \infty$. For very large distance from the perfect conductor, the energy shift has to vanish. Also, in this limit, one can replace the squares of the trigonometric functions by their average values, $\sin^2(k_zz) \to 1/2$ and $\cos^2(k_zz) \to 1/2$, arguing that deviations from the average values could be damped away by a convergent factor $\exp(-\eta\, z)$ for infinitesimal η. We can thus implement the renormalization by the replacements

$$\sin^2(k_z\, z) \to \sin^2(k_z\, z) - \tfrac{1}{2}, \tag{5.142a}$$

$$\cos^2(k_z\, z) \to \cos^2(k_z\, z) - \tfrac{1}{2} = \tfrac{1}{2} - \sin^2(k_z\, z) = -\left(\sin^2(k_z\, z) - \frac{1}{2}\right). \tag{5.142b}$$

So, after renormalization and the transition to the continuum limit of k vectors, one obtains with the help of Eq. (5.131),

$$\begin{aligned}\mathcal{E}_1(z) &= \frac{\hbar}{4\epsilon_0 V}\sum_{\vec{k}\lambda}\alpha(\omega)\,\omega\,\cos(2k_z z)\left[\left(\hat{\epsilon}_{\lambda\,x}(\vec{k})\right)^2+\left(\hat{\epsilon}_{\lambda\,y}(\vec{k})\right)^2-\left(\hat{\epsilon}_{\lambda\,z}(\vec{k})\right)^2\right]\\ &= \frac{\hbar}{4\epsilon_0 V}\sum_{\vec{k}}\alpha(\omega)\,\omega\,\cos(2k_z z)\left[\frac{k_y^2+k_z^2}{k^2}+\frac{k_x^2+k_z^2}{k^2}-\frac{k_x^2+k_y^2}{k^2}\right]\\ &= \frac{\hbar}{2\epsilon_0 V}\sum_{\vec{k}}\alpha(\omega)\,\omega\,\cos(2k_z z)\,\frac{k_z^2}{k^2} = \frac{\hbar}{2\epsilon_0 V}\,\frac{V}{8\pi^3}\int \mathrm{d}^3k\,\alpha(\omega)\,\omega\,\cos(2k_z z)\,\frac{k_z^2}{k^2}\,.\end{aligned} \tag{5.143}$$

In the last step, we have used Eq. (2.81) one more time. If we now use the relation $\omega = c\,k$ and transform to spherical coordinates, then the z-dependent Casimir energy is obtained as

$$\begin{aligned}\mathcal{E}_1(z) &= \frac{\hbar}{2\epsilon_0}\,\frac{1}{8\pi^3}\,2\pi c\int_0^\pi \mathrm{d}\theta\,\sin\theta\cos^2\theta\int_0^\infty \mathrm{d}k\,\alpha(c\,k)\,k^3\,\cos(2kz\cos\theta)\\ &= \frac{\hbar c}{8\pi^2\epsilon_0}\int_0^\infty \mathrm{d}k\,\alpha(c\,k)\,k^3\int_0^\pi \mathrm{d}\theta\,\sin\theta\cos^2\theta\cos(2kz\cos\theta)\\ &= \lim_{\eta\to 0}\frac{\hbar c}{8\pi^2\epsilon_0}\int_0^\infty \mathrm{d}\omega\alpha(\omega)\exp\left(\frac{2\mathrm{i}\omega z}{c}\right)\left(\frac{\mathrm{i}}{4cz^3}+\frac{\omega}{2c^2z^2}-\frac{\mathrm{i}\omega^2}{2c^3z}\right)\mathrm{e}^{-\eta\,\omega/c}+\text{c.c.}\\ &= -\frac{\hbar}{16\pi^2\epsilon_0 z^3}\int_0^\infty \mathrm{d}\omega\,\alpha(\mathrm{i}\omega)\,\exp\left(-\frac{2\omega z}{c}\right)\left[1+2\frac{\omega z}{c}+2\left(\frac{\omega z}{c}\right)^2\right]\\ &= \begin{cases} -\dfrac{\hbar}{16\pi^2\epsilon_0 z^3}\displaystyle\int_0^\infty \mathrm{d}\omega\,\alpha(\mathrm{i}\omega) & (z\to 0)\\ -\dfrac{3\hbar c\,\alpha(\omega=0)}{32\pi^2\,\epsilon_0\,z^4}, & (z\to\infty)\end{cases}\,.\end{aligned} \tag{5.144}$$

In the last two lines in this equation, we assume that the reference state is the ground state, so that the Wick rotation can be performed without problems. Considerable additional complexity is encountered in the analysis of excited-state interactions [229, 230]. We have used a convergence generating factor $\exp(-\eta\,\omega/c)$. In the calculation of the short- and long-distance limits in Eq. (5.144), the meaning of the retardation expansion is very obvious. Namely, for $c \to \infty$, we can simply replace

$$\exp\left(-\frac{2\omega z}{c}\right)\left[1+2\frac{\omega z}{c}+2\left(\frac{\omega z}{c}\right)^2\right]\to 1\,,\qquad c\to\infty\,, \tag{5.145}$$

which is the appropriate expansion for the short-distance limit. In the non-retardation limit, the speed of light can be approximated as being infinite, and one can let $1/c \to 0$. By contrast, for $z \to \infty$, the quantity $\omega\,z/c$ is not small, and thus all the terms in the retardation expansion [i.e., the expansion of the factor (5.145)] in terms of its argument $\omega\,z/c$ are relevant. In that limit, the magnitude of the quantity c/z relevant to the integration region over ω in Eq. (5.144) is so tiny that it can be neglected when compared to an atomic excitation frequency, and thus, we can approximate $\alpha(\omega) \to \alpha(\omega = 0)$.

5.5.3 *Multipole Interactions*

It is very instructive to generalize the result (5.144) to higher multipole orders. The higher multipoles correspond to corrections to the excitation dynamics due to the small changes in the phases of the virtual electromagnetic field quanta on length scales commensurate with the Bohr radius (radius of the atom). The 2^ℓ-pole dynamic polarizability α_ℓ is the generalization of the dipole ($\ell = 1$) expression in Eq. (3.53) to higher multipole orders; these are called dipole ($2^{\ell=1} = 2$), quadrupole ($2^{\ell=2} = 4$), octupole ($2^{\ell=3} = 8$), and hexadecupole ($2^{\ell=4} = 16$) corrections.

In order to approach the problem of a consistent definition of the multipole oscillator strength, let us make contact with the literature and consult Ref. [231], which contains a number of useful formulas for quantities of universal interest in atomic physics. We start from Eq. (27) of Ref. [231], where the Einstein A coefficient is defined for spontaneous emission from an upper level, denoted as level 2, to a lower level 1,

$$A_{21} = \frac{e^2 \omega_{21}^3}{3\pi\epsilon_0 \hbar c^3} \sum_{m_1} |\langle 1\, m_1 | \vec{r} | 2\, m_2 \rangle|^2 . \tag{5.146}$$

A comparison with Eq. (3.37), which was written in natural units, reveals that

$$A_{21} = W_{fi} = \gamma_{21} , \tag{5.147}$$

where $\gamma_{21} = 1/\tau_{21}$ is the contribution to the decay rate of the upper level 2 due to electric-dipole (E1) decay, to the lower level 1, due to spontaneous emission of a resonant photon. By τ_{21}, we denote the lifetime of level 2 against spontaneous electric-dipole decay to level 1. If there is more than one decay channel for the spontaneous transition $2 \to 1$, then the corresponding contributions to the decay rate have to be added to the electric-dipole decay rate $\gamma_{21} = A_{21}$. Additional decay channels are common. For example, in the case of the $nS \to 1S$ decay in hydrogenlike ions, the two most important contributions to the total decay rate are the sum of two-photon decay (2E1, see Refs. [232,233]) and one-photon magnetic-dipole (M1) decay. Two-photon decay $nS \to 1S$ decay proceeds via a virtual intermediate $n'P$ state, where n' can be smaller or larger than n. For $nS \to 1S$ decay, with $n \geq 3$, it is an interesting problem to separate the coherent contribution due to two-photon decay from the cascade process $nS \to n'P \to 1S$, where, for the cascade process, one has $n > n' > 1$ (see Refs. [183, 196, 197, 234]). For $P \to S$ transition, the (dominant) electric-dipole decay rate receives corrections due to more complex processes like two-photon E1M1 and E1M2 decay (see Refs. [235, 236]).

For clarity and ease of understanding, we remark that in Eq. (27) of Ref. [231], the trivial relationship $2/(3h) = 1/(3\pi\hbar)$ is used. The (dipole) oscillator strength, according to Eq. (22) of Ref. [231], is defined as

$$f_{21} = -\frac{1}{3} A_{21}\, \gamma_{\rm cl}^{-1} , \tag{5.148}$$

where the classical decay rate for a single-electron oscillator at frequency ω_{21} is

$$\gamma_{\rm cl} = \frac{e^2\, \omega_{21}^2}{6\pi\epsilon_0 m c^3} . \tag{5.149}$$

Here, ω_{21} has the interpretation of the (angular) transition frequency, in atomic units. Note that the oscillator strength defined according to Eq. (5.148) is a manifestly dimensionless quantity, obtained as the ratio of the spontaneous (atomic) dipole decay rate to the classical decay rate.

According to Eq. (23) of Ref. [231], the so-called gf value for a transition is given as

$$g_1 f_{12} = -g_2 f_{21} \equiv gf \,. \tag{5.150}$$

Furthermore, according to Eq. (24) of Ref. [231], the oscillator strength f_{12} (as opposed to f_{21}) for the transition $1 \to 2$ (lower to upper level) is given as

$$f_{12} = -\frac{g_2}{g_1} f_{21} = \frac{g_2}{g_1} A_{21} \frac{2\pi\epsilon_0 m c^3}{e^2 \omega_{21}^2} = \frac{2}{3} \frac{g_2}{g_1} \frac{m}{\hbar} \omega_{21} \sum_{m_1} |\langle 1\, m_1 | \vec{r} | 2\, m_2 \rangle|^2 \,. \tag{5.151}$$

Here, g_1 and g_2 are the multiplicities of levels 1 and 2 (due to magnetic sublevels). The sum over m_1 ensures that f_{12} is independent of m_2. In Eq. (5.151), the sum over m_1 and the division by g_1 correspond to an averaging over the magnetic projections of level 1. If $|1 m_1\rangle$ is the $1S$ ground state of a hydrogenlike system, then one can alternatively write

$$f_{12} = \frac{2}{3} \frac{m}{\hbar} \omega_{21} \sum_{m_2} |\langle 1S | \vec{r} | 2\, m_2 \rangle|^2 \,, \tag{5.152}$$

where the factor $1/g_1$ cancels the sum over m_1, and the factor g_2 is alternatively written as a sum over m_2.

The above considerations will now be applied to the dynamic polarizability of an atom, which we assume has a spherically symmetric ground state. We recall the definition of the ground-state dynamic polarizability $\alpha(\omega)$ from Eq. (3.53),

$$\alpha(\omega) = \frac{e^2}{3} \sum_{\pm} \left\langle 1S \left| \vec{r} \, \frac{1}{H_S - E_{1S} \pm \hbar\omega} \, \vec{r} \right| 1S \right\rangle , \tag{5.153}$$

where H_S is the Schrödinger–Coulomb Hamiltonian. (For a many-electron atom, one needs to sum over the coordinates of all electrons, as described by the substitution $\vec{r} \to \sum_i \vec{r}_i$.) Inserting a complete set of virtual states $|k\rangle$, one easily shows that

$$\alpha(\omega) = \sum_k \frac{F_{1k}}{E_{k1}^2 - (\hbar\omega)^2} \,, \tag{5.154}$$

where the sum over k comprises the continuous spectrum, and F_{1k} is a dimensionful oscillator strength defined in Eq. (5.155). Here, $E_{k1} = E_k - E_1 = \hbar\omega_{k1}$ is the transition energy for the transition $|1\rangle \to |k\rangle$, where $|1\rangle$ is the reference $1S$ ground state and $|k\rangle \equiv |k\, m_k\rangle$ is the virtual state which is connected to the ground state by an allowed electric-dipole transition and has magnetic projection m_k. We should add that we here do not consider the infinitesimal displacement of the poles (as a function of ω) in Eq. (5.154). For a comprehensive discussion of this question, see

Sec. 12.2.2. Furthermore, F_{1k} is the dimensionful (dipole) oscillator strength for the transition,

$$F_{1k} = \frac{2}{3} e^2 E_{k1} \sum_{m_k} |\langle 1S|\vec{r}|k\, m_k\rangle|^2 = e^2\, a_0^2\, E_h\, f_{1k}\,, \tag{5.155}$$

where f_{1k} is the dimensionless oscillator strength for the transition $|1\rangle \to |k\rangle$, defined according to Eq. (5.152). The prefactors $e^2\, a_0^2\, E_h$ imply that the dimensionful oscillator strength is equal to its value expressed in atomic units (we recall that E_h is the Hartree energy).

The question now is how to generalize the concept of the oscillator strength to multipole transition. To this end, we first write the dimensionful dipole oscillator strength as

$$\begin{aligned} F_{1k} &= \frac{2}{3} e^2 E_{kn} \sum_{m_k} |\langle 1S|\vec{r}|k\, m_k\rangle|^2 \\ &= 2\, e^2 \frac{1}{3} E_{kn} \sum_{m} \sum_{m_k} \left| \langle 1S| \sqrt{\frac{4\pi}{3}}\, r\, Y_{1m}(\hat{r}) |k\, m_k\rangle \right|^2 . \end{aligned} \tag{5.156}$$

For the argument of the spherical harmonics, we have used a more compact notation, denoting the arguments of the spherical harmonic as

$$Y_{\ell m}(\hat{r}) \equiv Y_{\ell m}(\theta, \varphi)\,, \tag{5.157}$$

observing that the polar and azimuth angles θ and φ define $\hat{r}$ uniquely, and vice versa. The 2^ℓ-pole polarizability can alternatively be written as

$$\alpha_\ell(\omega) = \frac{1}{2\ell+1} \sum_m \left\langle 1S \left| Q_{\ell m} \left(\frac{1}{H - E_{1S} - \hbar\omega} + \frac{1}{H - E_{1S} + \hbar\omega} \right) Q^*_{\ell m} \right| 1S \right\rangle , \tag{5.158}$$

where the $Q_{\ell m}$ tensor is given as

$$Q_{\ell m} = e \sqrt{\frac{4\pi}{2\ell + 1}}\, r\, Y_{\ell m}(\hat{r}_i)\,. \tag{5.159}$$

The physical dimension of the 2^ℓ-pole polarizability is charge squared, times energy multiplied by length to the power of 2ℓ. Therefore, the multipole corrections to the dipole-induced energy shift $\mathcal{E}_1(z)$ given in Eq. (5.144) have more inverse powers of z in order to compensate for the larger number of atomic coordinates in the multipole polarizabilities. The generalization of $\mathcal{E}_1(z)$ given in Eq. (5.140), which we recall for convenience as $\mathcal{E}_1(z) = -\frac{1}{2} \sum_{\vec{k}\lambda} \alpha(\omega) \left|\vec{E}_{\vec{k}\lambda}(\vec{r}_z)\right|^2$, to the quadrupole and octupole interactions, is as follows. For the quadrupole term ($\ell = 2$), one finds

$$\mathcal{E}_{\ell=2} = -\frac{1}{12} \sum_{\vec{k}\lambda} \sum_{ij} \alpha_2(\omega) \left| \left\{ \nabla^j E^i_{\vec{k}\lambda}(\vec{r}) \right\} \right|_{\ell=2} \Big|^2 , \tag{5.160}$$

where we reserve the index i for the component of the electric field. Here,

$$T^{ij}\Big|_{\ell=2} = \frac{1}{2} \left(T^{ij} + T^{ji} \right) - \frac{1}{3} \delta^{ij}\, \mathrm{Tr}(T) \tag{5.161}$$

is the traceless quadrupole component of a general tensor T^{ij} (see Sec. 6.2.1). The octupole energy shift ($\ell = 3$) is

$$\mathcal{E}_3 = -\frac{1}{180} \sum_{\vec{k}\lambda} \sum_{ijk} \alpha_3(\omega) \left| \{\nabla^j \nabla^k E^i_{\vec{k}\lambda}(\vec{r})\}\big|_{\ell=3} \right|^2 . \tag{5.162}$$

The $\ell = 3$ component can be extracted from a rank-3 tensor T^{ijk} as follows [237]. One first decomposes T^{ijk} into a totally symmetric part S^{ijk}, a totally skew-symmetric part A^{ijk}, and a remainder term N^{ijk},

$$S^{ijk} = T^{(ijk)} = \frac{1}{6}\left(T^{ijk} + T^{jki} + T^{kij} + T^{jik} + T^{kji} + T^{ikj}\right), \tag{5.163a}$$

$$A^{ijk} = T^{[ijk]} = \frac{1}{6}\left(T^{ijk} + T^{jki} + T^{kij} - T^{jik} - T^{kji} - T^{ikj}\right), \tag{5.163b}$$

$$N^{ijk} = T^{[ijk]} - S^{ijk} - A^{ijk} . \tag{5.163c}$$

The matrix S^{ijk} contains both $\ell = 1$ as well as $\ell = 3$ components. One then forms a vector α^k which one promotes to a matrix K^{ijk},

$$\alpha^k = \delta^{ij}\, S^{ijk} = \frac{1}{9}\left(\delta^{ij}\, T^{ijk} + \delta^{ij}\, T^{ikj} + \delta^{ij}\, T^{kij}\right),$$
$$K^{ijk} = \frac{1}{5}\left(\alpha^i\, \delta^{jk} + \alpha^j\, \delta^{ik} + \alpha^k\, \delta^{ij}\right). \tag{5.164}$$

Finally, the $\ell = 3$ components is extracted as follows,

$$T^{ijk}\big|_{\ell=3} = S^{ijk} - K^{ijk} . \tag{5.165}$$

From Eqs. (5.160) and (5.162), one derives the following results for the energy shifts in the case of a perfectly conducting surface. For $\ell = 2$ (quadrupole case), the energy shift is found as follows,

$$\mathcal{E}_2(z) = -\frac{\hbar}{16\pi^2\epsilon_0 z^5} \int_0^\infty \mathrm{d}\omega\, \alpha_2(\mathrm{i}\omega)\, \exp\left(-\frac{2\omega z}{c}\right) \times \left[\frac{3}{4} + \frac{3}{2}\frac{\omega z}{c} + \frac{4}{3}\left(\frac{\omega z}{c}\right)^2 + \frac{2}{3}\left(\frac{\omega z}{c}\right)^3 + \frac{1}{6}\left(\frac{\omega z}{c}\right)^4\right]. \tag{5.166}$$

For the octupole term, we have

$$\mathcal{E}_3(z) = -\frac{\hbar}{16\pi^2\epsilon_0 z^7} \int_0^\infty \mathrm{d}\omega\, \alpha_3(\mathrm{i}\omega) \mathrm{e}^{-2\omega z/c} \left[\frac{69}{320} + \frac{69}{320}\frac{\omega z}{c} + \frac{2863}{7200}\left(\frac{\omega z}{c}\right)^2 + \frac{793}{3600}\left(\frac{\omega z}{c}\right)^3 + \frac{71}{900}\left(\frac{\omega z}{c}\right)^4 + \frac{47}{2700}\left(\frac{\omega z}{c}\right)^5 + \frac{1}{675}\left(\frac{\omega z}{c}\right)^6\right]. \tag{5.167}$$

From the above formulas, it becomes clear that, for large distances, the multipole polarizabilities are suppressed by higher powers of z.

5.5.4 *Interactions with a Dielectric Surface*

It is very instructive and useful to generalize the above results, which were obtained for atom-surface interactions with a perfect conductor, to a nontrivial frequency-dependent dielectric function $\epsilon(\omega)$. In this case, the reflection and transmission at the boundary (plane $z = 0$) of the upper half space ($z > 0$, outside of the dielectric) and the lower half space $z < 0$ (inside the dielectric) need to be considered. In principle, one needs to enlarge the reference volume $V = L^2 L_z$ used in the previous considerations to include a region filled by the dielectric, and to calculate the field modes in the enlarged reference volume. In practice, it is possible to simplify the calculation by considering the field modes only in the region above the surface, $z > 0$, as a superposition of incident and reflected waves. If the incident wave has a wave vector $\vec{k} = (k_x, k_y, k_z)$, then the wave vector of the reflected wave reads as $\vec{k}^{\mathrm{R}} = (k_x, k_y, -k_z)$. There is no change in the wavelength; we have $|\vec{k}| = |\vec{k}^{\mathrm{R}}|$. However, the amplitude of the reflected wave is less than that of the incident wave, due to a nonvanishing transmission coefficient. One then has to consider the so-called TE and TM modes separately, where TE in this case stands for incident waves whose electric field is transverse to the plane of incidence, whereas TM stands for waves whose electric field is in the plane of incidence.

The calculation is described in some detail in Ref. [238]. For given wave vector $\vec{k}$, one has only one polarization vector characterizing the TE mode, and one for the TM mode. Alternatively, one can average over the polarizations, which corresponds to an average over a conical region of incident waves. The weight factor in formulating the superposition of incident and reflected waves is $1/\sqrt{2}$, which preserves the normalization of the modes given in Eq. (5.135) in the limit of a perfect conductor. With these ideas in mind, it is relatively easy to rederive the following result, originally derived in Ref. [214] for a dipole polarizable particle in contact with a dielectric surface,

$$\mathcal{E}_1(z) = -\frac{\hbar}{8\pi^2 \epsilon_0 c^3} \int_0^\infty \mathrm{d}\omega\, \omega^3 \alpha_1(\mathrm{i}\omega) \int_1^\infty \mathrm{d}\xi \exp\left(-\frac{2\xi\omega z}{c}\right) \mathrm{H}(\xi, \epsilon(\mathrm{i}\omega)), \tag{5.168a}$$

$$\mathrm{H}(\xi,\epsilon) = (1 - 2\xi^2)\frac{\sqrt{\xi^2 + \epsilon - 1} - \epsilon\xi}{\sqrt{\xi^2 + \epsilon - 1} + \epsilon\xi} + \frac{\sqrt{\xi^2 + \epsilon - 1} - \xi}{\sqrt{\xi^2 + \epsilon - 1} + \xi}. \tag{5.168b}$$

It is useful to investigate the function

$$K(\epsilon, z) = \int_1^\infty \mathrm{d}\xi \exp\left(-\frac{2\xi\omega z}{c}\right) \mathrm{H}(\xi, \epsilon). \tag{5.169}$$

For $\epsilon \to \infty$, we have

$$\mathrm{H}(\xi,\epsilon) \sim 2\xi^2, \qquad K(\epsilon, z) \sim \frac{1}{2}\left(\frac{c}{z\omega}\right)^3 \left[1 + 2\frac{z\omega}{c} + 2\left(\frac{z\omega}{c}\right)^2\right], \qquad \epsilon \to \infty, \tag{5.170}$$

which shows that the formula (5.168a) is consistent with the result given in Eq. (5.144) for the dipole interaction with a perfect conductor. For $z \to 0$, we have, on the other hand,

$$K(\epsilon, z) = \frac{1}{2}\left(\frac{c}{z\,\omega}\right)^3 \frac{\epsilon - 1}{\epsilon + 1} - \frac{1}{2}\left(\frac{c}{z\,\omega}\right) \frac{(\epsilon - 1)\,(3\,\epsilon + 1)}{(\epsilon + 1)^2} + \mathcal{O}(z^0)\,, \qquad z \to 0\,. \tag{5.171}$$

Inserting (5.171) into (5.168a), we have

$$\mathcal{E}_1(z) = -\frac{\hbar}{16\pi^2\,\epsilon_0\,z^3} \int_0^\infty \mathrm{d}\omega\,\alpha_1(\mathrm{i}\omega)\,\frac{\epsilon(\mathrm{i}\omega) - 1}{\epsilon(\mathrm{i}\omega) + 1}\,, \qquad z \to 0\,, \qquad \epsilon \to \infty\,. \tag{5.172}$$

The generalization to the quadrupole polarizability reads as follows,

$$\mathcal{E}_2(z) = -\frac{\hbar}{16\pi^2\,\epsilon_0\,c^5} \int_0^\infty \mathrm{d}\omega\,\omega^5\,\alpha_2(\mathrm{i}\omega) \int_1^\infty \mathrm{d}\xi\,\mathrm{e}^{-2\xi\omega z/c} \left(\frac{\xi^2}{2} - \frac{1}{3}\right) \mathrm{H}(\xi, \epsilon(\mathrm{i}\omega))\,. \tag{5.173}$$

In the perfect conductor limit ($\epsilon \to \infty$), this result is in agreement with the previously derived result given in Eq. (5.166). The octupole energy shift reads

$$\begin{aligned} \mathcal{E}_3(z) = &-\frac{\hbar}{16\pi^2\,\epsilon_0\,c^7} \int_0^\infty \mathrm{d}\omega\,\omega^7\,\alpha_3(\mathrm{i}\omega) \int_1^\infty \mathrm{d}\xi\,\mathrm{e}^{-2\xi\omega z/c} \\ &\times \left(\frac{23}{1200}\,\xi^4 - \frac{121}{5400}\,\xi^2 + \frac{17}{3600}\right) \mathrm{H}(\xi, \epsilon(\mathrm{i}\omega))\,. \end{aligned} \tag{5.174}$$

In the perfect conductor limit ($\epsilon \to \infty$), the result given in Eq. (5.167) is recovered. For completeness, we should mention that the calculation of the dielectric response function for imaginary input frequencies is a nontrivial problem. For a typical solid, it involves the application of the KRAMERS[3]–KRONIG[4] relation to tabulated data for the frequency-dependent dielectric function of a particular material. A comprehensive compilation of pertinent data can be found in Ref. [239]. The reader should be aware, though, that some of the data compiled in Ref. [239] needs to be seen as outdated from today's point of view.

5.6 Further Thoughts

Here are some suggestions for further thought.

(1) **Ion–Ion and Ion–Atom Interactions.** In Sec. 5.4.2, we had observed that, if the two nuclei of the hydrogenlike systems have charge numbers other than unity, then the two-photon exchange energy is dominated by either the electrostatic interaction among the ions, or the ion–atom interaction. Calculate the leading functional form of both interactions.

(2) **Expressions for the Propagator Denominators.** Derive the propagator denominators listed in Eq. (5.67), one by one.

(3) **Interpolating Formula.** Derive Eq. (5.117) for the interpolating formula, in between the Van-der-Waals and Casimir–Polder regimes, on the basis of collecting all terms listed in Eq. (5.67).

[3]Hendrik Anthony Kramers (1894–1952).
[4]Ralph de Laer Kronig (1904–1995).

(4) **Symmetrized Integrand.** The asymmetric (in k_1 and k_2) form of the integrand given in Eq. (5.106) allows us to carry out the k_2-integration as a principal-value integral [see Eqs. (5.97), (5.106) and (5.108)]. Convince yourself, conceivably with the help of a computer algebra system, that the symmetrized form of the integrand $\sum_{i=1}^{12}[f_i(k_1,k_2)]^{-1}$ is equal to the symmetrized form of the integrand given in Eq. (5.106), and try to complete the calculation of the long-range energy shift, with the aim of verifying Eq. (5.117).

(5) **Long-Range Tail for Excited States.** Derive the oscillating tail of the long-range interaction for excited reference states, by matching the energy shift with the corresponding S-matrix element (see Sec. 12.2.2). Contrast with the treatment in Sec. 5.4.5. The matching of the S-matrix element with the energy shift must be used for excited states, in order to correctly assign the position of poles in the propagator denominators, in the complex plane. Consult Refs. [208, 209, 240] for additional inspiration.

(6) **Atom-Surface Interactions.** Calculate the curl of the vector potentials given in Eq. (5.125) and show that the normal component of the magnetic field (the component in the direction of the normal to the boundary surface) vanishes on the boundaries. Furthermore, show that this is necessary, in order to fulfill Faraday's law.

(7) **Atom-Surface Interactions.** Generalize the vector potentials $\vec{A}(\vec{r})$ and the vector field operator given in Eq. (5.136) given in Eq. (5.125) to an interaction-picture vector-field operator $\vec{A}(\vec{r},t)$. Write the electric-field and magnetic-field operators $\vec{E}(\vec{r},t)$ and $\vec{B}(\vec{r},t)$ and show that these are Hermitian. Give a physical argument why these operators should be Hermitian even under the presence of boundary conditions corresponding to a box.

(8) **Two-Photon Decays.** Study the separation of the coherent two-photon decay rate correction from cascades (see Refs. [183, 196, 197, 234]). Generalize the treatment to forbidden transitions of current interest for high-precision spectroscopy.

(9) **Oscillator Strengths.** Calculate the dimensionful oscillator strength, defined according to Eq. (5.155), in the nonrelativistic approximation for a hydrogen atom in the limit of infinite nuclear mass, for a $1S \to nP$ transition. *Hint:* You might obtain a result similar to

$$F_{1n} = \frac{256\, n^5}{3\,(n^2-1)^4} \left(\frac{n-1}{n+1}\right)^{2n} e^2\, a_0^2\, E_h\,. \tag{5.175}$$

Compare your result to Ref. [241].

(10) **Incident and Reflected Waves.** Try to derive Eq. (5.168a), based on the incident and reflected waves off of a dielectric surface. *Hint:* Consult Ref. [238].

(11) **Excited States.** How would the formalism outlined in Sec. 5.5 have to be altered for excited states? *Hint:* Consult Refs. [229, 230]. Would you still be able to use the *ansatz* given in Eq. (5.140), for excited states?

Chapter 6

Racah–Wigner Algebra

6.1 Overview

The most important global symmetry of an atom is the rotational symmetry of the Coulomb potential, which is generated by the nucleus. Therefore, atomic physics is intimately connected with the group of rotations in three dimensions, which is the special orthogonal group $SO(3)$. This symmetry can be augmented, in the case of the hydrogen atom, to $SO(4)$, as described in detail in Chap. 4. The theory of the rotation group has been perfectioned over the years by generations of physicists, starting from the work of RACAH[1] and WIGNER;[2] it is our goal here to present an intuitively accessible, concise chapter on angular integration methods in atomic physics which involve the Clebsch–Gordan coefficients. Emphasis is laid on an intuitive understanding, so that more specialized literature can be digested by the interested reader with ease. We refer to the treatises [14, 15, 242, 243] for general understanding, as well as to the monumental handbook [16] for use in research, and we also mention Ref. [23], which is very useful in the analysis of more complex systems. All of these books are self-contained and voluminous in nature. However, a student interested in gaining a working knowledge of the field, or a working scientist interested in refreshing one's knowledge on related topics, sometimes does not have the time to work the way through an entire book in order to familiarize oneself, or re-familiarize oneself with the topic. In general, the Racah–Wigner algebra associated with a group (in this case, the group of rotations) is generally understood as the set of algebraic manipulations concerning the coupling and recoupling coefficients that pertain to the irreducible representations of the group.

6.2 Clebsch–Gordan Coefficients

6.2.1 *Expansions and Clebsch–Gordan Coefficients*

The motivation for the definition of Clebsch–Gordan coefficients is twofold: first, to find the expansion coefficients of a tensor of higher rank as it is composed out of

[1]Giulio (Yoel) Racah (1909–1965).
[2]Eugene Paul Wigner (1902–1995).

elements of tensors of lower rank, and second, to formalize a certain proportionality phenomenon that affects matrix elements of tensors of specific rank, all of which can be written in terms of a so-called reduced matrix element and an expansion factor, which is a Clebsch–Gordan coefficient. We start with the first motivation.

As outlined in Chaps. 5 and 8 of Ref. [81], the matrix-valued elements of the Lie algebra of the rotation group can be chosen as

$$(\mathbb{S}^k)^{ij} = -\mathrm{i}\epsilon^{ijk}\,, \qquad \epsilon^{123} = 1\,, \tag{6.1}$$

where ϵ^{ijk} is the totally antisymmetric Levi-Cività tensor. Here, k denotes the Cartesian component of the vector of $\mathbb{S}$ matrices. The rotation matrices fulfill the same algebraic relations as the components of the angular momentum operator,

$$[\,\mathbb{S}^i, \mathbb{S}^j\,] = \mathrm{i}\epsilon^{ijk}\,\mathbb{S}^k\,, \qquad [L^i, L^j] = \mathrm{i}\,\epsilon^{ijk}\,L^k\,, \tag{6.2}$$

where the angular momentum operator $\vec{L}$ has been defined in Eq. (4.2) and the EINSTEIN[3] summation convention is used. The angular momentum operator can be written as

$$\vec{L} = -\mathrm{i}\,\vec{r}\times\vec{\nabla}\,, \qquad L^i = -\mathrm{i}\epsilon^{ijk}\,x^j\,\frac{\partial}{\partial x^k}\,. \tag{6.3}$$

Here, we use natural units with $\hbar = c = \epsilon_0 = 1$, which simplifies the notation considerably. For all calculations reported here, SI mksA units can be restored easily. The three-dimensional rotation matrix $\mathbb{R}$ is obtained as the matrix exponential

$$\mathbb{R} = \exp\left(\mathrm{i}\sum_{k=1}^{3}\varphi^k\,\mathbb{S}^k\right) = \exp\left(\mathrm{i}\,\vec{\varphi}\cdot\vec{\mathbb{S}}\right)\,, \qquad \vec{r}\,' = \mathbb{R}\cdot\vec{r}\,. \tag{6.4}$$

Here, $\vec{\varphi}$ is a vector of rotation angles about the three axes. For small $\vec{\varphi} = \hat{n}\,\delta\varphi$, one can expand

$$\left[\exp\left(\mathrm{i}\vec{\varphi}\cdot\vec{\mathbb{S}}\right)\right]^{ij} \approx \left[\mathbb{1} + \mathrm{i}\vec{\varphi}\cdot\vec{\mathbb{S}}\right]^{ij} \approx \delta^{ij} + \varphi^k\,\epsilon^{kij}\,. \tag{6.5}$$

The components of a vector $\vec{v}$ thus transform as follows,

$$\left[\exp\left(\mathrm{i}\vec{\varphi}\cdot\vec{\mathbb{S}}\right)\right]^{ij}\,v^j \approx \left(\delta^{ij} + \varphi^k\,\epsilon^{kij}\right)v^j = v^i - \left(\vec{\varphi}\times\vec{v}\right)^i\,. \tag{6.6}$$

This is just the representation of an infinitesimal rotation by an angle $\vec{\varphi}$, in the case of a passive interpretation of the rotation. That means that the coordinate system turns by an angle $\vec{\varphi}$, and the coordinate vector $\vec{r}\to\vec{r}\,'$ needs to turn by an angle $-\vec{\varphi}$ with respect to the coordinate system, with the net result that $\vec{r}\,'$ and $\vec{r}$ represent the same vector in real space.

Under a passive rotation, a scalar function transforms as

$$f'(\vec{r}\,') = f(\vec{r}) \tag{6.7}$$

and therefore

$$\begin{aligned} f'(\vec{r}) &\approx f(\mathbb{R}^{-1}\cdot\vec{r}) = f(x^i - \varphi^k\,\epsilon^{kij}\,x^j) \approx f(\vec{r}) - \varphi^k\,\epsilon^{kij}\,x^j\,\frac{\partial}{\partial x^i}f(\vec{r}) \\ &= f(\vec{r}) + \mathrm{i}\varphi^k\left(-\mathrm{i}\,\epsilon^{kij}\,x^i\,\frac{\partial}{\partial x^j}\right)f(\vec{r}) = f(\vec{r}) + \mathrm{i}\,\vec{\varphi}\cdot\vec{L}\,f(\vec{r})\,. \end{aligned} \tag{6.8}$$

[3] Albert Einstein (1879–1955).

For finite rotations, we may thus summarize the transformation law of a vector $\vec{r}$ and of a scalar function as

$$\vec{r}' = \exp(\mathrm{i}\vec{\varphi}\cdot\vec{S})\cdot\vec{r}\,, \qquad f'(\vec{r}) = \exp(\mathrm{i}\,\vec{\varphi}\cdot\vec{L})\,f(\vec{r})\,. \tag{6.9}$$

We see that the angular momentum operator generates rotations. To fix ideas, let us study the case $\varphi = 90^\circ = \pi/2$, with

$$\mathbb{R}_{\pi/2} = \begin{pmatrix} 0 & 1 & 0 \\ -1 & 0 & 0 \\ 0 & 0 & 1 \end{pmatrix}, \qquad \mathbb{R}_{\pi/2}\cdot\begin{pmatrix} u_x \\ u_y \\ u_z \end{pmatrix} = \begin{pmatrix} u_y \\ -u_x \\ u_z \end{pmatrix}. \tag{6.10}$$

The rotated vector is $\vec{u}' = \mathbb{R}_{\pi/2}\cdot\vec{u}$, where $\vec{u}$ has the Cartesian components as indicated. It is easy to show that the modulus of a vector or the scalar product of two vectors $\vec{u}$ and $\vec{v}$ is invariant under rotations,

$$\vec{u}'\cdot\vec{v}' = \left(\mathbb{R}_{\pi/2}\cdot\vec{u}\right)\cdot\left(\mathbb{R}_{\pi/2}\cdot\vec{v}\right) = u_x v_x + u_y v_y + u_z v_z\,. \tag{6.11}$$

We shall now attempt to convince ourselves that scalar functions, which depend only on the radial variable, vectors, and the components of the quadrupole tensor, transform separately under the rotation group, i.e., they do not mix. We consider a second-rank tensor

$$\mathbb{T} = \vec{u}\otimes\vec{v} = \begin{pmatrix} u_x\,v_x & u_x\,v_y & u_x\,v_z \\ u_y\,v_x & u_y\,v_y & u_y\,v_z \\ u_z\,v_x & u_z\,v_y & u_z\,v_z \end{pmatrix}, \tag{6.12}$$

which can be decomposed as follows,

$$\mathbb{T} = \mathbb{T}|_{\ell=0} + \mathbb{T}|_{\ell=1} + \mathbb{T}|_{\ell=2}\,, \tag{6.13}$$

$$\mathbb{T}|_{\ell=0} = \frac{\mathrm{Tr}(\mathbb{T})}{3}\,\mathbb{1}_{3\times3}\,, \qquad \mathrm{Tr}(\mathbb{T}) = u_x\,v_x + u_y\,v_y + u_z\,v_z\,, \tag{6.14}$$

$$\mathbb{T}|_{\ell=1} = \frac{1}{2}\left(\mathbb{T} - \mathbb{T}^{\mathrm{T}}\right) = \begin{pmatrix} 0 & \dfrac{u_x\,v_y - u_y\,v_x}{2} & -\dfrac{u_z\,v_x - u_x\,v_z}{2} \\ -\dfrac{u_x\,v_y - u_y\,v_x}{2} & 0 & \dfrac{u_y\,v_z - u_z\,v_y}{2} \\ \dfrac{u_z\,v_x - u_x\,v_z}{2} & -\dfrac{u_y\,v_z - u_z\,v_y}{2} & 0 \end{pmatrix}. \tag{6.15}$$

A comparison of Eq. (6.14) with Eq. (6.11) shows that the $(\ell = 0)$-component is invariant under rotations.

The entries of the matrix $\mathbb{T}|_{\ell=1}$ can be identified as the component of the vector product of $\vec{u}$ and $\vec{v}$, which will be discussed in the following. Finally, the

($\ell = 2$)-component reads as

$$\mathbb{T}|_{\ell=2} = \frac{1}{2}\left(\mathbb{T} + \mathbb{T}^{\mathrm{T}}\right) - \frac{\mathrm{trc}(\mathbb{T})}{3}\,\mathbb{1}_{3\times 3} \tag{6.16}$$

$$= \begin{pmatrix} \frac{2u_x v_x - u_y v_y - u_z v_z}{3} & \frac{u_x v_y + u_y v_x}{2} & \frac{u_x v_z + u_z v_x}{2} \\ \frac{u_x v_y + u_y v_x}{2} & \frac{2u_y v_y - u_x v_x - u_z v_z}{3} & \frac{u_y v_z + u_z v_y}{2} \\ \frac{u_x v_z + u_z v_x}{2} & \frac{u_y v_z + u_z v_y}{2} & \frac{2u_z v_z - u_x v_x - u_y v_y}{3} \end{pmatrix}.$$

As already anticipated, we can order the components of $\mathbb{T}|_{\ell=1}$ into a vector, which is known as the vector product of $\vec{u}$ and $\vec{v}$,

$$\vec{u} \times \vec{v} = \begin{pmatrix} u_y v_z - u_z v_y \\ u_z v_x - u_x v_z \\ u_x v_y - u_y v_x \end{pmatrix}, \qquad (\vec{u} \times \vec{v})^i = \epsilon^{ijk} u^j v^k . \tag{6.17}$$

On the example of our rotation by 90° about the z axis, the fact that $\vec{u}\times\vec{v}$ transforms as a (pseudo-)vector can be verified as follows,

$$\vec{u}' \times \vec{v}' = \left(\mathbb{R}_{\pi/2} \cdot \vec{u}\right) \times \left(\mathbb{R}_{\pi/2} \cdot \vec{v}\right) = \begin{pmatrix} u_z v_x - u_x v_z \\ -(u_y v_z - u_z v_y) \\ u_x v_y - u_y v_x \end{pmatrix} = \mathbb{R}_{\pi/2} \cdot (\vec{u} \times \vec{v}) . \tag{6.18}$$

With the help of Eq. (6.5), we can generalize this statement to arbitrary rotations,

$$\begin{aligned} \vec{u}' \times \vec{v}' &= (\vec{u} + \vec{\varphi} \times \vec{u}) \times (\vec{v} + \vec{\varphi} \times \vec{v}) \\ &\approx \vec{u} \times \vec{v} + (\vec{\varphi} \times \vec{u}) \times \vec{v} + \vec{u} \times (\vec{\varphi} \times \vec{v}) = \vec{u} \times \vec{v} + \vec{\varphi} \times (\vec{u} \times \vec{v}) , \end{aligned} \tag{6.19}$$

where we expand to first order in φ. It is an instructive exercise to show the identity $(\vec{\varphi} \times \vec{u}) \times \vec{v} + \vec{u} \times (\vec{\varphi} \times \vec{v}) = \vec{\varphi} \times (\vec{u} \times \vec{v})$, which can be done component-wise, using the ϵ tensor.

Indeed, $\mathbb{T}|_{\ell=1}$ transforms as a (pseudo-)vector; it is possible to construct a (pseudo-)vector from products of vector components. In Cartesian components, the "coupling coefficients" can directly be read off from Eq. (6.17). However, the canonical formalism employs the spherical basis. The z component of the angular momentum operator has a particularly simple form in spherical coordinates,

$$L_z = -\mathrm{i}\left(x\,\frac{\partial}{\partial y} - y\,\frac{\partial}{\partial x}\right) = -\mathrm{i}\,\frac{\partial}{\partial \varphi} . \tag{6.20}$$

The spherical basis of vector components is chosen to generate an explicit φ-dependence of the form $\exp(im\varphi)$, i.e., it consists of eigenfunctions of L_z. The components are denoted as x_{+1}, x_0, and x_{-1} in the spherical basis and read as

$$x_{+1} = -\frac{1}{\sqrt{2}}(x + \mathrm{i}\,y) = -\frac{1}{\sqrt{2}}|\vec{r}|\,\sin\theta\,\mathrm{e}^{\mathrm{i}\phi} = \sqrt{\frac{4\pi}{3}}\,|\vec{r}|\,Y_{11}(\theta,\varphi) , \tag{6.21a}$$

$$x_0 = z = |\vec{r}|\,\cos\theta = \sqrt{\frac{4\pi}{3}}\,|\vec{r}|\,Y_{10}(\theta,\varphi) , \tag{6.21b}$$

$$x_{-1} = \frac{1}{\sqrt{2}}(x - \mathrm{i}\,y) = \frac{1}{\sqrt{2}}|\vec{r}|\,\sin\theta\,\mathrm{e}^{-\mathrm{i}\phi} = \sqrt{\frac{4\pi}{3}}\,|\vec{r}|\,Y_{1-1}(\theta,\varphi) , \tag{6.21c}$$

where we have the luxury to write $r = |\vec{r}| = \sqrt{x^2 + y^2 + z^2}$ explicitly. The emergence of phase factors of the form $\exp(\mathrm{i}\, m\, \varphi)$ is evident. The spherical components are complemented by spherical basis vectors as follows,

$$\vec{e}_{+1} = -\frac{1}{\sqrt{2}}\left(\hat{\mathrm{e}}_x + \mathrm{i}\,\hat{\mathrm{e}}_y\right), \qquad \vec{e}_0 = \hat{\mathrm{e}}_z\,, \qquad \vec{e}_{-1} = \frac{1}{\sqrt{2}}\left(\hat{\mathrm{e}}_x - \mathrm{i}\,\hat{\mathrm{e}}_y\right). \tag{6.22}$$

The coordinate vector can easily be expanded into the spherical basis,

$$\vec{r} = \sum_q (-1)^q\, x_{-q}\, \vec{e}_q\,, \qquad \vec{e}_q \cdot \vec{e}_{q'} = (-1)^q\, \delta_{q\,-q'}\,, \tag{6.23}$$

but the spherical basis vectors are not normalized to unity. Components can be extracted by calculating the scalar product of the coordinate vector $\vec{r}$ with a spherical basis vector,

$$\vec{r} \cdot \vec{e}_q = \sum_{q'} (-1)^{q'}\, x_{-q'}\, \vec{e}_{q'} \cdot \vec{e}_q = \sum_{q'} (-1)^{2q'}\, x_{-q'}\, \delta_{q\,-q'} = x_q\,, \tag{6.24}$$

where the sum over q runs over the indices $q = -1, 0, 1$.

The paradigm of the vector addition, or Clebsch–Gordan coefficients, is that one obtains a tensor component of magnetic quantum number m for a tensor of rank j by coupling two tensors of ranks j_1 and j_2 as follows,

$$w(jm) = \sum_{m_1 m_2} C^{jm}_{j_1 m_1 j_2 m_2}\, u(j_1\, m_1)\, v(j_2\, m_2)\,. \tag{6.25}$$

Here, $u(j_1\, m_1)$ and $v(j_2\, m_2)$ are two distinct tensors. In our example, we couple two tensors of rank one (the vectors $\vec{u}$ and $\vec{v}$ with spherical components u_{+1}, u_0 and u_{-1}, as well as v_{+1}, v_0 and v_{-1}), to a tensor of rank one, which is the vector product $\vec{w} = \vec{u} \times \vec{v}$. So, we have $j_1 = j_2 = j = 1$ for our example. The Clebsch–Gordan coefficients are tabulated and nowadays implemented in most modern computer algebra systems [151]. Using tabulated values, one finds

$$w_q = \sum_{q'q''} C^{1q}_{1q'\,1q''}\, u_{q'}\, v_{q''}\,, \tag{6.26a}$$

$$w_{+1} = \frac{u_{+1}\, v_0 - u_0\, v_{+1}}{\sqrt{2}} = \frac{\mathrm{i}}{\sqrt{2}} \left\{ -\frac{1}{\sqrt{2}} \left[(\vec{u} \times \vec{v})_x + \mathrm{i}\, (\vec{u} \times \vec{v})_y \right] \right\}, \tag{6.26b}$$

$$w_0 = \frac{u_{+1}\, v_{-1} - u_{-1}\, v_{+1}}{\sqrt{2}} = \frac{\mathrm{i}}{\sqrt{2}}\, (\vec{u} \times \vec{v})_z\,, \tag{6.26c}$$

$$w_{-1} = \frac{u_0\, v_{-1} - u_{-1}\, v_0}{\sqrt{2}} = \frac{\mathrm{i}}{\sqrt{2}} \left\{ \frac{1}{\sqrt{2}} \left[(\vec{u} \times \vec{v})_x - \mathrm{i}\, (\vec{u} \times \vec{v})_y \right] \right\}, \tag{6.26d}$$

$$\vec{w} = \frac{\mathrm{i}}{\sqrt{2}}\, (\vec{u} \times \vec{v})\,, \tag{6.26e}$$

where the subscripts x, y, and z denote the Cartesian components of the vector product, i.e., $(u \times v)_z = u_x\, v_y - u_y\, v_x$. The components of $\vec{w}$ are equal to those obtained by writing $\vec{u} \times \vec{v}$ in the spherical basis up to a prefactor $\mathrm{i}/\sqrt{2}$.

One can form two further linear combinations of the spherical basis vectors $\vec{e}_q$ which are of interest,

$$\vec{e}_q \times \vec{e}_{q'} = \mathrm{i}\sqrt{2}\sum_{\lambda} C^{1\lambda}_{1q1q'}\,\vec{e}_\lambda\,, \qquad \vec{r} = -\sqrt{3}\sum_{qq'} C^{00}_{1q1q'}\,x_q\,\vec{e}_{q'}\,. \tag{6.27}$$

The first of these illustrates that the vector product of two basis vectors in the spherical basis again is a vector; the second clarifies that the coordinate vector $\vec{r}$ actually is a scalar under rotations, obtained as the scalar combination (tensor of rank zero, component number zero) composed out of the spherical coordinates x_q and the spherical basis vectors $\vec{e}_q$. Of course, the vector of the x_q is not a scalar. Here, however, we consider the scalar product of the vector composed of the x_q with the vector composed of the $\vec{e}_q$. Upon rotation, this scalar product is equal to the coordinate vector obtained using the new coordinates x'_q, but multiplied with the rotated basis vectors $\vec{e}\,'_q$, thus corresponding to a passive interpretation of the rotation which leaves the physical coordinate vector $\vec{r} = \vec{r}\,'$ invariant.

Let us now investigate the ($\ell = 2$)-component given in Eq. (6.16). It is symmetric and traceless and thus has five independent components. This leaves three off-diagonal and two diagonal components to be determined; the third component on the diagonal is fixed by the condition $\mathrm{trc}\,(\mathbb{T}|_{\ell=2}) = 0$. A scalar ($\ell = 0$) has one component, a vector ($\ell = 1$) has three magnetic components which depend on φ as $\exp(im\varphi)$ with $m = -1, 0, 1$, and a quadrupole tensor has five magnetic components [which depend on φ as $\exp(im\varphi)$ with $m = -2, -1, 0, 1, 2$]. The generalization calls for $2\ell + 1$ magnetic components for a tensor of rank ℓ. Again, using tabulated values for Clebsch–Gordan coefficients of the form $C^{2q}_{1q'\,1q''}$, with $q', q'' = -1, 0, 1$ and $q = -2, -1, 0, 1, 2$, we have

$$t_q = \sum_{q'q''} C^{2q}_{1q'\,1q''}\,u_{q'}\,v_{q''}\,, \tag{6.28a}$$

$$t_{+2} = u_{+1}\,v_{+1}\,, \qquad t_{+1} = \frac{u_{+1}\,v_0 + u_0\,v_{+1}}{\sqrt{2}}\,, \tag{6.28b}$$

$$t_0 = \frac{3\,u_0\,v_0 - \vec{u}\cdot\vec{v}}{\sqrt{6}}\,, \tag{6.28c}$$

$$t_{-1} = \frac{u_{-1}\,v_0 + u_0\,v_{-1}}{\sqrt{2}}\,, \qquad t_{-2} = u_{-1}\,v_{-1}\,, \tag{6.28d}$$

where the spherical components u_{-1}, u_0, and u_{+1} are defined in Eq. (6.21). If the vectors $\vec{u}$ and $\vec{v}$ "live" in the coordinate spaces of different particles 1 and 2, then the functions t_q are eigenfunctions of the operator $\vec{L}^2 = (\vec{L}_1 + \vec{L}_2)^2$ with eigenvalue $\vec{L}^2 \to \ell(\ell+1) = 6$ and of the z component $L_z = L_{1z} + L_{2z}$ with eigenvalue $L_z \to m = q$. This can be seen easily by expressing the components of t_q in terms of spherical harmonics, in the variables θ_1 and φ_1 (for the vector u), and in the variables θ_2 and φ_2 (for the vector v).

6.2.2 *Matrix Elements and Clebsch–Gordan Coefficients*

A separate foundation for the interest in Clebsch–Gordan coefficients lies in the fact that matrix elements, which have the same radial symmetry but different angular symmetry, can be unified in terms of reduced matrix elements. This is known as the WIGNER[4]–ECKART[5] theorem and will be discussed below, by way of example. Let us consider the Cartesian components of the momentum operator, which are $p^i = -\mathrm{i}\,\partial/\partial x^i$. We have the following spherical components,

$$p_{+1} = -\frac{\mathrm{i}}{\sqrt{2}}\left(\frac{\partial}{\partial x} + \mathrm{i}\frac{\partial}{\partial y}\right), \qquad p_0 = -\mathrm{i}\,\partial_z\,, \qquad p_{-1} = \frac{\mathrm{i}}{\sqrt{2}}\left(\frac{\partial}{\partial x} - \mathrm{i}\frac{\partial}{\partial y}\right), \tag{6.29}$$

while the corresponding components of the position vector are given in Eq. (6.21). Let us investigate both matrix elements of the position as well as momentum operator "sandwiched" in between the ground state and the nonrelativistic $2P$ state of hydrogen ($Z = 1$) with angular momentum projection quantum number $m = 0$,

$$\langle 1S|\, x_{+1}\, |2P(m=0)\rangle = \langle 1S|\, p_{+1}\, |2P(m=0)\rangle = 0\,, \tag{6.30a}$$

$$\langle 1S|\, x_{-1}\, |2P(m=0)\rangle = \langle 1S|\, p_{-1}\, |2P(m=0)\rangle = 0\,, \tag{6.30b}$$

$$\langle 1S|\, x_0\, |2P(m=0)\rangle = \frac{8\mathrm{i}a_0^2}{3}\,\langle 1S|\, p_0\, |2P(m=0)\rangle = \frac{128\,\sqrt{2}}{243}\,a_0\,. \tag{6.30c}$$

Here, a_0 is the Bohr radius, and we remember that we use units with $\hbar = c = \epsilon_0 = 1$. The angular part of the $2P(m = 0)$ state, given by the spherical harmonic $Y_{10}(\theta, \varphi)$, is oriented along the z axis (see Fig. 2.2 of Ref. [81]). Physically, the fact that the only nonvanishing transition matrix element is the one with x_0 or p_0 means that the dipole radiation emitted during the $2P(m = 0) \to 1S$ transition has a dipole pattern aligned with the z axis. If the "incoming" $2P$ state has magnetic projection $m = 1$, then we have

$$\langle 1S|\, x_{+1}\, |2P(m=1)\rangle = \langle 1S|\, p_{+1}\, |2P(m=1)\rangle = 0\,, \tag{6.31a}$$

$$\langle 1S|\, x_{-1}\, |2P(m=1)\rangle = \frac{8\mathrm{i}a_0^2}{3}\,\langle 1S|\, p_{-1}\, |2P(m=1)\rangle = -\frac{128\,\sqrt{2}}{243}\,a_0\,, \tag{6.31b}$$

$$\langle 1S|\, x_0\, |2P(m=1)\rangle = \langle 1S|\, p_0\, |2P(m=1)\rangle = 0\,. \tag{6.31c}$$

Of course, the $1S$ state always has $m = 0$. The matrix elements vanish unless the magnetic quantum number of the rightmost state ($2P$ state) and the magnetic quantum number of the transition operator (position or momentum operator) add up to zero. Physically, this means that the light emitted during the transition $2P(m = 1) \to 1S$ has to be circularly polarized. Finally, for an incoming state $2P$ with magnetic projection $m = -1$, we have the following matrix elements:

$$\langle 1S|\, x_{+1}\, |2P(m=-1)\rangle = \frac{8\mathrm{i}a_0^2}{3}\,\langle 1S|\, p_{+1}\, |2P(m=-1)\rangle = -\frac{128\,\sqrt{2}}{243}\,a_0\,, \tag{6.32a}$$

$$\langle 1S|\, x_{-1}\, |2P(m=-1)\rangle = \langle 1S|\, p_{-1}\, |2P(m=-1)\rangle = 0\,, \tag{6.32b}$$

$$\langle 1S|\, x_0\, |2P(m=-1)\rangle = \langle 1S|\, p_0\, |2P(m=-1)\rangle = 0\,. \tag{6.32c}$$

An inspection of Eqs. (6.30), (6.31), and Eq. (6.32) reveals a certain proportionality phenomenon among the matrix elements. Namely, either the matrix elements

[4]Eugene Paul Wigner (1902–1995).
[5]Carl Henry Eckart (1902–1973).

$\langle 1S|\, x_q\, |2P(m = m_z)\rangle$ and $\langle 1S|\, p_q\, |2P(m = m_z)\rangle$ with $q = -1, 0, 1$ and $m_z = -1, 0, 1$ both vanish, or they are proportional to each other with proportionality constant $8\mathrm{i}a_0^2/3$. The same phenomenon can be observed in matrix elements of other vectors "sandwiched" between states of the same symmetry.

This observation can be formalized in terms of the Wigner–Eckart theorem. Let us consider states with quantum numbers γ, j and m, and γ', j' and m', where j and j' are angular momenta and m and m' are angular momentum projections, whereas γ and γ' are supplementary quantum numbers. When these are sandwiched with an operator of the form T_{kq}, which represents a spherical tensor of rank k with magnetic projection q, we have

$$\langle \gamma\, j\, m | T_{kq} | \gamma'\, j'\, m'\rangle = \frac{C^{jm}_{j'\, m'\, k\, q}}{\sqrt{2j+1}} \langle \gamma\, j || \vec{T}(k)\, || \gamma'\, j'\rangle\,, \tag{6.33}$$

where $\langle \gamma\, j || \vec{T}(k)\, || \gamma'\, j'\rangle$ is the so-called reduced matrix element, and the Clebsch–Gordan coefficient $C^{jm}_{j'\, m'\, k\, q}$ is the same as the one that enters Eq. (6.25). The vector arrow over $\vec{T}_k$ in the reduced matrix element indicates that the reduced elements are a genuine property of the tensor of rank k, not of its components. The functional dependence given in Eq. (6.33) is commonly referred to as the Wigner–Eckart theorem. Using tabulated values of Clebsch–Gordan coefficients, we can then reduce Eqs. (6.30), (6.31), and (6.32) to two matrix elements, appropriately called "reduced matrix elements", which read as follows,

$$\langle 1S || \vec{r}\, || 2P\rangle = -\frac{128}{81}\sqrt{\frac{2}{3}}\, a_0\,, \qquad \langle 1S || \vec{p}\, || 2P\rangle = \frac{16}{27}\sqrt{\frac{2}{3}}\, \frac{\mathrm{i}}{a_0}\,. \tag{6.34}$$

Using the Wigner–Eckart theorem, the nine equations given in the sub-equations of (6.30), (6.31), and Eq. (6.32) have been reduced to just two equations given in Eq. (6.34). The supplementary quantum numbers γ and γ' are often suppressed in writing down the Wigner–Eckart theorem (6.33).

6.3 Coefficients and Rotations

6.3.1 *Vector Addition or Clebsch–Gordan Coefficients*

The Clebsch–Gordan coefficients are commonly referred to as vector addition coefficients, as evident from the discussion in Sec. 6.2.1. In the following, for general angular momentum eigenvectors, we use the convention $u(j\, m)$ or $u(\gamma\, j\, m)$, where j is the angular momentum and m is the angular momentum projection (supplementary quantum numbers are summarized into the multi-index γ). It is customary not to set a comma between the arguments to avoid an increased workload when writing down long expressions, except when necessary for clarity.

In order to make contact with the literature, we should recall that some authors (e.g., Ref. [244]) use the alternative notation

$$C^{jm}_{j_1\, m_1\, j_2\, m_2} = (j_1\, m_1\, j_2\, m_2 | j_1\, j_2\, j\, m) \tag{6.35}$$

for the Clebsch–Gordan coefficients. The notation $(j_1 m_1 j_2 m_2 | j_1 j_2 j m)$ is inspired by the Dirac bra-ket notation for scalar products and by the fact that the resulting vector $v(j_1 j_2 j m) = |j_1 j_2 j m\rangle$ carries a "memory" of how it was constructed. In full generality, the definition of the Clebsch–Gordan coefficients is given by the following equation,

$$w(\gamma\, j_1\, j_2\, j\, m) = \sum_{m_1 m_2} C^{jm}_{j_1 m_1 j_2 m_2}\, u(\gamma_1\, j_1\, m_1)\, v(\gamma_2\, j_2\, m_2)\,, \tag{6.36}$$

where γ stands for the combined multi-index $\gamma_1\,\gamma_2$. Upon interchanging indices, Clebsch–Gordan coefficients typically acquire a phase factor, as in the relation

$$C^{c\gamma}_{a\alpha b\beta} = (-1)^{a+b-c}\, C^{c\gamma}_{b\beta a\alpha}\,. \tag{6.37}$$

According to Eq. (5.4.1) of Ref. [14], Clebsch–Gordan coefficients fulfill a number of symmetry relations,

$$\begin{aligned} C^{j_3 m_3}_{j_1 m_1\, j_2 m_2} &= (-1)^{j_1+j_2-j_3}\, C^{j_3 m_3}_{j_2 m_2\, j_1 m_1} = (-1)^{j_1+j_2-j_3}\, C^{j_3\, -m_3}_{j_1\, -m_1\, j_2\, -m_2} \\ &= (-1)^{j_1-m_1}\sqrt{\frac{2j_3+1}{2j_2+1}}\, C^{j_2 m_2}_{j_3\, m_3\, j_1\, -m_1} = (-1)^{j_2+m_2}\sqrt{\frac{2j_3+1}{2j_1+1}}\, C^{j_1 m_1}_{j_2\, -m_2\, j_3\, m_3}\,. \end{aligned} \tag{6.38}$$

In general, Clebsch–Gordan coefficients vanish unless $j_1 + j_2 + j_3$ is integer. Two of the j_i's ($i = 1, 2, 3$) can be half-integer valued, but then the last remaining j has to be integer rather than half-integer. Variations of the Wigner–Eckart theorem include various forms,

$$\begin{aligned} \langle j\, m|T_{kq}|j'\, m'\rangle &= \frac{C^{jm}_{j'\, m'\, k\, q}}{\sqrt{2j+1}}\, \langle j\|\vec{T}(k)\|j'\rangle = (-1)^{k-j+j'}\, \frac{C^{jm}_{k\, q\, j'\, m'}}{\sqrt{2j+1}}\, \langle j\|\vec{T}(k)\|j'\rangle \\ &= (-1)^{k-j+j'}\, \frac{C^{j\, -m}_{j'\, -m'\, k\, -q}}{\sqrt{2j+1}}\, \langle j\|\vec{T}(k)\|j'\rangle = (-1)^{j'-m'}\, \frac{C^{kq}_{j\, m\, j'\, -m'}}{\sqrt{2k+1}}\, \langle j\|\vec{T}(k)\|j'\rangle \\ &= (-1)^{k+q}\, \frac{C^{j'\, m'}_{k\, -q\, j\, m}}{\sqrt{2j'+1}}\, \langle j\|\vec{T}(k)\|j'\rangle\,. \end{aligned} \tag{6.39}$$

The first representation on the right-hand side has the advantage over the other representations that no phase factors or minus signs have to be introduced in the projections m, q and m'. The normalization of the reduced matrix element is chosen so that

$$\sum_{m\, q\, m'} |\langle j\, m|T_{kq}|j'\, m'\rangle|^2 = \left|\langle j\|\vec{T}(k)\|j'\rangle\right|^2\,. \tag{6.40}$$

Some example values for some reduced matrix elements have already been given in Eq. (6.34). Other examples are as follows. For the vector $\vec{\sigma}$ of Pauli spin matrices, we have

$$\left\langle \tfrac{1}{2}\, \|\vec{\sigma}\|\, \tfrac{1}{2} \right\rangle = \sqrt{6}\,. \tag{6.41}$$

This formula is relevant to the evaluation of matrix elements of the Pauli matrices in the basis of the states

$$\chi_m = |\tfrac{1}{2}\, m\rangle\,, \qquad |\tfrac{1}{2}\, \tfrac{1}{2}\rangle = |\uparrow\rangle = \begin{pmatrix} 1 \\ 0 \end{pmatrix}, \qquad |\tfrac{1}{2}\, -\tfrac{1}{2}\rangle = |\downarrow\rangle = \begin{pmatrix} 0 \\ 1 \end{pmatrix}, \tag{6.42}$$

which are the "spin up" and "spin down" states. The unit vector operator has the reduced matrix element

$$\langle \ell_1 \, \| \hat{r} \| \, \ell_2 \rangle = \sqrt{2\ell_2 + 1}\, C^{\ell_1 0}_{\ell_2 0 1 0} \,. \tag{6.43}$$

The matrix element is supposed to be evaluated among spherical harmonics, i.e.,

$$\int \mathrm{d}\Omega \, Y^*_{\ell_1 m_1}(\theta,\varphi)\, \hat{r}_q Y_{\ell_2 m_2}(\theta,\varphi) = \frac{C^{\ell_1 m_1}_{\ell_2 m_2 1 q}}{\sqrt{2\ell_1 + 1}} \, \langle \ell_1 \, \| \hat{r} \| \, \ell_2 \rangle \,, \tag{6.44}$$

where $\hat{r}_q = x_q / r$ [see Eq. (6.21)].

6.3.2 *Wigner 3j, 6j and 9j Symbols*

The Wigner $3j$ symbol is optimized for maximum symmetry and is related to the Clebsch–Gordan coefficients,

$$\begin{pmatrix} j_1 & j_2 & j_3 \\ m_1 & m_2 & m_3 \end{pmatrix} = \frac{(-1)^{j_1 - j_2 - m_3}}{\sqrt{2 j_3 + 1}} \, C^{j_3 \, -m_3}_{j_1 \, m_1 \, j_2 \, m_2} \,. \tag{6.45}$$

The $3j$ symbol is invariant under an even permutation of columns, whereas an odd permutation of the columns leads to a multiplicative factor $(-1)^{j_1+j_2+j_3}$. The Clebsch–Gordan coefficient is converted into a $3j$ symbol as follows,

$$C^{j_3 m_3}_{j_1 m_1 j_2 m_2} = (-1)^{j_2 - j_1 - m_3} \sqrt{2 j_3 + 1} \begin{pmatrix} j_1 & j_2 & j_3 \\ m_1 & m_2 & -m_3 \end{pmatrix} . \tag{6.46}$$

In terms of the Wigner $3j$ symbol, the Wigner–Eckart theorem can be given in the form

$$\langle j\, m | T_{kq} | j' \, m' \rangle = (-1)^{j' - k + m} \begin{pmatrix} j' & k & j \\ m' & q & -m \end{pmatrix} \langle j \| \vec{T}(k) \, \| j' \rangle \,. \tag{6.47}$$

Upon three interchanges of columns, we can rewrite this as

$$\langle j\, m | T_{kq} | j' \, m' \rangle = (-1)^{j - m + 2k} \begin{pmatrix} j & k & j' \\ -m & q & m' \end{pmatrix} \langle j \| \vec{T}(k) \, \| j' \rangle \,. \tag{6.48}$$

If k is integer, then we can simplify the prefactor to read $(-1)^{j-m}$. In general, one has to be extremely careful with the phase factors in Clebsch–Gordan coefficients and Wigner $3j$ symbols if one or two of the j's are half-integer. For the $3j$ symbol not to vanish, the sum $J = j + k + j'$ has to be integer-valued (half-integer would not be acceptable), in which case $(-1)^{j+k+j'} = (-1)^{-j-k-j'}$. An important special formula concerns the $3j$ symbol with three zero magnetic quantum numbers reads (see Appendix 2 on page 125 of Ref. [14])

$$\begin{aligned} \begin{pmatrix} j_1 & j_2 & j_3 \\ 0 & 0 & 0 \end{pmatrix} = (-1)^{J/2} & \left(\frac{(J - 2j_1)!(J - 2j_2)!(J - 2j_3)!}{(J+1)!} \right)^{1/2} \\ & \times \frac{(J/2)!}{(J/2 - j_1)!(J/2 - j_2)!(J/2 - j_3)!} \, \delta_{\mathrm{mod}(J,2),0} \,, \end{aligned} \tag{6.49}$$

where $J = j_1 + j_2 + j_3$ and $\mathrm{mod}(J,2)$ is defined to be the remainder of the division of J by two, i.e., $\mathrm{mod}(J,2)$ is zero if J is even, otherwise it is one. Note that the expression vanishes for odd J, because of the factor $\delta_{\mathrm{mod}(J,2),0}$.

The reduced matrix element of the unit vector operator, in terms of the Wigner $3j$ symbol, reads as

$$\langle \ell_1 \| \hat{r} \| \ell_2 \rangle = (-1)^{\ell_1} \, \Pi_{\ell_1 \ell_2} \begin{pmatrix} \ell_1 & 1 & \ell_2 \\ 0 & 0 & 0 \end{pmatrix}. \tag{6.50}$$

Here, an important shorthand notation is used for a mathematical expression that often occurs in angular momentum algebra (see Chap. 12 of Ref. [16]),

$$\Pi_{j_a\, j_b\, j_c \cdots j_f} = \sqrt{(2j_a + 1)\,(2j_b + 1)\,(2j_c + 1) \cdots (2j_f + 1)}\,. \tag{6.51}$$

Because of the complex nature of angular momentum algebra, the notation is not always unique. E.g., the Π symbol with three arguments (rather than indices) is commonly used to denote the test function for an even sum of the arguments,

$$\Pi(j_1, j_2, j_3) = \begin{cases} 1 & \text{if } j_1 + j_2 + j_3 \text{ even} \\ 0 & \text{otherwise} \end{cases}. \tag{6.52}$$

This is not to be confused with the Δ symbol of three arguments which is defined as a test function for the triangularity condition,

$$\Delta(j_a, j_b, j_c) = \begin{cases} 1 & \text{if } |j_a - j_b| \le j_c \le |j_a + j_b| \\ 0 & \text{otherwise} \end{cases}. \tag{6.53}$$

Vector coupling coefficients among irreducible representations whose j values do not fulfill the triangularity condition, vanish. Note that in Sec. 9.8 of Ref. [16], the test function for the triangularity condition is denoted differently, namely as $\{j_a\; j_b\; j_c\} \equiv \Delta(j_a, j_b, j_c)$, but this notation is not used here.

The Wigner $6j$ symbols are the vector recoupling coefficients that switch between different j_1's and j_2's which can be used to construct the same j and m. This ominous definition can be illustrated as follows. Let us consider the re-coupling of angular momenta. Let us consider three angular momenta j_1, j_2, and j_3 which are coupled to some resulting J. We may either couple j_1 and j_2 to j_{12}, before coupling j_{12} and j_3 to J, or we may couple j_2 and j_3 to j_{23}, before coupling j_{23} and j_1 to J. The resulting vectors are different,

$$\begin{aligned} w((j_1 j_2) j_{12}, j_3, JM) &= \sum_{m_{12} m_3} C^{JM}_{j_{12} m_{12} j_3 m_3} \, v((j_1 j_2) j_{12} m_{12}) \, u(j_3 m_3) \\ &= \sum_{m_{12} m_2 m_3} C^{JM}_{j_{12} m_{12} j_3 m_3} \, C^{j_{12} m_{12}}_{j_1 m_1 j_2 m_2} \, u(j_1 m_1) \, u(j_2 m_2) \, u(j_3 m_3) \end{aligned} \tag{6.54}$$

whereas

$$\begin{aligned} w(j_1, (j_2 j_3) j_{23}, JM) &= \sum_{m_1 m_{23}} C^{JM}_{j_1 m_1 j_{23} m_{23}} \, u(j_1 m_1) \, v((j_2 j_3) j_{23} m_{23}) \\ &= \sum_{m_{12} m_2 m_3} C^{JM}_{j_1 m_1 j_{23} m_{23}} \, C^{j_{23} m_{23}}_{j_2 m_2 j_3 m_3} \, u(j_1 m_1) \, u(j_2 m_2) \, u(j_3 m_3)\,. \end{aligned} \tag{6.55}$$

The recoupling coefficients, which are independent of M, are proportional to the overlap integral of the two w vectors given above, and can be written according Eqs. (6.1.4) and (6.1.5) of Ref. [14],

$$\begin{aligned}
&\langle (j_1 j_2) j_{12}, j_3, J | j_1, (j_2 j_3) j_{23}, J \rangle \\
&= \sum_{m_1\, m_2\, m_3} \sum_{m_{12}\, m_{23}} C^{JM}_{j_{12} m_{12} j_3 m_3} \, C^{j_{12} m_{12}}_{j_1 m_1 j_2 m_2} \, C^{JM}_{j_1 m_1 j_{23} m_{23}} \, C^{j_{23} m_{23}}_{j_2 m_2 j_3 m_3} \\
&= \sqrt{(2j_{12}+1)(2j_{23}+1)}\, (-1)^{j_1+j_2+j_3+J} \begin{Bmatrix} j_1 & j_2 & j_{12} \\ j_3 & J & j_{23} \end{Bmatrix} .
\end{aligned} \tag{6.56}$$

The latter equality merely constitutes a definition of the Wigner $6j$ symbol. The $6j$ symbols are invariant under any permutation of the columns, whether even or odd. We have, for instance,

$$\begin{Bmatrix} j_1 & j_2 & j_3 \\ j_4 & j_5 & j_6 \end{Bmatrix} = \begin{Bmatrix} j_2 & j_1 & j_3 \\ j_5 & j_4 & j_6 \end{Bmatrix} . \tag{6.57}$$

They are also invariant under a simultaneous exchange of two entries in each of the rows,

$$\begin{Bmatrix} j_1 & j_2 & j_3 \\ j_4 & j_5 & j_6 \end{Bmatrix} = \begin{Bmatrix} j_4 & j_5 & j_3 \\ j_1 & j_2 & j_6 \end{Bmatrix} . \tag{6.58}$$

The $6j$ symbols have many applications which go far beyond their original definition, one of which concerns decoupled angular momenta. A particularly useful result can be found in Eq. (7.1.7) of Ref. [14], and is relevant for the case of a reduced matrix element of a coupled state with angular momenta $\gamma j_1 j_2 J$, evaluated on an operator $\vec{T}_k$ which acts only on j_1, not j_2. The formula reads

$$\langle \gamma j_1' j_2 J' \| \vec{T}(k) \| \gamma j_1 j_2 J \rangle = (-1)^{j_1'+j_2+J+k}\, \Pi_{JJ'} \begin{Bmatrix} j_1' & J' & j_2 \\ J & j_1 & k \end{Bmatrix} \langle \gamma j_1' \| \vec{T}(k) \| \gamma j_1 \rangle , \tag{6.59}$$

where the Π symbol has been defined in Eq. (6.51). Let us also consider the case of two decoupled angular momenta. If a tensor X of rank k with spherical basis components q is composed of tensors T and U, as in

$$X = T \otimes U, \quad X_{kq} = \sum_{q_1 q_2} C^{kq}_{k_1 q_1 k_2 q_2}\, T_{k_1 q_1}\, U_{k_2 q_2} , \tag{6.60}$$

where T acts on j_1 and U acts on j_2 only, then a $9j$ symbol is incurred [see Eq. (7.1.7) of Ref. [14]] ,

$$\langle \gamma j_1 j_2 j \| \vec{X}(k) \| \gamma' j_1' j_2' j' \rangle = \Pi_{jj'k} \begin{Bmatrix} j_1 & j_1' & k_1 \\ j_2 & j_2' & k_2 \\ j & j' & k \end{Bmatrix} \sum_{\gamma''} \langle \gamma j_1 \| \vec{T}(k_1) \| \gamma'' j_1' \rangle \, \langle \gamma'' j_2 \| \vec{U}(k_2) \| \gamma' j_2' \rangle , \tag{6.61}$$

where the sum over γ'' extends over all virtual states necessary to complete the spectrum. The entity in curly brackets in Eq. (6.61) is known as a Wigner $9j$ symbol, and $\Pi_{jj'k}$ is defined in Eq. (6.51).

6.3.3 *Gaunt Coefficients*

Finally, it is perhaps not surprising that different conventions for vector coupling coefficients have been proposed in the literature; it is unclear up to this point if the most appropriate notation is the one commonly adopted today. Still, it represents the consensus of the community. For example, the so-called Racah V and W coefficients are not used very often today; still, it is useful to know that they exist and are related by conversion formulas (which we do not discuss in any further detail) to the Clebsch–Gordan coefficients.

One example of an alternative convention which we briefly mention, concerns the so-called GAUNT[6] coefficients. Let us recall the definition of the spherical harmonic from Ref. [81],

$$Y_{\ell m}(\theta,\varphi) = (-1)^m \left[\frac{(2\ell+1)\,(\ell-m)!}{4\pi(\ell+m)!}\right]^{1/2} P_\ell^m(\cos\theta)\,\mathrm{e}^{\mathrm{i}\,m\varphi}\,, \tag{6.62}$$

$$Y^*_{\ell m}(\theta,\varphi) = (-1)^m\, Y_{\ell\,-m}(\theta,\varphi)\,. \tag{6.63}$$

In the notation $Y_{\ell-m}$, there is no comma to separate the two indices. The Gaunt coefficient is defined as an integral over three spherical harmonics,

$$\int \mathrm{d}\Omega\, Y^*_{l_1 m_1}(\theta,\varphi)\, Y_{l_2 m_2}(\theta,\varphi)\, Y_{l_3 m_3}(\theta,\varphi) = (-1)^{m_1}\, Y^{\ell_1\ell_2\ell_3}_{-m_1\,m_2\,m_3}\,, \tag{6.64}$$

where

$$Y^{\ell_1\ell_2\ell_3}_{m_1 m_2 m_3} = \sqrt{\frac{(2\ell_1+1)(2\ell_2+1)(2\ell_3+1)}{4\pi}} \begin{pmatrix} \ell_1 & \ell_2 & \ell_3 \\ 0 & 0 & 0 \end{pmatrix} \begin{pmatrix} \ell_1 & \ell_2 & \ell_3 \\ m_1 & m_2 & m_3 \end{pmatrix}. \tag{6.65}$$

This formula involves a $3j$ symbol with three zero magnetic projections which has been indicated in Eq. (6.49). The literature is not always consistent in terms of the definitions used; e.g., in Eq. (107.14) of Chap. 107 of Ref. [245], the formula is a little different because different phase conventions for the spherical harmonics are used.

According to Eq. (6.45), one may reformulate the $3j$ symbols in terms of Clebsch–Gordan coefficients. Conversely, it is very instructive to reformulate the Gaunt coefficient in terms of Clebsch–Gordan coefficients, in order to illustrate the general formalism of Racah–Wigner algebra. Using the fact that an even permutation of columns leaves a $3j$ symbol invariant, one may apply Eq. (6.45) to Eq. (6.64) in the following manner,

$$\begin{pmatrix} \ell_1 & \ell_2 & \ell_3 \\ -m_1 & m_2 & m_3 \end{pmatrix} = \begin{pmatrix} \ell_2 & \ell_3 & \ell_1 \\ m_2 & m_3 & -m_1 \end{pmatrix} = \frac{(-1)^{\ell_2-\ell_3+m_1}}{\sqrt{2\ell_1+1}}\, C^{\ell_1\,m_1}_{\ell_2\,m_2\,\ell_3\,m_3}\,. \tag{6.66}$$

In terms of Clebsch–Gordan coefficients, the integral (6.64) may thus be

[6] John Arthur Gaunt (1904–1944).

expressed as

$$\begin{aligned}\int \mathrm{d}\Omega\, Y^*_{\ell_1 m_1}(\theta,\varphi)\, Y_{\ell_2 m_2}(\theta,\varphi)\, Y_{\ell_3 m_3}(\theta,\varphi) \\ = (-1)^{\ell_2-\ell_3}\sqrt{\frac{(2\ell_2+1)\,(2\ell_3+1)}{4\pi}} \begin{pmatrix} \ell_1 & \ell_2 & \ell_3 \\ 0 & 0 & 0 \end{pmatrix} C^{\ell_1 m_1}_{\ell_2 m_2 \ell_3 m_3} \\ = \sqrt{\frac{(2\ell_2+1)\,(2\ell_3+1)}{(2\ell_1+1)\,4\pi}}\, C^{\ell_1 0}_{\ell_2 0 \ell_3 0} C^{\ell_1 m_1}_{\ell_2 m_2 \ell_3 m_3} \,. \end{aligned} \tag{6.67}$$

This representation illustrates that the integral (6.64) is related to the Clebsch–Gordan coefficient for the coupling of $|j_2\, m_2\rangle$ and $|j_3\, m_3\rangle$ to $|j_1\, m_1\rangle$, but not exactly equal (due to nontrivial prefactors). On the other hand, one may also interpret the integral (6.67) using the Wigner–Eckart theorem given in Eq. (6.39) as

$$\begin{aligned}\langle \ell_1 m_1 | Y_{\ell_2 m_2} | \ell_3 m_3 \rangle &= \int \mathrm{d}\Omega\, Y^*_{\ell_1 m_1}(\theta,\varphi)\, Y_{\ell_2 m_2}(\theta,\varphi)\, Y_{\ell_3 m_3}(\theta,\varphi) \\ &= (-1)^{\ell_2-\ell_3+\ell_1} \frac{C^{\ell_1 m_1}_{\ell_2 m_2 \ell_3 m_3}}{\sqrt{2\ell_1+1}} \left\langle \ell_1 || \vec{Y}(\ell_2) || \ell_3 \right\rangle . \end{aligned} \tag{6.68}$$

A comparison of (6.68) to the second line of Eq. (6.67) leads to the highly symmetric and beautiful expression,

$$\langle \ell_1 || \vec{Y}(\ell_2) || \ell_3 \rangle = (-1)^{\ell_1} \sqrt{\frac{(2\ell_1+1)\,(2\ell_2+1)\,(2\ell_3+1)}{4\pi}} \begin{pmatrix} \ell_1 & \ell_2 & \ell_3 \\ 0 & 0 & 0 \end{pmatrix}, \tag{6.69}$$

for the reduced matrix element of the spherical harmonic tensor of rank ℓ_2. Let us set $\ell_2 = 1$ and use

$$Y_{1q}(\theta,\varphi) = \sqrt{\frac{3}{4\pi}}\, x_q \,, \tag{6.70}$$

where x_q is the qth spherical basis component of the unit vector $\hat{r}$ [see Eq. (6.21)]. Multiplying both sides of Eq. (6.69) by $\sqrt{(4\pi)/3}$, we see that formula (6.69) is consistent with Eq. (6.50) for the case $\ell_2 = 1$, where ℓ_2 in this case refers to the angular momentum ℓ_2 as given in Eq. (6.69).

6.3.4 *Representation of Finite Rotations*

We have shown that the Clebsch–Gordan coefficients enable us to construct higher-rank tensors out of lower-rank ones, and to decompose tensors into individual components that transform separately under rotations, i.e., they do not mix. This is called the reduction of the rotation group into irreducible representations. Within the irreducible representations, the effect of a rotation can be represented in terms of so-called Wigner D matrices, to be discussed below; these do not mix tensors of different ℓ. Suppose we wish to rotate a tensor of rank j by a finite rotation defined by the Euler angles α, β and γ. Here, the Euler angles assume their usual definition, as defined in Ref. [246] or in Eq. (1.3.1) of Ref. [14]. According to the fundamental paradigm underlying irreducible representations, a rotation transforms tensor components of rank j into components of the same rank.

However, one may ask how this is done in practice, precisely. To answer the question, one may conjecture a linear dependence of the rotated components on the original components of the tensor, and then, find the matrix, within each irreducible representation of the rotation group, which describes this linear mapping. We have according to Eq. (4.1.11) of Ref. [14], for the action of a rotation by the Euler angles α, β and γ onto a spherical tensor of rank j and magnetic projection m,

$$D(\alpha\,\beta\,\gamma)\,|jm\rangle = \sum_{m'} |jm'\rangle\,\langle jm'|D(\alpha\,\beta\,\gamma)\,|jm\rangle = \sum_{m'} |jm'\rangle\,\mathcal{D}^{(j)}_{m'm}(\alpha\,\beta\,\gamma)\,, \tag{6.71}$$

where $\mathcal{D}^{(j)}_{m'm}(\alpha\,\beta\,\gamma)$ is the element with indices m' and m of the representation of the rotation $D(\alpha\,\beta\,\gamma)$ in the space of tensors of rank j. The relation (4.270) serves as a definition of the matrix $\mathcal{D}^{(j)}_{m'm}(\alpha\,\beta\,\gamma)$. An analogous relation holds for general tensors [see Eq. (5.2.1) of Ref. [14]],

$$D(\alpha\,\beta\,\gamma)T(k\,q)D^{-1}(\alpha\,\beta\,\gamma) = \sum_{q'=-k}^{k} T(k\,q')\,\mathcal{D}^{(k)}_{q'q}(\alpha\,\beta\,\gamma)\,. \tag{6.72}$$

Because the rotation about the Euler angles involves two rotations about (mutually tilted) z axes, which only add a phase, we have [see Eq. (4.1.12) of Ref. [14]]

$$\mathcal{D}^{(j)}_{m'm}(\alpha\,\beta\,\gamma) = \exp(\mathrm{i}\,m'\gamma)\,d^{(j)}_{m'm}(\beta)\,\exp(\mathrm{i}\,m\,\alpha)\,. \tag{6.73}$$

The rotation $d^{(j)}_{m'm}(\beta)$ can finally be represented as [see Eq. (4.1.23) of Ref. [14]]

$$d^{(j)}_{m'm}(\beta) = \sqrt{\frac{(j+m')!(j-m')!}{(j+m)!(j-m)!}}\left(\cos\frac{\beta}{2}\right)^{m'+m}\left(\sin\frac{\beta}{2}\right)^{m'-m} P^{(m'-m,m'+m)}_{j-m'}(\cos\beta)\,. \tag{6.74}$$

Here, P is a Jacobi polynomial, which carries three indices and one argument, not to be confused with the Legendre polynomial or associated Legendre polynomial, which is also denoted as P but carries a lesser number of indices. If desired, the integration over the Euler angles can be performed using

$$\begin{aligned}&\int_0^{2\pi} \mathrm{d}\alpha \int_0^{\pi} \mathrm{d}\beta\,\sin\beta \int_0^{2\pi} \mathrm{d}\gamma\,\mathcal{D}^{(j_1)}_{m_1m_1'}(\alpha,\beta,\gamma)\,\mathcal{D}^{(j_2)*}_{m_2m_2'}(\alpha,\beta,\gamma)\\ &\quad= \frac{8\pi^2}{2j_1+1}\,\delta_{j_1j_2}\,\delta_{m_1m_2}\,\delta_{m_1'm_2'}\,.\end{aligned} \tag{6.75}$$

Products of rotation matrices can be decomposed using Clebsch–Gordan coefficients,

$$\mathcal{D}^{(j_1)}_{m_1m_1'}(\alpha,\beta,\gamma)\,\mathcal{D}^{(j_2)}_{m_2m_2'}(\alpha,\beta,\gamma) = \sum_{j=|j_1-j_2|}^{|j_1+j_2|}\sum_{mm'} C^{jm}_{j_1m_1j_2m_2}\,C^{jm'}_{j_1m_1'j_2m_2'}\,\mathcal{D}^{(j)}_{mm'}(\alpha,\beta,\gamma)\,, \tag{6.76}$$

where the sum over j encompasses all angular momenta j that fulfill the triangularity condition. They have a symmetry property,

$$\mathcal{D}^{(j)*}_{\mu\mu'}(\alpha,\beta,\gamma) = (-1)^{\mu'-\mu}\,\mathcal{D}^{(j)}_{-\mu-\mu'}(\alpha,\beta,\gamma)\,, \tag{6.77}$$

which can be applied as follows (example calculation),

$$\begin{aligned}\mathcal{D}^{(j_1)}_{m_1\lambda}\,\mathcal{D}^{(j_2)*}_{m_2\lambda} &= (-1)^{\lambda-m_2}\,\mathcal{D}^{(j_1)}_{m_1\lambda}\,\mathcal{D}^{(j_2)}_{-m_2-\lambda} \\ &= (-1)^{\lambda-m_2}\sum_{jmm'} C^{jm}_{j_1m_1j_2-m_2}\,C^{jm'}_{j_1\lambda j_2-\lambda}\,\mathcal{D}^{(j)}_{mm'}(\alpha,\beta,\gamma) \\ &= (-1)^{\lambda-m_2}\sum_{jm} C^{jm}_{j_1m_1j_2-m_2}\,C^{j0}_{j_1\lambda j_2-\lambda}\,\mathcal{D}^{(j)}_{m0}(\alpha,\beta,\gamma)\,.\end{aligned} \tag{6.78}$$

Here, the sum over m' could be carried out because the magnetic projections λ and $-\lambda$ have to add up to zero. Finally, let us point out that a large collection of formulas, symmetries, and useful relations can be found in a number of works, notably, Refs. [14] and [16].

6.4 Composed Tensors of Higher Order

6.4.1 *Construction of the Spin-Angular Function*

A spinor amplitude (with two components) is obtained by angular coupling of an orbital angular momentum to a spin. The two-component spinor wave functions are used to characterize eigenfunctions of the total angular momentum of an electron, composed of the orbital angular momentum and the spin. An electron carries spin-1/2, and orbital angular momentum ℓ. One may construct the spinor amplitude $|j\ell m\rangle$ by coupling the orbital angular momentum to the spin,

$$w(j\,m) = \sum_{m_1\,m_2} C^{j\,m}_{\ell\,m_1\,1/2\,m_2}\,u(\ell\,m_1)\,v(\tfrac{1}{2}\,m_2)\,. \tag{6.79}$$

In coordinate space, one has $u(\ell\,m_1) = Y_{\ell_1 m_1}(\theta,\varphi)$ and $v(\frac{1}{2}\,m_2) = \chi_{m_2}$, where χ_{m_2} is given in Eq. (6.42). The two-component spin-angular function $\chi_{\kappa\,\mu}(\hat{r})$ is obtained by assembling the total angular momentum j of the electron from the orbital angular momentum ℓ and the spin-1/2, with the help of vector-coupling (Clebsch–Gordan) coefficients,

$$\chi_{\kappa\,\mu}(\hat{r}) = \sum_m C^{j\,\mu}_{\ell\,\mu-m\,1/2\,m}\,Y_{\ell\mu-m}(\hat{r})\,\chi_m\,, \tag{6.80}$$

where $j = j(\kappa) = |\kappa| - 1/2$ and $\ell = \ell(\kappa) = |\kappa + 1/2| - 1/2$. Here and in the following, on occasion, we use the compact notation $Y_{\ell m}(\hat{r}) \equiv Y_{\ell m}(\theta,\varphi)$, observing that the polar and azimuth angles θ and φ define $\hat{r}$ uniquely, and vice versa [see also Eq. (5.157)].

The Dirac angular momentum operator K and its eigenvalue $-\kappa$ (by convention) are defined as follows,

$$K = \vec{\sigma}\cdot\vec{L} + 1\,, \qquad K\,\chi_{\kappa\,\mu}(\hat{r}) = -\kappa\,\chi_{\kappa\,\mu}(\hat{r})\,. \tag{6.81}$$

We recall from Eq. (4.348) that the Dirac angular quantum number κ fulfills

$$\kappa = (-1)^{j+\ell+1/2}\left(j + \frac{1}{2}\right) = \begin{cases} -(j+1/2) & \text{for} \quad j = \ell + 1/2 \\ (j+1/2) & \text{for} \quad j = \ell - 1/2 \end{cases}\,. \tag{6.82}$$

As such, the Dirac quantum number κ encodes information about both ℓ and j. Specifically, the Dirac quantum number κ is negative if the spin is added to the orbital angular momentum, and positive if the spin is subtracted from the orbital angular momentum. For $\ell = 0, 1, 2, \ldots$ and $j = 1/2, 3/2, \ldots$, the Dirac quantum number κ attains all possible positive and negative integer values except for zero. The orbital angular momentum $\ell = \ell(\kappa)$ and the total electron angular momentum j can be written in terms of κ, as follows,

$$\ell(\kappa) = \left|\kappa + \frac{1}{2}\right| - \frac{1}{2}, \qquad j = |\kappa| - \frac{1}{2}. \tag{6.83}$$

The first cases of special interest are as follows. For $S_{1/2}$ states, we have $\ell = 0$ and $j = 1/2$, so $\kappa = -1$. For $\ell = 1$ and $j = 1/2$ ($P_{1/2}$ state), we have $\kappa = 1$. For a $P_{3/2}$ state, we have $\ell = 1$ and $j = 3/2$, so $\kappa = -2$. For $D_{3/2}$ symmetry, one has $\ell = 2$ and $j = 3/2$, so $\kappa = 2$. Upon using explicit formulas for the Clebsch–Gordan coefficients, one finds the exact form for the two-component spin-angular function $\chi_{\kappa\,\mu}(\hat{r})$,

$$\chi_{\kappa\,\mu}(\hat{r}) = \begin{pmatrix} -\dfrac{\kappa}{|\kappa|} \left(\dfrac{\kappa + 1/2 - \mu}{2\kappa + 1} \right)^{1/2} Y_{\ell\,\mu-1/2}(\hat{r}) \\[2ex] \left(\dfrac{\kappa + 1/2 + \mu}{2\kappa + 1} \right)^{1/2} Y_{\ell\,\mu+1/2}(\hat{r}) \end{pmatrix}, \tag{6.84}$$

where $\ell = \ell(\kappa)$ is given in Eq. (6.83). The following identities, found in Eqs. (7.2.5.23) and (7.2.5.26) of Ref. [16], are quite important,

$$(\vec{\sigma} \cdot \hat{r})\, \chi_{\kappa\,\mu}(\hat{r}) = -\chi_{-\kappa\,\mu}(\hat{r}), \qquad r(\vec{\sigma} \cdot \vec{\nabla})\, \chi_{\kappa\,\mu}(\hat{r}) = -(1 + \kappa)\, \chi_{-\kappa\,\mu}(\hat{r}). \tag{6.85}$$

Note that a change $\kappa \to -\kappa$ preserves $j = |\kappa| - 1/2$ but entails a change $\ell \to \overline{\ell} = 2j - \ell$ for ℓ. If $\ell = j \pm 1/2$, then $\overline{\ell} = j \mp 1/2$.

It is somewhat hard to find a pertinent proof of the formula (6.85) in the literature. Let us try to indicate a possible proof for the first identity, $\vec{\sigma} \cdot \hat{r}\, \chi_{\kappa\mu}(\hat{r}) = -\chi_{-\kappa\mu}(\hat{r})$. First, one shows that the operator $\vec{\sigma} \cdot \hat{r}$ commutes with the total angular momentum $\vec{J}$,

$$[\vec{J}, \vec{\sigma} \cdot \hat{r}] = 0, \qquad [\vec{J}^2, \vec{\sigma} \cdot \hat{r}] = 0. \tag{6.86}$$

Because $|\kappa| = j + 1/2$, we can conclude that the function $[\vec{\sigma} \cdot \hat{r}\, \chi_{\kappa\mu}(\hat{r})]$ must belong to the same $\vec{J}^2 = j(j+1)$ manifold as the original function $\chi_{\kappa\mu}(\hat{r})$.

Furthermore, because $[\vec{J}, \vec{\sigma} \cdot \hat{r}] = 0$, one has a relation of the form

$$\vec{\sigma} \cdot \hat{r}\, \chi_{\kappa\mu}(\hat{r}) = c_1\, \chi_{\kappa\mu}(\hat{r}) + c_2\, \chi_{-\kappa\mu}(\hat{r}), \tag{6.87}$$

where the coefficients c_1 and c_2 remain to be determined. On both sides of Eq. (6.87), we have the same magnetic projection μ and the same total angular momentum j in the spin-angular functions. The orbital angular momenta, $\ell = |\kappa + \frac{1}{2}| - \frac{1}{2}$, are different, though, under the replacement $\kappa \to -\kappa$. Form Eq. (6.84), we can immediately infer that $\langle \chi_{\kappa\mu} | \vec{\sigma} \cdot \hat{r} | \chi_{\kappa\mu} \rangle = 0$ and thus $c_1 = 0$. In order to determine c_2, we observe that

$$\|\vec{\sigma} \cdot \hat{r}\, \chi_{\kappa\mu}(\hat{r})\|^2 = \left\langle \chi_{\kappa\mu} \middle| (\vec{\sigma} \cdot \hat{r})^2 \middle| \chi_{\kappa\mu} \right\rangle = \langle \chi_{\kappa\mu} | \chi_{\kappa\mu} \rangle = 1 = |c_2|^2. \tag{6.88}$$

This means that c_2 can only be a phase factor, $c_2 = \mathrm{e}^{\mathrm{i}\delta}$, where δ remains to be determined. It can be determined as follows. Let us investigate the value of the spin-angular function at $\theta = 0$, or, equivalently, $\hat{r} = \hat{\mathrm{e}}_z$. We have

$$Y_{\ell m}(\hat{\mathrm{e}}_z) = \left(\frac{2\ell+1}{4\pi}\right)^{1/2} \delta_{m0} \,. \tag{6.89}$$

We recall that [see Eq. (6.80)]

$$\chi_{\kappa\mu}(\hat{r}) = \sum_m C^{j\mu}_{\ell\,\mu-m\,1/2\,m}\, Y_{\ell\,\mu-m}(\hat{r})\,\chi_m\,, \tag{6.90}$$

$$\chi_{-\kappa\mu}(\hat{r}) = \sum_m C^{j\mu}_{\overline{\ell}\,\mu-m\,1/2\,m}\, Y_{\overline{\ell}\,\mu-m}(\hat{r})\,\chi_m\,, \tag{6.91}$$

$$\ell = |\kappa + \tfrac{1}{2}| - \tfrac{1}{2}\,, \qquad \overline{\ell} = |-\kappa + \tfrac{1}{2}| - \tfrac{1}{2}\,. \tag{6.92}$$

Specialized to the case $\theta = 0$, which is equivalent to $\hat{r} = \hat{\mathrm{e}}_z$, one has

$$\chi_{\kappa\mu}(\hat{\mathrm{e}}_z) = \sum_m C^{j\mu}_{\ell\,\mu-m\,1/2\,m} \left(\frac{2\ell+1}{4\pi}\right)^{1/2} \delta_{(\mu-m)0}\,\chi_m = C^{j\mu}_{\ell\,0\,1/2\,\mu} \left(\frac{2\ell+1}{4\pi}\right)^{1/2} \chi_\mu\,. \tag{6.93}$$

We have considered the equation $\vec{\sigma}\cdot\hat{r}\,\chi_{\kappa\mu}(\hat{r}) = c_2\,\chi_{-\kappa\mu}(\hat{r})$ for $\hat{r} = \hat{\mathrm{e}}_z$. The derivation of the relation $\vec{\sigma}\cdot\hat{\mathrm{e}}_z\,\chi_{\kappa\mu}(\hat{\mathrm{e}}_z) = 2\mu\,\chi_{\kappa\mu}(\hat{\mathrm{e}}_z)$ is left as an exercise to the reader. We arrive at the relationship

$$(2\mu)\,C^{j\mu}_{\ell\,0\,1/2\,\mu} \left(\frac{2\ell+1}{4\pi}\right)^{1/2} = c_2\,C^{j\mu}_{\overline{\ell}\,0\,1/2\,\mu} \left(\frac{2\overline{\ell}+1}{4\pi}\right)^{1/2} . \tag{6.94}$$

If $j = \ell + 1/2$, then $\overline{\ell} = j + \frac{1}{2} = \ell + 1$, while if $j = \ell - 1/2$, then $\overline{\ell} = j - \frac{1}{2} = \ell - 1$. In both cases, an explicit evaluation of the Clebsch–Gordan coefficient reveals that $c_2 = -1$, completing the proof of Eq. (6.85).

Let $\vec{r}_1$ and $\vec{r}_2$ be two vectors, r_1 and r_2 be their magnitudes, and

$$\xi = \cos\gamma = \frac{\vec{r}_2\cdot\vec{r}_1}{r_2\,r_1} = \cos\theta_2\,\cos\theta_1 + \sin\theta_2\,\sin\theta_1\,\cos(\varphi_2 - \varphi_1) \tag{6.95}$$

be the direction cosine, where (θ_1, φ_1) and (θ_2, φ_2) are the polar and azimuthal angles of $\vec{r}_1$ and $\vec{r}_2$, respectively. The sum over magnetic quantum numbers leads to the following explicit formula,

$$\pi_\kappa(\hat{r}_1, \hat{r}_2) = \sum_\mu \chi_{\kappa\,\mu}(\hat{r}_1) \otimes \chi^+_{\kappa\,\mu}(\hat{r}_2) = \frac{|\kappa|}{4\pi} \left\{ \mathbb{1}_{2\times 2}\,P_{\ell(\kappa)}(\xi) + \frac{\mathrm{i}}{\kappa}\,\vec{\sigma}\cdot(\hat{r}_1 \times \hat{r}_2)\;P'_{\ell(\kappa)}(\xi) \right\}, \tag{6.96}$$

where $\chi^+ = \chi^{\mathrm{T}*}$. The result (6.96) generalizes the well-known addition theorem for spherical harmonics,

$$\sum_{m=-\ell}^{\ell} Y_{\ell m}(\hat{r}_1)\,Y^*_{\ell m}(\hat{r}_2) = \frac{2\ell+1}{4\pi}\,P_\ell(\hat{r}_1\cdot\hat{r}_2)\,. \tag{6.97}$$

The Schrödinger wave function of a bound state has been given as

$$\phi_{n\ell m}(\vec{r}) = R_{n\ell}(r)\,Y_{\ell m}(\hat{r})\,. \tag{6.98}$$

The Schrödinger–Pauli two-component nonrelativistic wave function adds the spin and reads as

$$\phi_{n\ell j\mu}(\vec{r}) = R_{n\ell}(r)\,\chi_{\kappa\,\mu}(\hat{r})\,, \tag{6.99}$$

where κ is given in Eq. (6.82). As we shall discuss in more detail later, one may approximate the Dirac wave function, which is a four-component wave function, by the bispinor ("two-spinor", i.e., four-component) amplitude

$$\psi(\vec{r}) \approx \begin{pmatrix} \left(1 - \dfrac{\vec{p}^{\,2}}{2m}\right) \phi_{n\ell j\mu}(\vec{r}) \\ \dfrac{\vec{\sigma}\cdot\vec{p}}{2\,m}\,\phi_{n\ell j\mu}(\vec{r}) \end{pmatrix}. \tag{6.100}$$

This needs to be contrasted with the fully relativistic expression for the wave function, which reads as (see the discussion in Chap. 8)

$$\psi(\vec{r}) = \begin{pmatrix} f_{n\kappa}(r)\,\chi_{\kappa\,\mu}(\hat{r}) \\ \mathrm{i}\,g_{n\kappa}(r)\,\chi_{-\kappa\,\mu}(\hat{r}) \end{pmatrix}. \tag{6.101}$$

Here, $f_{n\kappa}(r)$ and $g_{n\kappa}(r)$ are radial components of the wave function which will be calculated in Sec. 8.2.2 below. The replacement $\kappa \to -\kappa$ implies that, in particular, the "lower component" (in the sense of two-component spinors) of a $P_{1/2}$ state is an $S_{1/2}$ state, whereas the lower component of a $P_{3/2}$ state is a $D_{3/2}$ state. Using the formulas given above, it is not only possible to show the consistency of Eqs. (6.100) and (6.101), but also, to give an approximation for the radial part of the lower component of the wave function,

$$\begin{aligned} -\mathrm{i}\vec{\sigma}\cdot\vec{\nabla}\,\phi_{n\ell j\mu}(\vec{r}) &= -\,\mathrm{i}\left[(\vec{\sigma}\cdot\vec{\nabla})R_{n\ell}(r)\right]\chi_{\kappa\,\mu}(\hat{x}) - \mathrm{i}R_{n\ell}(r)\,(\vec{\sigma}\cdot\vec{\nabla})\chi_{\kappa\,\mu}(\hat{r}) \\ &= -\,\mathrm{i}\,(\vec{\sigma}\cdot\hat{r})R'_{n\ell}(r)\,\chi_{\kappa\,\mu}(\hat{r}) + \mathrm{i}\,(1+\kappa)R_{n\ell}(r)\chi_{-\kappa\,\mu}(\hat{r}) \\ &= \mathrm{i}\left[R'_{n\ell}(x) + (1+\kappa)R_{n\ell}(x)\right]\chi_{-\kappa\,\mu}(\hat{r})\,. \end{aligned} \tag{6.102}$$

So, the lower component of the Dirac wave function given in Eq. (6.100) has the right symmetry, and a comparison of Eq. (6.101) to (6.102), with the help of Eq. (6.85) and the identity $\vec{p} = -\mathrm{i}\,\vec{\nabla}$, reveals that

$$f_{n\kappa}(r) \approx R_{n\ell}(r)\,, \qquad\qquad g_{n\kappa}(r) \approx R'_{n\ell}(r) + (1+\kappa)R_{n\ell}(r)\,. \tag{6.103}$$

In many cases, the approximations given in Eq. (6.103) are suitable approximations for the radial components of the full Dirac–Coulomb wave function given in Eq. (6.101).

6.4.2 *Construction of the Vector Spherical Harmonic*

The coupling of a spin to an orbital angular momentum is not the only fundamentally important case of a higher-order tensor which is relevant to experiments. Let us consider the coupling of a polarization vector, i.e., of a unit vector $\vec{e}_q$ given in the spherical basis, to an orbital angular momentum, described by a spherical harmonic. The resultant object is a vector spherical harmonic $\vec{Y}^L_{JM}(\hat{x})$ and reads as (see Sec. 5.4.2 of Ref. [81])

$$\vec{Y}^L_{JM}(\hat{r}) = \sum_{mq} C^{JM}_{Lm\,1q}\, Y_{Lm}(\hat{r})\, \vec{e}_q\,, \tag{6.104}$$

where the basis vectors $\vec{e}_q$ have been given in Eq. (6.22). Vector spherical harmonics are only nonvanishing when the triangularity condition $\Delta(L,1,J) = 1$ is fulfilled [see Eq. (6.53)], i.e., when $J = L-1, L, L+1$. It is useful to consider the space spanned by vector spherical harmonics with a given value of total angular momentum J. We can transform to a different basis spanned by the vector spherical harmonics $\vec{Y}^{(1)}_{JM}(\theta,\varphi)$, $\vec{Y}^{(0)}_{JM}(\theta,\varphi)$ and $\vec{Y}^{(-1)}_{JM}(\theta,\varphi)$. The transformation reads as follows,

$$\vec{Y}^{J-1}_{JM}(\hat{r}) = \sqrt{\frac{J+1}{2J+1}}\, \vec{Y}^{(1)}_{JM}(\hat{r}) + \sqrt{\frac{J}{2J+1}}\, \vec{Y}^{(-1)}_{JM}(\hat{r})\,, \tag{6.105a}$$

$$\vec{Y}^{J+1}_{JM}(\hat{r}) = \sqrt{\frac{J}{2J+1}}\, \vec{Y}^{(1)}_{JM}(\hat{r}) - \sqrt{\frac{J+1}{2J+1}}\, \vec{Y}^{(-1)}_{JM}(\hat{r})\,, \tag{6.105b}$$

$$\vec{Y}^{J}_{JM}(\hat{r}) = \vec{Y}^{(0)}_{JM}(\hat{r})\,. \tag{6.105c}$$

In spherical coordinates, the gradient operator can be decomposed into a radial and an angular part as follows,

$$\vec{\nabla} = \hat{e}_r \frac{\partial}{\partial r} + \frac{1}{r}\vec{\nabla}_\Omega\,, \qquad \vec{\nabla}_\Omega = \hat{e}_\theta \frac{\partial}{\partial \theta} + \hat{e}_\varphi \frac{1}{\sin\theta}\frac{\partial}{\partial \varphi}\,, \tag{6.106}$$

where the $\hat{e}_r$, $\hat{e}_\theta$ and $\hat{e}_\varphi$ are unit vectors in the radial, polar and azimuthal directions. When acting on a function $f = f(\theta,\varphi) = f(\hat{r})$ defined on the unit sphere, the radial derivative vanishes, and we have $\vec{\nabla} f(\hat{r}) = \vec{\nabla}_\Omega f(\hat{r})$. So, for example, one has $\vec{\nabla}_\Omega Y_{JM}(\theta,\varphi) = \vec{\nabla} Y_{JM}(\theta,\varphi)$. The following relations illustrate the relation of the alternative basis of the vector spherical harmonics $\vec{Y}^{(i)}_{JM}(\theta,\varphi)$ to the spherical harmonics $Y_{JM}(\theta,\varphi)$,

$$\vec{Y}^{(1)}_{JM}(\theta,\varphi) = \frac{1}{\sqrt{J\,(J+1)}}\, \vec{\nabla}\, Y_{JM}(\theta,\varphi)\,, \tag{6.107a}$$

$$\vec{Y}^{(0)}_{JM}(\theta,\varphi) = \frac{1}{\sqrt{J\,(J+1)}}\, \vec{L}\, Y_{JM}(\theta,\varphi)\,, \tag{6.107b}$$

$$\vec{Y}^{(-1)}_{JM}(\theta,\varphi) = \hat{r}\, Y_{JM}(\theta,\varphi)\,. \tag{6.107c}$$

A comparison with formulas given in Chap. 5 of Ref. [81] reveals that $\vec{Y}^{(1)}_{JM}(\theta,\varphi)$ is related to the longitudinal component of emitted radiation, as its curl vanishes. Furthermore, $\vec{Y}^{(0)}_{JM}(\theta,\varphi)$ is related to the electric field encountered in magnetic multipole components of the radiation. Its divergence vanishes. Finally, $\vec{Y}^{(-1)}_{JM}(\theta,\varphi)$ is related to the electric field encountered in electric multipole components of the radiation [see Eq. (5.164) of Ref. [81]].

Bethe's gradient formula reads as follows,

$$\vec{\nabla}\left(u(r)\,Y_{\ell m}(\theta,\varphi)\right) = \sqrt{\frac{\ell}{2\ell+1}}\left(\frac{\mathrm{d}}{\mathrm{d}r} + \frac{\ell+1}{r}\right) u(r)\,\vec{Y}^{\ell-1}_{\ell m}(\theta,\varphi) - \sqrt{\frac{\ell+1}{2\ell+1}}\left(\frac{\mathrm{d}}{\mathrm{d}r} - \frac{\ell}{r}\right) u(r)\,\vec{Y}^{\ell+1}_{\ell m}(\theta,\varphi)\,. \tag{6.108}$$

It applies to any function which can be represented as $u(r)\,Y_{\ell m}(\theta,\varphi)$ and can be used in order to simplify the angular algebra involved in calculations with the Schrödinger–Coulomb propagator, significantly. The two terms on the right-hand side of Eq. (6.108) involve spherical harmonics of rank $\ell - 1$ and $\ell + 1$, respectively. There is a subtlety, here. Under rotations, the vector $\vec{\nabla}$ itself, seen as a physical entity, is a scalar according to Eq. (6.27). That is because according to Eq. (6.27), any vector can be written as a tensor of rank 0, with magnetic projection 0, composed of vector components with magnetic projection q, and basis vectors with magnetic projection q', coupled with a Clebsch–Gordan coefficient $C^{00}_{1q\,1q'}$. While the physical operator $\vec{\nabla}$ is a scalar, its components and the unit vectors in the basis transform as contravariant and covariant components, respectively. Therefore, the total angular momentum $J = \ell$ of the vector spherical harmonics on the right-hand side of Eq. (6.108) still is $J = \ell$, while the purely angular part involves spherical harmonics of rank $\ell - 1$ and $\ell + 1$, respectively.

6.4.3 *Spherical Biharmonic*

In Sec. 6.4.1, we have coupled the electron spin to its orbital angular momentum, to obtain the standard spin-angular functions, and in Sec. 6.4.2, the coupling of a unit vector to an orbital momentum has been described. This leads to the vector spherical harmonics. The third possibility of coupling angular momenta, also of fundamental importance, concerns two individual orbital angular momenta (say, of two electrons) coupled together. The total angular momentum operator for the two electrons is given by

$$\vec{L} = \vec{\ell}_1 + \vec{\ell}_2\,, \tag{6.109}$$

and the coupled angular momenta are described by a spherical biharmonic,

$$\mathcal{Y}^{\ell_1\,\ell_2}_{L\,M}(\theta_1,\varphi_1,\theta_2,\varphi_2) = \sum_{m_1 m_2} C^{LM}_{\ell_1\,m_1\,\ell_2\,m_2} Y_{\ell_1\,m_1}(\theta_1,\varphi_1)\,Y_{\ell_2\,m_2}(\theta_2,\varphi_2)\,. \tag{6.110}$$

Again, using the Clebsch–Gordan coefficients, the two individual angular momenta ℓ_1 and ℓ_2 are added to produce the total angular momentum L. The spherical biharmonic has the properties

$$\vec{L}^{\,2}\,\mathcal{Y}^{\ell_1\,\ell_2}_{L\,M}(\hat{r}_1,\hat{r}_2) = L\,(L+1)\,\mathcal{Y}^{\ell_1\,\ell_2}_{L\,M}(\hat{r}_1,\hat{r}_2)\,, \qquad L_z\,\mathcal{Y}^{\ell_1\,\ell_2}_{L\,M}(\hat{r}_1,\hat{r}_2) = M\,\mathcal{Y}^{\ell_1\,\ell_2}_{L\,M}(\hat{r}_1,\hat{r}_2)\,. \tag{6.111}$$

The function $\mathcal{Y}^{\ell_1\,\ell_2}_{L\,M}(\hat{r}_1,\hat{r}_2)$ scalar-valued, rather than a vector (as in the case of the vector spherical harmonic).

6.5 Applications of Racah–Wigner Algebra

6.5.1 *Tensor Decomposition of the Light Shift*

Let us consider the following physical situation: A z-polarized laser illuminates an atom which has been prepared in a specific substate of a hyperfine manifold. In Secs. 3.3.1, 3.3.2, and 3.3.3, we have already encountered three alternative derivations of the dynamic Stark effect. Yet, on the level of nonrelativistic quantum mechanics, we have certainly neglected the electron spin, and certainly neglected the nuclear spin which defines the hyperfine manifold. The task then is to investigate the effect of the dynamic Stark shift operator on a specific substate of the hyperfine manifold. But the question is how to resolve this for a particular hyperfine level.

We start by defining the zz, or "33", component of the ac Stark shift operators (light shift operator) as follows,

$$Q^{33} = e^2 \, z \left(\frac{1}{E - H - \omega_{\rm L}} + \frac{1}{E - H + \omega_{\rm L}} \right) z \, . \tag{6.112}$$

In our discussion of the dynamic Stark shift in Sec. (3.3.1), we had already encountered the following matrix element of the reference state, given in Eq. (3.53),

$$\langle \phi_n | Q^{33} | \phi_n \rangle = e^2 \sum_{\pm} \left\langle \phi_n \left| z \left(\frac{1}{H_0 - E_n + \omega_{\rm L}} + \frac{1}{H_0 - E_n - \omega_{\rm L}} \right) z \right| \phi_n \right\rangle , \tag{6.113}$$

where we ignore prefactors which are irrelevant for the following discussion, and specialize the matrix element to a z-polarized laser. The Cartesian components of the light shift operator read as follows,

$$Q^{ij} = x^i \left(\frac{1}{E - H - \omega} + \frac{1}{E - H + \omega} \right) x^j \, . \tag{6.114}$$

For a spherically symmetric reference state $|\phi_n\rangle$, we can write

$$\begin{aligned} e^2 \, \langle \phi_n | Q^{33} | \phi_n \rangle &= \frac{e^2}{3} \, \langle \phi_n | Q^{11} + Q^{22} + Q^{33} | \phi_n \rangle \\ &= \frac{e^2}{3} \sum_{\pm} \left\langle \phi_n \left| x^i \, \frac{1}{H_0 - E_n \pm \omega_{\rm L}} \, x^i \right| \phi_n \right\rangle = \alpha(\omega_{\rm L}) \, , \end{aligned} \tag{6.115}$$

where $\omega_{\rm L}$ is the laser frequency and $\alpha(\omega_{\rm L})$ is the dynamic dipole polarizability of the reference state $|\phi_n\rangle$,

$$\alpha(\omega_{\rm L}) = \frac{e^2}{3} \sum_{\pm} \left\langle \phi_n \left| x^i \, \frac{1}{H_0 - E_n \pm \omega_{\rm L}} \, x^i \right| \phi_n \right\rangle . \tag{6.116}$$

We also recall that by the summation convention, the Cartesian coordinates x^i are summed over (for a complex atom, one has to sum over the coordinates x^i_a of all electrons labeled by the index a). For a spherically symmetric atom, according to Eq. (3.54), the dynamic Stark shift of the reference state can be expressed as

$$\Delta E_n = -\frac{I_{\rm L}}{2 \, \epsilon_0 \, c} \, \alpha(\omega_{\rm L}) \, , \tag{6.117}$$

but for a z-polarized laser field and a reference state which is not spherically symmetric (such as a specific magnetic substate within the hyperfine manifold), one has to manifestly evaluate

$$\Delta E_n = -\frac{I_{\mathrm{L}}}{2\,\epsilon_0\,c}\,\langle\phi_n|Q^{33}|\phi_n\rangle\,. \tag{6.118}$$

The task then is to evaluate the matrix element $\langle\phi_n|Q^{33}|\phi_n\rangle$ in terms of the quantum numbers, possibly within a hyperfine manifold.

According to the discussion in Sec. 6.2.1, one may decompose any tensor with two indices into an $\ell = 0$, three $\ell = 1$, and five $\ell = 2$ components. Our task is to decompose a diagonal matrix element (in an arbitrary reference state) of the light shift operator into irreducible tensor components. We would first like to put the decomposition given in Sec. 6.2.1 on a more solid footing, using the totally antisymmetric tensor. For a general tensor with Cartesian components T^{ij}, we write

$$T^{ij} = (T|_{\ell=0})^{ij} + (T|_{\ell=1})^{ij} + (T|_{\ell=2})^{ij}\,. \tag{6.119}$$

In particular, we have, from Eqs. (6.14)–(6.16),

$$(T|_{\ell=0})^{ij} = \frac{1}{3}\,T^{kk}\,\delta^{ij}\,, \tag{6.120a}$$

$$(T|_{\ell=1})^{ij} = \frac{1}{2}\left(T^{ij} - T^{ji}\right) = \frac{1}{2}\,\epsilon^{ijk}\,\epsilon^{k\ell m}\,T^{\ell m}\,, \tag{6.120b}$$

$$(T|_{\ell=2})^{ij} = \frac{1}{2}\left(T^{ij} + T^{ji}\right) - \frac{1}{3}\,T^{kk}\,\delta^{ij}\,. \tag{6.120c}$$

Let us apply the Wigner–Eckart theorem in the form given in Eq. (6.48), where it is given in terms of the $3j$ symbol,

$$\langle j\,m|T_{kq}|j'\,m'\rangle = (-1)^{j-m+2k}\begin{pmatrix} j & k & j' \\ -m & q & m' \end{pmatrix}\langle j||\vec{T}(k)||j'\rangle\,. \tag{6.121}$$

For the components of the T tensor, this means that

$$\left\langle L'\,0\left|T_{(\ell=k)(q=0)}\right|L''\,0\right\rangle = (-1)^{L'}\begin{pmatrix} L' & k & L'' \\ 0 & 0 & 0 \end{pmatrix}\left\langle L'\left\|\vec{T}(\ell=k)\right\|L''\right\rangle\,, \tag{6.122}$$

where $\ell = k$ assumes the values $\ell = 0$, $\ell = 1$, or $\ell = 2$. The above equation needs an explanation. On the left-hand side, we have an expression derived from a Cartesian representation of the tensor, with the $\ell = k$ component being isolated. The expressions for the particular components are given in Eq. (6.120). Note that, in this sense, the indices ij on the left-hand side of Eqs. (6.120a), (6.120b), and (6.120c) are not independent Cartesian indices, but merely represent the running Cartesian indices relevant to the $\ell = k$ tensor component matrices given in Eq. (6.120).

For the $\ell = k = 2$ component of a Cartesian tensor T^{ij}, *a priori*, we have nine components $(i,j = 1,2,3)$. However, $(T|_{\ell=2})^{ij}$ is a symmetric and traceless matrix, which reduces the number of independent components to $9-3-1 = 5$, which is exactly the number of available quantum number $q = -2,-1,0,1,2$ in the spherical basis. In order to apply the theorem (6.122), it is still necessary to convert the Cartesian to the spherical components. This is illustrated in some detail in Sec. 2.3.7 of Ref. [81]. In some cases, the conversion from the Cartesian to the spherical basis is particularly easy. For example, it is immediate to show that the Cartesian component $(T|_{\ell=2})^{33}$ is equal to the $q=0$ quadrupole component of the same tensor,

$$(T|_{\ell=2})^{33} = \frac{2}{3}T^{33} - \frac{1}{3}T^{11} - \frac{1}{3}T^{22} = (T_{\ell=2})_{q=0}\,. \tag{6.123}$$

The proof can be done on the basis of an inspection of the spherical harmonic $Y_{20}(\theta,\varphi)$, which constitutes a tensor element of a second-rank spherical quadrupole tensor, as follows. One observes that

$$\begin{aligned} r^2 Y_{20}(\theta,\varphi) &= r^2\sqrt{\frac{5}{16\pi}}\left(3\cos^2\theta - 1\right) = \sqrt{\frac{5}{16\pi}}\left(3z^2 - r^2\right) \\ &= 3\sqrt{\frac{5}{16\pi}}\left(\frac{2}{3}z^2 - \frac{1}{3}x^2 - \frac{1}{3}y^2\right) = 3\sqrt{\frac{5}{16\pi}}\,(R_{\ell=2})^{33}\,, \end{aligned} \tag{6.124}$$

where $R^{ij} = x^i\,x^j$ is the second-rank tensor composed of the products of the Cartesian coordinates, and $(R_{\ell=2})^{ij}$ is the matrix describing its quadrupole component. Because the relation of the Cartesian components to the spherical components of a particular tensor is the same for any tensor of the same rank, Eq. (6.123) is obvious.

One may thus identify the light-shift operator Q^{33} as a linear superposition of the $k=0$, $q=0$ and $k=2$, $q=0$ tensor components,

$$Q^{33} = \underbrace{\frac{1}{3}\left(Q^{11} + Q^{22} + Q^{33}\right)}_{Q_{0\,0}} + \underbrace{\frac{2}{3}Q^{33} - \frac{1}{3}Q^{11} - \frac{1}{3}Q^{22}}_{Q_{2\,0}} = Q_{0\,0} + Q_{2\,0}\,. \tag{6.125}$$

Using ideas outlined in Sec. 4.3.4, it is possible to evaluate the matrix elements $\left\langle\phi_{n\ell(m=0)}|Q_{0\,0}|\phi_{n\ell(m=0)}\right\rangle$ and $\left\langle\phi_{n\ell(m=0)}|Q_{2\,0}|\phi_{n\ell(m=0)}\right\rangle$ by explicit calculations. Here, $\phi_{n\ell(m=0)}$ is a nonrelativistic state with quantum numbers n, ℓ and $m=0$. Using Eq. (6.122), these elements may be related to the reduced matrix elements via the relations

$$\begin{aligned} \left\langle\phi_{n\ell(m=0)}|Q_{0\,0}|\phi_{n\ell(m=0)}\right\rangle &= (-1)^\ell \begin{pmatrix} \ell & 0 & \ell \\ 0 & 0 & 0 \end{pmatrix} \langle n\ell\|\vec{Q}(0)\|n\ell\rangle \\ &= \frac{1}{\sqrt{2\ell+1}}\,\langle n\ell\|\vec{Q}(0)\|n\ell\rangle\,, \end{aligned} \tag{6.126a}$$

$$\begin{aligned} \langle n\ell(m=0)|Q_{2\,0}|n\ell(m=0)\rangle &= (-1)^\ell \begin{pmatrix} \ell & 2 & \ell \\ 0 & 0 & 0 \end{pmatrix} \langle n\ell\|\vec{Q}(2)\|n\ell\rangle \\ &= -\frac{\sqrt{2\ell(\ell+1)}}{\sqrt{(2\ell-1)(2\ell+1)(2\ell+2)(2\ell+3)}}\,\langle n\ell\|\vec{Q}(2)\|n\ell\rangle\,. \end{aligned} \tag{6.126b}$$

These two relations enable us to find the reduced matrix elements $\langle n\ell\|\vec{Q}(0)\|n\ell\rangle$ and $\langle n\ell\|\vec{Q}(2)\|n\ell\rangle$ from the two elements $\left\langle\phi_{n\ell(m=0)}|Q_{0\,0}|\phi_{n\ell(m=0)}\right\rangle$ and

$\langle \phi_{n\ell(m=0)} | Q_{2\,0} | \phi_{n\ell(m=0)} \rangle$, which may be explicitly calculated. The factor $1/\sqrt{2\ell+1}$ for the tensor of rank zero finds a natural explanation in terms of Eq. (6.40).

There is an additional problem, however. In ultra-high precision spectroscopy, we are addressing particular hyperfine-structure levels. We must find a way of expressing the light shift for each hyperfine level individually, although the light-shift operator is independent of the nuclear spin as well as the electron spin. In our case, the electron orbital angular momentum ℓ and its spin S first couple to the total electron angular momentum J. Then, the nuclear angular momentum I couples to J, resulting in the total angular momentum F, which is the sum of the electron and nuclear angular momenta.

We start from the outermost level and consider the hyperfine level with quantum numbers n, F, and magnetic projection m_F. Using Eq. (6.121), we have

$$\langle n\, F\, m_F | Q_{2\,0} | n\, F\, m_F \rangle = (-1)^{F-m_F} \begin{pmatrix} F & 2 & F \\ -m_F & 0 & m_F \end{pmatrix} \langle n\, F \| \vec{Q}(2) \| n\, F \rangle \,. \tag{6.127}$$

We here use a shorthand notation where the quantum numbers alone denote the state. The hyperfine level $|n\, F\, m_F\rangle$ qualifies itself as $|n\, F\, m_F\rangle = |nJIF\, m_F\rangle$, because J and I couple to F. We may now apply (6.59) in order to reformulate $\langle nF \| \vec{Q}(2) \| nF \rangle$, because the light shift operator acts only on J, not on I, and obtain

$$\langle nJIF \| \vec{Q}(2) \| nJIF \rangle = (-1)^{J+I+F} (2F+1) \begin{Bmatrix} J & F & I \\ F & J & 2 \end{Bmatrix} \langle nJ \| \vec{Q}(2) \| nJ \rangle \,. \tag{6.128}$$

Now, the state $|nJ\rangle$ qualifies itself again as $|nJ\rangle = |n\ell SJ\rangle$, because ℓ and S couple to J. So, applying (6.59) one more time, we obtain

$$\begin{aligned} \langle nJ \| \vec{Q}(2) \| nJ \rangle &= \langle n\ell SJ \| \vec{Q}(2) \| n\ell SJ \rangle \\ &= (-1)^{\ell+S+J} (2J+1) \begin{Bmatrix} \ell & J & S \\ J & \ell & 2 \end{Bmatrix} \langle n\ell \| \vec{Q}(2) \| n\ell \rangle \,. \end{aligned} \tag{6.129}$$

The two symmetry properties (6.57) and (6.58) allow us to reformulate

$$\begin{Bmatrix} J & F & I \\ F & J & 2 \end{Bmatrix} = \begin{Bmatrix} J & J & 2 \\ F & F & I \end{Bmatrix} = \begin{Bmatrix} J & 2 & J \\ F & I & F \end{Bmatrix} . \tag{6.130}$$

Also, we have

$$\begin{Bmatrix} \ell & J & S \\ J & \ell & 2 \end{Bmatrix} = \begin{Bmatrix} \ell & \ell & 2 \\ J & J & S \end{Bmatrix} = \begin{Bmatrix} \ell & 2 & \ell \\ J & S & J \end{Bmatrix} . \tag{6.131}$$

The forms on the right-hand sides are more symmetric; they bring the rank (two) of the Q tensor into the middle column. Using Eqs. (6.128), (6.129), (6.130), and (6.131), we now have

$$\begin{aligned} \langle nF \| \vec{Q}(2) \| nF \rangle &= (-1)^{I+F+2J+\ell+S} (2J+1)(2F+1) \\ &\quad \times \begin{Bmatrix} J & 2 & J \\ F & I & F \end{Bmatrix} \begin{Bmatrix} \ell & 2 & \ell \\ J & S & J \end{Bmatrix} \langle n\ell \| \vec{Q}(2) \| n\ell \rangle \,. \end{aligned} \tag{6.132}$$

Together with (6.127), we finally have the relation

$$\langle nFm_F|Q_{20}|nFm_F\rangle = (-1)^{I+2F+2J+\ell+S-m_F}\,(2J+1)(2F+1) \times \begin{pmatrix} F & 2 & F \\ -m_F & 0 & m_F \end{pmatrix} \begin{Bmatrix} J & 2 & J \\ F & I & F \end{Bmatrix} \begin{Bmatrix} \ell & 2 & \ell \\ J & S & J \end{Bmatrix} \langle n\,\ell\|\vec{Q}(2)\|n\,\ell\rangle\,. \tag{6.133}$$

For the component of rank zero, the analog of Eq. (6.127) is

$$\begin{aligned}\langle nFm_F|Q_{00}|nFm_F\rangle &= (-1)^{F-m_F}\begin{pmatrix} F & 0 & F \\ -m_F & 0 & m_F \end{pmatrix}\langle n\,F\|\vec{Q}(0)\|n\,F\rangle \\ &= (-1)^{F-m_F}\frac{(-1)^{-F+m_F}}{\sqrt{2F+1}}\langle n\,F\|\vec{Q}(0)\|n\,F\rangle \\ &= \frac{1}{\sqrt{2F+1}}\langle n\,F\|\vec{Q}(0)\|n\,F\rangle\,. \end{aligned} \tag{6.134}$$

For the analogs of Eqs. (6.128) and (6.129), we can use the following relations for specific $6j$ symbols as they enter into the analogs of Eqs. (6.130) and (6.131),

$$\begin{Bmatrix} J & 0 & J \\ F & I & F \end{Bmatrix} = \frac{(-1)^{-F-I-J}}{\sqrt{(2F+1)(2J+1)}}\,, \tag{6.135a}$$

$$\begin{Bmatrix} L & 0 & L \\ J & S & J \end{Bmatrix} = \frac{(-1)^{-J-L-S}}{\sqrt{(2J+1)(2L+1)}}\,. \tag{6.135b}$$

In the analog of (6.132), the factors of $2J+1$ and $2F+1$ as well as the powers of -1 thus cancel; for the factor $2F+1$, this can be seen in combining the corresponding factors in Eq. (6.134) and (6.135a). Finally, we have

$$\langle n\,F\,m_F|Q_{00}|n\,F\,m_F\rangle = \frac{1}{\sqrt{2\ell+1}}\langle n\,\ell\|\vec{Q}(0)\|n\,\ell\rangle\,. \tag{6.136}$$

We can now use Eqs. (6.133) and (6.136) to write the matrix element of the light shift operator as

$$\begin{aligned}\langle nFm_F|Q^{33}|nFm_F\rangle &= \langle nFm_F|Q_{00}|nFm_F\rangle + \langle nFm_F|Q_{20}|nFm_F\rangle \\ &= \frac{1}{\sqrt{2L+1}}\langle n\ell\|\vec{Q}(0)\|n\ell\rangle + (-1)^{I+2F+2J+\ell+S-m_F}(2J+1)(2F+1) \\ &\quad\times\begin{pmatrix} F & 2 & F \\ -m_F & 0 & m_F \end{pmatrix} \begin{Bmatrix} J & 2 & J \\ F & I & F \end{Bmatrix} \begin{Bmatrix} \ell & 2 & \ell \\ J & S & J \end{Bmatrix} \langle n\,\ell\|\vec{Q}(2)\|n\,\ell\rangle\,. \end{aligned} \tag{6.137}$$

We remember that the absolute normalization of the reduced matrix elements is given by Eqs. (6.126a) and (6.126b).

Our result in Eq. (6.137) is consistent with the result given originally in Eq. (12) of Ref. [247] and with the analysis presented in Ref. [117]; it is of considerable importance for practical questions that arise in the analysis of high-precision spectroscopic experiments. Because the quadrupole term $\langle \phi_{nS} \| \vec{Q}(2) \| \phi_{nS} \rangle$ vanishes for $\ell = 0$ (i.e., for S states), the light shift is independent of the hyperfine structure components for S states (only the term proportional to $\langle n\ell \| \vec{Q}(0) \| n\ell \rangle$ survives). For states of a different angular symmetry, say, D states, the situation is different. The result (6.137) implies, in particular, that the light shift of D states does not only depend on the hyperfine component, as manifest in the total angular momentum quantum number F, but also on the projection m_F. This dependence is numerically not very large but significant in the analysis of high-precision spectroscopic experiments [117].

In our treatment so far, we have assumed that the light shift operator describes a z polarized laser, and that the hyperfine state also is aligned along the z axis, which is the quantization axis. This raises the question of how to rotate the z-polarized laser into the quantization axis of the atomic total angular momentum via the relation (6.71). We have to use a Wigner D matrix in order to describe the rotation,

$$D(\alpha\,\beta\,\gamma)\,Q_{20}D^{-1}(\alpha\,\beta\,\gamma) = \sum_{q'} Q_{2\,q'}\,\mathcal{D}^{(2)}_{q'0}(\alpha\,\beta\,\gamma)\,. \tag{6.138}$$

This leads to the following, modified formula for the matrix element of the rotated light shift operator,

$$\begin{aligned}\langle nFm_F|Q'^{33}|nFm_F\rangle &= \frac{1}{\sqrt{2L+1}}\,\langle nL\|\vec{Q}(0)\|nL\rangle + \sum_{q'}(-1)^{I+2F+2J+\ell+S-m_F} \\ &\quad\times(2J+1)\,(2F+1)\begin{pmatrix} F & 2 & F \\ -m_F & q' & m_F \end{pmatrix}\mathcal{D}^{(2)}_{q'0}(\alpha\,\beta\,\gamma) \\ &\quad\times\begin{Bmatrix} J & 2 & J \\ F & I & F \end{Bmatrix}\begin{Bmatrix} L & 2 & L \\ J & S & J \end{Bmatrix}\langle nL\|\vec{Q}(2)\|nL\rangle\,. \end{aligned} \tag{6.139}$$

The $3j$ symbol in this equation vanishes except for the case $q' = 0$, and so

$$\begin{pmatrix} F & 2 & F \\ -m_F & q' & m_F \end{pmatrix} = \begin{pmatrix} F & 2 & F \\ -m_F & 0 & m_F \end{pmatrix}\delta_{q'0}\,. \tag{6.140}$$

Furthermore, we have $\mathcal{D}^{(2)}_{00}(\alpha\,\beta\,\gamma) = d^{(2)}_{00}(\beta)$ according to Eq. (6.74), and thus

$$\begin{aligned}\left\langle nFm_F|Q'^{33}|nFm_F\right\rangle &= \frac{1}{\sqrt{2\ell+1}}\,\left\langle n\,\ell|\vec{Q}(0)|n\,\ell\right\rangle + (-1)^{I+2F+2J+\ell+S-m_F} \\ \times(2J+1)(2F+1)&\begin{pmatrix} F & 2 & F \\ -m_F & 0 & m_F \end{pmatrix}\begin{Bmatrix} J & 2 & J \\ F & I & F \end{Bmatrix}\begin{Bmatrix} \ell & 2 & \ell \\ J & S & J \end{Bmatrix} d^{(2)}_{00}(\beta)\,\left\langle n\,\ell\|\vec{Q}(2)\|n\,\ell\right\rangle\,. \end{aligned} \tag{6.141}$$

Here, β is the angle of the laser polarization axis to the quantization axis of the total angular momentum. The tensor element $d^{(2)}_{00}(\beta)$ can be expressed as [see Eq. (4.1.26) of Ref. [14]]

$$d^{(2)}_{00}(\beta) = P_2^{(0,0)}(\beta) = P_2(\beta) = \frac{1}{2}\left(3\cos^2\beta - 1\right) = 1 - \frac{3}{2}\beta^2 + \mathcal{O}(\beta^4)\,. \tag{6.142}$$

This result can be used in order to estimate the effect of a small tilt angle θ of the laser polarization axis against the angular momentum quantization axis.

6.5.2 *Tensorial Decomposition of a Dipole Transition*

We have treated the decay rate of an excited atomic state against one-photon dipole decay in Sec. 3.2.3 on the basis of Fermi's golden rule, and in Sec. 4.7.1 on the basis of the imaginary part of the one-loop self-energy, identifying the decay width as a quantity proportional to the imaginary part of the energy shift. In this treatment, the excited state was treated as a manifestly nonrelativistic state, without fine structure. Our task now is to decompose the transition matrix element for a dipole transition including fine-structure levels.

In the length gauge, the dipole transition operator is proportional to x_q, where x_q is the qth component of the position vector in spherical coordinates [see Eq. (6.21)]. We have $q = 0$ for the case of a z-polarized laser, and $q = \pm 1$ for circular polarization [see Eq. (6.22)]. Taking into account the fine-structure (total angular momentum quantum number J), the matrix element to decompose is

$$\langle n' \ell' S' J' \mu' | x_q | n \ell S J \mu \rangle \,, \tag{6.143}$$

where in the following we manifestly assume that $S' = S$ (for a one-electron state, we have $S = 1/2$). The Wigner–Eckart theorem in the form (6.121) yields

$$\langle n' \ell' S J' \mu' | x_q | n \ell S J \mu \rangle = (-1)^{J'-\mu'} \begin{pmatrix} J' & 1 & J \\ -\mu' & q & \mu \end{pmatrix} \langle n' \ell' S J' \| \vec{r} \| n \ell S J \rangle \,. \tag{6.144}$$

We recall (6.59) in the following form [with $\vec{T}(k) = \vec{r}$ a tensor of rank $k = 1$]

$$\langle n' j_1' j_2' J' \| \vec{r} \| n j_1 j_2 J \rangle = (-1)^{j_1' + j_2 + J + 1} \, \Pi_{JJ'} \begin{Bmatrix} j_1' & J' & j_2 \\ J & j_1 & 1 \end{Bmatrix} \langle n' j_1' \| \vec{r} \| n j_1 \rangle \,. \tag{6.145}$$

For a dipole transition, we identify $j_1' = \ell'$, $j_1 = \ell$, $j_2' = j_2 = S$. Only the electron orbital angular momentum couples to the position operator, and so

$$\langle n' \ell' S J' \| \vec{r} \| n \ell S J \rangle = (-1)^{\ell' + S + J + 1} \, \Pi_{JJ'} \begin{Bmatrix} \ell' & J' & S \\ J & \ell & 1 \end{Bmatrix} \langle n' \ell' \| \vec{r} \| n \ell \rangle \,. \tag{6.146}$$

We have finally reduced the dipole transition element to a reduced matrix element involving only the angular momentum states with quantum numbers n, ℓ, n', ℓ',

$$\langle n' \ell' S J' \mu' | x_q | n \ell S J \mu \rangle = (-1)^{J' + J + \ell' + S + 1 - \mu'} \times \Pi_{JJ'} \begin{pmatrix} J' & 1 & J \\ -\mu' & q & \mu \end{pmatrix} \begin{Bmatrix} \ell' & J' & S \\ J & \ell & 1 \end{Bmatrix} \langle n' \ell' \| \vec{r} \| n \ell \rangle \,. \tag{6.147}$$

For a one-electron state, we can now go a step further and use the fact that $S = 1/2$. Furthermore, we take into account that the reduced matrix element $\langle n' \ell' \| \vec{r} \| n \ell \rangle$ needs to be evaluated among nonrelativistic states of the form $\phi_{n\ell m}(\vec{r}) = R_{n\ell}(r) \, Y_{\ell m}(\hat{r})$. One may use the following result, given previously in Eq. (6.43) for the reduced matrix element of the unit vector operator, sandwiched between spherical harmonics $Y_{\ell' m'}(\hat{r})$ and $Y_{\ell m}(\hat{r})$,

$$\langle \ell' \| \hat{r} \| \ell \rangle = (-1)^{\ell'} \, \Pi_{\ell \ell'} \begin{pmatrix} \ell' & 1 & \ell \\ 0 & 0 & 0 \end{pmatrix} . \tag{6.148}$$

Then,

$$\langle n'\,\ell'\|\vec{r}\|n\,\ell\rangle = (-1)^{\ell'}\,\Pi_{\ell\ell'}\begin{pmatrix}\ell' & 1 & \ell\\ 0 & 0 & 0\end{pmatrix}\mathcal{R}^{(3)}_{n'\ell' n\ell}\,,\qquad \mathcal{R}^{(j)}_{n'\ell' n\ell} = \int_0^\infty \mathrm{d}r\, r^j\, R_{n'\ell'}(r)\, R_{n\ell}(r)\,. \tag{6.149}$$

Putting everything together, one finds

$$\langle n'\,\ell'\,\tfrac{1}{2}\,J'\,\mu'|x_q|n\,\ell\,\tfrac{1}{2}\,J\,\mu\rangle = (-1)^{J+J'-\mu'-\frac{1}{2}}\,\Pi_{JJ'\ell\ell'} \times\begin{pmatrix}J' & 1 & J\\ -\mu' & q & \mu\end{pmatrix}\begin{pmatrix}\ell' & 1 & \ell\\ 0 & 0 & 0\end{pmatrix}\begin{Bmatrix}\ell' & J' & \frac{1}{2}\\ J & \ell & 1\end{Bmatrix}\mathcal{R}^{(3)}_{n'\ell' n\ell}\,. \tag{6.150}$$

Again, we can describe the effect of a rotation of the laser polarization axis into the quantization axis of the total atomic angular momentum via the relation (6.73). Namely, we have

$$D(\alpha\,\beta\,\gamma)\,x_q\,D^{-1}(\alpha\,\beta\,\gamma) = \sum_{q'} x_{q'}\,\mathcal{D}^{(1)}_{q'q}(\alpha\,\beta\,\gamma)\,, \tag{6.151}$$

leading to the following matrix element for the rotated vector component $x'_q = D(\alpha\,\beta\,\gamma)\,x_q\,D^{-1}(\alpha\,\beta\,\gamma)$,

$$\langle n'\,\ell'\,\tfrac{1}{2}\,J'\,\mu'|\,x'_q\,|n\,\ell\,\tfrac{1}{2}\,J\,\mu\rangle = \sum_{q'}(-1)^{J+J'-\mu'-1/2}\,\Pi_{JJ'\ell\ell'}\,\mathcal{D}^{(1)}_{q'q}(\alpha\,\beta\,\gamma) \times\begin{pmatrix}J' & 1 & J\\ -\mu' & q' & \mu\end{pmatrix}\begin{pmatrix}\ell' & 1 & \ell\\ 0 & 0 & 0\end{pmatrix}\begin{Bmatrix}\ell' & J' & \frac{1}{2}\\ J & \ell & 1\end{Bmatrix}\mathcal{R}^{(3)}_{n'\ell' n\ell}\,. \tag{6.152}$$

There is no immediate, general simplifying formula available for the $3j$ symbols in this equation. However, in a concrete case, with all quantum numbers fixed, one may use this formula in order to estimate the effect of a slight tilt of the laser axis against the quantization axis of the angular momentum.

6.6 Rydberg Electron and Hydrogenlike Core

6.6.1 *Physical Foundation*

A great deal can be learned about the physics of atoms by considering the interaction of a single Rydberg electron, in a highly excited state, with a hydrogenlike core (where there is an attractive force, even if the hydrogenlike core is electrically neutral). Let us consider a two-electron ion, with the nucleus having charge number Z. One of the electrons is in the $1s$ ground state, the other electron is in an excited state with principal quantum number n and orbital angular momentum quantum number ℓ. Again, we use natural units with $\hbar = c = \epsilon_0 = 1$. We label the inner electron by the subscript 1, and the outer electron by the subscript 2. The Hamiltonian reads as

$$H = \frac{\vec{p}_1^{\,2}}{2m} + \frac{\vec{p}_2^{\,2}}{2m} - \frac{Z\alpha}{r_1} - \frac{Z\alpha}{r_2} + \frac{\alpha}{|\vec{r}_1 - \vec{r}_2|}\,. \tag{6.153}$$

The unperturbed Hamiltonian H_0 is the sum of an unscreened Hamiltonian (Coulomb potential with charge number Z) for the inner electron, and a fully screened Coulomb potential (charge number $Z-1$) for the outer electron,

$$H_0 = \frac{\vec{p}_1^{\,2}}{2m} + \frac{\vec{p}_2^{\,2}}{2m} - \frac{Z\alpha}{r_1} - \frac{(Z-1)\,\alpha}{r_2} . \tag{6.154}$$

The perturbation is just the difference $\Delta V = H - H_0$ and reads as

$$\Delta V = -\frac{\alpha}{r_2} + \frac{\alpha}{|\vec{r}_1 - \vec{r}_2|} . \tag{6.155}$$

From the multipole expansion of the electrostatic interaction (see Chap. 2 of Ref. [81]), we know that

$$\frac{1}{|\vec{r}_2 - \vec{r}_1|} = \sum_{\ell=0}^{\infty} \sum_{m=-\ell}^{\ell} \frac{4\pi}{2\ell+1} \frac{r_<^{\ell}}{r_>^{\ell+1}} Y_{\ell m}(\theta_1, \varphi_1)\, Y^*_{\ell m}(\theta'_2, \varphi'_2) . \tag{6.156}$$

For our situation, we have $r_< = r_1$ and $r_> = r_2$. The first two multipoles thus read as

$$\frac{1}{|\vec{r}_2 - \vec{r}_1|} = \frac{1}{r_2} + \frac{4\pi}{3} \frac{r_1}{r_2^2} \sum_{m=-1}^{1} Y_{1\,m}(\theta_1, \varphi_1)\, Y^*_{1\,m}(\theta_2, \varphi_2) + \mathcal{O}\left(\frac{1}{r_2^3}\right) . \tag{6.157}$$

We now enter with (6.157) into (6.155), we obtain for the sum of the first two terms in the multipole expansion of ΔV, written here very explicitly as

$$\Delta V = -\frac{\alpha}{r_2} + \frac{\alpha}{r_2} + \frac{4\pi\,\alpha}{3} \frac{r_1}{r_2^2} \sum_{m=-1}^{1} Y_{1\,m}(\theta_1, \varphi_1)\, Y^*_{1\,m}(\theta_2, \varphi_2) + \mathcal{O}\left(\frac{1}{r_2^3}\right) . \tag{6.158}$$

The monopole term cancels, and the first nonvanishing interaction between the outer electron and the core (beyond the screened Coulomb interaction) consists of a dipole interaction. The unperturbed eigenstates of H_0 consist of an electron in the $|1s\rangle$ ground state, and a second one in an $|n\ell\rangle$ excited Rydberg state. The two-electron state is characterized by the quantum numbers $|1s, n\ell\rangle$. In general, the eigenstates of H_0 are $|n_1\ell_1, n_2\ell_2\rangle$ with one electron in state $|n_1\ell_1\rangle$ and the other in the state $|n_2\ell_2\rangle$. We recall that H_0 does not couple the electrons. The second-order energy shift reads as

$$\begin{aligned} \Delta E &= \left\langle 1s, n\ell \left| \Delta V \left(\frac{1}{E^{(0)}_{1s,n\ell} - H_0} \right)' \Delta V \right| 1s, n\ell \right\rangle \\ &= \sum_{n_1, n_2, \pm} \frac{\langle 1s, n\ell | \Delta V | n_1\, p, n_2\, (\ell \pm 1) \rangle \, \langle n_1\, p, n_2\, (\ell \pm 1) | \Delta V | 1s, n\ell \rangle}{E^{(0)}_{1s,n\ell} - E^{(0)}_{n_1\, p, n_2\, (\ell\pm 1)}} , \end{aligned} \tag{6.159}$$

where $E^{(0)}_{n_1\ell_1, n_2\ell_2}$ is the energy eigenvalue of the state $|n_1\ell_1, n_2 l_2\rangle$ with respect to the unperturbed Hamiltonian H_0 as given in Eq. (6.154). It may be expressed as

$$E^{(0)}_{n_1\ell_1, n_2\ell_2} = E_{n_1\,\ell_1}(Z) + E_{n_2\,\ell_2}(Z-1) , \qquad E_{n\ell}(Z) = E_n(Z) = -\frac{(Z\alpha)^2 m}{2n^2} . \tag{6.160}$$

The dipole selection rule has been used. Both electrons undergo a dipole transition during the interaction. Therefore, the first electron, initially in the ground state, gets excited into a p state, and the second electron, initially in a state with orbital angular momentum ℓ, makes a transition into a state with angular momentum $\ell - 1$, or $\ell + 1$. So, the second-order energy shift reads as

$$\Delta E = \frac{16\pi^2\alpha^2}{9} \sum_{n_1, n_2, \pm} \sum_{m=-1}^{1} \sum_{m'=-1}^{1} \frac{1}{E_{1s}(Z) - E_{n_1 p}(Z) + E_{n\ell}(Z-1) - E_{n_2\, \ell\pm 1}(Z-1)}$$
$$\times \langle 1s | r_1 Y_{1m}(\theta_1, \varphi_1) | n_1 p \rangle \langle n_1 p | r_1 Y^*_{1m'}(\theta_1, \varphi_1) | 1s \rangle$$
$$\times \left\langle n\ell \left| \frac{1}{r_2^2} Y^*_{1m}(\theta_2, \varphi_2) \right| n_2\, (\ell \pm 1) \right\rangle \left\langle n_2\, (\ell \pm 1) \left| \frac{1}{r_2^2} Y_{1m'}(\theta_2, \varphi_2) \right| n\ell \right\rangle . \tag{6.161}$$

The decisive observation is now as follows: the state $|n\ell\rangle$ is a Rydberg state, and therefore its radial overlap with energetically lower states is small. Therefore, the largest contributions to the sum over intermediate states n_2 come from states with $n_2 \approx n$. Since Rydberg states are very close energetically, we may approximate

$$E_{1s}(Z) - E_{n_1 p}(Z) + E_{n\ell}(Z-1) - E_{n_2\, \ell\pm 1}(Z-1) \approx E_{1s}(Z) - E_{n_1 p}(Z) , \tag{6.162}$$

essentially ignoring the Rydberg-state energy difference. In that case, the propagator denominator becomes independent of the Rydberg electron, and the energy shift reads as

$$\Delta E = \frac{16\pi^2\alpha^2}{9} \sum_{n_1, n_2, \pm} \sum_{m=-1}^{1} \sum_{m'=-1}^{1} \frac{\langle 1s | r_1 Y_{1m}(\theta_1, \varphi_1) | n_1 p \rangle \langle n_1 p | r_1 Y^*_{1m'}(\theta_1, \varphi_1) | 1s \rangle}{E_{1s}(Z) - E_{n_1 p}(Z)}$$
$$\times \left\langle n\ell \left| \frac{1}{r_2^2} Y^*_{1m}(\theta_2, \varphi_2) \right| n_2\, (\ell \pm 1) \right\rangle \left\langle n_2\, (\ell \pm 1) \left| \frac{1}{r_2^2} Y_{1m'}(\theta_2, \varphi_2) \right| n\ell \right\rangle . \tag{6.163}$$

The sum over n_1 and n_2 also entails the integral over the continuous part of the spectrum.

6.6.2 *Angular Algebra*

From now on, without loss of generality, we assume that the Rydberg state has a zero magnetic quantum number. We wish to carry out the angular integration in full detail, assuming the structure $\langle \vec{r} | n\ell \rangle = R_{n\ell}(r)\, Y_{\ell m}(\hat{r})$ for all unperturbed orbitals. Under this assumption, the energy shift may be separated into radial and angular parts as follows,

$$\Delta E = \sum_{n_1} \sum_{n_2} \sum_{m_2=-1}^{1} \sum_{m_2'=-1}^{1} \sum_{m_3=-1}^{1} \frac{16\pi^2\alpha^2}{9} \frac{\mathcal{T}_{\ell+1} + \mathcal{T}_{\ell-1}}{E_{1s}(Z) - E_{n_1}(Z)} \left(\mathcal{R}^{(3)}_{1s n_1 p}\right)^2$$
$$\times \left(\int \mathrm{d}\Omega_1\, Y^*_{00}(\theta_1, \varphi_1)\, Y_{1m_2}(\theta_1, \varphi_1)\, Y_{1m_3}(\theta_1, \varphi_1) \right)$$
$$\times \left(\int \mathrm{d}\Omega_1\, Y^*_{1m_3}(\theta_1, \varphi_1)\, (-1)^{m_2'}\, Y_{1\,-m_2'}(\theta_1, \varphi_1)\, Y_{00}(\theta_1, \varphi_1) \right) . \tag{6.164}$$

The sums over m_3 and m_3' that are seen explicitly in the above equation have been implicit in Eq. (6.163). The terms $\mathcal{T}_{\ell\pm 1}$ are given by

$$\mathcal{T}_{\ell+1} = \sum_{m_3'=-(\ell+1)}^{\ell+1} \left(\int \mathrm{d}\Omega_2 \, Y^*_{\ell 0}(\theta_2,\varphi_2) \, (-1)^{m_2} Y_{1\,-m_2}(\theta_2,\varphi_2) \, Y_{(\ell+1)\,m_3'}(\theta_2,\varphi_2) \right)$$
$$\times \left(\int \mathrm{d}\Omega_2 \, Y^*_{(\ell+1)m_3'}(\theta_2,\varphi_2) \, Y_{1\,m_2'}(\theta_2,\varphi_2) \, Y_{\ell 0}(\theta_2,\varphi_2) \right) \left(\mathcal{R}^{(0)}_{n\ell n_2(\ell+1)} \right)^2 , \quad (6.165)$$
$$\mathcal{T}_{\ell-1} = \sum_{m_3'=-(\ell-1)}^{\ell-1} \left(\int \mathrm{d}\Omega_2 \, Y^*_{\ell 0}(\theta_2,\varphi_2) \, (-1)^{m_2} Y_{1\,-m_2}(\theta_2,\varphi_2) \, Y_{(\ell-1)m_3'}(\theta_2,\varphi_2) \right)$$
$$\times \left(\int \mathrm{d}\Omega_2 \, Y^*_{(\ell-1)m_3'}(\theta_2,\varphi_2) \, Y_{1\,m_2'}(\theta_2,\varphi_2) \, Y_{\ell 0}(\theta_2,\varphi_2) \right) \left(\mathcal{R}^{(0)}_{n\ell n_2(\ell-1)} \right)^2 . \quad (6.166)$$

The radial integrals have been defined in Eq. (6.149). We now use the result (6.65) in order to express an integral over three spherical harmonics in terms of $3j$ symbols, with the result

$$\Delta E = \sum_{n_1} \sum_{n_2} \sum_{m_2=-1}^{1} \sum_{m_2'=-1}^{1} \sum_{m_3=-1}^{1} \frac{16\pi^2\alpha^2}{9} \frac{1}{E_{1s}(Z) - E_{n_1}(Z)} \left(\mathcal{R}^{(3)}_{1sn_1p} \right)^2$$
$$\times \frac{\Pi_{011}}{\sqrt{4\pi}} \begin{pmatrix} 0 & 1 & 1 \\ 0 & 0 & 0 \end{pmatrix} \begin{pmatrix} 0 & 1 & 1 \\ 0 & m_2 & m_3 \end{pmatrix} (-1)^{m_3+m_2'} \frac{\Pi_{110}}{\sqrt{4\pi}} \begin{pmatrix} 1 & 1 & 0 \\ 0 & 0 & 0 \end{pmatrix} \begin{pmatrix} 1 & 1 & 0 \\ -m_3 & -m_2' & 0 \end{pmatrix}$$
$$\times \left[\sum_{m_3'=-(\ell+1)}^{\ell+1} (-1)^{m_2} \frac{\Pi_{\ell 1 (\ell+1)}}{\sqrt{4\pi}} \begin{pmatrix} \ell & 1 & \ell+1 \\ 0 & 0 & 0 \end{pmatrix} \begin{pmatrix} \ell & 1 & \ell+1 \\ 0 & -m_2 & m_3' \end{pmatrix} \right.$$
$$\times (-1)^{m_3'} \frac{\Pi_{(\ell+1)1\ell}}{\sqrt{4\pi}} \begin{pmatrix} \ell+1 & 1 & \ell \\ 0 & 0 & 0 \end{pmatrix} \begin{pmatrix} \ell+1 & 1 & \ell \\ -m_3' & m_2' & 0 \end{pmatrix} \left(\mathcal{R}^{(0)}_{n\ell n_2(\ell+1)} \right)^2$$
$$+ \sum_{m_3'=-(\ell-1)}^{\ell-1} (-1)^{m_2} \frac{\Pi_{\ell 1 \ell-1}}{\sqrt{4\pi}} \begin{pmatrix} \ell & 1 & \ell-1 \\ 0 & 0 & 0 \end{pmatrix} \begin{pmatrix} \ell & 1 & \ell-1 \\ 0 & -m_2 & m_3' \end{pmatrix}$$
$$\left. \times (-1)^{m_3'} \frac{\Pi_{(\ell-1)1\ell}}{\sqrt{4\pi}} \begin{pmatrix} \ell-1 & 1 & \ell \\ 0 & 0 & 0 \end{pmatrix} \begin{pmatrix} \ell-1 & 1 & \ell \\ -m_3' & m_2' & 0 \end{pmatrix} \left(\mathcal{R}^{(0)}_{n\ell n_2(\ell-1)} \right)^2 \right] . \quad (6.167)$$

This formula is not as frightening as it looks. The following special cases must now be considered, $\Pi_{011} \begin{pmatrix} 0 & 1 & 1 \\ 0 & 0 & 0 \end{pmatrix} = -\sqrt{3}$, and $\Pi_{\ell 1 (\ell+1)} \begin{pmatrix} \ell & 1 & \ell+1 \\ 0 & 0 & 0 \end{pmatrix} = (-1)^{\ell+1} \sqrt{3}\sqrt{\ell+1}$ as well as $\Pi_{\ell 1 (\ell-1)} \begin{pmatrix} \ell & 1 & \ell-1 \\ 0 & 0 & 0 \end{pmatrix} = (-1)^{\ell} \sqrt{3}\sqrt{\ell}$. After some algebra, we can rewrite ΔE as

$$\Delta E = \mathcal{U}_{\ell+1} + \mathcal{U}_{\ell-1} , \quad (6.168a)$$

where

$$\mathcal{U}_{\ell+1} = \sum_{n_1 n_2} \frac{\alpha^2}{E_{1s}(Z) - E_{n_1}(Z)} \left(\mathcal{R}^{(3)}_{1sn_1p}\right)^2 \left(\mathcal{R}^{(0)}_{n\ell n_2(\ell+1)}\right)^2$$

$$\times \sum_{m_2\, m_2'\, m_3\, m_3'} \left[(\ell+1)\,(-1)^{m_2+m_2'+m_3+m_3'} \begin{pmatrix} 1 & 0 & 1 \\ m_3 & 0 & m_2 \end{pmatrix} \begin{pmatrix} 1 & \ell & \ell+1 \\ -m_2 & 0 & m_3' \end{pmatrix}\right.$$

$$\left. \times \begin{pmatrix} \ell+1 & \ell & 1 \\ -m_3' & 0 & m_2' \end{pmatrix} \begin{pmatrix} 1 & 0 & 1 \\ -m_2' & 0 & -m_3 \end{pmatrix}\right] \tag{6.168b}$$

and

$$\mathcal{U}_{\ell-1} = +\sum_{n_1 n_2} \frac{\alpha^2}{E_{1s}(Z) - E_{n_1}(Z)} \left(\mathcal{R}^{(3)}_{1sn_1p}\right)^2 \left(\mathcal{R}^{(0)}_{n\ell n_2(\ell-1)}\right)^2$$

$$\times \sum_{m_2\, m_2'\, m_3\, m_3'} \left[\ell\,(-1)^{m_2+m_2'+m_3+m_3'} \begin{pmatrix} 1 & 0 & 1 \\ m_3 & 0 & m_2 \end{pmatrix} \begin{pmatrix} 1 & \ell & \ell-1 \\ -m_2 & 0 & m_3' \end{pmatrix}\right.$$

$$\left. \times \begin{pmatrix} \ell-1 & \ell & 1 \\ -m_3' & 0 & m_2' \end{pmatrix} \begin{pmatrix} 1 & 0 & 1 \\ -m_2' & 0 & -m_3 \end{pmatrix}\right]. \tag{6.168c}$$

In order to carry out the summation over the magnetic projections, we can now apply the following formula, which is found on page 455 of Ref. [16],

$$\sum_{\psi\kappa\rho\sigma} (-1)^{p-\psi+q-\kappa+r-\rho+s-\sigma} \begin{pmatrix} p & a & q \\ \psi & \alpha & -\kappa \end{pmatrix} \begin{pmatrix} q & b & r \\ \kappa & \beta & -\rho \end{pmatrix} \begin{pmatrix} r & c & s \\ \rho & \gamma & -\sigma \end{pmatrix} \begin{pmatrix} s & d & p \\ \sigma & \delta & -\psi \end{pmatrix}$$

$$= (-1)^{s-a-d-q} \sum_{x\xi} (-1)^{x-\xi} (2x+1) \begin{pmatrix} a & x & d \\ \alpha & -\xi & \delta \end{pmatrix} \begin{pmatrix} b & x & c \\ \beta & \xi & \gamma \end{pmatrix} \begin{Bmatrix} a & x & d \\ s & p & q \end{Bmatrix} \begin{Bmatrix} b & x & c \\ s & r & q \end{Bmatrix}. \tag{6.169}$$

For Eq. (6.168b), this formula implies that

$$\sum_{m_2, m_2', m_3, m_3'} (-1)^{\ell} (-1)^{m_2+m_2'+m_3+m_3'} \begin{pmatrix} 1 & 0 & 1 \\ m_3 & 0 & m_2 \end{pmatrix} \begin{pmatrix} 1 & \ell & \ell+1 \\ -m_2 & 0 & m_3' \end{pmatrix}$$

$$\times \begin{pmatrix} \ell+1 & \ell & 1 \\ -m_3' & 0 & m_2' \end{pmatrix} \begin{pmatrix} 1 & 0 & 1 \\ -m_2' & 0 & -m_3 \end{pmatrix}$$

$$= \begin{pmatrix} 0 & 0 & 0 \\ 0 & 0 & 0 \end{pmatrix} \begin{pmatrix} \ell & 0 & \ell \\ 0 & 0 & 0 \end{pmatrix} \begin{Bmatrix} 0 & 0 & 0 \\ 1 & 1 & 1 \end{Bmatrix} \begin{Bmatrix} \ell & 0 & \ell \\ 1 & \ell+1 & 1 \end{Bmatrix} = \frac{(-1)^{\ell}}{3\,(2\ell+1)}. \tag{6.170}$$

For the term in Eq. (6.168c), this formula means that

$$\sum_{m_2, m_2', m_3, m_3'} (-1)^{\ell} (-1)^{m_2+m_2'+m_3+m_3'} \begin{pmatrix} 1 & 0 & 1 \\ m_3 & 0 & m_2 \end{pmatrix} \begin{pmatrix} 1 & \ell & \ell-1 \\ -m_2 & 0 & m_3' \end{pmatrix}$$

$$\times \begin{pmatrix} \ell-1 & \ell & 1 \\ -m_3' & 0 & m_2' \end{pmatrix} \begin{pmatrix} 1 & 0 & 1 \\ -m_2' & 0 & -m_3 \end{pmatrix} = \frac{(-1)^{\ell}}{3\,(2\ell+1)}. \tag{6.171}$$

Finally, we obtain a compact expression for ΔE,

$$\Delta E = \frac{\alpha^2}{3} \sum_{n_1} \sum_{n_2} \frac{1}{E_{1s}(Z) - E_{n_1}(Z)} \left(\mathcal{R}^{(3)}_{1sn_1p}\right)^2$$

$$\times \left[\frac{\ell+1}{2\ell+1} \left(\mathcal{R}^{(0)}_{n\ell n_2(\ell+1)}\right)^2 + \frac{\ell}{2\ell+1} \left(\mathcal{R}^{(0)}_{n\ell n_2(\ell-1)}\right)^2\right]. \tag{6.172}$$

One may observe that

$$\sum_{n_2}\left[\frac{\ell+1}{2\ell+1}\left(\mathcal{R}^{(0)}_{n\ell n_2(\ell+1)}\right)^2+\frac{\ell}{2\ell+1}\left(\mathcal{R}^{(0)}_{n\ell n_2(\ell-1)}\right)^2\right]=\left\langle n\ell\left|\frac{1}{r_2^4}\right|n\ell\right\rangle . \tag{6.173}$$

A remark might be in order. One might think that the right-hand side is divergent for S states ($\ell = 0$), while the left-hand side of the equation features only convergent matrix elements. However, the divergence of the left-hand side is relegated, in this case, to the divergence introduced when one sums over all virtual states. Finally, one rewrites the energy perturbation as

$$\Delta E=-\frac{1}{2}\left\langle n\ell\left|\frac{\alpha^2}{r_2^4}\right|n\ell\right\rangle\left(\frac{2}{3}\sum_{i=1}^{3}\sum_{m=-1}^{1}\frac{\left|\left\langle 1s\left|x_1^i\right|n_1pm\right\rangle\right|^2}{E_{n_1}-E_{1s}}\right), \tag{6.174}$$

where we have re-introduced the sum over the Cartesian coordinates x_1^i with $i = 1, 2, 3$ in order to make the connection to the polarizability (see Chap. 4) explicit. Indeed, we identify the scaled static polarizability of the hydrogenlike core [see Eqs. (4.154) and (5.75a)] as

$$\alpha_{1s}(\omega=0)=\frac{2e^2}{3}\sum_{i=1}^{3}\sum_{m=-1}^{1}\frac{\left|\left\langle 1s\left|x_1^i\right|n_1pm\right\rangle\right|^2}{E_{n_1}-E_{1s}}=\frac{9e^2}{2\alpha^4 m^3}, \tag{6.175}$$

where we have used the fact that $Z = 1$ for the hydrogenlike core. The interaction energy can be written as an expectation value, proportional to the core polarizability $\alpha(\omega = 0)$ and given in terms of the $1/r^4$ matrix element, evaluated on the Rydberg state,

$$\Delta E=\langle n\ell|\Delta V|n\ell\rangle\,,\qquad \Delta V\approx-\frac{\alpha(\omega=0)}{2\,e^2}\frac{\alpha^2}{r^4}. \tag{6.176}$$

This energy shift should be added to the unperturbed energy $E^{(0)}_{1s,n\ell} = E_{1s}(Z) + E_{n\ell}(Z-1)$, as given in Eq. (6.160). The total energy thus is the sum of the inner ground-state electron, the outer Rydberg electron (which sees a screened charge $Z - 1$), and the interaction term,

$$E_{1s,n\ell}=-\frac{Z^2\alpha^2 m}{2}-\frac{(Z-1)^2\alpha^2 m}{2n^2}-\frac{\alpha(\omega=0)}{2\,e^2}\left\langle n\ell\left|\frac{\alpha^2}{r^4}\right|n\ell\right\rangle . \tag{6.177}$$

If desired, an explicit expression for the matrix element of the $1/r^4$ operator can be found in Eq. (4.346). An obvious generalization of Eq. (6.177) holds for general states with a Rydberg electron orbiting a neutral, polarizable core, and for a general long-range ion–atom interaction.

6.7 Further Thoughts

Here are some suggestions for further thought.

(1) **Noether Theorem.** Let us remember the physical foundation for the angular momentum being a conserved quantity. Let us study the Lagrangian

$$\mathcal{L} = \frac{1}{2} m \vec{v}^{\,2} - V(|\vec{r}|) , \tag{6.178}$$

on the level of classical mechanics. It does not depend on the angles $\vec{\varphi}$. We identify the conserved quantities as

$$\frac{\partial \mathcal{L}}{\partial \dot{q}} \frac{\partial q}{\partial \varphi^m} \qquad \to \qquad Q^m = \sum_{i=1}^{3} \frac{\partial \mathcal{L}}{\partial v^i} \frac{\partial x^i}{\partial \varphi^m} . \tag{6.179}$$

Show that the rotation by a small vector of angles $\vec{\varphi}$ can be accounted for by the change of coordinates

$$\delta x^i = -(\vec{\varphi} \times \vec{r})^i = -\epsilon^{ijk} \varphi^j x^k . \tag{6.180}$$

Furthermore, identify the conserved quantity as the angular momentum, e.g., as follows,

$$Q^m = \frac{\partial \mathcal{L}}{\partial v^\ell} \frac{\partial x^\ell}{\partial \varphi^m} = m v^\ell \left(-\epsilon^{\ell m k} x^k\right) = -\epsilon^{m k \ell} x^k m v^\ell = -(\vec{x} \times \vec{p})^m = -L^m . \tag{6.181}$$

(2) **Properties of Spin-Angular Functions.** Verify that Eq. (6.94) implies that $c_2 = -1$ for both signs in the relation $j = \ell \pm 1/2$. Then, prove the second identity listed in Eq. (6.85).

(3) **Tensorial Decomposition of the Light Shift Operator.** Repeat the analysis presented in Sec. 6.5.1 for the xx component of the light-shift operator.

(4) **Gauge-Invariance of the Light Shift.** Carry out the calculation in Sec. 6.5.1 in the velocity as opposed to the length gauge. Show gauge invariance of the angle-resolved matrix elements of the light shift operator.

(5) **Static Polarizability.** Show Eq. (6.175) based on Eqs. (4.154) and (5.75a).

(6) **Rydberg Electron and Core.** Exercise: Convert Eq. (6.177), which is given in natural units ($\hbar = \epsilon_0 = c = 1$), to atomic units and show that all factors of α disappear, i.e., that

$$E_{1s,n\ell}\big|_{\text{a.u.}} = -\frac{Z^2}{2} - \frac{(Z-1)^2}{2n^2} - \frac{\alpha(\omega = 0)|_{\text{a.u.}}}{2} \left\langle n\ell \left| \frac{1}{r^4} \right| n\ell \right\rangle \bigg|_{\text{a.u.}} , \tag{6.182}$$

where "a.u." denotes a quantity, expressed in atomic units. Find the obvious generalization for general states with a Rydberg electron orbiting a neutral, polarizable core.

(7) **Multi-Electron Core.** Generalize the derivation outlined in Sec. 6.6 to the interaction of a Rydberg electron with a general, multi-electron, core, by adding the interactions with all core electrons. Show, explicitly, as anticipated, that the general result is still equivalent to Eq. (6.177), with an obvious generalization of $\alpha(\omega = 0)$ to the core polarizability.

Chapter 7

Free Dirac Equation

7.1 Overview

The Dirac equation plays a special role in the analysis of free and interacting spin-1/2 particles, and in particular, in bound-state problems. The free Dirac equation is naturally obtained as the linearized Klein–Gordon equation, i.e., as a relativistically invariant wave equation describing spin-1/2 particles involving only first-order derivatives. The Dirac equation is separately charge-conjugation ($\mathcal{C}$) invariant, parity ($\mathcal{P}$) invariant, and invariant under time reversal ($\mathcal{T}$); this generic statement will be supported by concrete calculations in the following. The Dirac equation is Lorentz-covariant, but showing the covariance is a little nontrivial. Namely, one first has to clarify how a four-component bispinor wave function behaves under Lorentz transformations.

While the Dirac equation was originally proposed in Refs. [82, 83], its solutions were found later. Good quantum numbers are either the plane-wave momentum and the spin projection in the rest frame, or the plane-wave momentum and the helicity, which measures the spin direction of the particle with respect to its propagation direction. We here start with a brief discussion of the properties of the Dirac equation (Sec. 7.2) and then go on to study the plane-wave solutions in Sec. 7.3.1, in the helicity basis. In the plane-wave and the angular-momentum basis, we find the solutions in Sec. 7.3. Finally, propagators are found in Sec. 7.4.

7.2 Properties of the Free Dirac Equation

7.2.1 *Dirac Equation as the Linearized Klein–Gordon Equation*

At this stage, it is useful to recall that we use units with $\hbar = c = \epsilon_0 = 1$. We start from the Schrödinger equation, calculate its conserved probability current, and then proceed to investigate the Klein–Gordon equation which is a relativistic Lorentz-invariant wave equation for spinless particles. The known difficulty of negative-probability states of the Klein–Gordon equation is briefly discussed. The Dirac equation is motivated by the necessity to obtain a first-order time translation operator (Hamiltonian) whose eigenstates nevertheless fulfill a Lorentz-invariant dispersion relation.

The free Schrödinger equation is obtained from the nonrelativistic dispersion relation $E = \vec{p}^2/(2m)$, via the identifications $E \to \mathrm{i}\partial_t$ and $\vec{p} \to -\mathrm{i}\vec{\nabla}$. The correspondence principle implies that we should identify classical physical quantities with quantum mechanical operators. Indeed, the time-dependent Schrödinger equation reads

$$\mathrm{i}\frac{\partial}{\partial t}\phi(t,\vec{r}) = -\frac{\vec{\nabla}^2}{2m}\phi(t,\vec{r}), \tag{7.1}$$

where $\phi(t,\vec{r})$ is the time-dependent wave function in the coordinate representation. The charge and current densities ρ and $\vec{J}$ read as

$$\rho(t,\vec{r}) = \phi^*(t,\vec{r})\,\phi(t,\vec{r}) = |\phi(t,\vec{r})|^2\,, \qquad \vec{J}(t,\vec{r}) = -\frac{\mathrm{i}}{2m}\,\phi^*(t,\vec{r})\,\overset{\leftrightarrow}{\vec{\nabla}}\,\phi(t,\vec{r}), \tag{7.2}$$

where the antisymmetric differential operator acts as follows,

$$a(\vec{r})\,\overset{\leftrightarrow}{\vec{\nabla}}\,b(\vec{r}) = a(\vec{r})\vec{\nabla}b(\vec{r}) - b(\vec{r})\vec{\nabla}a(\vec{r})\,. \tag{7.3}$$

The Schrödinger equation implies the conservation law

$$\frac{\partial}{\partial t}\rho(t,\vec{r}) + \vec{\nabla}\cdot\vec{J}(t,\vec{r}) = 0\,, \tag{7.4}$$

which is a continuity equation, so that we can interpret the positive definite quantity $\rho(t,\vec{r}) = \phi^*(t,\vec{r})\,\phi(t,\vec{r})$ as a probability density.

Generalizing the treatment to the relativistic domain, in view of the relativistic dispersion relation $E^2 = \vec{p}^2 + m^2$, one is immediately led to the Klein–Gordon equation

$$\left(\frac{\partial^2}{\partial t^2} - \vec{\nabla}^2 + m^2\right)\phi(t,\vec{r}) = 0\,, \qquad (\partial^\mu\partial_\mu + m^2)\,\phi = 0\,, \tag{7.5}$$

where the latter form is obtained in a relativistic notation (see Chap. 8 of Ref. [81], $\mu = 0,1,2,3$ is a Lorentz index). The four-vector current $J^\mu = (J^0,\vec{J})$ has the components

$$J^0(t,\vec{r}) = \mathrm{i}\,\phi^*(t,\vec{r})\,\overset{\leftrightarrow}{\partial_t}\,\phi(t,\vec{r})\,, \qquad \vec{J}(t,\vec{r}) = -\mathrm{i}\,\phi^*(t,\vec{r})\,\overset{\leftrightarrow}{\vec{\nabla}}\,\phi(t,\vec{r})\,, \tag{7.6}$$

where

$$a(t)\,\overset{\leftrightarrow}{\partial_t}\,b(t) = a(t)\,\partial_t b(t) - b(t)\,\partial_t a(t)\,. \tag{7.7}$$

The continuity equation then reads as

$$\frac{\partial}{\partial t}\rho(t,\vec{r}) + \vec{\nabla}\cdot\vec{J}(t,\vec{r}) = 0\,, \qquad \partial_\mu J^\mu(x) = 0\,. \tag{7.8}$$

One may easily realize that there is a problem connected with "negative probabilities" if we take into account that the Klein–Gordon equation has solutions proportional to ($E = \sqrt{\vec{p}^2 + m^2} > 0$)

$$\exp\left[\mathrm{i}(E\,t - \vec{k}\cdot\vec{r})\right] = \exp(\mathrm{i}k\cdot x)\,, \qquad \exp\left[-\mathrm{i}(E\,t - \vec{k}\cdot\vec{r})\right] = \exp(-\mathrm{i}k\cdot x)\,. \tag{7.9}$$

Here, $E = \sqrt{\vec{k}^2 + m^2}$ is the modulus of the energy of the particle, and $k \cdot x = k^\mu x_\mu = E t - \vec{k} \cdot \vec{r}$. The first factor in Eq. (7.9) gives rise to a negative energy eigenvalue $-E$ when the time derivative operator (Hamilton operator) $\mathrm{i}\,\partial_t$ acts on it, i.e., one would have

$$J^0(t, \vec{r}) = \mathrm{i}\,\phi^*(t, \vec{r})\, \overleftrightarrow{\partial_t}\, \phi(t, \vec{r}) < 0\,, \qquad \phi \propto \exp\left[\mathrm{i}(E t - \vec{k} \cdot \vec{r})\right]. \tag{7.10}$$

As a consequence, solutions proportional to $\exp(\mathrm{i}k\cdot x)$ correspond to negative probability (negative spatial integral of J^0), which is physically unacceptable and suggests that the "probability" should be reinterpreted as a physical quantity proportional to the charge density of antiparticles (which carry the opposite sign of the charge as compared to particles).

The Dirac equation was invented because of the necessity to unify the advantages of the Schrödinger equation (namely, positive-definite probability density and existence of a well-defined first-order time translation operator) with the advantages of the Klein–Gordon equation, namely, the relativistic invariance. Let us consider the quantity

$$\begin{aligned} X &= \gamma^0 E - \gamma^x p^x - \gamma^y p^y - \gamma^z p^z \\ &= \gamma^0 E - \gamma^1 p^1 - \gamma^2 p^2 - \gamma^3 p^3 = \gamma^0 E - \vec{\gamma} \cdot \vec{p} = \gamma^\mu p_\mu\,, \end{aligned} \tag{7.11}$$

where $p^0 = E$, and γ^μ may be matrix-valued (and hence noncommutative under multiplication). We also distinguish from here on between the covariant (lower) and contravariant (upper) indices. Let us also postulate that the square of X should be equal to

$$X^2 = (E^2 - \vec{p}^{\,2})\, \mathbb{1}_{n\times n}\,, \tag{7.12}$$

where we are pedantic in noting that the unit matrix has to be obtained as an overall factor (the dimension n is fixed at $n = 4$, as we shall see in the following). By inspection, we realize that the following anticommutator relations [see Eq. (2.49)],

$$\{\gamma^\mu, \gamma^\nu\} = 2\, g^{\mu\nu}\,, \qquad g_{\mu\nu} = \mathrm{diag}(1, -1, -1, -1)\,, \qquad \mu, \nu = 0, 1, 2, 3\,, \tag{7.13}$$

are sufficient for the expression given in Eq. (7.11) to fulfill Eq. (7.12). By elementary calculations, one can show that the set of matrices (the so-called "Dirac representation")

$$\gamma^0 = \begin{pmatrix} \mathbb{1}_{2\times2} & 0 \\ 0 & -\mathbb{1}_{2\times2} \end{pmatrix}, \qquad \vec{\gamma} = \begin{pmatrix} 0 & \vec{\sigma} \\ -\vec{\sigma} & 0 \end{pmatrix}, \tag{7.14}$$

fulfill the anticommutator relations (7.13). The Pauli spin matrices have been given in Eq. (4.343). For completeness, we also mention the γ^5 matrix,

$$\gamma^5 = \mathrm{i}\,\gamma^0 \gamma^1 \gamma^2 \gamma^3 = \begin{pmatrix} 0 & \mathbb{1}_{2\times2} \\ \mathbb{1}_{2\times2} & 0 \end{pmatrix}, \qquad \{\gamma^5, \gamma^\mu\} = 0\,. \tag{7.15}$$

For reasons to be discussed below, the γ^5 matrix is sometimes called the "fifth current." It anticommutes with all four other Dirac "current" matrices γ^μ (where

$\mu = 0,1,2,3$) and has the property $(\gamma^5)^2 = 1$. This situation raises a pertinent question: The upper index of the Dirac γ matrices is a Lorentz index. So, the vector of γ matrices has to transform like a Lorentz vector under Lorentz transformations. However, one might also counter-argue that the entire formalism is questionable if the explicit form of the γ matrices should change under a Lorentz transformation. One might otherwise ask what the "choice" means if it needs to change upon a change of the Lorentz frame. The solution is that under Lorentz transformation, the Dirac matrices not only transform with regard to their Lorentz index, but also with regard to the internal spinor structure. The combined action of the Lorentz transformation on the vector index and on the spinor structure then leads to the invariance of the Dirac matrices under Lorentz transformations.

The Dirac equation is finally written down as

$$(\mathrm{i}\gamma^\mu\,\partial_\mu - m)\,\psi(x) = 0\,, \qquad x = (t, \vec{r})\,. \tag{7.16}$$

The price to pay for a linearized relativistically invariant wave equation is the introduction of a multi-component wave function, i.e., of an "internal spinor space" of the particle. Namely, the wave function ψ necessarily has to be formulated in terms of a four-component entity, which we will later identify as a bispinor (composed of two spinors with two components each).

Multiplication by γ^0 from the left brings the Dirac equation (7.17) into non-covariant, Hamiltonian form,

$$\mathrm{i}\frac{\partial}{\partial t}\psi(t,\vec{r}) = \left(-\mathrm{i}\vec{\alpha}\cdot\vec{\nabla} + \beta\,m\right)\psi(t,\vec{r}) = H_{\rm FD}\,\psi(t,\vec{r})\,, \tag{7.17}$$

where the free Dirac (FD) Hamiltonian is given as

$$H_{\rm FD} = -\mathrm{i}\vec{\alpha}\cdot\vec{\nabla} + \beta\,m = \vec{\alpha}\cdot\vec{p} + \beta\,m\,, \tag{7.18}$$

with

$$\beta = \gamma^0 = \begin{pmatrix} \mathbb{1}_{2\times 2} & 0 \\ 0 & \mathbb{1}_{2\times 2} \end{pmatrix}, \qquad \alpha^i = \gamma^0\,\gamma^i = \begin{pmatrix} 0 & \sigma^i \\ \sigma^i & 0 \end{pmatrix}, \qquad i = 1,2,3\,. \tag{7.19}$$

The free Dirac Hamiltonian $H_{\rm FD}$ is matrix-valued, and $\vec{p} = -\mathrm{i}\vec{\nabla}$ is the momentum operator. From this representation, it is clear that the momentum operator $\vec{p}$ commutes with the Hamiltonian. We denote the momentum eigenvalue as $\vec{k}$. However, a specification of $\vec{k}$ is not sufficient in order to characterize the eigenstate. In view of the matrix-valued character of the Dirac Hamiltonian, one needs at least one additional quantum number, which is connected with the spin. One may characterize the spin by its asymptotic value for a particle at rest (see Chap. 2 of Ref. [2]). This is commonly referred to as the rest-mass basis.

The helicity operator $\widehat{\Sigma}$ for a relativistic particle is the normalized projection of the particle's spin onto the propagation direction. We write the helicity operator $\widehat{\Sigma}$ in terms of the (4×4)-spin matrices $\vec{\Sigma}$,

$$\widehat{\Sigma} = \frac{\vec{\Sigma}\cdot\vec{p}}{|\vec{p}|}\,, \qquad \vec{\Sigma} = \gamma^5\,\gamma^0\,\vec{\gamma} = \begin{pmatrix} \vec{\sigma} & 0 \\ 0 & \vec{\sigma} \end{pmatrix}. \tag{7.20}$$

The eigenvalues of the $\widehat{\Sigma}$ operator are the helicities $\sigma = \pm 1$. The vector of spin matrices $\vec{\Sigma}$ finds the following covariant generalization in terms of the $\sigma^{\mu\nu}$ matrices,

$$\sigma^{\mu\nu} = \frac{\mathrm{i}}{2}\left[\gamma^\mu, \gamma^\nu\right], \qquad \sigma^{ij} = \epsilon^{ijk}\,\Sigma^k, \qquad \left[\gamma^i, \gamma^j\right] = -2\,\mathrm{i}\,\epsilon^{ijk}\,\Sigma^k. \tag{7.21}$$

The helicity operator commutes with the Hamiltonian $H_{\rm FD}$. This can be seen as follows. Since the momentum operator (all components) commutes with itself, we have to consider the matrix commutators of the Σ^i with the matrices α^j and β, which occur in the Dirac Hamiltonian $H_{\rm FD}$. The commutator $[|\vec{p}|\widehat{\Sigma}, \vec{\alpha}\cdot\vec{p}]$ reads as follows,

$$\begin{aligned}[\Sigma^i\,p^i, \alpha^j\,p^j] &= \left(\gamma^5\,\gamma^0\,\gamma^i\,\gamma^0\,\gamma^j - \gamma^0\,\gamma^j\,\gamma^5\,\gamma^0\,\gamma^i\right) p^i\,p^j \\ &= \left(-\gamma^5\,\gamma^i\,\gamma^j - \gamma^j\,\gamma^5\,\gamma^i\right) p^i\,p^j = 2\mathrm{i}\gamma^5\,\epsilon^{ijk}\,\Sigma^k\,p^i\,p^j = 0\,. \end{aligned} \tag{7.22}$$

Here, the identities

$$\{\gamma^5, \gamma^\mu\} = 0\,, \qquad \{\gamma^5, \gamma^0\} = \{\gamma^5, \gamma^i\} = 0\,, \tag{7.23}$$

have been used repeatedly. Furthermore, one can show the identity

$$[\Sigma^i\,p^i, \beta] = [\gamma^5\,\gamma^0\,\gamma^i, \gamma^0]\,p^i = 0\,. \tag{7.24}$$

Using the results (7.22) and Eq. (7.24), it follows immediately that

$$\left[\widehat{\Sigma}, H_{\rm FD}\right] = \frac{1}{|\vec{p}|}\left[\vec{\Sigma}\cdot\vec{p},\, \vec{\alpha}\cdot\vec{p} + \beta\,m\right] = 0\,. \tag{7.25}$$

A complete set of good quantum numbers is thus identified as the helicity $\sigma = \pm 1$ and the momentum $\vec{k} \in \mathbb{R}^3$. One can uniquely characterize states according to these quantum numbers.

7.2.2 *Spinor Lorentz Transformation*

As outlined in Chap. 8 of Ref. [81], a Lorentz transformation involves the generator matrices $(\mathbb{M}_{\alpha\beta})^\mu{}_\nu$ where α and β designate the matrix, while μ and ν characterize the components,

$$(\mathbb{M}_{\alpha\beta})^\mu{}_\nu = g^\mu{}_\alpha\,g_{\nu\beta} - g^\mu{}_\beta\,g_{\nu\alpha}\,. \tag{7.26}$$

The Lorentz transformation $\Lambda^\mu{}_\nu$, which has the defining property $g_{\mu\nu}\Lambda^\mu{}_\nu\Lambda^\nu{}_\delta = g_{\rho\delta}$, is obtained as the matrix exponential

$$\Lambda^\mu{}_\nu = \exp\left[\frac{1}{2}\,\omega^{\alpha\beta}\,(\mathbb{M}_{\alpha\beta})^\mu{}_\nu\right] = \left(\exp\left[\frac{1}{2}\,\omega^{\alpha\beta}\,\mathbb{M}_{\alpha\beta}\right]\right)^\mu{}_\nu\,, \tag{7.27}$$

so that, symbolically, one has $\Lambda = \exp\left[\frac{1}{2}\,\omega^{\alpha\beta}\,\mathbb{M}_{\alpha\beta}\right]$. Here, the expression $\exp\left[\frac{1}{2}\,\omega^{\alpha\beta}\,(\mathbb{M}_{\alpha\beta})^\mu{}_\nu\right]$ needs to be interpreted as follows. One first multiplies the generalized rotation angles $\omega^{\alpha\beta}$ with the generator matrices $\mathbb{M}_{\alpha\beta}$, where the components of each individual $\mathbb{M}_{\alpha\beta}$ are taken to be the Lorentz components of upper index μ and lower index ν. One then exponentiates the resultant matrix with components $\frac{1}{2}\,\omega^{\alpha\beta}\,(M_{\alpha\beta})^\mu{}_\nu$ and obtains a Lorentz transformation matrix which can be multiplied with a Lorentz vector component x^ν, to obtain the transformed vector component x'^μ.

The symbols $\omega^{\alpha\beta} = -\omega^{\beta\alpha}$ denote the generalized rotation angles in four-dimensional space-time. The generator matrices fulfill the following commutator relations,

$$[\mathbb{M}_{\mu\nu}, \mathbb{M}_{\kappa\lambda}] = g_{\mu\lambda}\, \mathbb{M}_{\nu\kappa} + g_{\nu\kappa}\, \mathbb{M}_{\mu\lambda} - g_{\mu\kappa}\, \mathbb{M}_{\nu\lambda} - g_{\nu\lambda}\, \mathbb{M}_{\mu\kappa}\,. \tag{7.28}$$

A Lorentz vector with components X^ν transforms as follows,

$$X'^\mu = \Lambda^\mu{}_\nu\, X^\nu\,. \tag{7.29}$$

In order to formulate the spinor Lorentz transformation, one simply exchanges the Lorentz matrices $\mathbb{M}_{\alpha\beta}$ with spin matrices $[-(\mathrm{i}/2)\sigma_{\alpha\beta}]$,

$$S(\Lambda) = \exp\left(-\frac{\mathrm{i}}{4}\,\omega^{\alpha\beta}\,\sigma_{\alpha\beta}\right), \qquad \sigma_{\alpha\beta} = \frac{\mathrm{i}}{2}\,[\gamma_\alpha, \gamma_\beta]\,. \tag{7.30}$$

The σ matrices fulfill the commutation relations

$$\left[\tfrac{1}{2}\sigma_{\mu\nu}, \tfrac{1}{2}\sigma_{\kappa\lambda}\right] = \mathrm{i}\left(g_{\mu\lambda}\tfrac{1}{2}\sigma_{\nu\kappa} + g_{\nu\kappa}\tfrac{1}{2}\sigma_{\mu\lambda} - g_{\mu\kappa}\tfrac{1}{2}\sigma_{\nu\lambda} - g_{\nu\lambda}\tfrac{1}{2}\sigma_{\mu\kappa}\right), \tag{7.31}$$

and it follows that the algebraic relations fulfilled by the matrices $[-(\mathrm{i}/2)\sigma_{\alpha\beta}]$ are analogous to those listed in Eq. (7.28). For infinitesimal rotation angles $\omega^\mu{}_\nu$, the Lorentz transformation matrix elements for vectors and spinors therefore read as

$$\Lambda^\mu{}_\nu = g^\mu{}_\nu + \omega^\mu{}_\nu\,, \qquad S(\Lambda) = 1 - \frac{\mathrm{i}}{4}\,\omega^{\alpha\beta}\,\sigma_{\alpha\beta}\,. \tag{7.32}$$

Under a spinor Lorentz transformation, the four-component wave function $\psi(x)$ does not simply transform as a scalar, which would otherwise imply that $\psi'(x') \stackrel{?}{=} \psi(x)$, but rather transforms as

$$\psi'(x') = S(\Lambda)\,\psi(x)\,, \tag{7.33}$$

where $S(\Lambda)$ is the additional spinor transformation factor. From here on, we will use the shorthand notation $x = (x^\mu) = (t, \vec{r})$ in order to denote the argument of the Dirac bispinor wave function, and in other contexts as well. In view of the identity $\gamma^0\,(\gamma^\mu)^+\,\gamma^0 = \gamma^\mu$, it is easy to check that the Dirac adjoint $\overline{\psi}(x)$ transforms as follows,

$$\overline{\psi}(x) = \psi^+(x)\,\gamma^0\,, \qquad \overline{\psi}'(x') = \overline{\psi}(x)\,S(\Lambda)^{-1}\,. \tag{7.34}$$

The net result is that under Lorentz transformations, the Dirac matrix γ^μ receives two contributions under a Lorentz transformation, one from the spinor index, and one from the Lorentz index,

$$\gamma'^\mu = \Lambda^\mu{}_\nu\,S(\Lambda)\,\gamma^\nu\,S(\Lambda)^{-1} = \gamma^\mu\,. \tag{7.35}$$

We here anticipate the result that the matrix actually is shape-invariant under the full Lorentz transformation, $\gamma'^\mu = \gamma^\mu$. This result is perhaps surprising and relies on two additional identities, which are shown here as an exercise in spinor algebra. First, the commutator of the γ_μ matrix with the spin matrix is calculated as follows,

$$\begin{aligned}[\gamma_\mu, \gamma_\alpha \gamma_\beta - \gamma_\beta \gamma_\alpha] &= \gamma_\mu \gamma_\alpha \gamma_\beta + \gamma_\beta \gamma_\alpha \gamma_\mu - \gamma_\alpha \gamma_\beta \gamma_\mu - \gamma_\mu \gamma_\beta \gamma_\alpha \\ &= \{\gamma_\mu, \gamma_\alpha\} \gamma_\beta - \gamma_\alpha \gamma_\mu \gamma_\beta + \gamma_\beta \{\gamma_\alpha, \gamma_\mu\} - \gamma_\beta \gamma_\mu \gamma_\alpha \\ &\quad - \gamma_\alpha \{\gamma_\beta, \gamma_\mu\} + \gamma_\alpha \gamma_\mu \gamma_\beta - \{\gamma_\mu, \gamma_\beta\} \gamma_\alpha + \gamma_\beta \gamma_\mu \gamma_\alpha \\ &= 4\, g_{\mu\alpha} \gamma_\beta - 4\, g_{\mu\beta} \gamma_\alpha \,. \end{aligned} \tag{7.36}$$

It follows that

$$[\gamma_\mu, \sigma_{\alpha\beta}] = \left[\gamma_\mu, \frac{\mathrm{i}}{2} (\gamma_\alpha \gamma_\beta - \gamma_\beta \gamma_\alpha)\right] = 2\,\mathrm{i}\, g_{\mu\alpha} \gamma_\beta - 2\,\mathrm{i}\, g_{\mu\beta} \gamma_\alpha \,. \tag{7.37}$$

Armed with this result, we can easily calculate the transformation of the Dirac matrix under an infinitesimal Lorentz transformation,

$$\begin{aligned}\gamma'^\mu(x) &= \Lambda^\mu{}_\nu\, S(\Lambda)\, \gamma^\nu(x)\, S(\Lambda)^{-1} \approx \Lambda^\mu{}_\nu \left(1 - \frac{\mathrm{i}}{4} \omega^{\alpha\beta} \sigma_{\alpha\beta}\right) \gamma^\nu \left(1 + \frac{\mathrm{i}}{4} \omega^{\gamma\delta} \sigma_{\gamma\delta}\right) \\ &= (g^\mu{}_\nu + \omega^\mu{}_\nu) \left(\gamma^\nu - \frac{\mathrm{i}}{4} \omega^{\alpha\beta} [\sigma_{\alpha\beta}, \gamma^\nu]\right) \\ &= (g^\mu{}_\nu + \omega^\mu{}_\nu) \left(\gamma^\nu + \frac{\mathrm{i}}{4} \omega^{\alpha\beta} \left(2\,\mathrm{i}\, g^\nu{}_\alpha \gamma_\beta - 2\,\mathrm{i}\, g^\nu{}_\beta \gamma_\alpha\right)\right) \\ &= (g^\mu{}_\nu + \omega^\mu{}_\nu) \left(\gamma^\nu - \frac{1}{2} \omega^{\nu\beta} \gamma_\beta + \frac{1}{2} \omega^{\alpha\nu} \gamma_\alpha\right) = (g^\mu{}_\nu + \omega^\mu{}_\nu) \left(\gamma^\nu - \omega^\nu{}_\beta \gamma^\beta\right) = \gamma^\mu \,. \end{aligned} \tag{7.38}$$

In each step, we have expanded to first order in the infinitesimal parameters $\omega^{\mu\nu}$. Furthermore, we have used the fact that $\omega^{\nu\mu} = -\omega^{\mu\nu}$. We can finally supply a well-defined current for the Dirac equation,

$$J^\mu(x) = \overline{\psi}(x)\, \gamma^\mu\, \psi(x) \,, \qquad J'^\mu(x') = \Lambda^\mu{}_\nu J^\nu(x) \,, \qquad \partial_\mu J^\mu(x) = 0 \,, \tag{7.39}$$

which transforms as a Lorentz vector. The γ^μ matrices extract the components of the current J^μ.

7.2.3 *Discrete Symmetries*

It is interesting to investigate the transformation properties of the Dirac equation under charge conjugation, parity transformation, and time reversal. We start with brief discussions of the transformations associated with parity and time reversal, and then describe the effects of the three transformations on the Dirac equation.

Because the Dirac equation entails a coupling to the electromagnetic field (or, electromagnetic four-vector potential), we need to start by investigating the transformation properties of $A_\mu(x)$ under parity and time reversal. We recall that the electric field $\vec{E}$ and the magnetic induction field $\vec{B}$ are derived as follows:

$$\vec{E} = -\vec{\nabla}\Phi - \partial_t \vec{A} \,, \qquad \vec{B} = \vec{\nabla} \times \vec{A} \,. \tag{7.40}$$

Under parity, $\vec{E} \to -\vec{E}$ (as it is a vector), while $\vec{B}$ is invariant (being a pseudovector). Because $\vec{\nabla}$ is a vector and changes sign under parity, we must have $\vec{A} \to -\vec{A}$

under parity, while Φ is invariant. Denoting the quantities transformed under parity with a superscript $\mathcal{P}$, we can formalize this as follows,

$$\Phi^{\mathcal{P}}(x) = \Phi(x_{\mathcal{P}}) , \qquad \vec{A}^{\mathcal{P}}(x_{\mathcal{P}}) = -\vec{A}(x_{\mathcal{P}}) , \qquad x_{\mathcal{P}} = (t, -\vec{r}) . \tag{7.41}$$

For time reversal, we consider the Lorentz force law,

$$\vec{F} = q\,(\vec{E} + \vec{v} \times \vec{B}). \tag{7.42}$$

Nonrelativistically, one has $\vec{F} = m\,\vec{a}$, where $\vec{a} = \dot{\vec{v}}$ is the acceleration. The derivative of $\vec{v}$, which is the acceleration, thus is time-reversal invariant. One might otherwise conjecture that since a trajectory flips under time reversal, so should the acceleration. However, one might consider the time reversal of a trajectory which describes an object falling in the gravitational field. Under time reversal, one would see an object shooting upward, but with decreasing upward velocity. This is consistent with the acceleration due to gravity being time-reversal invariant. Hence, under time reversal, $\vec{E} \to \vec{E}$ and since $\vec{v} \to -\vec{v}$ under time reversal, one has $\vec{B} \to -\vec{B}$. Using Eq. (7.40), one may formalize these considerations as follows ($\mathcal{T}$ denoted the quantities transformed under time reversal),

$$\Phi^{\mathcal{T}}(x) = \Phi(x_{\mathcal{T}}) , \qquad \vec{A}^{\mathcal{T}}(x_{\mathcal{T}}) = -\vec{A}(x_{\mathcal{T}}) , \qquad x_{\mathcal{T}} = (-t, \vec{r}) . \tag{7.43}$$

We begin our discussion of the transformations of the Dirac equation with charge conjugation. From nonrelativistic quantum mechanics, we know that the electromagnetic coupling can be described by the covariant coupling $\vec{p} \to \vec{p} - e\,\vec{A}$, where $\vec{A}$ is the vector potential of the electromagnetic field. The electromagnetically coupled Dirac equation reads as

$$\left[\gamma^\mu \left(\mathrm{i}\partial_\mu - e\,A_\mu\right) - m\right] \psi = 0 , \tag{7.44}$$

where the four-vector potential has components $A^\mu = (\Phi, \vec{A})$ with the scalar potential Φ and the vector potential $\vec{A}$ (see Chap. 8 of Ref. [81]). Under transposition and complex conjugation, one obtains

$$\psi^+ \left((\gamma^\mu)^+ \left(-\mathrm{i}\overleftarrow{\partial}_\mu - e\,A_\mu \right) - m \right) = 0 , \tag{7.45}$$

where the differential operator acts on the ψ to the left. Insertion of the γ^0 matrix leads to

$$(\psi^+\gamma^0)\,\gamma^0 \left((\gamma^\mu)^+ \left(-\mathrm{i}\overleftarrow{\partial}_\mu - e\,A_\mu \right) - m \right) \gamma^0 = 0 . \tag{7.46}$$

One now uses the familiar identity $\gamma^0\,(\gamma^\mu)^+\,\gamma^0 = \gamma^\mu$ to bring the equation into the form

$$\bar{\psi} \left(\gamma^\mu \left(-\mathrm{i}\overleftarrow{\partial}_\mu - e\,A_\mu \right) - m \right) = 0 . \tag{7.47}$$

Under an additional transposition, one obtains

$$\left((\gamma^\mu)^{\mathrm{T}} \left(-\mathrm{i}\partial_\mu - e\,A_\mu \right) - m \right) \bar{\psi}^{\mathrm{T}} = 0 . \tag{7.48}$$

One introduces the charge conjugation matrix C with the defining property

$$C\,(\gamma^\mu)^{\mathrm{T}}\,C^{-1} = -\gamma^\mu , \tag{7.49}$$

so that the charge conjugated spinor is $\psi^{\mathcal{C}}(x) = C\,\bar{\psi}^{\mathrm{T}}(x)$. The charge conjugated Dirac equation fulfilled by $\psi^{\mathcal{C}}(x)$ thus reads as

$$[\gamma^{\mu}\,(\mathrm{i}\partial_{\mu} + e\,A_{\mu}) - m]\;\psi^{\mathcal{C}}(x) = 0\,. \tag{7.50}$$

It differs from the ordinary Dirac equation (7.44) only in the sign of the physical charge e of the particle. In the Dirac representation, a possible choice is $C = \mathrm{i}\,\gamma^2\,\gamma^0$. Charge conjugation dictates that the sign of this charge, too, flips under charge conjugation, so we can write the charge-conjugated Dirac equation in terms of $e^{\mathcal{C}} = -e$ as follows,

$$\left[\gamma^{\mu}\left(\mathrm{i}\partial_{\mu} - e^{\mathcal{C}}\,A_{\mu}\right) - m\right]\;\psi^{\mathcal{C}}(x) = 0\,. \tag{7.51}$$

This form of the equation is formally identical to the original Dirac equation, but formulated in terms of the charge-conjugated quantities, and hence shows the charge-conjugation invariance of the Dirac equation.

In order to discuss parity, we start from the Dirac equation,

$$[\gamma^{\mu}(\mathrm{i}\partial_{\mu} - e\,A_{\mu}) - m]\;\psi(x) = \left[\gamma^0(\mathrm{i}\partial_0 - e\,\Phi) + \gamma^i(\mathrm{i}\partial_i - e\,A_i) - m\right]\;\psi(x) = 0\,, \tag{7.52}$$

where $i = 1, 2, 3$ is a spatial index. Under parity, $x \to x_{\mathcal{P}} = (t, -\vec{r})$, while $\partial_i \to -\partial_i$, and so the parity-transformed wave function $\psi^{\mathcal{P}}(x)$ needs to fulfill the equation

$$\left[\gamma^0(\mathrm{i}\partial_0 - e\,\Phi^{\mathcal{P}}(x)) + \gamma^i(-\mathrm{i}\partial_i + e\,A_i^{\mathcal{P}}(x)) - m\right]\;\psi^{\mathcal{P}}(x) = 0\,. \tag{7.53}$$

One introduces a parity transformation matrix P, with the property

$$P\gamma^0 P^{-1} = \gamma^0\,, \qquad P\gamma^i P^{-1} = -\gamma^i\,, \qquad \psi^{\mathcal{P}}(x) = P\psi(x_{\mathcal{P}})\,, \tag{7.54}$$

where again $i = 1, 2, 3$ is spatial. In the Dirac representation, we may choose $P = \gamma^0$ with $P^{-1} = P$. The Dirac equation may thus be parity-transformed as follows,

$$P\left[\gamma^0(\mathrm{i}\partial_0 - e\,\Phi^{\mathcal{P}}(x)) + \gamma^i(-\mathrm{i}\partial_i + e\,A_i^{\mathcal{P}}(x)) - m\right]\,P^{-1}\,P\,\psi(x_{\mathcal{P}}) = 0\,. \tag{7.55}$$

The parity-transformed bispinor $\psi^{\mathcal{P}}(x) = P\,\psi(x_{\mathcal{P}})$ fulfills the parity transformed Dirac equation, which in view of Eq. (7.41) reads as

$$\left[\gamma^0(\mathrm{i}\partial_0 - e\,\Phi(x_{\mathcal{P}})) + \gamma^i(\mathrm{i}\partial_i - e\,A_i(x_{\mathcal{P}})) - m\right]\,P\,\psi(x_{\mathcal{P}}) = 0\,, \tag{7.56}$$

which can be written as

$$[\gamma^{\mu}(\mathrm{i}\partial_{\mu} - e\,A_{\mu}(x_{\mathcal{P}})) - m]\;\psi^{\mathcal{P}}(x) = 0\,. \tag{7.57}$$

The Dirac equation is thus seen to be $\mathcal{P}$ invariant.

The time reversal of the Dirac equation is treated next. Again, we start from the ordinary Dirac equation,

$$[\mathrm{i}\gamma^{\mu}(\partial_{\mu} - e\,A_{\mu}) - m]\;\psi(x) = 0\,. \tag{7.58}$$

Transposition and complex conjugation leads to

$$\psi^{+}(x)\left((\gamma^{\mu})^{+}\left(-\mathrm{i}\overleftarrow{\partial}_{\mu} - e\,A_{\mu}\right) - m\right) = 0\,. \tag{7.59}$$

The insertion of the γ^0 matrix leads to the Dirac adjoint [see Eq. (7.34)],

$$\left(\psi^{+}(x)\,\gamma^0\right)\left[\gamma^0\left(\left(\gamma^\mu\right)^{+}\left(-\mathrm{i}\overleftarrow{\partial}_\mu - e\,A_\mu\right) - m\right)\gamma^0\right] = 0\,, \tag{7.60}$$

which is equivalent to

$$\overline{\psi}(x)\left[\gamma^\mu(-\mathrm{i}\overleftarrow{\partial}_\mu - e\,A_\mu) - m\right] = 0\,. \tag{7.61}$$

We here use the fact that a proper formulation of time reversal must respect the fact that the in and out states interchange under $\mathcal{T}$, so that it becomes indispensable to consider the adjoint of Dirac equation (see Sec. 4.4.2 of Ref. [248]). Again, we transpose and write

$$\left[\left(\gamma^\mu\right)^{\mathrm{T}}\left(-\mathrm{i}\partial_\mu - e\,A_\mu(x)\right) - m\right]\bar{\psi}^{\mathrm{T}}(x) = 0\,. \tag{7.62}$$

We introduce the time reversal matrix T, with the properties ($\mathcal{T}$ means time reversal, T is the time reversal matrix, and T denotes the transpose),

$$T\left(\gamma^0\right)^{\mathrm{T}}T^{-1} = \gamma^0\,, \qquad T\left(\gamma^i\right)^{\mathrm{T}}T^{-1} = -\gamma^i\,, \qquad \psi^{\mathcal{T}}(x) = T\,\bar{\psi}^{\mathrm{T}}(x_{\mathcal{T}})\,, \tag{7.63}$$

where the time-reversed bispinor is denoted as $\psi^{\mathcal{T}}(x)$. We now carry out an explicit time-reversal operation with $x_{\mathcal{T}} = (-t, \vec{r})$ on Eq. (7.62), use the transformation $\partial_t \to -\partial_t$ (under time reversal) and write

$$T\left[\left(\gamma^0\right)^{\mathrm{T}}\left(\mathrm{i}\partial_t - e\,\Phi^{\mathcal{T}}(x)\right) + \left(\gamma^i\right)^{\mathrm{T}}\left(-\mathrm{i}\partial_i - e\,A_i^{\mathcal{T}}(x)\right) - m\right]T^{-1}\,T\bar{\psi}^{\mathrm{T}}(x_{\mathcal{T}}) = 0\,. \tag{7.64}$$

Using Eq. (7.43), one obtains

$$\left[\gamma^0\left(\mathrm{i}\partial_t - e\,\Phi(x_{\mathcal{T}})\right) + (-\gamma^i)\left(-\mathrm{i}\partial_i + e\,A_i(x_{\mathcal{T}})\right) - m\right]\,T\bar{\psi}^{\mathrm{T}}(x_{\mathcal{T}}) = 0\,, \tag{7.65}$$

and thus

$$\left[\gamma^\mu\left(\mathrm{i}\partial_\mu - e\,A_\mu(x_{\mathcal{T}})\right) - m\right]\,\psi^{\mathcal{T}}(x) = 0\,, \tag{7.66}$$

which is formally equal to the original Dirac equation. Thus, we derive the $\mathcal{T}$ invariance. In the Dirac representation, we may choose $T = \mathrm{i}\,\gamma^5\,\gamma^2$. It is extremely instructive to calculate the time reversal of the fifth current, based exclusively on the property (7.63),

$$\begin{aligned} T\left(\gamma^5\right)^{\mathrm{T}}T^{-1} &= T\,\mathrm{i}\left(\gamma^3\right)^{\mathrm{T}}\left(\gamma^2\right)^{\mathrm{T}}\left(\gamma^1\right)^{\mathrm{T}}\left(\gamma^0\right)^{\mathrm{T}}T^{-1} \\ &= \mathrm{i}T\left(\gamma^3\right)^{\mathrm{T}}T^{-1}T\left(\gamma^2\right)^{\mathrm{T}}T^{-1}T\left(\gamma^1\right)^{\mathrm{T}}T^{-1}T\left(\gamma^0\right)^{\mathrm{T}}T^{-1} \\ &= \mathrm{i}\left(-\gamma^3\right)\left(-\gamma^2\right)\left(-\gamma^1\right)\left(+\gamma^0\right) = -\mathrm{i}\gamma^3\gamma^2\gamma^1\gamma^0 = -\mathrm{i}\gamma^0\gamma^1\gamma^2\gamma^3 = -\gamma^5\,. \end{aligned} \tag{7.67}$$

The literature is not always consistent in regard to the minus sign, and it is instructive to clarify its origin.

7.2.4 *Overview of the Symmetry Properties*

It is instructive to present an overview. We recall the charge conjugation matrix C, the parity transformation matrix P, and the time reversal matrix T,

$$C = \mathrm{i}\gamma^2\gamma^0\,, \qquad C\,(\gamma^\mu)^{\mathrm{T}}\,C^{-1} = -\gamma^\mu\,, \tag{7.68a}$$

$$P = \gamma^0\,, \qquad P\gamma^0 P^{-1} = \gamma^0\,, \qquad P\gamma^i P^{-1} = -\gamma^i\,, \tag{7.68b}$$

$$T = \mathrm{i}\gamma^5\gamma^2\,, \qquad T\left(\gamma^0\right)^{\mathrm{T}} T^{-1} = \gamma^0\,, \qquad T\left(\gamma^i\right)^{\mathrm{T}} T^{-1} = -\gamma^i\,. \tag{7.68c}$$

The matrices C, P and T fulfill the relations $C^{\mathrm{T}} = C^{-1}$, $P^{\mathrm{T}} = P^{-1}$, and $T^{\mathrm{T}} = T^{-1}$, where we denote the transpose by T. In view of the property $\{\gamma^\mu, \gamma^5\} = 0$, their product can be written as

$$CPT = -\gamma^5\,. \tag{7.69}$$

This implies that

$$(CPT)\gamma^\mu(CPT)^{-1} = -\gamma^\mu \tag{7.70}$$

[see page 239 of Ref. [5] and Eq. (40.48) of Ref. [249]]. The transformation of the wave function reads as follows,

$$\psi^{\mathcal{C}}(x) = C\,\bar{\psi}^{\mathrm{T}}(x)\,, \qquad \psi^{\mathcal{P}}(x) = P\,\psi(x_{\mathcal{P}})\,, \qquad \psi^{\mathcal{T}}(x) = T\,\bar{\psi}^{\mathrm{T}}(x_{\mathcal{T}})\,, \tag{7.71}$$

where $x_{\mathcal{T}} = -x_{\mathcal{P}} = (-t, \vec{r})$. This implies that $x_{\mathcal{PT}} = x_{\mathcal{CPT}} = -x$, because charge conjugation does not affect the coordinate.

Based on the result

$$\psi^{\mathcal{PT}}(x) = \mathrm{i}\gamma^0\,\gamma^5\gamma^2\gamma^0\psi^*(x_{\mathcal{PT}}) = \mathrm{i}\gamma^5\gamma^2\psi^*(x_{\mathcal{PT}})\,, \tag{7.72}$$

where the first γ^0 is due to $\mathcal{P}$, the second due to $\mathcal{T}$, one can show that

$$\begin{aligned}
\psi^{\mathcal{CPT}}(x) &= \left(\mathrm{i}\gamma^2\gamma^0\right)\gamma^0\left(-\mathrm{i}\gamma^5\left(\gamma^2\right)^*\,\psi(x_{\mathcal{CPT}})\right) \\
&= \gamma^2\gamma^5(-\gamma^2)\psi(x_{\mathcal{CPT}}) = \left(\gamma^2\right)^2\gamma^5\,\psi(x_{\mathcal{CPT}}) \\
&= -\gamma^5\,\psi(x_{\mathcal{CPT}}) = -\gamma^5\,\psi(-x)\,.
\end{aligned} \tag{7.73}$$

The combined action of $\mathcal{CPT}$ involves two complex conjugations.

The $\mathcal{CPT}$ invariance of the Dirac equation can thus be shown as follows. We first carry out the transformation $x \to -x = x_{\mathcal{PT}}$ in the Dirac equation,

$$\left[\gamma^\mu(-\mathrm{i}\partial_\mu - e\,A_\mu^{\mathcal{PT}}(x)) - m\right]\psi(-x) = 0\,. \tag{7.74}$$

We then identify, according to Eqs. (7.41) and (7.43), $A_\mu^{\mathcal{PT}}(x) = A_\mu(-x)$, and $e = -e^{\mathcal{C}} = -e^{\mathcal{CPT}}$, in view of the charge conjugation, observe that the coordinate does not change under charge conjugation, and insert a γ^5 matrix,

$$\gamma^5\left[\gamma^\mu(-\mathrm{i}\partial_\mu + e^{\mathcal{CPT}}\,A_\mu(x_{\mathcal{CPT}})) - m\right]\gamma^5\,(-\gamma^5)\psi(-x) = 0\,. \tag{7.75}$$

Because γ^5 anti-commutes with all γ^μ matrices, we have

$$\left[\gamma^\mu(\mathrm{i}\partial_\mu - e^{\mathcal{CPT}}\,A_\mu(x_{\mathcal{CPT}})) - m\right]\psi^{\mathcal{CPT}}(x) = 0\,. \tag{7.76}$$

This shows the $\mathcal{CPT}$ invariance of the Dirac equation and explains the minus sign in Eq. (7.70).

Furthermore, it is instructive to recall the transformation properties of the fifth current,

$$C\left(\gamma^5\right)^{\mathrm{T}} C^{-1} = \gamma^5\,, \qquad P\gamma^5 P^{-1} = -\gamma^5\,, \qquad T\left(\gamma^5\right)^{\mathrm{T}} T^{-1} = -\gamma^5\,. \tag{7.77}$$

We note that the transformation property for time reversal depends on our precise formulation of the time reversal operation as detailed in Sec. 7.2.3. It is also useful to remark that all above considerations need to be adjusted when formulated in terms of the (second-quantized) Dirac field operator as opposed to the (first-quantized) Dirac wave function ψ, with potential confusion because both quantities are usually denoted by the same letter ψ. Regarding this point, we also refer to Chaps. 2 and 3 of Ref. [2] and Chap. 7 of Ref. [250], and to Ref. [251], as well as Sec. 7.5. Roughly speaking, it is interesting to observe that it is only due to the anticommutativity of the second-quantized field operators that the overall sign of the Dirac Lagrangian is restored under $\mathcal{CPT}$. Otherwise, the minus sign in Eq. (7.70) would lead to a counter-intuitive reversal of the overall sign of the Dirac Lagrangian under $\mathcal{CPT}$.

7.3 Solutions of the Free Dirac Equation

7.3.1 *Plane-Wave Solutions of the Dirac Equation*

The Lorentz-invariant action for the Dirac particle reads as

$$S = \int \mathrm{d}^4 x \;\overline{\psi}(x)\,\left(\mathrm{i}\gamma^\mu \partial_\mu - m\right)\,\psi(x)\,, \tag{7.78}$$

with the Lagrangian density

$$\mathcal{L}(x) = \overline{\psi}(x)\,\left(\mathrm{i}\gamma^\mu \partial_\mu - m\right)\,\psi(x)\,. \tag{7.79}$$

Variation with respect to $\overline{\psi}(x)$ leads to the Dirac equation $(\mathrm{i}\gamma^\mu \partial_\mu - m)\,\psi(x)$. Upon using a plane-wave *ansatz* of the form $\psi(x) = U(\vec{k})\exp(-ik\cdot x)$, the free Dirac equation is transformed into the form of an algebraic equation, $(\not{k} - m)\,U(\vec{k}) = 0$, where the Feynman slash notation is introduced as $\not{k} = \gamma^\mu\,k_\mu$. Here, $U(\vec{k})$ is an (as yet) unspecified four-component spinor. One observes the algebraic identity

$$(\not{k} - m)\,(\not{k} + m) = k^2 - m^2 = 0\,, \tag{7.80}$$

where

$$k = (k^0, \vec{k}) = (\sqrt{\vec{k}^2 + m^2}, \vec{k}) \tag{7.81}$$

is the momentum four-vector. The Dirac equation is solved by $U(\vec{k}) = (\not{k} + m)\,\chi$, where χ is any constant four-component vector. Canonically, one constructs a representative set of solutions by selecting for χ the vectors which have one of the four entries equal to one, and the other entries equal to zero. This is described in Eqs. (2.36) and (2.37) of Ref. [2]. However, the use of this approach leads to a somewhat unclear situation with regard to the interpretation of the spin degree of freedom. Namely, while the solutions are "boosted" from the rest frame by the factor $(\not{k}+m)$, the interpretation of the spin index of the solution in the laboratory frame becomes obscure.

In the helicity basis, solutions of the free Dirac equation are labeled by the plane-wave momentum vector $\vec{k}$ and the helicity quantum number $\sigma = \pm 1$. We explore the action of the helicity operator $\widehat{\Sigma} = \vec{\Sigma}\cdot\vec{p}/|\vec{p}|$, defined in Eq. (7.20). The helicity spinors are given by

$$a_+(\vec{k}) = \begin{pmatrix} \cos\left(\frac{\theta}{2}\right) \\ \sin\left(\frac{\theta}{2}\right) \mathrm{e}^{\mathrm{i}\varphi} \end{pmatrix}, \qquad a_-(\vec{k}) = \begin{pmatrix} -\sin\left(\frac{\theta}{2}\right) \mathrm{e}^{-\mathrm{i}\varphi} \\ \cos\left(\frac{\theta}{2}\right) \end{pmatrix}, \tag{7.82}$$

where θ and φ are the polar and azimuth angles of the vector $\vec{k}$. The (2×2)-version of the helicity operator acts on the helicity spinors as follows,

$$\frac{\vec{\sigma}\cdot\vec{k}}{|\vec{k}|}\, a_\sigma(\vec{k}) = \sigma\, a_\sigma(\vec{k}), \qquad \sigma = \pm 1. \tag{7.83}$$

One observes the property

$$\frac{\vec{\sigma}\cdot\vec{p}}{|\vec{p}|} a_\sigma(\vec{k})\, \mathrm{e}^{\pm\mathrm{i}\vec{k}\cdot\vec{r}} = \pm\frac{\vec{\sigma}\cdot\vec{k}}{|\vec{k}|}\, a_\sigma(\vec{k})\, \mathrm{e}^{\pm\mathrm{i}\vec{k}\cdot\vec{r}} = \pm\sigma a_\sigma(\vec{k})\, \mathrm{e}^{\pm\mathrm{i}\vec{k}\cdot\vec{r}}, \tag{7.84}$$

where $\vec{p} = \pm\vec{k}$. Particle states are proportional to $\exp(-\mathrm{i}k\cdot x) = \exp(-\mathrm{i}k^0 t)\, \exp(\mathrm{i}\vec{k}\cdot\vec{r})$, while antiparticle waves are proportional to $\exp(\mathrm{i}k\cdot x) = \exp(\mathrm{i}k^0 t)\, \exp(-\mathrm{i}\vec{k}\cdot\vec{r})$. Hence, the helicity quantum number σ of $a_\sigma(\vec{k})$ is equal to the normalized projection of the momentum operator onto the spin of the particle, for positive-energy (particle) states, while a minus sign is incurred in the corresponding relation for negative-energy (anti-particle) states. After invoking the reinterpretation principle, which will be discussed in more detail in Sec. 7.4, one sees that σ is equal to the normalized projection of the *physical* momentum onto the spin, for both particles and antiparticles. The helicity spinors fulfill the following relations,

$$\sum_\sigma a_\sigma(\vec{k}) \otimes a_\sigma^+(\vec{k}) = \mathbb{1}_{2\times 2}, \qquad \sum_\sigma \sigma\, a_\sigma(\vec{k}) \otimes a_\sigma^+(\vec{k}) = \frac{\vec{\sigma}\cdot\vec{k}}{|\vec{k}|}, \tag{7.85}$$

where the sum in each case runs over the possible values $\sigma = \pm 1$. The fundamental spinor solutions to the Dirac equation read as

$$\psi(x) = \mathcal{U}_\pm(\vec{k})\, \exp(-\mathrm{i}k\cdot x), \qquad \phi(x) = \mathcal{V}_\pm(\vec{k})\, \exp(\mathrm{i}k\cdot x), \tag{7.86}$$

where the four-vector k is of the structure given in Eq. (7.81). Here, $\mathcal{U}_\pm$ describes a positive-energy solution, whereas $\mathcal{V}_\pm$ describes a negative-energy solution. In view of the Dirac equation, the following algebraic relations have to be fulfilled by the bispinor amplitudes $\mathcal{U}_\pm(\vec{k})$ and $\mathcal{V}_\pm(\vec{k})$,

$$\left(\not{k} - m\right) \mathcal{U}_\pm(\vec{k}) = 0, \qquad \left(\not{k} + m\right) \mathcal{V}_\pm(\vec{k}) = 0. \tag{7.87}$$

The positive-energy solutions can be projected out using the identity $(\not{k}-m)(\not{k}+m) = k^2 - m^2 = 0$. Let us define the massless bispinors

$$u_+(\vec{k}) = \begin{pmatrix} \frac{a_+(\vec{k})}{\sqrt{2}} \\ \frac{a_+(\vec{k})}{\sqrt{2}} \end{pmatrix}, \quad u_-(\vec{k}) = \begin{pmatrix} \frac{a_-(\vec{k})}{\sqrt{2}} \\ -\frac{a_-(\vec{k})}{\sqrt{2}} \end{pmatrix}, \quad v_+(\vec{k}) = -u_+(\vec{k}), \quad v_-(\vec{k}) = -u_-(\vec{k}). \tag{7.88}$$

Denoting the energy as E, we can then write

$$\mathcal{U}_+(\vec{k}) = \sqrt{\frac{E}{m}} \frac{(\not{k} + m)\, u_+(\vec{k})}{\sqrt{(E - |\vec{k}|)^2 + m^2}} = \begin{pmatrix} \sqrt{\frac{E+m}{2m}}\, a_+(\vec{k}) \\ \sqrt{\frac{E-m}{2m}}\, a_+(\vec{k}) \end{pmatrix}, \tag{7.89a}$$

$$\mathcal{U}_-(\vec{k}) = \sqrt{\frac{E}{m}} \frac{(\not{k} + m)\, u_-(\vec{k})}{\sqrt{(E - |\vec{k}|)^2 + m^2}} = \begin{pmatrix} \sqrt{\frac{E+m}{2m}}\, a_-(\vec{k}) \\ -\sqrt{\frac{E-m}{2m}}\, a_-(\vec{k}) \end{pmatrix}. \tag{7.89b}$$

The negative-energy eigenstates are given as

$$\mathcal{V}_+(\vec{k}) = \sqrt{\frac{E}{m}} \frac{(m - \not{k})\, v_+(\vec{k})}{\sqrt{(E - |\vec{k}|)^2 + m^2}} = \begin{pmatrix} -\sqrt{\frac{E-m}{2m}}\, a_+(\vec{k}) \\ -\sqrt{\frac{E+m}{2m}}\, a_+(\vec{k}) \end{pmatrix}, \tag{7.90a}$$

$$\mathcal{V}_-(\vec{k}) = \sqrt{\frac{E}{m}} \frac{(m - \not{k})\, v_-(\vec{k})}{\sqrt{(E - |\vec{k}|)^2 + m^2}} = \begin{pmatrix} -\sqrt{\frac{E-m}{2m}}\, a_-(\vec{k}) \\ \sqrt{\frac{E+m}{2m}}\, a_-(\vec{k}) \end{pmatrix}. \tag{7.90b}$$

The normalization conditions are

$$\overline{\mathcal{U}}_\sigma(\vec{k})\, \mathcal{U}_\sigma(\vec{k}) = 1\,, \qquad \overline{\mathcal{V}}_\sigma(\vec{k})\, \mathcal{V}_\sigma(\vec{k}) = -1 \tag{7.91}$$

and

$$\overline{\mathcal{U}}_\sigma(\vec{k})\, \gamma^0 \mathcal{U}_{\sigma'}(\vec{k}) = \frac{E}{m}\,, \qquad \overline{\mathcal{V}}_\sigma(\vec{k})\, \gamma^0 \mathcal{V}_{\sigma'}(\vec{k}) = -\frac{E}{m}\,. \tag{7.92}$$

The Dirac adjoint has been defined in Eq. (7.34). The following projector sum formulas are instrumental in the derivation of the propagator (see Sec. 7.4),

$$\sum_\sigma \mathcal{U}_\sigma(\vec{k}) \otimes \overline{\mathcal{U}}_\sigma(\vec{k}) = \frac{\not{k} + m}{2m}\,, \qquad \sum_\sigma \mathcal{V}_\sigma(\vec{k}) \otimes \overline{\mathcal{V}}_\sigma(\vec{k}) = \frac{\not{k} - m}{2m}\,. \tag{7.93}$$

A massless particle is described by the Dirac equation $\mathrm{i}\gamma^\mu \partial_\mu \psi(x) = 0$. The free massless Dirac Hamiltonian reads as

$$H = \vec{\alpha} \cdot \vec{p} = \gamma^5 \left(\gamma^5 \vec{\alpha} \cdot \vec{p}\right) = \gamma^5 \left(\vec{\Sigma} \cdot \vec{p}\right) = |\vec{p}|\, \gamma^5\, \widehat{\Sigma}\,, \tag{7.94}$$

where $\widehat{\Sigma}$ is the helicity projection operator defined in Eq. (7.20). The fundamental bispinor solutions to the massless Dirac equation read as

$$\psi(x) = u_\pm(\vec{k}) \exp(-\mathrm{i}k \cdot x)\,, \qquad \phi(x) = v_\pm(\vec{k}) \exp(\mathrm{i}k \cdot x)\,, \tag{7.95}$$

where we refer to Eq. (7.88). The dispersion relation for the massless equation is $E = k^0 = |\vec{p}| = |\vec{k}|$. In order for the energy to be negative, the eigenvalues of chirality (γ^5 matrix) and helicity ($\widehat{\Sigma}$ operator) have to have opposite sign, while the same sign is incurred for positive energy. As an example, one may observe that the solution $v_-(\vec{k})\exp(\mathrm{i}k\cdot x)$ has negative energy and negative chirality, but a positive eigenvalue of the helicity operator $\widehat{\Sigma}$, while it is still composed of the fundamental helicity spinors $a_-(\vec{k})$ which have negative eigenvalue in regard to the projection operator $\vec{\sigma}\cdot\vec{k}/|\vec{k}|$. One may thus conclude that the quantum number σ in the fundamental bispinors u_σ and v_σ is a quantum number which is equal to the (physical) helicity for particles, and equal to the negative of the (physical) helicity for antiparticles. This observation is crucial in recent attempts to formulate a consistent quantum theory for tachyonic neutrinos, in which case a certain set of curious identities helps one to quantize the theory so that neutrinos of negative helicity, and antineutrinos of positive helicity, are suppressed in view of Gupta–Bleuler conditions [252, 253].

The massless solutions (7.95) fulfill two projector sum formulas, which constitute analogues of those given in Eq. (7.93),

$$\sum_\sigma 2\,|\vec{k}|\,u_\sigma(\vec{k})\otimes\overline{u}_\sigma(\vec{k}) = \not{k}\,, \qquad \sum_\sigma 2\,|\vec{k}|\,v_\sigma(\vec{k})\otimes\overline{v}_\sigma(\vec{k}) = \not{k}\,, \tag{7.96}$$

as well as

$$\sum_\sigma 2\,|\vec{k}|\,(-\sigma)\,u_\sigma(\vec{k})\otimes\overline{u}_\sigma(\vec{k})\,\gamma^5 = \not{k}\,, \qquad \sum_\sigma 2\,|\vec{k}|\,(-\sigma)\,v_\sigma(\vec{k})\otimes\overline{v}_\sigma(\vec{k})\,\gamma^5 = \not{k}\,. \tag{7.97}$$

In contrast to the normalization conditions (7.91), one has $\overline{u}_\sigma(\vec{k})\,u_\sigma(\vec{k}) = \overline{v}_\sigma(\vec{k})\,v_\sigma(\vec{k}) = 0$ for the massless case.

7.3.2 *Dirac Angular Quantum Number*

In order to discuss the angular-momentum resolved eigenstates of the free Dirac equation, we recall from Eq. (4.348) the Dirac angular quantum number $\varkappa$, which unifies the orbital angular momentum ℓ and the total angular momentum j. The eigenfunctions of the free Dirac Hamiltonian are neither eigenfunctions of the orbital angular momentum $\vec{L}$ nor of the spin angular momentum $\vec{S} = \vec{\Sigma}/2$, because neither of these operators commutes with the Hamiltonian (7.18). Indeed, one has

$$[H_{\mathrm{FD}}, \vec{L}] = -\mathrm{i}\,(\vec{\alpha}\times\vec{p}) = -[H_{\mathrm{FD}}, \tfrac{1}{2}\vec{\Sigma}]\,. \tag{7.98}$$

The total angular momentum operator $\vec{J} = \vec{L} + \frac{1}{2}\vec{\Sigma}$ thus constitutes a good quantum number,

$$[H_{\mathrm{FD}}, \vec{J}] = [H_{\mathrm{FD}}, \vec{L} + \tfrac{1}{2}\vec{\Sigma}] = 0\,. \tag{7.99}$$

Alternatively, we (re-)define the Dirac angular operator K as the (4×4)-matrix generalization of the operator originally given in Eq. (6.81) of Sec. 6.4.1,

$$K = \beta\,(\vec{\Sigma}\cdot\vec{L} + 1)\,. \tag{7.100}$$

It is crucial to observe that K commutes with the Hamiltonian (7.18), and with any Hamiltonian in which a spherically symmetric potential is added to the Dirac

Hamiltonian. However, the proof is a little tricky; it is helpful to decompose the commutator $[H_{\rm FD}, K]$ into 2×2 matrix structures,

$$\begin{aligned}
-[H_{\rm FD}, K] &= -\left[\vec{\alpha}\cdot\vec{p},\, \beta\,(\vec{\Sigma}\cdot\vec{L}+1)\right] \\
&= \beta\,\vec{\alpha}\cdot\vec{p}\,(\vec{\Sigma}\cdot\vec{L}+1) + (\vec{\Sigma}\cdot\vec{L}+1)\,\beta\,\vec{\alpha}\cdot\vec{p} \\
&= \begin{pmatrix} 0 & \{\vec{\sigma}\cdot\vec{p},\vec{\sigma}\cdot\vec{L}\} + 2\,\vec{\sigma}\cdot\vec{p} \\ -\{\vec{\sigma}\cdot\vec{p},\vec{\sigma}\cdot\vec{L}\} - 2\,\vec{\sigma}\cdot\vec{p} & 0 \end{pmatrix} \\
&= \begin{pmatrix} 0 & -2\,\vec{\sigma}\cdot\vec{p} + 2\,\vec{\sigma}\cdot\vec{p} \\ 2\,\vec{\sigma}\cdot\vec{p} - 2\,\vec{\sigma}\cdot\vec{p} & 0 \end{pmatrix} = 0\,.
\end{aligned} \tag{7.101}$$

We have used the relation $\{\vec{\sigma}\cdot\vec{p},\vec{\sigma}\cdot\vec{L}\} = -2\,\vec{\sigma}\cdot\vec{p}$ which deserves a further explanation, given in Eqs. (7.103)–(7.105). The Pauli spin matrices [Eq. (4.343)] fulfill the relation

$$\sigma^i\,\sigma^j = \delta^{ij}\,\mathbb{1}_{2\times 2} + \mathrm{i}\,\epsilon^{ijk}\,\sigma^k\,. \tag{7.102}$$

The anticommutator $\{\vec{\sigma}\cdot\vec{p},\vec{\sigma}\cdot\vec{L}\}$ is easily calculated,

$$\begin{aligned}
\{\vec{\sigma}\cdot\vec{p},\vec{\sigma}\cdot\vec{L}\} &= \sigma^i\,p^i\,\sigma^j\,L^j + \sigma^j\,L^j\,\sigma^i\,p^i \\
&= p^i\,L^i + L^i\,p^i + \mathrm{i}\,\epsilon^{ijk}\,\sigma^k\,p^i\,L^j + \mathrm{i}\,\epsilon^{jik}\,\sigma^k\,L^j\,p^i \\
&= \epsilon^{ijk}\,x^j\,p^i\,p^k + \epsilon^{ijk}\,x^j\,p^k\,p^i + \mathrm{i}\,\epsilon^{ijk}\,\sigma^k\,L^i\,p^j - \mathrm{i}\,\epsilon^{ijk}\,\sigma^k\,p^j\,L^i \\
&= \mathrm{i}\,\epsilon^{ijk}\,\sigma^k\,(L^i\,p^j - p^j\,L^i)\,.
\end{aligned} \tag{7.103}$$

For the commutator of the angular momentum and the spatial momentum operator, one obtains

$$[L^i, p^\ell] = [\epsilon^{ijk}\,x^j\,p^k, p^\ell] = \epsilon^{ijk}\,[x^j, p^\ell]\,p^k = \mathrm{i}\,\epsilon^{i\ell k}\,p^k\,. \tag{7.104}$$

Finally, we obtain the relation

$$\{\vec{\sigma}\cdot\vec{L},\vec{\sigma}\cdot\vec{p}\} = \mathrm{i}\,\epsilon^{ijk}\,\sigma^k\,\mathrm{i}\,\epsilon^{ij\ell}\,p^\ell = -2\,\delta^{k\ell}\,\sigma^k\,p^\ell = -2\,\vec{\sigma}\cdot\vec{p}\,, \tag{7.105}$$

which proves that $\{\vec{\sigma}\cdot\vec{L},\vec{\sigma}\cdot\vec{p}\} = -2\,\vec{\sigma}\cdot\vec{p}$. In order to fix ideas, we recall that the eigenvalues of the Dirac operator K are denoted as $-\kappa$.

One may thus construct the bispinor (four-component) solutions of the Dirac equation in the angular-momentum basis from two two-component spinors, which carry the same total angular momentum j but are constructed from orbital angular momenta $\ell = j - 1/2$ and $\ell = j + 1/2$, adding or subtracting the electron spin to or from the orbital angular momentum. The spin-angular functions $\chi_{\kappa\,\mu}(\vec{r})$, which had originally been defined in Eq. (6.80), read as follows,

$$(\vec{\sigma}\cdot\vec{L}+1)\,\chi_{\kappa\,\mu}(\hat{r}) = -\kappa\,\chi_{\kappa\,\mu}(\hat{r})\,, \qquad \chi_{\kappa\mu}(\hat{r}) = \sum_m C^{j\mu}_{\ell\,\mu-m\,\frac{1}{2}\,m}\,Y_{\ell\,\mu-m}(\hat{r})\,|\tfrac{1}{2}\,m\rangle\,. \tag{7.106}$$

Here, $|\frac{1}{2}\,m\rangle$ is a fundamental spinor pointing up or down [see Eq. (6.42)], and we use $\hat{r}$ instead of θ and φ as the arguments of $\chi_{\kappa\mu}$. The Dirac quantum number κ

summarizes both the orbital quantum number ℓ as well as the spin and total angular momentum quantum number j into one single integer. We recall Eq. (6.82),

$$\kappa = (-1)^{j+\ell+1/2}\left(j+\frac{1}{2}\right) = \begin{cases} -(j+1/2) & \text{for} \quad j=\ell+1/2 \\ (j+1/2) & \text{for} \quad j=\ell-1/2 \end{cases}. \tag{7.107}$$

The identification of κ for the atomic states with the lowest angular momenta is as follows,

$$S_{1/2}: \quad \ell=0, \quad j=\tfrac{1}{2}, \quad \kappa=-1, \tag{7.108a}$$
$$P_{1/2}: \quad \ell=1, \quad j=\tfrac{1}{2}, \quad \kappa=1, \tag{7.108b}$$
$$P_{3/2}: \quad \ell=1, \quad j=\tfrac{3}{2}, \quad \kappa=-2, \tag{7.108c}$$
$$D_{3/2}: \quad \ell=2, \quad j=\tfrac{3}{2}, \quad \kappa=2, \tag{7.108d}$$
$$D_{5/2}: \quad \ell=2, \quad j=\tfrac{5}{2}, \quad \kappa=-3, \tag{7.108e}$$
$$F_{5/2}: \quad \ell=3, \quad j=\tfrac{5}{2}, \quad \kappa=3, \tag{7.108f}$$

and so on. The Dirac angular quantum number κ assumes all positive and negative integer values except zero.

7.3.3 *Angular Momenta and Massless Dirac Equation*

In Eqs. (7.95) and (7.88), we had already obtained the solutions of the massless Dirac equation, in the helicity basis. However, for a number of problems, e.g., the radiative orbital electron capture by a nucleus [254], it is advantageous to formulate the bispinor solution of the massless Dirac equation in an angular-momentum-resolved, spherical basis. The bispinor *ansatz* for the wave function reads as follows,

$$\psi(\vec{r}) = \begin{pmatrix} f(r)\,\chi_{\kappa\,\mu}(\hat{r}) \\ \mathrm{i}g(r)\,\chi_{-\kappa\,\mu}(\hat{r}) \end{pmatrix}. \tag{7.109}$$

It is convenient to introduce the imaginary unit as a prefactor in the lower component, because of another imaginary unit present in the momentum operator which couples upper and lower components. The action of the free massless Dirac Hamiltonian $H_{\mathrm{FMD}} = \vec{\alpha}\cdot\vec{p}$ on the wave function can be decomposed into 2×2 matrices as follows,

$$H_{\mathrm{FMD}}\,\psi(\vec{r}) = \vec{\alpha}\cdot\vec{p}\,\psi(\vec{r}) = \begin{pmatrix} 0 & \vec{\sigma}\cdot\vec{p} \\ \vec{\sigma}\cdot\vec{p} & 0 \end{pmatrix} \begin{pmatrix} f(r)\,\chi_{\kappa\,\mu}(\hat{r}) \\ \mathrm{i}g(r)\,\chi_{-\kappa\,\mu}(\hat{r}) \end{pmatrix}. \tag{7.110}$$

We now need to decompose the scalar product $\vec{\sigma}\cdot\vec{p}$ into angular and radial parts. This is achieved by writing

$$\begin{aligned} \vec{\sigma}\cdot\vec{p} &= \frac{1}{r^2}\,(\vec{\sigma}\cdot\vec{r})\,(\vec{\sigma}\cdot\vec{r})\,\vec{\sigma}\cdot\vec{p} = \frac{1}{r^2}\,(\vec{\sigma}\cdot\vec{r})\left(\vec{r}\cdot\vec{p}+\mathrm{i}\,\epsilon^{ijk}\,\sigma^k\,x^i\,p^j\right) \\ &= (\vec{\sigma}\cdot\hat{r})\left(-\mathrm{i}\,\frac{\partial}{\partial r}+\mathrm{i}\,\frac{\vec{\sigma}\cdot\vec{L}}{r}\right). \end{aligned} \tag{7.111}$$

We now use several properties of the spinors discussed in Sec. 6.4.1 and take advantage of Eq. (7.106), to rewrite the action of the free Hamiltonian $H = \vec{\alpha} \cdot \vec{p}$ as follows,

$$
\begin{aligned}
H_{\mathrm{FMD}} \psi(\vec{r}) &= \begin{pmatrix} 0 & (\vec{\sigma} \cdot \hat{r}) \left(-\mathrm{i} \dfrac{\partial}{\partial r} + \mathrm{i} \dfrac{\vec{\sigma} \cdot \vec{L}}{r} \right) \\ (\vec{\sigma} \cdot \hat{r}) \left(-\mathrm{i} \dfrac{\partial}{\partial r} + \mathrm{i} \dfrac{\vec{\sigma} \cdot \vec{L}}{r} \right) & 0 \end{pmatrix} \begin{pmatrix} f(r) \, \chi_{\kappa\mu}(\hat{r}) \\ \mathrm{i} g(r) \, \chi_{-\kappa\mu}(\hat{r}) \end{pmatrix} \\
&= \begin{pmatrix} (\vec{\sigma} \cdot \hat{r}) \left(\dfrac{\partial}{\partial r} g(r) - \dfrac{1}{r} g(r) \, (\kappa - 1) \right) \chi_{-\kappa\mu}(\hat{r}) \\ -\mathrm{i} \, (\vec{\sigma} \cdot \hat{r}) \left(\dfrac{\partial}{\partial r} f(r) - \dfrac{1}{r} f(r) \, (-\kappa - 1) \right) \chi_{\kappa\mu}(\hat{r}) \end{pmatrix} \\
&= \begin{pmatrix} \left(-\dfrac{\partial}{\partial r} + \dfrac{1}{r} \, (\kappa - 1) \right) g(r) \, \chi_{\kappa\mu}(\hat{r}) \\ \mathrm{i} \left(\dfrac{\partial}{\partial r} + \dfrac{1}{r} \, (\kappa + 1) \right) f(r) \, \chi_{-\kappa\mu}(\hat{r}) \end{pmatrix} = E \begin{pmatrix} f(r) \, \chi_{\kappa\mu}(\hat{r}) \\ \mathrm{i} g(r) \, \chi_{-\kappa\mu}(\hat{r}) \end{pmatrix} ,
\end{aligned}
\tag{7.112}
$$

where the last equality is postulated if $\psi(\vec{r})$ is an eigenstate of the Hamiltonian. For the angular part, we have used the identities $(\vec{\sigma} \cdot \vec{L}) \chi_{\kappa\mu}(\hat{r}) = [(\vec{\sigma} \cdot \vec{L} + 1) - 1] \chi_{\kappa\mu}(\hat{r}) = (-\kappa - 1) \, \chi_{\kappa\mu}(\hat{r})$ and $(\vec{\sigma} \cdot \hat{r}) \, \chi_{\kappa\mu}(\hat{r}) = -\chi_{-\kappa\mu}(\hat{r})$ from Sec. 6.4.1. The radial eigenvalue equations are thus given by

$$
\left(\frac{\partial}{\partial r} + \frac{1 + \kappa}{r} \right) f(r) = E \, g(r) , \qquad \left(\frac{\partial}{\partial r} + \frac{1 - \kappa}{r} \right) g(r) = -E \, f(r) . \tag{7.113}
$$

Given a total angular momentum j, we can choose either one of two possibilities for the Dirac angular quantum number κ, namely, $\kappa = \pm(j + 1/2)$. For negative κ, the solution to the free Dirac equation for massless particles is given as follows,

$$
f_{E\kappa}(r) = \frac{E}{\sqrt{\pi}} j_{j-1/2}(E r) , \quad g_{E\kappa}(r) = -\frac{E}{\sqrt{\pi}} j_{j+1/2}(E r) , \quad \kappa = -(j + 1/2) < 0 , \tag{7.114a}
$$

whereas for positive $\kappa = (j + 1/2)$,

$$
f_{E\kappa}(r) = \frac{E}{\sqrt{\pi}} j_{j+1/2}(E r) , \quad g_{E\kappa}(r) = \frac{E}{\sqrt{\pi}} j_{j-1/2}(E r) , \quad \kappa = (j + 1/2) > 0 . \tag{7.114b}
$$

These formulas are easily combined as follows,

$$
f_{E\kappa}(r) = \frac{E}{\sqrt{\pi}} j_{|\kappa| - 1/2 + \mathrm{sgn}(\kappa)/2)}(E r) , \quad g_{E\kappa}(r) = \mathrm{sgn}(\kappa) \, \frac{E}{\sqrt{\pi}} j_{|\kappa| - 1/2 - \mathrm{sgn}(\kappa)/2}(E r) , \tag{7.115}
$$

where $\mathrm{sgn}(\kappa) = \kappa / |\kappa|$ is the sign function.

The defining differential equation for spherical Bessel functions reads as

$$
\left(\frac{\partial^2}{\partial x^2} + \frac{2}{x} \frac{\partial}{\partial x} - \frac{\ell \, (\ell + 1)}{x^2} + 1 \right) j_\ell(x) = 0 . \tag{7.116}
$$

The recursion relations are

$$j_{\ell-1}(x) + j_{\ell+1}(x) = \frac{2\ell+1}{x}\, j_\ell(x)\,, \tag{7.117a}$$

$$\ell\, j_{\ell-1}(x) - (\ell+1)\, j_{\ell+1}(x) = (2\ell+1)\, j'_\ell(x)\,. \tag{7.117b}$$

Alternatively, one can express the derivative as follows,

$$\begin{aligned} j'_\ell(x) &= j_{\ell-1}(x) - \frac{\ell+1}{x}\, j_\ell(x) = -j_{\ell+1}(x) + \frac{\ell}{x}\, j_\ell(x) \\ &= \frac{1}{2}\,\left(j_{\ell+1}(x) + j_{\ell-1}(x)\right) - \frac{1}{2x}\, j_\ell(x)\,. \end{aligned} \tag{7.118}$$

Let us verify, in the spirit of Eq. (7.113), the solutions of the massless Dirac equation for $\kappa = -(j+1/2) < 0$. Using Eq. (7.118), one verifies two relations,

$$\frac{\partial}{\partial r} f_{E\kappa}(r) = \frac{E^2}{\sqrt{\pi}}\, j'_{j-1/2}(E\,r) = -\frac{E^2}{\sqrt{\pi}}\, j_{j+1/2}(E\,r) + \frac{E^2}{\sqrt{\pi}}\,\frac{j-1/2}{E\,r}\, j_{j-1/2}(E\,r)\,, \tag{7.119a}$$

$$\frac{\kappa+1}{r}\, f_{E\kappa}(r) = -\frac{E^2}{\sqrt{\pi}}\,\frac{j-1/2}{E\,r}\, j_{j-1/2}(E\,r)\,, \qquad \kappa = -(j+\tfrac{1}{2})\,. \tag{7.119b}$$

Adding these, one obtains

$$\frac{\partial}{\partial r} f_{E\kappa}(r) + \frac{\kappa+1}{r}\, f_{E\kappa}(r) = -E\,\frac{E}{\sqrt{\pi}}\, j_{j+1/2}(k\,r) = E\, g_{E\kappa}(r)\,, \tag{7.120}$$

verifying the first entry in Eq. (7.113). Again, using Eq. (7.118), the differentiation of the radial function $g_{E\kappa}(r)$ proceeds as follows,

$$-\frac{\partial}{\partial r} g_{E\kappa}(r) = \frac{E^2}{\sqrt{\pi}}\, j'_{j+1/2}(E\,r) = \frac{E^2}{\sqrt{\pi}}\, j_{j-1/2}(E\,r) - \frac{E^2}{\sqrt{\pi}}\,\frac{j+3/2}{E\,r}\, j_{j+1/2}(E\,r)\,, \tag{7.121a}$$

$$\frac{\kappa-1}{r}\, g_{E\kappa}(r) = \frac{E^2}{\sqrt{\pi}}\,\frac{j+3/2}{E\,r}\, j_{j+1/2}(E\,r)\,, \qquad \kappa = -(j+1/2)\,. \tag{7.121b}$$

So, one verifies the second entry in Eq. (7.113),

$$-\frac{\partial}{\partial r} g_{E\kappa}(r) + \frac{\kappa-1}{r}\, g_{E\kappa}(r) = \frac{E^2}{\sqrt{\pi}}\, j_{j-1/2}(E\,r) = E\left(\frac{E}{\sqrt{\pi}}\, j_{j-1/2}(E\,r)\right) = E\, f_{E\kappa}(r)\,. \tag{7.122}$$

This concludes the verification of the solution of the massless Dirac equation for negative κ, as given in Eq. (7.114a).

The normalization integrals are given as follows,

$$\int_0^\infty \mathrm{d}r\, r^2 \left(\frac{E}{\sqrt{\pi}}\, j_{j-1/2}(E\,r)\right)\left(\frac{E'}{\sqrt{\pi}}\, j_{j-1/2}(E'\,r)\right) = \frac{1}{2}\,\delta(E-E')\,, \tag{7.123a}$$

$$\int_0^\infty \mathrm{d}r\, r^2 \left(\frac{E}{\sqrt{\pi}}\, j_{j+1/2}(E\,r)\right)\left(\frac{E'}{\sqrt{\pi}}\, j_{j+1/2}(E'\,r)\right) = \frac{1}{2}\,\delta(E-E')\,. \tag{7.123b}$$

These relations follow from the orthogonality properties of the spherical Bessel functions, which are discussed in detail in Sec. 2.4.3 of Ref. [81]. The bispinor wave function given in Eq. (7.109) with the radial components given in Eq. (7.114), is normalized to the energy scale,

$$\psi_{E\kappa\mu}(\vec{r}) = \begin{pmatrix} f_{E\kappa}(r)\,\chi_{\kappa\,\mu}(\hat{r}) \\ \mathrm{i}g_{E\kappa}(r)\,\chi_{-\kappa\,\mu}(\hat{r}) \end{pmatrix}, \tag{7.124a}$$

$$\langle E'\kappa'\mu'|E\kappa\mu\rangle = \int \mathrm{d}^3r\,\psi^+_{E'\kappa'\mu'}(\vec{r})\,\psi_{E\kappa\mu}(\vec{r}) = \delta(E-E')\,\delta_{\kappa\kappa'}\,\delta_{\mu\mu'}\,. \tag{7.124b}$$

A final remark is in order. In principle, the wave function given in Eq. (7.109) can be used for the description of a Dirac particle in a process in which its mass is negligible. However, a slight generalization is necessary for neutrinos. Because of the properties of the weak interaction, neutrinos are born in helicity and chirality eigenstates within the zero-mass approximation. The massless Dirac Hamiltonian $H_{\mathrm{FMD}} = \vec{\alpha}\cdot\vec{p}$ commutes with the chirality operator γ^5. Furthermore, any simultaneous eigenstate of H_{FMD} and γ^5 also is an eigenstate of the helicity operator $\widehat{\Sigma}$, because the latter is proportional to $\gamma^5\,\vec{\alpha}\cdot\vec{p}$. In the helicity basis, the appropriate eigenstates have been given in Eqs. (7.88). The situation is a little more complicated in the angular-momentum-resolved basis. It helps to observe that for given j, one may choose two possible values of κ, namely, $\kappa_1 = -(j+1/2)$ and $\kappa_2 = (j+1/2)$, and form the following linear combination of solutions,

$$\begin{aligned}\psi^{(\nu)}_{Ej\mu}(\vec{r}) &= \frac{1}{\sqrt{2}}\begin{pmatrix} f_{E\kappa_1}(r)\,\chi_{\kappa_1\mu}(\hat{r}) + \mathrm{i}f_{E\kappa_2}(r)\,\chi_{\kappa_2\mu}(\hat{r}) \\ \mathrm{i}\,[g_{E\kappa_1}(r)\,\chi_{-\kappa_1\mu}(\hat{r}) + \mathrm{i}\,g_{E\kappa_2}(r)\,\chi_{-\kappa_2\mu}(\hat{r})] \end{pmatrix} \\ &= \frac{E}{\sqrt{2\pi}}\begin{pmatrix} j_{j-1/2}(Er)\,\chi_{-(j+1/2)\,\mu}(\hat{r}) + \mathrm{i}\,j_{j+1/2}(Er)\,\chi_{(j+1/2)\,\mu}(\hat{r}) \\ -\mathrm{i}\,j_{j+1/2}(Er)\,\chi_{(j+1/2)\,\mu}(\hat{r}) - j_{j-1/2}(Er)\,\chi_{-(j+1/2)\,\mu}(\hat{r}) \end{pmatrix}. \end{aligned} \tag{7.125}$$

This wave function is an eigenstate of chirality, and has negative helicity, as required for neutrinos,

$$\psi^{(\nu)}_{Ej\mu}(\vec{r}) = \begin{pmatrix} \nu_{Ej\mu}(\vec{r}) \\ -\nu_{Ej\mu}(\vec{r}) \end{pmatrix}, \tag{7.126a}$$

$$\nu_{Ej\mu}(\vec{r}) = \frac{E}{\sqrt{2\pi}}\left(j_{j-1/2}(Er)\,\chi_{-(j+1/2)\,\mu}(\hat{r}) + \mathrm{i}\,j_{j+1/2}(Er)\,\chi_{(j+1/2)\,\mu}(\hat{r})\right). \tag{7.126b}$$

The normalization integral for the neutrino wave function is given as follows,

$$\int \mathrm{d}^3r\,\psi^{(\nu)+}_{Ej\mu}(\vec{r})\,\psi^{(\nu)}_{E'j'\mu'}(\vec{r}) = \delta(E-E')\,\delta_{jj'}\,\delta_{\mu\mu'}\,. \tag{7.127}$$

This wave function is normalized to the energy scale, just like the one given in Eq. (7.124).

7.3.4 *Angular Momenta and Free Dirac Equation*

The free massive Dirac Hamiltonian reads as $H = \vec{\alpha}\cdot\vec{p} + \beta m$, where βm is the mass term. In the angular-momentum representation, the action of the free Hamiltonian can be written as

$$
\begin{aligned}
H\psi(\vec{r}) &= \begin{pmatrix} m & (\vec{\sigma}\cdot\hat{r})\left(-\mathrm{i}\dfrac{\partial}{\partial r} + \mathrm{i}\dfrac{\vec{\sigma}\cdot\vec{L}}{r}\right) \\ (\vec{\sigma}\cdot\hat{r})\left(-\mathrm{i}\dfrac{\partial}{\partial r} + \mathrm{i}\dfrac{\vec{\sigma}\cdot\vec{L}}{r}\right) & -m \end{pmatrix} \begin{pmatrix} f(r)\,\chi_{\kappa\,\mu}(\hat{r}) \\ \mathrm{i}g(r)\,\chi_{-\kappa\,\mu}(\hat{r}) \end{pmatrix} \\
&= \begin{pmatrix} \left[\left(-\dfrac{\partial}{\partial r} + \dfrac{\kappa-1}{r}\right) g(r) + m\,f(r)\right]\chi_{\kappa\,\mu}(\hat{r}) \\ \mathrm{i}\left[\left(\dfrac{\partial}{\partial r} + \dfrac{\kappa+1}{r}\right) f(r) - m\,g(r)\right]\chi_{-\kappa\,\mu}(\hat{r}) \end{pmatrix} = E\begin{pmatrix} f(r)\,\chi_{\kappa\,\mu}(\hat{r}) \\ \mathrm{i}\,g(r)\,\chi_{-\kappa\,\mu}(\hat{r}) \end{pmatrix}.
\end{aligned}
\tag{7.128}
$$

The radial equations differ from Eq. (7.113) by the inclusion of the mass term,

$$\left(\frac{\partial}{\partial r} + \frac{1-\kappa}{r}\right) g(r) = (m-E)\,f(r)\,, \tag{7.129a}$$

$$\left(\frac{\partial}{\partial r} + \frac{1+\kappa}{r}\right) f(r) = (m+E)\,g(r)\,. \tag{7.129b}$$

The solutions of the free Dirac equation follow as a generalization of the massless case [Eq. (7.114a)], with prefactors adjusted to account for the mass. For negative κ, one has

$$f_{E\kappa}(r) = \frac{(E^2-m^2)^{1/4}}{\sqrt{\pi}}\,\sqrt{E+m}\;j_{j-1/2}(\sqrt{E^2-m^2}\,r)\,, \tag{7.130a}$$

$$g_{E\kappa}(r) = -\,\frac{(E^2-m^2)^{1/4}}{\sqrt{\pi}}\,\sqrt{E-m}\;j_{j+1/2}(\sqrt{E^2-m^2}\,r)\,, \qquad \kappa = -(j+1/2)\,. \tag{7.130b}$$

For positive κ, the solution is given as

$$f_{E\kappa}(r) = \frac{(E^2-m^2)^{1/4}}{\sqrt{\pi}}\,\sqrt{E+m}\;j_{j+1/2}(\sqrt{E^2-m^2}\,r)\,, \tag{7.131a}$$

$$g_{E\kappa}(r) = \frac{(E^2-m^2)^{1/4}}{\sqrt{\pi}}\,\sqrt{E-m}\;j_{j-1/2}(\sqrt{E^2-m^2}\,r)\,, \qquad \kappa = j+1/2\,. \tag{7.131b}$$

The normalization integral is finally found as follows,

$$\psi_{E\kappa\mu}(\vec{r}) = \begin{pmatrix} f_{E\kappa}(r)\,\chi_{\kappa\,\mu}(\hat{r}) \\ \mathrm{i}g_{E\kappa}(r)\,\chi_{-\kappa\,\mu}(\hat{r}) \end{pmatrix}, \qquad \int \mathrm{d}^3r\,\psi^+_{E'\kappa'\mu'}(\vec{r})\,\psi_{E\kappa\mu}(\vec{r}) = \delta(E-E')\,\delta_{\kappa\kappa'}\,\delta_{\mu\mu'}\,. \tag{7.132}$$

The bispinor solutions, for both the massive as well as the massless free Dirac equation, combine an upper spinor component of orbital angular momentum $\ell = j \mp 1/2$ with a lower spinor component of orbital angular momentum $\ell = j \pm 1/2$. The solutions $\psi_{k\tilde{\kappa}\mu}(\vec{r})$, normalized to the momentum scale,

$$\psi_{k\tilde{\kappa}\mu}(\vec{r}) = \begin{pmatrix} f_{k\,\kappa}(r)\,\chi_{\kappa\,\mu}(\hat{r}) \\ \mathrm{i}g_{k\,\kappa}(r)\,\chi_{-\kappa\,\mu}(\hat{r}) \end{pmatrix}, \tag{7.133}$$

are easily obtained on the basis of the formula

$$\delta(k - k') = \frac{\mathrm{d}E}{\mathrm{d}k}\,\delta(E - E') = \frac{\sqrt{E^2 - m^2}}{E}\,\delta(E - E')\,, \tag{7.134}$$

where $E = \sqrt{k^2 + m^2}$. Hence, the wave functions normalized to the momentum scale are obtained from those normalized to the energy scale [see Eqs. (7.130) and (7.131)], via multiplication by a factor $(E^2 - m^2)^{1/4}/E^{1/2}$, i.e.,

$$f_{k\kappa}(r) = \frac{(E^2 - m^2)^{1/4}}{E^{1/2}}\,f_{E\kappa}(r)\,, \tag{7.135}$$

$$g_{k\kappa}(r) = \frac{(E^2 - m^2)^{1/4}}{E^{1/2}}\,g_{E\kappa}(r)\,. \tag{7.136}$$

The normalization integral for the latter solutions is

$$\psi_{k\kappa\mu}(\vec{r}) = \begin{pmatrix} f_{k\kappa}(r)\,\chi_{\kappa\,\mu}(\hat{r}) \\ \mathrm{i}g_{k\kappa}(r)\,\chi_{-\kappa\,\mu}(\hat{r}) \end{pmatrix}, \qquad \int \mathrm{d}^3 r\,\psi^{+}_{k'\kappa'\mu'}(\vec{r})\,\psi_{k\kappa\mu}(\vec{r}) = \delta(k - k')\,\delta_{\kappa\kappa'}\,\delta_{\mu\mu'}\,. \tag{7.137}$$

For the massless case, one has $E = k$, and so the distinction between the two normalizations is not necessary.

7.4 Quantized Dirac Field and Propagators

7.4.1 *Free Dirac Progator in Feynman's Formulation*

In the calculation of classical electromagnetic fields, the use of the retarded, Feynman and advanced propagator depends on the physical interpretation of positive- and negative-frequency components of the incoming signal, and is discussed in detail in Chap. 4 of Ref. [81]. If we interpret the charge and current distributions as sources of radiation, and the fields are to be calculated, then the most adequate representation involves the retarded Green function. The physical interpretation is that both positive-frequency as well as negative-frequency components of the oscillating charges and currents act as sources of radiation. However, if we wish to calculate the emission and absorption of virtual photons within quantum field theory (quantum electrodynamics), then the most adequate representation involves the use of the Feynman Green function, which is naturally obtained from the time-ordered products of field operators. Here, negative-frequency components are interpreted as sinks of radiation; this corresponds to the time-ordered product of the electromagnetic field operators.

In the case of fermion fields, only one interpretation is possible. Namely, the negative-energy solutions of the Dirac theory have to be (re-)interpreted as positive-energy solutions, using the following procedure: One writes the propagator in such a way that negative-energy solutions are automatically propagated into the past, and reinterpreted as positive-energy solutions propagating into the future, which describe the trajectories of antiparticles. Let us suppose that t'' is the time at the endpoint of a trajectory, and t' is the time at the "beginning." The antiparticle phase factor is $\exp[\mathrm{i}E\,(t''-t')]$; if the particle propagates into the past, with $t''=t_<$, and $t'=t_>$, then the accumulated phase factor is $\exp[\mathrm{i}E\,(t_<-t_>)]=\exp[-\mathrm{i}E\,(t_>-t_<)]$, and is equal to the phase factor accumulated during a positive-energy trajectory, propagating into the future, from $t_<$ to $t_>$. So, in a Feynman diagram, we can describe an incoming positron by an outgoing line, propagating into the past, decorated with a four-momentum which is the opposite of the physical, incoming positron momentum. In particular, the energy is negative. This is equivalent to an incoming positron with a positive energy, but with the arrow of propagation in space-time reversed.

To put these statements into perspective, a positron falling to the ground in a gravitational field [255] is interpreted as a positron falling upward, but backward in time, so that, effectively, it falls down just like any other particle. In our formulas, we have already taken the reinterpretation principle into account, labeling the solution by $\mathcal{V}_\sigma(\vec{k})$ where $\vec{k}$ is the physical momentum along the antiparticle trajectory, even if the phase factor is $\exp(\mathrm{i}k\cdot x)=\exp(\mathrm{i}Et-\mathrm{i}\vec{k}\cdot\vec{r})$ with a manifestly positive energy variable E. The reinterpretation principle, which reinterprets trajectories propagating into the past, is key in abandoning the hole theory, which would otherwise necessitate us to think about antiparticles as "holes" in a Dirac "sea" of filled negative-energy states.

Let us illustrate these statements by calculating the time-ordered vacuum expectation value of the field operators of a spin-1/2 theory, which is supposed to yield the time-ordered (Feynman) propagator. In the momentum representation, it is equal to the inverse of the Hamiltonian (upon multiplication with γ^0). The Lorentz-invariant integration measure used in its derivation is

$$\frac{\mathrm{d}^3k}{(2\pi)^3}\,\frac{1}{2E}=\int\frac{\mathrm{d}^4k}{(2\pi)^4}\,\delta(k^2-m^2)\Big|_{k^0>0}\,, \tag{7.138}$$

where

$$E=\sqrt{\vec{k}^2+m^2-\mathrm{i}\,\epsilon}\,, \tag{7.139}$$

and the energy-momentum four-vector is $k^\mu=(E,\vec{k})$. We start from the field operator

$$\psi(x)=\int\frac{\mathrm{d}^3k}{(2\pi)^3}\,\frac{m}{E}\sum_{\sigma=\pm}\left\{b_\sigma(k)\,\mathcal{U}_\sigma(\vec{k})\,\mathrm{e}^{-\mathrm{i}\,k\cdot x}+d^+_\sigma(k)\,\mathcal{V}_\sigma(\vec{k})\,\mathrm{e}^{\mathrm{i}\,k\cdot x}\right\}\,, \tag{7.140}$$

where b_σ annihilates particles and d^+_σ creates antiparticles. This operator is denoted by the symbol ψ although it is manifestly distinct from the first-quantized Dirac wave function ψ, which is usually denoted by the same symbol. The fundamental bispinors $\mathcal{U}_\sigma(\vec{k})$ and $\mathcal{V}_\sigma(\vec{k})$ have been given in Eqs. (7.89) and (7.90). The Dirac adjoint of the field operator is

$$\overline{\psi}(x)=\int\frac{\mathrm{d}^3k}{(2\pi)^3}\,\frac{m}{E}\sum_{\rho=\pm}\left[b^+_\rho(k)\,\overline{\mathcal{U}}_\rho(\vec{k})\,\mathrm{e}^{\mathrm{i}\,k\cdot x}+d_\rho(k)\,\overline{\mathcal{V}}_\rho(\vec{k})\,\mathrm{e}^{-\mathrm{i}\,k\cdot x}\right]. \tag{7.141}$$

Formally, these formulas are consistent with the following reinterpretation,

$$b_\sigma(-k) \to d^+_\sigma(k)\,, \qquad b^+_\sigma(-k) \to d_\sigma(k)\,, \tag{7.142}$$

for negative-energy states, in agreement with the above reinterpretation principle. The anticommutators of annihilation and creation operators vanish,

$$\{b_\sigma(k), b_\rho(k')\} = \left\{b^+_\sigma(k), b^+_\rho(k')\right\} = 0\,, \tag{7.143a}$$

$$\{d_\sigma(k), d_\rho(k')\} = \left\{d^+_\sigma(k), d^+_\rho(k')\right\} = 0\,. \tag{7.143b}$$

However, anticommutators of creation and annihilation operators do not vanish,

$$\left\{b_\sigma(k), b^+_\rho(k')\right\} = (2\pi)^3\,\frac{E}{m}\,\delta^{(3)}(\vec{k}-\vec{k}')\,\delta_{\sigma\rho}\,, \tag{7.144a}$$

$$\left\{d_\sigma(k), d^+_\rho(k')\right\} = (2\pi)^3\,\frac{E}{m}\,\delta^{(3)}(\vec{k}-\vec{k}')\,\delta_{\sigma\rho}\,. \tag{7.144b}$$

We have the following expression for the time-ordered field propagator, with $x_1 = (t_1, \vec{r}_1)$, $x_2 = (t_2, \vec{r}_2)$, and $k'^\mu = (E', \vec{k}')$,

$$\begin{aligned}&\left\langle 0 \left| T\,\psi_\xi(x_1)\,\overline{\psi}_{\xi'}(x_2) \right| 0\right\rangle \\ &= \Theta(t_1-t_2)\,\left\langle 0 \left| \psi_\xi(x_1)\,\overline{\psi}_{\xi'}(x_2) \right| 0\right\rangle - \Theta(t_2-t_1)\,\left\langle 0 \left| \overline{\psi}_{\xi'}(x_2)\,\psi_\xi(x_1) \right| 0\right\rangle\,.\end{aligned} \tag{7.145}$$

Here, $|0\rangle$ denotes the vacuum ground state of the underlying field theory. The negative sign in front of the second term is introduced for fermion field operators, by convention, for the time-ordered product of fermionic field operators. The indices in the expressions $\psi_\xi(x_1)$ and $\overline{\psi}_{\xi'}(x_2)$ denote the spinor components. The time-ordered product of fermion field operators is a tensor product in spinor space, $\left\langle 0 \left| T\,\psi(x_1)\,\overline{\psi}(x_2) \right| 0\right\rangle \equiv \left\langle 0 \left| T\,\psi(x_1) \otimes \overline{\psi}(x_2) \right| 0\right\rangle$, of which we calculate the component with spinor indices ξ, ξ'.

We can go on to calculate

$$\begin{aligned}&\left\langle 0 \left| T\,\psi_\xi(x_1)\,\overline{\psi}_{\xi'}(x_2) \right| 0\right\rangle \\ &= \Theta(t_1-t_2)\,\Bigg\langle 0 \Bigg| \int \frac{d^3k}{(2\pi)^3}\,\frac{m}{E} \sum_{\sigma=\pm} \left[b_\sigma(k)\,\mathcal{U}_{\sigma,\xi}(\vec{k})\,e^{-i\,k\cdot x_1} + d^+_\sigma(k)\,\mathcal{V}_{\sigma,\xi}(\vec{k})\,e^{i\,k\cdot x_1}\right] \\ &\quad\times \int \frac{d^3k'}{(2\pi)^3}\,\frac{m}{E'} \sum_{\rho=\pm} \left[b^+_\rho(k')\,\overline{\mathcal{U}}_{\rho,\xi'}(\vec{k}')\,e^{i\,k'\cdot x_2} + d_\rho(k')\,\overline{\mathcal{V}}_{\rho,\xi'}(\vec{k}')\,e^{-i\,k'\cdot x_2}\right] \Bigg| 0 \Bigg\rangle \\ &\quad - \Theta(t_2-t_1)\,\Bigg\langle 0 \Bigg| \int \frac{d^3k'}{(2\pi)^3}\,\frac{m}{E'} \sum_{\sigma=\pm} \left[b^+_\sigma(k')\,\overline{\mathcal{U}}_{\sigma,\xi'}(\vec{k}')\,e^{i\,k'\cdot x_2} + d_\sigma(k')\,\overline{\mathcal{V}}_{\sigma,\xi'}(\vec{k}')\,e^{-i\,k'\cdot x_2}\right] \\ &\quad\times \int \frac{d^3k}{(2\pi)^3}\,\frac{m}{E} \sum_{\rho=\pm} \left[b_\rho(k)\,\mathcal{U}_{\rho,\xi}(\vec{k})\,e^{-i\,k\cdot x_1} + d^+_\rho(k)\,\mathcal{V}_{\rho,\xi}(\vec{k})\,e^{i\,k\cdot x_1}\right] \Bigg| 0 \Bigg\rangle\,.\end{aligned} \tag{7.146}$$

It is very instructive to leave the spinor indices in place. Noting that annihilation operators lead to a vanishing contribution when acting on the vacuum, we can ascertain that the expression (7.146) becomes

$$
\begin{aligned}
\langle 0 | T\, \psi_\xi(x_1)\, \overline{\psi}_{\xi'}(x_2) | 0 \rangle = {} & \Theta(t_1 - t_2) \int \frac{d^3k}{(2\pi)^3} \frac{m}{E} \int \frac{d^3k'}{(2\pi)^3} \frac{m}{E'} \\
& \times \sum_{\sigma=\pm} \sum_{\rho=\pm} \langle 0 | b_\sigma(k)\, b_\rho^+(k') | 0 \rangle \big[\mathcal{U}_\sigma(\vec{k}) \otimes \overline{\mathcal{U}}_\rho(\vec{k}') \big]_{\xi\xi'}\, e^{-i\,k\cdot x_2 + i\,k'\cdot x_2} \\
& - \Theta(t_2 - t_1) \int \frac{d^3k}{(2\pi)^3} \frac{m}{E} \int \frac{d^3k'}{(2\pi)^3} \frac{m}{E'} \\
& \times \sum_{\sigma=\pm} \sum_{\rho=\pm} \langle 0 | d_\sigma(k')\, d_\rho^+(k) | 0 \rangle \big[\mathcal{V}_\rho(\vec{k}) \otimes \overline{\mathcal{V}}_\sigma(\vec{k}') \big]_{\xi\xi'}\, e^{i\,k\cdot x_2 - i\,k'\cdot x_2} .
\end{aligned}
\tag{7.147}
$$

Here,

$$
\big[\mathcal{U}_\sigma(\vec{k}) \otimes \overline{\mathcal{U}}_\rho(\vec{k}') \big]_{\xi\xi'} = \mathcal{U}_{\sigma,\xi}(\vec{k})\, \overline{\mathcal{U}}_{\rho,\xi'}(\vec{k}') , \tag{7.148}
$$

$$
\big[\mathcal{V}_\rho(\vec{k}) \otimes \overline{\mathcal{V}}_\sigma(\vec{k}') \big]_{\xi\xi'} = \mathcal{V}_{\rho,\xi}(\vec{k})\, \overline{\mathcal{V}}_{\sigma,\xi'}(\vec{k}') . \tag{7.149}
$$

Using the anticommutators given in Eq. (7.144) and the sum rule (7.93) in order to convert the double integral into a single one, one may write a much more compact expression,

$$
\begin{aligned}
\langle 0 | T\, \psi(x_1)\, \overline{\psi}(x_2) | 0 \rangle = {} & \Theta(t_1 - t_2) \int \frac{d^3k}{(2\pi)^3} \frac{m}{E}\, e^{-ik\cdot(x_1 - x_2)}\, \frac{\gamma^0 E - \vec{\gamma}\cdot\vec{k} + m}{2m} \\
& - \Theta(t_2 - t_1) \int \frac{d^3k}{(2\pi)^3} \frac{m}{E}\, e^{ik\cdot(x_1 - x_2)}\, \frac{\gamma^0 E - \vec{\gamma}\cdot\vec{k} - m}{2m} ,
\end{aligned}
\tag{7.150}
$$

where the tensor product in spinor space is understood, and $k \cdot x = E t - \vec{k}\cdot\vec{r}$.

The following contour integral representations are introduced for the Heaviside step functions,

$$
\Theta(t_1 - t_2)\, f(E)\, e^{-iE(t_1 - t_2)} = i \int \frac{dk_0}{2\pi} e^{-ik_0\,(t_1 - t_2)}\, \frac{1}{k_0 - E + i\epsilon}\, f(k_0) , \tag{7.151a}
$$

$$
\Theta(t_2 - t_1)\, f(E)\, e^{iE(t_1 - t_2)} = -i \int \frac{dk_0}{2\pi} e^{-ik_0\,(t_1 - t_2)}\, \frac{1}{k_0 + E - i\epsilon}\, f(-k_0) . \tag{7.151b}
$$

The disturbing relative minus sign in Eq. (7.150) is eliminated, and we obtain

$$
\begin{aligned}
& \langle 0 | T \psi(x_1) \overline{\psi}(x_2) | 0 \rangle \\
& = i \int \frac{d^4k}{(2\pi)^4} \frac{m}{E} e^{-ik_0\,(t_1 - t_2) + i\vec{k}\cdot(\vec{r}_1 - \vec{r}_2)}\, \frac{\gamma^0 k_0 - \vec{\gamma}\cdot\vec{k} + m}{2m}\, \frac{1}{k_0 - E + i\epsilon} \\
& \quad + i \int \frac{d^4k}{(2\pi)^4} \frac{m}{E} e^{-ik_0\,(t_1 - t_2) + i\vec{k}\cdot(\vec{r}_1 - \vec{r}_2)}\, \frac{-\gamma^0 k_0 + \vec{\gamma}\cdot\vec{k} - m}{2m}\, \frac{1}{k_0 + E - i\epsilon} .
\end{aligned}
\tag{7.152}
$$

Here, we have replaced $\vec{k} \to -\vec{k}$ in the second term, without affecting the result of the integration (the integral extends over all $\vec{k} \in \mathbb{R}^3$). The integration symbol is $d^4k = dk^0\, d^3k$. There is, actually, a (tacitly assumed) convention [2, 256] that in any integrals $\int d^3k$, the component k_0 is set equal to E, as given in Eq. (7.139), wherever it occurs, but if the integral is over the full d^4k, then the integration interval is the full $k_0 \in (-\infty, \infty)$. With this notion in mind, we have

$$\begin{aligned}
\langle 0 | T\, \psi(x_1)\, \overline{\psi}(x_2) | 0 \rangle &= \mathrm{i} \int \frac{\mathrm{d}^4k}{(2\pi)^4} \mathrm{e}^{-\mathrm{i}k\cdot(x_1-x_2)} \frac{\not{k}+m}{2m} \frac{m}{E} \frac{1}{k_0 - E + \mathrm{i}\epsilon} \\
&\quad + \mathrm{i} \int \frac{\mathrm{d}^4k}{(2\pi)^4} \mathrm{e}^{-\mathrm{i}k\cdot(x_1-x_2)} \frac{-\not{k}-m}{2m} \frac{m}{E} \frac{1}{k_0 + E - \mathrm{i}\epsilon} \\
&= \mathrm{i} \int \frac{\mathrm{d}^4k}{(2\pi)^4} \mathrm{e}^{-\mathrm{i}k\cdot(x_1-x_2)} \frac{\not{k}+m}{2} \frac{1}{E} \left(\frac{1}{k_0 - E + \mathrm{i}\epsilon} - \frac{1}{k_0 + E - \mathrm{i}\epsilon} \right) \\
&= \mathrm{i} \int \frac{\mathrm{d}^4k}{(2\pi)^4} \mathrm{e}^{-\mathrm{i}k\cdot(x_1-x_2)} \frac{\not{k}+m}{k^2 - m^2 + \mathrm{i}\epsilon} = \mathrm{i}\, S_F(x_1, x_2)\,. \qquad (7.153)
\end{aligned}$$

In the last line, we implicitly define the Feynman propagator S_F. In energy-momentum space, we can read off the representation of the Feynman propagator as follows,

$$S_F(k) = \frac{1}{\not{k} - m + \mathrm{i}\epsilon} = \frac{\not{k}+m}{k^2 - m^2 + \mathrm{i}\epsilon}\,. \qquad (7.154)$$

This is, of course, a well-known, classic result.

In addition to the propagator, it is extremely instructive to also calculate the field anticommutator $\{\psi(x_1), \overline{\psi}(x_2)\}$, which is of course equal to its own vacuum expectation value. Simplifying the notation by leaving out the spinor index as in Eq. (7.147), and using the anticommutator (7.144), we have

$$\begin{aligned}
\{\psi(x_1), \overline{\psi}(x_2)\} &= \int \frac{\mathrm{d}^3k}{(2\pi)^3} \frac{m}{E} \sum_\sigma \big[\mathrm{e}^{-\mathrm{i}k\cdot(x_1-x_2)}\, \mathcal{U}_\sigma(\vec{k}) \otimes \overline{\mathcal{U}}_\sigma(\vec{k}) \\
&\quad + \mathrm{e}^{\mathrm{i}k\cdot(x_1-x_2)}\, \mathcal{V}_\sigma(\vec{k}) \otimes \overline{\mathcal{V}}_\sigma(\vec{k}) \big]\,. \qquad (7.155)
\end{aligned}$$

With the help of the spin sum (7.93), it follows that

$$\begin{aligned}
\{\psi(x_1), \overline{\psi}(x_2)\} &= \int \frac{\mathrm{d}^3k}{(2\pi)^3} \frac{m}{E} \left(\mathrm{e}^{-\mathrm{i}k\cdot(x_1-x_2)} \frac{\not{k}+m}{2m} + \mathrm{e}^{\mathrm{i}k\cdot(x_1-x_2)} \frac{\not{k}-m}{2m} \right) \\
&= \int \frac{\mathrm{d}^3k}{(2\pi)^3} \frac{m}{E} \left(\frac{\mathrm{i}\not{\partial}_1 + m}{2m} \mathrm{e}^{-\mathrm{i}k\cdot(x_1-x_2)} + \frac{-\mathrm{i}\not{\partial}_1 - m}{2m} \mathrm{e}^{\mathrm{i}k\cdot(x_1-x_2)} \right) \\
&= \left(\mathrm{i}\not{\partial}_1 + m \right) \int \frac{\mathrm{d}^3k}{(2\pi)^3} \frac{1}{2E} \left(\mathrm{e}^{-\mathrm{i}k\cdot(x_1-x_2)} - \mathrm{e}^{\mathrm{i}k\cdot(x_1-x_2)} \right) \\
&= \left(\mathrm{i}\not{\partial}_1 + m \right) \mathrm{i}\,\Delta(x_1 - x_2)\,, \qquad (7.156)
\end{aligned}$$

where $\not{\partial}_1 = \gamma^\mu\, \partial/\partial x_1^\mu$. Finally, $\Delta(x_1 - x_2)$ is the expression [see Eqs. (3.56) and (3.77) of Ref. [2]],

$$\mathrm{i}\,\Delta(x_1 - x_2) = \int \frac{\mathrm{d}^3k}{(2\pi)^3} \frac{1}{2E} \left(\mathrm{e}^{-\mathrm{i}k\cdot(x_1-x_2)} - \mathrm{e}^{\mathrm{i}k\cdot(x_1-x_2)} \right)\,. \qquad (7.157)$$

The time derivative of the Δ distribution at equal times reads as [we denote $x_1^\mu = (t_1, \vec{r}_1)$ and $x_2^\mu = (t_2, \vec{r}_2)$]

$$\begin{aligned} \frac{\partial}{\partial t_1}\Delta(x_1 - x_2)\bigg|_{t_1=t_2} &= -\mathrm{i}\frac{\partial}{\partial t_1}\int \frac{\mathrm{d}^3k}{(2\pi)^3}\frac{1}{2E}\left(\mathrm{e}^{-\mathrm{i}k\cdot(x_1-x_2)} - \mathrm{e}^{\mathrm{i}k\cdot(x_1-x_2)}\right)\bigg|_{x_0=y_0} \\ &= -\int \frac{\mathrm{d}^3k}{(2\pi)^3}\frac{1}{2}\left(\mathrm{e}^{\mathrm{i}\vec{k}\cdot(\vec{r}_1-\vec{r}_2)} + \mathrm{e}^{\mathrm{i}\vec{k}\cdot(\vec{r}_1-\vec{r}_2)}\right) = -\delta^{(3)}(\vec{r}_1 - \vec{r}_2)\,. \end{aligned} \tag{7.158}$$

For $x_0 = y_0$, the only contribution to Eq. (7.156) is the one from the time derivative, and we have

$$\{\psi(x_1), \overline{\psi}(x_2)\}\big|_{t_1=t_2} = -\gamma^0\,\partial_0\,\Delta(x_1 - x_2) = \gamma^0\,\delta^{(3)}(\vec{r}_1 - \vec{r}_2)\,. \tag{7.159}$$

This result guarantees the locality of the theory. Furthermore, considering the Lagrangian (7.79), one can actually show that the conjugate field momentum is

$$\pi(x) \equiv \frac{\partial\mathcal{L}}{\partial(\partial_0\psi)} = \mathrm{i}\,\overline{\psi}(x)\,\gamma^0 = \mathrm{i}\,\psi^+(x). \tag{7.160}$$

Hence, Eq. (7.159) can also be interpreted as follows,

$$\{\psi(x_1), \pi(x_2)\}\big|_{t_1=t_2} = \mathrm{i}\,\delta^{(3)}(\vec{r}_1 - \vec{r}_2)\,, \tag{7.161}$$

which is the canonical equal-time anticommutation relation for fermionic fields. This relation should be compared to Eq. (10.282).

7.4.2 *Feynman Propagator and Green Function*

Green functions for the electron-positron Dirac field have, in general, many uses, including the calculation of time propagation and the calculation of propagator denominators in perturbative diagram calculations. In consequence, it seems that the Dirac Green function appears, in the literature, in at least three slightly different forms. All three versions are adapted, in each case, to the application one has in mind.

In the following, we will attempt to summarize the formulas for a generic Dirac Hamiltonian

$$H = \vec{\alpha}\cdot(\vec{p} - e\,\vec{A}) + e\,A^0 + \beta m\,, \tag{7.162a}$$

$$\vec{p} = -\mathrm{i}\vec{\nabla}\,, \qquad \vec{A} = \vec{A}(\vec{r})\,, \qquad A^0 = \Phi = \Phi(\vec{r})\,. \tag{7.162b}$$

For the Coulomb field, one often denotes $V(\vec{r}) = e\,\Phi(\vec{r})$ as the potential. It is useful to distinguish

- the time evolution Green function $G_F = G_F(t_1 - t_2, \vec{r}_1, \vec{r}_2)$,
- the Feynman propagator $S_F = S_F(t_1 - t_2, \vec{r}_1, \vec{r}_2)$,
- the perturbative Green function $\mathcal{G}_F = \mathcal{G}_F(t_1 - t_2, \vec{r}_1, \vec{r}_2)$.

Here, we have used the fact that the static nature of the Hamiltonian in Eq. (7.162) allows us to express the Green functions in terms of the time difference $t_1 - t_2$ only. However, for nontrivial scalar and vector potentials Φ and $\vec{A}$, spatial translation invariance could be lost, and we need to indicate the two arguments $\vec{r}_1$ and $\vec{r}_2$ separately.

Let us assume that a complete set of eigenfunctions exists,

$$H\,\psi_n(\vec{r}) = E_n\,\psi_n(\vec{r})\,, \tag{7.163}$$

which may include a continuum, and let us also define the summation over eigenfunctions to be generalized to include the continuum [see also the remarks following Eq. (3.45)]. The completeness of the spectrum is summarized in the formula

$$\sum_n \psi_n(\vec{r}_1) \otimes \psi_n^+(\vec{r}_2) = \mathbb{1}_{4\times 4}\,\delta^{(3)}(\vec{r}_1 - \vec{r}_2)\,. \tag{7.164}$$

Here, the outer product is denoted as $\otimes$ and is taken in the four-component bispinor space. One might ask how one can understand the completeness relation in the nonrelativistic limit. This is because the Dirac equation has "two spectra" in this limit, one for the particle, and one for the antiparticle solutions. In the nonrelativistic limit, the particle solutions represent two-component Schrödinger–Pauli spinors which will provide the "upper" components of the completeness relation, i.e., a term $\mathbb{1}_{2\times 2}\,\delta^{(3)}(\vec{r}_1 - \vec{r}_2)$, while the antiparticle solution will provide the "lower" completeness, also proportional to $\mathbb{1}_{2\times 2}\,\delta^{(3)}(\vec{r}_1 - \vec{r}_2)$. Their sum gives the unit matrix $\mathbb{1}_{4\times 4}$ in bispinor space.

The time evolution Green function $G_F = G_F(t_1 - t_2, \vec{r}_1, \vec{r}_2)$ fulfills the differential equation

$$\left(\mathrm{i}\frac{\partial}{\partial t} - H\right) G_F(t_1 - t_2, \vec{r}_1, \vec{r}_2) = \mathrm{i}\,\delta(t_1 - t_2)\,\delta^{(3)}(\vec{r}_1 - \vec{r}_2)\,, \tag{7.165}$$

with boundary conditions chosen so that positive-energy states propagate into the future, which negative-energy state propagator into the past. The function G_F can be written as

$$\begin{aligned} G_F(t_1 - t_2, \vec{r}_1, \vec{r}_2) &= \Theta(t_1 - t_2) \sum_{E_n > 0} \psi_n(\vec{r}_1) \otimes \psi_n^+(\vec{r}_2)\,\mathrm{e}^{-\mathrm{i}\,[E_n - \mathrm{i}\epsilon]\,(t_1 - t_2)} \\ &\quad - \Theta(t_2 - t_1) \sum_{E_m < 0} \psi_m(\vec{r}_1) \otimes \psi_m^+(\vec{r}_2)\,\mathrm{e}^{-\mathrm{i}\,[E_m + \mathrm{i}\epsilon]\,(t_1 - t_2)}\,, \end{aligned} \tag{7.166a}$$

$$\begin{aligned} G_F(E, \vec{r}_1, \vec{r}_2) &= \int_{-\infty}^{\infty} \mathrm{d}\tau\, G(\tau, \vec{r}_1, \vec{r}_2)\,\mathrm{e}^{\mathrm{i}\,E\,\tau} = \frac{1}{\mathrm{i}} \sum_n \frac{\psi_n(\vec{r}_1) \otimes \psi_n^+(\vec{r}_2)}{E_n(1 - \mathrm{i}\epsilon) - E} \\ &= \left\langle \vec{r}_1 \left| \frac{1}{\mathrm{i}} \frac{1}{H(1 - \mathrm{i}\epsilon) - E} \right| \vec{r}_2 \right\rangle\,. \end{aligned} \tag{7.166b}$$

Here, we have introduced infinitesimal imaginary parts, chosen in such a way that the time integrations separately converge for both subintervals $0 < \tau = t_1 - t_2 < \infty$ and $-\infty < \tau = t_1 - t_2 < 0$. In the Fourier transform, the sign of the imaginary parts is absorbed in the multiplicative correction $E_n \to E_n(1 - \mathrm{i}\epsilon)$ in the Fourier representation, which is subsequently absorbed in the multiplicative correction to the Hamiltonian.

One might ask how one can intuitively understand the emergence of the energy differences $E_n(1-\mathrm{i}\epsilon) - E \to E_n - E$ in the propagator denominators in the Fourier representation. The answer can be understood if one considers the Fourier representation of a state, proportional to $\psi_k(\vec{r})\exp(-\mathrm{i}E_k t)$. It is proportional to $\delta(E-E_k)$, and thus peaked near the reference-state energy. The propagator denominator in Eq. (7.166b) also is peaked near the resonance energies. Its structure emphasizes the fact that, in Fourier space, the probability of the propagation of wave packets is peaked near the resonance energies of the Hamiltonian.

Let us illustrate some properties of G_F by way of example. First, we observe that, in the space of positive-energy reference states, $G_F(t_1-t_2,\vec{r}_1,\vec{r}_2)$ takes the role of the space-time representation of the time evolution operator $U(t_1-t_2) \to \exp[-\mathrm{i}H(t_1-t_2)]$, i.e., that $G_F(t_1-t_2,\vec{r}_1,\vec{r}_2) \to \langle\vec{r}_1|U(t_1-t_2)|\vec{r}_2\rangle$. Note that $|\psi(t_1)\rangle = U(t_1-t_2)\,|\psi(t_2)\rangle$ for positive-energy states, without further integration over t_2. In coordinate space, this means that if

$$\psi(t_2,\vec{r}_2) = \sum_{E_n>0} c_n\,\psi_n(\vec{r}_2) \tag{7.167}$$

and $t_1 > t_2$, then

$$\psi(t_1,\vec{r}_1) = \int \mathrm{d}^3 r_2\, G_F(t_1-t_2,\vec{r}_1,\vec{r}_2)\,\psi(t_2,\vec{r}_2)\,, \tag{7.168}$$

without any further integration over t_2.

It is still very instructive to consider what would happen if we were to integrate over t_2. Let us assume that $\psi(t_2,\vec{r}_2) = \psi_k(\vec{r}_2)\exp(-\mathrm{i}E_k t_2)$ is the wave function at time t_2, where $\psi_k(\vec{r}_2)$ is one of the eigenfunctions of H (where $E_k > 0$). The function obtained by convolution is

$$\begin{aligned}\phi_k(t_1,\vec{r}_1) &= \int \mathrm{d}^3 r_2\,\mathrm{d}t_2\, G_F(t_1-t_2,\vec{r}_1,\vec{r}_2)\,\psi_k(t_2,\vec{r}_2)\\ &= 2\pi\delta(0)\,\psi_k(\vec{r}_1)\,\exp(-\mathrm{i}E_k t_1)\,.\end{aligned} \tag{7.169}$$

We have set $\int \mathrm{d}t_2 = 2\pi\delta(0)$, for obvious reasons, where the Dirac-δ, as usual, denotes the corresponding distribution. Since Eq. (7.169) denotes a convolution with respect to the time t_2, which is equivalent to multiplication in Fourier space, we can alternatively multiply in Fourier (frequency) space and then, transform back into the time domain. Hence, we investigate the function

$$\begin{aligned}\phi_k(E,\vec{r}_1) &= \int \mathrm{d}^3 r_2\, G_F(E,\vec{r}_1,\vec{r}_2)\,\psi_k(E,\vec{r}_2)\,,\\ \psi_k(E,\vec{r}_2) &= 2\pi\,\delta(E-E_k)\,\psi_k(\vec{r}_2)\,.\end{aligned} \tag{7.170}$$

Hence,

$$\begin{aligned}\phi_k(E,\vec{r}_1) &= \int \mathrm{d}^3 r_2\, \frac{1}{\mathrm{i}}\sum_n \frac{\psi_n(\vec{r}_1)\otimes\psi_n^+(\vec{r}_2)}{E_n(1-\mathrm{i}\epsilon)-E}\, 2\pi\,\delta(E-E_k)\,\psi_k(\vec{r}_2)\\ &= \frac{1}{\mathrm{i}}\,\frac{1}{E_k - E - \mathrm{i}\epsilon}\, 2\pi\,\delta(E-E_k)\,\psi_k(\vec{r}_1)\,,\end{aligned} \tag{7.171}$$

where we assume that $E_k > 0$. The Fourier backtransform of $\phi_k(E, \vec{r}_1)$ should confirm the result (7.169). Closing the integration contour in the lower half of the complex plane, we find that

$$
\begin{aligned}
\phi_k(t_1, \vec{r}_1) &= \int \frac{\mathrm{d}E}{2\pi} \exp(-\mathrm{i}Et_1)\, \phi_k(E, \vec{r}_1) \\
&= \int \frac{\mathrm{d}E}{2\pi} \exp(-\mathrm{i}Et_1) \left(-\frac{1}{\mathrm{i}} \frac{1}{E - (E_k - \mathrm{i}\epsilon)} \right) 2\pi\, \delta(E - E_k)\, \psi_k(\vec{r}_1) \\
&= \int \frac{\mathrm{d}E}{2\pi} \exp(-\mathrm{i}Et_1) \left[-\frac{1}{\mathrm{i}} \left(-2\pi\mathrm{i}\delta(E - E_k)\right) \right] 2\pi\, \delta(E - E_k)\, \psi_k(\vec{r}_1) \\
&= 2\pi\delta(0)\, \psi_k(\vec{r}_1) \exp(-\mathrm{i}E_k t_1)\,.
\end{aligned} \tag{7.172}
$$

This confirms the result (7.169).

The Green function $\mathcal{G}_F$, adapted to perturbative calculations, is defined so that its Fourier transform yields the expression $1/(H - E)$, where, in writing $1/(H - E)$, we ignore infinitesimal imaginary parts. The expression is slightly different from Eq. (7.166a) and reads as follows,

$$
\begin{aligned}
\mathcal{G}_F(t_1 - t_2, \vec{r}_1, \vec{r}_2) &= \mathrm{i}\Theta(t_1 - t_2) \sum_{E_n > 0} \psi_n(\vec{r}_1) \otimes \psi_n^+(\vec{r}_2)\, \mathrm{e}^{-\mathrm{i}[E_n - \mathrm{i}\epsilon](t_1 - t_2)} \\
&\quad - \mathrm{i}\Theta(t_2 - t_1) \sum_{E_m < 0} \psi_m(\vec{r}_1) \otimes \psi_m^+(\vec{r}_2)\, \mathrm{e}^{-\mathrm{i}[E_m + \mathrm{i}\epsilon](t_1 - t_2)}\,,
\end{aligned} \tag{7.173a}
$$

$$
\begin{aligned}
\mathcal{G}_F(E, \vec{r}_1, \vec{r}_2) &= \int_{-\infty}^{\infty} \mathrm{d}\tau\, \mathcal{G}_F(\tau, \vec{r}_1, \vec{r}_2)\, \mathrm{e}^{\mathrm{i}E\tau} = \sum_n \frac{\psi_n(\vec{r}_1) \otimes \psi_n^+(\vec{r}_2)}{E_n(1 - \mathrm{i}\epsilon) - E} \\
&= \left\langle \vec{r}_1 \left| \frac{1}{H(1 - \mathrm{i}\epsilon) - E} \right| \vec{r}_2 \right\rangle .
\end{aligned} \tag{7.173b}
$$

Finally, the Feynman propagator S_F is defined so that it transforms covariantly under Lorentz transformations, i.e., it is constructed from the bispinor wave functions $\psi_n(\vec{r}_1)$ and their Dirac adjoints $\overline{\psi}_n(\vec{r}_2)$, and with a sign convention so that, in the case of the free Hamiltonian H_{FD}, one recovers the result (7.154). Again, the formulas are slightly different as compared to Eqs. (7.166a) and (7.173), and read as follows,

$$
\begin{aligned}
S_F(t_1 - t_2, \vec{r}_1, \vec{r}_2) &= -\mathrm{i}\Theta(t_1 - t_2) \sum_{E_n > 0} \psi_n(\vec{r}_1) \otimes \overline{\psi}_n(\vec{r}_2)\, \mathrm{e}^{-\mathrm{i}[E_n - \mathrm{i}\epsilon](t_1 - t_2)} \\
&\quad + \mathrm{i}\Theta(t_2 - t_1) \sum_{E_m < 0} \psi_m(\vec{r}_1) \otimes \overline{\psi}_m(\vec{r}_2)\, \mathrm{e}^{-\mathrm{i}[E_m + \mathrm{i}\epsilon](t_1 - t_2)}\,,
\end{aligned} \tag{7.174a}
$$

while the spectral decomposition requires a careful distinction between positive-energy and negative-energy virtual states,

$$\begin{aligned}
S_F(E,\vec{r}_1,\vec{r}_2) &= \int_{-\infty}^{\infty} \mathrm{d}\tau\, S_F(\tau,\vec{r}_1,\vec{r}_2)\, \mathrm{e}^{\mathrm{i}\,E\,\tau} = \sum_n \frac{\psi_n(\vec{r}_1)\otimes\psi_n^+(\vec{r}_2)}{E - E_n(1-\mathrm{i}\epsilon)}\,\gamma^0 \\
&= \left\langle \vec{r}_1 \left| \frac{1}{\gamma^0 E - \gamma^0 H + \mathrm{i}\gamma^0 H\epsilon} \right| \vec{r}_2 \right\rangle \\
&= \left\langle \vec{r}_1 \left| \frac{1}{\gamma^0 E - \vec{\gamma}\cdot(\vec{p} - e\vec{A}) - \gamma^0 e\,\Phi - m + \mathrm{i}\epsilon} \right| \vec{r}_2 \right\rangle \\
&= \left\langle \vec{r}_1 \left| \frac{1}{\gamma^\mu (p_\mu - e\,A_\mu) - m + \mathrm{i}\epsilon} \right| \vec{r}_2 \right\rangle = \left\langle \vec{r}_1 \left| \frac{1}{\not{p} - e\not{A} - m + \mathrm{i}\epsilon} \right| \vec{r}_2 \right\rangle .
\end{aligned} \tag{7.174b}$$

The latter representation is valid if we identify $p^\mu = (E, \vec{p})$ with a four-vector composed of the energy argument E and the momentum vector operator $\vec{p}$, which is part of the Hamiltonian operator. We recall that the Feynman prescription for the propagator denominator can alternatively be replaced by the specification of an appropriate integration contour for the energy variable E.

The following conversion formulas are relevant,

$$G_F = \mathrm{i} S_F \gamma^0\,, \qquad S_F = -\mathrm{i} G_F\,\gamma^0\,, \tag{7.175a}$$

$$\mathcal{G}_F = \mathrm{i} G_F\,, \qquad G_F = -\mathrm{i}\mathcal{G}_F\,, \tag{7.175b}$$

$$S_F = -\,\mathcal{G}_F \gamma^0\,, \qquad \mathcal{G}_F = -S_F\,\gamma^0\,. \tag{7.175c}$$

We will use all conventions in the following.

7.4.3 *Free Dirac Propagator in the Angular Momentum Basis*

In order to calculate the fully relativistic free Dirac Green function, we work in mixed coordinate-frequency space. Because the free Dirac Hamiltonian is a (4×4)-matrix in spinor space, the corresponding propagator must also be of the same structure. We thus investigate the inverse of the free Dirac Hamiltonian H_{FD} [see Eq. (7.18)], in the spirit of the Green function $\mathcal{G}_F$ in Sec. 7.4.2 and in analogy to Eq. (4.82). The free Dirac Green function $\mathcal{F}$ fulfills the equation

$$(H_{\mathrm{FD}} - E)\,\mathcal{F}(\vec{r}_1,\vec{r}_2,E) = \delta^{(3)}(\vec{r}_1 - \vec{r}_2)\;\mathbb{1}_{4\times 4}\,, \tag{7.176}$$

where $H_{\mathrm{FD}} = -\mathrm{i}\vec{\alpha}\cdot\vec{\nabla}_1 + \beta\,m$ acts on $\vec{r}_1$. If we choose the infinitesimal imaginary parts of the virtual-state energies as in Eq. (7.166), then $\mathcal{F}$ becomes identical to the Feynman Green function $\mathcal{G}_F$ defined in Eq. (7.173b). In the (2×2)-submatrix formalism, we have

$$\begin{pmatrix} (m-E)\;\mathbb{1}_{2\times 2} & -\mathrm{i}\vec{\sigma}\cdot\vec{\nabla}_{r_1} \\ -\mathrm{i}\vec{\sigma}\cdot\vec{\nabla}_{r_1} & (-m-E)\;\mathbb{1}_{2\times 2} \end{pmatrix} \mathcal{F}(\vec{r}_1,\vec{r}_2,E) = \delta^{(3)}(\vec{r}_1 - \vec{r}_2)\;\mathbb{1}_{4\times 4}\,. \tag{7.177}$$

The free Dirac Green function has an expansion in terms of angular-momentum components, which are characterized by the Dirac angular momentum quantum number κ defined in Eq. (6.81). According to Eq. (6.82), it is constructed from the orbital angular momentum and spin as follows,

$$\kappa = (-1)^{j+\ell+1/2}\left(j+\frac{1}{2}\right) = \begin{cases} -(j+1/2) & \text{for} \quad j=\ell+1/2 \\ (j+1/2) & \text{for} \quad j=\ell-1/2 \end{cases}. \tag{7.178}$$

The angular-momentum expansion of the Dirac–Coulomb Green function reads as

$$\begin{aligned} &\mathcal{F}(\vec{r}_1,\vec{r}_2,E) \\ &= \sum_{\kappa\mu}\begin{pmatrix} \mathcal{F}^{11}_\kappa(r_1,r_2,E)\chi_{\kappa\mu}(\hat{r}_1)\otimes\chi^+_{\kappa\mu}(\hat{r}_2) & -\mathrm{i}\mathcal{F}^{12}_\kappa(r_1,r_2,E)\chi_{\kappa\mu}(\hat{r}_1)\otimes\chi^+_{-\kappa\mu}(\hat{r}_2) \\ \mathrm{i}\mathcal{F}^{21}_\kappa(r_1,r_2,E)\chi_{-\kappa\mu}(\hat{r}_1)\otimes\chi^+_{\kappa\mu}(\hat{r}_2) & \mathcal{F}^{22}_\kappa(r_1,r_2,E)\chi_{-\kappa\mu}(\hat{r}_1)\otimes\chi^+_{-\kappa\mu}(\hat{r}_2) \end{pmatrix} \\ &= \sum_{\kappa}\begin{pmatrix} \mathcal{F}^{11}_\kappa(r_1,r_2,E)\,\pi_\kappa(\hat{r}_1,\hat{r}_2) & \mathcal{F}^{12}_\kappa(r_1,r_2,E)\,\mathrm{i}\vec{\sigma}\cdot\hat{r}_1\,\pi_{-\kappa}(\hat{r}_1,\hat{r}_2) \\ -\mathcal{F}^{21}_\kappa(r_1,r_2,E)\,\mathrm{i}\vec{\sigma}\cdot\hat{r}_1\,\pi_\kappa(\hat{r}_1,\hat{r}_2) & \mathcal{F}^{22}_\kappa(r_2,r_1,E)\,\pi_{-\kappa}(\hat{r}_1,\hat{r}_2) \end{pmatrix}. \end{aligned} \tag{7.179}$$

Here, the sum over κ runs over the positive and negative integers, with the exception of zero, and the sum over the magnetic projections runs over $\mu = -j,\dots,j$, where $j = |\kappa| - 1/2$. We have used the identity $\vec{\sigma}\cdot\hat{r}\,\chi_{\kappa\mu}(\hat{r}) = -\chi_{-\kappa\mu}(\hat{r})$ in the reformulation of the angular projection [see Eq. (6.85)]. The angular part involves the projection operator π_κ, originally defined in Eq. (6.96),

$$\pi_\kappa(\hat{r}_1,\hat{r}_2) = \sum_\mu \chi_{\kappa\mu}(\hat{r}_1)\otimes\chi^+_{\kappa\mu}(\hat{r}_2) = \frac{|\kappa|}{4\pi}\left\{\mathbb{1}_{2\times 2}\,P_{\ell(\kappa)}(\xi) + \frac{\mathrm{i}}{\kappa}\vec{\sigma}\cdot(\hat{r}_1\times\hat{r}_2)\,P'_{\ell(\kappa)}(\xi)\right\}, \tag{7.180}$$

where $\ell(\kappa) = \left|\kappa+\frac{1}{2}\right| - \frac{1}{2}$, and $\xi = \cos\gamma = \hat{r}_1\cdot\hat{r}_2$. We note the useful identities

$$\pi_\kappa(\hat{r}_1,\hat{r}_2) = \sum_\mu \chi_{\kappa\mu}(\hat{r}_1)\otimes\chi^+_{\kappa\mu}(\hat{r}_2), \tag{7.181a}$$

$$\vec{\sigma}\cdot\hat{r}_1\,\pi_{-\kappa}(\hat{r}_1,\hat{r}_2) = -\sum_\mu \chi_{\kappa\mu}(\hat{r}_1)\otimes\chi^+_{-\kappa\mu}(\hat{r}_2), \tag{7.181b}$$

$$\vec{\sigma}\cdot\hat{r}_1\,\pi_{\kappa}(\hat{r}_1,\hat{r}_2) = -\sum_\mu \chi_{-\kappa\mu}(\hat{r}_1)\otimes\chi^+_{\kappa\mu}(\hat{r}_2), \tag{7.181c}$$

$$\pi_{-\kappa}(\hat{r}_1,\hat{r}_2) = \sum_\mu \chi_{-\kappa\mu}(\hat{r}_1)\otimes\chi^+_{-\kappa\mu}(\hat{r}_2). \tag{7.181d}$$

From our discussion in Sec. 4.3, it is clear that a suitable representation for the Dirac–Coulomb Green functions can be found if one orders the arguments into

$$r_> = \max(r_1,r_2), \qquad r_< = \min(r_1,r_2). \tag{7.182}$$

The details of the derivation are analogous to the other Green functions discussed in Chap. 4, but with one important difference which is mentioned below. In agreement with Eq. (A.20) in Ref. [257], one finds for $r_1 = r_< = \min(r_1,r_2)$ and $r_2 = r_> = \max(r_1,r_2)$

$$\mathcal{F}^{11}_\kappa(r_<,r_>,E) = -(E+m)\,C\,j_{\ell(\kappa)}(\mathrm{i}C\,r_<)\,h^{(1)}_{\ell(\kappa)}(\mathrm{i}C\,r_>), \tag{7.183a}$$

$$\mathcal{F}^{12}_\kappa(r_<,r_>,E) = -\mathrm{i}\,C^2\,\frac{\kappa}{|\kappa|}\,j_{\ell(\kappa)}(\mathrm{i}C\,r_<)\,h^{(1)}_{\ell(-\kappa)}(\mathrm{i}C\,r_>), \tag{7.183b}$$

$$\mathcal{F}^{21}_\kappa(r_<,r_>,E) = -\mathrm{i}\,C^2\,\frac{\kappa}{|\kappa|}\,j_{\ell(-\kappa)}(\mathrm{i}C\,r_<)\,h^{(1)}_{\ell(\kappa)}(\mathrm{i}C\,r_>), \tag{7.183c}$$

$$\mathcal{F}^{22}_\kappa(r_<,r_>,E) = -(E-m)\,C\,j_{\ell(-\kappa)}(\mathrm{i}C\,r_<)\,h^{(1)}_{\ell(-\kappa)}(\mathrm{i}C\,r_>), \tag{7.183d}$$

where

$$C = \sqrt{m^2 - E^2}\,, \tag{7.183e}$$

and the $h_\ell^{(1)}(x) = j_\ell(x) + \mathrm{i}\, y_\ell(x)$ are spherical Hankel function of the first kind [159]. The definitions of the $\mathcal{F}$ functions are completed by the symmetry relations

$$\mathcal{F}_\kappa^{11}(r_>, r_<, E) = \mathcal{F}_\kappa^{11}(r_<, r_>, E)\,, \tag{7.183f}$$

$$\mathcal{F}_\kappa^{12}(r_>, r_<, E) = \mathcal{F}_\kappa^{21}(r_<, r_>, E)\,, \tag{7.183g}$$

$$\mathcal{F}_\kappa^{21}(r_>, r_<, E) = \mathcal{F}_\kappa^{12}(r_<, r_>, E)\,, \tag{7.183h}$$

$$\mathcal{F}_\kappa^{22}(r_>, r_<, E) = \mathcal{F}_\kappa^{22}(r_<, r_>, E)\,. \tag{7.183i}$$

Note that the functions $\mathcal{F}_\kappa^{11}(r_1, r_2, E)$ and $\mathcal{F}_\kappa^{22}(r_1, r_2, E)$ are continuous at the cusp $r_1 = r_2$, while the functions $\mathcal{F}_\kappa^{12}(r_1, r_2, E)$ and $\mathcal{F}_\kappa^{21}(r_1, r_2, E)$ have a kink (discontinuity). We recall from the derivation of the Schrödinger–Coulomb Green function [the result has been given in Eq. (4.108)] that the radial component is continuous at the cusp $r_1 = r_2$. The same applies to the photon Green functions discussed in Chap. 4 of Ref. [81]. Here, we encounter radial Green functions which are discontinuous at the cusp. This is due to the fact that the Dirac Hamiltonian, as opposed to the Schrödinger Hamiltonian, actually is a first-order differential operator, when expressed in coordinate space. The presence of the discontinuity is required by the necessity to generate a Dirac-δ, by a single differentiation. We recall the variable C defined in Eq. (7.183e). Due to the continuum spectrum, the free Dirac Green function has cuts for $E > m$ and $E < -m$; these are encoded in the variable $C = \sqrt{m^2 - E^2}$. The Feynman prescription can either be implemented by lending to E an infinitesimal imaginary part along a suitable integration contour, or, by the substitution $m \to m - \mathrm{i}\epsilon$. The latter ensures that the branch cuts occur for $E = \mathrm{Re}E - \mathrm{i}\epsilon$, for $\mathrm{Re}E > m$, and $E = \mathrm{Re}E + \mathrm{i}\epsilon$, for $\mathrm{Re}E < -m$. The branch cut of the square root is chosen along the negative real axis, the square root is calculated on the principal branch, which results in

$$\mathrm{Re}(C) = \mathrm{Re}(\sqrt{m^2 - E^2}) > 0\,. \tag{7.184}$$

The sum over κ can be performed for the free Dirac Green function, with the result

$$\mathcal{F}(\vec{r}_1, \vec{r}_2, E) = \left[\left(\frac{C}{r_{12}} + \frac{1}{r_{12}^2}\right)\mathrm{i}\,\vec{\alpha}\cdot\vec{r}_{12} + \beta m + E\right]\frac{\mathrm{e}^{-C\,r_{12}}}{4\pi r_{12}}\,, \tag{7.185}$$

where $r_{12} = |\vec{r}_1 - \vec{r}_2|$. However, the radial decomposition (7.179) is more useful in many practical calculations.

Higher powers of the free Dirac Green Function $\mathcal{F}^2$, which are sometimes needed in bound-state calculations, can be generated upon differentiation. E.g., one finds that

$$\mathcal{F}^2(\vec{r}_1, \vec{r}_2, E) = \left\langle \vec{r}_1 \left| \left(\frac{1}{H - E}\right)^2 \right| \vec{r}_2 \right\rangle = \frac{\partial}{\partial E}\left\langle \vec{r}_1 \left| \frac{1}{H - E} \right| \vec{r}_2 \right\rangle = \frac{\partial}{\partial E}\mathcal{F}(\vec{r}_1, \vec{r}_2, E)\,. \tag{7.186}$$

The angular-momentum decomposition reads as

$$\mathcal{F}^2(\vec{r}_1, \vec{r}_2, z) = \sum_{\kappa\mu} \begin{pmatrix} (\mathcal{F}^2)^{11}_\kappa(r_1, r_2, E)\, \pi_\kappa(\hat{r}_1, \hat{r}_2) & (\mathcal{F}^2)^{12}_\kappa(r_1, r_2, E)\, \mathrm{i}\vec{\sigma}\cdot\hat{r}_1\, \pi_{-\kappa}(\hat{r}_1, \hat{r}_2) \\ -(\mathcal{F}^2)^{21}_\kappa(r_1, r_2, E)\, \mathrm{i}\vec{\sigma}\cdot\hat{r}_1\, \pi_\kappa(\hat{r}_1, \hat{r}_2) & (\mathcal{F}^2)^{22}_\kappa(r_2, r_1, E)\, \pi_{-\kappa}(\hat{r}_1, \hat{r}_2) \end{pmatrix}. \tag{7.187}$$

Explicit expressions for the radial components can be derived by explicit differentiation of the functions given in Eq. (7.183) with respect to the energy. The following result is then easily obtained,

$$(\mathcal{F}^2)^{11}_\kappa(r_1, r_2, E) = \frac{m + 2E\kappa}{C^2}\mathcal{F}^{11}_\kappa - \frac{E}{m-E}\left(r_2\,\mathcal{F}^{12}_\kappa + r_1\,\mathcal{F}^{21}_\kappa\right), \tag{7.188a}$$

$$(\mathcal{F}^2)^{12}_\kappa(r_1, r_2, E) = -\frac{E}{m-E} r_1\,\mathcal{F}^{22}_\kappa - \frac{E}{m+E} r_2\,\mathcal{F}^{11}_\kappa, \tag{7.188b}$$

$$(\mathcal{F}^2)^{21}_\kappa(r_1, r_2, E) = -\frac{E}{m-E} r_2\,\mathcal{F}^{22}_\kappa - \frac{E}{m+E} r_1\,\mathcal{F}^{11}_\kappa, \tag{7.188c}$$

$$(\mathcal{F}^2)^{22}_\kappa(r_1, r_2, E) = -\frac{m + 2E\kappa}{C^2}\mathcal{F}^{22}_\kappa - \frac{E}{m+E}\left(r_2\,\mathcal{F}^{21}_\kappa + r_1\,\mathcal{F}^{12}_\kappa\right). \tag{7.188d}$$

We here suppress the arguments of the components on the right-hand sides. This result is easily obtained by direct differentiation of the coordinate space representation (7.183), and by using Eq. (7.118).

Often, in bound-state calculations, we need to extract divergences connected with the insertion of a Coulomb interaction into the propagator. This amounts to a replacement $\vec{\alpha}\cdot\vec{p} + \beta m \to \vec{\alpha}\cdot\vec{p} + \beta m + V$ in the propagator denominator, or, to the expansion $\mathcal{F} \to \mathcal{G} = \mathcal{F} - \mathcal{F}\,V\,\mathcal{F} + \mathcal{F}\,V\,\mathcal{F}\,V\,\mathcal{F} + \cdots$ for the Dirac–Coulomb Green function G. In many cases, the divergent terms can be described as we commute the potential outside of the free Green function, which, in turn, leads to a much simpler expression for the Green function that needs to be used in typical subtraction terms. Indeed, it is often useful to define

$$\mathcal{G}_V = \mathcal{F} - \frac{1}{2}\left\{V, \mathcal{F}^2\right\}, \qquad V = -\frac{Z\alpha}{r}. \tag{7.189}$$

Here, recall that the coordinate-space representation of the term $V\,\mathcal{F}^2$ is $V(\vec{r}_1)\mathcal{F}^2(\vec{r}_1, \vec{r}_2)$, while that of $\mathcal{F}^2\,V$ is $\mathcal{F}^2(\vec{r}_1, \vec{r}_2)\,V(\vec{r}_2)$; the two terms are not equal, and so V does not commute with $\mathcal{F}^2$. In the angular-momentum decomposition, one has

$$\mathcal{G}_V(\vec{r}_1, \vec{r}_2, z) = \sum_{\kappa\mu} \begin{pmatrix} \mathcal{G}^{11}_{V,\kappa}(r_1, r_2, E)\, \pi_\kappa(\hat{r}_1, \hat{r}_2) & \mathcal{G}^{12}_{V,\kappa}(r_1, r_2, E)\, \mathrm{i}\vec{\sigma}\cdot\hat{r}_1\, \pi_{-\kappa}(\hat{r}_1, \hat{r}_2) \\ -\mathcal{G}^{12}_{V,\kappa}(r_1, r_2, E)\, \mathrm{i}\vec{\sigma}\cdot\hat{r}_1\, \pi_\kappa(\hat{r}_1, \hat{r}_2) & \mathcal{G}^{22}_{V,\kappa}(r_1, r_2, E)\, \pi_{-\kappa}(\hat{r}_1, \hat{r}_2) \end{pmatrix}. \tag{7.190}$$

The components of $\mathcal{G}_V$ read as follows [we suppress the arguments (r_1, r_2, E)],

$$\mathcal{G}^{11}_{V,\kappa} = \mathcal{F}^{11}_{\kappa} + (Z\alpha)\,\frac{r_1 + r_2}{2\,r_1\,r_2\,C^2}\,\left\{(m + 2\,E\,\kappa)\,\mathcal{F}^{11}_{\kappa} - E\,(E + m)\,\left[r_2\,\mathcal{F}^{12}_{\kappa} + r_1\,\mathcal{F}^{21}_{\kappa}\right]\right\}\,, \tag{7.191a}$$

$$\mathcal{G}^{12}_{V,\kappa} = \mathcal{F}^{12}_{\kappa} + (Z\alpha)\,\frac{r_1 + r_2}{2\,r_1\,r_2\,C^2}\,\left\{E\,(E - m)\,r_2\,\mathcal{F}^{11}_{\kappa} - E\,(E + m)\,r_1\,\mathcal{F}^{22}_{\kappa}\right\}\,, \tag{7.191b}$$

$$\mathcal{G}^{21}_{V,\kappa} = \mathcal{F}^{21}_{\kappa} + (Z\alpha)\,\frac{r_1 + r_2}{2\,r_1\,r_2\,C^2}\,\left\{E\,(E - m)\,r_1\,\mathcal{F}^{11}_{\kappa} - E\,(E + m)\,r_2\,\mathcal{F}^{22}_{\kappa}\right\}\,, \tag{7.191c}$$

$$\mathcal{G}^{22}_{V,\kappa} = \mathcal{F}^{22}_{\kappa} + (Z\alpha)\,\frac{r_1 + r_2}{2\,r_1\,r_2\,C^2}\,\left\{-(m + 2\,E\,\kappa)\,\mathcal{F}^{22}_{\kappa} + E\,(E - m)\,\left[r_2\,\mathcal{F}^{21}_{\kappa} + r_1\,F^{12}_{\kappa}\right]\right\}\,. \tag{7.191d}$$

The structure for the angular-momentum expansion is the same as in Eq. (7.179). A Green function related to the one given in Eq. (7.191) has been calculated in Eq. (5.4) of Ref. [258].

7.5 Further Thoughts

Here are some suggestions for further thought.

(1) **Lorentz Group.** In view of Eqs. (7.27) and (7.30), explain the correspondence

$$-\frac{\mathrm{i}}{4}\,\sigma_{\alpha\beta} \leftrightarrow \frac{1}{2}\,M_{\alpha\beta} \tag{7.192}$$

on the basis of the commutator relations fulfilled by the matrices.

(2) **Lorentz Transformation and Dirac Adjoint.** Verify Eq. (7.34) by an explicit calculation.

(3) **Dirac Current.** Show Eq. (7.39), i.e., convince yourself that

$$J^\mu(x) = \overline{\psi}(x)\,\gamma^\mu\,\psi(x)\,, \qquad J'^\mu(x') = \Lambda^\mu{}_\nu\,J^\nu(x)\,, \qquad \partial_\mu J^\mu(x) = 0\,. \tag{7.193}$$

(4) **CPT Transform.** Verify all intermediate steps in Eq. (7.73) in detail.

(5) **C Transform.** Show that

$$\overline{\psi^{\mathcal{C}}}(x) = \overline{\psi}^{\mathcal{C}}(x)\,. \tag{7.194}$$

Convince yourself that the Dirac current is invariant under charge conjugation,

$$J^\mu(x) = e\overline{\psi}(x)\,\gamma^\mu\,\psi(x) = e\overline{\psi^{\mathcal{C}}}(x)\,\gamma^\mu\,\psi^{\mathcal{C}}(x)\,. \tag{7.195}$$

This result is counter-intuitive because the charge-conjugated current characterizes antiparticles, which have the opposite charge as compared to particles. In the second-quantized formalism, $\psi(x)$ denotes the field operator (7.140). Let us temporarily denote the second-quantized field operator by the symbol $\hat{\psi}(x)$. Show that, in the second-quantized formulation, one has

$$\hat{J}^\mu(x) = e\,\hat{\overline{\psi}}(x)\,\gamma^\mu\,\hat{\psi}(x) = -e\,\hat{\overline{\psi}}^{\mathcal{C}}(x)\,\gamma^\mu\,\hat{\psi}^{\mathcal{C}}(x)\,, \tag{7.196}$$

i.e., that the second-quantized current operator changes sign, in accordance with intuition. In the calculation, pay close attention to the sequence of the anticommuting field operators.

(6) **PT Transform.** On the basis of Eq. (7.71), show that

$$\psi^{\mathcal{PT}}(x) = \mathrm{i}\,\gamma^5\,\gamma^2\,\psi^*(x_{\mathcal{PT}})\,, \qquad \psi^{\mathcal{C}}(x) = \mathrm{i}\,\gamma^2\,\psi^*(x)\,, \tag{7.197}$$

so that

$$\psi^{\mathcal{PT}}(x) = \gamma^5\,\psi^{\mathcal{C}}(-x)\,. \tag{7.198}$$

Knowing that $\psi^{\mathcal{C}}(x)$ fulfills the charge-conjugated Dirac equation, attempt to determine whether $\psi^{\mathcal{PT}}(x)$ fulfills the charge-conjugated, or the original, Dirac equation.

(7) **Dirac Algebra.** Show Eq. (7.80).

(8) **Dirac Field and Anticommutators.** Consider the canonical anticommutation relations given in Eq. (7.144) and the quantized Lagrangian given in Eq. (7.78). From the quantized Lagrangian (7.78), derive the field momentum, and show that the equal-time anticommutator given in Eq. (7.159) is compatibility with the canonical anticommutation relation of the field and its conjugate momentum. For the anticommutator, consult Eqs. (7.161) and (10.282).

(9) **Radial Dirac Equation for a Massless Particle.** Consider the radial equations governing a Dirac eigenstate, given in Eq. (7.113). Solve the first of these equations for $g(r)$, and insert the result into the second equation. You will find a second-order differential equation for $f(r)$. Also, solve the second of the given differential equations for $f(r)$, and insert the result into the first equation. You should find a second-order differential equation for $g(r)$. Verify that both obtained second-order differential equations are compatible with the explicit solutions given in Eq. (7.114), and with Eq. (7.116).

(10) **Neutrino Solutions of the Radial Dirac Equation.** Consider Eqs. (7.110), (7.117), (7.118) and (7.126). Show, by explicit calculation, that

$$H_{\mathrm{FMD}}\,\psi^{(\nu)}_{Ej\mu}(\vec{r}) = E\,\psi^{(\nu)}_{Ej\mu}(\vec{r}) = \begin{pmatrix} 0 & \vec{\sigma}\cdot\vec{p} \\ \vec{\sigma}\cdot\vec{p} & 0 \end{pmatrix}\psi^{(\nu)}_{Ej\mu}(\vec{r})\,. \tag{7.199}$$

Show that, in view of the structure

$$\psi^{(\nu)}_{Ej\mu}(\vec{r}) = \begin{pmatrix} \nu_{Ej\mu}(\vec{r}) \\ -\nu_{Ej\mu}(\vec{r}) \end{pmatrix}, \tag{7.200}$$

the property (7.199) implies that $\vec{\Sigma}\cdot\vec{p}\,\psi^{(\nu)}_{Ej\mu}(\vec{r}) = -|\vec{p}|\,\psi^{(\nu)}_{Ej\mu}(\vec{r})$, i.e., that $\psi^{(\nu)}_{Ej\mu}(\vec{r})$ is a negative-helicity eigenstate.

(11) **Antineutrino Wavefunction.** Write the analogy of Eq. (7.126) for an antineutrino solution. Observe that antineutrinos have negative energy, negative chirality, but positive helicity.

(12) **Discrete Symmetries and Second Quantization.** Consider the Lagrangian for the Dirac field, given, in Eq. (7.79),

$$\mathcal{L}(x) = \overline{\psi}(x)\,(\mathrm{i}\,\gamma^\mu\partial_\mu - m)\,\psi(x)\,, \tag{7.201}$$

but this time interpreting the variables $\psi(x)$ and $\overline{\psi}(x)$ as second-quantized field operators, in the sense of Eqs. (7.140) and (7.141). Show that, upon subsequent parity ($\mathcal{P}$), time-reversal($\mathcal{T}$), and charge-conjugation ($\mathcal{C}$) transformations, the Lagrangian restores its original sign, because of an exchange in the sequence of field operators. In view of the anticommutators given in Eqs. (7.143) and (7.144), show that various minus signs encountered in the derivation of the CPT transformation in the first-quantized theory are compensated, and the overall sign of Lagrangian is fully restored. Show that this conclusion is consistent with arguments presented in Chap. 7 of Ref. [250], and, under the inclusion of gravitational coupling, with the conclusions of Ref. [251].

(13) **Dirac Propagator.** Perform an explicit Fourier transformation of Eq. (7.150) with respect to time, and show that you obtain Eq. (7.153) after formally introducing the Fourier backtransformation.

(14) **Tachyon Propagator.** Study pertinent literature references [252, 253, 259] and repeat the derivation in Sec. 7.4.1 for the tachyonic spin-1/2 field operators.

(15) **Field Anticommutator.** In regard to Eq. (7.159), one might ask if the spatial derivatives in the expression $\not{\partial}\Delta(x-y)$ contribute to the field anticommutator at $x^0 = y^0$. Convince yourself that for $x_0 = y_0$, the expression $\mathrm{e}^{-\mathrm{i}k\cdot(x-y)} - \mathrm{e}^{\mathrm{i}k\cdot(x-y)} = 2\mathrm{i}\,\sin(\vec{k}\cdot\vec{r})$ generates a factor proportional to $k^i\,\cos(\vec{k}\cdot\vec{r})$ upon differentiation with respect to x^i, and the integral vanishes upon symmetrization with respect to the directions of $\vec{k}$.

(16) **Green Function and Physical Dimension.** Perform a quick check of the physical dimension of Eq. (7.183). Convince yourself that in natural units, the physical dimension of F in Eq. (7.176) has to be equal to the mass squared, because the Hamiltonian on the left-hand side of Eq. (7.176) has the dimension of mass, whereas the Dirac-δ on the right-hand side has the dimension of mass to the power of three. Then, convince yourself that the expressions in Eq. (7.183) have exactly this mass dimension.

(17) **Green Function and Physical Dimension.** Show Eqs. (7.188) and (7.191).

Chapter 8

Dirac Equation for Bound States, Lasers and Gravity

8.1 Overview

In the current chapter, we explore the Dirac equation in situations where the particle is coupled to electromagnetic fields, or, is moving in a curved space-time. The unifying element here is the concept of the covariant derivative. Namely, the application of a local gauge transformation multiplies the wave function by an unobservable phase factor, which can be different at any point in space-time. The application of the differential operator yields two terms, one of which amounts to the derivative, multiplied by said phase factor, while the other is a residual term which is absorbed in a gauge-transformed four-vector potential (electromagnetic coupling) or a gauge-transformed connection matrix (gravitational coupling). The former gauge transformation has been discussed, e.g., in Chap. 1 of Ref. [81], and is recalled in Sec. 8.2.1. Dirac particles in Coulomb fields are discussed in Sec. 8.2. The laser-dressed propagator is discussed in Sec. 8.3. Gravitational coupling is discussed in Sec. 8.4. In the latter context, a gauge transformation amounts to a change of the local Lorentz frame.

For Dirac particles in strong, plane-wave laser fields (see Sec. 8.3), we discuss a case of electromagnetic coupling, yet different from Sec. 8.2, in that the coupling concerns a time-varying electromagnetic (plane-wave) field. There is a certain analogy between the Dirac–Coulomb propagator (see Sec. 8.2.5) and the Dirac–Volkov propagator (see Sec. 8.3.2). Note that the corresponding problem of evaluating the dressed propagator, for gravitational coupling, has not yet been solved yet. Throughout this chapter, the Dirac wave function ψ should be taken as the wave function coming from the first-quantized theory, not as the second-quantized field operator.

8.2 Dirac Equation and Coulomb Field

8.2.1 *Electromagnetic Covariant Derivative*

The four-vector potential is $A^\mu = (\Phi, \vec{A})$. Under a gauge transformation of the electromagnetic scalar and vector potential (see Chap. 1 of Ref. [81]), they transform as follows,

$$\vec{A}' = \vec{A} + \vec{\nabla}\Lambda\,, \qquad \Phi' = \Phi - \partial_t\Lambda\,, \qquad A'^{\mu} = A^{\mu} - \partial^{\mu}\Lambda\,, \tag{8.1}$$

where again, throughout this entire chapter, we use natural units with $\hbar = c = \epsilon_0 = 1$. The gauge transformation of the wave function proceeds as follows,

$$\psi' = \exp(\mathrm{i}\,e\,\Lambda)\,\psi\,. \tag{8.2}$$

One defines the covariant derivative $\mathcal{D}_\mu$ with the properties,

$$\mathcal{D}_\mu = \partial_\mu + \mathrm{i}e\,A_\mu\,. \tag{8.3}$$

$$\begin{aligned}\mathcal{D}'_\mu\,\psi' &= \left(\partial_\mu + \mathrm{i}e\,A'_\mu\right)\psi' = \left(\partial_\mu + \mathrm{i}e\,A_\mu - \mathrm{i}e\,\partial_\mu\Lambda\right)\,\exp(\mathrm{i}\,e\,\Lambda)\,\psi \\ &= \exp(\mathrm{i}\,e\,\Lambda)\,\left(\partial_\mu + \mathrm{i}e\,A_\mu\right)\,\psi = \left(\mathcal{D}_\mu\,\psi\right)'\,.\end{aligned} \tag{8.4}$$

The covariant derivative of the gauge-transformed wave function is equal to the gauge transform of the covariant derivative. Introducing the gauge transformation $G(\Lambda)$, we can write this property as

$$G(\Lambda) = \exp\left(\mathrm{i}\,e\,\Lambda\right)\,, \qquad G(\Lambda)^{-1}\,\mathcal{D}'_\mu\,G(\Lambda)\,\psi = \mathcal{D}_\mu\,\psi\,. \tag{8.5}$$

For the Coulomb potential, one has

$$A^\mu = (\Phi, \vec{A}) = \left(-\frac{Z\,e}{4\pi r}, \vec{0}\right)\,, \qquad eA^0 = -\frac{Z\,e^2}{4\pi r} = -\frac{Z\,\alpha}{r}\,. \tag{8.6}$$

If $\vec{A} = \vec{0}$, then the electromagnetically coupled Dirac equation (7.44), which we recall for convenience,

$$\left[\gamma^\mu\,(\mathrm{i}\partial_\mu - e\,A_\mu) - m\right]\,\psi = 0\,, \tag{8.7}$$

can thus be brought into the following form [see Eq. (7.17)]

$$\gamma^0\left[\gamma^0\,(\mathrm{i}\partial_0 - e\,A_0) - \mathrm{i}\vec{\gamma}\cdot\vec{\nabla} - m\right]\,\psi = 0\,, \tag{8.8}$$

from which the Dirac–Coulomb Hamiltonian, which has the property $\mathrm{i}\partial_t\psi = H_{\mathrm{DC}}\,\psi$, can easily be derived,

$$H_{\mathrm{DC}} = \vec{\alpha}\cdot\vec{p} + \beta\,m + V\,, \qquad V = -\frac{Z\alpha}{r}\,, \qquad H_{\mathrm{DC}}\,\psi = E\psi\,. \tag{8.9}$$

The Dirac–Coulomb Hamiltonian is extremely important for bound-state problems.

8.2.2 *Dirac–Coulomb Bound-State Wave Functions*

The Dirac–Coulomb Hamiltonian, which we have just encountered in Eq. (8.9), should still be recalled at the start of the current discussion,

$$H_{\mathrm{DC}} = \vec{\alpha}\cdot\vec{p} + \beta\,m + V\,, \qquad V = -\frac{Z\alpha}{r}\,. \tag{8.10}$$

The $\vec{\alpha}$ and β matrices have been defined in Eq. (7.19). We now need to find the result of the Hamiltonian operator acting onto the standard solution with upper component $f(r)\,\chi_{\kappa\,\mu}(\hat{r})$ and lower component $\mathrm{i}g(r)\,\chi_{-\kappa\,\mu}(\hat{r})$, where we recall the discussion in Sec. 6.4.1 and Sec. 7.3.3 [see Eq. (7.111)]. Our standard *ansatz* is

$$\psi = \begin{pmatrix} f(r)\,\chi_{\kappa\mu}(\theta,\varphi) \\ \mathrm{i}\,g(r)\,\chi_{-\kappa\mu}(\theta,\varphi) \end{pmatrix} . \tag{8.11}$$

We give this derivation in great detail, following our earlier treatment of the massive but free Dirac equation,

$$\begin{aligned} H_{\mathrm{DC}}\psi(\vec{r}) &= \begin{pmatrix} m - \dfrac{Z\alpha}{r} & (\vec{\sigma}\cdot\hat{r})\left(-\mathrm{i}\dfrac{\partial}{\partial r} + \mathrm{i}\dfrac{\vec{\sigma}\cdot\vec{L}}{r}\right) \\ (\vec{\sigma}\cdot\hat{r})\left(-\mathrm{i}\dfrac{\partial}{\partial r} + \mathrm{i}\dfrac{\vec{\sigma}\cdot\vec{L}}{r}\right) & -m - \dfrac{Z\alpha}{r} \end{pmatrix} \begin{pmatrix} f(r)\,\chi_{\kappa\,\mu}(\hat{r}) \\ \mathrm{i}g(r)\,\chi_{-\kappa\,\mu}(\hat{r}) \end{pmatrix} \\ &= \begin{pmatrix} \left[\left(-\dfrac{\partial}{\partial r} + \dfrac{1}{r}\,(\kappa - 1)\right) g(r) + \left(m - \dfrac{Z\alpha}{r}\right) f(r)\right] \chi_{\kappa\,\mu}(\hat{r}) \\ \mathrm{i}\left[\left(\dfrac{\partial}{\partial r} + \dfrac{1}{r}\,(\kappa + 1)\right) f(r) - \left(m + \dfrac{Z\alpha}{r}\right) g(r)\right] \chi_{-\kappa\,\mu}(\hat{r}) \end{pmatrix} = E\,\psi(\vec{r})\,. \end{aligned} \tag{8.12}$$

We have used Eq. (6.81). The radial Dirac–Coulomb equations are thus given as follows,

$$\left(-\frac{\partial}{\partial r} + \frac{1}{r}\,(\kappa - 1)\right) g(r) - \left(E - m + \frac{Z\alpha}{r}\right) f(r) = 0\,, \tag{8.13a}$$

$$\left(\frac{\partial}{\partial r} + \frac{1}{r}\,(\kappa + 1)\right) f(r) - \left(E + m + \frac{Z\alpha}{r}\right) g(r) = 0\,. \tag{8.13b}$$

In this form, the equations are amenable to further manipulation and simplification. Inserting one equation into the other, we could obtain a second-order differential equation for either $f(r)$ or $g(r)$. However, these equations do not have immediate analytic solutions. The radial wave functions $f(r)$ and $g(r)$ in the bound-state solutions are actually composed of two, not one, Laguerre polynomials in the terminating hypergeometric functions which describe the bound states. Therefore, we proceed in two steps. In the first, we perform a linear transformation on the $f(r)$ and $g(r)$ to write them in terms of two other functions $h_1(r)$ and $h_2(r)$. The linear transformation depends on a free parameter X. The parameter X is fixed so that the two resulting equations for h_1 and h_2 take a particularly simple form: One of the functions appears only with a simple prefactor, independent of the radial variable, while the other function has both a differential operator acting on it, as well as an r-dependent prefactor. From these two equations, one can immediately form two second-order equations which take the form of Whittaker's differential equations, and the solution can be written in compact form. We then find the criterion for the two Laguerre polynomials to terminate, which is equivalent to the relativistic quantization condition and leads to the relativistic Dirac formula for the energy levels. The second step involves the calculation of normalization integrals and will be discussed in the following.

We thus introduce a free parameter X, and write the functions $h_1(r)$ and $h_2(r)$ as follows,

$$\begin{pmatrix} h_1(r) \\ h_2(r) \end{pmatrix} = \begin{pmatrix} 1 & X \\ X & 1 \end{pmatrix} \begin{pmatrix} f(r) \\ g(r) \end{pmatrix}, \qquad \begin{pmatrix} f(r) \\ g(r) \end{pmatrix} = \frac{1}{1-X^2} \begin{pmatrix} 1 & -X \\ -X & 1 \end{pmatrix} \begin{pmatrix} h_1(r) \\ h_2(r) \end{pmatrix}, \tag{8.14}$$

where the latter equation is obtained by matrix inversion. Inserting these new functions into Eq. (8.13), and leaving X as a free parameter, we obtain

$$\left(\frac{\kappa - 1 + XZ\alpha}{r} + X(E-m) - \partial_r \right) h_2(r) + \left(m-E - \frac{Z\alpha + X(\kappa-1)}{r} + X\partial_r \right) h_1(r) = 0, \tag{8.15a}$$

$$\left(\frac{\kappa + 1 + XZ\alpha}{r} + X(E+m) + \partial_r \right) h_1(r) - \left(m+E + \frac{Z\alpha + X(\kappa+1)}{r} + X\partial_r \right) h_2(r) = 0. \tag{8.15b}$$

We now multiply both Eqs. (8.15a) and (8.15b) by X, solve them for the expressions $X\partial_r h_1(r)$ and $X\partial_r h_2(r)$, respectively, and re-insert these results into Eq. (8.15). The resulting are found as

$$\begin{aligned} &\left[(1-X^2)m - (1+X^2)E - \frac{(1+X^2)Z\alpha + 2X\kappa}{r} \right] h_1(r) \\ &\quad + \left[\frac{(1+X^2)\kappa + 2XZ\alpha - (1-X^2)}{r} + 2XE - (1-X^2)\frac{\partial}{\partial r} \right] h_2(r) = 0, \end{aligned} \tag{8.16a}$$

$$\begin{aligned} &\left[\frac{(1+X^2)\kappa + 2XZ\alpha + (1-X^2)}{r} + 2XE + (1-X^2)\frac{\partial}{\partial r} \right] h_1(r) \\ &\quad - \left[(1+X^2)E + (1-X^2)m + \frac{(1+X^2)Z\alpha + 2X\kappa}{r} \right] h_2(r) = 0. \end{aligned} \tag{8.16b}$$

Dividing both equations by $(1-X^2)$, one obtains two equations with somewhat simplified prefactors,

$$\begin{aligned} &\left[m - \frac{1+X^2}{1-X^2}E - \frac{(1+X^2)Z\alpha + 2X\kappa}{(1-X^2)r} \right] h_1(r) \\ &\quad + \left[\frac{\frac{1+X^2}{1-X^2}\kappa + \frac{2X}{1-X^2}Z\alpha - 1}{r} + \frac{2X}{1-X^2}E - \frac{\partial}{\partial r} \right] h_2(r) = 0, \end{aligned} \tag{8.17a}$$

$$\begin{aligned} &\left[\frac{\frac{1+X^2}{1-X^2}\kappa + \frac{2X}{1-X^2}Z\alpha + 1}{r} + \frac{2X}{1-X^2}E + \frac{\partial}{\partial r} \right] h_1(r) \\ &\quad - \left[\frac{1+X^2}{1-X^2}E + m + \frac{(1+X^2)Z\alpha + 2X\kappa}{(1-X^2)r} \right] h_2(r) = 0. \end{aligned} \tag{8.17b}$$

We now use our freedom to choose X so that $(1+X^2)Z\alpha + 2X\kappa = 0$. One of the two solutions of the quadratic equation for X reads as

$$X = \frac{1}{Z\alpha}(\gamma - \kappa) = -\frac{\kappa}{|\kappa|}\sqrt{\frac{\kappa-\gamma}{\kappa+\gamma}}, \qquad \gamma = \sqrt{\kappa^2 - (Z\alpha)^2}. \tag{8.18}$$

The variable X is positive when κ is negative, and negative when κ is positive. The following relations fulfilled by the variable X are not completely trivial to derive,

$$X^2 = \frac{2\kappa^2}{(Z\alpha)^2} - \frac{2\kappa\gamma}{(Z\alpha)^2} - 1\,, \qquad 1 - X^2 = \frac{2\gamma}{\kappa+\gamma}\,, \tag{8.19a}$$

$$\frac{1+X^2}{1-X^2} = \frac{\frac{2\kappa}{(Z\alpha)^2}(\kappa-\gamma)}{\frac{2\gamma}{\kappa+\gamma}} = \frac{\kappa(\kappa^2-\gamma^2)}{(Z\alpha)^2\gamma} = \frac{\kappa}{\gamma}\,, \tag{8.19b}$$

$$\frac{2X}{1-X^2} = \frac{\frac{2}{Z\alpha}(\gamma-\kappa)}{\frac{2\gamma}{\kappa+\gamma}} = \frac{\gamma^2-\kappa^2}{(Z\alpha)\gamma} = -\frac{Z\alpha}{\gamma}\,. \tag{8.19c}$$

With these useful relations, we can directly replace the corresponding terms in Eq. (8.17) and obtain

$$\left[m - \frac{\kappa}{\gamma}E\right] h_1(r) + \left[\frac{\kappa^2 - (Z\alpha)^2 - \gamma}{\gamma r} - \frac{Z\alpha}{\gamma}E - \frac{\partial}{\partial r}\right] h_2(r) = 0\,, \tag{8.20a}$$

$$\left[\frac{\kappa^2 - (Z\alpha)^2 + \gamma}{\gamma r} - \frac{Z\alpha}{\gamma}E + \frac{\partial}{\partial r}\right] h_1(r) - \left[\frac{\kappa}{\gamma}E + m\right] h_2(r) = 0\,. \tag{8.20b}$$

These relations can trivially be simplified upon observing that $\kappa^2 - (Z\alpha)^2 = \gamma^2$. As envisaged, the function $h_1(r)$ in Eq. (8.20a) carries only a constant prefactor, while the same is true for the function $h_2(r)$ in Eq. (8.20b). We can now solve Eq. (8.20a) for $h_1(r)$, with the result

$$h_1(r) = \left(m - \frac{\kappa}{\gamma}E\right)^{-1} \left[\frac{Z\alpha}{\gamma}E - \frac{(\gamma-1)}{r} + \frac{\partial}{\partial r}\right] h_2(r)\,, \tag{8.21}$$

and use the result in Eq. (8.20b). Likewise, we can solve Eq. (8.20b) for $h_2(r)$, with the result

$$h_2(r) = \left(m + \frac{\kappa}{\gamma}E\right)^{-1} \left[\frac{(\gamma+1)}{r} - \frac{Z\alpha}{\gamma}E + \frac{\partial}{\partial r}\right] h_1(r)\,, \tag{8.22}$$

and use the result in Eq. (8.20a). We thus find the uncoupled second-order differential equations fulfilled by $h_1(r)$ and $h_2(r)$ separately,

$$\begin{aligned}\left[m - \frac{\kappa}{\gamma}E\right] h_1(r) &+ \left(m + \frac{\kappa}{\gamma}E\right)^{-1} \left[-\frac{\partial^2}{\partial r^2} + \left(\frac{\gamma-1}{r} - \frac{(\gamma+1)}{r}\right)\frac{\partial}{\partial r}\right.\\ &\left.+ \frac{\gamma(\gamma+1)}{r^2} - \frac{Z\alpha}{r}2E + \left(\frac{Z\alpha}{\gamma}E\right)^2\right] h_1(r) = 0\,,\end{aligned} \tag{8.23a}$$

$$\begin{aligned}\left(m - \frac{\kappa}{\gamma}E\right)^{-1} &\left[\frac{\partial^2}{\partial r^2} + \left(\frac{\gamma+1}{r} - \frac{(\gamma-1)}{r}\right)\frac{\partial}{\partial r} - \frac{\gamma(\gamma-1)}{r^2}\right.\\ &\left.+ \frac{Z\alpha}{r}2E - \left(\frac{Z\alpha}{\gamma}E\right)^2\right] h_2(r) - \left[\frac{\kappa}{\gamma}E + m\right] h_2(r) = 0\,.\end{aligned} \tag{8.23b}$$

We have simplified the equations by recalling that $\gamma^2 = \kappa^2 - (Z\alpha)^2$. Furthermore, it is useful to multiply Eq. (8.23a) by $m + \frac{\kappa}{\gamma}E$ and Eq. (8.23b) by $m - \frac{\kappa}{\gamma}E$. This yields

$$\left[\frac{\partial^2}{\partial r^2}+\frac{2}{r}\frac{\partial}{\partial r}-\frac{\gamma(\gamma+1)}{r^2}\right.$$
$$\left.+\frac{Z\alpha}{r}2E-\left(\frac{Z\alpha}{\gamma}E\right)^2+\left(\frac{\kappa^2}{\gamma^2}E^2-m^2\right)\right]h_1(r)=0\,, \tag{8.24a}$$
$$\left[\frac{\partial^2}{\partial r^2}+\frac{2}{r}\frac{\partial}{\partial r}-\frac{\gamma(\gamma-1)}{r^2}\right.$$
$$\left.+\frac{Z\alpha}{r}2E-\left(\frac{Z\alpha}{\gamma}E\right)^2+\left(\frac{\kappa^2}{\gamma^2}E^2-m^2\right)\right]h_2(r)=0\,. \tag{8.24b}$$

Finally, one arrives at two very compact equations,

$$\left(\frac{\partial^2}{\partial r^2}+\frac{2}{r}\frac{\partial}{\partial r}-\frac{\gamma(\gamma+1)}{r^2}+\frac{Z\alpha}{r}2E-\left(m^2-E^2\right)\right)h_1(r)=0\,, \tag{8.25a}$$
$$\left(\frac{\partial^2}{\partial r^2}+\frac{2}{r}\frac{\partial}{\partial r}-\frac{\gamma(\gamma-1)}{r^2}+\frac{Z\alpha}{r}2E-\left(m^2-E^2\right)\right)h_2(r)=0\,. \tag{8.25b}$$

After all the manipulations, we have arrived at differential equations that can be seen as a generalization of the Schrödinger equation. We now transform $h_i(r)\to v_i(r)/r$ for $i=1,2$ in order to get rid of the first-order derivative. We additionally define, with Eq. (7.183e),

$$C=\sqrt{m^2-E^2}\,, \tag{8.26}$$

and find

$$\left(\frac{\partial^2}{\partial r^2}-\frac{\gamma(\gamma+1)}{r^2}+\frac{Z\alpha}{r}2E-C^2\right)v_1(r)=0\,, \tag{8.27a}$$
$$\left(\frac{\partial^2}{\partial r^2}-\frac{\gamma(\gamma-1)}{r^2}+\frac{Z\alpha}{r}2E-C^2\right)v_2(r)=0\,. \tag{8.27b}$$

We now rescale the variables according to the relations

$$\rho=2C\,r\,,\qquad k=Z\alpha\frac{E}{C}=\frac{Z\alpha\,E}{\sqrt{m^2-E^2}}\,. \tag{8.28}$$

The functions are transformed according to $v_i(r)=v_i(\rho/(2C))=\overline{v}_i(\rho)\equiv v_i(\rho)$. In the last step, we simply redefine $\overline{v}_i(\rho)$ to be equal to $v_i(\rho)$. The rescaling turns Eq. (8.27) into a dimensionless equation of the form

$$\left[\frac{\partial^2}{\partial\rho^2}-\frac{\gamma(\gamma+1)}{\rho^2}+\frac{k}{\rho}-\frac{1}{4}\right]v_1(\rho)=\left[\frac{\partial^2}{\partial\rho^2}-\frac{1}{4}+\frac{k}{\rho}+\frac{\frac{1}{4}-(\gamma+\frac{1}{2})^2}{\rho^2}\right]v_1(\rho)=0\,, \tag{8.29a}$$
$$\left[\frac{\partial^2}{\partial\rho^2}-\frac{\gamma(\gamma-1)}{\rho^2}+\frac{k}{\rho}-\frac{1}{4}\right]v_2(\rho)=\left[\frac{\partial^2}{\partial\rho^2}-\frac{1}{4}+\frac{k}{\rho}+\frac{\frac{1}{4}-(\gamma-\frac{1}{2})^2}{\rho^2}\right]v_2(\rho)=0\,. \tag{8.29b}$$

These equations are related to each other by the replacement $\gamma\to\gamma-1$. This replacement indirectly accounts for the fact that a state of defined j (total angular momentum of the electron, defined as the sum of the orbital angular momentum and the spin) is made up of a combination of ℓ and $\ell - 1$. The component with $j = \ell - 1/2$ is obtained by subtracting the spin, and the component with $j = \ell + 1/2$ is obtained by adding the spin.

The solution to Eq. (8.29a) above is given by a Whittaker function, which was originally discussed in Sec. 4.2.2,

$$v_1(\rho) = M_{k,\gamma+\frac{1}{2}}(\rho) = \mathrm{e}^{-\frac{1}{2}\rho}\,\rho^{\gamma+1}\,{}_1F_1(\gamma+1-k, 2\gamma+2, \rho)\,, \tag{8.30}$$

whereas the solution of Eq. (8.29b) is obtained after the replacement $\gamma \to \gamma - 1$, namely,

$$v_2(\rho) = M_{k,\gamma-\frac{1}{2}}(\rho) = \mathrm{e}^{-\frac{1}{2}\rho}\,\rho^{\gamma}\,{}_1F_1(\gamma-k, 2\gamma, \rho)\,. \tag{8.31}$$

If at least one of the solutions is supposed to terminate (we can set the other one equal to zero), then we must demand that the first argument of the hypergeometric functions be equal to zero, or to a negative integer n_r, where n_r could be interpreted as a relativistic principal quantum number. This already determines the relativistic spectrum of hydrogenlike ions,

$$-n_r = \gamma - k = \gamma - \frac{Z\alpha\,E}{\sqrt{m^2 - E^2}}\,, \qquad E = \frac{m}{\sqrt{1 + \dfrac{(Z\alpha)^2}{(n_r+\gamma)^2}}}\,, \tag{8.32}$$

with $n_r = 0, 1, 2, 3, \ldots$ and $\gamma = \sqrt{\kappa^2 - (Z\alpha)^2}$. In the special case n_r, we set v_1 equal to zero and convince ourselves that the energy of the ground state of hydrogenlike ions reads as

$$n_r = 0\,, \qquad \kappa = -1, \qquad E = m\sqrt{1 - (Z\alpha)^2}\,. \tag{8.33}$$

A Taylor expansion for $Z\alpha \to 0$ gives rise to the expansion $E \approx m - (Z\alpha)^2 m/2$, where the first term is the rest-mass energy and the second term is equal to the nonrelativistic (Schrödinger) binding energy. We can thus replace $k \to n_r + \gamma$ in the Whittaker functions (8.30) and (8.31),

$$v_1(\rho) = M_{n_r+\gamma,\gamma+\frac{1}{2}}(\rho) = \mathrm{e}^{-\frac{1}{2}\rho}\,\rho^{\gamma+1}\,\frac{\Gamma(n_r)\,\Gamma(2\gamma+2)}{\Gamma(n_r+2\gamma+1)}\,L_{n_r-1}^{2\gamma+1}(\rho)\,, \tag{8.34a}$$

$$v_2(\rho) = M_{n_r+\gamma,\gamma-\frac{1}{2}}(\rho) = \mathrm{e}^{-\frac{1}{2}\rho}\,\rho^{\gamma}\,\frac{\Gamma(n_r+1)\,\Gamma(2\gamma)}{\Gamma(n_r+2\gamma)}\,L_{n_r}^{2\gamma-1}(\rho)\,, \tag{8.34b}$$

where we convert to a notation with associated Laguerre polynomials. The functions $h_1(r)$ and $h_2(r)$ are obtained from $v_1(\rho)$ and $v_2(\rho)$ by a simple division through ρ. Recalling that $\rho = 2\,C\,r$, we have

$$h_1(r) = N_1\,\mathrm{e}^{-C\,r}\,(2Cr)^{\gamma}\,L_{n_r-1}^{2\gamma+1}(2Cr)\,, \tag{8.35a}$$

$$h_2(r) = N_2\,\mathrm{e}^{-C\,r}\,(2Cr)^{\gamma-1}\,L_{n_r}^{2\gamma-1}(2Cr)\,, \tag{8.35b}$$

where we recall that $C = \sqrt{m^2 - E^2}$. The Γ functions in Eqs. (8.34a) and (8.34b) are absorbed in the normalization constants N_1 and N_2 implicitly defined in Eqs. (8.35a) and (8.35b).

In the second step of the calculation, we now have to calculate the normalization constants. This is nontrivial, because we have already exhausted the freedom of choice with regard to the parameter X in the first step. Now, we have two free "relative" normalization constants N_1 and N_2 in front of the two independent solutions $h_1(r)$ and $h_2(r)$, and we also have an overall normalization constant in front of the $f(r)$ and $g(r)$, which, in turn, constitute linear combinations of the $h_1(r)$ and $h_2(r)$. In order to fix the "relative normalization constants", we have to go back to the defining first-order, coupled differential equations which link the $h_1(r)$ to the $h_2(r)$. This leads to a first relation which needs to be fulfilled by N_1 and N_2. The overall normalization condition for the bound-state wave function then leads to a condition which finally fixes the constants N_1 and N_2 to defined values.

We start by observing that an expansion of Eq. (8.22) about the point $r = 0$ leads to the condition

$$n_r(2\gamma + n_r)\, C\, N_1 = (E\,\kappa + m\,\gamma)\, N_2\,. \tag{8.36}$$

In view of the relation

$$f(r)^2 + g(r)^2 = \frac{1+X^2}{(1-X^2)^2}\left(h_1(r)^2 + h_2(r)^2 - \frac{4X}{1+X^2}\, h_1(r)\, h_2(r)\right), \tag{8.37}$$

we can write with the normalization integral with the help of Eq. (8.19),

$$\int \mathrm{d}^3 r\, |\psi(\vec{r})|^2 = \int_0^\infty \mathrm{d}r\, r^2\, \left[f(r)^2 + g(r)^2\right] \tag{8.38}$$

$$= \frac{\kappa\,(\kappa+\gamma)}{2\gamma^2} \int_0^\infty \mathrm{d}r\, r^2 \left[h_1(r)^2 + \frac{2Z\alpha}{\kappa}\, h_1(r)\, h_2(r) + h_2(r)^2\right]. \tag{8.39}$$

This integral has three terms, proportional to N_1^2, N_2^2, and $N_1\, N_2$, and evaluates to

$$\frac{\kappa\,(\kappa+\gamma)\,\Gamma(2\gamma+n_r)}{8\gamma\, n_r!\, C^3}\big[(\gamma+n_r)\,(2\gamma+n_r)\, n_r\, N_1^2 - \frac{2Z\alpha}{\kappa}(2\gamma+n_r)\, n_r\, N_1\, N_2 + (\gamma+n_r)\, N_2^2\big] = 1\,. \tag{8.40}$$

Let us do a dimensional analysis. All factors are dimensionless with the exception of the factor C^3 in the denominator, which carries a mass dimension of power three in natural units. This implies that N_1 and N_2 should have physical dimension of mass to the power 3/2. The two equations (8.36) and (8.40) can be solved for the two unknowns N_1 and N_2. However, the algebraic simplification is tricky. One needs two identities,

$$\frac{\gamma^2 C^2}{Z\alpha\,(m\,\gamma)^2 + E\kappa^2\,((n_r+\gamma)\,C - E\,Z\alpha)} = \frac{C^2}{Z\alpha\, m^2}\,, \tag{8.41}$$

$$\frac{(\kappa E - \gamma\, m)\,(\kappa E + \gamma\, m)}{C^2} = n_r\,(n_r + 2\,\gamma)\,. \tag{8.42}$$

Using these identities, one can finally bring the result for N_1 and N_2 into a compact form,

$$N_1 = \frac{\kappa}{|\kappa|} \frac{2C^2}{\sqrt{Z\alpha m}} \sqrt{\frac{(n_r - 1)!\,(E\,\kappa + \gamma\, m)}{\Gamma(2\gamma + n_r + 1)\,(\kappa + \gamma)}}\,, \tag{8.43a}$$

$$N_2 = \frac{2C^2}{\sqrt{Z\alpha m}} \sqrt{\frac{n_r!\,(E\,\kappa - \gamma\, m)}{\Gamma(2\gamma + n_r)\,(\kappa + \gamma)}}\,. \tag{8.43b}$$

Here, we might think that we have reached our goal, but, in fact, we have not. The above formulas for N_1 and N_2 are not universally applicable and have to be modified when $n_r = 0$.

The expression $\mathrm{e}^{-Cr}\,(2Cr)^\gamma\, L^{2\gamma+1}_{n_r-1}(2Cr)$ vanishes when $n_r = 0$. However, N_1 also becomes singular in this case, in view of the factor $(n_r - 1)!$ which tends to infinity for $n_r \to 0$. In this case, we are free to choose the normalization factor N_1, because the function multiplied by it vanishes anyway. Let us now investigate the case $n_r = 0$ in greater detail. In this case, we have

$$n_r = 0\,, \qquad E = \frac{m\gamma}{|\kappa|}\,, \qquad E\,\kappa + m\,\gamma = \gamma\, m\,\frac{\kappa + |\kappa|}{|\kappa|}\,. \tag{8.44}$$

The latter factor refers to the condition in Eq. (8.36). It implies that when $n_r = 0$, only negative values of κ are allowed. The normalization factors are then given as follows,

$$N_1 = 0\,, \qquad \kappa = -|\kappa|\,, \qquad E = \frac{m\gamma}{|\kappa|}\,, \qquad n_r = 0\,, \tag{8.45a}$$

$$N_2 = 4\gamma \left(\frac{Z\alpha m}{|\kappa|}\right)^{3/2} \frac{1}{\sqrt{\Gamma(2\gamma+1)\,\kappa\,(\kappa+\gamma)}}\,. \tag{8.45b}$$

The first few states for which $n_r = 0$ and $\kappa = -|\kappa|$ are as follows,

$$E(1S_{1/2}) = \sqrt{1-(Z\alpha)^2} = 1 - \frac{(Z\alpha)^2}{2} - \frac{(Z\alpha)^4}{8} + \mathcal{O}(Z\alpha)^6\,, \qquad n_r = 0\,, \qquad \kappa = -1\,, \tag{8.46}$$

$$E(2P_{3/2}) = \frac{\sqrt{4-(Z\alpha)^2}}{2} = 1 - \frac{(Z\alpha)^2}{8} - \frac{(Z\alpha)^4}{128} + \mathcal{O}(Z\alpha)^6\,, \qquad n_r = 0\,, \qquad \kappa = -2\,, \tag{8.47}$$

$$E(3D_{5/2}) = \frac{\sqrt{9-(Z\alpha)^2}}{3} = 1 - \frac{(Z\alpha)^2}{18} - \frac{(Z\alpha)^4}{648} + \mathcal{O}(Z\alpha)^6\,, \qquad n_r = 0\,, \qquad \kappa = -3\,. \tag{8.48}$$

Let us summarize once more the decisive formulas. The wave function has the form

$$\psi(\vec{r}) = \begin{pmatrix} f(r)\,\chi_{\kappa\mu}(\hat{r}) \\ \mathrm{i}\,g(r)\,\chi_{-\kappa\mu}(\hat{r}) \end{pmatrix}, \qquad \frac{1}{1-X^2} = \frac{\kappa+\gamma}{2\gamma}, \qquad \frac{X}{1-X^2} = -\frac{Z\alpha}{2\gamma}, \tag{8.49a}$$

$$f(r) = \frac{1}{1-X^2}\,h_1(r) - \frac{X}{1-X^2}\,h_2(r), \qquad g(r) = -\frac{X}{1-X^2}\,h_1(r) + \frac{1}{1-X^2}\,h_2(r), \tag{8.49b}$$

$$h_1(r) = N_1\,\mathrm{e}^{-Cr}\,(2Cr)^{\gamma}\,L_{n_r-1}^{2\gamma+1}(2Cr), \qquad h_2(r) = N_2\,\mathrm{e}^{-Cr}\,(2Cr)^{\gamma-1}\,L_{n_r}^{2\gamma-1}(2Cr). \tag{8.49c}$$

The parameters are given as

$$C = \sqrt{m^2 - E^2}, \qquad \gamma = \sqrt{\kappa^2 - (Z\alpha)^2}, \qquad \kappa = (-1)^{j+\ell+1/2}\left(j + \frac{1}{2}\right), \tag{8.49d}$$

$$E = m\,f(n,j), \qquad f(n,j) = \left(1 + \frac{(Z\alpha)^2}{(n_r+\gamma)^2}\right)^{-\frac{1}{2}}, \qquad n_r = n - |\kappa| = n - j - \tfrac{1}{2}. \tag{8.49e}$$

We have used Eq. (6.82), and defined n_r to be a "reduced" principal quantum number. The normalization constants N_1 and N_2 are given in Eqs. (8.43a) and (8.43b) for a general case and in Eqs. (8.45a) and (8.45b) for the case $n_r = 0$. The latter case is relevant only to the $1S$ state.

8.2.3 *Dirac–Coulomb Continuum-State Wave Functions*

The calculation of the continuum states of the Dirac equation, with $E > 0$, is an interesting task in view of the more involved nature of the eigenvalue problem, as compared to the nonrelativistic Schrödinger equation, for which we had discussed the continuum states in Sec. 4.2.5. We use the standard ansatz (8.11) for the Dirac bispinor wave function, eigenstate of the Hamiltonian given in Eq. (8.10), with the spin-angular function $\chi_{\kappa\mu}(\theta,\varphi)$ being defined in Eq. (6.84). Following the same steps as were used for the bound-state problem in Sec. 8.2.2, we obtain the radial equations previously given in Eq. (8.13), and write them in a slightly different form,

$$\frac{\partial}{\partial r} f(r) = -\frac{\kappa+1}{r} f(r) + (E + m - V)\,g(r), \tag{8.50a}$$

$$\frac{\partial}{\partial r} g(r) = \frac{\kappa-1}{r} g(r) - (E - m - V)\,f(r). \tag{8.50b}$$

One can rescale the problem by introducing the functions u_1 and u_2,

$$u_1(r) = r\,f(r), \qquad u_2(r) = r\,g(r). \tag{8.51}$$

The radial equations become

$$\frac{\partial}{\partial r}\begin{pmatrix} u_1(r) \\ u_2(r) \end{pmatrix} = \begin{pmatrix} -\kappa/r & E+m-V \\ -(E-m-V) & \kappa/r \end{pmatrix}\begin{pmatrix} u_1(r) \\ u_2(r) \end{pmatrix}. \tag{8.52}$$

For a continuum state, it is advantageous to set, following Chap. 5 of Ref. [17],

$$u_1 = (E+m)^{1/2}\,(\Phi_1 + \Phi_2)\,, \tag{8.53a}$$

$$u_2 = \mathrm{i}\,(E-m)^{1/2}\,(\Phi_1 - \Phi_2)\,. \tag{8.53b}$$

Here, we remember that $E > m$, as we are talking about a continuum state. These equations entail that the radial parts u_1 and u_2 are real rather than imaginary if we set

$$\Phi_1^* = \Phi_2\,. \tag{8.54}$$

We will be using this observation later. One defines the variable p so that in the asymptotic limit of a free particle, the quantity p would be the modulus of the momentum (in view of the free dispersion relation $E^2 = p^2 + m^2$). One also defines the imaginary radial variable $x = 2\mathrm{i}pr$. We summarize,

$$p = \sqrt{E^2 - m^2}\,, \qquad x = 2\,\mathrm{i}\,pr\,. \tag{8.55}$$

Defining the functions $\Phi_j(x) \equiv \Phi_j(r)$ ($j = 1,2$), we "overload" the symbols $\Phi_j(x)$ with an additional definition, which is consistent as we observe that x is a dimensionless variable (in natural units), while r is dimensionful. This implies that

$$\frac{\partial}{\partial r}\Phi_j(r) = 2\mathrm{i}p\frac{\partial}{\partial x}\Phi_j(x)\,, \qquad j = 1,2\,. \tag{8.56}$$

The radial equations (8.50) take the form

$$\frac{\partial}{\partial x}\Phi_1(x) = \left(\frac{1}{2} + \frac{\mathrm{i}Z\alpha E}{px}\right)\Phi_1(x) - \left(\frac{\kappa}{x} - \frac{\mathrm{i}Z\alpha}{px}\right)\Phi_2(x)\,, \tag{8.57a}$$

$$\frac{\partial}{\partial x}\Phi_2(x) = -\left(\frac{\kappa}{x} + \frac{\mathrm{i}Z\alpha}{px}\right)\Phi_1(x) - \left(\frac{1}{2} + \frac{\mathrm{i}Z\alpha E}{px}\right)\Phi_2(x)\,. \tag{8.57b}$$

The variable γ has been defined in Eq. (8.18). One solves Eq. (8.57a) for Φ_2, which expresses Φ_2 in terms of $\Phi_1(x)$ and $\partial\Phi_1(x)/\partial x$. Plugging in the resulting expression into Eq. (8.57b), one obtains a first-order, and a second-order derivative term, and writes the equation

$$\frac{\partial^2}{\partial x^2}\Phi_1 + \frac{1}{x}\frac{\partial}{\partial x}\Phi_1(x) - \left[\frac{1}{4} + \left(\frac{1}{2} + \frac{\mathrm{i}Z\alpha E}{p}\right)\frac{1}{x} + \frac{\gamma^2}{x^2}\right]\Phi_1(x) = 0\,. \tag{8.58}$$

With the substitution

$$\mathcal{M}(x) = x^{1/2}\,\Phi_1(x)\,, \tag{8.59}$$

one finds the differential equation

$$\frac{\partial^2}{\partial x^2}\mathcal{M}(x) + \left[-\frac{1}{4} - \left(\frac{1}{2} + \frac{\mathrm{i}Z\alpha E}{p}\right)\frac{1}{x} + \frac{\frac{1}{4} - \gamma^2}{x^2}\right]\mathcal{M}(x) = 0\,. \tag{8.60}$$

This matches the Whittaker differential equation, given in Eq. (4.34), which we recall for convenience,

$$\frac{\mathrm{d}^2 w}{\mathrm{d}\rho^2} + \left(-\frac{1}{4} + \frac{\kappa}{\rho} + \frac{\frac{1}{4} - \mu^2}{\rho^2} \right) w = 0 \,, \tag{8.61}$$

with the identification

$$\kappa = -\tfrac{1}{2} - \mathrm{i}y \,, \qquad y = \frac{Z\alpha E}{p} \,, \qquad \mu = \gamma \,. \tag{8.62}$$

Here, we note the difference of the generic κ parameter in Whittaker's differential equation, and the Dirac angular quantum number $\varkappa$ in the spin-angular function. The solution of Eq. (8.61) regular at the origin is given as

$$w = M_{\kappa,\mu}(x) = \mathrm{e}^{-x/2}\, x^{\mu+1/2}\, {}_1F_1(\mu + \tfrac{1}{2} - \kappa, 2\mu + 1, x) \,. \tag{8.63}$$

For the parameters given in Eq. (8.62), it takes the form

$$\mathcal{M}(x) = M_{-1/2-\mathrm{i}y,\gamma}(x) = \mathrm{e}^{-x/2}\, x^{\gamma+1/2}\, {}_1F_1(\gamma + 1 + \mathrm{i}y, 2\gamma + 1, x) \,. \tag{8.64}$$

Our solution $\Phi_1(x)$ for the radial part of the Dirac–Coulomb continuum wave function therefore has to be equal to $x^{-1/2}\,\mathcal{M}(x)$. Restoring the radial variable r according to $x = 2\mathrm{i}pr$, and decorating the function with a normalization factor N and some other convenient prefactors, as well as a phase factor $\exp(\mathrm{i}\eta)$, one has

$$\Phi_1(x) = \Phi_1(r) = N(\gamma + \mathrm{i}y)\, \mathrm{e}^{\mathrm{i}\eta}\, (2pr)^{\gamma}\, \mathrm{e}^{-\mathrm{i}pr}\, {}_1F_1(\gamma + 1 + \mathrm{i}y, 2\gamma + 1, 2\mathrm{i}pr) \,. \tag{8.65}$$

It is useful to define the function $\Phi(r)$,

$$\Phi(r) = r^{\gamma}\, \mathrm{e}^{-\mathrm{i}pr}\, {}_1F_1(\gamma + 1 + \mathrm{i}y, 2\gamma + 1, 2\mathrm{i}pr) \,, \tag{8.66}$$

so that

$$\Phi_1(x) = N(\gamma + \mathrm{i}y)\, \mathrm{e}^{\mathrm{i}\eta}\, (2p)^{\gamma}\, \Phi(r) \,. \tag{8.67}$$

The complex conjugates of the derivatives of the radial wave functions $\Phi_j(x)$ take the form

$$\left(\frac{\partial}{\partial x} \Phi_j(x) \right)^{*} = \frac{\partial \Phi_j(x^*)}{\partial x^*} = -\frac{\partial \Phi_j^*(x)}{\partial x} \,, \qquad j = 1, 2 \,, \tag{8.68}$$

because $x^* = -x$. So, the complex-conjugated radial equations are

$$\frac{\partial}{\partial x} \Phi_2^*(x) = \left(\frac{1}{2} + \frac{\mathrm{i}Z\alpha E}{px} \right) \Phi_2^*(x) - \left(\frac{\varkappa}{x} - \frac{\mathrm{i}Z\alpha}{px} \right) \Phi_1^*(x) \,, \tag{8.69a}$$

$$\frac{\partial}{\partial x} \Phi_1^*(x) = -\left(\frac{\varkappa}{x} + \frac{\mathrm{i}Z\alpha}{px} \right) \Phi_2^*(x) - \left(\frac{1}{2} + \frac{\mathrm{i}Z\alpha E}{px} \right) \Phi_1^*(x) \,. \tag{8.69b}$$

The equation systems (8.57) and (8.69) are identical upon the replacement

$$\Phi_1 \to \Phi_2^* \,, \tag{8.70}$$

justifying the *ansatz* (8.54). Using the property (8.70), Eq. (8.57a) therefore becomes

$$\frac{\partial}{\partial x} \Phi_1(x) = \left(\frac{1}{2} + \frac{\mathrm{i}Z\alpha E}{px} \right) \Phi_1(x) - \left(\frac{\varkappa}{x} - \frac{\mathrm{i}Z\alpha}{px} \right) \Phi_1^*(x) \,. \tag{8.71}$$

With the help of the definitions (8.65) and (8.66), this equation can be rewritten in terms of a relation for the phase factor $\exp(-2i\eta)$,

$$\exp(-2i\eta) = -\frac{\gamma + iy}{\gamma - iy} \frac{r}{\kappa - iy/E} \left[\frac{1}{\Phi^*} \frac{\partial \Phi}{\partial r} - ip \left(1 + \frac{y}{pr} \right) \frac{\Phi}{\Phi^*} \right] . \tag{8.72}$$

Quite surprisingly, one can show that the right-hand side is independent of r. To this end, one writes the complex conjugate of the function $\Phi(r)$ as

$$\Phi^*(r) = r^\gamma \, e^{ipr} \, {}_1F_1(\gamma + 1 - iy, 2\gamma + 1, -2ipr) = r^\gamma \, e^{-ipr} \, {}_1F_1(\gamma + iy, 2\gamma + 1, 2ipr) , \tag{8.73}$$

where the Kummer transformation (see page 253 of Ref. [154]) has been used,

$${}_1F_1(a, b, z) = \exp(z) \, {}_1F_1(b - a, b, -z). \tag{8.74}$$

In the numerator of Eq. (8.72), one generates confluent hypergeometric functions according to the formula

$$\frac{\partial}{\partial z} {}_1F_1(a, b, z) = \frac{a}{b} \, {}_1F_1(a + 1, b + 1, z) . \tag{8.75}$$

Both the a as well as the b parameters have to be lowered in order to lead to a cancelation of the r-dependent terms in Eq. (8.72), i.e., to a cancelation of the hypergeometric functions in the numerator and denominator. One achieves this by the use of the contiguous relations (see page 254 of Ref. [154])

$$b \, {}_1F_1(a + 1, b, z) - b \, {}_1F_1(a, b, z) - z \, {}_1F_1(a + 1, b + 1, z) = 0 , \tag{8.76}$$

$$(a - b) \, {}_1F_1(a - 1, b, z) + a \, {}_1F_1(a + 1, b, z) + (b - 2a - z) \, {}_1F_1(a + 1, b + 1, z) = 0 . \tag{8.77}$$

The first of these equations, given in (8.76), can be used in order to lower both the a as well as b parameters, when applied to the derivative of the Kummer function given in Eq. (8.75). The second equation, given in (8.77), can be used in order to lower the a parameter of the function $\Phi(r)$ given in Eq. (8.66), in the sense of the consecutive replacements $\gamma + 2 + iy \to \gamma + 1 + iy \to \gamma + iy$. These operations eventually lead to a complete elimination of the r-dependence in Eq. (8.72). One finally arrives at the following formula

$$e^{2i\eta} = -\frac{\kappa - iy/E}{\gamma + iy} . \tag{8.78}$$

Assembling the information collected in Eqs. (8.51), (8.53) and (8.65), one arrives at the following formulas for the radial components,

$$r \, f(r) = (E + m)^{1/2} \, N \, (2pr)^\gamma \left[(\gamma + iy) \, e^{-ipr + i\eta} \, {}_1F_1(\gamma + 1 + iy, 2\gamma + 1, 2ipr) + \text{c.c.} \right] , \tag{8.79a}$$

$$r \, g(r) = i \, (E - m)^{1/2} \, N \, (2pr)^\gamma \left[(\gamma + iy) \, e^{-ipr + i\eta} \, {}_1F_1(\gamma + 1 + iy, 2\gamma + 1, 2ipr) - \text{c.c.} \right] . \tag{8.79b}$$

We now need to calculate the normalization factor N, with the notion of normalizing the radial wave functions to a Dirac-δ on the energy scale. To this end, we need to study the asymptotic behavior for $r \to \infty$, of the wave function. The relevant formula for the ${}_1F_1$ function is as follows [see page 278 of Ref. [154]],

$$ {}_1F_1(a,b,z) \to \frac{\Gamma(b)}{\Gamma(a)} x^{a-c} \, \mathrm{e}^{x} \,, \qquad x \to \infty \,. \tag{8.80} $$

Using the recurrence relation of the Γ function, $(\gamma + \mathrm{i}y)/\Gamma(\gamma + 1 + \mathrm{i}y) = 1/\Gamma(\gamma + \mathrm{i}y)$, as well as the formula $\Gamma(\gamma + \mathrm{i}y) = |\Gamma(\gamma + \mathrm{i}y)| \exp(\mathrm{i} \arg(\Gamma(\gamma + \mathrm{i}y)))$, one finds

$$ (\gamma + \mathrm{i}y)\, \mathrm{e}^{\mathrm{i}\eta} \, {}_1F_1(\gamma + 1 + \mathrm{i}y, 2\gamma + 1, 2\mathrm{i}pr) \to (2pr)^{-\gamma} \frac{\Gamma(2\gamma+1)}{|\Gamma(\gamma + \mathrm{i}y)|} \mathrm{e}^{-\pi y/2} \, \mathrm{e}^{\mathrm{i}\delta} \exp(2\mathrm{i}pr) \,, \tag{8.81} $$

for large r, where the phase factor is

$$ \delta = -\tfrac{1}{2}\pi\gamma - \arg[\Gamma(\gamma + \mathrm{i}y)] + y \ln(2pr) + \eta \,. \tag{8.82} $$

Its dependence on r is only logarithmic. The asymptotics of the radial upper and lower component wave functions take the following forms,

$$ r\,f(r) \to 2\,(E+m)^{1/2} \, \mathrm{e}^{-\pi y/2} \, N \frac{\Gamma(2\gamma+1)}{|\Gamma(\gamma + \mathrm{i}y)|} \cos(pr + \delta) = \sqrt{E+m}\,\mathcal{C} \cos(pr + \delta) \,, \tag{8.83a} $$

$$ r\,g(r) \to -2\,(E-m)^{1/2} \, \mathrm{e}^{-\pi y/2} \, N \frac{\Gamma(2\gamma+1)}{|\Gamma(\gamma + \mathrm{i}y)|} \sin(pr + \delta) = -\sqrt{E-m}\,\mathcal{C} \sin(pr + \delta) \,, \tag{8.83b} $$

with an obvious definition of $\mathcal{C}$,

$$ \mathcal{C} = 2\,\mathrm{e}^{-\pi y/2} \, N \frac{\Gamma(2\gamma+1)}{|\Gamma(\gamma + \mathrm{i}y)|} \,. \tag{8.84} $$

From these asymptotics, we can derive the normalization integral of two Dirac eigenstates $|1\rangle$ and $|2\rangle$, composed of radial wave functions f_1 and f_2, and g_1 and g_2, of different energies E_1 and E_2, and in consequence, different asymptotic free momenta p_1 and p_2, in the limit $E_1 \to E_2$, as follows,

$$ \begin{aligned} \langle 1|2\rangle &= \int \mathrm{d}r \, [(r\,f_1(r))(r\,f_2(r)) + (r\,g_1(r))(r\,g_2(r))] \\ &\to \mathcal{C}^2 \int_0^\infty \mathrm{d}r \, [(E_1 + m) \cos(p_1 r + \delta) \cos(p_2 r + \delta)] \\ &\quad + [(E_1 - m) \sin(p_1 r + \delta) \sin(p_2 r + \delta)] \,. \end{aligned} \tag{8.85} $$

The radial integral with the cosines is

$$ \begin{aligned} \int_0^\infty \mathrm{d}r \, \cos(p_1 r + \delta) \cos(p_2 r + \delta) &= \int_0^\infty \mathrm{d}r \, \frac{\mathrm{e}^{\mathrm{i}(p_1 r + \delta)} + \mathrm{e}^{-\mathrm{i}(p_1 r + \delta)}}{2} \frac{\mathrm{e}^{\mathrm{i}(p_2 r + \delta)} + \mathrm{e}^{-\mathrm{i}(p_2 r + \delta)}}{2} \\ &\to \frac{1}{4} \int_0^\infty \mathrm{d}r \left[\mathrm{e}^{\mathrm{i}(p_1 r - p_2 r)} + \mathrm{e}^{-\mathrm{i}(p_1 r - p_2 r)} \right] \\ &= \frac{1}{2} \int_{-\infty}^\infty \mathrm{d}r \, \mathrm{e}^{\mathrm{i}(p_1 r - p_2 r)} = \pi \delta(p_1 - p_2) \,. \end{aligned} \tag{8.86} $$

Analogously, for the sine functions, we have

$$\int_0^\infty \mathrm{d}r \, \sin(p_1 r + \delta) \, \sin(p_2 r + \delta) = \pi \delta(p_1 - p_2) \,. \tag{8.87}$$

Hence,

$$\langle 1|2 \rangle = 2\pi \, C^2 \, E_1 \, \delta(p_1 - p_2) \equiv \delta(E_1 - E_2) \,, \tag{8.88}$$

where the last equality holds if the wave functions are normalized to the energy scale. Using known properties of the Dirac-δ function, one can easily show that

$$\delta(p_1 - p_2) = \frac{1}{|\mathrm{d}p/\mathrm{d}E|}\bigg|_{E=E_1} \delta(E_1 - E_2) = \frac{p_1}{E_1} \, \delta(E_1 - E_2) \,, \tag{8.89}$$

where we recall Eq. (8.55). Hence, we must demand that

$$2\pi \, C^2 = \frac{1}{p} \,, \tag{8.90}$$

for a general state with asymptotic free momentum p and energy E. Using Eq. (8.84), we have

$$C = 2 \, \mathrm{e}^{-\pi y/2} \, N \, \frac{\Gamma(2\gamma + 1)}{|\Gamma(\gamma + \mathrm{i}\, y)|} \equiv \frac{1}{\sqrt{2\pi \, p}} \,. \tag{8.91}$$

So, the normalization factor has to be chosen as

$$N = \frac{1}{2\sqrt{2\pi \, p}} \, \mathrm{e}^{\pi y/2} \, \frac{|\Gamma(\gamma + \mathrm{i}\, y)|}{\Gamma(2\gamma + 1)} \,. \tag{8.92}$$

With Eq. (8.79), it means that the final result for the radial components of the wave functions is as follows,

$$r \, f(r) = (E + m)^{1/2} \, \frac{(2pr)^\gamma \, \mathrm{e}^{\pi y/2}}{2\sqrt{2\pi \, p}} \, \frac{|\Gamma(\gamma + \mathrm{i}\, y)|}{\Gamma(2\gamma + 1)} \times \left[(\gamma + \mathrm{i}y) \, \mathrm{e}^{-\mathrm{i}pr + \mathrm{i}\eta} \, {}_1F_1(\gamma + 1 + \mathrm{i}\, y, 2\gamma + 1, 2\mathrm{i}pr) + \mathrm{c.c.} \right] , \tag{8.93a}$$

$$r \, g(r) = \mathrm{i} \, (E - m)^{1/2} \, \frac{(2pr)^\gamma \, \mathrm{e}^{\pi y/2}}{2\sqrt{2\pi \, p}} \, \frac{|\Gamma(\gamma + \mathrm{i}\, y)|}{\Gamma(2\gamma + 1)} \times \left[(\gamma + \mathrm{i}y) \, \mathrm{e}^{-\mathrm{i}pr + \mathrm{i}\eta} \, {}_1F_1(\gamma + 1 + \mathrm{i}\, y, 2\gamma + 1, 2\mathrm{i}pr) - \mathrm{c.c.} \right] . \tag{8.93b}$$

These wave functions are normalized to the energy scale [see Eq. (8.88)].

8.2.4 *Dirac–Coulomb Virial Theorem*

We had discussed the virial theorem for the Schrödinger–Coulomb Hamiltonian in Sec. 4.2.4. For the Dirac–Coulomb case, the Hamiltonian is given by Eq. (8.9),

$$H_{\mathrm{DC}} = \vec{\alpha} \cdot \vec{p} + \beta \, m + V \,, \qquad V = -\frac{Z\alpha}{r} \,, \qquad \vec{\alpha} = \mathrm{i} \, [H_{\mathrm{DC}}, \vec{r}] \,. \tag{8.94}$$

For an eigenstate of the Dirac–Coulomb Hamiltonian, $(H_{\mathrm{DC}} - E)|\psi\rangle = 0$, we have

$$[H_{\mathrm{DC}} - E, A] \, |\psi\rangle = (H_{\mathrm{DC}} - E) A |\psi\rangle \,. \tag{8.95}$$

One of the virial theorems for the Dirac–Coulomb Hamiltonian, which reads as

$$\langle \vec{\alpha} \cdot \vec{p} \rangle = - \langle V \rangle \,, \tag{8.96}$$

can be derived as follows,

$$\begin{aligned} \langle \vec{\alpha} \cdot \vec{p} \rangle &= \langle \mathrm{i} \, [H_{\mathrm{DC}} - E, \vec{r}\,] \cdot \vec{p} \rangle = - \langle \mathrm{i} \, \vec{r} \cdot (H_{\mathrm{DC}} - E) \, \vec{p} \rangle \\ &= - \langle \mathrm{i} \, \vec{r} \cdot [H_{\mathrm{DC}} - E, \vec{p}\,] \rangle = \langle \mathrm{i} \, \vec{r} \cdot [\vec{p}, V] \rangle = \left\langle \vec{r} \cdot \vec{\nabla}(V) \right\rangle . \end{aligned} \tag{8.97}$$

For an eigenstate of the Dirac–Coulomb Hamiltonian, with $V = -Z\alpha/r$, one has

$$\left\langle \vec{r} \cdot \vec{\nabla}(V) \right\rangle = \left\langle \vec{r} \cdot \vec{\nabla} \left(-\frac{Z\alpha}{r} \right) \right\rangle = \left\langle \vec{r} \cdot \frac{(Z\alpha)\vec{r}}{r^3} \right\rangle = \left\langle \frac{Z\alpha}{r} \right\rangle = - \langle V \rangle \,. \tag{8.98}$$

This result is consistent with Eq. (7) of Ref. [260]. A consequence is

$$E = \langle \vec{\alpha} \cdot \vec{p} \rangle + \langle V \rangle + \langle \beta m \rangle = \langle \beta m \rangle \,. \tag{8.99}$$

This virial theorem reflects certain symmetries of the Schrödinger–Coulomb, and Dirac–Coulomb Hamiltonian.

Let us briefly explore the calculation of matrix elements by the Hellmann–Feynman theorem [261, 262], applying scaling arguments to the calculation of two important matrix elements. Only two parameters enter the Dirac Hamiltonian H_{DC}, and these are the nuclear charge number Z and the electron mass m. We thus consider the eigenfunction $\psi = \psi(Z,m)$ formally as a function of Z and m and differentiate

$$\begin{aligned} & Z \frac{\partial}{\partial Z} \langle \psi(Z,m) \, | H_{\mathrm{DC}}(Z,m) | \, \psi(Z,m) \, \rangle \\ &= 2 \left\langle Z \, \partial_Z \psi(Z,m) \middle| H_{\mathrm{DC}}(Z,m) \middle| \psi(Z,m) \right\rangle + \left\langle \psi(Z,m) \middle| Z \partial_Z H_{\mathrm{DC}}(Z,m) \middle| \psi(Z,m) \right\rangle \\ &= E(Z,m) \, Z \frac{\partial}{\partial Z} \left\langle \psi(Z,m) \middle| \psi(Z,m) \right\rangle + \left\langle \psi(Z,m) \middle| \left(-\frac{Z\alpha}{r} \right) \middle| \psi(Z,m) \right\rangle \\ &= \left\langle \psi(Z,m) \middle| \left(-\frac{Z\alpha}{r} \right) \middle| \psi(Z,m) \right\rangle = \langle \psi(Z,m) \, | \, V \, | \, \psi(Z,m) \rangle \,. \end{aligned} \tag{8.100}$$

This derivation is akin to the canonical derivation of the Hellmann–Feynman theorem [261, 262]. So, we have

$$Z \frac{\partial}{\partial Z} E(Z,m) = \langle \psi(Z,m) \, | \, V \, | \, \psi(Z,m) \rangle \,, \tag{8.101}$$

to all orders in the $Z\alpha$ expansion. The normalization $\langle \psi(Z,m) | \psi(Z,m) \rangle = 1$ holds independent of Z and m. Let us now differentiate with respect to m,

$$\begin{aligned} & \frac{\partial}{\partial m} \langle \, \psi(Z,m) \, | H_{\mathrm{DC}}(Z,m) | \, \psi(Z,m) \, \rangle \\ &= 2 \left\langle \frac{\partial}{\partial m} \psi(Z,m) \middle| H_{\mathrm{DC}}(Z,m) \middle| \psi(Z,m) \right\rangle + \left\langle \psi(Z,m) \middle| \frac{\partial}{\partial m} H_{\mathrm{DC}}(Z,m) \middle| \psi(Z,m) \right\rangle \\ &= E(Z,m) \, \frac{\partial}{\partial m} \left\langle \psi(Z,m) \middle| \psi(Z,m) \right\rangle + \langle \psi(Z,m) \, | \beta | \, \psi(Z,m) \rangle \\ &= \langle \psi(Z,m) \, | \beta | \, \psi(Z,m) \rangle \,. \end{aligned}$$

So, we have

$$\frac{\partial}{\partial m} E(Z,m) = \frac{E(Z,m)}{m} = \langle \psi(Z,m) \,|\, \beta \,|\, \psi(Z,m) \rangle \tag{8.102}$$

to all orders in the $Z\alpha$ expansion. Using the explicit form of the Dirac–Hamiltonian, this identity is seen to be equivalent to the virial theorem Eq. (8.97) for the Dirac–Coulomb equation.

The last equality can independently be derived from the fact that the Dirac energy $E(Z,m)$ actually is directly proportional to the electron mass m, i.e. it has the functional form

$$E(Z,m) = m\, f(Z)\,. \tag{8.103}$$

In order to show this, we consider the Dirac–Coulomb Hamiltonian once more and write it as

$$\begin{aligned} H_{\rm DC} &= \vec{\alpha} \cdot \vec{p} + \beta\, m - \frac{Z\alpha}{r} = -\mathrm{i}\, \vec{\alpha} \cdot \frac{\partial}{\partial \vec{r}} + \beta\, m - \frac{Z\alpha}{r} \\ &= m \left(-\mathrm{i}\, \vec{\alpha} \cdot \frac{\partial}{\partial (m\vec{r})} + \beta - \frac{Z\alpha}{mr} \right) = m \left(-\mathrm{i}\, \vec{\alpha} \cdot \frac{\partial}{\partial \vec{\rho}} + \beta - \frac{Z\alpha}{\rho} \right). \end{aligned} \tag{8.104}$$

Here, we have defined $\vec{\rho} = m\vec{r}$. The electron mass scales out as a prefactor, and the scaled Hamiltonian $-\mathrm{i}\, \vec{\alpha} \cdot \vec{\nabla}_{\vec{\rho}} + \beta - Z\alpha/\rho$ is free of the mass. Further relativistic sum rules and integral properties of the Dirac equation have been discussed by Goldman[1] and Drake[2] in Ref. [263]. Useful generalizations of virial relations have been derived by Shabaev[3] in Refs. [264, 265]. Additional relevant considerations, also in regard to the Hellmann–Feynman theorem, can be found in Ref. [266].

8.2.5 *Dirac–Coulomb Propagator*

The calculation of the Dirac–Coulomb propagator follows essentially the same lines as the calculation of the Schrödinger–Coulomb Green function in Sec. 4.3. However, because the Dirac–Coulomb Hamiltonian is a (4×4)-matrix in spinor space, the Dirac–Coulomb propagator also must be of the same structure. We thus investigate the inverse of the Dirac–Coulomb Hamiltonian $H_{\rm DC} = \vec{\alpha} \cdot \vec{p} + \beta\, m - Z\alpha/r$. The Green function has the following structure [see Eqs. (7.173) and (7.176)]

$$\mathcal{G}_{\rm DC}(\vec{r}_1, \vec{r}_2, E) = \left\langle \vec{r}_1 \left| \frac{1}{H_{\rm DC} - E} \right| \vec{r}_2 \right\rangle, \tag{8.105a}$$

$$(H_{\rm DC} - E)\, \mathcal{G}_{\rm DC}(\vec{r}_1, \vec{r}_2, E) = \delta^{(3)}(\vec{r}_1 - \vec{r}_2)\, \mathbb{1}_{4\times4}\,. \tag{8.105b}$$

A remark is in order. When using the Feynman prescription for the displacement of the poles, the expression for the Green function gets modified into $\langle \vec{r}_1 \,|1/[H_{\rm DC}(1 - \mathrm{i}\epsilon) - E]|\, \vec{r}_2 \rangle$, i.e., the energies of virtual positive-energy states obtain an infinitesimal negative imaginary part, while the energies of virtual

[1]Samuel Pedro Goldman (b. 1949).
[2]Gordon William Frederic Drake (b. 1944).
[3]Vladimir Moiseevich Shabaev (b. 1959).

negative-energy states obtain an infinitesimal positive imaginary part. In bound-state calculations, the energy variable E typically is equal to a reference-state energy, minus the energy of one or more virtual photons [257, 258, 267–270]. As is well known, one can implement the Feynman prescription either by infinitesimally displacing the virtual-state poles in the Dirac–Coulomb propagator, or, by a judicious choice of the photon integration contours [257, 258, 267–270]. In the latter case, which is relevant to practical calculations [257, 258, 267–270], it would be redundant to also displace the poles infinitesimally, and this approach is implemented in the following. If one were to insist on the former approach, then one would need to separate the contributions of positive-energy, and negative-energy states, to the Dirac–Coulomb Green function. This separation is immediately accomplished in a finite-basis-set representation (see Sec. 2.2 of Ref. [270]). If one were to attempt a corresponding separation of the contributions for the analytic approach below, inspiration could be drawn from Ref. [271]. Let us also point out that Ref. [271] treats the calculation of the reduced Dirac–Coulomb Green function.

Returning to the solution of Eq. (8.105), we observe that these equations constitute the generalization of Eq. (4.82) to the Dirac–Coulomb case. In (2×2)-matrix notation, we have

$$\begin{pmatrix} \left(-\dfrac{Z\alpha}{r_1} + m - E\right) \mathbb{1}_{2\times 2} & -\mathrm{i}\vec{\sigma}\cdot\vec{\nabla}_{r_1} \\ -\mathrm{i}\vec{\sigma}\cdot\vec{\nabla}_{r_1} & \left(-\dfrac{Z\alpha}{r_1} - m - E\right) \mathbb{1}_{2\times 2} \end{pmatrix} \mathcal{G}_{\rm DC}(\vec{r}_1, \vec{r}_2, E) = \delta^{(3)}(\vec{r}_1 - \vec{r}_2)\, \mathbb{1}_{4\times 4}\,. \tag{8.106}$$

In full analogy to the Schrödinger–Coulomb Green function, the Dirac–Coulomb Green function also has an expansion in terms of angular-momentum components, which are characterized by the Dirac angular momentum quantum number κ defined in Eq. (6.82), as $\kappa = (-1)^{j+\ell+1/2}\left(j+\frac{1}{2}\right)$. In analogy to the free Dirac Green function (7.179), the expansion of the Dirac–Coulomb Green function has the expansion

$$\begin{aligned} &\mathcal{G}_{\rm DC}(\vec{r}_1, \vec{r}_2, E) \\ &= \sum_\kappa \begin{pmatrix} \mathcal{G}^{11}_{\mathrm{DC},\kappa}(r_1, r_2, E)\pi_\kappa(\hat{r}_1, \hat{r}_2) & \mathcal{G}^{12}_{\mathrm{DC},\kappa}(r_1, r_2, E)\mathrm{i}\vec{\sigma}\cdot\hat{r}_1\pi_{-\kappa}(\hat{r}_1, \hat{r}_2) \\ -\mathcal{G}^{21}_{\mathrm{DC},\kappa}(r_1, r_2, E)\mathrm{i}\vec{\sigma}\cdot\hat{r}_1\pi_\kappa(\hat{r}_1, \hat{r}_2) & \mathcal{G}^{22}_{\mathrm{DC},\kappa}(r_2, r_1, E)\pi_{-\kappa}(\hat{r}_1, \hat{r}_2) \end{pmatrix}. \end{aligned} \tag{8.107}$$

Here, the sum over κ runs over the positive and negative integers, with the exception of zero, and the sum over the magnetic projections runs over $\mu = -(|\kappa| - 1/2), \ldots, |\kappa| - 1/2$. We have used the identity $\vec{\sigma}\cdot\hat{r}\,\chi_{\kappa\mu}(\hat{r}) = -\chi_{-\kappa\mu}(\hat{r})$ in the reformulation of the angular projection. Otherwise, the angular part involves the projection operator, originally defined in Eq. (6.96). The angular decomposition of the Dirac–Coulomb Green function follows the same pattern as the free Green function, given in Eq. (7.179). The result is given in Eq. (A.16) of Ref. [257]. A compact representation of the radial Dirac–Coulomb Green functions can be given in the following functional form (for $r_1 = r_< = \min(r_1, r_2)$ and $r_2 = r_> = \max(r_1, r)2)$,

$$\mathcal{G}^{11}_{\kappa}(r_<, r_>, E) = Q\,(1+z)\,M_-(\lambda,\nu,2cr_<)\,W_+(\lambda,\nu,2cr_>)\,, \tag{8.108a}$$
$$\mathcal{G}^{12}_{\kappa}(r_<, r_>, E) = Q\,c\,M_-(\lambda,\nu,2cr_<)\,W_-(\lambda,\nu,2cr_>)\,, \tag{8.108b}$$
$$\mathcal{G}^{21}_{\kappa}(r_<, r_>, E) = Q\,c\,M_+(\lambda,\nu,2cr_<)\,W_+(\lambda,\nu,2cr_>)\,, \tag{8.108c}$$
$$\mathcal{G}^{22}_{\kappa}(r_<, r_>, E) = Q\,(1-z)\,M_+(\lambda,\nu,2cr_<)\,W_-(\lambda,\nu,2cr_>)\,. \tag{8.108d}$$

The symmetry relations are completely analogous to Eq. (7.183f),

$$\mathcal{G}^{11}_{\kappa}(r_>, r_<, E) = \mathcal{G}^{11}_{\kappa}(r_<, r_>, E)\,, \tag{8.109a}$$
$$\mathcal{G}^{12}_{\kappa}(r_>, r_<, E) = \mathcal{G}^{21}_{\kappa}(r_<, r_>, E)\,, \tag{8.109b}$$
$$\mathcal{G}^{21}_{\kappa}(r_>, r_<, E) = \mathcal{G}^{12}_{\kappa}(r_<, r_>, E)\,, \tag{8.109c}$$
$$\mathcal{G}^{22}_{\kappa}(r_>, r_<, E) = \mathcal{G}^{22}_{\kappa}(r_<, r_>, E)\,. \tag{8.109d}$$

The functions denoted as $M_\pm$ and $W_\pm$ are related to linear combinations of two Whittaker functions,

$$M_\pm(\lambda,\nu,y) = (\lambda-\nu)\;M_{\nu-1/2,\lambda}(y) \pm p_-\;M_{\nu+1/2,\lambda}(y)\,, \tag{8.110a}$$
$$W_\pm(\lambda,\nu,y) = p_+\;W_{\nu-1/2,\lambda}(y) \pm W_{\nu+1/2,\lambda}(y)\,, \tag{8.110b}$$

where we recall the definition of the Whittaker functions given in Eqs. (4.35a) and (4.35b). Various symbols enter these expressions; these are given as follows,

$$Q = \frac{1}{4\,c^2\,(r_1\,r_2)^{3/2}}\,\frac{\Gamma(\lambda-\nu)}{\Gamma(1+2\,\lambda)}\,, \qquad c = \sqrt{m^2-E^2}\,, \qquad p_\pm = \kappa \pm \frac{Z\alpha}{c}\,, \tag{8.111a}$$
$$\lambda = \sqrt{\kappa^2-(Z\alpha)^2} = |\kappa| - \zeta\,, \qquad \zeta = \frac{(Z\alpha)^2}{|\kappa|+\lambda}\,, \qquad \nu = \frac{(Z\alpha)\,E}{\sqrt{m^2-E^2}}\,. \tag{8.111b}$$

In practical calculations [257,258,267], the energy variable E may assume manifestly complex values. For practical calculations, it is sometimes useful to rewrite the Whittaker functions M and W in terms of more fundamental hypergeometric U and ${}_1F_1$ functions, profiting from the isolation of the integer part of λ as given in Eq. (8.111). We have the following formula for the M function,

$$\begin{aligned} M_{\nu\pm1/2,\lambda}(y) &= \mathrm{e}^{-\frac{1}{2}y}\,y^{\lambda+\frac{1}{2}}\;{}_1F_1\left(\frac{1}{2}\mp\frac{1}{2}+\lambda-\nu, 2\,\lambda+1, y\right) \\ &= \mathrm{e}^{-\frac{1}{2}y}\,y^{|\kappa|+\frac{1}{2}-\zeta}\;{}_1F_1\left(|\kappa|+1-\delta_\pm-\zeta-\nu, 2\,|\kappa|+1-2\,\zeta, y\right)\,, \end{aligned} \tag{8.112}$$

whereas the W functions can be rewritten as follows,

$$\begin{aligned} W_{\nu\pm1/2,\lambda}(y) &= \mathrm{e}^{-\frac{1}{2}y}\,y^{\lambda+\frac{1}{2}}\;U\left(\frac{1}{2}\mp\frac{1}{2}+\lambda-\nu, 2\,\lambda+1, y\right) \\ &= \mathrm{e}^{-\frac{1}{2}y}\,y^{|\kappa|+\frac{1}{2}-\zeta}\,U\left(|\kappa|+1-\delta_\pm-\zeta-\nu, 2\,|\kappa|+1-2\,\zeta, y\right)\,. \end{aligned} \tag{8.113}$$

We here define the variable $\delta_\pm$ as

$$\delta_\pm = \delta_{\pm1,1} = \begin{cases} 1 \text{ for the case } + \\ 0 \text{ for the case } - \end{cases}\,. \tag{8.114}$$

For the W function, it is possible to rewrite the U function as a ${}_2F_0$ hypergeometric,

$$W_{\nu\pm 1/2,\lambda}(y) = \mathrm{e}^{-\frac{1}{2}y}\, y^{\nu\pm 1/2}\, {}_2F_0\left(\frac{1}{2} \mp \frac{1}{2} + \lambda - \nu, \frac{1}{2} \mp \frac{1}{2} - \lambda - \nu, -\frac{1}{y}\right)$$
$$= \mathrm{e}^{-\frac{1}{2}y}\, y^{\nu\pm 1/2}\, {}_2F_0\left(|\kappa| + 1 - \delta_\pm - \zeta - \nu, -|\kappa| + 1 - \delta_\pm + \zeta - \nu, -\frac{1}{y}\right). \tag{8.115}$$

A last representation useful for small y is the decomposition of the W function in terms of two confluent hypergeometric functions,

$$W_{\nu\pm 1/2,\lambda}(y) = \mathrm{e}^{-\frac{1}{2}y}\, y^{\lambda+\frac{1}{2}} \left[\frac{\Gamma(-2\lambda)}{\Gamma\left(\frac{1}{2} \mp \frac{1}{2} - \lambda - \nu\right)}\, {}_1F_1\left(\frac{1}{2} \mp \frac{1}{2} + \lambda - \nu, 2\lambda + 1, y\right)\right.$$
$$\left. + \frac{\Gamma(2\lambda)}{\Gamma\left(\frac{1}{2} \mp \frac{1}{2} + \lambda - \nu\right)}\, {}_1F_1\left(\frac{1}{2} \mp \frac{1}{2} - \lambda - \nu, 1 - 2\lambda, y\right)\right]$$
$$= \mathrm{e}^{-\frac{1}{2}y}\, y^{|\kappa|+\frac{1}{2}-\zeta} \left[\frac{\Gamma(-2|\kappa| + 2\zeta)}{\Gamma\left(-|\kappa| + 1 - \delta_\pm - \nu + \zeta\right)}\, {}_1F_1\left(|\kappa| + 1 - \delta_\pm - \nu - \zeta, 2|\kappa| + 1 - 2\zeta, y\right)\right.$$
$$\left. + \frac{\Gamma(2|\kappa| - 2\zeta)}{\Gamma\left(|\kappa| + 1 - \delta_\pm - \nu - \zeta\right)}\, {}_1F_1\left(-|\kappa| + 1 - \delta_\pm - \nu + \zeta, -2|\kappa| + 1 + 2\zeta, x\right)\right]. \tag{8.116}$$

This concludes our discussion of the Dirac–Coulomb Green function.

8.3 Dirac–Volkov Equation for Laser Fields

8.3.1 *Dirac–Volkov Solutions for Laser Fields*

In the presence of electromagnetic background fields, characterized by a vector potential $A^\mu(\phi)$, the Dirac equation takes the form

$$(\mathrm{i}\gamma^\mu\, \partial_\mu - e\gamma^\mu\, A_\mu - m)\, \psi(x) = 0\,, \tag{8.117}$$

where we assume minimal coupling in the form of the replacement $p_\mu = \mathrm{i}\,\partial_\mu \to p_\mu - e\, A_\mu$. A particularly interesting case is obtained when A_μ assumes the form of a vector potential for a plane-wave field, encompassing both linear as well as circular polarization.

The Dirac–Volkov wave functions $\psi(x)$ which fulfill the Dirac equation with an external potential can be written in the form (we use the Feynman slash notation $\not{k} = \gamma^\mu\, k_\mu$)

$$\left(\mathrm{i}\not{\partial} - e\not{A}(\phi) - m\right) \psi(x) = 0\,, \qquad \phi = k \cdot x = \omega\, t - \vec{k} \cdot \vec{r}\,. \tag{8.118}$$

In general, Dirac–Volkov solutions can be characterized by a constant four-vector p, which is equal to the asymptotic momentum of the Dirac–Volkov state in the absence of the laser field, and by a spin projection σ. The Dirac–Volkov solutions differ from the solutions of the free Dirac equation in a matrix-valued phase-dependent prefactor, and in a laser-phase factor which is obtained by integrating over the laser-field vector potential [3]. We have

$$\psi_{p,\sigma}(x) = \sqrt{\frac{m}{QV}} \left(1 + \frac{e\not{k}\not{A}(\phi)}{2\,(k \cdot p)}\right) \mathcal{U}_\sigma(p)\, \exp\left(\mathrm{i}\, S_p\, [\phi]\right) \equiv \sqrt{\frac{m}{QV}}\, E(p, x)\, \mathcal{U}_\sigma(p)\,. \tag{8.119}$$

Here, $E(p,x)$ is a laser-dressed prefactor which still needs to be evaluated more precisely and expanded into plane waves, and $Q = q^0$ is the timelike component of the effective momentum within the laser field [see also Eq. (8.122) below]. The bispinor $\mathcal{U}_\sigma(p)$ is defined for an electron in the helicity basis, according to Eq. (7.89). The normalization volume is V. The phase of the laser field is given as

$$\phi = k \cdot x = \omega t - \vec{k} \cdot \vec{r} \,. \tag{8.120}$$

The phase factor in the exponential is given as follows,

$$S_p[\phi] = -p \cdot x - \int_0^\phi \mathrm{d}\phi' \left(e \frac{p \cdot A(\phi')}{k \cdot p} - e^2 \frac{A^2(\phi')}{2\,(k \cdot p)} \right) . \tag{8.121}$$

After averaging over a laser period, one obtains the effective four-momentum q for the electron in the laser field,

$$q = p + \frac{e^2 \, |\overline{a}^2|}{2k \cdot p} k \,, \qquad \overline{a}^2 = \frac{1}{2\pi} \int_0^{2\pi} A^2(\phi') \, \mathrm{d}\phi' < 0 \,. \tag{8.122}$$

The quantity $\overline{a}^2$ is the average of the square of the vector potential over a laser period, and it is negative for a space-like vector potential (radiation gauge). Here, we write the modulus $|\overline{a}^2| = -\overline{a}^2 > 0$. Because $k^2 = 0$ for the laser photon, we can define the effective mass of the electron in the laser field as the four-vector product of q with itself,

$$q^2 = m^2 + e^2 \, |\overline{a}^2| = m_*^2 \,. \tag{8.123}$$

Observing that

$$q^2 = (q^0)^2 - \vec{q}^{\,2} = Q^2 - \vec{q}^{\,2} = m_*^2 \,, \tag{8.124}$$

we end up with a modified relativistic energy-momentum relation.

The solution (8.119) is not written in terms of plane waves, which are otherwise useful in order to carry out the integrations over time in S-matrix calculations of laser-dressed processes. In order to be able to carry out these integrations, we thus need to expand the Dirac–Volkov state into plane waves. We first turn our attention to linear polarization, in which case the vector potential reads as

$$A^\mu(\phi) = a^\mu \cos\phi \,. \tag{8.125}$$

One could otherwise define the vector potential as $A^\mu(\phi) = a\,\epsilon^\mu \cos\phi$, where a is the amplitude and ϵ^μ is a polarization vector, but the above notation is more compact. Usually, in laser-physics problems, one works in the radiation gauge, where $\epsilon^0 = 0$, and the four-vector is $\epsilon = (0, \vec{\epsilon}_i)$ (for $i = 1, 2$). Of course, the four-vector square is $\epsilon^2 = \epsilon^\mu \, \epsilon_\mu = -1$. The four-vector square of the four-vector potential of the laser is given as

$$A^\mu(\phi) A_\mu(\phi) = a^\mu \, a_\mu \cos^2\phi = a^2 \cos^2\phi = -|a^2| \cos^2\phi \,. \tag{8.126}$$

After integrating out the laser phase, the phase factor evaluates to

$$\mathrm{i} S_p[\phi] = -\mathrm{i} \left(p + \frac{e^2 \, |a^2|}{4(k \cdot p)} k \right) \cdot x - \mathrm{i} \frac{e \, a \cdot p}{k \cdot p} \sin\phi - \mathrm{i} \frac{e^2 \, |a^2|}{8\,(k \cdot p)} \sin(2\phi) \,. \tag{8.127}$$

The first term in round brackets is easily identified as the laser-dressed momentum in the linearly polarized field,

$$q = p + \frac{e^2 |a^2|}{4(k \cdot p)} k \,, \qquad q^2 = p^2 + \frac{e^2 |a^2|}{2} = m^2 + \frac{e^2 |a^2|}{2} = m_*^2 \,. \tag{8.128}$$

We also define the parameters

$$\alpha = \frac{e\, a \cdot p}{k \cdot p} \,, \qquad \beta = -\frac{e^2 |a^2|}{8\,(k \cdot p)} \,. \tag{8.129}$$

We need to expand the phase factor $\exp(\mathrm{i}\, S_p\,[\phi])$. When comparing with the literature, one has to be very careful about the phase conventions used. Also, we should point out one little subtlety. Namely, in Eq. (50) of Ref. [272], we have the expression $y = \frac{e^2 a^2}{8\,(k \cdot p)}$, but the quantity denoted as a^2 in Ref. [272] is our $-|a^2|$. There, we have $A^\mu = a^\mu \cos\phi$ and not $A^\mu = a\,\epsilon^\mu \cos\phi$ [consult the text before Eq. (47) of Ref. [272]]. The definitions of α here and of the corresponding variable in Ref. [272] are in mutual agreement. The phase factor in Eq. (8.121) therefore evaluates to

$$\mathrm{i} S_p\,[\phi] = -q \cdot x - \mathrm{i}\,\alpha\, \sin\phi + \mathrm{i}\,\beta\, \sin(2\phi) \,. \tag{8.130}$$

In order to expand this phase factor into plane waves, we have to define generalized Bessel functions which are given as

$$\begin{aligned} A_0(s,\alpha,\beta) &= \frac{1}{2\pi} \int_{-\pi}^{\pi} \exp[\mathrm{i}\, s\,\theta - \mathrm{i}\,\alpha\, \sin(\theta) + \mathrm{i}\,\beta\, \sin(2\theta)]\, \mathrm{d}\theta \\ &= \frac{1}{2\pi} \int_{-\pi}^{\pi} \cos[s\,\theta - \alpha\, \sin(\theta) + \beta\, \sin(2\theta)]\, \mathrm{d}\theta \,. \end{aligned} \tag{8.131}$$

The general theorem of Fourier decomposition then implies that

$$\exp\left(-\mathrm{i}\alpha \sin\phi + \mathrm{i}\beta \sin(2\phi)\right) = \sum_{s=-\infty}^{\infty} A_0(s,\alpha,\beta)\, \mathrm{e}^{-\mathrm{i}s\phi} \,. \tag{8.132}$$

It is useful to also define the following generalized A functions,

$$\exp\left(-\mathrm{i}\alpha \sin\phi + \mathrm{i}\beta \sin(2\phi)\right) = \sum_{s=-\infty}^{\infty} A_0(s,\alpha,\beta)\, \mathrm{e}^{-\mathrm{i}s\phi} \,, \tag{8.133a}$$

$$\cos^L \phi \, \exp\left(-\mathrm{i}\alpha \sin\phi + \mathrm{i}\beta \sin(2\phi)\right) = \sum_{s=-\infty}^{\infty} A_L^c(s,\alpha,\beta)\, \mathrm{e}^{-\mathrm{i}s\phi} \,, \tag{8.133b}$$

$$\sin^L \phi \, \exp\left(-\mathrm{i}\alpha \sin\phi + \mathrm{i}\beta \sin(2\phi)\right) = \sum_{s=-\infty}^{\infty} A_L^s(s,\alpha,\beta)\, \mathrm{e}^{-\mathrm{i}s\phi} \,. \tag{8.133c}$$

The generalized Bessel functions with cosine and sine prefactors, given by $A_L^c(1,\alpha,\beta)$ and $A_L^s(1,\alpha,\beta)$, fulfill the following relations,

$$A_0(s,\alpha,\beta) = A_0^c(s,\alpha,\beta) = A_0^s(s,\alpha,\beta) \,, \tag{8.134a}$$

$$A_L^c(s,\alpha,\beta) = \tfrac{1}{2}\left(A_{L-1}^c(s+1,\alpha,\beta) + A_{L-1}^c(s-1,\alpha,\beta)\right) , \tag{8.134b}$$

$$A_L^s(s,\alpha,\beta) = \tfrac{1}{2\mathrm{i}}\left(A_{L-1}^s(s+1,\alpha,\beta) - A_{L-1}^s(s-1,\alpha,\beta)\right) , \tag{8.134c}$$

as is easily shown by inserting the parameters $s+1$ and $s-1$ into the right-hand side of Eqs. (8.133b) and (8.133c). The exponentials in the Dirac–Volkov state (8.119) can be expanded into Bessel functions in view of the Jacobi–Anger expansions,

$$\exp(\mathrm{i}z\cos\phi) = \sum_{s=-\infty}^{\infty} \mathrm{i}^s\, J_s(z)\, \mathrm{e}^{\mathrm{i}s\phi}, \qquad \exp(\mathrm{i}z\sin\phi) = \sum_{s=-\infty}^{\infty} J_s(z)\, \mathrm{e}^{\mathrm{i}s\phi}. \tag{8.135}$$

We have the following relations and sum rules,

$$A_0(s,\alpha,\beta) = \sum_{n=-\infty}^{\infty} J_{2n+s}(\alpha)\, J_n(\beta), \qquad \sum_{s=-\infty}^{\infty} A_0(s,\alpha,\beta) = \sum_{s=-\infty}^{\infty} [A_0(s,\alpha,\beta)]^2 = 1. \tag{8.136}$$

The Volkov solutions for *linear polarization* can thus be expanded into plane waves, with expansion coefficients being given by generalized Bessel functions,

$$\begin{aligned} \psi_{p,\sigma}(x) &= \sqrt{\frac{m}{QV}} \sum_{s=-\infty}^{\infty} \left(A_0(s,\alpha,\beta) + \frac{e\not{k}\not{a}}{2(k\cdot p)} A_1^c(s,\alpha,\beta) \right) \\ &\quad \times \mathcal{U}_\sigma(p) \exp[-\mathrm{i}(q+s\,k)\cdot x] \equiv \sqrt{\frac{m}{QV}}\, F(p,x)\, \mathcal{U}_\sigma(p). \end{aligned} \tag{8.137}$$

Here, $F(p,x)$ is a prefactor which constitutes a 4×4 matrix, and q is given by Eq. (8.128). The factor $F(p,x)$ constitutes a specialization of the prefactor $E(p,x)$ given in Eq. (8.119). The four-vector potential for a circularly polarized plane-wave field can be given in a unified notation as

$$A^\mu(\phi) = a_1^\mu \cos\phi + a_2^\mu \sin\phi, \qquad a_1^\mu = \widetilde{a}\,\epsilon_1^\mu, \qquad a_2^\mu = \widetilde{a}\,\epsilon_2^\mu. \tag{8.138}$$

The two polarization four-vectors ϵ_1 and ϵ_2 have zero timelike components $\epsilon_\lambda^0 = 0$ (for $\lambda = 1,2$), i.e., $\epsilon_\lambda = (0,\vec{\epsilon}_\lambda)$, and $\vec{k}\cdot\vec{\epsilon}_\lambda = 0$, so that $\epsilon_\lambda\cdot\epsilon_\rho = -\delta_{\lambda\rho}$ $(\lambda,\rho = 1,2)$. For circular polarization, we have $A^2(\phi) = A^\mu(\phi)A_\mu(\phi) - \widetilde{a}^2$ independent of ϕ $(\widetilde{a}^2 > 0)$. The phase factor evaluates to

$$S_p[\phi] = -\left(p + \frac{e^2\widetilde{a}^2}{2(k\cdot p)} k \right)\cdot x - \frac{e\,a_1\cdot p}{k\cdot p}\sin\phi + \frac{e\,a_2\cdot p}{k\cdot p}\cos\phi, \tag{8.139}$$

where the first term in round brackets in easily identified as the laser-dressed momentum in the circularly polarized field,

$$q = p + \frac{e^2\widetilde{a}^2}{2(k\cdot p)} k, \qquad q^2 = m^2 + e^2\widetilde{a}^2 = m_*^2. \tag{8.140}$$

We can thus write the argument in the exponential in terms of the two parameters

$$\xi = e\widetilde{a}\,\frac{q\cdot\epsilon_1}{k\cdot q}, \qquad \eta = e\widetilde{a}\,\frac{q\cdot\epsilon_2}{k\cdot q}, \tag{8.141}$$

so that

$$S_p[\phi] = -q\cdot x - \xi\sin\phi + \eta\cos\phi. \tag{8.142}$$

Hence, we can easily expand the exponentials into plane waves, as follows,

$$\exp(-\mathrm{i}\xi\sin\phi + \mathrm{i}\eta\cos\phi) = \sum_{s=-\infty}^{+\infty} B_0(s,\xi,\eta)\, \mathrm{e}^{-\mathrm{i}s\phi}. \tag{8.143}$$

Here, we may define the generalized Bessel functions as Fourier components of the function which needs to be expanded,

$$B_0(s,\xi,\eta) = \frac{1}{2\pi}\int_0^{2\pi} \exp\left(\mathrm{i}\,s\,\theta - \mathrm{i}\xi\sin\theta + \mathrm{i}\eta\cos\theta\right)\mathrm{d}\theta\,. \tag{8.144}$$

With the help of the Jacobi–Anger expansions (8.135), we can write

$$B_0(s,\xi,\eta) = \sum_{n=-\infty}^{\infty} \mathrm{i}^n\, J_{s+n}(\xi)\, J_n(\eta)\,. \tag{8.145}$$

It is useful to also define the following generalized B functions,

$$\exp\left(-\mathrm{i}\xi\sin\phi + \mathrm{i}\eta\cos\phi\right) = \sum_{s=-\infty}^{\infty} B_0(s,\alpha,\beta)\,\mathrm{e}^{-\mathrm{i}s\phi}\,, \tag{8.146a}$$

$$\cos^L\phi\,\exp\left(-\mathrm{i}\xi\sin\phi + \mathrm{i}\eta\cos\phi\right) = \sum_{s=-\infty}^{\infty} B_L^c(s,\alpha,\beta)\,\mathrm{e}^{-\mathrm{i}s\phi}\,, \tag{8.146b}$$

$$\sin^L\phi\,\exp\left(-\mathrm{i}\xi\sin\phi + \mathrm{i}\eta\cos\phi\right) = \sum_{s=-\infty}^{\infty} B_L^s(s,\alpha,\beta)\,\mathrm{e}^{-\mathrm{i}s\phi}\,. \tag{8.146c}$$

The generalized Bessel functions with cosine and sine prefactors, given by $B_L^c(s,\alpha,\beta)$ and $B_L^s(s,\alpha,\beta)$, fulfill the following relations,

$$B_0(s,\xi,\eta) = B_0^c(s,\xi,\eta) = B_0^s(s,\xi,\eta)\,, \tag{8.147a}$$

$$B_L^c(s,\xi,\eta) = \tfrac{1}{2}\left(B_{L-1}^c(s+1,\xi,\eta) + B_{L-1}^c(s-1,\xi,\eta)\right), \tag{8.147b}$$

$$B_L^s(s,\xi,\eta) = \tfrac{1}{2\mathrm{i}}\left(B_{L-1}^s(s+1,\xi,\eta) - B_{L-1}^s(s-1,\xi,\eta)\right), \tag{8.147c}$$

in full analogy with Eq. (8.134).

The Volkov solutions for *circular polarization* are thus expanded into so-called generalized Bessel functions, as follows,

$$\begin{aligned}\psi_{p,\sigma}(x) &= \sqrt{\frac{m}{QV}}\sum_{s=-\infty}^{\infty}\left(B_0\left(s,\xi,\eta\right) + \frac{e\not{k}}{2\left(k\cdot p\right)}\left(\not{a}_1\,B_1^c\left(s,\xi,\eta\right) + \not{a}_2\,B_1^s\left(s,\xi,\eta\right)\right)\right)\\ &\quad\times\mathcal{U}_\sigma\left(p\right)\exp\left[-\mathrm{i}(q+s\,k)\cdot x\right]\\ &\equiv \sqrt{\frac{m}{QV}}\,G(p,x)\,\mathcal{U}_\sigma(p).\end{aligned} \tag{8.148}$$

The last line implicitly defines the (4×4)-matrix $G(p,x)$, which tends to the unit matrix as the laser field is being switched off. In the circular case, we should point out that a simplification of the generalized Bessel function is possible, which leads to a full reduction to ordinary Bessel functions. Graf's addition theorem (see Sec. 7.6.1 of Ref. [156]) states that

$$\sum_{n=-\infty}^{\infty} J_{s+n}(x)\,J_n(y)\,\mathrm{e}^{\mathrm{i}n\varphi} = J_s(z)\left(\frac{x-y\,\mathrm{e}^{-\mathrm{i}\varphi}}{x-y\,\mathrm{e}^{\mathrm{i}\varphi}}\right)^{s/2}, \tag{8.149a}$$

$$z = \left(x^2+y^2-2xy\cos\varphi\right)^{1/2} = \left[\left(x-y\,\exp(-\mathrm{i}\varphi)\right)\left(x-y\,\exp(\mathrm{i}\varphi)\right)\right]^{1/2}. \tag{8.149b}$$

This is valid for $|y\,\mathrm{e}^{\pm\mathrm{i}\varphi}| < |x|$. If we apply this formula to B_s, we get

$$B_s(0,\xi,\eta) = J_{-s}\left(\sqrt{\xi^2+\eta^2}\right)\left(\frac{-\xi-\mathrm{i}\,\eta}{-\xi+\mathrm{i}\,\eta}\right)^{s/2}. \tag{8.150}$$

Using well-known properties of Bessel functions, it is easy to prove that $B_s{}^*(0,\xi,\eta) = B_s(0,\xi,-\eta)$ and $B_{-s}(0,\xi,\eta) = B_s(0,-\xi,\eta)$. Similar relations also hold for $B_s^c(0,\xi,\eta)$ and $B_s^s(0,\xi,\eta)$.

8.3.2 *Dirac–Volkov Propagator*

We start the discussion by recalling the Dirac–Volkov solution, which has been given in Eq. (8.119) as follows,

$$\psi_{p,\sigma}(x) = \sqrt{\frac{m}{QV}} \left(1 + \frac{e \not{k} \not{A}(\phi)}{2\,(k\cdot p)}\right) \mathcal{U}_\sigma(p) \exp\left(\mathrm{i}\, S_p\,[\phi]\right) \equiv \sqrt{\frac{m}{QV}}\, E(p,x)\, \mathcal{U}_\sigma(p)\,. \tag{8.151}$$

Here, p is the asymptotic energy-momentum four-vector of the electron, outside the laser field (it fulfills the relation $p^2 = m^2$), and the bispinor $\mathcal{U}_\sigma(p)$ is defined for an electron in the helicity basis, according to Eq. (7.89). The matrix $E(p,x)$ can be given as

$$E(p,x) = \left[1 + \frac{e \not{k} \not{A}(\phi)}{2k\cdot p}\right] \exp\left(-\mathrm{i} p\cdot x - \mathrm{i}\int_0^\phi \mathrm{d}\phi' \left[\frac{e\, p\cdot A(\phi')}{k\cdot p} - \frac{e^2\, A^2(\phi')}{2(k\cdot p)}\right]\right). \tag{8.152}$$

Assuming that $A^\mu(\phi)$ is a real rather than complex field, we define the adjoint by

$$\begin{aligned} \overline{E}(p,x) &= \gamma^0\, E^+(p,x)\, \gamma^0 \\ &= \left[1 + \frac{e \not{A}(\phi)\, \not{k}}{2k\cdot p}\right] \exp\left(\mathrm{i} p\cdot x + \mathrm{i}\int_0^\phi \mathrm{d}\phi' \left[\frac{e\, p\cdot A(\phi')}{k\cdot p} - \frac{e^2\, A^2(\phi')}{2k\cdot p}\right]\right). \end{aligned} \tag{8.153}$$

Here, $E^+(p,x)$ denotes the Hermitian conjugate of the matrix $E(p,x)$. By the standard rules for Dirac γ matrices, this means that in $\overline{E}$, one forms the complex conjugate of all scalar quantities and reverses the order of all Dirac matrices.

We generalize the E matrix, which initially has been defined for on-mass-shell momenta ($p^2 = m^2$), to momenta off the mass shell. The E matrices can be shown to have the following properties,

$$\left(\mathrm{i}\not{\partial} - e\not{A}(\phi)\right) E(p,x) = E(p,x)\, \not{p}, \tag{8.154}$$

$$\overline{E}(p,x) \left(-\mathrm{i}\overleftarrow{\not{\partial}} - e\not{A}(\phi)\right) = \not{p}\, \overline{E}(p,x)\,. \tag{8.155}$$

They also fulfill the following orthogonality properties,

$$\int \frac{\mathrm{d}^4 x}{(2\pi)^4} \overline{E}(p,x) E(p',x) = \delta^{(4)}(p-p')\, \mathbb{1}_{4\times 4}, \tag{8.156}$$

$$\int \frac{\mathrm{d}^4 p}{(2\pi)^4} E(p,x) \overline{E}(p,x') = \delta^{(4)}(x-x')\, \mathbb{1}_{4\times 4}\,. \tag{8.157}$$

We recall that both E as well as $\overline{E}$ are 4×4 matrices, and that the products $\overline{E}(p,x)E(p',x)$ as well as $E(p,x)\overline{E}(p,x')$ include matrix multiplication operations.

From the properties (8.154) and (8.157), we can easily find the form of the laser-dressed Green function [273–275],

$$\begin{aligned} G(x_1,x_2) &= \int \frac{\mathrm{d}^4p}{(2\pi)^4} E(p,x_1)\,\frac{\not p+m}{p^2-m^2+\mathrm{i}\epsilon}\,\overline{E}(p,x_2) \\ &= \int \frac{\mathrm{d}^4p}{(2\pi)^4}\left[1+\frac{e\not k\not A(\phi_1)}{2\,(k\cdot p)}\right]\frac{\not p+m}{p^2-m^2+\mathrm{i}\epsilon}\left[1+\frac{e\not A(\phi_2)\not k}{2\,(k\cdot p)}\right] \\ &\quad\times\exp\left(-\mathrm{i}p\cdot(x_1-x_2)-\mathrm{i}\int_{\phi_2}^{\phi_1}\mathrm{d}\phi'\left[\frac{e p\cdot A(\phi')}{k\cdot p}-\frac{e^2 A^2(\phi')}{2\,(k\cdot p)}\right]\right), \end{aligned} \tag{8.158}$$

where $\phi_1 = k\cdot x_1$ and $\phi_2 = k\cdot x_2$. Note that this Green function is the generalization of the Feynman propagator $S_F(x_1,x_2)$, given in Eq. (7.153), to the laser-dressed setting. Per our discussion in Chap. 7, the Feynman propagator differs from the time-ordered expectation value of the Dirac field operators by an overall prefactor which is equal to the imaginary unit. The Dirac–Volkov propagator fulfills the time-dependent equation

$$\left(\mathrm{i}\not\partial_1 - e\not A(\phi_1) - m\right)G(x_1,x_2) = \delta^{(4)}(x-x')\,\mathbb{1}_{4\times4}\,, \tag{8.159}$$

where $\not\partial_1 = \gamma^\mu\,\partial/\partial x_1^\mu$, and $\phi_1 = k\cdot x_1 = \omega t_1 - \vec{k}\cdot\vec{r}_1$. In contrast to Eqs. (7.176) and (8.105), one cannot eliminate the time dependence which is inherent to the oscillating laser field. One can show Eq. (8.159) by acting with the differential operator on the left-hand side of Eq. (8.158), using Eq. (8.154) and the property $(\not p - m)(\not p + m) = p^2 - m^2$, canceling the propagator denominator in Eq. (8.158), and finally carrying out the integral over the momenta, using Eq. (8.157).

From now on, we assume linear polarization, where we recall the Dirac–Volkov solution from Eq. (8.137) as follows,

$$\begin{aligned} \psi_{p,\sigma}(x) &= \sqrt{\frac{m}{QV}}\sum_{s=-\infty}^{\infty}\left(A_0(s,\alpha,\beta)+\frac{e\not k\not a}{2\,(k\cdot p)}A_1^c(s,\alpha,\beta)\right) \\ &\quad\times\mathcal{U}_\sigma(p)\exp\left[-\mathrm{i}\,(q+s\,k)\cdot x\right] \equiv \sqrt{\frac{m}{QV}}\,F(p,x)\,\mathcal{U}_\sigma(p)\,. \end{aligned} \tag{8.160}$$

In this case, the matrix $E(p,x)$ specializes to the matrix $F(p,x)$ defined in Eq. (8.137). Equation (8.158) can be rewritten using the expansion into generalized Bessel functions, assuming linear polarization $A^\mu = a^\mu\cos\phi$, as

$$\begin{aligned} G(x_1,x_2) &= \int\frac{\mathrm{d}^4p}{(2\pi)^4}\sum_{s,s'}\exp\left[-\mathrm{i}\left(p+\frac{e^2\,|a^2|}{4k\cdot p}k\right)\cdot(x_1-x_2)-\mathrm{i}k\cdot(sx_1-s'x_2)\right] \\ &\times\left[A_0(s,\alpha,\beta)+\frac{e\not k\not a}{2k\cdot p}A_1^c(s,\alpha,\beta)\right]\frac{\not p+m}{p^2-m^2+\mathrm{i}\epsilon}\left[A_0(s',\alpha,\beta)+\frac{e\not a\not k}{2k\cdot p}A_1^c(s',\alpha,\beta)\right], \end{aligned} \tag{8.161}$$

where the sums over s and s' extend from $-\infty$ to ∞ and we recall from Eq. (8.129) that $\alpha = \frac{e\,a\cdot p}{k\cdot p}$, and $\beta = -\frac{e^2\,|a^2|}{8k\cdot p}$. The Green function can thus be expressed in terms of the free propagator, inserted between the Volkov-like matrix-valued functions E and $\overline{E}$.

In Eq. (8.161), the dependence on the integration variable p is such that the Fourier components of the propagator are not uniquely identified. It would be better if one could write the propagator as an expansion over the Fourier components $\exp(-\mathrm{i}\, q \cdot x)$, i.e., in terms of

$$q = p + \frac{e^2\,|a^2|}{4k \cdot p} k. \tag{8.162}$$

The Jacobian of this transformation is

$$\begin{aligned} J &= \det\left(\frac{\partial q^\mu}{\partial p^\nu}\right) = \det\left(\delta^{\mu\nu} - \frac{e^2\,|a^2|\,k^\nu\,k_\alpha\delta^{\alpha\mu}}{4(k\cdot p)^2}\right) = \epsilon^{\mu\nu\alpha\beta}\,\frac{\partial q^0}{\partial p^\mu}\,\frac{\partial q^1}{\partial p^\nu}\,\frac{\partial q^2}{\partial p^\alpha}\,\frac{\partial q^3}{\partial p^\beta} \\ &= \epsilon^{\mu\nu\alpha\beta}\left(\delta^{0\mu} - \frac{e^2\,a^2\,k^\mu k_\gamma\delta^{\gamma 0}}{4(k\cdot p)^2}\right)\left(\delta^{1\nu} - \frac{e^2\,a^2\,k^\nu\,k_\gamma\delta^{\gamma 1}}{4(k\cdot p)^2}\right) \\ &\quad\times\left(\delta^{2\alpha} - \frac{e^2a^2k^\alpha k_\gamma\delta^{\gamma 2}}{4\,(k\cdot p)^2}\right)\left(\delta^{3\beta} - \frac{e^2\,a^2\,k^\beta\,k_\gamma\,\delta^{\gamma 3}}{4\,(k\cdot p)^2}\right) = 1. \end{aligned} \tag{8.163}$$

Here, $\epsilon^{\mu\nu\alpha\beta}$ is the usual anti-symmetric symbol with $\epsilon^{0123} = 1$. The last step can be seen most easily by choosing an explicit coordinate system, e.g., by letting $k = (\omega, \omega, 0, 0)$. We then have

$$\mathrm{d}p^0\mathrm{d}p^1\mathrm{d}p^2\mathrm{d}p^3 = J^{-1}\mathrm{d}q^0\mathrm{d}q^1\mathrm{d}q^2\mathrm{d}q^3 = \mathrm{d}q^0\mathrm{d}q^1\mathrm{d}q^2\mathrm{d}q^3. \tag{8.164}$$

Written in the new integration variable, and noting that $k\cdot p = k\cdot q$ and $a\cdot p = a\cdot q$, the expression for the propagator reads

$$\begin{aligned} G(x_1, x_2) &= \int \frac{\mathrm{d}^4 q}{(2\pi)^4} \sum_{s,s'=-\infty}^{\infty} \left(A_0(s,\alpha,\beta) + \frac{e\,\not{k}\,\not{a}}{2k\cdot q}\,A_1^c(s,\alpha,\beta)\right) \frac{\not{q} - \frac{e^2\,|a^2|}{4\,(k\cdot q)}\not{k} + m}{q^2 - m_*^2 + \mathrm{i}\epsilon} \\ &\times \left(A_0(s',\alpha,\beta) + \frac{e\,\not{a}\,\not{k}}{2\,(k\cdot q)} A_1^c(s',\alpha,\beta)\right)\exp\left(-\mathrm{i}q\cdot(x_1 - x_2) - \mathrm{i}k\cdot(s\,x_1 - s'\,x_2)\right). \end{aligned} \tag{8.165}$$

We see that the poles are shifted to the effective mass shell, which for linear polarization reads as follows [see Eq. (8.128)]

$$p^2 = m_*^2 = m^2 + \frac{e^2\,|a^2|}{2}. \tag{8.166}$$

Finally, for circular polarization, we have from Eq. (8.148),

$$\begin{aligned} \psi_{p,\sigma}(x) &= \sqrt{\frac{m}{QV}} \sum_{s=-\infty}^{\infty} \exp\left[-\mathrm{i}(q + s\,k)\cdot x\right] \\ &\quad\times\left(B_0\,(s,\xi,\eta) + \frac{e\,\not{k}}{2\,(k\cdot p)}\,(\not{a}_1\,B_1^c\,(s,\xi,\eta) + \not{a}_2\,B_1^s\,(s,\xi,\eta))\right)\mathcal{U}_\sigma\,(p) \\ &\equiv \sqrt{\frac{m}{QV}}\,H(p,x)\,\mathcal{U}_\sigma(p), \end{aligned} \tag{8.167}$$

which defines the matrix-valued function $H(p,x)$. We recall the definitions $\xi = e\,\widetilde{a}\,\frac{q\cdot\epsilon_1}{k\cdot q}$ and $\eta = e\,\widetilde{a}\,\frac{q\cdot\epsilon_2}{k\cdot q}$ from Eq. (8.141), and the laser-dressed mass in the circularly polarized field,

$$m_*^2 = m^2 + e^2\,|\widetilde{a}^2|\,, \qquad |\widetilde{a}^2| = \widetilde{a}^2 > 0\,. \tag{8.168}$$

The propagator thus reads as

$$G(x_1,x_2) = \int \frac{\mathrm{d}^4p}{(2\pi)^4} H(p,x_1)\,\frac{\not{p}+m}{p^2-m^2+i\epsilon}\,\overline{H}(p,x_2) \tag{8.169}$$

and can be expressed as follows,

$$\begin{aligned} G(x_1,x_2) = \int \frac{\mathrm{d}^4p}{(2\pi)^4} \sum_{s,s'=-\infty}^{\infty} &\left(B_0\,(s,\xi,\eta) + \frac{e\,\not{k}}{2\,(k\cdot p)}\,(\not{\epsilon}_1 B_1^c\,(s,\xi,\eta) + \not{\epsilon}_2 B_1^s\,(s,\xi,\eta))\right) \\ &\times \frac{\not{p}+m}{p^2-m^2+\mathrm{i}\epsilon}\left(B_0\,(s,\xi,\eta) + (\not{\epsilon}_1\,B_1^c\,(s,\xi,\eta) + \not{\epsilon}_2\,B_1^s\,(s,\xi,\eta))\,\frac{e\,\not{k}}{2\,(k\cdot p)}\right) \\ &\times \exp\left[-\mathrm{i}\left(p + \frac{e^2\,|\widetilde{a}^2|}{4k\cdot p}k\right)\cdot(x_1-x_2) - \mathrm{i}k\cdot(s\,x_1 - s'\,x_2)\right]. \end{aligned} \tag{8.170}$$

After the change of variable $p \to q$ in the Green function, governed by Eqs. (8.162) and (8.163), we can express $G(x_1,x_2)$ as follows,

$$\begin{aligned} G(x_1,x_2) = \int \frac{\mathrm{d}^4q}{(2\pi)^4} \sum_{s,s'=-\infty}^{\infty} &\left(B_0\,(s,\xi,\eta) + \frac{e\,\not{k}}{2\,(k\cdot q)}\,(\not{\epsilon}_1\,B_1^c\,(s,\xi,\eta) + \not{\epsilon}_2\,B_1^s\,(s,\xi,\eta))\right) \\ &\times \frac{\not{q} - \frac{e^2\,|\widetilde{a}^2|}{4\,(k\cdot q)}\not{k} + m}{q^2-m_*^2+\mathrm{i}\epsilon}\left(B_0\,(s,\xi,\eta) + (\not{\epsilon}_1 B_1^c\,(s,\xi,\eta) + \not{\epsilon}_2 B_1^s\,(s,\xi,\eta))\,\frac{e\not{k}}{2(k\cdot q)}\right) \\ &\times \exp\left[-\mathrm{i}q\cdot(x_1-x_2) - \mathrm{i}k\cdot(s\,x_1 - s'\,x_2)\right]. \end{aligned} \tag{8.171}$$

In the limit of a vanishing laser field, the Volkov solution asymptotically tends to $\psi_{p,\sigma}(x) \to \sqrt{m/(EV)}u_\sigma(p)\exp(-\mathrm{i}p\cdot x)$, which is a the solution of the free Dirac equation, in view of the property $B_0(s,0,0) = \delta_{s0}$. Also, in this limit, we have $H(p,x) \to \exp(-\mathrm{i}p\cdot x)$, which is the laser-field-free expression. For circular polarization, one can simplify the amplitudes as described in Ref. [276], using identities related to Eq. (8.149).

Furthermore, in the limit $a \to 0$, the effective four-momentum (8.122) takes the same value as the free momentum, $q^\mu = p^\mu$. Obviously, the arguments of the generalized Bessel functions (8.141) are equal to zero in this case. One can then show that the laser-dressed electron propagator approaches the expression

$$G(x_1,x_2) \xrightarrow{a\to 0} \int \frac{\mathrm{d}^4p}{(2\pi)^4}\,\frac{\mathrm{e}^{-\mathrm{i}\,p\cdot(x_1-x_2)}}{\not{p}-m+\mathrm{i}\epsilon}\,, \tag{8.172}$$

i.e., the free electron propagator.

8.4 Dirac Equation with Coupling to Gravitational Fields

8.4.1 *Metric and Covariant Derivative*

In Sec. 4.2.1, we had already encountered the flat-space-time metric in "West-Coast" conventions, $\{\gamma^\mu, \gamma^\nu\} = 2\,g^{\mu\nu}$, where $g^{\mu\nu} = \mathrm{diag}(1,-1,-1,-1)$, as adopted in the modern particle physics literature [2,256] [see also Eq. (2.49)]. In general relativity, the metric tensor can locally be chosen to be identical to the flat-space metric, but globally, space-time curvature induces a deviation from the flat-space metric, according to the replacement $g^{\mu\nu} \to \overline{g}^{\mu\nu}(x)$. In order to describe the coupling of Dirac particles to curved space-times, an interesting and quite sophisticated formalism will need to be discussed in detail.

The presence of the coordinate-dependent metric implies that the partial derivative needs to be replaced by a covariant derivative. For tensors, the definition of the covariant derivative ∇_μ is well established [see Exercise 8.4 on page 211 of Ref. [277]]. Let us still include a pertinent discussion. From differential geometry, we know that Christoffel symbols enter the definition of the covariant derivative. For a position-dependent metric $\overline{g}_{\mu\nu}(x)$, the covariant components $A_\mu(x)$ are connected to the contravariant components $A^\mu(x)$ of a vector field by the relations

$$\begin{aligned} x_\mu &= \overline{g}_{\mu\nu}(x)\, x^\nu\,, & x^\mu &= \overline{g}^{\mu\nu}(x)\, x_\nu\,, \\ A_\mu(x) &= \overline{g}_{\mu\nu}(x)\, A^\nu(x)\,, & A^\mu(x) &= \overline{g}^{\mu\nu}(x)\, A_\nu(x)\,. \end{aligned} \tag{8.173}$$

From now on, we suppress the dependence of the metric on the coordinates and just write $\overline{g}_{\mu\nu}$ for $\overline{g}_{\mu\nu}(x)$ for the curved-space metric, in order to distinguish it from the flat-space metric given in Eq. (2.49).

Let us briefly motivate the definition of the covariant derivative. First of all, we recall that a vector $\vec{A}$ actually is independent of a coordinate system. A local basis consists of the set ("tetrad" basis) of unit vectors $\vec{e}_\mu$,

$$\vec{A} = A^\mu\, \vec{e}_\mu\,, \qquad \vec{e}_\mu \cdot \vec{e}_\nu = \overline{g}_{\mu\nu}\,, \qquad \vec{e}^{\,\mu} \cdot \vec{e}_\nu = \overline{g}^\mu{}_\nu = \delta^\mu{}_\nu\,. \tag{8.174}$$

The basis vectors $\vec{e}_\mu$ are non-constant, and an additional term emerges as one calculates the derivative of the vector $\vec{A}$ with respect to a coordinate,

$$\begin{aligned} \frac{\partial}{\partial x^\nu}\vec{A} &= \partial_\nu \vec{A} = (\partial_\nu A^\mu)\,\vec{e}_\mu + A^\mu\,(\partial_\nu \vec{e}_\mu) = (\partial_\nu A^\mu)\,\vec{e}_\mu + A^\mu\,\vec{e}_\lambda\,\overline{g}^{\lambda\rho}\,(\vec{e}_\rho \cdot \partial_\nu \vec{e}_\mu) \\ &= (\partial_\nu A^\mu)\,\vec{e}_\mu + A^\mu\,\vec{e}_\lambda\,\overline{g}^{\lambda\rho}\,\Gamma_{\rho\,\mu\,\nu} = (\partial_\nu A^\mu)\,\vec{e}_\mu + A^\mu\,\vec{e}_\lambda\,\Gamma^\lambda_{\mu\,\nu}\,, \end{aligned} \tag{8.175}$$

where $\Gamma_{\rho\,\mu\,\nu} = \vec{e}_\rho \cdot \partial_\nu \vec{e}_\mu$ is a Christoffel symbol, and the Einstein summation convention has been used (all the Greek indices are summed from zero to three). We now swap the dummy indices (summation) in the second term according to $(\lambda \leftrightarrow \mu)$ and identify the components of the covariant derivative,

$$\partial_\nu \vec{A} = \vec{e}_\mu\,\nabla_\nu A^\mu\,, \qquad \nabla_\nu A^\mu = \partial_\nu A^\mu + \Gamma^\mu_{\lambda\,\nu}\,A^\lambda\,. \tag{8.176}$$

While the Christoffel symbols $\Gamma^\mu_{\lambda\,\nu}$ are not tensors, one can still raise the first index with the metric,

$$\Gamma_{\rho\mu\nu} = \vec{e}_\rho \cdot \frac{\partial \vec{e}_\mu}{\partial x^\nu}\,, \qquad \Gamma^\beta_{\mu\nu} = \overline{g}^{\beta\rho}\,\Gamma_{\rho\mu\nu}\,. \tag{8.177}$$

A rather straightforward calculation shows that

$$\Gamma_{\beta\mu\nu} = \frac{1}{2}\left(\frac{\partial\overline{g}_{\beta\mu}}{\partial x^\nu} + \frac{\partial\overline{g}_{\beta\nu}}{\partial x^\mu} - \frac{\partial\overline{g}_{\mu\nu}}{\partial x^\beta}\right). \tag{8.178}$$

By inspection, one can easily derive the symmetry properties

$$\Gamma_{\beta\mu\nu} = \Gamma_{\beta\nu\mu}, \qquad \Gamma^\beta_{\mu\nu} = \Gamma^\beta_{\nu\mu}. \tag{8.179}$$

The covariant derivative of a vector field $A^\mu(x)$ has the properties,

$$\nabla_\nu A^\mu = \partial_\nu A^\mu + \Gamma^\mu_{\nu\beta}\, A^\beta, \qquad \partial_\nu \equiv \frac{\partial}{\partial x^\nu}, \tag{8.180a}$$

$$\nabla_\nu A_\mu = \partial_\nu A_\mu - \Gamma^\beta_{\nu\mu}\, A_\beta, \tag{8.180b}$$

$$\overline{g}^{\beta\nu}\,\nabla_\mu A_\nu = \nabla_\mu\left(\overline{g}^{\beta\nu}A_\nu\right) = \nabla_\mu A^\beta, \tag{8.180c}$$

i.e., the covariant derivative transforms constitutes a tensor (unlike the partial derivative), and its indices can therefore be raised and lowered with the metric, both before and after the covariant differentiation.

The property (8.180b) can be shown as follows,

$$\begin{aligned}
\frac{\partial}{\partial x^\nu}\vec{A} &= \partial_\nu\vec{A} = (\partial_\nu A_\mu)\,\vec{e}^{\,\mu} + A_\mu\,(\partial_\nu\vec{e}^{\,\mu}) = (\partial_\nu A_\mu)\,\vec{e}^{\,\mu} + A_\mu\,\partial_\nu\left(\overline{g}^{\mu\rho}\,\vec{e}_\rho\right) \\
&= (\partial_\nu A_\mu)\,\vec{e}^{\,\mu} + A_\mu\,(\partial_\nu\overline{g}^{\mu\rho})\,\overline{g}_{\rho\sigma}\,\vec{e}^{\,\sigma} + A^\rho\,\partial_\nu\vec{e}_\rho \\
&= (\partial_\nu A_\mu)\,\vec{e}^{\,\mu} - A_\mu\,\overline{g}^{\mu\rho}\left(\partial_\nu\overline{g}_{\rho\sigma}\right)\vec{e}^{\,\sigma} + A^\rho\,\vec{e}^{\,\sigma}\,(\vec{e}_\sigma\cdot\partial_\nu\vec{e}_\rho) \\
&= (\partial_\nu A_\mu)\,\vec{e}^{\,\mu} - A^\rho\left(\partial_\nu\overline{g}_{\rho\sigma}\right)\vec{e}^{\,\sigma} + A^\rho\,\vec{e}^{\,\sigma}\,\partial_\nu\underbrace{(\vec{e}_\sigma\cdot\vec{e}_\rho)}_{=\overline{g}_{\sigma\rho}} - A^\rho\,\vec{e}^{\,\sigma}\,(\vec{e}_\rho\cdot\partial_\nu\vec{e}_\sigma) \\
&= (\partial_\nu A_\mu)\,\vec{e}^{\,\mu} - A^\rho\,\vec{e}^{\,\sigma}\,\Gamma_{\rho\sigma\nu} = (\partial_\nu A_\mu)\,\vec{e}^{\,\mu} - A_\mu\,\Gamma^\mu_{\sigma\nu}\,\vec{e}^{\,\sigma}.
\end{aligned} \tag{8.181}$$

We can now use the property (8.179) and write

$$\nabla_\nu A_\mu = \partial_\nu A_\mu - \Gamma^\beta_{\nu\mu}\,A_\beta. \tag{8.182}$$

The equations (8.176) and (8.182) are complementary.

The Riemann curvature tensor $R^\rho{}_{\sigma\mu\nu}$ is of rank four and can be expressed in terms of the Christoffel symbols,

$$R^\rho{}_{\sigma\mu\nu} = \partial_\mu\Gamma^\rho_{\nu\sigma} - \partial_\nu\Gamma^\rho_{\mu\sigma} + \Gamma^\rho_{\mu\lambda}\,\Gamma^\lambda_{\nu\sigma} - \Gamma^\rho_{\nu\lambda}\,\Gamma^\lambda_{\mu\sigma}. \tag{8.183}$$

For the Ricci tensor $R_{\mu\nu}$, one contracts the first and the third index of the Riemann curvature tensor,

$$R_{\mu\nu} = R^\alpha{}_{\mu\alpha\nu}, \qquad R = \overline{g}^{\mu\nu}\,R_{\mu\nu} = R^\mu{}_\mu. \tag{8.184}$$

The scalar curvature R of space-time is obtained as the contraction of the Ricci tensor. The Einstein–Hilbert action S_{EH} is given by

$$S_{EH} = \int \mathrm{d}^4x\,\sqrt{-\det\overline{g}}\,R, \qquad \frac{\delta S_{EH}}{\delta\overline{g}^{\alpha\beta}} = \sqrt{-\det\overline{g}}\left(R_{\alpha\beta} - \frac{1}{2}\,\overline{g}_{\alpha\beta}\,R\right), \tag{8.185}$$

and the variational derivative of the action with respect to the metric tensor leads to the condition $R_{\alpha\beta} - \frac{1}{2}\overline{g}_{\alpha\beta} R = 0$, which is the Einstein–Hilbert equation for a region of the Universe, in which there is no matter and no other fields which give rise to a nonvanishing energy-momentum tensor. The variational condition $\delta S_{EH} = 0$ is equivalent to saying that the global integral over the curvature of space-time R acts like a soap film whose surface tension integral is being minimized by the shape assumed by space-time geometry, which is four-dimensional but might be embedded in a higher-dimensional manifold which would defy our intuitive understanding. The integration measure $\int d^4x \sqrt{-\det \overline{g}}$ is Lorentz-invariant. The coupling of the space-time curvature to the matter density in the Universe can be written in terms of a sum of the Einstein–Hilbert action S_{EH} and the matter action S_M, where the latter can be expressed in terms of the matter Lagrangian L_M,

$$S_{GR} = \frac{1}{16\pi G} S_{EH} + S_M \,, \qquad S_M = \int d^4x \sqrt{-\det \overline{g}}\, L_M \,, \tag{8.186a}$$

$$T_{\mu\nu} = -\frac{2}{\sqrt{-\det g}} \frac{\delta S_M}{\delta \overline{g}^{\mu\nu}} = -\frac{2}{\sqrt{-\det g}} \frac{\partial(\sqrt{-\det g}\, L_M)}{\partial \overline{g}^{\mu\nu}} = -2\frac{\partial L_M}{\partial \overline{g}^{\mu\nu}} + \overline{g}_{\mu\nu} L_M \,. \tag{8.186b}$$

The latter functional derivative acts as a definition of the energy-momentum tensor $T_{\mu\nu}$. The form given in Eq. (8.186b) is valid if the matter Lagrangian does not depend on derivatives of the metric. Examples of the energy-momentum tensor include a collection of particles,

$$T^{\mu\nu} = \sum_a \int d\tau_a \frac{\delta^{(4)}(x - x(\tau_a))}{\sqrt{-\det \overline{g}}}\, m_a\, v_a^\mu\, v_a^\nu \,. \tag{8.187}$$

Here, the v_a^μ are the four-vector components of the velocity v_a of particle a, and τ_a is the proper time of the trajectory of particle a. For an electromagnetic field, the stress-energy tensor is given as follows (see Ref. [81]),

$$T^{\mu\nu} = -\left(F^{\mu\alpha}\, F^{\nu}{}_{\alpha} - \frac{1}{4} \overline{g}^{\mu\nu}\, F^{\alpha\beta}\, F_{\alpha\beta} \right) , \tag{8.188}$$

where we recall that $\epsilon_0 = \mu_0 = 1$ for the current discussion. Under the inclusion of the energy-momentum tensor, the variational equations read as follows,

$$R^{\alpha\beta} - \frac{1}{2} R\, \overline{g}^{\alpha\beta} = 8\pi G\, T^{\alpha\beta} \,. \tag{8.189}$$

These equations have all been derived under a variation of the action with respect to the metric. They still do not clarify how the Maxwell equations need to be generalized to curved space-time, and this will be explored next.

It is not hard to guess that for a free electromagnetic field, one simply has to use the Lorentz-invariant measure $\int d^4x \sqrt{-\det \overline{g}}$ in the action integral for the electromagnetic field,

$$S_{EM} = \int d^4x \sqrt{-\det \overline{g}} \left(-\frac{1}{4} F^{\mu\nu}\, F_{\mu\nu} \right) . \tag{8.190}$$

The field-strength tensor, in the curved space-time, is given as follows,

$$\nabla_\mu A_\nu - \nabla_\nu A_\mu = \partial_\mu A_\nu - \partial_\nu A_\mu \,. \tag{8.191}$$

Here, we have used the definition (8.180) of the covariant derivative and the symmetry properties (8.179) of the Christoffel symbols. Variation with respect to the metric leads to the energy-momentum tensor already given in Eq. (8.188), modulo some total divergence. However, the variational principle also dictates that the action S_{EM} should be invariant under a variation of the four-vector field A^μ. The corresponding Euler–Lagrange equation gives rise to the condition

$$\frac{\partial}{\partial x^\mu}\left(F^{\mu\nu}\sqrt{-\det \overline{g}}\right) = 0\,, \qquad \nabla_\mu F^{\mu\nu} = 0\,, \tag{8.192}$$

where ∇_μ is the covariant derivative. It might be useful to remind the reader of the identity

$$\frac{1}{\sqrt{-\det \overline{g}}}\frac{\partial}{\partial x^\mu}\left(F^{\mu\nu}\sqrt{-\det \overline{g}}\right) = \nabla_\mu F^{\mu\nu}\,, \tag{8.193}$$

which follows from $\partial_\mu \ln(\sqrt{-|g|}) = \frac{1}{2} g^{\alpha\beta}\partial_\mu g_{\alpha\beta} = \Gamma^\kappa_{\mu\kappa}$ (we recall the Einstein summation convention one more time) and the fact that $F^{\mu\nu}$ is antisymmetric. The covariant derivative of the field-strength tensor can be expressed in terms of Christoffel symbols,

$$\nabla_\alpha F^{\beta\gamma} = \partial_\alpha F^{\beta\gamma} + \Gamma^\beta_{\alpha\mu}\, F^{\mu\gamma} + \Gamma^\gamma_{\alpha\rho}\, F^{\beta\rho}\,, \tag{8.194a}$$

$$\nabla_\mu F_{\alpha\beta} = \partial_\mu F_{\alpha\beta} - \Gamma^\rho_{\alpha\mu}\, F_{\rho\beta} - \Gamma^\rho_{\beta\mu}\, F_{\alpha\rho}\,. \tag{8.194b}$$

If we consider a source term, then Eq. (8.190) gets modified to read

$$S_{EM} = \delta \int \mathrm{d}^4x \sqrt{-\det \overline{g}}\left(-\frac{1}{4} F^{\mu\nu}\, F_{\mu\nu} - J_\mu\, A^\mu\right)$$

$$\frac{\delta S_{EM}}{\delta A_\mu} = \frac{\partial}{\partial x^\nu}\left(F^{\nu\mu}\sqrt{-\det \overline{g}}\right) - J^\mu \sqrt{-\det \overline{g}} = 0\,. \tag{8.195}$$

In component form, we have

$$\frac{1}{\sqrt{-\det \overline{g}}}\frac{\partial}{\partial x^\nu}\left(F^{\nu\mu}\sqrt{-\det \overline{g}}\right) = J^\mu\,, \qquad \nabla_\mu F^{\mu\nu} = J^\nu\,, \tag{8.196}$$

where ∇_μ is the covariant derivative. The equations (8.196) are the generalizations of the inhomogeneous Maxwell equations (see Ref. [81]), namely, $\partial_\mu F^{\mu\nu} = J^\nu$. Finally, we should remark that the covariant derivative of the metric is obtained as a generalization of Eq. (8.194b),

$$\nabla_\mu \overline{g}_{\alpha\beta} = \overline{g}_{\alpha\beta,\mu} - \Gamma^\lambda_{\alpha\mu}\, \overline{g}_{\lambda\beta} - \Gamma^\lambda_{\beta\mu}\, \overline{g}_{\alpha\lambda} = \overline{g}_{\alpha\beta,\mu} - \Gamma_{\beta\alpha\mu} - \Gamma_{\alpha\beta\mu} = 0\,. \tag{8.197}$$

It vanishes according to Eq. (8.178).

8.4.2 *Tetrad Basis and Affine Connection Matrix*

One might think that the Dirac Hamiltonian could be coupled to a central gravitational field by simply adding the Newtonian gravitational potential to the free Dirac Hamiltonian, leading to the expression

$$H_G = \vec{\alpha} \cdot \vec{p} + \beta\, m - G \frac{mM}{r} . \tag{8.198}$$

One might argue that, in the same right, we had simply added the static Coulomb potential to the free Dirac Hamiltonian, which leads to the Dirac–Coulomb Hamiltonian given in Eq. (8.9).

However, the Hamiltonian given in Eq. (8.198) does not constitute a valid description of a Dirac particle in a central gravitational field. The reason is that the Coulomb potential is not added to the free Dirac Hamiltonian just by chance. Rather, this additional term results from the minimal coupling

$$\mathrm{i}\,\partial_\mu \to \mathrm{i}\,\partial_\mu - e\,A_\mu \tag{8.199}$$

which is applicable to electromagnetic gauge theory (see Sec. 8.2.1). Setting $\mu = 0$ and $e\,A_0 = V = -Z\alpha/r$, one obtains the Dirac–Coulomb Hamiltonian. No such $U(1)$-gauge theory inspired prescription is available for gravitational coupling. The gravitational potential of a planet cannot be inserted at face value into the Dirac equation, like the Coulomb potential. Rather, the full formalism of general relativity needs to be employed, and the gravitational coupling involves both a redefinition of the algebra of Dirac matrices, adjusted to curved space-time. The coupling is formulated covariantly under local spinor Lorentz transformations; the covariance principle is derived from general relativity, not from electromagnetic $U(1)$ gauge theory.

We recall the Dirac equation in flat space according to Eq. (7.17),

$$(\mathrm{i}\,\gamma^\mu\,\partial_\mu - m)\,\psi(x) = 0\,, \qquad x = (t, \vec{r})\,. \tag{8.200}$$

The γ^μ matrices fulfill the (anti-)commutation relations $\{\gamma^\mu, \gamma^\nu\} = 2\,g^{\mu\nu}$, where $g^{\mu\nu} = \mathrm{diag}(1,-1,-1,-1)$ is the flat-space-time metric in "West-Coast" conventions adopted in the modern particle physics literature [2, 256]. In general relativity, the metric tensor can locally be chosen to be identical to the flat-space metric, but globally, space-time curvature induces a deviation from the flat-space metric, according to the replacement $g^{\mu\nu} \to \overline{g}^{\mu\nu}(x)$. The curved-space-time Dirac matrices fulfill the commutation relations

$$\{\overline{\gamma}^\mu(x), \overline{\gamma}^\nu(x)\} = 2\,\overline{g}^{\mu\nu}(x)\,, \tag{8.201}$$

where the local curved-space metric is $\overline{g}^{\mu\nu}(x)$ with $\mu, \nu = 0, 1, 2, 3$. Just to fix ideas, we should stress that the Dirac equation in curved space-time cannot simply be obtained by replacing the γ^μ matrices in Eq. (8.200) by the equivalent $\overline{\gamma}^\mu(x)$ matrices. This perhaps surprising fact has already been observed in Refs. [278–280].

We here follow the conventions used in Refs. [255, 281] for the flat-space and curved-space Dirac gamma matrices. The flat-space-time Dirac gamma matrices are denoted with a tilde ($\widetilde{\gamma}$) while the curved-space-time Dirac gamma matrices are written with an overline ($\overline{\gamma}$). *This means that the* $\widetilde{\gamma}$ *matrices used in the following are the same as the* γ *matrices (without tilde) used for the flat-space Dirac equation.* The notation for the curved-space-time matrices is inspired by the covariant structure of their anticommutator, expressed in Eq. (8.201), and the tensor ("vector") structure is denoted by the overline. By contrast, from the point of view of general relativity, the flat-space matrices can be regarded as "modified" γ matrices, *hence the tilde.*

Via integration by parts, we can modify the Lagrangian of the free Dirac field, given by the Lagrangian (7.79),

$$\mathcal{L}(x) = \overline{\psi}(x)\,(\mathrm{i}\gamma^\mu\partial_\mu - m)\,\psi(x)\,, \tag{8.202}$$

as follows,

$$\mathcal{S}_0 = \int \mathrm{d}^4x\,\overline{\psi}(x)\left(\frac{\mathrm{i}}{2}\gamma^\mu\overleftrightarrow{\partial}_\mu - m\right)\psi(x)\,, \tag{8.203}$$

where

$$A(x)\,\overleftrightarrow{\partial}_\mu\,B(x) \equiv A(x)\,\partial_\mu\,B(x) - B(x)\,\partial_\mu\,A(x)\,. \tag{8.204}$$

The generalization of the Dirac action (8.203) to curved space-time involves two steps: *(i)* an obvious generalization of the anticommutator relations (8.201) to curved space, $\{\overline{\gamma}^\mu(x), \overline{\gamma}^\mu(x)\} = 2\,\overline{g}^{\mu\nu}(x)$, and *(ii)* the introduction of a covariant derivative, $\partial_\mu \to \nabla_\mu$, which is defined as

$$\nabla_\mu = \partial_\mu - \Gamma_\mu(x)\,, \tag{8.205}$$

where ∇_μ is the covariant derivative and $\Gamma_\mu(x)$ is a (4×4)-matrix that specifies the affine connection. The action for the Dirac particle in curved space-time is obtained as a generalization of Eq. (7.78)

$$S = \int \mathrm{d}^4x\,\sqrt{-\det\,\overline{g}}\;\overline{\psi}(x)\,(\mathrm{i}\overline{\gamma}^\mu(x)\nabla_\mu - m)\,\psi(x)\,, \tag{8.206}$$

where $\det\overline{g} < 0$ is the determinant of the local space-time metric, and $\sqrt{-\det\overline{g}}\,\mathrm{d}^4x$ is a Lorentz-invariant measure. Because the local Dirac γ matrices fulfill the relation $\{\overline{\gamma}^\mu(x), \overline{\gamma}^\nu(x)\} = 2\,\overline{g}^{\mu\nu}(x)$, it is natural to postulate that the covariant derivative of a single curved-space Dirac matrix $\overline{\gamma}^\mu(x)$ needs to vanish, individually,

$$\nabla_\nu\,\overline{\gamma}_\mu(x) = 0\,. \tag{8.207}$$

Under the assumption (8.207), we can symmetrize Eq. (8.206) as follows,

$$S = \int \mathrm{d}^4x\,\sqrt{-\det\overline{g}}\;\overline{\psi}(x)\left(\frac{\mathrm{i}}{2}\,\overline{\gamma}^\mu(x)\,\overleftrightarrow{\nabla}_\mu - m\right)\psi(x)\,. \tag{8.208}$$

By variation, the gravitationally coupled Dirac equation is obtained as

$$(\mathrm{i}\gamma^\mu\,\nabla_\mu - m)\,\psi(x) = [\mathrm{i}\,\gamma^\mu\,(\partial_\mu - \Gamma_\mu) - m]\,\psi(x) = 0\,. \tag{8.209}$$

For a Lorentz vector T^μ, we recall that according to Eq. (8.182),

$$\nabla_\nu T_\mu = \partial_\nu T_\mu - \Gamma^\lambda_{\nu\mu}\,T_\lambda\,, \tag{8.210}$$

where the $\Gamma^{\mu}_{\rho\sigma}(x)$ are the Christoffel symbols defined according to Eq. (8.178). Based on Eq. (8.210), we write the covariant derivative [278–280] of the Dirac matrix $\overline{\gamma}_{\mu}(x)$ in terms of the Christoffel symbols $\Gamma^{\rho}_{\mu\nu}(x)$ and a commutator with the affine connection matrices $\Gamma^{\nu}(x)$,

$$\nabla_{\nu}\overline{\gamma}_{\mu}(x) = \partial_{\nu}\overline{\gamma}_{\mu}(x) - \Gamma^{\rho}_{\nu\mu}(x)\,\overline{\gamma}_{\rho}(x) - [\Gamma_{\nu}(x), \overline{\gamma}_{\mu}(x)] = 0\,. \tag{8.211}$$

The reason for this *ansatz* will be discussed in the following.

We need to find a representation of the curved-space Dirac algebra $\{\overline{\gamma}^{\mu}(x), \overline{\gamma}^{\nu}(x)\} = 2\,\overline{g}^{\mu\nu}(x)$. This is most easily done in a "tetrad" formalism, where one writes the curved-space $\overline{\gamma}_{\rho}(x)$ in terms of the free-space $\widetilde{\gamma}^{A}$ matrices, multiplied by a "square root" of the metric, namely, the tetrad (or *vielbein*) [see also Eqs. (8.174) and (8.218)],

$$\overline{\gamma}_{\rho}(x) = e^{A}_{\rho}(x)\,\widetilde{\gamma}_{A}\,, \qquad \widetilde{\gamma}_{A} = e^{\alpha}_{A}(x)\,\overline{\gamma}_{\alpha}(x)\,, \tag{8.212a}$$

$$\overline{\gamma}^{\alpha}(x) = e^{\alpha}_{A}(x)\,\widetilde{\gamma}^{A}\,, \qquad \widetilde{\gamma}^{A} = e^{A}_{\rho}(x)\,\overline{\gamma}^{\rho}(x)\,, \qquad e^{\mu}_{A}(x)\,e^{A}_{\rho}(x) = \delta^{\mu}{}_{\rho}\,, \tag{8.212b}$$

where

$$\{\widetilde{\gamma}^{A}, \widetilde{\gamma}^{B}\} = 2\widetilde{g}^{AB}\,. \tag{8.213}$$

We recall that we use the "West-Coast" convention for the flat-space metric, which we denote as $\widetilde{g}^{AB} = \widetilde{g}_{AB} = \mathrm{diag}(1,-1,-1,-1)$.

In our notation, we draw inspiration from the book [282] and denote space-time indices related to a local, freely falling Lorentz frame ("anholonomic basis") with capital Latin indices $A, B, C, \ldots = 0, 1, 2, 3$. By contrast, lowercase Latin indices starting with $i, j, k, \ldots = 1, 2, 3$ are reserved for "spatial" global coordinates. Our notation of the Dirac matrices addresses some ambiguities which could otherwise result from other approaches [278–280, 283–285]. For example, unless the Dirac matrices are distinguished by overlining or tildes, the explicit expression γ^{1} could be associated with a flat-space matrix $\widetilde{\gamma}^{I=1}$ or with a curved-space matrix $\overline{\gamma}^{i=1}$. The metric is recovered as

$$\{\overline{\gamma}_{\rho}(x), \overline{\gamma}_{\sigma}(x)\} = e^{A}_{\rho}(x)\,e^{B}_{\sigma}(x)\,\{\widetilde{\gamma}_{A}, \widetilde{\gamma}_{B}\} = 2\,\widetilde{g}_{AB}\,e^{A}_{\rho}(x)\,e^{B}_{\sigma}(x) = 2\,\overline{g}_{\rho\sigma}(x)\,. \tag{8.214}$$

In case we raise the indices, the result is

$$\{\overline{\gamma}^{\rho}(x), \overline{\gamma}^{\sigma}(x)\} = e^{\rho}_{A}(x)\,e^{\sigma}_{B}(x)\,\{\widetilde{\gamma}^{A}, \widetilde{\gamma}^{B}\} = 2\,\widetilde{g}^{AB}\,e^{\rho}_{A}(x)\,e^{\sigma}_{B}(x) = 2\,\overline{g}^{\rho\sigma}(x)\,. \tag{8.215}$$

The matrix with components $\overline{g}^{\rho\sigma}(x)$ is the inverse of the matrix with entries $\overline{g}_{\rho\sigma}(x)$, and the matrix with components $e^{\mu}_{A}(x)$ is the inverse of the matrix with components $e^{A}_{\mu}(x)$. From now on, we will suppress the dependence of the curved-space-time symbols on the space-time coordinate x.

A remark on our notation is in order. We have denoted the flat-space Dirac matrices $\widetilde{\gamma}^{A}$ and the curved-space matrices as $\overline{\gamma}^{\mu}$. An alternative notation would be to denote all the indices from the non-coordinate basis in brackets, so that vector components in a non-coordinate basis would be denoted as $V^{(A)}$ (such a basis is used for a local Lorentz frame). This is in contrast to a vector component in a *global coordinate (holonomic) basis, which would read V^μ. This would imply an alternative notation for the metric too; one might leave out the overline and denote the curved-space metric as $g_{\mu\nu} = \overline{g}_{\mu\nu}$, while $g_{(A)(B)} = g_{(AB)} = \widetilde{g}_{AB}$. However, all things considered, we found that the additional brackets to be supplemented for the flat-space metric would probably rather clutter the notation, while the differentiation between the curved- and flat-space matrices using the overline and tilde requires no extra space.*

8.4.3 *Ricci Rotation Coefficients*

Just as much as the Christoffel symbols are naturally obtained upon a differentiation of a vector field, in the holonomic basis spanned by the $\vec{e}_\mu$, with

$$\vec{e}_\mu \cdot \vec{e}_\nu = \overline{g}_{\mu\nu}\,, \tag{8.216}$$

we can define a local, orthonormal basis spanning the "local Lorentz frame" as follows,

$$\hat{e}_A \cdot \hat{e}_B = \widetilde{g}_{AB}\,. \tag{8.217}$$

Here, the $\hat{e}_A$ may be chosen locally; their dependence on x is not explicitly indicated. Of course,

$$\vec{e}^{\,\mu} = e^\mu_A\, \hat{e}^A\,, \qquad \hat{e}^A = e^A_\mu\, \vec{e}^{\,\mu}\,. \tag{8.218}$$

One can expand a Lorentz vector, in the orthonormal basis, as follows,

$$V = V^A\, \hat{e}_A\,. \tag{8.219}$$

In analogy with Eq. (8.175), one may differentiate a vector, as follows,

$$\begin{aligned}
\partial_\nu \vec{V} &= \partial_\nu \left[V^A\, \hat{e}_A\right] = \left(\partial_\nu V^A\right) \hat{e}_A + V^A\, \partial_\nu \hat{e}_A \\
&= \left(\partial_\nu V^B\right) \hat{e}_B + V^A\, \partial_\nu \left(e^\mu_A\, \hat{e}_\mu\right). \\
&= \left(\partial_\nu V^B\right) \hat{e}_B + V^A\, e^\mu_A \left(\partial_\nu \hat{e}_\mu\right) + V^A \left(\partial_\nu e^\mu_A\right) \hat{e}_\mu\,.
\end{aligned} \tag{8.220}$$

The second term on the right-hand side can be rewritten in terms of a Christoffel symbol, while the third term is re-expanded in the orthonormal basis,

$$\begin{aligned}
\partial_\nu \vec{V} &= \left(\partial_\nu V^B\right) \hat{e}_B + \underbrace{V^A\, e^\mu_A}_{=V^\mu} \underbrace{\left[\hat{e}^\rho \cdot \partial_\nu \hat{e}_\mu\right]}_{=\Gamma^\rho_{\nu\mu}} \hat{e}_\rho + V^A\, e^B_\mu \left(\partial_\nu e^\mu_A\right) \hat{e}_B \\
&= \left(\partial_\nu V^B\right) \hat{e}_B + V^\mu\, \Gamma^\rho_{\nu\mu}\, e^B_\rho\, \hat{e}_B + V^A \left(e^B_\mu \partial_\nu e^\mu_A\right) \hat{e}_B \\
&= \left(\partial_\nu V^B + \left(e^B_\mu \partial_\nu e^\mu_A\right) V^A + e^B_\mu\, \Gamma^\mu_{\nu\rho}\, e^\rho_A\, V^A\right) \hat{e}_B \\
&= \left(\partial_\nu V^B + \left(e^B_\mu \nabla_\nu e^\mu_A\right) V^A\right) \hat{e}_B = \left(\nabla_\nu V^B + \omega_\nu{}^B{}_A V^A\right) \hat{e}_B = \left(\mathcal{D}_\nu V^B\right) \hat{e}_B\,.
\end{aligned} \tag{8.221}$$

Here $\mathcal{D}_\nu V^B$ is a covariant derivative for vectors,

$$\mathcal{D}_\nu V^B = \nabla_\nu V^B + \omega_{\nu\,A}^{\,B} V^A\,, \tag{8.222}$$

with respect to local Lorentz transformations, and constitutes a definition. The connection 1-form is given as follows,

$$\omega_\mu^{AB} = e_\nu^A \, \nabla_\mu \, e^{\nu B}\,. \tag{8.223}$$

The Ricci rotation coefficient is obtained via multiplication by an additional tetrad coefficient,

$$\omega^{ABC} = e^{\mu C} \, \omega_\mu^{AB}\,. \tag{8.224}$$

However, we shall use the terminology of a spin-connection 1-form and a Ricci rotation coefficient interchangeably, here. In particular, indices in the orthonormal basis can be raised and lowered with the flat-space metric $\widetilde{g}^{AB}$.

8.4.4 *Covariant Derivative of a Spinor*

Let us consider a change of the tetrad basis, which alternatively can be interpreted as a Lorentz transformation in the "internal" flat space,

$$e'^{\mu A} = \Lambda^A{}_B \, e^{\mu B}\,. \tag{8.225}$$

We define

$$e_{\mu A} = e_A^\nu \, \overline{g}_{\nu\mu} = e_\mu^B \, \widetilde{g}_{BA}\,, \qquad e^{\mu A} = e_\nu^A \, \overline{g}^{\nu\mu} = e_B^\mu \, \widetilde{g}^{BA}\,. \tag{8.226}$$

We can then use $\widetilde{g}_{CA}\,\Lambda^A{}_B\,\widetilde{g}^{BD} = (\Lambda^{-1})^D{}_C$ (which is a consequence of the fact that Lorentz transformations preserve the four-dimensional inner product of vectors) and multiply this expression by e_D^μ to find

$$e'^\mu_C = e_D^\mu \left(\Lambda^{-1}\right)^D{}_C\,. \tag{8.227}$$

We now turn our attention to the curved space Dirac γ matrices, defined in (8.212). These carry a Lorentz index and spinor indices.

The transformation of $\widetilde{\gamma}^A$ can be inferred by recognizing that A is a Lorentz index, and that the matrix additionally has two spinor indices which also transform. Hence, just like in free space, one has

$$\widetilde{\gamma}'^A = \Lambda^A{}_B \, S(\Lambda) \, \widetilde{\gamma}^B \, S(\Lambda^{-1})\,. \tag{8.228}$$

Just like in free space, the spinor transformation is defined such as to reverse the effect of the Lorentz index transformation,

$$S(\Lambda) \, \widetilde{\gamma}^B \, S(\Lambda^{-1}) = (\Lambda^{-1})^B{}_C \, \widetilde{\gamma}^C\,. \tag{8.229}$$

So, in agreement of Eq. (7.35), we have

$$\widetilde{\gamma}'^A = \Lambda^A{}_B \, S(\Lambda) \, \widetilde{\gamma}^B \, S(\Lambda^{-1}) = \Lambda^A{}_B \, (\Lambda^{-1})^B{}_C \, \widetilde{\gamma}^C = \widetilde{\gamma}^A\,. \tag{8.230}$$

It means that just like in free space, we can choose one particular representation of the Clifford algebra, and then calculate all effects in that representation. The flat-space limit is reached smoothly.

A Lorentz transform will thus alter the tetrad, and in consequence, the $\overline{\gamma}'^{\mu}$ are different from the $\overline{\gamma}^{\mu}$,

$$\overline{\gamma}'^{\mu} = e'^{\mu}_{A}\,\widetilde{\gamma}^{A} = e^{\mu}_{A}\left(\Lambda^{-1}\right)^{A}{}_{B}\,\widetilde{\gamma}^{B} = e^{\mu}_{A}\,S(\Lambda)\,\widetilde{\gamma}^{A}\,S(\Lambda)^{-1} = S(\Lambda)\,\overline{\gamma}^{\mu}\,S(\Lambda)^{-1}\,. \tag{8.231}$$

The spinor Lorentz transformation $S(\Lambda)$ has been defined in Eq. (8.233b). Note that in comparison to Eq. (7.38), the Lorentz index of the Dirac γ matrix is not being transformed, hence the change $\overline{\gamma}^{\mu} \to \overline{\gamma}'^{\mu} \neq \overline{\gamma}^{\mu}$. We can then show that although the curved–space Dirac γ matrices change under a Lorentz transformation, the metric $\overline{g}$ remains invariant, i.e.,

$$\begin{aligned}\overline{g}'^{\mu\nu} &= \frac{1}{2}\left\{\overline{\gamma}'^{\mu}, \overline{\gamma}'^{\nu}\right\} = \frac{1}{2}\left\{e^{\mu}_{A}\left(\Lambda^{-1}\right)^{A}{}_{B}\,\widetilde{\gamma}^{B}, e^{\nu}_{C}\left(\Lambda^{-1}\right)^{C}{}_{D}\,\widetilde{\gamma}^{D}\right\} \\ &= e^{\mu}_{A}\,e^{\nu}_{C}\left(\Lambda^{-1}\right)^{A}{}_{B}\,\frac{1}{2}\left\{\widetilde{\gamma}^{B}, \widetilde{\gamma}^{D}\right\}\left(\Lambda^{-1}\right)^{C}{}_{D} \\ &= e^{\mu}_{A}\,e^{\nu}_{C}\left(\Lambda^{-1}\right)^{A}{}_{B}\,\widetilde{g}^{BD}\left(\Lambda^{-1\,\mathrm{T}}\right)^{D}{}_{C} = e^{\mu}_{A}\,e^{\nu}_{C}\,\widetilde{g}^{AC} = \overline{g}^{\mu\nu}\,.\end{aligned} \tag{8.232}$$

We can formulate the local vector and spinor Lorentz transformations as follows [see Eqs. (7.27) and (7.30)],

$$\Lambda^{C}{}_{D} = \left(\exp\left[\frac{1}{2}\epsilon^{AB}\,\mathbb{M}_{AB}\right]\right)^{C}{}_{D} \approx \delta^{C}{}_{D} + \epsilon^{C}{}_{D}\,, \tag{8.233a}$$

$$S(\Lambda) = \exp\left(-\frac{\mathrm{i}}{4}\widetilde{\sigma}^{AB}\,\epsilon_{AB}\right) \approx 1 - \frac{\mathrm{i}}{4}\widetilde{\sigma}^{AB}\,\epsilon_{AB}\,, \tag{8.233b}$$

where the matrices $\mathbb{M}_{CD}$ have been defined in Eq. (7.26) [see also Eq. (8.301)], and the spin matrices $\widetilde{\sigma}^{AB}$ have been defined in Eq. (7.30). Furthermore, using the relation (7.31), we have already shown in Eq. (7.38) that the $\widetilde{\gamma}^{A}$ matrices are shape-invariant under Lorentz transformations, provided one also transforms the Lorentz index. The (conceivably infinitesimal) generators $\epsilon_{AB} = -\epsilon_{BA}$ are formulated in terms of upper-case Latin instead of lower-case Greek letters, in order to distinguish them from the Ricci rotation coefficients.

Under a local Lorentz transformation (change of the tetrad), the Lagrangian density for the gravitationally coupled Dirac particle transforms as follows [see Eqs. (7.79), (8.206) and (8.208)],

$$\mathcal{L} = \overline{\psi}\left(\mathrm{i}\overline{\gamma}^{\mu}\,\nabla_{\mu} - m\right)\psi \to \mathcal{L}' = \overline{\psi}'\left(\mathrm{i}\overline{\gamma}'^{\mu}\,\nabla'_{\mu} - m\right)\psi'\,. \tag{8.234}$$

We suppress the arguments here, and recall that

$$\psi' = S(\Lambda)\,\psi\,, \qquad \overline{\psi}' = \overline{\psi}\,S(\Lambda)^{-1}\,, \tag{8.235}$$

in which case (8.234) becomes

$$\begin{aligned}\mathcal{L}' &= \overline{\psi}\,S(\Lambda)^{-1}\left[\mathrm{i}\left(S(\Lambda)\,\overline{\gamma}^{\mu}\,S(\Lambda)^{-1}\right)\nabla'_{\mu} - m\right]S(\Lambda)\,\psi \\ &= \overline{\psi}\left[\mathrm{i}\overline{\gamma}^{\mu}\left(S(\Lambda)^{-1}\,\nabla'_{\mu}\,S(\Lambda)\right) - m\right]\psi\,.\end{aligned} \tag{8.236}$$

Then, for the curved-space Lagrangian density to be Lorentz invariant, we must require that

$$S(\Lambda)^{-1}\,\nabla'_\mu\,S(\Lambda)\psi = \nabla_\mu\,\psi\,, \tag{8.237}$$

i.e., that the covariant derivative of a spinor after Lorentz transformation is equal to the Lorentz-transformed covariant derivative,

$$\nabla'_\mu\,\psi' = \nabla'_\mu\,S(\Lambda)\,\psi = S(\Lambda)\,\nabla_\mu\,\psi = (\nabla_\mu\,\psi)'\,. \tag{8.238}$$

Furthermore, we do not want the local change of the tetrad basis to alter the fundamental structure of the covariant derivative acting on a spinor, i.e.,

$$\nabla'_\mu = \partial_\mu - \Gamma'_\mu\,. \tag{8.239}$$

So,

$$\begin{aligned}\left(\partial_\mu - \Gamma'_\mu\right) S(\Lambda)\psi &= \left(\partial_\mu\,S(\Lambda)\right)\psi + S(\Lambda)\,\partial_\mu\,\psi - \Gamma'_\mu\,S(\Lambda)\,\psi \\ &= S(\Lambda)\left[\partial_\mu + S(\Lambda)^{-1}\,\partial_\mu\,S(\Lambda) - S(\Lambda)^{-1}\,\Gamma'_\mu\,S(\Lambda)\right]\psi\,.\end{aligned} \tag{8.240}$$

Comparing this result with the right-hand side of Eq. (8.238), we conclude that

$$\Gamma_\mu = S(\Lambda)^{-1}\,\Gamma'_\mu\,S(\Lambda) - S(\Lambda)^{-1}\,\partial_\mu\,S(\Lambda)\,, \tag{8.241}$$

or alternatively

$$\Gamma'_\mu = S(\Lambda)\,\Gamma_\mu\,S(\Lambda)^{-1} + \left[\partial_\mu\,S(\Lambda)\right]\,S(\Lambda)^{-1}\,. \tag{8.242}$$

The spinor Lorentz transformation has been given in Eq. (8.233b). Also, we have $\widetilde{\sigma}_{AB} = \frac{\mathrm{i}}{2}\left[\widetilde{\gamma}_A, \widetilde{\gamma}_B\right]$. In this case, one can reformulate Eqs. (8.241) and (8.242) in terms of a simplification of the second term on the right-hand side,

$$\Gamma_\mu = S(\Lambda)^{-1}\,\Gamma'_\mu\,S(\Lambda) + \frac{\mathrm{i}}{4}\left(\partial_\mu\,\epsilon^{AB}\right)\widetilde{\sigma}_{AB}\,, \tag{8.243a}$$

$$\Gamma'_\mu = S(\Lambda)\,\Gamma_\mu\,S(\Lambda)^{-1} - \frac{\mathrm{i}}{4}\left(\partial_\mu\,\epsilon^{AB}\right)\widetilde{\sigma}_{AB}\,. \tag{8.243b}$$

The generators ϵ^{AB} are antisymmetric, and despite the fact that the connection matrix Γ_μ is changed by a Lorentz transformation, one might conjecture that their overall mathematical structure, including the symmetry properties with respect to the anholonomic indices (A and B) and Lorentz indices (μ), should be conserved. This suggests the *ansatz*

$$\Gamma_\mu = \frac{\mathrm{i}}{4} C^{AB}_\mu\,\widetilde{\sigma}_{AB}\,, \tag{8.244}$$

where C^{AB}_μ is antisymmetric in A and B and carries another Lorentz index μ. Furthermore, because the covariant derivative of a spinor is formulated only in regard to *local* Lorentz transformations (a spinor does not constitute a vector), we may conjecture that the C^{AB}_μ coefficients might be connected with the Ricci rotation coefficients defined in Eq. (8.223), rather than the Christoffel symbols.

8.4.5 *Covariant Derivative of the Dirac Matrices*

A last preparatory step in the calculation of the Γ_μ matrix is the motivation of the covariant derivative of the curved-space $\overline{\gamma}^\mu$ matrix that we already encountered in Eq. (8.211),

$$\nabla_\mu \overline{\gamma}^\nu = \partial_\mu \overline{\gamma}^\nu + \Gamma^\nu_{\mu\rho} \overline{\gamma}^\rho - [\Gamma_\mu, \overline{\gamma}^\nu] \,. \tag{8.245}$$

We begin with the Dirac Lagrangian density in curved space, which we recall from Eq. (8.234),

$$\mathcal{L} = \overline{\psi} \left[\mathrm{i}\overline{\gamma}^\mu \left(\partial_\mu - \Gamma_\mu \right) - m \right] \psi \,. \tag{8.246}$$

Its adjoint reads as

$$\mathcal{L}^+ = \psi^+ \left[-\mathrm{i} \left(\overleftarrow{\partial}_\mu - \Gamma^+_\mu \right) \left(\overline{\gamma}^\mu \right)^+ - m \right] \overline{\psi}^+ \,. \tag{8.247}$$

We define $\overline{\psi}$ as

$$\overline{\psi} \equiv \psi^+ \widetilde{\gamma}^0 \,, \tag{8.248}$$

with the flat-space $\widetilde{\gamma}^0$, ensuring that $\overline{\psi}$ transforms with the inverse of the local Lorentz transform. With this in mind, and the fact that $(\widetilde{\gamma}^0)^2 = 1$, Eq. (8.247) becomes

$$\begin{aligned} \mathcal{L}^+ &= \psi^+ \widetilde{\gamma}^0 \widetilde{\gamma}^0 \left[-\mathrm{i} \left(\overleftarrow{\partial}_\mu - \Gamma^+_\mu \right) \left(\overline{\gamma}^\mu \right)^+ - m \right] \widetilde{\gamma}^0 \psi \\ &= \overline{\psi} \left[-\mathrm{i} \left(\overleftarrow{\partial}_\mu - \widetilde{\gamma}^0 \, \Gamma^+_\mu \, \widetilde{\gamma}^0 \right) \widetilde{\gamma}^0 \, \left(\overline{\gamma}^\mu \right)^+ \, \widetilde{\gamma}^0 - m \right] \psi \,. \end{aligned} \tag{8.249}$$

Since $\widetilde{\gamma}^0 (\widetilde{\gamma}^\mu)^+ \widetilde{\gamma}^0 = \widetilde{\gamma}^\mu$, it is trivial to show that $\widetilde{\gamma}^0 (\overline{\gamma}^\mu)^+ \widetilde{\gamma}^0 = \overline{\gamma}^\mu$.

We assume for Γ_μ, the *ansatz* given in Eq. (8.244), and write

$$\widetilde{\gamma}^0 \left(\Gamma_\mu \right)^+ \widetilde{\gamma}^0 = \widetilde{\gamma}^0 \left(\frac{\mathrm{i}}{4} C^{AB}_\mu \, \widetilde{\sigma}_{AB} \right)^+ \widetilde{\gamma}^0 = -\frac{\mathrm{i}}{4} C^{AB}_\mu \, \widetilde{\gamma}^0 \, \left(\widetilde{\sigma}_{AB} \right)^+ \widetilde{\gamma}^0 \,. \tag{8.250}$$

One can show that

$$\widetilde{\gamma}^0 \, \left(\widetilde{\sigma}_{AB} \right)^+ \, \widetilde{\gamma}^0 = \widetilde{\sigma}_{AB} \,. \tag{8.251}$$

Plugging this into (8.250), we find

$$\widetilde{\gamma}^0 \, \Gamma^+_\mu \, \widetilde{\gamma}^0 = -\frac{\mathrm{i}}{4} \, C^{AB}_\mu \, \widetilde{\sigma}_{AB} = -\Gamma_\mu \,. \tag{8.252}$$

Thus, Eq. (8.249) becomes

$$\mathcal{L}^+ = \overline{\psi} \left[-\mathrm{i} \left(\overleftarrow{\partial}_\mu + \Gamma_\mu \right) \overline{\gamma}^\mu - m \right] \psi \,. \tag{8.253}$$

The Lagrange density has to be Hermitian; hence $\mathcal{L} = \mathcal{L}^+$ and

$$\overline{\psi} \left[\mathrm{i} \overline{\gamma}^\mu \left(\partial_\mu - \Gamma_\mu \right) - m \right] \psi = \overline{\psi} \left[-\mathrm{i} \left(\overleftarrow{\partial}_\mu + \Gamma_\mu \right) \overline{\gamma}^\mu - m \right] \psi \,. \tag{8.254}$$

This can be reformulated as follows,

$$\overline{\psi} \, \overline{\gamma}^\mu \left(\partial_\mu \psi \right) + \left(\partial_\mu \overline{\psi} \right) \overline{\gamma}^\mu \, \psi = \overline{\psi} \, \left(\overline{\gamma}^\mu \, \Gamma_\mu - \Gamma_\mu \, \overline{\gamma}^\mu \right) \psi \,. \tag{8.255}$$

Adding the term $\overline{\psi}\,(\partial_\mu\,\overline{\gamma}^\mu)\,\psi$ to both sides of this equation, we have

$$\partial_\mu\left(\overline{\psi}\,\overline{\gamma}^\mu\,\psi\right)=\overline{\psi}\,(\partial_\mu\,\overline{\gamma}^\mu)\,\psi-\overline{\psi}\,[\Gamma_\mu,\overline{\gamma}^\mu]\,\psi\,. \tag{8.256}$$

The curved-space probability current is given as

$$J^\mu=\overline{\psi}\,\overline{\gamma}^\mu\,\psi\,,\qquad \nabla_\mu\,J^\mu=0\,. \tag{8.257}$$

Thus,

$$\partial_\mu\,J^\mu+\Gamma^\mu_{\mu\rho}\,J^\rho=\partial_\mu\left(\overline{\psi}\,\overline{\gamma}^\mu\,\psi\right)+\Gamma^\mu_{\mu\rho}\left(\overline{\psi}\,\overline{\gamma}^\rho\,\psi\right)=0\,. \tag{8.258}$$

In this expression, we replace the term $\overline{\psi}\,\overline{\gamma}^\mu\,\psi$ with the help of Eq. (8.256), yielding

$$\overline{\psi}\,(\partial_\mu\,\overline{\gamma}^\mu)\,\psi-\overline{\psi}\,[\Gamma_\mu,\overline{\gamma}^\mu]\,\psi+\overline{\psi}\,\Gamma^\mu_{\mu\rho}\,\overline{\gamma}^\rho\,\psi=0\,. \tag{8.259}$$

Introducing, artificially, a Kronecker symbol $\delta_\nu{}^\mu=\overline{g}_\nu{}^\mu$, one has

$$\overline{\psi}\,\overline{g}_\nu{}^\mu\left((\partial_\mu\,\overline{\gamma}^\nu)+\Gamma^\nu_{\mu\rho}\,\overline{\gamma}^\rho-[\Gamma_\mu,\overline{\gamma}^\nu]\right)\psi=\overline{\psi}\,\left(\overline{g}_\nu{}^\mu\,\nabla_\mu\,\overline{\gamma}^\nu\right)\psi=0\,, \tag{8.260}$$

where we identify the covariant derivative $\nabla_\mu\,\overline{\gamma}^\nu$ of a Dirac γ matrix in accordance with Eq. (8.211).

8.4.6 *Spin Connection*

It remains to calculate the explicit form of the C^{AB}_μ in Eq. (8.244). We impose the restriction that $\nabla_\mu\overline{\gamma}^\nu=0$, use the tetrad basis ($\overline{\gamma}^\mu=e^\mu_A\,\widetilde{\gamma}^A$), and find that

$$\nabla_\mu\overline{\gamma}^\nu=\left(\partial_\mu\,e^\nu_B+\Gamma^\nu_{\mu\rho}\,e^\rho_B\right)\widetilde{\gamma}^B-e^\nu_B\left[\Gamma_\mu,\widetilde{\gamma}^B\right]=\left(\nabla_\mu\,e^\nu_B\right)\widetilde{\gamma}^B-e^\nu_B\left[\Gamma_\mu,\widetilde{\gamma}^B\right]=0\,. \tag{8.261}$$

We recall that the Ricci rotation coefficient [the "spin connection" of Eq. (8.223)] is

$$\omega^{AB}_\mu=e^A_\nu\,\nabla_\mu\,e^{\nu B}\,, \tag{8.262}$$

and multiply Eq. (8.261) by e^A_ν to obtain

$$e^A_\nu\,\left(\nabla_\mu\,e^\nu_B\right)\widetilde{\gamma}^B-e^A_\nu\,e^\nu_B\left[\Gamma_\mu,\widetilde{\gamma}^B\right]=\omega^{AB}_\mu\,\widetilde{\gamma}_B-\left[\Gamma_\mu,\widetilde{\gamma}^A\right]=0\,. \tag{8.263}$$

In order to proceed, we use our *ansatz* for Γ_μ given in Eq. (8.244), and Eq. (7.37), to find

$$\begin{aligned}\left[\Gamma_\mu,\widetilde{\gamma}^A\right]&=\frac{\mathrm{i}}{4}C^{BC}_\mu\left[\widetilde{\sigma}_{BC},\widetilde{\gamma}^A\right]=\frac{\mathrm{i}}{4}C^{BC}_\mu\left[2\,\mathrm{i}\left(g^A{}_C\,\widetilde{\gamma}_B-g^A{}_B\,\widetilde{\gamma}_C\right)\right]\\&=\frac{1}{2}\left(C^{AB}_\mu\,\widetilde{\gamma}_B+C^{AC}_\mu\,\widetilde{\gamma}_C\right)=C^{AB}_\mu\,\widetilde{\gamma}_B\,,\end{aligned} \tag{8.264}$$

where we used the assumption that C^{AB}_μ is antisymmetric ($C^{AB}_\mu=-C^{BA}_\mu$). Combining this result with (8.263), we obtain that quite simply, C^{AB}_μ is a rotation coefficient,

$$C^{AB}_\mu=\omega^{AB}_\mu\,, \tag{8.265}$$

and

$$\Gamma_\mu=\frac{\mathrm{i}}{4}\,\omega^{AB}_\mu\,\widetilde{\sigma}_{AB}\,,\qquad \omega^{AB}_\mu=e^A_\nu\,\nabla_\mu\,e^{\nu B}\,. \tag{8.266}$$

Finally, we can write the covariant derivative operating on a spinor as

$$\nabla_\mu\psi=\left(\partial_\mu-\Gamma_\mu\right)\psi=\left(\partial_\mu-\frac{\mathrm{i}}{4}\omega^{AB}_\mu\,\widetilde{\sigma}_{ab}\right)\psi\,. \tag{8.267}$$

In principle, this result is well known [255, 278, 280–282, 284, 286–289], including a missing prefactor in Ref. [278], as pointed out in Ref. [281]. However the derivations in the literature are not very detailed, and it is thus useful to recall the basis of the formalism here.

8.4.7 *Transformation Properties of Rotation Coefficients*

The transformation properties of Ricci rotation coefficients, or connection 1-forms, deserve a special discussion. We have, in the unprimed frame, according to Eq. (8.223),

$$\omega_\mu^{AB} = e_\rho^A \, \nabla_\mu \, e^{\rho B} \,. \tag{8.268}$$

This translates into the primed frame as follows,

$$\omega_\mu^{\prime AB} = e_\rho^{\prime A} \, \nabla_\mu \, e^{\prime \rho B} \,. \tag{8.269}$$

According to Eqs. (8.233a) and (8.233b), infinitesimal transformations read as follows,

$$\Lambda^A{}_B \approx g^A{}_B + \epsilon^A{}_B \,, \qquad S(\Lambda) = \mathbb{1}_{4\times 4} - \frac{\mathrm{i}}{4} \epsilon^{AB} \, \widetilde{\sigma}_{AB} \,. \tag{8.270}$$

The transformation properties of the Ricci rotation coefficients under infinitesimal local Lorentz transformations can be inferred directly,

$$\begin{aligned} \omega_\mu^{\prime AB} &\approx \left(g^A{}_C + \epsilon^A{}_C\right) e_\rho^C \, \nabla_\mu \left[\left(g^B{}_D + \epsilon^B{}_D\right) e^{\rho D}\right] \\ &= \left(e_\rho^A + \epsilon^A{}_C \, e_\rho^C\right) \nabla_\mu \, e^{\rho B} + e_\rho^A \, \nabla_\mu \left(\epsilon^B{}_D \, e^{\rho D}\right) \\ &= \omega_\mu^{AB} + \epsilon^A{}_C \, \omega_\mu^{CB} + \left(\nabla_\mu \, \epsilon^B{}_D\right) \widetilde{g}^{AD} + \epsilon^B{}_D \, e_\rho^A \, \nabla_\mu \, e^{\rho D} \\ &= \omega_\mu^{AB} + \epsilon^A{}_C \, \omega_\mu^{CB} + \nabla_\mu \, \epsilon^{BA} + \epsilon^B{}_D \, \omega_\mu^{AD} \,. \end{aligned} \tag{8.271}$$

Summarizing, one has

$$\omega_\mu^{\prime AB} \approx \omega_\mu^{AB} - \partial_\mu \, \epsilon^{AB} + \epsilon^A{}_C \, \omega_\mu^{CB} - \omega_\mu^{AD} \, \epsilon_D{}^B \,. \tag{8.272}$$

One particular observation is very important. Because ϵ^{AB} is invariant under global Lorentz transformations, we have $\nabla_\mu \, \epsilon^{AB} = \partial_\mu \, \epsilon^{AB}$, commensurate with the transformation properties of a Lorentz scalar. From $\epsilon^{AC} = -\epsilon^{CA}$, one gets the relation $\epsilon^A{}_B = -\epsilon_B{}^A$, via multiplication by $\widetilde{g}_{CB} = \widetilde{g}_{BC}$.

On the other hand, the transformation property of the spin connection matrix has been given in Eq. (8.243b),

$$\Gamma'_\mu = S(\Lambda) \, \Gamma_\mu \, S(\Lambda)^{-1} + \left[\partial_\mu S(\Lambda)\right] S(\Lambda)^{-1} \,, \qquad \Gamma_\mu = \frac{\mathrm{i}}{4} \, \omega_\mu^{AB} \, \widetilde{\sigma}_{AB} \,. \tag{8.273}$$

This formula has to be compatible with the result (8.272) [see also Eq. (8.266)]. This is because we can otherwise calculate the transformation properties of the Γ_μ matrices separately. Indeed, Γ'_μ has to read as follows,

$$\Gamma'_\mu = \frac{\mathrm{i}}{4} \, \omega_\mu^{\prime AB} \, \widetilde{\sigma}'_{AB} \,, \tag{8.274}$$

with Lorentz-transformed coefficients $\omega_\mu^{\prime AB}$ and Lorentz-transformed matrices $\widetilde{\sigma}'_{AB}$. However, in view of Eq. (8.230), one has

$$\widetilde{\sigma}'_{AB} = \frac{\mathrm{i}}{2} [\widetilde{\gamma}'^A, \widetilde{\gamma}'^B] = \frac{\mathrm{i}}{2} [\widetilde{\gamma}^A, \widetilde{\gamma}^B] = \widetilde{\sigma}_{AB} \,, \tag{8.275}$$

and so

$$\Gamma'_\mu = \frac{\mathrm{i}}{4}\,\omega'^{AB}_\mu\,\widetilde{\sigma}_{AB}\,. \tag{8.276}$$

We recall from Eq. (8.270) the relation

$$S(\Lambda)^{-1} \approx \mathbb{1}_{4\times 4} + \frac{\mathrm{i}}{4}\epsilon^{AB}\,\widetilde{\sigma}_{AB}\,. \tag{8.277}$$

So, the derivative term in Eq. (8.273) can be expressed as follows,

$$[\partial_\mu S(\Lambda)]\;S(\Lambda)^{-1} = -\frac{\mathrm{i}}{4}\partial_\mu\epsilon^{AB}\,\widetilde{\sigma}_{AB}\,. \tag{8.278}$$

It remains to calculate the following expression, to first order in the ϵ parameters,

$$\begin{aligned}
S(\Lambda)\,\Gamma_\mu\,S(\Lambda)^{-1} &\approx \left(\mathbb{1}_{4\times 4} - \frac{\mathrm{i}}{4}\epsilon^{AB}\,\widetilde{\sigma}_{AB}\right)\frac{\mathrm{i}}{4}\omega^{CD}_\mu\,\widetilde{\sigma}_{CD}\left(\mathbb{1}_{4\times 4} + \frac{\mathrm{i}}{4}\epsilon^{EF}\,\widetilde{\sigma}_{EF}\right) \\
&\approx \frac{\mathrm{i}}{4}\omega^{CD}_\mu\,\widetilde{\sigma}_{CD} + \left(-\frac{\mathrm{i}}{4}\epsilon^{AB}\,\widetilde{\sigma}_{AB}\right)\left(\frac{\mathrm{i}}{4}\omega^{CD}_\mu\,\widetilde{\sigma}_{CD}\right) \\
&\quad + \left(\frac{\mathrm{i}}{4}\omega^{CD}_\mu\,\widetilde{\sigma}_{CD}\right)\left(\frac{\mathrm{i}}{4}\epsilon^{EF}\,\widetilde{\sigma}_{EF}\right) \\
&= \frac{\mathrm{i}}{4}\omega^{CD}_\mu\,\widetilde{\sigma}_{CD} - \frac{\mathrm{i}}{4}\epsilon^{AB}\,\widetilde{\sigma}_{AB}\,\frac{\mathrm{i}}{4}\omega^{CD}_\mu\,\widetilde{\sigma}_{CD} + \frac{\mathrm{i}}{4}\omega^{CD}_\mu\,\widetilde{\sigma}_{CD}\,\frac{\mathrm{i}}{4}\epsilon^{AB}\,\widetilde{\sigma}_{AB} \\
&= \frac{\mathrm{i}}{4}\omega^{CD}_\mu\,\widetilde{\sigma}_{CD} + \frac{1}{16}\,\epsilon^{AB}\,\omega^{CD}_\mu\,[\widetilde{\sigma}_{AB},\widetilde{\sigma}_{CD}]\,.
\end{aligned} \tag{8.279}$$

Using the commutator relation [see Eq. (7.31) and Chap. 8 of Ref. [81]]

$$[\widetilde{\sigma}_{AB},\widetilde{\sigma}_{CD}] = 2\mathrm{i}\,(\widetilde{g}_{AD}\,\widetilde{\sigma}_{BC} + \widetilde{g}_{BC}\,\widetilde{\sigma}_{AD} - \widetilde{g}_{AC}\,\widetilde{\sigma}_{BD} - \widetilde{g}_{BD}\,\widetilde{\sigma}_{AC})\,, \tag{8.280}$$

one can derive the result

$$\frac{1}{16}\,\epsilon^{AB}\,\omega^{CD}_\mu\,[\widetilde{\sigma}_{AB},\widetilde{\sigma}_{CD}] = \frac{\mathrm{i}}{4}\left(\epsilon^{A}{}_{C}\,\omega^{CB}_\mu - \omega^{AD}_\mu\,\epsilon_{D}{}^{B}\right)\widetilde{\sigma}_{AB}\,. \tag{8.281}$$

In the derivation of Eq. (8.281), the identity

$$f^{AB}\,\widetilde{\sigma}_{AB} = \frac{1}{2}\,(f^{AB} - f^{BA})\,\widetilde{\sigma}_{AB} \tag{8.282}$$

may be useful. Adding, now, the results given in Eqs. (8.278) and (8.279), one easily obtains

$$\begin{aligned}
\Gamma'_\mu &= S(\Lambda)\,\Gamma_\mu\,S(\Lambda)^{-1} + [\partial_\mu S(\Lambda)]\;S(\Lambda)^{-1} \\
&= \frac{\mathrm{i}}{4}\omega^{AB}_\mu\,\widetilde{\sigma}_{AB} - \frac{\mathrm{i}}{4}\partial_\mu\epsilon^{AB}\,\widetilde{\sigma}_{AB} + \frac{\mathrm{i}}{4}\left(\epsilon^{A}{}_{C}\,\omega^{CB}_\mu - \omega^{AD}_\mu\,\epsilon_{D}{}^{B}\right)\widetilde{\sigma}_{AB} \\
&= \frac{\mathrm{i}}{4}\left(\omega^{AB}_\mu - \partial_\mu\epsilon^{AB} + \epsilon^{A}{}_{C}\,\omega^{CB}_\mu - \omega^{AD}_\mu\,\epsilon_{D}{}^{B}\right)\widetilde{\sigma}_{AB} \approx \frac{\mathrm{i}}{4}\omega'^{AB}_\mu\,\widetilde{\sigma}_{AB}\,,
\end{aligned} \tag{8.283}$$

compatible with Eq. (8.272).

8.5 Applications of Gravitational Coupling

8.5.1 *Dirac–Schwarzschild Hamiltonian*

For a space-time geometry independent of time, it is possible to invoke a Hamiltonian formalism. The Schwarzschild metric for a non-rotating black hole in isotropic coordinates reads (see Sec. 43 of Chap. 3 of Ref. [290]),

$$\overline{g}_{\mu\nu} = \mathrm{diag}\left(w^2(r), -v^2(r), -v^2(r), -v^2(r)\right), \tag{8.284}$$

and the coefficients are given as

$$w(r) = \left(1 - \frac{r_s}{4r}\right)\left(1 + \frac{r_s}{4r}\right)^{-1}, \qquad v(r) = \left(1 + \frac{r_s}{4r}\right)^2. \tag{8.285}$$

The Schwarzschild radius is given as $r_s = 2GM$, where G is Newton's gravitational constant, and M is the mass of the black hole (or planet). The gravitationally coupled Dirac equation is written out as follows,

$$\left[\mathrm{i}\overline{\gamma}^\mu \left(\partial_\mu - \Gamma_\mu\right) - m\right]\psi = \left[\mathrm{i}\,\overline{\gamma}^0\,\partial_0 + \mathrm{i}\overline{\gamma}^i\,\partial_i - \mathrm{i}\overline{\gamma}^\mu\,\Gamma_\mu - m\right]\psi = 0. \tag{8.286}$$

The Hamiltonian form thus is

$$\mathrm{i}\left(\overline{\gamma}^0\right)^2 \partial_0\psi = \left(-\mathrm{i}\,\overline{\gamma}^0\,\overline{\gamma}^i\,\partial_i + \mathrm{i}\overline{\gamma}^0\,\overline{\gamma}^\mu\,\Gamma_\mu + \overline{\gamma}^0 m\right)\psi, \tag{8.287}$$

with $\overline{\gamma}^0 = \widetilde{\gamma}^0/w(r)$ and $\overline{\gamma}^i = \widetilde{\gamma}^i/v(r)$ for $i = 1, 2, 3$. An explicit calculation establishes that

$$\overline{\gamma}^0\,\overline{\gamma}^\mu\,\Gamma_\mu = -\frac{\widetilde{\gamma}^0\,\widetilde{\vec{\gamma}}\cdot\hat{r}}{v(r)\,w(r)}\,G(r), \qquad G(r) = \frac{2\,v'(r)\,w(r) + v(r)\,w'(r)}{2\,v(r)\,w(r)}, \tag{8.288}$$

where $\hat{r}$ is the position unit vector. In the Schwarzschild-type metric, we have

$$\overline{\gamma}^0 = \frac{\widetilde{\gamma}^0}{w(r)}, \qquad \overline{\gamma}^i = \frac{\widetilde{\gamma}^i}{v(r)}, \qquad \overline{\gamma}^0\,\overline{\gamma}^i = \frac{\widetilde{\gamma}^0\,\widetilde{\gamma}^i}{w(r)\,v(r)} = \frac{\alpha^i}{w(r)\,v(r)}, \tag{8.289}$$

where $\alpha^i = \widetilde{\gamma}^0\,\widetilde{\gamma}^i$. By convention, the Dirac matrix β is denoted as $\beta = \widetilde{\gamma}^0$ in the Hamiltonian formalism, and thus

$$\mathrm{i}\partial_t\psi = \left(\frac{w(r)}{v(r)}\,\vec{\alpha}\cdot\vec{p} - \mathrm{i}\frac{w(r)}{v(r)}\,\vec{\alpha}\cdot\hat{r}\,G(r) + w(r)\,\beta\,m\right)\psi. \tag{8.290}$$

Using the explicit result for the G function from Eq. (8.288), one can identify the gravitationally coupled Dirac Hamiltonian H for a static space-time metric as follows,

$$\mathrm{i}\partial_t\psi = H\,\psi, \qquad H = \frac{w}{v}\,\vec{\alpha}\cdot\vec{p} - \frac{\mathrm{i}}{2v}\,\vec{\alpha}\cdot\vec{\nabla}w - \frac{\mathrm{i}w}{v^2}\,\vec{\alpha}\cdot\vec{\nabla}v + \beta m w. \tag{8.291}$$

We now scale space according to the scaling

$$\psi' = v^{3/2}\,\psi, \qquad H' = v^{3/2}\,H\,v^{-3/2}. \tag{8.292}$$

This leads to a Hermitian gravitational Hamiltonian, which acts on the Hilbert space of square-integrable functions with the scalar product $\langle\phi, \psi\rangle = \int \mathrm{d}^3r\,\phi^*(\vec{r})\,\psi(\vec{r})$. It reads as [281, 287]

$$H_{\mathrm{GR}} = \frac{1}{2}\,\{\vec{\alpha}\cdot\vec{p}, \mathcal{F}\} + \beta m w, \qquad \mathcal{F} = \frac{w}{v}. \tag{8.293}$$

In order to write this equation in terms of the radial components of an energy eigenstate ψ, one uses an *ansatz* analogous to Eq. (7.109),

$$\psi(\vec{r}) = \begin{pmatrix} f(r)\,\chi_{\kappa\,\mu}(\hat{r}) \\ \mathrm{i}g(r)\,\chi_{-\kappa\,\mu}(\hat{r}) \end{pmatrix}, \qquad H_{\mathrm{GR}}\psi(\vec{r}) = E_{\mathrm{GR}}\,\psi(\vec{r})\,. \tag{8.294}$$

Using various identities fulfilled by the fundamental spin-angular functions $\chi_{\kappa\,\mu}(\hat{r})$, as discussed in Sec. 6.4.1, together with the identity (7.111), one finally arrives at the following set of radial equations,

$$\left(\frac{\partial}{\partial r} + \frac{1-\kappa}{r} + \frac{2\,v'(r)\,w(r) + v(r)\,w'(r)}{2\,v(r)\,w(r)}\right) g(r) = v(r)\left(m - \frac{E_{\mathrm{GR}}}{w(r)}\right) f(r)\,, \tag{8.295a}$$

$$\left(\frac{\partial}{\partial r} + \frac{1+\kappa}{r} + \frac{2\,v'(r)\,w(r) + v(r)\,w'(r)}{2\,v(r)\,w(r)}\right) f(r) = v(r)\left(m + \frac{E_{\mathrm{GR}}}{w(r)}\right) g(r)\,. \tag{8.295b}$$

Because of the rather involved structure of the w and v functions for the Schwarzschild metric, as given in Eq. (8.285), the exact calculation of the bound and continuum solutions of Eq. (8.295) still constitutes an open problem, even for the relatively simple Schwarzschild geometry of space-time. The particle-antiparticle symmetry

$$E \leftrightarrow -E, \qquad f(r) \leftrightarrow g(r)\,, \qquad \kappa \leftrightarrow -\kappa\,, \tag{8.296}$$

which leaves the radial equations (8.295) invariant, ensures that both particles as well as antiparticles are attracted by the gravitational field, and establishes the equivalence principle for antiparticles.

For the Schwarzschild metric in isotropic coordinates, given in Eq. (8.285), we have to first order in the Schwarzschild radius r_s,

$$w = \left(1 - \frac{r_s}{4r}\right)\left(1 + \frac{r_s}{4r}\right)^{-1} = \frac{4r - r_s}{4r + r_s} \approx 1 - \frac{r_s}{2r}\,, \tag{8.297a}$$

$$v = \left(1 + \frac{r_s}{4r}\right)^2 \approx 1 + \frac{r_s}{2r}\,, \tag{8.297b}$$

$$\frac{w}{v} = \frac{16\,r^2\,(4r - r_s)}{(4r + r_s)^3} \approx 1 - \frac{r_s}{r}\,. \tag{8.297c}$$

The Schwarzschild radius is given as $r_s = 2\,G\,M$, where G is Newton's gravitational constant, and M is the mass of the planet. To first order in r_s, we have to analyze the Dirac–Schwarzschild Hamiltonian H_{DS}, which is given by

$$H_{\mathrm{DS}} = \frac{1}{2}\left\{\vec{\alpha}\cdot\vec{p}, \left(1 - \frac{r_s}{r}\right)\right\} + \beta m \left(1 - \frac{r_s}{2r}\right). \tag{8.298}$$

It is amenable to a Foldy–Wouthuysen transformation (see Chap. 11).

8.5.2 *Dirac Adjoint for Curved Space-Times*

Let us try to derive the particle-antiparticle symmetry of the gravitationally and electromagnetically coupled Dirac equation on the basis of the Lagrangian of the field. To this end, we first need to generalize the concept of the Dirac adjoint to curved space-times. We recall that the Dirac adjoint transforms with the inverse of the Lorentz transform as compared to the original Dirac spinor. A general spinor Lorentz transformation $S(\Lambda)$ is given as follows,

$$S(\Lambda) = \exp\left(-\frac{\mathrm{i}}{4}\,\epsilon^{AB}\,\widetilde{\sigma}_{AB}\right), \qquad \widetilde{\sigma}_{AB} = \frac{\mathrm{i}}{2}\left[\widetilde{\gamma}^{A}, \widetilde{\gamma}^{B}\right]. \tag{8.299}$$

Note that the generator parameters ϵ^{AB}, for local Lorentz transformations, can be coordinate-dependent. The spin matrices $\widetilde{\sigma}_{AB}$ are, however, the flat-space spin matrices. The spin matrices fulfill the commutation relations [see also Eq. (7.31)]

$$\left[\tfrac{1}{2}\widetilde{\sigma}^{CD}, \tfrac{1}{2}\widetilde{\sigma}^{EF}\right] = \mathrm{i}\left(\widetilde{g}^{CF}\,\tfrac{1}{2}\widetilde{\sigma}^{DE} + \widetilde{g}^{DE}\,\tfrac{1}{2}\widetilde{\sigma}^{CF} - \widetilde{g}^{CE}\,\tfrac{1}{2}\widetilde{\sigma}^{DF} - \widetilde{g}^{DF}\,\tfrac{1}{2}\widetilde{\sigma}^{CE}\right). \tag{8.300}$$

It is highly instructive to compare these relations to those fulfilled by the matrices $\mathbb{M}_{AB}$ that generate (four-)vector Lorentz transformations,

$$\left(\mathbb{M}_{AB}\right)^{C}{}_{D} = \widetilde{g}^{C}_{A}\,\widetilde{g}_{DB} - \widetilde{g}^{C}_{B}\,\widetilde{g}_{DA}\,. \tag{8.301}$$

The Lorentz transformation $\Lambda^{C}{}_{D}$ is obtained as the matrix exponential

$$\Lambda^{C}{}_{D} = \left(\exp\left[\frac{1}{2}\,\epsilon^{AB}\,\mathbb{M}_{AB}\right]\right)^{C}{}_{D}. \tag{8.302}$$

The algebra fulfilled by the $\mathbb{M}$ matrices is well known to be

$$\left[\mathbb{M}^{CD}, \mathbb{M}^{EF}\right] = \widetilde{g}^{CF}\,\mathbb{M}^{DE} + \widetilde{g}^{DE}\,\mathbb{M}^{CF} - \widetilde{g}^{CE}\,\mathbb{M}^{DF} - \widetilde{g}^{DF}\,\mathbb{M}^{CE}. \tag{8.303}$$

The two algebraic relations (7.31) and (8.303) are equivalent if one replaces

$$\mathbb{M}^{CD} \to -\frac{\mathrm{i}}{2}\,\sigma^{CD}, \tag{8.304}$$

which exactly leads from Eq. (8.299) to Eq. (8.302). Under a local Lorentz transformation, a Dirac spinor transforms as

$$\psi'(x') = S(\Lambda)\,\psi(x)\,. \tag{8.305}$$

In order to write the Lagrangian, one needs to define the Dirac adjoint in curved space-time. In order to address this question, one has to remember that in flat-space-time, the Dirac adjoint $\overline{\psi}(x)$ is defined in such a way that it transforms with the inverse of the spinor Lorentz transform as compared to $\psi(x)$,

$$\overline{\psi}'(x') = \overline{\psi}(x)\,S(\Lambda^{-1}) = \overline{\psi}(x)\,[S(\Lambda)]^{-1}\,. \tag{8.306}$$

The problem of the definition of $\overline{\psi}(x)$ in curved space-time is sometimes treated in the literature in a rather cursory fashion [278]. Let us see if in curved space-time, we can use the *ansatz*

$$\overline{\psi}(x) = \psi^{+}(x)\,\widetilde{\gamma}^{0}\,, \tag{8.307}$$

with the same flat-space $\widetilde{\gamma}^0$ as in the flat-space Dirac adjoint. In this case,

$$\overline{\psi}'(x') = \psi^+(x')\, S^+(\Lambda)\,\widetilde{\gamma}^0 = \left(\psi^+(x')\,\widetilde{\gamma}^0\right)\left[\widetilde{\gamma}^0\, S^+(\Lambda)\,\widetilde{\gamma}^0\right]. \tag{8.308}$$

To first order in the Lorentz generators ϵ_{AB}, we have indeed,

$$\widetilde{\gamma}^0\, S^+(\Lambda)\,\widetilde{\gamma}^0 = 1 + \frac{\mathrm{i}}{4}\,\epsilon^{AB}\,\widetilde{\gamma}^0\,\widetilde{\sigma}^+_{AB}\,\widetilde{\gamma}^0 = 1 + \frac{\mathrm{i}}{4}\,\epsilon^{AB}\,\widetilde{\sigma}_{AB} = [S(\Lambda)]^{-1}, \tag{8.309}$$

where we have used the identity

$$\begin{aligned}\widetilde{\sigma}^+_{AB} &= -\frac{\mathrm{i}}{2}\,[\widetilde{\gamma}^+_B, \widetilde{\gamma}^+_A] = -\frac{\mathrm{i}}{2}\,\widetilde{\gamma}^0\,[\widetilde{\gamma}^0\widetilde{\gamma}^+_B\widetilde{\gamma}^0,\, \widetilde{\gamma}^0\widetilde{\gamma}^+_A\widetilde{\gamma}^0]\,\widetilde{\gamma}^0 \\ &= -\frac{\mathrm{i}}{2}\,\widetilde{\gamma}^0\,[\widetilde{\gamma}_B,\, \widetilde{\gamma}_A]\,\widetilde{\gamma}^0 = -\widetilde{\gamma}^0\,\widetilde{\sigma}_{BA}\,\widetilde{\gamma}^0 = \widetilde{\gamma}^0\,\widetilde{\sigma}_{AB}\,\widetilde{\gamma}^0\,.\end{aligned} \tag{8.310}$$

It is easy to show that Eq. (8.309) generalizes to all orders in the ϵ^{AB} parameters, which justifies our *ansatz* given in Eq. (8.307).

8.5.3 *Lagrangian and Charge Conjugation*

Equipped with an appropriate form of the Dirac adjoint in curved space-time, we start from the Lagrangian density

$$\mathcal{L} = \overline{\psi}(x)\,[\overline{\gamma}^\mu\,\{\mathrm{i}\,(\partial_\mu - \Gamma_\mu) - e\,A_\mu\} - m]\,\psi(x)\,, \tag{8.311}$$

and attempt to derive the particle-antiparticle symmetry on the level of a transformation of the Lagrangian. The Lagrangian is Hermitian, and so

$$\mathcal{L} = \mathcal{L}^+ = \psi^+(x)\left[(\overline{\gamma}^\mu)^+\left\{-\mathrm{i}\overleftarrow{\partial}_\mu - e\,A_\mu\right\} - (-\mathrm{i})\,(\Gamma_\mu)^+\,(\overline{\gamma}^\mu)^+ - m\right]\left[\overline{\psi}(x)\right]^+. \tag{8.312}$$

An insertion of $\widetilde{\gamma}^0$ matrices and use of the identity $(\widetilde{\gamma}^0)^2 = 1$ leads to the relation

$$\begin{aligned}\mathcal{L}^+ &= \psi^+(x)\,\widetilde{\gamma}^0\left[\widetilde{\gamma}^0\,(\overline{\gamma}^\mu)^+\,\widetilde{\gamma}^0\left\{-\mathrm{i}\overleftarrow{\partial}_\mu - e\,A_\mu\right\}\right. \\ &\quad \left. + \mathrm{i}\left\{\widetilde{\gamma}^0\,(\Gamma_\mu)^+\,\widetilde{\gamma}^0\right\}\widetilde{\gamma}^0\,(\overline{\gamma}^\mu)^+\,\widetilde{\gamma}^0 - m\right]\widetilde{\gamma}^0\left[\overline{\psi}(x)\right]^+.\end{aligned} \tag{8.313}$$

Also, we recall that $\widetilde{\gamma}^0\,(\Gamma_\mu)^+\,\widetilde{\gamma}^0 = -\Gamma_\mu$, because

$$\Gamma^+_\mu = -\frac{\mathrm{i}}{4}\,\omega_\mu^{AB}\,\widetilde{\sigma}^+_{AB} = -\frac{\mathrm{i}}{4}\,\omega_\mu^{AB}\,\widetilde{\gamma}^0\,\widetilde{\sigma}_{AB}\,\widetilde{\gamma}^0 = -\widetilde{\gamma}^0\,\Gamma_\mu\,\widetilde{\gamma}^0\,. \tag{8.314}$$

So,

$$\mathcal{L}^+ = \psi^+(x)\,\widetilde{\gamma}^0\left[\overline{\gamma}^\mu\left\{-\mathrm{i}\overleftarrow{\partial}_\mu - e\,A_\mu\right\} - \mathrm{i}\,\Gamma_\mu\,\overline{\gamma}^\mu - m\right]\widetilde{\gamma}^0\left[\overline{\psi}(x)\right]^+. \tag{8.315}$$

Now, we use the relations $\psi^+(x)\,\widetilde{\gamma}^0 = \overline{\psi}(x)$ and $\widetilde{\gamma}^0\left[\overline{\psi}(x)\right]^+ = \psi(x)$, and arrive at the form

$$\mathcal{L}^+ = \overline{\psi}(x)\left[\overline{\gamma}^\mu\left\{-\mathrm{i}\overleftarrow{\partial}_\mu - e\,A_\mu\right\} - \mathrm{i}\,\Gamma_\mu\,\overline{\gamma}^\mu - m\right]\psi(x)\,. \tag{8.316}$$

Because $\mathcal{L}$ is a scalar, a transposition again does not change the Lagrangian, and we have

$$(\mathcal{L}^+)^{\mathrm{T}} = \psi^{\mathrm{T}}(x)\left[(\overline{\gamma}^\mu)^{\mathrm{T}}\left\{-\mathrm{i}\overrightarrow{\partial}_\mu - e\,A_\mu\right\} - \mathrm{i}\,(\overline{\gamma}^\mu)^{\mathrm{T}}\,(\Gamma_\mu)^{\mathrm{T}} - m\right]\left[\overline{\psi}(x)\right]^{\mathrm{T}}. \tag{8.317}$$

An insertion of the charge conjugation matrix $C = \mathrm{i}\widetilde{\gamma}^2\,\widetilde{\gamma}^0$ leads to

$$\begin{aligned}(\mathcal{L}^+)^{\mathrm{T}} = \psi^{\mathrm{T}}(x)\,C^{-1}\Big[C\,(\overline{\gamma}^{\mu})^{\mathrm{T}}\,C^{-1}\Big\{-\mathrm{i}\,\overrightarrow{\partial}_{\mu} - e\,A_{\mu}\Big\} \\ -\mathrm{i}\,C\,(\overline{\gamma}^{\mu})^{\mathrm{T}}\,C^{-1}\,C\,\Gamma_{\mu}^{\mathrm{T}}\,C^{-1} - m\Big]\,C\,\left[\overline{\psi}(x)\right]^{\mathrm{T}}.\end{aligned} \tag{8.318}$$

We have used the identities $C\,(\overline{\gamma}^{\mu})^{\mathrm{T}}\,C^{-1} = -\overline{\gamma}^{\mu}$, and $C\,(\Gamma_{\mu})^{\mathrm{T}}\,C^{-1} = -\Gamma_{\mu}$. The latter of these can be shown as follows,

$$C\,\Gamma_{\mu}^{\mathrm{T}}\,C^{-1} = \frac{\mathrm{i}}{4}\left\{\frac{\mathrm{i}}{2}\omega_{\mu}^{AB}\,C\left[\widetilde{\gamma}_B^{\mathrm{T}},\widetilde{\gamma}_A^{\mathrm{T}}\right]C^{-1}\right\} = \frac{\mathrm{i}}{4}\left\{\frac{\mathrm{i}}{2}\omega_{\mu}^{AB}\left[-\widetilde{\gamma}_B,-\widetilde{\gamma}_A\right]\right\} = -\Gamma_{\mu}\,. \tag{8.319}$$

The result is the expression

$$(\mathcal{L}^+)^{\mathrm{T}} = \psi^{\mathrm{T}}(x)\,C^{-1}\Big[(-\overline{\gamma}^{\mu})\Big\{-\mathrm{i}\,\overrightarrow{\partial}_{\mu} - e\,A_{\mu}\Big\} - \mathrm{i}\,(-\overline{\gamma}^{\mu})\,(-\Gamma_{\mu}) - m\Big]\,C\,\left[\overline{\psi}(x)\right]^{\mathrm{T}}. \tag{8.320}$$

Now we express the result in terms of the charge-conjugated spinor and its adjoint (further remarks on this point are presented in the following),

$$\psi^{\mathcal{C}}(x) = C\left[\overline{\psi}(x)\right]^{\mathrm{T}}, \qquad \overline{\psi^{\mathcal{C}}(x)} = -\psi^{\mathrm{T}}(x)\,C^{-1}, \tag{8.321}$$

to write

$$\begin{aligned}\mathcal{L} = (\mathcal{L}^+)^{\mathrm{T}} &= -\overline{\psi^{\mathcal{C}}(x)}\Big[\overline{\gamma}^{\mu}\Big\{\mathrm{i}\,\overrightarrow{\partial}_{\mu} + e\,A_{\mu}\Big\} - \mathrm{i}\,\overline{\gamma}^{\mu}\,\Gamma_{\mu} - m\Big]\,\psi^{\mathcal{C}}(x) \\ &= -\,\overline{\psi^{\mathcal{C}}(x)}\,\left[\overline{\gamma}^{\mu}\,\{\mathrm{i}(\partial_{\mu} - \Gamma_{\mu}) + e\,A_{\mu}\} - m\right]\psi^{\mathcal{C}}(x)\,.\end{aligned} \tag{8.322}$$

The Lagrangian (8.322) differs from (8.311) only with respect to the sign of electric charge, as is to be expected, and with respect to the replacement of the Dirac spinor $\psi(x)$ by its charge conjugation $\psi^{\mathcal{C}}(x)$. The overall minus sign is a consequence of the fact that we here investigate the Lagrangian on the level of the (first-quantized) relativistic quantum theory, but not on the level of the (second-)quantized quantum field theory, for which we refer the reader to Ref. [251]. We had already seen that the charge conjugation of the Dirac current only yields the intuitive result in the second-quantized formalism [see Eqs. (7.195) and (7.196)]. The gravitational covariant derivative $\partial_{\mu} - \Gamma_{\mu}$ has retained its form in going from (8.312) to (8.322), in agreement with the particle-antiparticle symmetry of the gravitational interaction.

We should now supplement the derivation of Eq. (8.321). Very explicitly, the charge-conjugated spinor is given by

$$\psi^{\mathcal{C}}(x) = C\left[\overline{\psi}(x)\right]^{\mathrm{T}} = C\left[\psi^+(x)\,\widetilde{\gamma}^0\right]^{\mathrm{T}} = C\,\widetilde{\gamma}^0\,\psi^*(x)\,. \tag{8.323}$$

Hence, we have the relation

$$\begin{aligned}\overline{\psi^{\mathcal{C}}(x)} &= \left(C\,\widetilde{\gamma}^0\,\psi^*(x)\right)^+\,\widetilde{\gamma}^0 = [\psi^*(x)]^+\,\widetilde{\gamma}^0\,C^+\,\widetilde{\gamma}^0 \\ &= \psi^{\mathrm{T}}(x)\,\widetilde{\gamma}^0\,C^+\,\widetilde{\gamma}^0 = \psi^{\mathrm{T}}(x)\,C = -\psi^{\mathrm{T}}(x)\,C^{-1}\,,\end{aligned} \tag{8.324}$$

where the two identities *(i)* $\widetilde{\gamma}^0\,C^+\,\widetilde{\gamma}^0 = C$ and *(ii)* $C^{-1} = -C$ have been used. These will be derived in the following. The explicit form of the $\widetilde{\gamma}^2$ matrix in the Dirac representation implies that $\left(\widetilde{\gamma}^2\right)^+ = -\widetilde{\gamma}^2$. Based on this relation, we can easily show that

$$C^+ = \left(\mathrm{i}\,\widetilde{\gamma}^2\,\widetilde{\gamma}^0\right)^+ = -\mathrm{i}\,\widetilde{\gamma}^0\,\left(\widetilde{\gamma}^2\right)^+ = \mathrm{i}\,\widetilde{\gamma}^0\,\widetilde{\gamma}^2 = -\mathrm{i}\,\widetilde{\gamma}^2\,\widetilde{\gamma}^0 = -C\,. \tag{8.325}$$

The first identity $\widetilde{\gamma}^0\,C^+\,\widetilde{\gamma}^0 = C$ can now be shown as follows,

$$\widetilde{\gamma}^0\,C^+\,\widetilde{\gamma}^0 = \widetilde{\gamma}^0\,\left[-\mathrm{i}\,\widetilde{\gamma}^2\,\widetilde{\gamma}^0\right]\,\widetilde{\gamma}^0 = -\mathrm{i}\,\widetilde{\gamma}^0\,\widetilde{\gamma}^2 = \mathrm{i}\,\widetilde{\gamma}^2\,\widetilde{\gamma}^0 = C\,. \tag{8.326}$$

Furthermore, one has

$$C\,C^+ = C\,(-C) = \mathrm{i}\,\widetilde{\gamma}^2\,\widetilde{\gamma}^0\,\mathrm{i}\,\widetilde{\gamma}^0\,\widetilde{\gamma}^2 = -\left(\widetilde{\gamma}^2\right)^2 = -\left(-\mathbb{1}_{4\times4}\right) = \mathbb{1}_{4\times4}\,, \tag{8.327}$$

so that

$$C^{-1} = C^+ = -C\,, \tag{8.328}$$

which proves, in particular, that $C^{-1} = -C$. One consequence of the derivation outlined here is that the equivalence principle holds for antimatter [251, 291]. This conclusion is supported by preliminary experimental results [292].

8.6 Further Thoughts

Here are some suggestions for further thought.

(1) **Dirac Ground State.** We investigate whether or not it is possible to "guess" the exact functional form of the $1S_{1/2}$ ground state wave function of the hydrogen atom within the fully relativistic formalism. To this end, we start from Eq. (8.12),

$$H_{\rm DC}\psi(\vec r) = \begin{pmatrix} m - \dfrac{Z\alpha}{r} & (\vec\sigma\cdot\hat r)\left(-\mathrm{i}\dfrac{\partial}{\partial r} + \mathrm{i}\dfrac{\vec\sigma\cdot\vec L}{r}\right) \\ (\vec\sigma\cdot\hat r)\left(-\mathrm{i}\dfrac{\partial}{\partial r} + \mathrm{i}\dfrac{\vec\sigma\cdot\vec L}{r}\right) & -m - \dfrac{Z\alpha}{r} \end{pmatrix} \begin{pmatrix} f(r)\chi_{\kappa\,\mu}(\hat r) \\ \mathrm{i}g(r)\chi_{-\kappa\,\mu}(\hat r) \end{pmatrix}$$
$$= \begin{pmatrix} \left[\left(-\dfrac{\partial}{\partial r} + \dfrac{1}{r}\,(\kappa-1)\right) g(r) + \left(m - \dfrac{Z\alpha}{r}\right) f(r)\right] \chi_{\kappa\,\mu}(\hat r) \\ \mathrm{i}\left[\left(\dfrac{\partial}{\partial r} + \dfrac{1}{r}\,(\kappa+1)\right) f(r) - \left(m + \dfrac{Z\alpha}{r}\right) g(r)\right] \chi_{-\kappa\,\mu}(\hat r) \end{pmatrix} = E\,\psi(\vec r) \tag{8.329}$$

where one has used

$$\sigma\cdot\vec L = K - 1\,, \qquad K\,\chi_{\kappa\mu} = -\kappa\,\chi_{\kappa\mu}\,, \qquad \vec\sigma\cdot\hat r\,\chi_{\kappa\,\mu}(\hat r) = -\chi_{\kappa\,\mu}(\hat r)\,. \tag{8.330}$$

The upper and lower component equations are

$$\left(m - \frac{Z\alpha}{r} - E\right) f(r) + \left(-\frac{\partial}{\partial r} + \frac{1}{r}\,(\kappa-1)\right) g(r) = 0\,, \tag{8.331}$$

$$\left(\frac{\partial}{\partial r} + \frac{1}{r}\,(\kappa+1)\right) f(r) - \left(m + \frac{Z\alpha}{r} + E\right) g(r) = 0\,. \tag{8.332}$$

Our *ansatz* for the wave function is

$$f(r) = A r^{\gamma} \exp(-\beta r) , \qquad g(r) = B r^{\gamma} \exp(-\beta r) . \tag{8.333}$$

Comparing coefficients of different powers of r, one should obtain the system of equations

$$(m - E) A + \beta B = 0 , \tag{8.334a}$$

$$(-2 - \gamma) B - Z \alpha A = 0 , \tag{8.334b}$$

$$(E + m) B + A \beta = 0 , \tag{8.334c}$$

$$-Z \alpha B + A \gamma = 0 . \tag{8.334d}$$

Furthermore, one may ascertain that all of these equations are satisfied if we set

$$E = m\sqrt{1 - (Z\alpha)^2} , \qquad \gamma = \sqrt{1 - (Z\alpha)^2} - 1 ,$$

$$\beta = Z\alpha m , \qquad B = \frac{\gamma A}{Z\alpha} = \frac{\sqrt{1 - (Z\alpha)^2} - 1}{Z\alpha} A . \tag{8.335}$$

Finally, try to determine if the overall normalization factor A could read as

$$A = \left(\frac{2^{2\sqrt{1-(Z\alpha)^2}-1} \, m^{2\sqrt{1-(Z\alpha)^2}+1} \, (Z\alpha)^{2\sqrt{1-(Z\alpha)^2}+3}}{\sqrt{1 - (Z\alpha)^2} \sqrt{1 - \sqrt{1 - (Z\alpha)^2}} \, \Gamma(2\sqrt{1 - (Z\alpha)^2})} \right)^{1/2} . \tag{8.336}$$

(2) **Dirac Continuum and Bound States.** Generalize the treatment, inspired by the second-order differential equation given in (8.58) for the continuum wave functions (see Sec. 8.2.3), to the case of bound-state wave functions (see Sec. 8.2.2). Instead of Eq. (8.53), try the *ansatz*

$$u_1 = (m + E)^{1/2} (\Phi_1 + \Phi_2) , \tag{8.337a}$$

$$u_2 = (m - E)^{1/2} (\Phi_1 - \Phi_2) . \tag{8.337b}$$

Compare with Eqs. (18) and (19) of Ref. [293].

(3) **Momentum Normalization for Continuum States.** Work out the Dirac continuum wave functions in the momentum normalization [as opposed to the energy normalization, see Eq. (8.88)]

$$\langle 1|2 \rangle = \delta(p_1 - p_2) . \tag{8.338}$$

Go through all intermediate steps of the derivation and convince yourself of the validity of the calculation of the normalization integral.

(4) **Positron Continuum States.** The derivation in Sec. 8.2.3 concerned electron continuum solutions. Work out the positron continuum solutions in explicit form.

(5) **Nonrelativistic Limit.** Calculate the nonrelativistic limit of the relativistic bound-state wave functions (see Sec. 8.2.2) and compare with the derivation of the nonrelativistic wave function calculated in Sec. 4.2.3. Which parameters

are relevant to the expansion, and which ones have to be kept constant? On the basis of the considerations, or otherwise, show the validity of the approximation given in Eq. (6.100), whose leading term can be found in Eq. (6.99). Then, attempt to do the same for the continuum wave functions discussed in Sec. 8.2.3 and compare to Sec. 4.2.5.

(6) **Virial Theorems.** Study the formalism introduced in Ref. [264]. Investigate whether or not the formalism allows one to derive closed-form analytic expressions for the correction to the wave function induced by an external magnetic field. These would be relevant for numerical calculations of radiative corrections to the bound-electron g factor, e.g., in Ref. [39].

(7) **Maxwell–Dirac–Mohr Equations.** A very interesting work by MOHR[4] concerns candidate wave functions for the photon [294]. We shall dwell on related questions in terms of potential further thoughts that might be invested into this intriguing problem, and start with the spin operators of the photon which are given by the matrices $\mathbb{S}^k$ $(k = 1, 2, 3)$ [see Eq. (5.139) on page 194 of Ref. [81]],

$$(\mathbb{S}^k)^{ij} = -\mathrm{i}\,\epsilon^{kij}\,, \tag{8.339}$$

where $\epsilon_{ijk} = \epsilon^{ijk}$ is the Levi-Cività tensor. The explicit representation for $k = 1, 2, 3$ reads as

$$\mathbb{S}^1 = \begin{pmatrix} 0 & 0 & 0 \\ 0 & 0 & -\mathrm{i} \\ 0 & \mathrm{i} & 0 \end{pmatrix} \qquad \mathbb{S}^2 = \begin{pmatrix} 0 & 0 & \mathrm{i} \\ 0 & 0 & 0 \\ -\mathrm{i} & 0 & 0 \end{pmatrix}, \qquad \mathbb{S}^3 = \begin{pmatrix} 0 & -\mathrm{i} & 0 \\ \mathrm{i} & 0 & 0 \\ 0 & 0 & 0 \end{pmatrix}. \tag{8.340}$$

These spin operators follow the conventions proposed in Ref. [295] for the generators of the global $SU(2)$ symmetry in the basis of the $\pi^{0,\pm}$ mesons. The commutation relation is

$$[\,\mathbb{S}^i, \mathbb{S}^j\,] = \mathrm{i}\,\epsilon^{ijk}\,\mathbb{S}^k\,. \tag{8.341}$$

For the discussion of hyperfine effects for spin-1 particles, e.g., the deuteron nucleus in the context of hyperfine splitting in deuterium, it is often advantageous to define a basis where the τ^3 matrix is diagonal, i.e.,

$$\tau^3 = \begin{pmatrix} 1 & 0 & 0 \\ 0 & 0 & 0 \\ 0 & 0 & -1 \end{pmatrix}, \tag{8.342}$$

supplemented by

$$\tau^1 = \frac{1}{\sqrt{2}} \begin{pmatrix} 0 & 1 & 0 \\ 1 & 0 & 1 \\ 0 & 1 & 0 \end{pmatrix}, \qquad \tau^2 = \frac{\mathrm{i}}{\sqrt{2}} \begin{pmatrix} 0 & -1 & 0 \\ 1 & 0 & -1 \\ 0 & 1 & 0 \end{pmatrix}. \tag{8.343}$$

These conventions are adopted in Ref. [294]. (In Ref. [294], the electric and magnetic fields are used in the spherical basis, but we here advocate a modified representation which allows for Cartesian coordinates to be used for the

[4]Peter Joseph Mohr (b. 1943).

description of the fields.) We recall the basis vectors in the spherical basis from Chap. 6,

$$\vec{e}_{+1} = -\frac{1}{\sqrt{2}}\left(\hat{e}_x + \mathrm{i}\,\hat{e}_y\right) = -\frac{1}{\sqrt{2}}\begin{pmatrix}1\\ \mathrm{i}\\ 0\end{pmatrix}, \qquad \vec{e}_0 = \hat{e}_z\,, \qquad \vec{e}_{-1} = \frac{1}{\sqrt{2}}\left(\hat{e}_x - \mathrm{i}\,\hat{e}_y\right). \tag{8.344}$$

In the spherical basis, $\mathbb{S}^3$ is diagonal, i.e., one has

$$\mathbb{S}^3\,\vec{e}_{+1} = \vec{e}_{+1}\,, \qquad \mathbb{S}^3\,\vec{e}_0 = \vec{0}\,, \qquad \mathbb{S}^3\,\vec{e}_{-1} = -\vec{e}_{-1}\,. \tag{8.345}$$

For an arbitrary vector field $\vec{B}$, one can show that

$$\mathrm{i}\left(\vec{\mathbb{S}}\cdot\vec{\nabla}\right)\vec{B} = -\vec{\nabla}\times\vec{B}\,, \tag{8.346}$$

where $\vec{B} = B_x\hat{e}_x + B_y\hat{e}_y + B_z\,\hat{e}_z$, given in the Cartesian basis. One may observe that, taking the divergence of the Ampere–Maxwell law, one obtains the time derivative of Gauss's law. The law of the absence of magnetic monopoles, $\vec{\nabla}\cdot\vec{B} = 0$, can be interpreted as an intrinsic property of the magnetic induction field.

Hence, the only Maxwell equations which need to be addressed, are the Ampere–Maxwell law, and the Faraday law. In the source-free case, with $\epsilon_0 = \mu_0 = c = 1$, they read as

$$\mathrm{i}\left(\vec{\mathbb{S}}\cdot\vec{\nabla}\right)\vec{B} + \frac{\partial}{\partial t}\vec{E} = 0\,, \tag{8.347}$$

$$\mathrm{i}\left(\vec{\mathbb{S}}\cdot\vec{\nabla}\right)\vec{E} - \frac{\partial}{\partial t}\vec{B} = 0\,, \tag{8.348}$$

which can be grouped as

$$\begin{pmatrix} \mathbb{1}_{2\times 2}\dfrac{\partial}{\partial t} & \vec{\mathbb{S}}\cdot\vec{\nabla} \\ -\vec{\mathbb{S}}\cdot\vec{\nabla} & -\mathbb{1}_{2\times 2}\dfrac{\partial}{\partial t}\end{pmatrix}\begin{pmatrix}\vec{E}\\ \mathrm{i}\vec{B}\end{pmatrix} = \begin{pmatrix}\vec{0}\\ \vec{0}\end{pmatrix}. \tag{8.349}$$

One may define modified Dirac–Mohr matrices $\widehat{\gamma}$ according to the following prescription,

$$\widehat{\gamma}^0 = \begin{pmatrix}\mathbb{1}_{2\times 2} & 0\\ 0 & -\mathbb{1}_{2\times 2}\end{pmatrix}, \qquad \widehat{\gamma}^i = \begin{pmatrix}0 & \mathbb{S}^i\\ -\mathbb{S}^i & 0\end{pmatrix}, \tag{8.350}$$

and write the Maxwell–Dirac–Mohr equation (8.349) as

$$\widehat{\gamma}^\mu\,\partial_\mu\,\Psi = 0\,, \tag{8.351}$$

where $\Psi = (\vec{E}, \mathrm{i}\vec{B})^{\mathrm{T}}$ is a candidate wave function of the photon. One of the advantages of Eq. (8.351) is that it is possible to define a Hamilton operator for the photon,

$$H = \vec{\widehat{\alpha}}\cdot\vec{p}\,, \qquad \vec{p} = -\mathrm{i}\vec{\nabla}\,, \qquad \widehat{\alpha}^i = \widehat{\gamma}^0\,\widehat{\gamma}^i\,. \tag{8.352}$$

Longitudinal photons can be identified as zero-energy states of the photon field [294].

A good exercise is to study the formalism introduced in Ref. [294]. Also, in Ref. [294], the Green functions of the electromagnetic field are calculated based on a first-order differential equation, $\widetilde{\gamma}^\mu \partial_\mu \Psi = \Xi(x)$, where $\Xi(x)$ is a six-component source term $\Xi(x) = (-\vec{J}(x), \vec{0})^{\rm T}$ [see Eq. (52) of Ref. [294]]. The formalism introduced in Ref. [294] has been used in Ref. [296]. One advantage of the formalism introduced in Ref. [294] is that the Maxwell Green function, given in Chap. 9 of Ref. [294], directly connects the sources to the fields, without any intermediate recourse to potentials. The advantages of the formalism introduced in Ref. [294] for practical applications still remain to be analyzed, and one is encouraged to expand on the formalism introduced in Ref. [296].

(8) **Numerical Evaluation of Generalized Bessel Functions.** For the numerical evaluation of an array of (ordinary) Bessel functions of increasing index, one may use recursive methods, described in Refs. [258,297,298] and in Sec. III A of Ref. [272]. Key to the solution of the numerical problem is the transformation of a five-term recurrence relation fulfilled by the generalized Bessel functions, given in Eq. (11) of Ref. [272], into a four-term, and a three-term recurrence relation, given in Eqs. (24)–(28) of Ref. [272]. The numerical behavior of the expansion coefficients of a Dirac–Volkov state, as a function of the index, is analyzed in Sec. V B of Ref. [272]. In this context, one considers the asymptotic behavior of the expansion coefficients, which constitute generalized Bessel functions, as illustrated in Fig. 5 of Ref. [272]. Investigate to which extent the observed behavior allows us to establish a classical-quantum correspondence relation for the Dirac–Volkov states.

(9) **Covariant Derivative.** Consider the analogy of Eqs. (8.4) and (8.238),

$$\mathcal{D}'_\mu \psi' = (D_\mu \psi)' \quad \Leftrightarrow \quad \nabla'_\mu \psi' = (\nabla_\mu \psi)' . \tag{8.353}$$

Furthermore, consider the analogy, on the basis of Eqs. (8.3) and (8.239),

$$\mathrm{i}\, e\, A_\mu \quad \Leftrightarrow \quad -\Gamma_\mu . \tag{8.354}$$

Then, on the basis of the analogy, interpret the identity (8.242),

$$\Gamma'_\mu = S(\Lambda)\, \Gamma_\mu\, S(\Lambda)^{-1} + [\partial_\mu S(\Lambda)]\, S(\Lambda)^{-1} , \tag{8.355}$$

in terms of the gauge transformation of the electromagnetic four-vector potential (8.1), i.e., in terms of the relation of A'_μ to A_μ.

(10) **Christoffel Symbols.** Show Eq. (8.178) and verify all intermediate steps in Eq. (8.181).

(11) **Lorentz Transformation of a Dirac Matrix.** Show, in analogy to Eq. (8.230), that the Dirac matrix with covariant index, $\widetilde{\gamma}_A$, is invariant under local Lorentz transformations, i.e., that $\widetilde{\gamma}'_A = \widetilde{\gamma}_A$.

(12) **Basic Properties of the Tetrad.** Show that

$$\left(e^{\rho}_{\ B}\right)^{-1} = e^{B}_{\ \rho}\,, \quad \left(e^{B}_{\ \rho}\right)^{-1} = e^{\rho}_{\ B}\,, \quad e^{\mu}_{\ A} e^{A}_{\ \nu} = \delta^{\mu}_{\nu}\,, \quad e^{A}_{\ \mu} e^{\mu}_{\ B} = \delta^{A}_{B}\,. \tag{8.356}$$

Using the properties

$$e_{\mu A} = e^{\nu}_{\ A}\, \overline{g}_{\nu\mu} = e^{B}_{\ \mu}\, \widetilde{g}_{BA}\,, \qquad e^{\mu A} = e^{A}_{\ \nu}\, \overline{g}^{\nu\mu} = e^{\mu}_{\ B}\, \widetilde{g}^{BA}\,, \tag{8.357}$$

convince yourself that

$$e^{A}_{\ \mu}\, e_{\nu A} = \overline{g}_{\mu\nu}\,, \quad e^{\mu}_{\ A}\, e^{\nu A} = \overline{g}^{\mu\nu}\,, \tag{8.358}$$

$$e^{\mu}_{\ A}\, e_{\mu B} = \widetilde{g}_{AB}\,, \quad e^{A}_{\ \mu}\, e^{\mu B} = \widetilde{g}^{AB}\,. \tag{8.359}$$

(13) **Lorentz Group.** Attempt to show the commutator relation (8.280) using the Clifford algebra of the flat-space Dirac matrices.

(14) **Comparison with Other Literature.** Study the formalism introduced in Ref. [289], as well as in Ref. [299], and Ref. [300], and, last, but not least, in Ref. [284]. Compare with Refs. [278, 280, 301–304], and try to decide if the formalism introduced in the cited references is compatible with our formulas for the gravitational coupling of a Dirac particle.

(15) **Spin Connection: Alternative Form.** Consider the following alternative form of the affine connection matrix [278, 281],

$$\Gamma_{\rho}(x) = -\frac{\mathrm{i}}{4}\, \overline{g}_{\mu\alpha}(x) \left[\frac{\partial e^{B}_{\ \nu}}{\partial x^{\rho}}\, e^{\alpha}_{\ B} - \Gamma^{\alpha}_{\nu\rho} \right] \overline{\sigma}^{\mu\nu}\,, \tag{8.360}$$

where the local spin matrix is $\overline{\sigma}^{\mu\nu} = \overline{\sigma}^{\mu\nu}(x) = \frac{\mathrm{i}}{2}\left[\overline{\gamma}^{\mu}(x), \overline{\gamma}^{\nu}(x)\right]$. Show the equivalence of Eq. (8.360) and Eq. (8.266). Then, use the equation

$$\left[\overline{\gamma}^{\mu}(x), \overline{\sigma}^{\alpha\beta}(x)\right] = 2\,\mathrm{i}\, \overline{g}^{\mu\alpha}(x)\, \overline{\gamma}^{\beta}(x) - 2\,\mathrm{i}\, \overline{g}^{\mu\beta}(x)\, \overline{\gamma}^{\alpha}(x)\,, \tag{8.361}$$

in order to verify that the functional form (8.360) fulfills Eq. (8.211).

(16) **Spin Connection: Lorentz Transformation.** The Ricci rotation coefficient, or "spin connection", was defined in Eq. (8.223) as

$$\omega^{AB}_{\mu} = e^{A}_{\ \nu}\, \nabla_{\mu}\, e^{\nu B}\,. \tag{8.362}$$

So, it is intuitively clear that under a Lorentz transformation of the local Lorentz frame [see Eq. (8.225)],

$$e'^{\mu A} = \Lambda^{A}_{\ \ B}\, e^{\mu B}\,, \tag{8.363}$$

one has

$$\omega'^{AB}_{\mu} = e'^{A}_{\nu}\, \nabla_{\mu}\, e'^{\nu B}\,. \tag{8.364}$$

Verify that this transformation property is consistent with Eq. (8.242),

$$\Gamma'_{\mu} = S(\Lambda)\, \Gamma_{\mu}\, S(\Lambda)^{-1} + \left[\partial_{\mu}\, S(\Lambda)\right]\, S(\Lambda)^{-1} \tag{8.365}$$

where, in contrast to the discussion of Sec. 8.4.7, one uses full, exponentiated Lorentz transformations,

$$\Lambda^{A}_{\ \ B} = \left(\exp\left[\epsilon^{CD}\, \mathbb{M}_{CD}\right]\right)^{A}_{\ \ B}\,, \tag{8.366a}$$

$$S(\Lambda) = \exp\left(-\frac{\mathrm{i}}{4}\, \widetilde{\sigma}^{AB}\, \epsilon_{AB}\right). \tag{8.366b}$$

(17) **Dirac–Schwarzschild Propagator.** Try to generalize the treatment of the Dirac–Coulomb propagator, outlined in Sec. (8.2.5), to the Dirac–Schwarzschild Hamiltonian given in Eq. (8.298). One needs to solve the equation

$$(H_{\rm DS} - E)\, G(\vec{r}_1, \vec{r}_2, E) = \delta^{(3)}\,(\vec{r}_1 - \vec{r}_2)\; \mathbb{1}_{4\times 4}\,, \tag{8.367}$$

where $H_{\rm DS}$ is given in Eq. (8.298). Show how the particle-antiparticle symmetry is reflected in the spin-angular decomposition of the propagator.

Chapter 9

Electromagnetic Field and Photon Propagators

9.1 Overview

In the current chapter, we investigate the photon propagator from different angles. Initially, we had discussed the quantization of the electromagnetic field in Sec. 2.4. However, we restricted the discussion to the transverse components of the electric field.

We proceed as follows. In the previous chapter, our focus was laid on the Feynman propagator of the fermion field, which was obtained as the time-ordered product of the fermion field operators, as we recall Eq. (7.153),

$$\left\langle 0 \left| T\, \psi(x_1)\, \overline{\psi}(x_2) \right| 0 \right\rangle = \mathrm{i}\, S_F(x_1, x_2)\,. \tag{9.1}$$

Here, we recall that, because of the bispinor structure of the Dirac field operator, the notation adopted in Eq. (9.1) is a short-hand notation for a matrix-valued function, where on the left-hand side, we suppress the spinor indices of the fermion field operators.

In the case of the photon field, we shall encounter the time-ordered product of field operators [see Eq. (9.120)]

$$\langle 0 | \mathrm{T}\; A^\mu(x)\, A^\nu(x') | 0 \rangle = \mathrm{i} D_F^{\mu\nu}(x - x') = \mathrm{i}\, g^{\mu\nu}\, G_F(x - x')\,, \tag{9.2}$$

where $D_F^{\mu\nu}(x - x')$ is the Feynman propagator of the photon field, and $G_F(x - x')$ is the scalar Green function of the wave equation, which fulfills the equation

$$\left(\frac{\partial}{\partial t^2} - \vec{\nabla}^2 \right) G_F(t - t', \vec{r} - \vec{r}\,') = \delta(t - t)\, \delta^{(3)}(\vec{r} - \vec{r}\,')\,. \tag{9.3}$$

This is discussed in greater detail in Sec. 9.2.1. Natural units with $\hbar = c = \epsilon_0 = 1$ are used throughout this chapter.

The Green function of the wave equation has been discussed at a number of places in the literature (see, e.g., Chap. 4 of Ref. [81]). We note that the retarded, advanced and Feynman Green functions all fulfill the wave equation (9.3), while being subject to different boundary conditions. Here, we derive a few properties of the Green functions of the wave equation, by a careful consideration of the Dirac-δ distributions entering the expressions in both coordinate as well as momentum space. This is done in Sec. 9.2.

As we had seen in Chap. 2, the timelike component of the four-vector potential (the so-called "scalar potential") is not quantized in Coulomb gauge, due to the lack of a conjugate field momentum. However, the spatial components of the vector-field operator are indeed quantized in Coulomb gauge. In general, the four-vector field $A^\mu = (\Phi, \vec{A})$ composed of the scalar potential Φ and the vector potential $\vec{A}$, fulfills different gauge conditions in different gauges. In turn, the time-ordered product of the photon field operators enters the expression for the photon propagator in Eq. (9.2). Gauge conditions such as

$$\vec{\nabla} \cdot \vec{A} = 0 \qquad \text{(Coulomb Gauge)} \tag{9.4}$$

[see Eq. (2.43)] and

$$\partial_\mu A^\mu = 0 \qquad \text{(Lorenz Gauge)} \tag{9.5}$$

[see Eq. (9.75)] are primarily imposed on the *classical* four-vector potential, while the expression (9.2) is constructed with the *quantized* vector-field operators. We shall learn that the gauge conditions imposed on the classical potentials can often be imposed on the quantized field operators only in the sense of expectation values of the physical states of the fields. Still, it becomes clear that the explicit form of the photon propagator will depend on the gauge being used.

In Coulomb gauge, we shall discuss the photon propagator in Sec. 9.3. The calculation is made more interesting by the fact that the "scalar potential" is not quantized in Coulomb gauge, per our discussion in Chap. 2. However, in the Hamiltonian of the electromagnetically coupled currents, one finds a term [see Eq. (9.50)] which contains the timelike component of the current J^0 and needs to be matched against the D^{00} component of the photon propagator [see Eq. (9.60)]. This matching is done against the second-order (as opposed to first-order) S matrix element, which defines the spatial components D^{ij} of the photon propagator [see Eq. (9.58)]. The result of the matching is the photon propagator in Coulomb gauge [see Eq. (9.73)], obtained as a result of adding the instantaneous, non-quantized Coulomb interaction (which couples to the scalar component of the current) to the retarded interaction which couples to the spatial components of the current.

In Lorenz gauge (see Sec. 9.4), one can "solve" the problem connected with the absence of a canonical momentum conjugate to A^0, by introducing a gauge-fixing term [see Eq. (9.79)]. This procedure enables us to write down a nonvanishing, canonical field momentum Π^0, conjugate to the scalar potential [see Eq. (9.102)], resulting in the photon propagator in Lorenz gauge [see Eq. (9.121)]. However, the artificial "rabbit drawn from the hat" (the nonvanishing Π^0), comes at a price: In writing the modified Lagrangian (9.79), we have introduced, at will, a term proportional to $(\partial \cdot A)^2$ which vanishes in Lorenz gauge (in the classical theory), but not necessarily in other gauges. Furthermore, a closer inspection reveals that $\partial \cdot A = \partial_\mu A^\mu$ actually does not vanish in the quantized theory, i.e., if A^μ is a field operator as opposed to a classical field. Hence, we have introduced nonphysical degrees of freedom, which will turn out to be scalar and longitudinal photons.

Physical states of the electromagnetic field cannot contain unphysical degrees of freedom. The intuitive feeling is that if we have introduced artificial degrees of freedom, then we somehow need to eliminate them. This is done by imposing the Lorenz-gauge condition $\partial \cdot A = 0$ not as an operator identity, but in terms of a condition $\partial \cdot A^{(+)}|\Psi\rangle = 0$ which is fulfilled for any physical state $|\Psi\rangle$ of the photon field. Here, $A^{(+)}$ contains the positive-frequency (annihilation) contribution to the photon-field operator. This is called the Gupta–Bleuler formalism (see Sec. 9.5.3), even if it could be characterized as a "Lorenz gauge condition in the mean." The transition to other gauges is discussed in Sec. 9.5.

Finally, in Sec. 9.6, we explore the interactions of quantized and classical fields. The notion is that the quantized description allows for the description of field quanta, while a classical description is adequate for strong fields for which we can neglect the influence of quantum fluctuations. For example, if (Dirac) currents are quantized and electromagnetic fields are not, then we may describe (virtual) fermion pair creation processes. This formalism is used in Sec. 18.2 in order to derive the effective Heisenberg–Euler Lagrangian, which is valid for very strong electromagnetic fields. Conversely, if we employ a quantized photon field and classical currents, then we may describe photon creation processes. This latter formalism is applied in Sec. 9.6.1. The way of calculation essentially relies on the Wick theorem, which is given (in various, equivalent forms) in Eqs. (9.163), (9.164), (9.165) and (9.166). Quantizing only one of the two involved fields leads us to a much more manageable theoretical description.

9.2 Time Orderings, Field Commutators and Green Functions

9.2.1 *Miscellaneous Fundamental Relations for Green Functions*

We write the defining equation (9.3) for the Green function of the wave equation as

$$\Box\, G(x) = \delta^{(4)}(x)\,, \qquad \Box = \frac{\partial}{\partial t^2} - \vec{\nabla}^2\,. \tag{9.6}$$

Here, $\Box$ is the "quabla" operator. We use a relativistically covariant formulation. The coordinates are [see also Eq. (2.58)]

$$x^0 = t, \quad x^1 = x, \quad x^2 = y, \quad x^3 = z. \tag{9.7}$$

As is common, we use Latin indices for spatial vectors ($i, j \in \{1, 2, 3\}$), whereas Greek indices are used for space-time four-vectors, $\mu, \nu \in \{0, 1, 2, 3\}$. The space-time metric is $g^{\mu\nu} = \mathrm{diag}(1, -1, -1, -1)$. We work in flat Minkowski space. The derivatives with respect to the coordinates are $\partial^\mu = \partial/(\partial x_\mu)$ and $\partial_\mu = \partial/(\partial x^\mu)$. The scalar product of four-vectors is $v \cdot w = v^0 w^0 - \vec{v} \cdot \vec{w} = v^\mu\, w_\mu$. For the spatial components of a general four-vector, we will sometimes use the notation $v^1 = v^x, v^2 = v^y, v^3 = v^z$. A somewhat nonstandard notation used here concerns the spatial components of the gradient operator, which we recall from Eq. (2.56),

$$\nabla^i \equiv \frac{\partial}{\partial x^i} \tag{9.8}$$

with upper indices.

In momentum space, with $k^2 = k^\mu k_\mu$, we have for the retarded, advanced and Feynman Green functions,

$$G_R(k) = -\frac{1}{(k^0 + \mathrm{i}\,\epsilon)^2 - \vec{k}^2}\,, \quad G_A(k) = -\frac{1}{(k^0 - \mathrm{i}\,\epsilon)^2 - \vec{k}^2}\,, \quad G_F(k) = -\frac{1}{k^2 + \mathrm{i}\,\epsilon}\,. \tag{9.9}$$

All three Green functions G_R, G_A and G_F fulfill the relation

$$\Box\, G_R(x) = \Box G_A(x) = \Box G_F(x) = \delta^{(4)}(x)\,. \tag{9.10}$$

A summary of important properties of the retarded, advanced, and so-called Feynman solutions can be found in Sec. 4.1.5 of Ref. [81], but we here wish to advance a little further. The poles of the retarded Green function are located at $k^0 = \pm|\vec{k}| - \mathrm{i}\,\epsilon$, i.e., the lower half of the complex plane. In the Fourier backtransform integral, the contour for the k^0 integration needs to be closed in the upper half of the complex plane when $t < 0$. Hence, the retarded Green function vanishes for negative t. Analogously (see Chap. 4 of Ref. [81]), one can show that $G_A(x)$ vanishes for $t > 0$. A special role is taken by the Feynman Green function G_F, whose poles are at $k^0 = \pm\sqrt{\vec{k}^2 - \mathrm{i}\,\epsilon}$, positive-frequency (positive-k^0) poles are picked up for $t > 0$, whereas negative-frequency poles are relevant for $t < 0$.

As explained in detail in Chap. 4 of Ref. [81], a Fourier transformation to coordinate space leads to the following formulas. One has for the retarded Green function,

$$\begin{aligned} G_R(x) &= -\int \frac{\mathrm{d}^4 k}{(2\pi)^4} \frac{\mathrm{e}^{-\mathrm{i}k\cdot x}}{(k^0 + \mathrm{i}\,\epsilon)^2 - \vec{k}^2} \\ &= \frac{\Theta(t)}{4\pi r}\,[\delta(t-r) - \delta(t+r)] \approx \frac{1}{2\pi}\,\Theta(x^0)\,\delta(x^2)\,. \end{aligned} \tag{9.11}$$

Here, $x = (t, \vec{r})$, with $x^0 = t$ and $r = |\vec{r}|$. The representations indicated after the "$\approx$" sign indicate a form obtained after we ignore the overlap term of the step function Θ and the representation of the Dirac-δ function. For the advanced Green function, one finds

$$\begin{aligned} G_A(x) &= -\int \frac{\mathrm{d}^4 k}{(2\pi)^4} \frac{\mathrm{e}^{-\mathrm{i}k\cdot x}}{(k^0 - \mathrm{i}\,\epsilon)^2 - \vec{k}^2} \\ &= -\frac{\Theta(-t)}{4\pi r}\,[\delta(t-r) - \delta(t+r)] \approx \frac{1}{2\pi}\,\Theta(-x^0)\,\delta(x^2)\,. \end{aligned} \tag{9.12}$$

Finally, the Feynman Green function reads as follows, in space-time coordinates,

$$G_F(x) = -\int \frac{\mathrm{d}^4 k}{(2\pi)^4} \frac{\mathrm{e}^{-\mathrm{i}k\cdot x}}{k^2 + \mathrm{i}\,\epsilon} = -\frac{\mathrm{i}}{4\pi^2} \frac{1}{x^2 - \mathrm{i}\epsilon}\,. \tag{9.13}$$

The derivation of this latter result will be discussed in more detail in the following.

For completeness, we now consider the Fourier transformations with respect to time, to obtain a mixed photon energy-coordinate representation which is convenient especially for bound-state calculations,

$$G_R(\omega,\vec{r}) = \int_{-\infty}^{\infty} \mathrm{d}t\, \mathrm{e}^{\mathrm{i}\omega t} G_R(x) = \frac{\mathrm{e}^{\mathrm{i}\,\omega r}}{4\pi\, r}, \tag{9.14}$$

$$G_A(\omega,\vec{r}) = \int_{-\infty}^{\infty} \mathrm{d}t\, \mathrm{e}^{\mathrm{i}\omega t} G_A(x) = \frac{\mathrm{e}^{-\mathrm{i}\,\omega r}}{4\pi\, r}, \tag{9.15}$$

$$G_F(\omega,\vec{r}) = \int_{-\infty}^{\infty} \mathrm{d}t\, \mathrm{e}^{\mathrm{i}\omega t} G_F(x) = \frac{\mathrm{e}^{\mathrm{i}\,\sqrt{\omega^2+\mathrm{i}\epsilon}\, r}}{4\pi\, r}, \qquad \mathrm{Im}\sqrt{\omega^2+\mathrm{i}\epsilon} > 0\,. \tag{9.16}$$

The condition in Eq. (9.16) ensures that the modulus of the Feynman Green function remains bounded in the entire complex ω plane.

9.2.2 *Distributions and Fourier Transforms*

In dealing with Green functions, one often has to treat expressions which contain infinitesimal quantities, Dirac-δ functions and other singular distributions. Let us try to compile a few important formulas and in particular, investigate in detail the derivation of the Fourier backtransform (9.13). We start with the following formula, which is relevant to the splitting of an integral into a principal-value term and a pole term, in the distributional sense,

$$\frac{1}{z - \mathrm{i}\epsilon} = (\mathrm{P.V.})\,\frac{1}{z} + \mathrm{i}\,\pi\,\delta(z)\,. \tag{9.17}$$

Here, (P.V.) denotes the principal value, and ϵ denotes an infinitesimal imaginary part. We now investigate a slight generalization of Eq. (9.17), where

$$\eta = (\epsilon, 0, 0, 0) \tag{9.18}$$

is a time-like infinitesimal four-vector.

Then, we have

$$(x - \mathrm{i}\eta)^2 = x^2 - 2\mathrm{i}x^0\epsilon = x^2 - \mathrm{i}\epsilon\,\mathrm{sgn}(x^0)\,, \tag{9.19}$$

where we have absorbed $2|x^0|$ into a redefined ϵ, which we denote as ϵ. The sign function is denoted $\mathrm{sgn}(t) = t/|t|$; it equals unity for positive argument and -1 for negative argument. So, we can write

$$\frac{1}{(x - \mathrm{i}\,\eta)^2} = (\mathrm{P.V.})\frac{1}{x^2} + \mathrm{i}\,\pi\,\mathrm{sgn}(x^0)\,\delta(x^2)\,. \tag{9.20}$$

We shall use this result in Eq. (9.26).

On another issue, collecting the zeros of the argument of a Dirac-δ, using the well-known formula

$$\delta(f(z)) = \sum_i \frac{1}{|f'(z_i)|}\,\delta(z - z_i)\,, \qquad f(z_i) = 0\,, \tag{9.21}$$

we can write

$$\delta(x^2) = \delta((x^0)^2 - \vec{r}^{\,2}) = \frac{1}{2|\vec{r}|}\left[\delta(x^0 - |\vec{r}|) + \delta(x^0 + |\vec{r}|)\right], \tag{9.22}$$

and

$$\Theta(x^0)\,\delta(x^2) \approx \frac{\Theta(x^0)}{2|\vec{r}|}\delta(x^0 - |\vec{r}|)\,, \tag{9.23}$$

where we ignore a "dangerous" term proportional to $\Theta(x^0)\,\delta(x^0 + |\vec{r}|)$ [see the discussion following Eq. (4.35) of Ref. [81] and Eqs. (9.11) and (9.12)]. The latter term can be important if we take into account the infinitesimal overlap of the representation of the Dirac-δ function with the step function Θ. This overlap can be neglected in most calculations but one should be aware of exceptions [305].

A Lorentz-invariant integration measure is given as

$$\frac{\mathrm{d}^4k}{(2\pi)^4}\Theta(k^0)\delta(k^2) = \frac{\mathrm{d}^3k}{(2\pi)^3}\frac{1}{2|\vec{k}|}\,, \tag{9.24}$$

where the integration proceeds on the full $\mathbb{R}^4$ for the left-hand side, while it is restricted to $\mathbb{R}^3$ for the right-hand side.

The following integral is of rather fundamental importance,

$$\int\frac{\mathrm{d}^3k}{(2\pi)^3}\frac{1}{2|\vec{k}|}\mathrm{e}^{\mathrm{i}k\cdot x} = -\frac{1}{4\pi^2}\frac{1}{(x+\mathrm{i}\eta)^2}\,, \qquad \int\frac{\mathrm{d}^3k}{(2\pi)^3}\frac{1}{2|\vec{k}|}\mathrm{e}^{-\mathrm{i}k\cdot x} = -\frac{1}{4\pi^2}\frac{1}{(x-\mathrm{i}\eta)^2}\,. \tag{9.25}$$

The derivation is a little tricky, which is why we present it in some detail [we use the convention that $k = |\vec{k}|$, $r = |\vec{r}|$, the relation $k\cdot x = \omega t - \vec{k}\cdot\vec{r}$, and a convergent factor $\exp(-\epsilon k)$],

$$\begin{aligned}
\int\frac{\mathrm{d}^3k}{(2\pi)^3}\frac{1}{|\vec{k}|}\mathrm{e}^{\mathrm{i}\,k\cdot x} &= \frac{1}{(2\pi)^2}\int_{-1}^{1}\mathrm{d}u\int_0^\infty \mathrm{d}k\,k\,\mathrm{e}^{\mathrm{i}\,(k\,t - k\,r\,u)} \\
&= \lim_{\epsilon\to 0}\frac{1}{(2\pi)^2}\frac{1}{\mathrm{i}r}\int_0^\infty \mathrm{d}k\,\left(\mathrm{e}^{\mathrm{i}\,k\,(t+r+\mathrm{i}\epsilon)} - \mathrm{e}^{\mathrm{i}\,k\,(t-r+\mathrm{i}\epsilon)}\right) \\
&= \lim_{\epsilon\to 0}\frac{1}{(2\pi)^2}\frac{1}{\mathrm{i}r}\left(\frac{1}{-\mathrm{i}\,(t+r+\mathrm{i}\epsilon)} - \frac{1}{-\mathrm{i}\,(t-r+\mathrm{i}\epsilon)}\right) \\
&= \frac{1}{(2\pi)^2 r}\left((\mathrm{P.V.})\frac{2r}{r^2-t^2} - \mathrm{i}\pi\,\delta(t+r) + \mathrm{i}\pi\,\delta(t-r)\right) \\
&= -\frac{1}{2\pi^2}\left[(\mathrm{P.V.})\frac{1}{x^2} - \mathrm{i}\,\pi\,\mathrm{sgn}(t)\,\delta(x^2)\right] = -\frac{1}{2\pi^2}\frac{1}{(x+\mathrm{i}\eta)^2}\,.
\end{aligned} \tag{9.26}$$

We have used the identity

$$\mathrm{sgn}(t)\,\delta(x^2) = \mathrm{sgn}(t)\,\delta(t^2 - r^2) = \frac{\mathrm{sgn}(t)}{2r}\left[\delta(t+r) - \delta(t-r)\right]\,, \tag{9.27}$$

and infinitesimal convergence-generating factors. Analogously, one can derive the results

$$\int\frac{\mathrm{d}^3k}{(2\pi)^3}\frac{1}{2|\vec{k}|}\mathrm{e}^{-\mathrm{i}k\cdot x} = -\frac{1}{4\pi^2}\frac{1}{(x-\mathrm{i}\eta)^2}\,, \tag{9.28a}$$

$$\int\frac{\mathrm{d}^3k}{(2\pi)^3}\frac{1}{2|\vec{k}|}\left(\mathrm{e}^{\mathrm{i}k\cdot x} + \mathrm{e}^{-\mathrm{i}k\cdot x}\right) = -\frac{1}{2\pi^2}(\mathrm{P.V.})\frac{1}{x^2}\,, \tag{9.28b}$$

$$\int\frac{\mathrm{d}^3k}{(2\pi)^3}\frac{1}{2|\vec{k}|}\left(\mathrm{e}^{\mathrm{i}k\cdot x} - \mathrm{e}^{-\mathrm{i}k\cdot x}\right) = \frac{\mathrm{i}}{2\pi}\mathrm{sgn}(x^0)\,\delta(x^2)\,. \tag{9.28c}$$

As anticipated, we now consider the Fourier backtransform (9.13) of the Feynman propagator, using the identity

$$\frac{1}{k^2 + \mathrm{i}\epsilon} = \frac{1}{k_0^2 - \vec{k}^2 + \mathrm{i}\epsilon} = \frac{1}{2|\vec{k}|}\left(\frac{1}{k_0 - |\vec{k}| + \mathrm{i}\epsilon} - \frac{1}{k^0 + |\vec{k}| - \mathrm{i}\epsilon}\right). \tag{9.29}$$

Starting from the momentum representation $G_F(k) = -1/(k^2 + \mathrm{i}\epsilon)$, and using Eq. (9.29), one obtains

$$G_F(x) = -\int \frac{\mathrm{d}^3k}{(2\pi)^3}\frac{\mathrm{e}^{\mathrm{i}\vec{k}\cdot\vec{x}}}{2|\vec{k}|}\frac{1}{2\pi}\left\{\Theta(x^0)\mathrm{e}^{-\mathrm{i}|\vec{k}|x^0}(-2\pi\mathrm{i}) - \Theta(-x^0)\mathrm{e}^{\mathrm{i}|\vec{k}|x^0}(+2\pi\mathrm{i})\right\}, \tag{9.30}$$

where one closes the contour either in the lower complex half-plane (for $x^0 > 0$), leading to the term proportional to $\Theta(x^0)$, or in the upper complex half-plane (for $x^0 < 0$), leading to the term proportional to $\Theta(-x^0)$. Because one integrates over d^3k, the two terms can be summarized as follows,

$$\begin{aligned} G_F(x) &= \mathrm{i}\int \frac{\mathrm{d}^3k}{(2\pi)^3}\frac{1}{2|\vec{k}|}\,\mathrm{e}^{-\mathrm{i}|\vec{k}|\,|x^0|+\mathrm{i}\vec{k}\cdot\vec{x}} \\ &= \frac{\mathrm{i}}{2}\,\Theta(x^0)\int \frac{\mathrm{d}^3k}{(2\pi)^3}\frac{1}{|\vec{k}|}\,\mathrm{e}^{-\mathrm{i}k\cdot x} + \frac{\mathrm{i}}{2}\,\Theta(-x^0)\int \frac{\mathrm{d}^3k}{(2\pi)^3}\frac{1}{|\vec{k}|}\,\mathrm{e}^{\mathrm{i}k\cdot x} \\ &= \frac{\mathrm{i}}{2}\,\Theta(x^0)\left[-\frac{1}{2\pi^2}\frac{1}{(x-\mathrm{i}\eta)^2}\right] + \frac{\mathrm{i}}{2}\,\Theta(-x^0)\left[-\frac{1}{2\pi^2}\frac{1}{(x+\mathrm{i}\eta)^2}\right], \end{aligned} \tag{9.31a}$$

where use has been made of Eq. (9.25). Finally, using Eq. (9.20), one obtains

$$\begin{aligned} G_F(x) &= -\frac{\mathrm{i}}{4\pi^2}\left[\Theta(x^0)\,(\mathrm{P.V.})\frac{1}{x^2} + \mathrm{i}\,\pi\,\mathrm{sgn}(x^0)\,\Theta(x^0)\,\delta(x^2)\right. \\ &\qquad \left. + \Theta(-x^0)\,(\mathrm{P.V.})\frac{1}{x^2} - \mathrm{i}\,\pi\,\mathrm{sgn}(x^0)\,\Theta(-x^0)\,\delta(x^2)\right] \\ &= -\frac{\mathrm{i}}{4\pi^2}\left[(\mathrm{P.V.})\frac{1}{x^2} + \mathrm{i}\pi\delta(x^2)\right] = -\frac{\mathrm{i}}{4\pi^2}\frac{1}{x^2 - \mathrm{i}\epsilon}. \end{aligned} \tag{9.31b}$$

The final result is surprisingly compact. For the transformations to the mixed frequency-coordinate representation, $(t, \vec{r}) \to (\omega, \vec{r})$, we use the identity

$$\frac{1}{t^2 - \vec{r}^{\,2} - \mathrm{i}\epsilon} = \frac{1}{2r}\left(\frac{1}{t - r - \mathrm{i}\epsilon} - \frac{1}{t + r + \mathrm{i}\epsilon}\right). \tag{9.32}$$

The Fourier transformation (9.16) of $G_F(x)$ can thus be calculated easily,

$$\begin{aligned} G_{\mathrm{F}}(\omega, \vec{r}) &= \int_{-\infty}^{\infty} \mathrm{d}t\,\mathrm{e}^{\mathrm{i}\,\omega\,t}\,G_F(x) = \int_{-\infty}^{\infty} \mathrm{d}t\,\mathrm{e}^{\mathrm{i}\,\omega\,t}\frac{1}{4\pi^2\mathrm{i}}\frac{1}{t^2 - \vec{r}^{\,2} - \mathrm{i}\epsilon} \\ &= \frac{1}{4\pi^2\mathrm{i}}\int_{-\infty}^{\infty} \mathrm{d}t\,\mathrm{e}^{\mathrm{i}\,\omega\,t}\frac{1}{2r}\left(\frac{1}{t - r - \mathrm{i}\epsilon} - \frac{1}{t + r + \mathrm{i}\epsilon}\right) \\ &= \frac{1}{4\pi^2\mathrm{i}}\left\{\frac{(2\mathrm{i}\pi)\,\mathrm{e}^{\mathrm{i}\omega\,r}}{2r}\Theta(\omega) - \frac{(-2\mathrm{i}\pi)\,\mathrm{e}^{-\mathrm{i}\omega\,r}}{2r}\Theta(-\omega)\right\} = \frac{\mathrm{e}^{\mathrm{i}|\omega|r}}{4\pi r}. \end{aligned} \tag{9.33}$$

This derivation holds for real ω, i.e., $\mathrm{Im}\,\omega = 0$. For complex ω, as discussed in Chap. 4 of Ref. [81], one has

$$G_F(\omega, \vec{r} - \vec{r}\,') = \frac{\exp\left(\mathrm{i}\sqrt{\omega^2 + \mathrm{i}\epsilon}\;r\right)}{4\pi\,r} = \frac{\exp(-b\,r)}{4\pi\,r}, \qquad b = -\mathrm{i}\sqrt{\omega^2 + \mathrm{i}\epsilon}. \tag{9.34}$$

The fulfillment of the condition $\mathrm{Re}(b) \geq 0$ is achieved by laying the branch cut of the square root function along the positive real axis, thus ensuring that one has $\mathrm{Re}(b) > 0$, irrespective of the complex phase of ω.

Some final remarks are in order. *(i)* For the backtransform of the Feynman propagator, we can state two equivalent prescriptions,

$$G_F(x) = -\int \frac{\mathrm{d}^4k}{(2\pi)^4} \frac{\mathrm{e}^{-ik\cdot x}}{k^2 + \mathrm{i}\,\epsilon} = -\int_{C_\mathrm{F}} \frac{\mathrm{d}^4k}{(2\pi)^4} \frac{\mathrm{e}^{-ik\cdot x}}{k^2}, \tag{9.35}$$

where C_F is the Feynman contour, which provides for the poles with $\mathrm{Re}\,\omega < 0$ to be encircled in the mathematically positive direction, while poles with $\mathrm{Re}\,\omega > 0$ are encircled in the mathematically negative direction, i.e., the pole with $\mathrm{Re}\,\omega > 0$ is infinitesimally below the real axis (see Chap. 4 of Ref. [81]). The indication of both an infinitesimal part $\mathrm{i}\epsilon$ and the Feynman contour is a tautology; one should decide for either one. Here, unless otherwise indicated, we always choose the explicit indication of the infinitesimal imaginary part in the denominator. *(ii)* The distributions $G_R(x)$ and $G_A(x)$ are centered on the light cone and vanish outside of it, but $G_F(x)$ is a distribution which is manifestly nonvanishing outside of the light cone. *(iii)* For $t = 0$, our result for the Feynman Green function simplifies to

$$G_\mathrm{F}(t=0,\vec{r}) = \frac{\mathrm{i}}{(2\pi)^2 r^2}, \tag{9.36}$$

in agreement with Eq. (1.180) of Ref. [2]. *(iv)* We have stressed that all Green functions we discussed are solutions of Eq. (9.10), while their difference is a solution of the homogeneous equation. The distribution $G_-(x)$, which is the difference of the retarded and advanced Green functions,

$$G_-(x) = G_R(x) - G_A(x) \approx \frac{1}{2\pi}\left[\Theta(x^0) - \Theta(-x^0)\right]\delta(x^2) = \frac{1}{2\pi}\,\mathrm{sgn}(x^0)\,\delta(x^2), \tag{9.37}$$

fulfills the homogeneous equation $\Box G_-(x) = 0$. The retarded and advanced Green functions can be reconstructed according to

$$G_R(x) = \Theta(x^0)\,G_-(x), \qquad G_A(x) = -\Theta(-x^0)\,G_-(x). \tag{9.38}$$

9.3 Photon Propagator in Coulomb Gauge

9.3.1 *Legendre Transformation and Hamiltonian*

As already discussed in Sec. 2.4.1, one of the main difficulties in the quantization of the electromagnetic field lies in the absence of a canonical momentum for the zeroth component A^0 of the four-vector potential, namely, the scalar potential. One can overcome the difficulties along two lines of argument. In Coulomb gauge (see Sec. 2.4.1), one simply refrains from quantizing the scalar potential. This implies that the scalar potential A^0 enters the QED Lagrangian via two terms, namely, the so-called longitudinal part of the field energy, $\frac{1}{2}\vec{E}_\parallel^2$, which is related to the longitudinal component $\vec{E}_\parallel$ of the electric field, $\vec{E}_\parallel = -\vec{\nabla}A^0$, and also, by the explicit term $-J_0\, A^0$, which is part of the covariant expression $-J_\mu\, A^\mu$. After the Legendre transformation, the QED Hamiltonian contains a non-quantized, nonretarded, term due to the scalar potential, which can be matched against the scattering amplitude generated by the quantized, spatial components of the photon propagator.

A complementary approach is based on gauge-fixing terms which can be added to the QED Lagrangian; these generate a nonvanishing canonical momentum Π^0 which allows us to quantize the theory. However, one pays a price: The alternative quantization procedure adds unphysical degrees of freedom which later have to be eliminated from the physical states of the photon field. The latter elimination proceeds by imposing additional conditions, e.g., the Gupta–Bleuler condition discussed in Sec. 9.5.3.

Here and in the following, we use the notations $\Phi \equiv A^0$ and $J_0 = \rho$ interchangeably, and set $x = (t, \vec{r})$ as well as $x' = (t', \vec{r}')$. We recall the Coulomb gauge condition (2.43) from Sec. 2.4.1,

$$\vec{\nabla}\cdot\vec{A}(t,\vec{r}) = 0\,, \qquad \vec{A}\,(t,\vec{r}) = \vec{A}_\perp\,(t,\vec{r})\,, \qquad \vec{A}_\parallel\,(t,\vec{r}) = 0\,, \tag{9.39}$$

and the potentials are coupled to the sources as follows, [see Eq. (2.44)],

$$\vec{\nabla}^2\Phi\,(t,\vec{r}) = -\,\rho\,(t,\vec{r})\,, \tag{9.40a}$$

$$\left(\frac{\partial^2}{\partial t^2} - \vec{\nabla}^2\right)\vec{A}_\perp\,(t,\vec{r}) = \vec{J}_\perp\,(t,\vec{r})\,, \tag{9.40b}$$

$$\vec{\nabla}\frac{\partial}{\partial t}\Phi\,(t,\vec{r}) = \vec{J}_\parallel(t,\vec{r})\,. \tag{9.40c}$$

The action-at-a-distance solution to Eq. (9.40a),

$$\Phi\,(t,\vec{r}) = \int \mathrm{d}^3r'\,\frac{1}{4\pi|\vec{r}-\vec{r}'|}\,\rho\,(t,\vec{r}') + \Phi_{\mathrm{hom}}(t,\vec{r})\,, \tag{9.41}$$

could be interpreted as being at variance with the causality principle; however, according to Chap. 4 of Ref. [81], the constraint (9.40c) actually forces the homogeneous solution $\Phi_{\mathrm{hom}}(t,\vec{r})$ (solution of the Laplace equation) to be chosen so that the full $\Phi\,(t,\vec{r})$ is connected to the sources only via retarded Green functions, within the classical theory. Furthermore, we shall show that the contribution of the homogeneous solution to the Hamiltonian density is not physically significant.

This can be seen as follows. We start from the QED Lagrangian, from which we exclude, for simplicity, the free Dirac density $\mathcal{L}_D = \overline{\psi}(\mathrm{i}\gamma^\mu\,\partial_\mu - m)\psi$ (reverting to flat space, $\gamma^\mu \equiv \widetilde{\gamma}^\mu$, in comparison to Chap. 8) and interpret $J_\mu = e\overline{\psi}\gamma_\mu\psi$ as a classical current distribution. The resulting Lagrangian is

$$\begin{aligned}\mathcal{L} &= -\frac{1}{4}\,F^{\mu\nu}\,F_{\mu\nu} - J_\mu\,A^\mu \\ &= \frac{1}{2}\left(\vec{E}_\parallel^2 + \vec{E}_\perp^2 - \vec{B}^2\right) - J_0 A^0 + \vec{J}\cdot\vec{A}\,,\end{aligned} \tag{9.42}$$

where we have split the electric field into longitudinal and transverse parts, according to Chap. 1 of Ref. [81]. In Coulomb gauge, one has

$$\vec{E}_\parallel = -\vec{\nabla}\Phi\,, \qquad \vec{E}_\perp = -\frac{\partial}{\partial t}\vec{A}_\perp\,. \tag{9.43}$$

The conjugate variable of the vector potential is the transverse electric field $\vec{E}_\perp(t,\vec{r})$ (up to a minus sign),

$$\vec{\Pi}(t,\vec{r}) = \frac{\partial}{\partial t}\vec{A}_\perp(t,\vec{r}) = -\vec{E}_\perp(t,\vec{r})\,. \tag{9.44}$$

Here, we define the $\vec{\Pi}$ vector to be composed of the conjugate variables of the $\vec{A}$ vector, in the sense of the remarks included after Eq. (2.49). We had seen in Eq. (2.74) that

$$\left[E^i_\perp(t,\vec{r}), A^j_\perp(t,\vec{r}\,')\right] = \mathrm{i}\,\delta^{\perp,ij}(\vec{r}-\vec{r}\,')\,, \tag{9.45}$$

where we recall the transverse Dirac-δ from Eq. (2.54),

$$\delta^{\perp,ij}(\vec{r}-\vec{r}\,') = \int \frac{\mathrm{d}^3k}{(2\pi)^3}\left(\delta^{ij} - \frac{k^i\,k^j}{\vec{k}^2}\right)\mathrm{e}^{\mathrm{i}\,\vec{k}\cdot(\vec{r}-\vec{r}\,')}\,. \tag{9.46}$$

For the Legendre transformation of the Lagrangian density, though, we have to use the formula

$$\mathcal{H} = \Pi_k\,\frac{\partial A^k}{\partial t} - \mathcal{L} = \frac{1}{2}\left(\vec{E}_\perp^2 + \vec{B}^2\right) - \frac{1}{2}\,\vec{E}_\parallel^2 + J_0 A^0 - \vec{J}\cdot\vec{A}\,, \tag{9.47}$$

in order to obtain the Hamiltonian. We now investigate the space integral

$$\begin{aligned}
H_\parallel(t) &= \int \mathrm{d}^3r\left(-\frac{1}{2}\,\vec{E}_\parallel^2(t,\vec{r}) + J_0(t,\vec{r})\,A^0(t,\vec{r})\right)\\
&= \int \mathrm{d}^3r\left(-\frac{1}{2}\left[\vec{\nabla}\Phi(t,\vec{r})\right]^2 + \rho(t,\vec{r})\,\Phi(t,\vec{r})\right)\\
&= \int \mathrm{d}^3r\left(\frac{1}{2}\,\Phi(t,\vec{r})\,\vec{\nabla}^2\Phi(t,\vec{r}) + \rho(t,\vec{r})\,\Phi(t,\vec{r})\right)\\
&= \frac{1}{2}\int \mathrm{d}^3r\,\rho(t,\vec{r})\,\Phi(t,\vec{r}) = \frac{1}{2}\int \mathrm{d}^3r\,J_0(t,\vec{r})\,A^0(t,\vec{r})\,.
\end{aligned} \tag{9.48}$$

Now, let Φ_0 be a particular solution of the inhomogeneous equation $\vec{\nabla}^2\Phi_0 = -\rho$. Then, the general solution to Eq. (9.40a) can be written as $\Phi = \Phi_0 + \Phi_{\mathrm{hom}}$, where Φ_{hom} is a solution to the homogeneous equation, and so, after integrations by parts, one has

$$\begin{aligned}
\int \mathrm{d}^3r\,J_0(t,\vec{r})\,A^0(t,\vec{r}) &= \int \mathrm{d}^3r\,\rho(t,\vec{r})\left(\Phi_0(t,\vec{r}) + \Phi_{\mathrm{hom}}\right)\\
&= \int \mathrm{d}^3r\left[-\vec{\nabla}^2\Phi_0(t,\vec{r})\right]\left(\Phi_0(t,\vec{r}) + \Phi_{\mathrm{hom}}\right)\\
&= -\int \mathrm{d}^3r\,\Phi_0(t,\vec{r})\,\vec{\nabla}^2\left(\Phi_0(t,\vec{r}) + \Phi_{\mathrm{hom}}\right)\\
&= \int \mathrm{d}^3r\,\Phi_0(t,\vec{r})\,\rho(t,\vec{r})\,.
\end{aligned} \tag{9.49}$$

In the end, $H_\parallel(t)$ finds a much simpler form,

$$H_\parallel(t) = \frac{1}{2}\int \mathrm{d}^3r\int \mathrm{d}^3r'\,J_0(t,\vec{r})\,\frac{1}{4\pi|\vec{r}-\vec{r}\,'|}\,J_0(t,\vec{r}\,')\,. \tag{9.50}$$

This implies that we can replace

$$-\frac{1}{2}\vec{E}_{\parallel}^{2}(t,\vec{r}) + J_0(t,\vec{r})\,A^0(t,\vec{r}) \to \frac{1}{2}\int \mathrm{d}^3r'\,J_0(t,\vec{r})\,\frac{1}{4\pi|\vec{r}-\vec{r}'|}\,J_0(t,\vec{r}') \equiv \mathcal{H}_{\parallel}(t,\vec{r}) \tag{9.51}$$

in Eq. (9.47), where $H_{\parallel}(t) = \int \mathrm{d}^3r\,\mathcal{H}_{\parallel}(t,\vec{r})$. Here, $\mathcal{H}_{\parallel}(t,\vec{r})$ is a suitable expression for the electrostatic self-energy density of the charge distribution. Furthermore, we write

$$\mathcal{H} = \mathcal{H}_0 + \mathcal{H}_{\mathrm{int}}\,, \qquad \mathcal{H}_0 = \frac{1}{2}\left(\vec{E}_{\perp}^2 + \vec{B}^2\right), \qquad \mathcal{H}_{\mathrm{int}} = \mathcal{H}_{\parallel} - \vec{J}\cdot\vec{A}\,, \tag{9.52}$$

where all entities are functions of t and $\vec{r}$.

9.3.2 *Matching and Photon Propagator*

In Chap. 2, we had discussed the quantization of $\mathcal{H}_0$, which resulted in the quantized-field Hamiltonian given in Eq. (2.71),

$$\vec{A}(t,\vec{r}) = \sum_{\lambda=1}^{2}\int \frac{\mathrm{d}^3k}{\sqrt{2\omega\,(2\pi)^3}}\,\hat{\epsilon}_{\lambda}(\vec{k})\,\left(a_{\lambda}(\vec{k})\,\mathrm{e}^{-\mathrm{i}\,k\cdot x} + a_{\lambda}^{+}(\vec{k})\,\mathrm{e}^{\mathrm{i}\,k\cdot x}\right), \tag{9.53}$$

with the fundamental field commutators given in Eq. (2.70),

$$[a_{\lambda}(\vec{k}), a_{\lambda'}^{+}(\vec{k}')] = \delta^{(3)}(\vec{k}-\vec{k}')\,\delta_{\lambda\lambda'}\,. \tag{9.54}$$

We now calculate the photon propagator in Coulomb gauge by matching the first-order and second-order contributions to the S matrix, generated by the Hamiltonian (9.52). From $H_{\parallel}(t)$, as given in Eq. (9.50), one has

$$H_{\parallel}(t) = \frac{1}{2}\int \mathrm{d}^3r\int \mathrm{d}^3r'\,J_0(t,\vec{r})\,\frac{1}{4\pi|\vec{r}-\vec{r}'|}\,J_0(t,\vec{r}')\,. \tag{9.55}$$

The first-order contribution to the S matrix, still of second order in the currents, thus is given by

$$\begin{aligned} S^{[1]} &= -\mathrm{i}\int \mathrm{d}t H_C(t) = -\frac{\mathrm{i}}{2}\int \mathrm{d}t\int \mathrm{d}^3r\int \mathrm{d}^3r'\,J_0(x)\,\frac{1}{4\pi|\vec{r}-\vec{r}'|}\,J_0(x') \\ &= -\frac{\mathrm{i}}{2}\int \mathrm{d}^4x\int \mathrm{d}^4x'\,J_0(x)\,\frac{\delta(t-t')}{4\pi|\vec{r}-\vec{r}'|}\,J_0(x')\,. \end{aligned} \tag{9.56}$$

The second-order contribution to the S matrix contains the term

$$\begin{aligned} S^{[2]} &= \frac{(-\mathrm{i})^2}{2!}\int \mathrm{d}^4x\int \mathrm{d}^4x'\,J_i(x)\,\mathrm{T}\langle 0|A^i(x)A^j(x')|0\rangle\,J_j(x') \\ &= -\frac{1}{2}\int \mathrm{d}^4x\int \mathrm{d}^4x'\,J_i(x)\,\left(\mathrm{i}D_C^{ij}(x-x')\right)\,J_j(x') \end{aligned} \tag{9.57}$$

where we identify the Feynman (time-ordered) propagator (spatial components) in Coulomb gauge as

$$\mathrm{i}D_C^{ij}(x-x') = \mathrm{T}\langle 0|A^i(x)A^j(x')|0\rangle\,. \tag{9.58}$$

The sum $S^{[1]} + S^{[2]}$ can be matched against the covariant form $[(ij) \to (\mu\nu)]$,

$$S^{[2]} = -\frac{1}{2} \int d^4x \int d^4x' \, J_\mu(x) \left(iD_C^{\mu\nu}(x - x') \right) J_\nu(x') , \tag{9.59}$$

by writing the expression for $S^{[1]}$ as

$$\begin{aligned} S^{[1]} &= -\frac{i}{2} \int d^4x \int d^4x' \, J_0(x) \, \frac{\delta(t - t')}{4\pi|\vec{r} - \vec{r}'|} \, J_0(x') \\ &= -\frac{1}{2} \int d^4x \int d^4x' \, J_0(x) \left(iD_C^{00}(x - x') \right) J_0(x') , \end{aligned} \tag{9.60}$$

where

$$D_C^{00}(x - x') = \frac{\delta(t - t')}{4\pi|\vec{r} - \vec{r}'|} . \tag{9.61}$$

Transformation to momentum space yields

$$\begin{aligned} D_C^{00}(k) &= \int d^4(x - x') \, e^{ik\cdot(x - x')} \, D_C^{00}(x - x') \\ &= \int d^3(r - r') \, \frac{e^{-i\vec{k}\cdot(\vec{r} - \vec{r}')}}{4\pi|\vec{r} - \vec{r}'|} = \frac{1}{\vec{k}^2} . \end{aligned} \tag{9.62}$$

It remains to perform the calculation of the time-ordered product of field operators in Eq. (9.58). To this end, we recall Eq. (2.71),

$$\vec{A}(t, \vec{r}) = \sum_{\lambda=1}^{2} \int \frac{d^3k}{\sqrt{2\omega\,(2\pi)^3}} \, \hat{\epsilon}_\lambda(\vec{k}) \left(a_\lambda(\vec{k}) \, e^{-i\,k\cdot x} + a_\lambda^+(\vec{k}) \, e^{i\,k\cdot x} \right) , \tag{9.63}$$

where $\omega = |\vec{k}|$, with the fundamental field commutator (2.70),

$$[a_\lambda(\vec{k}), a_{\lambda'}^+(\vec{k}')] = \delta^{(3)}(\vec{k} - \vec{k}') \, \delta_{\lambda\lambda'} , \tag{9.64}$$

as well as Eq. (9.46) for the transverse Dirac-δ function,

$$\delta^{\perp,ij}(\vec{r} - \vec{r}') = \int \frac{d^3k}{(2\pi)^3} \left(\delta^{ij} - \frac{k^i \, k^j}{\vec{k}^2} \right) e^{i\,\vec{k}\cdot(\vec{r} - \vec{r}')} = \left(\delta^{ij} - \frac{\nabla^i \, \nabla^j}{\vec{\nabla}^2} \right) \delta^{(3)}(\vec{r} - \vec{r}') . \tag{9.65}$$

For the sum over polarizations, we have according to Eq. (2.63),

$$\hat{\epsilon}_\lambda(\vec{k}) \cdot \hat{\epsilon}_{\lambda'}(\vec{k}) = \delta_{\lambda\lambda'} , \qquad \vec{k} \cdot \hat{\epsilon}_\lambda(\vec{k}) = 0 , \qquad \sum_{\lambda=1}^{2} \epsilon_\lambda^i(\vec{k}) \, \epsilon_{\lambda'}^j(\vec{k}) = \delta^{ij} - \frac{k^i \, k^j}{\vec{k}^2} . \tag{9.66}$$

The vacuum expectation value of Coulomb-gauge field operators for the vector potential reads as follows,

$$\begin{aligned} iD^{ij}(x - x') &= \langle 0|A^i(t, \vec{r}) A^j(t', \vec{r}')|0\rangle \qquad (9.67) \\ &= \sum_{\lambda,\lambda'=1}^{2} \int d^3k \int d^3k' \frac{\epsilon_\lambda^i(\vec{k}) \epsilon_{\lambda'}^j(\vec{k}') \langle 0|a_\lambda(\vec{k}) a_{\lambda'}^+(\vec{k})|0\rangle e^{-i\,k\cdot(x - x')}}{\sqrt{2\omega(2\pi)^3} \sqrt{2\omega'(2\pi)^3}} \\ &= \sum_{\lambda=1}^{2} \int \frac{d^3k}{2\omega\,(2\pi)^3} \, \epsilon_\lambda^i(\vec{k}) \, \epsilon_\lambda^j(\vec{k}) \, e^{-i\,k\cdot(x - x')} \\ &= \left(\delta^{ij} - \frac{\nabla^i \, \nabla^j}{\vec{\nabla}^2} \right) \int d^3k \frac{e^{i\,k\cdot(x' - x)}}{2\omega\,(2\pi)^3} = -\frac{1}{4\pi^2} \left(\delta^{ij} - \frac{\nabla^i \, \nabla^j}{\vec{\nabla}^2} \right) \frac{1}{(x - x' - i\eta)^2} , \end{aligned}$$

where we have used Eqs. (9.25) and (9.54). For the time-ordered product of field operators, one obtains

$$\begin{aligned}
\mathrm{i}D_C^{ij}(x-x') &= \langle 0|\mathrm{T}\, A^i(t,\vec{r})A^j(t',\vec{r}')|0\rangle \\
&= \Theta(t-t')\,\langle 0|\mathrm{T}\, A^i(t,\vec{r})A^j(t',\vec{r}')|0\rangle + \Theta(t'-t)\,\langle 0|\mathrm{T}\, A^j(t',\vec{r}')A^i(t,\vec{r})|0\rangle \\
&= \Theta(t-t')\,\mathcal{D}^{ij}(x-x') + \Theta(t'-t)\,\mathcal{D}^{ij}(x'-x)\,, \qquad (9.68)
\end{aligned}$$

and thus

$$\begin{aligned}
\mathrm{i}D_C^{ij}(x-x') &= -\frac{1}{4\pi^2}\left(\delta^{ij} - \frac{\nabla^i\,\nabla^j}{\vec{\nabla}^2}\right)\left[\frac{\Theta(t-t')}{(x-x'-\mathrm{i}\eta)^2} + \frac{\Theta(t'-t)}{(x'-x-\mathrm{i}\eta)^2}\right] \\
&= -\frac{1}{4\pi^2}\left(\delta^{ij} - \frac{\nabla^i\,\nabla^j}{\vec{\nabla}^2}\right)\Bigg[\Theta(t-t')\left\{(\mathrm{P.V.})\frac{1}{(x-x')^2}\right. \\
&\qquad \left. + \,\mathrm{i}\,\pi\,\mathrm{sgn}(t-t')\,\delta((x-x')^2)\right\} + (x \leftrightarrow x')\Bigg] \\
&= -\frac{1}{4\pi^2}\left(\delta^{ij} - \frac{\nabla^i\,\nabla^j}{\vec{\nabla}^2}\right)\frac{1}{(x-x')^2 - \mathrm{i}\,\epsilon} \\
&= -\,\mathrm{i}\left(\delta^{ij} - \frac{\nabla^i\,\nabla^j}{\vec{\nabla}^2}\right)G_F(x-x')\,, \qquad (9.69)
\end{aligned}$$

where we have used the result (9.13),

$$G_F(x) = -\frac{\mathrm{i}}{4\pi^2}\left[(\mathrm{P.V.})\frac{1}{x^2} + \mathrm{i}\pi\delta(x^2)\right] = -\frac{\mathrm{i}}{4\pi^2}\,\frac{1}{x^2 - \mathrm{i}\epsilon}\,. \qquad (9.70)$$

The time-ordered product of vector potential field operators in coordinate space (Coulomb gauge, spatial components) thus reads as

$$D_C^{ij}(x-x') = -\left(\delta^{ij} - \frac{\nabla^i\,\nabla^j}{\vec{\nabla}^2}\right)G_F(x-x')\,. \qquad (9.71)$$

Using the momentum representation $G_F(k) = -1/(k^2 + \mathrm{i}\epsilon)$, one finds

$$D_C^{ij}(k) = \frac{1}{k^2 + \mathrm{i}\epsilon}\left(\delta^{ij} - \frac{k^i\,k^j}{\vec{k}^2}\right)\,. \qquad (9.72)$$

So, in the Coulomb gauge, we can express the time-order expectation value of the components of the four-vector potential as follows,

$$\begin{aligned}
\mathrm{i}D_C^{\mu\nu}(x-x') &= \langle 0|\mathrm{T}\, A^\mu(t,\vec{r})A^\mu(t',\vec{r}')|0\rangle\,, \\
D_C^{\mu\nu}(x-x') &= \int \frac{\mathrm{d}^4k}{(2\pi)^4}\, D_C^{\mu\nu}(k)\, \mathrm{e}^{-\mathrm{i}k\cdot(x-x')}\,, \\
D_C^{\mu\nu}(k) &= \begin{cases} \dfrac{1}{\vec{k}^2} & \mu = \nu = 0 \\[2ex] \dfrac{1}{k^2 + \mathrm{i}\epsilon}\left(\delta^{ij} - \dfrac{k^i\,k^j}{\vec{k}^2}\right) & \mu = i, \nu = j \\[2ex] 0 & \mu \neq \nu \end{cases}\,. \qquad (9.73)
\end{aligned}$$

We reemphasize that the result (9.73) has been obtained without formally quantizing the Coulomb interaction, thereby remedying the lack of a suitable canonical momentum Π_0, to supplement the scalar potential A^0, which persists in Coulomb gauge. The matching procedure described here leads to a successful quantization of the theory in Coulomb gauge and constitutes an alternative to the covariant procedure which will be outlined in the following.

9.4 Photon Propagator in Lorenz Gauge

9.4.1 *Gauge Invariance and Mass of Photon*

As a prelude to the calculation of the photon propagator in Lorenz gauge, let us briefly recall a few aspects of classical electrodynamics, with an emphasis on the properties of the Lorenz (covariant) gauge. We had briefly discussed gauge transformations in Sec. 4.6.2 and Sec. 8.2.1, in the context of the covariant derivative. A gauge transformation Λ changes the scalar and vector potentials as follows [see Eq. (8.1)],

$$\vec{A} \to \vec{A}' = \vec{A} + \vec{\nabla}\Lambda \,, \qquad A^0 \to (A^0)' = A^0 - \frac{\partial}{\partial t}\Lambda \,, \qquad A'^{\mu} = A^{\mu} - \partial^{\mu}\Lambda \,. \tag{9.74}$$

Here, Λ is an arbitrary gauge function of time and space, and $\mu = 0, 1, 2, 3$ is a space-time index. The Lorenz gauge condition is

$$\vec{\nabla} \cdot \vec{A} + \frac{\partial}{\partial t} A^0 = \partial_{\mu} A^{\mu} = 0 \,. \tag{9.75}$$

Let us briefly mention why the gauge invariance of QED protects the masslessness of the photon. Without sources, the Maxwell Lagrangian reads as

$$\mathcal{L} = -\frac{1}{4} F^{\mu\nu} F_{\mu\nu} \,, \qquad F_{\mu\nu} = \partial_{\mu} A_{\nu} - \partial_{\nu} A_{\mu} \,. \tag{9.76}$$

The variational field equations $\Box A_{\mu} - \partial_{\mu}(\partial \cdot A) = \partial_{\nu} F^{\nu\mu} = 0$ are invariant under a gauge transformation, $A^{\mu} \to A^{\mu} - \partial^{\mu}\Lambda$. The gauge invariance of the field equations can be traced to the trivial identity $\Box\partial_{\mu}\Lambda - \partial_{\mu}(\partial^{\nu}\partial_{\nu}\Lambda) = 0$. However, if we assume the field to be massive,

$$\mathcal{L} = -\frac{1}{4} F^{\mu\nu} F_{\mu\nu} + \frac{1}{2}\mu^2 A^2 \,, \tag{9.77}$$

then the field equations become

$$\Box\, A^{\mu} - \partial^{\mu}(\partial \cdot A) = \partial_{\nu} F^{\nu\mu} = -\mu^2 A^{\mu} \,. \tag{9.78}$$

While the field-strength tensor $F_{\nu\mu}$ is gauge invariant, the expression $-\mu^2 A_{\mu}$ is not; if the photon were massive then gauge invariance would be lost.

9.4.2 *Gauge-Fixing Term and Quantization*

Let us briefly outline the quantization procedure for the electromagnetic field in Lorenz gauge. The main problem is the lack of a canonical momentum conjugate to the scalar potential, as outlined in Eq. (2.48). In the first step, we add a term $-(\lambda/2)\,(\partial \cdot A)^2$ to the Lagrangian, with a Lagrange multiplier λ. This leads to a nonvanishing canonical momentum Π^0, complementing the time-like component A^0 of the four-vector potential, and makes it possible to formulate the canonical commutation relations.

These relations, for the fundamental creation and annihilation operators (in Lorenz gauge), will be different from those for the Coulomb gauge. The reason lies in our somewhat artificial introduction of a nonvanishing canonical momentum, conjugate to A^0. It implies the introduction of a quantized representation (which does not vanish) for the quantity $\partial \cdot A$ which, according to the classical formulation, should vanish (but does not do so in the fully quantized formalism). In the course of the calculation, we have introduced two "unphysical" degrees of freedom of the photon field: the scalar and longitudinal polarizations, augmenting the two polarizations orthogonal to the propagation vector, to four.

One might ask: Why do we have to deal with two additional degrees of freedom, not one? Finally, we have only introduced the canonical variables A^0 and Π^0 into the formalism, which should give us one, not two, additional polarizations. The reason is a subtle difference of the four-vector potential operators in the Lorenz gauge. In Coulomb gauge, our operator $\vec{A}$ given in Eq. (2.71) fulfills the gauge condition $\vec{\nabla} \cdot \vec{A} = 0$, even in the fully quantized setting (because of the properties of the polarization vectors contained therein). This gauge condition reduces the number of available photon polarizations from three to two, within Coulomb gauge. The quantized Lorenz gauge operators which we shall use do not fulfill the Lorenz gauge condition $\partial \cdot A = 0$. As already mentioned, we have introduced two additional degrees of freedom, the first for the introduction of the A^0 (and Π^0) into the quantization procedure, the second for the lack of a gauge condition which would be fulfilled by the quantized operators. In fact, in the quantized formalism, the Lorenz gauge condition is implemented in a subtle way, as discussed in the following.

The addition of the two "unphysical" degrees of freedom allows us to carry out the quantization of the electromagnetic field in a fully covariant form. However, we now have to deal with the two unphysical photon polarizations. In fact, the two problems are intertwined. Namely, it will turn out that the canonical momentum Π^0 is proportional to $\partial \cdot A$, so that the equation $\partial \cdot A = 0$ cannot be fulfilled as an operator equation in the quantized theory anyway. We simply create "too many" dynamical variables, some of which belong to "unphysical" degrees of freedom (these are "unphysical" scalar and longitudinal photons, specifically). Some of the states turn out to have a negative norm, leading to a Fock space with an indefinite metric. A natural idea is to demand that if the condition $\partial \cdot A = 0$ cannot be fulfilled as an operator identity, then at least, it should be fulfilled in terms of an expectation value. Or, we could postulate that $\partial \cdot A$ should vanish when acting on the physical states of the electromagnetic field, i.e., $\partial \cdot A^{(+)}|\Psi\rangle = 0$ for any physical state $|\Psi\rangle$ which contains real photons. Here, $A^{(+)}$ is the "annihilation part" of the four-vector potential operator, to be discussed in Eq. (9.113). This latter condition is called the Gupta–Bleuler condition and is discussed in Sec. 9.5.3.

Let us now cast this program into mathematical form. We start from the following Lagrangian for the source-free photon field with a gauge-fixing term,

$$\mathcal{L} = -\frac{1}{4} F^{\mu\nu} F_{\mu\nu} - \frac{\lambda}{2}(\partial \cdot A)^2 . \tag{9.79}$$

This Lagrangian is not gauge-invariant any more; upon a transformation $A^\mu \to A^\mu - \partial^\mu \Lambda$, it changes as

$$\mathcal{L} \to \mathcal{L} - \frac{\lambda}{2}\left(-2\,(\Box\Lambda)\;(\partial \cdot A) + (\Box\Lambda)^2\right) . \tag{9.80}$$

The Lagrangian (9.79) thus is gauge-invariant only within the family of Lorenz gauges; any gauge transformation Λ which fulfills $\Box\Lambda = 0$ is still permissible in the sense that a "gauge always shoots twice" (see Sec. 1.2.4 of Ref. [81]). A variation with respect to λ of (9.79) yields back the gauge condition $\partial \cdot A = 0$. However, we shall rather interpret λ as a coupling constant whose value can be fixed in order to cast the field equations into a desired form. In fact, we shall arrange for λ to be chosen so that the field equations have the same form as in Lorenz gauge, or, more precisely, the same as we would obtain if we were to impose the Lorenz gauge condition on the four-vector potential, after the variation of the action. Still, the formalism, in the first place, knows nothing about the Lorenz gauge condition, and treats A^μ as a normal four-vector quantum field with all available degrees of freedom and polarizations at one's disposal. Fixing λ to the appropriate value required to reproduce the Lorenz gauge field equations is referred to as the "Feynman gauge" while strictly speaking, it is not a gauge condition imposed on the four-vector potential, but a condition on an artificially introduced coupling parameter in the Lagrangian (9.79), fixed to cast the field equations into a desired form.

The variational equations read as follows,

$$\partial_\nu \left(\frac{\partial \mathcal{L}}{\partial(\partial_\nu A_\mu)} \right) - \frac{\partial \mathcal{L}}{\partial A_\mu} = 0 . \tag{9.81}$$

The only nonvanishing partial derivatives are

$$\frac{\partial \mathcal{L}}{\partial(\partial_\nu A_\mu)} = -\left(\partial^\nu A^\mu - \partial^\mu A^\nu\right) - \lambda\, g^{\nu\mu}\,(\partial \cdot A) , \tag{9.82}$$

where we use the representation

$$\partial \cdot A = \partial_\rho\, A^\rho = g^{\nu\mu} \partial_\nu\, A_\mu . \tag{9.83}$$

The variational equations, in the source-free case, therefore read as follows,

$$\begin{aligned} \partial_\nu \left(\frac{\partial \mathcal{L}}{\partial(\partial_\nu A_\mu)} \right) &= -\left(\partial^\nu \partial_\nu A^\mu - \partial^\mu \partial_\nu A^\nu\right) - \lambda\, \partial^\mu (\partial \cdot A) \\ &= -\left[\Box A^\mu - (1 - \lambda)\, \partial^\mu (\partial \cdot A)\right] = 0 . \end{aligned} \tag{9.84}$$

The resulting equation,

$$\Box A^\mu - (1-\lambda)\,\partial^\mu(\partial\cdot A) = 0 = \partial_\nu F^{\nu\mu} + \lambda\,\partial^\mu(\partial\cdot A)\,, \tag{9.85}$$

turns into the Lorenz gauge source-free Maxwell equations for $\lambda = 1$. Having calculated (9.82), we now evaluate the canonically conjugate momenta

$$\Pi^\rho = \frac{\partial\mathcal{L}}{\partial(\partial_0 A_\rho)} = -\left(\partial^0 A^\rho - \partial^\rho A^0\right) - \lambda\, g^{0\rho}\,(\partial\cdot A) = F^{\rho 0} - \lambda\, g^{\rho 0}\,(\partial\cdot A)\,. \tag{9.86}$$

The canonical momentum Π^0 conjugate to A_0 reads as

$$\Pi^0 = -\lambda\,(\partial\cdot A)\,. \tag{9.87}$$

From now on, we shall assume that

$$\lambda = 1\,, \qquad \Pi^0 = -\partial\cdot A\,, \qquad \Pi^i = F^{i0} = E^i\,, \tag{9.88}$$

where $\vec{E}$ is the electric-field operator. We now postulate the following canonical quantization relations at equal times,

$$[\Pi^\mu(t,\vec{r}), A_\nu(t,\vec{r}\,')] = -\mathrm{i}\, g^\mu{}_\nu\, \delta^{(3)}(\vec{r}-\vec{r}\,')\,, \tag{9.89}$$

which is consistent with the indefinite metric of the Hilbert space implied by the commutation relation Eq. (9.92) discussed in the following. We note that these commutation relations are valid at equal times. When specialized to the spatial components, Eq. (9.89) reads as follows,

$$\left[\Pi^i(t,\vec{r}), A_j(t,\vec{r}\,')\right] = -\mathrm{i}\, g^i{}_j\, \delta^{(3)}(\vec{r}-\vec{r}\,')\,, \tag{9.90}$$

where $g^i{}_j = \delta^{ij}$ is equal to the Kronecker delta. We notice that $\Pi_i = -E_i$, not Π^i, is the canonically conjugate variable corresponding to A^i. (The correspondence to the quantum mechanical relation $[p^i, x^j] = -\mathrm{i}\,\delta^{ij}$ is maintained.) It is instructive to compare Eq. (9.89) with Eq. (2.59) in Sec. 2.4.1. In the latter case, the transverse delta function is obtained, in momentum space, after a summation over only two polarizations, not four.

The covariant photon vector potential operator reads as

$$A^\mu(x) = \sum_{\lambda=0}^{3} \int \frac{\mathrm{d}^3k}{(2\pi)^3}\, \frac{1}{2|\vec{k}|}\, \epsilon^\mu_\lambda(\vec{k})\, \left\{a_{\vec{k}\lambda}\, \mathrm{e}^{-\mathrm{i}\,k\cdot x} + a^+_{\vec{k}\lambda}\, \mathrm{e}^{\mathrm{i}\,k\cdot x}\right\}, \tag{9.91}$$

where we work in a fully relativistic setting and aim to use the Lorentz-invariant integration measure (9.24). This necessitates a change in the normalizations of the fundamental field operators, as already indicated in Sec. 2.5.2. We therefore distinguish the "relativistically" transformed operators by their subscripts $a_{\vec{k}\lambda}$ from the "nonrelativistic" operators $a_\lambda(\vec{k})$. In accordance with Eq. (2.117), we postulate the following commutation relations,

$$[a_{\vec{k}\lambda}, a^+_{\vec{k}'\lambda'}] = -g_{\lambda\lambda'}\,(2|\vec{k}|)\,(2\pi)^3\,\delta^{(3)}(\vec{k}-\vec{k}')\,. \tag{9.92}$$

Here, $g_{\lambda\lambda'}$ is determined by the photon polarizations, which we augment by scalar and longitudinal photons. The symbol $g_{\lambda\lambda'}$ would otherwise indicate a space-time metric, which we use in the West-Coast conventions $g_{\mu\nu} = \mathrm{diag}(1,-1,-1,-1)$, but its meaning in Eq. (9.92) really comes from the indefinite metric associated with the Fock space of photons, where the two "additional" photon polarizations are suppressed because their state acquire a negative norm, as per the commutation relation (9.92). Details are discussed in the following. The sum over polarizations is carried out according to

$$\sum_{\lambda=0}^{3} \sum_{\lambda'=0}^{3} \epsilon^{\mu}_{\lambda}(\vec{k})\, \epsilon^{\nu}_{\lambda'}(\vec{k})\, g_{\lambda\lambda'} = g^{\mu\nu} \,. \tag{9.93}$$

This formula is in need of an explanation and illustration. We first augment the two transverse photons by scalar and longitudinal photons, to arrive at four virtual photon polarizations. For the polarization with $\lambda = 0$ (scalar photons), the polarization vector is parallel to the time-like unit vector

$$\hat{n} = (1,0,0,0)\,, \qquad \epsilon_0(\vec{k}) = \hat{n}\,, \qquad \epsilon_0(\vec{k}) \cdot \epsilon_0(\vec{k}) = 1\,. \tag{9.94}$$

The physically allowed polarizations (for real photons) are $\lambda = 1, 2$. We have a vanishing time-like component $\epsilon^0_{1,2}(\vec{k}) = 0$ and $\vec{k} \cdot \vec{\epsilon}_{1,2}(\vec{k}) = 0$. Of course, just like for any two orthogonal space-like vectors normalized to unity, we have

$$\epsilon_i(\vec{k}) \cdot \epsilon_j(\vec{k}) = -\delta^{ij}\,, \qquad i,j = 1,2\,. \tag{9.95}$$

The third polarization vector (longitudinal photons) is in the propagation direction of the photon, but has a vanishing time-like component. In terms of the wave vector k^μ, it can be written as follows,

$$\nu^\mu = k^\mu - (k \cdot n)\, n^\mu = (0, \vec{k})\,, \qquad \epsilon^\mu_3(\vec{k}) = \frac{\nu^\mu}{\sqrt{-\nu \cdot \nu}}\,, \qquad \epsilon_3(\vec{k}) \cdot \epsilon_3(\vec{k}) = -1\,. \tag{9.96}$$

We convince ourselves that

$$\nu \cdot \nu = k^2 - 2(k \cdot n)^2 + (k \cdot n)^2 = \omega^2 - \vec{k}^2 - \omega^2 = -\vec{k}^2\,. \tag{9.97}$$

Furthermore,

$$k \cdot \epsilon_0(\vec{k}) = |\vec{k}| = -k \cdot \epsilon_3(\vec{k})\,, \qquad k \cdot \epsilon_{1,2}(\vec{k}) = 0\,. \tag{9.98}$$

For a photon propagating in the z direction, with $\vec{k} = |\vec{k}|\, \hat{\mathrm{e}}_z$, we have

$$\epsilon_0(\vec{k}) = \begin{pmatrix} 1 \\ 0 \\ 0 \\ 0 \end{pmatrix}, \qquad \epsilon_1(\vec{k}) = \begin{pmatrix} 0 \\ 1 \\ 0 \\ 0 \end{pmatrix}, \qquad \epsilon_2(\vec{k}) = \begin{pmatrix} 0 \\ 0 \\ 1 \\ 0 \end{pmatrix}, \qquad \epsilon_3(\vec{k}) = \begin{pmatrix} 0 \\ 0 \\ 0 \\ 1 \end{pmatrix}. \tag{9.99}$$

For the case $\vec{k} = |\vec{k}|\, \hat{\mathrm{e}}_z$, we can now verify Eq. (9.93) as follows, even by inspection,

$$\epsilon_\lambda(|\vec{k}|\, \hat{\mathrm{e}}_z) \cdot \epsilon_{\lambda'}(|\vec{k}|\, \hat{\mathrm{e}}_z) = g_{\lambda\lambda'}\,, \qquad \sum_{\lambda=0}^{3} \sum_{\lambda'=0}^{3} \epsilon^{\mu}_{\lambda}(|\vec{k}|\, \hat{\mathrm{e}}_z)\, \epsilon^{\nu}_{\lambda'}(|\vec{k}|\, \hat{\mathrm{e}}_z) g_{\lambda\lambda'} = g^{\mu\nu}\,. \tag{9.100}$$

Let us verify the compatibility of the field operator, in the convention used in Eq. (9.91), with the explicit form of the commutators given in Eq. (9.92), and the canonical commutation relation (9.89). One has according to Eq. (9.91),

$$A^\mu(x) = \sum_{\lambda=0}^{3} \int \frac{\mathrm{d}^3 k}{(2\pi)^3}\, \frac{1}{2\,|\vec{k}|}\, \epsilon^\mu_\lambda(\vec{k})\, \left\{ a_{\vec{k}\lambda}\, \mathrm{e}^{-\mathrm{i}\, k \cdot x} + a^+_{\vec{k}\lambda}\, \mathrm{e}^{\mathrm{i}\, k \cdot x} \right\}, \tag{9.101}$$

so that

$$\Pi^0(x) = -\partial\cdot A = \sum_{\lambda=0}^{3}\int \frac{\mathrm{d}^3k}{(2\pi)^3}\,\frac{1}{2|\vec{k}|}\,(\mathrm{i}k\cdot\epsilon_\lambda(\vec{k}))\,\left\{a_{\vec{k}\lambda}\,\mathrm{e}^{-\mathrm{i}\,k\cdot x} - a^{+}_{\vec{k}\lambda}\,\mathrm{e}^{\mathrm{i}\,k\cdot x}\right\}$$

$$= \mathrm{i}\int \frac{\mathrm{d}^3k}{(2\pi)^3}\,\frac{1}{2|\vec{k}|}\,|\vec{k}|\,\left\{(a_{\vec{k}0} - a_{\vec{k}3})\,\mathrm{e}^{-\mathrm{i}\,k\cdot x} - (a^{+}_{\vec{k}0} - a^{+}_{\vec{k}3})\,\mathrm{e}^{\mathrm{i}\,k\cdot x}\right\}, \tag{9.102}$$

where Eq. (9.98) has been used. Also, the expression for A^0 may be simplified,

$$A^0(x) = \int \frac{\mathrm{d}^3k}{(2\pi)^3}\,\frac{1}{2|\vec{k}|}\,\left\{a_{\vec{k}0}\,\mathrm{e}^{-\mathrm{i}\,k\cdot x} + a^{+}_{\vec{k}0}\,\mathrm{e}^{\mathrm{i}\,k\cdot x}\right\} \tag{9.103}$$

in view of $\epsilon^0_0(\vec{k}) = 1$, and $\epsilon^0_\lambda(\vec{k}) = 0$, for $\lambda = 1,2,3$. From now on, within the discussion of the field commutators, we shall assume that

$$t = t', \tag{9.104}$$

so that $x = (t,\vec{r})$ and $x' = (t,\vec{r}\,')$. The commutator is

$$[\Pi^0(x), A_0(x')] = \mathrm{i}\int \frac{\mathrm{d}^3k}{(2\pi)^3}\,\frac{1}{2|\vec{k}|}\int \frac{\mathrm{d}^3k'}{(2\pi)^3}\,\frac{1}{2|\vec{k}'|}\,|\vec{k}|\,\Big[\Big\{(a_{\vec{k}0} - a_{\vec{k}3})\,\mathrm{e}^{-\mathrm{i}\,k\cdot x}$$

$$- (a^{+}_{\vec{k}0} - a^{+}_{\vec{k}3})\,\mathrm{e}^{\mathrm{i}\,k\cdot x}\Big\}, \Big\{a_{\vec{k}'0}\,\mathrm{e}^{-\mathrm{i}\,k'\cdot x'} + a^{+}_{\vec{k}'0}\,\mathrm{e}^{\mathrm{i}\,k'\cdot x'}\Big\}\Big]. \tag{9.105}$$

According to Eq. (9.92),

$$[a_{\vec{k}0}, a^{+}_{\vec{k}'0}] = -(2|\vec{k}|)\,(2\pi)^3\,\delta^{(3)}(\vec{k}-\vec{k}'), \tag{9.106}$$

while $[a_{\vec{k}0}, a^{+}_{\vec{k}'3}]$ and other commutators vanish. Then,

$$[\Pi^0(x), A_0(x')] = \mathrm{i}\int \frac{\mathrm{d}^3k}{(2\pi)^3}\,\frac{1}{2|\vec{k}|}\int \frac{\mathrm{d}^3k'}{(2\pi)^3}\,\frac{1}{2|\vec{k}'|}\,|\vec{k}|$$

$$\times\left(\left[a_{\vec{k}0}, a^{+}_{\vec{k}'0}\right]\mathrm{e}^{-\mathrm{i}\,k\cdot x+\mathrm{i}k'\cdot x'} - \left[a^{+}_{\vec{k}0}, a_{\vec{k}'0}\right]\mathrm{e}^{\mathrm{i}\,k\cdot x-\mathrm{i}k'\cdot x'}\right)$$

$$= -\frac{\mathrm{i}}{2}\int \frac{\mathrm{d}^3k}{(2\pi)^3}\left[\mathrm{e}^{-\mathrm{i}\,k\cdot x+\mathrm{i}k\cdot x'} + \mathrm{e}^{\mathrm{i}\,k\cdot x-\mathrm{i}k\cdot x'}\right] = -\mathrm{i}g^0{}_0\,\delta^{(3)}(\vec{r}-\vec{r}\,'). \tag{9.107}$$

It remains to calculate $[\Pi^i(x), A^j(x')]$. We have

$$\Pi^i(x) = \partial^i A^0 - \partial^0 A^i = -\partial_i A^0 - \partial_0 A^i$$

$$= -\int \frac{\mathrm{d}^3k}{(2\pi)^3}\,\frac{1}{2|\vec{k}|}\,(+\mathrm{i}k^i)\,\left\{a_{\vec{k}0}\,\mathrm{e}^{-\mathrm{i}\,k\cdot x} - a^{+}_{\vec{k}0}\,\mathrm{e}^{\mathrm{i}\,k\cdot x}\right\}$$

$$-\sum_{\lambda=1}^{3}\int \frac{\mathrm{d}^3k}{(2\pi)^3}\,\frac{1}{2|\vec{k}|}\,(-\mathrm{i}|\vec{k}|)\,\epsilon^i_\lambda(\vec{k})\,\left\{a_{\vec{k}\lambda}\,\mathrm{e}^{-\mathrm{i}\,k\cdot x} - a^{+}_{\vec{k}\lambda}\,\mathrm{e}^{\mathrm{i}\,k\cdot x}\right\}, \tag{9.108}$$

while

$$A^j(x) = \sum_{\lambda=1}^{3}\int \frac{\mathrm{d}^3k}{(2\pi)^3}\,\frac{1}{2|\vec{k}|}\,\epsilon^j_\lambda(\vec{k})\,\left\{a_{\vec{k}\lambda}\,\mathrm{e}^{-\mathrm{i}\,k\cdot x} + a^{+}_{\vec{k}\lambda}\,\mathrm{e}^{\mathrm{i}\,k\cdot x}\right\}, \tag{9.109}$$

where we carefully take note of the summation limits, which extend from $\lambda = 1$ to $\lambda = 3$. So, the commutator among the spatial components evaluates to

$$\begin{aligned}
[\Pi^i(x), A_j(x')] &= -[\Pi^i(x), A^j(x')] = -\left(T^{ij} + U^{ij}\right), \\
T^{ij} &= \left[-\partial_i A^0, A^j\right] = 0, \\
U^{ij} &= \left[\sum_{\lambda=1}^{3} \int \frac{d^3k}{(2\pi)^3} \frac{1}{2|\vec{k}|} i|\vec{k}|\, \epsilon^i_\lambda(\vec{k}) \left\{a_{\vec{k}\lambda}\, e^{-i\,k\cdot x} - a^+_{\vec{k}\lambda}\, e^{i\,k\cdot x}\right\}, \right. \\
&\quad \left. \sum_{\lambda'=1}^{3} \int \frac{d^3k'}{(2\pi)^3} \frac{1}{2|\vec{k}'|} \epsilon^j_{\lambda'}(\vec{k}') \left\{a_{\vec{k}'\lambda'}\, e^{-i\,k'\cdot x'} + a^+_{\vec{k}'\lambda'}\, e^{i\,k'\cdot x'}\right\}\right] \\
&= \sum_{\lambda,\lambda'=1}^{3} \int \frac{d^3k}{(2\pi)^3} \frac{1}{2|\vec{k}|} \int \frac{d^3k'}{(2\pi)^3} \frac{1}{2|\vec{k}'|} i|\vec{k}|\, \epsilon^i_\lambda(\vec{k})\, \epsilon^j_{\lambda'}(\vec{k}')\, 2|\vec{k}|(2\pi)^3 \\
&\quad \times \left(\delta_{\lambda\lambda'}\, \delta^{(3)}(\vec{k} - \vec{k}')\, e^{-i\,k\cdot x + ik'\cdot x'} + \delta_{\lambda\lambda'}\, \delta^{(3)}(\vec{k} - \vec{k}')\, e^{i\,k\cdot x - ik'\cdot x'}\right) \\
&= i\delta^{ij}\, \delta^{(3)}(\vec{r} - \vec{r}').
\end{aligned} \tag{9.110}$$

We have used the polarization sum formula [appealing to Eq. (9.99)]

$$\sum_{\lambda=1}^{3} \epsilon^i_\lambda(\vec{k})\, \epsilon^j_\lambda(\vec{k}) = \delta^{ij}, \tag{9.111}$$

which is, notably, different from Eq. (2.63) because of the addition of the third polarization. The calculations described in Eqs. (9.101)–(9.111) are different from the Coulomb gauge [see Eq. (2.4.2)] because of the different normalizations of the fundamental commutators [Eq. (9.89) versus Eq. (2.70)] and because of the addition of the scalar and longitudinal photons.

9.4.3 *Representations of the Photon Propagator*

It is instrumental to calculate the vacuum expectation value of the product of two field operators,

$$\begin{aligned}
i\mathcal{D}^{\mu\nu}(x - x') &= \langle 0|A^\mu(x)\, A^\nu(x')|0\rangle \\
&= \sum_{\lambda,\lambda'=0}^{3} \int \frac{d^3k}{(2\pi)^3} \frac{1}{2|\vec{k}|} \int \frac{d^3k'}{(2\pi)^3} \frac{1}{2|\vec{k}'|} \epsilon^\mu_\lambda(\vec{k})\, \epsilon^\nu_{\lambda'}(\vec{k}') \\
&\quad \times \left\langle 0\left|a_{\vec{k}\lambda}\, e^{-i\,k\cdot x}\, a^+_{\vec{k}'\lambda'}\, e^{i\,k'\cdot x'}\right|0\right\rangle \\
&= -g^{\mu\nu} \int \frac{d^3k}{(2\pi)^3} \frac{1}{2|\vec{k}|} e^{-i\,k\cdot(x - x')} = \frac{g^{\mu\nu}}{4\pi^2} \frac{1}{(x - x' - i\eta)^2},
\end{aligned} \tag{9.112}$$

where we have used Eqs. (9.25), (9.92), and (9.93), and recall that η is an infinitesimal time-like vector, with $\eta = (\epsilon, 0, 0, 0)$. The introduction of the imaginary unit in the definition of the photon propagator follows the conventions used in Eqs. (7.153) and (7.154) for the fermionic case. We now decompose the field operator $A^\mu(x)$

into a negative-frequency, creation part $A^{(-)\mu}(x)$ and positive-frequency, annihilation part $A^{(+)\mu}(x)$,

$$A^\mu(x) = A^{(+)\mu}(x) + A^{(-)\mu}(x)\,, \qquad A^{(+)\mu}(x) = \sum_{\lambda=0}^{3} \int \frac{\mathrm{d}^3k}{(2\pi)^3} \frac{1}{2|\vec{k}|} \epsilon^\mu_\lambda(\vec{k})\, a_{\vec{k}\lambda}\, \mathrm{e}^{-\mathrm{i}\,k\cdot x}\,. \tag{9.113}$$

Of course, one has $A^{(-)\mu}(x) = A^\mu(x) - A^{(+)\mu}(x) = [A^{(+)\mu}(x)]^+$, illustrating the fact that $A^\mu(x)$ is a Hermitian operator. Because the sign of a frequency (of an energy) does not change under a Lorentz transformation, the separation into $A^{(+)}$ and $A^{(-)}$ is Lorentz-invariant. This implies that we can write $\mathcal{D}^{\mu\nu}(x-x')$ as follows,

$$\begin{aligned} \mathrm{i}\mathcal{D}^{\mu\nu}(x-x') &= \left\langle 0 \left| A^{(+)\mu}(x)\, A^{(-)\nu}(x') \right| 0 \right\rangle = \left\langle 0 \left| \left[A^{(+)\mu}(x), A^{(-)\nu}(x') \right] \right| 0 \right\rangle \qquad (9.114) \\ &= -\left[A^{(-)\mu}(x'), A^{(+)\nu}(x) \right] = -g^{\mu\nu} \int \frac{\mathrm{d}^4k}{(2\pi)^4}\, \mathrm{e}^{-\mathrm{i}\,k\cdot(x-x')}\, \Theta(k^0)\, \delta(k^2)\,, \end{aligned}$$

where we have used the results (9.112) and (9.24). The commutator itself is a complex number, free of operators, and therefore equal to its own vacuum expectation value. Thus, the vacuum expectation value of the field commutator, $\mathrm{i}\mathcal{D}^{\mu\nu}_-(x-x') = \langle 0 | [A^\mu(x), A^\nu(x')] | 0 \rangle$, actually is centered on the light cone,

$$\begin{aligned} \mathrm{i}\mathcal{D}^{\mu\nu}_-(x-x') &= \langle 0 | [A^\mu(x), A^\nu(x')] | 0 \rangle \qquad (9.115) \\ &= \mathrm{i}\mathcal{D}^{\mu\nu}(x-x') - \mathrm{i}\mathcal{D}^{\nu\mu}(x'-x) = \mathrm{i}\, g^{\mu\nu}\, \frac{\mathrm{sgn}(t-t')}{2\pi}\, \delta((x-x')^2)\,. \end{aligned}$$

Also, in the Lorenz gauge, we have

$$\mathcal{D}^{\mu\nu}_-(x-x') = g^{\mu\nu}\, G_-(x-x')\,, \tag{9.116}$$

where $G_-(x-x')$ is the Green function defined in Eq. (9.37). The retarded and advanced propagators are centered on the light cone; they can be obtained from the field commutator as follows,

$$\begin{aligned} D^{\mu\nu}_R(x-x') &= \Theta(t-t')\, \mathcal{D}^{\mu\nu}_-(x-x') \\ &= g^{\mu\nu}\, \frac{\Theta(t-t')}{2\pi}\, \delta((x-x')^2) = g^{\mu\nu}\, G_R(x-x')\,. \qquad (9.117) \end{aligned}$$

For the advanced propagator, we have

$$\begin{aligned} D^{\mu\nu}_A(x-x') &= -\,\Theta(t'-t)\, \mathcal{D}^{\mu\nu}_-(x-x') = -g^{\mu\nu}\, \Theta(t'-t)\, G_-(x-x') \\ &= -\,g^{\mu\nu}\, \frac{\Theta(t'-t)}{2\pi}\, \delta((x-x')^2) = g^{\mu\nu}\, G_A(x-x')\,. \qquad (9.118) \end{aligned}$$

Here, we have used the result (9.38) in the form

$$G_A(x-x') = -\Theta(-(t-t'))\, G_-(x-x') = \frac{\Theta(-(t-t'))}{2\pi}\, \delta((x-x')^2)\,. \tag{9.119}$$

Finally, the time-ordered vacuum expectation value of field operators leads to the Feynman propagator of the photon field,

$$\begin{aligned} \mathrm{i}D^{\mu\nu}_F(x-x') &= \langle 0 | \mathrm{T}\, A^\mu(x)\, A^\nu(x') | 0 \rangle \\ &= \Theta(t-t')\ \langle 0 | A^\mu(x)\, A^\nu(x') | 0 \rangle + \Theta(t'-t)\ \langle 0 | A^\nu(x')\, A^\mu(x) | 0 \rangle \\ &= \Theta(t-t')\ \mathcal{D}^{\mu\nu}(x-x') + \Theta(t'-t)\ \mathcal{D}^{\nu\mu}(x'-x) \\ &= \frac{g^{\mu\nu}}{4\pi^2} \frac{1}{(x-x')^2 - \mathrm{i}\,\epsilon} = \mathrm{i}\, g^{\mu\nu}\, G_F(x-x')\,, \qquad (9.120) \end{aligned}$$

where we encounter the same combination of infinitesimal parts as in Eq. (9.31). In the last step, the relation (9.13) has been used. Using Eq. (9.9), we can find the form

$$D_F^{\mu\nu}(k) = -\frac{g^{\mu\nu}}{k^2 + \mathrm{i}\epsilon} \tag{9.121}$$

for the Feynman propagator of the photon field in momentum space.

9.5 Photon Propagator in General Gauges

9.5.1 *Construction Principle*

Starting from Eq. (9.120), we can define the photon propagator $D^{\mu\nu}(x - x')$ as the vacuum expectation value of the time-ordered product of four-vector field operators,

$$\langle 0 | \mathrm{T}\, A^\mu(x)\, A^\nu(x') | 0 \rangle = \mathrm{i} D^{\mu\nu}(x - x'), \tag{9.122}$$

where $D^{\mu\nu}(x - x')$, in the Lorenz gauge, is given in Eq. (9.120). In Coulomb gauge, the space-time expressions are given in Eqs. (9.61) and (9.71), while in momentum space, the formulas can be found in Eqs. (9.62), (9.72), and (9.73).

In order to derive the photon propagator in an arbitrary gauge, one proceeds in the following steps. First, one takes a step back and matches the second-order S matrix element, calculated using classical current distributions but with quantized field operators, against the first-order S matrix element, calculated using the same classical current distribution and the classical (retarded) four-vector potential generated by the same current distribution. This leads to a relation of the time-ordered product of field operators and the retarded Green function. One concludes that, in order to derive the photon propagator in an arbitrary gauge, it is sufficient to establish the equations which couple the *classical* four-vector potential A^μ to the four-current density J^μ, then, invert these relations, and place the poles of the photon propagator according to the prescriptions pertinent to the Feynman, retarded, or advanced propagators.

Let us start with the first step. In second-order perturbation theory, with a classical current distribution $J^\mu(x)$, the S-matrix element for forward scattering of the vacuum state of the electromagnetic field can be expressed in terms of the retarded Green function,

$$\begin{aligned} S^{[2]} &= \frac{(-\mathrm{i})^2}{2!} \int \mathrm{d}^4x \int \mathrm{d}^4x'\, J_\mu(x)\, \mathrm{T}\langle 0|A^\mu(x) A^\nu(x')|0\rangle\, J_\nu(x') \\ &= -\frac{1}{2} \int \mathrm{d}^4x \int \mathrm{d}^4x'\, J_\mu(x)\, \big(\Theta(t-t')\, \langle 0|A^\mu(x) A^\nu(x')|0\rangle \\ &\quad + \Theta(t'-t)\, \langle 0|A^\nu(x') A^\mu(x)|0\rangle\big)\, J_\nu(x') \\ &= -\int \mathrm{d}^4x \int \mathrm{d}^4x' J_\mu(x) \Theta(t-t') \langle 0|A^\mu(x) A^\nu(x')|0\rangle J_\nu(x') \\ &= -\mathrm{i} \int \mathrm{d}^4x \int \mathrm{d}^4x' J_\mu(x)\, D_R^{\mu\nu}(x - x')\, J_\nu(x') \equiv -\mathrm{i} \int \mathrm{d}^4x J_\mu(x)\, A_R^\mu(x)\,. \end{aligned} \tag{9.123}$$

So, the matching of the retarded Green function with the second-order S-matrix is successful if the retarded photon propagator is given by

$$\mathrm{i}D_R^{\mu\nu}(x-x') = \Theta(t-t')\langle 0|A_\mu(x)A_\nu(x')|0\rangle\,, \tag{9.124}$$

and in the classical sense

$$A_R^\mu(x) = \int \mathrm{d}^4x'\, D_R^{\mu\nu}(x-x')\, J_\nu(x')\,. \tag{9.125}$$

Because the latter equation can be given in momentum space as

$$A_R^\mu(k) = D_R^{\mu\nu}(k)\, J_\nu(k)\,, \tag{9.126}$$

we may derive the photon propagator in various gauges by considering the coupling of the vector potential to its source,

$$\Box\, A^\mu(x) - \partial^\mu(\partial\cdot A(x)) = J^\mu(x)\,. \tag{9.127}$$

After imposing gauge conditions, these equations generally take a simpler form, e.g., in the Lorenz gauge where one would neglect the term $\partial\cdot A(x)$. This would entail the second step outlined above. One then transforms this equation to energy-momentum space, and inverting the matrix to solve for A^μ. In Lorenz gauge, for example, one would impose $\partial\cdot A(x) = 0$, resulting in

$$-k^2 A^\mu(k) = J^\mu(k)\,, \qquad A^\mu(k) = D_R^{\mu\nu}(k)\, J_\nu(k)\,, \tag{9.128a}$$

$$D_R^{\mu\nu}(k) = -\frac{g^{\mu\nu}}{(k^0+\mathrm{i}\epsilon)-\vec{k}^2} \to -\frac{g^{\mu\nu}}{k^2+\mathrm{i}\epsilon}\,, \tag{9.128b}$$

where the second equation illustrates the transition from the retarded to the Feynman Green function via a simple reassignment of the positions of the poles. One thus verifies Eq. (9.121). Indeed, one should remember that the solution of the equation (9.126) does not yet specify the prescription to place the poles; these must be placed differently for the Feynman and retarded propagators, as discussed.

9.5.2 *Most General Form and Weyl Gauge*

An inspection of Eqs. (9.127) and (9.128) shows that one can add arbitrary terms of the form

$$D_R^{\mu\nu} \to D_R^{\mu\nu} + \frac{1}{2k^2}(f^\mu k^\nu + f^\nu k^\mu) \tag{9.129}$$

to the retarded or Feynman photon propagators, without changing the physical fields. The continuity equation for the current paves the way in the derivation. E.g., the transition from the Feynman to the Coulomb gauge is accomplished using

$$f^0 = \frac{k^0}{\vec{k}^2}\,, \qquad f^i = -\frac{k^i}{\vec{k}^2}\,. \tag{9.130}$$

With a gauge transformation

$$\Lambda = \int \mathrm{d}t\, A^0\,, \tag{9.131}$$

the transformed scalar potential can be made to vanish,

$$(A^0)' = A^0 - \frac{\partial}{\partial t}\Lambda = A^0 - A^0 = 0\,, \qquad \vec{A}' = \vec{A} + \vec{\nabla}\Lambda = \vec{A} + \int \mathrm{d}t\,\vec{\nabla}A^0\,, \tag{9.132}$$

at the expense of adding an additional term to the vector potential. This is called the temporal or WEYL[1] gauge. In this gauge, one has

$$D_F^{00}(k) = 0\,, \qquad D_F^{ij}(k) = \frac{1}{k^2 + \mathrm{i}\epsilon}\left(\delta_{ij} - \frac{k_i\,k_j}{\omega^2}\right)\,. \tag{9.133}$$

Other choices for the functions f^μ in Eq. (9.129) lead to the so-called Landau and Fried–Yennie gauges (see also Sec. 9.7).

9.5.3 *Gupta–Bleuler Condition*

We now have to carry out the program sketched in Sec. 9.4.2 and reestablish the consistency of the quantization for the electromagnetic field in the covariant formalism, by implementing the condition $\partial \cdot A \to 0$ in a "weak" sense, as a condition on any state which contains real photons. This reestablishes the Lorenz gauge condition equation fulfilled by the physical state vectors. The "physical" Hilbert space then is assured to have a positive definite metric.

Let us consider a state $|s\rangle$ which constitutes a superposition of scalar photon states,

$$|s\rangle = \int \mathrm{d}^3k\,f(\vec{k})\,a_0^+(\vec{k})\,|0\rangle\,. \tag{9.134}$$

Its norm is

$$\begin{aligned}\langle s|s\rangle &= \int \mathrm{d}^3k' \int \mathrm{d}^3k\,f^*(\vec{k}')\,f(\vec{k})\,\langle 0|[a_0(\vec{k}'), a_0^+(\vec{k})]|0\rangle \\ &= -\,g_{00}\int \mathrm{d}^3k\,f^*(\vec{k})\,f(\vec{k}) = -\int \mathrm{d}^3k\,\left|f(\vec{k})\right|^2 < 0\,,\end{aligned} \tag{9.135}$$

exemplifying the Fock space with negative norm (or, "indefinite norm" because the other polarizations actually have positive norm).

Let us first investigate the consequences of postulating that the physical states of the electromagnetic field should fulfill a variation of the condition $\partial_\mu A^\mu = 0$ in the mean, e.g., in the sense of quantum mechanical expectation values. This would otherwise imply that

$$\langle \Psi'|\partial_\mu A^\mu|\Psi\rangle = 0 \tag{9.136}$$

for any physical states $|\Psi\rangle$ and $|\Psi'\rangle$ of the photon field. Decomposing this condition into creation and annihilation parts, we see that the above equation is equivalent to

$$\langle \Psi'|\partial_\mu A^{(+)\,\mu}|\Psi\rangle + \langle \Psi'|\partial_\mu A^{(-)\,\mu}|\Psi\rangle = 0\,, \tag{9.137}$$

[1]Hermann Klaus Hugo Weyl (1885–1955).

where the "positive-frequency" (annihilation) part [see Eq. (9.113)] has the divergence

$$\partial_\mu A^{(+)\,\mu}(t,\vec{x}) = -\mathrm{i} \sum_{\lambda=0}^{3} \int \frac{\mathrm{d}^3k}{(2\pi)^3} \frac{1}{2|\vec{k}|} |\vec{k}| \left[a_0(\vec{k}) - a_3(\vec{k})\right] \mathrm{e}^{-\mathrm{i}\,k\cdot x} . \tag{9.138}$$

In view of Eqs. (9.94), (9.97) and (9.98), only scalar and longitudinal photons ($\lambda = 0, 3$) contribute to the result. We now formulate the Gupta–Bleuler condition in a way which is a little stricter than Eq. (9.136), namely, as the operator equation

$$\partial_\mu A^{(+)\,\mu} |\Psi\rangle = 0 \tag{9.139}$$

for all physical states. This means that

$$\left[a_0(\vec{k}) - a_3(\vec{k})\right] |\Psi\rangle = 0 \tag{9.140}$$

for all $\vec{k}$. The Gupta–Bleuler condition (9.140) is sufficient to ensure that there are no scalar and no longitudinal photons present in $|\Psi\rangle$. An intuitive argument, valid for states with a finite number of photons, can be given as follows. The operators $a_0(\vec{k})$ and $a_3(\vec{k})$ annihilate different photon polarizations. Hence, the occupation numbers for scalar and longitudinal photons in the states $a_0(\vec{k})|\Psi\rangle$ and $a_3(\vec{k})|\Psi\rangle$ are different. Because $a_0(\vec{k})\,|\Psi\rangle$ has to be equal to $a_3(\vec{k})\,|\Psi\rangle$, the only way to fulfill Eq. (9.140) is for $|\Psi\rangle$ not to contain any scalar or longitudinal photons in the first place.

More formally, the proof is as follows. We first denote the allowed wave vectors of the electromagnetic field as $\vec{k}_1, \vec{k}_2, \ldots$. Then, we decompose $|\Psi\rangle$ in an occupation number representation of Fock space in terms of states of the form

$$|\Psi\rangle = |n_0(\vec{k}_1), n_0(\vec{k}_2), \ldots, n_1(\vec{k}_1), n_1(\vec{k}_2), \ldots, n_2(\vec{k}_1), n_2(\vec{k}_2), \ldots, n_3(\vec{k}_1), n_3(\vec{k}_2), \ldots\rangle, \tag{9.141}$$

which enumerates the occupation numbers of the photon states with polarizations $\lambda = 0, 1, 2, 3$ (subscript of the occupation number n) and with the allowed wave vectors $\vec{k}_j$ ($j \in \mathbb{N}$). Here, the state with occupation number $n_\lambda(\vec{k})$ is lifted from the vacuum using the photon creation operators,

$$|0, \ldots, n_\lambda(\vec{k}), \ldots\rangle = \frac{1}{\sqrt{n_\lambda(\vec{k})!}} \left(a_\lambda^+(\vec{k})\right)^n |0, \ldots, 0, \ldots\rangle . \tag{9.142}$$

The application of the operator $a_0(\vec{k}) - a_3(\vec{k})$ (with an arbitrary allowable wave vector labeled $\vec{k}$) leads to the equation

$$a_0(\vec{k})\,|\Psi\rangle = \sqrt{n_0(\vec{k})}\,|\ldots, n_0(\vec{k}) - 1, \ldots\rangle = a_3(\vec{k})\,|\Psi\rangle = \sqrt{n_3(\vec{k})}\,|\ldots, n_3(\vec{k}) - 1, \ldots\rangle . \tag{9.143}$$

We notice the reduction by one of the relevant occupation numbers. This equation involves two states with different occupation numbers for longitudinal and scalar modes with wave vector $\vec{k}$. Hence, the only possibility for it to be fulfilled is if

$$n_0(\vec{k}) = n_3(\vec{k}) = 0 . \tag{9.144}$$

The Gupta–Bleuler condition (9.139) or alternatively (9.140) thus implies that there are no longitudinal or scalar photons present at all in the physical state $|\Psi\rangle$. Let us figure an analogy to the reasoning just presented: One could say that (9.140), when applied to a box full of "matchbox" cars, says that after one "green" car (provided one is present) has been taken away, the resulting state is the same as the one obtained when one takes one "blue" car away (the color denotes the polarization). If there is only a finite number of cars, then the only way that could happen is if there were no blue and no green cars present in the box in the first place. And that is precisely the Gupta–Bleuler condition.

9.6 Wick Theorem and Applications

9.6.1 *Time Ordering and Wick Theorem*

The Wick theorem states, essentially, that the time-ordered product of field operators is equal to the normal-ordered product, plus all possible contractions. For the photon field, we can derive the Wick theorem by investigating a number of identities, which pertain to exponentiated field operators and Hamiltonians, where the classical current acts as a source for the quantized electromagnetic field.

This relatively little-known set of identities is the subject of the current derivation. We begin by deriving an important identity which is valid for any time-ordered exponential. For a Hamiltonian $H = H_0 + H_1$, the interaction Hamiltonian in the interaction picture (this is not redundant) acquires a time dependence according to Eq. (2.108),

$$H_I(t) = \exp(-\mathrm{i}H_0\,t)\,H_1\,\exp(\mathrm{i}H_0\,t)\,. \tag{9.145}$$

The time-ordered exponential of the interaction Hamiltonian is

$$\mathrm{T}\exp\left[-\mathrm{i}\int_{t_i}^{t_f}\mathrm{d}t\,H_I(t)\right] = \sum_{n=0}^{\infty}\frac{(-\mathrm{i})^n}{n!}\int_{t_i}^{t_f}\mathrm{d}t_1\cdots\int_{t_i}^{t_f}\mathrm{d}t_n\,\mathrm{T}\left[H_I(t_1)\ldots H_I(t_n)\right]\,. \tag{9.146}$$

Our purpose will be to rewrite the time-ordered exponential into a potentially easier form. We introduce a tiling of the time interval (t_i, t_f), which for $N \to \infty$ reads as

$$\Delta t = \frac{t_f - t_i}{N}\,, \qquad t_k = t_i + \left(k - \tfrac{1}{2}\right)\Delta t\,, \qquad k \in (1,\ldots,N)\,, \tag{9.147}$$

so that

$$\mathrm{T}\exp\left[-\mathrm{i}\int_{t_i}^{t_f}\mathrm{d}t\,H_I(t)\right] = \mathrm{e}^{-\mathrm{i}\,\Delta t\,H_I(t_N)}\,\mathrm{e}^{-\mathrm{i}\,\Delta t\,H_I(t_{N-1})}\cdots\mathrm{e}^{-\mathrm{i}\,\Delta t\,H_I(t_1)}\,. \tag{9.148}$$

The right-hand side of (9.148) contains products of exponentials which can be transformed using the Campbell–Baker–Hausdorff identity

$$\mathrm{e}^{A}\,\mathrm{e}^{B} = \mathrm{e}^{A+B+[A,B]/2}\,, \tag{9.149}$$

which is valid if $[A,[A,B]] = [B,[A,B]] = 0$. We shall assume in the following that the commutator of two interaction Hamiltonians is a complex number, possibly a function of space and time, and commutes with the interaction Hamiltonian itself, which is supposed to contain no derivative operators. We begin by considering the product of the two leftmost factors in Eq. (9.148),

$$\begin{aligned} e^{-i\,\Delta t\, H_I(t_N)}\, e^{-i\,\Delta t\, H_I(t_{N-1})} &= e^{-i\,\Delta t\,[H_I(t_N)+H_I(t_N)]}\, e^{\frac{1}{2}[-i\,\Delta t\, H_I(t_N),-i\,\Delta t\, H_I(t_{N-1})]} \\ &= e^{-i\,\Delta t\,[H_I(t_N)+H_I(t_{N-1})]}\, e^{-\frac{1}{2}\Delta t^2\,[H_I(t_N),H_I(t_{N-1})]}\,. \end{aligned} \tag{9.150}$$

This consideration generalizes to three temporal arguments as follows,

$$\begin{aligned} & e^{-i\,\Delta t\, H_I(t_N)}\, e^{-i\,\Delta t\, H_I(t_{N-1})}\, e^{-i\,\Delta t\, H_I(t_{N-2})} \\ &\quad = e^{-i\,\Delta t\,[H_I(t_N)+H_I(t_{N-1})]}\, e^{-i\,\Delta t\, H_I(t_{N-2})} e^{-\frac{1}{2}\Delta t^2\,[H_I(t_N),H_I(t_{N-1}]} \\ &\quad = e^{-i\,\Delta t\,[H_I(t_N)+H_I(t_{N-1})+H_I(t_{N-2})]} \\ &\qquad \times e^{-\frac{1}{2}\Delta t^2\,([H_I(t_N),H_I(t_{N-1})]+[H_I(t_{N-1}),H_I(t_{N-2})]+[H_I(t_N),H_I(t_{N-2})])}\,. \end{aligned} \tag{9.151}$$

In the commutators on the right-hand side, the time argument in the left Hamiltonian is greater than the time in the right Hamiltonian. In the limit $N \to \infty$, we thus have

$$\begin{aligned} \mathrm{T}\,\exp\left[-i\int_{t_i}^{t_f} dt\, H_I(t)\right] &= \exp\left[-i\int_{t_i}^{t_f} dt\, H_I(t)\right] \\ &\quad \times \exp\left[-\tfrac{1}{2}\int_{t_i}^{t_f} dt \int_{t_i}^{t_f} dt'\,\Theta(t-t')[H_I(t),H_I(t')]\right]. \end{aligned} \tag{9.152}$$

In the following, we use the quantum electrodynamic interaction Hamiltonian in the form

$$H_I(t) = \int d^3x\, A^\mu(x)\, J_\mu(x) = \int d^3x\, A(x)\cdot J(x)\,, \tag{9.153}$$

where $J_\mu(x)$ is assumed to be a classical current distribution, not a quantum mechanical operator [see also Eq. (2.108)]. We recall that in our formulation of the photon field operator, we have induced the time-dependence by going to the interaction picture. The S-matrix is given by the time-ordered exponential

$$S = \mathrm{T}\,\exp\left[-i\int d^4x\, A(x)\cdot J(x)\right] = T\,\exp\left[-i\int_{-\infty}^{\infty} dt\, H_I(t)\right]. \tag{9.154}$$

In the following, we are inspired by the derivation following Eq. (4-60) of Ref. [2], but we found it useful to add a number of explanations for intermediate steps which make the derivation much easier to follow and put the considerations into their proper context. The commutator of the interaction Hamiltonians $H_I(t)$ and $H_I(t')$ is [see Eq. (9.116)]

$$\begin{aligned} [H_I(t), H_I(t')] &= \int d^3x \int d^3x'\, J_\mu(x)\, [A^\mu(x), A^\nu(x')]\, J_\nu(x') \\ &= i\, g^{\mu\nu} \int d^3x \int d^3x'\, J_\mu(x)\, G_-(x,x')\, J_\nu(x') \\ &= i \int d^3x \int d^3x'\, J_\mu(x)\, G_-(x,x')\, J^\mu(x')\,, \end{aligned} \tag{9.155}$$

where $G_-(x,x') = G_R(x,x') - G_A(x,x')$ according to Eq. (9.38). We note that since $H_I(t)$ is a Hermitian operator, the commutator $[H_I(t), H_I(t')]$ is antihermitian, and since we assume $J^\mu(x)$ to be a classical current distribution, the commutator can be expressed as $[H_I(t), H_I(t')] = \mathrm{i}\, f(t,t')$ where $f(t,t')$ is a real function. We conclude that

$$\begin{aligned} S &= \exp\left[-\mathrm{i} \int_{-\infty}^{\infty} \mathrm{d}t\, H_I(t)\right] \exp\left[-\tfrac{1}{2} \int_{-\infty}^{\infty} \mathrm{d}t \int_{-\infty}^{\infty} \mathrm{d}t'\, \Theta(t-t')[H_I(t), H_I(t')]\right] \\ &= \exp\left[-\mathrm{i} \int \mathrm{d}^4x\, A(x) \cdot J(x)\right] \exp\left[-\frac{\mathrm{i}}{2} \int \mathrm{d}^4x \int \mathrm{d}^4x' J_\mu(x) G_R(x,x') J^\mu(x')\right], \end{aligned} \tag{9.156}$$

where we have used the connection of the retarded Green function and the field commutator, namely $G_R(x,x') = \Theta(t-t')\, G_-(x,x')$.

The presence of the retarded Green function instead of the time-ordered Green function, or Feynman propagator, is disturbing. We can now proceed by carefully analyzing the term $\exp\left[-\mathrm{i} \int \mathrm{d}^4x\, A(x) \cdot J(x)\right]$. To this end, we first decompose the expression $\exp\left[-\mathrm{i} \int \mathrm{d}^4x\, A(x) \cdot J(x)\right]$ into creation and annihilation parts, which are labeled as positive-frequency (annihilation) components [superscript (+)] and negative-frequency (creation) components [superscript (−)],

$$\begin{aligned} &\exp\left[-\mathrm{i} \int \mathrm{d}^4x\, A(x) \cdot J(x)\right] = \exp\left[-\mathrm{i} \int \mathrm{d}^4x \left[A^{(+)}(x) + A^{(-)}(x)\right] \cdot J(x)\right] \\ &= \exp\left[-\mathrm{i} \int \mathrm{d}^4x\, A^{(-)}(x) \cdot J(x)\right] \exp\left[-\mathrm{i} \int \mathrm{d}^4x'\, A^{(+)}(x') \cdot J(x')\right] \\ &\quad \times \exp\left[\tfrac{1}{2} \int \mathrm{d}^4x \int \mathrm{d}^4x'\, J_\mu(x) \left[A^{(-)\,\mu}(x), A^{(+)\,\nu}(x')\right] J_\nu(x)\right]. \end{aligned} \tag{9.157}$$

Again, we have used the formula $\mathrm{e}^{A+B} = \mathrm{e}^A\, \mathrm{e}^B\, \mathrm{e}^{-[A,B]/2}$. We can now identify the expression

$$\exp\left[-\mathrm{i} \int \mathrm{d}^4x\, A^{(-)}(x) \cdot J(x)\right] \exp\left[-\mathrm{i} \int \mathrm{d}^4x\, A^{(+)}(x) \cdot J(x)\right] \equiv 1 + :S:, \tag{9.158}$$

where $:S:$ is the normal-ordered version of the S matrix defined in Eq. (9.154), because the annihilation operators contained in $A^{(+)}(x)$ stand to the right of the creation operators contained in $A^{(-)}(x)$. Note that the presence of the unit operator on the right-hand side is essential because the leading term in the expansion of the operator on the left-hand side of the above equation (in the limit $J^\mu(x) \to 0$) is just unity, and the normal-ordered variant of the unit operator vanishes. Putting everything together, we thus have proven the identity

$$\begin{aligned} S = (1 + :S:) \exp\Big[\tfrac{1}{2} \int \mathrm{d}^4x \int \mathrm{d}^4x'\, J_\mu(x) \Big(&\left[A^{(-)\,\mu}(x), A^{(+)\,\nu}(x')\right] \\ &- \Theta(t-t')\, [A^\mu(x), A^\nu(x')]\Big)\, J_\nu(x')\Big]. \end{aligned} \tag{9.159}$$

Here, we have rewritten the retarded Green function again as a field commutator, using the identity

$$\mathrm{i} J_\mu(x)\, G_R(x-x') J^\mu(x') = \Theta(t-t')\, J_\mu(x)\, [A^\mu(x), A^\nu(x')]\, J_\nu(x'). \tag{9.160}$$

The expression $\left[A^{(-)\,\mu}(x), A^{(+)\,\nu}(x')\right] - \Theta(t-t')\ [A^\mu(x), A^\nu(x')]$ is a field commutator and therefore equal to its vacuum expectation value. The time-ordered product emerges as follows,

$$\begin{aligned}
&\left\langle 0\left|\left[A^{(-)\,\mu}(x), A^{(+)\,\nu}(x')\right] - \Theta(t-t')\ [A^\mu(x), A^\nu(x')]\right|0\right\rangle \\
&= \left\langle 0\left|-A^{(+)\,\nu}(x')A^{(-)\,\mu}(x) - \Theta(t-t')A^\mu(x)A^\nu(x') + \Theta(t-t')A^\nu(x')A^\mu(x)\right|0\right\rangle \\
&= \langle 0|-A^\nu(x')\,A^\mu(x) + \Theta(t-t')\,A^\nu(x')\,A^\mu(x) - \Theta(t-t')\,A^\mu(x)\,A^\nu(x')|0\rangle \\
&= \langle 0|-[1-\Theta(t-t')]\,A^\nu(x')\,A^\mu(x) - \Theta(t-t')\,A^\mu(x)\,A^\nu(x')|0\rangle \\
&= \langle 0|-\Theta(t'-t)\,A^\nu(x')\,A^\mu(x) - \Theta(t-t')\,A^\mu(x)\,A^\nu(x')|0\rangle \\
&= -\langle 0|T\,A^\mu(x)\,A^\nu(x')|0\rangle\,.
\end{aligned} \tag{9.161}$$

Our end result is

$$S = (1+:S:)\,\exp\left[-\frac{1}{2}\int \mathrm{d}^4x \int \mathrm{d}^4x'\,J_\mu(x)\,\langle 0|T\,A^\mu(x)\,A^\nu(x')|0\rangle\,J_\nu(x')\right], \tag{9.162}$$

or, written explicitly,

$$\begin{aligned}
&\mathrm{T}\exp\left[-\mathrm{i}\int \mathrm{d}^4x\,A(x)\cdot J(x)\right] = \left(1+:\exp\left[-\mathrm{i}\int \mathrm{d}^4x\,A(x)\cdot J(x)\right]:\right) \\
&\times\exp\left[-\frac{1}{2}\int \mathrm{d}^4x \int \mathrm{d}^4x'\,J_\mu(x)\,\langle 0|\mathrm{T}\,A^\mu(x)\,A^\nu(x')|0\rangle\,J_\nu(x')\right].
\end{aligned} \tag{9.163}$$

This formula constitutes an "exponentiated" version of the Wick theorem. The classical current $J^\mu(x)$ acts as a source for the quantized electromagnetic field. Upon expansion in powers of the vector potential, mediated by functional differentiation of Eq. (9.163) with respect to $J^\mu(x)$, we recover the identities which make up the Wick theorem. When we act with $\delta^2/(\delta J^\mu(x)\delta J^\nu(x'))$ on both sides of Eq. (9.163), we see that

$$\mathrm{T}\,A^\mu(x)\,A^\nu(x') \;=\, :A^\mu(x)\,A^\nu(x'): + \langle 0|T\,A^\mu(x)\,A^\nu(x')|0\rangle\,. \tag{9.164}$$

The functional differentiation $\delta^3/(\delta J^\mu(x)\delta J^\nu(x')\delta J^\rho(x''))$ leads to the identity

$$\begin{aligned}
\mathrm{T}A^\mu(x)A^\nu(x')A^\rho(x'') \;&=\, :A^\mu(x)A^\nu(x')A^\rho(x''): + :A^\mu(x):\langle 0|\mathrm{T}A^\nu(x')A^\rho(x'')|0\rangle \\
&+ :A^\nu(x'):\langle 0|\mathrm{T}A^\mu(x)A^\rho(x'')|0\rangle + :A^\rho(x''):\langle 0|\mathrm{T}A^\mu(x)A^\nu(x')|0\rangle\,.
\end{aligned} \tag{9.165}$$

In a general order, we have

$$\begin{aligned}
&\mathrm{T}\,A^{\mu_1}(x_1)\cdots A^{\mu_n}(x_n) \;=\, :A^{\mu_1}(x_1)\cdots A^{\mu_n}(x_n): \\
&+\sum_{k<\ell} :A^{\mu_1}(x_1)\cdots \underline{A^{\mu_k}(x_k)}\cdots \underline{A^{\mu_l}(x_\ell)}\cdots A^{\mu_n}(x_n):\,\langle 0|\mathrm{T}A^{\mu_k}(x_k)\,A^{\mu_l}(x_\ell)|0\rangle \\
&+\sum_{Q=2}^{\infty}\;\sum_{k_1<k_2<\cdots<k_{2Q}} :A^{\mu_1}(x_1)\cdots \underline{A^{\mu_{k_1}}(x_{k_1})}\cdots \underline{A^{\mu_{k_{2Q}}}(x_{k_{2Q}})}\cdots A^{\mu_n}(x_n): \\
&\times\sum_P \left\langle 0\left|\mathrm{T}A^{\mu_{k_{P(1)}}}(x_{k_{P(1)}})\,A^{\mu_{k_{P(2)}}}(x_{k_{P(2)}})\right|0\right\rangle\cdots \\
&\langle 0|\mathrm{T}A^{\mu_{k_{P(2Q-1)}}}(x_{k_{P(2Q-1)}})\,A^{\mu_{k_{P(2Q)}}}(x_{k_{P(2Q)}})|0\rangle\,,
\end{aligned} \tag{9.166}$$

where operators left out from the normal-ordered product are underlined. This equation is to be read as follows. The permutations P transform the indices

$$\{1,\ldots,2Q\} \to \{P(1),\ldots,P(2Q)\}\,, \tag{9.167a}$$

$$\{k_1,\ldots,k_{2Q}\} \to \{k_{P(1)},\ldots,k_{P(2Q)}\}\,, \tag{9.167b}$$

$$\{x_{k_1},\ldots,x_{k_{2Q}}\} \to \{x_{k_{P(1)}},\ldots,x_{k_{P(2Q)}}\}\,, \tag{9.167c}$$

i.e., they act on the indices of the indices *(sic!)* of the space-time coordinates, and on the indices of the indices *(sic!)* of the space-time indices of the four-vector potential field operators. All possible permutations of $k_1,\ldots,k_{2Q}$ are summed over. The Wick theorem implies that the time ordering equals the normal ordering plus all contractions. We observe that the result given in Eq. (9.163), which is given in exponentiated form, is much more compact in comparison to the collection of formulas given in Eqs. (9.164), (9.165), and (9.166).

9.6.2 *Field Commutators and Current Distributions*

The full S matrix can only be evaluated under special circumstances, one of which concerns the generation of photons by a given, classical current distribution. The QED interaction Hamiltonian $H_I(t) = J(x)\cdot A(x)$ is split into a quantized part [four-vector potential $A^\mu(x)$] and a classical part [current distribution $J^\mu(x)$]. In the calculation, we shall assume that the electromagnetic field, initially, is in the vacuum state, and the currents $J^\mu(x)$ are switched on adiabatically in the distant past. We then calculate the vacuum persistence amplitude, i.e., the probability amplitude for the system to remain in the vacuum state (no photons are created), and also, the probability for creating n photons.

We split the four-vector potential field operator $A^\mu(x)$ into "in" and "out" fields, which are solutions of the homogeneous field equations and describe the asymptotic behavior as $t \to \mp\infty$,

$$\lim_{t\to-\infty} A^\mu(x) \to A^\mu_{\rm in}(x)\,, \qquad \lim_{t\to+\infty} A^\mu(x) \to A^\mu_{\rm out}(x)\,, \tag{9.168}$$

where $x^\mu = (t,\vec r)$. In Lorenz gauge, one has

$$\begin{aligned} A^\mu(x) &= A^\mu_{\rm in}(x) + \int \mathrm{d}^4x'\, G_R(x-x')\, J^\mu(x') \\ &= A^\mu_{\rm out}(x) + \int \mathrm{d}^4x'\, G_A(x-x')\, J^\mu(x')\,, \end{aligned} \tag{9.169}$$

where we assume that the sources are switched off adiabatically in the distant past and future, and according to Eq. (9.10), we have $\Box G_R(x) = \Box G_A(x) = \delta^{(4)}(x)$. The difference [see Eq. (9.37)]

$$A^\mu_{\rm out}(x) - A^\mu_{\rm in}(x) = \int \mathrm{d}^4x'\, G_-(x-x')\, J^\mu(x') \tag{9.170}$$

is a solution of the homogeneous equation. *A priori*, Eq. (9.169) only holds for classical four-vector potentials, and classical current distributions. In the fully quantized formalism, the current would be an operator involving quantized fermion fields.

However, one can show [see also Eq. (9.186)] that the formalism still holds and leads to a consistent S matrix if we replace $A^\mu_{\rm out}(x)$ and $A^\mu_{\rm in}(x)$ by quantized operators in the interaction picture. This is what we shall do in the following, in order to describe photon creation by a classical, given current distribution. We introduce the S matrix as the matrix which connects the in and out states, as follows,

$$|a\,{\rm out}\rangle = S^{-1}\,|a\,{\rm in}\rangle\,, \qquad \langle a\,{\rm out}| = \langle a\,{\rm in}|S\,, \tag{9.171a}$$

$$|a\,{\rm in}\rangle = S\,|a\,{\rm out}\rangle\,, \qquad \langle a\,{\rm in}| = \langle a\,{\rm out}|S^{-1}\,, \tag{9.171b}$$

for a general state $|a\rangle$. The usual interpretation of the S matrix as the time-ordered exponential, integrated from time equal to $-\infty$ to $+\infty$, proceeds from the formula

$$S_{ab} = \langle a\,{\rm out}|b\,{\rm in}\rangle = \langle a\,{\rm in}|S|b\,{\rm in}\rangle = \langle a\,{\rm out}|S|b\,{\rm out}\rangle\,. \tag{9.172}$$

The S matrix element can be evaluated either in the space of the in and out states.

The S matrix also relates the in and out fields,

$$A^\mu_{\rm out}(x) = S^{-1}\,A^\mu_{\rm in}(x)\,S\,, \qquad S^+ = S^{-1}\,. \tag{9.173}$$

The field operators $A^\mu_{\rm out}(x)$ and $A^\mu_{\rm in}(x)$ are assumed to retain their meaning in the Fock space of the asymptotic states, so that

$$\langle a\,{\rm out}|A^\mu_{\rm out}(x)|b\,{\rm out}\rangle = \langle a\,{\rm in}|A^\mu_{\rm in}(x)|b\,{\rm in}\rangle\,. \tag{9.174}$$

An important example is the vacuum persistence amplitude

$$S_{ab} = \langle a\,{\rm in}|S|b\,{\rm in}\rangle = |a\,{\rm in}\rangle = |b\,{\rm in}\rangle = |0\,{\rm in}\rangle\,, \tag{9.175}$$

which is

$$\langle 0\,{\rm out}|0\,{\rm in}\rangle = \langle 0\,{\rm in}|S|0\,{\rm in}\rangle = \langle 0\,{\rm out}|S|0\,{\rm out}\rangle\,. \tag{9.176}$$

The probability

$$p_0 = |\langle 0\,{\rm out}|0\,{\rm in}\rangle|^2 = |\langle 0\,{\rm in}|S|0\,{\rm in}\rangle|^2 \le 1 \tag{9.177}$$

gives a measure of the likelihood of the vacuum remaining in its ground state, without any photons present in the radiation field.

It is instructive to observe that Eq. (9.173) implies the relations

$$a^{\rm out}_{\vec k\lambda} = S^{-1}\,a^{\rm in}_{\vec k\lambda}\,S\,, \qquad \left(a^{\rm out}_{\vec k\lambda}\right)^+ = S^{-1}\left(a^{\rm in}_{\vec k\lambda}\right)^+ S \tag{9.178}$$

for the fundamental annihilation and creation operators. We take the opportunity to illustrate the action of the S matrix one more time, by way of example. In the sense of Eq. (2.87), a state

$$|1_{\vec k\lambda}\,{\rm in}\rangle = \left(a^{\rm in}_{\vec k\lambda}\right)^+ |0\rangle \tag{9.179}$$

is a one-photon state in the limit of $t \to -\infty$, while

$$|1_{\vec k'\lambda'}\,{\rm out}\rangle = \left(a^{\rm out}_{\vec k'\lambda'}\right)^+ |0\rangle \tag{9.180}$$

constitutes the same in the limit $t \to \infty$. The transition probability is

$$\langle 1_{\vec k'\lambda'}\,{\rm out}\,|\,1_{\vec k\lambda}\,{\rm in}\rangle = \langle 1_{\vec k'\lambda'}\,{\rm out}\,|S|1_{\vec k\lambda}\,{\rm out}\rangle = \langle 1_{\vec k'\lambda'}\,{\rm in}\,|S|1_{\vec k\lambda}\,{\rm in}\rangle\,, \tag{9.181}$$

and can be evaluated using the S matrix, within both the space of the in as well as the out states.

We now combine the definition of the S matrix as given in Eq. (9.173) with Eq. (9.169), to write

$$\begin{aligned} A^\mu_{\rm out}(x) = S^{-1} A^\mu_{\rm in}(x) S &= A^\mu_{\rm in}(x) + \int \mathrm{d}^4x'\, G_-(x-x')\, J^\mu(x') \\ &= A^\mu_{\rm in}(x) - \mathrm{i} \int \mathrm{d}^4x'\, \mathcal{D}^{\mu\nu}_-(x-x')\, J^\nu(x')\,, \end{aligned} \tag{9.182}$$

where we recall that $\mathcal{D}^{\mu\nu}_-(x-x') = [A^\mu(x), A^\nu(x')] = \mathrm{i}\, g^{\mu\nu}\, G_-(x-x')$. One may ask how the fields $A^\mu(x)$ and $A^\nu(x)$, which enter the definition of $\mathcal{D}^{\mu\nu}_-(x-x')$, qualify themselves in the context of the in and out fields. Using Eqs. (9.169) and (9.173), it is easily possible to show that

$$\mathcal{D}^{\mu\nu}_-(x-x') = [A^\mu(x), A^\nu(x')] = [A^\mu_{\rm in}(x), A^\nu_{\rm in}(x')] = [A^\mu_{\rm out}(x), A^\nu_{\rm out}(x')]\,. \tag{9.183}$$

So, we have

$$A^\mu_{\rm out}(x) = S^{-1} A^\mu_{\rm in}(x) S = A^\mu_{\rm in}(x) - \mathrm{i} \int \mathrm{d}^4x'\, [A^\mu_{\rm in}(x), A^\nu_{\rm in}(x')]\, J_\nu(x')\,. \tag{9.184}$$

From this expression, we can immediately find an exponentiated form of the S matrix operator, because

$$\mathrm{e}^{-X}\, B\, \mathrm{e}^{X} = B + [B, X] + \frac{1}{2!}[[B,X],X] + \frac{1}{3!}[[[B,X],X],X] + \cdots \tag{9.185}$$

reduces to $\mathrm{e}^{-X}\, B\, \mathrm{e}^{X} = B + [B, X]$ (exact identity) in the case where the higher-order commutators vanish. Because the field commutator is just a complex number, we can identify $X = -\mathrm{i} \int \mathrm{d}^4x'\, A_{\rm in}(x') \cdot J(x')$ and $B = A^\mu_{\rm in}(x)$, to identify $S = \exp(X)$ with the expression

$$S = \exp\left[-\mathrm{i} \int \mathrm{d}^4x'\, A_{\rm in}(x') \cdot J(x')\right]. \tag{9.186}$$

For a classical current distribution, this result differs from Eq. (9.156) only by an unobservable phase. We now decompose the field operator according to Eq. (9.113),

$$A^\mu_{\rm in}(x) = A^{(-)\,\mu}_{\rm in}(x) + A^{(+)\,\mu}_{\rm in}(x)\,, \tag{9.187}$$

into a negative-frequency, creation part $A^{(-)}_{\rm in}(x)$ and positive-frequency, annihilation part $A^{(+)}_{\rm in}(x)$. We recall the Campbell–Baker–Hausdorff identity (9.149) in the form $\mathrm{e}^{A+B} = \mathrm{e}^{A}\, \mathrm{e}^{B}\, \mathrm{e}^{-[A,B]/2}$, which implies that

$$\begin{aligned} S &= \exp\left[-\mathrm{i} \int \mathrm{d}^4x\, A^{(-)}_{\rm in}(x) \cdot J(x)\right] \exp\left[-\mathrm{i} \int \mathrm{d}^4x\, A^{(+)}_{\rm in}(x) \cdot J(x)\right] \\ &\quad \times \exp\left[\frac{1}{2} \int \mathrm{d}^4x \int \mathrm{d}^4x' \left[A^{(-)}_{\rm in}(x) \cdot J(x),\, A^{(+)}_{\rm in}(x') \cdot J(x')\right]\right] \\ &= (1 + :S:) \exp\left[\frac{1}{2} \int \mathrm{d}^4x \int \mathrm{d}^4x' \left[A^{(-)}_{\rm in}(x) \cdot J(x),\, A^{(+)}_{\rm in}(x') \cdot J(x')\right]\right], \end{aligned} \tag{9.188}$$

where we have used Eq. (9.158). The vacuum persistence amplitude thus reads as

$$\sqrt{p_0} = \langle 0\,\mathrm{in}|S|0\,\mathrm{in}\rangle = \exp\left[\frac{1}{2}\int \mathrm{d}^4x \int \mathrm{d}^4x' \left[A_{\mathrm{in}}^{(-)}(x)\cdot J(x), A_{\mathrm{in}}^{(+)}(x')\cdot J(x')\right]\right]. \tag{9.189}$$

With the help of Eq. (9.114), one obtains

$$[A_{\mathrm{in}}^{(-)\mu}(x), A_{\mathrm{in}}^{(+)\nu}(x')] = g^{\mu\nu}\int \frac{\mathrm{d}^4k}{(2\pi)^4}\,\mathrm{e}^{\mathrm{i}\,k\cdot(x-x')}\,\Theta(k^0)\,\delta(k^2)\,. \tag{9.190}$$

Let us define the Fourier transformation of the current distribution in the convention

$$J^\mu(k) = \int \mathrm{d}^4x\, J^\mu(x)\,\mathrm{e}^{\mathrm{i}k\cdot x}\,, \qquad J^\mu(x) = \int \frac{\mathrm{d}^4k}{(2\pi)^4}\,J^\mu(k)\,\mathrm{e}^{-\mathrm{i}k\cdot x}\,. \tag{9.191}$$

It is then rather straightforward to evaluate the convolution of the current distribution and the field operators,

$$\begin{aligned}
&\frac{1}{2}\int \mathrm{d}^4x \int \mathrm{d}^4x' \left[A_{\mathrm{in}}^{(-)}(x)\cdot J(x), A_{\mathrm{in}}^{(+)}(x')\cdot J(x')\right] \\
&= \frac{1}{2}\int \mathrm{d}^4x \int \mathrm{d}^4x'\, J^\mu(x)\,J_\mu(x') \int \frac{\mathrm{d}^4k}{(2\pi)^4}\,\mathrm{e}^{\mathrm{i}\,k\cdot(x-x')}\,\Theta(k^0)\,\delta(k^2) \\
&= \frac{1}{2}\int \frac{\mathrm{d}^4k}{(2\pi)^4}\,J^{*\,\mu}(k)\,J_\mu(k)\,\Theta(k^0)\,\delta(k^2) = \frac{1}{2}\int \frac{\mathrm{d}^3k}{(2\pi)^3}\,\frac{1}{2\,|\vec{k}|}\,J^{*\,\mu}(k)\,J_\mu(k)\big|_{k^0=+|\vec{k}|}\,,
\end{aligned} \tag{9.192}$$

where we use $J_\mu(-k) = J^*_\mu(k)$ in view of the reality of the current in space-time. We now decompose the current $J^\mu(k)$ into a longitudinal component $J_\parallel(k)$ and transverse components $J^\mu_\perp(k)$, with $k_\mu\, J^\mu_\perp(k) = 0$,

$$J^\mu(k) = k^\mu\, J_\parallel(k) + J^\mu_\perp(k) \tag{9.193}$$

where we recall the general decomposition outlined in Chap. 1 of Ref. [81] and observe that the introduction of the scalar function $J_\parallel(k)$ is useful in other contexts as well [see the related analysis of the scalar potential in Coulomb gauge, described in Sec. 4.2.4 of Ref. [81]].

For real photons, one has $k^0 = |\vec{k}|$, and only the transverse component of the current contributes, because

$$J^*(k)\cdot J(k) = J(-k)\cdot J(k) = -k^2 J^2_\parallel(k) + J_\perp(-k)\cdot J_\perp(k) = J^*_\perp(k)\cdot J_\perp(k)\,. \tag{9.194}$$

Any vector field perpendicular to the direction of $\vec{k}$ can be decomposed into two transverse components as follows [see also Eq. (9.95)],

$$\begin{aligned}
&J^\mu_\perp(k) = -\sum_{\lambda=1}^{2}\epsilon^\mu_\lambda(k)\,J_{\perp,\lambda}(k)\,, \quad \epsilon_\lambda(k)\cdot\epsilon_{\lambda'}(k) = -\delta_{\lambda\lambda'}\,, \\
&J_{\perp,\lambda}(k) = \epsilon_\lambda(k)\cdot J(k)\,, \quad J^*_\perp(k)\cdot J_\perp(k) = -\left(|J_{\perp,1}(k)|^2 + |J_{\perp,2}(k)|^2\right)\,.
\end{aligned} \tag{9.195}$$

We finally have the result for the vacuum persistence amplitude,

$$\sqrt{p_0} = \langle 0\,\mathrm{in}|S|0\,\mathrm{in}\rangle = \exp\left[-\frac{1}{2}\int \frac{\mathrm{d}^3k}{(2\pi)^3}\,\frac{1}{2\,|\vec{k}|}\left(|J_{\perp,1}(k)|^2 + |J_{\perp,2}(k)|^2\right)\right] \le 1\,, \tag{9.196}$$

which implies that only the transverse components of the currents produce photons. The probability for the electromagnetic field to remain in the vacuum state is p_0, and it is obtained by squaring the result given in Eq. (9.196).

In order to clarify how many photons are produced, we investigate the n-photon state

$$|k_1\lambda_1, \ldots, k_n\lambda_n\rangle = a^+_{\lambda_1}(\vec{k}_1) \cdots a^+_{\lambda_n}(\vec{k}_n)\,|0\rangle\,. \tag{9.197}$$

Here, the $a_\lambda(k)$ are normalized as in Secs. 9.4.2 and 9.3.2, i.e.,

$$[a_\lambda(\vec{k}), a^+_{\lambda'}(\vec{k}')] = \delta_{\lambda\lambda'}\,\delta^{(3)}(\vec{k} - \vec{k}') \qquad (\lambda = 1, 2)\,. \tag{9.198}$$

For the discussion of the n-photon production amplitude, this particular normalization is advantageous. The n-photon states are normalized according to the condition

$$\begin{aligned}&\langle k_1\lambda_1, \ldots, k_n\lambda_n | k'_1\lambda'_1, \ldots, k'_n\lambda'_n\rangle \\ &= \sum_P \delta_{\lambda_1\lambda'_{P(1)}} \cdots \delta_{\lambda_1\lambda'_{P(n)}}\,\delta^{(3)}(\vec{k}_1 - \vec{k}_{P(1)}) \cdots \delta^{(3)}(\vec{k}_1 - \vec{k}_{P(n)})\,,\end{aligned} \tag{9.199}$$

where the sum runs over all permutations P of $\{1, \ldots, n\}$, just like in Eq. (9.167). The projector over the n-photon state involves a combinatorial factor,

$$P_n = \frac{1}{n!}\int \mathrm{d}^3k_1 \cdots \int \mathrm{d}^3k_n \sum_{\lambda_i=1,2} |k_1\lambda_1, \ldots, k_n\lambda_n\rangle\,\langle k_1\lambda_1, \ldots, k_n\lambda_n|\,. \tag{9.200}$$

The insertion of the projector P_n leaves the overlap matrix element between two n-photon states invariant,

$$\langle k_1\lambda_1, \ldots, k_n\lambda_n | k'_1\lambda'_1, \ldots, k'_n\lambda'_n\rangle = \langle k_1\lambda_1, \ldots, k_n\lambda_n | P_n | k'_1\lambda'_1, \ldots, k'_n\lambda'_n\rangle\,. \tag{9.201}$$

Now, we calculate the probability that $|0\,\mathrm{in}\rangle$ becomes an out state with n photons, which as we emphasize is only possible in a fully quantized framework. To this end, we employ the S-matrix formalism,

$$\begin{aligned}&\langle k_1\lambda_1, \ldots, k_n\lambda_n\,\mathrm{out}\,|\,0\,\mathrm{in}\rangle = \langle k_1\lambda_1, \ldots, k_n\lambda_n\,\mathrm{in}\,|\,S\,|\,0\,\mathrm{in}\rangle \\ &= \sqrt{p_0}\,\left\langle k_1\lambda_1, \ldots, k_n\lambda_n\,\mathrm{in}\left|\exp\left[-\mathrm{i}\int \mathrm{d}^4x\,A^{(-)}_{\mathrm{in}}(x)\cdot J(x)\right]\right|0\,\mathrm{in}\right\rangle\,,\end{aligned} \tag{9.202}$$

where p_0 is given in (9.196) and we have used Eq. (9.188) in order to factor out the vacuum part. We write down the photon field operator as follows, in accordance with our normalization $[a_\lambda(\vec{k}), a^+_{\lambda'}(\vec{k}')] = \delta_{\lambda\lambda'}\,\delta^{(3)}(\vec{k} - \vec{k}')$ of the fundamental field operators, and with Sec. 9.3.2,

$$A^\mu(x) = \sum_{\lambda=1}^{2}\int \frac{\mathrm{d}^3k}{\sqrt{2\omega\,(2\pi)^3}}\,\epsilon^\mu_\lambda(\vec{k})\,\left[a_\lambda(\vec{k})\,\exp(-\mathrm{i}\,k\cdot x) + a^+_\lambda(\vec{k})\,\exp(\mathrm{i}\,k\cdot x)\right]\,. \tag{9.203}$$

The space-time integral of the creation part of the photon field and the classical, photon-generating current can be reduced to the transverse components of the current, in momentum space,

$$\begin{aligned}\int \mathrm{d}^4x\, A_{\mathrm{in}}^{(-)}(x)\cdot J(x) &= \sum_{\lambda=1}^{2}\int \frac{\mathrm{d}^3k}{\sqrt{2|\vec{k}|\,(2\pi)^3}}\,\epsilon_\lambda^\mu(\vec{k})\,a_\lambda^+(\vec{k})\int \mathrm{d}^4x\,\mathrm{e}^{\mathrm{i}\,k\cdot x}\,J_\mu(x)\\ &= \sum_{\lambda=1}^{2}\int \frac{\mathrm{d}^3k}{\sqrt{2|\vec{k}|\,(2\pi)^3}}\,\epsilon_\lambda^\mu(\vec{k})\,a_\lambda^+(\vec{k})\,J_\mu(k)\\ &= \sum_{\lambda=1}^{2}\int \frac{\mathrm{d}^3k}{\sqrt{2\omega\,(2\pi)^3}}\,a_\lambda^+(\vec{k})\,J_{\perp,\lambda}(k)\,. \end{aligned}\tag{9.204}$$

We have used the fact that according to Eq. (9.195), $\epsilon_\lambda^\mu(\vec{k})\cdot J_\mu(k) = J_{\perp,\lambda}(k)$. Now, the amplitude for the generation of n photons can be expressed as follows,

$$\begin{aligned}&\left\langle k_1\lambda_1,\ldots,k_n\lambda_n,\,\mathrm{in}\left|\exp\left[-\mathrm{i}\int \mathrm{d}^4x\,A_{\mathrm{in}}^{(-)}(x)\cdot J(x)\right]\right|0\,\mathrm{in}\right\rangle\\ &= \sum_{\lambda_1\ldots\lambda_n}\frac{(-\mathrm{i})^n}{n!}\left\langle k_1\lambda_1,\ldots,k_n\lambda_n,\,\mathrm{in}\right|\int\frac{\mathrm{d}^3q_1}{\sqrt{2|\vec{q}_1|\,(2\pi)^3}}\cdots\frac{\mathrm{d}^3q_n}{\sqrt{2|\vec{q}_n|\,(2\pi)^3}}\\ &\quad\times a^+_{\vec{k}_1\,\lambda_1}(\vec{q}_1)\,J_{\perp,\lambda_1}(q_1)\cdots a^+_{\vec{k}_n\,\lambda_n}(\vec{q}_n)\,J_{\perp,\lambda_n}(q_n)\big|0\,\mathrm{in}\big\rangle\\ &= (-\mathrm{i})^n\,\frac{J_{\perp,\lambda_1}(k_1)}{\sqrt{2|\vec{k}_1|\,(2\pi)^3}}\cdots\frac{J_{\perp,\lambda_n}(k_n)}{\sqrt{2|\vec{k}_n|\,(2\pi)^3}}\,,\end{aligned}\tag{9.205}$$

where the combinatorial factor $1/n!$ is canceled by the permutations over the quantum numbers of the photons, all of which lead to the same contribution. Finally, one finds that

$$\langle k_1\lambda_1,\ldots,k_n\lambda_n\,\mathrm{in}\,|\,S\,|\,0\,\mathrm{in}\rangle = \sqrt{p_0}\,(-\mathrm{i})^n\,\frac{J_{\perp,\lambda_1}(k_1)}{\sqrt{2|\vec{k}_1|\,(2\pi)^3}}\cdots\frac{J_{\perp,\lambda_n}(k_n)}{\sqrt{2|\vec{k}_n|\,(2\pi)^3}}\,.\tag{9.206}$$

Assuming that the states $|k_1\lambda_1,\ldots,k_n\lambda_n\rangle$ in Eq. (9.200) are out states, we now put everything together,

$$\begin{aligned}p_n &= \langle 0\,\mathrm{in}|P_n|0\,\mathrm{in}\rangle\\ &= \frac{1}{n!}\int \mathrm{d}^3k_1\cdots\mathrm{d}^3k_n\sum_{\lambda_i=1,2}\langle 0\,\mathrm{in}\,|\,S\,|\,k_1\lambda_1,\ldots,k_n\lambda_n\,\mathrm{in}\rangle\,\langle k_1\lambda_1,\ldots,k_n\lambda_n\,\mathrm{in}\,|\,S\,|\,0\,\mathrm{in}\rangle\\ &= \frac{p_0}{n!}\int\frac{\mathrm{d}^3k_1}{2|\vec{k}_1|(2\pi)^3}\cdots\int\frac{\mathrm{d}^3k_n}{2|\vec{k}_n|(2\pi)^3}\sum_{\lambda_i=1,2}|J_{\perp,\lambda_1}(k_1)|^2\cdots|J_{\perp,\lambda_n}(k_n)|^2\\ &= \frac{p_0}{n!}\left[\int\frac{\mathrm{d}^3k}{2|\vec{k}|(2\pi)^3}\left(|J_{\perp,1}(k)|^2+|J_{\perp,2}(k)|^2\right)\right]^n\,.\end{aligned}\tag{9.207}$$

The intermediate step here is nontrivial; one has to convince oneself that the sum over all intermediate polarizations generates all the terms relevant to the exponential. This is done best by writing down the expressions for the first few n explicitly, say, $n = 2$ and $n = 3$. For $n = 3$, e.g., one has the identity

$$\int \frac{d^3 k_1}{2|\vec{k}_1|(2\pi)^3} \int \frac{d^3 k_2}{2|\vec{k}_2|(2\pi)^3} \sum_{\lambda_1=1,2} \sum_{\lambda_2=1,2} |J_{\perp,\lambda_1}(k_1)|^2 \, |J_{\perp,\lambda_2}(k_2)|^2$$
$$= \int \frac{d^3 k_1}{2|\vec{k}_1|(2\pi)^3} (|J_{\perp,1}(k_1)|^2 + |J_{\perp,2}(k_1)|^2) \int \frac{d^3 k_2}{2|\vec{k}_2|(2\pi)^3} (|J_{\perp,1}(k_2)|^2 + |J_{\perp,2}(k_2)|^2)$$
$$= \left[\int \frac{d^3 k}{2|\vec{k}|(2\pi)^3} (|J_{\perp,1}(k)|^2 + |J_{\perp,2}(k)|^2) \right]^2 . \tag{9.208}$$

We find the mean number $\bar{n}$ of produced photons as follows,

$$\bar{n} = \int \frac{d^3 k}{2|\vec{k}|(2\pi)^3} \left(|J_{\perp,1}(k)|^2 + |J_{\perp,2}(k)|^2 \right) , \tag{9.209}$$

so that $p_0 = \exp(-\bar{n})$ and

$$p_n = \frac{\bar{n}^n}{n!} \exp(-\bar{n}) , \qquad \sum_{n=0}^{\infty} p_n = 1 . \tag{9.210}$$

The distribution of emitted photons follows Poisson statistics. We identify $\bar{n}$ as the mean number of produced photons. The unitarity of the S matrix is expressed by the fact that the n-photon production probabilities add up to one. In a perturbative treatment, we would have found a probability of $\bar{n}$ for the creation of one and only one photon. However, the probability actually is $\bar{n} \exp(-\bar{n})$. The correction due to the exponential factor is not small, it is known as a "unitarity correction" in the literature. Unitarity corrections [306] can only be obtained and understood via the nonperturbative, exact treatment presented in the current section.

9.7 Further Thoughts

Here are some suggestions for further thought.

(1) **Green Functions in the Mixed Representation.** Verify the formulas (9.14) and (9.15) by evaluating the Fourier transform, with respect to time, of Eqs. (9.11) and (9.12). For the space-time representation, use the "exact" forms which contain terms proportional to $\Theta(t)\,\delta(t+r)$ and $\Theta(-t)\,\delta(t-r)$. In order to evaluate the limit as $r \to 0$ of the Fourier transform integrals, replace $r \to \epsilon$ and study the limit $\epsilon \to 0$.

(2) **Longitudinal Electric Field and Field Hamiltonian.** Convince yourself that $H_\parallel(t)$ is equal to the formula for the self-energy of a charge distribution, both in the electrostatic limit, as well as in the sense of the integral, over all space, of the energy density $\frac{1}{2}\vec{E}_\parallel^2 = \frac{1}{2}(\vec{\nabla}\Phi)^2$.

(3) **Discrete Representation of the Vector Potential.** Repeat all calculations in Sec. 9.3.2 using the discrete representation of the $\vec{A}$ field, with the explicit normalization volume. Consult Sec. 2.4.3 and use the representation (2.89),

$$\vec{A}(t,\vec{r}) = \sum_{\vec{k}\lambda} \frac{1}{\sqrt{V}} \frac{1}{\sqrt{2\omega}} \hat{\epsilon}_\lambda(\vec{k}) \left(a_\lambda(\vec{k})\, e^{-i k\cdot x} + a_\lambda^+(\vec{k})\, e^{i k\cdot x} \right) , \tag{9.211}$$

with the field commutators given in Eq. (2.93),

$$[\mathrm{a}_\lambda(\vec{k}),\, \mathrm{a}^+_{\lambda'}(\vec{k}')] = \delta_{\vec{k}\vec{k}'}\, \delta_{\lambda\lambda'}\,. \tag{9.212}$$

(4) **Fried–Yennie and Landau Gauges.** Search the literature for the precise form of the FRIED[2]–YENNIE[3] and LANDAU[4] gauges of the photon propagator, and determine the respective gauge-change functions f^μ in the sense of Eq. (9.129), for the change of gauge from the Lorenz gauge to the Fried–Yennie, or Landau gauge.

(5) **Negative Norm and Fock States.** Investigate Eq. (9.135). On page 246 in Chap. 9 of Ref. [10], it is stressed that only states with an odd number of scalar photons actually acquire a negative norm. Investigate if this statement is true, by calculating the norm of the two-scalar-photon state $\int \mathrm{d}^3k \int \mathrm{d}^3k'\, f(\vec{k},\vec{k}')\, a^+_0(\vec{k})\, a^+_0(\vec{k}')\, |0\rangle$.

(6) **In and Out Forms of the Photon Propagator.** Show the equivalence of the in and out forms of $\mathcal{D}_-$ as given in Eq. (9.183). *Hint:* Use Eq. (9.173) and observe that the field commutator $[A^\mu(x), A^\nu(x')]$ is just a complex number.

---

[2]Herbert M. Fried (b. 1928).
[3]Donald Robert Yennie (1924–1993).
[4]Lev Davidovich Landau (1908–1968).

Chapter 10

Tree-Level and Loop Diagrams, and Renormalization

10.1 Overview

We now have all the tools at hand in order to carry out fully relativistic calculations in quantum electrodynamics. The QED Lagrangian reads as

$$\mathcal{L} = \bar{\psi}\,[\gamma^{\mu}\,(\mathrm{i}\partial_{\mu} - e\,A_{\mu}) - m]\,\psi - \frac{1}{4}\,F^{\mu\nu}\,F_{\mu\nu}\,, \tag{10.1}$$

and we are familiar with all its ingredients. The field strength tensor $F^{\mu\nu} = \partial^{\mu} A^{\nu} - \partial^{\nu} A^{\mu}$ is familiar from Chap. 6 of Ref. [81], and has been introduced in this treatise in Eq. (2.47) of Sec. 2.4.1. Furthermore, we have carried out the (second) quantization of the electromagnetic four-vector potential A^{μ} in Secs. 2.4, 9.3 and 9.4, in the Coulomb and Lorenz gauges, respectively. Finally, the free Dirac field operator has been discussed in Sec. 7.4, including its quantization and the calculation of the propagator. With the exception of Sec. 10.2.1, we shall assume, throughout the current chapter, that ψ denotes the (second-)quantized Dirac field operator rather than the (first-quantized) Dirac wave function [see Eq. (7.140)].

We can split the Lagrangian density (10.1) into a free part $\mathcal{L}_0$ and an interaction part $\mathcal{L}_I$. The free Lagrangian is given as follows,

$$\mathcal{L}_0 = \bar{\psi}\,(\mathrm{i}\,\gamma^{\mu}\partial_{\mu} - m)\,\psi - \frac{1}{4}\,F^{\mu\nu}\,F_{\mu\nu}\,, \tag{10.2}$$

and it is composed of a free electromagnetic field and a free Dirac field (describing the electrons and positrons). The interaction Lagrangian density is

$$\mathcal{L}_I = -e\,\bar{\psi}\,\gamma^{\mu}\,A_{\mu}\,\psi = -J^{\mu}\,A_{\mu}\,, \qquad J^{\mu} = e\,\bar{\psi}\,\gamma^{\mu}\,\psi\,. \tag{10.3}$$

The interaction Lagrangian density is equal to the negative of the QED interaction Hamiltonian (density),

$$\mathcal{H}_I = J^{\mu}\,A_{\mu}\,, \qquad J^{\mu} = e\,\bar{\psi}\,\gamma^{\mu}\,\psi\,. \tag{10.4}$$

Here J^{μ} is the current of the Dirac field. The interaction Hamiltonian density (10.4) has the right physical dimension: The current density has dimension charge per time per area, which is charge times velocity per volume, or, in terms of characteristic quantities, $e\,|\vec{v}|/V$. Then, the curl of $\vec{A}$ gives the magnetic induction field $\vec{B}$. The

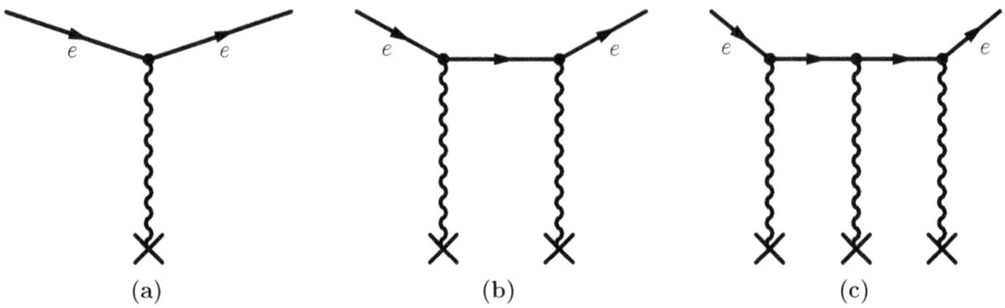

Fig. 10.1 Illustration of Rutherford scattering of an electron on an atomic nucleus, showing the single scattering term (a), the double-scattering term (b), and the triple-scattering event (c) from Eq. (10.6).

dimension of $\vec{A}$ is length times the dimension of $\vec{B}$. Let us denote the mass dimension of a physical quantity X as $[X]$. We find that the dimension of $J^\mu A_\mu$, in terms of characteristic quantities is

$$\left[\frac{e|\vec{v}|}{V}\,\ell\,|\vec{B}|\right] = \left[\frac{(e|\vec{v}|\,|\vec{B}|)\,\ell}{V}\right] = \left[\frac{|\vec{F}|\,\ell}{V}\right] = \left[\frac{E}{V}\right], \tag{10.5}$$

which is the right dimension for an energy density. Here, $|\vec{F}|$ is the Lorentz force, and E denotes a characteristic energy.

Finally, the S matrix formalism allows us to establish a connection of the quantized fields with dynamical processes and transition amplitudes; it has been introduced in Sec. 9.6.2. In Feynman diagrams with virtual loops, we shall encounter divergences which need to be regularized and renormalized. In the context of problems encountered in classical physics, these procedures have been introduced at other places in the literature; the reader is referred to Ref. [307] and Chap. 8 of Ref. [81]. In Chap. 8 of Ref. [81], the transition to quantum field theory has been anticipated on the example of the Casimir effect (i.e., the interaction between two perfectly conducting plates). Now we use all of the wisdom acquired previously, in order to calculate a number of basic processes in QED.

We start with a calculation of a tree-level process in Sec. 10.2. Our focus is on Rutherford scattering (see Sec. 10.2.1), which is the scattering of an electron in a central (Coulomb) field. We then proceed to the calculation of the one-loop vertex correction in QED, in Sec. 10.3, in both PAULI–VILLARS[1] and dimensional regularization, and supplement a calculation of the vacuum-polarization tensor in Sec. 10.4. The latter calculation is carried out in dimensional regularization. A brief discussion of the self-energy operator follows in Sec. 10.5, and the renormalization of QED is being discussed in Sec. 10.6.

[1]Felix Villars (1921–2002).

10.2 Tree-Level

10.2.1 *Rutherford Scattering*

We have already encountered the S matrix in Secs. 9.6.1 and 9.6.2 as the unitary transformation of the *in* to the *out* states in a scattering process. Here, we shall treat a rather basic problem on the basis of the S matrix, namely, the scattering of an incoming electron, described as a plane-wave, by a central Coulomb potential, and we shall do so including all relativistic corrections. The wisdom learned in the concrete problem will suggest the Feynman rules which describe general terms in the scattering matrix and in the corrections to propagators, which will be explored in the following. But first, we explore Rutherford scattering (scattering off of an atomic nucleus) at tree-level.

In order to understand the diagram in Fig. 10.1, we first take notice of the operator identity

$$\frac{\mathrm{i}}{X-Y} = \frac{1}{X} + \frac{\mathrm{i}}{X}(-\mathrm{i}Y)\frac{\mathrm{i}}{X} + \frac{\mathrm{i}}{X}(-\mathrm{i}Y)\frac{\mathrm{i}}{X}(-\mathrm{i}Y)\frac{\mathrm{i}}{X} + \frac{\mathrm{i}}{X}(-\mathrm{i}Y)\frac{\mathrm{i}}{X}(-\mathrm{i}Y)\frac{\mathrm{i}}{X}(-\mathrm{i}Y)\frac{\mathrm{i}}{X} + \cdots. \tag{10.6}$$

For Rutherford scattering, we set

$$X = \not{p} - m + \mathrm{i}\epsilon\,, \quad Y = e\not{A}, \quad \mathrm{i}S_F(p) = \frac{\mathrm{i}}{\not{p} - m + \mathrm{i}\epsilon}\,. \tag{10.7}$$

The formula for $S_F(p)$ is taken from Eq. (7.154).

In the spirit of Sec. 7.4.2 [see Eq. (7.174b)], the Feynman propagator in the presence of the external four-vector potential A^μ can be expressed as follows

$$\mathrm{i}\widetilde{S}_F = \frac{\mathrm{i}}{\not{p} - e\not{A} - m + \mathrm{i}\epsilon}\,. \tag{10.8}$$

An expansion on the basis of Eq. (10.6) yields

$$\begin{aligned}\mathrm{i}\widetilde{S}_F &= \frac{\mathrm{i}}{\not{p} - e\not{A} - m + \mathrm{i}\epsilon} = \frac{\mathrm{i}}{\not{p} - m + \mathrm{i}\epsilon} + \frac{\mathrm{i}}{\not{p} - m + \mathrm{i}\epsilon}\left(-\mathrm{i}e\not{A}\right)\frac{\mathrm{i}}{\not{p} - m + \mathrm{i}\epsilon} + \ldots \\ &= (\mathrm{i}S_F) + (\mathrm{i}S_F)\left(-\mathrm{i}e\not{A}\right)(\mathrm{i}S_F) + \ldots\,. \end{aligned} \tag{10.9}$$

The S-matrix element for the transition $|i\rangle \to |f\rangle$ is evaluated in the limit of $t' \to \infty$, as an overlap integral of the time-evolved state $|\widetilde{\psi}_i(t)\rangle$ with the final state $|\widetilde{\psi}_f(t)\rangle$,

$$S_{fi} = \lim_{t\to\infty} \int \mathrm{d}^3 r\, \widetilde{\psi}_f^+(t,\vec{r})\, \widetilde{\psi}_i(t,\vec{r})\,. \tag{10.10}$$

In the following, we will write expressions where it is understood that the limit $t \to \infty$ should be taken at the end of the calculation. The initial state $\widetilde{\psi}_i(t,\vec{r})$ at time $t \to \infty$, i.e., propagated from the past to the time t, is the result of a propagation from another time t', in the infinite past, via the Green function $\widetilde{G}_F = \mathrm{i}\widetilde{S}_F\,\gamma^0$.

We take

$$\widetilde{\psi}_f^+(t,\vec{r}) \to \psi_f^+(t,\vec{r}) \tag{10.11}$$

to be equal to an eigenstate of the free Hamiltonian, at the "future" time t, while

$$\widetilde{\psi}_i(t,\vec{r}) \to \int \mathrm{d}^3r' \left(\mathrm{i}\widetilde{S}_F(t,\vec{r},t',\vec{r}')\,\gamma^0\right) \psi_i(t',\vec{r}') \tag{10.12}$$

is a time-evolved state at time t, originating from an eigenstate ψ_i of the free Hamiltonian, via propagation, though, with the help of the full Green function $\widetilde{G}_F = \mathrm{i}\widetilde{S}_F\,\gamma^0$. So, the S matrix element can be interpreted as a probability amplitude for the transition of a defined eigenstate of the free Hamiltonian, under the perturbation, into a different eigenstate of the free Hamiltonian in the infinite future. The notion is that the perturbation will be adiabatically damped in the infinite future and past.

Hence, the S matrix element can be evaluated as

$$\begin{aligned}
S_{fi} &= \lim_{\substack{t\to\infty \\ t'\to-\infty}} \int \mathrm{d}^3r \int \mathrm{d}^3r'\, \psi_f^+(t,\vec{r}) \left(\mathrm{i}\,\widetilde{S}_F(t,\vec{r},t',\vec{r}')\,\gamma^0\right) \psi_i(t',\vec{r}') \\
&= \lim_{\substack{t\to\infty \\ t'\to-\infty}} \int \mathrm{d}^3r \int \mathrm{d}^3r'\, \psi_f^+(t,\vec{r}) \left(\mathrm{i}\,S_F(t,\vec{r},t',\vec{r}')\,\gamma^0\right) \psi_i(t',\vec{r}') \\
&\quad + \int \mathrm{d}^3r \int \mathrm{d}^3r' \int \mathrm{d}^3r'' \int \mathrm{d}t''\, \psi_f^+(t,\vec{r}) \left(\mathrm{i}\,S_F(t,\vec{r},t'',\vec{r}'')\gamma^0\right) \gamma^0 \\
&\quad \times \left(-\mathrm{i}\,e\not{A}(t'',\vec{r}'')\right) \left(\mathrm{i}\,S_F(t'',\vec{r}'',t',\vec{r}')\,\gamma^0\right) \psi_i(t',\vec{r}')\,.
\end{aligned} \tag{10.13}$$

In view of the identity $G_F = \mathrm{i}\,S_F\,\gamma^0$ [see Eq. (7.175a)], which clarifies that the free time-evolution Green function G_F is just equal to $\mathrm{i}\,S_F\,\gamma^0$, we can now replace as follows,

$$\psi_i(t'',\vec{r}'') = \int \mathrm{d}^3r \left(\mathrm{i}S_F(t'',\vec{r}'',t',\vec{r}')\,\gamma^0\right) \psi_i(t',\vec{r}')\,, \tag{10.14a}$$

$$\overline{\psi}_f(t'',\vec{r}'') = \psi_f^+(t'',\vec{r}'')\,\gamma^0 = \int \mathrm{d}^3r'\, \psi_f^+(t,\vec{r}) \left(\mathrm{i}\,S_F(t,\vec{r},t'',\vec{r}'')\gamma^0\right) \gamma^0\,. \tag{10.14b}$$

The S matrix element is given as

$$\begin{aligned}
S_{fi} &= \int \mathrm{d}^3r\, \psi_f^+(t,\vec{r})\, \psi_i(t,\vec{r}) - \mathrm{i} \int \mathrm{d}^3r'' \int \mathrm{d}t''\, \psi_f^+(t'',\vec{r}'')\,\gamma^0\, e\not{A}(t'',\vec{r}'')\, \psi_i(t'',\vec{r}'') \\
&= \delta_{fi} - \mathrm{i} \int \mathrm{d}t \int \mathrm{d}^3r\, \overline{\psi}_f(t,\vec{r})\, e\not{A}(t,\vec{r})\, \psi_i(t,\vec{r})\,,
\end{aligned} \tag{10.15}$$

where in the second step we have evaluated the first matrix element to be equal to δ_{fi}, and replaced $(t'',\vec{r}'') \to (t,\vec{r})$ for the integration variable in the second. Our *ansatz* for the bispinor solutions is as follows,

$$\psi_i(t,\vec{r}) = \sqrt{\frac{m}{E_i\,V}}\,\mathcal{U}_\sigma(\vec{k}_i)\,\mathrm{e}^{-\mathrm{i}\,k_i\cdot x}\,, \qquad \overline{\psi}_f(t,\vec{r}) = \sqrt{\frac{m}{E_f\,V}}\,\overline{\mathcal{U}}_\sigma(\vec{k}_f)\,\mathrm{e}^{\mathrm{i}\,k_f\cdot x}\,, \tag{10.16a}$$

where V is the normalization volume. Here, we use the bispinors $\mathcal{U}_\sigma(\vec{k})$, which are normalized according to Eqs. (7.91) and (7.92). Furthermore, we have the four-vectors $k_i^\mu = (k_i^0, \vec{k}_i)$ and $k_f = (k_f^0, \vec{k}_f)$ with both energies k_i^0 and k_f^0 positive quantities, and $k_i^0 > m$, $k_f^0 > m$. For arbitrary t, these solutions are thus normalized according to

$$\begin{aligned}\int \mathrm{d}^3r\, \psi_f^+(t,\vec{r})\, \psi_i(t,\vec{r}) &= \int \mathrm{d}^3r\, \overline{\psi}_f(t,\vec{r})\, \gamma^0\, \psi_i(t,\vec{r}) \\ &= \sqrt{\frac{m}{E_f V}}\sqrt{\frac{m}{E_i V}} \int \mathrm{d}^3r\, \mathcal{U}_{\sigma_f}^+(\vec{k}_f)\mathcal{U}_{\sigma_i}(\vec{k}_i)\, \mathrm{e}^{-\mathrm{i}\,(k_i^0-k_f^0)t}\, \mathrm{e}^{\mathrm{i}\,(\vec{k}_i-\vec{k}_f)\cdot\vec{r}} \\ &= \delta_{\vec{k}_i\vec{k}_f}\, \delta_{\sigma_i\,\sigma_f} \equiv \delta_{fi}\,. \end{aligned} \tag{10.16b}$$

We use the fact that, in our finite normalization volume, we have $\int \mathrm{d}^3r = V$. This explains the formula for the first term in Eq. (10.15). So, for $e\not{A} = -\gamma^0\, Z\alpha/r$, which describes the Coulomb potential,

$$\begin{aligned} S_{fi} - \delta_{fi} &= -\mathrm{i}\int \mathrm{d}t \int \mathrm{d}^3r\, \overline{\psi}_f(t,\vec{r})\, e\not{A}(t,\vec{r})\, \psi_i(t,\vec{r}) \\ &= -\mathrm{i}\int \mathrm{d}t \int \mathrm{d}^3r \sqrt{\frac{m}{E_f V}}\, \overline{\mathcal{U}}_{\sigma_f}(\vec{k}_f)\, \mathrm{e}^{\mathrm{i}\,k_f\cdot x}\, \gamma^0 \left(-\frac{Z\alpha}{r}\right) \sqrt{\frac{m}{E_i V}}\, \mathcal{U}_{\sigma_i}(\vec{k}_i)\, \mathrm{e}^{-\mathrm{i}\,k_i\cdot x} \\ &= \mathrm{i}\, \frac{m}{V\sqrt{E_f E_i}} \left(\overline{\mathcal{U}}_{\sigma_f}(\vec{k}_f)\, \gamma^0\, \mathcal{U}_{\sigma_i}(\vec{k}_i)\right) \int \mathrm{d}t\, \mathrm{e}^{\mathrm{i}\,(E_f-E_i)t} \int \mathrm{d}^3r\, \mathrm{e}^{\mathrm{i}\,(\vec{k}_f-\vec{k}_i)\cdot\vec{r}}\, \frac{Z\alpha}{r}\,. \end{aligned} \tag{10.17}$$

The time and space integrations for the Fourier integrals lead to

$$\int \mathrm{d}t\, \mathrm{e}^{\mathrm{i}\,(E_f-E_i)t} = 2\pi\, \delta(E_f - E_i)\,, \qquad \int \mathrm{d}^3r\, \mathrm{e}^{\mathrm{i}\,(\vec{k}_f-\vec{k}_i)\cdot\vec{r}}\, \frac{Z\alpha}{r} = \frac{4\pi(Z\alpha)}{(\vec{k}_f - \vec{k}_i)^2}\,, \tag{10.18}$$

so that

$$S_{fi} - \delta_{fi} = \mathrm{i}\, \frac{m}{V\sqrt{E_f E_i}} \left(\overline{\mathcal{U}}_{\sigma_f}(\vec{k}_f)\, \gamma^0\, \mathcal{U}_{\sigma_i}(\vec{k}_i)\right) 2\pi\, \delta(E_f - E_i)\, \frac{4\pi(Z\alpha)}{(\vec{k}_f - \vec{k}_i)^2}\,. \tag{10.19}$$

For the cross section, one has to sum over the number of available final states, and normalize to the flux of incoming particles. According to Eq. (3.20), the number of available final states is given as

$$\mathrm{d}n_f = \frac{V\, \mathrm{d}^3k_f}{(2\pi)^3}\,, \tag{10.20}$$

in the infinitesimal momentum element d^3k_f. In order to match the result with Fermi's golden rule, we start from Eq. (3.10) and set $\hbar = 1$, to obtain the time-dependent probability

$$\begin{aligned} P_f(t) &= \left|\langle\psi_f|H_1|\psi_i\rangle\right|^2 \left(4\frac{\sin^2[(E_f - E_i)\,t/2]}{(E_f - E_i)^2}\right) \\ &\to \left|\langle\psi_f|H_1|\psi_i\rangle\right|^2\, 2\pi\, \delta(E_f - E_i)\, T\,, \qquad t \to T\,. \end{aligned} \tag{10.21}$$

We have used Eq. (3.14). Here, T is the large time over which the process is being observed. Summing over the number $\mathrm{d}n_f$ of available final states, we have for the transition rate $\mathrm{d}W_{fi}$,

$$\mathrm{d}W_{fi} = \frac{P_f(T)}{T}\, \mathrm{d}n_f = \left|\langle\psi_f|H_1|\psi_i\rangle\right|^2\, 2\pi\, \delta(E_f - E_i)\, \mathrm{d}n_f\,. \tag{10.22}$$

Here, it is understood that the integration over the infinitesimal energy interval about the final state needs to be performed in order to catch the possible outcomes of the experiment. Let us try to define a differential cross-section $\mathrm{d}\sigma$ by postulating that the transition rate $i \to f$ should be obtained as the product of the incoming particle current density and the cross-section, as follows,

$$\mathrm{d}W_{fi} = J_{\mathrm{in}}\,\mathrm{d}\sigma = \frac{v_z}{V}\,\mathrm{d}\sigma\,, \qquad J_{\mathrm{in}} = \frac{\mathrm{d}n_{\mathrm{in}}}{\mathrm{d}t\,\mathrm{d}A} = \frac{v_z}{V}\,. \tag{10.23}$$

Here, $\mathrm{d}n_{\mathrm{in}}$ is the number of particles that pass into the scattering area, per time interval $\mathrm{d}t$ and cross-sectional area $\mathrm{d}A$ (if normalized to one particle in volume V). One can solve Eqs. (10.22) and (10.23) for $\mathrm{d}\sigma$,

$$\mathrm{d}\sigma = \frac{V}{v_z}\,|\langle\psi_f|H_1|\psi_i\rangle|^2\; 2\pi\,\delta(E_f - E_i)\,\mathrm{d}n_f\,. \tag{10.24}$$

We now have to match the matrix element $\langle\psi_f|H_1|\psi_i\rangle$ in Eq. (10.22) against the S matrix element $S_{fi} - \delta_{fi}$ given in Eq. (10.19).

In order to perform the matching, we go back to Eq. (10.15) and write

$$\begin{aligned} S_{fi} - \delta_{fi} &= -\mathrm{i}\int \mathrm{d}t \int \mathrm{d}^3r\, \overline{\psi}_f(t,\vec{r})\, e\not{A}(t,\vec{r})\, \psi_i(t,\vec{r}) \\ &= -\mathrm{i}\int \mathrm{d}t \int \mathrm{d}^3r\, \psi_f^{+}(\vec{r})\, \left[e\,A^0(\vec{r}) - e\,\vec{\alpha}\cdot\vec{A}(\vec{r})\right]\, \psi_i(\vec{r}) \\ &= -\mathrm{i}\,T \int \mathrm{d}^3r\, \psi_f^{+}(\vec{r})\, \left[e\,A^0(\vec{r}) - e\,\vec{\alpha}\cdot\vec{A}(\vec{r})\right]\, \psi_i(\vec{r})\,. \end{aligned} \tag{10.25}$$

Here, we have set $\int \mathrm{d}t = T$, assumed that all components of the four-vector potential are time-independent, and used the fact that the matrix element is nonvanishing only if the states ψ_i and ψ_f have the same energy. The interaction Hamiltonian, under these assumptions, reads as

$$H_1(\vec{r}) = e\,A^0(\vec{r}) - e\,\vec{\alpha}\cdot\vec{A}(\vec{r})\,. \tag{10.26}$$

We can match

$$S_{fi} - \delta_{fi} = -\mathrm{i}\,T\;\langle\psi_f\,|H_1|\,\psi_i\rangle\,. \tag{10.27}$$

The Dirac-δ in the energy, for $E_f \to E_i$, can be expressed as

$$2\pi\delta(E_f - E_i) = \int_{-T/2}^{T/2} \mathrm{d}t\,\mathrm{e}^{\mathrm{i}(E_f - E_i)t} \qquad \Rightarrow \qquad T = 2\pi\delta(0)\,. \tag{10.28}$$

We can thus match, for $E_f \to E_i$, in Eq. (10.24),

$$|\langle\psi_f|H_1|\psi_i\rangle|^2\; 2\pi\,\delta(E_f - E_i) = |\langle\psi_f|H_1|\psi_i\rangle|^2\; T = \frac{1}{T}\,|S_{fi} - \delta_{fi}|^2\,. \tag{10.29}$$

We now continue the discussion of the differential cross-section, based on Eqs. (10.19), (10.20), and (10.23), to obtain

$$\begin{aligned}
\mathrm{d}\sigma &= \frac{V}{v_z}\left(\frac{1}{T}\,|S_{fi}-\delta_{fi}|^2\right)\mathrm{d}n_f \\
&= \frac{m^2}{V^2\,(E_f\,E_i)}\left|\overline{\mathcal{U}}_{\sigma_f}(\vec{k}_f)\,\gamma^0\,\mathcal{U}_{\sigma_i}(\vec{k}_i)\right|^2\left[\delta(E_f-E_i)\,\frac{T}{2\pi}\right]\frac{64\pi^4(Z\alpha)^2}{(\vec{k}_f-\vec{k}_i)^4} \\
&\quad\times\frac{1}{T\,v_z/V}\;\frac{V\,k_f^2\,\mathrm{d}k_f\,\mathrm{d}\Omega_f}{(2\pi)^3} \\
&= \frac{m^2}{(E_f\,E_i)}\left|\overline{\mathcal{U}}_{\sigma_f}(\vec{k}_f)\,\gamma^0\,\mathcal{U}_{\sigma_i}(\vec{k}_i)\right|^2\left[\delta(E_f-E_i)\,\frac{1}{2\pi}\right]\frac{64\pi^4(Z\alpha)^2}{(\vec{k}_f-\vec{k}_i)^4}\,\frac{1}{v_z}\,\frac{k_f^2\,\mathrm{d}k_f\,\mathrm{d}\Omega_f}{(2\pi)^3} \\
&= \frac{m^2}{(E_f\,E_i)}\left|\overline{\mathcal{U}}_{\sigma_f}(\vec{k}_f)\,\gamma^0\,\mathcal{U}_{\sigma_i}(\vec{k}_i)\right|^2\delta(E_f-E_i)\,\frac{4\,(Z\alpha)^2}{(\vec{k}_f-\vec{k}_i)^4}\,\frac{1}{v_z}\,k_f^2\,\mathrm{d}k_f\,\mathrm{d}\Omega_f\,. \qquad (10.30)
\end{aligned}$$

The differential cross-section thus is

$$\frac{\mathrm{d}\sigma}{\mathrm{d}\Omega_f} = \left|\overline{\mathcal{U}}_{\sigma_f}(\vec{k}_f)\,\gamma^0\,\mathcal{U}_{\sigma_i}(\vec{k}_i)\right|^2\frac{4\,(Z\alpha)^2\,m^2}{(\vec{k}_f-\vec{k}_i)^4}\,\frac{\delta(E_f-E_i)\,k_f^2\,\mathrm{d}k_f}{E_f\,E_i\,v_z}\,. \qquad (10.31)$$

We now have a singular Dirac-δ peak in this expression and an infinitesimal integration interval $\mathrm{d}k_f$ for the integration over the final state momentum interval. From

$$E_f^2 = k_f^2 + m^2 \qquad\Rightarrow\qquad E_f\,\mathrm{d}E_f = k_f\,\mathrm{d}k_f\,, \qquad (10.32)$$

we have

$$\begin{aligned}
\frac{\delta(E_f-E_i)\,k_f^2\,\mathrm{d}k_f}{E_f\,E_i\,v_z} &= \frac{\delta(E_f-E_i)\,k_f\,E_f\,\mathrm{d}E_f}{E_f\,E_i\,v_z} = \frac{\delta(E_f-E_i)\,k_f\,\mathrm{d}E_f}{E_i\,v_z} \\
&= \frac{\delta(E_f-E_i)\,(m\,\gamma\,v_z)\,\mathrm{d}E_f}{(m\,\gamma)\,v_z} = \delta(E_f-E_i)\,\mathrm{d}E_f\,. \qquad (10.33)
\end{aligned}$$

We have used energy conservation here, as manifest in the peak of the Dirac-δ function $\delta(E_f - E_i)$, and the formulas $E_i = m\,\gamma$ and $k_f = k_i = m\,\gamma\,v_z$. After integration over the infinitesimal energy interval characterizing the final state, which can be done by the replacement $\delta(E_f - E_i)\,\mathrm{d}E_f \to 1$, the differential cross-section becomes

$$\frac{\mathrm{d}\sigma}{\mathrm{d}\Omega_f} = \left|\overline{\mathcal{U}}_{\sigma_f}(\vec{k}_f)\,\gamma^0\,\mathcal{U}_{\sigma_i}(\vec{k}_i)\right|^2\frac{4\,(Z\alpha)^2\,m^2}{(\vec{k}_f-\vec{k}_i)^4}\,. \qquad (10.34)$$

This cross-section, as written down, is spin-resolved. Let us continue with a general consideration. With Γ being a structure of Dirac matrices, we have

$$\begin{aligned}
\left(\overline{\mathcal{U}}_{\sigma_f}(\vec{k}_f)\,\Gamma\,\mathcal{U}_{\sigma_i}(\vec{k}_i)\right)^* &= \mathcal{U}^+_{\sigma_i}(\vec{k}_i)\,\Gamma^+\,\gamma^0\,\mathcal{U}_{\sigma_f}(\vec{k}_f) \\
&= \mathcal{U}^+_{\sigma_i}(\vec{k}_i)\,\gamma^0\,\left(\gamma^0\,\Gamma^+\,\gamma^0\right)\mathcal{U}_{\sigma_f}(\vec{k}_f) = \overline{\mathcal{U}}_{\sigma_i}(\vec{k}_i)\,\widetilde{\Gamma}\,\mathcal{U}_{\sigma_f}(\vec{k}_f) \qquad (10.35)
\end{aligned}$$

where $\widetilde{\Gamma} = \gamma^0 \, \Gamma^+ \, \gamma^0$. Here, the relation $\gamma^0 \, (\gamma^\mu)^+ \, \gamma^0 = \gamma^\mu$ has been used in order to establish that in $\widetilde{\Gamma}$, the Dirac matrices retain their meaning, but their sequence is reversed. The spin sum can be carried out using formula (7.93),

$$\sum_{\sigma_i} \mathcal{U}_{\sigma_i}(\vec{k}_i) \otimes \overline{\mathcal{U}}_{\sigma_i}(\vec{k}_i) = \frac{\not{k}_i + m}{2m}, \tag{10.36}$$

and analogously for σ_f and k_f. Finally, the summation over outgoing and averaging over incoming spins leads to the following projectors,

$$\frac{1}{2} \sum_{s_i \, s_f} \left(\overline{\mathcal{U}}_{\sigma_f}(\vec{k}_f) \, \Gamma \, \mathcal{U}_{\sigma_i}(\vec{k}_i) \right)^* \left(\overline{\mathcal{U}}_{\sigma_f}(\vec{k}_f) \, \Gamma \, \mathcal{U}_{\sigma_i}(\vec{k}_i) \right) = \frac{1}{2} \, \mathrm{Tr} \left[\frac{\not{k}_i + m}{2m} \, \widetilde{\Gamma} \, \frac{\not{k}_f + m}{2m} \, \Gamma \right], \tag{10.37}$$

where in our case simply $\Gamma = \widetilde{\Gamma} = \gamma^0$. We have used the fact that $E_i = E_f$. The following general rule will be used,

$$\mathrm{Tr} \left(\gamma^0 \, \not{k}_i \, \gamma^0 \, \not{k}_f \right) = 4 \left(E_i \, E_f + \vec{k}_i \cdot \vec{k}_f \right) . \tag{10.38}$$

Using $|\vec{k}_f - \vec{k}_i| = |\vec{q}| = 2 \, |\vec{k}| \, \sin(\frac{1}{2} \, \theta)$, and $E = E_i = E_f$, we have after some additional simplification:

$$\begin{aligned} \frac{\mathrm{d}\sigma}{\mathrm{d}\Omega_f} &= \frac{4 \, (Z\alpha)^2 \, m^2}{(\vec{k}_f - \vec{k}_i)^4} \left\{ \frac{1}{2} \, \mathrm{Tr} \left[\gamma^0 \left(\frac{\not{k}_i + m}{2m} \right) \gamma^0 \left(\frac{\not{k}_f + m}{2m} \right) \right] \right\} \\ &= \frac{4 \, (Z\alpha)^2 \, m^2}{\left[2 \, |\vec{k}| \, \sin(\frac{1}{2} \, \theta) \right]^4} \left\{ \frac{1}{2m^2} \left[E^2 + \vec{k}^{\,2} \left(1 - 2 \sin^2 \left(\tfrac{1}{2} \theta \right) \right) + m^2 \right] \right\} \\ &= \frac{4 \, (Z\alpha)^2}{16 \, |\vec{k}|^4 \, \sin^4(\frac{1}{2} \, \theta)} \left\{ \frac{1}{2} \left[E^2 + \vec{k}^{\,2} + m^2 - 2 \vec{k}^{\,2} \, \sin^2 \left(\tfrac{1}{2} \theta \right) \right] \right\} \\ &= \frac{(Z\alpha)^2}{8 \, |\vec{k}|^4 \, \sin^4(\frac{1}{2} \, \theta)} \left\{ 2 \, E^2 - 2 \, (\beta \, E)^2 \, \sin^2 \left(\tfrac{1}{2} \theta \right) \right\} \\ &= \frac{(Z\alpha)^2 \, E^2}{4 |\vec{k}|^4 \, \sin^4(\frac{1}{2} \, \theta)} \left\{ 1 - \beta^2 \, \sin^2 \left(\tfrac{1}{2} \theta \right) \right\} = \frac{(Z\alpha)^2}{4 \beta^2 \, |\vec{k}|^2 \, \sin^4(\frac{1}{2} \, \theta)} \left\{ 1 - \beta^2 \, \sin^2 (\tfrac{1}{2} \, \theta) \right\} . \end{aligned} \tag{10.39}$$

Here, we have used the fact that $|\vec{k}| = |\vec{k}_i| = |\vec{k}_f| = m \, \gamma \, v_z = m \, \gamma \, \beta = E \, \beta$, where $\beta = v_z = v$ is the velocity (divided by the speed of light c, which we set equal to unity). It is instructive to recall what measures had to be taken in order to convert the S matrix, or scattering matrix (transition matrix) element into a differential cross-section. We first obtained the S-matrix element as a physically dimensionless transition matrix element between incoming and outgoing states, after subtracting the forward scattering term δ_{fi}. We then identified the cross-section using the formula $\mathrm{d}W_{fi} = J_{\mathrm{in}} \, \mathrm{d}\sigma$, where J_{in} is the incoming particle current density, and related $\mathrm{d}W_{fi}$ to the S matrix element and the number of available outgoing states, by Fermi's golden rule. In the square of the S matrix element, we treated the square of the Dirac-δ function according to $\left[\delta(E_f - E_i) \right]^2 \to \delta(E_f - E_i) \, T/(2\pi)$, where T is the (long) observation time. Finally, an integration over the available energy interval according to $\delta(E_i - E_f) \, \mathrm{d}E_f \to 1$ counts the available final states.

10.2.2 *Feynman Rules*

From the previous discussion of Rutherford scattering, we have already gathered that there is a certain pattern in associating a specific QED process, which can be represented as a diagram, with an S matrix element. Indeed, Feynman rules establish the connection between S matrix elements and mathematical expressions. Yet, we should remember that there is an extra step involved in connecting the S matrix element with a physical quantity of interest. For example, as we have seen, it can be nontrivial to connect an S matrix element with a cross-section; this has been discussed in Sec. 10.2.1. In another case, we have to connect the S matrix element with a correction to a propagator, say, by identifying the vacuum polarization correction to a photon line. This involves a comparison of the contribution to the S matrix due to a tree-level diagram, to a loop diagram, and associating the difference with a correction to the photon propagator. Other calculations, say, of energy shifts, start with an expression for an S matrix element, and then associate the element with an energy shift for a reference state, after the element is evaluated with a specific reference state in mind. Sometimes, one has to adiabatically damp an energy-dependent perturbation, as already anticipated in Sec. 3.3.3 (derivation of the Gell-Mann–Low–Sucher theorem).

Different physical processes can be described by one and the same Feynman diagram, but differ in the details of the initial state considered. As an example, the annihilation diagram of two electrons into three photons can represent the orthopositronium decay width (if the initial state is a bound state of orthopositronium), but also the annihilation of a free electron-positron pair into three photons (the initial state in this case consists of two free fermions). The Feynman rules are prescriptions for the construction of S matrix elements, nothing more.

The *antiparticle* solutions are proportional to $\mathcal{V}(\vec{k})\exp(ik\cdot x)$. Here, we write $\mathcal{V}(\vec{k})$, because the energy $k^0 = E = \sqrt{\vec{p}^{\,2}+m^2}$ is chosen to be positive, and therefore, equal to the physical energy of the antiparticle. In this way, $p = p^\mu$ parameterizes the momentum flow in a diagram, for an outgoing particle. Indeed, in Table 10.1, p^μ is the momentum flow inside the diagram, as one calculates the amplitude. Formally, p^μ is an incoming momentum, even though the physical momentum $\vec{k}$ of the antiparticle moving forward in time is outgoing, with $\vec{p} = -\vec{k}$.

Our p^μ parameter in the antiparticle solution thus parameterizes the negative value of the physical momentum,

$$p = p_{\rm math} = p_{\rm diagram} = -k = -k_{\rm phys} \qquad \text{(antiparticle)}\,. \tag{10.40}$$

The mathematical momentum p of the antiparticle is what enters the momentum conservation of the Feynman diagram, and it is equal to the negative physical momentum k. In Table 10.1, this is manifest in the relations $\mathcal{V}_\sigma(\vec{k}) = \mathcal{V}_\sigma(-\vec{p})$ and $\overline{\mathcal{V}}_\sigma(\vec{k}) = \overline{\mathcal{V}}_\sigma(-\vec{p})$.

In a Feynman diagram, one writes the positron arrow in the "physically wrong" direction. Let the positron be incoming from the past. One describes it, in the diagrammatic language, by an outgoing positron spinor, $\overline{\mathcal{V}}_\sigma(\vec{k})$, which is multiplied by a phase factor $\exp(-\mathrm{i}k \cdot x)$, where $k \cdot x = Et - \vec{k} \cdot \vec{r}$, and k^μ is the physical four-momentum, and one has $k^0 = E = \sqrt{\vec{k}^2 + m^2} > 0$. By contrast, $p^\mu = -k^\mu$ is the four-momentum in the diagrammatic sense, i.e., the momentum flow through the diagram, with $k^\mu = -p^\mu$. Hence, one uses the bispinor $\overline{\mathcal{V}}_\sigma(\vec{k})$, where $\vec{k}$ is the momentum in the "physical" direction, i.e., with the same momentum, and in the same direction, in which one would have expected the positron to move. The correct sign for the time-dependent term in the exponential is determined by the interpretation of the positron as a necessarily positive-energy, incoming particle.

As a rather straightforward generalization of the considerations in Sec. 10.2.1, we may write down the Feynman rules for QED, which express of a correspondence of a mathematical expression and a diagram, as given in Table 10.1.

The imaginary units (factors "i") for the photon and fermion lines can be understood if we consider the time-ordered expectation values of the field operators, which are connected to the propagators by a relation that involves an imaginary unit, as in the relations [see Eq. (9.112)]

$$\langle 0|\mathrm{T}\, A^\mu(x)\, A^\nu(y)|0\rangle = \mathrm{i}\, D_F^{\mu\nu}(x-y) = \mathrm{i}\int \frac{\mathrm{d}^4k}{(2\pi)^4}\, D_F^{\mu\nu}(k)\, \mathrm{e}^{-\mathrm{i}k\cdot(x-y)}\,. \tag{10.41}$$

In Sec. 9.4.3, we have used the photon propagator in the Lorenz gauge, with the Feynman prescription for the poles in the complex energy plane, commensurate with the time ordering of the four-vector potential operators. The photon propagator has been given in Eq. (9.121), in Lorenz gauge, as

$$D_F^{\mu\nu}(k) = -\frac{g^{\mu\nu}}{k^2 + \mathrm{i}\epsilon}\,. \tag{10.42}$$

For the corresponding expressions in the Coulomb gauge, still, with the Feynman prescription for the pole terms, we refer to Eq. (9.73).

For the fermion field, we have [see Eqs. (7.153) and (7.154)],

$$\langle 0|\mathrm{T}\, \psi(x)\,\overline{\psi}(y)|0\rangle = \mathrm{i}\, S_F(x-y) = \mathrm{i}\int \frac{\mathrm{d}^4k}{(2\pi)^4}\, S_F(k)\, \mathrm{e}^{-\mathrm{i}k\cdot(x-y)}\,. \tag{10.43}$$

The Feynman propagator for the fermion field has been given in Eq. (7.154),

$$S_F(k) = \frac{1}{\not{k} - m + \mathrm{i}\epsilon}\,. \tag{10.44}$$

In the incoming positron from the past, the Dirac adjoint in $\overline{\mathcal{V}}_\sigma(\vec{k})$ is understandable because mathematically, it is an outgoing positron into the past. According to the reinterpretation principle, positrons always propagate into the past. The arrow of the momentum points from the future to the past. An incoming positron, according the Feynman's reinterpretation principle which replaces hole theory, is generated by the action of the operator $\overline{\psi}$ on the one-positron state $d_\sigma^+(\vec{k})|0\rangle$, which has been excited from the vacuum. The field operator $\overline{\psi}$ contains a term proportional to $d_\sigma^+(\vec{k})\,\overline{\mathcal{V}}_\sigma(\vec{k})\,\exp(-\mathrm{i}k\cdot x)$. That means that the four-momentum k is continuously flowing in the diagram, along the internal fermion line, from the future. Now, the

Table 10.1 Within the Feynman rules, the spin projection of the bispinors $\mathcal{U}_\sigma(\vec{k})$ and $\mathcal{V}_\sigma(\vec{k})$ is denoted as σ. The three-vector $\vec{k}$ is the physical momentum of the particle or antiparticle, which is completed to a four-vector by adding the timelike component $k^0 = \sqrt{\vec{k}^{\,2} + m^2} > 0$. The mathematical momentum, required for momentum conservation at the vertex, is being denoted as p. The dot on one end of external lines indicates the presence of an internal vertex. The polarization vectors for incoming and outgoing photons are defined according to Eq. (9.91) [see also Eqs. (9.99) and (9.100)]. While the polarization vectors can be chosen as real rather than complex, we indicate the complex conjugate of the polarization vector for an outgoing photon, in order to account for possible manifestly complex choices of the photon polarization vectors.

Incoming electron	$\mathcal{U}_\sigma(\vec{k}) = \mathcal{U}_\sigma(\vec{p})$	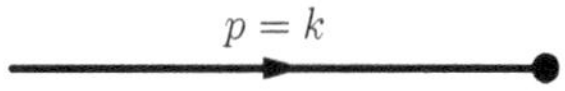
Outgoing electron	$\overline{\mathcal{U}}_\sigma(\vec{k}) = \overline{\mathcal{U}}_\sigma(\vec{p})$	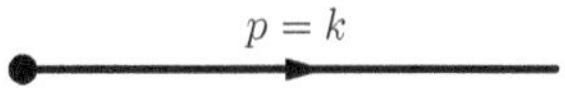
Incoming positron from the past	$\overline{\mathcal{V}}_\sigma(\vec{k}) = \overline{\mathcal{V}}_\sigma(-\vec{p})$	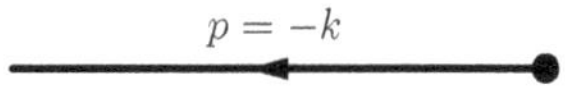
Outgoing positron into the future	$\mathcal{V}_\sigma(\vec{k}) = \mathcal{V}_\sigma(-\vec{p})$	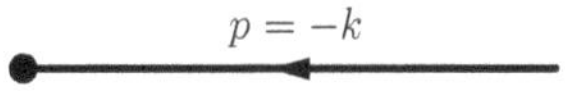
Incoming photon	$\epsilon^\mu_\lambda(\vec{k})$	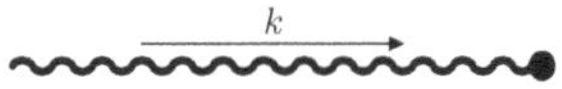
Outgoing photon	$[\epsilon^\mu_\lambda(\vec{k})]^*$	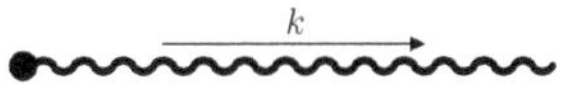
Photon line	$\mathrm{i}D_F^{\mu\nu}(p)$	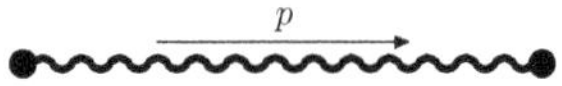
Fermion line	$\mathrm{i}S_F(p)$	
Fermion-photon vertex	$-\mathrm{i}\,e\,\gamma_\mu$	
Outgoing positron	(-1)	
Interchanged fermion field operator	(-1)	
Closed fermion loop	(-1)	

energy component of this four-vector is $k^0 = \sqrt{\vec{k}^{\,2} + m^2} > 0$. So, provided we draw the arrows for the momenta correctly in a Feynman diagram, we have to represent the position state by the bispinor $\overline{\mathcal{V}}_\sigma(\vec{p})$, where $\vec{p} = -\vec{k}$ is the momentum in the diagram flow, which is opposite to the physical momentum $\vec{k}$ of the incoming positron.

There are further overall phase factors to consider for Feynman diagram calculations. These are as follows,

- (-1) for an outgoing positron,
- (-1) for fermion field operator interchange,
- (-1) for a closed fermion loop.

The (-1) for the outgoing positron line is understandable because the outgoing positron reverses the sequence of the field operators $\psi(x)$ and $\overline{\psi}(x)$ in an internal line leading to the outgoing positron. That is to say, we observe that an outgoing positron is described by an expression $\langle 0|d_\sigma(k)$ in a diagram; internal lines are generated by instances of the interaction Lagrangian $-e\overline{\psi}(x)\,\gamma^\mu\,\psi(x)\,A_\mu(x)$. However, the positron annihilation operator $d_\sigma(k)$ which pairs off with the positron creation operator $d^+_\sigma(k)$, is to be found in $\psi(x)$, not $\overline{\psi}(x)$. Because of the negative sign in the time-ordered product (7.145), the required "flipping" of the sequence of fermionic field operators implies the presence of an additional minus sign, which is incurred for an outgoing positron. Once the sign flipping is done, no additional minus sign is incurred for incoming positrons. For a closed fermion loop, we can understand the minus sign by the relative minus sign between the two terms in time-ordered product (7.145). Namely, working in space and time coordinates, we see that a loop can only be closed if the time ordering of the last pair in the sequence $x_1 \to x_2 \to \ldots \to x_n \to x_1$ is inverted: $(x_n, x_1) \to (x_1, x_n)$. Again, considering the minus sign in Eq. (7.145), this leads to a factor (-1).

The factor (-1) for fermion field operator interchange is sometimes not obvious just by inspecting Feynman diagrams. The presence of the factor is clear when, say, it pertains to the interchange of fermions in the out channel of, e.g., Møller scattering $(e^- e^- \to e^- e^-)$, where it corresponds to an antisymmetrization of the wave function of the outgoing fermions. However, for Bhabha scattering $(e^+ e^- \to e^+ e^-)$, there is also a relative minus sign between the so-called t channel (photon exchange) and the so-called s channel (annihilation diagram). The presence of this latter relative minus sign in Bhabha scattering is less obvious and requires a careful consideration of the assignment of incoming and outgoing momenta, which is left as an exercise to the reader.

A coherent (amplitude) sum is formed over all diagrams which contribute to a given process in a given order of perturbation theory. This gives the S matrix element. The modulus square is then formed once the coherent sum over all amplitudes is taken.

10.3 Vertex Correction

10.3.1 *Vertex and Pauli–Villars Regularization*

In principle, the Feynman rules provide us with the necessary tools to translate diagrams into formulas. The vertex correction is shown in Fig. 10.2. The Feynman

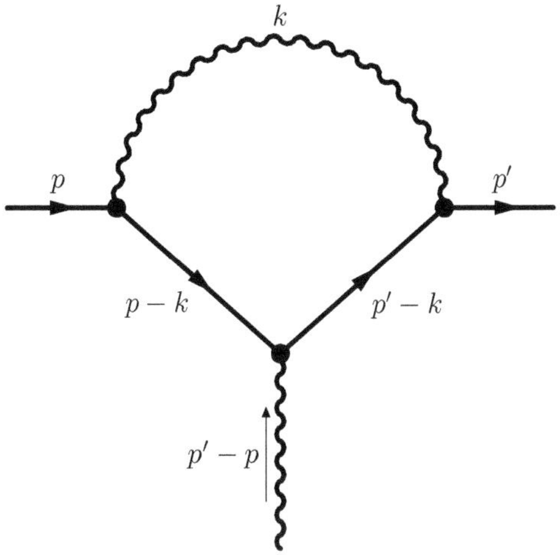

Fig. 10.2 Feynman diagram for the vertex correction with momentum flow.

rules, rather than being used in order to evaluate a cross section, here lead to a one-loop correction to the electron-photon vertex. One compares the simple factor $-\mathrm{i}e\gamma^\mu$, which is usually associated to a vertex (see Table 10.1), to the one-loop correction, which depends on the momenta p and p'. The replacement reads as follows,

$$-\mathrm{i}e\gamma^\nu \to -\mathrm{i}e\gamma^\nu - \mathrm{i}e\Lambda^\nu(p,p'), \tag{10.45a}$$

$$\begin{aligned}-\mathrm{i}e\Lambda^\nu(p,p') = \int \frac{\mathrm{d}^4k}{(2\pi)^4}\left(\mathrm{i}D_F^{\mu\rho}(k)\right)(-\mathrm{i}e\gamma_\mu) \\ \times\left(\mathrm{i}S_F(p'-k)\right)(-\mathrm{i}e\gamma^\nu)\left(\mathrm{i}S_F(p-k)\right)(-\mathrm{i}e\gamma_\rho).\end{aligned} \tag{10.45b}$$

The correction Λ^μ to the vertex factor Γ^μ, which needs to be added to the Dirac matrix γ^μ, is defined by

$$\Gamma^\nu(p,p') = \gamma^\nu + \Lambda^\nu(p,p'). \tag{10.46}$$

Using the expression in Eq. (10.44), the correction can be expressed as

$$\begin{aligned}\Lambda^\nu(p,p') &= \mathrm{i}e^2\int\frac{\mathrm{d}^4k}{(2\pi)^4}D_F^{\mu\rho}(k)\,\gamma_\mu\frac{1}{\not{p}'-\not{k}-m+\mathrm{i}\epsilon}\gamma^\nu\frac{1}{\not{p}-\not{k}-m+\mathrm{i}\epsilon}\gamma_\rho \\ &= \mathrm{i}e^2\int\frac{\mathrm{d}^4k}{(2\pi)^4}\,D_F^{\mu\rho}(k)\,\frac{\gamma_\mu(\not{p}'-\not{k}+m)\,\gamma^\nu(\not{p}-\not{k}+m)\,\gamma_\rho}{[(p'-k)^2-m^2+\mathrm{i}\epsilon]\,[(p-k)^2-m^2+\mathrm{i}\epsilon]}.\end{aligned} \tag{10.47}$$

In typical loop calculations, it is convenient to use the Feynman gauge,

$$D_F^{\mu\rho}(k) = -\frac{g^{\mu\rho}}{k^2+\mathrm{i}\epsilon}. \tag{10.48}$$

For $|k| \to \infty$, the integral diverges logarithmically. We can establish this observation by simply counting the powers of k in the Euclidean region where we can replace any k by its modulus $|k|$, for power counting arguments. This means that the integral (10.47) diverges logarithmically at large momenta.

The usual trick to still be able to evaluate such an integral is to introduce a regularization, i.e., we use an additional parameter (in this case denoted as Λ) which cuts off the divergence at large momenta. The integral no longer diverges, and we can take the limit $\Lambda \to \infty$ at the end of the calculation (i.e., after regularization and renormalization, see Sec. 8.4 of Ref. [81]). In Pauli–Villars regularization, we subtract from the photon propagator a term describing a photon with a very large mass,

$$D_F^{\mu\rho}(k) \to -\frac{g^{\mu\rho}}{k^2 + \mathrm{i}\epsilon} + \frac{g^{\mu\rho}}{k^2 - \Lambda^2 + \mathrm{i}\epsilon}. \tag{10.49}$$

Anticipating potential infrared divergences, we also introduce an infrared regulator (infinitesimal photon mass) λ as follows,

$$D_F^{\mu\rho}(k) \to -\frac{g^{\mu\rho}}{k^2 - \lambda^2 + \mathrm{i}\epsilon} + \frac{g^{\mu\rho}}{k^2 - \Lambda^2 + \mathrm{i}\epsilon} = \int_{\lambda^2}^{\Lambda^2} \mathrm{d}L \frac{g^{\mu\rho}}{(k^2 - L + \mathrm{i}\epsilon)^2} \tag{10.50}$$

where the limits $\lambda \to 0$ and $\Lambda \to \infty$ are tacitly assumed to be taken at the end of the calculation. The loop correction to the vertex can be written as

$$\Lambda^\nu(p,p') = 4\pi\mathrm{i}\alpha \int \frac{\mathrm{d}^4k}{(2\pi)^4} \int_{\lambda^2}^{\Lambda^2} \frac{\mathrm{d}L}{(k^2 - L + \mathrm{i}\epsilon)^2} \frac{\gamma^\mu(\not{p}' - \not{k} + m)\gamma^\nu(\not{p} - \not{k} + m)\gamma_\mu}{[(p'-k)^2 - m^2 + \mathrm{i}\epsilon][(p-k)^2 - m^2 + \mathrm{i}\epsilon]}, \tag{10.51}$$

where we have used the relation $e^2 = 4\pi\alpha$. For practical reasons, we focus on the matrix element of the vertex correction $\Lambda^\nu(p,p')$ between on-shell (OS) bispinors, which we write as

$$\Lambda^\nu_{\mathrm{OS}}(p,p') = \Lambda^\nu(p,p')|_{\not{p}=\not{p}'=m}, \tag{10.52}$$

where $\Lambda^\nu_{\mathrm{OS}}(p,p')$ is the on-shell vertex correction. In the evaluation of $\Lambda^\nu_{\mathrm{OS}}(p,p')$, we can replace any occurrence of a p^2 or p'^2 with the mass square, m^2. Also, because $\mathcal{U}_\sigma(\vec{p})$ fulfills the Dirac equation, we have $\not{p}\,\mathcal{U}_\sigma(\vec{p}) = m\mathcal{U}_\sigma(\vec{p})$. In the same way $\overline{\mathcal{U}}_{\sigma'}(\vec{p}')$ has to fulfill the conjugate Dirac equation, $\overline{\mathcal{U}}_{\sigma'}(\vec{p}')\not{p}' = \overline{\mathcal{U}}_{\sigma'}\, m$. It is understood that whenever $\not{p}$ in the numerator of $\Lambda^\nu_{\mathrm{OS}}(p,p')$ is brought to the right, or $\not{p}'$ is brought to the left, it can be replaced by the mass m, implementing the OS (on shell) condition. Using the on-shell condition, we can simplify the denominator and write

$$\Lambda^\nu_{\mathrm{OS}}(p,p') = 4\pi\mathrm{i}\alpha \int \frac{\mathrm{d}^4k}{(2\pi)^4} \int_{\lambda^2}^{\Lambda^2} \frac{\mathrm{d}L}{(k^2 - L + \mathrm{i}\epsilon)^2} \times \left. \frac{\gamma^\mu(\not{p}' - \not{k} + m)\gamma^\nu(\not{p} - \not{k} + m)\gamma_\mu}{[k^2 - 2p'\cdot k + \mathrm{i}\epsilon][k^2 - 2p\cdot k + \mathrm{i}\epsilon]}\right|_{\not{p}=\not{p}'=m}. \tag{10.53}$$

As a next step, we consider the algebra of Dirac matrices and recall that the anticommutation relations read as $\{\gamma^\mu, \gamma^\nu\} = 2g^{\mu\nu}$ with a diagonal flat-space metric $g^{\mu\nu} = \mathrm{diag}(1,-,1,-1,-1)$. In view of the Dirac matrices in Eq. (10.53), we now have to practice a little Dirac algebra. The first, rather trivial, relation reads as

$$\gamma^\mu\gamma_\mu = g_{\mu\nu}\gamma^\mu\gamma^\nu = \frac{1}{2}g_{\mu\nu}\{\gamma^\mu,\gamma^\nu\} = g_{\mu\nu}g^{\mu\nu} = d. \tag{10.54}$$

We here take the opportunity to supplement a generalization of this formula, valid for d space-time dimensions, which we evaluate, for later purposes, in d space-time dimensions. This result will become useful later, in the context of dimensional regularization, where one usually sets $d = 4 - 2\epsilon$, i.e., the number of dimensions is slightly lower than four. Using simple commutator relations, it is then easy to find formulas for slightly more complex Dirac structures,

$$\gamma^\mu \not{a} \gamma_\mu = \gamma^\mu \gamma^\nu \gamma_\mu a_\nu = -\gamma^\nu \gamma^\mu \gamma_\mu a_\nu + 2g^{\mu\nu} \gamma_\mu a_\nu = -d\gamma^\nu a_\nu + 2\gamma^\nu a_\nu = (2-d)\not{a} . \quad (10.55)$$

For two Feynman slashes sandwiched between Dirac γ^μ matrices, we find

$$\begin{aligned}\gamma^\mu \not{a}\not{b} \gamma_\mu &= \gamma^\mu \gamma^\nu \gamma^\kappa \gamma_\mu a_\nu b_\kappa = \gamma^\nu \gamma^\kappa \gamma^\mu \gamma_\mu a_\nu b_\kappa - 2\,\gamma^\nu g^{\mu\kappa} \gamma_\mu a_\nu b_\kappa + 2\,\gamma^\kappa \gamma^\nu a_\nu b_\kappa \\ &= d\,\gamma^\nu \gamma^\kappa a_\nu b_\kappa - 2\,\gamma^\nu \gamma^\kappa a_\nu b_\kappa - 2\,\gamma^\nu \gamma^\kappa a_\nu b_\kappa + 4\,g^{\kappa\nu} a_\nu b_\kappa \\ &= (d-4)\,\gamma^\nu \gamma^\kappa a_\nu b_\kappa + 4\,a^\kappa b_\kappa = \ (d-4)\,\not{a}\not{b} + 4a\cdot b . \end{aligned} \quad (10.56)$$

For three Feynman slashes sandwiched between Dirac γ^μ and γ_μ matrices, we commute the leftmost γ^μ sufficiently to the right before it "sees" a γ_μ, to obtain

$$\begin{aligned}\gamma^\mu \not{a}\not{b}\not{c} \gamma_\mu &= \gamma^\mu \gamma^\nu \gamma^\kappa \gamma^\rho \gamma_\mu a_\nu b_\kappa c_\rho = -\gamma^\nu \gamma^\mu \gamma^\kappa \gamma^\rho \gamma_\mu a_\nu b_\kappa c_\rho + 2\,g^{\mu\nu} \gamma^\kappa \gamma^\rho \gamma_\mu a_\nu b_\kappa c_\rho \\ &= -\,\gamma^\nu \gamma^\kappa \gamma^\rho \gamma^\mu \gamma_\mu a_\nu b_\kappa c_\rho + 2\,\gamma^\nu \gamma^\kappa g^{\mu\rho} \gamma_\mu a_\nu b_\kappa c_\rho - 2\,\gamma^\nu \gamma^\rho \gamma^\kappa a_\nu b_\kappa c_\rho \\ &\quad - 2\,\gamma^\rho \gamma^\kappa \gamma^\nu a_\nu b_\kappa c_\rho + 4\,g^{\kappa\rho} \gamma^\nu a_\nu b_\kappa c_\rho \\ &= -\,d\,\gamma^\nu \gamma^\kappa \gamma^\rho a_\nu b_\kappa c_\rho + 2\,\gamma^\nu \gamma^\kappa \gamma^\rho a_\nu b_\kappa c_\rho + 2\,\gamma^\nu \gamma^\kappa \gamma^\rho a_\nu b_\kappa c_\rho \\ &\quad - 4\,\gamma^\nu g^{\rho\kappa} a_\nu b_\kappa c_\rho - 2\,\gamma^\rho \gamma^\kappa \gamma^\nu a_\nu b_\kappa c_\rho + 4\,\gamma^\nu a_\nu b^\rho c_\rho \\ &= (4-d)\,\not{a}\not{b}\not{c} - 2\not{c}\not{b}\not{a} . \end{aligned} \quad (10.57)$$

We have used the relations

$$\gamma^\mu \gamma_\mu = d\,, \qquad \gamma^\rho \gamma^\kappa = 2g^{\rho\kappa} - \gamma^\kappa \gamma^\rho\,, \quad (10.58)$$

repeatedly. Let use now assume that $\not{a} = \not{p}'$ and that $\not{c} = \not{p}$. The Feynman slashes are commuted to the outside, but to the wrong side. We would like to have expressions like $\overline{\mathcal{U}}_{\sigma'}(p')\,\not{p}'$ and $\not{p}\,\mathcal{U}_\sigma(p)$, not $\not{p}'\,\mathcal{U}_\sigma(p)$ and $\overline{\mathcal{U}}_{\sigma'}(p')\,\not{p}$. However, one can move the $\not{a}$ to the left and the $\not{c}$ to the right by writing identities like

$$\begin{aligned}\not{c}\not{b}\not{a} &= \gamma^\rho \gamma^\kappa \gamma^\nu c_\rho b_\kappa a_\nu = (2g^{\rho\kappa} - \gamma^\kappa \gamma^\rho)\,\gamma^\nu b_\kappa c_\rho a_\nu \\ &= -\not{a}\not{b}\not{c} + 2\,(a\cdot b)\,\not{c} - 2\,(a\cdot c)\,\not{b} + 2\,(b\cdot c)\,\not{a} . \end{aligned} \quad (10.59)$$

After repeatedly commuting the Dirac γ matrices, one arrives at

$$\gamma^\mu \not{a}\not{b}\not{c} \gamma_\mu = (6-d)\,\not{a}\not{b}\not{c} - 4\,(a\cdot b)\,\not{c} + 4\,(a\cdot c)\,\not{b} - 4\,(b\cdot c)\,\not{a}\,, \quad (10.60)$$

where now all the slashed vectors keep their order. The numerator of the integrand in Eq. (10.53) can be rewritten as

$$\begin{aligned}\mathcal{N}^\nu &= \gamma^\mu (\not{p}' - \not{k} + m)\,\gamma^\nu\,(\not{p} - \not{k} + m)\gamma_\mu \\ &= \gamma^\mu (\not{p}' - \not{k})\gamma^\nu (\not{p} - \not{k})\gamma_\mu + \gamma^\mu (\not{p}' - \not{k})\gamma^\nu m \gamma_\mu + \gamma^\mu m \gamma^\nu (\not{p} - \not{k})\gamma_\mu + \gamma^\mu m \gamma^\nu m \gamma_\mu\,, \end{aligned} \quad (10.61)$$

where we have separated the slashed momenta and the masses. We can rewrite this expression using Eq. (10.60) for the first term (observing that $\not{b} = \gamma^\rho b_\rho = \gamma^\nu$ for $b_\rho = g^\nu{}_\rho$), Eq. (10.56) for the second and third, and finally Eq. (10.55) for the last term. So we find

$$\begin{aligned}\mathcal{N}^\nu &= (6-d)(\not{p}'-\not{k})\gamma^\nu(\not{p}-\not{k}) - 4(p'-k)^\nu(\not{p}-\not{k}) + 4\gamma^\nu(p'-k)\cdot(p-k)\\ &\quad - 4(p-k)^\nu(\not{p}'-\not{k}) + (d-4)m(\not{p}'-\not{k})\gamma^\nu + 4m(p'-k)^\nu + (d-4)m\gamma^\nu(\not{p}-\not{k})\\ &\quad + 4m(p-k)^\nu + (2-d)m^2\gamma^\nu\,. \end{aligned} \tag{10.62}$$

Using the on-shell condition, we now apply $\not{p}'$ and $\not{p}$ directly onto their respective bispinors. After some additional manipulation, we obtain finally

$$\mathcal{N}^\nu = -4m\,k^\nu + (d-6)\,\gamma^\nu k^2 + 4\not{k}\left[p'^\nu + p^\nu - \left(\frac{d}{2}-1\right)k^\nu\right] + 4\,\gamma^\nu\,(p'-k)\cdot(p-k)\,. \tag{10.63}$$

We shall now return to the case $d = 4$, where we have

$$\mathcal{N}^\nu = -4m\,k^\nu - 2\gamma^\nu k^2 + 4\not{k}\,[p'^\nu + p^\nu - k^\nu] + 4\,\gamma^\nu\,(p'-k)\cdot(p-k)\,. \tag{10.64}$$

Now, let us investigate the denominator of the integrand of the vertex correction $\Lambda^\mu_{\rm OS}(p,p')$. It contains the expressions $[k^2 - 2p'\cdot k + {\rm i}\epsilon]$ and $[k^2 - 2p\cdot k + {\rm i}\epsilon]$ which are not "spherically" symmetric in the four-vector k. However, we can enforce this symmetry, by joining denominators using so-called Feynman parameters, based on formulas like

$$\int_0^1 {\rm d}x \left(\frac{1}{A\,x + B\,(1-x)}\right)^2 = \frac{1}{A\,B}, \qquad \int_0^1 {\rm d}x \frac{6\,x\,(1-x)}{[A\,x + B\,(1-x)]^2} = \frac{1}{(A\,B)^2}, \tag{10.65}$$

which can be generated by differentiation of the integral $\int_0^1 {\rm d}x\,(A\,x + B\,(1-x))^{-2}$ with respect to parameters. After joining the denominators, one can simply shift the integration (loop) momentum, to obtain "spherically" symmetric integrals. We begin by combining the two denominators from the fermion propagators, writing

$$\frac{1}{(k^2 - 2p\cdot k + {\rm i}\epsilon)\,(k^2 - 2p'\cdot k + {\rm i}\epsilon)} = \int_0^1 {\rm d}y \frac{1}{(k^2 - 2p_y\cdot k + {\rm i}\epsilon)^2}\,, \tag{10.66}$$

where we have defined $p_y = y\,p + (1-y)p'$. The next step is to combine this denominator with the one from the photon propagator, using the second entry in Eq. (10.65),

$$\begin{aligned}&\int_{\lambda^2}^{\Lambda^2} {\rm d}L \frac{1}{(k^2 - L + {\rm i}\epsilon)^2} \int_0^1 {\rm d}y \frac{1}{(k^2 - 2p_y\cdot k + {\rm i}\epsilon)^2}\\ &= \int_{\lambda^2}^{\Lambda^2} {\rm d}L \int_0^1 {\rm d}z \int_0^1 {\rm d}y \frac{6z(1-z)}{\left(Q^2 - Lz - p_y^2(1-z)^2 + {\rm i}\epsilon\right)^4}\,,\end{aligned} \tag{10.67}$$

where we have completed the square to shift the integration variable according to

$$Q = k - p_y(1-z)\,, \tag{10.68}$$

using the identity $k^2 - 2p_y\cdot k(1-z) = [k - p_y(1-z)]^2 - p_y^2(1-z^2) = Q^2 - p_y^2(1-z^2)$. The integrand is now "spherically"' symmetric in the new integration momentum

Q. The shift leaves the momentum integration measure $\mathrm{d}^4k \to \mathrm{d}^4Q$ invariant, as we integrate over $\mathbb{R}^4$. We now use the identity $k = Q + p_y(1-z)$ in Eq. (10.63) in order to reformulate the numerator $\mathcal{N}^\nu$, very explicitly, as follows,

$$\begin{aligned}\mathcal{N}^\nu = &-4m\,Q^\nu - 4m(1-z)p_y^\nu - 2\,\gamma^\nu Q^2 - 4\,\gamma^\nu(1-z)\,Q\cdot p_y - 2\,\gamma^\nu(1-z)^2 p_y^2 \\ &+ 4\not{Q}\,\left[p'^\nu + p^\nu - (1-z)p_y^\nu\right] - 4\not{Q}\,Q^\nu - 4(1-z)\not{p}_y Q^\nu \\ &+ 4(1-z)\not{p}_y\left[p'^\nu + p^\nu - (1-z)p_y^\nu\right] - 4\gamma^\nu Q\cdot\left[p + p'^\rho - 2(1-z)p_y\right] + 4\gamma^\nu Q^2 \\ &+ 4\,\gamma^\nu\,(p' - (1-z)p_y)\cdot(p-(1-z)p_y)\,. \end{aligned} \tag{10.69}$$

We now distinguish terms quadratic in Q, terms linear in Q and terms without Q. Linear terms vanish after integration over Q because of the $\mathbb{R}^4$ symmetry (after the Wick rotation, to be discussed below). The only contributing terms are summarized in a redefined expression $\mathcal{N}^\nu$,

$$\begin{aligned}\mathcal{N}^\nu = &\,2\,\gamma^\nu Q^2 - 4\not{Q}Q^\nu - 4m(1-z)p_y^\nu \\ &- 2\,\gamma^\nu(1-z)^2 p_y^2 + 4(1-z)\not{p}_y\left[p'^\nu + p^\nu - (1-z)p_y^\nu\right] \\ &+ 4\,\gamma^\nu\,\left(p'\cdot p - (1-z)p'\cdot p_y - (1-z)p\cdot p_y + (1-z)^2\,p_y^2\right)\,. \end{aligned} \tag{10.70}$$

The two first expressions are quadratic in Q, and we replace them as follows,

$$2\,\gamma^\nu Q^2 - 4\not{Q}Q^\nu = 2\,\gamma^\nu Q^2 - 4\gamma_\mu\,(Q^\mu Q^\nu) \to 2\,\gamma^\nu Q^2 - 4\gamma_\mu\left(\frac{1}{4}\,g^{\mu\nu}Q^2\right) = \gamma^\nu Q^2\,. \tag{10.71}$$

We have carried out a replacement based on the formula

$$\int \mathrm{d}^dQ\;Q^\mu Q^\nu\;f(Q^2) = \frac{g^{\mu\nu}}{d}\int \mathrm{d}^dQ\,Q^2 f(Q^2)\,, \tag{10.72}$$

applied to the case $d = 4$. This relation follows because any general second-rank tensor $B^{\mu\nu}$ in four-dimensional space-time has a decomposition, much in analogy to Eq. (6.13),

$$B^{\mu\nu} = \frac{g^{\mu\nu}}{d}\,T + A^{\mu\nu} + S^{\mu\nu}\,, \tag{10.73}$$

where T is its trace, A an antisymmetric tensor ($A^{\mu\nu} = -A^{\nu\mu}$), and S is a traceless, symmetric tensor ($S^{\mu\nu} = S^{\nu\mu}$, and $g_{\mu\nu}\,S^{\mu\nu} = 0$). These tensors do not mix under Lorentz transformations. When integrated with a function symmetric in Q^2 over all space-time, the anti-symmetric part of the tensor $Q^\mu Q^\nu$ vanishes due to symmetry. The same is true of the symmetric, traceless part. Only the scalar part (trace) yields a nonvanishing contribution to the integral, as expressed in Eq. (10.72).

The part of $\mathcal{N}^\mu$ without Q can be further simplified with the help of the Dirac equation, which allows the replacement $\not{p}_y \to m$, because $\not{p}$ and $\not{p}'$ act to the right and left. Using this identity, expressing $p_y = yp + (1-y)p'$, and replacing $p^2 \to m^2$, $p'^2 \to m^2$, as well as $p'\cdot p \to m^2 - q^2/2$, with

$$q = p' - p\,, \tag{10.74}$$

we get the result that $\mathcal{N}^\nu$ can be equivalently replaced under the integral sign, by

$$\begin{aligned}\mathcal{N}^\nu = &\,\gamma^\nu Q^2 + 4m(1-z)yq^\nu - 4m(1-z)^2(1-y)\,q^\nu - 2\,\gamma^\nu m^2\,(1-2z-z^2) \\ &+ 4z\,(1-z)\,mp^\nu + 2\,\gamma^\nu q^2\left[y^2\,(1-z)^2 - y\,(1-z)^2 - z\right]. \end{aligned} \tag{10.75}$$

We now recall that in calculating the numerator structure $\mathcal{N}^\nu$, we have always assumed that the numerator is "sandwiched" between bispinors $\overline{\mathcal{U}}_{\sigma'}(p')$ and $\mathcal{U}_\sigma(p)$ (see Sec. 7.3.1). We shall use the on-shell condition one more time, in order to get rid of the term proportional to p^ν in Eq. (10.75) and write it in terms a function of the momentum transfer $q = p' - p$. To this end, we need the Gordon identity,

$$\begin{aligned}
\overline{\mathcal{U}}_{\sigma'}(\vec{p}\,')\,\gamma^\nu\,\mathcal{U}_\sigma(\vec{p}) &= \frac{1}{2m}\overline{\mathcal{U}}_{\sigma'}(\vec{p}\,')\,[\not{p}'\gamma^\nu + \gamma^\nu\not{p}]\,\mathcal{U}_\sigma(\vec{p}) \\
&= \frac{1}{2m}\overline{\mathcal{U}}_{\sigma'}(\vec{p}\,')\Bigg[\frac{1}{2}\underbrace{\{\gamma^\mu,\gamma^\nu\}}_{=2g^{\nu\mu}}p'_\mu + \frac{1}{2}\underbrace{\{\gamma^\nu,\gamma^\mu\}}_{=2g^{\nu\mu}}p_\mu + \frac{1}{2}\underbrace{[\gamma^\mu,\gamma^\nu]}_{=2\mathrm{i}\sigma^{\nu\mu}}p'_\mu + \frac{1}{2}\underbrace{[\gamma^\nu,\gamma^\mu]}_{=-2\mathrm{i}\sigma^{\nu\mu}}p_\mu\Bigg]\mathcal{U}_\sigma(\vec{p}) \\
&= \overline{\mathcal{U}}_{\sigma'}(\vec{p}\,')\left[\frac{p^\nu + p'^\nu}{2m} + \frac{\mathrm{i}\sigma^{\nu\mu}q_\mu}{2m}\right]\mathcal{U}_\sigma(\vec{p})\,. \qquad (10.76)
\end{aligned}$$

Using (again) the identity $p' = q + p$, this leads to the relation

$$\overline{\mathcal{U}}_{\sigma'}(\vec{p}\,')\,p^\nu\,\mathcal{U}_\sigma(\vec{p}) = \overline{\mathcal{U}}_{\sigma'}(\vec{p}\,')\left[m\gamma^\nu - \frac{q^\nu}{2} - \frac{\mathrm{i}\sigma^{\nu\mu}q_\mu}{2}\right]\mathcal{U}_\sigma(\vec{p})\,, \qquad (10.77)$$

which allows to express the term proportional to p^ν in Eq. (10.75) by a matrix element of the Dirac γ function, and terms that depend on the momentum transfer q. We thus find that $\mathcal{N}^\nu$ can be equivalently replaced by

$$\begin{aligned}
\mathcal{N}^\nu &= \gamma^\nu Q^2 + 2mq^\nu(2y-1)(z-1)(z-2) - 2\gamma^\nu m^2(1 - 4z + z^2) \\
&\quad - 2z(1-z)m\mathrm{i}\sigma^{\nu\mu}q_\mu + 2\gamma^\nu q^2\left[y^2(1-z)^2 - y(1-z)^2 - z\right]. \qquad (10.78)
\end{aligned}$$

After all these simplifications to the numerator, we can finally proceed to the evaluation of the Q-integral. The integral over d^4Q is normally carried out in Minkowski space with a metric $g^{\mu\nu} = \mathrm{diag}(1,-1,-1,-1)$. However, if we rotate the integral of the timelike component $\mathrm{d}\ell^0 \to \mathrm{i}\,\mathrm{d}\ell^0_E$, then this so-called Wick rotation shifts the integration contour of the timelike component of the loop momentum integration by $\pi/2$, in the mathematically positive direction i.e. counterclockwise.

Let us remember that the denominator D of the integrand in Eq. (10.67) is of the form

$$D = Q^2 - X^2 + \mathrm{i}\epsilon\,, \qquad X^2 = Lz + p_y^2(1-z)^2 > 0\,. \qquad (10.79)$$

Here, we recall that the integration domain for z is the interval $z \in (0,1)$, and that $p_y = yp + (1-y)p'$, where $y \in (0,1)$. The next step is to combine the zeros of the denominator as given by the equation

$$Q^2 = X^2 - \mathrm{i}\epsilon\,, \qquad \omega_Q = \pm\sqrt{\vec{k}_Q^2 + X^2 - \mathrm{i}\epsilon}\,, \qquad Q^\mu = (\omega_Q, \vec{k}_Q)\,. \qquad (10.80)$$

Expanding in ϵ, and redefining ϵ, we find that the position of the poles in the ω_Q integration contour is thus given by (see also Fig. 10.3)

$$\omega_Q = \left|\sqrt{\vec{k}_Q^2 + X^2}\right| - \mathrm{i}\epsilon\,, \qquad \omega_Q = -\left|\sqrt{\vec{k}_Q^2 + X^2}\right| + \mathrm{i}\epsilon\,, \qquad (10.81)$$

i.e., in the second and fourth quadrant of the complex plane. The integrand in Eq. (10.67) is an analytic function, which can be continued to complex momenta

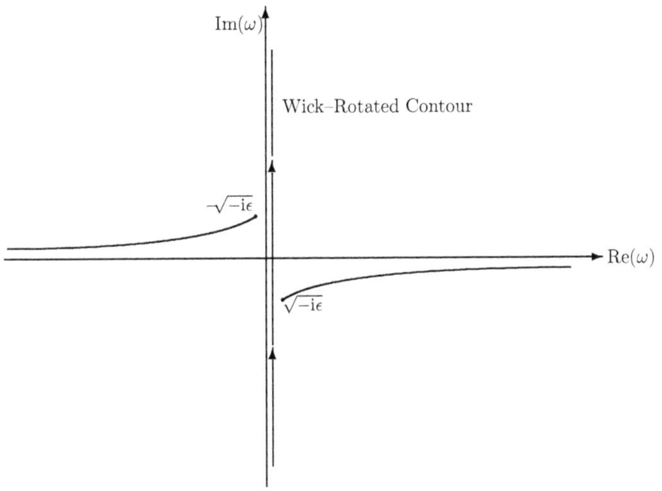

Fig. 10.3 The Wick-rotated integration contour for the vertex correction extends along the imaginary axis in the complex ω plane, where we understand that $\omega \equiv \omega_Q$ is the timelike component of the four-vector Q [see Eqs. (10.68) and (10.80)]. The denominator of the vertex function has poles at $\omega^2 - \vec{k}^2 - X^2 + \mathrm{i}\epsilon = 0$, i.e., at $\omega = \pm\sqrt{\vec{k}^2 + X^2 - \mathrm{i}\epsilon}$, where $\omega = \omega_Q$ and $\vec{k} = \vec{k}_Q$. The minimum value of X^2 in the relevant range of integration parameters is zero, and so, as we integrate over d^3k, a branch cut is generated, with branch points at $\pm\sqrt{X^2 - \mathrm{i}\epsilon}$ [see also Eq. (9.34)]. The curves in the figure display the geometric locus of the branch cuts for $X = 0$. We observe that the fermionic denominators, after multiplication with $\gamma^0 = \beta$, involve the free Dirac Hamiltonian $\vec{\alpha}\cdot\vec{p} + \beta m$ (see also Chap. 7). Let us include the cryptic remark that, for a free reference Dirac state, there is no further branch cut from the fermion propagator, because the reference state is an eigenstate of the momentum operator, and also, the Dirac Hamiltonian in the denominator only involves the momentum operator.

and falls off sufficiently fast for large $|\omega_Q|$. We can thus actively shift the ω_Q integration contour from "horizontal" to "vertical" in the complex plane, changing the integration measure from Minkowskian to Euclidean,

$$\mathrm{d}\omega_Q \to \mathrm{i}\mathrm{d}\omega_{Q,E}\,, \qquad \vec{k}_Q \to \vec{k}_{Q,E}\,, \qquad \mathrm{d}^4Q \to \mathrm{i}\,\mathrm{d}^4Q_E\,, \tag{10.82}$$

$$Q^2 \to -Q_E^2 = -\omega_{Q,E}^2 - \vec{k}_{Q,E}^2\,. \tag{10.83}$$

The integral over d^4Q_E also extends over the entire space $\mathbb{R}^4$. As is illustrated in Fig. 10.3, the Wick rotation is compatible with the Feynman prescription for the poles of the denominators, because of the active shift by an angle $+\pi/2$ in the complex plane. This implies that we can drop the infinitesimal imaginary part $\mathrm{i}\epsilon$ in the on-mass shell vertex function [see Eqs. (10.53), (10.67) and (10.78)],

$$\begin{aligned}\Lambda^{\nu}_{\mathrm{OS}}(p,p') &= 4\pi\mathrm{i}\alpha\int\frac{\mathrm{d}^4Q}{(2\pi)^4}\int_{\lambda^2}^{\Lambda^2}\mathrm{d}L\int_0^1\mathrm{d}z\int_0^1\mathrm{d}y\frac{6z(1-z)\,\mathcal{N}^{\nu}}{\left(Q^2 - Lz - p_y^2(1-z)^2\right)^4}\\ &= -4\pi\alpha\int\frac{\mathrm{d}^4Q_E}{(2\pi)^4}\int_{\lambda^2}^{\Lambda^2}\mathrm{d}L\int_0^1\mathrm{d}z\int_0^1\mathrm{d}y\frac{6z(1-z)\,\mathcal{N}^{\nu}}{\left(-Q_E^2 - Lz - p_y^2(1-z)^2\right)^4}\,.\end{aligned} \tag{10.84}$$

We have already transformed the integral into a spherically symmetric form. Now, it remains to separate off the angular and radial parts. For the angular part, we need the (generalized) surface are Ω_d of the (generalized) unit sphere embedded in d-dimensional space, in order to plug it into the formula

$$\int \mathrm{d}^d Q_E \, f(Q_E^2) = \int \mathrm{d}\Omega_d \int \mathrm{d}Q_E \, Q_E^{d-1} f(Q_E^2) \,, \tag{10.85}$$

which holds for any function $f(Q_E^2)$ of the square of the Euclidean momentum. The volume of the d-dimensional unit sphere is $\Omega_d = \int \mathrm{d}\Omega_d$. We here discuss the general result for non-integer dimension d, as it will become useful later.

One possible way to derive a formula for Ω_d is as follows. We consider the d-dimensional Gaussian integral of $\exp(-\vec{k}^2)$ and calculate it in two ways. First, we express $\vec{k}^2$ in terms of its d components and integrate each of them separately, with the result

$$\int_{\mathbb{R}^d} \frac{\mathrm{d}^d k}{(2\pi)^d} \mathrm{e}^{-\vec{k}^2} = \left[\int_{-\infty}^{\infty} \frac{\mathrm{d}k}{2\pi} \mathrm{e}^{-k^2}\right]^d = \left(\frac{\sqrt{\pi}}{2\pi}\right)^d = \frac{1}{2^d \, \pi^{d/2}} = \frac{\pi^{d/2}}{(2\pi)^d} \,. \tag{10.86}$$

On the other hand, if we single out the integral over the angular variable $\int \mathrm{d}\Omega_d$ and perform only a radial integral, then we have

$$\int_{\mathbb{R}^d} \frac{\mathrm{d}^d k}{(2\pi)^d} \mathrm{e}^{-\vec{k}^2} = \frac{1}{(2\pi)^d} \left[\int \mathrm{d}\Omega_d\right] \int_0^{\infty} \mathrm{d}k \, k^{d-1} \, \mathrm{e}^{-k^2} = \frac{\Gamma(d/2)}{2(2\pi)^d} \left[\int \mathrm{d}\Omega_d\right], \tag{10.87}$$

where the Γ function is defined as $\Gamma(x) = \int_0^\infty t^{x-1} \exp(-t) \, \mathrm{d}t$, for $x > 0$. So, we have

$$\Omega_d = \frac{2\pi^{d/2}}{\Gamma\left(\frac{d}{2}\right)} \,, \tag{10.88}$$

which yields $\Omega_4 = 2\pi^2$ in $d = 4$. This result could have been obtained by more elementary means for integer dimension, e.g., for the case $d = 4$, by integrating Eq. (4.196) over the angles. However, anticipating the discussion in Sec. 10.3.4, we need a more general expression for non-integer d. Putting together Eqs. (10.78) and (10.84), we have

$$\begin{aligned}
\Lambda^\nu_{\mathrm{OS}}(p,p') &= -4\pi\alpha \int \frac{\mathrm{d}^4 Q}{(2\pi)^4} \int_{\lambda^2}^{\Lambda^2} \mathrm{d}L \int_0^1 \mathrm{d}z \int_0^1 \mathrm{d}y \frac{6z(1-z)\,\mathcal{N}^\nu}{\left(-Q_E^2 - Lz - p_y^2(1-z)^2\right)^4} \\
&= -4\pi\alpha \int_0^1 \mathrm{d}z \int_0^1 \mathrm{d}y [6z(1-z)] [-I_2(y,z)\,\gamma^\nu + I_0(y,z) \\
&\quad \times \left\{2mq^\nu(2y-1)(z-1)(z-2) + \gamma^\nu \left(2\,q^2 \left[y^2(1-z)^2 - y(1-z)^2 - z\right]\right.\right. \\
&\quad \left.\left. - 2m^2(1-4z+z^2)\right) - 2z(1-z)mi\sigma^{\nu\mu}q_\mu\right\}] \,.
\end{aligned} \tag{10.89}$$

The $\mathrm{d}^4 Q_E$ integrals have been absorbed in $I_2(y,z)$ and $I_0(y,z)$. The integral $I_2(y,z)$ carries two powers of Q_E^2 in the numerator,

$$\begin{aligned}
I_2(y,z) &= \int \frac{\mathrm{d}^4 Q_E}{(2\pi)^4} \int_{\lambda^2}^{\Lambda^2} \mathrm{d}L \frac{Q_E^2}{\left(Q_E^2 + Lz + p_y^2(1-z)^2\right)^4} \\
&= \frac{1}{8\pi^2} \frac{1}{6z} \ln\left[\frac{p_y^2(1-z)^2 + z\Lambda^2}{p_y^2(1-z)^2 + z\lambda^2}\right] = \frac{1}{48\pi^2 z} \ln\left[\frac{z\Lambda^2}{p_y^2(1-z)^2}\right] \gamma^\mu \,.
\end{aligned} \tag{10.90}$$

We have used the identity $\int \mathrm{d}^4 Q_E = 2\pi^2 \int \mathrm{d}Q_E\, Q_E^3$, which is valid for radially symmetric integrands. This expression is ultraviolet divergent in the limit $\Lambda \to \infty$, while the infrared limit $\lambda \to 0$ is safe. This confirms the result of naive power counting. By contrast, the integral $I_0(y,z)$, by naive power counting, is infrared divergent but ultraviolet safe,

$$\begin{aligned} I_0(y,z) &= \int \frac{\mathrm{d}^4 Q_E}{(2\pi)^4} \int_{\lambda^2}^{\Lambda^2} \frac{\mathrm{d}L}{\left(Q_E^2 + Lz + p_y^2(1-z)^2\right)^4} = \int_{\lambda^2}^{\Lambda^2} \frac{\mathrm{d}L}{8\pi^2} \int_0^\infty \frac{\mathrm{d}Q_E\, Q_E^3}{\left(Q_E^2 + Lz + p_y^2(1-z)^2\right)^4} \\ &= \frac{1}{8\pi^2} \frac{\Lambda^2 - \lambda^2}{12(p_y^2(1-z)^2 + z\lambda^2)(p_y^2(1-z)^2 + z\Lambda^2)} = \frac{1}{96\pi^2 z (p_y^2(1-z)^2 + z\lambda^2)}. \end{aligned} \tag{10.91}$$

In the last step, the limit $\Lambda \to \infty$ is taken. With these results, we are now in a position to write for the full one-loop vertex correction $[p_y^2 = m^2 - q^2 y(1-y)]$

$$\begin{aligned} \Lambda^\nu_{\mathrm{OS}}(p,p') &= \frac{\alpha}{2\pi} \int_0^1 \mathrm{d}z \int_0^1 \mathrm{d}y \left\{ (1-z)\,\gamma^\nu \ln\left[\frac{z\Lambda^2}{(1-z)^2\,(m^2 - q^2 y(1-y))} \right] \right. \\ &\quad + \frac{(1-z)}{(1-z)^2 (m^2 - q^2 y(1-y)) + z\lambda^2} \Big(-m q^\nu (2y-1)(z-1)(z-2) \\ &\quad \left. + \gamma^\nu \left(q^2 \left[y\,(1-y)\,(1-z)^2 + z \right] + m^2 (1 - 4z + z^2) \right) + z(1-z) m \mathrm{i} \sigma^{\nu\mu} q_\mu \Big) \right\}. \end{aligned} \tag{10.92}$$

Based on symmetry properties, we can show that the term proportional to q^ν vanishes after integration over y. Namely, the numerator of this term is directly proportional to $2y-1$, which is antisymmetric in the interval $y \in [0,1]$, whereas the denominator can be written in terms of the variable $y\,(1-y)$, which is symmetric in the interval $y \in [0,1]$. So, $\Lambda^\nu_{\mathrm{OS}}(p,p')$ simplifies to

$$\begin{aligned} \Lambda^\nu_{\mathrm{OS}}(p,p') &= \frac{\alpha}{2\pi} \int_0^1 \mathrm{d}z \int_0^1 \mathrm{d}y \, \left\{ (1-z)\,\gamma^\nu \, \ln\left[\frac{z\Lambda^2}{(1-z)^2\,(m^2 - q^2 y(1-y))} \right] \right. \\ &\quad + \frac{(1-z)}{(1-z)^2\,(m^2 - q^2 y(1-y)) + z\lambda^2} \Big(\gamma^\nu m^2 (1 - 4z + z^2) \\ &\quad \left. + \gamma^\nu q^2 \left[y(1-y)(1-z)^2 + z \right] + z(1-z) m \mathrm{i} \sigma^{\nu\mu} q_\mu \Big) \right\}. \end{aligned} \tag{10.93}$$

This concludes our discussion of the regularized vertex function. We now need to take care of the renormalization, and calculate the renormalized vertex function, as well as the form factors of the electron.

10.3.2 *Vertex and Form Factors*

Intuitively, one would assume that the vertex correction $\Lambda^\nu_{\mathrm{OS}}(p,p')$ has to vanish for zero momentum transfer, i.e., for $q = 0$. This is because, according to Eq. (10.45), any correction to the "vertex factor" $-\mathrm{i}e\gamma^\nu$ at zero momentum transfer counts as a modification δe of the electron's charge, $-\mathrm{i}\, e\, \gamma^\nu \to -\mathrm{i}\,(e+\delta e)\,\gamma^\nu$. Since the e parameter in our theory parameterizes the physical electron charge, this implies that we should interpret the term δe as a perturbative, radiative contribution to the electron's charge which should be naturally absorbed in the physical value of e. The original parameter e of the theory is then reinterpreted as a bare parameter, to which we have to add a radiative correction. Variously, if we wish to keep the interpretation of e as the physical, renormalized electron charge, then we need to subtract the radiative contribution δe because it is already contained in the physical value of e; we just did not know, when we wrote down the original QED Lagrangian, that the parameter e that appears in it also contained a radiative correction. The general motivation and implementation of the renormalization program of QED will be discussed in Sec. 10.3.3, where we shall introduce the charge renormalization constant $Z_1 \equiv Z_1(\Lambda, \mu)$.

It follows that we should subtract from the expression (10.93) the vertex function at zero momentum transfer ($p = p'$),

$$\begin{aligned}\Lambda^\nu_{\mathrm{OS}}(p,p) &= \gamma^\nu \, \frac{\alpha}{2\pi} \int_0^1 \mathrm{d}z \left[(1-z) \left(\ln\left[\frac{z\Lambda^2}{m^2(1-z)^2} \right] + \frac{1-4z+z^2}{(1-z)^2 + z\frac{\lambda^2}{m^2}} \right) \right] \\ &= \gamma^\nu \, \frac{\alpha}{\pi} \left[\frac{1}{4} \ln\left(\frac{\Lambda^2}{m^2} \right) - \frac{1}{2} \ln\left(\frac{m^2}{\lambda^2} \right) + \frac{9}{8} + \mathcal{O}\left(\frac{\lambda^2}{m^2} \right) \right]. \end{aligned} \tag{10.94}$$

This function is conveniently proportional to γ^ν.

At zero momentum transfer, the unrenormalized vertex function is ultraviolet and infrared divergent (limits $\Lambda \to \infty$ and $\mu \to 0$). The on-shell, renormalized [subscript (R)] vertex function is obtained as

$$\begin{aligned}\Lambda^\nu_{\mathrm{R}}(p,p') &= \Lambda^\nu_{\mathrm{OS}}(p,p') - \Lambda^\nu_{\mathrm{OS}}(p,p) = \frac{\alpha}{2\pi} \int_0^1 \mathrm{d}z \int_0^1 \mathrm{d}y \Bigg\{ \gamma^\nu \, (1-z) \\ &\times \ln\left[\left(1 - \frac{q^2}{m^2} y(1-y) \right)^{-1} \right] + \frac{(1-z)}{(1-z)^2 \left[1 - \frac{q^2}{m^2} y(1-y) \right] + z\frac{\lambda^2}{m^2}} \Big(\gamma^\nu (1-4z+z^2) \\ &+ \gamma^\nu \frac{q^2}{m^2} \left[y(1-y)(1-z)^2 + z \right] + z(1-z) \mathrm{i}\sigma^{\nu\mu} \frac{q_\mu}{m} \Big) - \gamma^\nu (1-z) \frac{1-4z+z^2}{(1-z)^2 + z\frac{\lambda^2}{m^2}} \Bigg\}. \end{aligned} \tag{10.95}$$

The renormalized vertex function is still infrared divergent.

Let us define the renormalized on-shell vertex function $\Gamma^\nu_{\mathrm{R}}(q)$ to be the sum of the Dirac γ^ν matrix and the renormalized one-loop correction $\Lambda^\nu_{\mathrm{R}}(p,p')$. An inspection of Eq. (10.95) reveals that we can write $\Gamma^\nu(q)$ in terms of two form factors F_1 and F_2,

$$\Gamma^\nu_{\mathrm{R}}(q) = \gamma^\nu + \Lambda^\nu_{\mathrm{R}}(p,p') = F_1(q^2)\,\gamma^\nu + F_2(q^2) \frac{\mathrm{i}}{2m} \sigma^{\nu\mu} q_\mu \,. \tag{10.96}$$

We assume on-shell incoming and outgoing bispinors. A more general analysis of the structure of possible interactions between a fermion and gauge fields would otherwise lead to a form [see Eq. (1.2) of Ref. [308]]

$$\Gamma^{\nu}_{\rm R}(q) = F_1(q^2)\,\gamma^\nu + F_2(q^2)\frac{\rm i}{2m}\sigma^{\nu\rho}q_\rho + F_3(q^2)\,\frac{1}{2m}\sigma^{\nu\rho}\gamma^5 q_\rho + F_A(q^2)\,\frac{G_F\,a}{8\pi\sqrt{2}}\left(\gamma^\nu\gamma^5 q^2 - 2m\,\gamma^5 q^\nu\right). \tag{10.97}$$

Here, $\sigma^{\nu\rho} = {\rm i}\,[\gamma^\nu,\gamma^\rho]/2$, and a is the contribution of the anapole moment [309]. A nonvanishing value of $F_3(0)$ would otherwise describe an electric dipole moment of the electron [see Eq. (1.3) of Ref. [308]]. Within QED, one obtains nonvanishing contributions for the F_1 and F_2 form factors (which are also called the Dirac and the Pauli form factors, respectively). By inspection of Eq. (10.95), one infers the integral representations

$$F_1(q^2) = 1 + \frac{\alpha}{2\pi}\int_0^1 {\rm d}z \int_0^1 {\rm d}y\,(1-z)\left\{\ln\left[\left(1 - \frac{q^2}{m^2}y(1-y)\right)^{-1}\right] + \frac{(1-4z+z^2) + \frac{q^2}{m^2}\left[y(1-y)(1-z)^2 + z\right]}{(1-z)^2\left[1 - \frac{q^2}{m^2}y(1-y)\right] + z\frac{\lambda^2}{m^2}} - \frac{1-4z+z^2}{(1-z)^2 + z\frac{\mu^2}{m^2}}\right\}, \tag{10.98}$$

$$F_2(q^2) = \frac{\alpha}{\pi}\int_0^1 {\rm d}z \int_0^1 {\rm d}y\,\frac{(1-z)^2\,z}{(1-z)^2\left[1 - \frac{q^2}{m^2}y(1-y)\right] + z\frac{\lambda^2}{m^2}}. \tag{10.99}$$

For the infrared finite form factor F_2, we can set the photon mass $\lambda = 0$,

$$F_2(q^2) = \frac{\alpha}{\pi}\underbrace{\int_0^1 {\rm d}z\,z}_{=1/2}\int_0^1 {\rm d}y\,\frac{1}{1 - \frac{q^2}{m^2}y(1-y)} = \frac{\alpha}{2\pi}\left[1 + \frac{1}{6}\frac{q^2}{m^2} + \frac{1}{30}\frac{q^4}{m^4} + \mathcal{O}\left(\frac{q^6}{m^6}\right)\right]. \tag{10.100}$$

The integrals for $F_1(q^2) = F_1^{[1]}(q^2) + F_1^{[2]}(q^2)$ are more involved. The logarithmic part is

$$F_1^{[1]}(q^2) = \frac{\alpha}{2\pi}\underbrace{\int_0^1 {\rm d}z(1-z)}_{=1/2}\int_0^1 {\rm d}y\,\ln\left[\left(1 - \frac{q^2}{m^2}y(1-y)\right)^{-1}\right] = \frac{\alpha}{2\pi}\left[\frac{q^2}{m^2}\frac{1}{12} + \frac{q^4}{m^4}\frac{1}{120} + \mathcal{O}\left(\frac{q^6}{m^6}\right)\right]. \tag{10.101}$$

The nonlogarithmic term is expanded as follows,

$$F_1^{[2]}(q^2) = \frac{\alpha}{2\pi}\int_0^1 {\rm d}z\int_0^1 {\rm d}y\,(1-z)\left\{\frac{(1-4z+z^2) + \frac{q^2}{m^2}\left[y(1-y)(1-z)^2 + z\right]}{(1-z)^2\left[1 - \frac{q^2}{m^2}y(1-y)\right] + z\frac{\lambda^2}{m^2}} - \frac{1-4z+z^2}{(1-z)^2 + z\frac{\lambda^2}{m^2}}\right\} = \frac{\alpha}{2\pi}\left\{\frac{q^2}{m^2}\left[\frac{2}{3}\ln\left(\frac{m}{\lambda}\right) - \frac{1}{3}\right] + \frac{q^4}{m^4}\left[\frac{1}{10}\ln\left(\frac{m}{\lambda}\right) - \frac{1}{10}\right] + \mathcal{O}\left(\frac{q^6}{m^6}\right)\right\}. \tag{10.102}$$

The integration over the Feynman parameters is greatly simplified if one observes that the infrared regulator is only relevant in the region $z \approx 1$; hence, one can replace (twice) $z\lambda^2/m^2 \to \lambda^2/m^2$ in the integrand of Eq. (10.102). In the final result, we only keep terms which are divergent or finite in the limit $\lambda \to 0$. The F_1 form factor in one-loop order is the sum of $F_1^{[1]}(q^2)$ and $F_1^{[2]}(q^2)$, and of the tree-level leading term (unity) from Eq. (10.98),

$$F_1(q^2) = 1 + \frac{\alpha}{\pi}\frac{q^2}{m^2}\left[\frac{1}{3}\ln\left(\frac{m}{\lambda}\right) - \frac{1}{8}\right] + \frac{\alpha}{\pi}\frac{q^4}{m^4}\left[\frac{1}{20}\ln\left(\frac{m}{\lambda}\right) - \frac{11}{240}\right] + \mathcal{O}\left(\frac{q^6}{m^6}\right). \quad (10.103)$$

One cannot stress enough that this result is still infrared divergent; it contains the photon mass λ. That is to say, despite all our heroic efforts to remove the (ultraviolet) divergences by means of renormalization, there are still divergences (of infrared nature) left in the final result. This observation finds a natural explanation if one considers a compensating divergence in the soft bremsstrahlung correction.

Roughly speaking, the explanation is as follows. The infrared divergence in the result (10.103) describes the physical fact that it is infinitely easy for the electron to emit, and reabsorb, a virtual photon before and after the interaction with the external field (at the vertex). Hence, this infrared divergence needs to be cut off, at an energy scale commensurate with the photon mass λ.

If one considers the high-energy part of the bound-electron self-energy for the Lamb shift, then the infrared divergence finds a natural explanation: simply, this infrared divergence is cut off at the scale of the atomic binding energy, where the energy could not be radiated and absorbed any more in the form of virtual infrared photons of arbitrarily small energy, but rather, the bound-state spectrum should be taken into account. We anticipate that the infrared divergence enters the expressions for the high-energy part of the bound-electron self-energy in photon mass regularization [see Eq. (11.148)] as well as in photon energy regularization [see Eq. (11.150)], with the matching given in Eq. (11.151). The infrared divergence of the high-energy part is compensated when the low-energy part is added. The low-energy part is given in Eq. (11.147) (photon mass regularization) and in Eq. (4.325) (photon energy regularization). The total result for the self-energy shift of an atomic level is independent of the infrared regularization procedure.

One might now ask how all of this plays out when one considers bremsstrahlung corrections to a scattering process. In this case, the infrared divergence of the vertex correction is compensated by a corresponding logarithmic infrared divergence of the infrared bremsstrahlung correction, which describes the (almost collinear) emission of infrared photons by the outgoing electron. Infrared divergences of the vertex correction and of the bremsstrahlung correction to the total cross-section can be shown to cancel, and the so-called infrared catastrophe is avoided (see Chap. 7 of Ref. [2]). In this case, the infrared photon energy cutoff ϵ [see Eq. (11.150)] takes the role of the detector resolution $\Delta E = \epsilon$, where $\Delta E = \epsilon$ is the smallest photon energy detectable by the apparatus. The matching between the photon energy and photon mass regularizations is given in Eq. (11.151).

Incidentally, one further aspect is of utmost interest. While, at every order of perturbation theory, infrared divergences of vertex and bremsstrahlung corrections cancel, one might ask about the total contribution of the vertex-only corrections, i.e., of the processes in which manifestly no extra bremsstrahlung photons are emitted. This question is answered in the works of YENNIE,[2] FRAUTSCHI[3] and SUURA,[4] and SUDAKOV.[5] Roughly speaking, one can ascertain that the corrections to the cross section, once all orders of perturbation theory are added, behave, to logarithmic accuracy, just like the following infinite series,

$$\sum_{n=0}^{\infty} \frac{1}{n!} \left[-\frac{\alpha}{\pi} \ln\left(\frac{m}{\epsilon}\right) \right]^n = \exp\left[-\frac{\alpha}{\pi} \ln\left(\frac{m}{\epsilon}\right) \right] = \left(\frac{\epsilon}{m}\right)^{\alpha/\pi} \to 0 \,, \tag{10.104}$$

where ϵ is the infrared cutoff. The summation of all orders of perturbation theory leads to a vanishing of the infrared divergence. Therefore, in the limit $\epsilon \to 0$, this leaves only a finite, infrared safe correction to the entire process. The process of the cancellations is similar to the calculation of the exponential function of large negative argument,

$$\exp(-1000) = 1 - 1000 + \frac{1000^2}{2!} + \cdots. \tag{10.105}$$

Each single term on the right-hand side is very large, but their sum approaches $\exp(-1000)$, which is very close to zero.

10.3.3 *Vertex and Renormalization*

As our considerations regarding the vertex charge renormalization show [see the beginning of Sec. 10.3.2], the renormalization establishes a relation between the bare quantities, which are present in the "original" QED Lagrangian, and the renormalized, physical quantities, such as the renormalized, physical charge.

In the context of the charge renormalization, we recall Eq. (10.45), which at $p' = p$, or $q = 0$, and between on-shell bispinors, reads as

$$-\,\mathrm{i}e\gamma^\nu \to -\mathrm{i}e\gamma^\nu - \mathrm{i}e\Lambda^\nu_{\mathrm{OS}}(p,p)\,. \tag{10.106}$$

Furthermore, we have Eq. (10.94), which we write as follows,

$$\Lambda^\nu_{\mathrm{OS}}(p,p) = \gamma^\nu \, \frac{e^2}{16\pi^2} \left[\frac{1}{4} \ln\left(\frac{\Lambda^2}{m^2}\right) - \frac{1}{2} \ln\left(\frac{m^2}{\lambda^2}\right) + \frac{9}{8} + \mathcal{O}\left(\frac{\lambda^2}{m^2}\right) \right]. \tag{10.107}$$

The parameter e in the original QED Lagrangian is thus seen to be the bare charge e_0, which receives a perturbative correction

$$\delta e = \frac{e^3}{16\pi^2} \left[\frac{1}{4} \ln\left(\frac{\Lambda^2}{m^2}\right) - \frac{1}{2} \ln\left(\frac{m^2}{\lambda^2}\right) + \frac{9}{8} \right]. \tag{10.108}$$

---

[2]Donald Robert Yennie (1924–1993).
[3]Stephen Clark Frautschi (b. 1933).
[4]Hiroshi Suura (1925–1998).
[5]Vladimir Vasilievich Sudakov (1925–1995).

This correction is perturbatively small (of higher order in the coupling parameter, which is the electron charge $e = \sqrt{4\pi\alpha}$) and yet, divergent in the limits $\Lambda \to \infty$ and $\lambda \to 0$. Upon re-identifying the parameter e in the relations above as the bare electron charge, we can write the following relation between the physical, renormalized charge $e_{\rm phys} = e_{\rm R}$, and the bare charge e_0,

$$e_{\rm phys} = e_{\rm R} = e_0 + \delta e = e_0 \left\{ 1 + \frac{e_0^2}{16\pi^2} \left[\frac{1}{4} \ln\left(\frac{\Lambda^2}{m^2}\right) - \frac{1}{2} \ln\left(\frac{m^2}{\lambda^2}\right) + \frac{9}{8} \right] \right\} . \tag{10.109}$$

This equation suggests that we can express, order by order in perturbation theory, the physical charge of the electron, in terms of the bare charge.

The renormalization program is characterized by the following steps. *(i)* The original Lagrangian is identified as the bare Lagrangian, expressed in terms of the bare quantities. *(ii)* The perturbative loop corrections identify counterterms, i.e., terms which are naturally added to the Lagrangian. These are proportional to terms of the original Lagrangian. These so-called counterterms shift the bare quantities so that, upon addition of radiative corrections, finite results are obtained from the theory. The postulate of renormalizability then states that the physical, renormalized Lagrangian, expressed in terms of the physical quantities, should have the same functional form as the bare Lagrangian, expressed in terms of the bare quantities, up to the addition of the counterterms. It will turn out that the bare and physical quantities are related by multiplicative renormalization constants. *(iii)* One then rewrites the bare Lagrangian, originally expressed in terms of the bare quantities, in terms of the physical quantities. This operation is equivalent to the combined operation of re-interpreting the parameters in the original Lagrangian in terms of the physical, renormalized parameters, and adding the counterterms (in the sense of the relation $e_0 = e_{\rm R} - \delta e$). Here, δe is the radiative contribution to the electron charge. Added to the bare charge, it gives the renormalized, physical charge. On the other hand, if we add, to the original Lagrangian, the counter term $-\delta e$, and re-interpret the charge parameter in the original Lagrangian as the renormalized charge, then, upon switching the radiative corrections "back on", finite physical predictions can be obtained from the theory. Due to the multiplicative nature of renormalization, the addition of the radiative corrections (subtraction of the counter terms) amounts to a multiplication of the bare parameters by renormalization constants. The final result of all these operations is that the bare Lagrangian, written in terms of the bare quantities, is equal to the renormalized Lagrangian, written in terms of the renormalized (physical) quantities.

As we carry out the entire renormalization procedure on the example of quantum electrodynamics, we will soon realize that the charge renormalization considered above, in fact, is a so-called Z_1 renormalization at the vertex, which cancels with the Z_2 wave function renormalization by virtue of the WARD[6]–TAKAHASHI[7] identity. But let us not anticipate details of the second step before completing the first.

[6]John Clive Ward (1924–2000).
[7]Yasushi Takahashi (1924–2013).

Rather, let us start by realizing that the quantum field operators in Eq. (10.1) constitute the unrenormalized, bare quantities that are obtained when we neglect the radiative corrections that lead to the necessity of renormalization. The interaction Hamiltonian originally given in Eq. (10.3) is renormalized as follows,

$$\mathcal{L}_I = -e_0\,\overline{\psi}_0 \not{A}_0 \psi_0 = -Z_1 e_{\rm R}\,\overline{\psi}_{\rm R} \not{A}_{\rm R}\,\psi_{\rm R}\,. \tag{10.110}$$

For full clarity, all renormalized quantities will be indicated with the subscript R in the following discussion. Surprisingly, the bare Lagrangian, once all renormalizations are taken into account (charge, wave function and four-vector potential), is formally equal to the renormalized Lagrangian, but with the bare parameters replaced by their renormalized counterparts and with counterterms being added. We note that the first form of the Lagrangian (10.110) is the QED interaction Lagrangian in the interpretation *(i)* given above, whereas the second form actually corresponds to the interpretation *(iii)*. Indeed, the full renormalization program of QED, including the Ward–Takahashi identity, implies that the bare quantities are related to the physical quantities as follows,

$$\psi_0 = Z_2^{1/2}\psi_{\rm R}\,, \qquad e_0 = \frac{Z_1}{Z_2\,Z_3^{1/2}}\,e_{\rm R}\,, \qquad A_0 = Z_3^{1/2}\,A_{\rm R}\,, \qquad Z_1 = Z_2\,. \tag{10.111}$$

Concerning the multiplicative renormalizations of the electron in Eq. (10.111), we will soon realize that, for the electron charge, the Z_2 renormalizations of the wave function and the Z_1 vertex renormalization of the charge cancel, and the only surviving renormalization is the Z_3 renormalization of the charge due to a phenomenon known as vacuum polarization. For the entire interaction Lagrangian given in Eq. (10.110), the only surviving renormalization is the overall Z_1 renormalization of the charge at the vertex.

The renormalization constants Z_1, Z_2 and Z_3 relate the bare to the physical quantities. The physical interaction Lagrangian is

$$\mathcal{L}_{I,{\rm R}} = -e_{\rm R}\,\overline{\psi}_{\rm R} \not{A}_{\rm R}\,\psi_{\rm R}\,, \tag{10.112}$$

where all quantities correspond to the physical ones, and $\mathcal{L}_{I,{\rm R}}$ is renormalized. The bare Lagrangian is

$$\mathcal{L}_{I,0} = -e_0\,\overline{\psi}_0 \not{A}_0\,\psi_0\,. \tag{10.113}$$

As a result of our calculations, we obtain a relation of the form

$$\mathcal{L}_{I,0} + \mathcal{L}_{\rm RAD} = \mathcal{L}_{I,{\rm R}}\,, \qquad \mathcal{L}_{I,0} = \mathcal{L}_{I,{\rm R}} - \mathcal{L}_{\rm CT}\,, \qquad \mathcal{L}_{\rm CT} = \mathcal{L}_{\rm RAD}\,, \tag{10.114}$$

where $\mathcal{L}_{\rm RAD} = \mathcal{L}_{\rm CT}$ contains the radiative corrections to the physical quantities of the theory (only those radiative corrections that need to be absorbed into physical parameters). In order to obtain finite results, we need to subtract, from the Lagrangian $\mathcal{L}_{I,{\rm R}}$ formulated in terms of finite, physical quantities, the term $\mathcal{L}_{\rm CT}$ which contains the counter terms that we still have to identify. From Eqs. (10.110) and (10.111), we see that

$$\mathcal{L}_{\rm CT} = (Z_1 - 1)\,e_{\rm R}\,\overline{\psi}_{\rm R} \not{A}_{\rm R}\,\psi_{\rm R} = \left(1 - Z_1^{-1}\right)\,e_0\,\overline{\psi}_0 \not{A}_0 \psi_0 = -\left(Z_1^{-1} - 1\right) e_0\,\overline{\psi}_0 \not{A}_0 \psi_0\,. \tag{10.115}$$

In the framework of the renormalization program, we thus have to identify the counter term from the vertex correction as follows,

$$\mathcal{L}_{\rm CT} = -\left(Z_1^{-1} - 1\right) e_0 \overline{\psi}_0 \not{A}_0 \psi_0 \,. \tag{10.116}$$

Now, we add this term to the Lagrangian and convince ourselves one more time that the physical Lagrangian is obtained, as a function of the physical parameters,

$$\begin{aligned} \mathcal{L}_{I,0} + \mathcal{L}_{\rm CT} &= -e_0 \overline{\psi}_0 \not{A}_0 \psi_0 - e_0 \left(Z_1^{-1} - 1\right) \overline{\psi}_0 \not{A}_0 \psi_0 \\ &= -e_0 Z_1^{-1} \overline{\psi}_0 \not{A}_0 \psi_0 = -e_{\rm R} \overline{\psi}_{\rm R} \not{A}_{\rm R} \psi_{\rm R} \,, \end{aligned} \tag{10.117}$$

where we carefully distinguish between bare quantities (with a subscript zero) and physical quantities (with a subscript R), and *all* quantities have been replaced by their renormalized counterparts [see also Eq. (10.111)]. We should mention that, in the sense of Eq. (10.111), we have only taken into account the charge renormalization at the vertex, which is given by the Z_1 renormalization constant, so that the simplified relation of the renormalized and the bare charge reads as $e_{\rm R} = Z_1^{-1} e_0$.

In leading order in the QED perturbative expansion, the renormalization constants Z_1, Z_2, and Z_3 all have the expansion $1 + \mathcal{O}(\alpha)$. In our calculation of the vertex correction, we had found the decomposition [see Eqs. (10.45), (10.46), and (10.52)]

$$\Gamma^\nu_{\rm OS}(p,p') = \gamma^\nu + \Lambda^\nu_{\rm OS}(p,p') \,, \qquad \Lambda^\nu_{\rm OS}(p,p') = \Lambda^\nu(p,p')\big|_{\not{p}=\not{p}'=m} \,. \tag{10.118}$$

We thus have to write the unrenormalized on-shell vertex function $\Lambda^\nu_{\rm OS}(p,p')$ as follows,

$$\Lambda^\nu_{\rm OS}(p,p') = \gamma^\nu \left(Z_1^{-1} - 1\right) + Z_1^{-1} \Lambda^\nu_{\rm R}(p,p') \,, \qquad \Gamma^\nu_{\rm OS}(p,p') = Z_1^{-1} \Gamma^\nu_{\rm R}(p,p') \,. \tag{10.119}$$

We might note that this definition of the renormalized vertex function only works for on-shell momenta. The renormalized vertex function also exists for more general values of p and p', and has been investigated intensively in the latter kinematic region, in part as an input in multiloop calculations in QED bound-state problems [310, 311]. The renormalized on-shell vertex function $\Lambda^\nu_{\rm R}(p,p')$ vanishes at zero momentum transfer [see Eqs. (10.94) and (10.95)].

In this case, the unrenormalized vertex factor $-{\rm i}e_0 \gamma^\nu$, plus the radiative correction, $-{\rm i}e_0 \Lambda^\nu_{\rm OS}(p,p)$, at zero momentum transfer, finds the representation

$$-{\rm i}e_0 \gamma^\nu - {\rm i}e_0 \Lambda^\nu_{\rm OS}(p,p) = -{\rm i}e_0 \gamma^\nu - {\rm i}e_0 \gamma^\nu \left(Z_1^{-1} - 1\right) = -{\rm i}Z_1^{-1} e_0 \gamma^\nu = -{\rm i}\, e_{\rm R} \gamma^\nu \,. \tag{10.120}$$

Now, the question is how the separation into the renormalization (counter-)term $\gamma^\nu(Z_1^{-1} - 1)$ and the renormalized vertex function has to be accomplished. The answer to this question can be found as follows. One has to formulate renormalization conditions, which determine how the bare quantities are related to the physical ones. For the renormalized vertex function, the on-shell renormalization condition reads as follows,

$$\Gamma^\nu_{\rm R}(p,p)\big|_{\not{p}=m} = \gamma^\nu \,, \qquad \Lambda^\nu_{\rm R}(p,p) = 0 \,, \tag{10.121a}$$

$$\Lambda^\nu_{\rm OS}(p,p) = \gamma^\nu \left(Z_1^{-1} - 1\right) . \tag{10.121b}$$

This condition forces us to subtract the vertex at zero momentum transfer, i.e., at $q = p - p = 0$. Furthermore, because $Z_1 = 1 + \mathcal{O}(\alpha)$ is equal to unity in leading order in α, we have

$$\Lambda^\nu_{\rm OS}(p,p) = \gamma^\nu (Z_1^{-1} - 1) \approx \gamma^\nu (1 - Z_1)\,, \tag{10.122a}$$

$$\Lambda^\nu_{\rm R}(p,p') = Z_1 \Lambda^\nu_{\rm OS}(p,p') - \gamma^\nu (1 - Z_1) \approx \Lambda^\nu_{\rm OS}(p,p') - \Lambda^\nu_{\rm OS}(p,p)\,. \tag{10.122b}$$

This consideration finally justifies our subtraction, carried out in Eq. (10.95) to one-loop order, i.e., in the first order in the fine-structure constant α. If we keep Z_1 in the intermediate steps (no expansion in α), then we need to write the equations

$$Z_1\, \Lambda^\nu_{\rm OS}(p,p) = \gamma^\nu (1 - Z_1)\,, \tag{10.123a}$$

$$\Lambda^\nu_{\rm R}(p,p') = Z_1 \left[\Lambda^\nu_{\rm OS}(p,p') - \Lambda^\nu_{\rm OS}(p,p)\right]\,, \tag{10.123b}$$

again with a single subtraction at zero momentum transfer. We can now also understand why there is an additional factor Z_1^{-1} in front of the renormalized terms in Eq. (10.119). Namely, after the absorption of the first term, $\gamma^\mu\,(Z_1^{-1} - 1)$, into the physical electron charge, there is still a term left, which multiplies the unrenormalized charge e_0 in the bare Lagrangian, and which must be multiplied with Z_1^{-1}, to obtain the physical charge which multiplies the renormalized vertex function.

Let us now extend this program to the complete QED Lagrangian. The original QED Lagrangian $\mathcal{L} \equiv \mathcal{L}^{(0)}_{\rm QED}$, given in Eq. (10.1), is identified as the bare Lagrangian. We add a gauge-fixing term as well as a small photon mass, which as we saw comes in useful for the calculation of vertex corrections,

$$\begin{aligned}\mathcal{L}^{(0)}_{\rm QED} = & -\frac{1}{4}\left(\partial^\mu A_0^\nu - \partial^\nu A_0^\mu\right)\left(\partial_\mu A_{0\nu} - \partial_\nu A_{0\mu}\right) - \frac{\lambda_0}{2}\left(\partial_\mu A_0^\mu\right)^2 \\ & + \frac{\lambda_0^2}{2} A_0^2 + \overline{\psi}_0 \left(\mathrm{i}\not{\partial} - e_0 \not{A}_0 - m_0\right)\psi_0\,.\end{aligned} \tag{10.124}$$

This is the QED Lagrangian in the interpretation *(i)* given above: All terms are expressed in terms of bare quantities. The term proportional to $\lambda_0^2 A_0^2$ describes the infinitesimal photon mass.

We apologize to the reader for the similarity of the fonts used for the (bare) photon mass λ_0 and the (bare) gauge fixing parameter λ_0. In view of the presence of a number of variations of μ in this treatise, among them, μ for the reduced mass of a two-body bound system, and μ for the renormalization scale, we prefer λ for the photon mass.

Now, we can write the original Lagrangian $\mathcal{L} \equiv \mathcal{L}^{(0)}_{\rm QED}$, given in Eq. (10.1), in view of the relations (10.111), as follows,

$$\begin{aligned}\mathcal{L}^{(0)}_{\rm QED} = & -\frac{Z_3}{4}\left(\partial^\mu A_{\rm R}^\nu - \partial^\nu A_{\rm R}^\mu\right)\left(\partial_\mu A_{{\rm R}\,\nu} - \partial_\nu A_{{\rm R}\,\mu}\right) - \frac{\lambda_0\, Z_3}{2}\left(\partial_\mu A_{\rm R}^\mu\right)^2 \\ & + \frac{\lambda_0^2\, Z_3}{2} A_{\rm R}^2 + Z_2\, \overline{\psi}_{\rm R}\left(\mathrm{i}\not{\partial} - m_0\right)\psi_{\rm R} - Z_1\, e_{\rm R}\, \overline{\psi}_{\rm R} \not{A}_{\rm R}\, \psi_{\rm R}\,.\end{aligned} \tag{10.125}$$

This is the QED Lagrangian in the interpretation *(iii)* as discussed above. Moreover, Ward–Takahashi identities imply that we can consistently renormalize QED,

$$Z_1 = Z_2\,, \qquad \lambda_{\rm R} = Z_3\, \lambda_0\,, \qquad \lambda_{\rm R}^2 = Z_3\, \lambda_0^2\,. \tag{10.126}$$

The renormalization of the gauge-fixing parameter λ, and of the photon mass λ, is physically irrelevant. This is because the gauge-fixing parameter constitutes an auxiliary, variational parameter, and because the photon mass is infinitesimally small (and remains so after renormalization). The identity $Z_1 = Z_2$ is especially important because both renormalization constants Z_1 and Z_2 are gauge dependent: If they were not equal, then, by choosing a different gauge, we could assign a different multiplicative correction to the renormalized interaction Lagrangian $\mathcal{L}_I^{(R)}$ as opposed to the free fermionic term $\overline{\psi}_0 \left(\mathrm{i}\partial\!\!\!/ - m_0\right) \psi_0$. This would destroy the gauge invariance of the theory on the level of loop corrections. Finally, the Ward–Takahashi identities imply that, after the specification of the counterterms, the QED Lagrangian reads as

$$\begin{aligned}\mathcal{L}^{(0)}_{\mathrm{QED}} &= -\frac{1}{4}\left(\partial^\mu A^\nu_{\mathrm{R}} - \partial^\nu A^\mu_{\mathrm{R}}\right)\left(\partial_\mu A_{\mathrm{R}\nu} - \partial_\nu A_{\mathrm{R}\mu}\right) - \frac{\lambda}{2}\left(\partial_\mu A^\mu_{\mathrm{R}}\right)^2 \\ &\quad + \frac{\lambda}{2} A^2_{\mathrm{R}} + \overline{\psi}_{\mathrm{R}}\left(\mathrm{i}\partial\!\!\!/ - eA\!\!\!/_{\mathrm{R}} - m_{\mathrm{R}}\right)\psi_{\mathrm{R}} + \mathcal{L}_{\mathrm{CT}}\,, \end{aligned} \tag{10.127}$$

where all quantities are equal to the renormalized, physical entities. All counterterms in $\mathcal{L}_{\mathrm{CT}}$ will be specified in the following. The bare mass m_0 and the renormalized mass m_{R} are related as $m_0 = m_{\mathrm{R}} - \delta m$, where δm is the mass counter term.

In Eq. (10.124), we had expressed the unrenormalized (bare) Lagrangian in terms of the bare quantities. By contrast, in Eq. (10.127), the renormalized Lagrangian is expressed in terms of the physical fields. The observation is that the unrenormalized (bare) Lagrangian in terms of the bare quantities, has the same functional form as the renormalized Lagrangian, expressed in terms of the physical fields. Let us therefore define a mapping of the vector of renormalization constants and physical parameters onto the Lagrangian, which read as follows,

$$(\vec{Z} = (Z_1, Z_2, Z_3), \delta m, \vec{v} = (e, m, \psi, A, \lambda, \lambda)) \to \mathcal{L}^{(0)}_{\mathrm{QED}}(\vec{Z}, \delta m, \vec{v})\,. \tag{10.128}$$

Defining $\vec{1} = (1, 1, 1)$, one can write the relation

$$\mathcal{L}^{(0)}_{\mathrm{QED}}(\vec{1}, 0, \vec{v}_0) = \mathcal{L}^{(0)}_{\mathrm{QED}}(\vec{Z}, \delta m, \vec{v}_{\mathrm{R}}) = \mathcal{L}^{(0)}_{\mathrm{QED}}(\vec{1}, 0, \vec{v}_{\mathrm{R}}) + \mathcal{L}_{\mathrm{CT}}(\vec{Z}, \delta m, \vec{v}_{\mathrm{R}})\,, \tag{10.129}$$

where $\vec{v}_0$ is the input "vector" of bare parameters, and $\vec{v}_{\mathrm{R}}$ is the input "vector" of renormalized parameters,

$$\vec{v}_0 = (e_0, \mu_0, m_0, \psi_0, A_0, \lambda_0, \lambda_0), \qquad \vec{v}_{\mathrm{R}} = (e_{\mathrm{R}}, \mu_{\mathrm{R}}, m_{\mathrm{R}}, \psi_{\mathrm{R}}, A_{\mathrm{R}}, \lambda_{\mathrm{R}}, \lambda_{\mathrm{R}}). \tag{10.130}$$

This relation describes the renormalizability of the theory: The physical, renormalized Lagrangian $\mathcal{L}_{\mathrm{R}}$ of the theory, given as $\mathcal{L}_{\mathrm{R}} = \mathcal{L}^{(0)}_{\mathrm{QED}}(\vec{Z}, \delta m, \vec{v}_{\mathrm{R}})$ expressed in terms of the physical, renormalized quantities, has the same functional form as the bare Lagrangian, expressed in terms of the bare quantities. It is often stated that QED actually is defined at some ultraviolet scale and then evolved, using renormalization-group equations, into the infrared (low-energy) region. This statement finds an easy explanation: Namely, looking at a charged electron at very high momenta (very small distances), we can say that the vacuum-polarization charge cloud is gradually "taken away", and in fact we see the more of the "bare" charge of the electron, which is larger than the charge seen at low momenta (large distance). Bare parameters are relevant to high energy scales. This explains why we say that the theory is defined at high energy, while the renormalization conditions used here actually describe low-energy physics.

In order to derive the Ward–Takahashi identities, it is in fact very useful to take a step back and recall the vertex function at $q = 0$, i.e. $p' = p$, initially *without* assuming that the incoming and outgoing bispinors are on the mass shell,

$$\Lambda^\mu(p,p) = \mathrm{i}\,e^2 \int \frac{\mathrm{d}^4 k}{(2\pi)^4} D_F^{\nu\rho}(k)\,\gamma_\nu \frac{1}{\not{p} - \not{k} - m + \mathrm{i}\epsilon}\gamma^\mu \frac{1}{\not{p} - \not{k} - m + \mathrm{i}\epsilon}\gamma_\rho\,. \tag{10.131}$$

Using the operator identity

$$\frac{1}{\not{p} - \not{k} - m + \mathrm{i}\epsilon}\gamma^\mu \frac{1}{\not{p} - \not{k} - m + \mathrm{i}\epsilon} = -\frac{\partial}{\partial p_\mu}\frac{1}{\not{p} - \not{k} - m + \mathrm{i}\epsilon}\,, \tag{10.132}$$

where the sequence of terms matters, we can rewrite the one-loop vertex function at zero momentum transfer as follows,

$$\Lambda^\mu(p,p) = -\frac{\partial}{\partial p_\mu}\Sigma(p)\,, \qquad \Sigma(p) = \mathrm{i}e^2 \int \frac{\mathrm{d}^4 k}{(2\pi)^4} D_F^{\nu\rho}(k)\,\gamma_\nu \frac{1}{\not{p} - \not{k} - m + \mathrm{i}\epsilon}\gamma_\rho\,, \tag{10.133}$$

where $\Sigma(p)$ is the one-loop self-energy. We can even ascertain that the first equality in the relation (10.133) is general and holds to all loop orders. A handwaving argument is as follows: One considers an arbitrary multi-loop self-energy diagram with the derivative $\partial/\partial p_\mu$ acting on all the different internal fermion propagators due to the product rule. This differentiation accounts for all possibilities to insert a γ^μ matrix. The Ward–Takahashi identity is appropriately named because it was Takahashi who generalized this for arbitrary momentum transfer $q = p' - p$ to

$$(p'_\mu - p_\mu)\Gamma^\mu(p,p') = S_F^{-1}(p') - S_F^{-1}(p)\,, \tag{10.134}$$

where $S_F^{-1}(p)$ is the inverse of the full (up to all orders) Feynman propagator (including self-energy insertions), and we refer to Eq. (10.118) for a definition of the vertex function, which we use in an extended sense here, writing $\Gamma^\mu(p,p') = \gamma^\mu + \Lambda^\mu(p,p')$, where $\Lambda^\mu(p,p')$ is an all-order vertex correction. In full analogy with Eq. (10.119), we can write an expansion for the self-energy,

$$\Sigma(p) = \delta m - \left[Z_2^{-1} - 1\right](\not{p} - m) + Z_2^{-1}\,\Sigma_{\mathrm{R}}(p)\,, \tag{10.135}$$

where δm is the mass counterterm, i.e., $m = m_0 + \delta m$. With the help of the Ward–Takahashi identity, we can then write for zero momentum transfer and on mass shell ($\not{p} = m$), in view of the relation (10.135),

$$\Lambda^\mu(p,p)\big|_{\not{p}=m} = \Lambda^\mu_{\mathrm{OS}}(p,p) = -\frac{\partial}{\partial p_\mu}\Sigma(p)\big|_{\not{p}=m} = \gamma^\mu\left(Z_2^{-1} - 1\right)\,, \qquad Z_1 = Z_2\,. \tag{10.136}$$

Indeed, a quick look at Eq. (10.121a) then reveals that $Z_1 = Z_2$. According to Eqs. (10.94) and (10.107), the renormalization constants can be calculated as follows,

$$\begin{aligned}\Lambda^{\mu}_{\rm OS}(p,p) &= \gamma^{\nu}\left[Z_1^{-1} - 1\right] = \gamma^{\nu}\left[Z_2^{-1} - 1\right] \\ &= \gamma^{\nu}\,\frac{\alpha}{\pi}\left[\frac{1}{4}\ln\left(\frac{\Lambda^2}{m^2}\right) - \frac{1}{2}\ln\left(\frac{m}{\lambda^2}\right) + \frac{9}{8} + \mathcal{O}\left(\frac{\lambda^2}{m^2}\right)\right].\end{aligned} \tag{10.137}$$

For the renormalized vertex function we thus have the integral given in Eq. (10.95).

10.3.4 *Detour on Dimensional Regularization*

A very brief detour on dimensional regularization is in order. We follow Sec. 8.2.5 of Ref. [81]. Dimensional regularization proceeds by artificially modifying the dimension of space-time into non-integer values, the notion being that typical (logarithmic) ultraviolet as well as infrared divergences can be eliminated once the quantum-field theoretical integrals are continued analytically into a fictitious space with a non-integer value of dimensions. The analytic continuation is inspired by mathematical considerations which lead to an analytic continuation of the solid angle into an arbitrary number of dimensions. Pauli–Villars regularization respects radial symmetries of the problem; dimensional regularization does so automatically, and in addition, it also respects gauge invariance. Strictly speaking, the introduction of the photon mass in dimensional regularization breaks gauge invariance, albeit only in terms of a mathematical computation device. In compensation, in parity violating weak interactions, which involve the fifth current matrix γ^5, there are problems associated with dimensional regularization: By construction of the electroweak theory, the γ^5 matrix is required to anticommute with all other γ^μ matrices, a property required in order to consistently define chirality. However, in an arbitrary number of dimensions d, this concept is problematic; we shall not dwell into this matter here in too much detail but point out that remedies (including the so-called naive γ^5 scheme) exist (see also Sec. 10.7).

In order to proceed, we need the volume surface area Ω_D of the unit sphere embedded in D-dimensional space. The matching of the result for integer dimensions to the result for non-integer dimensions is recalled from Eq. (10.88) as follows,

$$\Omega_D = \int \mathrm{d}\Omega_D = \frac{2\,\pi^{D/2}}{\Gamma(D/2)}. \tag{10.138}$$

This result takes the following values for integer dimension,

$$\Omega_D = \begin{cases} 2 & (D = 1) \\ 2\pi & (D = 2) \\ 4\pi & (D = 3) \\ 2\pi^2 & (D = 4) \end{cases}. \tag{10.139}$$

One may ask about the origin of the result $\Omega_{D=1} = 2$. The reason is that we attempt to replace an integral over the entire D-dimensional space $\int_{\mathbb{R}^D} \mathrm{d}^D r$ with an integral of the form proportional to $\int_0^\infty \mathrm{d}r\, r^{D-1}$ over a radially symmetric function. Then, in one dimension, for a function $f(x) = f(-x)$, with $x \in \mathbb{R}$, we have a factor two because the "solid angle" in one dimension simply counts the number of endpoints of the line segment $|x| \leq 1$, which is the one-dimensional analog of a unit sphere. Regimes $x \in (-\infty, 0)$ and $x \in (0, \infty)$ have to be treated separately.

For a radially symmetric integrand $f(\vec{r}) = f(r)$, one has

$$\int_{\mathbb{R}^D} \mathrm{d}^D r\, f(\vec{r}) = \Omega_D \int_0^\infty \mathrm{d}r\, r^{D-1} f(r)\,, \tag{10.140}$$

in a general dimension D (where $r = |\vec{r}|$ and the function f is defined accordingly, depending on the scalar or vector character of its argument). In the following, we will assume a non-integer dimension D, slightly displaced from the $n = 3$ by 2ε,

$$D = 3 - 2\varepsilon\,, \tag{10.141}$$

and consider the continuation of the Coulomb potential to a non-integer number of dimensions. Note that we here denote the deviation from an integer number of dimensions by the letter ε, while energy cutoffs are denoted as ϵ (see Chap. 4). We also use an infinitesimal parameter $\epsilon > 0$ in propagator denominators. Our considerations here supplement other works (see Chap. 8 of Ref. [81] and Ref. [307]) where, within classical as opposed to quantum electrodynamics, electrostatic potentials have been evaluated using dimensional regularization.

In the following, we typically use the symbol $D = 3 - 2\varepsilon$ for the dimensionality of a regularized three-dimensional space, and

$$d = 4 - 2\varepsilon \tag{10.142}$$

for the dimensionality of a regularized four-dimensional space-time. Per Eq. (10.138), the surface of the unit sphere in $D = 3 - 2\varepsilon$ dimensions is $\Omega_{3-2\varepsilon} = 2\pi^{3-2\varepsilon}/\Gamma[(3-2\varepsilon)/2]$. The radial part of the Laplacian is

$$\vec{\nabla}^2 \sim \frac{\partial^2}{\partial r^2} + \frac{D-1}{r}\frac{\partial}{\partial r}\,. \tag{10.143}$$

Let us define the Coulomb potential (strictly speaking, the potential energy experienced by a bound particle of charge number Z) in D dimensions as the radially symmetric potential V that solves the equation

$$\left(\frac{\partial^2}{\partial r^2} + \frac{2-2\varepsilon}{r}\frac{\partial}{\partial r}\right) V(r) = 4\pi Z\alpha\, \mu^{3-D}\, \delta^{(D)}(\vec{r})\,. \tag{10.144}$$

Here, the mass scaling is as follows. The potential V has mass dimension one, the Laplacian has mass dimension two. The Dirac-δ in D dimensions, $\delta^{(D)}(\vec{r})$, has mass dimension D, because the dimension of the D-dimensional Dirac-δ is the inverse of the dimension of its argument. In order for the right-hand side of Eq. (10.144) to have the same mass dimension as the left-hand side, we need to insert a factor $\mu^{3-D} = \mu^{2\varepsilon}$, which is typical for calculations that involve dimensional regularization.

The right-hand side of Eq. (10.144) is zero almost everywhere, namely, except at the origin of D-dimensional space. Hence, we use an *ansatz* of the form

$$V = \frac{C}{r^{D-2}} = \frac{C}{r^{1-2\varepsilon}} \tag{10.145}$$

for the potential. Let us now return to the defining equation

$$\vec{\nabla}^2 V(r) = 4\pi Z\alpha\, \mu^{2\varepsilon}\, \delta^{(D)}(\vec{r})\,, \tag{10.146}$$

and try to determine the coefficient C in the *ansatz* given in Eq. (10.145). Certainly, the integral over $\int \mathrm{d}^D r$ of the right-hand side of Eq. (10.146) is equal to $4\pi Z\alpha\, \mu^{2\varepsilon}$, by a fundamental property of the D-dimensional Dirac-δ.

For the left-hand side of Eq. (10.146), one can integrate over the D-dimensional sphere S_R of radius R and obtain

$$\int_{S_R} \mathrm{d}^D r\, \vec{\nabla}^2 V(r) = \int_{\partial S_R} \mathrm{d}\vec{A} \cdot \vec{\nabla} V(r) = R^{D-1}\, \Omega_D \left. \left(\frac{\partial}{\partial r} V(r) \right) \right|_{r=R}\,, \qquad D = 3 - 2\varepsilon\,. \tag{10.147}$$

Here, GAUSS's[8] theorem has been used in D dimensions, and we have taken advantage of the radial symmetry of the potential. The surface of the D-dimensional sphere of radius R is denoted as ∂S_R. In view of the identity [see Eq. (10.145)]

$$\left. \left(\frac{\partial}{\partial r} V(r) \right) \right|_{r=R} = -C\,(D-2)\, R^{1-D}\,, \tag{10.148}$$

one can derive the result

$$C = -Z\alpha\, \mu^{2\varepsilon}\, \pi^{-1/2+\varepsilon}\, \Gamma\left(\frac{1}{2} - \varepsilon \right). \tag{10.149}$$

So, the Coulomb potential in $D = 3 - 2\varepsilon$ dimensions is

$$V(r) = -\frac{Z\alpha}{r^{1-2\varepsilon}}\, \mu^{2\varepsilon}\, \pi^{-1/2+\varepsilon}\, \Gamma\left(\frac{1}{2} - \varepsilon \right). \tag{10.150}$$

This relation confirms results given in Refs. [312] and [313].

10.3.5 *Detour on Feynman Parameterization*

Feynman parameters have been outlined in Eq. (10.65). We will need the appropriate generalization to m denominators, which reads as follows,

$$\frac{1}{A_1\, A_2 \cdots A_m} = (m-1)! \left(\prod_{j=1}^{m} \int_0^1 \mathrm{d}u_j \right) \frac{\delta(1 - u_1 - u_2 - \cdots - u_m)}{[A_1\, u_1 + A_2\, u_2 + \cdots + A_m\, u_m]^m}\,. \tag{10.151}$$

For $m = 2$, we recover Eq. (10.65). There are two general identities which are useful for integrating out Feynman parameters. The first of these is given as

$$\int_0^1 \mathrm{d}u_m\, \delta(1 - u_1 - u_2 - \cdots - u_m)\, f(u_1, u_2, \ldots, u_m)$$
$$= \Theta(1 - u_1 - u_2 - \cdots - u_{m-1})\, f(u_1, u_2, \ldots, 1 - u_1 - u_2 - \cdots - u_{m-1})\,. \tag{10.152}$$

[8] Johann Carl Friedrich Gauss (1777–1855).

This identity holds because, if the expression $u_1 + u_2 + \cdots + u_{m-1}$ turns out to be negative or greater than one, then the peak of the Dirac-δ function will be outside of the interval $u_m \in (0,1)$. If we did not assume that all the u_i with $i = 1, \ldots, m$ are positive, then we would have to indicate an extra Heaviside step function $\Theta(u_1 + u_2 + \cdots + u_{m-1})$ on the right-hand side. The second identity is

$$\begin{aligned}\int_0^1 & du_{m-1}\, \Theta(1 - u_1 - u_2 - \cdots - u_{m-1})\, f(u_1, u_2, \ldots, 1 - u_1 - u_2 - \cdots - u_{m-1}) \\ &= \Theta(1 - u_1 - u_2 - \cdots - u_{m-2}) \\ &\times \int_0^{1-u_1-\cdots-u_{m-2}} du_{m-1}\, f(u_1, u_2, \ldots, 1 - u_1 - u_2 - \cdots - u_{m-1}) \,. \end{aligned} \tag{10.153}$$

The upper integration limit on the right-hand side is $1 - u_1 - \cdots - u_{m-2}$. We need to indicate an extra Heaviside step function $\Theta(1 - u_1 - u_2 - \cdots - u_{m-2})$ on the right-hand side because, if the expression $u_1 + u_2 + \cdots + u_{m-2}$ is greater than one, then the integrand vanishes on the entire integration interval $u_{m-1} \in (0,1)$.

With the help of the identities (10.152) and (10.153), one can subsequently integrate out the Feynman parameters. For example, one has

$$\begin{aligned}\int_0^1 du_1 \int_0^1 du_2 \int_0^1 du_3\, \delta(1 - u_1 - u_2 - u_3)\, f(u_1, u_2, u_3) \\ = \int_0^1 du_1 \int_0^{1-u_1} du_2\, f(u_1, u_2, 1 - u_1 - u_2) \,, \end{aligned} \tag{10.154}$$

where we note the integration limits, and

$$\begin{aligned}\int_0^1 du_1 \int_0^1 du_2 \int_0^1 du_3 \int_0^1 du_4\, \delta(1 - u_1 - u_2 - u_3 - u_4)\, f(u_1, u_2, u_3, u_4) \\ = \int_0^1 du_1 \int_0^{1-u_1} du_2 \int_0^{1-u_1-u_2} du_3\, f(u_1, u_2, u_3, 1 - u_1 - u_2 - u_3) \,. \end{aligned} \tag{10.155}$$

We can thus rewrite Eq. (10.151) as follows,

$$\begin{aligned}\frac{1}{A_1\, A_2 \cdots A_m} &= (m-1)! \int_0^1 du_1 \int_0^{1-u_1} du_2 \cdots \int_0^{1-u_1-u_2-\cdots-u_{m-2}} du_{m-1} \\ &\times \frac{1}{[A_1\, u_1 + A_2\, u_2 + \cdots + A_{m-1}\, u_{m-1} + A_m\, (1 - u_1 - u_2 - \cdots - u_{m-1})]^m} \,. \end{aligned} \tag{10.156}$$

Using the substitution (with unit Jacobian)

$$v_1 = 1 - u_1 \,, \qquad v_2 = 1 - u_1 - u_2 \,, \tag{10.157}$$

$$v_3 = 1 - u_1 - u_2 - u_3 \,, \ldots \,, \tag{10.158}$$

one can show the validity of an alternative representation

$$\begin{aligned}\frac{1}{A_1\, A_2 \cdots A_m} &= (m-1)! \int_0^1 dv_1 \int_0^{v_1} dv_2 \cdots \int_0^{v_{m-2}} dv_{m-1} \\ &\times \frac{1}{[A_1 + v_1\, (A_2 - A_1) + \cdots + v_{m-1}\, (A_m - A_{m-1})]^m} \,, \end{aligned} \tag{10.159}$$

which is sometimes useful. By a suitable differentiation under the integral sign, one can show the more general formula

$$\frac{1}{A_1^{\alpha_1} \cdots A_m^{\alpha_m}} = \frac{\Gamma(\sum_{k=1}^m \alpha_k)}{\prod_{k=1}^m \Gamma(\alpha_k)} \int_0^1 du_1 \ldots \int_0^1 du_m \frac{\delta(1 - \sum_{k=1}^m u_k)\, u_1^{\alpha_1 - 1} \cdots u_m^{\alpha_m - 1}}{(\sum_{k=1}^m u_k\, A_k)^{\sum_{k=1}^m \alpha_k}} \,, \tag{10.160}$$

which is valid for general exponents $\alpha_1, \ldots, \alpha_m$.

10.3.6 *Vertex and Dimensional Regularization*

Let us now delve into a calculation to illustrate the method. We start from Eq. (10.47), generalized to d dimensions,

$$\Lambda^{\nu}(p,p') = \mathrm{i}e^2 \int \frac{\mathrm{d}^d k}{(2\pi)^d} D_F^{\mu\rho}(k) \frac{\gamma_\mu (\not{p}' - \not{k} + m)\gamma^\nu (\not{p} - \not{k} + m)\gamma_\rho}{[(p'-k)^2 - m^2 + \mathrm{i}\epsilon][(p-k)^2 - m^2 + \mathrm{i}\epsilon]}. \quad (10.161)$$

It is important to mention here that also in $d = 4 - 2\varepsilon$ dimensions [see Eq. (10.142)], we still have one time-like dimension and thus $d - 1$ space-like dimension. Consequently, the dimension of the Wick rotation, which affects only the time-like component of the photon four-momentum, remains unchanged. One immediately visible advantage of dimensional regularization, in comparison to Pauli–Villars regularization (see Sec. 10.3.1), is that we do not need to introduce a separate infrared regulator (photon mass λ). The integration over the L parameter, per Eq. (10.50), is thus unnecessary, and we can directly proceed to the combination of the propagator denominators via Feynman parameterization.

Evaluating Eq. (10.161) on mass shell, we find

$$\Lambda^{\nu}_{\mathrm{OS}}(p,p') = -\mathrm{i}e^2 \int \frac{\mathrm{d}^d k}{(2\pi)^d} \frac{g^{\mu\rho}}{k^2 + \mathrm{i}\epsilon} \frac{\gamma_\mu (\not{p}' - \not{k} + m)\gamma^\nu (\not{p} - \not{k} + m)\gamma_\rho}{[k^2 - 2p'\cdot k + \mathrm{i}\epsilon][k^2 - 2p\cdot k + \mathrm{i}\epsilon]}\bigg|_{\not{p}=\not{p}'=m}. \quad (10.162)$$

Here, we use the same definition of the on-mass-shell condition as in Eq. (10.53): whenever $\not{p}$ in the numerator of $\Lambda^{\nu}_{\mathrm{OS}}(p,p')$ is brought to the right, or $\not{p}'$ is brought to the left, it will be replaced by the mass m.

We can use our result from Eq. (10.63), where we already calculated the Dirac algebra in d dimensions, to write

$$\Lambda^{\nu}_{\mathrm{OS}}(p,p') = -\mathrm{i}e^2 \int \frac{\mathrm{d}^d k}{(2\pi)^d} \frac{\mathcal{N}^\nu}{[k^2 + \mathrm{i}\epsilon][k^2 - 2p'\cdot k + \mathrm{i}\epsilon][k^2 - 2p\cdot k + \mathrm{i}\epsilon]}, \quad (10.163\text{a})$$

$$\mathcal{N}^\nu = -4mk^\nu + (d-6)\gamma^\nu k^2 + 4\not{k}\left[p'^\nu + p^\nu - \left(\frac{d}{2} - 1\right)k^\nu\right] + 4\gamma^\nu (p'-k)\cdot(p-k). \quad (10.163\text{b})$$

In combining the denominators, we proceed as outlined in Sec. 10.3.4, use Eq. (10.156) for $m = 3$, to combine the denominators in Eq. (10.163), and find

$$\begin{aligned}\Lambda^{\nu}_{\mathrm{OS}}(p,p') &= -\mathrm{i}e^2\,(2!) \int_0^1 \mathrm{d}u_1 \int_0^1 \mathrm{d}u_2 \int_0^1 \mathrm{d}u_3\, \delta(1 - u_1 - u_2 - u_3) \\ &\quad \times \int_{C_F} \frac{\mathrm{d}^d k}{(2\pi)^d} \frac{\mathcal{N}^\nu}{[u_1 k^2 + u_2 k^2 - 2u_2 p'\cdot k + u_3 k^2 - 2u_3 p\cdot k]^3}, \\ &= -2\mathrm{i}e^2 \int_0^1 \mathrm{d}u_1 \int_0^1 \mathrm{d}u_2 \int_0^1 \mathrm{d}u_3\, \delta(1 - u_1 - u_2 - u_3) \\ &\quad \times \int_{C_F} \frac{\mathrm{d}^d k}{(2\pi)^d} \frac{\mathcal{N}^\nu}{[k^2 - 2k\cdot(u_2 p' + u_3 p)]^3},\end{aligned} \quad (10.164)$$

where we have suppressed the infinitesimal imaginary part $\mathrm{i}\epsilon$ in view of the specification of the Feynman contour C_F. The integrand could be rearranged with the help of the identity $u_1 + u_2 + u_3 = 1$. The square in the denominator is completed according to

$$k = \ell + u_2\, p' + u_3\, p\,, \tag{10.165a}$$

$$D = k^2 - 2k \cdot (u_2 p' + u_3 p) = \ell^2 - \Delta\,, \tag{10.165b}$$

$$\Delta = m^2\,(u_2 + u_3)^2 - q^2\, u_2\, u_3\,. \tag{10.165c}$$

We rewrite the loop momentum k in terms of ℓ, with unit Jacobian, and use the replacement

$$\ell^\mu \ell^\nu \to \frac{1}{d} g^{\mu\nu} \ell^2\,, \tag{10.166}$$

in the numerator in Eq. (10.164), motivated by the angular structure in d dimensions. Terms linear in ℓ in the numerator vanish after loop momentum integration, due to symmetry. We can thus replace the numerator structure in Eq. (10.164), originally defined in Eq. (10.63), as follows,

$$\begin{aligned}\mathcal{N}^\nu \to\ & \frac{(d-2)^2}{d} \gamma^\nu \ell^2 + 2\gamma^\nu\, m^2 \left\{2 - 2u_2 + \left(1 - \frac{d}{2}\right) u_2^2 - 2u_3 + (2-d)\, u_2 u_3 \right. \\ & \left. + \left(1 - \frac{d}{2}\right) u_3^2 \right\} + 2\gamma^\nu q^2 \left[u_2 + u_3 - 1 + \left(1 - \frac{d}{2}\right) u_2 u_3\right] \\ & - 4m\,\mathrm{i}\,\sigma^{\nu\rho} q_\rho \left[u_2 + \left(1 - \frac{d}{2}\right) u_2^2 + u_3 + 2\left(1 - \frac{d}{2}\right) u_2 u_3 + \left(1 - \frac{d}{2}\right) u_3^2\right].\end{aligned} \tag{10.167}$$

Here, we have left out a term proportional to q^ν on the right-hand side, which, per our discussion following Eq. (10.92), vanishes after integration. Another argument to show this is as follows. From the Ward–Takahashi identity, given in Eq. (10.134), which we use in the form $(p'_\nu - p_\nu)\, \Lambda^\nu(p,p') = S_F^{-1}(p') - S_F^{-1}(p)$, one can infer that any conceivable contribution to $\Lambda^\mu(p,p')$, proportional to q^ν, needs to vanish. This is because it would lead to a term $(p'_\nu - p_\mu)\Lambda^\nu(p,p') \to q_\nu q^\nu = q^2$, which cannot be reproduced by any on-mass-shell self-energy insertion on the right-hand side $S_F^{-1}(p') - S_F^{-1}(p)$, in view of the decomposition (10.135).

We are interested in the numerator $\mathcal{N}^\nu$ for $d = 4 - 2\varepsilon$ dimension, i.e., close to the ordinary four dimensions of space-time. As far as $\mathcal{N}^\nu$ is concerned, it turns out to be advantageous to set $d = 4 - 2\varepsilon$ everywhere except the first term, and we write

$$\begin{aligned}\mathcal{N}^\nu =\ & \frac{(d-2)^2}{d} \gamma^\nu \ell^2 + \gamma^\nu \left\{2m^2 \left[2 - 2u_2 + (\varepsilon - 1)\, u_2^2 - 2u_3 + 2\,(\varepsilon - 1)\, u_2 u_3 + (\varepsilon - 1)\, u_3^2\right]\right. \\ & \left. + 2q^2 \left[u_2 + u_3 - 1 + (\varepsilon - 1)\, u_2 u_3\right]\right\} \\ & - 2mi\,\sigma^{\nu\rho} q_\rho \left[u_2 + (\varepsilon - 1)\, u_2^2 + u_3 + 2\,(\varepsilon - 1)\, u_2 u_3 + (\varepsilon - 1)\, u_3^2\right] \\ =\ & \frac{(d-2)^2}{d} \gamma^\nu \ell^2 + \gamma^\nu\, \mathcal{N}_1 + \mathrm{i}\sigma^{\nu\rho} q_\rho\, \mathcal{N}_2\,.\end{aligned} \tag{10.168}$$

Here,

$$\begin{aligned}\mathcal{N}_1 =\ & 2m^2 \left[2 - 2u_2 + (\varepsilon - 1)\, u_2^2 - 2u_3 + 2\,(\varepsilon - 1)\, u_2 u_3 + (\varepsilon - 1)\, u_3^2\right] \\ & + 2q^2 \left[u_2 + u_3 - 1 + (\varepsilon - 1)\, u_2 u_3\right],\end{aligned} \tag{10.169a}$$

$$\mathcal{N}_2 = -2m \left[u_2 + (\varepsilon - 1)\, u_2^2 + u_3 + 2\,(\varepsilon - 1)\, u_2 u_3 + (\varepsilon - 1)\, u_3^2\right]. \tag{10.169b}$$

We can now put these simplified expressions for the numerator and denominator back into $\Lambda^{\nu}_{\rm OS}$.

However, there is an obstacle. Namely, the charge e in d dimensions cannot be dimensionless. In order to see this, let us consider the QED interaction Lagrangian in d dimensions [compare with Eq. (10.3) for the Lagrangian density],

$$\mathcal{L}_I = -e \int \mathrm{d}^d x \, \overline{\psi}(x) \, \gamma^{\mu} \, \psi(x) \, A_{\mu}(x) \,. \tag{10.170}$$

Let M temporarily denote a mass scale, which, in natural units, is equal to the inverse of a length scale. Then, if we denote the mass dimension of a physical quantity X by $[X]$,

$$[\mathrm{d}^d x] = M^{-d} \,, \qquad [\psi] = M^{(d-1)/2} \,, \qquad [A^{\mu}] = M^{(d-2)/2} \,. \tag{10.171}$$

The latter mass scaling can be derived based on the requirement that the free electromagnetic part of the Lagrangian (the Maxwell Lagrangian in d dimensions) be dimensionless. Namely, we have $\mathcal{L} = -(1/4) \int \mathrm{d}^d x \, F_{\mu\nu} \, F^{\mu\nu}$, where $F_{\mu\nu} = \partial_{\mu} A_{\nu} - \partial_{\nu} A_{\mu}$. The mass scaling of ψ can be derived from the requirement that the integral of the spatial probability density operator (absolute square of the field operator), given as $\int \mathrm{d}^{d-1} x \, \psi^{+}(x) \, \psi(x)$, must be dimensionless. Now, for $\mathcal{L}_I$ in Eq. (10.170) to be dimensionless, in view of the mass scalings in Eq. (10.171), $\mathcal{L}_I$ can be made dimensionless provided e acquires a mass dimension of

$$[e] = M^{2-d/2} = M^{\varepsilon} \,, \qquad d = 4 - 2\varepsilon \,. \tag{10.172}$$

We verify that if $d = 4$, the charge e is dimensionless. In a general number of $d = 4 - 2\varepsilon$ dimensions, it turns out that a favorable ansatz (known as "modified minimal subtraction", or $\overline{\rm MS}$) for the square of the charge is

$$e^2 = (4\pi)^{d/2-1} \, \alpha \, \mu^{4-d} \, \mathrm{e}^{\gamma_E (2-d/2)} = (4\pi)^{1-\varepsilon} \, \alpha \, \mu^{2\varepsilon} \mathrm{e}^{\gamma_E \varepsilon} \,, \tag{10.173}$$

where $\gamma_E = 0.57721\,56649\ldots$ is the Euler–Mascheroni constant, which we had already encountered in Eq. (4.272). Notice that we were forced to introduce an arbitrary mass parameter μ in order to keep the dimensions consistent. The introduction of μ is characteristic of dimensional regularization and has already appeared in Eq. (10.144). The parameter μ finds a natural interpretation as a renormalization scale. With the help of Eqs. (10.164) and (10.173), as well as Eq. (10.168), we find for $\Lambda_{\rm OS}(p, p')$,

$$\begin{aligned}\Lambda^{\nu}_{\rm OS}(p, p') = & -8\pi \mathrm{i} \alpha \mu^{2\varepsilon} \mathrm{e}^{\gamma_E \varepsilon} (4\pi)^{-\varepsilon} \int_0^1 \mathrm{d}u_1 \int_0^1 \mathrm{d}u_2 \int_0^1 \mathrm{d}u_3 \\ & \times \delta(1 - u_1 - u_2 - u_3) \int \frac{\mathrm{d}^d \ell}{(2\pi)^d} \frac{\frac{(d-2)^2}{d} \gamma^{\nu} \ell^2 + \gamma^{\nu} \mathcal{N}_1 + \mathrm{i} \sigma^{\nu\rho} q_{\rho} \, \mathcal{N}_2}{(\ell^2 - \Delta + \mathrm{i}\epsilon)^3} \,.\end{aligned} \tag{10.174}$$

Following the discussion in Sec. 10.3.1 [see the text surrounding Eq. (10.84)], we actively shift the integration contour along the time-like dimension to the imaginary axis. We proceed to do this in one time-like (Wick-rotated) dimension, and $3 - 2\varepsilon$ spatial dimensions,

$$\ell^0 \to \mathrm{i}\ell_E^0, \quad \vec{\ell} = \vec{\ell}_E, \quad \ell^2 \to -\ell_E^2, \qquad \int \mathrm{d}^d\ell \to \mathrm{i}\int \mathrm{d}^d\ell_E. \tag{10.175}$$

This results in

$$\begin{aligned}
\Lambda^\nu_{\mathrm{OS}}(p,p') &= 8\pi\alpha\mu^{2\varepsilon}\mathrm{e}^{\gamma_E\varepsilon}(4\pi)^{-\varepsilon}\int_0^1 \mathrm{d}u_1\int_0^1 \mathrm{d}u_2\int_0^1 \mathrm{d}u_3 \\
&\quad\times\delta(1-u_1-u_2-u_3)\int\frac{\mathrm{d}^d\ell_E}{(2\pi)^d}\left[\frac{\frac{(d-2)^2}{d}\gamma^\nu\ell_E^2}{(\ell_E^2+\Delta)^3}-\frac{\gamma^\nu\mathcal{N}_1}{(\ell_E^2+\Delta)^3}-\frac{\mathrm{i}\sigma^{\nu\rho}q_\rho\mathcal{N}_2}{(\ell_E^2+\Delta)^3}\right] \\
&= 8\pi\alpha\int_0^1 \mathrm{d}u_1\int_0^1 \mathrm{d}u_2\int_0^1 \mathrm{d}u_3\,\delta(1-u_1-u_2-u_3)\left[T_1^\nu(q^2)+T_2^\nu(q^2)+T_3^\nu(q^2)\right] \\
&= I_1^\nu(q^2)+I_2^\nu(q^2)+I_3^\nu(q^2),
\end{aligned} \tag{10.176}$$

where $q = p' - p$ and the terms $T_j^\nu(q^2)$ and $I_j^\nu(q^2)$ $(k = 1,2,3)$ are defined in the obvious way; they will be used in the following. We shall make use of the fact that the vertex correction can be expressed in terms of the momentum transfer q. We note the overall sign change in the denominator term $(\ell^2-\Delta)^3 \to -(\ell_E^2+\Delta)^3$. The term $T_1^\nu(q^2)$ with the Euclidean loop momentum square in the numerator is, very explicitly,

$$\begin{aligned}
T_1^\nu(q^2) &= \gamma^\nu\mu^{2\varepsilon}\mathrm{e}^{\gamma_E\varepsilon}(4\pi)^{-\varepsilon}\frac{(d-2)^2}{d}\int\frac{\mathrm{d}^d\ell_E}{(2\pi)^d}\frac{\ell_E^2}{(\ell_E^2+\Delta)^3} \\
&= \gamma^\nu\,\mu^{2\varepsilon}\mathrm{e}^{\gamma_E\varepsilon}(4\pi)^{-\varepsilon}\frac{(d-2)^2}{d}\frac{2\pi^{d/2}}{(2\pi)^d\,\Gamma(\frac{d}{2})}\int_0^\infty \mathrm{d}\ell_E\frac{\ell_E^{d+1}}{(\ell_E^2+\Delta)^3} \\
&= \gamma^\nu\,\mu^{2\varepsilon}\,\mathrm{e}^{\gamma_E\varepsilon}\frac{(1-\varepsilon)^2}{16\,\pi\,\Gamma(1-\varepsilon)}\frac{1}{\sin(\varepsilon\pi)}\frac{1}{\Delta^\varepsilon},
\end{aligned} \tag{10.177}$$

where we set $d = 4-2\varepsilon$ in the result, and we have used the result for the d-dimensional solid angle, $\Omega_d = (2\pi^{d/2})/\Gamma(\frac{d}{2})$. We also note that the dependence on the momentum transfer enters via the implicit dependence of Δ on q^2. We must restrict d to the range $-2 < d < 4$ in order for the integral over ℓ_E to converge. For the second and third integrals, using a similar procedure, we get

$$T_2^\nu(q^2)+T_3^\nu(q^2) = -\,\mu^{2\varepsilon}\mathrm{e}^{\gamma_E\varepsilon}(4\pi)^{d/2-2}\int\frac{\mathrm{d}^d\ell_E}{(2\pi)^d}\frac{\gamma^\nu\mathcal{N}_1^\nu+\mathrm{i}\sigma^{\nu\rho}\,q^\rho\,\mathcal{N}_2}{(\ell_E^2+\Delta)^3} \tag{10.178}$$

$$= \mu^{2\varepsilon}\,\mathrm{e}^{\gamma_E\varepsilon}\,\frac{1}{32\pi\,\Gamma(-\varepsilon)}\,\frac{1}{\sin(\varepsilon\pi)}\,\frac{\gamma^\nu\mathcal{N}_1+\mathrm{i}\sigma^{\nu\rho}\,q^\rho\,\mathcal{N}_2}{\Delta^{1+\varepsilon}}. \tag{10.179}$$

For $I_1^\nu(q^2)$, we find

$$\begin{aligned}
I_1^\nu(q^2) &= 8\pi\alpha\int_0^1 \mathrm{d}u_1\int_0^1 \mathrm{d}u_2\int_0^1 \mathrm{d}u_3\,\delta(1-u_1-u_2-u_3)\,T_1^\nu(q^2) \\
&= \gamma^\nu\frac{\alpha}{2}\frac{(1-\varepsilon)^2}{\Gamma(1-\varepsilon)}\frac{\mu^{2\varepsilon}\mathrm{e}^{\gamma_E\varepsilon}}{\sin(\varepsilon\pi)}\int_0^1 \mathrm{d}u_2\int_0^{1-u_2}\mathrm{d}u_3\,\frac{1}{(m^2(u_2+u_3)^2-q^2u_2u_3)^\varepsilon}.
\end{aligned} \tag{10.180}$$

We recall the definition of Δ in Eq. (10.165c). Because the calculation is already complicated as it stands, we elect to expand the integrand of I_1 in q^2 and obtain

$$I_1^\nu(q^2) = \gamma^\nu \frac{\alpha}{2} \frac{(1-\varepsilon)^2}{\Gamma(1-\varepsilon)} \frac{\mu^{2\varepsilon} e^{\gamma_E \varepsilon}}{\sin(\varepsilon\pi)} \frac{1}{[m(u_2+u_3)]^{2\varepsilon}} \times \int_0^1 du_2 \int_0^{1-u_2} du_3 \left(1 + \frac{q^2}{m^2} \frac{u_2 u_3 \varepsilon}{(u_2+u_3)^2} + \frac{q^4}{2m^4} \frac{u_2^2 u_3^2 \varepsilon}{(u_2+u_3)^4}\right), \tag{10.181}$$

where we ignore terms of order ε^2 in the integrand (it would otherwise be incurred in the term proportional to q^4). This happens as we anticipate that these vanish in the limit $\varepsilon \to 0$ in the final result for the form factors. We also note that ε acts as a suitable ultraviolet as well as infrared regulator: Simply, it regularizes the dimensionality of space-time, regularizing, simultaneously, infrared as well as ultraviolet divergences. We change variables as follows,

$$u_2 = \rho\,\omega\,, \quad u_3 = \rho\,(1-\omega)\,, \tag{10.182}$$

which allows to convert the integration limits in both variables to the interval $(0, 1)$, the Jacobian being equal to ρ. Hence, we can write $I_1^\nu(q^2)$ as

$$\begin{aligned} I_1^\nu(q^2) &= \gamma^\nu \frac{\alpha}{2} \frac{(1-\varepsilon)^2}{\Gamma(1-\varepsilon)} \left(\frac{\mu}{m}\right)^{2\varepsilon} \frac{e^{\gamma_E \varepsilon}}{\sin(\varepsilon\pi)} \\ &\quad \times \int_0^1 d\rho \int_0^1 d\omega\, \rho^{1-2\varepsilon} \left(1 + \frac{q^2\,\varepsilon}{m^2}\omega(1-\omega) + \frac{q^4 \varepsilon}{2m^4}\omega^2(1-\omega)^2\right) \\ &= \gamma^\nu \frac{\alpha}{4\pi}\left[\frac{1}{\varepsilon} - 1 + \ln\left(\frac{\mu^2}{m^2}\right) + \frac{1}{6}\frac{q^2}{m^2} + \frac{1}{60}\frac{q^4}{m^4} + \mathcal{O}\left(\frac{q^6}{m^6}\right)\right]. \end{aligned} \tag{10.183}$$

We now turn our attention to the infrared divergent integral $I_2^\nu(q^2)$, which contributes to the form factor $F_1(q^2)$, in contrast to the term $T_3^\nu(q^2)$, which contributes to the magnetic $F_2(q^2)$ form factor. One finds, after an expansion in powers of the momentum transfer,

$$\begin{aligned} I_2^\nu(q^2) &= 8\pi\alpha \int_0^1 du_1 \int_0^1 du_2 \int_0^1 du_3\, \delta(1-u_1-u_2-u_3)\, T_2^\nu(q^2) \\ &= \gamma^\nu \frac{\alpha}{4} \frac{1}{\Gamma(-\varepsilon)} \frac{\mu^{2\varepsilon} e^{\gamma_E \varepsilon}}{\sin(\varepsilon\pi)} \int_0^1 du_2 \int_0^{1-u_2} du_3 \frac{\mathcal{N}_1}{(m^2(u_2+u_3)^2 - q^2 u_2 u_3)^{1+\varepsilon}} \\ &= \gamma^\nu \frac{\alpha}{2\pi}\left\{\frac{1}{\varepsilon} + \frac{5}{2} + \ln\left(\frac{\mu^2}{m^2}\right) - \frac{1}{3}\frac{q^2}{m^2}\left[1 + \frac{1}{\varepsilon} + \ln\left(\frac{\mu^2}{m^2}\right)\right]\right. \\ &\quad \left. - \frac{1}{20}\frac{q^4}{m^4}\left[2 + \frac{1}{\varepsilon} + \ln\left(\frac{\mu^2}{m^2}\right)\right] + \mathcal{O}\left(\frac{q^6}{m^6}\right)\right\}. \end{aligned} \tag{10.184}$$

Similarly, we carry out the calculation of $I_3^\nu(q^2)$, with the result

$$\begin{aligned} I_3^\nu(q^2) &= 8\pi\alpha \int_0^1 du_1 \int_0^1 du_2 \int_0^1 du_3\, \delta(1-u_1-u_2-u_3)\, T_3^\nu(q^2) \\ &= i\sigma^{\nu\rho} q_\rho \frac{\alpha}{4} \frac{1}{\Gamma(-\varepsilon)} \frac{\mu^{2\varepsilon} e^{\gamma_E \varepsilon}}{\sin(\varepsilon\pi)} \int_0^1 du_2 \int_0^{1-u_2} du_3 \frac{\mathcal{N}_2}{(m^2(u_2+u_3)^2 - q^2 u_2 u_3)^{1+\varepsilon}} \\ &= \frac{i\sigma^{\nu\rho} q_\rho}{2m} \frac{\alpha}{2\pi}\left[1 + \frac{1}{6}\frac{q^2}{m^2} + \frac{1}{30}\frac{q^4}{m^4} + \mathcal{O}\left(\frac{q^6}{m^6}\right)\right]. \end{aligned} \tag{10.185}$$

For the anomalous magnetic moment term, one can actually set $\varepsilon = 0$ in all intermediate steps, since neither infrared nor ultraviolet divergences are being encountered. For amusement, it is actually quite entertaining to renounce on the expansion in q^2, do the same substitutions as for the integral $I_2^\nu(q^2)$, to obtain

$$\begin{aligned} I_3^\nu(q^2) &= \frac{\mathrm{i}\sigma^{\nu\rho} q_\rho}{2m} \frac{\alpha}{2\pi} \int_0^1 \int_0^1 \mathrm{d}\omega \left[1 - \frac{q^2}{m^2}\omega(1-\omega)\right]^{-1} \\ &= \frac{\mathrm{i}\sigma^{\nu\rho} q_\rho}{2m} \frac{\alpha}{2\pi} \frac{4}{\sqrt{q^2/m^2}\sqrt{4 - q^2/m^2}} \arctan\left(\sqrt{\frac{q^2/m^2}{4 - q^2/m^2}}\right), \end{aligned} \tag{10.186}$$

which confirms the corresponding terms in Eqs. (10.95), (10.99), and (10.100).

Based on Eqs. (10.176), (10.183), (10.184), and (10.185), we can write down the result for the unrenormalized, on-shell one-loop vertex correction $\Lambda^\nu_{\mathrm{OS}}(p,p')$, in dimensional regularization, as follows,

$$\begin{aligned} \Lambda^\nu_{\mathrm{OS}}(p,p') &= I_1^\nu(q^2) + I_2^\nu(q^2) + I_3^\nu(q^2) \\ &= \gamma^\nu \frac{\alpha}{\pi}\left[1 + \frac{3}{4\varepsilon} + \frac{3}{4}\ln\left(\frac{\mu^2}{m^2}\right)\right] + \gamma^\nu \frac{\alpha}{\pi}\left[\frac{q^2}{m^2}\left(-\frac{1}{8} - \frac{1}{6\varepsilon} + \frac{1}{6}\ln\left(\frac{m^2}{\mu^2}\right)\right)\right. \\ &\quad \left. + \frac{q^4}{m^4}\left(-\frac{11}{240} - \frac{1}{40\varepsilon} + \frac{1}{40}\ln\left(\frac{m^2}{\mu^2}\right)\right) + \mathcal{O}\left(\frac{q^6}{m^6}\right)\right] \\ &\quad + \frac{\mathrm{i}\sigma^{\nu\rho} q_\rho}{2m}\frac{\alpha}{2\pi}\left[1 + \frac{1}{6}\frac{q^2}{m^2} + \frac{1}{30}\frac{q^4}{m^4} + \mathcal{O}\left(\frac{q^6}{m^6}\right)\right]. \end{aligned} \tag{10.187}$$

The on-mass-shell renormalization condition can be implemented by subtracting the q^2-independent term, as outlined in Sec. 10.3.1 [see Eq. (10.95)], namely

$$\Lambda^\nu_{\mathrm{R}}(p,p') = \Lambda^\nu_{\mathrm{OS}}(p,p') - \Lambda^\nu_{\mathrm{OS}}(p,p). \tag{10.188}$$

Hence, $\Lambda^\nu_{\mathrm{R}}(p,p')$ differs from Eq. (10.163a) by the subtraction of the q^2-independent term, which also contains the logarithm $\ln(\mu^2/m^2)$. Following Eq. (10.96), we use the form factors F_1 and F_2,

$$\Gamma^\nu_{\mathrm{R}}(q) = \gamma^\nu + \Lambda^\nu_{\mathrm{R}}(p,p') = F_1(q^2)\,\gamma^\nu + F_2(q^2)\frac{\mathrm{i}}{2m}\sigma^{\nu\mu} q_\mu\,. \tag{10.189}$$

From Eq. (10.163a), we read off the results

$$\begin{aligned} F_1(q^2) &= 1 + \frac{\alpha}{\pi}\left[\frac{q^2}{m^2}\left(-\frac{1}{8} - \frac{1}{6\varepsilon} + \frac{1}{6}\ln\left(\frac{m^2}{\mu^2}\right)\right)\right. \\ &\quad \left. + \frac{q^4}{m^4}\left(-\frac{11}{240} - \frac{1}{40\varepsilon} + \frac{1}{40}\ln\left(\frac{m^2}{\mu^2}\right)\right) + \mathcal{O}\left(\frac{q^6}{m^6}\right)\right], \end{aligned} \tag{10.190a}$$

$$F_2(q^2) = \frac{\alpha}{2\pi}\left[1 + \frac{1}{6}\frac{q^2}{m^2} + \frac{1}{30}\frac{q^4}{m^4} + \mathcal{O}\left(\frac{q^6}{m^6}\right)\right]. \tag{10.190b}$$

If we set the renormalization mass scale equal to the electron mass, $\mu = m$, then the results specialize to

$$F_1(q^2) = 1 + \frac{\alpha}{\pi}\left[\frac{q^2}{m^2}\left(-\frac{1}{8} - \frac{1}{6\varepsilon}\right) + \frac{q^4}{m^4}\left(-\frac{11}{240} - \frac{1}{40\varepsilon}\right) + \mathcal{O}\left(\frac{q^6}{m^6}\right)\right], \tag{10.191a}$$

$$F_2(q^2) = \frac{\alpha}{2\pi}\left[1 + \frac{1}{6}\frac{q^2}{m^2} + \frac{1}{30}\frac{q^4}{m^4} + \mathcal{O}\left(\frac{q^6}{m^6}\right)\right]. \tag{10.191b}$$

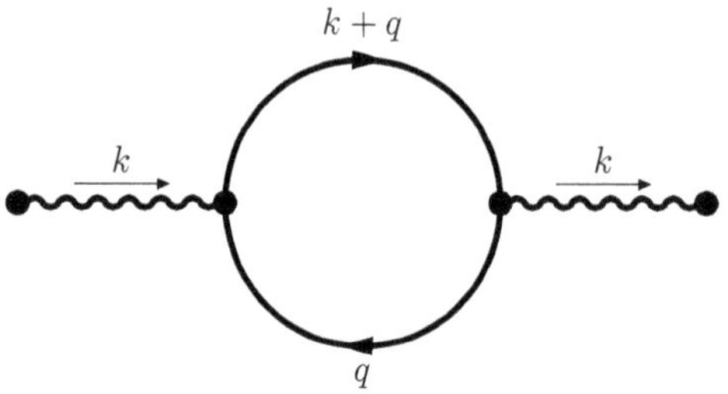

Fig. 10.4 The Feynman diagram for the vacuum polarization involves a closed fermion loop.

The result given in Eq. (10.191a) constitutes the conversion of the one given in Eq. (10.103) into dimensional regularization, while the result in Eq. (10.191b) confirms Eq. (10.100). In Eq. (10.191a), ε acts as an infrared regulator. Note that, in contrast to the result (10.103), the photon mass λ has been replaced by a dimensional infrared regulator. We anticipate that, as we will show in Sec. 11.4.4, the infrared divergence of the high-energy part of the bound-state self-energy (in dimensional regularization) will be compensated by an opposite divergent term (in $1/\varepsilon$) coming from the low-energy part of the Lamb shift. At this point, it is useful to remember that the on-shell condition states that $F_1(q^2 = 0) = 1$ (because of the charge renormalization), while the value $F_2(q^2 = 0)$ does not require any additional normalization. In fact, the result $F_2(q^2 = 0) = \alpha/(2\pi)$ reproduces Schwinger's expression [53] for the anomalous magnetic moment of the electron.

10.4 Vacuum Polarization

10.4.1 *Initial Considerations*

We consider the Feynman diagram in Fig. 10.4. The photon four-momentum k enters the loop, while the fermion momentum circulating inside the loop is q. According to Table 10.1, we need to assign the following expressions: $-\mathrm{i}\,e\,\gamma^\mu$ and $-\mathrm{i}\,e\,\gamma^\nu$ for the fermion-photon vertices, $\mathrm{i}S_F(q)$ and $\mathrm{i}S_F(k+q)$ for the fermion propagators, and $\mathrm{i}D_{\rho\mu}$ as well as $\mathrm{i}D_{\nu\sigma}$ for the photon propagators. A further factor (-1) is incurred for the closed fermion loop. Hence, the vacuum-polarization insertion into the photon propagator leads to the following expression,

$$\begin{aligned} \mathrm{i}D_{\rho\sigma}(k) \to (-1) \int \frac{\mathrm{d}^d q}{(2\pi)^d} \mathrm{i}D_{\rho\mu}(k)(-\mathrm{i}e\gamma^\mu)_{\alpha\beta}\, \mathrm{i}\,[S_F(q+k)]_{\beta\gamma} \\ \times (-\mathrm{i}e\gamma^\nu)_{\gamma\delta}\, \mathrm{i}\,[S_F(q)]_{\delta\alpha}\, \mathrm{i}D_{\nu\sigma}(k)\,. \end{aligned} \tag{10.192}$$

Here, α, β, γ and δ are spinor indices, which are summed over. Because the first and the last spinor index are paired off, we encounter the trace in spinor space. By contrast, ρ, μ, ν and σ are Lorentz indices; repeated Lorentz indices are also summed over.

This can be reformulated as

$$\mathrm{i}D_{\rho\sigma}(k) \to \mathrm{i}D_{\rho\mu}(k) \underbrace{\left\{(-1)\,e^2 \int \frac{\mathrm{d}^d q}{(2\pi)^d} \mathrm{Tr}\left[\gamma^\mu\, S_F(q+k)\,\gamma^\nu\, S_F(q)\right]\right\}}_{=\,\mathrm{i}\,\Pi^{\mu\nu}(k)} \mathrm{i}D_{\nu\sigma}(k)\,. \tag{10.193}$$

We write the sum of the tree-level photon propagator and the one-loop correction in terms of the vacuum-polarization tensor $\Pi^{\mu\nu}(k)$,

$$\mathrm{i}\mathcal{D}_{\rho\sigma}(k) = \mathrm{i}D_{\rho\sigma}(k) + \mathrm{i}D_{\rho\mu}(k)\,\left[\mathrm{i}\,\Pi^{\mu\nu}(k)\right]\,\mathrm{i}D_{\nu\sigma}(k) + \cdots, \tag{10.194}$$

where

$$\mathrm{i}\,\Pi^{\mu\nu}(k) = (-1)\,e^2 \int \frac{\mathrm{d}^d q}{(2\pi)^d} \mathrm{Tr}\left[\gamma^\mu\, S_F(q+k)\,\gamma^\nu\, S_F(q)\right]. \tag{10.195}$$

The following considerations are easiest when we consider the original photon propagator to be formulated in Landau gauge,

$$D_{\mu\nu}(k) = -\frac{1}{k^2 + \mathrm{i}\epsilon}\, P^{\mu\nu}(k)\,, \qquad P^{\mu\nu}(k) = g^{\mu\nu} - \frac{k^\mu k^\nu}{k^2}\,. \tag{10.196}$$

The projector $P^{\mu\nu}(k)$ has the property

$$P^{\mu\nu}(k)\; P_{\nu\rho}(k) = P^{\mu}{}_{\rho}(k)\,. \tag{10.197}$$

Let us have a look at the exact photon propagator and how it is affected by vacuum polarization. From Eq. (10.194), we have

$$\mathcal{D}_{\rho\sigma}(k) = D_{\rho\sigma}(k) - D_{\rho\mu}(k)\,\Pi^{\mu\nu}(k)\,D_{\nu\sigma}(k) + \cdots. \tag{10.198}$$

One can show (see also Sec. 10.7) that the vacuum-polarization tensor $\Pi^{\mu\nu}(k)$ has the following structure,

$$\Pi^{\mu\nu}(k) = k^2\, P^{\mu\nu}(k)\,\Pi(k^2)\,, \tag{10.199}$$

with a scalar vacuum polarization correction $\Pi(k^2)$. The modified photon propagator becomes

$$\begin{aligned}
\mathcal{D}_{\rho\sigma}(k^2) = & -\frac{1}{k^2+\mathrm{i}\epsilon}\, P_{\rho\sigma}(k) \\
& - \left(-\frac{1}{k^2+\mathrm{i}\epsilon}\, P_{\rho\mu}(k)\right) k^2\, P^{\mu\nu}(k)\,\Pi(k^2)\, \left(-\frac{1}{k^2+\mathrm{i}\epsilon}\, P_{\nu\sigma}(k)\right) + \cdots \\
= & -\frac{1}{k^2+\mathrm{i}\epsilon}\, P_{\rho\sigma}(k) - \left(\frac{1}{k^2+\mathrm{i}\epsilon}\right)^2 \left[k^2\Pi(k^2)\right]\, P_{\rho\sigma}(k) + \cdots \\
= & -\frac{1}{k^2\left[1-\Pi(k^2)\right]+\mathrm{i}\epsilon}\, P_{\rho\sigma}(k)\,.
\end{aligned} \tag{10.200}$$

The residue of the photon propagator at $k^2 = 0$ is thus shifted from unity, to a value

$$\frac{1}{1-\Pi(0)} \approx 1 + \Pi(0) = Z_3\,, \qquad Z_3 \equiv 1 + \Pi(0)\,. \tag{10.201}$$

In Eq. (10.295), we shall realize that the proper identification is $Z_3 \equiv 1 + \prod(0)$, even if, in one-loop order, the difference does not matter. If we wish to retain a unit residue, then we must add to the QED Lagrangian a counter term, to be specified in the following, so that the summation of diagrams due to the counter term results in the subtraction of the polarization operator insertion at $k^2 = 0$, to yield the renormalized vacuum-polarization insertion

$$\Pi_{\mathrm{R}}(k^2) = \Pi(k^2) - \Pi(k^2 = 0)\,. \tag{10.202}$$

A closer inspection reveals that, somewhat surprisingly, this can be achieved if one adds, to the QED Lagrangian, a counter term [see also Eq. (10.270c)]

$$\mathcal{L}_{\mathrm{CT}} = -\frac{1}{4}(Z_3 - 1)F^{\mu\nu}F_{\mu\nu}\,, \tag{10.203}$$

where Z_3 is defined in Eq. (10.201), and $F^{\mu\nu} = \partial^\mu A^\nu - \partial^\nu A^\mu$ is the field-strength tensor of QED. This is most easily seen by going into momentum space, where $F_{\mu\nu}(k) = -\mathrm{i}\,(k_\mu A_\nu(k) - k_\nu A_\mu(k))$, and considering all possible contractions of the field operators (see also Sec. 10.6.3).

One treats the counter term given in Eq. (10.203) perturbatively. A summation of diagrams involving, instead of the vacuum-polarization insertion given in Fig. 10.4, the counter term $\mathcal{L}_{\mathrm{CT}}$, then yields a modified photon propagator of the form

$$\mathcal{D}_{\mathrm{R},\rho\sigma}(k^2) = -\frac{1}{k^2 Z_3\left[1 - \Pi(k^2)\right] + \mathrm{i}\epsilon}\,P_{\mu\nu}(k) \tag{10.204}$$

where

$$Z_3\left[1 - \Pi(k^2)\right] \approx \left[1 + \Pi(0)\right]\left[1 - \Pi(k^2)\right] \approx 1 - \left[\Pi(k^2) - \Pi(0)\right] = 1 - \Pi_{\mathrm{R}}(k^2)\,. \tag{10.205}$$

Alternatively, we observe that a change in the residue of the photon propagator at $k^2 = 0$ is equivalent to a renormalization of the charge. Specifically, the physical charge e_{R} is obtained as $e_{\mathrm{R}} = \sqrt{Z_3}e_0$, where e_0 is the bare charge, i.e. the charge parameter that appears in the bare Lagrangian (see also Sec. 10.6.1).

10.4.2 *Vacuum Polarization and Dimensional Regularization*

After this initial discussion, let us turn to the actual calculation of the vacuum polarization diagram in 1-loop order. Using the well-known expressions for the propagator S_F [see Eq. (7.154)] in Eq. (10.195), we obtain

$$\mathrm{i}\,\Pi^{\mu\nu}(k^2) = -e^2 \int \frac{\mathrm{d}^d q}{(2\pi)^d}\,\mathrm{Tr}\left\{\frac{\gamma^\mu(\not{q} + \not{k} + m)\gamma^\nu(\not{q} + m)}{(q^2 - m^2 + \mathrm{i}\epsilon)\left[(q+k)^2 - m^2 + \mathrm{i}\epsilon\right]}\right\}, \tag{10.206}$$

where $d = 4 - 2\varepsilon$ is the dimension of space-time.

By naive power counting, the integral of the vertex correction, given in Eq. (10.47), is logarithmically divergent at large loop momentum k. This is because the photon propagator goes as $\propto 1/k^2$, while both fermionic propagators are proportional to $1/k$. Together with the integration measure $\int \mathrm{d}^4 k$, logarithmic divergence is observed.

For the integral (10.206), naive power counting results in a quadratic divergence for large q. We should note that the integration over $\int \mathrm{d}^d q$ in Eq. (10.206) is an integration over the four-momentum of a fermionic particle, so that the Pauli–Villars prescription of subtracting an auxiliary heavy photon propagator, as done in Sec. 10.3.1, requires a modification. Namely, one has to subtract the contribution from auxiliary heavy fermions (possibly, more than one) inside the fermion loop. Naively, one might think that it will be required to subtract the contribution of at least three auxiliary fermions in order to reduce the degree of divergence of the integral from quadratic to linear, and then, to logarithmic. Finally, with the third subtraction, convergence will be reached.

However, a more careful analysis, to be completed below, reveals that $\Pi^{\mu\nu}(k^2)$ has the form $\Pi^{\mu\nu}(k^2) = k^2 \left(g^{\mu\nu} - k^\mu k^\nu / k^2\right) \Pi(k^2)$, where $\Pi(k^2)$ has a finite limit as $k^2 \to 0$. The additional factor k^2 implies that the divergence of the integral given in Eq. (10.206) is only logarithmic, and the subtraction of the contribution of only one auxiliary heavy fermion is sufficient to ensure the convergence of the loop integral.

Still, the calculation of the integral in Eq. (10.206) is much easier in dimensional as opposed to Pauli–Villars regularization. One avoids any problems involved with the introduction of "auxiliary heavy fermions", and the subtraction procedure becomes much more clear. In addition, one can point out that the Pauli–Villars subtraction procedure only works if the mass M of the auxiliary fermion is much larger than the incoming photon momentum scale k. This issue is circumvented when using dimensional regularization.

In evaluating the Dirac trace in Eq. (10.206), we naturally encounter the problem of calculating traces of Dirac γ matrices in d dimensions. Some introductory notes on this topic are therefore required. The relations we are going to find have a strong connection to those that we encountered in Sec. 10.3.1.

The trace of the product of two γ matrices can be calculated with the help of the fundamental anticommutator relation,

$$\mathrm{Tr}\left[\gamma^\mu \gamma^\nu\right] = \mathrm{Tr}\left[2 g^{\mu\nu} \mathbb{1} - \gamma^\nu \gamma^\mu\right] = 8 g^{\mu\nu} - \mathrm{Tr}\left[\gamma^\nu \gamma^\mu\right] . \tag{10.207}$$

It follows that

$$\mathrm{Tr}\left[\gamma^\mu \gamma^\nu\right] = 4 g^{\mu\nu} , \qquad \mathrm{Tr}\left[\gamma_\mu \gamma^\mu\right] = 4 g^{\mu\nu} g_{\mu\nu} = 4d . \tag{10.208}$$

Despite working in d dimensions, we set

$$\mathrm{Tr}\, \mathbb{1} = 4 , \tag{10.209}$$

because the trace of the unit matrix is obtained as a universal prefactor in all traces and can be safely set equal to four right at the start of the calculation. The reason is that $d = 4 - 2\varepsilon$ differs from four only by a term of order ε. Because the term of order $1/\varepsilon$ of the vacuum-polarization function $\Pi(k^2)$ is momentum-independent (just a constant). An alternative convention like $\mathrm{Tr}\, \mathbb{1} = d$ would thus only add a constant term to $\Pi(k^2)$, which is absorbed in the renormalization anyway.

It is possible to show that the trace of an odd number of Dirac γ matrices vanishes,

$$\mathrm{Tr}\left[\gamma^{\mu_1} \gamma^{\mu_2} \cdots \gamma^{\mu_{2n+1}}\right] = 0 . \tag{10.210}$$

For the trace of four γ matrices, one has, based on the repeated application of relations akin to $\gamma^\nu\gamma^\kappa = \{\gamma^\nu, \gamma^\kappa\} - \gamma^\kappa\gamma^\nu$,

$$\mathrm{Tr}\left[\gamma^\mu\gamma^\nu\gamma^\kappa\gamma^\lambda\right] = \mathrm{Tr}\left[2g^{\mu\nu}\gamma^\kappa\gamma^\lambda\right] - \mathrm{Tr}\left[2g^{\mu\kappa}\gamma^\nu\gamma^\lambda\right] + \mathrm{Tr}\left[2g^{\mu\lambda}\gamma^\nu\gamma^\kappa\right] - \mathrm{Tr}\left[\gamma^\nu\gamma^\kappa\gamma^\lambda\gamma^\mu\right]. \tag{10.211}$$

Based on a cyclic permutation of the matrix product (argument of the last trace), and the property $\mathrm{Tr}\left[\gamma^\kappa\gamma^\lambda\right] = 4g^{\kappa\lambda}$ [see Eq. (10.207)], we have

$$\mathrm{Tr}\left[\gamma^\mu\gamma^\nu\gamma^\kappa\gamma^\lambda\right] = 4g^{\mu\nu}g^{\kappa\lambda} - 4g^{\mu\kappa}g^{\nu\lambda} + 4g^{\mu\lambda}g^{\nu\kappa}. \tag{10.212}$$

Further considerations on traces of Dirac γ matrices can be found in Sec. 11.6.

Using the derived formulas for the traces in Eqs. (10.207) and (10.212) we find for the numerator of Eq. (10.206),

$$\begin{aligned}\mathcal{N}^{\mu\nu} &= \mathrm{Tr}\left[\gamma^\mu(\not{q} + \not{k} + m)\gamma^\nu(\not{q} + m)\right] = \mathrm{Tr}\left[\gamma^\mu(\not{q} + \not{k})\gamma^\nu\not{q}\right] + m^2\mathrm{Tr}\left[\gamma^\mu\gamma^\nu\right] \\ &= g^{\mu\nu}\left[4m^2 - 4q\cdot(q+k)\right] + 4q^\mu(q+k)^\nu + 4q^\nu(q+k)^\mu,\end{aligned} \tag{10.213}$$

where in going from the first to the second line, we have used the fact that the trace of an odd number of γ-matrices vanishes.

The two denominators in Eq. (10.206) are combined using a Feynman parameter,

$$\frac{1}{AB} = \int_0^1 \mathrm{d}x \, \frac{1}{[Ax + B\,(1-x)]^2}, \tag{10.214}$$

which together with the above result for the trace leads to

$$\mathrm{i}\Pi^{\mu\nu}(k^2) = -4e^2 \int \frac{\mathrm{d}^d q}{(2\pi)^d} \int_0^1 \mathrm{d}x \, \frac{g^{\mu\nu}[m^2 - q\cdot(q+k)] + q^\mu(q+k)^\nu + q^\nu(q+k)^\mu}{[q^2 + 2q\cdot k(1-x) + k^2(1-x) - m^2]^2}. \tag{10.215}$$

In order to complete the square in the denominator, we shift the integration variable according to

$$\ell = q + k\,(1-x), \tag{10.216}$$

which yields for $\Pi^{\mu\nu}(k^2)$ the formula

$$\mathrm{i}\Pi^{\mu\nu}(k^2) = -4\mathrm{e}^2 \int \frac{\mathrm{d}^d \ell}{(2\pi)^d} \int_0^1 \mathrm{d}x \frac{\mathcal{N}^{\mu\nu}}{[\ell^2 + k^2 x(1-x) - m^2]^2}, \tag{10.217}$$

where the numerator is

$$\begin{aligned}\mathcal{N}^{\mu\nu} &= g^{\mu\nu}\left(m^2 - \ell^2 + k\cdot\ell(1-2x) + k^2x(1-x)\right) \\ &\quad + 2\ell^\mu\ell^\nu - (1-2x)(k^\mu\ell^\nu + k^\nu\ell^\mu) - 2k^\mu k^\nu x(1-x).\end{aligned} \tag{10.218}$$

Due to Lorentz invariance, all terms linear in ℓ vanish when integrated over the d-dimensional space. Furthermore, as described for the vertex correction, we have $\ell^\mu\ell^\nu \to \frac{1}{d}g^{\mu\nu}\ell^2$. The expression for $\Pi^{\mu\nu}(k^2)$ thus simplifies to

$$\mathrm{i}\Pi^{\mu\nu}(k^2) = -4e^2 \int \frac{\mathrm{d}^d \ell}{(2\pi)^d} \int_0^1 \mathrm{d}x \, \frac{g^{\mu\nu}\left[\left(\frac{2-d}{d}\right)\ell^2 + m^2 + k^2x(1-x)\right] - 2k^\mu k^\nu x(1-x)}{(\ell^2 - \Delta)^2}, \tag{10.219}$$

where

$$\Delta = m^2 - k^2 x(1-x)\,. \tag{10.220}$$

A suitable Wick rotation of the integration momentum, following $\ell^0 \to \mathrm{i}\ell^0_{\mathrm{E}}$ and $\ell^2 \to -\ell^2_{\mathrm{E}}$, results in

$$\mathrm{i}\Pi^{\mu\nu}(k^2) = -4\mathrm{i}e^2 \int \frac{\mathrm{d}^d \ell_{\mathrm{E}}}{(2\pi)^d} \int_0^1 \mathrm{d}x \, \frac{g^{\mu\nu}\left[\left(\frac{d-2}{d}\right)\ell^2_{\mathrm{E}} + m^2 + k^2 x(1-x)\right] - 2k^\mu k^\nu x(1-x)}{[\ell^2_{\mathrm{E}} + \Delta]^2}, \tag{10.221}$$

where e^2 in d dimensions is, according to Eq. (10.173),

$$e^2 = 4\pi\alpha\, \mu^{2\varepsilon} \mathrm{e}^{\gamma_E \varepsilon} (4\pi)^{-\varepsilon}\,. \tag{10.222}$$

It is now possible to carry out the d-dimensional integration over the loop momentum, in terms of integrals K_1 and K_2, to be defined according to

$$\begin{aligned}
\Pi^{\mu\nu}(k^2) &= -\,16\pi\alpha\, \mu^{2\varepsilon} \mathrm{e}^{\gamma_E \varepsilon} (4\pi)^{-\varepsilon} \int_0^1 \mathrm{d}x \int \frac{\mathrm{d}^d \ell_{\mathrm{E}}}{(2\pi)^d} \frac{1}{[\ell^2_{\mathrm{E}} + \Delta]^2} \\
&\quad \times \left[g^{\mu\nu} \left(\frac{d-2}{d}\right) \ell^2_{\mathrm{E}} + g^{\mu\nu}\left(m^2 + k^2 x(1-x)\right) - 2k^\mu k^\nu x(1-x) \right] \\
&= -\,16\pi\alpha \int_0^1 \mathrm{d}x \left\{ g^{\mu\nu} K_1 + \left[g^{\mu\nu}\left(m^2 + k^2 x(1-x)\right) - 2k^\mu k^\nu x(1-x)\right] K_2 \right\} \\
&= J_1^{\mu\nu}(k^2) + J_2^{\mu\nu}(k^2)\,.
\end{aligned} \tag{10.223}$$

The terms $J_1^{\mu\nu}(k^2)$ and $J_2^{\mu\nu}(k^2)$ are generated by the terms $K_1 = K_1(k^2)$ and $K_2 = K_2(k^2)$ in the integrand. The K_1 term reads as follows,

$$\begin{aligned}
K_1(k^2) &= \mu^{2\varepsilon} \mathrm{e}^{\gamma_E \varepsilon} (4\pi)^{-\varepsilon} \frac{d-2}{d} \int \frac{\mathrm{d}^d \ell_E}{(2\pi)^d} \frac{\ell^2_E}{(\ell^2_E + \Delta)^2} \\
&= -\,\mu^{2\varepsilon}\, \mathrm{e}^{\gamma_E \varepsilon} \frac{1}{16\,\pi\,\Gamma(1-\varepsilon)} \frac{1}{\sin(\varepsilon\pi)} \Delta^{1-\varepsilon}\,,
\end{aligned} \tag{10.224}$$

while K_2 is given as

$$\begin{aligned}
K_2(k^2) &= \mu^{2\varepsilon} \mathrm{e}^{\gamma_E \varepsilon} (4\pi)^{-\varepsilon} \int \frac{\mathrm{d}^d \ell_E}{(2\pi)^d} \frac{1}{(\ell^2_E + \Delta)^2} \\
&= \mu^{2\varepsilon}\, \mathrm{e}^{\gamma_E \varepsilon} \frac{1}{16\,\pi\,\Gamma(1-\varepsilon)} \frac{1}{\sin(\varepsilon\pi)} \frac{1}{\Delta^\varepsilon}\,.
\end{aligned} \tag{10.225}$$

The dependence on k^2 and x is due to the denominator structure Δ defined in Eq. (10.220). We have

$$J_1^{\mu\nu}(q^2) = \alpha\, g^{\mu\nu} \frac{\mu^{2\varepsilon}\, \mathrm{e}^{\gamma_E \varepsilon}}{\Gamma(1-\varepsilon)\, \sin(\varepsilon\pi)} \int_0^1 \mathrm{d}x\, \Delta^{1-\varepsilon}\,, \tag{10.226}$$

$$J_2^{\mu\nu}(q^2) = -\alpha \frac{\mu^{2\varepsilon}\, \mathrm{e}^{\gamma_E \varepsilon}}{\Gamma(1-\varepsilon)\, \sin(\varepsilon\pi)} \int_0^1 \mathrm{d}x\, \Delta^{-\varepsilon} \left[g^{\mu\nu}\left[m^2 + k^2 x(1-x)\right] - 2k^\mu k^\nu x(1-x)\right], \tag{10.227}$$

(overall minus sign for $J_2^{\mu\nu}(q^2)$). In the sum of $J_1^{\mu\nu}(q^2)$ and $J_2^{\mu\nu}(q^2)$, using Eq. (10.220), we can find the tensor structure

$$\Pi^{\mu\nu}(k^2) = \left(k^2\, g^{\mu\nu} - k^\mu\, k^\nu\right) \Pi(k^2)\,. \tag{10.228}$$

The scalar function $\Pi(k^2)$ is calculated as

$$\begin{aligned}\Pi(k^2) &= -2\alpha \frac{\mu^{2\varepsilon}\, \mathrm{e}^{\gamma_E \varepsilon}}{\Gamma(1-\varepsilon)\,\sin(\varepsilon\pi)} \int_0^1 \mathrm{d}x\, x(1-x)\,\Delta^{-\varepsilon} \\ &= -\frac{\alpha}{3\pi}\left[\frac{1}{\varepsilon} + \ln\left(\frac{\mu^2}{m^2}\right)\right] + \frac{2\alpha}{\pi}\int_0^1 \mathrm{d}x\, x(1-x)\,\ln\left(1 - x(1-x)\,\frac{k^2}{m^2}\right) \\ &= \underbrace{-\frac{\alpha}{3\pi}\left[\frac{1}{\varepsilon} + \ln\left(\frac{\mu^2}{m^2}\right)\right]}_{=\Pi(k^2=0)} + \Pi_{\mathrm{R}}(k^2)\,,\end{aligned} \tag{10.229}$$

where

$$\begin{aligned}\Pi_{\mathrm{R}}(k^2) &= \frac{2\alpha}{\pi}\int_0^1 \mathrm{d}x\, x(1-x)\,\ln\left(1 - x(1-x)\,\frac{k^2}{m^2}\right) \\ &= \frac{\alpha}{\pi}\left\{-\frac{12 + 5k^2/m^2}{9k^2/m^2} + \frac{2(2 + k^2/m^2)}{3(k^2/m^2)}\sqrt{\frac{4 - k^2/m^2}{k^2/m^2}}\,\arctan\left(\sqrt{\frac{k^2/m^2}{4 - k^2/m^2}}\right)\right\}.\end{aligned} \tag{10.230}$$

We encounter the same variable

$$\xi(k^2) = \arctan\left(\sqrt{\frac{k^2/m^2}{4 - k^2/m^2}}\right) \tag{10.231}$$

as in Eq. (10.186).

The asymptotic limits are as follows,

$$\Pi_{\mathrm{R}}(k^2) = \begin{cases} -\dfrac{\alpha}{\pi}\left[\dfrac{1}{15}\dfrac{k^2}{m^2} + \dfrac{1}{140}\left(\dfrac{k^2}{m^2}\right)^2 + \dfrac{1}{945}\left(\dfrac{k^2}{m^2}\right)^3 + \cdots\right] & k^2 \ll m^2 \\ \dfrac{\alpha}{\pi}\left[\dfrac{1}{3}\ln\left(-\dfrac{k^2}{m^2}\right) - \dfrac{5}{9} - \dfrac{2m^2}{k^2} + \cdots\right] & k^2 \gg m^2 \end{cases}. \tag{10.232}$$

For $k^2/m^2 \gg 1$, the incoming virtual photon can produce a pair, which is manifest in the imaginary part due to the logarithm.

10.4.3 *Vacuum Polarization and Coulomb Potential*

The Coulomb potential is due to the exchange of space-like virtual photons with four-momenta $k^\mu = (0, \vec{k})$,

$$V(\vec{k}) = -\frac{4\pi Z\alpha}{\vec{k}^2}\,. \tag{10.233}$$

Here, Z is the nuclear charge number, while α is the fine-structure constant. Vacuum polarization modifies the Coulomb potential to

$$V(\vec{k}) = -\frac{4\pi\alpha}{\vec{k}^2[1 - \Pi_{\rm R}(-\vec{k}^2)]} \approx -\frac{4\pi\alpha}{\vec{k}^2}\,[1 + \Pi_R(-\vec{k}^2)]\,. \tag{10.234}$$

The result is a small modification of the Coulomb potential, which is especially prevalent for large $|\vec{k}|^2$, i.e. small distances. Indeed, in coordinate space, vacuum polarization yields a Dirac-δ modification. Let us evaluate the leading correction to the Coulomb potential, which reads as

$$\begin{aligned} V(\vec{r}) &\approx -4\pi Z\alpha \int \frac{\mathrm{d}^3k}{(2\pi)^3} \mathrm{e}^{\mathrm{i}\vec{k}\cdot\vec{r}} \left(\frac{1}{|\vec{k}|^2} + \frac{\alpha}{15\pi m^2}\right) \\ &= -\frac{Z\alpha}{r} - \frac{4}{15}\frac{\alpha\,(Z\alpha)}{m^2}\delta^{(3)}\,(\vec{r})\,. \end{aligned} \tag{10.235}$$

In hydrogen atoms, the consequence is that, in leading order, only S states receive a correction through vacuum polarization because only these states have an overlap with the core (for a point-like core at least).

It is very instructive, and a good exercise in bound-state calculations, to derive a concise integral representation for $V(\vec{r})$, including the vacuum-polarization correction, based on complex contour integration. The Cauchy residue theorem will be of importance. We start from ($k \equiv |\vec{k}|$)

$$\begin{aligned} V(\vec{r}) &= -\int \frac{\mathrm{d}^3k}{(2\pi)^3} \mathrm{e}^{\mathrm{i}\vec{k}\cdot\vec{r}} \frac{4\pi Z\alpha}{|\vec{k}|^2\left(1 - \Pi_R(-|\vec{k}|^2)\right) - \mathrm{i}\epsilon}\,. \\ &= -\frac{4\pi Z\alpha}{(2\pi)^2}\int_0^\infty \mathrm{d}k\,k^2\frac{\mathrm{e}^{\mathrm{i}kr} - \mathrm{e}^{-\mathrm{i}kr}}{\mathrm{i}kr}\frac{1}{k^2\left(1 - \Pi_R(-k^2)\right) - \mathrm{i}\epsilon} \\ &\approx \frac{\mathrm{i}Z\alpha}{\pi r}\int_{-\infty}^\infty \mathrm{d}k\,k\frac{\mathrm{e}^{\mathrm{i}kr}}{k^2 - \mathrm{i}\epsilon}\left(1 + \Pi_R(-k^2)\right)\,. \end{aligned} \tag{10.236}$$

We have used the fact that, due to radial symmetry, we can set $\vec{k}\cdot\vec{r} = kr\cos(\theta)$, carry out the integration over the azimuth angle, and then, do the polar angle integration. In the second and third lines of Eq. (10.236), we have changed the notation to $k \equiv |\vec{k}|$.

We must close the contour in Eq. (10.236) in the positive half-plane. Let us set $k = \mathrm{i}q$. The expression in Eq. (10.236) has a pole at $k \approx \pm\sqrt{\mathrm{i}\epsilon}$. However, because the argument of the logarithm in $\Pi_R(-k^2)$ becomes negative for $-k^2 = q^2 > 4m^2$, the integrand in Eq. (10.236) has a branch cut starting at $k = \pm 2m\mathrm{i}$, extending along the imaginary axes in the complex k-plane. We now write

$$\begin{aligned} \frac{\mathrm{i}Z\alpha}{\pi r}\oint_C \mathrm{d}k\,k\frac{\mathrm{e}^{\mathrm{i}kr}}{k^2 - \mathrm{i}\epsilon}(1+\Pi_R(-k^2)) &= \frac{\mathrm{i}Z\alpha}{\pi r}\left(\int_{-\infty}^\infty + \int_{C_H} + \int_{C_B}\right)\mathrm{d}k\,k\frac{\mathrm{e}^{\mathrm{i}kr}}{k^2 - \mathrm{i}\epsilon}(1+\Pi_R(-k^2)) \\ &= \frac{\mathrm{i}Z\alpha}{\pi r}(2\pi\mathrm{i})\,\underset{k=\sqrt{\mathrm{i}\epsilon}}{\mathrm{Res}}\left\{k\frac{\mathrm{e}^{\mathrm{i}kr}}{k^2 - \mathrm{i}\epsilon}\left(1 + \Pi_{\rm R}(-k^2)\right)\right\}. \end{aligned} \tag{10.237}$$

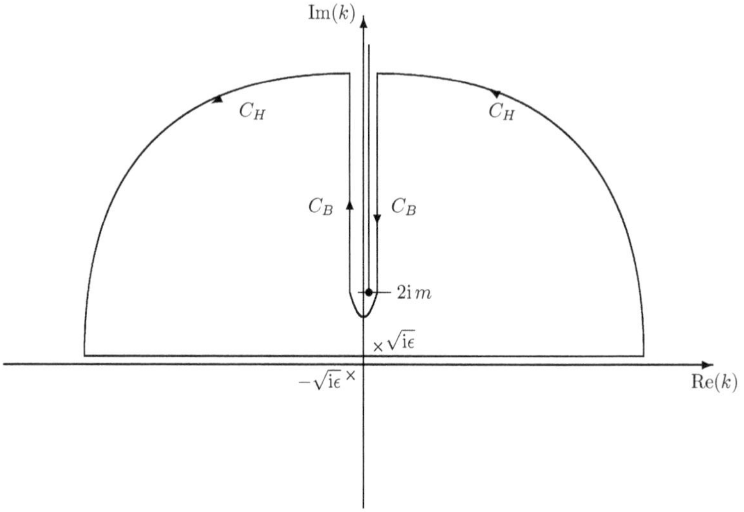

Fig. 10.5 The integration contours used in the representation (10.237) of the vacuum polarization involve an integration around the branch cut of the vacuum-polarization insertion from $k = |\vec{k}| = 2\mathrm{i}m$.

The integration contours are shown in Fig. 10.5. The integral $\int_{C_H}$ vanishes because of the exponential suppression of the integrand. Then, we have

$$V(\vec{r}) = \underbrace{-\frac{2Z\alpha}{r} \operatorname*{Res}_{k=\sqrt{\mathrm{i}\epsilon}} \left\{ k \frac{\mathrm{e}^{\mathrm{i}kr}}{k^2 - \mathrm{i}\epsilon} \left(1 + \Pi_{\mathrm{R}}(-k^2)\right) \right\}}_{=V_C(r)} \underbrace{- \frac{\mathrm{i}Z\alpha}{\pi r} \int_{C_B} \mathrm{d}k\, k \frac{\mathrm{e}^{\mathrm{i}kr}}{k^2 - \mathrm{i}\epsilon} \left(1 + \Pi_{\mathrm{R}}(-k^2)\right)}_{=V_R(r)}, \tag{10.238}$$

where we define the Coulomb term $V_C(r)$ and the radiative correction $V_R(r)$. In order to evaluate the pole term, one writes the identity

$$k^2 - \mathrm{i}\epsilon = \left(k + \sqrt{\mathrm{i}\epsilon}\right)\left(k - \sqrt{\mathrm{i}\epsilon}\right). \tag{10.239}$$

From the residue term at $k = \sqrt{\mathrm{i}\epsilon}$, we have

$$\begin{aligned} V_C(\vec{r}) &= -\frac{2Z\alpha}{r} \operatorname*{Res}_{k=\sqrt{\mathrm{i}\epsilon}} \left\{ k \frac{\mathrm{e}^{\mathrm{i}kr}}{k^2 - \mathrm{i}\epsilon} \left(1 + \Pi_{\mathrm{R}}(-k^2)\right) \right\} \\ &= -\frac{2Z\alpha}{r} \times \frac{1}{2} \times \mathrm{e}^{\mathrm{i}\sqrt{\mathrm{i}\epsilon}r} \left(1 + \Pi_{\mathrm{R}}(0)\right) = -\frac{Z\alpha}{r}. \end{aligned} \tag{10.240}$$

This term reproduces the Coulomb potential. The second term contributing to the potential is due to the branch cut, taken along the contour C_B, with the branch cut extending from pair-production threshold to infinity,

$$\begin{aligned} V_R(r) &= -\frac{\mathrm{i}Z\alpha}{\pi r} \int_{C_B} \frac{\mathrm{d}k\, k\, \mathrm{e}^{\mathrm{i}kr}}{k^2 - \mathrm{i}\epsilon} \left(1 + \Pi_{\mathrm{R}}(-k^2)\right) = +2\frac{\mathrm{i}Z\alpha}{\pi r} \int_{k=2m\mathrm{i}}^{\mathrm{i}\infty} \frac{\mathrm{d}k\, k\, \mathrm{e}^{\mathrm{i}kr}}{k^2 - \mathrm{i}\epsilon} \left\{\mathrm{i}\, \mathrm{Im}\Pi_{\mathrm{R}}(-k^2)\right\} \\ &= -\frac{Z\alpha}{r}\frac{2}{\pi} \int_{2m}^{\infty} \mathrm{d}q\, \frac{\mathrm{e}^{-qr}}{q}\, \mathrm{Im}\left[\Pi_{\mathrm{R}}(q^2 - \mathrm{i}\epsilon)\right]. \end{aligned} \tag{10.241}$$

The form of the integral now forces us to find the imaginary part of Π_R if we want to find the one loop result for the correction to the Coulomb potential. According to Eq. (10.230), we have

$$\mathrm{Im}\,\Pi_R(q^2 - \mathrm{i}\epsilon) = \frac{2\alpha}{\pi} \int_0^1 \mathrm{d}x\, x(1-x)\,\mathrm{Im}\left[\ln\left(1 - x(1-x)\,\frac{q^2}{m^2} + \mathrm{i}\epsilon\right)\right]. \qquad (10.242)$$

We thus have to evaluate, as a function of q^2, the imaginary part of the integral. The imaginary part of the logarithm is $+\pi$ when the real part of the argument of the logarithm is negative. The imaginary part of the argument of the logarithm is infinitesimally positive. We thus need to evaluate the branch points, as solutions of the equation

$$1 - \frac{q^2}{m^2} x(1-x) = 0\,, \qquad x = x_\pm \equiv \frac{1}{2} \pm \frac{1}{2}\sqrt{1 - \frac{4m^2}{q^2}}\,. \qquad (10.243)$$

Hence,

$$\mathrm{Im}\big[\Pi_R(q^2 - \mathrm{i}\epsilon)\big] = \frac{2\alpha}{\pi}\,(+\pi) \int_{x_-}^{x_+} \mathrm{d}x\; x\,(1-x) = \frac{\alpha}{3}\sqrt{1 - \frac{4m^2}{q^2}}\left(1 + \frac{2m^2}{q^2}\right). \qquad (10.244)$$

The radiative correction to the Coulomb potential is finally evaluated as

$$V_R(r) = -\frac{2\alpha(Z\alpha)}{3\pi r} \int_{2m}^{\infty} \mathrm{d}q\, \frac{\mathrm{e}^{-qr}}{q} \sqrt{1 - \frac{4m^2}{q^2}}\left(1 + \frac{2m^2}{q^2}\right). \qquad (10.245)$$

We have a found a compact integral representation of the one loop vacuum polarization correction to the Coulomb potential, that is, the so-called UEHLING[9] potential.

10.4.4 *Vacuum Polarization and Asymptotics*

It is extremely useful to explore the asymptotics of the Uehling potential, in the limits $r \gg 1/m$ and $r \ll 1/m$. For $r \gg 1/m$, the exponential suppression in Eq. (10.245) is dominant. We substitute $t = q - 2m$, which leads to

$$\begin{aligned} V_R(r) &= -\frac{2\alpha(Z\alpha)}{3\pi r} \int_0^\infty \mathrm{d}t\, \frac{\mathrm{e}^{-(2m+t)r}}{(2m+t)^3} \sqrt{\frac{t(4m+t)}{(2m+t)^2}}\left(t^2 + 4mt + 6m^2\right) \\ &= -\frac{\alpha\,(Z\alpha)}{2\pi r} \int_0^\infty \mathrm{d}t\, \frac{\mathrm{e}^{-(2m+t)r}}{m} \sqrt{\frac{t}{m}}\left(1 - \frac{29}{24}\frac{t}{m} + \frac{445}{384}\frac{t^2}{m^2} - \frac{3037}{3072}\frac{t^3}{m^3} + \cdots\right) \\ &= -\frac{\alpha}{r}\,\frac{Z\alpha}{4\sqrt{\pi}}\,\frac{\mathrm{e}^{-2mr}}{(mr)^{3/2}}\left(1 - \frac{29}{16}\frac{1}{mr} + \frac{2225}{512}\frac{1}{(mr)^2} - \frac{106295}{8192}\frac{1}{(mr)^3} + \cdots\right). \end{aligned} \qquad (10.246)$$

For hydrogen, we can use the above formula as an improvement over Eq. (10.235).

The determination of the asymptotic behavior for $r \to 0$ requires the use of another trick, which is an application of the method of the overlapping parameter

[9]Edwin Albrecht Uehling (1901–1985).

(see Sec. 4.5.2) or, a special case for the method of regions outlined in Sec. 16.3.2. A first substitution $q = 2mu$ brings the integral (10.245) into the form

$$V_{\rm R}(r) = \frac{\alpha}{\pi}\left[-\frac{Z\alpha}{r}\right]\int_1^\infty du\, e^{-2mur}\,\frac{\sqrt{u^2-1}\,(2u^2+1)}{3u^4}. \tag{10.247}$$

A second substitution $u = (1-v^2)^{-1/2}$ gives an alternative and perhaps even more economical representation [314],

$$V_{\rm R}(r) = \frac{\alpha}{\pi}\left[-\frac{Z\alpha}{r}\right]\int_0^1 dv\,\exp\left[-\frac{2mr}{\sqrt{1-v^2}}\right]\frac{v^2\left(1-\frac{v^2}{3}\right)}{1-v^2}. \tag{10.248}$$

The advantage of the representation in Eq. (10.248) above is that we have projected the infinite integration regime into a finite and very simple interval $[0,1]$. A naive expansion of the integrand in Eq. (10.248) in powers r leads to problems in the range $v \approx 1$, due to the expression $\sqrt{1-v^2}$ in the denominator. We thus split the integral into

$$V_{\rm R}(\vec{r}) = \lim_{\epsilon\to 0}\left[V_L(r) + V_H(r)\right], \tag{10.249a}$$

$$V_L(r,\epsilon) = \frac{\alpha}{\pi}\left[-\frac{Z\alpha}{r}\right]\int_0^{1-\epsilon} dv\,\exp\left[-\frac{2mr}{\sqrt{1-v^2}}\right]\frac{v^2\left(1-\frac{v^2}{3}\right)}{1-v^2}, \tag{10.249b}$$

$$V_H(r,\epsilon) = \frac{\alpha}{\pi}\left[-\frac{Z\alpha}{r}\right]\int_{1-\epsilon}^1 dv\,\exp\left[-\frac{2mr}{\sqrt{1-v^2}}\right]\frac{v^2\left(1-\frac{v^2}{3}\right)}{1-v^2}. \tag{10.249c}$$

In $V_L(r)$, we can expand the integrand (including the exponential) in r, without problems, because we avoid the problematic region $v \approx 1$. A subsequent expansion in ϵ leads to the result

$$V_L(r,\epsilon) = \frac{\alpha}{\pi}\left[\frac{2\sqrt{2}Z\alpha m}{3\sqrt{\epsilon}} - \frac{\pi}{2}Z\alpha m + \frac{5}{9}\frac{Z\alpha}{r} + \frac{Z\alpha}{3r}\ln\left(\frac{\epsilon}{2}\right) + \mathcal{O}(r)\right]. \tag{10.250}$$

For the high-v part V_H of the vacuum polarization correction, we set $v^2 = 1-h$, so that the integration variable is small. Hence, the integral for the upper part takes the form

$$V_H(r,\epsilon) = \frac{\alpha}{\pi}\left[-\frac{Z\alpha}{r}\right]\int_0^{2\epsilon-\epsilon^2} dh\,\exp\left[-\frac{2mr}{\sqrt{h}}\right]\frac{2-h-h^2}{6h\sqrt{1-h}}. \tag{10.251}$$

Because of the exponential suppression factor, the relevant r-range is of the same order as $\sqrt{h}$, and therefore, an expansion in h of the non-exponential terms corresponds to an expansion in r. The result is

$$V_H(r,\epsilon) = \frac{\alpha}{\pi}\left\{-\frac{2\sqrt{2}Z\alpha m}{3\sqrt{\epsilon}} + \frac{Z\alpha}{3r}\left[2\gamma_E + \ln\left(\frac{2mr}{\epsilon}\right)\right] + \mathcal{O}(r)\right\}, \tag{10.252}$$

where γ_E, again, is the Euler gamma constant. The addition of V_L and V_H leads to the cancelation of the ϵ parameter, and gives the short-range asymptotics of $V_R(r)$,

$$V_{\rm R}(r) = \frac{\alpha}{\pi}\frac{Z\alpha}{r}\left[\frac{1}{3}\left(\ln(mr) + 2\gamma_E\right) - \frac{\pi}{2}mr + \frac{5}{9}\right] + \mathcal{O}(r). \tag{10.253}$$

It is important to mention that the argument of the logarithm is, in fact, dimensionless: We recall that in the natural unit system, the mass is given in inverse units of length. Moreover, the leading term $\sim \ln(mr)/r$ constitutes an attractive potential for distances $r < 1/m$ since the logarithm is negative if the argument is smaller than unity.

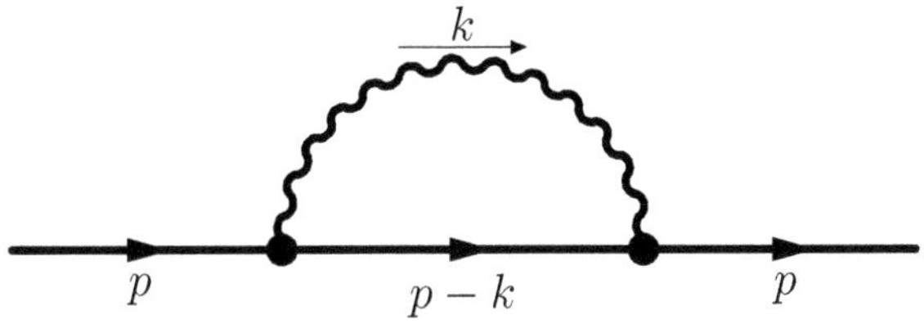

Fig. 10.6 The four-vector p of the incoming fermion splits into the photon loop momentum k, and the fermion loop momentum $p-k$, in the one-loop self-energy diagram.

10.5 Self–Energy Operator

10.5.1 *Self–Energy and Pauli–Villars Regularization*

We have discussed, in great detail, the evaluation of the vertex correction (Sec. 10.3) as well as the vacuum polarization correction to the photon propagator (Sec. 10.4). In the evaluation of the quantum electrodynamic self-energy (see Fig. 10.6), we will thus concentrate on the results.

In Eq. (10.45), we had derived the vertex correction based on the addition to the vertex factor $-\mathrm{i}e\gamma^\nu$. For the vacuum-polarization diagram, we had considered the insertion of the fermionic loop into the photon propagator in Eq. (10.192). Based on the Feynman rules of QED, the self-energy diagram (see Fig. 10.6) leads to the following insertion into the fermionic propagator,

$$\begin{aligned}\mathrm{i}S_F(p) &\to \mathrm{i}S_F(p)\\ &+\int\frac{\mathrm{d}^4k}{(2\pi)^4}\,\mathrm{i}S_F(p)(\mathrm{i}D_{\mu\nu}(k))(-\mathrm{i}e\gamma^\mu)(\mathrm{i}S_F(p-k))(-\mathrm{i}e\gamma^\nu)\mathrm{i}S_F(p)+\cdots.\end{aligned} \tag{10.254}$$

On the other hand, if we can show that the self-energy insertion can be formulated in terms of an operator $\Sigma(p)$ as

$$\mathrm{i}S_F(p) \to \mathrm{i}S_F(p) + \mathrm{i}S_F(p)\Sigma(p)\,S_F(p)+\cdots, \tag{10.255}$$

then the infinite series of insertions can be summed into the propagator denominator, as follows,

$$\begin{aligned}\mathrm{i}\frac{1}{\not p-m+\mathrm{i}\epsilon} &\to \mathrm{i}\left[\frac{1}{\not p-m+\mathrm{i}\epsilon}+\frac{1}{\not p-m+\mathrm{i}\epsilon}\,\Sigma(p)\,\frac{1}{\not p-m+\mathrm{i}\epsilon}+\cdots\right]\\ &=\mathrm{i}\left[\frac{1}{\not p-m-\Sigma(p)+\mathrm{i}\epsilon}\right].\end{aligned} \tag{10.256}$$

We appeal to Eq. (7.154) for the functional form of the Feynman propagator in momentum space. A comparison of Eqs. (10.254) and (10.256) reveals that

$$\Sigma(p) = \mathrm{i}e^2\int\frac{\mathrm{d}^4k}{(2\pi)^4}\,D_{\mu\nu}(k)\,\gamma^\mu\,S_F(p-k)\gamma^\nu\,. \tag{10.257}$$

In Feynman gauge [see Eq. (9.121)], one can write this as follows,

$$\begin{aligned}\Sigma(p) &= e^2 \int \frac{d^4k}{(2\pi)^4 i} \frac{g_{\mu\nu}}{k^2 + i\epsilon} \gamma^\mu \frac{1}{\not p - \not k - m + i\epsilon} \gamma^\nu \\ &= e^2 \int \frac{d^4k}{(2\pi)^4 i} \left(\frac{1}{k^2 - \Lambda^2 + i\epsilon} - \frac{1}{k^2 - \lambda^2 + i\epsilon} \right) \gamma^\mu \frac{1}{\not p - \not k - m + i\epsilon} \gamma_\mu . \end{aligned} \tag{10.258}$$

In the second step, we have introduced a photon mass and a Pauli–Villars regularization term Sec. 10.3.1. As we can see in Eq. (10.256), the self-energy insertion $\Sigma(p)$ is an additive insertion in the propagator denominator, to be added to the particle's mass. Because the mass of a particle is fixed by experiment, the term δm in the expansion

$$\begin{aligned}\Sigma(p) &= \delta m - \left(Z_2^{-1} - 1\right)(\not p - m - \delta m) + Z_2^{-1} \Sigma_{\rm R}(p) \\ &= Z_2^{-1}\, \delta m - \left(Z_2^{-1} - 1\right)(\not p - m) + Z_2^{-1} \Sigma_{\rm R}(p)\end{aligned} \tag{10.259}$$

constitutes an electromagnetic radiative contribution to the particle's mass, which, according to modern understanding, should be added to the mass term which is otherwise generated by the Higgs[10] mechanism. In the spirit of Sec. 10.3.3, we thus have to interpret δm as a correction to the mass which has to absorbed in the physical parameters of the theory. The summed and self-energy corrected propagator reads as follows,

$$\frac{1}{\not p - m - \Sigma(p) + i\epsilon} = \frac{Z_2}{\not p - m - \delta m - \Sigma_{\rm R}(p) + i\epsilon} . \tag{10.260}$$

This will be discussed in further detail in Sec. 10.6.2; the physical mass is $m + \delta m$, so that the initial parameter m in the theory must be interpreted as the bare mass. Alternatively, if we wish to keep the interpretation of m as the physical mass of the electron, then we need to subtract a so-called mass counter term, proportional to δm, from the Lagrangian. A calculation using Pauli–Villars regularization reveals that

$$\delta m = m \frac{3\alpha}{4\pi} \left[\ln\left(\frac{\Lambda^2}{m^2} \right) + \frac{1}{2} \right] . \tag{10.261}$$

Note that, in leading order in α, we have $Z_2 = 1 + \mathcal{O}(\alpha)$, so that $Z_2 \delta m = \delta m + \mathcal{O}(\alpha^2)$. The result for the renormalization constant $Z_2 = Z_1$ can be found in Eq. (10.137), per Eq. (10.134). In Pauli–Villars regularization, just as for the vertex correction, we use a photon mass λ in order to regularize infrared divergences, which can persist even in the renormalized expressions (see Sec. 10.3.2). Indeed, the renormalized self-energy function is found as ($p^2 < m^2$)

$$\begin{aligned}\Sigma_{\rm R}(p) &= \frac{\alpha m}{2\pi} \left[-1 + \frac{2(m^2 - p^2)}{p^2} \ln\left(1 - \frac{p^2}{m^2}\right) \right] \\ &\quad + \frac{\alpha \not p}{2\pi} \left[\frac{3}{2} - \frac{m^2}{2p^2} - \frac{m^4 - (p^2)^2}{2\,(p^2)^2} \ln\left(1 - \frac{p^2}{m^2}\right) \right] + \frac{\alpha}{2\pi} (\not p - m) \ln\left(\frac{\lambda^2}{m^2}\right) .\end{aligned} \tag{10.262}$$

Here, λ is the photon mass.

[10] Peter Ware Higgs CH FRS (b. 1929).

10.5.2 *Self–Energy and Dimensional Regularization*

We use the dimensionally regularized Dirac algebra outlined in Sec. 10.3.1, and the dimensionally regularized electron charge as given in Eq. (10.173) (inspired by the $\overline{\text{MS}}$ scheme). The renormalization procedure is the same as in Sec. 10.5.1, and the renormalization constants in the dimensional regularization are listed in Table 10.2.

Using a judicious choice of the integration methods, it is possible to derive the following closed-form result for the self-energy insertion,

$$\Sigma_{\rm R}(p) = \frac{\alpha m}{2\pi}\left[-1 + \frac{2(m^2 - p^2)}{p^2}\ln\left(1 - \frac{p^2}{m^2}\right)\right]$$
$$\times \frac{\alpha \not{p}}{2\pi}\left[\frac{3}{2} - \frac{m^2}{2\,p^2} - \frac{m^4 - (p^2)^2}{2\,(p^2)^2}\ln\left(1 - \frac{p^2}{m^2}\right)\right] + \frac{\alpha}{2\pi}(\not{p} - m)\left[\frac{1}{\epsilon} + \ln\left(\frac{\mu^2}{m^2}\right)\right]. \tag{10.263}$$

Note that the above result is obtained from the result in Pauli–Villars regularization by the substitution

$$\ln\left(\frac{\lambda^2}{m^2}\right) \to \frac{1}{\epsilon} + \ln\left(\frac{\mu^2}{m^2}\right). \tag{10.264}$$

The proof is left as an exercise to the reader.

10.6 Renormalization of QED

10.6.1 *Bare and Renormalized Lagrangian*

We are now in the position to expand on the arguments presented in Sec. 10.3.3. For simplicity and clarity, we shall exclude the gauge-fixing term, and the photon mass, from the discussion. The bare QED Lagrangian, which is identified as the "original" Lagrangian, i.e., the one which we use in order to start constructing the quantum field theory, is

$$\mathcal{L}_0 = -\frac{1}{4}F_0^2 + \overline{\psi}_0\left(\mathrm{i}\not{\partial} - m_0\right)\psi_0 - e_0\,\overline{\psi}_0 \not{A}_0\,\psi_0\,, \tag{10.265}$$

where

$$F_0^2 = F_{0,\mu\nu}\,F_0^{\mu\nu} = \left(\partial_\mu A_{0,\nu} - \partial_\nu A_{0,\nu}\right)\left(\partial^\mu A_0^\nu - \partial^\nu A_0^\nu\right). \tag{10.266}$$

The unrenormalized, bare physical quantities carry a subscript zero. This Lagrangian can describe all physical processes at tree level. Furthermore, at tree level, the unrenormalized quantities are equal to the physical ones (there is no renormalization necessary at tree level). The "renormalized QED Lagrangian", expressed in terms of renormalized quantities (denoted by the subscript R), is

$$\mathcal{L}_0 = \mathcal{L}_{\rm R} = -\frac{1}{4}Z_3\,F_{\rm R}^2 + Z_2\,\overline{\psi}_{\rm R}\left(\mathrm{i}\not{\partial} - m_{\rm R}\right)\psi_{\rm R} + Z_2\,\overline{\psi}_{\rm R}\,\delta m_{\rm R}\,\psi_{\rm R} - Z_1\,e\,\overline{\psi}_{\rm R}\not{A}_{\rm R}\,\psi_{\rm R}\,, \tag{10.267}$$

where

$$F_{\rm R}^2 = F_{{\rm R},\mu\nu}\, F_{\rm R}^{\mu\nu} = (\partial_\mu A_{{\rm R},\nu} - \partial_\nu A_{{\rm R},\nu})\, (\partial^\mu A_{\rm R}^\nu - \partial^\nu A_{\rm R}^\nu)\,. \tag{10.268}$$

The renormalized (physical) and bare parameters are related as follows,

$$\psi_0 = \sqrt{Z_2}\,\psi_{\rm R}\,, \qquad m_0 = m_{\rm R} - \delta m\,, \tag{10.269a}$$

$$A_0 = \sqrt{Z_3}\,A_{\rm R}\,, \qquad e_0 = Z_1\, Z_2^{-1}\, Z_3^{-1/2}\, e_{\rm R}\,. \tag{10.269b}$$

The renormalized Lagrangian can be written as follows [see also Eq. (10.129)],

$$\mathcal{L}_{\rm R} = \mathcal{L}_1 + \mathcal{L}_2\,, \tag{10.270a}$$

$$\mathcal{L}_1 = -\frac{1}{4}\, F_{\rm R}^2 + \overline{\psi}_{\rm R}\, ({\rm i}\not\partial - m_{\rm R})\, \psi_{\rm R} - e_{\rm R}\, \overline{\psi}_{\rm R} \not{A}_{\rm R}\, \psi_{\rm R}\,, \tag{10.270b}$$

$$\begin{aligned} \mathcal{L}_2 = &- (Z_3 - 1)\, \frac{1}{4}\, F_{\rm R}^2 + (Z_2 - 1)\, \overline{\psi}_{\rm R}\, ({\rm i}\not\partial - m_{\rm R})\, \psi_{\rm R} - (Z_1 - 1)\, e_{\rm R}\, \overline{\psi}_{\rm R} \not{A}_{\rm R}\, \psi_{\rm R} \\ &+ Z_2 \overline{\psi}_{\rm R}\, \delta m_{\rm R}\, \psi_{\rm R}\,. \end{aligned} \tag{10.270c}$$

Here, the functional form of $\mathcal{L}_1$ is equal to that of the original Lagrangian (10.265), with the bare parameters replaced by the renormalized ones, while $\mathcal{L}_2$ contains the counter terms which render the physical Green functions finite.

We may add the following two general summary statements: *(i)* The unrenormalized Lagrangian contains bare quantities but is void of counterterms. *(ii)* The renormalized Lagrangian contains physical quantities and all counter terms. The renormalization program will be illustrated in the following.

10.6.2 *Renormalization of Vertex and Self–Energy*

The renormalization program sketched above will now be illustrated. From Sec. 10.5, we recall that the self-energy has the expansion

$$\Sigma = Z_2^{-1}\, \delta m - (Z_2^{-1} - 1)\, (\not p - m_0) + Z_2^{-1}\, \Sigma_{\rm R}\,, \tag{10.271}$$

this time, suppressing the argument of Σ, in order to simplify the notation. Unlike in Eq. (10.259), we denote the initial mass parameter of the theory, which has been identified as the bare mass m_0, with the corresponding subscript zero. Through an iterative insertion of self-energy diagrams, the unrenormalized self-energy insertion Σ is summed into the propagator denominator as follows,

$$\frac{1}{\not p - m_0} + \frac{1}{\not p - m_0}\, \Sigma\, \frac{1}{\not p - m_0} + \frac{1}{\not p - m_0}\, \Sigma\, \frac{1}{\not p - m_0}\, \Sigma\, \frac{1}{\not p - m_0} + \cdots = \frac{1}{\not p - m_0 - \Sigma}\,. \tag{10.272}$$

In view of the relation of the unrenormalized and renormalized quantities (we carefully distinguish m_0 and $m_{\rm R}$), one can write the relation

$$\begin{aligned} \not p - m_0 - \Sigma &= \not p - m_0 - \delta m + (Z_2^{-1} - 1)\, (\not p - m_{\rm R}) - Z_2^{-1}\, \Sigma_{\rm R} \\ &= \not p - m_{\rm R} + (Z_2^{-1} - 1)\, (\not p - m_{\rm R}) - Z_2^{-1}\, \Sigma_{\rm R} \\ &= Z_2^{-1} (\not p - m_{\rm R} - \Sigma_{\rm R})\,, \end{aligned} \tag{10.273}$$

where $m_{\rm R} = m_0 + \delta m$. The summation of self-energy insertions thus leads to the following modification of the fermionic propagator,

$$S_F^{(0)} = \frac{1}{\not{p} - m_0 - \Sigma} = \frac{Z_2}{\not{p} - m_{\rm R} - \Sigma_{\rm R}}. \tag{10.274}$$

The unrenormalized theory, after the identification of the mass counter term, leads to a propagator whose residue at $\not{p} = m_{\rm R}$ is not equal to unity any more.

Now, let us consider the corresponding terms in the renormalized Lagrangian (10.267), and check whether they produce finite results,

$$\mathcal{L} \sim Z_2 \overline{\psi}(\mathrm{i}\not{\partial} - m_{\rm R})\psi + Z_2\, \delta m\, \overline{\psi}\, \psi. \tag{10.275}$$

Upon quantization, these terms create a propagator (indeed) of the functional form

$$S_F' = \frac{1}{Z_2} \frac{1}{\not{p} - m_{\rm R}}, \tag{10.276}$$

where the prime only denotes the modified form (no differentiation). We recall that the propagator, in the momentum representation, has to be equal to the inverse of the kinetic operator, in this case, the inverse of the operator sandwiched in between $\overline{\psi}$ and ψ.

Let us also consider the one-loop self-energy insertion derived from the renormalized Lagrangian. The self-energy has two vertices and one propagator. The vertices acquire a factor Z_1, while the propagator acquires a factor $1/Z_2$. Altogether, one has the factor

$$Z_1 \frac{1}{Z_2} Z_1 = Z_2\,, \qquad Z_1 = Z_2\,, \tag{10.277}$$

where advantage has been taken of the Ward–Takahashi identity [see Eq. (10.136)].

So, using the renormalized Lagrangian, taking into account the mass counter term perturbatively in each order of perturbation theory, the summation of self-energy insertions gives

$$\begin{aligned} S_F'' &= \frac{1}{Z_2(\not{p} - m_{\rm R})} + \frac{1}{Z_2(\not{p} - m_{\rm R})} \left(Z_2\, \Sigma - Z_2\, \delta m\right) \frac{1}{Z_2(\not{p} - m_{\rm R})} + \cdots \\ &= \frac{1}{Z_2(\not{p} - m_{\rm R} + \delta m - \Sigma)}, \end{aligned} \tag{10.278}$$

the mass counter term ending up being subtracted from $m_{\rm R}$. Again, we emphasize that the double prime on S_F'' denotes the modified form of the fermionic Feynman propagator obtained from the renormalized Lagrangian (no differentiation). But then, the expression in the denominator of S_F'' can be reformulated as follows,

$$\begin{aligned} Z_2(\not{p} - m_{\rm R} + \delta m - \Sigma) &= Z_2 \left[\not{p} - m_{\rm R} + \delta m - \delta m + (Z_2^{-1} - 1)\,(\not{p} - m_{\rm R}) - Z_2^{-1}\, \Sigma_{\rm R}\right] \\ &= Z_2 \left[\not{p} - m_{\rm R} + (Z_2^{-1} - 1)\,(\not{p} - m_{\rm R}) - Z_2^{-1}\, \Sigma_{\rm R}\right] \\ &= (\not{p} - m_{\rm R} - \Sigma_{\rm R})\,. \end{aligned} \tag{10.279}$$

We are using Eq. (10.259) in the form

$$\Sigma(p) = Z_2^{-1}\, \delta m - \left[Z_2^{-1} - 1\right](\not{p} - m_R) + Z_2^{-1} \Sigma_{\rm R}(p)\,. \tag{10.280}$$

So, we finally obtain the result

$$S''_F = \frac{1}{\not{p} - m_\mathrm{R} - \Sigma_\mathrm{R}}, \tag{10.281}$$

which is finite and has unit residue at $\not{p} = m_\mathrm{R}$. Consistent renormalized results have been obtained from the renormalized Lagrangian (10.270).

A further remark is in order. Based on a general consideration, we have justified, in the text following Eq. (10.276), that the Lagrangian $\mathcal{L} = Z_2 \overline{\psi}_\mathrm{R} (\mathrm{i}\not{\partial} - m) \psi_\mathrm{R} + \cdots$ generates a term $Z_2^{-1} (\not{p} - m)^{-1}$ in the propagator. This can be further justified as follows. First, one quantizes the theory by postulating the following equal-time anticommutation relation of the field ψ_R and its conjugate momentum π_R,

$$\{\psi_\mathrm{R}(t,\vec{r}), \pi_\mathrm{R}(t,\vec{r}\,')\} = \mathrm{i}\,\delta^{(3)}(\vec{r} - \vec{r}\,'). \tag{10.282}$$

From $\mathcal{L}$, the canonical momentum is

$$\pi_\mathrm{R} = \frac{\partial L}{\partial(\partial_t \psi)} = \mathrm{i}\, Z_2 \overline{\psi}_\mathrm{R} \gamma^0 = \mathrm{i}\, Z_2\, \psi^+_\mathrm{R}. \tag{10.283}$$

This expression differs from the usual field momentum given in Sec. 7.4.1 [see Eq. (7.160)] only in the multiplicative factor Z_2. If we impose the anticommutator condition

$$\{\psi_\mathrm{R}(t,\vec{r}), \pi_\mathrm{R}(t,\vec{r}\,')\} = \mathrm{i}\, Z_2 \{\psi_\mathrm{R}(t,\vec{r}), \psi^+_\mathrm{R}(t,\vec{r}\,')\} = \mathrm{i}\,\delta^{(3)}(\vec{r} - \vec{r}\,'), \tag{10.284}$$

then it become clear that this condition can be fulfilled if we choose for ψ_R and ψ^+_R field operators which formally differ from the usual definitions only in a multiplicative factor $1/\sqrt{Z_2}$. This observation finally justifies the form of the propagator as derived from the renormalized Lagrangian. Because the Feynman propagator is obtained as the time-ordered product of two field operators, the factor Z_2 in Eq. (10.276) is consistently obtained.

10.6.3 *Renormalization of Vacuum Polarization*

Let us supplement the considerations outlined in Sec. 10.4.1, regarding the renormalization of the vacuum-polarization insertions, by additional considerations, which are also connected to the renormalized Lagrangian given in Eqs. (10.267) and (10.270).

Let us first treat the counter terms in Eq. (10.270c) perturbatively. A quantization of the electromagnetic field based on the Lagrangian (10.270b) leads to the photon propagator [in Landau gauge, see Eq. (10.196)]. Let us also remember Eq. (10.41), which in Landau gauge reads as

$$\langle 0|\mathrm{T}\, A^\mu_\mathrm{R}(x)\, A^\nu_\mathrm{R}(z)|0\rangle = \mathrm{i} \int \frac{\mathrm{d}^4 k}{(2\pi)^4} D^{\mu\nu}_F(k)\, \mathrm{e}^{-\mathrm{i}k\cdot(x-z)} = \mathrm{i} D^{\mu\nu}_F(x-z), \tag{10.285a}$$

$$\mathrm{i} D^{\mu\nu}_F(k) = -\frac{\mathrm{i}}{k^2 + \mathrm{i}\epsilon} P^{\mu\nu}(k), \qquad P^{\mu\nu}(k) = g^{\mu\nu} - \frac{k^\mu k^\nu}{k^2}. \tag{10.285b}$$

The counter term is

$$\mathcal{L}_2 \sim (Z_3 - 1) \left\{ -\frac{1}{4} F_{\mathrm{R},\mu\nu} F_{\mathrm{R}}^{\mu\nu} \right\} \tag{10.286}$$

from Eq. (10.270c). We treat it perturbatively. One insertion of the perturbative counter term leads to

$$\begin{aligned} \mathrm{i}\Delta D^{\mu\nu} &= (Z_3 - 1)\,\mathrm{i} \int \mathrm{d}^4 y \left\{ -\frac{1}{4} \langle 0 | \mathrm{T}\, A_{\mathrm{R}}^{\mu}(x)\, [\partial_\rho A_{\mathrm{R},\sigma}(y) - \partial_\sigma A_{\mathrm{R},\rho}(y)] \right. \\ &\quad \left. \times \left[\partial^\rho A_{\mathrm{R}}^{\sigma}(y) - \partial^\sigma A_{\mathrm{R}}^{\rho}(y) \right] A_{\mathrm{R}}^{\nu}(z) | 0 \rangle \right\} \sim -\frac{Z_3 - 1}{4}\,\mathrm{i} \\ &\quad \times \int \mathrm{d}^4 y \,\langle 0 | \mathrm{T}\, A_{\mathrm{R}}^{\mu}(x) \left[\partial_\rho A_{\mathrm{R},\sigma}(y)\, \partial^\rho A_{\mathrm{R}}^{\sigma}(y) + \partial_\sigma A_{\mathrm{R},\rho}(y)\, \partial^\sigma A_{\mathrm{R}}^{\rho}(y) \right] A_{\mathrm{R}}^{\nu}(z) | 0 \rangle \,, \end{aligned} \tag{10.287}$$

where we have selected, in the second step, after the $\sim$ sign, the terms with paired-off Lorentz indices in the derivative operators. The other terms would generate terms proportional to $\partial_\rho D_F^{\rho\sigma}(x-y) \sim k_\rho\, D_F^{\rho\sigma}(k) = 0$ (in momentum space). We are considering only those contractions in Eq. (10.287) which lead to connected diagrams, i.e., the operators $A_{\mathrm{R}}^{\mu}(x)$ and $A_{\mathrm{R}}^{\nu}(z)$ both get contracted with four-vector potentials at y. One finally gets four equivalent terms, from the contractions of each of $A_{\mathrm{R}}^{\mu}(x)$ and $A_{\mathrm{R}}^{\nu}(z)$ with one of the two $A_{\mathrm{R}}^{\sigma}(y)$ [and $A_{\mathrm{R}}^{\rho}(y)$], resulting in

$$\begin{aligned} \mathrm{i}\Delta D^{\mu\nu} &\sim -\frac{Z_3 - 1}{2}\,\mathrm{i} \int \mathrm{d}^4 y \left\{ \left[\frac{\partial}{\partial y^\rho} \langle 0 | \mathrm{T}\, A_{\mathrm{R}}^{\mu}(x)\, A_{\mathrm{R},\sigma}(y) | 0 \rangle \, \frac{\partial}{\partial y_\rho} \langle 0 | \mathrm{T}\, A_{\mathrm{R}}^{\sigma}(y)\, A_{\mathrm{R}}^{\nu}(z) | 0 \rangle \right. \right. \\ &\quad \left. \left. + \frac{\partial}{\partial y^\sigma} \langle 0 | \mathrm{T}\, A_{\mathrm{R}}^{\mu}(x)\, A_{\mathrm{R},\rho}(y) | 0 \rangle \, \frac{\partial}{\partial y_\sigma} \langle 0 | \mathrm{T}\, A_{\mathrm{R}}^{\rho}(y)\, A_{\mathrm{R}}^{\nu}(z) | 0 \rangle \right] \right\} \\ &= -(Z_3 - 1)\,\mathrm{i} \int \mathrm{d}^4 y \left[-\frac{\partial}{\partial y^\rho} \frac{\partial}{\partial y_\rho} \Big(\mathrm{i}\, D_F{}^{\mu}{}_{\sigma}(x-y) \Big) \Big(\mathrm{i}\, D_F^{\sigma\nu}(y-z) \Big) \right], \end{aligned} \tag{10.288}$$

where in the last step, we have performed a partial integration. In momentum space, we can replace convolution by multiplication, and so,

$$-\frac{\partial}{\partial y^\rho} \frac{\partial}{\partial y_\rho} \to -(-\mathrm{i}k_\rho)(-\mathrm{i}k^\rho) = k^2 \,. \tag{10.289}$$

So, the one-loop correction becomes

$$\begin{aligned} \mathrm{i}\Delta D^{\mu\nu} &\sim -(Z_3 - 1)\,\mathrm{i}k^2 \Big(\mathrm{i} D_F{}^{\mu}{}_{\sigma}(k) \Big) \Big(\mathrm{i} D_F^{\sigma\nu}(k) \Big) \\ &= (Z_3 - 1)\,\mathrm{i}k^2 \left(-\frac{1}{k^2 + \mathrm{i}\epsilon} P^{\mu}{}_{\sigma}(k) \right) \left(-\frac{1}{k^2 + \mathrm{i}\epsilon} P^{\sigma\nu}(k) \right) \\ &= (Z_3 - 1)\,\mathrm{i} \frac{1}{k^2 + \mathrm{i}\epsilon} P^{\mu\nu}(k) \,. \end{aligned} \tag{10.290}$$

Hence, the series of perturbative insertions of the counter term given in Eq. (10.286) leads to

$$\mathrm{i}D_{\mu\nu}(k) + \mathrm{i}\Delta D_{\mu\nu}(k) + \cdots = -\frac{\mathrm{i}}{k^2 + \mathrm{i}\epsilon} P^{\mu\nu}(k) + (Z_3 - 1) \frac{\mathrm{i}}{k^2 + \mathrm{i}\epsilon} P^{\mu\nu}(k) + \cdots . \tag{10.291}$$

The resulting series can be summed,

$$1-(Z_3-1)+(Z_3-1)^2+\cdots=\frac{1}{1+(Z_3-1)}=\frac{1}{Z_3}\,. \tag{10.292}$$

The same result would be obtained if we take the multiplicative character of the counter term in to account, and write the Lagrangian of the free photon field as

$$\mathcal{L}\sim-\frac{1}{4}\,Z_3\,F_{\mathrm{R},\mu\nu}\,F^{\mathrm{R},\mu\nu}\,, \tag{10.293}$$

by arguments analogous to those used in the discussion surrounding the passage between Eqs. (10.282) and (10.284).

We then sum all insertions of the counter term (10.286) and of the vacuum-polarization diagrams into the photon propagator, over all iterative insertions (including the mixed ones). This is in full analogy to the fermionic self-energy (see Sec. 10.5). This procedure leads to a modified propagator denominator,

$$\mathcal{D}_{\mu\nu}(k^2)=-\frac{1}{k^2\,[1+(Z_3-1)-\Pi(k^2)]+\mathrm{i}\epsilon}\,P_{\mu\nu}(k)\,. \tag{10.294}$$

Writing

$$\Pi(k^2)=\Pi(0)+\Pi_{\mathrm{R}}(k^2)\,,\qquad Z_3=1+\Pi(0)\,, \tag{10.295}$$

one obtains the result for Z_3.

We conclude with an illustrative remark. Let us return to Eq. (10.269), and consider the relation of the bare charge, and bare four-vector potential, to the renormalized counterparts. Strictly speaking, the counter term (10.286) is due to a renormalization of the four-vector potential (and consequently, of the electromagnetic field). Of course, it enters the field-strength tensor,

$$F_{0,\mu\nu}=\sqrt{Z_3}\,F_{\mathrm{R},\mu\nu}\,,\qquad A_{0,\mu}=\sqrt{Z_3}\,A_{\mathrm{R},\mu}\,. \tag{10.296}$$

This constitutes a renormalization of the photon field. However, per Eq. (10.267), the only renormalization which persists in the QED interaction Lagrangian is the Z_1 renormalization at the vertex,

$$\mathcal{L}\sim-Z_1\,e\,\overline{\psi}_{\mathrm{R}}\,\not{A}_{\mathrm{R}}\,\psi_{\mathrm{R}}\,. \tag{10.297}$$

This implies that, in view of the renormalization of the Dirac field $\psi_0=\sqrt{Z_2}\,\psi_{\mathrm{R}}$, and of the photon field $A_{0,\mu}=\sqrt{Z_3}\,A_{\mathrm{R},\mu}$, the electron charge has to be renormalized as

$$e_0=Z_1\,Z_2^{-1}\,Z_3^{-1/2}\,e_{\mathrm{R}}=Z_3^{-1/2}\,e_{\mathrm{R}}\,, \tag{10.298}$$

where we make use of the Ward–Takahashi identity (10.136), which implies that $Z_1=Z_2$. In other words, the renormalization of the photon field, per Eq. (10.296), implies a renormalization of the electron charge, according to Eq. (10.298).

Table 10.2 Summary of the one-loop renormalization constants using Pauli–Villars and dimensional regularization. Feynman gauge has been used throughout, although only $Z_1 = Z_2$ is gauge dependent. For the Pauli–Villars regularization scheme, we have included an additional artificial photon with a large mass Λ, and have also included a photon mass term λ to regulate infrared divergences [see Eq. (10.50)]. We see that $Z_1 = Z_2$ has both ultraviolet and infrared divergences. For dimensional regularization, we calculate in $d = 4-2\epsilon$ dimensions and see that ϵ regulates both ultraviolet and infrared divergences. The mass parameter μ is included to maintain dimensional consistency [see Eq. (10.173)].

One-Loop Renormalization Constants

	Pauli–Villars Regularization (λ is the photon mass)	Dimensional Regularization (μ is the renormalization scale)
$Z_1 = Z_2$	$1-\frac{\alpha}{\pi}\left[\frac{1}{4}\ln\left(\frac{\Lambda^2}{m^2}\right)-\frac{1}{2}\ln\left(\frac{m^2}{\lambda^2}\right)+\frac{9}{8}\right]$	$1-\frac{\alpha}{\pi}\left[\frac{3}{4\epsilon}+\frac{3}{4}\ln\left(\frac{\mu^2}{m^2}\right)+1\right]$
Z_3	$1-\frac{\alpha}{\pi}\left[\frac{1}{3}\ln\left(\frac{\Lambda^2}{m^2}\right)\right]$	$1-\frac{\alpha}{\pi}\left[\frac{1}{3\epsilon}+\frac{1}{3}\ln\left(\frac{\mu^2}{m^2}\right)\right]$
δm	$\frac{\alpha\, m}{\pi}\left[\frac{3}{4}\ln\left(\frac{\Lambda^2}{m^2}\right)+\frac{3}{8}\right]$	$\frac{\alpha\, m}{\pi}\left[\frac{3}{4\epsilon}+\frac{3}{4}\ln\left(\frac{\mu^2}{m^2}\right)+1\right]$

10.6.4 *Compilation of Renormalization Constants*

We would like to summarize the one-loop renormalization constants we have obtained in this chapter. We have employed two regularization schemes: Pauli–Villars and dimensional, and have used them to calculate the set of constants $Z_1 = Z_2$, Z_3, and δm. Our results are presented in Table 10.2.

The theory of QED is defined by the simple Lagrangian (10.1). However, for actual calculations a number of choices must be made, and these choices affect the character of a QED calculation in a profound way. The initial and unavoidable choice concerns the gauge for the quantization procedure. We have restricted ourselves to covariant gauges, and Feynman gauge in particular, in this chapter. In general, for calculations of cross sections and for renormalization related considerations, the covariant gauges have the great advantage of preserving Lorentz symmetry and allowing us to use arguments based on covariance to simplify calculations. On the other hand, the non-covariant Coulomb gauge is advantageous for the study of bound systems in QED. Once a gauge has been established, calculations beyond tree-level bring the need to evaluate loop integrals. As we have seen, these integrals often contain divergences: ultraviolet and infrared. Regularization procedures must be adopted in order to deal with these divergences. In this chapter, we have studied Pauli–Villars regularization, which involves the introduction of fictitious particles with large masses (that will eventually be taken to be infinitely large) coupled in such a way as to cancel the ultraviolet divergences that would otherwise be present. When using Pauli–Villars regularization, we made the choice to regulate the infrared by inserting an artificial small photon mass. The resulting photon propagator with Pauli–Villars regularization was given in Eq. (10.50):

$$D_F^{\mu\nu}(k) \to -\frac{g^{\mu\nu}}{k^2 - \lambda^2 + \mathrm{i}\epsilon} + \frac{g^{\mu\nu}}{k^2 - \Lambda^2 + \mathrm{i}\epsilon}, \tag{10.299}$$

where λ is the small photon mass and Λ is the large mass of the unphysical (note the sign of its propagator) additional "photon". We also illustrated the use of dimensional regularization, in which both ultraviolet and infrared divergences are regulated by working in a number of space-time dimensions different from four. The algebraic structure of QED is preserved with dimensional regularization since the propagators and vertices are unchanged, but tensor and spinor calculations and integrations must be done in d dimensions. When working in $d = 4 - 2\epsilon$ dimensions, the elementary charge acquires a dimension, and we write $e^2 = 4\pi\alpha(\mu^2 e^{\gamma_E}/(4\pi))^\epsilon$ where μ is a mass parameter that we are forced to introduce in order to keep dimensions consistent. It is convenient to introduce the terms involving the Euler–Mascheroni constant γ_E and π along with μ in order to minimize factors of γ_E and $\ln(4\pi)$ that would appear in our results, but this itself is a choice that is inspired by modified minimal subtraction ("MS-bar", $\overline{\mathrm{MS}}$) [see also Eqs. (10.173) and (16.145)].

Other choices of factors in the relation between e^2 and the dimensionless α are also possible. Moreover, other regularization schemes could have been chosen. Among them are "analytic regularization", in which propagator denominators are generalized from $k^2 - m^2 + i\epsilon$ to $(k^2 - m^2 + i\epsilon)^\alpha$. The diagrams become analytic functions of α in some region of the complex plane. Adjustment of α can be used to regulate ultraviolet and infrared divergences. Another regularization method involves the imposition of an explicit momentum-space cut-off. The cut-off is usually chosen to limit the Euclidean-space momentum magnitude to a finite value. These regularization procedures break gauge invariance and the equality $Z_1 = Z_2$. For that reason, they are seldom used in serious calculations. Dimensional regularization is the usual choice due to its favorable mathematical qualities and the fact that it minimizes the structural disruption suffered by the theory upon regularization.

When dealing with Coulombic bound states such as hydrogen, muonium, positronium, etc., choices about the procedure for dealing with the binding must also be made. Perhaps the most fundamental is whether to stick with traditional QED or move to an effective low-energy quantum field theory such as NRQED. An introduction to NRQED will be given in Chap. 17.

10.6.5 *Forest Formula*

In diagrams with more than one nested loop, one may encounter so-called nested divergences. The renormalization, in this case, becomes a little more complicated. One first has to subtract all divergences associated with the inner loops; i.e., in a

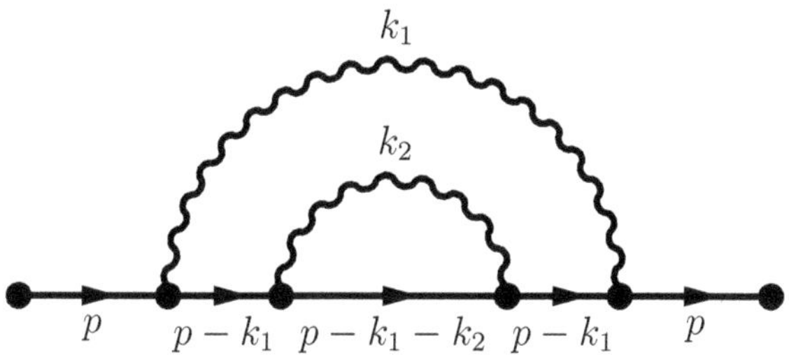

Fig. 10.7 Two consecutive photon emissions and absorptions constitute the "rainbow" nested-loop two-loop self-energy Feynman diagram.

two-loop self-energy diagram like the one shown in Fig. 10.7, one would subtract the mass counter term, and the wave function renormalization term, from the inner self-energy loop. The subtraction operations are done recursively. If γ_a is a divergent subgraph of a larger graph G, and $\mathcal{T}_{\gamma_a}$ is the appropriate subtraction information for the graph γ_a, $\mathcal{I}_G$ is the integral defining the graph G, and $\mathcal{I}_{G\backslash\{\gamma_1,\ldots,\gamma_n\}}$ is the integral with the divergent subgraphs excluded, then the so-called *forest formula* reads as

$$\overline{\mathcal{R}}_G = \mathcal{I}_G + \sum_{\substack{\{\gamma_1,\ldots,\gamma_n\} \\ \gamma_a \cap \gamma_b = \varnothing}} \mathcal{I}_{G\backslash\{\gamma_1,\ldots,\gamma_n\}} \prod_{a=1}^{n} \left(-\mathcal{T}_{\gamma_a} \overline{\mathcal{R}}_{\gamma_a}\right) , \tag{10.300}$$

$$\mathcal{R}_G = (1 - \mathcal{T}_G)\, \overline{\mathcal{R}}_G . \tag{10.301}$$

Here, $\mathcal{R}_G$ is the renormalized integral corresponding to the graph G, and it requires a further, overall subtraction $\mathcal{T}_G$, while $\overline{\mathcal{R}}_G$ is a partially subtracted graph, where all subgraphs have been renormalized, but an overall remaining subtraction has not yet been implemented.

For a two-loop vertex with the external field acting in the middle of the rainbow (see Fig. 10.7), with G being the whole diagram, and γ being the inner loop, this means that we can carry out the renormalization recursively. For the inner loop, we have

$$\overline{\mathcal{R}}_G = \mathcal{I}_G + \mathcal{I}_{G\backslash\gamma} \left(-\mathcal{T}_\gamma \overline{\mathcal{R}}_\gamma\right) = (1 - \mathcal{T}_\gamma)\, \mathcal{I}_G . \tag{10.302}$$

Then, for the outer loop, there is a further subtraction,

$$\mathcal{R}_G = (1 - \mathcal{T}_G)\, \overline{\mathcal{R}}_G = (1 - \mathcal{T}_G)\, (1 - \mathcal{T}_\gamma)\, \mathcal{I}_G . \tag{10.303}$$

The leftmost subtraction is the outer subtraction, the second one is the inner subtraction.

10.7 Further Thoughts

Here are some suggestions for further thought.

(1) **Form Factors and QED.** Consider Eq. (10.97). Show that nonvanishing contributions to F_3 and F_A would be incompatible with the discrete symmetries of QED.

(2) **An Integral.** Do the analytic evaluation of the integral given in the Eq. (10.94). Why can you replace $z\lambda^2/m^2 \to \lambda^2/m^2$ in the integrand, in the physically relevant limit $\lambda \to 0$?

(3) **Feynman Parameters.** Verify the derivations of relevant formulas concerning Feynman parameters, given in Eqs. (10.151), (10.156), (10.159), and (10.160).

(4) **Magnetic Form Factor and Anomalous Magnetic Moment.** Find the connection of the result for the magnetic (Pauli) form factor of the electron, given in Eqs. (10.100) and (10.191b), to the anomalous magnetic moment of the electron.

(5) **Two-Loop Form Factors.** Study the two-loop generalizations of the results given in Eq. (10.191) and their role in bound-state calculations. Consult Refs. [199, 312, 315].

(6) **Tensor Structure of the Vacuum-Polarization Insertion.** Consider Eq. (10.195):

$$\begin{aligned}\mathrm{i}\,\Pi^{\mu\nu}(k) &= (-1)\,e^2 \int \frac{\mathrm{d}^d q}{(2\pi)^d} \mathrm{Tr}\left[\gamma^\mu\, S_F(q+k)\,\gamma^\nu\, S_F(q)\right] \\ &= (-1)\,e^2 \int \frac{\mathrm{d}^d q}{(2\pi)^d} \mathrm{Tr}\left[\gamma^\mu \frac{1}{\not{q} - \not{k} - m + \mathrm{i}\epsilon}\gamma^\nu \frac{1}{\not{q} - m + \mathrm{i}\epsilon}\right].\end{aligned} \tag{10.304}$$

In order to show that the vacuum-polarization has the tensor structure (10.199), one does not necessarily have to go through the derivation outlined in Sec. 10.4.2, which leads to the explicit result (10.228). it suffices to show that $k_\mu\,\Pi^{\mu\nu}(k) = 0$. Show that

$$k_\mu\,\Pi^{\mu\nu}(k) = 0\,, \tag{10.305}$$

by writing $k_\mu\gamma^\mu = \not{k} = -(\not{q} - \not{k} - m + \mathrm{i}\epsilon) + (\not{q} - m + \mathrm{i}\epsilon)$, canceling denominators in Eq. (10.304), and using the cyclic invariance of the trace.

(7) **Self-Energy Operator.** Repeat the calculation of the self-energy operator, discussed in Sec. 10.5 in Pauli–Villars regularization, using dimensional regularization. Be inspired by Secs. 10.3.6 and 10.4.2. In particular, find the one-loop mass counter term in dimensional regularization [see Eq. (10.261)].

(8) **Two-Loop Renormalization.** Consider Ref. [311]. Convince yourself that the renormalized fermion propagator given in Eq. (2.14) of Ref. [311] is consistent with Eq. (10.281). Then, continue to study the two-loop renormalization constants on the basis of Ref. [311], possibly, extending the treatment to a unified regularization of both infrared as well as ultraviolet divergences through dimensional regularization.

(9) **Trace Technology for an Odd Number of Dirac Matrices.** Investigate why the trace of an odd number of Dirac γ^μ matrices vanishes. Consider the case of four space-time dimensions, and consider the definition of γ^5 given in Eq. (7.15). Then, one considers the series of transformations ($2n+1$ is odd, where n is an integer),

$$\begin{aligned}\mathrm{Tr}\left[\gamma^{\mu_1}\gamma^{\mu_2}\ldots\gamma^{\mu_{2n+1}}\right] &= \mathrm{Tr}\left[\left(\gamma^5\right)^2\gamma^{\mu_1}\left(\gamma^5\right)^2\gamma^{\mu_2}\ldots\left(\gamma^5\right)^2\gamma^{\mu_{2n+1}}\right] \\ &= \mathrm{Tr}\left[\left(\gamma^5\gamma^{\mu_1}\gamma^5\right)\left(\gamma^5\gamma^{\mu_2}\gamma^5\right)\cdots\left(\gamma^5\gamma^{\mu_{2n+1}}\gamma^5\right)\right] \\ &= \mathrm{Tr}\left[\left(-\gamma^{\mu_1}\right)\left(-\gamma^{\mu_2}\right)\cdots\left(-\gamma^{\mu_{2n+1}}\right)\right] \\ &= (-1)^{2n+1}\,\mathrm{Tr}\left[\gamma^{\mu_1}\gamma^{\mu_2}\ldots\gamma^{\mu_{n+1}}\right] \\ &= -\,\mathrm{Tr}\left[\gamma^{\mu_1}\gamma^{\mu_2}\ldots\gamma^{\mu_{2n+1}}\right], \end{aligned} \tag{10.306}$$

from which the theorem follows. Clarify where the properties $\left(\gamma^5\right)^2 = \mathbb{1}$, $\left\{\gamma^\mu,\gamma^5\right\} = 0$, and the invariance of the trace under a cyclic permutation, have been used in the above transformations. Now, consider what happens in d dimensions. We cannot define γ^5 consistently in d dimensions, as we convinced ourselves. However, in the case of three γ^μ matrices, we can use the fourth current, not contained in any of the three matrices, in place of the γ^5. Also, because the case $d = 4$ has to be approached smoothly, any deviation from the result for $d = 4$ could only be of order ϵ.

(10) **Fifth Current and Dimensional Regularization.** Investigate the possibility of defining a γ^5 matrix in d dimensions. Using the equality $\gamma^\mu\gamma_\mu = d\,\mathbb{1}$, one can derive the following relation, under repeated use of the invariance of the trace under cyclic permutations of the matrices in the argument of the trace, as well as the assumed anticommutativity of the γ^5 matrix with all γ^μ matrices,

$$\begin{aligned} d\,\mathrm{Tr}\left[\gamma^5\gamma^\rho\gamma^\sigma\gamma^\kappa\gamma^\lambda\right] &= \mathrm{Tr}\left[\gamma^5\gamma_\mu\gamma^\mu\gamma^\rho\gamma^\sigma\gamma^\kappa\gamma^\lambda\right] \\ &= \mathrm{Tr}\left[\gamma^\mu\gamma^\rho\gamma^\sigma\gamma^\kappa\gamma^\lambda\gamma^5\gamma_\mu\right] = -\mathrm{Tr}\left[\gamma^\mu\gamma^\rho\gamma^\sigma\gamma^\kappa\gamma^\lambda\gamma_\mu\gamma^5\right] \\ &= -\,\mathrm{Tr}\left[\gamma^5\gamma^\mu\gamma^\rho\gamma^\sigma\gamma^\kappa\gamma^\lambda\gamma_\mu\right] \\ &= -\,\mathrm{Tr}\left[\gamma^5\gamma^\mu\gamma^\rho\gamma^\sigma\gamma^\kappa\{\gamma^\lambda,\gamma_\mu\}\right] + \mathrm{Tr}\left[\gamma^5\gamma^\mu\gamma^\rho\gamma^\sigma\gamma^\kappa\gamma_\mu\gamma^\lambda\right] \\ &= -\,2\,\mathrm{Tr}\left[\gamma^5\gamma^\lambda\gamma^\rho\gamma^\sigma\gamma^\kappa\right] + \mathrm{Tr}\left[\gamma^5\gamma^\mu\gamma^\rho\gamma^\sigma\gamma^\kappa\gamma_\mu\gamma^\lambda\right] \\ &= \ldots \\ &= -\,(d-8)\,\mathrm{Tr}\left[\gamma^5\gamma^\rho\gamma^\sigma\gamma^\kappa\gamma^\lambda\right]. \end{aligned} \tag{10.307}$$

Here, we denote missing intermediate steps using dots; the corresponding transformations start with the use of the equality $\gamma^\kappa\gamma_\mu = \{\gamma^\kappa,\gamma_\mu\} - \gamma_\mu\gamma^\kappa$. One still has to permute the arguments cyclically a few times, and use the anticommutativity of the γ^5 matrix a few times. One ends up with the relation

$$d\mathrm{Tr}\left[\gamma^5\gamma^\rho\gamma^\sigma\gamma^\kappa\gamma^\lambda\right] = -(d-8)\,\mathrm{Tr}\left[\gamma^5\gamma^\rho\gamma^\sigma\gamma^\kappa\gamma^\lambda\right]. \tag{10.308}$$

Alternatively, one can formulate this as follows,

$$(d-4)\,\mathrm{Tr}\left[\gamma^5\gamma^\rho\gamma^\sigma\gamma^\kappa\gamma^\lambda\right] = 0. \tag{10.309}$$

If $d \neq 4$, then this implies that the trace $\mathrm{Tr}\left[\gamma^5 \gamma^\rho \gamma^\sigma \gamma^\kappa \gamma^\lambda\right]$ needs to vanish, and cannot be proportional to $\epsilon^{\rho\sigma\kappa\lambda}$, as we would otherwise wish.

(11) Investigate further properties of the fifth current, in particular, the trace of the γ^5 matrix in d dimensions. Convince yourself of the validity of all intermediate steps in the transformation

$$d\mathrm{Tr}\left[\gamma^5\right] = \mathrm{Tr}\left[\gamma^5 \gamma_\mu \gamma^\mu\right] = \mathrm{Tr}\left[\gamma^\mu \gamma^5 \gamma_\mu\right] = -\mathrm{Tr}\left[\gamma^\mu \gamma_\mu \gamma^5\right] = -d\,\mathrm{Tr}\left[\gamma^5\right], \quad (10.310)$$

from which it follows that

$$d\,\mathrm{Tr}\left[\gamma^5\right] = 0. \quad (10.311)$$

Then, consider the next more complicated trace containing γ^5, which is

$$\begin{aligned} d\mathrm{Tr}\left[\gamma^5\gamma^\rho\gamma^\sigma\right] &= \mathrm{Tr}\left[\gamma^5\gamma_\mu\gamma^\mu\gamma^\rho\gamma^\sigma\right] = \mathrm{Tr}\left[\gamma^\mu\gamma^\rho\gamma^\sigma\gamma^5\gamma_\mu\right] = -\mathrm{Tr}\left[\gamma^\mu\gamma^\rho\gamma^\sigma\gamma_\mu\gamma^5\right] \\ &= -\,\mathrm{Tr}\left[\gamma^5\gamma^\mu\gamma^\rho\gamma^\sigma\gamma_\mu\right] \\ &= -\,\mathrm{Tr}\left[\gamma^5\gamma^\mu\gamma^\rho\{\gamma^\sigma,\gamma_\mu\}\right] + \mathrm{Tr}\left[\gamma^5\gamma^\mu\gamma^\rho\gamma_\mu\gamma^\sigma\right] \\ &= \dots \\ &= -\,(d-4)\,\mathrm{Tr}\left[\gamma^5\gamma^\rho\gamma^\sigma\right]. \end{aligned} \quad (10.312)$$

Here, the same transformation has been used as for the trace of a γ^5 matrix and four Dirac γ matrices, in Eq. (10.307). From the relation $(d-2)\mathrm{Tr}\left[\gamma^5\gamma^\rho\gamma^\sigma\right] = 0$, near $d = 4$ dimensions, one finds that the trace $\mathrm{Tr}\left[\gamma^5\gamma^\rho\gamma^\sigma\right]$ vanishes.

(12) **Fifth Current and Traces.** In four space-time dimensions, the trace of a γ^5 matrix and four Dirac γ matrices has to be proportional to the Levi-Cività symbol $\epsilon^{\rho\sigma\kappa\lambda}$ (where $\epsilon^{0123} = +1$). The constant can be determined by recalling that $\left(\gamma^0\right)^2 = \mathbb{1}$ and $\left(\gamma^i\right)^2 = -\mathbb{1}$, for $i = 1,2,3$ (no summation over i). Considering the definition (7.15) and the anticommutativity of the Dirac γ matrices, show that

$$\mathrm{Tr}\left[\gamma^5\gamma^\rho\gamma^\sigma\gamma^\kappa\gamma^\lambda\right] = -4\,\mathrm{i}\,\epsilon^{\rho\sigma\kappa\lambda}. \quad (10.313)$$

Chapter 11

Foldy–Wouthuysen Transformation and Lamb Shift

11.1 Overview

The Foldy–Wouthuysen transformation is an essential tool in the analysis of bound states, and interacting particles in general. The idea is the following: The free Dirac Hamiltonian, given in Eq. (7.18),

$$H_{\rm FD} = \begin{pmatrix} m\mathbb{1}_{2\times 2} & \vec{\sigma}\cdot\vec{p} \\ \vec{\sigma}\cdot\vec{p} & -m\,\mathbb{1}_{2\times 2} \end{pmatrix}, \tag{11.1}$$

couples upper and lower components of the Dirac bispinor. For zero momentum, we can ignore the kinetic term, and identify the upper Dirac spinor (in the bispinor) as the particle wave function, while the lower spinor is identified as an antiparticle wave function, but this is impossible for nonvanishing momenta. The aim of the (unitary) Foldy–Wouthuysen transformation is to separate the upper (particle) from the lower (antiparticle) degrees of freedom in higher orders in the momentum expansion.

It is rather common wisdom that the basic relativistic corrections to the hydrogen spectrum, induced by the relativistic correction to the free-particle dispersion relation, the *zitterbewegung* term and the spin-orbit coupling, can be derived from the Dirac–Coulomb Hamiltonian, by means of a Foldy–Wouthuysen transformation (a particularly instructive derivation is presented in Ref. [316]). Yet, it is our goal here to go further, and to derive higher-order correction terms. In considering the Foldy–Wouthuysen transformation, we should remember that we apply it to the Dirac equation, which in principle is a one-particle equation. This equation is valid, *a priori*, only in the limit of infinite nuclear mass. The momentum of a bound particle in the Coulomb field of a heavy nucleus is of order $Z\alpha m$, where Z is the nuclear charge number, α is the fine-structure constant, and m is the mass of the orbiting particle. Hence, the expansion in higher powers of the momenta is, in fact, an expansion in powers of $Z\alpha$.

We proceed as follows. In Sec. 11.2, we derive the leading relativistic corrections to the Foldy–Wouthuysen transformed Dirac Hamiltonian, for general electric and magnetic background fields. We include the anomalous magnetic moment into the transformation. Applications to the Coulomb field, and to magnetic-field coupling, and to the fermionic transition current, are discussed in Sec. 11.3. All of these calculations take into account the anomalous-magnetic-moment correction, given by the Pauli form factor, but not the Dirac form factor corrections. The latter aspect is treated in Sec. 11.4, including the high-energy part of the Lamb shift which can be derived on the basis of the transformed Hamiltonian. Further applications are discussed in Sec. 11.5.

11.2 Leading Relativistic Corrections

11.2.1 *Unitary Transformation and Hamiltonian*

Let us study the transformation of the Hamiltonian operator under a general unitary transformation

$$U = \exp(\mathrm{i}S)\,, \qquad S = S^+\,, \tag{11.2}$$

where U may be time-dependent and S is Hermitian (not to be confused with the S matrix, of course). The transformed Hamiltonian acts on the transformed wave function $\psi' = \exp(\mathrm{i}S)\,\psi$, for which

$$\begin{aligned}\mathrm{i}\partial_t\psi' &= \mathrm{i}\partial_t \exp(\mathrm{i}S)\,\psi = \exp(\mathrm{i}S)(H - \partial_t S)\,\psi \\ &= \left[\exp(\mathrm{i}S)[H - \mathrm{i}\partial_t]\,\exp(-\mathrm{i}S)\right]\psi'\,.\end{aligned} \tag{11.3}$$

We thus write the Foldy–Wouthuysen (FW) transformed Hamiltonian as follows,

$$H_{\mathrm{FW}} = \exp(\mathrm{i}S)\,(H - \mathrm{i}\,\partial_t)\exp(-\mathrm{i}S)\,. \tag{11.4}$$

In typical cases, the operator S carries some power of the momenta, which is typically small in atomic systems, and it is thus possible to expand in nested commutators as follows,

$$\begin{aligned}H_{\mathrm{FW}} &= H + [\mathrm{i}\,S, H - \mathrm{i}\,\partial_t] + \frac{1}{2!}\,[\mathrm{i}S,[\mathrm{i}\,S, H - \mathrm{i}\,\partial_t]] \\ &\quad + \frac{1}{3!}\,[\mathrm{i}S,[\mathrm{i}S,[\mathrm{i}\,S, H - \mathrm{i}\,\partial_t]]] + \frac{1}{4!}\,[\mathrm{i}S,[\mathrm{i}S,[\mathrm{i}S,[\mathrm{i}\,S, H - \mathrm{i}\,\partial_t]]]] + \cdots.\end{aligned} \tag{11.5}$$

The order required for the latter transformation depends on the problem under study. However, in general, we can say that S will be of the same order as the momenta in the problem. In many cases, the momenta define a convenient coupling parameter for a bound-state problem.

In the case of an explicit time dependence of the fields, we can formulate the first-order correction as follows,

$$\delta H^{(1)} = [\mathrm{i}S, H - \mathrm{i}\partial_t] = [\mathrm{i}S, H] - [-\mathrm{i}\,\partial_t(\mathrm{i}\,S)] = \mathrm{i}[S,H] - \partial_t S\,. \tag{11.6}$$

The next higher-order corrections can be calculated iteratively as follows,

$$\delta H^{(2)} = \frac{1}{2}[\mathrm{i}S, \delta H^{(1)}]\,, \quad \delta H^{(3)} = \frac{1}{3}[\mathrm{i}S, \delta H^{(2)}]\,, \quad \delta H^{(n+1)} = \frac{1}{n+1}[\mathrm{i}S, \delta H^{(n)}]\,. \tag{11.7}$$

It is only in special cases that the Foldy–Wouthuysen transformation can be carried out to all orders in the coupling parameter.

11.2.2 *Free Dirac Particle*

It is natural to apply the idea of the Foldy–Wouthuysen transformation first to the example of the free Dirac Hamiltonian. Indeed, in Chaps. 7 and 8, we have studied the relativistic theory of the electron, on the basis of the Dirac equation. The particle wave function (positive energy) and antiparticle wave functions (negative energy) were obtained within the fully relativistic theory, without approximations. The solutions of the free Dirac equation were given in Chap. 7, and those of the Dirac–Coulomb problem were discussed in Chap. 8. All of these solutions have nonvanishing entries for the upper and lower spinors, i.e., all four entries of the bispinor wave functions are populated. The operator $\vec{\alpha}\cdot\vec{p}$ in the Dirac Hamiltonian mixes the upper and lower components. Indeed, in the free Dirac Hamiltonian $H_{\rm FD} = \vec{\alpha}\cdot\vec{p} + \beta\, m$ given in Eq. (7.18), the kinetic term $\vec{\alpha}\cdot\vec{p}$ mixes the upper and lower spinor components.

The Hamiltonian for a particle at rest, which reads as $\beta\, m$, is diagonal in bispinor space, with an energy eigenvalue $+m$ for the particle solutions, and energy $-m$ for the antiparticle solutions. The Foldy–Wouthuysen transformation serves to accomplish a separation of the total Hamiltonian into an effective upper-component Hamiltonian, and an effective lower-component Hamiltonian, perturbatively, in successively higher orders of the momentum operator.

The Foldy–Wouthuysen transformation is best explained by way of example [317]. For illustrative purposes, let us therefore consider the free Dirac Hamiltonian defined in Eq. (7.18),

$$H_{\rm FD} = \vec{\alpha}\cdot\vec{p} + \beta m \equiv H_1 \,, \tag{11.8}$$

where the subscript "1" denotes the first iteration of the Foldy–Wouthuysen transformation. For the first Foldy–Wouthuysen transformation, we choose S_1, a Hermitian operator, to be proportional to $-{\rm i}\beta\,\mathcal{O}_1$, where $\mathcal{O}_1$ is the odd part (in bispinor space) of the Hamiltonian H_1. Specifically, we have

$$S_1 = -\frac{\rm i}{2m}\,\beta\,\mathcal{O}_1 \,, \qquad \mathcal{O}_1 = \vec{\alpha}\cdot\vec{p} \,. \tag{11.9}$$

In view of the lack of an explicit time dependence of S_1 [one compares with Eq. (11.4)], the transformation toward the Hamiltonian H_2 is calculated in terms of commutators as follows,

$$\begin{aligned} H_2 &= {\rm e}^{{\rm i}S_1}\, H_1\, {\rm e}^{-{\rm i}S_1} = H_1 + {\rm i}\,[S_1, H_1] + \frac{{\rm i}^2}{2!}\,[S_1,[S_1,H_1]] \\ &\quad + \frac{{\rm i}^3}{3!}\,[S_1,[S_1,[S_1,H_1]]] + \frac{{\rm i}^4}{4!}\,[S_1,[S_1,[S_1,[S_1,H_1]]]] + \cdots . \end{aligned} \tag{11.10}$$

For a calculation in fourth order in the momenta, one needs all terms given in Eq. (11.10). The result reads as follows,

$$H_2 = \beta\left(m + \frac{\vec{p}^{\,2}}{2m} - \frac{\vec{p}^{\,4}}{8m^3}\right) - \frac{(\vec{\alpha}\cdot\vec{p})^3}{3m^2} + \mathcal{O}(p^5) \,, \tag{11.11}$$

where $\mathcal{O}(p^5)$ denotes a generic term, possibly involving Dirac matrices, of fifth power in momentum operator $\vec{p}$. For the second Foldy–Wouthuysen transformation, we choose the odd operator to be the operator derived from the Hamiltonian H_2,

$$S_2 = -\frac{\mathrm{i}}{2m}\,\beta\,\mathcal{O}_2\,, \qquad \mathcal{O}_2 = -\frac{(\vec{\alpha}\cdot\vec{p})^3}{3m^2}\,. \tag{11.12}$$

The second transformation

$$H_{\mathrm{FW}} = \mathrm{e}^{\mathrm{i}S_2}\,H_2\,\mathrm{e}^{-\mathrm{i}S_2} \tag{11.13}$$

eliminates the term proportional to $(\vec{\alpha}\cdot\vec{p})^3$ and leads to the final form,

$$H_3 = \beta\left(m + \frac{\vec{p}^{\,2}}{2m} - \frac{\vec{p}^{\,4}}{8m^3} + \mathcal{O}(p^6)\right). \tag{11.14}$$

Eventually, extending the method to arbitrary order in the momenta, one arrives at the Hamiltonian (see also Sec. 11.6)

$$H_\infty = \beta\sqrt{\vec{p}^{\,2}+m^2} = \begin{pmatrix} \sqrt{\vec{p}^{\,2}+m^2}\cdot\mathbb{1}_{2\times 2} & 0 \\ 0 & -\sqrt{\vec{p}^{\,2}+m^2}\cdot\mathbb{1}_{2\times 2} \end{pmatrix}, \tag{11.15}$$

where (2×2)-matrix notation has been used. The particle degrees of freedom (spin-up and spin-down, upper components) have been separated from the antiparticle degrees of freedom (lower components). Furthermore, because an antiparticle solution with eigenvalue $-E$ of the Hamiltonian is interpreted in terms of a physical energy $+E$, one verifies the dispersion relation

$$E = \sqrt{\vec{p}^{\,2}+m^2} \tag{11.16}$$

for electrons and positrons. Generically, it is customary in the literature to refer to the Foldy–Wouthuysen transformed Hamiltonian as the particle (not antiparticle) term, i.e., as the (2×2)-upper component of H_3. For the case of a free particle, one obtains

$$H_{\mathrm{FW}} = \sqrt{\vec{p}^{\,2}+m^2}\cdot\mathbb{1}_{2\times 2} = \mathbb{1}_{2\times 2}\left(m + \frac{\vec{p}^{\,2}}{2m} - \frac{\vec{p}^{\,4}}{8m^3} + \mathcal{O}(p^6)\right). \tag{11.17}$$

The free Dirac Hamiltonian (11.8) is invariant under parity and so is the transformed Hamiltonian (11.17). It is instructive to observe that parity invariance is maintained in every step of the transformation. In view of Eq. (7.54), the first transformation S_1 given in Eq. (11.9) transforms as follows,

$$S_1 = -\frac{\mathrm{i}}{2m}\,\beta\,\vec{\alpha}\cdot\vec{p} \;\xrightarrow{\;\mathcal{P}\;}\; \beta\left[-\frac{\mathrm{i}}{2m}\,\beta\,\vec{\alpha}\cdot(-\vec{p})\right]\beta = \frac{\mathrm{i}}{2m}\,\vec{\alpha}\cdot\vec{p}\,\beta = S_1\,. \tag{11.18}$$

In fact, it is known that spurious terms can be generated in variants of the Foldy–Wouthuysen program when parity invariance is broken in intermediate steps of the calculation [286]. In particular, attempts to short-cut the procedure are prone to generating spurious terms. For example, it is known that parity-breaking, still unitary transformation give rise to a deceptively simple diagonal form (in bispinor space), but feature spurious terms [255, 286, 287, 318–321]. The main difficulty concerned with the separation of the upper and lower components cannot be avoided in general cases. The Foldy–Wouthuysen transformation eliminates odd terms one after the other, perturbatively in the momenta. It is a tedious but unavoidable endeavor.

11.2.3 *Transformation in the General Case*

Let us dwell a little more on the central idea the Foldy–Wouthuysen transformation, and anticipate a few aspects which will be important in more complex applications. The generalized Dirac Hamiltonians considered in Chaps. 7 and 8 have the structure

$$H = \mathcal{E} + \mathcal{O}\,, \tag{11.19a}$$

$$\mathcal{E} = \tfrac{1}{2}\,(H + \beta\,H\,\beta) = \beta m + \cdots\,, \tag{11.19b}$$

$$\mathcal{O} = \tfrac{1}{2}\,(H - \beta\,H\,\beta) = \vec{\alpha}\cdot\vec{p} + \cdots\,. \tag{11.19c}$$

Here, $\mathcal{E}$ is the even part in bispinor space, whereas $\mathcal{O}$ is the odd part, where even and odd are understood with respect to the separation of the Dirac matrix into 2×2 submatrices. For example, $\beta\,m$ is even, while $\vec{\alpha}\cdot\vec{p}$ is odd,

$$\beta\,m = \begin{pmatrix} m\,\mathbb{1}_{2\times2} & 0 \\ 0 & -m\,\mathbb{1}_{2\times2} \end{pmatrix}, \qquad \vec{\alpha}\cdot\vec{p} = \begin{pmatrix} 0 & \vec{\sigma}\cdot\vec{p} \\ \vec{\sigma}\cdot\vec{p} & 0 \end{pmatrix}. \tag{11.20}$$

In typical cases, the dominant term in the even part consists of the term $\beta\,m$, which has eigenvalues $\pm\,m$ and describes a Dirac particle at rest. The dominant term in the odd contribution is the kinetic term $\vec{\alpha}\cdot\vec{p}$. The central idea of the Foldy–Wouthuysen transformation is based on the identity

$$[\beta\,\mathcal{O}, \beta m] = -2m\,\mathcal{O}\,, \tag{11.21}$$

which holds for a general odd term $\mathcal{O}$, because $\beta\,m$ is a matrix with constant entries in space and time. It is for this reason that one defines the Hermitian operator S and the unitary operator $U = \exp(\mathrm{i}S)$ as follows,

$$S = -\mathrm{i}\,\beta\,\frac{\mathcal{O}}{2m}\,, \qquad U = \mathrm{e}^{\mathrm{i}S}\,. \tag{11.22}$$

One defines a perturbative expansion parameter, usually, a parameter connected with the average velocity of the quantum-mechanical wave packet, or with the spatial dimension of the system, and calculates all the terms of the Foldy–Wouthuysen transformation, to be defined in detail below, perturbatively, in the respective parameters. The calculation of multiple nested commutators of S and H, to describe the unitary transformation induced by U on H, then proceeds until further nested commutators yield higher-order terms beyond the order of interest. In many cases, the Foldy–Wouthuysen transformation needs to be iterated in order to yield the desired result.

11.2.4 *Radiatively Corrected Dirac Hamiltonian*

Before we discuss the Foldy–Wouthuysen transformation of a general electromagnetically coupled Dirac Hamiltonian, we should dwell on the radiatively corrected Dirac equation, because it is very useful to incorporate at least some radiative effects directly into the Foldy–Wouthuysen transformation. Our considerations rely on an understanding of the free Dirac Hamiltonian [see Eq. (7.18)] and of the Dirac–Coulomb Hamiltonian [see Eq. (8.9)]. Furthermore, in Eq. (10.96), we had already investigated the vertex function Λ^μ. Results for the Dirac form factor F_1, and for the Pauli form factor F_2, were given in Eqs. (10.103) and (10.100), respectively.

In a general four-vector potential $A^\mu(x)$, the Dirac Hamiltonian is

$$H_{\rm D} = \vec{\alpha} \cdot \vec{\pi} + \beta\, m + e\, A^0\,, \tag{11.23}$$

where e is the electron charge, and $\vec{\pi}$ is the canonical momentum,

$$\vec{\pi} = \vec{p} - e\,\vec{A}\,. \tag{11.24}$$

The Dirac form factor F_1, and the Pauli form factor F_2, are usually formulated as functions of the momentum transfer q^2. However, the general Dirac Hamiltonian (11.23) is formulated in terms of differential operators, and multiplication operators, expressed in space-time coordinates. Let us assume static fields and define

$$H_{\rm D}(\vec{r}) = \vec{\alpha} \cdot \left[-{\rm i}\vec{\nabla} - e\,\vec{A}(\vec{r})\right] + \beta\, m + e\, A^0(\vec{r})\,. \tag{11.25}$$

Following the paradigm that multiplication in coordinate space is equivalent to convolution in momentum space, and differentiation in coordinate space is equivalent to multiplication in momentum space, we can formulate the Dirac Hamiltonian in momentum space as

$$H_{\rm D}(\vec{p}\,',\vec{p}) = \vec{\alpha} \cdot \left[\delta^{(3)}(\vec{p}\,' - \vec{p})\vec{p} - e\,\vec{A}(\vec{p}\,' - \vec{p})\right] + \delta^{(3)}(\vec{p}\,' - \vec{p})\,\beta\, m + e\, A^0(\vec{p}\,' - \vec{p})\,. \tag{11.26}$$

This is consistent with our earlier considerations for the Schrödinger–Coulomb problem [Eqs. (4.171)–(4.173)]. One finally has for the action of the Hamiltonian $H_{\rm D}$ in momentum space,

$$\left[H_{\rm D}\,\psi\right](\vec{p}\,') = \int \frac{{\rm d}^3 p}{(2\pi)^3}\, H_{\rm D}(\vec{p}\,',\vec{p})\;\psi(\vec{p}). \tag{11.27}$$

An example is the Coulomb potential

$$e\,A^0(\vec{r}) = V(\vec{r})\,, \qquad V(\vec{q}) = -\frac{4\pi Z\alpha}{\vec{q}^{\,2}}\,, \qquad \vec{q} = \vec{p}\,' - \vec{p}\,. \tag{11.28}$$

A hypothetical multiplication of $V(\vec{q})$ by a function $F(\vec{q})$ of $\vec{q}$ would modify the potential. By Fourier transformation, one sees that the action of $F(-\vec{q}^{\,2})\,V(\vec{q})$ in momentum space is thus equivalent to the action of $F(\vec{\nabla}^2)\,V(\vec{r})$ in coordinate space, where the gradient operator only acts on the potential, but not on the following wave function. Thus, we replace, according to Eqs. (10.45), (10.118) and (10.189),

$$\begin{aligned}
\gamma^\mu\, A_\mu(\vec{q}) &\to \gamma^\mu F_1(q^2) A_\mu(\vec{q}) + \frac{{\rm i}\,\sigma^{\mu\nu} q_\nu}{2m} F_2(q^2) A_\mu(\vec{q}) \\
&\to \gamma^\mu F_1(-\vec{q}^{\,2}) A_\mu(\vec{q}) + \frac{{\rm i}\,\sigma^{\mu\nu} q_\nu}{2m} F_2(-\vec{q}^{\,2}) A_\mu(\vec{q}) \\
&\to \gamma^\mu F_1(\vec{\nabla}^2) A_\mu(\vec{r}) + \frac{{\rm i}\,\sigma^{\mu\nu}}{2m} F_2(\vec{\nabla}^2)\left\{{\rm i}\,\partial_\nu A_\mu(\vec{r})\right\}.
\end{aligned} \tag{11.29}$$

The Laplacian operators in the arguments of the form factors act only on the argument of the four-vector potentials. We use the identity

$$\sigma^{\mu\nu}\partial_\nu A_\mu(\vec{r}) = -\sigma^{\mu\nu}\partial_\mu A_\nu(\vec{r}) = -\frac{1}{2}\left(\sigma^{\mu\nu} - \sigma^{\nu\mu}\right)\partial_\mu A_\nu(\vec{r})$$
$$= -\frac{1}{2}\sigma^{\mu\nu}\left[\partial_\mu A_\nu(\vec{r}) - \partial_\nu A_\mu(\vec{r})\right] = -\frac{1}{2}\sigma^{\mu\nu} F_{\mu\nu}\,. \qquad (11.30)$$

Hence, the replacement rule is

$$\gamma^\mu\, A_\mu(\vec{q}) \to \gamma^\mu F_1(\vec{\nabla}^2) A_\mu(\vec{r}) + \frac{F_2(\vec{\nabla}^2)}{4m}\,\sigma^{\mu\nu}\, F_{\mu\nu}\,. \qquad (11.31)$$

We recall that the covariant form of the Dirac equation (11.23) is

$$\left[\gamma^\mu \pi_\mu - m\right]\Psi = \left[\gamma^\mu p_\mu - e\,\gamma^\mu A_\mu - m\right]\Psi = 0\,. \qquad (11.32)$$

We can thus reformulate the covariant form of the Dirac equation, with the interaction term given in Eq. (11.31), in the static case, as

$$\left[\gamma^\mu p_\mu - e\,F_1(\vec{\nabla}^2)\,\gamma^\mu\, A_\mu(\vec{r}) - \frac{e}{4m}\,F_2(\vec{\nabla}^2)\,\sigma^{\mu\nu}\, F_{\mu\nu} - m\right]\Psi = 0\,. \qquad (11.33)$$

We now use the identity

$$\sigma^{0i} F_{0i} = \frac{\mathrm{i}}{2}\left[\gamma^0, \gamma^i\right] E^i = \mathrm{i}\,\beta\,\vec{\gamma}\cdot\vec{E}\,, \qquad \vec{\gamma} = \begin{pmatrix} 0 & \vec{\sigma} \\ -\vec{\sigma} & 0 \end{pmatrix}. \qquad (11.34)$$

The term $\sigma^{i0}F_{i0}$ leads to an equal term which has to be added. The magnetic interaction is written as

$$\sigma^{ij}\, F_{ij} = \epsilon^{ijk}\,\Sigma^k\,(-\epsilon^{ij\ell}\, B^\ell) = -2\,\vec{\Sigma}\cdot\vec{B}\,, \qquad \vec{\Sigma} = \begin{pmatrix} \vec{\sigma} & 0 \\ 0 & \vec{\sigma} \end{pmatrix}. \qquad (11.35)$$

Finally, we have

$$F_2(\vec{\nabla}^2)\,\frac{e}{4\,m}\,\sigma^{\mu\nu}\, F_{\mu\nu} = F_2(\vec{\nabla}^2)\,\frac{e}{2\,m}\left(\mathrm{i}\,\beta\,\vec{\gamma}\cdot\vec{E} - \vec{\Sigma}\cdot\vec{B}\right). \qquad (11.36)$$

Converting Eq. (11.33) to Hamiltonian form, we thus have the radiatively corrected Dirac Hamiltonian H_{R} as

$$H_{\mathrm{R}}(\vec{r}) = \vec{\alpha}\cdot\left[\vec{p} - e\,F_1(\vec{\nabla}^2)\,\vec{A}(\vec{r})\right] + \beta\, m + F_1(\vec{\nabla}^2)\,eA^0(\vec{r})$$
$$+ F_2(\vec{\nabla}^2)\,\frac{e}{2\,m}\left[\mathrm{i}\,\vec{\gamma}\cdot\vec{E}(\vec{r}) - \beta\,\vec{\Sigma}\cdot\vec{B}(\vec{r})\right]. \qquad (11.37)$$

In the following, we will use the approximations [see Eqs. (10.100) and (10.103)],

$$F_1(\vec{\nabla}^2) \approx F_1(0) = 1\,, \qquad F_2(\vec{\nabla}^2) \approx F_2(0) \equiv \kappa \approx \frac{\alpha}{2\pi}\,, \qquad (11.38)$$

where the g factor of the electron is expressed as $g = 2(1+\kappa)$.

Furthermore, under the approximations given in Eq. (11.38), the form factors become independent of time. In this case, we can relax again the assumptions on the time-independence of the vector and scalar potentials and write the radiatively corrected Dirac Hamiltonian as

$$H_{\mathrm{R}}(\vec{r},t) = \vec{\alpha}\cdot\left[\vec{p} - e\,\vec{A}(\vec{r},t)\right] + \beta\, m + eA^0(\vec{r},t)$$
$$+ \frac{e\,\kappa}{2\,m}\left[\mathrm{i}\,\vec{\gamma}\cdot\vec{E}(\vec{r},t) - \beta\,\vec{\Sigma}\cdot\vec{B}(\vec{r},t)\right]. \qquad (11.39)$$

The possible explicit time dependence will need to be taken into account as we discuss the Foldy–Wouthuysen transformation.

11.2.5 General Electromagnetic Coupling

We need to generalize the Foldy–Wouthuysen transformation, carried out at this point only for a free Dirac particle, to the case of an electromagnetic coupling. It is useful to also include leading radiative corrections, corresponding to the anomalous magnetic moment, as given in the effective Dirac Hamiltonian (11.37). We shall consider only terms linear in the electron anomaly κ defined in Eq. (1.9).

With reference to Eqs. (11.24) and (11.39), we investigate the radiatively corrected Hamiltonian

$$H_{\rm R} = \vec{\alpha}\cdot\vec{\pi} + \beta m + e\,A^0 + \frac{e\kappa}{2m}\left(\mathrm{i}\beta\,\vec{\alpha}\cdot\vec{E} - \beta\,\vec{\Sigma}\cdot\vec{B}\right), \tag{11.40}$$

where we take into account the possibility of an explicit time dependence of the vector and scalar potentials, and the fields.

We shall carry out the transformation of the Hamiltonian (11.40) up to eighth order in the momenta, under the premise of the following scaling,

$$\vec{\pi}\sim\alpha\,, \qquad e\vec{A}\sim\alpha\,, \qquad e\,\vec{B}\sim\alpha^2\,, \qquad e\,\vec{E} = -\vec{\nabla}V\sim\alpha^3\,, \qquad e\,\partial_t\vec{E}\sim\alpha^5\,. \tag{11.41}$$

In the context of the general method outlined in Sec. 11.2.1, we find that the time derivative term in Eq. (11.4) is decisive in ensuring that the Foldy–Wouthuysen transformed Hamiltonian contains the complete electric field, and not only the transverse or longitudinal components separately. In Coulomb gauge, these components are naturally separated as follows,

$$\vec{E}_{\parallel} = -\vec{\nabla}A^0\,, \qquad \vec{E}_{\perp} = -\partial_t\vec{A}\,. \tag{11.42}$$

The result of the iterative eighth-order standard Foldy–Wouthuysen transformation [286], carried out up to order α^8, can be written in terms of

$$\mathcal{H}_{\rm FW} = \mathcal{H}^{[0]} + \mathcal{H}^{[2]} + \mathcal{H}^{[3]} + \mathcal{H}^{[4]} + \mathcal{H}^{[5]} + \mathcal{H}^{[6]} + \mathcal{H}^{[7]} + \mathcal{H}^{[8]}\,. \tag{11.43}$$

The superscript denotes the power of the coupling parameter at which the term becomes relevant. The coupling parameter is usually denoted as α. From the zeroth to third order in α, the terms read as follows,

$$\mathcal{H}^{[0]} = \beta m\,, \tag{11.44}$$

$$\mathcal{H}^{[2]} = \beta\frac{1}{2m}(\vec{\Sigma}\cdot\vec{\pi})^2 + V\,, \tag{11.45}$$

$$\mathcal{H}^{[3]} = -\frac{e\,\kappa}{2m}\,\beta\,\vec{\Sigma}\cdot\vec{B}\,. \tag{11.46}$$

The α^4 and α^5 terms can be expressed very compactly, and are found in agreement with the literature [316, 322],

$$\mathcal{H}^{[4]} = -\beta\,\frac{1}{8m^3}(\vec{\Sigma}\cdot\vec{\pi})^4 - \frac{\mathrm{i}\,e}{8m^2}[\,\vec{\Sigma}\cdot\vec{\pi},\vec{\Sigma}\cdot\vec{E}\,]\,, \tag{11.47}$$

$$\mathcal{H}^{[5]} = -\frac{\mathrm{i}\,e\,\kappa}{4m^2}\,[\,\vec{\Sigma}\cdot\vec{\pi},\vec{\Sigma}\cdot\vec{E}\,] + \beta\,\frac{e\,\kappa}{16m^3}\,\{\vec{\Sigma}\cdot\vec{\pi},\{\,\vec{\Sigma}\cdot\vec{\pi},\vec{\Sigma}\cdot\vec{B}\}\}\,. \tag{11.48}$$

Here, $\{A,B\} = A\,B + B\,A$ denotes the anticommutator.

For the α^6 terms, we indicate two alternative representations. The straightforward application of the method described in Sec. (11.2.1) leads to the result

$$\mathcal{H}^{[6]} = \beta \frac{1}{16m^5}(\vec{\Sigma}\cdot\vec{\pi})^6 - \frac{5ie}{128m^4}[\vec{\Sigma}\cdot\vec{\pi},[\vec{\Sigma}\cdot\vec{\pi},[\vec{\Sigma}\cdot\vec{\pi},\vec{\Sigma}\cdot\vec{E}]]] + \frac{ie}{8m^4}\{(\vec{\Sigma}\cdot\vec{\pi})^2,[\vec{\Sigma}\cdot\vec{\pi},\vec{\Sigma}\cdot\vec{E}]\} + \beta\frac{e^2\vec{E}^2}{8m^3}. \tag{11.49}$$

Using the identity

$$[\vec{\Sigma}\cdot\vec{\pi},[\vec{\Sigma}\cdot\vec{\pi},[\vec{\Sigma}\cdot\vec{\pi},\vec{\Sigma}\cdot\vec{E}]]] = 2\left\{(\vec{\Sigma}\cdot\vec{\pi})^2,[\vec{\Sigma}\cdot\vec{\pi},\vec{\Sigma}\cdot\vec{E}]\right\} - \left[(\vec{\Sigma}\cdot\vec{\pi})^2,\{\vec{\Sigma}\cdot\vec{E},\vec{\Sigma}\cdot\vec{\pi}\}\right], \tag{11.50}$$

the result for the sixth-order terms can alternatively be written as

$$\mathcal{H}^{[6]} = \beta \frac{1}{16m^5}(\vec{\Sigma}\cdot\vec{\pi})^6 + \frac{5ie}{128m^4}\left[(\vec{\Sigma}\cdot\vec{\pi})^2,\{\vec{\Sigma}\cdot\vec{E},\vec{\Sigma}\cdot\vec{\pi}\}\right] + \frac{3ie}{64m^4}\{(\vec{\Sigma}\cdot\vec{\pi})^2,[\vec{\Sigma}\cdot\vec{\pi},\vec{\Sigma}\cdot\vec{E}]\} + \beta\frac{e^2\vec{E}^2}{8m^3}. \tag{11.51}$$

The α^7 terms contain the anomalous magnetic moment, proportional to κ. As we recall, the electron g factor is related to κ by the relations $g = 2(1+\kappa)$. The result, obtained using the method outlined in Sec. 11.2.1, is

$$\begin{aligned}\mathcal{H}^{[7]} &= \beta\frac{e^2\kappa}{8m^3}\vec{E}^2 - \frac{e\kappa}{16m^3}\beta\{\vec{\Sigma}\cdot\vec{\pi},\vec{\Sigma}\cdot\partial_t\vec{E}\} + \frac{ie\kappa}{16m^4}\{(\vec{\Sigma}\cdot\vec{\pi})^2,[\vec{\Sigma}\cdot\vec{\pi},\vec{\Sigma}\cdot\vec{E}]\}\\ &\quad - \beta\frac{e\kappa}{32m^5}[\vec{\Sigma}\cdot\vec{\pi},[\vec{\Sigma}\cdot\vec{\pi},\{\vec{\Sigma}\cdot\vec{\pi},\{\vec{\Sigma}\cdot\vec{\pi},\vec{\Sigma}\cdot\vec{B}\}\}]]\\ &\quad - \beta\frac{3e\kappa}{256m^5}\{\vec{\Sigma}\cdot\vec{\pi},\{\vec{\Sigma}\cdot\vec{\pi},\{\vec{\Sigma}\cdot\vec{\pi},\{\vec{\Sigma}\cdot\vec{\pi},\vec{\Sigma}\cdot\vec{B}\}\}\}\}\\ &\quad + \frac{ie^2\kappa}{16m^4}[\vec{\Sigma}\cdot\vec{E},\{\vec{\Sigma}\cdot\vec{\pi},\vec{\Sigma}\cdot\vec{B}\}].\end{aligned} \tag{11.52}$$

We here supplement a few terms which are relevant for strong magnetic fields, and which have not been taken into account in the literature. One notes that, in the seventh order in α, one cannot avoid the emergence of terms proportional to the time derivative of electric fields.

In the somewhat non-standard approach chosen in Ref. [322], the time derivative terms already appear in the order α^6. Specifically, according to Eq. (18) of Ref. [322], one encounters a term $-[e/(16m^3)]\{\sigma\cdot\vec{\pi},\sigma\cdot\dot{\vec{E}}\}$. The terms involving the time-derivative of the electric field were advocated to be eliminated by an additional unitary transformation given in Eq. (19) of Ref. [323]. In the standard Foldy–Wouthuysen approach, as we show here, the time derivative of the electric field appears only in the order α^7.

In the order α^8, it is not necessary to consider the anomalous magnetic moment, provided we neglect terms proportional to κ^2. After a somewhat elaborate calculation, using computer algebra techniques [151] in order to keep the complexity of intermediate expressions under control, one arrives at the following terms,

$$\begin{aligned}
\mathcal{H}^{[8]} &= -\beta\frac{5}{128m^7}(\vec{\Sigma}\cdot\vec{\pi})^8 - \frac{\mathrm{i}\,e^2}{32m^4}[\vec{\Sigma}\cdot\vec{E},\vec{\Sigma}\cdot\partial_t\vec{E}] + \frac{7\,e^2}{192m^5}\beta[\vec{\Sigma}\cdot\vec{\pi},\vec{\Sigma}\cdot\vec{E}][\vec{\Sigma}\cdot\vec{\pi},\vec{\Sigma}\cdot\vec{E}] \\
&\quad - \frac{3\,e^2}{64m^5}\,\beta\,\{\vec{\Sigma}\cdot\vec{\pi},\vec{\Sigma}\cdot\vec{E}\}\,\{\vec{\Sigma}\cdot\vec{\pi},\vec{\Sigma}\cdot\vec{E}\} - \frac{e^2}{24m^5}\,\beta\,[\vec{\Sigma}\cdot\vec{\pi},[\vec{\Sigma}\cdot\vec{\pi},(\vec{\Sigma}\cdot\vec{E})^2]] \\
&\quad + \frac{e}{48m^5}\,\beta\{(\vec{\Sigma}\cdot\vec{\pi})^3,\vec{\Sigma}\cdot\partial_t\vec{E}\} \\
&\quad - \frac{5\,\mathrm{i}\,e}{1024m^6}[\vec{\Sigma}\cdot\vec{\pi},[\vec{\Sigma}\cdot\vec{\pi},[\vec{\Sigma}\cdot\vec{\pi},[\vec{\Sigma}\cdot\vec{\pi},[\vec{\Sigma}\cdot\vec{\pi},\vec{\Sigma}\cdot\vec{E}]]]]] \\
&\quad - \frac{\mathrm{i}\,e}{32m^6}\{\vec{\Sigma}\cdot\vec{\pi},\{\vec{\Sigma}\cdot\vec{\pi},[\vec{\Sigma}\cdot\vec{\pi},[\vec{\Sigma}\cdot\vec{\pi},[\vec{\Sigma}\cdot\vec{\pi},\vec{\Sigma}\cdot\vec{E}]]]\}\} \\
&\quad - \frac{\mathrm{i}\,e}{48m^6}\{\vec{\Sigma}\cdot\vec{\pi},\{\vec{\Sigma}\cdot\vec{\pi},\{\vec{\Sigma}\cdot\vec{\pi},\{\vec{\Sigma}\cdot\vec{\pi},[\vec{\Sigma}\cdot\vec{\pi},\vec{\Sigma}\cdot\vec{E}]\}\}\}\}. \qquad (11.53)
\end{aligned}$$

The α^8 terms have been obtained by mapping the algebra of the commutators onto a computer symbolic program [151].

One can check that the terms given in Eqs. (11.44)–(11.53) fulfill the particle-antiparticle symmetry relations, i.e., they are invariant under the application of the following operations: *(i)* multiplication by an overall factor -1, *(ii)* replacement $\beta \to -\beta$, *(iii)* replacement $\vec{\pi} \to -\vec{\pi}$, and $\partial_t \to -\partial_t$, *(iv)* replacement $\vec{\Sigma} \to -\vec{\Sigma}$, and *(v)* replacements $e \to -e$, $V \to -V$ and $\vec{E} \to -\vec{E}$. All of these replacements are dictated by the reinterpretation principle of quantum field theory.

11.2.6 *General Particle Hamiltonians*

Let us concentrate on the upper left 2×2 submatrix of $\mathcal{H}_{\rm FW}$, which is the particle Hamiltonian. It is well known that the Dirac Hamiltonian describes particle and antiparticle states simultaneously, and that the lower right 2×2 submatrix of $\mathcal{H}_{\rm FW}$ describes the corresponding antiparticle. In principle, the particle Hamiltonian can be obtained from the results given in Eqs. (11.45)–(11.53) by simply replacing $\vec{\Sigma} \to \vec{\sigma}$, and $\beta \to \mathbb{1}_{2\times 2}$, but it is still instructive to give the results separately.

Because the rest mass term βm given in Eq. (11.44) is a physically irrelevant constant, we write the general particle Hamiltonian H under the presence of the external electric and magnetic fields as

$$H_{\rm FW} = H^{[2]} + H^{[3]} + H^{[4]} + H^{[5]} + H^{[6]} + H^{[7]} + H^{[8]}\,, \qquad (11.54)$$

where we take into account up to eighth-order terms. One finds

$$H^{[2]} = \frac{1}{2m}(\vec{\sigma}\cdot\vec{\pi})^2 + V\,, \qquad H^{[3]} = -\frac{e\,\kappa}{2m}\,\vec{\sigma}\cdot\vec{B}\,. \qquad (11.55)$$

In regard to Eq. (11.55), the following relation is of extreme importance, which is why we present the derivation in great detail,

$$\begin{aligned}
(\vec{\sigma}\cdot\vec{\pi})^2 &= \sigma^i\,\sigma^j\,\pi^i\,\pi^j = \left(\delta^{ij} + \mathrm{i}\,\epsilon^{ijk}\,\sigma^k\right)\pi^i\,\pi^j \\
&= \vec{\pi}^{\,2} + \frac{1}{2}\mathrm{i}\,\epsilon^{ijk}\,\sigma^k[\pi^i,\pi^j] = \vec{\pi}^{\,2} + \frac{1}{2}\mathrm{i}\,\epsilon^{ijk}\,\sigma^k\left[-\mathrm{i}\nabla^i(-e\,A^j) + \mathrm{i}\nabla^j(-e\,A^i)\right] \\
&= \vec{\pi}^{\,2} + \frac{1}{2}\,\epsilon^{ijk}\,\sigma^k\left\{2\nabla^i(-e\,A^j)\right\} = \vec{\pi}^{\,2} - e\,\epsilon^{kij}\,\sigma^k\nabla^i A^j = \vec{\pi}^{\,2} - e\,\vec{\sigma}\cdot\vec{B}\,. \qquad (11.56)
\end{aligned}$$

The fourth-order terms find the compact representation

$$H^{[4]} = -\frac{1}{8m^3}(\vec{\sigma}\cdot\vec{\pi})^4 - \frac{\mathrm{i}\,e}{8m^2}[\,\vec{\sigma}\cdot\vec{\pi},\vec{\sigma}\cdot\vec{E}\,], \tag{11.57}$$

while the fifth-order terms contain the anomalous magnetic-moment,

$$H^{[5]} = -\frac{\mathrm{i}\,e\,\kappa}{4m^2}\,[\,\vec{\sigma}\cdot\vec{\pi},\vec{\sigma}\cdot\vec{E}\,] + \frac{e\,\kappa}{16m^3}\,\{\,\vec{\sigma}\cdot\vec{\pi},\{\,\vec{\sigma}\cdot\vec{\pi},\vec{\sigma}\cdot\vec{B}\,\}\}. \tag{11.58}$$

The α^6 terms, in the general setting, are given as

$$H^{[6]} = \frac{(\vec{\sigma}\cdot\vec{\pi})^6}{16m^5} + \frac{5\mathrm{i}e}{128m^4}\left[(\vec{\sigma}\cdot\vec{\pi})^2,\{\vec{\sigma}\cdot\vec{E},\vec{\sigma}\cdot\vec{\pi}\}\right] + \frac{3\mathrm{i}\,e}{64m^4}\,\{\,(\vec{\sigma}\cdot\vec{\pi})^2,[\,\vec{\sigma}\cdot\vec{\pi},\vec{\sigma}\cdot\vec{E}\,]\} + \frac{e^2\vec{E}^{\,2}}{8m^3}. \tag{11.59}$$

In the above form, the sixth-order terms in the Hamiltonian are compatible with those used in Ref. [324] for spin-1/2 particles [see Eqs. (36)–(38) of Ref. [324]]. The α^6 terms listed in Eq. (11.51) are also equal to those obtained by applying the unitary transformation outlined in Eq. (19) of Ref. [323] to the Hamiltonian given in Eq. (15) of Ref. [323], i.e., to the Hamiltonian obtained by adding the terms given in Eqs. (15) and (20) of Ref. [323]. Also, the result in Eq. (11.51) is equal to the Hamiltonian considered in Eq. (7) of Ref. [325], which in turn has been derived from the NRQED approach outlined in Ref. [326]. The α^7 terms attain the following structure,

$$\begin{aligned} H^{[7]} = {} & \frac{e^2\,\kappa}{8m^3}\,\vec{E}^{\,2} - \frac{e\,\kappa}{16m^3}\,\{\,\vec{\sigma}\cdot\vec{\pi},\vec{\sigma}\cdot\partial_t\vec{E}\,\} + \frac{\mathrm{i}\,e\,\kappa}{16m^4}\,\{\,(\vec{\sigma}\cdot\vec{\pi})^2,[\,\vec{\sigma}\cdot\vec{\pi},\vec{\sigma}\cdot\vec{E}\,]\} \\ & - \frac{e\,\kappa}{32m^5}\,[\vec{\sigma}\cdot\vec{\pi},[\vec{\sigma}\cdot\vec{\pi},\{\,\vec{\sigma}\cdot\vec{\pi},\{\,\vec{\sigma}\cdot\vec{\pi},\vec{\sigma}\cdot\vec{B}\,\}\}\,]] \\ & - \frac{3e\,\kappa}{256m^5}\,\{\vec{\sigma}\cdot\vec{\pi},\{\vec{\sigma}\cdot\vec{\pi},\{\,\vec{\sigma}\cdot\vec{\pi},\{\,\vec{\sigma}\cdot\vec{\pi},\vec{\sigma}\cdot\vec{B}\,\}\}\,\}\} + \frac{\mathrm{i}\,e^2\,\kappa}{16\,m^4}\,[\,\vec{\sigma}\cdot\vec{E},\{\vec{\sigma}\cdot\vec{\pi},\vec{\sigma}\cdot\vec{B}\}]. \end{aligned} \tag{11.60}$$

Finally, the α^8 terms are derived from Eq. (11.53) and read as follows,

$$\begin{aligned} H^{[8]} = {} & -\frac{5}{128m^7}(\vec{\sigma}\cdot\vec{\pi})^8 - \frac{\mathrm{i}\,e^2}{32m^4}[\vec{\sigma}\cdot\vec{E},\vec{\sigma}\cdot\partial_t\vec{E}] + \frac{7\,e^2}{192m^5}[\vec{\sigma}\cdot\vec{\pi},\vec{\sigma}\cdot\vec{E}][\vec{\sigma}\cdot\vec{\pi},\vec{\sigma}\cdot\vec{E}] \\ & - \frac{3\,e^2}{64m^5}\{\vec{\sigma}\cdot\vec{\pi},\vec{\sigma}\cdot\vec{E}\}\{\vec{\sigma}\cdot\vec{\pi},\vec{\sigma}\cdot\vec{E}\} - \frac{e^2}{24m^5}[\vec{\sigma}\cdot\vec{\pi},[\vec{\sigma}\cdot\vec{\pi},(\vec{\sigma}\cdot\vec{E})^2]] \\ & + \frac{e}{48m^5}\beta\{(\vec{\sigma}\cdot\vec{\pi})^3,\vec{\sigma}\cdot\partial_t\vec{E}\} \\ & - \frac{5\mathrm{i}e}{1024m^6}[\vec{\sigma}\cdot\vec{\pi},[\vec{\sigma}\cdot\vec{\pi},[\vec{\sigma}\cdot\vec{\pi},[\vec{\sigma}\cdot\vec{\pi},[\vec{\sigma}\cdot\vec{\pi},\sigma\cdot\vec{E}]]]]] \\ & - \frac{\mathrm{i}\,e}{32m^6}\{\vec{\sigma}\cdot\vec{\pi},\{\vec{\sigma}\cdot\vec{\pi},[\vec{\sigma}\cdot\vec{\pi},[\vec{\sigma}\cdot\vec{\pi},[\vec{\sigma}\cdot\vec{\pi},\sigma\cdot\vec{E}]]]\}\} \\ & - \frac{\mathrm{i}\,e}{48m^6}\{\vec{\sigma}\cdot\vec{\pi},\{\vec{\sigma}\cdot\vec{\pi},\{\vec{\sigma}\cdot\vec{\pi},\{\vec{\sigma}\cdot\vec{\pi},[\vec{\sigma}\cdot\vec{\pi},\sigma\cdot\vec{E}]\}\}\}\}. \end{aligned} \tag{11.61}$$

Again, we stress that we restrict the discussion to terms linear in κ. Also, while we indicate general results for the FW transformed Hamiltonians to order α^8, we restrict the discussion to order α^6 in the applications discussed below.

In the following, we shall discuss four applications of the Foldy–Wouthuysen Hamiltonian given in Eq. (11.54) in the regime of bound states. The first of these will concern a plain Coulomb field, in which case the electric field is static and its time derivative vanishes. We shall derive explicit expressions, with comments on the applicability to bound states in regard to reduced-mass corrections, in each order in the $Z\alpha$-expansion. A second application will concern magnetic-field coupling, where we shall put special emphasis on the spin-dependent terms. A third application concerns the extraction of the transition current, including relativistic corrections, from the Foldy–Wouthuysen transformed Hamiltonian in the presence of (conceivably quantized) fields. Finally, a fourth application goes beyond the treatment restricted to the F_2 form factor, and includes form-factor derivatives as well as the Dirac F_1 form factor.

11.3 Applications

11.3.1 *Coulomb Field Coupling*

One of the important applications of the Hamiltonian (11.54) concerns Coulombic bound states. In this case, one has a situation where all fields are purely classical,

$$e\,A^0 = V = -\frac{Z\alpha}{r}\,, \qquad \vec{A} = \vec{0}\,, \qquad \vec{B} = \vec{0}\,, \qquad e\,\vec{E} = -e\,\vec{\nabla}A^0 = -\vec{\nabla}V\,, \qquad \vec{\pi} = \vec{p}\,. \tag{11.62}$$

One ends up with the following leading term,

$$H^{[2]} = \frac{\vec{p}^{\,2}}{2m} + V = \frac{\vec{p}^{\,2}}{2m} - \frac{Z\alpha}{r}\,, \tag{11.63}$$

which is the Schrödinger–Coulomb Hamiltonian given in Eq. (4.1), in the limit of an infinite nuclear mass. We recall that the Dirac formalism is inherently a one-particle formalism and does not allow for the inclusion of reduced-mass effects, which have to be treated on the basis of a manifestly two-body formalism.

For the evaluation of the fourth-order corrections given by Eq. (11.57), one needs the identities,

$$e\,\vec{E} = -\vec{\nabla}V = \vec{\nabla}\frac{Z\alpha}{r} = -\frac{Z\alpha\,\vec{r}}{r^3}\,, \qquad \mathrm{i}e\,[\vec{\sigma}\cdot\vec{p}, \vec{\sigma}\cdot\vec{E}] = -\vec{\nabla}^2 V - 2\,\vec{\sigma}\cdot(\vec{\nabla}V\times\vec{p})\,. \tag{11.64}$$

According to Eq. (11.57), the leading relativistic corrections in the Foldy–Wouthuysen Hamiltonian in this case reads as follows,

$$\begin{aligned} H^{[4]} &= -\frac{\vec{p}^{\,4}}{8m^3} + \frac{1}{8m^2}\vec{\nabla}^2 V + \frac{1}{4m^2}\vec{\sigma}\cdot(\vec{\nabla}V\times\vec{p}) \\ &= -\frac{\vec{p}^{\,4}}{8m^3} + \frac{\pi\,(Z\alpha)}{2m^2}\,\delta^{(3)}(\vec{r}) + \frac{Z\alpha}{4m^2\,r^3}\,\vec{\sigma}\cdot\vec{L}\,. \end{aligned} \tag{11.65}$$

This Hamiltonian contains the familiar *zitterbewegung* term (Darwin term), proportional to a Dirac-δ, and the spin-orbit coupling, both of which receive anomalous magnetic-moment corrections. From Eq. (11.58), it follows that the fifth-order (in $Z\alpha$) terms reads as follows,

$$\begin{aligned} H^{[5]} &= 2\kappa \left[\frac{1}{8m^2} \vec{\nabla}^2 V + \frac{1}{4m^2} \vec{\sigma} \cdot (\vec{\nabla} V \times \vec{p}) \right] \\ &= 2\kappa \left[\frac{\pi\,(Z\alpha)}{2m^2}\, \delta^{(3)}(\vec{r}) + \frac{Z\,\alpha}{4m^2\,r^3}\, \vec{\sigma} \cdot \vec{L} \right]. \end{aligned} \tag{11.66}$$

The anomalous magnetic-moment corrections simply multiply the potential terms in Eq. (11.65) by an additional factor 2κ.

For the sixth-order corrections, we note the following useful identity which follows from Eq. (11.62),

$$e\{\vec{\sigma} \cdot \vec{p}, \vec{\sigma} \cdot \vec{E}\} = -\{\vec{\sigma} \cdot \vec{p}, \vec{\sigma} \cdot \vec{\nabla} V\} = -\mathrm{i}[\vec{p}^{\,2}, V]\,. \tag{11.67}$$

Thus, the sixth-order corrections attain the following form,

$$\begin{aligned} H^{[6]} &= \frac{\vec{p}^{\,6}}{16m^5} - \frac{3}{64m^4} \left\{ \vec{p}^{\,2}, \vec{\nabla}^2 V \right\} - \frac{3}{32m^4} \left\{ \vec{p}^{\,2}, \vec{\sigma} \cdot (\vec{\nabla} V \times \vec{p}) \right\} \\ &\quad + \frac{5}{128m^4} \left[\vec{p}^{\,2}, [\vec{p}^{\,2}, V] \right] + \frac{(\vec{\nabla} V)^2}{8m^3}\,. \end{aligned} \tag{11.68}$$

This result has to be compared to the one used in Ref. [312],

$$\overline{H}^{[6]} = \frac{\vec{p}^{\,6}}{16\,m^5} - \frac{3}{64\,m^4} \{\vec{p}^{\,2}, \vec{\nabla}^{\,2} V\} - \frac{3}{32\,m^4} \{\vec{p}^{\,2}, \vec{\sigma} \cdot (\vec{\nabla} V \times \vec{p})\} + \frac{1}{128\,m^4} [\vec{p}^{\,2}, [\vec{p}^{\,2}, V]]\,. \tag{11.69}$$

We use the Schrödinger equation given Eq. (4.1) in the non-recoil limit, where the reduced mass is equal to the electron mass,

$$H_S = \frac{\vec{p}^2}{2m} + V\,, \qquad H_S\,|\phi\rangle = E\,|\phi\rangle\,, \qquad (H_S - E)\,|\phi\rangle = 0\,. \tag{11.70}$$

One can show the following identity fulfilled by diagonal matrix elements of the Schrödinger eigenstates,

$$\langle\phi|[\vec{p}^2, [\vec{p}^2, V]]|\phi\rangle = -4m\,\langle\phi|\langle(\vec{\nabla} V)^2|\phi\rangle\,. \tag{11.71}$$

One thus has the relation

$$\begin{aligned} &\frac{5}{128m^4} \langle\phi| \left[\vec{p}^{\,2}, [\vec{p}^{\,2}, V]\right] |\phi\rangle + \frac{1}{8m^3} \langle\phi|(\vec{\nabla} V)^2|\phi\rangle \\ &= \frac{1}{128\,m^4} \langle\phi|\, [\vec{p}^{\,2}, [\vec{p}^{\,2}, V]]\, |\phi\rangle = -\frac{1}{32m^3} \langle\phi|\langle(\vec{\nabla} V)^2|\phi\rangle\,. \end{aligned} \tag{11.72}$$

So, with reference to Eqs. (11.68) and (11.69), one can show that, on the level of diagonal matrix element, the Hamiltonians $H^{[6]}$ and $\overline{H}^{[6]}$ are equivalent,

$$\langle\phi|H^{[6]}|\phi\rangle = \langle\phi|\overline{H}^{[6]}|\phi\rangle\,. \tag{11.73}$$

On the level of the α^6 corrections, the direct approach for the Foldy–Wouthuysen transformation chosen here, and the modified approach used in Refs. [312, 322], are thus equivalent.

A few remarks are in order. For hydrogen and positronium, and other two-particle bound systems, the evaluation of the *zitterbewegung* terms Eqs. (11.65) and (11.66) (those proportional to Dirac-δ) is immediate. However, if one generalizes the approach to helium, the evaluation of the diagonal matrix element of the *zitterbewegung* term can be far less trivial, as has been noticed in Ref. [327]. This fact has inspired the development [328] of a scheme where the expectation value of a singular Dirac-δ operator is replaced by less singular momentum operators. It is interesting to observe that the identities derived in Ref. [328] can be traced to the relation

$$4\pi(Z\alpha)\delta^{(3)}(\vec{r}) = [\vec{p},[\vec{p},V]] = -\vec{\nabla}^2 V = \{\vec{p}^2, V\} - 2\vec{p}\,V\,\vec{p}\,. \tag{11.74}$$

On the left-hand side of this equation, we have a Dirac-δ, on the right-hand side, some standard momentum operators and potentials. Taking the expectation value of the expression on the right-hand side, and using the Hermiticity of the momentum operators, one arrives at less singular expressions. This approach, though, meets some limitations in higher-order. This and similar approaches are commonly referred to as "Drachmanization" in view of the author of Ref. [328].

Some additional remarks on the sixth-order corrections are in order. Let us consider the expectation value of the second term on the right-hand side of Eq. (11.68),

$$\langle\phi|\left\{\vec{p}^{\,2}, \vec{\nabla}^{\,2} V\right\}|\phi\rangle\,. \tag{11.75}$$

When evaluated on a Schrödinger eigenstate ψ, one can replace $\vec{p}^{\,2} \to 2m(E - V)$, where E is the reference-state energy. For $V = -Z\,\alpha/r$, this leads to a matrix element proportional to $\langle\phi|(1/r)\,\delta^{(3)}(\vec{r})|\phi\rangle$, which is divergent for S states which have a nonvanishing probability density at the origin. The application of relations such as those given in Eq. (11.74) cannot cure the divergence. Indeed, for the exact Coulomb potential $V = -Z\,\alpha/r$, several matrix elements of $H^{[6]}$ become singular for reference S states. A more far-reaching approach is needed. One possibility, as shown in Ref. [44], consists in the regularization (at short range) of the Coulomb potential,

$$V = -\frac{Z\alpha}{r} \to -\frac{Z\alpha}{r}\left(1 - \mathrm{e}^{-\lambda r}\right). \tag{11.76}$$

The original Coulomb potential is obtained in the limit $\lambda \to \infty$, while for any finite λ, the divergence of the Coulomb potential at $r = 0$ is regularized and made finite. Using this approach, one can obtain (see Ref. [44]) the $(Z\alpha)^6$ contributions to the energies of the bound states, denoted as $E^{[6]}$, on the basis of the formula

$$E^{[6]} = \left\langle H^{[4]}\left(\frac{1}{E-H}\right)' H^{[4]}\right\rangle + \left\langle H^{[6]}\right\rangle. \tag{11.77}$$

Here, both the Hamiltonians $H^{[4]}$ as well as $H^{[6]}$ are understood to be formulated with the regularized potential given in Eq. (11.76). The idea is to recover the full $(Z\alpha)^6$ result in the limit $\lambda \to \infty$, the limit being taken after all integrations are complete. In particular, the corrections to the wave function, due to the correction term proportional to $\mathrm{e}^{-\lambda r}$, need to be taken into account consistently (see Ref. [44]). The energy at $\mathcal{O}(Z\alpha)^6$ can also be regularized dimensionally, as was done for positronium in Ref. [329].

11.3.2 *Magnetic Field Coupling and Electron Moment*

The Hamiltonian (11.54) has a rich structure. A second illustrative application concerns an otherwise free electron in a homogeneous, spatially and temporally constant magnetic field $\vec{B}$. A constant magnetic field is described by the vector potential $\vec{A} = \frac{1}{2}(\vec{B} \times \vec{r})$. We summarize the following relations,

$$\vec{A} = \tfrac{1}{2}(\vec{B} \times \vec{r}), \qquad eA^0 = V = 0, \qquad \vec{E} = \vec{0}, \qquad \vec{\pi} = \vec{p} - \frac{e}{2}(\vec{B} \times \vec{r}). \tag{11.78}$$

Furthermore, one has the relation [see also Eq. (11.56)]

$$(\vec{\sigma} \cdot \vec{\pi})^2 = \vec{\pi}^2 - e\vec{\sigma} \cdot \vec{B} = \vec{p}^2 - e\vec{\sigma} \cdot \vec{B} - e\vec{L} \cdot \vec{B} + \frac{e^2}{4}(\vec{B} \times \vec{r})^2. \tag{11.79}$$

The term proportional to $(\vec{r} \times \vec{B})^2$ describes the leading diamagnetic interaction. The sum of the terms $H^{[2]}$ and $H^{[3]}$ from the Foldy–Wouthuysen Hamiltonian (11.54) becomes, ignoring higher-order corrections in the momenta,

$$H_{\rm FW} \approx H^{[2]} + H^{[3]} = \frac{\vec{p}^2}{2m} - \frac{e}{2m}\left\{(1+\kappa)\,\vec{\sigma} \cdot \vec{B} + \vec{L} \cdot \vec{B}\right\} + \frac{e^2}{8m}\left(\vec{r} \times \vec{B}\right)^2. \tag{11.80}$$

We note that the Pauli $\vec{\sigma}$ matrices, defined in Eq. (4.343), are proportional to the electron spin operator $\vec{s} = \vec{\sigma}/2$. The 2×2-matrices $\vec{s}$ describe the electron spin angular momentum [see also Eq. (7.102)], and they fulfill the commutator relations $[s^i, s^j] = \mathrm{i}\epsilon^{ijk} s^k$ which characterize the Lie algebra of the rotation group. From Eq. (11.80), it is obvious that only the coupling of the spin to the magnetic field (as opposed to the coupling of the orbital angular momentum to the magnetic field) receives a correction due to the anomalous magnetic moment.

The electron spin g factor is related to the spin-flip frequency (or spin-flip energy) of a free electron in a homogeneous and constant magnetic field. In order to clarify the definition of the g factor, in Eq. (11.80), one isolates the electron-spin dependent term, and writes the energy shift incurred upon a spin flip, as a function of the magnetic field strength. There is a little subtlety involved in the sign of the electron g factor, in view of its negative charge. This subtlety has given rise to some confusion, and, in particular, to proposals of redefining the sign of the electron's g factor [330]. Let us go into detail and isolate the spin-dependent terms in Eq. (11.80),

$$H_{\rm FW} = -2(1+\kappa)\,\frac{e}{2m}\,\vec{s} \cdot \vec{B} \equiv -g\,\frac{e}{2m}\,\vec{s} \cdot \vec{B}, \qquad \vec{s} = \frac{1}{2}\vec{\sigma}. \tag{11.81}$$

Here, $e = -|e|$ is the electron charge, and the expression following the "$\equiv$" sign serves as a definition of the electron g factor. The spin operator is $\vec{s} = \vec{\sigma}/2$. Note that a matching with Eq. (11.81) leads to the identification

$$g = 2(1+\kappa) \tag{11.82}$$

for the electron g factor. This result should be compared to Eqs. (1.10a), (1.10b) and (1.11). It is intuitively preferable to define the electron g factor as a positive quantity, and this desire is fulfilled in Eq. (11.82) as well as in the literature, in general (see, however, Ref. [330]). We must now compare the definition given in Eq. (11.82) to the well-known interaction energy of a magnetic dipole with magnetic moment $\vec{\mu}$ in an external magnetic field; the latter reads as $E = -\langle \vec{\mu} \cdot \vec{B} \rangle$. This formalism is consistent with Eq. (11.82) if we identify, for the electron,

$$E = -\langle \vec{\mu} \cdot \vec{B} \rangle \,, \qquad \vec{\mu} = g\,(-\mu_B)\,\vec{s}\,, \qquad \mu_B = \frac{|e|}{2m} = -\frac{e}{2m}\,. \tag{11.83}$$

Here, μ_B is the Bohr magneton, which, in SI mksA units, is equal to $|e|\,\hbar/(2mc)$, where $|e|$ is the positron charge, and m is the electron mass. Note that, in order for the g factor of the electron to come out positive, one has to define the relationship of the magnetic moment $\vec{\mu}$, the spin operator, and the Bohr magneton μ_B, by introducing a minus sign, as explicitly written in the relation $\vec{\mu} = -g\,\mu_B\,\vec{s}$, where $\vec{s}$ is the spin matrix. If one were to insist that $\vec{\mu}$ be equal to $g\,\mu_B\,\vec{s}$ for the electron, then the sign of the electron g factor would be negative.

11.3.3 *Magnetic Fields and Electrostatic Potentials*

Now, we turn our attention to the problem of an electron coupled to a constant, uniform magnetic field, which is simultaneously bound to an electrostatic field. The canonical momentum is given as

$$\vec{\pi} = \vec{p} - \frac{e}{2}(\vec{B} \times \vec{r})\,. \tag{11.84}$$

Furthermore, because the electric field is static, and thus equal to the gradient of a potential, we have the relation

$$e\vec{E} = e\vec{E}_{\parallel} = -\vec{\nabla} V\,, \qquad V = e\,A^0\,. \tag{11.85}$$

In the combined fields, the sum of the terms $H^{[2]}$ and $H^{[3]}$ is just

$$H^{[2]} + H^{[3]} = \frac{\vec{p}^{\,2}}{2m} + V - \frac{e}{2m}\,\vec{L} \cdot \vec{B} - \frac{e}{2m}\,(1+\kappa)\,\vec{\sigma} \cdot \vec{B} + \frac{e^2}{8m}\,\left(\vec{B} \times \vec{r}\right)^2\,. \tag{11.86}$$

The expression for $H^{[4]}$ simplifies as follows,

$$\begin{aligned} H^{[4]} = &-\frac{\vec{\pi}^{\,4}}{8m^3} + \frac{e\vec{\pi}^{\,2}}{4m^3}\vec{\sigma} \cdot \vec{B} - \frac{e^2}{8m^3}\,\vec{B}^2 + \frac{1}{8m^2}\vec{\nabla}^2 V \\ &+ \frac{1}{4m^2}\vec{\sigma} \cdot (\vec{\nabla} V \times \vec{p}) - \frac{e}{8m^2}\vec{\sigma} \cdot (\vec{\nabla} V \times (\vec{B} \times \vec{r}))\,, \end{aligned} \tag{11.87}$$

where, using Eq. (11.79), one derives the relation

$$\vec{\pi}^4 = \vec{p}^{\,4} - 2e\vec{L} \cdot \vec{B}\,\vec{p}^{\,2} + \frac{e^2}{4}\left\{\vec{p}^{\,2}, \left(\vec{B} \times \vec{r}\right)^2\right\} + e^2(\vec{L} \cdot \vec{B})^2 + \frac{e^4}{16}(\vec{B} \times \vec{r})^4\,. \tag{11.88}$$

We recall that, in the current calculation, we assume a strong magnetic field, of the order given in Eq. (11.41). Hence, we keep higher powers of the magnetic field, commensurate with the desired order of the expansion. E.g., the $\vec{\pi}^{\,4}$ term in Eq. (11.87) contains terms of fourth order in the magnetic fields.

For the fifth-order Hamiltonian $H^{[5]}$, we need the following relation, which is valid for a constant, uniform magnetic field,

$$\{\vec{\sigma}\cdot\vec{\pi},\{\vec{\sigma}\cdot\vec{\pi},\vec{\sigma}\cdot\vec{B}\}\} = 4\vec{B}\cdot\vec{p}\,\vec{\sigma}\cdot\vec{p} + 2e\left((\vec{\sigma}\times\vec{r})\cdot\vec{B}\right)(\vec{B}\cdot\vec{p})\,. \tag{11.89}$$

This implies that

$$\begin{aligned} H^{[5]} &= \frac{\kappa}{4m^2}\vec{\nabla}^2 V + \frac{\kappa}{2m^2}\vec{\sigma}\cdot(\vec{\nabla}V\times\vec{p}) - \frac{e\kappa}{8m^2}\vec{\sigma}\cdot(\vec{\nabla}V\times(\vec{B}\times\vec{r})) \\ &\quad + \frac{e\kappa}{4m^3}\vec{B}\cdot\vec{p}\,\vec{\sigma}\cdot\vec{p} + \frac{e\kappa}{8m^3}[(\vec{\sigma}\times\vec{r})\cdot\vec{B}]\,\vec{B}\cdot\vec{p}\,. \end{aligned} \tag{11.90}$$

Again, we stress that, in view of the assumed scaling relations given in Eq. (11.41), we keep terms quadratic in the magnetic fields. In the combined magnetic and electrostatic fields, the sixth-order corrections summarized in $H^{[6]}$ evaluate as follows,

$$\begin{aligned} H^{[6]} &= \frac{(\vec{\sigma}\cdot\vec{\pi})^6}{16m^5} + \frac{5\mathrm{i}e}{128m^4}\left[(\vec{\sigma}\cdot\vec{\pi})^2,\{\vec{\sigma}\cdot\vec{E},\vec{\sigma}\cdot\vec{\pi}\}\right] \\ &\quad + \frac{3\mathrm{i}\,e}{64m^4}\{(\vec{\sigma}\cdot\vec{\pi})^2,[\vec{\sigma}\cdot\vec{\pi},\vec{\sigma}\cdot\vec{E}]\} + \frac{(\vec{\nabla}V)^2}{8m^3}\,. \end{aligned} \tag{11.91}$$

From Eq. (11.56), one can derive the following relation,

$$(\vec{\sigma}\cdot\vec{\pi})^6 = (\vec{\pi}^2 - e\vec{\sigma}\cdot\vec{B})^3 = \vec{\pi}^{\,6} + 3e^2\,\vec{B}^2\,\vec{\pi}^2 - 3\,e\,\vec{\sigma}\cdot\vec{B}\,\vec{\pi}^{\,4} - e^3\,\vec{B}^2\,\vec{\sigma}\cdot\vec{B}\,. \tag{11.92}$$

For the higher-order term $\vec{\pi}^6$ (without Pauli matrices), one finds

$$\vec{\pi}^{\,6} = \left[\vec{p}^{\,2} - e\vec{L}\cdot\vec{B} + \frac{e^2}{4}(\vec{B}\times\vec{r})^2\right]^3\,. \tag{11.93}$$

Two further useful relations can be given which enable us to simplify the sixth-order terms. The first of these is as follows,

$$e\,[(\vec{\sigma}\cdot\vec{\pi})^2,\{\vec{\sigma}\cdot\vec{E},\vec{\sigma}\cdot\vec{\pi}\}] = -\mathrm{i}e\left[\vec{p}^{\,2} - e\vec{L}\cdot\vec{B} + \frac{e^2}{4}(\vec{B}\times\vec{r})^2,\left[\vec{p}^{\,2} - e\vec{L}\cdot\vec{B},V\right]\right]. \tag{11.94}$$

We also have the following relation,

$$\begin{aligned} e\,\{(\vec{\sigma}\cdot\vec{\pi})^2,[\vec{\sigma}\cdot\vec{\pi},\vec{\sigma}\cdot\vec{E}]\} &= \mathrm{i}\Bigg\{\vec{p}^{\,2} - e\,(\vec{\sigma}+\vec{L})\cdot\vec{B} + \frac{e^2}{4}(\vec{B}\times\vec{r})^2, \\ &\qquad \vec{\nabla}^2 V + 2\vec{\sigma}\cdot\vec{\nabla}V\times\left[\vec{p} - \frac{e}{2}\left(\vec{B}\times\vec{r}\right)\right]\Bigg\}\,. \end{aligned} \tag{11.95}$$

Hence, the result for the sixth-order terms of the Foldy–Wouthuysen transformed Hamiltonian, in the combined magnetic and electrostatic fields, reads as follows,

$$\begin{aligned} H^{[6]} &= \frac{1}{16m^5}\left(\vec{\pi}^{\,6} + 3e^2\,\vec{B}^2\,\vec{\pi}^2 - 3\,e\,\vec{\sigma}\cdot\vec{B}\,\vec{\pi}^{\,4} - e^3\,\vec{B}^2\,\vec{\sigma}\cdot\vec{B}\right) + \frac{(\vec{\nabla}V)^2}{8m^3} \\ &\quad + \frac{5}{128m^4}\left[\vec{p}^{\,2} - e\vec{L}\cdot\vec{B} + \frac{e^2}{4}(\vec{B}\times\vec{r})^2,\left[\vec{p}^{\,2} - e\vec{L}\cdot\vec{B},V\right]\right] \\ &\quad - \frac{3}{64m^4}\left\{\vec{p}^{\,2} - e\,(\vec{\sigma}+\vec{L})\cdot\vec{B} + \frac{e^2}{4}(\vec{B}\times\vec{r})^2, \vec{\nabla}^2 V + 2\vec{\sigma}\cdot\vec{\nabla}V\times\left[\vec{p} - \frac{e}{2}\left(\vec{B}\times\vec{r}\right)\right]\right\}. \end{aligned} \tag{11.96}$$

Expressions for the seventh-order and eighth-order terms in the combined magnetic and electrostatic fields. can be obtained from Eqs. (11.60) and (11.61) under the substitutions $\vec{\pi}\to\vec{p} - (e/2)(\vec{B}\times\vec{r})$, and $e\vec{E}\to-\vec{\nabla}V$.

11.3.4 *Transition Current*

We now investigate the question of how to define the transition current operator J^μ, which is to be extracted from the Foldy–Wouthuysen transformed Hamiltonian (11.54). The answer to this question can be derived from a consideration of the Lagrangian of the classical electromagnetic field, coupled to a classical four-vector current distribution J^μ, as given in Eq. (8.84) of Ref. [81],

$$\mathcal{L} = -\frac{1}{4} F^{\mu\nu} F_{\mu\nu} - J^\mu A_\mu \,. \tag{11.97}$$

For the fully quantized theory, we have encountered the Lagrangian in Eq. (10.1). It can be written as the sum of the free Lagrangian $\mathcal{L}_0$ of the quantized electromagnetic and Dirac fields, given in Eq. (10.2), and the interaction Lagrangian $\mathcal{L}_I$, given in Eq. (10.3). Let us remember that, according to Eq. (10.4), the quantum electrodynamic interaction Hamiltonian has the structure

$$H_I = J^\mu A_\mu = e\, j^\mu A_\mu = e\, j^0 A^0 - e\vec{j} \cdot \vec{A} \,, \tag{11.98}$$

where in the latter form, we write out the time-like and spatial components explicitly. We define the current operator $j^\mu = J^\mu / e$, where e is the electron charge.

In the context of the Foldy–Wouthuysen transformation, we should remember that the momentum and position operators in Eq. (11.54) act on the first-quantized wave function describing the electron. In the second-quantized formalism, the current becomes an operator, i.e., $j^\mu = \overline{\psi} \gamma^\mu \psi$, where ψ and $\overline{\psi} = \psi^+ \gamma^0$ are the field operators of the Dirac field. The functional form of the total Hamiltonian given in Eq. (11.54) does not change if we assume that the components A^μ of the vector potential are quantized rather than classical fields (see also Chap. 2). If we assume that the electron wave functions are taken in the first-quantized formalism, then we can extract, from the interaction part H_I of the Foldy–Wouthuysen Hamiltonian (11.54), the time-like and spatial components j^0 and $\vec{j}$ of the current operator by the identification

$$H_{I\gamma} = e\, j^0 A^0_\gamma - e\,\vec{j} \cdot \vec{A}_\gamma \,, \tag{11.99}$$

where the additional subscripts γ in $H_{I\gamma}$ and A^μ_γ refer to the four-vector potentials of a quantized field, describing a photon. The reason for singling out the four-vector potentials of the photon field, distinguished by the subscript γ, will become apparent in the following. We will see that in many cases, the elements of the current, j^0 and $\vec{j}$, are in fact operator-valued in the nonrelativistic limit (they contain momentum operators). Now, one has to be careful: The momentum and gradient operators in j^0 and $\vec{j}$ could either act on the quantum mechanical wave functions between which the interaction Hamiltonian H_I is "sandwiched", or, on the four-vector potential A^μ. Indeed, we shall encounter both cases.

Let us now discuss how to extract the current from the interaction Hamiltonian. To this end, we consider a typical transition matrix element of the form $\langle \phi_f, 0 | H_{I\gamma} | \phi_i, 1_{\vec{k}\lambda} \rangle = e \langle \phi_f | j^0 | \phi_i \rangle \langle 0 | A^0_\gamma | 1_{\vec{k}\lambda} \rangle - e \langle \phi_f | \vec{j} | \phi_i \rangle \cdot \langle 0 | \vec{A}_\gamma | 1_{\vec{k}\lambda} \rangle$, which describes a photon annihilation process, namely, the annihilation of one photon in the state with wave vector $\vec{k}$ and polarization λ, with a concomitant transition of the electron from the initial state ϕ_i to the final state ϕ_f. The physical interpretation is that the electron-wave-function part of the given matrix element describes the current which couples to the four-vector potential. These considerations imply that the current operator is the coefficient of the linear terms in A^μ in the interaction Hamiltonian.

Some additional remarks on the choice of gauge are in order. For the discussion of bound-state problems, the Coulomb gauge is known to be very practical. This fact is being exploited in the following Chap. 12, when we derive higher-order corrections to the interaction potential between quantum mechanical Dirac particles. It is also well known that the Coulomb gauge is suitable for the discussion of the low-energy part of the self-energy (see Chap. 4). As we saw in Chap. 2, the quantization of the electromagnetic field in the Coulomb gauge proceeds only for the spatial components of the vector potential operator. However, we saw in Sec. 9.3 that the $(\mu = 0)(\nu = 0)$-component (henceforth denoted as the 00-component) of the photon propagator is still nonvanishing in the Coulomb gauge. For us, it means that we cannot ignore the time-like part j^0 of the current operator.

The Foldy–Wouthuysen transformed Hamiltonian given in Eq. (11.54) contains the canonical momentum $\vec{\pi} = \vec{p} - e\,\vec{A}$, which in turn depends on the spatial components of the vector potential $\vec{A}$. We shall use the Hamiltonian given in Eq. (11.54) under the following kinematic conditions,

$$\vec{\pi} = \vec{p} - e\,\vec{A}_{\rm ext} - e\,\vec{A}_\gamma = \vec{\pi}_{\rm ext} - e\,\vec{A}_\gamma\,, \tag{11.100a}$$

$$e\,A^0 = e\,A^0_{\rm ext} + e\,A^0_\gamma = V + e\,A^0_\gamma\,, \qquad V = e\,A^0_{\rm ext}\,. \tag{11.100b}$$

Furthermore, we assume that the action of the gradient operator on the four-vector potential describing the photon, generates a photon wave vector whose magnitude is of the order of $\nabla^i A^\mu_\gamma \sim (Z\alpha)^2\, m\, A^\mu_\gamma$. That is, we allow for an additional external four-vector potential $A^\mu_{\rm ext}$, while the quantized electromagnetic field is described by the four-vector potential A^μ_γ. The action of the gradient operator on the photon four-vector potential results in a prefactor proportional to the photon wave vector, which, for a typical atomic transition, is of order $(Z\alpha)^2$. Here, Z is the nuclear charge number. Alternatively, we can say that, for an optical photon, one has $k = |\vec{k}| \sim (Z\alpha)^2$. The magnetic field operator therefore of the order of $\vec{B} \sim \vec{k} \times \vec{A} \sim (Z\alpha)^2\,\vec{A}$. In Eq. (11.100), we implicitly define the quantities $\vec{\pi}_{\rm ext}$ and V.

For the nuclear Coulomb field, we have, as an example,

$$A^0_{\rm ext} = -\frac{Ze}{4\pi\epsilon_0 r} = -\frac{Ze}{4\pi r}\,, \qquad V = e\,A^0_{\rm ext} = -\frac{Ze^2}{4\pi r} = -\frac{Z\alpha}{r}\,. \tag{11.101}$$

The total electric field $\vec{E} = \vec{E}_{\rm tot}$ is a sum of a longitudinal component, due to the scalar potential of both the external field and the photon field, and a transverse component, due to the time derivative of the photon field,

$$e\,\vec{E}_{\rm tot} = -\,\vec{\nabla} V - e\vec{\nabla} A^0_\gamma - e\frac{\partial}{\partial t}\vec{A}_\gamma = e\,\vec{E}_{\rm ext} + e\,\vec{E}_\gamma\,, \tag{11.102a}$$

$$e\,\vec{E}_{\rm ext} = -\,\vec{\nabla} V\,, \tag{11.102b}$$

$$\vec{E}_\gamma = -\,\vec{\nabla} A^0_\gamma - \frac{\partial}{\partial t}\vec{A}_\gamma = \vec{E}^{[1]}_\gamma + \vec{E}^{[2]}_\gamma\,, \tag{11.102c}$$

where $\vec{E}^{[1]}_\gamma = -\vec{\nabla} A^0_\gamma$ and $\vec{E}^{[2]}_\gamma = -\frac{\partial}{\partial t}\vec{A}_\gamma$. Fields with a subscript γ denote photon fields. The total magnetic field $\vec{B}_{\rm tot}$ is

$$\vec{B} = \vec{B}_{\rm tot} = \vec{\nabla}\times\vec{A}_{\rm ext} + \vec{\nabla}\times\vec{A}_\gamma = \vec{B}_{\rm ext} + \vec{B}_\gamma\,, \tag{11.103}$$

where we implicitly define $\vec{B}_{\rm ext}$ and $\vec{B}_\gamma$. We use the Coulomb gauge. The fields A^μ_γ enter the relation (9.73),

$$\langle 0|{\rm T}\,A^\mu_\gamma(t,\vec{r})A^\nu_\gamma(t',\vec{r}\,')|0\rangle = {\rm i}D^{\mu\nu}_C(x-x')\,. \tag{11.104}$$

Under the kinematic conditions summarized in Eq. (11.100), we can write the Foldy–Wouthuysen Hamiltonian given in Eq. (11.54) as follows (we keep terms up to the fourth-order terms in the momenta),

$$\begin{aligned}
H_{\rm FW} &= H^{[2]} + H^{[3]} + H^{[4]} \\
&= \frac{(\vec{\pi}_{\rm ext} - e\,\vec{A}_\gamma)^2}{2m} + V + e\,A^0_\gamma - \frac{e}{2m}(1+\kappa)\,\vec{\sigma}\cdot\vec{B}_{\rm tot} - \frac{(\vec{\pi}_{\rm ext} - e\,\vec{A}_\gamma)^4}{8m^3} \\
&\quad - \frac{e}{8m^2}(1+2\kappa)\left[\vec{\nabla}\cdot\vec{E}_{\rm tot} + \vec{\sigma}\cdot\left(\vec{E}_{\rm tot}\times(\vec{\pi}_{\rm ext} - e\vec{A}_\gamma) - (\vec{\pi}_{\rm ext} - e\vec{A}_\gamma)\times\vec{E}_{\rm tot}\right)\right] \\
&= H^{\rm (R)}_{\rm FW} + e\,j^0\,A^0_\gamma - e\,\vec{j}\cdot\vec{A}_\gamma = H^{\rm (R)}_{\rm FW} + H^{(00)}_{I\gamma} + H^{(ij)}_{I\gamma}\,.
\end{aligned} \tag{11.105}$$

Here, $H^{\rm (R)}_{\rm FW}$ is the rest term, which contains terms without A^γ_μ and the seagull term $e^2\vec{A}^2_\gamma/(2m^2)$ which of second order in the quantized field. Also, we define the following Hamiltonians $H^{(00)}_{I\gamma}$ and $H^{(ij)}_{I\gamma}$,

$$H^{(00)}_{I\gamma} = e\,j^0\,A^0_\gamma\,, \qquad H^{(ij)}_{I\gamma} = -e\,\vec{j}\cdot\vec{A}_\gamma\,. \tag{11.106}$$

Of course, the superscript (ij) does not indicate a tensor structure, but rather, the origin of the Hamiltonian from the spatial components of the current, which are usually denoted by Latin as opposed to Greek indices. We shall, in the following, attempt to extract the components j^μ ($\mu = 0,1,2,3$) of the transition current, up to order $(Z\alpha)^3$.

For $\mu = 0$, we obtain the timelike component of the current, which we extract from the Hamiltonian

$$\begin{aligned}
H^{(00)}_{I\gamma} &= e\,A^0_\gamma - \frac{e}{8m^2}(1+2\kappa)\left[\vec{\nabla}\cdot\vec{E}^{[1]}_\gamma + \vec{\sigma}\cdot\left(\vec{E}^{[1]}_\gamma\times\vec{\pi}_{\rm ext} - \vec{\pi}_{\rm ext}\times\vec{E}^{[1]}_\gamma\right)\right] \\
&= e\,A^0_\gamma + \frac{e\,(1+2\kappa)}{8m^2}\,\vec{\nabla}^2 A^0_\gamma - \frac{e\,(1+2\kappa)}{4m^2}\,\vec{\sigma}\cdot\left[(-\vec{\nabla} A^0_\gamma)\times\vec{\pi}_{\rm ext}\right] \\
&= e\left[A^0_\gamma + \frac{1+2\kappa}{8m^2}\,\vec{\nabla}^2 A^0_\gamma + \frac{1+2\kappa}{4m^2}\,\vec{\sigma}\cdot\left(\vec{\nabla} A^0_\gamma\times\vec{\pi}_{\rm ext}\right)\right] \\
&= e\left[1 + \frac{1+2\kappa}{8m^2}\,\vec{\nabla}^2 - \frac{1+2\kappa}{4m^2}\,\vec{\sigma}\cdot\left(\vec{\pi}_{\rm ext}\times\vec{\nabla}\right)\right]A^0_\gamma = e\,j^0\,A^0_\gamma\,,
\end{aligned} \tag{11.107}$$

where the latter expression reflects the definition given in Eq. (11.106), and we appeal to Eq. (11.102c). There is a small subtlety in the formulation of the current operator j^0. Namely, it contains gradient operators which act on the time-like component A^0_γ of the photon field. By contrast, the momentum operator which is part of $\vec{\pi}_{\rm ext} = \vec{p} - e\,\vec{A}$, acts on the quantum mechanical wave function, between which the interaction Hamiltonian is being sandwiched. With this understanding, we can write

$$j^0 = 1 + \frac{1+2\kappa}{8m^2}\,\vec{\nabla}^2 - \frac{1+2\kappa}{8m^2}\,\vec{\sigma}\cdot\left(\vec{\pi}_{\rm ext}\times\vec{\nabla}\right). \tag{11.108}$$

If we transform the vector potential into momentum space, then we can use the correspondence $\vec{k} \Leftrightarrow -{\rm i}\vec{\nabla}$, and thus $\vec{\nabla} \Leftrightarrow {\rm i}\vec{k}$, where $\vec{k}$ is the photon momentum. This replacement is valid only for a photon annihilation process. In order to see this, let us recall Eq. (2.71) for the transverse vector potential in Coulomb gauge,

$$\vec{A}(t,\vec{r}) = \sum_{\lambda=1}^{2}\int \frac{{\rm d}^3k}{\sqrt{2\omega\,(2\pi)^3}}\,\hat{\epsilon}_\lambda(\vec{k})\,\left(a_\lambda(\vec{k})\,{\rm e}^{-{\rm i}\,k\cdot x} + a^+_\lambda(\vec{k})\,{\rm e}^{{\rm i}\,k\cdot x}\right), \tag{11.109}$$

where $\omega = |\vec{k}|$, with the fundamental field commutator (2.70),

$$[a_\lambda(\vec{k}), a^+_{\lambda'}(\vec{k}')] = \delta^{(3)}(\vec{k}-\vec{k}')\,\delta_{\lambda\lambda'}\,. \tag{11.110}$$

For a photon annihilation process, the vector potential therefore is proportional to

$$\vec{A}(t,\vec{r}) \propto \hat{\epsilon}_\lambda(\vec{k})\,{\rm e}^{-{\rm i}\,k\cdot x} = \hat{\epsilon}_\lambda(\vec{k})\,{\rm e}^{-{\rm i}\,\omega\,t+{\rm i}\vec{k}\cdot\vec{r}}\,, \qquad \omega = |\vec{k}|\,. \tag{11.111}$$

In the following, we shall assume that the vector potential is of the form (11.111), and evaluate correction terms for an expansion of $\vec{A}(t,\vec{r})$ about the origin ($\vec{r} = \vec{0}$). This naturally gives rise to an expansion of $\vec{j}$ in terms of $\vec{k}$. One observes that the Fourier transform of the current only depends on the spatial components of the photon momentum $\vec{k}$, so that we can write

$$j^0 = 1 + \frac{{\rm i}(1+2\kappa)}{4m^2}\,\vec{\sigma}\cdot\left(\vec{k}\times\vec{\pi}_{\rm ext}\right) - \frac{1+2\kappa}{8m^2}\,\vec{k}^2\,. \tag{11.112}$$

This result is important because, e.g., for the electron-electron interaction, the timelike ($\mu = \nu = 0$) component of the photon propagator (in Coulomb gauge) leads to a nonvanishing and important contribution (see Chap. 12).

Let us now turn our attention to the spatial components of the current operator. We first consider the problem in lower order and approximate the Hamiltonian (11.105) as follows,

$$H_{\rm FW} \approx \frac{(\vec{\pi}_{\rm ext} - e\,\vec{A}_\gamma)^2}{2m} - \frac{e}{2m}(1+\kappa)\,\vec{\sigma}\cdot\vec{B}_{\rm tot}\,. \tag{11.113}$$

We here omit the $(Z\alpha)^3$ terms from Eq. (11.119) and (11.126), and keep only the $(Z\alpha)^2$ terms. In Eq. (11.113), the relevant replacement for the extraction of the current from the canonical momentum square is $(\vec{\pi}_{\rm ext} - e\,\vec{A}_\gamma)^2 \to -2e\vec{A}_\gamma\cdot\vec{\pi}_{\rm ext}$. In Coulomb gauge, we may ignore terms proportional to $[\vec{p},\vec{A}_\gamma] \propto \vec{\nabla}\cdot\vec{A}_\gamma$. With reference to Eq. (11.106), one therefore has the following result for the interaction Hamiltonian $H_I^{(ij)}$, as derived from the spatial components of the current,

$$\begin{aligned} H_{I\gamma}^{(ij)} &\approx -\frac{e}{m}\vec{A}_\gamma \cdot \vec{\pi}_{\rm ext} - \frac{e}{2m}(1+\kappa)\,\vec{\sigma}\cdot(\vec{\nabla}\times\vec{A}_\gamma) \\ &= -e\left(\frac{\vec{\pi}_{\rm ext}}{m} + \frac{1+\kappa}{2m}\,\vec{\sigma}\times\vec{\nabla}\right)\cdot\vec{A}_\gamma = -e\,\vec{j}\cdot\vec{A}_\gamma\,. \end{aligned} \tag{11.114}$$

In coordinate space, this implies that

$$\vec{j} = \frac{\vec{\pi}_{\rm ext}}{m} + \frac{1+\kappa}{2m}\,\vec{\sigma}\times\vec{\nabla}\,. \tag{11.115}$$

The addition of the anomalous-magnetic-moment term (correction of order κ) generates part of the α^3 correction. One might argue that if *one* term of order α^3 is included, then one should include *all* terms in this order. However, we get this term "for free", namely, as an additional multiplicative (radiative) correction, and thus include it here. Finally, after a Fourier transformation, the current is given as follows,

$$\vec{j} = \frac{\vec{\pi}_{\rm ext}}{m} + \frac{\mathrm{i}\,(1+\kappa)}{2m}\,(\vec{\sigma}\times\vec{k})\,. \tag{11.116}$$

Now, let us analyze the corrections to the leading-order term given in Eq. (11.115). To this end, we recall Eq. (11.105),

$$\begin{aligned} H_{\rm FW} &= \frac{(\vec{\pi}_{\rm ext} - e\,\vec{A}_\gamma)^2}{2m} + V + e\,A_\gamma^0 - \frac{e}{2m}\,(1+\kappa)\,\vec{\sigma}\cdot\vec{B}_{\rm tot} - \frac{(\vec{\pi}_{\rm ext} - e\,\vec{A}_\gamma)^4}{8m^3} \\ &\quad - \frac{e}{8m^2}(1+2\kappa)\left[\vec{\nabla}\cdot\vec{E}_{\rm tot} + \vec{\sigma}\cdot\left(\vec{E}_{\rm tot}\times(\vec{\pi}_{\rm ext} - e\vec{A}_\gamma) - (\vec{\pi}_{\rm ext} - e\vec{A}_\gamma)\times\vec{E}_{\rm tot}\right)\right]. \end{aligned} \tag{11.117}$$

The relevant interaction terms for the extraction of the spatial current $\vec{j}$, in the Hamiltonian $H_{I\gamma}^{(ij)}$, read as follows,

$$\begin{aligned} H_{I\gamma}^{(ij)} &= -\frac{e\,\{\vec{A}_\gamma, \vec{\pi}_{\rm ext}\}}{2m} + e\frac{\{\vec{A}_\gamma, \vec{\pi}_{\rm ext}\,\vec{\pi}^{\,2}_{\rm ext}\}}{4\,m^3} - \frac{e}{2m}(1+\kappa)\,\vec{\sigma}\cdot(\vec{\nabla}\times\vec{A}_\gamma) \\ &\quad + (1+2\kappa)\frac{e}{4\,m^2}\,\vec{\sigma}\cdot\left(\frac{\partial\vec{A}_\gamma}{\partial t}\times\vec{\pi}_{\rm ext}\right) - (1+2\kappa)\frac{e}{4\,m^2}\,\vec{\sigma}\cdot\left(\vec{\nabla}V\times\vec{A}_\gamma\right) \\ &= -e\vec{j}\cdot\vec{A}_\gamma\,. \end{aligned} \tag{11.118}$$

The first of the terms in the second line of the above relation involves $\vec{E}_\gamma^{[2]} = -\partial\vec{A}_\gamma/\partial t$ as defined in Eq. (11.102c), while the second term in the second line of the above relation involves $\vec{E}_{\rm ext}$, also according to the definition given in Eq. (11.102), and the term $-e\,\vec{A}_\gamma$ from the canonical momentum inside the cross product. We now remember that to lowest order in α, we can replace the factors $\exp(\pm\mathrm{i}\vec{k}\cdot\vec{r})$ in the field operator (11.109) by unity, so that $\vec{A}_\gamma$ commutes with $\vec{\pi}_{\rm ext}$. Finally, we find for the spatial components of the current,

$$\vec{j} = \frac{\vec{\pi}_{\rm ext}}{m} + \frac{1+\kappa}{2m}\vec{\sigma}\times\vec{\nabla} - \frac{\vec{\pi}_{\rm ext}\,\vec{\pi}^{\,2}_{\rm ext}}{2\,m^3} + \frac{\vec{\sigma}\times\vec{\pi}_{\rm ext}}{4\,m^2}\frac{\partial}{\partial t} + \frac{\vec{\sigma}\times\vec{\nabla}V}{4\,m^2}\,. \tag{11.119}$$

Here, the momentum operator $\vec{\pi}_{\rm ext}$ acts on the atomic degrees of freedom, while the differential operators (gradient operator and time derivative) act on the vector potential. The terms on the right-hand side of Eq. (11.119) are, respectively, of order α and α^2, while the last three terms are of order α^3. After a transformation into Fourier space, one has

$$\vec{j} = \frac{\vec{\pi}_{\rm ext}}{m} + \frac{{\rm i}(1+\kappa)}{2m}\vec{\sigma}\times\vec{k} - \frac{\vec{\pi}_{\rm ext}\,\vec{\pi}^{\,2}_{\rm ext}}{2\,m^3} - \frac{{\rm i}\omega}{4\,m^2}\,\vec{\sigma}\times\vec{\pi}_{\rm ext} + \frac{\vec{\sigma}\times\vec{\nabla}V}{4\,m^2}\,, \tag{11.120}$$

where ω is the photon frequency.

11.3.5 *Nonrelativistic Transition Current and Multipoles*

One might think that, at this point, we have derived all formulas necessary for a convenient evaluation of transition current matrix elements in the nonrelativistic approximation. This is, however, not the case. Namely, our considerations up to this point ignore the fact that the phase factors in the vector potential operator lead to multipole corrections to the transition current, which cannot be ignored. Let us consider, once more, the Hamiltonians given in Eq. (11.106),

$$H^{(00)}_{I\gamma} = e\,j^0\,A^0_\gamma\,, \qquad H^{(ij)}_{I\gamma} = -e\,\vec{j}\cdot\vec{A}_\gamma\,. \tag{11.121}$$

Let us also remember that the vector potential operator carries exponential factors according to Eq. (11.109), which we recall for convenience,

$$\vec{A}(t,\vec{r}) = \sum_{\lambda=1}^{2}\int\frac{{\rm d}^3k}{\sqrt{2\omega\,(2\pi)^3}}\,\hat{\epsilon}_\lambda(\vec{k})\,\left(a_\lambda(\vec{k})\,{\rm e}^{-{\rm i}\,k\cdot x} + a^+_\lambda(\vec{k})\,{\rm e}^{{\rm i}\,k\cdot x}\right). \tag{11.122}$$

The timelike and spatial current operators are given in Eqs. (11.112) and (11.120), respectively,

$$\begin{aligned} j^0 &= 1 - \frac{1+2\kappa}{8m^2}\,\vec{k}^2 + \frac{{\rm i}(1+2\kappa)}{4m^2}\,\vec{\sigma}\cdot\left(\vec{k}\times\vec{\pi}_{\rm ext}\right)\,,\\ \vec{j} &= \frac{\vec{\pi}_{\rm ext}}{m} + \frac{{\rm i}(1+\kappa)}{2m}\vec{\sigma}\times\vec{k} - \frac{\vec{\pi}_{\rm ext}\,\vec{\pi}^{\,2}_{\rm ext}}{2\,m^3} - \frac{{\rm i}\omega}{4\,m^2}\,\vec{\sigma}\times\vec{\pi}_{\rm ext} + \frac{\vec{\sigma}\times\vec{\nabla}V}{4\,m^2}\,. \end{aligned} \tag{11.123}$$

The characteristic exponential factor in the (annihilation part of the) vector potential operator is

$$\vec{A}_\gamma \propto {\rm e}^{-{\rm i}\,k\cdot x} = {\rm e}^{-{\rm i}\,\omega\,t+{\rm i}\vec{k}\cdot\vec{r}}\,, \qquad \omega = |\vec{k}|\,. \tag{11.124}$$

Let us define a nonrelativistic current $\vec{j}_{\rm NR}$ by the matching condition

$$\vec{j}\cdot\vec{A}_\gamma(\vec{r}) = \vec{j}_{\rm NR}\cdot\vec{A}_\gamma(\vec{r}=\vec{0})\,{\rm e}^{-{\rm i}\omega t}\,, \tag{11.125}$$

so that $\vec{j}_{\rm NR}$ contains the effect of the exponential factor ${\rm e}^{{\rm i}\vec{k}\cdot\vec{r}}$. For atomic-physics calculations, the vector potential of the photon, given in Eq. (11.111), can be expanded in powers of $\vec{k}\cdot\vec{r}$. This is because $\vec{r}$, which measures the electron-nucleus distance, is of the order of the Bohr radius, or, of the order of $1/(Z\alpha)$, when expressed in terms of the electron Compton wavelength (i.e., in units with $\hbar = c = \epsilon_0 = 1$). Also, $\vec{k}$ is of the order of an optical transition photon, and thus, of the order $(Z\alpha)^2$ in natural units. We aim to expand the transition current $\vec{j}_{\rm NR}$ up to order $(Z\alpha)^3$. The ith component of the transition current can thus be written as

$$j^i_{\rm NR} = \frac{\pi^i_{\rm ext}}{m} + \frac{1}{2}\left\{\frac{\pi^i_{\rm ext}}{m}, {\rm i}\vec{k}\cdot\vec{r}\right\} - \frac{1}{4}\left\{\frac{\pi^i_{\rm ext}}{m}, (\vec{k}\cdot\vec{r})^2\right\} - \frac{\pi^i_{\rm ext}\,\vec{\pi}^{\,2}_{\rm ext}}{2\,m^3}$$
$$+ \frac{\rm i}{2\,m}\left(\vec{\sigma}\times\vec{k}\right)^i\left(1 + {\rm i}\vec{k}\cdot\vec{r}\right) - \frac{{\rm i}\omega}{4\,m^2}\left(\vec{\sigma}\times\pi_{\rm ext}\right)^i + \frac{1}{4\,m^2}\left(\vec{\sigma}\times\vec{\nabla}V\right)^i . \quad (11.126)$$

The interpretation of the terms is as follows. The first three terms summarize multipole corrections to the transition current p^i/m, the fourth term is a relativistic kinetic correction, the fifth term is a magnetic interaction, with retardation effects, while the sixth and seventh terms describe the covariant coupling of the spin-orbit interaction to the quantized electromagnetic field of the photon, the latter being described by the vector field $\vec{A}_\gamma$. The analogous expression for the time-like nonrelativistic current is

$$j^0_{\rm NR} = 1 + {\rm i}\vec{k}\cdot\vec{r} - \frac{1}{2}\left(\vec{k}\cdot\vec{r}\right)^2 - \frac{1+2\kappa}{8m^2}\vec{k}^2 + \frac{{\rm i}(1+2\kappa)}{4m^2}\vec{\sigma}\cdot\left(\vec{k}\times\vec{\pi}_{\rm ext}\right). \quad (11.127)$$

A final remark is in order. One might ask if the anticommutators in the current operator (11.126) are really necessary. Specifically, one might ask what would happen if one were to write these terms as $\left(1 + {\rm i}\vec{k}\cdot\vec{r} - \frac{1}{2}(\vec{k}\cdot\vec{r})^2\right)p^i/m$, i.e., if one were to commute the momentum operator to the right. Starting from Eq. (11.109), the additional term would involve the commutator

$$\left[p^i, A^j(\vec{r})\right] = k^i \sum_{\lambda=1}^{2}\int \frac{{\rm d}^3k}{\sqrt{2\omega\,(2\pi)^3}}\,\epsilon^j_\lambda(\vec{k})\left(a_\lambda(\vec{k})\,{\rm e}^{-{\rm i}\,k\cdot x} - a^+_\lambda(\vec{k})\,{\rm e}^{{\rm i}\,k\cdot x}\right). \quad (11.128)$$

The factor k^i vanishes when contracted with the Coulomb-gauge photon propagator, because the latter is proportional to $\delta^{ij} - k^i\,k^j/\vec{k}^2$ [see Eq. (9.73)]. For a self-energy calculation in Coulomb gauge, therefore, $\vec{p}$ and $\vec{A}$ effectively commute, and the ordering of p^i/m and $\left(1 + {\rm i}\vec{k}\cdot\vec{r} - \frac{1}{2}(\vec{k}\cdot\vec{r})^2\right)$ in Eq. (11.126) does not matter. We note that it was precisely the assumption of Coulomb gauge which let us indicate the results given in Eqs. (11.112) and (11.120); otherwise, one would have had to indicate the anticommutator of the canonical momentum $\vec{\pi}_{\rm ext}$ and $\vec{A}_\gamma$.

The corrections to the current, listed in Eq. (11.126), are exactly those terms that have to be used in the low-energy part of the self-energy (Lamb shift), to supplement the leading-order term $\pi^i_{\rm ext}/m$ [see Eqs. (4.302) and (4.314)].

11.4 Dirac Form Factor and Bound-State Radiative Energy Shifts

11.4.1 *Foldy–Wouthuysen Transformation and Form Factors*

So far, we have considered the anomalous-magnetic-moment corrections in the FW-transformed Hamiltonians, which can be traced to the Pauli form factor. As we turn our attention to bound-state energy shifts, we need to generalize the treatment and consider, in addition, corrections due to the Dirac form factor. We thus take a step back and recall the radiatively corrected Dirac Hamiltonian given in Eq. (11.37),

$$H_{\rm R}(\vec{r}) = \vec{\alpha}\cdot\left[\vec{p} - e\,F_1(\vec{\nabla}^2)\,\vec{A}(\vec{r})\right] + \beta m + F_1(\vec{\nabla}^2)\,eA^0(\vec{r}) \\ + F_2(\vec{\nabla}^2)\,\frac{e}{2\,m}\left[\mathrm{i}\,\vec{\gamma}\cdot\vec{E}(\vec{r}) - \beta\,\vec{\Sigma}\cdot\vec{B}(\vec{r})\right]. \tag{11.129}$$

In Eq. (11.38), we made the following approximations,

$$F_1(\vec{\nabla}^2) \approx F_1(0) = 1\,, \qquad F_2(\vec{\nabla}^2) \approx F_2(0) \equiv \kappa \approx \frac{\alpha}{2\pi}\,. \tag{11.130}$$

For the discussion of the high-energy part of the bound-electron self-energy, the approximation $F_1(\vec{\nabla}^2) \approx F_1(0) = 1$ is not sufficient. We recall at this stage the result given in Eq. (10.103),

$$F_1(\vec{\nabla}^2) = 1 + \frac{\alpha}{\pi}\left[\frac{1}{3}\ln\left(\frac{m}{\lambda}\right) - \frac{1}{8}\right]\frac{\vec{\nabla}^2}{m^2} + \mathcal{O}\left(\frac{\vec{\nabla}^4}{m^4}\right). \tag{11.131}$$

In dimensional regularization, one has according to Eq. (10.190a),

$$F_1(\vec{\nabla}^2) = 1 + \frac{\alpha}{\pi}\left[-\frac{1}{8} - \frac{1}{6\varepsilon} + \frac{1}{6}\ln\left(\frac{m^2}{\mu^2}\right)\right]\frac{\vec{\nabla}^2}{m^2} + \mathcal{O}\left(\frac{\vec{\nabla}^4}{m^4}\right). \tag{11.132}$$

According to Eqs. (10.100) and (10.190b), the F_2 form factor is infrared convergent,

$$F_2(q^2) = \frac{\alpha}{2\pi}\int_0^1 \mathrm{d}y\,\frac{1}{1 - \frac{q^2}{m^2}y(1-y)} = \frac{\alpha}{2\pi}\left[1 + \frac{1}{6}\frac{q^2}{m^2} + \frac{1}{30}\frac{q^4}{m^4} + \mathcal{O}\left(\frac{q^6}{m^6}\right)\right]. \tag{11.133}$$

The form factors may be expanded in powers of their argument $\vec{\nabla}^2/m^2$, because an expansion in the gradient operator corresponds to an expansion in $Z\alpha$, in the sense that $\vec{\nabla}^2 \sim (Z\alpha)^2$. Under the kinematic conditions listed in Eq. (11.62), with $\kappa \equiv F_2(0)$, the new Hamiltonian takes the form

$$H_{\rm FF} = \frac{\vec{p}^{\,2}}{2\,m} - \frac{\vec{p}^{\,4}}{8\,m^3} + [1 + F_1'(0)]\,\vec{\nabla}^2 V \\ + \frac{1}{8\,m^2}\,(1 + 2\,\kappa)\left[\vec{\nabla}^2 V + 2\,\vec{\sigma}\cdot\left(\vec{\nabla}V \times \vec{p}\right)\right], \tag{11.134}$$

where we keep the potential V in general form. Of course, we shall consider the Coulomb potential $V = -Z\alpha/r$ in the following. We here restrict the discussion to terms up to order α^5. We should be well aware that the form-factor treatment has its limits, but it is sufficient to calculate the high-energy part of the bound-electron self-energy. Additional higher-order corrections are described in Refs. [199, 312], and explored further in Chap. 15.

11.4.2 *High-Energy Part in Photon Mass Regularization*

Let us isolate, from Eq. (11.134), those operators which are relevant to the high-energy part (HEP) of the electron self-energy in the order α^5,

$$H_{\rm HEP} = F_1'(0)\,\vec{\nabla}^2 V + \frac{\kappa}{4\,m^2}\left[\vec{\nabla}^2 V + 2\vec{\sigma}\cdot\left(\vec{\nabla}V \times \vec{p}\right)\right] = \Delta V_1 + \Delta V_2\,, \tag{11.135}$$

where we define the effective potentials $\Delta V_1 = F_1'(0)\,\vec{\nabla}^2 V$ and, with reference to Sec. 4.6.5,

$$\Delta V_2 = \frac{\kappa}{4m^2}\left[\vec{\nabla}^2 V + 2\vec{\sigma}\cdot\left(\vec{\nabla}V \times \vec{p}\right)\right]$$
$$= \frac{\alpha}{\pi}\frac{1}{2}\frac{\pi\,(Z\alpha)\,\delta^{(3)}(\vec{r})}{m^2} + \frac{\alpha}{\pi}\frac{(Z\alpha)\,\vec{\sigma}\cdot\vec{L}}{4m^2\,r^3}. \tag{11.136}$$

The one-loop value for κ is $\kappa = \alpha/(2\pi)$, according to Eq. (11.130). Here, in going from the first to the second line, we use properties of the Coulomb potential $V = -Z\alpha/r$. It is perhaps surprising that the effective potential ΔV_2 does not capture the reduced-mass dependence correctly, and in fact, one needs to redefine ΔV_2 as follows,

$$\Delta V_2 \to \Delta V_2 \equiv \frac{\alpha}{\pi}\frac{1}{2}\frac{\pi\,(Z\alpha)\,\delta^{(3)}(\vec{r})}{m^2} + \frac{\alpha}{\pi}\frac{(Z\alpha)\,\vec{\sigma}\cdot\vec{L}}{4m\,\mu\,r^3}. \tag{11.137}$$

We here correct the reduced-mass dependence on account of the fact that the self-energy actually is derived from the two-body BREIT[1] Hamiltonian, to be discussed in more detail in Sec. 12.5.3 (a keyword to remember in this context is the proton's convection current). We remind the reader of our notation for the reduced mass μ of the two-body bound system and for photon mass λ. With the proper reduced-mass dependence, the result in Eq. (11.137) confirms the expression previously indicated in Eq. (4.342).

Recalling the formalism outlined in Sec. 4.6.5 for the evaluation of matrix elements of Schrödinger–Pauli wave functions, we confirm a result previously indicated in Eq. (4.347), namely,

$$\Delta E_2 = \langle \Delta V_2 \rangle = \frac{\alpha}{\pi}\frac{(Z\alpha)^4\,m}{n^3}\left[\left(\frac{\mu}{m}\right)^3\frac{\delta_{\ell 0}}{2} + \left(-\frac{1}{2\kappa\,(2\ell+1)}\right)\left(\frac{\mu}{m}\right)^2(1-\delta_{\ell 0})\right]. \tag{11.138}$$

Now, we consider the challenging part of the calculation, which is connected to the effective potential

$$\Delta V_1 = F_1'(0)\,\vec{\nabla}^2 V = 4\left[m^2\,F_1'(0)\right]\frac{\pi\,Z\alpha\,\delta^{(3)}(\vec{r})}{m^2}, \tag{11.139a}$$

$$\Delta E_1 = \langle \Delta V_1 \rangle = 4\left[m^2\,F_1'(0)\right]\frac{(Z\alpha)^4 m}{n^3}\left(\frac{\mu}{m}\right)^3\delta_{\ell 0}, \tag{11.139b}$$

where we appeal to the definition of the "standard potential" in Eq. (4.344) and use the same conventions for the bound-state quantum numbers as in Chap. 4. According to Eqs. (10.103) and (11.131), one has

$$m^2\,F_1'(0) = \frac{\alpha}{3\pi}\left[\ln\left(\frac{m}{\lambda}\right) - \frac{3}{8}\right] \tag{11.140}$$

in photon mass regularization. When comparing to the low-photon-energy contribution indicated in Eq. (4.325), we are faced with a problem. The form-factor

[1]Gregory Breit, transcribed from Russian: Grigory Alfredovich Breit-Shneider (1899–1981).

slope indicated in Eq. (11.140) is given as a function of the photon mass λ, which in this case acts as an infrared regulator. By contrast, it will turn out that the photon energy cutoff ϵ in Eq. (11.146) acts as an ultraviolet regulator. In principle, one should assume that the dependence on the infrared and ultraviolet regulators should cancel at the end of the calculation, but how should the quantities ϵ and λ be matched against each other? For the record, we recall that ε in Eq. (11.159) indicates the regularized dimension according to $d = 4 - 2\varepsilon$, ϵ in Eq. (11.146) denotes the (noncovariant) photon energy cutoff, and λ is the photon mass, while μ is the reduced mass of the two-body system for which the Lamb shift is being calculated.

In order to compare ϵ and λ, one may recalculate the Dirac form factor slope using photon energy regularization, as outlined in Chap. 7 of Ref. [2]. This leads to a conversion of the high-energy potential ΔV_1 from photon mass to photon energy regularization. However, these is an easier way, which we would like to describe in the following. Namely, we can alternatively concentrate on the low-energy part of the bound-electron self-energy and reevaluate it in photon mass regularization. Let us consider the penultimate step in Eq. (4.311), modified for a massive photon. Based on Eq. (12b) and the discussion in the Appendix of Ref. [331], one can show that the appropriate formula reads as follows,

$$\Delta E^{(\lambda)}_{\rm LEP} = \frac{e^2}{m^2} \int \frac{\mathrm{d}^3 k}{(2\pi)^3} \frac{\delta^{ij} - \frac{k^i k^j}{\vec{k}^2 + \lambda^2}}{2\sqrt{\vec{k}^2 + \lambda^2}} \left\langle \phi \left| p^i \frac{1}{E - H_S - \sqrt{\vec{k}^2 + \lambda^2}} p^j \right| \phi \right\rangle . \tag{11.141}$$

Here, the quantity $\sqrt{\vec{k}^2 + \lambda^2}$ is the energy of the massive photon, and the Schrödinger Hamiltonian H_S is given in Eq. (4.1). In view of the presence of an infrared regulator in Eq. (11.141), with the condition that

$$m \gg \lambda \gg (Z\alpha)^2 m \,, \tag{11.142}$$

we can actually expand the expression

$$\frac{1}{E - H_S - \sqrt{\vec{k}^2 + \lambda^2}} = -\frac{1}{\sqrt{\vec{k}^2 + \lambda^2}} + \frac{H_S - E}{\vec{k}^2 + \lambda^2} + \mathcal{O}\left(\frac{1}{(\sqrt{\vec{k}^2 + \lambda^2})^{3/2}} \right) , \tag{11.143}$$

and use the second term in Eq. (11.141). The first term is absorbed in a mass renormalization inherent to the low-energy part (see Chap. 4). With the result

$$\lim_{\lambda \to 0} \int_0^\epsilon \frac{\mathrm{d}^3 k}{(2\,\pi)^3} \frac{\delta^{ij} - \frac{k^i k^j}{k^2 + \lambda^2}}{2\sqrt{k^2 + \lambda^2}} \frac{1}{k^2 + \lambda^2} = \frac{2}{3} \frac{\delta^{ij}}{4\,\pi^2} \left[\ln\left(\frac{2\,\epsilon}{\lambda} \right) - \frac{5}{6} \right] , \tag{11.144}$$

we can conclude that

$$\begin{aligned} \Delta E^{(\lambda)}_{\rm LEP} &= \frac{2}{3} \frac{e^2}{m^2} \frac{\delta^{ij}}{4\,\pi^2} \left[\ln\left(\frac{2\,\epsilon}{\lambda} \right) - \frac{5}{6} \right] \left\langle p^i \, (H_S - E) \, p^j \right\rangle \\ &= \frac{1}{3} \frac{\alpha}{\pi m^2} \left[\ln\left(\frac{2\,\epsilon}{\lambda} \right) - \frac{5}{6} \right] \left\langle \vec{\nabla}^2 V \right\rangle \\ &= \frac{4\alpha}{3\pi} \frac{(Z\alpha)^4 m}{n^3} \left(\frac{\mu}{m} \right)^3 \left[\ln\left(\frac{2\,\epsilon}{\lambda} \right) - \frac{5}{6} \right] \delta_{\ell 0} \,. \end{aligned} \tag{11.145}$$

This result has to be compared to Eq. (4.325), which we recall for convenience,

$$\Delta E^{(\epsilon)}_{\rm LEP} = \frac{4\,\alpha}{3\,\pi}\,\frac{(Z\alpha)^4\,m}{n^3}\,\left(\frac{\mu}{m}\right)^3 \left[\ln\left(\frac{2\epsilon}{(Z\alpha)^2\,\mu}\right)\delta_{\ell 0} - \ln k_0(n\ell)\right]. \tag{11.146}$$

We can argue as follows. The result given in Eq. (11.145) depends on both ϵ, which takes the role of an ultraviolet regulator, and the photon mass λ, which is an infrared regulator. This is natural, because the divergence for large ϵ is inherent to the formulation of the low-energy part of the self-energy and cannot be cured by an infrared regulator. The low-energy part can be converted to photon mass regularization, though, on account of the following consideration. The high-energy effective potential given in Eq. (11.139), in photon mass regularization, contains the entire interval $\omega \in (0, \infty)$, which is made possible by the fact that the infrared divergence is cut off at the photon mass scale. By contrast, the low-energy part given in Eq. (11.146) comprises the photon energy interval $\omega \in (0, \epsilon)$. Hence, if one were to add the contributions given in Eq. (11.139) and Eq. (11.146), then one would double-count the integration interval $\omega \in (0, \epsilon)$. This double counting is avoided if one subtracts, from the low-energy part given in Eq. (11.146), the result of an integration over the integration interval $\omega \in (0, \epsilon)$, with the latter one being evaluated in photon mass regularization — the latter quantity is precisely given in Eq. (11.145). Hence, the low-energy part, in photon mass regularization, is given as follows,

$$\begin{aligned}\Delta E_{\rm LEP} &= \Delta E^{(\epsilon)}_{\rm LEP} - \Delta E^{(\lambda)}_{\rm LEP}\\ &= \frac{4\,\alpha}{3\,\pi}\,\frac{(Z\alpha)^4\,m}{n^3}\,\left(\frac{\mu}{m}\right)^3 \left\{\left[\ln\left(\frac{\lambda}{(Z\alpha)^2\,\mu}\right) + \frac{5}{6}\right]\delta_{\ell 0} - \ln k_0(n\ell)\right\}. \end{aligned} \tag{11.147}$$

In view of Eqs. (11.139b) and (11.140), one has

$$\Delta E_1 = \frac{4\alpha}{3\pi}\left[\ln\left(\frac{m}{\lambda}\right) - \frac{3}{8}\right]\frac{(Z\alpha)^4 m}{n^3}\left(\frac{\mu}{m}\right)^3 \delta_{\ell 0}\,, \tag{11.148}$$

where λ is the photon mass and μ is the reduced mass of the two-body system. With ΔE_2 being given in Eq. (11.138), and on account of the algebraic identity $5/6 - 3/8 + (3/4) \times (1/2) = 5/6$, the total result for the self-energy shift is

$$\begin{aligned}\Delta E &= \Delta E_1 + \Delta E_2 + \Delta E_{\rm LEP}\\ &= \frac{4\,\alpha}{3\,\pi}\,\frac{(Z\alpha)^4\,m}{n^3}\,\left(\frac{\mu}{m}\right)^3 \left\{\left[\ln\left(\frac{m}{(Z\alpha)^2\,\mu}\right) + \frac{5}{6}\right]\delta_{\ell 0}\right.\\ &\quad \left. + \left(-\frac{1}{2\kappa\,(2\,\ell + 1)}\right)\frac{m}{\mu}\,(1 - \delta_{\ell 0}) - \ln k_0(n\ell)\right\}. \end{aligned} \tag{11.149}$$

For S states, this expression reduces to the result given in Eq. (4.352), while for non-S states, one obtains the result given in Eq. (4.353).

Note that the matching indicated in Eq. (11.147) was a very difficult problem in the early days of quantum electrodynamics, as evident from a footnote on page 777 of [332]. The complete correction in this order was obtained first in the years

1948 and 1949 (see Refs. [332–335]). The full reduced mass dependence (which we indicate in all detail) was clarified much later in Ref. [336] by SAPIRSTEIN[2] and YENNIE.[3] The matching of the low-energy part in photon energy and photon mass regularization in higher orders has been discussed in Refs. [312,337]. Indeed, higher-order terms in the $Z\alpha$-expansion of the self-energy were evaluated over the past decades. The correction term in the order $\alpha(Z\alpha)^5 m$ was found in Refs. [338–340], and the higher-order binding corrections in the order $\alpha(Z\alpha)^6 m$ were calculated in Refs. [163, 173, 337, 341–352].

11.4.3 *High-Energy Part in Photon Energy Regularization*

The result given in Eq. (11.147) has been obtained by converting the low-energy part of the bound-electron self-energy to photon mass regularization. As evident from the discussion in Chap. 4, the most natural way to calculate the low-energy part of the self-energy proceeds via photon energy regularization. One can alternatively convert the high-energy part given in Eq. (11.139b) to photon energy regularization as follows. One first observes that the high-energy part, in photon-mass regularization, comprises all of the interval $\omega \in (0, \infty)$. However, if one uses photon energy regularization, then the contribution given in Eq. (11.145), which precisely describes the integration interval $\omega \in (0, \epsilon)$, should be subtracted, because the high-energy part, in photon energy regularization, corresponds to the integration interval $\omega \in (\epsilon, \infty)$.

Hence, $\Delta E_1^{(\epsilon)}$, in photon energy regularization, reads as

$$\Delta E_1^{(\epsilon)} = \Delta E_1 - \Delta E_{\rm LEP}^{(\lambda)} = \frac{4\alpha}{3\pi}\frac{(Z\alpha)^4 m}{n^3}\left(\frac{\mu}{m}\right)^3\left[\ln\left(\frac{m}{2\,\epsilon}\right) + \frac{11}{24}\right]\delta_{\ell 0}\,, \tag{11.150}$$

which precisely reproduces the result given in Eq. (4.341). The quantity $\Delta E_{\rm LEP}^{(\lambda)}$ is given in Eq. (11.145), while ΔE_1 is given in Eq. (11.148). Adding ΔE_2 and the result for the low-energy part given in Eqs. (4.325) and (11.146), one easily reproduces the total result for the self-energy shift given in Eq. (11.149). The latter result is independent of the regularization procedure and demonstrates that the physical energy shift does not depend on whether photon energy or photon mass regularization is used for a calculation.

Note that, as one converts ΔE_1 given in Eq. (11.148) to $\Delta E_1^{(\epsilon)}$ given in Eq. (11.150), one can establish the following equivalence,

$$\ln\left(\frac{m}{\lambda}\right) \leftrightarrow \ln\left(\frac{m}{2\,\epsilon}\right) + \frac{5}{6}\,, \tag{11.151}$$

which provides the matching between the infrared photon mass cutoff λ and the noncovariant photon energy cutoff ϵ. The precise form of this matching, though, depends on the process under study and the nonlogarithmic term changes as one considers, e.g., higher-order multipole corrections to the Bethe logarithms.

[2]Jonathan Robert Sapirstein (b. 1951).
[3]Donald Robert Yennie (1924–1993).

11.4.4 Self–Energy in Dimensional Regularization

As a last step, beyond photon mass and photon energy regularization, we discuss dimensional regularization, which has already been used in various contexts in this treatise (see Secs. 10.3.4, 10.3.6, and 10.4.2). Note that the slope of the Dirac form factor, in dimensional regularization, is given in Eq. (11.159).

We will ignore reduced-mass corrections and set the reduced mass equal to the electron mass. We work in $D = 3 - 2\varepsilon$ space dimensions [see Eq. (10.141)] and one time dimension, so that the dimension of space-time is $d = 4 - 2\varepsilon$ [see Eq. (10.142)]. Also, we recall that, according to Eq. (10.173),

$$e^2 = (4\pi)^{1-\varepsilon}\,\alpha\,\mu^{2\varepsilon} e^{\gamma_E \varepsilon} . \tag{11.152}$$

Using the symmetry of the integrand in D dimensions, we start from the penultimate step in Eq. (4.311), and use some obvious symmetry relations of the integrand, to show that

$$\begin{aligned}\Delta E_{\rm LEP} &= e^2 \int_{\mathbb{R}^D} \frac{\mathrm{d}^D k}{(2\pi)^D\, 2k} \left(\delta^{ij} - \frac{k^i\, k^j}{k^2}\right) \left\langle \frac{p^i}{m} \frac{1}{E - k - H} \frac{p^j}{m} \right\rangle \\ &= e^2 \int \frac{\mathrm{d}^D k}{(2\pi)^D\, 2k} \left(\delta^{ij} - \frac{k^i\, k^j}{k^2}\right) \frac{\delta^{ij}}{D} \left\langle \frac{p^k}{m} \frac{1}{E - k - H} \frac{p^k}{m} \right\rangle \\ &= e^2 \frac{D-1}{D} \frac{\Omega_D}{(2\pi)^D} \frac{1}{2} \int_0^\infty \mathrm{d}k\, k^{D-2} \left\langle \frac{p^k}{m} \frac{1}{E - k - H} p^k \right\rangle ,\end{aligned} \tag{11.153}$$

where Ω_D is given in Eq. (10.138). We now introduce a separation parameter Λ, which cancels at the end of the calculation, to separate the small-momentum from the large-momentum regions. The key to the calculation shall be that dimensional regularization is only required for the integration region $k \in (\Lambda, \infty)$,

$$\begin{aligned}\Delta E_{\rm LEP} &= e^2 \frac{D-1}{D} \frac{\Omega_D}{(2\pi)^D} \frac{1}{2} \int_\Lambda^\infty \mathrm{d}k\, k^{D-2} \left\langle \frac{p^k}{m} \frac{1}{E - k - H} \frac{p^k}{m} \right\rangle \\ &\quad + \frac{2\alpha}{3\pi} \int_0^\Lambda \mathrm{d}k\, k \left\langle \frac{p^k}{m} \frac{1}{E - k - H} \frac{p^k}{m} \right\rangle \\ &= e^2 \frac{D-1}{D} \frac{\Omega_D}{(2\pi)^D} \frac{1}{2} \int_\Lambda^\infty \mathrm{d}k\, k^{D-2} \left\langle \frac{p^k}{m} \left(-\frac{1}{k} + \frac{H-E}{k^2} - \frac{(H-E)^2}{k^3} \right) \frac{p^k}{m} \right\rangle \\ &\quad + \frac{2\alpha}{3\pi} \int_0^\Lambda \mathrm{d}k\, k \left\langle \frac{p^k}{m} \frac{1}{E - k - H} \frac{p^k}{m} \right\rangle .\end{aligned} \tag{11.154}$$

Using Eq. (10.173), we can show that

$$e^2 \frac{D-1}{D} \frac{\Omega_D}{(2\pi)^D} \frac{1}{2} = \frac{2}{3} + \varepsilon \left[\frac{10}{9} + \frac{4}{3} \ln\left(\frac{\mu}{2}\right) \right] . \tag{11.155}$$

We can now use the following results ($D = 3 - 2\varepsilon$), ignoring terms generated by the upper limit at $k = \infty$ in the spirit of dimensional regularization,

$$\int_\Lambda^\infty \mathrm{d}k\, k^{D-3} = -\Lambda + \mathcal{O}(\varepsilon)\,, \qquad \int_\Lambda^\infty \mathrm{d}k\, k^{D-4} = \frac{1}{2\varepsilon} - \ln(\Lambda) + \mathcal{O}(\varepsilon)\,, \tag{11.156a}$$

$$\int_\Lambda^\infty \mathrm{d}k\, k^{D-5} = \frac{1}{\Lambda} + \mathcal{O}(\varepsilon)\,, \qquad \int_\Lambda^\infty \mathrm{d}k\, k^{D-6} = \frac{1}{2\Lambda^2} + \mathcal{O}(\varepsilon)\,, \tag{11.156b}$$

and so on. We see that higher-order terms for large k in the expansion of $1/(E-H-k)$ give rise to vanishing contributions in the limit of large Λ. We now specialize the discussion to the limit of infinite nuclear mass, where the reduced mass is equal to the electron mass m, in order to be able to concentrate on the essential elements of the discussion. The intermediate expression for the low-energy part is found as follows,

$$\begin{aligned}\Delta E_{\rm LEP} &= \frac{\alpha}{\pi}\left\{\frac{2}{3} + \varepsilon\left[\frac{10}{9} + \frac{4}{3}\ln\left(\frac{\mu}{2}\right)\right]\right\}\left[\left(\frac{1}{2\varepsilon} - \ln(\Lambda)\right)\left\langle p^k\,(H-E)\,p^k\right\rangle + \Lambda\left\langle \vec{p}^{\,2}\right\rangle\right] \\ &\quad - \frac{2\alpha}{3\pi}\Lambda\left\langle \vec{p}^{\,2}\right\rangle + \frac{4\alpha}{3\pi}\frac{(Z\alpha)^4 m}{n^3}\left(\ln\left(\frac{2\Lambda}{(Z\alpha)^2 m}\right) - \ln k_0(n,\ell)\right).\end{aligned} \tag{11.157}$$

For the definition of the Bethe logarithm $\ln k_0(n,\ell)$, we refer to the corresponding discussion in Chap. 4. Using $\left\langle p^k\,(H-E)\,p^k\right\rangle = \left\langle 2\pi Z\alpha\,\delta^{(D)}(\vec{r})\right\rangle \to 2(Z\alpha)^4/n^3$ in the limit $D \to 3$, we can now see that Λ cancels in the sum of the two above contributions (both the linear as well as the logarithmic term), and that also the $\ln(2)$ terms cancel. We finally obtain

$$\Delta E_{\rm LEP} = \frac{4\alpha}{3\pi}\frac{(Z\alpha)^4 m}{n^3}\left[\left(\frac{1}{2\varepsilon} + \frac{5}{6} + \ln\left(\frac{\mu}{(Z\alpha)^2 m}\right)\right)\delta_{\ell 0} - \ln k_0(n,\ell)\right]. \tag{11.158}$$

This result is strictly of order $\alpha(Z\alpha)^4 m$ and does not contain any higher-order terms, in contrast to photon energy regularization, which was discussed in Sec. 4.7.3. This behavior is due to a separation of the kinematic variables (the powers of $Z\alpha$) from the divergences (i.e., from the divergence for large virtual photon energy). In photon energy regularization, the two mechanism for the generation of divergences are intertwined, and therefore the nonrelativistic dipole term generates divergent contributions in arbitrarily high order in the $(Z\alpha)$-expansion. Another advantage of the dimensional regularization is that all $\ln(Z\alpha)$ terms come from the low-energy part, to which they are physically associated.

Let us recall the Dirac form-factor slope in dimensional regularization from Eq. (11.132) [see also Eq. (10.190a)],

$$F_1'(0) = \frac{\alpha}{\pi m^2}\left(-\frac{1}{8} - \frac{1}{6\varepsilon} + \frac{1}{6}\ln\left(\frac{m^2}{\mu^2}\right)\right). \tag{11.159}$$

From Eq. (10.190b), we have $F_2(0) = \alpha/(2\pi)$.

We write $\Delta E_{\rm HEP} = \Delta E_1 + \Delta E_2$ as the sum of two contributions. According to Eqs. (11.139a) and (11.159), the high-energy potential ΔV_1 and its associated energy shift are given by (again, in the limit of infinite nuclear mass)

$$\Delta V_1 = \frac{4\alpha}{\pi}\left[-\frac{1}{8} - \frac{1}{6\varepsilon} + \frac{1}{6}\ln\left(\frac{m^2}{\mu^2}\right)\right]\frac{\pi Z\alpha \left\langle \delta^{(3)}(\vec{r})\right\rangle}{m^2}, \tag{11.160a}$$

$$\Delta E_1 = \frac{4}{3}\left[-\frac{3}{8} - \frac{1}{2\varepsilon} + \frac{1}{2}\ln\left(\frac{m^2}{\mu^2}\right)\right]\frac{\alpha}{\pi}\frac{(Z\alpha)^4 m}{n^3}\,\delta_{\ell 0}\,. \tag{11.160b}$$

After adding the results in Eqs. (11.158) and (11.160b), *both the dependence on the ε parameter as well as the dependence on the renormalization scale μ cancel.* For the record, let us consider the energy shift ΔE_2 listed in Eq. (11.138), in the limit of infinite nuclear mass,

$$\Delta E_2 = \frac{\alpha}{\pi}\frac{(Z\alpha)^4 m}{n^3}\left[\frac{\delta_{\ell 0}}{2} + \left(-\frac{1}{2\kappa\,(2\ell+1)}\right)(1-\delta_{\ell 0})\right]. \tag{11.161}$$

Adding the results from Eqs. (11.158), (11.160b) and (11.161), one obtains for the self-energy shift $\Delta E_{\rm SE}$,

$$\begin{aligned}\Delta E_{\rm SE} &= \Delta E_{\rm HEP} + \Delta E_{\rm LEP}\\ &= \frac{4\,\alpha}{3\,\pi}\frac{(Z\alpha)^4 m}{n^3}\left\{\left[\ln(Z\alpha)^{-2} + \frac{5}{6}\right]\delta_{\ell 0} + \left(-\frac{1}{2\kappa\,(2\ell+1)}\right)(1-\delta_{\ell 0}) - \ln k_0(n\ell)\right\},\end{aligned} \tag{11.162}$$

which is precisely equal to the result given in Eq. (11.149) in the limit of infinite nuclear mass. We have avoided the discussion of the nuclear-mass dependence in the context of dimensional regularization in order to keep formulas compact. It is a good exercise to restore this dependence in intermediate steps.

11.4.5 *Vacuum Polarization*

For S states, there is a further correction which is determined by the vacuum polarization correction to the Coulomb potential, which generates an effective potential [see Eq. (10.235)]

$$\Delta V_{\rm VP} = -\frac{4}{15}\,\alpha(Z\alpha)\,\frac{\delta^{(3)}(\vec{r})}{m^2} \equiv \Delta V_3\,, \tag{11.163}$$

to supplement ΔV_1 given in Eq. (11.139a) and ΔV_2 given in Eq. (11.137). The vacuum-polarization potential (11.163) leads to an energy shift

$$\Delta E_{\rm VP} = \langle n\ell|\Delta V_3|n\ell\rangle = -\frac{4}{15}\frac{\alpha}{\pi}\frac{(Z\alpha)^4 m}{n^3}\left(\frac{\mu}{m}\right)^3\delta_{l0} \approx -\frac{4}{15}\frac{\alpha}{\pi}\frac{(Z\alpha)^4 m}{n^3}\delta_{l0}\,, \tag{11.164}$$

where the latter form is valid in the limit of infinite nuclear mass. Adding this result to the self-energy shift (4.352) for S states, we obtain for the sum of self-energy and vacuum-polarization corrections, for S states,

$$\Delta E_{\rm SE+VP}(n{\rm S}) = \frac{\alpha}{\pi}\frac{(Z\alpha)^4 m}{n^3}\left(\frac{4}{3}\ln\left[(Z\alpha)^{-2}\right] + \frac{38}{45} - \frac{4}{3}\ln k_0(n,\ell)\right) \tag{11.165}$$

as the total Lamb shift in the order $\alpha(Z\alpha)^4 m$ (again, in the non-recoil approximation). For states with $\ell \neq 0$, the result (4.353) is already complete in the order $\alpha(Z\alpha)^4 m$. The reduced-mass dependence adds an overall multiplicative factor $(\mu/m)^3$, according to Eqs. (4.352) and (11.164).

11.5 Foldy–Wouthuysen Transformation and Gravity

We recall the Dirac–Schwarzschild (DS) Hamiltonian $H_{\rm DS}$ from Eq. (8.298),

$$H_{\rm DS} = \frac{1}{2}\left\{\vec{\alpha}\cdot\vec{p}, \left(1 - \frac{r_s}{r}\right)\right\} + \beta m\left(1 - \frac{r_s}{2r}\right), \tag{11.166}$$

where $r_s = 2GM$. This Hamiltonian describes the coupling of a Dirac particle to the gravitational field surrounding a black hole. Below, we recall a transformation [255] which describes the expansions to first order in $\chi = r_s/r$ and to higher orders order in $\xi = \vec{p}/m$ or $\xi = 1/(mr)$. Terms are calculated up to order ξ^4 if there is no gravitational interaction, and up to order $\xi^3\,\chi$ if there is a gravitational interaction. Terms of order χ^2 (second order in the gravitational interaction) are ignored. The Foldy–Wouthuysen transformed Hamiltonian is found as follows [255],

$$\begin{aligned} H_{\rm FW} &= \beta\left(m + \frac{\vec{p}^{\,2}}{2m} - \frac{\vec{p}^{\,4}}{8m^3}\right) - \beta\,\frac{m\,r_s}{2\,r} \\ &\quad + \beta\left(-\frac{3r_s}{8m}\left\{\vec{p}^{\,2}, \frac{1}{r}\right\} + \frac{3\pi r_s}{4m}\delta^{(3)}(\vec{r}) + \frac{3r_s}{8m}\,\frac{\vec{\Sigma}\cdot\vec{L}}{r^3}\right). \end{aligned} \tag{11.167}$$

As pointed out in Ref. [255], the very last term on the right-hand side of Eq. (11.167) describes the gravitational spin-orbit coupling. The classical result for the interaction of a spinning classical particle with the gravitational field is otherwise known as the geodesic precession or Fokker precession [353–355], or de Sitter precession. On the classical level, it has been verified in the Gravity Probe B experiment [356]. When comparing the spin-orbit term to Eq. (26) of Ref. [355], one might notice a prefactor of 3/2 instead of 3/8. However, one should take note that the spin operator used in Ref. [355] carries a factor one half ($\vec{S} = \frac{1}{2}\vec{\Sigma}$), and that there is an additional factor 2 in the definition of the Schwarzschild radius. The prefactor β in Eq. (11.167) describes the particle-antiparticle symmetry in the gravitational interactions, which cannot be obtained based on classical considerations [353–355]. In particular, gravitationally coupled relativistic Dirac theory makes the unique prediction that the equivalence principle holds for antimatter [251, 291, 357, 358].

Finally, we quote the gravitational corrections to the transition current from Ref. [255],

$$\begin{aligned} j^i_{\rm FW} &= \frac{p^i}{m} + \frac{1}{2}\left\{\frac{p^i}{m}, \mathrm{i}\vec{k}\cdot\vec{r}\right\} - \frac{1}{4}\left\{\frac{p^i}{m}, (\vec{k}\cdot\vec{r})^2\right\} - \frac{p^i\,\vec{p}^{\,2}}{2m^3} + \frac{\mathrm{i}}{2\,m}\left(\vec{\sigma}\times\vec{k}\right)^i\left(1 + \mathrm{i}\,\vec{k}\cdot\vec{r}\right) \\ &- \frac{3}{4}\left\{\frac{p^i}{m}, \frac{r_s}{r}\right\} + \frac{r_s}{2r}\,\frac{(\vec{\sigma}\times\vec{r})^i}{m r^2} - \frac{1}{2}\left\{\mathrm{i}\vec{k}\cdot\vec{r}, \left\{\frac{p^i}{m}, \frac{r_s}{r}\right\}\right\} - \frac{3\mathrm{i}r_s}{4r}\,\frac{(\vec{\sigma}\times\vec{k})^i}{m} + \frac{1}{4}\left\{\frac{r_s}{r}(\mathrm{i}\vec{k}\cdot\vec{r}), \frac{p^i}{m}\right\}. \end{aligned} \tag{11.168}$$

The well-known corrections to the relativistic transition current, given in Eqs. (11.112) and (11.120), are supplemented by the gravitational kinetic correction, and gravitational corrections to the magnetic coupling.

11.6 Further Thoughts

Here are some suggestions for further thought.

(1) **Alternative Foldy–Wouthuysen Transformations.** Consider the free Dirac particle. Verify that the modified Foldy–Wouthuysen transformation $U = \exp(\mathrm{i}S)$, given as

$$U = \mathrm{e}^{\mathrm{i}S}\,, \qquad S = -\frac{\mathrm{i}\beta}{2m}\left(\vec{\alpha}\cdot\vec{p} - \frac{(\vec{\alpha}\cdot\vec{p})^3}{3m^2}\right) = S_1 + S_2\,, \tag{11.169}$$

transforms $H_{\rm FD}$, given in Eq. (11.8), directly into $H_{\rm FW}$, given in Eq. (11.17), and captures the relativistic corrections up to and including fourth order in the momenta. In addition, show, using the Campbell–Baker–Hausdorff formula, that if

$$U = \mathrm{e}^{\mathrm{i}(S_1+S_2)}\,, \qquad S_1 = -\frac{\mathrm{i}\beta}{2m}\vec{\alpha}\cdot\vec{p}\,, \qquad S_2 = \frac{\mathrm{i}\beta}{6m^3}(\vec{\alpha}\cdot\vec{p})^3\,, \tag{11.170}$$

then

$$U = \mathrm{e}^{\mathrm{i}(S_1+S_2)} = U_1\,U_2 = \mathrm{e}^{\mathrm{i}S_1}\,\mathrm{e}^{\mathrm{i}S_2}\,. \tag{11.171}$$

Hint: This calculation is still perturbative in the momentum operators; you may discard higher-order terms in intermediate steps.

(2) **One-Step Transformations.** Consider H_∞ as given in Eq. (11.15), and $H_{\rm FD}$ as given in Eq. (11.8). Find a one-step unitary transformation S_∞ for which

$$H_\infty = \mathrm{e}^{\mathrm{i}S_\infty}\,H_{\rm FD}\,\mathrm{e}^{-\mathrm{i}S_\infty}\,. \tag{11.172}$$

Hint: Try the *ansatz*

$$\exp(\mathrm{i}S) = \exp[\beta\,\vec{\alpha}\cdot\vec{p}\,\theta(\vec{p})] = \cos[|\vec{p}|\,\theta(\vec{p})] + \frac{\beta\,\vec{\alpha}\cdot\vec{p}}{|\vec{p}|}\,\sin[|\vec{p}|\,\theta(\vec{p})]\,. \tag{11.173}$$

Can this approach be easily generalized easily to more complicated scenarios?

(3) **Antiparticle Hamiltonians.** Show that the effective antiparticle Hamiltonian, corresponding to the physical system described by Eq. (11.54), can be obtained from Eq. (11.54) by the replacement $e \to -e$.

(4) **Foldy–Wouthuysen Transformation.** Compare the result given in Eq. (11.54) to the treatment in Chap. 4 of Ref. [316]. In particular, show the identity

$$\vec{E}\times\vec{\pi} - \vec{\pi}\times\vec{E} = 2\vec{E}\times\vec{\pi} + \mathrm{i}\vec{\nabla}\times\vec{E} \tag{11.174}$$

and compare our result to the one given in Eq. (4.5) of Ref. [316]. *Hint:* You may have to consider the limit $\kappa \to 0$.

(5) **Comparison to the Literature.** In Ref. [359], a Foldy–Wouthuysen transformation is being used with the operator

$$U = \exp(\mathrm{i}S)\,, \qquad S = S_0 + \delta S\,, \tag{11.175a}$$

$$S_0 = -\frac{\mathrm{i}}{2m}\left\{\beta\,\vec{\alpha}\cdot\vec{\pi} - \frac{1}{3m^2}\beta\,(\vec{\alpha}\cdot\vec{\pi})^3 + \frac{e}{2m}\,\vec{\alpha}\cdot\vec{E}\right\}, \tag{11.175b}$$

$$\delta S = -\frac{\mathrm{i}}{2m}\left\{\frac{e\kappa}{2m}\,\mathrm{i}\vec{\alpha}\cdot\vec{E} - \frac{e\kappa}{4m^2}\left[\vec{\alpha}\cdot\vec{\pi}, \beta\vec{\Sigma}\cdot\vec{B}\,\right]\right\}. \tag{11.175c}$$

In Eq. (2.8) of Ref. [312], it is precisely the Foldy–Wouthuysen transformation in the form given in Eq. (11.175) which is being used in order to derive the α^6 and α^7 Hamiltonians.

In Ref. [322], a Foldy–Wouthuysen transformation is being used which is a little different from the one used in Refs. [359] and [312] and is given as

$$S = -\frac{\mathrm{i}}{2m}\left\{\beta\,\vec{\alpha}\cdot\vec{\pi} - \frac{1}{3m^2}\,\beta\,(\vec{\alpha}\cdot\vec{\pi})^3 + \frac{1}{2m}\,[\vec{\alpha}\cdot\vec{\pi}\,,\, e\,A^0 - i\,\partial_t] + \delta Y\right\}, \tag{11.176a}$$

$$\begin{aligned}\delta Y = {} & \frac{\beta}{5\,m^4}\,(\vec{\alpha}\cdot\vec{\pi})^5 - \frac{\beta\,e}{4\,m^2}\,\vec{\alpha}\cdot\dot{\vec{E}} + \frac{\mathrm{i}\,e}{24\,m^3}\left[\vec{\alpha}\cdot\vec{\pi}, [\vec{\alpha}\cdot\vec{\pi}, \vec{\alpha}\cdot\vec{E}]\right] \\ & - \frac{\mathrm{i}\,e}{3\,m^3}\left\{(\vec{\alpha}\cdot\vec{\pi})^2\,\vec{\alpha}\cdot\vec{E} + \vec{\alpha}\cdot\vec{E}\,(\vec{\alpha}\cdot\vec{\pi})^2\right\}.\end{aligned} \tag{11.176b}$$

In terms of the $(Z\alpha)$-expansion, one has $\delta Y \sim (Z\alpha)^5$. The breakdown of individual contributions and iterative transformations is being discussed in Ref. [322]. The corresponding sixth-order Hamiltonian is

$$\begin{aligned}H^{[6]} = {} & \frac{(\vec{\sigma}\cdot\vec{\pi})^6}{16\,m^5} - \frac{e}{16\,m^3}\left\{\vec{\sigma}\cdot\vec{\pi},\, \vec{\sigma}\cdot\dot{\vec{E}}\right\} - \frac{\mathrm{i}\,e}{128\,m^4}\,[\vec{\sigma}\cdot\vec{\pi}, [\vec{\sigma}\cdot\vec{\pi}, [\vec{\sigma}\cdot\vec{\pi}, \vec{\sigma}\cdot\vec{E}]]] \\ & + \frac{\mathrm{i}\,e}{16\,m^4}\left\{(\vec{\sigma}\cdot\vec{\pi})^2, [\vec{\sigma}\cdot\vec{\pi}, \vec{\sigma}\cdot\vec{E}]\right\}.\end{aligned} \tag{11.177}$$

One notes, for a Coulomb field, the identities given in Eqs. (11.64) and (11.67). The sixth-order terms in Eq. (11.177) start with the term proportional to $(\vec{\sigma}\cdot\vec{\pi})^6$. Our result, given in Eq. (11.59), differs in the sense that the term proportional to $\left\{\vec{\sigma}\cdot\vec{\pi},\, \vec{\sigma}\cdot\dot{\vec{E}}\right\}$ is absent, and also, the term proportional to $[\vec{\sigma}\cdot\vec{\pi}, [\vec{\sigma}\cdot\vec{\pi}, [\vec{\sigma}\cdot\vec{\pi}, \vec{\sigma}\cdot\vec{E}]]]$ is absent, while the term proportional to $\left\{(\vec{\sigma}\cdot\vec{\pi})^2, [\vec{\sigma}\cdot\vec{\pi}, \vec{\sigma}\cdot\vec{E}]\right\}$ carries a different prefactor.

Compare the following α^6 Hamiltonians: *(i)* the Hamiltonian given in Eq. (11.177), together with the modified FW transformation from Eq. (11.176), *(ii)* the Hamiltonian given in Ref. [312], used in its Coulomb-field form in Eq. (11.69), together with the transformation given in Eq. (11.175), *(iii)* our result (11.59), obtained by the straightforward application of the iterative FW method, where subsequently, one eliminates the odd operators in the transformed Dirac Hamiltonians. Consider both diagonal as well as off-diagonal matrix elements.

(6) **Gauge Invariance and Hamiltonians.** Strictly speaking, the Hamiltonian (11.54) is not gauge invariant, as it contains the gauge-dependent scalar and vector potential. However, this observation is equally true of the original Hamiltonian (11.40) which contains the canonical momentum $\vec{\pi} = \vec{p} - e\vec{A}$ (it is gauge covariant, but not gauge invariant). Show, however, that energy eigenvalues will be gauge invariant. *Hint:* Consider the equation $H\psi = E\psi$ and investigate if the gauge-transformed wave function ψ' [see Eq. (8.2)] fulfills $H'\psi' = E\psi'$, where H' is the gauge-transformed Hamiltonian.

(7) **Bound-Electron g Factor.** Consider the bound-electron g factor on the basis of Eq. (11.54), for the $1S$ state of a hydrogenlike bound state, in which case you can replace the orbital angular momentum operator $\vec{L}$ by zero. Treat all terms which contribute to order $(Z\alpha)^6$. Repeat the exercise in the length gauge, i.e., with the form of the interaction Hamiltonian obtained after a Power–Zienau transformation.

(8) **Corrections to the Transition Current.** Consider additional corrections to the photon transition current, induced by the addition of an additional vector potential $\vec{A}_{\mathrm{hfs}}$. Use Eq. (22) of Ref. [360] as an inspiration. *Hint:* A nonrelativistic motivation can be derived by replacing the Hamiltonian

$$\frac{(\vec{p} - e\vec{A} - e\,A_{\mathrm{hfs}})^2}{2m} \to -e\vec{A}\cdot\frac{\vec{p}}{m} - e\vec{A}\cdot\frac{|e|\vec{A}_{\mathrm{hfs}}}{m}, \tag{11.178}$$

and matching with the coefficient of the vector potential $\vec{A}$.

(9) **Exponential Tail and Local Potential.** Match Eq. (11.163) with the exponential factors in Eq. (10.246).

Chapter 12

Relativistic Interactions for Many-Particle and Compound Systems

12.1 Overview

In the current chapter, we shall discuss the derivation of time derivative operators (Hamiltonians), and time evolution operators, for systems consisting of many particles. At first, one might conjecture that it is next-to-impossible, from a relativistic point of view, to calculate the effective Hamiltonian for a two-body, or three-body system, and for quantum mechanical systems or more than three particles, because the proper time cannot be uniquely defined for this system. Each of the two (or more) particles has its own proper time. It might be that the only way to define a Hamiltonian in a relativistically invariant way would involve a time derivative with respect to the proper time of each particle in question. How can this question be solved? The key is to observe that we are not calculating an energy level of a particular, isolated particle, but energy levels of a compound system consisting of many particles. We can first formulate a Hamiltonian for the compound system, i.e., a Hamiltonian that contains terms for all involved particles. Then, we go into the center-of-mass frame of the compound system, and define the Hamiltonian as the time derivative operator with respect to the proper time in the center-of-mass system. This approach will prove to be extremely useful.

We shall proceed as follows. In Sec. 12.3, we describe the derivation of the many-body Breit Hamiltonian based on the matching of the forward scattering amplitude with the effective Hamiltonian, in the no-retardation (zero-frequency) limit of the photon. The derivation is heavily based on previous work carried out in Chap. 11. Namely, from the Foldy–Wouthuysen transform of the single-particle Dirac Hamiltonian, we shall first derive the transition current operators which we need in order to write down the two-particle, forward scattering amplitude, which is formulated as an expectation value of the many-particle wave function. The calculation culminates in a general formula, given in Eq. (12.64), for the two-particle interaction Hamiltonian of atomic electrons, including the anomalous-magnetic-moment corrections. The atomic nucleus (of arbitrary spin) is treated in Sec. 12.4.2. An orbiting particle of arbitrary spin is treated in Sec. 12.4.3. In these treatments, we need to generalize the Pauli spin matrices; this is accomplished in detail for the spin-1 case. Applications of the derived formalism to two-body systems will allow us to derive the exact reduced-mass dependence of a number of tiny, but important energy shifts for bound systems (see Sec. 12.5). Finally, in Sec. 12.6, we shall discuss the derivation of Hamiltonians applicable to more complicated scenarios, such as the interaction of a compound system with an external magnetic field.

12.2 Interatomic Interactions in Covariant Formalism

12.2.1 *General Paradigm of the Matching*

Let us discuss the general paradigm underlying the matching of an effective Hamiltonian with the S matrix. First, it needs to be observed that the S matrix describes the time evolution of the system from the infinite past to the infinite future. In turn, of course, the time evolution of a quantum system is described by its Hamiltonian (time derivative operator). Hence, there must be a connection.

The S matrix is given as the time-ordered exponential [see Eq. (3.41)]

$$S = \mathrm{T}\exp\left[-\mathrm{i}\int_{-\infty}^{\infty} \mathrm{d}t'\, H_I(t')\right] \approx 1 - \mathrm{i}\int_{-\infty}^{\infty} \mathrm{d}t'\, H_I(t')\,. \tag{12.1}$$

Here, $H_I(t)$ is the interaction Hamiltonian in the interaction representation [see Eq. (2.108)],

$$H_I(t,\vec{r}) = \mathrm{e}^{\mathrm{i}\,H_0\,t}\, H_I(\vec{r})\, \mathrm{e}^{-\mathrm{i}\,H_0\,t}\,, \qquad H_I \equiv H_I(t=0)\,. \tag{12.2}$$

Now, let $H_I = H_I(\vec{r})$ be an effective, time-independent potential. The S matrix element describes the scattering of an initial state $|i\rangle$ to a final state $|f\rangle$. Following the discussion in Chap. 3, we assume that $E_f = E_i = E$ and write

$$\begin{aligned} S_{fi} &= \left\langle f\left|1 - \mathrm{i}\int_{-\infty}^{\infty} \mathrm{d}t\, \mathrm{e}^{\mathrm{i}\,H_0\,t}\, H_I\, \mathrm{e}^{-\mathrm{i}\,H_0\,t}\right|i\right\rangle \qquad (12.3)\\ &= \langle f|i\rangle - \mathrm{i}\int_{-\infty}^{\infty} \mathrm{d}t\,\left\langle f\left|\mathrm{e}^{\mathrm{i}\,E\,t}\, H_I\, \mathrm{e}^{-\mathrm{i}\,E\,t}\right|i\right\rangle = \delta_{fi} - \mathrm{i}T\int \mathrm{d}^3r\, \phi_f(\vec{r})\, H_I(\vec{r})\, \phi_i(\vec{r})\,. \end{aligned}$$

The matching relation thus is as follows,

$$S_{fi} = \delta_{fi} - \mathrm{i}T\int \mathrm{d}^3r\, \phi_f(\vec{r})\, H_I(\vec{r})\, \phi_i(\vec{r})\,. \tag{12.4}$$

Or, if we ignore the Kronecker-δ term, and focus on the first-order contribution to the S matrix (the Kronecker-δ would be the zeroth-order term), we can write, in a somewhat shorthand notation,

$$\langle S(\vec{r})\rangle^{[1]} = -\mathrm{i}T\,\langle U(\vec{r})\rangle\,, \tag{12.5}$$

where, in a typical case, some interaction potential $H_I(\vec{r}) = U(\vec{r})$ is matched against the S matrix element.

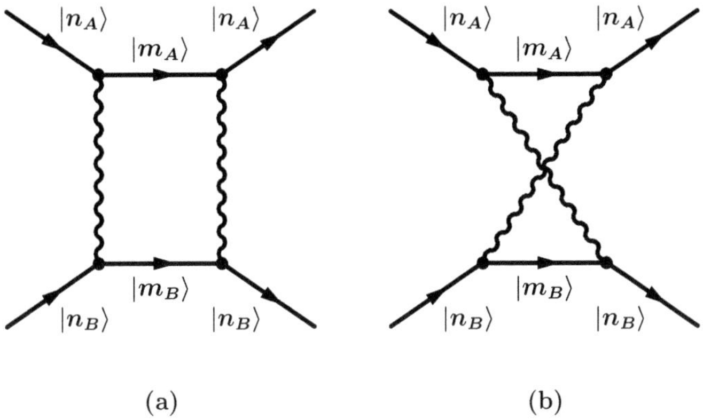

Fig. 12.1 We investigate the two-photon exchange between two active electrons in two atoms, labeled A and B. Diagram (a) is the ladder exchange, while diagram (b) constitutes the crossed-ladder photon exchange. Atom A makes a virtual transition from the reference states $|n_A\rangle$ to the virtual state $|m_A\rangle$, and back. Atom B undergoes an analogous transition.

12.2.2 *Scattering Amplitudes and Interatomic Interactions*

In order to illustrate the considerations outlined in Sec. 12.2.1, let us consider a possible application of the propagator formalism introduced in Chaps. 7 and 9, to the calculation of the interatomic interaction, which we had considered in the Schrödinger picture in Sec. 5.4. The problem with the treatment in Sec. 5.4 is that no time variable is introduced in the description of photon exchange. Hence, it is difficult to correctly assess the placement of the propagator poles in the denominators. It is very interesting to realize that the calculation performed in Sec. 5.4 becomes a lot easier when one resorts to time-ordered perturbation theory and just calculates the forward scattering amplitude for the two-atom system, rather than calculating the Hamiltonian on the basis of time-independent, Schrödinger-picture field operators, as in Sec. 5.4. Following Ref. [208], we here illustrate, briefly, how to derive the starting expression for the self-energy and the Casimir–Polder interaction. We are also inspired by the derivation outlined in Chap. 85 of Ref. [3].

The calculation described in Ref. [208] roughly proceeds as follows. One considers two atoms in states $\psi_A(\vec{r}_A)$ and $\psi_B(\vec{r}_B)$ which scatter into states $\psi'_A(\vec{r}_A)$ and $\psi'_B(\vec{r}_B)$ under the action of a potential $U(|\vec{r}_A - \vec{r}_B|)$. If we denote the initial state by i and the final state by the subscript f, then the corresponding S-matrix element reads as follows,

$$
\begin{aligned}
S_{fi} &= -\mathrm{i} \int \mathrm{d}^3 r_A \int \mathrm{d}^3 r_B \, \psi'^*_A(\vec{r}_A)\, \psi'^*_B(\vec{r}_B)\, U(|\vec{r}_A - \vec{r}_B|)\, \psi_A(\vec{r}_A)\, \psi_B(\vec{r}_B) \\
&\quad \times \int \mathrm{d}t \, \exp[-\mathrm{i}\,(E_1 + E_2 - E'_1 - E'_2)\, t] \\
&= -\mathrm{i} T \int \mathrm{d}^3 r_A \int \mathrm{d}^3 r_B \, \psi'^*_A(\vec{r}_A)\, \psi'^*_B(\vec{r}_B)\, U(\vec{r}_A - \vec{r}_B)\, \psi_A(\vec{r}_A)\, \psi_B(\vec{r}_B) \,, \qquad (12.6)
\end{aligned}
$$

where we have assumed energy conservation ($E_1 + E_2 = E'_1 + E'_2$).

One uses time-dependent quantum electrodynamic (QED) perturbation theory, where the interaction is formulated in the interaction picture [2, 116]. This means that the second-quantized operators in the interaction Hamiltonian have a time dependence which is generated by the action of the free Hamiltonian [112]. The designation of an interaction Hamiltonian being formulated in the interaction picture is not redundant [112]. The two relevant Feynman diagrams in Fig. 12.1 (as opposed to the twelve diagrams in Fig. 5.1) describe all possible time orderings.

We use, specifically, the temporal gauge for the photon propagator, outlined in Eq. (9.133). We recall that the 00 component of the photon propagator vanishes in the temporal (Weyl) gauge,

$$D_F^{00}(k) = 0\,, \qquad D_F^{ij}(k) = \frac{1}{k^2 + \mathrm{i}\epsilon}\left(\delta_{ij} - \frac{k_i\,k_j}{\omega^2}\right). \tag{12.7}$$

Because the scalar potential vanishes, the interaction Hamiltonian can then be restricted to the dipole interaction with the (spatial components of the) electric field,

$$V(t) = -\vec{E}(\vec{r}_A,t)\cdot\vec{d}_A(t) - \vec{E}(\vec{r}_B,t)\cdot\vec{d}_B(t) \approx -\vec{E}(\vec{R}_A,t)\cdot\vec{d}_A(t) - \vec{E}(\vec{R}_B,t)\cdot\vec{d}_B(t)\,, \tag{12.8}$$

where $\vec{d}_i = e\,\vec{r}_i$ is the dipole operator for atom i (for atoms with more than one electron, one has to sum over all the electrons in the atoms $i = A, B$). By $\vec{R}_A$ and $\vec{R}_B$, we denote the positions of the atomic nuclei.

We use the relation (9.122) in order to relate the time-ordered product of field operators to the Hamiltonian. This differs, by an overall minus sign, from the conventions used in Eq. (10) of Ref. [208]. In the mixed frequency-position representation, one has

$$D^{ik}(\omega,\vec{r}) = -\left(\delta^{ik} + \frac{\nabla^i\,\nabla^k}{\omega^2}\right)\frac{\mathrm{e}^{\mathrm{i}|\omega|r}}{4\pi r}\,, \tag{12.9}$$

where $|\omega| \equiv \sqrt{\omega^2 + \mathrm{i}\epsilon}$, and the branch cut of the square-root function chosen along the positive real axis, so that $\mathrm{Im}(\sqrt{\omega^2 + \mathrm{i}\epsilon}) > 0$ in the spirit of Eq. (9.34). We carry out the differentiations with the result,

$$\nabla^i\,\nabla^k\frac{\mathrm{e}^{\mathrm{i}|\omega|r}}{4\pi r} = \omega^2\,\delta^{ik}\left(-\frac{1}{\omega^2 r^2} + \frac{\mathrm{i}}{|\omega|\,r}\right)\frac{\mathrm{e}^{\mathrm{i}|\omega|r}}{4\pi r} + \omega^2\,\frac{r^i\,r^k}{r^2}\left(\frac{3}{\omega^2\,r^2} - \frac{3\mathrm{i}}{|\omega|\,r} - 1\right)\frac{\mathrm{e}^{\mathrm{i}|\omega|r}}{4\pi r}\,. \tag{12.10}$$

The temporal-gauge photon propagator in the mixed representation becomes

$$D^{ik}(\omega,\vec{r}) = -\left[\alpha^{ik} + \beta^{ik}\left(\frac{\mathrm{i}}{|\omega| r} - \frac{1}{\omega^2\,r^2}\right)\right]\frac{\mathrm{e}^{\mathrm{i}|\omega|r}}{4\pi r}\,,$$

$$\alpha^{ik} = \delta^{ik} - \frac{r^i\,r^k}{r^2}\,, \qquad \beta^{ik} = \delta^{ik} - 3\frac{r^i\,r^k}{r^2}\,. \tag{12.11}$$

Throughout this chapter, we shall depart from the notation (2.58) and use r^i rather than x^i for the ith component of the position operator $\vec{r}$. This is necessitated by the fact that, for composite systems, a differentiation of the position vector $\vec{r}$ and the center-of-mass coordinate $\vec{x}$ becomes convenient [see Eq. (12.140)]. The photon propagator, which is the propagator for the vector potential $\vec{A}$, can be translated into the propagator for the electric field by differentiation with respect to the time,

$$\begin{aligned} D_E^{ik}(x_1 - x_2) &= \frac{\partial}{\partial t_1}\frac{\partial}{\partial t_2}\left\langle 0\left|\mathrm{T}\left[A^i(\vec{r}_A, t_1)\, A^j(\vec{r}_B, t_2)\right]\right|0\right\rangle \\ &= \frac{\partial}{\partial(t_1 - t_2)}\frac{\partial}{\partial(t_2 - t_1)}\left(\mathrm{i}\, D^{ik}(x_1 - x_2)\right) = -\mathrm{i}\,\frac{\partial^2}{\partial t_1^2}\, D^{ik}(x_1 - x_2)\,. \end{aligned} \tag{12.12}$$

If we work in the mixed representation, we can implement the differentiation with respect to time in the Fourier integral as follows,

$$\left\langle 0\left|\mathrm{T}\left[E^i(\vec{r}_A, t_1)\, E^j(\vec{r}_B, t_2)\right]\right|0\right\rangle = D_E^{ik}(x_1 - x_2) = \mathrm{i}\int \frac{\mathrm{d}\omega}{2\pi}\,\omega^2\, D^{ik}(\omega, \vec{r})\,\mathrm{e}^{-\mathrm{i}\omega(t_1 - t_2)}\,. \tag{12.13}$$

This result will prove useful in the following.

The fourth-order contribution to the S-matrix is

$$S^{(4)} = \frac{(-\mathrm{i})^4}{4!}\int \mathrm{d}t_1 \int \mathrm{d}t_2 \int \mathrm{d}t_3 \int \mathrm{d}t_4\, \mathrm{T}[V(t_1)V(t_2)V(t_3)V(t_4)]\,. \tag{12.14}$$

According to the Wick theorem, the time-ordered product is equal to the normal ordered product, plus all contractions. So,

$$S^{(4)} = \frac{(-\mathrm{i})^4}{4!} \times 3 \times \prod_{\ell=1}^{4}\int \mathrm{d}t_\ell\, \langle\phi|(\langle 0|\mathrm{T}[V(t_1)V(t_2)|0\rangle\, \langle 0|\mathrm{T}[V(t_3)V(t_4)|0\rangle)|\phi\rangle\,. \tag{12.15}$$

We denote by $|\phi\rangle$ the reference state of the atoms, and by $|0\rangle$ the vacuum of the electromagnetic field. Now,

$$\begin{aligned} \langle 0|\mathrm{T}[V(t_1)V(t_2)|0\rangle \sim{}& \left\langle 0\left|\mathrm{T}\left[\left(\vec{E}(\vec{r}_A, t_1)\cdot\vec{d}_A(t_1)\right)\left(\vec{E}(\vec{r}_B, t_2)\cdot\vec{d}_B(t_2)\right)\right]\right|0\right\rangle \\ &+ \left\langle 0\left|\mathrm{T}\left[\left(\vec{E}(\vec{r}_B, t_1)\cdot\vec{d}_B(t_1)\right)\left(\vec{E}(\vec{r}_A, t_2)\cdot\vec{d}_A(t_2)\right)\right]\right|0\right\rangle\,, \end{aligned} \tag{12.16}$$

where by $\sim$ we denote the omission of operators which pertain to the self-energy of the two atoms (these would otherwise contain two A's or two B's in the indices), and keep only the terms relevant for the interaction energy. The time ordering of the electric-field operators results in the expression,

$$\begin{aligned} \langle 0|\mathrm{T}[V(t_1)V(t_2)|0\rangle \sim{}& \left\langle 0\left|\mathrm{T}\left[E^i(\vec{r}_A, t_1)\, E^j(\vec{r}_B, t_2)\right]\right|0\right\rangle d_A^i(t_1)\, d_B^j(t_2) \\ &+ \left\langle 0\left|\mathrm{T}\left[E^i(\vec{r}_B, t_1)\, E^j(\vec{r}_A, t_2)\right]\right|0\right\rangle d_B^i(t_1)\, d_A^j(t_2)\,, \end{aligned} \tag{12.17}$$

and analogously for $t_1 \to t_3$, and $t_2 \to t_4$, in the last line of Eq. (12.15). After the time ordering of the atomic dipole operators [the T symbol in Eq. (12.16) refers to both field as well as dipole operators], for the times t_1, t_2, t_3 and t_4, we end up with four contributions,

$$C_1 \equiv \langle n_A|\mathrm{T}\, d^i_A(t_1)\, d^k_A(t_3)|n_A\rangle\, \langle n_B|\mathrm{T}\, d^j_B(t_2)\, d^\ell_B(t_4)|n_B\rangle\,, \tag{12.18a}$$

$$C_2 \equiv \langle n_A|\mathrm{T}\, d^i_A(t_1)\, d^\ell_A(t_4)|n_A\rangle\, \langle n_B|\mathrm{T}\, d^j_B(t_2)\, d^k_B(t_3)|n_B\rangle\,, \tag{12.18b}$$

$$C_3 \equiv \langle n_B|\mathrm{T}\, d^i_B(t_1)\, d^\ell_B(t_4)|n_B\rangle\, \langle n_A|\mathrm{T}\, d^j_A(t_2)\, d^k_A(t_3)|n_A\rangle\,, \tag{12.18c}$$

$$C_4 \equiv \langle n_B|\mathrm{T}\, d^i_B(t_1)\, d^k_B(t_3)|n_B\rangle\, \langle n_A|\mathrm{T}\, d^j_A(t_2)\, d^\ell_A(t_4)|n_A\rangle\,. \tag{12.18d}$$

Contributions C_2 and C_4 correspond to the crossed-ladder diagram (in the language of Feynman diagrams), whereas C_1 and C_3 correspond to the two-photon ladder exchange (see also Fig. 12.1). The contributions of atoms A and B to the atomic reference state are denoted as $|n_A\rangle$ and $|n_B\rangle$, respectively (see Fig. 12.1). With all terms C_1, C_2, C_3, and C_4 leading to equivalent contributions, we finally arrive at

$$S^{(4)} = \frac{1}{2} \prod_{\ell=1}^{4} \int \mathrm{d}t_\ell \left\langle 0\left|\mathrm{T}\left[E^i(\vec r_A, t_1)\, E^j(\vec r_B, t_2)\right]\right|0\right\rangle \left\langle 0\left|\mathrm{T}\left[E^k(\vec r_A, t_3)\, E^\ell(\vec r_B, t_4)\right]\right|0\right\rangle$$
$$\times \left\langle n_A\left|\mathrm{T}\, d^i_A(t_1)\, d^k_A(t_3)\right|n_A\right\rangle \left\langle n_B\left|\mathrm{T}\, d^j_B(t_2)\, d^\ell_B(t_4)\right|n_B\right\rangle\,. \tag{12.19}$$

The time-ordered product of electric-field operators can be evaluated as follows,

$$D^{ik}_E(x_1 - x_2) = \left\langle 0\left|\mathrm{T}\left[E^i(\vec r_A, t_1)\, E^k(\vec r_B, t_2)\right]\right|0\right\rangle\,, \tag{12.20}$$

where $D^{ik}_E(x_1 - x_2)$ is given in Eq. (12.12).

Now, let us proceed to the time-ordered product of dipole operators, which is given as follows (for atom A),

$$\alpha_{A,ik}(t_1 - t_2) = \mathrm{i}\left\langle n_A\left|\mathrm{T}(d^i_A(t_1)\, d^k_A(t_2)\right|n_A\right\rangle\,, \qquad \alpha^{ik}_A(t) = \int_{-\infty}^{\infty} \frac{\mathrm{d}\omega}{2\pi}\, \mathrm{e}^{-\mathrm{i}\,\omega\, t}\, \alpha^{ik}_A(\omega)\,, \tag{12.21}$$

so that

$$\left\langle n_A\left|\mathrm{T}\, d^i_A(t_1)\, d^k_A(t_2)\right|n_A\right\rangle = -\mathrm{i}\,\alpha^{ik}_A(t_1 - t_2) = -\mathrm{i}\int_{-\infty}^{\infty} \frac{\mathrm{d}\omega}{2\pi}\, \mathrm{e}^{-\mathrm{i}\,\omega\,(t_1 - t_2)}\, \alpha^{ik}_A(\omega)\,. \tag{12.22}$$

The time-ordered product of dipole operators can be evaluated in terms of the polarizability of the atom, with the pole being displaced according to the Feynman prescription,

$$\begin{aligned}
\alpha^{ik}_A(\omega) &= \int_{-\infty}^{\infty} \mathrm{d}t\, \mathrm{e}^{\mathrm{i}\omega t}\, \alpha^{ik}_A(t) = \mathrm{i}\int_{-\infty}^{\infty} \mathrm{d}t\, \mathrm{e}^{\mathrm{i}\omega t} \left\langle n_A\left|\mathrm{T}(d^i_A(t)\, d^k_A(0)\right|n_A\right\rangle \\
&= \mathrm{i}\int_0^{\infty} \mathrm{d}t\, \mathrm{e}^{\mathrm{i}\omega t} \left\langle n_A\left|d^i_A(t)\, d^k_A(0)\right|n_A\right\rangle + \mathrm{i}\int_{-\infty}^{0} \mathrm{d}t\, \mathrm{e}^{\mathrm{i}\omega t} \left\langle n_A\left|d^k_A(0)\, d^i_A(t)\right|n_A\right\rangle \\
&= \sum_{m_A} \left(\frac{\left\langle n_A\left|d^i_A\right|m_A\right\rangle \left\langle m_A\left|d^k_A\right|n_A\right\rangle}{E_{mn,A} - \omega - \mathrm{i}\epsilon} + \frac{\left\langle n_A\left|d^i_A\right|m_A\right\rangle \left\langle m_A\left|d^k_A\right|n_A\right\rangle}{E_{mn,A} + \omega - \mathrm{i}\epsilon} \right),
\end{aligned} \tag{12.23}$$

where $E_{mn,A} = E_{m_A} - E_{n_A}$. We recognize the polarizability, with poles displaced according to the Feynman prescription. The attentive reader will have noticed that the sequence of the indicated i and k in the last term has been inverted in going from the second to the third line of Eq. (12.23). This requires an explanation. For a nonrelativistic atomic state, one can show that, taking into account the summation over the m_A, and using the formalism of reduced matrix elements outlined Chap. 6,

$$\frac{\langle n_A | d_A^k | m_A \rangle \langle m_A | d_A^i | n_A \rangle}{E_{mn,A} + \omega - \mathrm{i}\epsilon} = \frac{\langle n_A | d_A^i | m_A \rangle \langle m_A | d_A^k | n_A \rangle}{E_{mn,A} + \omega - \mathrm{i}\epsilon} . \tag{12.24}$$

Details are left as an exercise to the reader. The positive-energy poles occur at $\omega = E_{mn,A} - \mathrm{i}\epsilon$, while the negative-energy poles are at $\omega = -E_{mn,A} + \mathrm{i}\epsilon$. If the virtual state is displaced toward lower energy, i.e., $E_{mn,A} < 0$, then the pole at $\omega = -E_{mn,A} + \mathrm{i}\epsilon$ migrates into the first quadrant of the complex plane. Note that the last transformation in Eq. (12.23) holds provided the matrix elements in the numerator are real, in which case we can formulate the Hermitian adjoint.

We now reformulate Eq. (12.15), with the help of Eq. (12.22), and approximate $\vec{r}_A \approx \vec{R}_A$, and $\vec{r}_B \approx \vec{R}_B$, as well as $\vec{R} = \vec{R}_A - \vec{R}_B$, for the interaction in the dipole approximation,

$$\begin{aligned} S^{(4)} &= \frac{1}{2} \prod_{\ell=1}^{4} \int \mathrm{d}t_\ell \left\langle 0 \left| \mathrm{T}\left[E^i(\vec{R}_A, t_1) \, E^j(\vec{R}_B, t_2) \right] \right| 0 \right\rangle \left\langle 0 \left| \mathrm{T}\left[E^k(\vec{R}_A, t_3) \, E^\ell(\vec{R}_B, t_4) \right] \right| 0 \right\rangle \\ &\quad \times \left\langle n_A \left| \mathrm{T} \, d_A^i(t_1) \, d_A^k(t_3) \right| n_A \right\rangle \left\langle n_B \left| \mathrm{T} \, d_B^j(t_2) \, d_B^\ell(t_4) \right| n_B \right\rangle \\ &= \frac{1}{2} \prod_{\ell=1}^{4} \int \mathrm{d}t_\ell \prod_{s=1}^{4} \int \frac{\mathrm{d}\omega_s}{2\pi} \, \omega_1^2 \, \omega_2^2 \, D^{ij}(\omega_1, \vec{R}) \, D^{k\ell}(\omega_2, \vec{R}) \, \alpha_A^{ik}(\omega_3) \, \alpha_B^{j\ell}(\omega_4) \\ &\quad \times \mathrm{e}^{-\mathrm{i}\omega_1(t_1 - t_2) - \mathrm{i}\omega_2(t_3 - t_4) - \mathrm{i}\omega_3(t_1 - t_3) - \mathrm{i}\omega_4(t_2 - t_4)} . \end{aligned} \tag{12.25}$$

One now carries out the $\mathrm{d}t_i$ integration one after the other, with the prescriptions $\int \mathrm{d}t_2 \to 2\pi\delta(\omega_1 - \omega_4)$, then $\int \mathrm{d}t_3 \to 2\pi\delta(\omega_2 - \omega_3)$, and $\int \mathrm{d}t_4 \to 2\pi\delta(\omega_2 + \omega_4)$. As a result, the condition $\omega_1 = \omega_4 = -\omega_2 = -\omega_3$ is implemented in the final result,

$$\begin{aligned} S^{(4)} &= \frac{1}{2} \int \mathrm{d}t_1 \int \frac{\mathrm{d}\omega_1}{2\pi} \, \omega_1^2 \, (-\omega_1)^2 \, D^{ij}(\omega_1, \vec{R}) \, D^{k\ell}(-\omega_1, \vec{R}) \, \alpha^{ik}(-\omega_1) \, \alpha^{j\ell}(\omega_1) \\ &= \frac{T}{2} \int \frac{\mathrm{d}\omega}{2\pi} \, \omega^4 \, D^{ij}(\omega, \vec{R}) \, D^{k\ell}(\omega, \vec{R}) \, \alpha_A^{ik}(\omega) \, \alpha_B^{j\ell}(\omega) . \end{aligned} \tag{12.26}$$

The matching of the effective potential with the scattering amplitude according to Eq. (12.5) results in the interaction energy

$$\begin{aligned} \Delta E &= \frac{\mathrm{i}}{2} \int_{-\infty}^{\infty} \frac{\mathrm{d}\omega}{2\pi} \, \omega^4 \, D^{ij}(\omega, \vec{R}) \, D^{k\ell}(\omega, \vec{R}) \, \alpha_A^{ik}(\omega) \, \alpha_B^{j\ell}(\omega) \\ &= \mathrm{i} \int_0^{\infty} \frac{\mathrm{d}\omega}{2\pi} \, \omega^4 \, D^{ij}(\omega, \vec{R}) \, D^{k\ell}(\omega, \vec{R}) \, \alpha_A^{ik}(\omega) \, \alpha_B^{j\ell}(\omega) , \end{aligned} \tag{12.27}$$

where the latter equality surprisingly holds because of perfect $\omega \leftrightarrow -\omega$ symmetry for the integrand; both the photon propagator as well as the atomic polarizability have that symmetry. We note that the Feynman prescription is crucial in establishing the perfect symmetry of $\alpha_{A,ik}(\omega)$ under the transformation $\omega \leftrightarrow -\omega$.

The polarizabilities in Eq. (12.27) are used according to the Feynman prescription,

$$\alpha_A^{ij}(\omega) = \sum_{m_A} \left(\frac{\langle n_A | d_A^i | m_A \rangle \langle m_A | d_A^j | n_A \rangle}{E_{m,A} - E_{n,A} - \omega - \mathrm{i}\epsilon} + \frac{\langle n_A | d_A^i | m_A \rangle \langle m_A | d_A^j | n_A \rangle}{E_{m,A} - E_{n,A} + \omega - \mathrm{i}\epsilon} \right), \quad (12.28\text{a})$$

$$\alpha_B^{ij}(\omega) = \sum_{m_B} \left(\frac{\langle n_B | d_B^i | m_B \rangle \langle m_B | d_B^j | n_B \rangle}{E_{q,B} - E_{n,B} - \omega - \mathrm{i}\epsilon} + \frac{\langle n_B | d_B^i | m_B \rangle \langle m_B | d_B^j | n_B \rangle}{E_{q,B} - E_{n,B} + \omega - \mathrm{i}\epsilon} \right). \quad (12.28\text{b})$$

Here, n_A and n_B are the atomic reference states, and m_A and m_B are the virtual states (which can be of lower energy). The tensor structure in Eq. (12.27) amounts to

$$D^{ij}(\omega, \vec{R})\, D^{ji}(\omega, \vec{R}) = \frac{2\, \mathrm{e}^{2\mathrm{i}|\omega| r}}{r^2} \left(1 + \frac{2\mathrm{i}}{|\omega| R} - \frac{5}{(\omega R)^2} - \frac{6\mathrm{i}}{(|\omega| R)^3} + \frac{3}{(\omega R)^4} \right). \quad (12.29)$$

For ground-state atoms, with $\alpha_A^{ij}(\omega) = \delta^{ij}\, \alpha_A(\omega)$ and $\alpha_B^{ij}(\omega) = \delta^{ij}\, \alpha_B(\omega)$, a Wick rotation of the expression (12.27) leads to

$$\Delta E = -\int_0^\infty \mathrm{d}\omega\, \frac{\mathrm{e}^{-2\omega R}\, \omega^4}{R^2} \left(1 + \frac{2}{\omega R} + \frac{5}{(\omega R)^2} + \frac{6}{(\omega R)^3} + \frac{3}{(\omega R)^4} \right) \alpha_A(\mathrm{i}\omega)\, \alpha_B(\mathrm{i}\omega), \quad (12.30)$$

confirming the result in Eq. (5.117). For excited-state atoms, the Feynman method dictates the positioning of the poles, which leads to long-range tails [209, 212, 361]. We have demonstrated the power of the Feynman formalism for the calculation of the effective interaction between atoms.

12.3 From the One-Body to the Two-Body Hamiltonian

12.3.1 *Relevant Hamiltonians*

Now, we apply the matching formalism (introduced in Sec. 12.2.1) to the interaction between two charged particles. We start from the effective Hamiltonian obtained by adding the terms $H^{[2]}$ and $H^{[3]}$ given in Eq. (11.55), and the terms $H^{[4]}$ and $H^{[5]}$ obtained in Eqs. (11.57) and (11.58), respectively. Using the relation (11.56) and assuming a spatially slowly varying magnetic field, which justifies the approximation

$$-\frac{(\vec{\sigma} \cdot \vec{\pi})^4}{8m^3} + \frac{e\kappa}{16m^3} \{\vec{\sigma} \cdot \vec{\pi}, \{\vec{\sigma} \cdot \vec{\pi}, \vec{\sigma} \cdot \vec{B}\}\} \approx -\frac{\vec{\pi}^4}{8m^3} + \frac{e}{8m^3} \left(\{\vec{\sigma} \cdot \vec{B}, \vec{\pi}^2\} + \kappa \{\vec{\pi} \cdot \vec{B}, \vec{\sigma} \cdot \vec{\pi}\} \right), \quad (12.31)$$

one can write the following general Hamiltonian, which contains all the α^4 terms as well as the α^5 corrections due to the anomalous magnetic moment,

$$\begin{aligned} H_{\mathrm{FW}} = {} & \frac{\vec{\pi}^2}{2m} + e\, A^0 - \frac{e}{2m}(1 + \kappa)\, \vec{\sigma} \cdot \vec{B} - \frac{\vec{\pi}^4}{8m^3} \\ & - \frac{e}{8m^2}(1 + 2\kappa) \left[\vec{\nabla} \cdot \vec{E} + \vec{\sigma} \cdot \left(\vec{E} \times \vec{\pi} - \vec{\pi} \times \vec{E} \right) \right] \\ & + \frac{e}{8m^3} \left(\{\vec{\sigma} \cdot \vec{B}, \vec{\pi}^2\} + \kappa\, \{\vec{\pi} \cdot \vec{B}, \vec{\pi} \cdot \vec{\sigma}\} \right). \end{aligned} \quad (12.32)$$

The $\vec{E}$ and $\vec{B}$ fields are related to the scalar and vector potentials as follows,

$$\begin{aligned}\vec{E} &= -\vec{\nabla}A^0 - \partial_t \vec{A}\,,\\ \vec{B} &= \vec{\nabla}\times\vec{A}\,.\end{aligned} \tag{12.33}$$

We take into account the lack of an explicit time dependence in Eq. (12.32), except for the electromagnetic fields. The interaction Hamiltonian is used in the representation given in Eq. (11.98),

$$H_I = e\,j^\mu\,A_\mu = e\,j^\mu(\vec{r},\mathrm{i}\partial_{t'},-\mathrm{i}\vec{\nabla}')\,A_\mu(t',\vec{r}')\big|_{\vec{r}'=\vec{r}} = e\,j^\mu(\vec{r},\mathrm{i}\partial_{x'})\,A_\mu(x')\big|_{x'=x}\,. \tag{12.34}$$

Our notation in Eq. (12.34) is a little more precise than in Eq. (11.98). Namely, we indicate the fact that the current operators in Eqs. (11.108) and (11.119) contain differential operators acting on the photon vector potential. After the differentiation with respect to the arguments of the vector potential A_μ, the locality of the current dictates that we set the primed coordinates equal to the unprimed ones. The differential operators with respect to the arguments of A_μ are denoted as $(\mathrm{i}\partial_{t'},-\mathrm{i}\vec{\nabla}')$, and summarized in the expression $\mathrm{i}\partial'_x$. It is, nevertheless, necessary to distinguish the differentiation with respect to the primed coordinates from the unprimed ones, because the details of the functional dependence determine the behavior in momentum space [see Eq. (12.37)].

12.3.2 *Matching the Two-Particle Interaction Hamiltonian*

The matching of the scattering amplitude for the interaction of two electrons, labeled a and b, against the effective Hamiltonian, is a little harder to derive than one might otherwise expect. We start from a second-order matrix element and consider the transition of electron a from the initial state $|i_a\rangle$ to the final state $|f_a\rangle$, and the transition of electron b from the initial state $|i_b\rangle$ to the final state $|f_b\rangle$. The procedure is slightly complicated by the fact that the Foldy–Wouthuysen transformed current operators contain momentum operators acting on the four-vector potential of the photon.

Following Eq. (12.34), the interaction Hamiltonian is (*nota bene* the distinction of J^μ versus j^μ, and the fact that the differential operator $\mathrm{i}\partial_{x'}$ denotes the corresponding four-vector with components $\mathrm{i}\partial/\partial x'_\rho$, where $\rho = 0,1,2,3$)

$$\mathcal{H} = J^\mu_a(x,\mathrm{i}\partial_{x'})A_\mu(x') + J^\mu_b(x,\mathrm{i}\partial_{x'})A_\mu(x')\big|_{x'=x}\,. \tag{12.35}$$

The transition currents read as follows,

$$\begin{aligned}J^\mu_a(x,\mathrm{i}\partial_{x'}) &= e\,\phi^+_{f_a}(x)\,j^\mu(\vec{r},\mathrm{i}\partial_{x'})\,\phi_{i_a}(x)\\ &= e\,\mathrm{e}^{-\mathrm{i}(E_{i_a}-E_{f_a})t}\;\phi^+_{f_a}(\vec{r})\,j^\mu(\vec{r},\mathrm{i}\partial_{x'})\,\phi_{i_a}(\vec{r})\,,\end{aligned} \tag{12.36a}$$

$$\begin{aligned}J^\mu_b(x,\mathrm{i}\partial_{x'}) &= e\,\phi^+_{f_b}(x)\,j^\mu(\vec{r},\mathrm{i}\partial_{x'})\,\phi_{i_b}(x)\\ &= e\,\mathrm{e}^{-\mathrm{i}(E_{i_b}-E_{f_b})t}\;\phi^+_{f_b}(\vec{r})\,j^\mu(\vec{r},\mathrm{i}\partial_{x'})\,\phi_{i_b}(\vec{r})\,.\end{aligned} \tag{12.36b}$$

Once more, we understand that the primed operators act on the four-vector potential, before setting the primed coordinates equal to the unprimed ones.

One considers the emergence of two equivalent terms because of the two terms in the interaction Hamiltonian, and writes the second-order S matrix element as follows,

$$\begin{aligned}
S_{fi} &= 2 \times \left\{ -\frac{1}{2!} \int \mathrm{d}^4 x_a \int \mathrm{d}^4 x_b \, \mathrm{T} \left[J_a^\mu(x_a, \mathrm{i}\partial_{x'_a}) \, J_b^\mu(x_a, \mathrm{i}\partial_{x'_b}) \right] \right. \\
&\quad \left. \times \langle 0 | \mathrm{T}[A_\mu(x'_a) \, A_\nu(x'_b)] | 0 \rangle \Big|_{x'_a = x_a, x'_b = x_b} \right\} \\
&= -\int \mathrm{d}^4 x_a \int \mathrm{d}^4 x_b \int \frac{\mathrm{d}^4 k}{(2\pi)^4} \mathrm{T} \left[J_a^\mu(x_a, k) \, J_b^\nu(x_b, -k) \right] \, \mathrm{i} D_{\mu\nu}(k) \, \mathrm{e}^{-\mathrm{i}k(x_a - x_b)} \\
&= -\mathrm{i} \int \frac{\mathrm{d}^4 k}{(2\pi)^4} D_{\mu\nu}(k) \, Q^{\mu\nu}(k) \,, \qquad (12.37)
\end{aligned}$$

where we have used Eq. (9.120) in the form

$$\langle 0 | \mathrm{T} \left[A_\mu(x_a) \, A_\nu(x_b) \right] | 0 \rangle = \mathrm{i} D_{\mu\nu}(x_a - x_b) = \mathrm{i} \int \frac{\mathrm{d}^4 k}{(2\pi)^4} D_{\mu\nu}(k) \, \mathrm{e}^{-\mathrm{i}k \cdot (x_a - x_b)} \,, \qquad (12.38)$$

and we define

$$Q^{\mu\nu}(k) = \int \mathrm{d}^4 x_a \int \mathrm{d}^4 x_b \, \mathrm{e}^{-\mathrm{i}k(x_a - x_b)} \, \mathrm{T} \left[J_a^\mu(x_a, k) \, J_b^\nu(x_b, -k) \right] . \qquad (12.39)$$

We write the four-vectors as $x_a^\mu = (t_a, \vec{r}_a)$ and $x_b^\mu = (t_b, \vec{r}_b)$. Then,

$$\begin{aligned}
Q^{\mu\nu}(k) &= \Theta(t_a - t_b) e^2 \int \mathrm{d}t_a \int \mathrm{d}t_b \int \mathrm{d}^3 r_a \int \mathrm{d}^3 r_b \phi^+_{f_a}(\vec{r}_a) j^\mu(\vec{r}_a, k) \mathrm{e}^{\mathrm{i}\vec{k}\cdot\vec{r}_a} \phi_{i_a}(\vec{r}_a) \\
&\quad \times \phi^+_{f_b}(\vec{r}_b) j^\mu(\vec{r}_b, -k) \mathrm{e}^{-\mathrm{i}\vec{k}\cdot\vec{r}_b} \phi_{i_b}(\vec{r}_b) \mathrm{e}^{-\mathrm{i}(E_{i_a} - E_{f_a})t_a} \, \mathrm{e}^{-\mathrm{i}(E_{i_b} - E_{f_b})t_b} \, \mathrm{e}^{-\mathrm{i}\omega(t_a - t_b)} \\
&\quad + \Theta(t_b - t_a) \, e^2 \int \mathrm{d}t_a \int \mathrm{d}t_b \int \mathrm{d}^3 r_a \int \mathrm{d}^3 r_b \phi^+_{f_b}(\vec{r}_b) j^\nu(\vec{r}_b, -k) \mathrm{e}^{-\mathrm{i}\vec{k}\cdot\vec{r}_b} \phi_{i_b}(\vec{r}_b) \\
&\quad \times \phi^+_{f_a}(\vec{r}_a) j^\mu(\vec{r}_a, k) \mathrm{e}^{\mathrm{i}\vec{k}\cdot\vec{r}_a} \phi_{i_a}(\vec{r}_a) \mathrm{e}^{-\mathrm{i}(E_{i_a} - E_{f_a})t_a} \mathrm{e}^{-\mathrm{i}(E_{i_b} - E_{f_b})t_b} \mathrm{e}^{-\mathrm{i}\omega(t_a - t_b)} \\
&= T \int \mathrm{d}^3 r_a \int \mathrm{d}^3 r_b \, \frac{\phi^+_{f_a}(\vec{r}_a) \, j^\mu(x_a, k) \phi_{i_a}(\vec{r}_a) \phi^+_{f_b}(\vec{r}_b) j^\nu(\vec{r}_b, -k) \phi_{i_b}(\vec{r}_b)}{\mathrm{i}(E_{i_a} - E_{f_a} + \omega - \mathrm{i}\epsilon)} \mathrm{e}^{\mathrm{i}\vec{k}\cdot\vec{r}_{ab}} \\
&\quad + T \int \mathrm{d}^3 r_a \int \mathrm{d}^3 r_b \, \frac{\phi^+_{f_b}(\vec{r}_b) j^\nu(\vec{r}_b, -k) \phi_{i_b}(\vec{r}_b) \phi^+_{f_a}(\vec{r}_a) j^\mu(x_a, k) \phi_{i_a}(\vec{r}_a)}{-\mathrm{i}(E_{i_a} - E_{f_a} + \omega + \mathrm{i}\epsilon)} \mathrm{e}^{\mathrm{i}\vec{k}\cdot\vec{r}_{ab}} \,,
\end{aligned} \qquad (12.40)$$

where $\omega = k^0$. Furthermore, we define $\vec{r}_{ab} = \vec{r}_a - \vec{r}_b$. The Jacobian of the substitutions

$$(t_a, t_b) \to (\tau, t_a - t_b) \,, \qquad (t_a, t_b) \to (\tau, t_b - t_a) \,, \qquad (12.41)$$

is just equal to unity. The long time interval is $\int \mathrm{d}\tau = T$ (as opposed to the time ordering operator T). We have used the fact that, in order to generate the factor T which we need for the matching with the effective Hamiltonian, we need the relations

$$E_{f_a} - E_{i_a} = E_{i_b} - E_{f_b} \,, \qquad E_{f_a} + E_{f_b} = E_{i_a} + E_{i_b} \,. \qquad (12.42)$$

Recalling Eq. (12.37), one matches with the effective Hamiltonian H_{ab} proceeds using the formula,

$$S_{fi} = -\mathrm{i} T \, \langle \phi_f | H_{ab} | \phi_i \rangle \,, \qquad (12.43)$$

and obtains

$$\langle\phi_f|H_{ab}|\phi_i\rangle = \frac{\mathrm{i}}{T}\left(-\mathrm{i}e^2\int\frac{\mathrm{d}^4k}{(2\pi)^4}\,D_{\mu\nu}(k)\,Q^{\mu\nu}(k)\right) = -\mathrm{i}e^2\int\mathrm{d}^3r_a\int\mathrm{d}^3r_b$$
$$\times\int\frac{\mathrm{d}^4k}{(2\pi)^4}\,D_{\mu\nu}(k)\,\mathrm{e}^{\mathrm{i}\vec{k}\cdot\vec{r}_{ab}}\left[\frac{\phi^+_{f_a}(\vec{r}_a)\,j^\mu(\vec{r}_a,k)\,\phi_{i_a}(\vec{r}_a)\,\phi^+_{f_b}(\vec{r}_b)\,j^\nu(\vec{r}_b,-k)\,\phi_{i_b}(\vec{r}_b)}{\omega-(E_{f_a}-E_{i_a})-\mathrm{i}\epsilon}\right.$$
$$\left.-\frac{\phi^+_{f_b}(\vec{r}_b)\,j^\nu(\vec{r}_b,-k)\,\phi_{i_b}(\vec{r}_b)\,\phi^+_{f_a}(\vec{r}_a)\,j^\mu(\vec{r}_a,k)\,\phi_{i_a}(\vec{r}_a)}{\omega-(E_{f_a}-E_{i_a})+\mathrm{i}\epsilon}\right]. \tag{12.44}$$

One now closes the integration contour over the photon energy in the upper half of the complex plane, and ignores the pole due to the photon propagator which otherwise leads to a self-energy type effect contributing to the recoil correction

Furthermore, one neglects the resonance frequency $E_{f_a} - E_{i_a}$ of the transition. This can be justified as follows. Picking up the poles given in Eq. (12.44) results in photon propagator expressions of the form $1/[(E_{f_a} - E_{i_a})^2 - \vec{k}^2]$. The Coulomb interaction is mediated by virtual photons with a momentum $|\vec{k}| \sim (Z\alpha)\,m$, while atomic-state energy difference are of the order of $E_{f_a} - E_{i_a} \sim (Z\alpha)^2\,m$ (in natural units with $\hbar = c = \epsilon_0 = 1$). Hence, we can ignore $(E_{f_a} - E_{i_a})^2 \ll \vec{k}^2$ to the order of approximation we are interested in. This corresponds to the approximation of a static interaction, and leads to the Breit Hamiltonian [see also task (1) in Sec. 12.7].

The result, in view of the residue theorem, is

$$\langle\phi_f|H_{ab}|\phi_i\rangle = \int\mathrm{d}^3r_a\int\mathrm{d}^3r_b\,\phi^+_{f_a}(\vec{r}_a)\,\phi^+_{f_b}(\vec{r}_b)$$
$$\times\left\{e^2\int\frac{\mathrm{d}^3k}{(2\pi)^3}D_{\mu\nu}(k)\mathrm{e}^{\mathrm{i}\vec{k}\cdot\vec{r}_{ab}}\,j^\mu(\vec{r}_a,k)\,j^\nu(\vec{r}_b,-k)\right\}\Bigg|_{k^0=0}\phi_{i_a}(\vec{r}_a)\phi_{i_b}(\vec{r}_b). \tag{12.45}$$

We can read off the effective two-body interaction Hamiltonian

$$H_{ab} = e^2\int\frac{\mathrm{d}^3k}{(2\pi)^3}\,D_{\mu\nu}(\vec{k})\,j^\mu_a(\vec{k})\,\mathrm{e}^{\mathrm{i}\,\vec{k}\cdot(\vec{r}_a-\vec{r}_b)}\,j^\nu_b(-\vec{k})\Bigg|_{k^0=0}, \tag{12.46}$$

where we use the short-hand notations defined in Eqs. (11.112) and (11.120). This expression will be the basis for our evaluation of the interelectronic and electron-nucleus interaction.

12.3.3 *Photon Exchange and Breit Hamiltonian*

In order to proceed with the evaluation of H_{ab} given in Eq. (12.46), we use the photon propagator in Coulomb gauge, and recall Eq. (9.73) for convenience,

$$D_{\mu\nu}(k) = \begin{cases} \dfrac{1}{\vec{k}^2} & \mu=\nu=0 \\ \dfrac{1}{\omega^2-\vec{k}^2+\mathrm{i}\,\epsilon}\left(\delta_{ij}-\dfrac{k_i\,k_j}{\vec{k}^2}\right) & \mu=i,\ \nu=j \\ 0 & \mu\neq\nu\,. \end{cases} \tag{12.47}$$

We employ the definition $\vec{r}_{ab} = \vec{r}_a - \vec{r}_b$. The Fourier transform

$$D_{\mu\nu}(\vec{r}_{ab}) = \int \frac{\mathrm{d}^3 k}{(2\pi)^3}\, D_{\mu\nu}(\vec{k})\, \mathrm{e}^{\mathrm{i}\,\vec{k}\cdot\vec{r}_{ab}} \tag{12.48}$$

is found as follows,

$$D_{\mu\nu}(\vec{r}_{ab}) = \begin{cases} \dfrac{1}{4\pi r_{ab}} & \mu = \nu = 0 \\ -\dfrac{1}{8\pi r_{ab}} \left(\delta^{ij} - \dfrac{r_a^i\, r_b^j}{r_{ab}^2}\right) & \mu = i, \quad \nu = j \\ 0 & \mu \neq \nu \end{cases} . \tag{12.49}$$

We recall, from Eq. (11.112), the time-like component of the current,

$$j^0(\vec{k}) = 1 + \frac{\mathrm{i}\,(1+2\kappa)}{4m^2}\, \vec{\sigma}\cdot(\vec{k}\times\vec{\pi}) - \frac{1+2\kappa}{8m^2}\, \vec{k}^2 . \tag{12.50}$$

In comparison to Eq. (11.112), we have replaced $\vec{\pi}_{\rm ext} \to \vec{\pi}$, identifying the quantity $\vec{\pi}$ with a canonical momentum that contains a vector potential from external sources. Finally, one obtains the matrix element of the effective Hamiltonian,

$$\begin{aligned} H_{ab}^{(00)} &= e^2 \int \frac{\mathrm{d}^3 k}{(2\pi)^3}\, D_{00}(\vec{k})\, j_a^0(\vec{k})\, \mathrm{e}^{\mathrm{i}\,\vec{k}\cdot(\vec{r}_a - \vec{r}_b)}\, j_b^0(-\vec{k}) \\ &= e^2 \int \frac{\mathrm{d}^3 k}{(2\pi)^3} \left[\frac{\mathrm{e}^{\mathrm{i}\,\vec{k}\cdot(\vec{r}_a - \vec{r}_b)}}{\vec{k}^2} + \frac{\mathrm{i}\,(1+2\kappa)}{4m^2}\, \vec{\sigma}_a\cdot(\vec{k}\times\vec{\pi}_a) \frac{\mathrm{e}^{\mathrm{i}\,\vec{k}\cdot(\vec{r}_a - \vec{r}_b)}}{\vec{k}^2} \right. \\ &\quad \left. + \frac{\mathrm{e}^{\mathrm{i}\,\vec{k}\cdot(\vec{r}_a - \vec{r}_b)}}{\vec{k}^2} \frac{\mathrm{i}\,(1+2\kappa)}{4m^2}\, \vec{\sigma}_b\cdot(-\vec{k}\times\vec{\pi}_b) - \frac{1+2\kappa}{4m^2}\, \mathrm{e}^{\mathrm{i}\,\vec{k}\cdot(\vec{r}_a - \vec{r}_b)} \right] \\ &= 4\pi\alpha \left[\frac{1}{4\pi|\vec{r}_a - \vec{r}_b|} + \frac{\mathrm{i}\,(1+2\kappa)}{4m^2}\, \vec{\sigma}_a\cdot(-\mathrm{i}\vec{\nabla}_a\times\vec{\pi}_a) \frac{1}{4\pi|\vec{r}_a - \vec{r}_b|} \right. \\ &\quad \left. + \frac{\mathrm{i}\,(1+2\kappa)}{4m^2}\, \vec{\sigma}_b\cdot(\mathrm{i}\vec{\nabla}_a\times\vec{\pi}_b) \frac{1}{4\pi|\vec{r}_a - \vec{r}_b|} - \frac{1+2\kappa}{4m^2}\, \delta^{(3)}(\vec{r}_a - \vec{r}_b) \right] . \end{aligned} \tag{12.51}$$

We have repeatedly used the correspondence $\vec{k} \Leftrightarrow -\mathrm{i}\,\vec{\nabla}$, and the convention that the operators $\vec{\nabla}_a = \partial/\partial\vec{r}_a$ act on the explicit $\vec{r}_a$-dependence of the potential factors, while the operators $\vec{\pi}_a$ and $\vec{\pi}_b$ act on the electron (wave function) coordinates. One may consult Eqs. (12.34) and (11.108) for the distinction between the "primed" gradient operators (here, the $\vec{\nabla}_a$, which act on the potentials) and the "unprimed" ones (here, the $\vec{\pi}_a$, which act on the wave functions). Using, again, the definition $\vec{r}_{ab} = \vec{r}_a - \vec{r}_b$, we have

$$H_{ab}^{(00)} = \frac{\alpha}{r_{ab}} - \frac{\pi\alpha}{m^2}\,(1+2\kappa)\,\delta^{(3)}(\vec{r}_{ab}) + \frac{\alpha(1+2\kappa)}{4m^2 r_{ab}^3} \left[\vec{\sigma}_b\cdot(\vec{r}_{ab}\times\vec{\pi}_b) - \vec{\sigma}_a\cdot(\vec{r}_{ab}\times\vec{\pi}_a)\right] . \tag{12.52}$$

We recall the spatial components of the current from Eq. (11.120),

$$\vec{j}(\vec{k}) = \frac{\vec{\pi}}{m} + \frac{\mathrm{i}\,(1+\kappa)}{2m}\,(\vec{\sigma}\times\vec{k}) . \tag{12.53}$$

These components generate an effective Hamiltonian $H_{ab}^{(ij)}$,

$$
\begin{aligned}
H_{ab}^{(ij)} &= e^2 \int \frac{\mathrm{d}^3 k}{(2\pi)^3}\, D_{ij}(\vec{k})\, j_a^i(\vec{k})\, \mathrm{e}^{\mathrm{i}\vec{k}\cdot(\vec{r}_a-\vec{r}_b)}\, j_b^j(-\vec{k}) \\
&= e^2 \int \frac{\mathrm{d}^3 k}{(2\pi)^3} \left(\frac{\pi_a^i}{m} + \frac{\mathrm{i}\,(1+\kappa)}{2m}\,(\vec{\sigma}\times\vec{k})^i\right)\left[-\frac{\mathrm{e}^{\mathrm{i}\vec{k}\cdot(\vec{r}_a-\vec{r}_b)}}{\vec{k}^2}\left(\delta^{ij} - \frac{k^i\,k^j}{\vec{k}^2}\right)\right] \\
&\quad \times \left(\frac{\pi_b^j}{m} + \frac{\mathrm{i}\,(1+\kappa)}{2m}\,(\vec{\sigma}\times(-\vec{k}))^j\right) = \mathcal{T}_1 + \mathcal{T}_2 + \mathcal{T}_3 + \mathcal{T}_4\,. \qquad (12.54)
\end{aligned}
$$

The expression for $\mathcal{T}_1$ is obtained immediately, while the terms $\mathcal{T}_2$, $\mathcal{T}_3$ and $\mathcal{T}_4$ require a more elaborate treatment of the Fourier (back-)transformation from momentum to coordinate space,

$$
\begin{aligned}
\mathcal{T}_1 &= e^2 \frac{\pi_a^i}{m}\left[-\frac{1}{2r_{ab}}\left(\delta^{ij} - \frac{r_a^i\,r_b^j}{r_{ab}^2}\right)\right]\frac{\pi_b^j}{m}\,, \\
\mathcal{T}_2 &= e^2 \int \frac{\mathrm{d}^3 k}{(2\pi)^3}\left(\frac{\mathrm{i}\,(1+\kappa)}{2m}\,(\vec{\sigma}_a\times\vec{k})^i\right)\left[-\frac{\mathrm{e}^{\mathrm{i}\vec{k}\cdot(\vec{r}_a-\vec{r}_b)}}{\vec{k}^2}\delta^{\perp,ij}\right]\frac{\pi_b^j}{m}\,, \\
\mathcal{T}_3 &= e^2 \int \frac{\mathrm{d}^3 k}{(2\pi)^3}\,\frac{\pi_a^i}{m}\left[-\frac{\mathrm{e}^{\mathrm{i}\vec{k}\cdot(\vec{r}_a-\vec{r}_b)}}{\vec{k}^2}\delta^{\perp,ij}\right]\left(\frac{\mathrm{i}\,(1+\kappa)}{2m}\,(\vec{\sigma}_b\times(-\vec{k}))^j\right), \\
\mathcal{T}_4 &= e^2 \int \frac{\mathrm{d}^3 k}{(2\pi)^3}\left(\frac{\mathrm{i}\,(1+\kappa)}{2m}\,(\vec{\sigma}_a\times\vec{k})^i\right)\left[-\frac{\mathrm{e}^{\mathrm{i}\vec{k}\cdot(\vec{r}_a-\vec{r}_b)}}{\vec{k}^2}\delta^{\perp,ij}\right] \\
&\quad \times \left(\frac{\mathrm{i}\,(1+\kappa)}{2m}\,(\vec{\sigma}_b\times(-\vec{k}))^j\right). \qquad (12.55)
\end{aligned}
$$

Here, $\delta^{\perp,ij} = \delta^{ij} - k^i\,k^j/\vec{k}^2$. A useful integral for the calculation of $\mathcal{T}_2$ is

$$
I = \int \frac{\mathrm{d}^3 k}{(2\pi)^3}\,(\vec{\sigma}_a\times\vec{k})^i\left[-\frac{\mathrm{e}^{\mathrm{i}\vec{k}\cdot(\vec{r}_a-\vec{r}_b)}}{\vec{k}^2}\left(\delta^{ij} - \frac{k^i\,k^j}{\vec{k}^2}\right)\right]\pi_b^j = -\mathrm{i}\frac{(\vec{\sigma}_a\times\vec{r}_{ab})\cdot\vec{\pi}_b}{4\pi r_{ab}^3}. \qquad (12.56)
$$

The corresponding evaluation, adapted to the term $\mathcal{T}_3$, reads as follows,

$$
J = \int \frac{\mathrm{d}^3 k}{(2\pi)^3}\pi_a^i\left[-\frac{\mathrm{e}^{\mathrm{i}\vec{k}\cdot(\vec{r}_a-\vec{r}_b)}}{\vec{k}^2}\left(\delta^{ij} - \frac{k^i\,k^j}{\vec{k}^2}\right)\right](\vec{\sigma}_b\times(-\vec{k}))^j = \mathrm{i}\frac{(\vec{\sigma}_b\times\vec{r}_{ab})\cdot\vec{\pi}_a}{4\pi r_{ab}^3}. \qquad (12.57)
$$

For $\mathcal{T}_4$, we find it useful to calculate

$$
\begin{aligned}
K &= \int \frac{\mathrm{d}^3 k}{(2\pi)^3}\,(\vec{\sigma}_a\times\vec{k})^i\left[-\frac{\mathrm{e}^{\mathrm{i}\vec{k}\cdot(\vec{r}_a-\vec{r}_b)}}{\vec{k}^2}\left(\delta^{ij} - \frac{k^i\,k^j}{\vec{k}^2}\right)\right](\vec{\sigma}_b\times(-\vec{k}))^j \\
&= -\int \frac{\mathrm{d}^3 k}{(2\pi)^3}\left(\vec{\sigma}_a\cdot\vec{\sigma}_b\,\vec{k}^2 - \vec{\sigma}_a\cdot\vec{k}\,\vec{\sigma}_b\cdot\vec{k}\right)\left[-\frac{\mathrm{e}^{\mathrm{i}\vec{k}\cdot(\vec{r}_a-\vec{r}_b)}}{\vec{k}^2}\right] \\
&= \vec{\sigma}_a\cdot\vec{\sigma}_b\,\delta^{(3)}(\vec{r}_{ab}) + \left(\vec{\sigma}_a\cdot\vec{\nabla}_a\,\vec{\sigma}_b\cdot\vec{\nabla}_a\right)\frac{1}{4\pi r_{ab}} \\
&= \frac{2}{3}\,\vec{\sigma}_a\cdot\vec{\sigma}_b\,\delta^{(3)}(\vec{r}_{ab}) - \frac{\sigma_a^i\sigma_b^j}{4\pi}\left(\frac{\delta^{ij}}{r_{ab}^3} - 3\frac{r_{ab}^i\,r_{ab}^j}{r_{ab}^5}\right). \qquad (12.58)
\end{aligned}
$$

Armed with the results for I and J, and K, we can calculate the "spatial" contribution to the Breit Hamiltonian as follows,

$$H_{ab}^{(ij)} = 4\pi\alpha \frac{\pi_a^i}{m}\left[-\frac{1}{8\pi r_{ab}}\left(\delta^{ij} + \frac{r_a^i r_b^j}{r_{ab}^2}\right)\right]\frac{\pi_b^j}{m} + \frac{\mathrm{i}(1+\kappa)}{2m^2} I + \frac{\mathrm{i}(1+\kappa)}{2m^2} J + \frac{-(1+\kappa)^2}{4m^2} K . \tag{12.59}$$

So, $H_{ab}^{(ij)}$ is eventually found as follows,

$$\begin{aligned} H_{ab}^{(ij)} &= -\frac{\alpha}{2m^2 r_{ab}}\, \pi_a^i \left(\delta^{ij} + \frac{r_{ab}^i r_{ab}^j}{r_{ab}^2}\right)\pi_b^j + \frac{\alpha(1+\kappa)}{2m^2 r_{ab}^3}\left[\vec{\sigma}_a \times \vec{r}_{ab}\cdot\vec{\pi}_b - \vec{\sigma}_b \times \vec{r}_{ab}\cdot\vec{\pi}_a\right] \\ &\quad - \frac{2\pi(1+\kappa)^2}{3m^2}\,\vec{\sigma}_a\cdot\vec{\sigma}_b\,\delta^{(3)}(\vec{r}_{ab}) + \frac{\alpha}{4m^2}\frac{\sigma_a^i\sigma_b^j}{r_{ab}^3}\left(\delta^{ij} - 3\frac{r_{ab}^i r_{ab}^j}{r_{ab}^2}\right), \end{aligned} \tag{12.60}$$

which is purely an α^4 contribution. Together with $H_{ab}^{(00)}$ from Eq. (12.52), which we recall for convenience,

$$H_{ab}^{(00)} = \frac{\alpha}{r_{ab}} - \frac{\pi\alpha}{m^2}(1+2\kappa)\,\delta^{(3)}(\vec{r}_{ab}) + \frac{\alpha(1+2\kappa)}{4m^2 r_{ab}^3}\left[\vec{\sigma}_b\cdot(\vec{r}_{ab}\times\vec{\pi}_b) - \vec{\sigma}_a\cdot(\vec{r}_{ab}\times\vec{\pi}_a)\right], \tag{12.61}$$

we obtain H_{ab} as the sum of $H_{ab}^{(00)}$ and $H_{ab}^{(ij)}$. Finally, the Breit Hamiltonian for a many-electron system is obtained as the sum of the one-particle, and the many-particle Hamiltonians,

$$H_{\mathrm{BP}} = \sum_a H_a + \sum_{a>b} H_{ab} . \tag{12.62}$$

The single-particle Hamiltonian is obtained from Eq. (12.32). However, we perform two adjustments in order to allow for a coupling to an external field, by replacing $e\,A^0 \to V_a + e\,A_a^0 = -Z\alpha/r_a + e\,A_a^0$, where $V_a = -Z\alpha/r_a$ is the Coulomb potential, A_a^0 is an external vector potential, evaluated at $\vec{r}_a$, and $e\vec{E} \to -\vec{\nabla}_a V + e\,\vec{E}_a$. Using the electron coordinate $\vec{r}_a$ of the ath electron, one obtains

$$\begin{aligned} H_a &= \frac{\vec{\pi}_a^{\,2}}{2m} - \frac{Z\alpha}{r_a} + e\,A_a^0 - \frac{e}{2m}(1+\kappa)\,\vec{\sigma}_a\cdot\vec{B}_a - \frac{\vec{\pi}_a^{\,4}}{8m^3} + \frac{\pi(Z\alpha)}{2m^2}(1+2\kappa)\,\delta^{(3)}(\vec{r}_a) \\ &\quad + \frac{Z\alpha(1+2\kappa)}{4m^2}\vec{\sigma}_a\cdot\left(\frac{\vec{r}_a}{r_a^3}\times\vec{\pi}_a\right) - \frac{e(1+2\kappa)}{8m^2}\left[\vec{\nabla}\cdot\vec{E}_a + \sigma_a\cdot\left(\vec{E}_a\times\vec{\pi}_a - \vec{\pi}_a\times\vec{E}_a\right)\right] \\ &\quad + \frac{e}{8m^3}\left(\left\{\vec{\sigma}_a\cdot\vec{B}_a, \vec{\pi}_a^{\,2}\right\} + \kappa\left\{\vec{\pi}_a\cdot\vec{B}_a, \vec{\pi}_a\cdot\vec{\sigma}_a\right\}\right). \end{aligned} \tag{12.63}$$

The two-particle interaction Hamiltonian is the sum of Eqs. (12.60) and (12.61),

$$\begin{aligned} H_{ab} &= \frac{\alpha}{r_{ab}} - \frac{\alpha}{2m^2}\pi_a^i\left(\frac{\delta^{ij}}{r_{ab}} + \frac{r_{ab}^i r_{ab}^j}{r_{ab}^3}\right)\pi_b^j - \frac{\pi\,\alpha}{m^2}(1+2\kappa)\,\delta^{(3)}(\vec{r}_{ab}) \\ &\quad - \frac{2\pi\alpha}{3m^2}(1+\kappa)^2\,\vec{\sigma}_a\cdot\vec{\sigma}_b\,\delta^{(3)}(\vec{r}_{ab}) + \frac{\alpha}{4m^2}(1+\kappa)^2\frac{\sigma_a^i\sigma_b^j}{r_{ab}^3}\left(\delta^{ij} - 3\frac{r_{ab}^i r_{ab}^j}{r_{ab}^3}\right) \\ &\quad + \frac{\alpha}{4m^2 r_{ab}^3}\left[2(1+\kappa)\left(\vec{\sigma}_a\cdot\vec{r}_{ab}\times\vec{\pi}_b - \vec{\sigma}_b\cdot\vec{r}_{ab}\times\vec{\pi}_a\right)\right. \\ &\quad \left. + (1+2\kappa)\left(\vec{\sigma}_b\cdot\vec{r}_{ab}\times\vec{\pi}_b - \vec{\sigma}_a\cdot\vec{r}_{ab}\times\vec{\pi}_a\right)\right]. \end{aligned} \tag{12.64}$$

These terms are identified and interpreted as follows. First, we have a plain Coulomb repulsion and a magnetic photon exchange, then, a *zitterbewegung* term corrected by the anomalous magnetic moment, a magnetic photon exchange, a spin-spin contact interaction, a higher-multipole correction to the spin-spin interaction, and finally, the spin-other-orbit and spin-orbit term. Note that we have kept the canonical momenta $\vec{\pi}_a$ in the most general form possible. One might argue that the vector potential in $\vec{\pi}_a = \vec{p}_a - e\,\vec{A}(\vec{r}_a)$ in Eq. (12.32) was absorbed in the photon propagator in Eq. (12.37). The idea is that $\vec{A}$ could be written as the sum of two terms, one of which describes the quantized field, while the second vector potential in $\vec{\pi}_a$ might correspond to additional external fields $\vec{A}_a$ (e.g., constant magnetic fields) which contribute to $\vec{\pi}_a$.

12.3.4 *Breit Hamiltonian for Unequal Particles*

With the same conventions as used in Sec. 12.3, we start from Eq. (12.46) and use the photon propagator in Coulomb gauge [see Eq. (12.47)]. However, unlike Eq. (12.50), we distinguish, in the timelike component of the current, the anomalous magnetic moment κ_a and the mass m_a,

$$j_a^0(\vec{k}) = 1 + \frac{\mathrm{i}\,(1+2\kappa_a)}{4m_a}\,\vec{\sigma}\cdot(\vec{k}\times\vec{\pi}) - \frac{1+2\kappa_a}{8m_a^2}\,\vec{k}^2\,, \tag{12.65}$$

and analogously for $a \to b$. The Hamiltonian will be denoted by $\mathcal{H}_{ab}$ instead of H_{ab}. After going through analogous steps as the ones discussed in Sec. 12.3.3, the effective Hamiltonian is calculated as follows,

$$\begin{aligned}
\mathcal{H}_{ab}^{(00)} &= e_a\,e_b \int \frac{\mathrm{d}^3k}{(2\pi)^3}\, D_{00}(\vec{k})\, j_a^0(\vec{k})\,\mathrm{e}^{\mathrm{i}\,\vec{k}\cdot(\vec{r}_a-\vec{r}_b)}\, j_b^0(-\vec{k}) \\
&= e_a\,e_b \int \frac{\mathrm{d}^3k}{(2\pi)^3} \left[\frac{\mathrm{e}^{\mathrm{i}\,\vec{k}\cdot(\vec{r}_a-\vec{r}_b)}}{\vec{k}^2} + \frac{\mathrm{i}\,(1+2\kappa_a)}{4m_a^2}\,\vec{\sigma}_a\cdot(\vec{k}\times\vec{\pi}_a)\frac{\mathrm{e}^{\mathrm{i}\,\vec{k}\cdot(\vec{r}_a-\vec{r}_b)}}{\vec{k}^2} \right. \\
&\quad \left. + \frac{\mathrm{e}^{\mathrm{i}\,\vec{k}\cdot(\vec{r}_a-\vec{r}_b)}}{\vec{k}^2}\frac{\mathrm{i}\,(1+2\kappa_b)}{4m_b^2}\,\vec{\sigma}_b\cdot(-\vec{k}\times\vec{\pi}_b) - \left(\frac{1+2\kappa_a}{8m_a^2} + \frac{1+2\kappa_b}{8m_b^2}\right)\mathrm{e}^{\mathrm{i}\,\vec{k}\cdot(\vec{r}_a-\vec{r}_b)} \right].
\end{aligned} \tag{12.66}$$

We use the convention

$$\alpha_{ab} = \frac{e_a\,e_b}{4\pi}\,, \tag{12.67}$$

and note that, e.g., for a negatively charged orbiting particle of charge e, and a nucleus of charge $-Z\,e$, one has

$$e_a = e\,, \qquad e_b = -Z\,e\,, \qquad \alpha_{ab} = -Z\alpha\,. \tag{12.68}$$

After a Fourier transformation, one obtains

$$\begin{aligned}
\mathcal{H}_{ab}^{(00)} &= \frac{\alpha_{ab}}{r_{ab}} - \frac{\alpha_{ab}\,(1+2\kappa_a)}{4m_a^2\,r_{ab}^3}\,\vec{\sigma}_a\cdot(\vec{r}_{ab}\times\vec{\pi}_a) \\
&\quad + \frac{\alpha_{ab}\,(1+2\kappa_b)}{4m_b^2\,r_{ab}^3}\,\vec{\sigma}_b\cdot(\vec{r}_{ab}\times\vec{\pi}_b) - \frac{\pi\alpha_{ab}}{2}\left(\frac{1+2\kappa_a}{m_a^2} + \frac{1+2\kappa_b}{m_b^2}\right)\delta^{(3)}(\vec{r}_{ab})\,.
\end{aligned} \tag{12.69}$$

Hence,

$$\mathcal{H}_{ab}^{(00)} = \frac{\alpha_{ab}}{r_{ab}} + \frac{\alpha_{ab}}{4r_{ab}^3}\left[(1+2\kappa_b)\frac{\vec{\sigma}_b\cdot(\vec{r}_{ab}\times\vec{\pi}_b)}{m_b^2} - (1+2\kappa_a)\frac{\vec{\sigma}_a\cdot(\vec{r}_{ab}\times\vec{\pi}_a)}{m_a^2}\right] - \frac{\pi\alpha_{ab}}{2}\left(\frac{1+2\kappa_a}{m_a^2}+\frac{1+2\kappa_b}{m_b^2}\right)\delta^{(3)}(\vec{r}_{ab}). \tag{12.70}$$

The spatial components of the current are given by Eq. (12.53), but with the distinction for the ath particle mass m_a and its anomalous magnetic moment κ_a,

$$\vec{j}_a(\vec{k}) = \frac{\vec{\pi}}{m_a} + \frac{\mathrm{i}(1+\kappa_a)}{2m_a}(\vec{\sigma}\times\vec{k}), \tag{12.71}$$

and analogously for $a \to b$. The effective Hamiltonian from the spatial elements of the current reads as

$$\begin{aligned}\mathcal{H}_{ab}^{(ij)} &= e_a\, e_b \int \frac{\mathrm{d}^3k}{(2\pi)^3} D_{ij}(\vec{k})\, j_a^i(\vec{k})\, \mathrm{e}^{\mathrm{i}\vec{k}\cdot(\vec{r}_a-\vec{r}_b)}\, j_b^j(-\vec{k}) \\ &= e_a\, e_b \int \frac{\mathrm{d}^3k}{(2\pi)^3}\left(\frac{\pi_a^i}{m_a} + \frac{\mathrm{i}(1+\kappa_a)}{2m_a}(\vec{\sigma}\times\vec{k})^i\right) \\ &\quad\times\left[-\frac{\mathrm{e}^{\mathrm{i}\vec{k}\cdot(\vec{r}_a-\vec{r}_b)}}{\vec{k}^2}\left(\delta^{ij}-\frac{k^i k^j}{\vec{k}^2}\right)\right]\left(\frac{\pi_b^j}{m_b} + \frac{\mathrm{i}(1+\kappa_b)}{2m_b}(\vec{\sigma}\times(-\vec{k}))^j\right).\end{aligned} \tag{12.72}$$

Using Eqs. (12.56), (12.57) and (12.58), one can obtain $\mathcal{H}_{ab}^{(ij)}$ as

$$\begin{aligned}\mathcal{H}_{ab}^{(ij)} &= -\frac{\alpha_{ab}}{2m_a m_b r_{ab}}\pi_a^i\left(\delta^{ij}+\frac{r_a^i r_b^j}{r_{ab}^2}\right)\pi_b^j - \frac{2\pi(1+\kappa_a)(1+\kappa_b)\alpha_{ab}}{3m_a m_b}\vec{\sigma}_a\cdot\vec{\sigma}_b\delta^{(3)}(\vec{r}_{ab}) \\ &\quad+ \frac{\alpha_{ab}}{4m_a m_b}(1+\kappa_a)(1+\kappa_b)\frac{\sigma_a^i\sigma_b^j}{r_{ab}^3}\left(\delta^{ij}-3\frac{r_{ab}^i r_{ab}^j}{r_{ab}^2}\right) \\ &\quad+ \frac{\alpha_{ab}}{4m_a m_b r_{ab}^3}2[(1+\kappa_a)\vec{\sigma}_a\cdot\vec{r}_{ab}\times\vec{\pi}_b - (1+\kappa_b)\vec{\sigma}_b\cdot\vec{r}_{ab}\times\vec{\pi}_a].\end{aligned} \tag{12.73}$$

We recall the result for $\mathcal{H}_{ab}^{(00)}$ from Eq. (12.70),

$$\mathcal{H}_{ab}^{(00)} = \frac{\alpha_{ab}}{r_{ab}} + \frac{\alpha_{ab}}{4r_{ab}^3}\left[(1+2\kappa_b)\frac{\vec{\sigma}_b\cdot(\vec{r}_{ab}\times\vec{\pi}_b)}{m_b^2} - (1+2\kappa_a)\frac{\vec{\sigma}_a\cdot(\vec{r}_{ab}\times\vec{\pi}_a)}{m_a^2}\right] - \frac{\pi\alpha_{ab}}{2}\left(\frac{1+2\kappa_a}{m_a^2}+\frac{1+2\kappa_b}{m_b^2}\right)\delta^{(3)}(\vec{r}_{ab}). \tag{12.74}$$

The full interaction Hamiltonian $\mathcal{H}_{ab}$ is the sum of $\mathcal{H}_{ab}^{(00)}$ and $\mathcal{H}_{ab}^{(ij)}$.

Because we treat all interactions on the same footing, we do not single out the electron-nucleus interaction as in Eq. (12.63), but rather, the single-particle Hamiltonian is a "free" Hamiltonian for the particle, plus coupling terms, to the external $\vec{E}$ and $\vec{B}$ fields, and to the external vector potential. It is obtained from

Eq. (12.63) by setting $Z\alpha \to 0$, and using masses and anomalous magnetic moments specific to particle a,

$$\begin{aligned}\mathcal{H}_a = & \frac{\vec{\pi}_a^{\,2}}{2m_a} - \frac{\vec{\pi}_a^{\,4}}{8m_a^3} + e\,A_a^0 - \frac{e}{2m_a}\,(1+\kappa_a)\,\vec{\sigma}_a\cdot\vec{B}_a \\ & - \frac{e}{8m_a^2}\,(1+2\kappa_a)\,\left[\vec{\nabla}\cdot\vec{E}_a + \sigma_a\cdot\left(\vec{E}_a\times\vec{\pi}_a - \vec{\pi}_a\times\vec{E}_a\right)\right] \\ & + \frac{e}{8m_a^3}\,\left(\left\{\vec{\sigma}_a\cdot\vec{B}_a, \vec{\pi}_a^2\right\} + \kappa_a\,\left\{\vec{\pi}_a\cdot\vec{B}_a, \vec{\pi}_a\cdot\vec{\sigma}_a\right\}\right) . \end{aligned} \tag{12.75}$$

The two-particle interaction Hamiltonian is the sum of $\mathcal{H}_{ab}^{(00)}$ and $\mathcal{H}_{ab}^{(ij)}$,

$$\begin{aligned}\mathcal{H}_{ab} = & \frac{\alpha_{ab}}{r_{ab}} - \frac{\alpha_{ab}\,\pi_a^i}{2m_a m_b r_{ab}}\left(\delta^{ij} + \frac{r_{ab}^i\, r_{ab}^j}{r_{ab}^2}\right)\pi_b^j - \frac{\pi\alpha_{ab}}{2}\left(\frac{1+2\kappa_a}{m_a^2} + \frac{1+2\kappa_b}{m_b^2}\right)\delta^{(3)}(\vec{r}_{ab}) \\ & - \frac{2\pi\alpha_{ab}}{3m_a m_b}\,(1+\kappa_a)(1+\kappa_b)\,\vec{\sigma}_a\cdot\vec{\sigma}_b\,\delta^{(3)}(\vec{r}_{ab}) \\ & + \frac{\alpha_{ab}}{4m_a m_b}\,(1+\kappa_a)\,(1+\kappa_b)\,\frac{\sigma_a^i\,\sigma_b^j}{r_{ab}^3}\left(\delta^{ij} - 3\,\frac{r_{ab}^i\, r_{ab}^j}{r_{ab}^2}\right) \\ & + \frac{\alpha_{ab}}{4\,r_{ab}^3}\left[\frac{2}{m_a\,m_b}\,\left((1+\kappa_a)\,\vec{\sigma}_a\cdot\vec{r}_{ab}\times\vec{\pi}_b - (1+\kappa_b)\,\vec{\sigma}_b\cdot\vec{r}_{ab}\times\vec{\pi}_a\right)\right. \\ & \left. + (1+2\kappa_b)\frac{\vec{\sigma}_b\cdot\vec{r}_{ab}\times\vec{\pi}_b}{m_b^2} - (1+2\kappa_a)\frac{\vec{\sigma}_a\cdot\vec{r}_{ab}\times\vec{\pi}_a}{m_a^2}\right] . \end{aligned} \tag{12.76}$$

This interaction Hamiltonian is valid for arbitrary spin-1/2 particles, with masses m_a and m_b, and anomalous magnetic moments describes by κ_a and κ_b.

12.4 Applications and Generalizations for Many-Particle Systems

12.4.1 *Application to Many-Electron Systems*

In Sec. 12.3.3, we have discussed an approach where we would describe the energy levels of an atom by the sum of "one-particle" Hamiltonians H_a given in Eq. (12.63), which describe the electron–nucleus interaction (in the non-recoil limit), and the electron-electron interaction, which is given by the Hamiltonian (12.64). In Sec. 12.3.4, for a general N-particle system with unequal particles $a, b = 1, \dots, N$, we add the free one-particle Hamiltonians given in Eq. (12.75) to the interactions given in Eq. (12.76). The latter approach includes the mass dependence of the many-body system exactly. However, one might ask as well where the connection is, i.e., in which limit the two approaches become identical.

To this end, we investigate an N-particle system with one nucleus of charge $-Z\,e$ and $N-1$ electrons of charge e (typically, $N-1 = Z$ for a neutral atom). This system can be described, according to the approach described in Sec. 12.3.4, by "one-particle" Hamiltonians (for the electrons) which are the sum of the single-particle Hamiltonians $\mathcal{H}_a$ given in Eq. (12.75), the electron-nucleus Hamiltonians $\mathcal{H}_{an}$ (with

$b = n$), and electron-electron interaction Hamiltonians $\mathcal{H}_{ab}$, which are obtained from Eq. (12.76) for $m_a = m_b = m$, $\alpha_{ab} = \alpha$, and $\kappa_a = \kappa_b = \kappa$, where $\kappa \approx \alpha/(2\pi)$ describes the anomalous contribution to the electron g factor [see Eqs. (1.11) and (11.38)].

We should mention that the latter approach from Sec. 12.3.4 is, strictly speaking, only valid if the nucleus is a spin-1/2 particle, because we have derived our general Hamiltonian for the interaction of two particles with masses m_a and m_b and g factors $2(1+\kappa_a)$ and $2(1+\kappa_b)$ under the assumptions of a spin-1/2 character.

It is interesting to verify the consistency of the approach outlined in Sec. 12.3.4 with the approach from Sec. 12.3.3, and with the single-particle formalism outlined in Sec. 11.3.1. First, one verifies, easily, that the interaction Hamiltonian (12.76), for the case $m_a = m_b = m$, $\alpha_{ab} = \alpha$, and $\kappa_a = \kappa_b = \kappa$, reproduces the interaction Hamiltonian in Eq. (12.64). The (somewhat trivial) details are left as an exercise to the reader (see Sec. 12.7). The next step is to calculate the limit of infinite nuclear mass ($m_n \to \infty$) of the Hamiltonian given in Eq. (12.76), under the conditions

$$\begin{aligned} &e_a \to e\,, \qquad e_b \to -Ze\,, \qquad \alpha_{ab} \to -Z\alpha\,, \\ &\kappa_a \to \kappa\,, \qquad \vec{r}_{ab} = \vec{r}_{an} \to \vec{r}_a\,, \qquad m_b = m_n \to \infty\,. \end{aligned} \tag{12.77}$$

One obtains

$$\lim_{m_n\to\infty} \mathcal{H}_{an} = -\frac{Z\alpha}{r_a} + \frac{\pi Z\alpha}{2m^2}\,(1+\kappa)\,\delta^{(3)}(\vec{r}_a) + \frac{Z\alpha}{4m^2}(1+2\kappa)\left[\vec{\sigma}_a \cdot \left(\frac{\vec{r}_a}{r_a^3} \times \vec{\pi}_a\right)\right]. \tag{12.78}$$

The second term in Eq. (12.78) corresponds to the term

$$-\frac{\pi\alpha_{ab}}{2}\left(\frac{1+\kappa_a}{m_a^2} + \frac{1+\kappa_b}{m_b^2}\right)\delta^{(3)}(\vec{r}_{ab}) \tag{12.79}$$

in Eq. (12.76). While the third term in Eq. (12.78) corresponds to the spin-orbit coupling term $\frac{\alpha_{ab}}{4\,r_{ab}^3}\left[\left((1+2\kappa_b)\frac{\vec{\sigma}_b\cdot\vec{r}_{ab}\times\vec{\pi}_b}{m_b^2} - (1+2\kappa_a)\frac{\vec{\sigma}_a\cdot\vec{r}_{ab}\times\vec{\pi}_a}{m_a^2}\right)\right]$ in Eq. (12.76), the latter taken under the replacements given in Eq. (12.77). Furthermore, the Hamiltonian (12.78) is equal to the sum of the terms $H^{[2]}+H^{[4]}+H^{[5]}$ given in Eqs. (11.63), (11.65) and (11.66), under the kinematic conditions given in Eq. (11.62). The latter fact ramifies the general wisdom that Dirac theory, augmented by the Foldy–Wouthuysen method, constitutes a single-particle theory and cannot, in principle, account for reduced-mass corrections.

The new single-particle Hamiltonian H_a is the sum of the single-particle Hamiltonian $\mathcal{H}_a$ given in Eq. (12.75), which contains the interactions with the external fields and the electron-nucleus Hamiltonian H_{an} taken from Eq. (12.78),

$$\begin{aligned} H_a = {}& \frac{\vec{\pi}_a^{\,2}}{2m_a} - \frac{\vec{\pi}_a^{\,4}}{8m_a^3} + e\,A_a^0 - \frac{e}{2m}\,(1+\kappa)\,\vec{\sigma}_a\cdot\vec{B}_a - \frac{e}{8m^2}\,(1+2\kappa)\,\big[\vec{\nabla}\cdot\vec{E}_a \\ & + \sigma_a\cdot\big(\vec{E}_a\times\vec{\pi}_a - \vec{\pi}_a\times\vec{E}_a\big)\big] + \frac{e}{8m^3}\left(\{\vec{\sigma}_a\cdot\vec{B}_a, \vec{\pi}_a^{\,2}\} + \kappa\,\{\vec{\pi}_a\cdot\vec{B}_a, \vec{\pi}_a\cdot\vec{\sigma}_a\}\right) \\ & - \frac{Z\alpha}{r_a} + \frac{\pi Z\alpha}{2m^2}\,(1+\kappa)\,\delta^{(3)}(\vec{r}_a) + \frac{Z\alpha}{4m^2}(1+2\kappa)\left[\vec{\sigma}_a\cdot\left(\frac{\vec{r}_a}{r_a^3}\times\vec{\pi}_a\right)\right]. \end{aligned} \tag{12.80}$$

This has to be compared to the result obtained for the single-particle Hamiltonian (12.63). A closer inspection reveals that the Hamiltonians (12.63) and (12.80) are identical. Reassuringly, the Hamiltonian given in Eq. (12.63), if one sets the external fields equal to zero ($A_a^0 \to 0$, $\vec{E}_a \to \vec{0}$, $\vec{B}_a \to \vec{0}$, $\vec{\pi}_a \to \vec{p}_a$) becomes identical to the Foldy–Wouthuysen Hamiltonian for the Dirac–Coulomb problem given in Eqs. (11.63), (11.65) and (11.66). Hence, we have clarified that the approaches outlined in Secs. 12.3.3 and 12.3.4 are in mutual agreement, in the non-recoil limit of infinite nuclear mass $m_N \to \infty$.

The approach outlined in Sec. 12.3.4 treats the atomic nucleus on a more equal footing as compared to the approach described in Sec. 12.3. Let us illustrate the advantage of the latter approach by deriving the so-called mass polarization contribution to the spectrum of helium. We have three particles, the electrons labeled as 1 and 2, and the helium nucleus N. We take the nonrelativistic kinetic-energy contributions from the two electrons and the nucleus, in the spirit of the approach taken in Sec. 12.3.4,

$$H_{\rm NR} = \frac{\vec{p}_1^{\,2}}{2m} + \frac{\vec{p}_2^{\,2}}{2m} + \frac{\vec{p}_n^{\,2}}{2m_n}, \tag{12.81}$$

which really is the sum over $a = 1, 2$ and $a = N$ of the free+field Hamiltonian given in Eq. (12.75), taking the first term under the replacement $\vec{\pi}_a^{\,2} \to \vec{p}_a^{\,2}$, assuming a vanishing external vector potential. In the centre-of-mass frame, one has $\vec{p}_N = -\vec{p}_1 - \vec{p}_2$, and thus

$$H_{\rm NR} = \frac{\vec{p}_1^{\,2}}{2m} + \frac{\vec{p}_2^{\,2}}{2m} + \frac{(-\vec{p}_1 - \vec{p}_2)^2}{2m_N} = \frac{\vec{p}_1^{\,2}}{2\mu} + \frac{\vec{p}_2^{\,2}}{2\mu} + \frac{\vec{p}_1 \cdot \vec{p}_2}{2m_N}. \tag{12.82}$$

Here, $\mu = m\, m_N/(m + m_N)$ is the reduced mass of the electron-nucleus system. The last term in Eq. (12.82) is just the mass-polarization term.

12.4.2 *Single-Particle Hamiltonian for Arbitrary Spin*

It is very interesting to consider the question of how to generalize the Hamiltonian (12.75) to the case of arbitrary nuclear spin. An example would be an atomic nucleus such as the deuteron, which constitutes a spin-1 particle. Let s^i be the spin matrices of the appropriate structure. For $s = \frac{1}{2}$, we can choose $s^i = \frac{1}{2}\sigma^i$, where the σ^i are the Pauli matrices [see Eq. (4.343)]. The vector $\vec{\mathbb{S}}$ of spin matrices for $s = 1$ reads as follows (see Chap. 5 of Ref. [81]),

$$\left(\mathbb{S}^i\right)^{jk} = -\mathrm{i}\,\epsilon^{ijk}, \tag{12.83}$$

where $\mathbb{S}^i$ is the spin matrix for the ith Cartesian direction, and its components are labeled as jk. An explicit representation has been given in Eq. (8.340), and we recall it here for convenience,

$$\mathbb{S}^1 = \begin{pmatrix} 0 & 0 & 0 \\ 0 & 0 & -\mathrm{i} \\ 0 & \mathrm{i} & 0 \end{pmatrix} \qquad \mathbb{S}^2 = \begin{pmatrix} 0 & 0 & \mathrm{i} \\ 0 & 0 & 0 \\ -\mathrm{i} & 0 & 0 \end{pmatrix}, \qquad \mathbb{S}^3 = \begin{pmatrix} 0 & -\mathrm{i} & 0 \\ \mathrm{i} & 0 & 0 \\ 0 & 0 & 0 \end{pmatrix}. \tag{12.84}$$

The commutator of the spin-1 matrices has been given in Eq. (8.341) as $[\,\mathbb{S}^i, \mathbb{S}^j\,] = \mathrm{i}\,\epsilon^{ijk}\,\mathbb{S}^k$, where the Einstein summation convention is used for the sum over $k = 1, 2, 3$ on the right-hand side. The eigenvalues and eigenvectors λ of $\mathbb{S}^3$ are manifest in the spherical basis vectors $\vec{e}_\lambda$, which fulfill $\mathbb{S}^3\,\vec{e}_\lambda = \lambda\,\vec{e}_\lambda$ and are given by

$$\vec{e}_{+1} = -\frac{1}{\sqrt{2}}\,(\hat{e}_x + \mathrm{i}\,\hat{e}_y) = \begin{pmatrix} -1/\sqrt{2} \\ -\mathrm{i}/\sqrt{2} \\ 0 \end{pmatrix}, \qquad \vec{e}_0 = \hat{e}_z = \begin{pmatrix} 0 \\ 0 \\ 1 \end{pmatrix}, \tag{12.85a}$$

$$\vec{e}_{-1} = \frac{1}{\sqrt{2}}\,(\hat{e}_x - \mathrm{i}\,\hat{e}_y) = \begin{pmatrix} 1/\sqrt{2} \\ -\mathrm{i}/\sqrt{2} \\ 0 \end{pmatrix}. \tag{12.85b}$$

A number of useful formulas can be found in Ref. [16] (see also Chap. 6).

We go even one step further and denote a general spin matrix as

$$\vec{\$}_a = \begin{cases} \frac{1}{2}\,\vec{\sigma}_a & s_a = 1/2 \\ \vec{\mathbb{S}}_a & s_a = 1 \end{cases}. \tag{12.86}$$

We define the g factor for particle a, denoted as g_a, as follows,

$$g_a \equiv 2(1 + \kappa_a)\,. \tag{12.87}$$

This definition deserves some comments. One substitutes the expression for the ath particle, in Eq. (12.75) as follows,

$$-\frac{e}{2m_a}\,(1 + \kappa_a)\,\vec{\sigma}_a \cdot \vec{B}_a \to -g_a\,\frac{e_a}{2m_a}\,\vec{\$}_a \cdot \vec{B}_a\,. \tag{12.88}$$

Here, we recall that e_a is the physical charge of the ath particle, and m_a is its mass. We had included some discussion of this convention around Eqs. (11.81) and (11.82), and has remarked that, with this definition, the g factor of the electron (and, by the way, also that of the positron), comes out as positive (cf. Ref. [330]). We had also remarked that the positive sign of the electron g factor hinges upon the negative sign in the relation that connects the electron's magnetic moment with the Bohr magneton, as in Eq. (11.83).

According to the substitution (12.88), the leading-order interaction of the ath particle's spin with the magnetic field is given by the Hamiltonian

$$H_a = -g_a\,\frac{e_a}{2m_a}\,\vec{\$}_a \cdot \vec{B}_a\,. \tag{12.89}$$

For electrons (and muons), this Hamiltonian is in agreement with the discussion surrounding Eqs. (11.81) and (11.82), and with the usual conventions from the literature. Let us now investigate if the g factor defined according to Eq. (12.87) is in agreement with the usual conventions used for nuclei. To this end, let us first recall the definition of the Bohr magneton μ_B and the nuclear magneton μ_N,

$$\mu_B = \frac{|e|}{2m}\,, \qquad \mu_N = \frac{|e|}{2m_p}\,, \tag{12.90}$$

where m is the electron mass, m_p is the proton mass, and $|e|$ is the positron (proton) charge.

Let us now assume that $a = p$ is a proton. Then, $e_a = |e|$, and because the proton carries a spin-1/2, the spin-magnetic interaction Hamiltonian becomes

$$H_p = -g_p \frac{|e|}{2m_p} \vec{\mathbb{s}}_p \cdot \vec{B}_a \,, \qquad \vec{\mathbb{s}}_p = \tfrac{1}{2}\vec{\sigma}_p \,. \tag{12.91}$$

The usual convention of the nuclear g factor, for a nucleus N, connects the nuclear magnetic moment $\vec{\mu}_N$ with the nuclear spin operator $\vec{\mathbb{s}}_N$ via the relation

$$\vec{\mu}_N = \mathbf{g}_N \, \mu_{\mathrm{N}} \, \vec{\mathbb{s}}_N \,, \qquad \vec{\mathbb{s}}_N = \vec{I}_N \,, \tag{12.92}$$

where $\vec{\mathbb{s}}_N = \vec{I}_N$ is the nuclear spin operator for the nucleus N. This leads to the spin-magnetic Hamiltonian

$$H_N = -\vec{\mu}_N \cdot \vec{B} = -\mathbf{g}_N \, \mu_{\mathrm{N}} \, \vec{\mathbb{s}}_N \cdot \vec{B} \,. \tag{12.93}$$

Here, $\mathbf{g}_N$ is the nuclear g factor in the usual convention. With the help of Eqs. (12.90) and (12.92), one can easily show that the Hamiltonians H_N and H_p given in Eqs. (12.93) and (12.91) are in agreement with the general Hamiltonian H_a given in Eq. (12.89), for $a = p$ being a proton.

The situation changes for nuclei other than the proton. Here, the relation (12.92) is still valid in the commonly accepted conventions for the nuclear g factor $\mathbf{g}_N$, but the nucleon's mass is not necessarily equal (or close) to the proton mass. In order to use the general formulas given below for nuclei other than the proton, one needs to equate

$$g_N \frac{e_N}{2m_N} = \mathbf{g}_N \, \mu_{\mathrm{N}} = \mathbf{g}_N \frac{|e|}{2m_p} \,, \tag{12.94}$$

where $a = N$ is a nucleus, $\mathbf{g}_N$ is the nuclear g factor in the usual conventions, while g_N is the g factor according to the definition (12.87), and e_N is the nucleon charge. For the neutron, $N = n$, the nuclear g factor, according to the definition (12.92), reads as $\mathbf{g}_n = -3.826\,085\,45(90)$ [60, 61]. The neutron, though, carries a charge $e_n = 0$. Hence, in order to apply any of the formulas given below to the neutron, one has to assume an infinitesimal neutron charge $e_n = \epsilon$. The g_n factor of the neutron assumes the large value $g_n = \mathbf{g}_n \, (|e|/\epsilon) \, (m_n/m_p)$, and one keeps the product $g_n \, \epsilon = \mathbf{g}_n \, |e| \, (m_n/m_p)$ constant throughout the calculation. In the end, one takes the limit $\epsilon \to 0$, keeping the value of the product $g_n \, \epsilon$ constant. For the deuteron $N = d$, the nuclear g factor, according to the definition (12.92), reads as $\mathbf{g}_d = 0.847\,438\,2338(22)$ [60,61]. while the g factor defined according to Eq. (12.87) reads as $g_d = \mathbf{g}_d(m_d/m_p) = 1.714\ldots$. It is disputable which definition of the g factor is the most appropriate one for the neutron and deuteron, but the above discussion clarifies the potential ambiguities.

After this detour on the conventions of the g factor, we substitute in Eq. (12.75), for a particle of general spin, as follows,

$$-\frac{e}{2m_a}(1+\kappa_a)\,\vec{\sigma}_a\cdot\vec{B}_a \to -g_a\,\frac{e_a}{2m_a}\,\vec{\mathbb{s}}_a\cdot\vec{B}_a\,, \tag{12.95a}$$

$$-\frac{e}{8m_a^2}(1+2\kappa_a)\,\vec{\nabla}\cdot\vec{E}_a$$
$$\to -\frac{e}{8m_a^2}(g_a-1)\,\delta_{s_a,1/2}\,\vec{\nabla}\cdot\vec{E}_a - \frac{e}{6}\,\langle r_E^2\rangle_a\,\vec{\nabla}\cdot\vec{E}_a\,, \tag{12.95b}$$

$$-\frac{e}{8m_a^2}(1+2\kappa_a)\sigma_a\cdot\left(\vec{E}_a\times\vec{\pi}_a-\vec{\pi}_a\times\vec{E}_a\right)$$
$$\to -\frac{e}{4m_N^2}(g_a-1)\,\vec{\mathbb{s}}_a\cdot\left(\vec{E}_a\times\vec{\pi}_a-\vec{\pi}_a\times\vec{E}_a\right). \tag{12.95c}$$

Here, we use the fact that according to Ref. [362], the so-called *zitterbewegung* term, proportional to $\delta_{s_a,1/2}\,\vec{\nabla}\cdot\vec{E}_a$, vanishes for a spin-1 nucleus. It has been a topic of discussion of whether to include, or exclude, the Darwin–Foldy term in the nuclear radius. Specifically, in Refs. [205, 363], the exclusion of this term from the nuclear radius is advocated, while in Ref. [364], the authors advocate to include this term. If this term were included, then for a point particle, we would have $\langle r_E^2\rangle = 3/(4m_N^2)$, reproducing the *zitterbewegung* term which is otherwise only present for a spin-1/2 nucleus. However, since this term is absent for a spin-1 particle [362], we should better exclude the Darwin term from the definition of the nuclear radius, in accordance with Refs. [205, 363]. Note that the mean-square charge radius $\langle r_E^2\rangle_a$ is defined as the slope of the Sachs G_E form factor [see Eq. (11) of Ref. [205]].

One can also replace, in Eq. (12.75),

$$\frac{e}{8m_a^3}\left(\left\{\vec{\sigma}_a\cdot\vec{B}_a,\vec{\pi}_a^2\right\}+\kappa\left\{\vec{\pi}_a\cdot\vec{B}_a,\vec{\pi}_a\cdot\vec{\sigma}_a\right\}\right)$$
$$\to \frac{e}{8m_a^3}\left[2\left\{\vec{\mathbb{s}}_a\cdot\vec{B}_a,\vec{\pi}_a^2\right\}+(g_a-2)\left\{\vec{\pi}_a\cdot\vec{B}_a,\vec{\pi}_a\cdot\vec{\mathbb{s}}_a\right\}\right]. \tag{12.95d}$$

For a particle with an internal structure, there can be an additional Hamiltonian term (12.96) due to a quadrupole moment, where

$$H_Q = -\frac{e}{6}\,(\mathbb{s}_a^i\,\mathbb{s}_a^j)^{(2)}\,Q_{Ea}\,(E_a)^i_{,j}\,,\qquad (E_a)^i_{,j} \equiv \left.\frac{\partial(E_a)^i}{\partial r^j}\right|_{\vec{r}=\vec{r}_a}\,, \tag{12.96}$$

and Q_{Ea} is the electric quadrupole constant of the ath particle. The traceless quadrupole spin tensor reads as follows,

$$(\mathbb{s}_a^i\,\mathbb{s}_a^j)^{(2)} = \frac{\mathbb{s}_a^i\,\mathbb{s}_a^j+\mathbb{s}_a^j\,\mathbb{s}_a^i}{2} - \frac{\delta^{ij}}{3}\,\vec{\mathbb{s}}_a\cdot\vec{\mathbb{s}}_a\,. \tag{12.97}$$

For completeness, we also add an interaction quadratic in the magnetic fields,

$$H_M = -\frac{e_a^2}{2}\,B^i\,B^j\left[\alpha_a^M\,\delta^{ij}+\gamma_a^M\,(\mathbb{s}_a^i\,\mathbb{s}_a^j)^{(2)}\right], \tag{12.98}$$

where α_a^M is the magnetic susceptibility of the ath particle, while γ_a^M is the magnetic quadrupole constant.

The superscript (2) denotes the selection of the ($\ell = 2$)-component of the interaction. Our final effective Hamiltonian for an arbitrary-spin particle (interacting with external fields) is obtained as a simplified version of the Hamiltonian given in Eqs. (1)–(4) of Ref. [324]; the Hamiltonian in Ref. [324] also comprises sixth-order terms in the momenta. Our Hamiltonian

$$\begin{aligned}\mathcal{H}'_a &= \frac{\vec{\pi}_a^2}{2m_a} + e_a A_a^0 - \frac{\vec{\pi}_a^4}{8m_a^3} - \frac{e_a}{2m_a} g_a \,\vec{\mathbb{s}}_a \cdot \vec{B}_a \\ &\quad - \frac{e_a}{8m_a^2} (g_a - 1)\, \delta_{s_a,1/2} \vec{\nabla} \cdot \vec{E}_a - \frac{e_a}{6} \left\langle r_E^2 \right\rangle_a \vec{\nabla} \cdot \vec{E}_a \\ &\quad - \frac{e_a}{4m_a^2} (g_a - 1)\, \vec{\mathbb{s}}_a \cdot \left(\vec{E}_a \times \vec{\pi}_a - \vec{\pi}_a \times \vec{E}_a \right) - \frac{e_a}{6} (\mathbb{s}_a^i \, \mathbb{s}_a^j)^{(2)} Q_{Ea} \, (E_a)^i_{,j} \\ &\quad + \frac{e_a}{8m_N^3} \left[2 \left\{ \vec{\mathbb{s}}_a \cdot \vec{B}_a, \vec{\pi}_a^2 \right\} + (g_a - 2) \left\{ \vec{\pi}_a \cdot \vec{B}_a, \vec{\pi}_a \cdot \vec{\mathbb{s}}_a \right\} \right] \\ &\quad - \frac{e_a^2}{2} B_a^i B_a^j \left[\alpha_a^M \delta^{ij} + \gamma_a^M (\mathbb{s}_a^i \, \mathbb{s}_a^j)^{(2)} \right], \end{aligned} \tag{12.99}$$

still has quite a rich structure, which we will explore in the following.

12.4.3 *Interaction Hamiltonian for General Spin*

In addition to the single-particle Hamiltonian (12.99), the generalization of the interaction Hamiltonian given in Eq. (12.76) to particles of general spin is equally interesting. It is obvious that we should replace, in Eq. (12.76), $\sigma_a^i \to 2\,\mathbb{s}_a^i$, as well as $2(1+\kappa_a) \to g_a$, to describe the spin vector and g factor of particle a. Using substitutions akin to those applied in Eq. (12.95), we derive the interaction Hamiltonian (12.64) for particles of a general spin quantum number as follows,

$$\begin{aligned}\mathcal{H}'_{ab} &= \frac{\alpha_{ab}}{r_{ab}} - \frac{\alpha_{ab}}{2m_a m_b} \pi_a^i \left(\frac{\delta^{ij}}{r_{ab}} - \frac{r_{ab}^i r_{ab}^j}{r_{ab}^3} \right) \pi_b^j - \frac{2\pi\alpha_{ab}}{3} \left(\langle \vec{r}_E^{\,2} \rangle_a + \langle \vec{r}_E^{\,2} \rangle_b \right) \delta^{(3)}(\vec{r}_{ab}) \\ &\quad - \frac{\pi\alpha_{ab}}{2m_a^2} (g_a - 1)\, \delta_{s_a,1/2} \delta^{(3)}(\vec{r}_{ab}) - \frac{\pi\alpha_{ab}}{2m_b^2} (g_b - 1)\, \delta_{s_b,1/2} \delta^{(3)}(\vec{r}_{ab}) \\ &\quad - \frac{2\pi\alpha_{ab} g_a g_b}{3m_a m_b} \vec{\mathbb{s}}_a \cdot \vec{\mathbb{s}}_b \, \delta^{(3)}(\vec{r}_{ab}) + \frac{\alpha_{ab} g_a g_b}{4m_a m_b r^3} \mathbb{s}_a^i \left(\delta^{ij} - 3 \frac{r_{ab}^i r_{ab}^j}{r_{ab}^2} \right) \mathbb{s}_b^j \\ &\quad + \frac{\alpha_{ab}}{2 r_{ab}^3} \left[\frac{g_a}{m_a m_b} \vec{\mathbb{s}}_a \cdot (\vec{r}_{ab} \times \vec{\pi}_b) - \frac{g_b}{m_a m_b} \vec{\mathbb{s}}_b \cdot (\vec{r}_{ab} \times \vec{\pi}_a) \right. \\ &\quad \left. + \frac{g_b - 1}{m_b^2} \vec{\mathbb{s}}_b \cdot (\vec{r}_{ab} \times \vec{\pi}_b) - \frac{g_a - 1}{m_a^2} \vec{\mathbb{s}}_a \cdot (\vec{r}_{ab} \times \vec{\pi}_a) \right] \\ &\quad + \frac{\alpha_{ab}}{6} \left[Q_{Ea} \, (\mathbb{s}_a^i \, \mathbb{s}_a^j)^{(2)} Q_{Eb} \, (\mathbb{s}_b^i \, \mathbb{s}_b^j)^{(2)} \right] \frac{3\, r_{ab}^i r_{ab}^j - \delta^{ij} \, \vec{r}_{ab}^{\,2}}{r_{ab}^5} \Bigg\} . \end{aligned} \tag{12.100}$$

This Hamiltonian is not quite equivalent to the ones given in Eq. (3) of Ref. [365] and in Eq. (17) of Ref. [366]. Namely, a closer inspection reveals that the Hamiltonian given in Eq. (3) of Ref. [365] and in Eq. (17) of Ref. [366] uses the radii $\langle \vec{r}_E^{\,2} \rangle_a$ and

$\langle \vec{r}_E^{\,2} \rangle_b$ in the conventions of Ref. [364], i.e., with the Darwin term included in the definition of the nuclear radius. However, the Darwin term actually is excluded from the nuclear radius in atomic-physics conventions, and also, canonically, in the analysis of scattering experiments, as explained in detail in Ref. [205]. The Hamiltonian given in Eq. (12.100) is compatible with a definition of the radii in terms of the derivative of the Sachs G_E form factor [see Eq. (11) of Ref. [205]]. This is the accepted definition, and it is implemented in Eq. (12.100).

In Eq. (12.100), the electric quadrupole terms are relevant only for spin-1 particles ($Q_E = 0$ for spin-1/2). Once more, we reemphasize that the radii of the spin-1/2 particles are defined in terms of the slope of the Sachs G_E form factor [see Eq. (11) of Ref. [205]].

12.5 Application to Two-Body Bound Systems

12.5.1 *General Aspects*

Considerable simplifications of the outlined formalism are possible for two-body systems such as hydrogen. We investigate a system of two spin-1/2 particles, of masses m_1 and m_2, without any external fields. Indeed, in some sense, atomic hydrogen constitutes the most imminent application of the formalism outlined in Sec. 12.3.4, in general. The total Hamiltonian is the sum of the single-particle Hamiltonians given in Eq. (12.75), but without coupling to external fields, i.e., $A_a^0 \to 0$, $\vec{E}_a \to \vec{0}$, $\vec{B}_a \to \vec{0}$, $\vec{\pi}_a \to \vec{p}_a$, for $a = 1, 2$, and of the interaction Hamiltonian (12.76), evaluated for $a = 1$, $b = 2$. Also, we assume that

$$e_1 = e\,, \qquad e_2 = -Ze\,, \qquad \alpha_{12} = -Z\alpha\,, \qquad \vec{r}_{12} = \vec{r}\,, \tag{12.101}$$

where e is the electron charge. Typically, we will choose m_1 to be the mass of the lighter ("orbiting") particle, while m_2 is the mass of the heavier particle. Then,

$$\mathcal{H} = \mathcal{H}_1 + \mathcal{H}_2 + \mathcal{H}_{12} = \mathcal{T}_1 + \mathcal{T}_2\,, \tag{12.102}$$

and it is useful to group the terms as follows. The term $\mathcal{T}_1$ contains the spin-independent relativistic effects,

$$\begin{aligned} \mathcal{T}_1 = {} & \left\{ \frac{\vec{p}_1^{\,2}}{2m_1} + \frac{\vec{p}_2^{\,2}}{2m_2} - \frac{Z\alpha}{r} \right\} - \frac{\vec{p}_1^{\,4}}{8m_1^3} - \frac{\vec{p}_2^{\,4}}{8m_2^3} + \frac{\pi Z\alpha}{2} \left(\frac{1+\kappa_1}{m_1^2} + \frac{1+\kappa_2}{m_2^2} \right) \delta^{(3)}(\vec{r}) \\ & + \frac{Z\alpha}{2m_1 m_2 r}\, p_1^i \left(\delta^{ij} + \frac{r^i r^j}{r^2} \right) p_2^j\,, \end{aligned} \tag{12.103}$$

while $\mathcal{T}_2$ contains the spin-dependent terms,

$$\begin{aligned} \mathcal{T}_2 = {} & + \frac{Z\alpha}{4r^3} (1 + 2\kappa_1) \frac{\vec{\sigma}_1 \cdot \vec{r} \times \vec{p}_1}{m_1^2} - \frac{Z\alpha}{4r^3} (1 + 2\kappa_2) \frac{\vec{\sigma}_2 \cdot \vec{r} \times \vec{p}_2}{m_2^2} \\ & + \frac{Z\alpha}{2m_1 m_2 r^3} \left[(1+\kappa_2) \vec{\sigma}_2 \cdot \vec{r} \times \vec{p}_1 - (1+\kappa_1) \vec{\sigma}_1 \cdot \vec{r} \times \vec{p}_2 \right] \\ & + \frac{Z\alpha}{4m_1 m_2} (1+\kappa_1)(1+\kappa_2) \left[\frac{8\pi}{3}\, \vec{\sigma}_1 \cdot \vec{\sigma}_2\, \delta^{(3)}(\vec{r}) - \frac{\sigma_1^i \sigma_2^j}{r^3} \left(\delta^{ij} - 3\,\frac{r^i r^j}{r^2} \right) \right]. \end{aligned} \tag{12.104}$$

Now, with reference to suggestion 1 in Sec. 4.8, as well as Secs. 13.2.2 and 16.2.1, we go into the center-of-mass frame, with $\vec{p}_1 = -\vec{p}_2 = \vec{p}$ and $\vec{P} = \vec{p}_1 + \vec{p}_2 = \vec{0}$, and introduce the orbital angular momentum $\vec{L} = \vec{r} \times \vec{p}$,

$$\begin{aligned}\mathcal{H} = &\left\{\frac{\vec{p}^{\,2}}{2\mu} - \frac{Z\alpha}{r}\right\} - \frac{\vec{p}^{\,4}}{8m_1^3} - \frac{\vec{p}^{\,4}}{8m_2^3} + \frac{\pi Z\alpha}{2}\left(\frac{1+\kappa_1}{m_1^2} + \frac{1+\kappa_2}{m_2^2}\right)\delta^{(3)}(\vec{r}) \\ &- \frac{Z\alpha}{2m_1m_2r}\,p^i\left(\delta^{ij} + \frac{r^i r^j}{r^2}\right)p^j + \frac{Z\alpha}{4r^3}\left(\frac{1+2\kappa_1}{m_1^2}\vec{\sigma}_1\cdot\vec{L} + \frac{1+2\kappa_2}{m_2^2}\vec{\sigma}_2\cdot\vec{L}\right) \\ &+ \frac{Z\alpha}{2m_1m_2r^3}\left[(1+\kappa_1)\vec{\sigma}_1\cdot\vec{L} + (1+\kappa_2)\vec{\sigma}_2\cdot\vec{L}\right] \\ &+ \frac{Z\alpha}{4m_1m_2}(1+\kappa_1)(1+\kappa_2)\left[\frac{8\pi}{3}\,\vec{\sigma}_1\cdot\vec{\sigma}_2\,\delta^{(3)}(\vec{r}) - \frac{\sigma_1^i\sigma_2^j}{r^3}\left(\delta^{ij} - 3\,\frac{r^i r^j}{r^2}\right)\right]. \end{aligned} \tag{12.105}$$

This Hamiltonian can be written as the sum

$$\mathcal{H} = H_S + H_{\rm rel} + H_{\rm fs} + H_{\rm hfs}\,, \tag{12.106a}$$

where H_S is the Schrödinger Hamiltonian, and the other terms read as follows. The nonrelativistic Schrödinger–Coulomb term H_S and the spin-independent relativistic corrections are

$$H_S = \frac{\vec{p}^{\,2}}{2\mu} - \frac{Z\alpha}{r}\,, \tag{12.106b}$$

$$H_{\rm rel} = -\frac{\vec{p}^{\,4}}{8m_1^3} - \frac{\vec{p}^{\,4}}{8m_2^3} + \frac{\pi Z\alpha}{2}\left(\frac{1+\kappa_1}{m_1^2} + \frac{1+\kappa_2}{m_2^2}\right)\delta^{(3)}(\vec{r}) - \frac{Z\alpha}{2m_1m_2r}\,p^i\left(\delta^{ij} + \frac{r^i r^j}{r^2}\right)p^j. \tag{12.106c}$$

Fine-structure and hyperfine-structure effects are described by the Hamiltonians $H_{\rm fs}$ and $H_{\rm hfs}$,

$$\begin{aligned} H_{\rm fs} &= \frac{Z\alpha}{4m_1^2r^3}(1+2\kappa_1)\vec{\sigma}_1\cdot\vec{L} + \frac{Z\alpha}{2m_1m_2r^3}(1+\kappa_1)\vec{\sigma}_1\cdot\vec{L} \\ &= \left[1 + 2\kappa_1 + 2\,\frac{m_1}{m_2}(1+\kappa_1)\right]\frac{Z\alpha\,\vec{\sigma}_1\cdot\vec{L}}{4m_1^2r^3}\,, \end{aligned} \tag{12.106d}$$

$$\begin{aligned} H_{\rm hfs} = &\left[1 + \kappa_2 + \frac{1}{2}\,\frac{m_1}{m_2}(1+2\kappa_2)\right]\frac{Z\alpha\,\vec{\sigma}_2\cdot\vec{L}}{2\,m_1\,m_2\,r^3} \\ &+ \frac{Z\alpha}{4m_1m_2}(1+\kappa_1)(1+\kappa_2)\left[\frac{8\pi}{3}\,\vec{\sigma}_1\cdot\vec{\sigma}_2\,\delta^{(3)}(\vec{r}) - \frac{\sigma_1^i\sigma_2^j}{r^3}\left(\delta^{ij} - 3\,\frac{r^i r^j}{r^2}\right)\right]. \end{aligned} \tag{12.106e}$$

Here, H_S is the Schrödinger–Coulomb Hamiltonian (which involves the reduced mass $\mu = m_1m_2/(m_1+m_2)$). In particular, the calculation clarifies one more time that the reduced mass should be used in the evaluation of the spectrum of the Schrödinger–Coulomb Hamiltonian. Moreover, $H_{\rm rel}$ describes the spin-independent relativistic correction terms (these do not contribute to the fine-structure). The

Hamiltonian $H_{\rm fs}$ contains all the terms which contain the spin $\vec{\sigma}_1$ of the orbiting particle; these terms contribute to the fine-structure (e.g., of hydrogenic P states, i.e., to the splitting of $P_{1/2}$ and $P_{3/2}$ states). Furthermore, $H_{\rm fs}$ enables us to calculate the reduced-mass dependence of the terms exactly, i.e., to all orders in the mass ratio (yet, we remember that the Breit Hamiltonian is valid only up to order α^4, or, with anomalous-magnetic-moment corrections, one might point out that a part of the order-α^5 corrections are also contained in it. Finally, $H_{\rm hfs}$ contains the nuclear-spin-dependent (or "other-particle-spin-dependent") terms (operator $\vec{\sigma}_2$), and it also includes the complete reduced-mass dependence.

12.5.2 *Definition of the Lamb Shift*

In order to calculate the Lamb shift $\mathcal{L}$ of atomic energy levels in a two-body bound systems, it is necessary to have a precise definition of the effect. This question alone is non-trivial, because in higher orders, even the spin of the nucleus plays a decisive role in calculating the spectrum [362]. Let us write the energy levels of a two-body system as

$$E = \mu\,[f(n,j) - 1] - \frac{\mu^2}{2(m_1 + m_2)}[f(n,j) - 1]^2 + \mathcal{L} + E_{\rm hfs}\,. \tag{12.107}$$

Here, $\mu = m_1 m_2/(m_1 + m_2)$ is the reduced mass of the system. This definition was given by Sapirstein and Yennie in Ref. [336] and has been adopted in many later works, e.g., in Eq. (67) of Ref. [173]. It is very instructive (and nontrivial) to investigate the origin of the idea behind Eq. (12.107).

The dimensionless Dirac energy $f(n,j)$ determines the energy levels of a two-body bound system, in the non-recoil limit, according to Eq. (8.49e) as

$$E = m\, f(n,j)\,, \qquad f(n,j) = \left(1 + \frac{(Z\alpha)^2}{[n_r + \sqrt{\kappa^2 - (Z\alpha)^2}]^2}\right)^{-\frac{1}{2}}, \tag{12.108}$$

$$n_r = n - |\kappa| = n - j - \tfrac{1}{2}\,. \tag{12.109}$$

Here, n_r is the "reduced" principal quantum number. The analytic form of the result (12.107) is motivated as follows: By expansion of $f(n,j)$ in $Z\alpha$, one obtains the Schrödinger spectrum as follows,

$$f(n,j) - 1 \approx \left(1 + \frac{(Z\alpha)^2}{(n - |\kappa| + |\kappa|)^2}\right)^{-\frac{1}{2}} - 1 = \left(1 + \frac{(Z\alpha)^2}{n^2}\right)^{-\frac{1}{2}} - 1 \approx -\frac{(Z\alpha)^2}{2\,n^2}\,. \tag{12.110}$$

The inclusion of the reduced mass in the Schrödinger spectrum has been discussed in Sec. 4.2.2. However, in the context of the Dirac equation, it is actually impossible to include reduced-mass effects from first principles, because the Dirac–Coulomb Hamiltonian (8.9), in contrast to the Breit Hamiltonian, constitutes a one-particle operator. Still, one can use the Breit Hamiltonian given in Eq. (12.106) in order to rewrite the reduced-mass dependence of the energy levels of the two-particle system into a unified formula, as far as possible.

Let us explore to which extent this is possible. First, we observe that hyperfine effects are described by the radiatively corrected hyperfine Breit Hamiltonian given in Eq. (12.106e) and therefore, are absorbed in $E_{\rm hfs}$ in Eq. (12.107). Let us write Eqs. (12.106c) and (12.106e) without radiative corrections, i.e., in the limit $\kappa_1 \to 0$, and $\kappa_2 \to 0$. This leads to the total Hamiltonian $\widetilde{H}$,

$$\widetilde{H} = H_S + \widetilde{H}_{\rm rel} + \widetilde{H}_{\rm fs}\,, \tag{12.111a}$$

where H_S is the Schrödinger Hamiltonian [see Eq. (4.1)], and

$$\widetilde{H}_{\rm rel} = -\frac{\vec{p}^{\,4}}{8m_1^3} - \frac{\vec{p}^{\,4}}{8m_2^3} + \frac{\pi Z\alpha}{2}\left(\frac{1}{m_1^2} + \frac{1}{m_2^2}\right)\delta^{(3)}(\vec{r}) - \frac{Z\alpha}{2m_1 m_2 r}\, p^i \left(\delta^{ij} + \frac{r^i r^j}{r^2}\right) p^j\,, \tag{12.111b}$$

$$\widetilde{H}_{\rm fs} = \left(1 + 2\,\frac{m_1}{m_2}\right)\frac{Z\alpha\,\vec{\sigma}_1\cdot\vec{L}}{4\,m_1^2\,r^3}\,. \tag{12.111c}$$

One can show that, up to order $(Z\alpha)^4$, the energy levels of the Hamiltonian (12.111) are given by the formula

$$\begin{aligned} E = {} & \mu\,[f(n,j) - 1] - \frac{\mu^2}{2(m_1 + m_2)}[f(n,j) - 1]^2 \\ & + \frac{(Z\alpha)^4}{2n^3 m_2^2}\left(\frac{1}{j + 1/2} - \frac{1}{\ell + 1/2}\right)(1 - \delta_{\ell 0})\,, \end{aligned} \tag{12.112}$$

including the complete reduced-mass dependence. Again, the formula (12.112) includes all the $(Z\alpha)^4$ terms, as well as a part of the higher-order corrections. The notion behind the decomposition in Eq. (12.112) is that all terms generated by the Breit Hamiltonian (12.111) which respect the (n,j) degeneracy, are still grouped together and written in terms of the dimensionless Dirac energy $f(n,j)$, while the terms in Eq. (12.112) which break the (n,j) degeneracy are known as the Barker–Glover corrections (see Ref. [367]) and read as follows,

$$H_{\rm BG} \sim \frac{(Z\alpha)^4}{2n^3 m_2^2}\left(\frac{1}{j + 1/2} - \frac{1}{\ell + 1/2}\right)(1 - \delta_{\ell 0})\,, \tag{12.113}$$

and thus are part of the Lamb shift $\mathcal{L}$, according to the definition (12.107). These considerations explain the motivation behind the definition (12.107).

12.5.3 *Application to Hydrogenlike Ions*

One application which is particularly important in the discussion of the hydrogen spectrum (but actually more general) concerns the contribution of the anomalous magnetic moment of the electron to the Lamb shift, and the reduced-mass dependence of the terms.

For hydrogenlike ions (and related systems, the discussion actually is more general), we set

$$m_1 = m\,, \qquad m_2 = m_N\,, \qquad \vec{\sigma}_1 \to \vec{\sigma}_e\,, \qquad \vec{\sigma}_2 \to \vec{\sigma}_N\,, \qquad \kappa_1 = \kappa \approx \frac{\alpha}{2\pi}\,, \tag{12.114}$$

and assume that $m \ll m_N$. However, we keep the nuclear charge number Z as a variable. Let us isolate, from $H_{\rm fs}$ defined in Eq. (12.106d),

$$H_{\rm fs} \sim \left[1 + 2\kappa + 2\,\frac{m}{m_N}\,(1+\kappa)\right] \frac{Z\,\alpha\,\vec{\sigma}_e \cdot \vec{L}}{4\,m^2\,r^3}\,, \tag{12.115}$$

those terms which contribute in the order $\alpha(Z\alpha)^4$. One obtains

$$\begin{aligned} H_A^{[5]} &\sim \left(2\kappa + 2\,\frac{m}{m_N}\kappa\right) \frac{Z\alpha\,\vec{\sigma}_e \cdot \vec{L}}{4\,m^2\,r^3} = \left(1 + \frac{m}{m_N}\right) \kappa\,\frac{Z\alpha\,\vec{\sigma}_e \cdot \vec{L}}{2\,m^2\,r^3} \\ &= \left(\frac{m + m_N}{m_N\,m}\right) \kappa\,\frac{Z\alpha\,\vec{\sigma}_e \cdot \vec{L}}{2\,m\,r^3} = \kappa\,\frac{Z\alpha\,\vec{\sigma}_e \cdot \vec{L}}{2\,m\,\mu\,r^3}\,. \end{aligned} \tag{12.116}$$

We use the relation

$$\begin{aligned} \left\langle \frac{Z\alpha\,\vec{\sigma}_e \cdot \vec{L}}{r^3} \right\rangle &= \left\langle \vec{\sigma}_e \cdot \vec{L} \right\rangle \left\langle \frac{Z\alpha}{r^3} \right\rangle \\ &= \frac{[j(j+1) - \ell(\ell+1) - s(s+1)]}{\ell\,(2\ell+1)\,(\ell+1)}\,\frac{2(Z\alpha)^4\mu^3}{n^3}\,(1 - \delta_{\ell 0}) \\ &= \frac{(Z\alpha)^4\,\mu^3}{n^3} \left(-\frac{2}{\varkappa(2\ell+1)}\right) (1 - \delta_{\ell 0})\,, \end{aligned} \tag{12.117}$$

for a hydrogenic state of symmetry $n^{2S+1}L_j$. Here, $\varkappa$ is not the anomalous contribution to the electron's magnetic moment, which, otherwise, we denote as $\kappa \neq \varkappa$ [see Eq. (1.9)], but the Dirac quantum number $\varkappa = (-1)^{j+\ell+1/2}(j+1/2)$, defined in Eq. (4.348). From Eqs. (12.116) and (12.117), we can derive the anomalous magnetic moment correction to the Lamb shift,

$$\begin{aligned} E_\kappa = \langle H_A^{[5]} \rangle &= \frac{\kappa}{2m\mu}\,\frac{(Z\alpha)^4\,\mu^3}{n^3} \left(-\frac{2}{\varkappa(2\ell+1)}\right) (1 - \delta_{\ell 0}) \\ &= \frac{\alpha}{\pi} \left(\frac{\mu}{m}\right)^2 \frac{(Z\alpha)^4 m}{n^3} \left(-\frac{1}{2\varkappa(2\ell+1)}\right) (1 - \delta_{\ell 0})\,, \end{aligned} \tag{12.118}$$

where we approximate $\kappa \approx \alpha/(2\pi)$ [see Eqs. (1.11) and (11.38)].

Here, an important observation is made: The dependence on the reduced mass of the anomalous-magnetic-moment contribution to the self-energy for states with $\ell \neq 0$ is given by the factor $(\mu/m)^2$, not $(\mu/m)^3$, as it would otherwise be expected for typical contributions to the Lamb shift (see Sec. 11.4). The correct reduced-mass dependence of the anomalous-magnetic moment correction to the Lamb shift can only be found within a two-body formalism; this consideration finally explains the replacement made in Eq. (11.137).

In Refs. [173, 336], the peculiar reduced-mass dependence is ascribed to "a consequence of the proton's convection current". In Ref. [336], we refer to a remark on page 569, in between Eqs. (2.4) and (2.5a). How should this remark be interpreted? Let us consider Eqs. (12.104) and (12.116). One realizes that a term proportional to $\vec{\sigma}_1 \cdot \vec{r} \times \vec{p}_2$ is the origin of the anomalous magnetic-moment correction to the Lamb shift; it leads to the peculiar reduced-mass dependence described by the factor $(\mu/m)^2$. This term, in turn, contains the proton's momentum operator $\vec{p}_2$ (as opposed to $\vec{p}_1$), which corresponds to the convection current as opposed to the spin current. In the center-of-mass frame with $\vec{p}_1 = -\vec{p}_2$, of course, the distinction between the electron's and the proton's convection current becomes unnecessary.

Now, we isolate, from Eq. (12.106c), specifically, from the term

$$H_{\rm rel} \sim \frac{\pi Z\alpha}{2} \left(\frac{1 + 2\kappa}{m^2} \right) \delta^{(3)}(\vec{r}) \,, \tag{12.119}$$

an anomalous-magnetic moment corrections which contributes to S states,

$$H_B^{[5]} = \pi Z\alpha \, \frac{\kappa}{m^2} \, \delta^{(3)}(\vec{r}) \,. \tag{12.120}$$

The energy correction induced by $H'_{\alpha 5}$ is

$$E'_\kappa = \langle H_B^{[5]} \rangle = \pi Z\alpha \, \frac{\kappa}{m^2} \, \langle \delta^{(3)}(\vec{r}) \rangle = \frac{\alpha}{\pi} \frac{1}{2} \left(\frac{\mu}{m} \right)^3 \frac{(Z\alpha)^4 m}{n^3} \, \delta_{\ell 0} \,. \tag{12.121}$$

For $\kappa = -1$ (which is the Dirac angular quantum number for S states), and $\mu = m$ (infinite nuclear mass limit), the formulas (12.118) reduce to (12.121); however, for a finite nuclear mass, one needs to distinguish between S and non-S states.

Two remarks are in order: *(i)* In regard to the discussion of the effective Dirac equation (with form factors) from Eq. (11.37), we should mention that there is a slight danger of double counting the anomalous magnetic-moment contributions to the Lamb shift. One needs to realize that the radiatively corrected Breit Hamiltonian describes the same effect as the $F_2(0)$ term in Eq. (11.37). Double counting should be avoided; once the anomalous magnetic moment has been taken into account on the level of the two-particle formalism, one cannot invoke it in addition, on the basis of a single-particle Foldy–Wouthuysen transformation. *(ii)* One might wonder about the term corresponding to Eq. (12.119), but with the proton mass as opposed to the electron mass in the denominator. This term is present in Eq. (12.105) but it forms part of the nuclear self-energy and is thus omitted here.

For hydrogenlike systems, the hyperfine splitting operator from Eq. (12.106e) reads as

$$\begin{aligned} H_{\rm hfs} = {} & \left[1 + \kappa_N + \frac{1}{2} \frac{m}{m_N} (1 + 2\kappa_N) \right] \frac{Z\alpha \, \vec{\sigma}_N \cdot \vec{L}}{2 \, m \, m_N \, r^3} \\ & + \frac{Z\alpha}{4 \, m \, m_N} (1 + \kappa)(1 + \kappa_N) \left[\frac{8\pi}{3} \, \vec{\sigma}_e \cdot \vec{\sigma}_N \, \delta^{(3)}(\vec{r}) - \frac{\sigma_e^i \, \sigma_N^j}{r^3} \left(\delta^{ij} - 3 \frac{r^i \, r^j}{r^2} \right) \right] . \end{aligned} \tag{12.122}$$

This operator can be brought into a much more familiar form by using the identity

$$Z \, (1 + \kappa_N) \frac{\alpha}{2m_N} \, \vec{\sigma}_N = \frac{1}{4\pi} \, 2Z(1 + \kappa_N) \frac{e^2}{2m_N} \frac{\vec{\sigma}_N}{2} = \frac{|e|}{4\pi} (1 + \kappa_N) \frac{Z|e|}{2m_N} \, \vec{s}_N = \frac{|e| \, \vec{\mu}_N}{4\pi} \,, \tag{12.123}$$

where $\vec{s}_N = \vec{\sigma}_N/2$ is the proton spin operator, and g_N is the proton g factor. We shall now explain the reasoning which leads to the identity (12.123) in greater detail. To this end, we first observe that, by definition, the nuclear magnetic moment is related to the nuclear spin operator as follows [see the discussion surrounding Eq. (12.90)],

$$\vec{\mu}_N = Z(1 + \kappa_N)\,\frac{|e|}{2m_N}\,\vec{s}_N \equiv g_N\,\frac{|e|}{2m_p}\,\frac{\vec{\sigma}_N}{2} = g_N\,\mu_N\,\frac{\vec{\sigma}_N}{2}\,. \tag{12.124}$$

We had initially calculated the hyperfine Hamiltonian, from the assumption that the nucleus is a point-like spin-1/2 particle, of charge $|q| = Z|e|$. Thus, *a priori*, its magnetic moment would be $\vec{\mu} = (1 + \kappa_N)\,|q|/(2m_N)\vec{s}_N$, according to our definition of the magnetic component of the current in Eq. (12.71). This would be valid for an internally structureless nucleus. Indeed, if the proton or other nucleus were a point-like, spin-1/2 particle of charge $Z\,|e|$, then we would have $\kappa_N = Z\kappa$, and this term, indeed, figures in a properly defined nuclear self-energy for S states, according to Eq. (28) of Ref. [205]. However, nuclear magnetic moments are, by definition, measured in terms of the nuclear magneton μ_N, which involves the proton mass, and this is the definition adopted in Eq. (12.124).

The formula

$$\vec{\mu}_N = g_N\,\mu_N\,\frac{\vec{\sigma}_N}{2}\,, \tag{12.125}$$

which measures nuclear magnetic moments in terms of the nuclear spin operator $\vec{s}_N = \vec{\sigma}_N/2$, also defines the nuclear g factor, which we denote as g_N. With these definitions, the general hyperfine Hamiltonian (12.106e) for an orbiting electron and a nucleus with magnetic moment vector $\vec{\mu}_N$ takes the form

$$\begin{aligned} H_{\rm hfs} = \frac{m|e|\,\vec{\mu}_N}{4\pi}\cdot\Bigg[&\left(1 + \frac{g_N - 1}{g_N}\,\frac{m}{m_N}\right)\frac{\vec{L}}{m^2\,r^3} \\ &+ (1+\kappa)\frac{4\pi}{3}\,\vec{\sigma}_e\,\frac{\delta^{(3)}(\vec{r})}{m^2} + \frac{1+\kappa}{2m^2r^3}\left(\frac{\vec{r}\,(\vec{\sigma}_e\cdot\vec{r})}{r^2} - \vec{\sigma}_e\right)\Bigg]. \end{aligned} \tag{12.126}$$

Let us assume that the nucleus is a proton ($N = p$), and that the quantization axis is the z axis. The matrix element of the z component of the proton's magnetic moment is

$$\langle\pm|\vec{\mu}_p\cdot\hat{\rm e}_z|\pm\rangle = \pm\frac{g_p\,\mu_N}{2}\,, \tag{12.127}$$

where $|\pm\rangle$ are states with the up or down projections of spin, onto the z axis. Irrespective of the spin orientation, the magnitude of the proton's magnetic moment amounts to

$$|\langle\pm|\vec{\mu}_p|\pm\rangle| = \frac{g_p\,\mu_N}{2} \approx 1.4106\times 10^{-26}\,\frac{\rm J}{\rm T}\,, \tag{12.128}$$

while the nuclear magneton is

$$\mu_N = \frac{|e|}{2m_p} \approx 5.0507\times 10^{-27}\,\frac{\rm J}{\rm T}\,, \tag{12.129}$$

leading to a proton g factor of

$$g_p = 2\,\frac{|\langle\pm|\vec{\mu}_p|\pm\rangle|}{\mu_N} \approx 5.58569\,. \qquad (12.130)$$

One notes the factor 2, which is somewhat non-obvious. For S states, the Hamiltonian given in Eq. (12.126) can be reduced to only one term (a contact interaction),

$$\begin{aligned}\langle H_{\rm hfs}\rangle &= \frac{m|e|}{4\pi}\,(1+\kappa)\frac{4\pi}{3}\left\langle\vec{\mu}_N\cdot\vec{\sigma}_e\,\frac{\delta^{(3)}(\vec{r})}{m^2}\right\rangle \\ &= (1+\kappa)\,\frac{|e|}{6m}\,g_N\,\mu_N\,\left\langle\vec{\sigma}_N\cdot\vec{\sigma}_e\,\delta^{(3)}(\vec{r})\right\rangle\,. \qquad (12.131)\end{aligned}$$

Now, for S states, the total angular momentum is $\vec{F} = \vec{s}_e + \vec{s}_N$, i.e., the sum of the spin angular momenta,

$$\langle\vec{\sigma}_N\cdot\vec{\sigma}_e\rangle = 4\,\langle\vec{s}_N\cdot\vec{s}_e\rangle = 2\,\left(\vec{F}^2 - \vec{s}_N^{\,2} - \vec{s}_e^{\,2}\right)\,. \qquad (12.132)$$

For $F = 1$ states, one has

$$X_1 = \vec{F}^2 - \vec{s}_N^{\,2} - \vec{s}_e^{\,2} = F(F+1) - s_N(s_N+1) - s_e(s_e+1) = 2 - \tfrac{3}{4} - \tfrac{3}{4} = \tfrac{1}{2}\,. \qquad (12.133)$$

For $F = 0$ states, one has

$$X_2 = \vec{F}^2 - \vec{s}_N^{\,2} - \vec{s}_e^{\,2} = F(F+1) - s_N(s_N+1) - s_e(s_e+1) = 0 - \tfrac{3}{4} - \tfrac{3}{4} = -\tfrac{3}{2}\,. \qquad (12.134)$$

The difference between $F = 1$ and $F = 0$ states thus is $X_1 - X_2 = 2$. Denoting the hyperfine-structure difference of the sublevels of the ground state of hydrogenlike ions by the symbol $\langle\langle\cdot\rangle\rangle$, we have

$$\langle\langle\vec{\sigma}_N\cdot\vec{\sigma}_e\rangle\rangle = 2\,(X_1 - X_2) = 4\,. \qquad (12.135)$$

Then,

$$\begin{aligned}\langle\langle H_{\rm hfs}\rangle\rangle &= (1+\kappa)\,\frac{|e|}{6m}\,g_N\,\mu_N\,\left\langle\left\langle\vec{\sigma}_N\cdot\vec{\sigma}_e\,\delta^{(3)}(\vec{r})\right\rangle\right\rangle \\ &= (1+\kappa)\,\frac{2|e|}{3m}\,g_N\,\mu_N\,|\phi(\vec{0})|^2\,. \qquad (12.136)\end{aligned}$$

This is the familiar result [see Eq. (12) of Ref. [37]]. It includes the leading correction due to the anomalous magnetic moment of the electron. The reduced-mass dependence is captured in the formula

$$\begin{aligned}\langle\langle H_{\rm hfs}\rangle\rangle &= (1+\kappa)\,\frac{e^2}{3\,m\,m_p}\,g_N\,\frac{(Z\alpha\mu)^3}{\pi n^3} \\ &= (1+\kappa)\left(\frac{\mu}{m}\right)^3 g_N\,\frac{4\alpha\,(Z\alpha)^3}{3n^3}\,\frac{m^2}{m_p}\,. \qquad (12.137)\end{aligned}$$

This formula describes the so-called 21 cm line that one observes in astrophysical spectra; it corresponds to the $F = 0$ to $F = 1$ transition among the hyperfine sublevels of the hydrogen ground state. The derivation has been given in great detail, in order to elucidate the origin of all terms, including the anomalous magnetic moment, and the reduced-mass dependence.

12.6 Compound System in a Homogeneous Magnetic Field

12.6.1 *Power–Zienau Transformation for a Magnetic Field*

In Secs. 12.3.3 and 12.3.4, we have set up the formalism for the treatment of a system of particles which interact both with external electromagnetic fields, as well as with each other. The ultimate aim of the endeavor, of course, is to be able to accurately describe the properties of compound systems, consisting of electrons and nuclei. We are now in the position to apply this formalism to the calculation of atomic interactions with external fields. In this, we are inspired by Refs. [365,366,368–370]. The entire setting is a nontrivial problem because one has to separate the interaction of the compound system with the external fields, from the internal interactions. Here, we will concentrate on the case of an external magnetic field, which we assume to be constant and uniform over the scale of the quantum system under study, and analyze its interaction with a compound system, as an example of the interaction of a compound system with an external field.

We assume arbitrary spin symmetries of the constituent particles and start from the Hamiltonian

$$\mathcal{H} = \sum_a \mathcal{H}'_a + \sum_{a,b>a} \mathcal{H}'_{ab} \,. \tag{12.138}$$

Here, $\mathcal{H}'_a$ is given by Eq. (12.99), in the sense of Eq. (12.86)]. We assume that there are no external electric fields ($\vec{E}_a \to \vec{0}$), while we have a uniform magnetic field ($\vec{B}_a = \vec{B}$). Furthermore, $\mathcal{H}'_{ab}$ is given by Eq. (12.100).

Our aim is to find a suitable unitary transformation ϕ, so that the transformed Hamiltonian

$$H = \mathrm{e}^{-\mathrm{i}\phi}\, \mathcal{H}\, \mathrm{e}^{\mathrm{i}\phi} \tag{12.139}$$

is brought into a form suitable for further calculations of the magnetic-field interaction of the compound system, with the external field. Let us go into the center-of-mass frame, with

$$\vec{x}_a = \vec{r}_a - \vec{R}\,, \qquad \vec{R} = \sum_a \frac{m_a}{M}\, \vec{r}_a\,, \tag{12.140a}$$

$$\vec{q}_a = \vec{p}_a - \frac{m_a}{M}\, \vec{P}\,, \qquad \vec{P} = \sum_a \vec{p}_a\,, \tag{12.140b}$$

where $M = \sum_a m_a$. The letter q here is reserved for the relative momenta. The covariantly coupled total momentum of the compound system is

$$\vec{\Pi} = \sum_a \vec{\pi}_a = \sum_a \left[\vec{p}_a - e_a \vec{A}(\vec{R})\right] = \vec{P} - Q\,\vec{A}(\vec{R})\,, \qquad Q = \sum_a e_a\,, \tag{12.141}$$

where Q is the total charge. In the following, we assume an exclusively spatially dependent vector potential, restricting the discussion to a uniform external magnetic field. Effects connected with external electric fields like the Röntgen term (see Ref. [371]) are excluded from the discussion.

In the following, it is very important to realize that we shall denote the components of $\vec{r}_a$ as r_a^k, while the components of $\vec{x}_a$ are labeled x_a^k (we had used different conventions at other places in this book). The commutators are easily calculated as follows (see also Sec. 12.7),

$$[x_a^i, q_b^j] = [x_a^i, p_b^j] = \mathrm{i}\,\delta^{ij}\left(\delta_{ab} - \frac{m_b}{M}\right), \tag{12.142a}$$

$$[R^i, P^j] = \mathrm{i}\,\delta^{ij}, \tag{12.142b}$$

$$[x_a^i, P^j] = [R^i, q_a^j] = 0. \tag{12.142c}$$

We now employ a gauge transformation inspired by the Power–Zienau method [176],

$$\phi = \sum_a e_a \int_0^1 \mathrm{d}u\, \vec{x}_a \cdot \vec{A}(\vec{R} + u\,\vec{x}_a), \tag{12.143}$$

$$= \sum_a e_a \left(x_a^i\, A^i(\vec{R}) + \frac{1}{2!}\, x_a^i\, x_a^j\, A^i_{,j}(\vec{R}) + \frac{1}{3!}\, x_a^i\, x_a^j\, x_a^k\, A^i_{,jk}(\vec{R}) + \cdots \right). \tag{12.144}$$

Here, the partial derivatives are

$$A^i_{,j}(\vec{R}) = \left.\frac{\partial}{\partial r^j} A^i\right|_{\vec{r}=\vec{R}}. \tag{12.145}$$

For a constant $\vec{B}$ field, one has for general argument $\vec{r}$,

$$\vec{A}(\vec{r}) = \frac{1}{2}(\vec{B} \times \vec{r}), \qquad A^i = \frac{1}{2}\epsilon^{i\ell m} B^\ell r^m, \qquad A^i_{,j} = \frac{1}{2}\epsilon^{i\ell j} B^\ell. \tag{12.146}$$

For a homogeneous magnetic field, the gauge phase ϕ can be formulated as follows, by directly inserting the expression $\vec{A} = \frac{1}{2}(\vec{B} \times \vec{r})$ into the formula for ϕ,

$$\begin{aligned}
\phi &= \sum_a e_a \int_0^1 \mathrm{d}u\, \vec{x}_a \cdot \vec{A}(\vec{R} + u\,\vec{x}_a) = \sum_a e_a \frac{1}{2} \int_0^1 \mathrm{d}u\, \vec{x}_a \cdot \vec{B} \times (\vec{R} + u\,\vec{x}_a) \\
&= \sum_a e_a \frac{1}{2} \int_0^1 \mathrm{d}u\, \vec{x}_a \cdot \vec{B} \times \vec{R} + \sum_a e_a \frac{1}{2} \int_0^1 \mathrm{d}u\, u\, \vec{x}_a \cdot \vec{B} \times \vec{x}_a \\
&= \sum_a e_a \frac{1}{2}\, \vec{x}_a \cdot \vec{B} \times \vec{R} + 0 = \frac{1}{2} \sum_a e_a\, \vec{x}_a \cdot \vec{B} \times \vec{R}.
\end{aligned} \tag{12.147}$$

Other important results are

$$\begin{aligned}
\frac{\partial x_c^i}{\partial r_a^k} &= \frac{\partial}{\partial r_a^k}\left(r_c^i - \sum_b \frac{m_b}{M} r_b^i \right) = \delta^{ik}\,\delta_{ac} - \sum_b \frac{m_b}{M}\,\delta^{ik}\,\delta_{ab} \\
&= \delta^{ik}\,\delta_{ac} - \frac{m_a}{M}\,\delta^{ik} = \delta^{ik}\left(\delta_{ac} - \frac{m_a}{M}\right),
\end{aligned} \tag{12.148a}$$

$$\frac{\partial R^i}{\partial r_a^k} = \frac{\partial}{\partial r_a^k} \sum_b \frac{m_b}{M} r_b^i = \sum_b \frac{m_b}{M} \delta^{ik}\,\delta_{ab} = \delta^{ik}\,\frac{m_a}{M}. \tag{12.148b}$$

It is helpful to calculate the commutator $[\frac{\partial}{\partial r_a^k}, \phi]$,

$$
\begin{aligned}
\frac{\partial}{\partial r_a^k}\phi &= \frac{1}{2}\sum_c e_c \frac{\partial x_c^i}{\partial r_a^k}\epsilon^{i\ell m} B^\ell R^m + \frac{1}{2}\sum_c e_c x_c^i \epsilon^{i\ell m} B^\ell \frac{\partial R^m}{\partial r_a^k} \\
&= \frac{1}{2}\sum_c e_c \delta^{ik}\left(\delta_{ac} - \frac{m_a}{M}\right)\epsilon^{i\ell m} B^\ell R^m + \frac{1}{2}\sum_c e_c x_c^i \epsilon^{i\ell m} B^\ell \frac{\partial R^m}{\partial r_a^k} \\
&= \frac{1}{2}\sum_c \left(e_c \delta^{ik}\delta_{ac}\epsilon^{i\ell m} B^\ell R^m - \frac{m_a}{M}\frac{e_c}{2}\delta^{ik}\epsilon^{i\ell m} B^\ell R^m + \frac{e_c}{2} x_c^i \epsilon^{i\ell m} B^\ell \delta^{mk}\frac{m_a}{M}\right) \\
&= \frac{e_a}{2}(\vec{B}\times\vec{R})^k - \frac{Q}{2}\frac{m_a}{M}(\vec{B}\times\vec{R})^k + \frac{1}{2}\frac{m_a}{M}(\vec{D}\times\vec{B})^k .
\end{aligned}
\tag{12.149}
$$

Here,

$$
\vec{D} = \sum_a e_a \vec{x}_a \tag{12.150}
$$

is the dipole moment of the charge distribution, measured in reference to its center-of-mass.

The Power–Zienau transformation serves to eliminate the vector potentials in favor of the fields (we need to differentiate the summation indices b and c from the particle index a for the momentum $\vec{\pi}_a$ under investigation). One finds for $\mathrm{e}^{-\mathrm{i}\phi}\,\pi_a^k\,\mathrm{e}^{\mathrm{i}\phi} = \pi_a'^k$ where $\vec{\pi}_a = -\mathrm{i}\vec{\nabla}_a - e_a\vec{A}(\vec{r}_a)$,

$$
\begin{aligned}
\pi_a'^k &= \exp(-\mathrm{i}\phi)\left(-\mathrm{i}\frac{\partial}{\partial r_a^k} - e_a A^k(\vec{r}_a)\right)\exp(\mathrm{i}\phi) = -\mathrm{i}\,\nabla_a^k - e_a A^k(\vec{x}_a + \vec{R}) + \left[\frac{\partial}{\partial r_a^k}, \phi\right] \\
&= p_a^k - e_a \frac{1}{2}(\vec{B}\times\vec{x}_a)^k - \frac{Q}{2}\frac{m_a}{M}(\vec{B}\times\vec{R})^k + \frac{1}{2}\frac{m_a}{M}(\vec{D}\times\vec{B})^k \\
&= \underbrace{\left\{p_a^k - \frac{m_a}{M}P^k\right\}}_{=q_a^k} + \frac{m_a}{M}\left[P^k - \frac{Q}{2}(\vec{B}\times\vec{R})^k\right] + \frac{1}{2}\left[\left(e_a\,\vec{x}_a + \frac{m_a}{M}\vec{D}\right)\times\vec{B}\right]^k \\
&= \underbrace{q_a^k + \frac{1}{2}\left[\left(e_a\,\vec{x}_a + \frac{m_a}{M}\vec{D}\right)\times\vec{B}\right]^k}_{\equiv\underline{\pi}_a^k} + \frac{m_a}{M}\underbrace{\left[P^k - \frac{Q}{2}(\vec{B}\times\vec{R})^k\right]}_{\equiv\Pi^k} = \underline{\pi}_a^k + \frac{m_a}{M}\Pi^k .
\end{aligned}
\tag{12.151}
$$

We have just established the result

$$
\mathrm{e}^{-\mathrm{i}\phi}\,\vec{\pi}_a\,\mathrm{e}^{\mathrm{i}\phi} = \vec{\underline{\pi}}_a + \frac{m_a}{M}\vec{\Pi}, \qquad \vec{\underline{\pi}}_a = \vec{q}_a + \frac{1}{2}\vec{\mathcal{D}}_a\times\vec{B}, \tag{12.152a}
$$

$$
\vec{\mathcal{D}}_a = e_a\,\vec{x}_a + \frac{m_a}{M}\vec{D}, \tag{12.152b}
$$

where $\vec{D}$ has been defined in Eq. (12.150), and $\vec{q}_a$ in Eq. (12.140b). With the help of Eq. (12.152), the kinetic energy transforms as follows (see also Sec. 12.7),

$$
\mathrm{e}^{-\mathrm{i}\phi}\left(\sum_a \frac{\vec{\pi}_a^2}{2m_a}\right)\mathrm{e}^{\mathrm{i}\phi} = \sum_a \frac{\vec{\underline{\pi}}_a^2}{2m_a} + \frac{\vec{\Pi}^2}{2M} + \frac{\vec{\Pi}}{M}\cdot\left(\vec{D}\times\vec{B}\right). \tag{12.153}
$$

The following identity has been used,

$$\begin{aligned}\sum_a \vec{\underline{\pi}}_a \cdot \vec{\Pi} &= \sum_a \left(\vec{q}_a + \frac{1}{2}\left(e_a\,\vec{x}_a + \frac{m_a}{M}\vec{D}\right)\times\vec{B}\right)\cdot\vec{\Pi} \\ &= \left(\sum_a \vec{q}_a\right)\cdot\vec{\Pi} + \left[\frac{1}{2}\left(\sum_a e_a\,\vec{x}_a\right)\times\vec{B}\right]\cdot\vec{\Pi} \\ &\quad + \frac{1}{2}\sum_a \frac{m_a}{M}\left(\vec{D}\times\vec{B}\right)\cdot\vec{\Pi} = \left(\vec{D}\times\vec{B}\right)\cdot\vec{\Pi}\,.\end{aligned} \tag{12.154}$$

One finds that the Hamiltonian $\mathcal{H}$, defined in Eq. (12.138), of the compound system, interacting with the external magnetic field, can be split into three contributions: *(i)* the nonrelativistic term H_0, *(ii)* the relativistic interaction, still in terms of the internal coordinates of the system H_R, and *(iii)* the overall kinetic term H_Π, which describes the interaction of the compound system with the external magnetic field. The terms read as follows [see Eq. (12.139)],

$$H = \mathrm{e}^{-\mathrm{i}\phi}\,\mathcal{H}\,\mathrm{e}^{-\mathrm{i}\phi} = H_0 + H_R + H_\Pi\,, \tag{12.155a}$$

$$H_0 = \sum_a \frac{\vec{\underline{\pi}}_a^{\,2}}{2m_a} + \sum_{a>b,b}\frac{\alpha_{ab}}{r_{ab}} - \sum_a \frac{e_a}{2m_a}\,g_a\,\vec{\mathbb{s}}_a\cdot\vec{B} + \frac{\vec{\Pi}^2}{2M} + \frac{\vec{\Pi}}{M}\cdot\left(\vec{D}\times\vec{B}\right)\,, \tag{12.155b}$$

$$\begin{aligned}H_R = \sum_a \Bigg\{ &-\frac{\vec{\underline{\pi}}_a^{\,4}}{2m_a^3} + \frac{e_a}{8m_a^3}\left[4\,\vec{\underline{\pi}}_a^{\,2}\,\vec{\mathbb{s}}_a\cdot\vec{B} + (g_a-2)\left\{\vec{\underline{\pi}}_a\cdot\vec{B},\vec{\underline{\pi}}_a\cdot\vec{\mathbb{s}}_a\right\}\right] \\ &-\frac{e_a^2}{2}\,B^i\,B^j\left[\alpha_a^M\,\delta^{ij} + \gamma_a^M\,(\mathbb{s}_a^i\,\mathbb{s}_a^j)^{(2)}\right]\Bigg\} + \sum_{a>b}\widetilde{\mathcal{H}}'_{ab}\,.\end{aligned} \tag{12.155c}$$

We recall the definition of α_{ab} from Eq. (12.67). Here, $\widetilde{\mathcal{H}}'_{ab}$ is obtained from $\mathcal{H}'_{ab}$ given in Eq. (12.100) under the replacement $\vec{\pi}_a \to \vec{\underline{\pi}}_a$. For a spin-1/2 particle, it reduces to the Hamiltonian given in Eq. (12.76). So, we have, specifically,

$$\begin{aligned}\widetilde{\mathcal{H}}'_{ab} = &\frac{\alpha_{ab}}{r_{ab}} - \frac{\alpha_{ab}}{2m_a m_b}\,\underline{\pi}_a^i\left(\frac{\delta^{ij}}{r_{ab}} - \frac{r_{ab}^i\,r_{ab}^j}{r_{ab}^3}\right)\underline{\pi}_b^j - \frac{2\pi\alpha_{ab}}{3}\left(\langle\vec{r}_E^{\,2}\rangle_a + \langle\vec{r}_E^{\,2}\rangle_b\right)\delta^{(3)}(\vec{r}_{ab}) \\ &- \frac{\pi\alpha_{ab}}{2m_a^2}\,(g_a-1)\,\delta_{s_a,1/2}\,\delta^{(3)}(\vec{r}_{ab}) - \frac{\pi\alpha_{ab}}{2m_b^2}\,(g_b-1)\,\delta_{s_b,1/2}\,\delta^{(3)}(\vec{r}_{ab}) \\ &- \frac{2\pi\alpha_{ab}\,g_a\,g_b}{3m_a\,m_b}\,\vec{\mathbb{s}}_a\cdot\vec{\mathbb{s}}_b\,\delta^{(3)}(\vec{r}_{ab}) + \frac{\alpha_{ab}\,g_a\,g_b}{4m_a\,m_b\,r^3}\,\mathbb{s}_a^i\left(\delta^{ij} - 3\,\frac{r_{ab}^i\,r_{ab}^j}{r_{ab}^2}\right)\mathbb{s}_b^j \\ &+ \frac{\alpha_{ab}}{2\,r_{ab}^3}\Bigg[\frac{g_a}{m_a\,m_b}\,\vec{\mathbb{s}}_a\cdot\left(\vec{r}_{ab}\times\vec{\underline{\pi}}_b\right) - \frac{g_b}{m_a\,m_b}\,\vec{\mathbb{s}}_b\cdot\left(\vec{r}_{ab}\times\vec{\underline{\pi}}_a\right) \\ &+ \frac{g_b-1}{m_b^2}\,\vec{\mathbb{s}}_b\cdot\left(\vec{r}_{ab}\times\vec{\underline{\pi}}_b\right) - \frac{g_a-1}{m_a^2}\,\vec{\mathbb{s}}_a\cdot\left(\vec{r}_{ab}\times\vec{\underline{\pi}}_a\right)\Bigg] \\ &+ \frac{\alpha_{ab}}{6}\left[Q_{Ea}\,(\mathbb{s}_a^i\mathbb{s}_a^j)^{(2)}Q_{Eb}\,(\mathbb{s}_b^i\mathbb{s}_b^j)^{(2)}\right]\frac{3\,r_{ab}^i\,r_{ab}^j - \delta^{ij}\,\vec{r}_{ab}^{\,2}}{r_{ab}^5}\,.\end{aligned} \tag{12.156}$$

One obtains the following initial expression for H_Π as defined in Eq. (12.155a) [see Eq. (32) of Ref. [366]]

$$\begin{aligned} H_\Pi = & -\frac{1}{8M^3}\vec{\Pi}^4 - \frac{1}{2M^3}\vec{\Pi}^2\,\vec{\Pi}\cdot(\vec{D}\times\vec{B}) - \frac{\vec{\Pi}^2}{2M^2}\left(\sum_a \frac{\vec{\underline{\pi}}_a^2}{2m_a} + \sum_{a>b,b}\frac{\alpha_{ab}}{r_{ab}}\right) \\ & - \frac{\Pi^i\,\Pi^j}{2M^2}\left(\sum_a \frac{\underline{\pi}_a^i\,\underline{\pi}_a^j}{2m_a} + \sum_{a>b,b}\frac{\alpha_{ab}\,r_{ab}^i\,r_{ab}^j}{r_{ab}^3}\right) - \sum_a \frac{1}{2Mm_a^2}\vec{\underline{\pi}}_a^2\,(\vec{\underline{\pi}}_a\cdot\vec{\Pi}) \\ & + \sum_a \frac{e_a}{4M^2 m_a}\left[2\vec{\Pi}^2\,\vec{\mathbb{s}}_a\cdot\vec{B} + (g_a-2)\,(\vec{\Pi}\cdot\vec{B})\,(\vec{\Pi}\cdot\vec{\mathbb{s}}_a)\right] + \sum_a \frac{e_a}{4Mm_a^2} \\ & \times\left[4(\vec{\underline{\pi}}_a\cdot\vec{\Pi})(\vec{\mathbb{s}}_a\cdot\vec{B}) + (g_a-2)(\vec{\Pi}\cdot\vec{B})(\vec{\underline{\pi}}_a\cdot\vec{\mathbb{s}}_a) + (g_a-2)\,(\vec{\Pi}\cdot\vec{\mathbb{s}}_a)(\vec{\underline{\pi}}_a\cdot\vec{B})\right] \\ & + \sum_{a>b,b}\alpha_{ab}\left[\frac{1}{2Mm_a r_{ab}^3}\,\vec{\mathbb{s}}_a\cdot\left(\vec{r}_{ab}\times\vec{\Pi}\right) - \frac{1}{2Mm_b}\Pi^i\left(\frac{\delta^{ij}}{r_{ab}} + \frac{r_{ab}^i\,r_{ab}^j}{r_{ab}^3}\right)\underline{\pi}_b^j\right]. \end{aligned} \tag{12.157}$$

However, after several unitary transformations, similar to those employed in Refs. [365, 366], one may simplify H_Π to read

$$\begin{aligned} H_\Pi = & -\frac{\vec{\Pi}^4}{8M^3} - \frac{\vec{\Pi}^2}{2M^2}\left(\sum_a \frac{\vec{\underline{\pi}}_a^2}{2m_a} + \sum_{a>b,b}\frac{\alpha_{ab}}{r_{ab}}\right) \\ & + \frac{Q}{2M^3}\left[\vec{\Pi}^2(\vec{J}\cdot\vec{B}) + \frac{g_J-2}{2}(\vec{J}\cdot\vec{\Pi})\,(\vec{B}\cdot\vec{\Pi})\right]. \end{aligned} \tag{12.158}$$

The angular momenta are defined as follows,

$$\frac{Q}{M}\,g_J\,\vec{J} = \sum_a\left[\frac{e_a}{m_a}\left(\vec{\ell}_a + g_a\vec{\mathbb{s}}_a\right)\right], \qquad \vec{J} = \sum_a(\vec{\ell}_a + \vec{\mathbb{s}}_a), \qquad \vec{\ell}_a = \vec{x}_a\times\vec{q}_a\,. \tag{12.159}$$

Here, $\vec{J}$ describes the sum of the orbital and spin angular momentum of the compound system. A closer inspection implies that our result in Eq. (12.158) and the result given in Eq. (33) of Ref. [366], are in agreement. Based on Eqs. (12.152) and (12.155b), and a few transformations (see Sec. 12.7), we can write the nonrelativistic Hamiltonian, inclusive of the magnetic interaction, in a frame of reference with $\vec{P} = \vec{0}$, and $\vec{R} = \vec{0}$, as follows,

$$H_0 = \sum_a \frac{\vec{p}_a^{\,2}}{2m_a} + \sum_{a>b,b}\frac{\alpha_{ab}}{r_{ab}} - \sum_a \frac{e_a}{2m_a}\left(\vec{\ell}_a + g_a\,\vec{\mathbb{s}}_a\right)\cdot\vec{B} + \sum_a \frac{1}{8\,m_a}\left(\vec{\mathcal{D}}_a\times\vec{B}\right)^2. \tag{12.160}$$

On the occasion, we recall Eqs. (12.67) and (12.152b). The total nonrelativistic magnetic interaction terms of the compound system are thus given by

$$\mathcal{H}_M = -\sum_a \frac{e_a}{2m_a}\left(\vec{\ell}_a + g_a\,\vec{\mathbb{s}}_a\right)\cdot\vec{B} + \sum_a \frac{1}{8\,m_a}\left(\vec{\mathcal{D}}_a\times\vec{B}\right)^2. \tag{12.161}$$

The term quadratic in the magnetic field describes the magnetic susceptibility.

$$\sum_a \frac{e_a^2}{8\,m_a}\left(\vec{x}_a\times\vec{B}\right)^2 + \frac{3}{8M}\left(\vec{D}\times\vec{B}\right)^2 = \sum_a \frac{1}{8\,m_a}\left(\vec{\mathcal{D}}_a\times\vec{B}\right)^2. \tag{12.162}$$

The term

$$\mathcal{H}_m = -\sum_a \frac{e_a}{2m_a}\left(\vec{\ell}_a + g_a\,\vec{\mathbb{s}}_a\right)\cdot\vec{B} = -\frac{Q}{2M}\,g_J\,\vec{J}\cdot\vec{B} \tag{12.163}$$

gives the magnetic moment including the finite-nuclear-mass corrections [we appeal to Eq. (12.159)].

12.6.2 Reduced-Mass Corrections to the Zeeman Effect

We now attempt to make contact with the literature. While the result given for $\mathcal{H}_M$ in Eq. (12.161) is entirely nonrelativistic, it incorporates the reduced-mass correction to the magnetic interaction, in the compound (many-particle) system. It is instructive to compare our result to Ref. [368], where an analogous calculation is carried out, but formulated in terms of relative coordinates of the electrons (with respect to the nucleus) and the nucleus is assumed to be spinless. We thus investigate the magnetic moment term (12.163). We assume an N-particle system, with the particles numbered as $a = 1, \ldots, N$, and the case $a = N$ represents the (spinless) nucleus. One writes the relative coordinates

$$\vec{r}_{ab} = \vec{r}_a - \vec{r}_b \,, \qquad \vec{p}_{ab} = -\mathrm{i}\frac{\partial}{\partial \vec{r}_{ab}} \,, \tag{12.164a}$$

$$\vec{L}' = \sum_a{}' \vec{r}_{aN} \times \vec{p}_{aN} \,, \qquad \vec{S}' = \sum_a{}' \vec{\mathfrak{s}}_a \,, \qquad \sum_a{}' \equiv \sum_{a=1}^{N-1} \,, \tag{12.164b}$$

where $\vec{r}_N$ is the coordinate of the atomic nucleus, and the summation index a runs over the electrons ($a = 1, \ldots, N-1$). One would like to include reduced-mass corrections of first order in m/M. We start from the orbital interaction terms in Eq. (12.163), which, for the spinless case under discussion, we denote as

$$\mathcal{H}_m^{(1)} = -\sum_a \frac{e_a}{2m_a} \vec{\ell}_a \cdot \vec{B} = -\sum_a \frac{e_a}{2m_a} (\vec{x}_a \times \vec{q}_a) \cdot \vec{B} \,, \tag{12.165a}$$

$$\mathcal{H}_m^{(1)} = \mathcal{H}_m'^{(1)} + \mathcal{H}_m^{(1),N} \,, \tag{12.165b}$$

where $\vec{\ell}_a = \vec{x}_a \times \vec{q}_a$ has been defined in Eq. (12.159), and $\vec{x}_a = \vec{r}_a - \vec{R}$ according to Eq. (12.140). Furthermore, the orbital coupling term (12.165a) will be split into an electron term $\mathcal{H}_m'^{(1)}$ and a nucleus term $\mathcal{H}_m^{(1),N}$, to be specified below.

The center-of-mass coordinate of the system is written as

$$\vec{R} = \sum_a{}' \frac{m}{M} \vec{r}_a + \frac{m_N}{M} \vec{r}_N \,, \tag{12.166}$$

where the primed sum is over the electron coordinates, m is the electron mass, and M is the total mass of the system, which is assumed to be neutral (the total charge is $Q = 0$). One then goes into a frame with [see Eqs. (12.140a) and (12.140b)]

$$\vec{R} = \vec{0} \,, \qquad \vec{P} = \sum_a \vec{p}_a = \vec{0} \,, \qquad \vec{q}_a = \vec{p}_a - \vec{P} = \vec{p}_a \,, \qquad \vec{x}_a = \vec{r}_a = \vec{r}_{aN} + \vec{r}_N \,, \tag{12.167}$$

where $a = N$ denotes the nucleus, and $\vec{r}_{aN} = \vec{r}_a - \vec{r}_N$.

Let us concentrate on the term $\mathcal{H}_m'^{(1)}$ that is exclusively due to the electrons. One can write it as follows,

$$\begin{aligned}
\mathcal{H}_m'^{(1)} &= -\sum_a{}' \frac{e}{2m} \left[\vec{r}_a - \sum_b{}' \frac{m}{M} \vec{r}_b - \frac{m_N}{M} \vec{r}_N \right] \times \vec{p}_a \cdot \vec{B} \\
&= -\sum_a{}' \frac{e}{2m} \left[\vec{r}_{aN} - \sum_b{}' \frac{m}{M} \vec{r}_{bN} - \sum_b{}' \frac{m}{M} \vec{r}_N + \left(1 - \frac{m_N}{M}\right) \vec{r}_N \right] \times \vec{p}_a \cdot \vec{B} \,, \\
&= -\sum_a{}' \frac{e}{2m} \left[\vec{r}_{aN} - \sum_b{}' \frac{m}{M} \vec{r}_{bN} \right] \times \vec{p}_a \cdot \vec{B} \,,
\end{aligned} \tag{12.168}$$

where we have used the relations $\vec{r}_b = \vec{r}_{bN} + \vec{r}_N$ and $M = m_N + \sum'_b m = m_N + Z\,m$. In the frame with $\vec{P} = \vec{0}$ and $\vec{R} = \vec{0}$ [see Eq. (12.167)], the nucleus term is

$$\begin{aligned} \mathcal{H}_m^{(1),N} &= -\frac{e_N}{2m_N}\left(\vec{r}_N - \vec{R}\right) \times \vec{p}_N \cdot \vec{B} = -\frac{e_N}{2m_N}\left(\vec{r}_N \times \vec{p}_N\right) \cdot \vec{B} \\ &= -\frac{e_N}{2M}\left(-\sum_b{}' \frac{m}{M}\vec{r}_b \times \vec{p}_N\right) \cdot \vec{B} = \frac{e_N}{2m}\left(\frac{m}{M}\right)^2 \sum_b{}' \left(\vec{r}_b \times \vec{p}_N\right) \cdot \vec{B} \,. \end{aligned} \tag{12.169}$$

This term is of relative order $(m/M)^2$ in comparison to the leading term and can thus be neglected. Finally, one has, with reference to Eq. (12.165b),

$$\mathcal{H}_m^{(1)} = \mathcal{H}_m'^{(1)} + \mathcal{H}_m^{(1),N} = -\sum_a{}' \frac{e}{2m}\left(\vec{r}_{aN} - \sum_b{}' \frac{m}{M}\vec{r}_{bN}\right) \times \vec{p}_a \cdot \vec{B} \,. \tag{12.170}$$

Under the variable change

$$\vec{r}_a \to \vec{r}_{aN} = \vec{r}_a - \vec{r}_N \,, \qquad \vec{r}_N \to \vec{r}_N \,, \tag{12.171}$$

the nuclear coordinate only acts as a constant shift, and one has

$$\frac{\partial}{\partial r_a^i} = \sum_b{}' \frac{\partial r_{bN}^j}{\partial r_a^i}\frac{\partial}{\partial r_{bN}^j} + \frac{\partial r_N^j}{\partial r_a^i}\frac{\partial}{\partial r_N^j} = \sum_b{}' \delta^{ij}\,\delta_{ab}\frac{\partial}{\partial r_{bN}^j} = \frac{\partial}{\partial r_{aN}^i} \,. \tag{12.172}$$

Hence, we can write

$$\begin{aligned} \mathcal{H}_m^{(1)} &= -\sum_a{}' \frac{e}{2m}\left[\left(\vec{r}_{aN} - \sum_b{}' \frac{m}{M}\vec{r}_{bN}\right) \times \vec{p}_{aN}\right] \cdot \vec{B} \\ &= -\frac{e}{2m}\sum_a{}'\left(1 - \frac{m}{M}\right)\left(\vec{r}_{aN} \times \vec{p}_{aN}\right) \cdot \vec{B} + \frac{e}{2M}\sum_{a\neq b,b}{}' \left(\vec{r}_{bN} \times \vec{p}_{aN}\right) \cdot \vec{B} \,. \end{aligned} \tag{12.173}$$

Under the addition of the spin term for the electrons and the nucleus,

$$\mathcal{H}_m^{(2)} = -g_a \sum_a \frac{e_a}{2m_a}\,\vec{\mathbb{s}}_a \cdot \vec{B} \,, \tag{12.174}$$

and appealing to the definitions in Eq. (12.164b), one obtains for the interaction given in Eq. (12.163),

$$\begin{aligned} \mathcal{H}_m &= \mathcal{H}_m^{(1)} + \mathcal{H}_m^{(2)} = -\sum_a \frac{e_a}{2m_a}\left(\vec{\ell}_a + g_a\,\vec{\mathbb{s}}_a\right) \cdot \vec{B} \\ &= -\frac{e}{2m}\left[\left(1 - \frac{m}{M}\right)\vec{L}' + g_S\,\vec{S}'\right] \cdot \vec{B} + \frac{e}{2M}\sum_{a\neq b,b}\left(\vec{r}_{aN} \times \vec{p}_{bN}\right) \cdot \vec{B} - g_N\frac{e_N}{2m_N}\vec{\mathbb{s}}_N \cdot \vec{B} \,. \end{aligned} \tag{12.175}$$

The $g_S = 2(1+\kappa)$ factor is the spin g factor for the electrons in the system. We have used the conventions $\vec{L}' = \sum'_a \vec{r}_{aN} \times \vec{p}_{aN}$ and $\vec{S}' = \sum'_a \vec{\mathbb{s}}_a$ outlined in Eq. (12.164). Our result, given in Eq. (12.175), generalizes a result given in the last (unnumbered) equation on page 1803 of Ref. [368], to an arbitrary nuclear spin, and clarifies that the summations over $i \neq j$ in the last (unnumbered) equation on page 1803 of Ref. [368], in the conventions used in Ref. [368], should be understood as summations over i and j, where i runs over all values $i > j$. The result for the reduced-mass correction (the reduction factor $1 - m/M$) reflects the fact that the orbital momentum of the nucleus partially compensates that of the electrons.

It is interesting to remark that the mixed term proportional to $(\vec{r}_{bN} \times \vec{p}_{aN})$ in $\mathcal{H}_M$ leads to a second-order energy shift

$$\Delta E = \left(\frac{e}{2M}\right)^2 \left\langle \left(\sum_{a\neq b,b}{}' (\vec{r}_{aN} \times \vec{p}_{bN}) \cdot \vec{B} \right) \left(\frac{1}{E-H}\right)' \left(\sum_{a\neq b,b}{}' (\vec{r}_{aN} \times \vec{p}_{bN}) \cdot \vec{B} \right) \right\rangle, \quad (12.176)$$

and contributes to the magnetic susceptibility. However, it is of second order in the electron-to-nucleus mass ratio.

From the sum of H_0 given in Eq. (12.155b) and H_R given in Eq. (12.155c), one may extract (still setting $\vec{\Pi} = 0$) the following result for the interaction Hamiltonian $\mathcal{H}_M$ of the compound system with the magnetic field, including relativistic corrections. One notes that in a weak magnetic field, it is possible to ignore terms quadratic in $\vec{B}$. One finds a generalization of a result reported in of Ref. [369], which was originally derived under the assumption of a spin-1/2 nucleus (see Ref. [366]),

$$\begin{aligned}
\mathcal{H}_M = & -\sum_a \frac{e_a}{2m_a} (\vec{\ell}_a + g_a \vec{\mathbb{s}}_a) \cdot \vec{B} + \sum_a \frac{1}{8m_a} (\vec{\mathcal{D}}_a \times \vec{B})^2 \\
& + \sum_a \frac{1}{4m_a^3} \left[\vec{q}_a^{\,2} \vec{\mathcal{D}}_a \cdot (\vec{q}_a \times \vec{B}) + 2e_a \vec{q}_a^{\,2} \vec{\mathbb{s}}_a \cdot \vec{B} + e_a (g_a - 2) (\vec{q}_a \cdot \vec{\mathbb{s}}_a)(\vec{q}_a \cdot \vec{B}) \right] \\
& + \sum_{a\neq b,b} \left[-\frac{\alpha_{ab}}{4m_a m_b} q_a^i \left(\frac{\delta^{ij}}{r_{ab}} + \frac{r_{ab}^i r_{ab}^j}{r_{ab}^3} \right) (\vec{\mathcal{D}}_b \times \vec{B})^j \right. \\
& \left. + \frac{\alpha_{ab}}{4r_{ab}^3} \frac{g_a}{m_a m_b} (\vec{\mathbb{s}}_a \times \vec{r}_{ab}) \cdot (\vec{\mathcal{D}}_b \times \vec{B}) - \frac{\alpha_{ab}}{4r_{ab}^3} \frac{g_a - 1}{m_a^2} (\vec{\mathbb{s}}_a \times \vec{r}_{ab}) \cdot (\vec{\mathcal{D}}_a \times \vec{B}) \right].
\end{aligned} \quad (12.177)$$

The reduced-mass-corrected dipole moment operator $\vec{\mathcal{D}}_a$ of particle a had been defined in Eq. (12.152b).

12.6.3 *Nuclear Magnetic Shielding*

It is instructive to consider, in Eq. (12.177), the term $a = N$ which describes the atomic nucleus. Indeed, nuclear magnetic resonance (NMR) spectroscopy investigates the spin-flip frequency of the nucleus, as modified by its environment, i.e., by atomic or molecular electrons. It is also known that NMR spectroscopy can be used as a diagnostic tool for the environmental conditions of an atom, e.g., within solvants, or in a chemical compound [372–374]. Hence, it is imperative to formulate an accurate theoretical description of the effect of the surrounding electrons on the nuclear magnetic moment.

The Hamiltonian (12.177) provides for a convenient basis to study this effect. One sets $a = N$ in Eq. (12.177) and isolates the nucleus term, including the nonrelativistic spin coupling. After an isolation of the terms which depend on the nuclear spin $\vec{\mathbb{s}}_N$, one may write the interaction Hamiltonian as follows,

$$\mathcal{H}_N^{(1)} = -\frac{e_N}{2m_N} g_N \vec{\mathbb{s}}_N \cdot \vec{B} + \frac{e_N}{4m_N^3} \left[2\vec{q}_N^{\,2} \vec{\mathbb{s}}_N \cdot \vec{B} + (g_N - 2)\,(\vec{q}_N \cdot \vec{\mathbb{s}}_N)\,(\vec{q}_N \cdot \vec{B})\right]$$
$$+ \sum_b{}' \frac{e_N\, e_b}{4\pi} \frac{\vec{\mathbb{s}}_N \times \vec{r}_{Nb}}{4r_{Nb}^3} \cdot \left[\frac{g_N}{m_N\, m_b}\left(\vec{\mathcal{D}}_b \times \vec{B}\right) - \frac{g_N - 1}{m_N^2}\left(\vec{\mathcal{D}}_N \times \vec{B}\right)\right]. \quad (12.178)$$

There is also a second-order energy shift which generates, surprisingly, a term linear (not quadratic) in the magnetic field. The first of these is the orbital coupling term

$$\mathcal{H}_N^{(2)} = -\sum_a \frac{e_a}{2m_a} \vec{\ell}_a \cdot \vec{B} \quad (12.179)$$

from Eq. (12.155c). We recall that according to Eq. (12.152), one has

$$\vec{\underline{\pi}}_a = \vec{q}_a + \frac{1}{2} \vec{\mathcal{D}}_a \times \vec{B}\,. \quad (12.180)$$

The last term in Eq. (12.178) is obtained by the replacement $\vec{\underline{\pi}}_a \to \frac{1}{2} \vec{\mathcal{D}}_a \times \vec{B}$ in the corresponding term in Eq. (12.177). However, there is another term generated if we replace $\vec{\underline{\pi}}_a \to \vec{q}_a$ in Eq. (12.177), and it reads

$$\mathcal{H}_N^{(3)} = \sum_b{}' \frac{e_N\, e_b}{4\pi} \frac{\vec{\mathbb{s}}_N \times \vec{r}_{Nb}}{2m_N\, r_{Nb}^3} \cdot \left[g_N \frac{\vec{q}_b}{m_b} - (g_N - 1)\frac{\vec{q}_N}{m_N}\right]. \quad (12.181)$$

The second-order energy shift generated by the combined action of $\mathcal{H}_N^{(2)}$ and $\mathcal{H}_N^{(3)}$ is proportional to $\vec{\mathbb{s}}_N \cdot \vec{B}$.

Combining all terms, one can derive the so-called shielding constant σ for the interaction of the nuclear magnetic moment with the external, homogenous magnetic field, as follows,

$$\Delta E = -g_N\, \mu_N\, \langle \vec{\mathbb{s}}_N \cdot \vec{B} \rangle\, (1 - \sigma)\,. \quad (12.182)$$

Here $1 - \sigma$ is an "attenuation factor" for the interaction of the nucleus with the external magnetic field, due to the orbiting electrons, and is commonly referred to as the shielding constant in the chemical literature [375], because it describes the way in which the nuclear spin is "shielded" by the atomic electrons.

There is a certain caveat, though, in our derivation up to this point. Namely, our derivation of the magnetic interaction terms, starting from Eq. (12.139), has been valid only under the assumption that the external $\vec{B}$ field generates the only nonvanishing vector potential in the physical system. A closer inspection reveals that in doing so, we risk overlooking the dominant contribution to the nuclear magnetic shielding. The nuclear magnetic moment itself [see Eq. (12.125)]

$$\vec{\mu}_N = g_N\, \mu_N\, \vec{\mathbb{s}}_N \quad (12.183)$$

generates, at the point $\vec{r}_a$ of the ath electron, a vector potential

$$\vec{A}_I(\vec{r}_a) = \frac{\vec{\mu}_N \times \vec{r}_{aN}}{4\pi r_{aN}^3}, \quad (12.184)$$

which exists in addition to the vector potential [see Eq. (12.146)]

$$\vec{A}_{\rm ext}(\vec{r}_a) = \frac{1}{2}(\vec{B} \times \vec{r}_{aN}), \tag{12.185}$$

which describes the constant external $\vec{B}$ field. Here, one might wonder why one uses the argument $\vec{r}_{aN}$ for the vector potential describing the constant external magnetic field, and not $\vec{r}_a$. The answer to this question is subtle. One can show that the only consistent way to define the vector potential corresponding to the constant external magnetic field is to adjust the origin of the coordinate system so that $\vec{A}_{\rm ext}(\vec{r}_a) = \frac{1}{2}(\vec{B} \times (\vec{r}_a - \vec{r}_0))$ where $\langle \vec{r}_a - \vec{r}_0 \rangle = \vec{0}$ in an electronic state. The latter condition is fulfilled for $\vec{r}_0 = \vec{r}_N$. From the seagull term $\vec{A}^2/(2m_a)$ in the original Hamiltonian (12.75) [see also Eqs. (12.99)], one has the mixed term

$$\mathcal{H}_S = \sum_a{}' e_a^2 \frac{(\vec{A}_I(\vec{r}_a) + \vec{A}_E(\vec{r}_a))^2}{2m_a} \to \sum_a{}' e_a^2 \frac{\vec{A}_I(\vec{r}_a) \cdot \vec{A}_E(\vec{r}_a)}{m_a}, \tag{12.186}$$

one obtains, for S states,

$$\begin{aligned} \langle \mathcal{H}_S \rangle &= \sum_a{}' \frac{e_a^2}{4\pi} \frac{g_N \mu_N}{2m_a} \left\langle \frac{1}{r_{aN}^3} (\vec{\$}_N \times \vec{r}_{aN}) \cdot (\vec{B} \times \vec{r}_{aN}) \right\rangle \\ &= \sum_a{}' \frac{e_a^2}{4\pi} \frac{g_N \mu_N}{2m_a} \left\langle \frac{1}{r_{aN}^3} \left[(\vec{\$}_N \cdot \vec{B}) \vec{r}_{aN}^{\,2} - (\vec{\$}_N \cdot \vec{r}_{aN})(\vec{B} \cdot \vec{r}_{aN}) \right] \right\rangle \\ &= \left(\sum_a{}' \frac{e_a^2}{4\pi} \frac{1}{3m_a} \left\langle \frac{1}{r_{aN}} \right\rangle \right) g_N \mu_N \langle \vec{\$}_N \cdot \vec{B} \rangle, \end{aligned} \tag{12.187}$$

implying that (in leading order)

$$\sigma = \sum_a{}' \frac{e_a^2}{4\pi} \frac{1}{3m_a} \left\langle \frac{1}{r_{aN}} \right\rangle. \tag{12.188}$$

Let us now consider S states in an atomic system with $e_a = e$ (electron charge) for all orbiting particles (hence, $e_a^2 = 4\pi\alpha$ for $a = 1, \dots, N-1$), and a nucleus $a = N$ with mass m_N. The general result for the shielding constant σ can be expressed, in relatively compact form, as (see Ref. [366])

$$\begin{aligned} \sigma = {}& \frac{\alpha}{3m_e} \left\langle \sum_a{}' \frac{1}{r_{aN}} \right\rangle + \frac{\langle \vec{p}_N^{\,2} \rangle}{m_N} \left(\frac{2 + 3g_N}{6g_N} \frac{1}{m_N} - \frac{g_N - 1}{3g_N} \frac{1}{Zm_e} \right) \\ & + \frac{\alpha}{3m_N} \left\langle \sum_{a,b \neq N, a \neq b} (\vec{r}_a \times \vec{p}_b) \left(\frac{1}{E - H} \right)' \sum_{c \neq N} \left[\frac{\vec{r}_c}{r_c^3} \times \left(\frac{\vec{p}_c}{m_e} - \frac{g_N - 1}{g_N} \frac{\vec{p}_N}{m_N} \right) \right] \right\rangle . \end{aligned} \tag{12.189}$$

This formula is valid in a system with $\vec{p}_N = -\sum_a{}' \vec{p}_a$, and $\vec{P} = \vec{0}$. The first term on the right-hand side of Eq. (12.189) is precisely Eq. (12.188). In deriving Eq. (12.189), one also notes the importance of second-order shifts generated by the seagull Hamiltonian $\mathcal{H}_S$ and the Breit Hamiltonian (12.75) for the electrons.

12.7 Further Thoughts

Here are some suggestions for further thought.

(1) **Coulomb Potential and Momentum Scales.** Show that the main integration region for expectation values of the Coulomb interaction in Schrödinger–Coulomb eigenstates is mediated by virtual photons with exchange momentum $|\vec{k}| \sim (Z\alpha)m$.
Hint: Consider the expectation value of the Coulomb potential,

$$\langle V \rangle \sim \int \mathrm{d}^3 r \, \phi^*(\vec{r}) \left(-\frac{Z\alpha}{r}\right) \phi(\vec{r}) \sim \alpha^2 \, m \,, \tag{12.190}$$

where $r \sim 1/(Z\alpha m)$ for distances of the order of the Bohr radius. Furthermore, consider the normalization integral

$$\int \mathrm{d}^3 r \, |\phi(\vec{r})|^2 = 1 \,, \tag{12.191}$$

which, since $\int \mathrm{d}^3 r \sim 1/(Z\alpha m)^3$, and $\int \mathrm{d}^3 k \sim (Z\alpha m)^3$, implies that $|\phi(\vec{r})|^2 \sim (Z\alpha)^3$, $|\phi(\vec{k})|^2 \sim (Z\alpha)^{-3}$. In momentum space, the expectation value of the Coulomb potential can be formulated as

$$\langle V \rangle \sim \int \mathrm{d}^3 k \int \mathrm{d}^3 k' \phi^*(\vec{k}) \left(-\frac{4\pi Z\alpha}{(\vec{k} - \vec{k}')^2}\right) \phi(\vec{k}') \sim (Z\alpha)^2 \, m \,. \tag{12.192}$$

Counting, now, the powers of $Z\alpha$, one has 6 powers from the $\mathrm{d}^3 k$ and $\mathrm{d}^3 k'$ integrations, and three inverse powers from the wave functions, and one explicit $Z\alpha$, result in a factor $(Z\alpha)^4$. These powers, divided by $(\vec{k} - \vec{k}')^2$, should reproduce the two powers of α on the right-hand side of Eq. (12.192). Show that this consideration justifies the order-of-magnitude estimate $|\vec{q}| = |\vec{k} - \vec{k}'| \sim Z\alpha$.

(2) **Difficult Integral.** Fill in the missing intermediate steps in the derivation (12.57). *Hint:* Why can you ignore the terms $k^i \, k^j / \vec{k}^2$ in the Coulomb-gauge photon propagator? Think about the properties of the ϵ tensor.

(3) **Another Difficult Integral.** Fill in the missing intermediate steps in the derivation (12.56). *Hint:* Extract the terms proportional to a Dirac-δ function using Gauss's theorem.

(4) **Interaction Hamiltonians.** Verify that the interaction Hamiltonian from Eq. (12.76), for the case $m_a = m_b = m$, $\alpha_{ab} = \alpha$, and $\kappa_a = \kappa_b = \kappa$, reproduces the interaction Hamiltonian in Eq. (12.64).

(5) **Nuclear Size Correction in Two-Body Systems.** Explain why the replacement

$$H \sim \frac{\pi Z\alpha}{2} \left(\frac{1}{m_1^2} + \frac{1}{m_2^2}\right) \delta^{(3)}(\vec{r}) \sim \frac{2Z\alpha}{3} \left(\frac{3}{4m_1^2} + \frac{3}{4m_2^2} + \langle r^2 \rangle_N\right) \pi \delta^{(3)}(\vec{r}) \tag{12.193}$$

in Eq. (12.111) is a good description for the effect of an extended nucleus. Clarify the role of the so-called Darwin term in the definition of the nuclear-size effect. Consult and compare Refs. [205, 364, 376].

(6) **Breit Hamiltonian and Muonic Hydrogen.** Study the application of the radiatively corrected Breit Hamiltonian from Eq. (12.106) to muonic hydrogen, where $m_1 = m_\mu$ and $m_2 = m_p$. Verify the correctness of the treatment of the anomalous-magnetic-moment correction to the Lamb shift in the literature, as discussed in Refs. [377] and [205]. Explore the inclusion of proton radius effects into the radiatively corrected Breit Hamiltonian according to Eq. (7) of Ref. [205]. Study the atomic-physics conventions for the definition of the proton radius, and the role of the so-called Darwin term and its relation to the terms

$$H \sim \frac{\pi Z\alpha}{2}\left(\frac{1+\kappa_1}{m_1^2}+\frac{1+\kappa_2}{m_2^2}\right)\delta^{(3)}(\vec{r}) \tag{12.194}$$

in Eq. (12.106). The proton radius excludes all effects which would be ascribed to the proton if it were a point-like spin-1/2 particle. Explain why the anomalous-magnetic-moment term with $\kappa = \alpha/(2\pi)$ should be included in the definition of the nuclear self-energy for S states, but not for P states, in muonic hydrogen. Consult Refs. [205, 376].

(7) **Numerical Data for Muonic Hydrogen.** Study once more the application of the radiatively corrected Breit Hamiltonian from Eq. (12.106) to muonic hydrogen, i.e., set $m_1 = m_\mu$ and $m_2 = m_p$. Derive the frequencies for the fine-structure, and hyperfine structure frequencies, of muonic hydrogen. In particular, supply the missing prefactors c and $\hbar$ in order to convert all formulas to the SI mksA unit system. Compare your results with Refs. [203, 204, 378–380].

(8) **Breit Hamiltonian and Positronium.** Apply the two-body Breit Hamiltonian (12.106) to positronium; so $m_1 = m_2 = m$ and $\kappa_1 = \kappa_2 = \kappa$. Derive the positronium hyperfine splitting to lowest order, with a radiative correction due to the anomalous magnetic moment. Study the literature (e.g., Chap. 85 of Ref. [3]) in order to find a proper way to include the annihilation diagram.

(9) **Hydrogen Hyperfine Transition.** Convince yourself that the formula given in Eq. (12.137) describes the transition wavelength among the hyperfine sublevels of the hydrogen ground state, including the reduced-mass dependence. Furthermore, calculate the theoretical prediction with and without the radiative correction of order α, and compare with the terms given in Eq. (7) of Ref. [39], and identify the term a_{00} in the conventions used in Ref. [39].

(10) **Breit Hamiltonian and Massive Photon.** Generalize the Breit Hamiltonian to an interaction with a massive photon. Thus, derive the relativistic corrections to the vacuum-polarization energy correction in muonic hydrogen. Consult Refs. [377, 381].

(11) **Hyperfine Structure and Nuclear Spin.** Generalize the hyperfine splitting Hamiltonian given in Eq. (12.136) to nuclei with spin $I = 1$, and $I = 3/2$. Derive, in leading order, the hyperfine frequencies of low-lying states of deuterium, and muonic deuterium.

(12) **Operators for External-Field Interactions.** Show the commutator relations in Eq. (12.142). *Hint:* Consider the following transformations,

$$\begin{aligned}[x^i_a, q^j_b] &= [x^i_a, p^j_b - \frac{m_b}{M} P^j] = [x^i_a, p^j_b] - \frac{m_b}{M} [x^i_a, P^j] \\ &= [x^i_a, p^j_b] = [r^i_a - \sum_c \frac{m_c}{M} r^i_c, p^j_b] = \mathrm{i}\delta^{ij}\,\delta_{ab} - \mathrm{i}\sum_c \frac{m_c}{M}\,\delta^{ij}\,\delta_{bc} \\ &= \mathrm{i}\delta^{ij}\left(\delta_{ab} - \frac{m_b}{M}\right). \end{aligned} \tag{12.195}$$

Verify that this consideration establishes that

$$[x^i_a, p^j_b] = \mathrm{i}\delta^{ij}\left(\delta_{ab} - \frac{m_b}{M}\right). \tag{12.196}$$

Then,

$$[x^i_a, P^j] = \sum_b [x^i_a, p^j_b] = \mathrm{i}\delta^{ij}\sum_b \left(\delta_{ab} - \frac{m_b}{M}\right) = \mathrm{i}\delta^{ij}\,(1-1) = 0. \tag{12.197}$$

Show the rest of the commutator relations in Eq. (12.142) in a similar fashion.

(13) **Transformation of the Kinetic Term.** Show Eq. (12.153). *Hint:* Maybe, you want to consider

$$\begin{aligned} \mathrm{e}^{-\mathrm{i}\phi}\left(\sum_a \frac{\vec{\pi}^2_a}{2m_a}\right)\mathrm{e}^{\mathrm{i}\phi} &= \sum_a \frac{\left(\vec{\underline{\pi}}_a + \frac{m_a}{M}\vec{\Pi}\right)^2}{2m_a} \\ &= \sum_a \frac{\vec{\underline{\pi}}^2_a}{2m_a} + \sum_a \frac{m_a}{2M^2}\vec{\Pi}^2 + \frac{1}{2M}\sum_a \left(\vec{\underline{\pi}}_a \cdot \vec{\Pi} + \vec{\Pi}\cdot\vec{\underline{\pi}}_a\right) \\ &= \sum_a \frac{\vec{\underline{\pi}}^2_a}{2m_a} + \frac{\vec{\Pi}^2}{2M} + \frac{\vec{\Pi}}{M}\cdot\left(\vec{D}\times\vec{B}\right). \end{aligned} \tag{12.198}$$

(14) **Nonrelativistic Hamiltonian with Magnetic Coupling.** Consider the steps which lead from Eq. (12.155b) to Eq. (12.160). *Hint:* You might proceed as follows. Profiting from the fact that in our frame of reference, we have $\sum_a \vec{p}_a = \vec{0}$, start from the relation

$$\begin{aligned} H_0 = {} & \sum_a \frac{\vec{p}^2_a}{2m_a} + \sum_a \frac{e^2_a}{8m^2_a}\left(\vec{x}_a\times\vec{B}\right)^2 + \underbrace{\sum_a \frac{m_a}{8M^2}\left(\vec{D}\times\vec{B}\right)^2}_{=T_1} \\ & + \sum_a \frac{e_a}{2m_a}(\vec{x}_a\times\vec{B})\cdot\vec{p}_a + \underbrace{\sum_a \frac{1}{4M}\left\{\vec{p}_a, \left(\vec{D}\times\vec{B}\right)\right\}}_{=0} \\ & + \underbrace{\sum_a \frac{e_a}{4M}\left(\vec{x}_a\times\vec{B}\right)\cdot\left(\vec{D}\times\vec{B}\right)}_{=T_2} + \sum_{a>b,b}\frac{\alpha_{ab}}{r_{ab}} - \sum_a \frac{e_a}{2m_a} g_a\,\vec{s}_a\cdot\vec{B}, \end{aligned} \tag{12.199}$$

where we implicitly define the terms T_1 and T_2. A very easy transformation should lead to the following expression,

$$
\begin{aligned}
H_0 = {} & \sum_a \frac{\vec{p}_a^{\,2}}{2m_a} + \sum_a \frac{e_a^2}{8m_a^2} \left(\vec{x}_a \times \vec{B}\right)^2 + \underbrace{\frac{3}{8M} \left(\vec{D} \times \vec{B}\right)^2}_{=T_1+T_2} \\
& - \sum_a \frac{e_a}{2m_a} \vec{\ell}_a \cdot \vec{B} + \sum_{a>b} \frac{e_a \, e_b}{4\pi \, r_{ab}} - \sum_a \frac{e_a}{2m_a} g_a \, \vec{s}_a \cdot \vec{B} \,.
\end{aligned} \tag{12.200}
$$

Finally, one should obtain

$$
\begin{aligned}
H_0 = {} & \sum_a \frac{\vec{p}_a^{\,2}}{2m_a} + \sum_a \frac{e_a^2}{8m_a^2} \left(\vec{x}_a \times \vec{B}\right)^2 + \frac{3}{8M} \left(\vec{D} \times \vec{B}\right)^2 \\
& - \sum_a \frac{e_a}{2m_a} \vec{\ell}_a \cdot \vec{B} + \sum_{a>b} \frac{e_a \, e_b}{4\pi \, r_{ab}} - \sum_a \frac{e_a}{2m_a} g_a \, \vec{s}_a \cdot \vec{B} \,.
\end{aligned} \tag{12.201}
$$

On the basis of the identity

$$
\sum_a \frac{e_a^2}{8\,m_a} \left(\vec{x}_a \times \vec{B}\right)^2 + \frac{3}{8M} \left(\vec{D} \times \vec{B}\right)^2 = \sum_a \frac{1}{8\,m_a} \left(\vec{\mathcal{D}}_a \times \vec{B}\right)^2 , \tag{12.202}
$$

this expression should easily be shown to be equivalent to the Hamiltonian given in Eq. (12.160).

(15) **Magnetic Shielding.** Study in some further detail the theory of magnetic shielding of the nucleus on the basis of Refs. [369, 382–385]. Attempt to derive at least the leading logarithmic QED correction, inspired by Refs. [375, 386, 387].

(16) **Electromagnetic Properties of Diatomic Molecules.** Attempt to extend the analysis of Sec. 12.6 to diatomic molecules. Be inspired by Refs. [384, 385].

Chapter 13

Fully Correlated Basis Sets and Helium

13.1 Overview

Helium is the simplest atom with more than one electron. It is a quantum three-body system for which no exact solutions are known, not even on the level of nonrelativistic quantum mechanics. The lack of exact solutions, however, does not prevent an accurate discussion of the spectrum of the helium atom, including the relativistic corrections which are so decisive for an accurate understanding of the spectrum of the system.

We proceed as follows. In Sec. 13.2, we review basic ingredients in the theoretical description of many-electron systems, starting from the spin wave functions in the nonrelativistic approximation. Relativistic and radiative corrections are discussed in Sec. 13.3. The derivation of the helium Lamb shift in Sec. 13.3.2 is rather tricky; it involves a combination of the self-energy of the bound electrons individually, as well as a radiative correction to the interelectron interaction (recoil correction). We conclude with a brief description of the actual, numerical calculation of the helium spectrum in Sec. 13.4. Within the Hartree–Fock approximation, the electron wave functions are modeled on the basis of a SLATER[1] determinant. The restrictions set forth by the Slater determinant are relaxed in the case of a fully correlated basis set, which takes into account the interelectron interaction to all orders right from the start of the calculation and does not rely on the specification of individual electronic orbitals. It is the latter approach which is relevant to helium.

13.2 Essential Ingredients for Many-Electron Systems

13.2.1 *Spin Wave Functions*

In helium, the spin-orbit coupling, which is of order of $(Z\alpha)^4 m$, is much weaker than the interelectron interaction, which is of the order of $(Z\alpha)^2 m$ (in natural units, $\hbar = c = \epsilon_0 = 1$, and m denotes the electron mass). The spin of the electron enters the spectrum only at the order of $(Z\alpha)^4 m$, and the spectrum as well as the properties of

[1] John Clarke Slater (1900–1976).

neutral helium are therefore spin-independent to a very good approximation. The decisive degree of freedom in helium therefore is the orbital angular momentum; the electron spins couple independently and interact only weakly with the environment. This is the regime of LS-coupling, where the spins couple independently of the angular momenta. First, the spins $\vec{s}_i$ couple to the total spin $\vec{S}$, the orbital angular momenta $\vec{\ell}_i$ couple to the total angular momentum $\vec{L}$, and then total spin couples to the total angular momentum. There is strong electron correlation, and weak binding to the nucleus.

The situation is reversed in jj coupling. First, the spin $\vec{s}_i$ of the ith individual electron couples to its orbital angular momentum $\vec{\ell}_i$, to form the individual total angular momentum $\vec{j}_i$, and then the total angular momenta $\vec{j}_i$ from the different electrons couple to each other. One canonically denotes the angular momenta of individual electrons by lowercase letters, whereas the combined momenta are denoted by uppercase letters. This explains the notation jj versus LS coupling. For weak electron correlation, the electrons are strongly bound to the nucleus. An example for jj coupling is the electronic configuration $(2p^2_{1/2})_{J=0}$, where both electrons are in states with a total angular momentum $j = 1/2$, and couple to a state with $J = 0$.

Let us now return to LS coupling. If the total spin wave function is symmetric, then, according to the Pauli exclusion principle, the orbital wave function has to be antisymmetric under particle exchange, and vice versa. This implies that the ground state of helium is a spin singlet. It has an anti-symmetric spin wave function. Within the approximation of individual atomic orbitals, we would say that the two electrons both occupy the $1s$ ground state, to yield the state $1s^2\,{}^1S_0$. This is referred to as para-helium. The metastable ground state of ortho-helium (with an $S = 1$ symmetric spin wave function) has an antisymmetric orbital wave function. The two electrons cannot occupy the same quantum state. The $1s2s\,{}^3S_1$ ground state of ortho-helium is energetically higher than the ground state of para-helium. The spin wave function of para-helium is given by

$$|S=0, M=0\rangle = \frac{1}{\sqrt{2}}\left(|\uparrow\downarrow\rangle - |\downarrow\uparrow\rangle\right), \tag{13.1}$$

whereas for triplet helium, the spin states are

$$|S=1, M=1\rangle = |\uparrow\uparrow\rangle, \qquad |S=1, M=-1\rangle = |\downarrow\downarrow\rangle, \tag{13.2a}$$

$$|S=1, M=0\rangle = \frac{1}{\sqrt{2}}\left(|\uparrow\downarrow\rangle + |\downarrow\uparrow\rangle\right). \tag{13.2b}$$

If we assume that the two electrons occupy independent orbitals, labeled 1 and 2, then the two-particle wave functions are given by the following general formulas,

$$\Psi_{S=0,M=0}(\vec{r}_1,\vec{r}_2) = \frac{1}{\sqrt{2}}\left(\Psi_1(\vec{r}_1)\,\Psi_2(\vec{r}_2) + \Psi_1(\vec{r}_2)\,\Psi_2(\vec{r}_1)\right)\frac{1}{\sqrt{2}}\left(|\uparrow\downarrow\rangle - |\downarrow\uparrow\rangle\right). \tag{13.3}$$

Here, we have split the wave function into an orbital and a spin part. For triplet states, we have

$$\Psi_{S=1,M=0}(\vec{r}_1,\vec{r}_2) = \frac{1}{\sqrt{2}}\left(\Psi_1(\vec{r}_1)\,\Psi_2(\vec{r}_2) - \Psi_1(\vec{r}_2)\,\Psi_2(\vec{r}_1)\right)\frac{1}{\sqrt{2}}\left(|\uparrow\downarrow\rangle + |\downarrow\uparrow\rangle\right). \tag{13.4}$$

Both of these wave functions are odd under particle exchange. They factorize into a spin and an orbital part. Because the electron spins interact only weakly and spin-flip induced transitions are slow, the ground-state of spin-1 helium (ortho-helium) actually is metastable. There are fundamental differences in the excitation spectra of para-helium and ortho-helium, which explains the names for the two different "kinds" of helium.

The limitations in the suitability of the two wave functions (13.3) and (13.4) for the description of electron correlation become clear when we look at the identity

$$\vec{L}\, f(|\vec{r}_1 - \vec{r}_2|) = (\vec{\ell}_1 + \vec{\ell}_2)\, f(|\vec{r}_1 - \vec{r}_2|) = 0\,, \tag{13.5}$$

which states that any function of the interelectron distance $r_{12} = |\vec{r}_1 - \vec{r}_2|$ belongs to a state of total angular momentum $L = 0$ (an S state). This suggests the presence of terms that involve $r_{12} = |\vec{r}_1 - \vec{r}_2|$ in the two-particle wave function; Eqs. (13.4) and (13.5) lack such terms. As we know from classical electrodynamics [81], an expression of the form $1/r_{12}$ gives rise to arbitrarily high multipoles proportional to $Y_{\ell m}(\vec{r}_1)\, Y^*_{\ell m}(\vec{r}_2)$ when $1/r_{12}$ is expanded in (inverse) powers of $r_< = \min(r_1, r_2)$ and $r_> = \max(r_1, r_2)$.

In order to correctly describe the electron correlation, one would thus have to describe a many-electron state in terms of a linear superposition of orbitals, each with a different orbital momentum, replacing, schematically,

$$\begin{aligned}\Psi_1(\vec{r}_1)\,\Psi_2(\vec{r}_2) + \Psi_1(\vec{r}_2)\,\Psi_2(\vec{r}_1) \to\ & a\,(\Psi_1(\vec{r}_1)\,\Psi_2(\vec{r}_2) + \Psi_1(\vec{r}_2)\,\Psi_2(\vec{r}_1)) \\ & + b\,(\Psi_3(\vec{r}_1)\,\Psi_4(\vec{r}_2) + \Psi_4(\vec{r}_2)\,\Psi_3(\vec{r}_1))\,, \end{aligned} \tag{13.6}$$

with suitable expansion coefficients a and b and different angular symmetries for the Ψ_j with $j = 1, 2, 3, 4$. A simple product *ansatz* consisting of individual atomic orbitals, as given by Eqs. (13.3) and (13.4), will not be sufficient. However, a wave function expanded over a suitable basis set which contains r_{12} is able to describe electron correlation correctly and to all orders. This is called a fully correlated basis set.

We recall that in LS coupling, the spins of the electrons first couple to the total spin S, the orbital angular momenta couple to the total orbital angular momentum L, and then L and S couple to J. A configuration is denoted as ${}^{2S+1}L_J$. There is strong electron correlation and a comparatively weak binding to the nucleus. This situation is relevant to all outer shells of neutral atoms, because the effective nuclear charge numbers for the outer electron shells are small. An example is the $1s^2\,{}^1S_0$ ground state of neutral helium. By contrast, in jj coupling, the spins s_i of the individual electrons first couple to the individual angular momenta ℓ_i, and then the total angular momenta j_i couple to each other. One can then couple the total electron angular momenta together, as in the ground-state configuration $(1s^2_{1/2})_0$ of heliumlike uranium. Here, the subscript denotes the total angular momentum of all electrons in the configuration ($J = 0$). In highly charged ions, there is weak electron correlation, and strong binding to the atomic nucleus. The transition region between LS and jj coupling can be estimated as follows. Let us consider the electron correlation term α/r_{12} in the Hamiltonian and compare it to the spin-orbit coupling term,

$$\frac{\alpha}{r} \sim Z\,\alpha^2\, m\,, \qquad \frac{Z\,\alpha}{4\,m^2\,r^3}\,\vec{\sigma}\cdot(\vec{r}\times\vec{p}) \sim (Z\alpha)^4 m\,. \tag{13.7}$$

The first term measures the strength of the electron-electron correlation, the second term measures the strength of the spin-orbit coupling, and we have used the fact that the dimension of the atom reads as $r \sim (Z\alpha m)^{-1}$. The transition occurs at a value of Z given by the condition

$$Z\,\alpha^2 = (Z\alpha)^4\,, \qquad Z \approx 26.57\,, \tag{13.8}$$

i.e., in between iron (Fe, $Z = 26$) and cobalt (Co, $Z = 27$) in the periodic table of elements.

13.2.2 *Two-Body Problem: Center-of-Mass Coordinates*

Before we dwell on the helium atom and its non-separability, and the mass polarization term, let us briefly recall how the two-body Schrödinger–Coulomb Hamiltonian can be separated into a free Hamiltonian describing the motion of the combined two-body system, and a Schrödinger–Coulomb Hamiltonian in the relative coordinate, and with the reduced mass. We denote the coordinates and momenta of the two particles by subscripts 1 and 2 and define the relative coordinate as $\vec{r} = \vec{r}_1 - \vec{r}_2$. The center-of-mass coordinates $\vec{r}$ and $\vec{R}$ are defined as follows,

$$\vec{R} = \frac{m_1}{M}\,\vec{r}_1 + \frac{m_2}{M}\,\vec{r}_2\,, \qquad \vec{r} = \vec{r}_1 - \vec{r}_2\,, \qquad M = m_1 + m_2\,, \tag{13.9}$$

where the subscripts 1 and 2 denote the "orbiting particle" and the "nucleus", respectively. (The discussion is general, and is also valid for, say, positronium with $m_1 = m_2$.) The coordinates of the individual particles can also be obtained from the center-of-mass coordinates,

$$\vec{r}_1 = \vec{R} + \frac{m_2}{M}\,\vec{r} = \vec{R} + \xi_2\,\vec{r}\,, \quad \vec{r}_2 = \vec{R} - \frac{m_1}{M}\,\vec{r} = \vec{R} - \xi_1\,\vec{r}\,, \quad \xi_1 = \frac{m_1}{M}\,, \quad \xi_2 = \frac{m_2}{M}\,. \tag{13.10}$$

The formalism with the ξ_i factors ($i = 1, 2$) becomes important in the discussion of the Bethe–Salpeter equation in Sec. 16.2.1. The total momentum of the system is denoted as $\vec{P}$, while $\vec{p}$ is the relative momentum,

$$\vec{P} = \vec{p}_1 + \vec{p}_2\,, \qquad \vec{p} = \frac{\mu}{m_1}\,\vec{p}_1 - \frac{\mu}{m_2}\,\vec{p}_2 = \xi_2\,\vec{p}_1 - \xi_1\,\vec{p}_2\,, \qquad \mu = \frac{m_1\,m_2}{m_1 + m_2}\,. \tag{13.11}$$

The individual momenta $\vec{p}_1$ and $\vec{p}_2$ can be obtained as follows,

$$\vec{p}_1 = \frac{m_1}{M}\,\vec{P} + \vec{p} = \xi_1\,\vec{P} + \vec{p}\,, \qquad \vec{p}_2 = \frac{m_2}{M}\,\vec{P} - \vec{p} = \xi_2\,\vec{P} - \vec{p}\,. \tag{13.12}$$

The commutation relations read as follows,

$$\left[P^i, R^j\right] = \left[p^i, r^j\right] = -\mathrm{i}\,\delta^{ij}\,, \qquad \left[P^i, r^j\right] = \left[p^i, R^j\right] = 0\,. \tag{13.13}$$

The Hamiltonian transforms as follows,

$$H = \frac{\vec{p}_1^{\,2}}{2m_1} + \frac{\vec{p}_2^{\,2}}{2m_2} - \frac{Z\alpha}{|\vec{r}_1 - \vec{r}_2|} = \frac{\vec{P}^2}{2M} + \frac{\vec{p}^{\,2}}{2\mu} - \frac{Z\alpha}{r}\,, \tag{13.14}$$

where of course $r = |\vec{r}| = |\vec{r}_1 - \vec{r}_2|$. The two-body problem is exactly separable in center-of-mass coordinates and there is no mass polarization term. The Hamiltonian is the sum of the kinetic energy term for the composite system, described by the kinetic term $\vec{P}^2/(2M)$, and the unperturbed Schrödinger–Coulomb Hamiltonian $H_S = \frac{\vec{p}^2}{2\mu} - \frac{Z\alpha}{r}$. This consideration provides for an *a posteriori* justification of the starting point for the Schrödinger–Coulomb Hamiltonian given in Eq. (2.1).

13.2.3 *Three-Body Problem: Mass Polarization*

For a general three-body system, we would have to introduce three masses m_1, m_2 and m_3. However, it is customary to restrict the discussion to a three-body system with two particles of mass m (typically, the electron mass), and a third particle of mass m_N (typically, the nucleus). The Hamiltonian of the three-body system is given as follows,

$$H = \frac{\vec{p}_1^{\,2}}{2m} + \frac{\vec{p}_2^{\,2}}{2m} + \frac{\vec{p}_N^{\,2}}{2m_N} - \frac{Z\alpha}{|\vec{r}_1 - \vec{r}_N|} - \frac{Z\alpha}{|\vec{r}_2 - \vec{r}_N|} + \frac{\alpha}{|\vec{r}_1 - \vec{r}_2|}\,. \tag{13.15}$$

If we are only interested in the "internal degrees of freedom" of the system, then we may actually obtain the three-body Hamiltonian using a simple trick. Namely, we work directly in the center-of-mass system, where the sum of the particle momenta vanishes and we have $\vec{p}_1 + \vec{p}_2 + \vec{p}_N = \vec{0}$, and the nucleus is at the origin of the coordinate system, $\vec{r}_N = \vec{0}$. In view of the relation $\vec{p}_N = -\vec{p}_1 - \vec{p}_2$, the kinetic term becomes

$$H = \vec{p}_1^{\,2}\left(\frac{1}{2m} + \frac{1}{2m_N}\right) + \vec{p}_2^{\,2}\left(\frac{1}{2m} + \frac{1}{2m_N}\right) + \frac{\vec{p}_1 \cdot \vec{p}_2}{m_N} = \frac{\vec{p}_1^{\,2}}{2\mu} + \frac{\vec{p}_2^{\,2}}{2\mu} + \frac{\vec{p}_1 \cdot \vec{p}_2}{m_N}\,. \tag{13.16}$$

The reduced mass μ corresponds to the two-body system of electron and nucleus,

$$\mu = \frac{m\, m_N}{m_N + m}\,. \tag{13.17}$$

The Hamiltonian of the three-body system, in the center-of-mass system, can therefore be written as

$$H = \frac{\vec{p}_1^{\,2}}{2\mu} + \frac{\vec{p}_2^{\,2}}{2\mu} - \frac{Z\alpha}{r_1} - \frac{Z\alpha}{r_2} + \frac{\alpha}{r_{12}} + \frac{\vec{p}_1 \cdot \vec{p}_2}{m_N}\,, \tag{13.18}$$

where $r_1 = |\vec{r}_1|$, $r_2 = |\vec{r}_2|$, and $r_{12} = |\vec{r}_1 - \vec{r}_2|$. The last term in Eq. (13.18) is the so-called mass polarization term. It is a consequence of the non-separability of the three-body Hamiltonian.

Let us repeat the derivation once more, but this time without going into the center-of-mass system. The total momentum of the system is $\vec{P} = \vec{p}_1 + \vec{p}_2 + \vec{p}_N$. The momenta $\vec{P}_1$ and $\vec{P}_2$ in this case have to be defined as

$$\begin{aligned}
\vec{P}_1 &= \frac{m_N + m}{m_N + 2m}\,\vec{p}_1 - \frac{m}{m_N + 2m}\,\vec{p}_2 - \frac{m}{m_N + 2m}\,\vec{p}_N\,, \\
\vec{P}_2 &= \frac{m_N + m}{m_N + 2m}\,\vec{p}_2 - \frac{m}{m_N + 2m}\,\vec{p}_1 - \frac{m}{m_N + 2m}\,\vec{p}_N\,.
\end{aligned} \tag{13.19}$$

The center-of-mass coordinates can be found as follows,

$$\vec{R}_1 = \vec{r}_1 - \vec{r}_N\,, \qquad \vec{R}_2 = \vec{r}_2 - \vec{r}_N\,, \qquad \vec{\rho} = \frac{m_N\,\vec{r}_N + m\,(\vec{r}_1 + \vec{r}_2)}{m_N + 2m}\,. \tag{13.20}$$

The commutation relations of the canonically conjugate variables follow almost immediately,

$$\begin{aligned}&\left[P_1^i, R_1^j\right] = \left[P_2^i, R_2^j\right] = \left[P^i, \rho^j\right] = -\mathrm{i}\,\delta^{ij}\,,\\ &\left[P_1^i, R_2^j\right] = \left[P_1^i, \rho^j\right] = \left[P_2^i, R_1^j\right] = \left[P_2^i, \rho^j\right] = \left[P^i, R_1^j\right] = \left[P^i, R_2^j\right] = 0\,.\end{aligned} \tag{13.21}$$

Here, $\vec{\rho}$ is the canonically conjugate variable of the total momentum $\vec{P}$. The momenta of the individual particles can be obtained in terms of canonically conjugate new variables as follows,

$$\vec{p}_1 = \frac{m}{m_N + 2m}\vec{P} + \vec{P}_1\,, \qquad \vec{p}_2 = \frac{m}{m_N + 2m}\vec{P} + \vec{P}_2\,, \tag{13.22a}$$

$$\vec{p}_N = \frac{m_N}{m_N + 2m}\vec{P} - \vec{P}_1 - \vec{P}_2\,. \tag{13.22b}$$

The kinetic term for the entire system can then be separated according to

$$\frac{\vec{p}_1^{\,2}}{2m} + \frac{\vec{p}_2^{\,2}}{2m} + \frac{\vec{p}_N^{\,2}}{2m_N} = \frac{\vec{P}^2}{2(m_N + 2m)} + \frac{\vec{P}_1^2}{2\mu} + \frac{\vec{P}_2^2}{2\mu} + \frac{\vec{P}_1 \cdot \vec{P}_2}{m_N}\,. \tag{13.23}$$

In the center-of-mass system, then, the helium Hamiltonian reads as

$$H = \frac{\vec{P}^2}{2(m_N + 2m)} + \frac{\vec{P}_1^2}{2\mu} - \frac{Z\alpha}{R_1} + \frac{\vec{P}_2^2}{2\mu} - \frac{Z\alpha}{R_2} + \frac{\alpha}{|\vec{R}_1 - \vec{R}_2|} + \frac{\vec{P}_1 \cdot \vec{P}_2}{m_N}\,. \tag{13.24}$$

Upon the identification $\vec{P}_1 \to \vec{p}_1$, $\vec{P}_2 \to \vec{p}_2$, $R_1 \to r_1$, $R_2 \to r_2$, $|\vec{R}_1 - \vec{R}_2| \to |\vec{r}_1 - \vec{r}_2| \equiv r$, this Hamiltonian is equivalent to the one given in Eq. (13.18). The mass polarization term $\vec{P}_1 \cdot \vec{P}_2/m_N$ is an essential ingredient to the description of the three-body system.

Finally, let us indicate the generalization of Eq. (13.18) to a bound system consisting of n electrons, and one nucleus,

$$H = \sum_{i=1}^{n}\left(\frac{\vec{p}_i^{\,2}}{2\mu} - \frac{Z\alpha}{r_i}\right) + \sum_{i<j}^{n}\left(\frac{\alpha}{r_{ij}} + \frac{\vec{p}_i \cdot \vec{p}_j}{m_N}\right), \tag{13.25}$$

where $r_i = |\vec{r}_i|$ measures the distance of the ith electron from the nucleus, and $r_{ij} = |\vec{r}_i - \vec{r}_j|$ is the relative distance of the ith and the jth electron. The derivation is straightforward in the center-of-mass system and left as an exercise to the reader.

The additional remarks are in order. (i) The reduced mass μ enters the helium Hamiltonian (13.18). However, if one would like to couple the electrons to the quantized electromagnetic field, then one has to do this individually. The interaction Hamiltonian for the individual electrons with the radiation field has to be written with the electron mass, not the reduced mass. (ii) The simple trick of writing all operators in the center-of-mass system, where the particle momenta add up to zero, i.e., $\vec{p}_1 + \vec{p}_2 + \cdots + \vec{p}_n = -\vec{p}_N$, has an important generalization. In higher orders of perturbation theory, it can be used in order to obtain effective Hamiltonians describing the higher-order terms such as Breit corrections. Going into the center-of-mass system means that we specialize the time evolution of the atomic energy levels to a particular reference system; this is manifestly noncovariant, and one might wonder about the implications of this approximation within special relativity. In order to address the issue, it is sufficient to recall that in a typical terrestrial environment like a thermal atomic beam, atoms move at manifestly nonrelativistic velocities. Corrections due to the Lorentz transformation of transition energies and/or frequencies into the center-of-mass system can then be taken into account on the basis of Doppler shifts and do not limit the accuracy to which experiment can be compared to theory.

13.3 Relativistic and Radiative Corrections

13.3.1 *Relativistic Corrections*

At face value, the nonrelativistic Hamiltonian (13.18) is spin-independent, but still, the electron spin has a tremendous influence on the spectrum. To good approximation, the electron spins couple to either a spin singlet or a spin triplet state, and the orbital angular momenta couple to a state of definite angular momentum. Within this approximation, the angular part of the helium wave function is given by a spherical biharmonic discussed in Sec. 6.4.3, with an appropriate addition for symmetrization or antisymmetrization. On the level of relativistic corrections, though, the situation is different. The relativistic operators contain manifestly spin-dependent terms. Let us consider a spin singlet state, which is antisymmetric under particle interchange [see also Eq. (13.1)], i.e., we have $|00\rangle = \frac{1}{\sqrt{2}}\,(|\uparrow\downarrow\rangle - |\downarrow\uparrow\rangle)$. Here and in the following, we denote the spin component of the singlet state as $|S\,M\rangle$ where S is the total spin and M is the projection onto the z axis. A simple calculation shows that the following expectation values hold for the spin-single state ($\sigma_{1,z}$ is the z component of the spin operator for particle 1),

$$\langle\uparrow\downarrow|\sigma_{1,z}|\uparrow\downarrow\rangle = 1\,, \qquad \langle\uparrow\downarrow|\sigma_{1,z}|\downarrow\uparrow\rangle = 0\,, \tag{13.26a}$$

$$\langle\downarrow\uparrow|\sigma_{1,z}|\downarrow\uparrow\rangle = -1\,, \qquad \langle\downarrow\uparrow|\sigma_{1,z}|\uparrow\downarrow\rangle = 0\,. \tag{13.26b}$$

The expectation values of both z spin components of both particles vanish,

$$\langle 00|\sigma_{1,z}|00\rangle = \langle 00|\sigma_{2,z}|00\rangle = 0\,. \tag{13.27}$$

Triplet states with spin wave functions given in Eq. (13.2) have the following expectation values,

$$\langle 11|\sigma_{1,z}|11\rangle = \langle 11|\sigma_{2,z}|11\rangle = 1\,. \tag{13.28}$$

A simple generalization of these statements implies that operators proportional to $(\vec{\sigma}_1 + \vec{\sigma}_2)/2$ vanish on a spin singlet, but they do not vanish on a spin triplet state. By contrast, operators proportional to $(\vec{\sigma}_1 - \vec{\sigma}_2)/2$ vanish on both spin singlets as well as spin triplets. This observation will help us in simplifying the expressions for the relativistic corrections in the helium atom.

Supplementing the discussion in Chap. 11 (for hydrogen), we should now consider the relativistic corrections for helium energy levels. The Breit Hamiltonian for helium is naturally split into three terms,

$$H^{[4]} = H_1^{[4]} + H_2^{[4]} + H_3^{[4]} . \tag{13.29}$$

The superscript denotes the corrections of order α^4 (in natural units). The first term, $H_1^{[4]}$, can be understood on the basis of the sum of the single-particle Hamiltonian listed in Eq. (11.3.1). One refers to Eq. (11.65) and consults the sum of relativistic corrections to the kinetic energy of both electrons, and *zitterbewegung* corrections to the electron-nucleus interactions, for both electrons,

$$H_1^{[4]} = -\frac{p_1^4 + p_2^4}{8m^3} + \frac{\pi(Z\alpha)}{2m^2}\left(\delta^{(3)}(\vec{r}_1) + \delta^{(3)}(\vec{r}_2)\right) . \tag{13.30}$$

The term $H_2^{[4]}$ summarizes the spin-independent terms in the two-particle Hamiltonian given in Eq. (12.76) (ignoring the spin-independent terms). Specifically, the term $H_2^{[4]}$ is given as the sum of a magnetic exchange correction to the electron-electron interaction, and electron-electron *zitterbewegung* corrections, and reads as follows,

$$H_2^{[4]} = -\frac{\alpha}{2m^2} p_1^i \left(\frac{\delta^{ij}}{r_{12}} + \frac{r_{12}^i r_{12}^j}{r_{12}^3}\right) p_2^j - \frac{\pi\alpha}{2m^2}\delta^{(3)}(\vec{r}_{12}) . \tag{13.31}$$

We recall that the vector $\vec{r}_{12}$ is given as $\vec{r}_{12} = \vec{r}_1 - \vec{r}_2$. Its Cartesian components are r_{12}^i, with $i = 1, 2, 3$.

The third contribution $H_3^{[4]}$ summarizes all spin-dependent terms. It is given as the sum of the spin-orbit coupling term given in Eq. (11.65), evaluated for both electrons, and of the spin-dependent terms in the two-particle Hamiltonian given in Eq. (12.76). It is instructive to decompose the Hamiltonian $H_3^{[4]}$ as follows [336,388],

$$H_3^{[4]} = H_{\rm so} + H'_{\rm soo} + H''_{\rm soo} + H_{\rm ss} + H_{\rm ssc} , \tag{13.32}$$

with the designation of the terms explained below. The term $H_{\rm so}$ simply is the sum of the spin-orbit interactions of the individual electrons, from Eq. (11.65),

$$H_{\rm so} = \frac{Z\alpha}{4m^2}\left(\frac{\vec{\sigma}_1\cdot(\vec{r}_1\times\vec{p}_1)}{r_1^3} + \frac{\vec{\sigma}_2\cdot(\vec{r}_2\times\vec{p}_2)}{r_2^3}\right) . \tag{13.33a}$$

The spin-other-orbit (soo) term is the sum of two terms, $H'_{\rm soo}$ and $H''_{\rm soo}$, the first of which reads as follows,

$$H'_{\rm soo} = -\frac{\alpha}{4m^2 r_{12}^3}\left[\vec{\sigma}_1\cdot(\vec{r}_{12}\times\vec{p}_1) + \vec{\sigma}_2\cdot(\vec{r}_{21}\times\vec{p}_2)\right] . \tag{13.33b}$$

Here, $\vec{r}_{12} = \vec{r}_1 - \vec{r}_2$, while $\vec{r}_{21} = \vec{r}_2 - \vec{r}_1$. The negative prefactor of the correction $H'_{\rm soo}$ becomes obvious if we consider that the position vector pointing toward particle 2 from particle 1 is given as $\vec{r}_{21}$, not $\vec{r}_{12}$. An additional spin-other-orbit term couples the spin of one particle to the momentum of the other,

$$H''_{\rm soo} = -\frac{\alpha}{2\,m^2\,r_{12}^3}\left[\vec{\sigma}_2\cdot(\vec{r}_{12}\times\vec{p}_1) + \vec{\sigma}_1\cdot(\vec{r}_{21}\times\vec{p}_2)\right] . \tag{13.33c}$$

There is also a spin-spin interaction,

$$H_{\rm ss} = \frac{\alpha}{4m^2}\left(\frac{\vec{\sigma}_1\cdot\vec{\sigma}_2}{r_{12}^3} - 3\,\frac{(\vec{\sigma}_1\cdot\vec{r}_{12})\,(\vec{\sigma}_2\cdot\vec{r}_{12})}{r_{12}^5}\right). \tag{13.33d}$$

Finally, the spin-spin-contact (ssc) term is proportional to a Dirac-δ in the relative coordinate,

$$H_{\rm ssc} = -\frac{2\pi\alpha}{3m^2}\,\vec{\sigma}_1\cdot\vec{\sigma}_2\,\delta^{(3)}(\vec{r}_{12})\,. \tag{13.33e}$$

For practical calculations, it is sometimes useful to write the sum of the spin-orbit and spin-other-orbit terms as follows,

$$H_{\rm so} + H'_{\rm soo} + H''_{\rm soo} = H_A + H_B\,, \tag{13.34}$$

where H_A and H_B involve the spin sums and differences $\frac{\vec{\sigma}_1+\vec{\sigma}_2}{2}$ and $\frac{\vec{\sigma}_1-\vec{\sigma}_2}{2}$. So, H_A contributes to triplet states, but both H_A and H_B vanish on singlet states,

$$H_A = \left[\frac{Z\alpha}{4m^2}\left(\frac{\vec{r}_1}{r_1^3}\times\vec{p}_1 + \frac{\vec{r}_2}{r_2^3}\times\vec{p}_2\right) - \frac{3\alpha}{4m^2}\left(\frac{\vec{r}_{12}}{r_{12}^3}\times(\vec{p}_1-\vec{p}_2)\right)\right]\cdot\frac{\vec{\sigma}_1+\vec{\sigma}_2}{2}\,, \tag{13.35}$$

$$H_B = \left[\frac{Z\alpha}{4m^2}\left(\frac{\vec{r}_1}{r_1^3}\times\vec{p}_1 - \frac{\vec{r}_2}{r_2^3}\times\vec{p}_2\right) + \frac{\alpha}{4m^2}\left(\frac{\vec{r}_{12}}{r_{12}^3}\times(\vec{p}_1+\vec{p}_2)\right)\right]\cdot\frac{\vec{\sigma}_1-\vec{\sigma}_2}{2}\,. \tag{13.36}$$

For the calculation of the relativistic corrections to the helium spectrum (corrections of order α^4), it is sufficient to first generate a good reference-state nonrelativistic wave function, and then, to evaluate the first-order perturbation

$$E^{(4)} = \langle H_1^{(4)} + H_2^{(4)}\rangle\,, \tag{13.37}$$

which gives the α^4 corrections to each individual state. For an informative account on the analytic and numerical intricacies of the helium fine-structure, we refer to Ref. [389].

13.3.2 *Derivation of the Helium Lamb Shift*

The derivation of the leading α^5 terms in the helium Lamb shift is more complicated than the corresponding calculation for hydrogen, described in Chap. 4. We use minimal coupling in order to couple the two electrons to the quantized electromagnetic field. The Hamiltonian reads as

$$\begin{aligned} H = {} & \frac{[\vec{p}_1 - e\,\vec{A}(\vec{r}_1)]^2}{2m} + \frac{[\vec{p}_2 - e\,\vec{A}(\vec{r}_2)]^2}{2m} + \frac{\vec{p}_N^{\,2}}{2m_N} \\ & - \frac{Z\alpha}{r_1} - \frac{Z\alpha}{r_2} + \frac{\alpha}{r_{12}} + \sum_{\lambda=1,2}\int \mathrm{d}^3k\,k\,a^+_\lambda(\vec{k})\,a_\lambda(\vec{k})\,, \end{aligned} \tag{13.38}$$

where the fundamental photon creation and annihilation operators are normalized so that $[a_\lambda(\vec{k}), a^+_{\lambda'}(\vec{k}')] = \delta^{(3)}(\vec{k}-\vec{k}')\,\delta_{\lambda\lambda'}$. We now go into the center-of-mass system, and ignore the mass polarization term [see Eq. (13.18)]. The unperturbed

Hamiltonian $\mathcal{H}_0$ consists of an atomic part, given for helium in Eq. (13.18), and the Hamiltonian of the free electromagnetic field,

$$\mathcal{H}_0 = \frac{\vec{p}_1^{\,2}}{2\mu} + \frac{\vec{p}_2^{\,2}}{2\mu} - \frac{Z\alpha}{r_1} - \frac{Z\alpha}{r_2} + \frac{\alpha}{r} + \sum_{\lambda=1,2} \int \mathrm{d}^3k\, k\, a_\lambda^+(\vec{k})\, a_\lambda(\vec{k}), \tag{13.39}$$

where μ is defined in Eq. (13.17). The perturbation is

$$H_1 = -\frac{e}{m}\vec{p}_1 \cdot \vec{A}(\vec{r}_1) - \frac{e}{m}\vec{p}_2 \cdot \vec{A}(\vec{r}_2) + \frac{\vec{A}^2(\vec{r}_1)}{2m} + \frac{\vec{A}^2(\vec{r}_2)}{2m}. \tag{13.40}$$

The electron mass m enters the interaction Hamiltonian, while the reduced mass μ figures in the unperturbed Hamiltonian. This is analogous to what we did for hydrogen. The vector potential operator is used in the dipole approximation, within the Schrödinger picture, as given in Eq. (2.71),

$$\vec{A}(\vec{r}) = \vec{A}_\perp(t=0,\vec{r}) \approx \vec{A}_\perp(t=0,\vec{r}=\vec{0}) = \sum_{\lambda=1}^{2} \int \frac{\mathrm{d}^3k}{\sqrt{2\omega\,(2\pi)^3}}\, \hat{\epsilon}_\lambda(\vec{k}) \left(a_\lambda(\vec{k}) + a_\lambda^+(\vec{k})\right). \tag{13.41}$$

One introduces an overlapping parameter ϵ, in order to calculate the low-energy part ΔE_{SL} of the self-energy, and sums over individual self-energies of the two bound electrons (labeled as $a = 1, 2$),

$$\begin{aligned}\Delta E_{SL} &= \sum_{a=1,2} \frac{e^2}{m^2} \langle\phi|\vec{p}_a \cdot \vec{A}(\vec{0}) \frac{1}{(E_0 - \mathcal{H}_0)'} \vec{p}_a \cdot \vec{A}(\vec{0})|\phi\rangle \\ &= \sum_{a=1,2} \frac{2e^2}{3m^2} \int_0^\epsilon \frac{\mathrm{d}^3k}{(2\pi)^3} \frac{1}{2k} \langle\phi|\vec{p}_a \frac{1}{E_0 - k - H_0} \vec{p}_a|\phi\rangle .\end{aligned} \tag{13.42}$$

Here, H_0 is the atomic part of $\mathcal{H}_0$, i.e., $H_0 = \mathcal{H}_0 - \sum_{\lambda=1,2} \int \mathrm{d}^3k\, k\, a_\lambda^+(\vec{k})\, a_\lambda(\vec{k})$. The two-electron reference state (assuming zero photons in the electromagnetic field) is denoted as $|\phi\rangle$, with unperturbed energy E_0. The photon energy integration can be carried out with the result,

$$\begin{aligned}\Delta E_{SL} &= \sum_{a=1,2} \frac{2\alpha}{3\pi m^2} \langle\phi|\vec{p}_a (H_0 - E_0) \ln\left(\frac{\epsilon\,\mu}{H_0 - E_0}\right) \vec{p}_a|\phi\rangle \\ &= -\sum_{a=1,2} \frac{2\alpha}{3\pi m^2} \langle\phi|\vec{p}_a (H_0 - E_0) \ln\left(\frac{H_0 - E_0}{(Z\alpha)^2\mu/2}\right) \vec{p}_a|\phi\rangle \\ &\quad + \left[\frac{4Z\alpha^2}{3m^2} \langle\delta^{(3)}(\vec{r}_1) + \delta^{(3)}(\vec{r}_2)\rangle - \frac{8\alpha^2}{3m^2} \langle\delta^{(3)}(\vec{r}_{12})\rangle\right] \ln\left(\frac{2\epsilon}{(Z\alpha)^2\mu}\right).\end{aligned} \tag{13.43}$$

Just like in Chap. 4, we have subtracted the low-energy part of the mass renormalization. In the evaluation of the logarithmically divergent term, we have used the relation

$$\begin{aligned}\sum_{a=1,2} [\vec{p}_a, [H_0 - E_0, \vec{p}_a]] &= \vec{\nabla}_1^2\left(-\frac{Z\alpha}{r_1}\right) + \vec{\nabla}_2^2\left(-\frac{Z\alpha}{r_2}\right) + \vec{\nabla}_1^2\left(\frac{\alpha}{r_{12}}\right) + \vec{\nabla}_2^2\left(\frac{\alpha}{r_{12}}\right) \\ &= 4\pi\left[Z\alpha\,\delta^{(3)}(\vec{r}_1) + Z\alpha\,\delta^{(3)}(\vec{r}_2)\right] - 8\pi\alpha\,\delta^{(3)}(\vec{r}_{12}),\end{aligned} \tag{13.44}$$

where $\vec{\nabla}_a \equiv \partial/\partial\vec{r}_a$. The logarithmically divergent terms from the electron-nucleus attraction and from the electron-electron repulsion carry different signs. The high-energy part of the one-electron Lamb shift was previously analyzed in Eq. (4.349). The term proportional to $-1/(2\kappa(2\ell+1))$ in Eq. (4.349) can be understood as a magnetic-moment correction to the spin-orbit interaction; for helium we will add this effect later on when we consider the radiative corrections to the Breit Hamiltonian. However, we immediately write down the contribution to the high-energy part of the self-energy from the spin-independent high-energy terms,

$$\Delta E_{SH} = \frac{Z\alpha^2}{m^2}\left[\frac{4}{3}\ln\left(\frac{m}{2\epsilon}\right) + \frac{10}{9} - \frac{4}{15}\right]\langle\delta^{(3)}(\vec{r}_1) + \delta^{(3)}(\vec{r}_2)\rangle - \frac{\alpha^2}{m^2}\left[2\times\frac{4}{3}\ln\left(\frac{m}{2\epsilon}\right) + 2\times\frac{10}{9} - \frac{4}{15}\right]\langle\delta^{(3)}(\vec{r}_{12})\rangle\,. \tag{13.45}$$

The first term is easy to understand as a generalization of the hydrogen result. In the second term, we multiply by two the contribution from both self-energies, but we also take into account that there is only one photon exchange diagram between the electrons in first order, which receives a correction from a vacuum-polarization insertion. The self-energy term $\Delta E_S = \Delta E_{SL} + \Delta E_{SH}$ then reads as

$$\Delta E_S = -\sum_{a=1,2}\frac{2\alpha}{3\pi m^2}\left\langle\phi\left|\vec{p}_a(H_0 - E_0)\ln\left(\frac{H_0 - E_0}{(Z\alpha)^2\mu/2}\right)\vec{p}_a\right|\phi\right\rangle + \frac{Z\alpha^2}{m^2}\langle\delta^{(3)}(\vec{r}_1) + \delta^{(3)}(\vec{r}_2)\rangle\left[\frac{4}{3}\ln\left(\frac{m}{(Z\alpha)^2\mu}\right) + \frac{10}{9} - \frac{4}{15}\right] - \frac{\alpha^2}{m^2}\langle\delta^{(3)}(\vec{r}_{12})\rangle\left[\frac{8}{3}\ln\left(\frac{m}{(Z\alpha)^2\mu}\right) + \frac{20}{9} - \frac{4}{15}\right]. \tag{13.46}$$

One might wonder why the Bethe logarithm term only has contributions of the form $\langle\vec{p}_1\cdots\vec{p}_1\rangle$ and $\langle\vec{p}_2\cdots\vec{p}_2\rangle$, but no contributions of the form $\langle(\vec{p}_1+\vec{p}_2)\cdots(\vec{p}_1+\vec{p}_2)\rangle$. One would otherwise expect the presence of such contributions based on the coupling of both electrons to the electromagnetic field. We shall find the answer to this question below. Indeed, in a two-electron system, the relativistic recoil correction (Salpeter correction, see Sec. 15.5), which is an order-α^5 correction beyond the Breit–Pauli Hamiltonian, needs to be added to the self-energy term, to obtain the complete result for the order-α^5 corrections to the Lamb shift.

For a two-body system (see Sec. 15.5), the relativistic recoil correction of order α^5 needs to be evaluated for the (repulsive) interaction of the electrons. The low-energy part of the Salpeter correction is due to the exchange of a transverse photon between the two electrons, beyond the Breit Hamiltonian,

$$\begin{aligned}\Delta E_L &= \sum_{a=1,2} \frac{2e^2}{3m^2} \int_0^{\epsilon\alpha/2} \frac{d^3k}{(2\pi)^3} \frac{1}{2k} \left\langle \phi | \vec{p}_a \frac{1}{E_0 - k - H_0} \vec{p}_{3-a} | \phi \right\rangle \\ &= \sum_{a=1,2} \frac{2\alpha}{3\pi m^2} \left\langle \phi | \vec{p}_a \, (H_0 - E_0) \ln\left(\frac{\epsilon\,\alpha\mu/2}{H_0 - E_0} \right) \vec{p}_{3-a} | \phi \right\rangle \\ &= -\sum_{a=1,2} \frac{2\alpha}{3\pi m^2} \left\langle \phi | \vec{p}_a \, (H_0 - E_0) \ln\left(\frac{2(H_0 - E_0)}{(Z\alpha)^2 \mu} \right) \vec{p}_{3-a} | \phi \right\rangle \\ &\quad + \frac{8}{3} \frac{\alpha^2}{m^2} \ln\left(\frac{\epsilon}{Z^2\,\alpha\,\mu} \right) \langle \delta^{(3)}(\vec{r}_{12}) \rangle .\end{aligned} \tag{13.47}$$

This term supplies the "mixed" terms of the form $\langle (\vec{p}_1 + \vec{p}_2) \dots (\vec{p}_1 + \vec{p}_2) \rangle$ in the Bethe logarithm for helium. For the remaining contributions, the calculation is similar to the one reported in Sec. 15.5 (see also Ref. [388]). Using the commutator relation

$$[\vec{p}_1, [(H_0 - E_0), \vec{p}_2]] = \left[\vec{p}_1, \left[\frac{\alpha}{r_{12}}, \vec{p}_2 \right] \right] = -\vec{p}_1 \cdot \vec{p}_2 \left(\frac{\alpha}{r} \right) = \vec{p}_1^{\,2} \left(\frac{\alpha}{r_{12}} \right) = 4\pi\alpha\, \delta^{(3)}(\vec{r}_{12}) , \tag{13.48}$$

the middle-energy part can be formulated as follows,

$$\Delta E_M = -\frac{2\alpha^5 m}{3\pi} \left\langle P\left(\frac{1}{(\alpha m r_{12})^3} \right) \right\rangle + \frac{8}{3} \frac{\alpha^2}{m^2} \left[\frac{4}{3} + \ln(2) - \ln\left(\frac{\epsilon}{m} \right) \right] \langle \delta^{(3)}(\vec{r}_{12}) \rangle . \tag{13.49}$$

We here employ the Araki–Sucher distribution P that was already discussed in Eq. (15.133).

Because of the transformation of the lower limit in the radial integration of the $1/r^3$ regularized matrix element, we have

$$(\alpha\mu)^3 P\left(\frac{1}{(\alpha\mu r_{12})^3} \right) = (\alpha m)^3 P\left(\frac{1}{(\alpha m r_{12})^3} \right) + 4\pi \delta^{(3)}(\vec{r}_{12}) \ln\left(\frac{\mu}{m} \right) . \tag{13.50}$$

Otherwise, the mass $\mu \to m$ scales out of the integral, but the logarithmically divergent term receives a correction. The double transverse part is given as follows,

$$\Delta E_D = -\frac{\alpha^5 m}{2\pi} \left\langle P\left(\frac{1}{(\alpha m r_{12})^3} \right) \right\rangle - \frac{2\alpha^2}{m^2} \left(\ln(\lambda) + \frac{4}{3} \ln(2) - \frac{17}{6} \right) \langle \delta^{(3)}(\vec{r}_{12}) \rangle . \tag{13.51}$$

Here, λ is a cutoff parameter for the ultraviolet region of the photon momentum, which multiplies the photon propagator (in whatever gauge) by $\Lambda^2/(\Lambda^2 + \vec{k}^2)$ where $\vec{k}^2$ is the momentum square of the exchanged photon, and $\Lambda = \lambda\,\alpha\,\mu$. We employ the same multiplicative regularization in Sec. 15.5. The high-energy part reads as

$$\Delta E_H = -\frac{\alpha^2}{m^2} \left(\langle \vec{\sigma}_1 \cdot \vec{\sigma}_2 \rangle + \frac{4}{3} - 2 \ln(\lambda\,\alpha) \right) \langle \delta^{(3)}(\vec{r}_{12}) \rangle , \tag{13.52}$$

with the same λ as before. The total recoil correction is easily obtained as follows,

$$\begin{aligned}\Delta E_R &= \Delta E_L + \Delta E_M + \Delta E_D + \Delta E_H \\ &= \frac{\alpha^2}{m^2} \left(\frac{71}{9} - \langle \vec{\sigma}_1 \cdot \vec{\sigma}_2 \rangle - \frac{16}{3} \ln(Z) - \frac{2}{3} \ln(\alpha) \right) \langle \delta^{(3)}(\vec{r}_{12}) \rangle \\ &\quad - \frac{14\alpha^5 m}{3} \left\langle \frac{1}{4\pi} P\left(\frac{1}{(\alpha\, m\, r_{12})^3} \right) \right\rangle .\end{aligned} \tag{13.53}$$

This result is independent any overlapping parameter, as it should be. The complete helium Lamb shift, which is the energy correction of order α^5, consists of the sum of the self-energy term E_S given in Eq. (13.46), the recoil term E_R [see Eq. (13.53)], and a magnetic-moment correction to the spin-dependent terms $H_2^{(4)}$ in the Breit–Pauli Hamiltonian,

$$E^{(5)} = \frac{\alpha}{\pi}\left\langle H_{\rm so} + H'_{\rm soo} + \frac{1}{2}H''_{\rm soo} + H_{\rm ss} + H_{\rm ssc}\right\rangle$$
$$+ \frac{\alpha^2}{m^2}\left[\frac{89}{15} + \frac{14}{3}\ln\alpha - \langle\vec{\sigma}_1\cdot\vec{\sigma}_2\rangle\right]\langle\delta^{(3)}(\vec{r}_{12})\rangle - \frac{14\alpha^5 m}{3(4\pi)}\left\langle P\left(\frac{1}{(\alpha\,\mu\, r_{12})^3}\right)\right\rangle$$
$$+ \left\{\left[\frac{4}{3}\ln\left(\frac{m}{(Z\alpha)^2\mu}\right) + \frac{38}{45}\right]\frac{Z\alpha^2}{m^2} - \frac{4\alpha}{3m^2}\ln k_0\right\}\left\langle\delta^{(3)}(\vec{r}_1) + \delta^{(3)}(\vec{r}_2)\right\rangle . \qquad (13.54)$$

All the spin-dependent terms in the Breit Hamiltonian receive an anomalous magnetic-moment correction [see also Eq. (13.31)]. When comparing to Eq. (12.76), one first recalls that $2\kappa = \alpha/\pi$ in one-loop order [see Eq. (1.11)]. Then, for the interaction between two electrons, in the spirit of Eq. (12.76), one has $(1+\kappa_a)(1+\kappa_b) = (1+\kappa)^2 = 1 + 2\kappa$, plus higher-order terms, which we ignore.

For neutral helium, one has $Z = 2$, while formula (13.54) is also valid for helium-like ions. Terms proportional to $\langle\delta^{(3)}(\vec{r}_{12})\rangle \ln(Z)$ vanish in the sum of $\Delta E_S + \Delta E_R$. As a sum of the pure self-energy and the recoil correction, we have the Bethe logarithm which contains the transition matrix elements with $\vec{p}_1 + \vec{p}_2$ and reads as

$$\ln k_0 = \frac{\left\langle(\vec{p}_1+\vec{p}_2)(H_0 - E_0)\ln\left(\frac{H_0 - E_0}{(Z\alpha)^2\mu/2}\right)(\vec{p}_1+\vec{p}_2)\right\rangle}{2\pi(Z\alpha)\left\langle\delta^{(3)}(\vec{r}_1) + \delta^{(3)}(\vec{r}_2)\right\rangle} . \qquad (13.55)$$

The last expression contains a definition of the Bethe logarithm $\ln k_0$. For numerical data, we refer to Table 13.1. It can be shown that, for heliumlike ions with charge number $Z > 2$, the quantity $\ln(k_0/Z^2)$ approximates a constant very well. The definition given in Eq. (13.55) is a natural generalization of the Bethe logarithm defined in Eq. (4.317). Indeed, the formula (4.317) can be obtained from Eq. (13.55) under the replacement

$$\pi(Z\alpha)\left\langle\delta^{(3)}(\vec{r}_1) + \delta^{(3)}(\vec{r}_2)\right\rangle \to \pi(Z\alpha)\left\langle\delta^{(3)}(\vec{r})\right\rangle = \frac{(Z\alpha)^4\mu^3}{n^3} . \qquad (13.56)$$

The indication of a result valid for general Z is useful for heliumlike ions.

Note that the definition (13.55) implies that the Bethe logarithms only depend on the orbital symmetry (total orbital quantum number L) and on the total spin angular quantum number S. Here,

$$\vec{L} = \vec{\ell}_1 + \vec{\ell}_2 = \vec{r}_1\times\vec{p}_1 + \vec{r}_2\times\vec{p}_2 , \qquad \vec{L}^2 = L(L+1) . \qquad (13.57)$$

The total spin angular momentum corresponds to the two-particle spin states given in Eqs. (13.1) and (13.2). The two-particle states are labeled as S states for $L = 0$, and as P states if $L = 1$. Furthermore, singlet states have $S = 0$, whereas for triplet states, one has $S = 1$.

Table 13.1 The Bethe logarithms for neutral helium (nuclear charge number $Z = 2$) are given for singlet-S, triplet-S, singlet-P and triplet-P states (data are taken from Ref. [390]). The 2^3S state is the ground state of ortho-helium. The definition (13.55) implies that the Bethe logarithms only depend on the orbital symmetry (total orbital quantum number L) and on the total spin quantum number S.

Bethe Logarithms for Low-Lying States of Helium

	n^1S	n^3S	n^1P	n^3P
$n = 1$	2.983 865 857(3)	–	–	–
$n = 2$	2.980 118 36(7)	2.977 742 46(1)	2.983 803 46(3)	2.983 690 84(2)
$n = 3$	2.982 870 510(8)	2.982 372 56(2)	2.984 001 39(9)	2.983 939 5(3)
$n = 4$	2.983 596 30(1)	2.983 429 08(5)	2.984 068 8(2)	2.984 039 80(5)
$n = 5$	2.983 857 3(1)	2.983 783 95(8)	2.984 096 18(4)	2.984 080 2(2)

The configurations can be addressed more clearly, in the sense that singlet-S states correspond to the configurations $1s\,ns\;{}^1S_0$, where the two-electron orbital configuration approximates the symmetrized combination of one electron in the $1S$ state, the other electron being in the ns state. The total angular momentum of orbital and spin angular momenta of both electrons is $J = 0$, and this is indicated in the spectroscopic notation $n^1S_{J=0}$. Correspondingly, n^1P states qualify themselves as $1s\,np\;{}^1P_1$, while n^1D states would be $1s\,nd\;{}^1D_2$, and n^1F states would be $1s\,nf\;{}^1F_3$. For triplet state, where the orbital wave function is antisymmetrized, one identifies the n^3S state configurations more precisely as $1s\,ns\;{}^3S_1$, while the n^3P configurations split up into $1s\,np\;{}^3P_{0,1,2}$ under the fine-structure splitting (with $J = 0, 1, 2$), and the n^3D configurations qualify themselves as $1s\,nd\;{}^3D_{1,2,3}$. Triplet-F configurations n^3F are qualified as $1s\,nf\;{}^3F_{2,3,4}$. In general, when we consider triplet states, then, for $L \geq 1$, the permissible values of J are $J = L - 1, L, L + 1$.

13.4 Numerical Calculation of the Helium Spectrum

13.4.1 *Fully Correlated Basis Sets*

The Hartree–Fock approximation still treats the electron wave function as being composed of individual orbitals and therefore is not suited for atomic systems with either a strong interelectronic correlation, or for systems where an exceptionally high degree of accuracy in the theoretical predictions is required. A paradigmatic example for a system of the latter kind is atomic helium, where experiments have meanwhile reached absolutely impressive accuracy [391].

In a fully correlated basis set, the functional dependence on the relative coordinate $r = r_{12} = |\vec{r}_1 - \vec{r}_2|$ is taken into account to all orders. The basic paradigm then consists in writing down a suitable set of basis ("trial") functions, and to evaluate the matrix elements of the helium Hamiltonian on the basis functions. In the limit of an infinite number of basis functions, the basis hopefully becomes complete and the calculated bound-state energies become accurate approximations of the true energy eigenvalues. This difficulty cannot be avoided because no exact, analytic eigenstate solutions of the three-body unperturbed helium Hamiltonian

$$H = \frac{\vec{p}_1^{\,2}}{2\mu} + \frac{\vec{p}_2^{\,2}}{2\mu} - \frac{Z\alpha}{r_1} - \frac{Z\alpha}{r_2} + \frac{\alpha}{r} \tag{13.58}$$

in the space of spin-symmetrized nonrelativistic wave functions are known. We recall that (13.58) is an approximation to (13.18), neglecting the mass polarization term.

In Sec. 6.4.3, we had introduced the spherical biharmonic,

$$\mathcal{Y}^{\ell_1\,\ell_2}_{L\,M}(\vec{r}_1, \vec{r}_2) = \sum_{m_1 m_2} C^{LM}_{\ell_1\,m_1\,\ell_2\,m_2} Y_{\ell_1\,m_1}(\hat{r}_1)\, Y_{\ell_2\,m_2}(\hat{r}_2)\,. \tag{13.59}$$

Here, the two individual angular momenta ℓ_1 and ℓ_2 are added to give the total angular momentum L. Singlet-state wave functions of total angular momentum L and magnetic projection M have the following structure,

$$\begin{aligned}\Psi_{S=0}(L, M, \vec{r}_1, \vec{r}_2) = \big[&R_{nL}(r_1, r_2, r_{12})\, \mathcal{Y}^{\ell_1\,\ell_2}_{L\,M}(\hat{r}_1, \hat{r}_2) \\ &+ R_{nL}(r_2, r_1, r_{12})\, \mathcal{Y}^{\ell_1\,\ell_2}_{L\,M}(\hat{r}_2, \hat{r}_1)\big]\ |00\rangle\,.\end{aligned} \tag{13.60}$$

Here, n is a multi-index for the approximate principal quantum numbers, which give the main admixtures to the state. The singlet ground state is $(1s^2)\,{}^1S_0$, and higher excited singlet states have the structure

$$\Psi_{S=0}(0, 0, \vec{r}_1, \vec{r}_2) = \frac{1}{4\pi}\,\left[R_{n0}(r_1, r_2, r_{12}) + R_{n0}(r_2, r_1, r_{12})\right] |00\rangle\,. \tag{13.61}$$

Triplet state wave functions of total angular momentum J and total magnetic projection M have the following structure,

$$\begin{aligned}\Psi_{S=1}(J, M, \vec{r}_1, \vec{r}_2) = \sum_{M'\,M''} C^{JM}_{L'\,M'\,1\,M''}\, \big[&R_{nL'}(r_1, r_2, r_{12})\, \mathcal{Y}^{\ell_1\,\ell_2}_{L'\,M'}(\hat{r}_1, \hat{r}_2) \\ &- R_{nL'}(r_2, r_1, r_{12})\, \mathcal{Y}^{\ell_1\,\ell_2}_{L'\,M'}(\hat{r}_2, \hat{r}_1)\big]\ |1\,M''\rangle\,.\end{aligned} \tag{13.62}$$

The triplet ground state is $1s\,2s\,{}^3S_1$, and triplet S configurations have the general structure

$$\Psi_{S=1}(1, M, \vec{r}_1, \vec{r}_2) = \left[R_{n0}(r_1, r_2, r_{12}) - R_{n0}(r_2, r_1, r_{12})\right] |1\,M\rangle\,. \tag{13.63}$$

Within the nonrelativistic approximation, the triplet ($S = 1$) state energies only depend on the total orbital angular momentum L', and are independent of the value of $J = L' - 1, L', L' + 1$ and of course independent of M. We reemphasize that the total angular momentum operator

$$\vec{L} = \vec{\ell}_1 + \vec{\ell}_2\,, \qquad \vec{L}\, R_{nL}(r_1, r_2, r_{12}) = 0\,, \tag{13.64}$$

has a zero eigenvalue when acting on any function of the radial coordinates $R_{nL}(r_1, r_2, r_{12})$.

In constructing the basis set, the following criteria are relevant: The basis set of trial functions must have sufficient mathematical structure in order to accommodate the particularities of the helium atom, but it must also be simple enough so that the integrals of the terms in the helium Hamiltonian, and in the relativistic corrections, can be easily calculated, preferentially in closed analytic form. Two basis sets have been used widely in calculations of properties of the helium atom, namely, the HYLLERAAS[2] [392–395] and the KOROBOV[3] [396–398] basis sets. Indeed, the different approaches for the treatment of the helium atom are distinguished based on the functional form of the radial component. The Hylleraas *ansatz* entails the functional form

$$R_{nL}(r_1, r_2, r_{12}) = \sum_{i=0}^{m} \sum_{j=0}^{m-i} \sum_{k=0}^{m-i-j} a_{ijk}\, (r_1)^i\, (r_2)^j\, (r_{12})^k\, \mathrm{e}^{-\alpha r_1 - \beta r_2}\,, \tag{13.65}$$

where of course the a_{ijk} depend on the principal quantum numbers. They are expansion parameters, to be determined in the diagonalization of the Hamiltonian matrix, and the α and β are global parameters which describe the exponential fall-off of the basis functions. The terms $(r_1)^i$, $(r_2)^j$ and $(r_{12})^k$ are powers of the radial coordinates, which we denote in a very explicit notation, in order to differentiate them from the Cartesian components r_1^i, r_2^j and $r_{12}^k = r_1^k - r_2^k$ as the kth component of $\vec{r}_{12} = \vec{r}_1 - \vec{r}_2$. The parameters α and β are treated as variational parameters in a typical calculation, which are optimized according to a variational principle. E.g., the best approximation to the ground-state energy, given a finite number of basis functions, is naturally determined as the trial function which, among the given set of basis functions, yields the minimum energy. In the Hylleraas basis, trial states are ordered into so-called Pekeris shells of order m with $i + j + k = m$.

A possible modification of the Hylleraas idea consists in a "double basis set" for the exponentials. The "double bases" are distinguished according to the distance from the nucleus which they model. Indeed, the double-basis Hylleraas *ansatz* for the helium wave functions of excited $1s\, ns$ states is given as

$$\begin{aligned} R_{nL}(r_1, r_2, r_{12}) = {} & \sum_{i=0}^{M} \sum_{j=0}^{M-i} \sum_{k=0}^{M-i-j} a_{ijk}\, r_1^i\, r_2^j\, r_{12}^k\, \mathrm{e}^{-\alpha_{\mathrm{I}} r_1 - \beta_{\mathrm{I}} r_2} \\ & + \sum_{i=0}^{M} \sum_{j=0}^{M-i} \sum_{k=0}^{M-i-j} b_{ijk}\, r_1^i\, r_2^j\, r_{12}^k\, \mathrm{e}^{-\alpha_{\mathrm{II}} r_1 - \beta_{\mathrm{II}} r_2}\,. \end{aligned} \tag{13.66}$$

In a typical calculation with the double basis set (see Refs. [399, 400]), the parameters α_{I} and β_{I} model the short-distance physics.

Sometimes, the Hylleraas *ansatz* is written in terms of the so-called Hylleraas variables

$$s = r_1 + r_2\,, \qquad t = r_1 - r_2\,, \qquad u = r_{12}\,. \tag{13.67}$$

[2]Egil Andersen Hylleraas (1898–1965).
[3]Vladimir Ivanovich Korobov (b. 1956).

Counting all orders in the expansion in s, t, and u, the same terms are generated as by the powers of r_1, r_2 and r_{12}. According to Refs. [401, 402], even negative powers of s have to be admitted. Matrix elements of operators in the Hylleraas basis are discussed in Ref. [399]. Calculations using Hylleraas basis sets are usually numerically quite stable, and the separation into a double basis can help to separate short-distance from long-distance physics [403]. For matrix elements in the Hylleraas basis, we refer to [404] and to the early compilation of recurrence relations in Ref. [405]. For the higher-lying Rydberg states, the double basis set has proven to be remarkably stable from a numerical standpoint [404].

In the Korobov basis set (exponential basis set), the radial functions are of the form

$$R_{nL}(r_1, r_2, r_{12}) = \sum_i a_i \, K(\alpha_i, \beta_i, \gamma_i, r_1, r_2, r_{12}),$$
$$K(\alpha_i, \beta_i, \gamma_i, r_1, r_2, r_{12}) = \exp\left(-\alpha_i \, r_1 - \beta_i \, r_2 - \gamma_i \, r_{12}\right). \tag{13.68}$$

The basis states are conveniently labeled by the parameters α, β and γ which define three regions in which these are optimized. The Korobov basis set is well suited for practical calculations because the integrals of the functions in the basis set assume a compact functional form. One basic integral [406] is

$$\int \mathrm{d}^3 r_1 \int \mathrm{d}^3 r_2 \, \frac{\mathrm{e}^{-\alpha r_1 - \beta r_2 - \gamma r_{12}}}{r_1 \, r_2 \, r_{12}} = \frac{16\pi^2}{(\alpha+\beta)(\beta+\gamma)(\gamma+\alpha)}. \tag{13.69}$$

The advantage of the exponential basis set is that the product of two wave functions in the basis set reproduces a mathematical expression of the same structure as a wave function within the basis set,

$$K(\alpha_1, \beta_1, \gamma_1, r_1, r_2, r_{12}) \, K(\alpha_2, \beta_2, \gamma_2, r_1, r_2, r_{12})$$
$$= K(\alpha_1 + \alpha_2, \beta_1 + \beta_2, \gamma_1 + \gamma_2, r_1, r_2, r_{12}). \tag{13.70}$$

Integrals of the form

$$\int \mathrm{d}^3 r_1 \int \mathrm{d}^3 r_2 \, r_1^k r_2^m r_{12}^n \, \frac{\mathrm{e}^{-\alpha r_1 - \beta r_2 - \gamma r_{12}}}{r_1 \, r_2 \, r_{12}}$$
$$= \left(\frac{\partial}{\partial \alpha}\right)^k \left(\frac{\partial}{\partial \beta}\right)^m \left(\frac{\partial}{\partial \gamma}\right)^n \frac{16\pi^2}{(\alpha+\beta)(\beta+\gamma)(\gamma+\alpha)} \tag{13.71}$$

can then be evaluated by repeated differentiation over parameters. $k, m, n \geq -1$; for -1 one has to replace differentiation by integration. Calculations made using the Korobov basis typically entail significant losses of numerical precision in intermediate steps, and extended-precision arithmetic [407–409] is often used.

13.4.2 *Matrix Elements and Nonorthogonal Basis*

For actual numerical calculations, it is customary to first transform to dimensionless variables, in a manner consistent with the length dimensions of the system. We start from the helium Hamiltonian given in Eq. (13.58),

$$H = -\frac{1}{2\mu} \vec{\nabla}_1^2 - \frac{Z\alpha}{r_1} - \frac{1}{2\mu} \vec{\nabla}_2^2 - \frac{Z\alpha}{r_2} + \frac{\alpha}{r_{12}}, \tag{13.72}$$

and scale the coordinate as follows,

$$r_1 \to \frac{\rho_1}{\alpha\,\mu}, \qquad r_2 \to \frac{\rho_2}{\alpha\,\mu}. \tag{13.73}$$

The Hamiltonian then becomes

$$\begin{aligned} H &= -\frac{(\alpha\,\mu)^2}{2\mu}\frac{\partial^2}{\partial\vec{\rho}_1^{\,2}} - \frac{(\alpha\,\mu)^2}{2\mu}\frac{\partial^2}{\partial\vec{\rho}_2^{\,2}} - \frac{Z\,\alpha^2\,\mu}{\rho_1} - \frac{Z\,\alpha^2\,\mu}{\rho_2} + \frac{\alpha^2\,\mu}{\rho_{12}} \\ &= \alpha^2\,\mu\left(-\frac{1}{2}\frac{\partial^2}{\partial\vec{\rho}_1^{\,2}} - \frac{Z}{\rho_1} - \frac{1}{2}\frac{\partial^2}{\partial\vec{\rho}_2^{\,2}} - \frac{Z}{\rho_2} + \frac{1}{\rho_{12}}\right), \end{aligned} \tag{13.74}$$

which enables us to scale the quantity $\alpha^2\,\mu$ out of the Hamiltonian term. The remaining term, after the redefinition $\vec{\rho}_i \to \vec{r}_i$, can be written as follows,

$$H_0 = -\frac{1}{2}\vec{\nabla}_1^2 - \frac{Z}{r_1} - \frac{1}{2}\vec{\nabla}_2^2 - \frac{Z}{r_2} + \frac{1}{r_{12}}. \tag{13.75}$$

Here, all quantities are dimensionless, and they assume numerical values of order unity for the helium atom. For definiteness, we shall now specialize to the evaluation of matrix elements of the unperturbed helium Hamiltonian H_0, for singlet S and triplet S states, in the Korobov basis, we have the matrix elements,

$$\mathcal{H}(\alpha_i,\beta_i,\gamma_i,\alpha_j,\beta_j,\gamma_j) = \langle K(\alpha_i,\beta_i,\gamma_i,r_1,r_2,r_{12})|H_0|K(\alpha_j,\beta_j,\gamma_j,r_1,r_2,r_{12})\rangle. \tag{13.76}$$

The matrix element of the Hamiltonian is obtained as

$$\begin{aligned} H(\alpha_i,\beta_i,\gamma_i,\alpha_j,\beta_j,\gamma_j) &= \mathcal{H}(\alpha_i,\beta_i,\gamma_i,\alpha_j,\beta_j,\gamma_j) \\ &\quad \pm(\beta_i \leftrightarrow \beta_j) \pm (\alpha_i \leftrightarrow \alpha_j) + (\alpha_i \leftrightarrow \alpha_j, \beta_i \leftrightarrow \beta_j) \end{aligned} \tag{13.77}$$

where the positive sign is to be used for spin singlet states and the negative sign is for spin triplet states. We work in a non-orthogonal basis and therefore define the overlap matrix (with multi-indices i and j which specify the states)

$$\mathcal{S}(\alpha_i,\beta_i,\gamma_i,\alpha_j,\beta_j,\gamma_j) = \langle K(\alpha_i,\beta_i,\gamma_i,r_1,r_2,r_{12})|K(\alpha_j,\beta_j,\gamma_j,r_1,r_2,r_{12})\rangle, \tag{13.78}$$

with

$$\begin{aligned} S(\alpha_i,\beta_i,\gamma_i,\alpha_j,\beta_j,\gamma_j) &= \mathcal{S}(\alpha_i,\beta_i,\gamma_i,\alpha_j,\beta_j,\gamma_j) \\ &\quad \pm(\beta_i \leftrightarrow \beta_j) \pm (\alpha_i \leftrightarrow \alpha_j) + (\alpha_i \leftrightarrow \alpha_j, \beta_i \leftrightarrow \beta_j). \end{aligned} \tag{13.79}$$

Here, S is properly (anti-)symmetrized. We can view the overlap matrix as a matrix which consists of the overlap integrals of the states $|\psi_i\rangle$ and $|\psi_j\rangle$,

$$S_{ij} = \langle\psi_i|\psi_j\rangle, \qquad \mathbb{1} = \sum_{ij} |\psi_i\rangle \left(S^{-1}\right)_{ij} \langle\psi_j|, \qquad \mathbb{S} = \mathbb{L}^T\,\mathbb{L}, \tag{13.80}$$

where the latter expression identifies the Cholesky decomposition. The above expression for the unit matrix is easily proven by writing out the explicit expression for the square $\mathbb{1}^2$, and using the fact that $S_{ij} = \langle\psi_i|\psi_j\rangle$. The lower-diagonal matrix $\mathbb{L}$ has the properties

$$\left(\mathbb{L}^{-1}\right)^T = \left(\mathbb{L}^T\right)^{-1}, \qquad \text{but} \qquad \mathbb{L}^{-1} \neq \mathbb{L}^T. \tag{13.81}$$

Using the fundamental decomposition of an eigenvector $|\psi\rangle = \sum_k c_k |\psi_k\rangle$, the eigenvalue problem can then be formulated as $H|\psi\rangle = E|\psi\rangle$. After inserting a unit matrix, this reads as

$$\mathbb{1}\, H \sum_k c_k |\psi_k\rangle = E \sum_i c_i |\psi_i\rangle . \tag{13.82}$$

Using the relation (13.80), we then have

$$\sum_i |\psi_i\rangle \left(\sum_{jk} \left(S^{-1}\right)_{ij} \langle \psi_j | H | \psi_k \rangle c_k \right) = \sum_i |\psi_i\rangle E c_i . \tag{13.83}$$

This has to be fulfilled for all i separately since the basis spanned by the $|\psi_i\rangle$ is non-singular. Thus,

$$\sum_{jk} \left(S^{-1}\right)_{ij} \langle \psi_j | H | \psi_k \rangle c_k = E c_i , \qquad \mathbb{S}^{-1} \mathbb{H} \mathbb{c} = E \mathbb{c} . \tag{13.84}$$

The latter equation is valid in terms of the column vector $\mathbb{c}$ spanned by the c_i coefficients and the matrix $\mathbb{H}$ spanned by the $\langle \psi_i | H | \psi_j \rangle$. This can be reformulated as $\mathbb{H}\,\mathbb{c} = E\,\mathbb{S}\,\mathbb{c}$. Inserting the unit operator as $\mathbb{1} = \mathbb{L}^{-1}\,\mathbb{L}$ on the left-hand side, and using the relation $\mathbb{S} = \mathbb{L}^T\,\mathbb{L}$ on the right-hand side, we obtain

$$\mathbb{H}\,\mathbb{L}^{-1}\,\mathbb{L}\,\mathbb{c} = E\,\mathbb{L}^T\,\mathbb{L}\,\mathbb{c} . \tag{13.85}$$

This can trivially be reformulated as

$$\left(\left(\mathbb{L}^T\right)^{-1} \mathbb{H}\,\mathbb{L}^{-1} \right) \mathbb{d} = E\,\mathbb{d} , \qquad \mathbb{d} = \mathbb{L}\,\mathbb{c} . \tag{13.86}$$

The Hamiltonian matrix which gives the eigenvalues thus is given by

$$\mathbb{H}_{\text{eff}} = \left(\mathbb{L}^T\right)^{-1} \mathbb{H}\,\mathbb{L}^{-1} , \qquad \mathbb{H}_{\text{eff}}\,\mathbb{d} = E\,\mathbb{d} , \qquad \mathbb{c} = \mathbb{L}^{-1}\,\mathbb{d} . \tag{13.87}$$

It is still symmetric, but it does not have a band structure.

13.4.3 *Numerical Results and Technical Issues*

Numerical calculations for helium have a long history. The starting point is the scaled Hamiltonian given in Eq. (13.75),

$$H_0 = -\frac{1}{2}\vec{\nabla}_1^2 - \frac{Z}{r_1} - \frac{1}{2}\vec{\nabla}_2^2 - \frac{Z}{r_2} + \frac{1}{r_{12}} . \tag{13.88}$$

Throughout the last two decades, it has been a challenge to determine the ground state of the helium atom to utmost accuracy. We give three recent results, obtained using fully correlated basis sets. The first result is taken from Ref. [410],

$$E_0 = -2.903\,724\,377\,034\,119\,598\,30(2) . \tag{13.89}$$

The second result is taken from a more recent article by Korobov [398],

$$E_0 = -2.903\,724\,377\,034\,119\,598\,311\,093\,1 . \tag{13.90}$$

The record in accuracy is still being held by Refs. [401, 402], and reads as

$$E_0 = -2.903\,724\,377\,034\,119\,598\,311\,159\,245\,194\,404\,446\,696\,925\,309 . \tag{13.91}$$

Here, E_0 refers to the energy of the ${}^1S_0$ state. The next higher excited singlet and triplet S state reference energies are given as follows [411],

$$E(1s\,2s\,{}^1S_0) = -2.145\,974\,046\,054\,426(18)\,, \tag{13.92a}$$

$$E(1s\,2s\,{}^3S_1) = -2.175\,229\,378\,236\,791\,23(10)\,. \tag{13.92b}$$

The numerical accuracy of these results exceeds that of the relativistic and radiative corrections, and is thus more than needed for a comparison of theory and experiment. Selected values for singlet and triplet states of the helium atom are given in Tables 13.2 and 13.3. For a recent compilation of level energies with relativistic and radiative corrections, one consults Refs. [412, 413]. For $1s\,nd$ states, one consults Ref. [414]. Progress toward the evaluation of all $\alpha^7\,m$ corrections in helium, with tremendous advances being reported for triplet $n = 2$ states, has recently been reported in Refs. [415–418].

One of the most important boundary conditions, to be integrated in the formalism, concerns the Kato cusp condition [419, 420]. We briefly generalize the discussion, return to a Hamiltonian formalism with dimensionful momenta and coordinates, and consider those terms in the Hamiltonian in a multi-electron atom which describe the kinetic energy of electrons i and j (or, say, charged particles i and j), and the electrostatic interaction between them,

$$H = \frac{\vec{p}_i^{\,2}}{2m_i} + \frac{\vec{p}_j^{\,2}}{2m_j} + \dots + \frac{q_i\,q_j}{4\pi\,r_{ij}} + \dots \quad (\text{no sum over } i \text{ or } j)\,. \tag{13.93}$$

We use natural units with $\hbar = c = \epsilon_0 = 1$. For the electron-electron interaction in the helium atom, we have $i = 1$ and $j = 2$, and $m_1 = m_2 = \mu$ as well as $q_1\,q_2 = e^2 = 4\pi\alpha$.

Table 13.2 The nonrelativistic energies of the singlet states ${}^1L_0$ of the helium atom are given for excitations with principal quantum numbers $n = 1, \dots, 6$, and total orbital angular momenta $L = 0, 1, 2, 3$. Data are taken from Refs. [403, 404]. The singlet states follow the same notation as was used in Table 13.1.

Nonrelativistic Binding Energies for Singlet States of Helium

	n^1S	n^1P	n^1D	n^1F
$n = 1$	-2.903 724 377 034	—	—	—
$n = 2$	-2.145 974 046 054	-2.123 843 086 498	—	—
$n = 3$	-2.061 271 989 741	-2.055 146 362 092	-2.055 620 732 852	—
$n = 4$	-2.033 586 717 031	-2.031 069 650 450	-2.031 279 846 179	-2.031 255 144 382
$n = 5$	-2.021 176 851 574	-2.019 905 989 901	-2.020 015 836 160	-2.020 002 937 158
$n = 6$	-2.014 563 098 447	-2.013 833 979 672	-2.013 898 227 424	-2.013 890 683 816

Table 13.3 The analog of Table 13.2 for nonrelativistic energies of triplet states ${}^3L_J$ of the helium atom involves excitations with principal quantum numbers $n = 1, \ldots, 6$, and total orbital angular momenta $L = 0, 1, 2, 3$. Data are taken from Refs. [403,404]. The same notation was used as in Table 13.1.

Nonrelativistic Binding Energies for Triplet States of Helium

	n^3S	n^3P	n^3D	n^3F
$n = 2$	-2.175 229 378 237	-2.133 164 190 779	—	—
$n = 3$	-2.068 689 067 472	-2.058 081 084 274	-2.055 636 309 453	—
$n = 4$	-2.036 512 083 098	-2.032 324 354 297	-2.031 288 847 502	-2.031 255 168 403
$n = 5$	-2.022 618 872 302	-2.020 551 187 256	-2.020 021 027 447	-2.020 002 957 377
$n = 6$	-2.015 377 452 993	-2.014 207 958 774	-2.013 901 415 454	-2.013 890 698 348

The eigenfunctions of H contain singularities at points where two charged particles collide, i.e., have zero relative distance, which results in an infinite Coulomb potential. The divergence in the potential energy must be compensated by a corresponding divergence in the kinetic energy. It is generally stated that an analysis of the two-particle collisions in a nonrelativistic Coulomb system leads to the Kato cusp condition, valid for fully correlated basis sets,

$$\frac{\partial \langle \psi \rangle}{\partial r_{ij}} = \mu_{ij} \, \frac{q_i \, q_j}{4\pi} \, \psi(r_{ij} = 0) \,, \qquad \mu_{ij} = \frac{m_i \, m_j}{m_i + m_j} \,. \tag{13.94}$$

Here, $\langle \psi \rangle$ is the wave function averaged over an infinitesimal sphere centered about $r_{ij} = 0$. Let us try to present a streamlined derivation of the Kato cusp condition. To this end, we write near the cusp,

$$\psi \sim a + b \, r_{ij} \,, \tag{13.95}$$

where a and b are constants with respect to r_{ij} (they can still depend on r_i and r_j, explicitly). In view of the relation

$$\vec{p}_i^{\,2} \, (r_{ij}) = -\vec{\nabla}_i^{\,2} \, (r_{ij}) = \vec{p}_j^{\,2} \, (r_{ij}) = -\vec{\nabla}_j^{\,2} \, (r_{ij}) = -\frac{2}{r_{ij}} \,. \tag{13.96}$$

The eigenvalues equation near the cusp reads as

$$H \, \psi \sim H(a + b \, r_{ij}) = E \psi = \left(-\frac{b}{m_i \, r_{ij}} - \frac{b}{m_j \, r_{ij}} + \frac{a \, q_1 \, q_2}{4\pi \, r_{ij}} + \mathcal{O}(r_{ij})^0 \right) . \tag{13.97}$$

The term proportional to $1/r_{ij}$ has to vanish if the eigenvalue condition is to hold near the cusp, and this leads to the relation

$$-\frac{b}{\mu_{ij} \, r_{ij}} + \frac{a \, q_i \, q_j}{4\pi \, r_{ij}} = 0 \,, \qquad \psi(r_{ij}) = a \,, \qquad \frac{\partial \langle \psi \rangle}{\partial r_{ij}} = b \,, \tag{13.98}$$

where we use the fact that $1/m_i + 1/m_j = 1/\mu_{ij}$. Equation (13.98) immediately leads to the Kato cusp condition (13.94). With trial wave functions that fail to obey the cusp condition, there will typically be some fluctuations around the exact eigenenergy as the size of the basis is increased. The wave functions fulfilling the cusp condition occupy a small portion of configuration space. Variational calculations of the helium spectrum will generally converge much more slowly unless one implements the Kato cusp condition; sometimes, the Kato cusp condition is implemented by hand in all trial wave functions in the basis.

A few final words are in order regarding the evaluation of matrix elements of singular operators (e.g., Dirac-δ's and relativistic corrections involving higher powers of the momentum operators), such as the ones appearing in the Breit Hamiltonian [see Eqs. (13.29), (13.30), (13.31) and (13.32)]. One of the most successful strategies consists in rewriting the singular operators in terms of other operators that probe the wave function over longer distances, a process which is sometimes called "Drachmanization" [see Ref. [421]]. Let us reproduce the derivation here. We start from a Hamiltonian

$$H = -\sum_i \frac{\vec{\nabla}_i^2}{2m_i} + V(\vec{r}_1, \dots, \vec{r}_N)\,, \qquad V(\vec{r}_1, \dots, \vec{r}_N) = \sum_i \frac{q_i\, q_Z}{4\pi r_i} + \sum_{i<j} \frac{q_i\, q_j}{4\pi r_{ij}}\,, \tag{13.99}$$

where typically $q_i\, q_Z = -4\pi Z\alpha$ and $q_i\, q_j = 4\pi\alpha$. The following relations are immediate,

$$\vec{\nabla}_i^2 \left(\frac{1}{r_{ij}}\right) = \vec{\nabla}_j^2 \left(\frac{1}{r_{ij}}\right) = -4\pi\delta^{(3)}(\vec{r}_{ij})\,. \tag{13.100}$$

Integrating, one obtains (no sum over i, j)

$$\int d^3r_1 \cdots \int d^3r_N\, \psi^2(\vec{r}_1, \dots, \vec{r}_N) \left(\frac{\vec{\nabla}_i^2}{2m_i} + \frac{\vec{\nabla}_j^2}{2m_j}\right)\left(\frac{1}{r_{ij}}\right) = -\frac{2\pi}{\mu_{ij}} \langle \delta^{(3)}(\vec{r}_{ij}) \rangle\,, \tag{13.101}$$

where $\mu_{ij} = m_i\, m_j/(m_i + m_j)$. It is very important to realize that we can actually add the "inert" kinetic energies of the remaining particle coordinates not contained in r_{ij}. After an integration by parts, this amounts to the expression

$$\int d^3r_1 \cdots \int d^3r_N \left(\frac{1}{r_{ij}}\right)\left(-\sum_{i=1}^{N} \frac{\vec{\nabla}_i^2}{2m_i}\right)\left(\psi^2(\vec{r}_1, \dots, \vec{r}_N)\right) = \frac{2\pi}{\mu_{ij}} \langle \delta^{(3)}(\vec{r}_{ij}) \rangle\,. \tag{13.102}$$

This relation involves the expression

$$-\frac{\vec{\nabla}_i^2}{2m_i}\psi^2 = \psi\left(-\frac{\vec{\nabla}_i^2}{m_i}\right)\psi - \frac{1}{m_i}\left(\vec{\nabla}_i \psi\right)^2\,, \tag{13.103}$$

where we assume the wave function to be real (this assumption can be dropped in view of the Hermiticity of the operators, the proof is left as an exercise to the reader). We are now able to rewrite the expectation value of the Dirac-δ operator in terms of a global integral ("Drachmanization", see Ref. [421]),

$$\int d^3r_1 \cdots \int d^3r_N \left(\frac{1}{r_{ij}}\right)\left[2(E - V)\,\psi^2 - \sum_i \frac{1}{m_i}\left(\vec{\nabla}_i \psi\right)^2\right] = \frac{2\pi}{\mu_{ij}} \langle \delta^{(3)}(\vec{r}_{ij}) \rangle\,, \tag{13.104}$$

where E is the energy that enters the eigenvalues condition $H\,\psi = E\,\psi$, with H being given in Eq. (13.99). Here, we suppress the arguments of the wave function $[\psi \equiv \psi(\vec{r}_1,\ldots,\vec{r}_N)]$. In order to rewrite expectation values of highly singular operators, we can also use other algebraic properties like the virial theorem. E.g.,

$$\vec{p}_1^{\,4} + \vec{p}_2^{\,4} = (\vec{p}_1^{\,2} + \vec{p}_2^{\,2})^2 - 2\,\vec{p}_1^{\,2}\,\vec{p}_2^{\,2} = 4\,T^2 - 2\,\vec{p}_1^{\,2}\,\vec{p}_2^{\,2} \tag{13.105}$$

where T is proportional to the kinetic energy of the two electrons whose value is related to the total energy, by the virial theorem. The derivation is described in Sec. 4.2.4 for the nonrelativistic hydrogen atoms; the generalization to the helium case is left as an exercise for the reader.

13.5 Further Thoughts

Here are some suggestions for further thought.

(1) **Virial Theorems.** Derive a virial theorem relating the kinetic energy of the electrons, to the potential energy and to the total energy of the bound system, for helium as opposed to hydrogen. Then, generalize the theorem to neutral lithium.

(2) **Spin-Dependent Terms.** Verify the consistency of the spin-dependent Hamiltonian terms in Eqs. (12.76) with the terms listed in Eq. (13.31). Then, calculate the anomalous-magnetic-moment correction to the spin-dependent terms based on an effective Hamiltonian.

(3) **Drachmanization.** Let us anticipate the treatment of the relativistic recoil correction, for the interaction of two particles with charges $e_1\,e_2 = -4\pi Z\alpha$. One can write Eq. (15.132a) as

$$J^{ij}\left[p^i,\left[p^j,-\frac{Z\alpha}{r}\right]\right] \sim -\frac{Z\alpha}{6\pi^2}\left[\frac{1}{r^3} + \left(1 - \gamma_E - \ln(Z\alpha\mu\eta\, r)\right)\frac{Z\alpha}{6\,\pi^2}\,\vec{\nabla}^2\left(\frac{1}{r}\right)\right]. \tag{13.106}$$

Attempt to evaluate the above matrix element for the ground state of a two-body system in terms of the "Drachmanization" discussed in Sec. 13.4.3. Is it possible that a closer inspection reveals that an integration by parts of the term $\vec{\nabla}^2\,(1/r)$ can only yield the logarithmically divergent term in η, obtained for the ground state obtained in Eq. (15.125), but misses the constant terms. What consequences result for helium?

(4) **Lithium Bethe Logarithms.** Consider Table I of Ref. [422]. It is found that Bethe logarithms in three-electron systems (Li and Be^+), as well as in two-electron systems, e.g., the lithium ion (Li^+) and in doubly ionized beryllium (Be^{++}), can be conveniently parameterized by the functional form

$$\beta = \beta^{(0)} + (\mu/M)\,\beta^{(1)} + \ln(Z^2\mu/M)\,, \tag{13.107}$$

where $\beta^{(0)}$ is almost independent of the quantum numbers of the excited states. Try to find a physical interpretation for this phenomenon.

(5) **Magnetic Moment of Triplet States.** Even though bound-state QED is in agreement with most current experiments on few-electron systems, a few discrepancies remain unexplained. One of these concerns the g factor of the $n = 2$ triplet states of helium. Despite a number of related experimental and theoretical efforts, this discrepancy [359, 423–427] remains unresolved. One might attempt to solve the discrepancy either by renewed theoretical or experimental efforts.

(6) **Higher-Order Corrections.** Consider the treatment of helium $\alpha^7 m$ corrections for triplet P states, as elucidated in a research group led by YEROKHIN[4] and PACHUCKI[5] in Refs. [417, 418]. Try to generalize the treatment to helium singlet states, and try to find the reason for the observed theoretical-experimental discrepancies for 2^3S–3^3D and 2^3P–3^3D transitions.

---

[4]Vladimir Anatolevich Yerokhin (b. 1968).
[5]Krzysztof Pachucki (b. 1963).

Chapter 14

Relativistic Many-Particle Calculations

14.1 Overview

The problem of the description of atomic many-particle systems (many-electron atoms) with a full account of the relativistic and quantum electrodynamic corrections is an active area of research. For the solution of the relativistic problem, we distinguish the configuration-interaction (CI) *ansatz* and the multiconfiguration Dirac–Hartree–Fock (MCDHF) approach. In a CI calculation, the orbitals are being held fixed, and the optimization of the solution (accomplished, e.g., by a minimization of the energy functional) takes place within a given, fixed set of functions included in the basis. In an MCDHF *ansatz*, the orbitals themselves are subject to optimization. Both CI as well as MCDHF codes have in common that the many-electron wave function is modeled as a Slater determinant of individual orbitals. In practice, one adds QED effects, after solving the relativistic (but not radiative) optimization problem, based on approximate treatments [428].

In Chap. 13, we had considered the helium system, based on fully correlated basis sets (Sec. 13.4.1). The Hylleraas *ansatz* given in Eq. (13.65) and the Korobov *ansatz* given in Eq. (13.68) are fully correlated in the sense that the wave functions depend on the relative coordinate r_{12} between the two electrons. However, once the wave functions are chosen, the only variational parameters are the coefficients a_{ijk} in Eqs. (13.65) and (13.68). In that sense, the fully correlated calculations reported in Chap. (12) rather correspond to CI than MCDHF calculations. However, because of the astute choice of the basis sets, exceptional numerical accuracy can be achieved.

Two further problems arise: The first is connected with the reduced-mass dependence of the energy levels and other atomic quantities under investigation. It is evident that a single-particle Dirac theory, and, for that matter, single-particle Dirac orbitals, cannot account for the reduced-mass dependence of the energies and other quantities of interest. This is because Dirac theory, a priori, is a single-particle theory: The Hamiltonian in Eq. (11.23) contains the electron mass, not the reduced mass. One can attempt to remedy the problem by an *ad hoc ansatz* based on the replacement of the electron mass by the reduced mass. However, it is clear that such an approach cannot capture the reduced-mass dependence of all

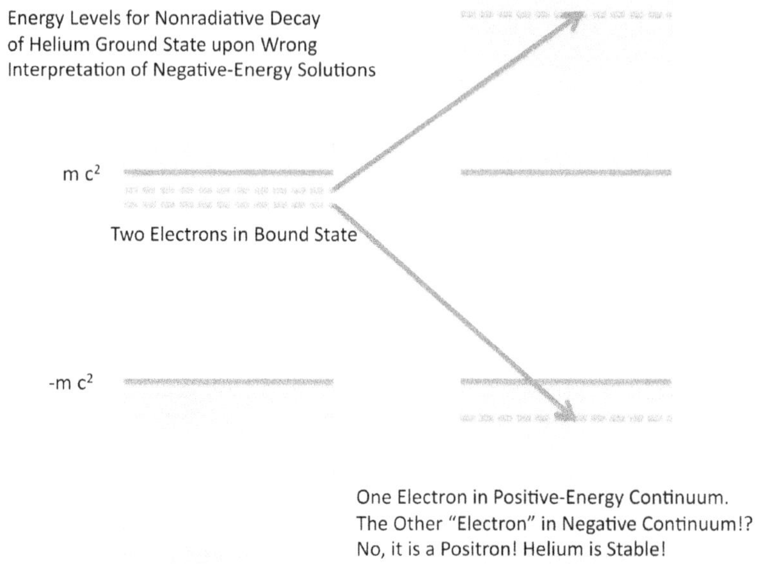

Fig. 14.1 If negative-energy states of the Dirac equation were not reinterpreted, then the ground-state of helium would become unstable against nonradiative decay into a state where one of the electron escapes into the positive-energy continuum, while the other electron escapes into the negative-energy continuum. The final state of the continuum dissolution process is energetically degenerate with respect to the initial state. This is depicted schematically in the panel.

corrections consistently. As we had seen in Sec. 11.4.2 (see the discussion following Eq. (11.137)), and in Sec. 12.5.3, in the discussion following Eq. (12.118), there are subtle reduced-mass effects which can only be captured if the problem is formulated as a many-body problem right from the start. The second problem is connected with the emergence of negative-energy solutions of the Dirac equation. The entire Chap. 11 is dedicated to a solution of related problems, in the sense that we separate the particle (positive-energy) and the antiparticle (negative-energy) Hamiltonians to be derived from the Dirac Hamiltonian. If we "personify" the computations done by a computer, we can say that numerical codes are unaware of the fact that the negative-energy solutions of the Dirac equation should be reinterpreted, physically, as antiparticle solutions. This fact has recently been stressed in other contexts (see Ref. [251]), and it was observed that the particle-antiparticle symmetry of the gravitationally coupled Dirac equation can be used in order to make predictions regarding the dynamics of antiparticles in gravitational fields [251,357]. The problem is that, numerically, the energy of an atomic system can be lowered (numerically) to arbitrarily large, negative values, if one allows for the atomic orbitals to include negative-energy solutions (see Fig. 14.1).

Still, because one would like to describe the substantial relativistic corrections experienced by electrons in the inner shells of an atom, we should use the Dirac formalism. The situation becomes even more involved because the nuclear charge is screened (by the core electrons) in the outer valence shells of an atom. With a small effective charge felt in the outer shells, the limitations of any *ansatz* based on individual atomic orbitals become ever more apparent. Based on the above considerations, it is clear that relativistic configuration-interaction (CI) multiconfiguration Dirac–Fock (MCDHF) codes will be best suited for the electrons in few-electron, highly charged ions, where the jj coupling prescription is most suitable. We remember, from Eqs. (13.7) and (13.8), that the transition between the LS and the jj coupling regimes occurs at an effective nuclear charge number of $Z \approx 26.57$.

The subject matter has been treated at length in a number of monographs [21–23, 429–431]. For the nonrelativistic Hartree–Fock problem, an authoritative reference is the book of FROESE FISCHER[1] [429] (see also Ref. [430] for a very informative introduction to the Hartree–Fock method). A very informative overview of the nonrelativistic multi-configuration calculations has recently been given in Ref. [432]. For the relativistic problem, we refer to JOHNSON[2] (Ref. [21]), LINDGREN[3] and Morrison (Refs. [23, 431]), and GRANT[4] (Ref. [22]) for authoritative references. Hence, in the current chapter, we attempt to give only a short digression on theoretical methods for many-particle systems (see Sec. 14.2). Furthermore, we illustrate in Sec. 14.3.5 that the relativistic, unretarded Breit interaction, which plays a crucial role in bound-state calculations [433, 434], gives rise to angular integrals which can be treated using the methods introduced in Chap. 6.

14.2 Many-Particle Systems and Hartree–Fock Approximation

14.2.1 *Nonrelativistic Hartree–Fock Approximation*

Within the approximation of defined, one-electron orbitals, one can devise a number of methods for the calculation of approximate energies of many-electron systems. We shall briefly dwell on the essential, basic idea of the Hartree–Fock approximation [435–437]. Let us imagine a simplified picture where two electrons are bound in a helium atom, each of which is in a defined orbital and spin state,

$$\phi_1(\vec{r}) = \psi_1(\vec{r})\,|s_1\rangle\,, \qquad \phi_2(\vec{r}) = \psi_2(\vec{r})\,|s_2\rangle\,. \tag{14.1}$$

Here, we can have $|s_1\rangle = |\uparrow\rangle$ or $|s_1\rangle = |\downarrow\rangle$, and likewise for $|s_2\rangle$. Obeying the Pauli exclusion principle, we write the appropriate *ansatz* for the two-particle wave function as follows,

$$\Psi(\vec{r}_1, \vec{r}_2) = \frac{1}{\sqrt{2}}\,[\phi_1(\vec{r}_1)\,\phi_2(\vec{r}_2) - \phi_1(\vec{r}_2)\,\phi_2(\vec{r}_1)]\,. \tag{14.2}$$

While the ground state of helium cannot be described by a Hartree–Fock *ansatz* to good accuracy, it is still instructive to study some approximations. If we set

$$\phi_1(\vec{r}) = \psi_{1S}(\vec{r})\,|\uparrow\rangle\,, \qquad \phi_2(\vec{r}) = \psi_{1S}(\vec{r})\,|\downarrow\rangle\,, \tag{14.3}$$

[1]Charlotte Froese Fischer (b. 1929).
[2]Walter Richard Johnson (b. 1929).
[3]Ingvar Lindgren (b. 1931).
[4]Ian Philip Grant (b. 1930).

then

$$\Psi(\vec{r}_1, \vec{r}_2) = \psi_{1S}(\vec{r}_1)\, \psi_{1S}(\vec{r}_2)\, \frac{1}{\sqrt{2}} \left(|\uparrow\downarrow\rangle - |\downarrow\uparrow\rangle \right) . \tag{14.4}$$

In the spin component of the wave function, we recognize the spin state $|S = 0, M = 0\rangle = \frac{1}{\sqrt{2}} \left(|\uparrow\downarrow\rangle - |\downarrow\uparrow\rangle \right)$ in the sense of Chap. 6. The total wave function (14.2) is antisymmetric under particle interchange, while for the wave function (14.4), the orbital component is symmetric, and the spin part is antisymmetric under interchange. The functional form of the state (14.4) is an acceptable approximation to the $S = 0$ (para) ground state of helium, and of heliumlike ions. If we set

$$\phi_1(\vec{r}) = \psi_{1S}(\vec{r})\, |\uparrow\rangle , \qquad \phi_2(\vec{r}) = \psi_{2S}(\vec{r})\, |\uparrow\rangle , \tag{14.5}$$

then we obtain

$$\Psi(\vec{r}_1, \vec{r}_2) = \frac{1}{\sqrt{2}} \left(\psi_{1S}(\vec{r}_1)\, \psi_{2S}(\vec{r}_2) - \psi_{1S}(\vec{r}_2)\, \psi_{2S}(\vec{r}_1) \right) |\uparrow\uparrow\rangle . \tag{14.6}$$

We recognize the spin state $|S = 1, M = 1\rangle = |\uparrow\uparrow\rangle$ in the sense of Chap. 6. The orbital wave function in Eq. (14.6) is antisymmetric under particle interchange, while the spin wave function is symmetric. The functional form of the state in Eq. (14.6) is an acceptable approximation to the ground state of ortho-helium, and of heliumlike ions.

Note that, in general, the use of an antisymmetrized combination of two (or more) single-particle orbitals works best when one has well-defined single-particle orbitals. This is the case when the spin-orbit interaction energy is larger than the electron-electron interaction energy, i.e., in cases where the orbital angular momentum ℓ is being coupled to the spin angular momentum s, to form the total angular momentum j, before the individual j are coupled together (jj coupling). For LS coupling, fully correlated basis sets (see Chap. 13) are better suited.

Let us resume the calculation. We set

$$H_0(\vec{r}_1) = -\frac{\vec{\nabla}_1^2}{2m} - \frac{Z\alpha}{r_1} , \qquad H_0(\vec{r}_2) = -\frac{\vec{\nabla}_2^2}{2m} - \frac{Z\alpha}{r_2} , \qquad U(\vec{r}_2, \vec{r}_1) = \frac{\alpha}{r_{12}} , \tag{14.7}$$

where we indicate the coordinate on which the Hamiltonian acts, as its argument. The energy expectation value is

$$\begin{aligned}
\mathcal{E} &\equiv \langle H_0 + U \rangle - \lambda_1 \left(|\phi_1|^2 - 1 \right) - \lambda_2 \left(|\phi_2|^2 - 1 \right) \\
&= \int \mathrm{d}^3 r_1\, \phi_1^+(\vec{r}_1)\, H_0(\vec{r}_1)\, \phi_1(\vec{r}_1) + \int \mathrm{d}^3 r_2\, \phi_2^+(\vec{r}_2)\, H_0(\vec{r}_2)\, \phi_2(\vec{r}_2) \\
&\quad + \frac{1}{2} \int \mathrm{d}^3 r_2 \left[\phi_1^+(\vec{r}_1)\, \phi_2^+(\vec{r}_2) - \phi_2^+(\vec{r}_1)\, \phi_1^+(\vec{r}_2) \right] U(\vec{r}_2, \vec{r}_1) \\
&\quad \times \left[\phi_1(\vec{r}_1)\, \phi_2(\vec{r}_2) - \phi_2(\vec{r}_1)\, \phi_1(\vec{r}_2) \right] \\
&\quad - \lambda_1 \left(\int \mathrm{d}^3 r_1\, |\phi_1(\vec{r}_1)|^2 - 1 \right) - \lambda_2 \left(\int \mathrm{d}^3 r_2\, |\phi_2(\vec{r}_2)|^2 - 1 \right) .
\end{aligned} \tag{14.8}$$

The last two terms describe the Lagrange multipliers for the normalization conditions. We evaluate the variation of the single-particle Hamiltonian term. For a general orbital+spin wave function $\phi = \phi(\vec{r})$, one has

$$\frac{\delta}{\delta\phi^+(\vec{r}\,')} \int \mathrm{d}^3r \, \phi^+(\vec{r}) \, H(\vec{r}) \, \phi(\vec{r}) = \int \mathrm{d}^3r \, \delta^{(3)}(\vec{r} - \vec{r}\,') \, H(\vec{r}) \, \phi(\vec{r}) = H(\vec{r}\,') \, \phi(\vec{r}\,') \,. \tag{14.9}$$

The variation of the interaction potential reads as follows,

$$\begin{aligned} &\frac{\delta}{\delta\phi_1^+(\vec{r}_1^{\,\prime})} \left[\frac{1}{2} \int \mathrm{d}^3r_1 \int \mathrm{d}^3r_2 \, (\phi_1^+(\vec{r}_1)\phi_2^+(\vec{r}_2) - \phi_2^+(\vec{r}_1)\phi_1^+(\vec{r}_2)) \, U(\vec{r}_2, \vec{r}_1) \right. \\ &\qquad \left. \times (\phi_1(\vec{r}_1)\phi_2(\vec{r}_2) - \phi_2(\vec{r}_1)\phi_1(\vec{r}_2)) \right] \\ &\qquad = \frac{1}{2} \int \mathrm{d}^3r_2 \, \phi_2^+(\vec{r}_2) \, U(\vec{r}_2, \vec{r}_1^{\,\prime}) \, (\phi_1(\vec{r}_1^{\,\prime}) \, \phi_2(\vec{r}_2) - \phi_2(\vec{r}_1^{\,\prime}) \, \phi_1(\vec{r}_2) \\ &\qquad + \phi_1(\vec{r}_1^{\,\prime}) \, \phi_2(\vec{r}_2) - \phi_2(\vec{r}_1^{\,\prime}) \, \phi_1(\vec{r}_2)) = \int \mathrm{d}^3r_2 \, \underbrace{\phi_2^+(\vec{r}_2) \, U(\vec{r}_2, \vec{r}_1^{\,\prime}) \, \phi_2(\vec{r}_2)}_{\text{direct term}} \, \phi_1(\vec{r}_1^{\,\prime}) \\ &\qquad - \int \mathrm{d}^3r_2 \, \underbrace{\phi_2^+(\vec{r}_2) \, U(\vec{r}_2, \vec{r}_1^{\,\prime}) \, \phi_1(\vec{r}_2)}_{\text{exchange term}} \, \phi_2(\vec{r}_1^{\,\prime}) \,. \end{aligned} \tag{14.10}$$

The variational equation for the term in the Lagrange multiplier is given as follows,

$$\frac{\delta}{\delta\phi_1^+(\vec{r}_1^{\,\prime})} \int \mathrm{d}^3r_1 \, |\phi_1(\vec{r}_1)|^2 = \phi_1(\vec{r}_1^{\,\prime}) \,. \tag{14.11}$$

Hence, setting the variational derivative $\delta\mathcal{E}/\delta\phi_1(\vec{r})$ equal to zero, one obtains

$$\begin{aligned} &H_0(\vec{r}_1) \, \phi_1(\vec{r}_1) + \int \mathrm{d}^3r_2 \, \underbrace{\phi_2^+(\vec{r}_2) \, U(\vec{r}_2, \vec{r}_1) \, \phi_2(\vec{r}_2)}_{\text{direct term}} \, \phi_1(\vec{r}_1) \\ &\quad - \int \mathrm{d}^3r_2 \, \underbrace{\phi_2^+(\vec{r}_2) \, U(\vec{r}_2, \vec{r}_1) \, \phi_1(\vec{r}_2)}_{\text{exchange term}} \, \phi_2(\vec{r}_1) = \lambda_1 \, \phi_1(\vec{r}_1) = \mathcal{E}_1 \, \phi_1(\vec{r}_1) \,. \end{aligned} \tag{14.12}$$

Interpreting $\lambda_1 = \mathcal{E}_1$ as the Hartree–Fock energy, we obtain the Hartree–Fock equations for a two-electron system.

The direct term describes the interaction of electron 1 with the electrostatic charge density of electron 2. For the exchange term to be nonvanishing, electrons 1 and 2 must have the same spin orientation. Using the Slater determinant *ansatz* in Eq. (14.2), one can achieve the antisymmetrization either by the spin part (singlet state) or by the orbital part of the wave function. If the spin part is antisymmetrized, then the orbital part must be symmetric, and the only way to achieve this using the *ansatz* (14.2) is by assuming the two electrons to be in the same state. So, excited singlet S state configurations of the form $1s\,ns\,{}^1S_0$ cannot be described by the *ansatz* (14.2). However, it is easy to enhance the Hartree–Fock formalism to accommodate for more complicated trial wave functions.

The generalization to N electrons is obvious. The Slater determinant for the n-electron system is

$$\Psi(\vec{r}_1,\ldots,\vec{r}_n) = \frac{1}{\sqrt{n!}}\,\det[\phi_{ij}(\vec{r}_1,\ldots,\vec{r}_n)]\,, \qquad \phi_{ij}(\vec{r}_1,\ldots,\vec{r}_n) = [\phi_i(\vec{r}_j)]_{i,j=1,\ldots,n}\,. \tag{14.13}$$

The wave function for the two-particle system can be found in Eq. (14.2). For three particles, the fully antisymmetrized, fermionic, orbital+spin wave function is

$$\begin{aligned}\Psi(\vec{r}_1,\vec{r}_2,\vec{r}_3) = \frac{1}{\sqrt{3!}}\,[&\phi_1(\vec{r}_1)\,\phi_2(\vec{r}_2)\,\phi_3(\vec{r}_3) + \phi_1(\vec{r}_2)\,\phi_2(\vec{r}_3)\,\phi_3(\vec{r}_1)\\ &+ \phi_1(\vec{r}_3)\,\phi_2(\vec{r}_1)\,\phi_3(\vec{r}_2) - \phi_1(\vec{r}_1)\,\phi_2(\vec{r}_3)\,\phi_3(\vec{r}_2)\\ &- \phi_1(\vec{r}_2)\,\phi_2(\vec{r}_1)\,\phi_3(\vec{r}_3) - \phi_1(\vec{r}_3)\,\phi_2(\vec{r}_2)\,\phi_3(\vec{r}_1)]\,.\end{aligned} \tag{14.14}$$

We label the reference orbital by the index a, and label the summation index by $n = 1,\ldots,N$, and write for the ath orbital,

$$\begin{aligned}&H_0(\vec{r}_1)\,\phi_a(\vec{r}_1) + \sum_{n\neq a}\int \mathrm{d}^3r_2\,\phi_n^+(\vec{r}_2)\,U(\vec{r}_2,\vec{r}_1)\,\phi_n(\vec{r}_2)\,\phi_a(\vec{r}_1)\\ &- \sum_{n\neq a}\int \mathrm{d}^3r_2\,\phi_n^+(\vec{r}_2)\,U(\vec{r}_2,\vec{r}_1)\,\phi_a(\vec{r}_2)\,\phi_n(\vec{r}_1) = \mathcal{E}_a\,\phi_a(\vec{r}_1)\,.\end{aligned} \tag{14.15}$$

We recall the definitions of H_0 and U from Eq. (14.7). Relativistic generalizations of the Hartree–Fock approximation will be discussed in the following.

14.2.2 *Implementation and Practical Aspects*

A practical implementation of the Hartree–Fock method is not as trivial as one might otherwise assume, and many variations exist. The traditional approach is to solve the problem iteratively. One starts with an approximate wavefunction for the state of interest. A Slater determinant of hydrogenic wave functions is good enough for this step in many cases, and the electron-electron interaction can be ignored. One then determines a mean-Coulomb field acting on the ith electron based on the wave functions of the other electron. One proceeds to solve the one-electron Schrödinger equation for each electron in its mean-Coulomb field to generate a new set of wave functions. These steps are repeated until convergence is reached, and the method is therefore commonly referred to as the self-consistent field (SCF) approach.

One important approach, somewhat beyond the original Hartree–Fock method, consists in the multi-configurational *ansatz*. In a first step, one constructs the Slater determinant from the wave functions (the spin-angular orbitals), ignoring the electron-electron interaction. The wave functions are eigenstates of the Hamiltonian without electron-electron interaction. The wave function resulting from this step is antisymmetric, but does not have defined values for the total combined angular momentum L and the total combined spin S of the atomic electrons. (We assume to be in a regime of LS coupling.)

In a second step, one combines the orbitals to defined L and S values, using the formalism outlined in Chap. 6. The orbital angular momenta of the electrons are subsequently coupled, adding the angular momenta of, say, electrons 1 and 2, then adding the angular momentum of electron 3, and so on [430]. The same procedure is used for the spin angular momenta. The resulting wave functions are the configuration state functions (CSFs). They are still eigenfunctions of the Hamiltonian without electron-electron interaction.

In a third step, it is recognized that different CSFs having the same L and S values need to be mixed in order to obtain a better approximation to the atomic wave function. The multi-configuration calculation with appropriate mixing coefficients based on the self-consistent field approximation yields the eigenstates of the Hartree–Fock equation. One determines the mixing coefficients and the eigenstate wave functions.

This method receives a few important simplifications in important cases. For example, when subshells are filled and an average is taken over all magnetic projections, one can simplify the exchange term in the Hartree–Fock equations. Let us approximate the wave functions nonrelativistically,

$$\phi_n(\vec{r}) = R_{n\ell_n}(r)\,\chi_{\kappa_n\,\mu_n}(\hat{r})\,, \tag{14.16}$$

where $\chi_{\kappa\mu}(\hat{r})$ is the spin-angular function discussed in Chap. 6, and $\ell_n = |\kappa_n + 1/2| - 1/2$ is the orbital angular momentum. Under the assumption of filled subshells, we can replace the exchange term, namely, the third term on the left-hand side of Eq. (14.15), by the expression [438, 439]

$$\begin{aligned}&-\sum_{n\neq a}\int \mathrm{d}^3 r_2\,\phi_n^+(\vec{r}_2)\,U(\vec{r}_2,\vec{r}_1)\,\phi_a(\vec{r}_2)\,\phi_n(\vec{r}_1)\\ &\to -\frac{1}{2}\sum_{n\neq a}\sum_{\ell}\begin{pmatrix}\ell_n & \ell & \ell_a\\ 0 & 0 & 0\end{pmatrix}^2 R^*_{n\ell_a}(r_1)\,\chi_{\kappa_a\ell_a}(\hat{r}_1)\int \mathrm{d}r_2 R^*_{n\ell_n}(r_2)\,\frac{r_<^\ell}{r_>^{\ell+1}}\,R^*_{a\ell_a}(r_2)\,.\end{aligned} \tag{14.17}$$

The occurrence of the expression $R^*_{n\ell_a}(r_1)\,\chi_{\kappa_a\ell_a}(\hat{r})$ in the above equation is not a typographical error. The spin-angular component with the same symmetry as the reference state $\phi_a(\vec{r})$ needs to be selected.

Many variations of the Hartree–Fock method exist. For example, ROOTHAAN[5] proposed to project the Hartree–Fock equations onto a basis [440], effectively transforming them into matrix equations. The same applies when one uses a radial lattice in order to project the differential equations [439]. Upon discretization, one obtains a difference operator and the eigenfunctions of the Hartree–Fock problems are obtained on the lattice.

Furthermore, one distinguishes the so-called Unrestricted Hartree–Fock method, where the eigenfunctions are allowed to depend on the magnetic quantum numbers, from the restricted Hartree–Fock method, where one averages over the magnetic

[5]Clemens Carel Johannes Roothaan (1918–2019).

projections. The use of the unrestricted method is imperative if one would like to describe hyperfine effects [441].

Many approaches exist to overcome the limitations set forth by the cursory model of a single Slater determinant for the atomic wave function. One possibility is to include virtual excitations of the atom on the basis of MØLLER[6]–PLESSET[7] perturbation theory. In many calculations, the Hartree–Fock problem is solved non-perturbatively, with the Coulomb interaction taken into account, but higher-order effects like the Breit interaction (Chap. 12) are taken into account only perturbatively. Effects which are modeled on the basis of given orbitals are known as configuration interactions (CI). Often, the higher-order effects are taken into account on the CI level, after the solution of the initial HF problem [430].

In problems with few active electrons, it can be very beneficial to model the system based on the perturbative treatment of single and double core excitations [442], which allows for a systematic treatment of the higher-order effects based on many-body perturbation theory (MBPT). From *ansatz* based on the frozen-core approximation, some very accurate theoretical calculations have been reported [443, 444].

14.3 Relativistic Hartree–Fock Methods

14.3.1 *Brown–Ravenhall Disease*

As has been shown in Chap. 11, the Dirac equation describes particles and antiparticles simultaneously. Historically, one may point out that, based on the Dirac equation, the existence of the positron was predicted, followed by its experimental detection in 1933, by ANDERSON[8] [445]. Anecdotally, one may point out that the helium atom would be unstable against decay into a state where the two electrons perform quantum jumps into continuum states (see Fig. 14.1 and Ref. [446]).

The Dirac equation has a negative-energy spectrum which belongs to the positron states. If one does naive variational perturbation theory, minimizing the energy, then the positron states are not automatically excluded. The energy, in other words, is not bounded from below. This leaves the energy to be unbounded from below, and calculations do not automatically converge to a physically sensible value. The "BROWN[9]–RAVENHALL[10] disease" leads to practical, numerical problems in MCDHF codes, where considerable numerical and analytic effort has been invested into a resolution of said problems, with the help of projection operators [22, 447–449].

[6]Christian Møller (1904–1980).
[7]Milton Spinoza Plesset (1908–1991).
[8]Carl David Anderson (1905–1991).
[9]Gerald Edward Brown (1926–2013).
[10]David Geoffrey Ravenhall (b. 1927).

One defines the projection operator

$$\Lambda_\pm(\vec{r}) = \tfrac{1}{2}\left(1 \pm H_{\rm FD}(\vec{r})/E_{\rm FD}\right), \tag{14.18}$$

$$H_{\rm FD}(\vec{r}) = \vec{\alpha}\cdot\vec{p} + \beta m = -\mathrm{i}\vec{\alpha}\cdot\vec{\nabla}_{\vec{r}} + \beta m, \tag{14.19}$$

$$E_{\rm FD}(\vec{r}) = \sqrt{\vec{p}^{\,2} + m^2} = \sqrt{-\vec{\nabla}^2 + m^2}. \tag{14.20}$$

Here, both $H_{\rm FD}$ as well as $E_{\rm FD}$ have to be understood as operators [see also Eq. (7.18)]. Because we are operating in a many-particle setting, it is useful to indicate the dependence on the individual electron coordinate $\vec{r}$. The positive square root is to be taken for $E_{\rm FD}$. The projectors $\Lambda_\pm$ have the following properties,

$$\Lambda_+\Lambda_- = \Lambda_-\Lambda_+ = 0, \qquad (\Lambda_\pm)^2 = \Lambda_\pm, \qquad \Lambda_+ + \Lambda_- = 1. \tag{14.21}$$

They project onto the positive-energy, and negative-energy, subspaces of the free Dirac Hamiltonian $H_{\rm FD}$. However, one can argue that, unless the nuclear charge number is excessively large [450], the positive- and negative-energy spectra do not mix when the electron-nucleus and the electron-electron interactions are added.

If one did not invoke the positive-energy projection operators in MCDHF codes, then, for a system as simple as helium, nonsensical results would be obtained. Namely, one of the electrons could undergo a quantum jump into the positive-energy continuum, the other, into the negative-energy continuum, with the sum of the energies of the two continuum states (final state of the two-electron non-radiative transition) being equal to the sum of the two bound-state energies of the helium atom from which the transition started [446, 451, 452]. The "Brown–Ravenhall disease" is of course addressed, on the theoretical level, by invoking the Dirac sea, or, alternatively, by invoking the reinterpretation principle and quantum field theory. Let us recall that the development of quantum field theory made it possible to reformulate the Dirac equation in a way that treats the antiparticles as "sea" particles rather than the absence of a particle from the "Dirac sea of negative-energy particles". Still, it is sometimes overlooked that the Dirac equation describes particles and antiparticles *simultaneously*, a numerical manifestation being "Brown–Ravenhall disease" (see Ref. [446]).

14.3.2 *Relativistic Frequency-Dependent Breit Interaction*

An important basis for all relativistic calculations of atomic structure is the relativistic frequency-dependent Breit interaction. Without taking into account this interaction, one could not possibly hope to accurately describe the inner electrons in systems with a high nuclear charge number Z. The derivation proceeds as a relativistic generalization of the considerations reported in Sec. 12.3.2. We have to consider the transition currents

$$j^\mu_a(x) = \psi^+_a(x)\,\alpha^\mu\,\psi_a(x), \qquad j^\mu_b(x) = \psi^+_b(x)\,\alpha^\mu\,\psi_b(x), \tag{14.22}$$

where the $\alpha^\mu = \gamma^0\gamma^\mu$ are Dirac matrices, and the wave functions $\psi_a(x)$ and $\psi_b(x)$ are four-component bispinors. We understand the four-vector as $x^\mu = (t, \vec{r})$. Then,

$$j^\mu_a(x) = \mathrm{e}^{-\mathrm{i}(E_{i_a}-E_{f_a})t}\,\psi^+_a(\vec{r})\,\alpha^\mu_a\,\psi_a(\vec{r}), \qquad j^\mu_b(x) = \mathrm{e}^{-\mathrm{i}(E_{i_b}-E_{f_b})t}\,\psi^+_b(\vec{r})\,\alpha^\mu_b\,\psi_b(\vec{r}). \tag{14.23}$$

The interaction Hamiltonian involves the currents of both interacting particles,

$$\mathcal{H} = e\, j_a^\mu(x) A_\mu(x) + e\, j_b^\mu(x) A_\mu(x)\,. \tag{14.24}$$

We start from the relativistic generalization of Eq. (12.37),

$$\begin{aligned} S_{fi} &= 2 \times \left\{ -\frac{e^2}{2!} \int \mathrm{d}^4 x_a \int \mathrm{d}^4 x_b\, \mathrm{T}\,[j_a^\mu(x_a)\, j_b^\nu(x_b)]\, \langle 0|\mathrm{T}[A_\mu(x_a)\, A_\nu(x_b)]|0\rangle \right\} \\ &= -e^2 \int \mathrm{d}^4 x_a \int \mathrm{d}^4 x_b \int \frac{\mathrm{d}^4 k}{(2\pi)^4}\, \mathrm{T}\,[j_a^\mu(x_a)\, j_b^\nu(x_b)]\;\, \mathrm{i} D_{\mu\nu}(k)\, \mathrm{e}^{-\mathrm{i}k(x_a - x_b)} \\ &= -\mathrm{i}e^2 \int \frac{\mathrm{d}^4 k}{(2\pi)^4}\, D_{\mu\nu}(k)\, Q^{\mu\nu}(k)\,, \end{aligned} \tag{14.25}$$

where we have used Eq. (9.120) in the form

$$\langle 0|\mathrm{T}\,[A_\mu(x_a)\, A_\nu(x_b)]|0\rangle = \mathrm{i} \int \frac{\mathrm{d}^4 k}{(2\pi)^4}\, D_{\mu\nu}(k)\, \mathrm{e}^{-\mathrm{i}k\cdot(x_a - x_b)}\,, \tag{14.26}$$

and we define the relativistic generalization of Eq. (12.39),

$$Q^{\mu\nu}(k) = \int \mathrm{d}^4 x_a \int \mathrm{d}^4 x_b\, \mathrm{e}^{-\mathrm{i}k(x_a - x_b)}\, \mathrm{T}\,[j_a^\mu(x_a)\, j_b^\nu(x_b)]\,. \tag{14.27}$$

We write $x_1 = (t_a, \vec{r}_a)$ and $x_2 = (t_b, \vec{r}_b)$. One can show that

$$\begin{aligned} Q^{\mu\nu} &= T \int \mathrm{d}^3 r_a \int \mathrm{d}^3 r_b\, \frac{\psi_{f_a}^+(\vec{r}_a)\, \alpha_a^\mu\, \psi_{i_a}(\vec{r}_a)\, \psi_{f_b}^+(\vec{r}_b)\, \alpha_b^\nu\, \psi_{i_b}(\vec{r}_b)}{\mathrm{i}(E_{i_a} - E_{f_a} + \omega - \mathrm{i}\epsilon)}\, \mathrm{e}^{\mathrm{i}\vec{k}\cdot\vec{r}_{ab}} \\ &+ T \int \mathrm{d}^3 r_a \int \mathrm{d}^3 r_b\, \frac{\psi_{f_b}^+(\vec{r}_b)\, \alpha_b^\nu\, \psi_{i_b}(\vec{r}_b)\, \psi_{f_a}^+(\vec{r}_a)\, \alpha_a^\mu\, \psi_{i_a}(\vec{r}_a)}{-\mathrm{i}(E_{i_a} - E_{f_a} + \omega + \mathrm{i}\epsilon)}\, \mathrm{e}^{\mathrm{i}\vec{k}\cdot\vec{r}_{ab}}\,. \end{aligned} \tag{14.28}$$

Here, $\omega = k^0$ is the photon energy. Just like in the nonrelativistic case (see Sec. 12.3.2), we use the fact that, in order to generate the factor $T = \int \mathrm{d}\tau$, which we need for the matching with the effective Hamiltonian, we need the energy conservation condition $E_{f_a} + E_{f_b} = E_{i_a} + E_{i_b}$. The Jacobian of the variable changes $(t_a, t_b) \to (\tau, t_a - t_b)$, and $(t_a, t_b) \to (\tau, t_b - t_a)$, is unity in both cases.

We recall Eq. (14.25) and the matching with the effective Hamiltonian H_{ab}, $S_{fi} = -\mathrm{i}T\,\langle \psi_f|H_{ab}|\psi_i\rangle$, which results in the relation

$$\begin{aligned} \langle \psi_f|H_{ab}|\psi_i\rangle &= \frac{\mathrm{i}}{T} \left(-\mathrm{i}e^2 \int \frac{\mathrm{d}^4 k}{(2\pi)^4}\, D_{\mu\nu}(k)\, Q^{\mu\nu}(k) \right) = -\mathrm{i}e^2 \int \mathrm{d}^3 r_a \int \mathrm{d}^3 r_b \\ &\times \int \frac{\mathrm{d}^4 k}{(2\pi)^4}\, D_{\mu\nu}(k) \mathrm{e}^{\mathrm{i}\vec{k}\cdot\vec{r}_{ab}} \left[\frac{\psi_{f_a}^+(\vec{r}_a)\, \alpha_a^\mu\, \psi_{i_a}(\vec{r}_a)\, \psi_{f_b}^+(\vec{r}_b)\, \alpha_b^\nu\, \psi_{i_b}(\vec{r}_b)}{\omega - (E_{f_a} - E_{i_a}) - \mathrm{i}\epsilon} \right. \\ &\left. - \frac{\psi_{f_b}^+(\vec{r}_b)\, \alpha_b^\nu\, \psi_{i_b}(\vec{r}_b)\, \psi_{f_a}^+(\vec{r}_a)\, \alpha_a^\mu\, \psi_{i_a}(\vec{r}_a)}{\omega - (E_{f_a} - E_{i_a}) + \mathrm{i}\epsilon} \right]. \end{aligned}$$

We use the photon propagator in Coulomb gauge, and recall that according to Eq. (9.73), $D_{00}(k) = 1/\vec{k}^2$ and $D_{ij}(k) = (\omega^2 - \vec{k}^2 + \mathrm{i}\,\epsilon)^{-1}\,(\delta_{ij} - k_i\, k_j/\vec{k}^2)$ for $\mu = i = 1, 2, 3$ and $\nu = j = 1, 2, 3$. Let us first consider the case $\mu = \nu = 0$. Then, there is only a pole from the fermion propagator, not from the photon propagator. One observes that the ω integral converges, provided both terms are added under the integral sign, and closes the ω integration in the upper half plane, with the following result for the matched effective Hamiltonian $H^{(00)}_{ab}$,

$$\begin{aligned}\langle\psi_f|H^{(00)}_{ab}|\psi_i\rangle &= e^2\int \mathrm{d}^3r_a\int \mathrm{d}^3r_b \int \frac{\mathrm{d}^3k}{(2\pi)^3}\,\frac{1}{\vec{k}^2}\,\mathrm{e}^{\mathrm{i}\vec{k}\cdot\vec{r}_{ab}}\,\psi^+_{f_a}(\vec{r}_a)\,\psi_{i_a}(\vec{r}_a)\,\psi^+_{f_b}(\vec{r}_b)\,\psi_{i_b}(\vec{r}_b)\\ &= \int \mathrm{d}^3r_a \int \mathrm{d}^3r_b\,\psi^+_{f_a}(\vec{r}_a)\,\psi_{i_a}(\vec{r}_a)\,\frac{e^2}{4\pi|\vec{r}_a-\vec{r}_b|}\,\psi^+_{f_b}(\vec{r}_b)\,\psi_{i_b}(\vec{r}_b)\,.\end{aligned} \tag{14.29}$$

This is just the Coulomb repulsion of the two electrons. The spatial components of the photon propagator lead to the Breit Hamiltonian,

$$\begin{aligned}\langle\psi_f|H^{(ij)}_{ab}|\psi_i\rangle &= -\mathrm{i}e^2\int \mathrm{d}^3r_a\int \mathrm{d}^3r_b\int\frac{\mathrm{d}^4k}{(2\pi)^4}\,D_{ij}(k)\,\mathrm{e}^{\mathrm{i}\vec{k}\cdot\vec{r}_{ab}}\\ &\times\left[\frac{\psi^+_{f_a}(\vec{r}_a)\,\alpha^i_a\,\psi_{i_a}(\vec{r}_a)\,\psi^+_{f_b}(\vec{r}_b)\,\alpha^j_b\,\psi_{i_b}(\vec{r}_b)}{\omega-(E_{f_a}-E_{i_a})-\mathrm{i}\epsilon}-\frac{\psi^+_{f_b}(\vec{r}_b)\,\alpha^\nu_b\,\psi_{i_b}(\vec{r}_b)\,\psi^+_{f_a}(\vec{r}_a)\,\alpha^\mu_a\,\psi_{i_a}(\vec{r}_a)}{\omega-(E_{f_a}-E_{i_a})+\mathrm{i}\epsilon}\right].\end{aligned} \tag{14.30}$$

Let us consider the first term in the spatial components of the photon propagator,

$$D_{ij}(k) = \frac{1}{\omega^2-\vec{k}^2+\mathrm{i}\,\epsilon}\left(\delta_{ij}-\frac{k_i\,k_j}{\vec{k}^2}\right)\to\frac{\delta_{ij}}{\omega^2-\vec{k}^2+\mathrm{i}\,\epsilon}\equiv D^{(1)}_{ij}(k)\,. \tag{14.31}$$

One obtains the first term,

$$\begin{aligned}\langle\psi_f|H^{(ij),1}_{ab}|\psi_i\rangle &= -\mathrm{i}e^2\int \mathrm{d}^3r_a\int \mathrm{d}^3r_b\int\frac{\mathrm{d}^4k}{(2\pi)^4}\,\frac{\mathrm{e}^{\mathrm{i}\vec{k}\cdot\vec{r}_{ab}}}{\omega^2-\vec{k}^2+\mathrm{i}\,\epsilon}\\ &\times\left[\frac{\psi^+_{f_a}(\vec{r}_a)\alpha^i_a\psi_{i_a}(\vec{r}_a)\,\psi^+_{f_b}(\vec{r}_b)\alpha^j_b\psi_{i_b}(\vec{r}_b)}{\omega-(E_{f_a}-E_{i_a})-\mathrm{i}\epsilon}-\frac{\psi^+_{f_b}(\vec{r}_b)\alpha^j_b\psi_{i_b}(\vec{r}_b)\,\psi^+_{f_a}(\vec{r}_a)\alpha^i_a\psi_{i_a}(\vec{r}_a)}{\omega-(E_{f_a}-E_{i_a})+\mathrm{i}\epsilon}\right].\end{aligned} \tag{14.32}$$

The photon propagator pole at $\omega = |\vec{k}|$ leads to a self-energy type effect, which is otherwise included in the relativistic recoil correction. The Breit Hamiltonian is generated by the pole of the fermion propagator at $\omega = (E_{f_a} - E_{i_a}) \pm \mathrm{i}\epsilon$. We close, just as for the case with $\mu = \nu = 0$, the contour in the upper half plane and write

$$\begin{aligned}\langle\psi_f|H^{(ij),1}_{ab}|\psi_i\rangle &= \int \mathrm{d}^3r_a\int \mathrm{d}^3r_b\,\psi^+_{f_a}(\vec{r}_a)\,\psi^+_{f_b}(\vec{r}_b)\\ &\times\left\{-\frac{e^2\cos[(E_{f_a}-E_{i_a})|\vec{r}_{ab}|]}{4\pi|\vec{r}_{ab}|}\,\vec{\alpha}_a\cdot\vec{\alpha}_b\right\}\psi_{i_a}(\vec{r}_a)\,\psi_{i_b}(\vec{r}_b)\,.\end{aligned} \tag{14.33}$$

We have used the result

$$\mathrm{P.V.}\int\frac{\mathrm{d}^3k}{(2\pi)^3}\,\frac{\mathrm{e}^{\mathrm{i}\vec{k}\cdot\vec{r}_{ab}}}{(E_{f_a}-E_{i_a})^2-\vec{k}^2+\mathrm{i}\,\epsilon} = -\frac{\cos[(E_{f_a}-E_{i_a})|\vec{r}_{ab}|]}{4\pi|\vec{r}_{ab}|}\,. \tag{14.34}$$

Let us consider now the second term in the spatial components of the photon propagator,

$$D_{ij}(k) = \frac{1}{\omega^2 - \vec{k}^2 + \mathrm{i}\epsilon}\left(\delta_{ij} - \frac{k_i\,k_j}{\vec{k}^2}\right) \to -\frac{k_i\,k_j}{\vec{k}^2(\omega^2 - \vec{k}^2 + \mathrm{i}\epsilon)} \equiv D^{(2)}_{ij}(k)\,. \tag{14.35}$$

One obtains the second term,

$$\langle\psi_f|H^{(ij),2}_{ab}|\psi_i\rangle = -\mathrm{i}e^2\int \mathrm{d}^3r_a\int \mathrm{d}^3r_b\int\frac{\mathrm{d}^4k}{(2\pi)^4}\left(\frac{\mathrm{i}k_i\,\mathrm{i}k_j\,\mathrm{e}^{\mathrm{i}\vec{k}\cdot\vec{r}_{ab}}}{\vec{k}^2(\omega^2 - \vec{k}^2 + \mathrm{i}\epsilon)}\right)$$
$$\times\left[\frac{\psi^+_{f_a}(\vec{r}_a)\,\alpha^i_a\,\psi_{i_a}(\vec{r}_a)\,\psi^+_{f_b}(\vec{r}_b)\,\alpha^j_b\,\psi_{i_b}(\vec{r}_b)}{\omega - (E_{f_a} - E_{i_a}) - \mathrm{i}\epsilon} - \frac{\psi^+_{f_b}(\vec{r}_b)\,\alpha^j_b\,\psi_{i_b}(\vec{r}_b)\,\psi^+_{f_a}(\vec{r}_a)\,\alpha^i_a\,\psi_{i_a}(\vec{r}_a)}{\omega - (E_{f_a} - E_{i_a}) + \mathrm{i}\epsilon}\right]. \tag{14.36}$$

As before, the contribution to the Breit Hamiltonian is generated by the pole of the fermion propagator at $\omega = (E_{f_a} - E_{i_a}) \pm \mathrm{i}\epsilon$. We close, just as for the case with $\mu = \nu = 0$, the contour in the upper half plane and obtain

$$\langle\psi_f|H^{(ij),2}_{ab}|\psi_i\rangle = \int \mathrm{d}^3r_a\int \mathrm{d}^3r_b\,\psi^+_{f_a}(\vec{r}_a)\,\psi^+_{f_b}(\vec{r}_b)$$
$$\times\left\{\vec{\alpha}_a\cdot\vec{\nabla}_a\,\vec{\alpha}_b\cdot\vec{\nabla}_b\frac{e^2\,\cos[(E_{f_a} - E_{i_a})|\vec{r}_{ab}|] - 1}{4\pi(E_{f_a} - E_{i_a})^2|\vec{r}_{ab}|}\right\}\psi_{i_a}(\vec{r}_a)\,\psi_{i_b}(\vec{r}_b)\,. \tag{14.37}$$

In order to ensure a certain symmetry of notation in the final result, we have replaced $\vec{\nabla}_a \to -\vec{\nabla}_b$. One uses the fact that

$$\mathrm{P.V.}\int\frac{\mathrm{d}^3k}{(2\pi)^3}\frac{1}{\vec{k}^2}\frac{\mathrm{e}^{\mathrm{i}\vec{k}\cdot\vec{r}_{ab}}}{(E_{f_a} - E_{i_a})^2 - \vec{k}^2 + \mathrm{i}\epsilon} = -\frac{\cos[(E_{f_a} - E_{i_a})|\vec{r}_{ab}|] - 1}{4\pi(E_{f_a} - E_{i_a})^2\,|\vec{r}_{ab}|} \to \frac{1}{8\pi}|\vec{r}_{ab}|\,. \tag{14.38}$$

Finally, one summarizes the contributions $H^{(00)}_{ab}$, $H^{(ij),1}_{ab}$, and $H^{(ij),2}_{ab}$,

$$H = H^{(00)}_{ab} + H^{(ij),1}_{ab} + H^{(ij),2}_{ab} = \frac{e^2}{4\pi|\vec{r}_a - \vec{r}_b|} \tag{14.39}$$
$$-\frac{e^2}{4\pi}\left\{\frac{\cos[(E_{f_a} - E_{i_a})|\vec{r}_{ab}|]}{|\vec{r}_{ab}|}\vec{\alpha}_a\cdot\vec{\alpha}_b + \vec{\alpha}_a\cdot\vec{\nabla}_a\vec{\alpha}_b\cdot\vec{\nabla}_b\frac{\cos[(E_{f_a} - E_{i_a})|\vec{r}_{ab}|] - 1}{(E_{f_a} - E_{i_a})^2|\vec{r}_{ab}|}\right\}.$$

This is the frequency-dependent Breit interaction.

14.3.3 *Relativistic Hartree–Fock Approximation*

The generalization of the approach outlined in Sec. 14.2.1 to the relativistic setting is accomplished as follows. The wave function, again, is a Slater determinant,

$$\psi(\vec{r}_1,\dots,\vec{r}_n) = \frac{1}{\sqrt{n!}}\det[\phi_{ij}(\vec{r}_1,\dots,\vec{r}_n)]\,,\qquad \phi_{ij}(\vec{r}_1,\dots,\vec{r}_n) = \phi_i(\vec{r}_j)\,, \tag{14.40}$$

with $\{i,j\} = 1,\dots,n$. The determinant is taken over the indices i and j, and the entry ij of the matrix whose determinant is being calculated, is just equal to the ith wave function ϕ_i being evaluated at the jth coordinate $\vec{r}_j$. In contrast to the nonrelativistic case, however, the $\phi_i(\vec{r}_j)$ are now relativistic wave functions with a defined spin-angular symmetry (see Sec. 8.2.2). One uses the single-particle no-pair Hamiltonian

$$H_0(\vec{r}) = \Lambda_+(\vec{r}) \left(\vec{\alpha} \cdot \vec{p} + \beta m - \frac{Z\alpha}{r} \right) \Lambda_+(\vec{r}) \,, \tag{14.41}$$

where r is the electron-nucleus distance and the projection operators are defined in Eq. (14.18). The electron-electron interaction Hamiltonian is written in the form

$$U(\vec{r}_1, \vec{r}_2) = U_C(\vec{r}_1, \vec{r}_2) + U_B(\vec{r}_1, \vec{r}_2) \,, \tag{14.42}$$

where $U_C(\vec{r}_1, \vec{r}_2)$ is the unretarded electron-electron Coulomb interaction,

$$U_C(\vec{r}_1, \vec{r}_2) = \Lambda_+(\vec{r}_1) \, \Lambda_+(\vec{r}_1) \, \frac{e^2}{4\pi r_{12}} \, \Lambda_+(\vec{r}_1) \, \Lambda_+(\vec{r}_2) \,, \tag{14.43}$$

with $r_{12} = |\vec{r}_1 - \vec{r}_2|$. We re-emphasize that all formulas are given in natural units (one has the relation $e^2 = 4\pi\alpha$, where α is the fine-structure constant).

The frequency-dependent Breit interaction is given by

$$\begin{aligned} U_B(\vec{r}_1, \vec{r}_2) &= \Lambda_+(\vec{r}_1)\Lambda_+(\vec{r}_2) \\ &\times \left[-\frac{e^2}{4\pi} \left\{ \frac{\cos(\omega|\vec{r}_{12}|)}{|\vec{r}_{12}|} \, \vec{\alpha}_1 \cdot \vec{\alpha}_2 - \vec{\alpha}_2 \cdot \vec{\nabla}_1 \vec{\alpha}_2 \cdot \vec{\nabla}_2 \frac{\cos(\omega|\vec{r}_{12}|) - 1}{\omega^2 |\vec{r}_{12}|} \right\} \right] \\ &\times \Lambda_+(\vec{r}_1)\Lambda_+(\vec{r}_2) \,, \end{aligned} \tag{14.44}$$

where ω is the energy difference, defined according to Eq. (14.39). The non-retarded Breit interaction is obtained as the zero-frequency limit of the former,

$$U_B(\vec{r}_1, \vec{r}_2) \stackrel{\omega \to 0}{\to} \Lambda_+(\vec{r}_1)\Lambda_+(\vec{r}_2) \left[-\frac{e^2}{4\pi} \left(\frac{\vec{\alpha}_1 \cdot \vec{\alpha}_2}{2 r_{12}} + \frac{\vec{\alpha}_1 \cdot \vec{r}_{12} \, \vec{\alpha}_2 \cdot \vec{r}_{12}}{2 \, r_{12}^3} \right) \right] \Lambda_+(\vec{r}_1)\Lambda_+(\vec{r}_2) \,, \tag{14.45}$$

where $r_{12} = |\vec{r}_{12}|$.

To summarize, the Dirac–Hartree–Fock (DHF) Hamiltonian is the sum of Dirac–Coulomb single-particle Hamiltonians with projection operators, and of the electron-electron interaction as well as the Breit interaction.

14.3.4 *Relativistic Multi–Configurational Approach*

In Refs. [22, 449, 453–456], a practical implementation of the Dirac–Fock approach is described. Other implementations exist (see Refs. [457, 458]). Let us provide for a mini-review on this subject. Following the conventions outlined in Sec. 8.2.2, we will denote single-particle wave functions as follows,

$$\psi_{n_a \kappa_a \mu_a}(\vec{r}) = \begin{pmatrix} f_{n_a \kappa_a}(r) \chi_{-\kappa_a \mu_a}(\hat{r}) \\ \mathrm{i} \, g_{n_a \kappa_a}(r) \chi_{\kappa_a \mu_a}(\hat{r}) \end{pmatrix} , \qquad \int_0^\infty \mathrm{d}r \, r^2 \left[f^2_{n_a \kappa_a}(r) + g^2_{n_a \kappa_a}(r) \right] = 1 \,. \tag{14.46}$$

A correlation orbital in the MCDHF procedure is referred to as a CSF (configuration state function). Each CSF has the structure $\psi^{(\mathrm{CSF})}_{JM,\{n_i\kappa_i\}}$, where $\{n_i\kappa_i\}$ is a multi-index characterizing the one-electron quantum numbers (i enumerates the electrons) used in the construction of the CSF, and JM are the total J and M quantum numbers of the orbital. E.g., one would use the notation $|(1s\,2s)_0\rangle$ for a CSF with principal quantum numbers $n_1 = 1$, $n_2 = 2$, and Dirac angular quantum numbers $\kappa_1 = \kappa_2 = -1$. These individual angular momenta are summarized in the multi-index $\{n_i\kappa_i\} = \{n_1, \kappa_1, n_2, \kappa_2\}$, to result in the total angular momenta $J = 0$, and of course $M = 0$. One can give this for two electrons,

$$\psi^{(\mathrm{CSF})}_{JM}(\vec{r}_1, \vec{r}_2) = \frac{1}{\sqrt{2}} \sum_{\mu_1\mu_2} C^{JM}_{j_1\mu_1 j_2\mu_2} \left[\psi_{n_1\kappa_1\mu_1}(\vec{r}_1)\, \psi_{n_2\kappa_2\mu_2}(\vec{r}_2) - \psi_{n_1\kappa_1\mu_1}(\vec{r}_2)\, \psi_{n_2\kappa_2\mu_2}(\vec{r}_1)\right] . \tag{14.47}$$

The final result of the MCDHF optimization procedure is an atomic state function (ASF). The ASF is a linear combination of CSFs, all of which have the same quantum numbers JM as the reference ASF. However, the different CSF contribute to the final ASF with different occupation numbers. Generally, one has

$$\psi^{(\mathrm{ASF})}_{JM} = \sum_{\{n_i\kappa_i\}} c_{\{n_i\kappa_i\}}\, \psi^{(\mathrm{CSF})}_{JM,\{n_i\kappa_i\}} . \tag{14.48}$$

As an example, the matrix element of the Breit interaction in an ASF is given by

$$\langle \mathrm{ASF}|H_{\mathrm{B}}|\mathrm{ASF}\rangle = \sum_{\{n_i\kappa_i\},\{n_j\kappa_j\}} c^*_{\{n_i\kappa_i\}}\, c_{\{n_j\kappa_j\}}\, \langle \psi^{(\mathrm{CSF})}_{JM,\{n_i\kappa_i\}} |H_{\mathrm{B}}| \psi^{(\mathrm{CSF})}_{JM,\{n_i\kappa_i\}} \rangle . \tag{14.49}$$

In practical implementations, we differentiate between CI calculations, where one incurs no perturbation of the orbitals, and only does a rediagonalization of the Hamiltonian matrix. In MCDHF calculations, the orbitals themselves are subject to optimization. In some cases, it is advantageous to use the self-consistent field approach for the Coulomb+unretarded Breit interactions, while the retarded Breit corrections are added within a second, CI-based step [22, 449, 453–456].

Some remarks on the implementation of the necessary positive-energy projection operators (see Sec. 14.3.1) are useful. The normalizable solutions of the Dirac equation happen to describe just the positive-energy bound states, and these states are exactly those that we are interested in. Assuming that any bound state can be written as a superposition of these normalizable solutions, it is therefore possible to write a CI calculation that automatically takes care of the negative-energy problem, because all functions in the basis set are normalizable. Likewise, if one does an optimization procedure for self-consistent equations only within the space spanned by normalizable wave functions, then one is sure to avoid the negative-energy problem altogether. However, in GRASP and in other implementations of MCDHF [22, 449, 453–458], the Hamiltonian is put on a lattice, implying a priori a discrete spectrum and normalizable, discrete solutions for both the positive-energy as well as the negative-energy domain. Here, one must either directly project out the negative-energy states in all intermediate stages of the calculations [459], or one must hope that the calculation converges sufficiently fast so that the negative-energy states play no role [22].

14.3.5 *Relativistic Breit Interaction and Spinors*

We consider the evaluation of the Coulomb interaction (14.43), and of the unretarded Breit interaction (14.45), between two Dirac bispinors of the form given in Eq. (14.46), $\psi_{n_a\kappa_a\mu_a}(\vec{r}_1) = [f_{n_a\kappa_a}(r_1)\chi_{-\kappa_a\mu_a}(\hat{r}_1), \mathrm{i}\, g_{n_a\kappa_a}(r_1)\chi_{\kappa_a\mu_a}(\hat{r}_1)]^{\mathrm{T}}$ and $\psi_{n_b\kappa_b\mu_b}(\vec{r}_2) = [f_{n_b\kappa_b}(r_2)\chi_{-\kappa_b\mu_b}(\hat{r}_2), \mathrm{i}\, g_{n_b\kappa_b}(r_2)\chi_{\kappa_b\mu_b}(\hat{r}_2)]^{\mathrm{T}}$. The Dirac α matrices have the structure

$$\vec{\alpha}_1 = \begin{pmatrix} 0 & \vec{\sigma}_1 \\ \vec{\sigma}_1 & 0 \end{pmatrix}, \qquad \vec{\alpha}_2 = \begin{pmatrix} 0 & \vec{\sigma}_2 \\ \vec{\sigma}_2 & 0 \end{pmatrix}, \tag{14.50}$$

for the two electrons. Upon invoking the multipole expansions of the interelectron interaction (see Chap. 2 of Ref. [81]), proportional to $1/r_{12}$, two characteristic angular integrals emerge, which we would like to discuss in greater detail.

The first of these concerns Dirac matrix elements with spherical harmonics, namely, the scalar product (in spinor space) $\chi^+_{\kappa_a\mu_a}(\hat{r})\,\chi_{\kappa_b\mu_b}(\hat{r})$, which only depends on the coordinates and can therefore be expanded on the spherical harmonics,

$$\chi^+_{\kappa_a\mu_a}(\hat{r})\,\chi_{\kappa_b\mu_b}(\hat{r}) = \sum_{LM} s^{ba}_{LM}\, C_L(\kappa_b,\kappa_a)\, Y_{LM}(\hat{r}), \tag{14.51}$$

with all coefficients on the right-hand side explained below. As we form the complex conjugate of this expression and use the orthogonality properties of the spherical harmonics, we obtain

$$\int \mathrm{d}\Omega\, \chi^+_{\kappa_a\mu_a}(\hat{r})\, Y_{LM}(\hat{r})\, \chi_{\kappa_b\mu_b}(\hat{r}) = s^{ab}_{LM}\, C_L(\kappa_a,\kappa_b). \tag{14.52}$$

The definition of the tensor s^{ab}_{LM} is inspired by Refs. [23, 460]. It reads

$$s^{ab}_{LM} = \frac{(-1)^{j_b-\mu_b}}{\sqrt{4\pi}}\, C^{LM}_{j_a\mu_a j_b-\mu_b} = \sqrt{\frac{2L+1}{4\pi}}\,(-1)^{j_a-\mu_b+M} \begin{pmatrix} j_a & j_b & L \\ \mu_a & -\mu_b & -M \end{pmatrix}. \tag{14.53}$$

Notice that s^{ab}_{JM} does not depend on the sign of the involved κ's, and can be seen as a multi-index depending on $a \equiv (j_a, \mu_a)$ and $b \equiv (j_b, \mu_b)$. Essentially, s^{ab}_{JM} is nothing but a Clebsch–Gordan coefficient, and its physical meaning in the context of the Wigner–Eckart theorem can be further clarified by either one of following forms,

$$s^{ab}_{LM} = \sqrt{\frac{2L+1}{4\pi}}\, \frac{C^{j_a\mu_a}_{j_b\mu_b LM}}{\sqrt{2j_a+1}} = \sqrt{\frac{2L+1}{4\pi}}\,(-1)^{j_a-\mu_a} \begin{pmatrix} j_a & L & j_b \\ -\mu_a & M & \mu_b \end{pmatrix}. \tag{14.54}$$

We realize that Eq. (14.52) has just the form of the Wigner–Eckart theorem [see Chap. 6, especially Eq. (6.33)] if we identify

$$C_L(\kappa_a,\kappa_b) = \sqrt{\frac{4\pi}{2L+1}}\,\left\langle \kappa_a \left\| \vec{Y}_L \right\| \kappa_b \right\rangle. \tag{14.55}$$

The reduced matrix element $C_L(\kappa_a,\kappa_b)$ may be expressed as follows,

$$\begin{aligned} C_L(\kappa_a,\kappa_b) &= (2j_a+1)^{1/2}\,(-1)^L\, C^{j_b\,1/2}_{j_a\,1/2\,L\,0}\,\Pi(l_a,l_b,L) \\ &= (-1)^{j_a+1/2}\,\Pi_{j_a,j_b} \begin{pmatrix} j_b & L & j_a \\ \frac{1}{2} & 0 & -\frac{1}{2} \end{pmatrix} \Pi(l_a,l_b,L), \end{aligned} \tag{14.56}$$

where $\Pi_{j_a,j_b} = \sqrt{(2j_a+1)(2j_b+1)}$ has been defined in Eq. (6.51). The symbol $\Pi(l_a, l_b, L) = 1$ is defined to be equal to unity if $l_a + l_b + L$ is even and zero otherwise,

$$\Pi(l_a, l_b, L) = \frac{(-1)^{l_a+l_b+L} + 1}{2}. \tag{14.57}$$

As the transformation $\kappa \to -\kappa$ leaves the j's invariant, we have the following symmetry properties,

$$C_L(\kappa_a, \kappa_b) = C_L(-\kappa_a, -\kappa_b) = (-1)^{j_a - j_b} C_L(\kappa_b, \kappa_a). \tag{14.58}$$

The algebra becomes a little more involved when σ matrices are involved, and Dirac matrix elements with spinor matrices are considered. When the vector of σ matrices is introduced between the spinor spherical harmonics, then we may expand the spinor-space scalar product on the vector spherical harmonics,

$$\chi^{+}_{\kappa_a \mu_a}(\hat{r})\, \vec{\sigma} \chi_{\kappa_b \mu_b}(\hat{r}) = \sum_{JLM} s^{ba}_{JM}\, S_{JL}(\kappa_a, \kappa_b)\, \vec{Y}^{L}_{JM}(\hat{r}), \tag{14.59}$$

from which we can immediately infer (by complex conjugation of the former and using the orthogonality properties of the vector spherical harmonics),

$$\int \mathrm{d}\Omega\, \chi^{+}_{\kappa_a \mu_a}(\hat{r})\, \vec{\sigma} \cdot \vec{Y}^{L}_{JM}(\hat{r})\, \chi_{\kappa_b \mu_b}(\hat{r}) = s^{ab}_{JM}\, S_{JL}(\kappa_b, \kappa_a). \tag{14.60}$$

The order of the indices in each case should be observed. It is not a typographical error. The s^{ab}_{JM} are real since we choose the Clebsch–Gordan coefficients to be real, and the same applies to the reduced matrix element $S_{JL}(\kappa_b, \kappa_a)$, which will be further explained in the following. We can identify s^{ab}_{JM}, as defined in (14.53), as $\sqrt{(2J+1)/(4\pi)}$ times the usual prefactor of the Wigner–Eckart theorem in front of the reduced matrix element, and thus

$$S_{JL}(\kappa_b, \kappa_a) = \sqrt{\frac{4\pi}{2J+1}} \left\langle \kappa_a \left\| \vec{\sigma} \cdot \vec{Y}_{JL} \right\| \kappa_b \right\rangle = \sqrt{\frac{4\pi}{2J+1}} \left\langle \kappa_a \left\| [Y_L \otimes \vec{\sigma}]_J \right\| \kappa_b \right\rangle. \tag{14.61}$$

Let us make a small detour and explain and verify the notation $[Y_L \otimes \vec{\sigma}]_J$. To this end, let us first recall that according to the definition of a tensor product, the JM component of $[Y_L \otimes \vec{\sigma}]_J$ fulfills

$$[Y_L \otimes \vec{\sigma}]_{JM} = \sum_{\mu q} C^{JM}_{L\mu 1q}\, Y_{L\mu}(\hat{r})\, \sigma_q, \tag{14.62}$$

where σ_q is the qth component of the vector σ in the spherical basis. Also in the spherical basis, we can express the Mth component of the scalar product $\vec{\sigma} \cdot \vec{Y}_{JL}$ as

$$\left(\vec{\sigma} \cdot \vec{Y}^{L}_{J}\right)_M = \sum_q (-1)^q \left(\vec{Y}^{L}_{JM}(\hat{r})\right)_q \sigma_{-q}. \tag{14.63}$$

We thus have to find the qth vector component of $\vec{Y}^{L}_{JM}(\hat{r})$ in the spherical basis. On one hand, any vector $\vec{a}$ fulfills the relation $\vec{a} = \sum_q (-1)^q\, a_{-q}\, \hat{e}_q$ by virtue of the definition of the spherical basis. On the other hand, we have

$$\vec{Y}^{L}_{JM}(\hat{r}) = \sum_{\mu q} C^{JM}_{L\mu 1q}\, Y_{L\mu}(\hat{r})\, \hat{e}_q, \tag{14.64}$$

which allows us to identify

$$(-1)^q \left(\vec{Y}^L_{JM}(\hat{r})\right)_{-q} = \sum_\mu C^{JM}_{L\mu 1q} Y_{L\mu}(\hat{r}), \tag{14.65}$$

so that, finally, $\left(\vec{Y}^L_{JM}(\hat{r})\right)_q$ can be determined as

$$\left(\vec{Y}^L_{JM}(\hat{r})\right)_q = (-1)^{-q} \sum_\mu C^{JM}_{L\mu 1-q} Y_{L\mu}(\hat{r}). \tag{14.66}$$

Using this relation in Eq. (14.63) gives the result

$$\left(\vec{\sigma}\cdot\vec{Y}_{JL}\right)_M = \sum_{\mu q} C^{JM}_{L\mu 1-q} Y_{L\mu}(\hat{r})\, \vec{\sigma}_{-q} = [Y_L \otimes \vec{\sigma}]_{JM}, \tag{14.67}$$

which proves the equivalence of the two representations in Eq. (14.61).

The only nonvanishing components of S_{JL} can be found for $J = L-1, L, L+1$ and are given by

$$S_{JJ+1}(\kappa_b,\kappa_a) = \sqrt{\frac{J+1}{2J+1}}\left(1 + \frac{\kappa_b+\kappa_a}{J+1}\right) C_J(-\kappa_a,\kappa_b), \tag{14.68a}$$

$$S_{JJ}(\kappa_b,\kappa_a) = \frac{\kappa_b-\kappa_a}{\sqrt{J(J+1)}} C_J(\kappa_a,\kappa_b), \tag{14.68b}$$

$$S_{JJ-1}(\kappa_b,\kappa_a) = \sqrt{\frac{J}{2J+1}}\left(-1 + \frac{\kappa_b+\kappa_a}{J}\right) C_J(-\kappa_a,\kappa_b). \tag{14.68c}$$

The symmetry properties are

$$S_{JL}(\kappa_a,\kappa_b) = (-1)^{J+L+1}(-1)^{j_b-j_a} S_{JL}(\kappa_b,\kappa_a). \tag{14.69}$$

14.3.6 *Subtleties of Nuclear Effects, QED and MCDHF*

Of particular interest in many-electron calculations are effects related to the region immediately surrounding the atomic origin. The reason is that the wave function of inner-shell electrons receives a large radiative correction. Also, the outer electrons press the inner ones toward the nucleus because of repulsion. So, one has to model the charge density at the nucleus with great care. This fact is stressed in Refs. [461, 462]. Possible searches for new interactions are based on the analysis of data obtained from the combination of isotope shift measurements, which refer to effects stemming from the region around the atomic nucleus (so-called King nonlinearities, see Ref. [463]). In analyzing the nuclear effects, one differentiates between volume shifts (so-called nuclear size corrections) and nuclear mass corrections (also known as mass shifts). We remember that the single-particle no-pair Dirac–Coulomb Hamiltonian (14.41) is formulated in the non-recoil limit. Recoil effects can be described based on a recoil operator [464]. For another instructive treatment of nuclear motional effects, one can consult Ref. [465].

In many-electron systems, there exists a further interesting correlation effect which is connected with QED and the behavior of wave functions near the origin. In Ref. [466], the effect of the Uehling potential (the one-loop vacuum polarization correction to the Coulomb interaction) is studied in the context of a many-electron system. It is found that s orbitals are "pulled in" by the attractive Uehling potential, and are drawn closer to the nucleus. Somewhat paradoxically, as a result, the screening of the nuclear charge by the inner s orbitals therefore becomes more efficient. As the low-lying core s orbitals are drawn closer to the nucleus, electrons further out feel a lesser effective nuclear charge, and their resultant energy shift is positive, even if the first-order perturbative shift of, say, p electrons (the expectation value of the Uehling potential, taken on a p state) is negative. As a result of the first-order perturbative shift, and the correlation effect, the enhanced screening remains dominant and leads to a reduced attraction of the electrons by the nucleus and to an increase in the energy of the outer electrons.

14.3.7 *Software Packages and Applications*

A very influential computer code based on the Hartree–Fock method is the COWAN[11] code [438, 467]. One important characteristic of the code is its ability to calculate transition probabilities and decay rates. This enhances the reliability of line identifications. In single-ionized Thulium (Tm II), wrongly assigned quantum numbers J of many energy levels were corrected based on the theoretical calculations [468, 469]. For clarity, we should remark that in physics, the charge state of an atom or ion is denoted by the charge state of the ionic core [470]. A neutral atom thus has the first spectrum of an element, denoted by the element symbol and the Roman letter I, while a single-ionized charge state is denoted by the Roman letter II. The Cowan code is still rather competitive today for nd- and nf-transition elements (lanthanides and actinides). Typically, ab initio multiconfiguration codes have difficulties for atomic systems with large electron-electron correlation effects. In typical cases, standard deviations $< 100\text{cm}^{-1}$ can be accomplished for a least-squared fitting on the basis of Cowan's codes, for the spectra of transition elements, including lanthanides and actinides. Cowan's code [438, 467] starts from a non-relativistic approximation, with relativistic corrections accounted as perturbations. Further developments of Cowan's code have been discussed in Refs. [471, 472], based on so-called orthogonal operators. Although the new code is similar to Cowan's, in that it uses a non-relativistic approximation with relativistic corrections accounted for as perturbations, it additionally accounts for small second-order interactions, which are omitted in Cowan's codes. Recent successful analyses of the Re III spectrum [473] and the Os III spectrum in Ref. [474], overcoming previous limitations [475] bear testimony to the topical interest in Cowan's code.

The MCHF and ATSP2K atomic structure packages are multiconfiguration methods (described in Refs. [476–480]) which have been used to good effect in the analysis of complex spectra of many atoms [432]. One particular advantageous aspect of the ATSP2K package [480] is that, in addition to spectral properties, line strengths and other physical quantities of interest can be calculated; this makes

[11] Robert Duane Cowan (1919–2018).

line identification much easier. Further independent computer codes include CIV3 (Ref. [481]), which is a nonrelativistic Hartree–Fock code with perturbative relativistic corrections.

Let us briefly discuss a curious development in the measurement and theory of the $3p^2P_{1/2}$ and $3p^2P_{3/2}$ levels of neutral sodium atoms (Na I), which also illustrates interesting interactions between theory and experiment. We reemphasize that an accurate understanding of level widths is of decisive importance for the identification of levels in more complex atoms.

A first experiment based on beam-gas-laser spectroscopy (BGLS), as reported in Ref. [482], resulted in a lifetime of (16.40 ± 0.03) ns, for the $3p^2P_{1/2}$ state in Na I (see Sec. V of Ref. [482]). Translated to a (electric dipole) line strength, this result translated into a value of $S = (37.04 \pm 0.08)\, e^2\, a_0^2$, where the numerical value is equal to the line strength in atomic units. For a precise definition of the (dipole) line strength, we refer to Eq. (31) of Ref. [231] and to Chap 5. The experimental result reported in Ref. [482] was at variance with a Hartree–Fock calculation [483] which included core-polarization and electron correlation, and gave a line strength of $S = 37.39\, e^2\, a_0^2$ (see the extrapolated value, to $n_{\rm max} = \infty$, in Fig. 2 of Ref. [483]). The discrepancy was puzzling because the spectrum of sodium ($Z = 11$) corresponds to an electronic configuration of a neon-like core and one single active valence electron, and one might thus expect it to be easily calculable. Hence, the discrepancy of theory and experiment remained extremely puzzling for a number of years. Later, three independent experimental groups (Refs. [483–486]) confirmed a shorter lifetime, corresponding to the larger theoretical value of the line strength, and thus confirming the theoretical prediction originally published in Ref. [483]. Specifically, the shorter lifetime of 16.299(21) ns for the $3p^2P_{1/2}$ was reported in Ref. [484].

In order to illustrate the power of modern Hartree–Fock methods, we give in Table 14.1 the theoretical predictions [487, 488] and NIST ASD data [470] for selected lines of neutral carbon. Note that the configuration $1s^2\, 2s^2\, 2p^2(^3_2P)\ ^3P_0$ is the ground-state configuration and is assigned zero energy. The concept of seniority, defined in Refs. [489–491], is used for the lower subscript in the assignment of a subshell; roughly speaking, the seniority is equal to the number of unpaired electrons. Because none of the $2p$ electrons in the term ^4_3S are spin-paired, the seniority is equal to three. One of the motivations for the investigations is the carbon abundance analysis in stars. For a re-analysis of the solar carbon abundance, we refer to Sec. 5 of Ref. [487].

Of particular importance is the GRASP2K computer code, which belongs to the most powerful multiconfiguration methods currently available [449, 453–456, 492, 493]. The current version is the result of decades-long research described in Refs. [447, 448, 494–498]. A good overview of the capabilities of the code is provided in Refs. [22, 492, 499].

Table 14.1 Theoretical predictions [487,488] for selected lines of neutral carbon atoms (C I) are obtained using a relativistic CI (RCI) calculation, following a Hartree–Fock optimization, and compared to the line data from the Atomic Spectra Database (ASD, see Ref. [470]) of the National Institute of Standards and Technology (NIST).

Species	State	E_{RCI} [cm^{-1}]	E_{NIST} [cm^{-1}]
C I	$1s^2\,2s^2\,2p^2(^3_2P)\ ^3P_0$	0	0
C I	$1s^2\,2s^2\,2p^2(^3_2P)\ ^3P_1$	16	16
C I	$1s^2\,2s^2\,2p^2(^3_2P)\ ^3P_2$	43	43
C I	$1s^2\,2s^2\,2p^2(^1_2D)\ ^1D_2$	10 275	10 193
C I	$1s^2\,2s^2\,2p^2(^1_0S)\ ^1S_0$	21 775	21 648
C I	$1s^2\,2s\ ^2S\,2p^3(^4_3S)\ ^5S^\circ_2$	33 859	33 735
C I	$1s^2\,2s^2\,2p\ ^2P\,3s\,^3P^\circ_0$	60 114	60 333
C I	$1s^2\,2s^2\,2p\ ^2P\,3s\ ^3P^\circ_1$	60 133	60 353
C I	$1s^2\,2s^2\,2p\ ^2P\,3s\ ^3P^\circ_2$	60 174	60 393
C I	$1s^2\,2s^2\,2p\ ^2P\,3s\ ^1P^\circ_1$	61 750	61 982
C I	$1s^2\,2s\ ^2S\,2p^3(^2_3D)\ ^3D^\circ_3$	65 014	64 087
C I	$1s^2\,2s\ ^2S\,2p^3(^2_3D)\ ^3D^\circ_1$	65 017	64 090
C I	$1s^2\,2s\ ^2S\,2p^3(^2_3D)\ ^3D^\circ_2$	65 018	64 091

The MCDFGME code [457,458,500] also constitutes a multiconfiguration Dirac–Fock code which can calculate many atomic properties in addition to the spectrum, such as hyperfine constants [441]. For the role of the Breit interaction in the code, we refer to Refs. [500, 501]. Superheavy elements have been analyzed with the MCDFGME code [502], opening the door to the theoretical description of additions to the periodic table. These calculations are motivated by possible additions to the periodic table of elements around a conceivable island of stability characterizing long-lived superheavy nuclei. Recently, the half-life of element 115 (with mass numbers 287, 288, and 289) has been measured to be around 140 and 190 milliseconds [503]. An early work on the subject is Ref. [504], where the question was investigated if Dirac theory breaks down when the product of nuclear charge number and fine-structure constant exceeds unity, i.e., $Z\alpha > 1$. Surprisingly, the spatial extent of the nucleus prevents this breakdown; the Coulomb potential is altered nonperturbatively. The conclusions of Ref. [504] indicated that one might eventually encounter new physics effects connected with the overlap of the most tightly bound electronic level with the Dirac sea when $Z = 173$, not $Z = 137$. In Ref. [505], despite the enormously large number of electrons, attempts were made to determine the electronic ground-state configurations, based on average level (AL) Dirac–Fock calculations, for a number of elements in the range $Z \approx 140$ and $Z \approx 170$. The ground state of the elements with $Z = 171$ was predicted to be the $[\mathrm{Rn}]5f^{14}6d^{10}7s^2 7p^6 8s^2 8p^6 7d^{10} 5g^{18} 6f^{14} 9s^2 9p$ configuration, with $J = 1/2$. For the element with $Z = 172$, a closely related configuration was found to have lowest energy, with $9s^2 9p$ replaced by $9s^2 9p^2$, and $J = 2$. For the $9s^2 9p^3$ ground-state configuration of the element with $Z = 173$, the authors of Ref. [505] find $J = 3/2$. The following best fit for the nuclear radii [506] was used in Ref. [505],

$$R = \left(r_0 + \frac{r_1}{A^{2/3}} + \frac{r_2}{A^{4/3}} \right) A^{1/3} , \tag{14.70}$$

with $r_0 = 0.891(2)\,\mathrm{fm}$, $r_1 = 1.52(3)\,\mathrm{fm}$, and $r_2 = -2.8(1)\,\mathrm{fm}$.

An independent ansatz based on the MCDHF procedure is given by the Flexible Atomic Code (FAC) described in Ref. [507]. Multireference (MR) Møller–Plesset (MP) perturbation theory is implemented in the MR–MP code described in Ref. [508]. The Jena Atomic Calculator (JAC) is described in Refs. [509, 510]. It is based on the JULIA language (see Ref. [511]), which combines the advantages of the high speed of a compiled computer language with the flexibility of modern computer algebra systems.

Finally, the QEDMOD package described in Ref. [428] represents the QED self-energy operator as an "almost" local potential, modeled as a Yukawa potential with a range equal to the Compton wavelength of the electron, and a prefactor adjusted to reproduce the known diagonal elements of the self-energy operator. It also contains expressions for Uehling and Wichmann–Kroll potentials and thus offers a reliable implementation of QED effects for the Dirac–Hartree–Fock orbitals.

14.4 Systems with a Large Number of Electrons

For systems with a large number of electrons, it is clear that further simplifications of the Hartree–Fock equations are desirable. Indeed, ideas to simplify the exchange term in the Hartree–Fock equations have been around for many decades; for example, in Eq. (6) of Ref. [512], it was proposed to approximate the exchange term with an average exchange energy, averaged over the charge density of the atom. In Eq. (14) of Ref. [512], a further simplification is proposed, based on the exchange term for a free electron gas, somewhat foreshadowing the development of density-functional theory [513, 514].

A few words on density-functional theory are thus indicated. It is sometimes being said that in density-functional theory, one writes the energy of a system exclusively in terms of the probability density $n(\vec{r}) = \sum_i n_i |\psi_i(\vec{r})|^2$ of the electrons, where i runs over all charged particles in the system, and n_i is the occupation number of the ith orbital. However, an inspection reveals that the kinetic term $E_{\rm kin}$ in the energy functional,

$$E_{\rm kin} = \sum_i n_i \int \mathrm{d}^3 r\, \psi_i^*(\vec{r}) \left[-\frac{\vec{\nabla}^2}{2m} \right] \psi_i(\vec{r}) \,, \tag{14.71}$$

cannot be written as a function of the density alone. In order to see this, one observes that any conceivable *ansatz* proportional to the volume integral of $\vec{\nabla}^2 |\psi_i(\vec{r})|^2$ vanishes, in view of the fact that $\vec{\nabla}^2 |\psi_i(\vec{r})|^2$ is the divergence of the gradient $\vec{\nabla} |\psi_i(\vec{r})|^2$. The divergence theorem then indicates that the volume integral vanishes. In fact, this problem may have motivated the investigations reported in Ref. [515]. It could be solved via a generalization of the functional derivative, or, via the introduction of a two-coordinate density $n(\vec{r}, \vec{r}\,') = \sum_i n_i\, \psi_i^*(\vec{r})\, \psi_i(\vec{r}\,')$ [see the unnumbered equation between Eqs. (5) and (6) of Ref. [515]].

According to Ref. [516], however, one can show that at least, a so-called density functional exists which could be used in place of the exchange term in the Hartree–Fock equations and still reproduce the spectrum of the fully interacting system. The project to formulate the entire system based on the probability density alone is relaxed in Ref. [516] via the introduction of one-particle orbitals ψ_i referred to as the KOHN[12]–SHAM[13] orbitals [Ref. [517]]. One arrives at equations of the type [Eq. (2.8) of Ref. [517]] called the Kohn–Sham equations

$$\left\{ -\frac{\vec{\nabla}^2}{2m} + V(\vec{r}) + \frac{e^2}{4\pi} \int \mathrm{d}^3 r' \, \frac{n(r')}{|\vec{r} - \vec{r}\,'|} + V_{\rm xc}(n(\vec{r})) \right\} \psi_i(\vec{r}) = \epsilon_i \, \psi_i(\vec{r}) \,. \tag{14.72}$$

The exchange term functional is $V_{\rm xc}(n(\vec{r}))$. Here, ϵ_i has the physical interpretation of a one-particle ("orbital") energy, and $V(\vec{r})$ is the external potential. The term

$$\frac{e^2}{4\pi} \int \mathrm{d}^3 r' \, \frac{n(r')}{|\vec{r} - \vec{r}\,'|} \tag{14.73}$$

is the direct term in the electron-electron interaction ("Hartree term"). The essential difference between Eq. (14.72) and the Hartree–Fock equations is that the exchange term is formulated in terms of a functional of the electron density $n(\vec{r})$, and does not require any knowledge of the wave functions. The computational effort is now only linearly related with the number of electrons in the system.

One might still hope to do away with the Kohn–Sham orbitals, a field of research now known as orbital-free density functional theory. The original idea came from the so-called Thomas–Fermi model, formulated in Refs. [518–520]. The assumption is that all energy levels in the interacting system up to the Fermi momentum are filled. This model has been augmented by VON WEIZSÄCKER[14] in Ref. [521]. Von Weizsäcker argued that the Thomas–Fermi model needs to be augmented because it

[12]Walter Kohn (1923–2016).
[13]Lu Jeu Sham (b. 1938).
[14]Carl Friedrich von Weizsäcker (1912–2007).

effectively assumes a sudden drop-off of the particle density at the Fermi momentum. Let us consider once more the kinetic-energy term,

$$\sum_i n_i \int \mathrm{d}^3r\, \psi_i^*(\vec{r}) \left(-\frac{\hbar^2\vec{\nabla}^2}{2m}\right) \psi_i(\vec{r}) = \sum_i \frac{\hbar^2}{2m} \int \mathrm{d}^3r\, |\vec{\nabla}\psi_i(\vec{r})|^2 . \tag{14.74}$$

The gradient of the wave function at the boundary of the nucleus, for which von Weizsäcker had developed his theory, enters the expression for the kinetic energy. In fact, Eq. (14.74) is the starting point [in an unnumbered equation] for the considerations of Ref. [521]. By assuming a more elaborate model for the particle wave functions as compared to the original Thomas–Fermi approximation, von Weizsäcker obtained the energy functional

$$\begin{aligned} E = {} & \frac{3h^2}{40m}\left(\frac{3}{\pi}\right)^{2/3} \int \mathrm{d}^3r\, [n(\vec{r})]^{5/3} + \frac{h^2}{32\pi^2 m} \int \mathrm{d}^3r\, \frac{[\vec{\nabla}n(\vec{r})]^2}{n(\vec{r})} \\ & + \int \mathrm{d}^3r\, n(\vec{r})\, V_N(\vec{r}) + \frac{e^2}{8\pi} \int \mathrm{d}^3r \int \mathrm{d}^3r' \frac{n(\vec{r})\, n(\vec{r}\,')}{|\vec{r}-\vec{r}\,'|} . \end{aligned} \tag{14.75}$$

The first two terms in this functional serve as approximations to the kinetic energy of the ensemble of particles. Although "orbital-free density functional theory" could in principle lead to a much simplified formulation, it still entails quite considerable approximations. A more elaborate *ansatz* involves Kohn–Sham quantum orbitals.

Implementations of density-functional theory include the GAUSSIAN package [522, 523], originally developed under the guidance of Pople,[15] which has been developed over a number of decades. The GAUSSIAN package uses atomic basis functions whose radial part is proportional to a functional form $\exp(-\alpha\, r^2)$, where α is a constant and r is the radial variable. These functions are different from Slater functions which are proportional to the functional form $\exp(-\zeta\, r)$. The use of the Gaussian *ansatz*, augmented by powers of the radial variable in order to correctly describe the wave functions near the atomic nuclei, has some computational advantages. GAUSSIAN can be regarded as one of the most powerful density-functional-theory inspired molecular structure packages available today [513, 514].

The DIRAC program [524, 525] also constitutes a very powerful tool for the analysis of molecular electronic structure. One feature of the DIRAC package is that relativistic basis sets for the notoriously problematic lanthanide series of elements are available [526] and have been used for the determination of ground-state spectroscopic constants for YbF [526] with good success. Also, the DIRAC package has been used to analyze the role of spin-orbit coupling in the notoriously problematic actinide cations, and evidence has been found to support the hypothesis that the x-ray photoelectron spectra (XPS) of ionic actinide materials may provide direct information about the angular momentum coupling within the $5f$ shell [527].

Finally, the General Atomic and Molecular Electronic Structure System (GAMESS) is a program for ab initio molecular quantum chemistry [528]. It has also evolved over a number of decades [529, 530]. A variety of atomic and molecular

[15]Sir John Anthony Pople (1925–2004).

properties, including dipole moments and frequency-dependent hyperpolarizabilities, may be computed. Finally, the AMBER package [531] is a relatively recent, highly powerful addition to the family of molecular structure packages currently available. One important focus of AMBER lies on biomolecules [532, 533].

14.5 Further Thoughts

Here are some suggestions for further thought.

(1) **Relativistic Breit Interaction.** Review the derivation of the relativistic Breit interaction (14.45) as a generalization of the approach used in Chap. 12, but with the propagator sandwiched between relativistic Dirac bispinors. Try to derive its nonrelativistic limit, and leading corrections, and compare with the treatment in Chap. 12.

(2) **Exchange Term for Filled Subshells.** Assume that subshells are filled and that an average is taken over all magnetic projections. Derive the simplification considered in Eq. (14.17),

$$-\sum_{n\neq a}\int \mathrm{d}^3r_2\,\phi_n^+(\vec r_2)\,U(\vec r_2,\vec r_1)\,\phi_a(\vec r_2)\,\phi_n(\vec r_1)$$
$$\to -\frac{1}{2}\sum_{n\neq a}\sum_{\ell}\begin{pmatrix}\ell_n & \ell & \ell_a\\ 0 & 0 & 0\end{pmatrix}^2 R^*_{n\ell_a}(r_1)\,\chi_{\kappa_a\ell_a}(\hat r_1)\int \mathrm{d}r_2 R^*_{n\ell_n}(r_2)\,\frac{r_<^\ell}{r_>^{\ell+1}}\,R^*_{a\ell_a}(r_2)\,, \tag{14.76}$$

and thus, confirm expressions obtained in Refs. [438, 439].

(3) **Interactions and Correlation Orbitals.** Apply the decompositions (14.52) and (14.59) to the calculation of matrix elements (reduction to radial integrals) between correlation orbitals of the form $\psi_{n_a\kappa_a\mu_a}(\vec r_1) = [f_{n_a\kappa_a}(r_1)\chi_{-\kappa_a\mu_a}(\hat r_1), \mathrm{i}\,g_{n_a\kappa_a}(r_1)\chi_{\kappa_a\mu_a}(\hat r_1)]^{\mathrm{T}}$ and $\psi_{n_b\kappa_b\mu_b}(\vec r_2) = [f_{n_b\kappa_b}(r_2)\chi_{-\kappa_b\mu_b}(\hat r_2), \mathrm{i}\,g_{n_b\kappa_b}(r_2)\chi_{\kappa_b\mu_b}(\hat r_2)]^{\mathrm{T}}$.

(4) **Hartree–Fock–Roothaan Equations.** Start from Eq. (14.15) and assume a basis set expansion of the orbital wave functions in a basis, e.g., in the form

$$\phi_n(\vec r) = \sum_\nu C_{\nu n}\,\widetilde\phi_\nu(\vec r)\,, \tag{14.77}$$

where the $\widetilde\phi_\nu(\vec r)$ are the basis functions ($\nu = 1, \dots, K$ where K is the size of the basis). Try to define a suitable one-particle operator h so that, with

$$h_{\mu\nu} = \int \mathrm{d}^3r\,\widetilde\phi_\mu^+(\vec r)\,h(\vec r)\,\widetilde\phi_\nu(\vec r)\,,$$
$$S_{\mu\nu} = \int \mathrm{d}^3r\,\widetilde\phi_\mu^+(\vec r)\,\widetilde\phi_\nu(\vec r)\,, \tag{14.78}$$

you can write the Hartree–Fock eigenvalue problem as

$$\sum_\nu h_{\mu\nu}\,C_{\nu i} = \sum_\nu S_{\mu\nu}\,C_{\nu i}\,\mathcal{E}_i\,, \qquad \mathbb{h}\,\mathbb{C} = \mathbb{S}\,\mathbb{C}\,\mathcal{E}\,. \tag{14.79}$$

Attempt to discuss similarities and differences to the approach outlined in Sec. 13.4.2, notably, Eq. (13.84), and attempt to clarify why the Hartree–Fock method is said to entail a so-called self-consistent field.

(5) **Tensors and Symmetries.** Use Eqs. (8.4.10) and (8.4.17) of Ref. [16] to establish the relation of s^{ba}_{JM} and s^{ab}_{JM}.

(6) **QED Effects and Multi-Electron Atoms.** Write a program implementing some of the ideas outlined in the current chapter, including QED effects. Be inspired by Refs. [22, 428, 449, 453–458].

Chapter 15

Beyond Breit Hamiltonian and On-Shell Form Factors

15.1 Overview

The energy corrections to atomic levels described so far could be understood based on the Foldy–Wouthuysen transformation and effective Dirac equations containing on-shell form factors (see Chaps. 10 and 11), and the Breit Hamiltonian (Chap. 12). Beyond this point, one has to resort to different and more sophisticated techniques for the description of bound-state energy levels. We will describe these in the following. In the current chapter, we shall thus dwell into the matching of scattering amplitudes with energy shifts in atoms. We have already learned that Feynman diagrams give us S matrix elements (see Secs. 9.5, 9.6 and 10.2.1), which can be matched against energy shifts [see Eq. (12.43)]. In the current chapter, this concept will be ramified and applied to the calculation of two paradigmatic and important energy shifts which affect bound-state energies in two-particle bound systems in higher order. Because the corrections are mediated by the binding Coulomb field, they are also referred to as "binding corrections".

We shall study these processes based on two examples. The first of these is the $\alpha(Z\alpha)^5$ two-Coulomb-vertex correction to the Lamb shift, originally evaluated in Ref. [340]. The second example, which entails a scattering amplitude, is a recoil correction to energy levels of order $(Z\alpha)^5$, which is a genuine two-particle effect and cannot, in principle, be derived from the Breit Hamiltonian or the Foldy–Wouthuysen transformation. The recoil correction entails a loop integration, and necessitates the evaluation of retardation effects which cannot be described by a static Hamiltonian. Alternatively, the recoil correction is an effect which we would like to refer to as the Salpeter correction [534]. Results for hydrogenic states with principal quantum numbers $n = 1$ and $n = 2$ were derived in Ref. [534], before the results were generalized for an arbitrary reference state in Refs. [346, 347, 535, 536]. The correction entails the evaluation of a matrix element of the form $1/r^3$, which is divergent when evaluated on a reference S state and thus, in need of a regularization. This regularization is accomplished by a distribution proposed in Refs. [537, 538]. The result for the recoil correction will also be relevant for our derivation of the helium Lamb shift in Chap. 13.

15.2 Furry Picture, Scattering Amplitude and Self–Energy

15.2.1 *Furry Picture*

In Chap. 10, we have studied quantum electrodynamics based on free states, with interactions happening "in the middle" of the diagram, while incoming and outgoing states are eigenstates of the free Dirac Hamiltonian. However, for bound-state calculations, it is often advantageous and in fact, appropriate, to start from a different formulation of QED, based on the FURRY[1] picture [539]. The Lagrangian of Furry-picture QED is

$$\mathcal{L} = \bar{\psi}\left[\gamma^\mu\left(\mathrm{i}\partial_\mu - e\,A_\mu\right) - \gamma^0 V - m\right]\psi - \frac{1}{4}F^{\mu\nu}F_{\mu\nu}\,. \tag{15.1}$$

In contrast to Eq. (10.1), we have incorporated the nuclear potential (Coulomb potential) $V = -Z\alpha/r$ as a classical potential in the Lagrangian. The four-vector potential A_μ is that of the quantized photon field. The free Lagrangian (10.2) is modified to read as follows,

$$\mathcal{L}_0 = \bar{\psi}\left(\mathrm{i}\gamma^\mu\partial_\mu - \gamma^0 V - m\right)\psi - \frac{1}{4}F^{\mu\nu}F_{\mu\nu}\,. \tag{15.2}$$

The eigenstates of the fermionic part of the corresponding Hamiltonian are the bound and continuum eigenstates of the Dirac–Coulomb Hamiltonian, discussed in Chap. 8. The appropriate quantum numbers to label the states are the principal quantum number n, the Dirac angular quantum number κ [see Eq. (4.348)], and the magnetic projection μ. One can of course equivalently exchange the quantum number n for the bound-state energy E. (Recall that κ summarizes the values of the total angular quantum number j and of the orbital quantum number ℓ into a single quantum number.) Given κ and μ, specification of n is tantamount to specifying E. In the Furry picture, the interaction Lagrangian is the same as in Eq. (10.3),

$$\mathcal{L}_I = -e\,\bar{\psi}\,\gamma^\mu\,A_\mu\,\psi\,. \tag{15.3}$$

In ordinary QED, we had encountered the fermionic field operators in Eqs. (7.140) and (7.141). In the Furry picture, one writes the field operator and its Dirac adjoint as a sum of contributions from positive-energy and negative-energy states,

$$\psi(x) = \sum_{(E>0)\kappa\mu} b_{E\kappa\mu}\,\psi_{E\kappa\mu}(\vec{r})\,\mathrm{e}^{-\mathrm{i}Et} + \sum_{(E'>0)\kappa'\mu'} d^{+}_{E'\kappa'\mu'}\,\psi_{-E'\kappa'\mu'}(\vec{r})\,\mathrm{e}^{\mathrm{i}E't}\,, \tag{15.4a}$$

$$\overline{\psi}(x) = \sum_{(E>0)\kappa\mu} b^{+}_{E\kappa\mu}\,\overline{\psi}_{E\kappa\mu}(\vec{r})\,\mathrm{e}^{\mathrm{i}Et} + \sum_{(E'>0)\kappa'\mu'} d_{E'\kappa'\mu'}\,\psi_{-E'\kappa'\mu'}(\vec{r})\,\mathrm{e}^{-\mathrm{i}E't}\,. \tag{15.4b}$$

The canonical conjugate field momentum is obtained from Eq. (15.1) as

$$\pi(x) = \mathrm{i}\,\overline{\psi}(x)\,\gamma^0\,. \tag{15.5}$$

[1]Wendell Hinkle Furry (1907–1984).

For two space-time points with equal time coordinates, $x_1^\mu = (t, \vec{r}_1)$ and $x_2^\mu = (t, \vec{r}_2)$, the canonical anticommutation relation (7.161),

$$\{\psi(x_1), \pi(x_2)\}|_{t_1=t_2} = \mathrm{i}\,\delta^{(3)}(\vec{r}_1 - \vec{r}_2)\,, \tag{15.6}$$

is fulfilled provided the eigenstates (see Sec. 8.2.2) are normalized as follows [see Eq. (8.88)]

$$\int \mathrm{d}^3 r\, \psi^+_{E_1\kappa_1\mu_1}(\vec{r})\, \psi_{E_2\kappa_2\mu_2}(\vec{r}) = \delta_{E_1 E_2}\, \delta_{\kappa_1\kappa_2}\, \delta_{\mu_1\mu_2}\,. \tag{15.7}$$

The interaction Lagrangian density in the Furry picture is

$$\mathcal{L}_I = -e\,\bar{\psi}\,\gamma^\mu\, A_\mu\, \psi = -J^\mu\, A_\mu\,, \qquad J^\mu = e\,\bar{\psi}\,\gamma^\mu\,\psi\,. \tag{15.8}$$

The Feynman rules in the Furry picture are given in Table 15.1. The double lines denote the electron bound inside the Coulomb field. The Feynman rule for the fermion propagator involves the Dirac–Coulomb Feynman propagator, which we had already encountered in Sec. 8.2.5,

$$S_F^{\mathrm{DC}} = \frac{1}{\not{p} - m + \gamma^0 V + \mathrm{i}\epsilon}\,, \tag{15.9}$$

in the slightly different form $\mathcal{G}_{\mathrm{DC}}$. The form $\mathcal{G}_{\mathrm{DC}}$ discussed in Sec. 8.2.5 does not specify the infinitesimal part in the denominator. The following conversion formulas are relevant in order to convert the Dirac–Coulomb Green function between the different conventions, in analogy to Eq. (7.175) for the free-electron case,

$$G_F^{\mathrm{DC}} = \mathrm{i} S_F^{\mathrm{DC}} \gamma^0\,, \qquad S_F^{\mathrm{DC}} = -\mathrm{i} G_F^{\mathrm{DC}}\,\gamma^0\,, \tag{15.10a}$$

$$\mathcal{G}_F^{\mathrm{DC}} = \mathrm{i} G_F^{\mathrm{DC}}\,, \qquad G_F^{\mathrm{DC}} = -\mathrm{i}\mathcal{G}_F^{\mathrm{DC}}\,, \tag{15.10b}$$

$$S_F^{\mathrm{DC}} = -\,\mathcal{G}_F^{\mathrm{DC}}\gamma^0\,, \qquad \mathcal{G}_F^{\mathrm{DC}} = -S_F^{\mathrm{DC}}\,\gamma^0\,. \tag{15.10c}$$

All conventions will be used in the following.

15.2.2 *Matching for the Bound-State Self–Energy*

In order to derive the starting expression for the bound-state self-energy shift, to all orders in the Coulomb field, we will carry out the matching in the Furry picture outlined above (see also Sec. 12.2 for details of the matching procedure). Let us consider the diagram in Fig. 15.1. From the Feynman rules in the Furry picture, one obtains the following forward-scattering S-matrix element,

$$\begin{aligned} S &= \int \mathrm{d}^4 x \int \mathrm{d}^4 x'\, \mathrm{i}\, D_{F\mu\nu}(x - x')\, \bar{\psi}_n(x)\, (-\mathrm{i}e\gamma^\mu)\, \mathrm{i}\, S_F^{\mathrm{DC}}(x, x')\, (-\mathrm{i}e\gamma^\nu)\, \psi_n(x') \\ &= e^2 \int \mathrm{d}^4 x \int \mathrm{d}^4 x'\, D_{F\mu\nu}(x - x')\, \bar{\psi}_n(x)\, \gamma^\mu S_F^{\mathrm{DC}}(x, x')\, \gamma^\nu\, \psi_n(x')\,. \end{aligned} \tag{15.11}$$

Recalling that $\psi_n(x) = \psi_n(\vec{r}) \exp(-\mathrm{i}\, E_n\, t)$ and separating the covariant four-dimensional integrals into their components, we can write

$$\begin{aligned} S &= e^2 \int \mathrm{d}t \int \mathrm{d}^3 r \int \mathrm{d}t' \int \mathrm{d}^3 r' \int \frac{\mathrm{d}^4 k}{(2\pi)^4} \\ &\quad \times \bar{\psi}_n(\vec{r})\, \mathrm{e}^{\mathrm{i}E_n t}\, \gamma^\mu \int \frac{\mathrm{d}E}{2\pi}\, S_F^{\mathrm{DC}}(E, \vec{r}, \vec{r}')\, \mathrm{e}^{-\mathrm{i}E(t-t')}\, \gamma^\nu\, \psi_n(\vec{r}')\, \mathrm{e}^{-\mathrm{i}E_n t'} \\ &\quad \times D_{F\mu\nu}(\omega, \vec{k})\, \mathrm{e}^{-\mathrm{i}\omega\cdot(t-t')}\, \mathrm{e}^{\mathrm{i}\vec{k}\cdot(\vec{r}-\vec{r}')}\,, \end{aligned} \tag{15.12}$$

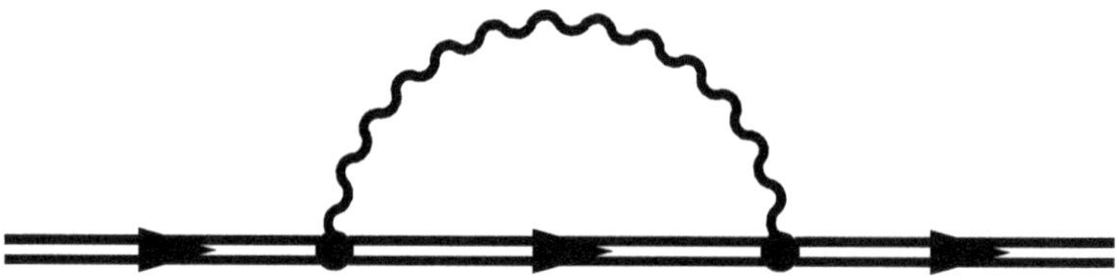

Fig. 15.1 In the Feynman diagram for the bound-electron self-energy in the Furry picture, the double line denotes a Dirac–Coulomb propagator given in Eq. (15.9).

where $\omega = k^0$. We change variables according to $(t, t') \to (t, \tau = t - t')$, with unit Jacobian, to obtain

$$\begin{aligned} S &= e^2 \int \mathrm{d}^3 r \int \mathrm{d}^3 r' \underbrace{\int \mathrm{d}t}_{=T} \int \mathrm{d}\tau \int \frac{\mathrm{d}^4 k}{(2\pi)^4} \int \frac{\mathrm{d}E}{2\pi} D_{F\mu\nu}(\omega, \vec{k}) \\ &\quad \times \bar{\psi}_n(\vec{r})\, \gamma^\mu\, S_F^{\mathrm{DC}}(E, \vec{r}, \vec{r}')\, \gamma^\nu\, \psi_n(\vec{r}')\, \mathrm{e}^{\mathrm{i}\vec{k}\cdot(\vec{r}-\vec{r}')} \mathrm{e}^{-\mathrm{i}(E-E_n+\omega)\tau} \\ &= e^2 T \int \mathrm{d}^3 r \int \mathrm{d}^3 r' \int \frac{\mathrm{d}^4 k}{(2\pi)^4} \int \frac{\mathrm{d}E}{2\pi} D_{F\mu\nu}(\omega, \vec{k}) \\ &\quad \times \bar{\psi}_n(\vec{r})\, \gamma^\mu\, S_F^{\mathrm{DC}}(E, \vec{r}, \vec{r}')\, \gamma^\nu\, \psi_n(\vec{r}')\, \mathrm{e}^{\mathrm{i}\vec{k}\cdot(\vec{r}-\vec{r}')}\, (2\pi)\, \delta(E - E_n + \omega)\,. \end{aligned} \tag{15.13}$$

Integrating out the Dirac-δ function, one obtains

$$\begin{aligned} S &= e^2 T \int \mathrm{d}^3 r \int \mathrm{d}^3 r' \int \frac{\mathrm{d}^4 k}{(2\pi)^4} D_{F\mu\nu}(\omega, \vec{k}) \mathrm{e}^{\mathrm{i}\vec{k}\cdot(\vec{r}-\vec{r}')} \\ &\quad \times \bar{\psi}_n(\vec{r})\, \gamma^\mu\, S_F^{\mathrm{DC}}(E_n - \omega, \vec{r}, \vec{r}')\, \gamma^\nu\, \psi_n(\vec{r}') = -\mathrm{i} T \, \langle \psi^+ | \Delta H_{\mathrm{SE}} | \psi \rangle\,, \end{aligned} \tag{15.14}$$

where the latter identification with the self-energy Hamiltonian H_{SE} is due to the matching given in Eq. (12.43). After all, the matching has resulted in the following equation for the self-energy Hamiltonian ΔH_{SE},

$$\begin{aligned} \Delta E_{\mathrm{SE}} &= \langle \psi^+ | \Delta H_{\mathrm{SE}} | \psi \rangle = \mathrm{i} e^2 \int \mathrm{d}^3 r \int \mathrm{d}^3 r' \int \frac{\mathrm{d}^4 k}{(2\pi)^4} D_{F\mu\nu}(\omega, \vec{k}) \\ &\quad \times \bar{\psi}_n(\vec{r})\, \gamma^\mu\, \mathrm{e}^{\mathrm{i}\vec{k}\cdot\vec{r}}\, S_F^{\mathrm{DC}}(E_n - \omega, \vec{r}, \vec{r}')\, \gamma^\nu\, \mathrm{e}^{-\mathrm{i}\vec{k}\cdot\vec{r}'}\, \psi_n(\vec{r}')\,. \end{aligned} \tag{15.15}$$

Using Eq. (15.10), in the form $\mathcal{G}_F^{\mathrm{DC}}(E_n - \omega, \vec{r}, \vec{r}') = -S_F^{\mathrm{DC}}(E_n - \omega, \vec{r}, \vec{r}')\, \gamma^0$, and defining the matrices $\alpha^\mu = \gamma^0 \gamma^\mu$, we can alternatively write

$$\begin{aligned} \Delta E_{\mathrm{SE}} &= \langle \psi^+ | \Delta H_{\mathrm{SE}} | \psi \rangle = -\mathrm{i} e^2 \int \frac{\mathrm{d}^4 k}{(2\pi)^4} D_{F\mu\nu}(\omega, \vec{k}) \int \mathrm{d}^3 r \int \mathrm{d}^3 r' \\ &\quad \times \psi_n^+(\vec{r})\, \alpha^\mu\, \mathrm{e}^{\mathrm{i}\vec{k}\cdot\vec{r}}\, \mathcal{G}_F^{\mathrm{DC}}(E_n - \omega, \vec{r}, \vec{r}')\, \alpha^\nu\, \mathrm{e}^{-\mathrm{i}\vec{k}\cdot\vec{r}'}\, \psi_n(\vec{r})\,. \end{aligned} \tag{15.16}$$

This latter representation is useful in numerical evaluations of the one-loop bound-state self-energy [257, 258, 267]. For the role of the mass counter term, we refer to Sec. 15.6.

Table 15.1 The Feynman rules of Furry-picture quantum electrodynamics can be expressed in terms of a correspondence of a mathematical expression and a diagram. The building blocks are given below; they are slightly different from those given in Table 10.1. The Coulomb potential is incorporated in the unperturbed Dirac propagator and breaks translational symmetry of wave functions and propagators. Energy is still conserved. Therefore, the Feynman propagator in the Furry picture, in momentum space, has the arguments $(p^0, \vec{p}_1, \vec{p}_2)$, or, in the mixed energy-coordinate representation, the arguments $(p^0, \vec{r}_1, \vec{r}_2)$. The mathematical energy variable, required for momentum conservation at the vertex, is being denoted as p^0, and it corresponds to the energy conservation at the vertex (denoted by a dot), when the energy flow is applied in the indicated direction. The energy variable E is understood to be positive. By contrast, the physical energy of a positron flows in the opposite direction. Symmetry factors for an outgoing positron, the interchange of a field operator, and a closed fermion loop, are the same as for ordinary quantum electrodynamics (see Table 10.1).

Incoming electron	$\psi_{E\kappa\mu}$	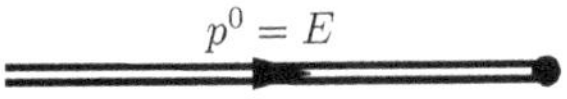
Outgoing electron	$\overline{\psi}_{E\kappa\mu}$	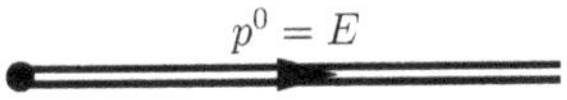
Incoming positron from the past	$\overline{\psi}_{-E\kappa\mu}$	
Outgoing positron into the future	$\psi_{-E\kappa\mu}$	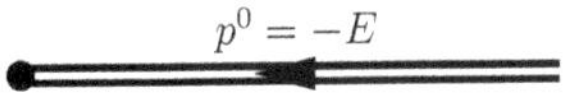
Incoming photon	$\epsilon^{\mu}_{k\lambda}$	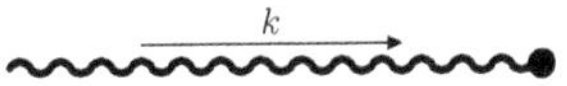
Outgoing photon	$\epsilon^{\mu}_{k\lambda}$	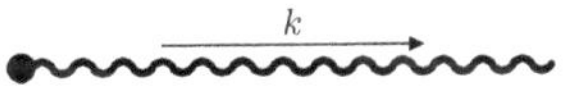
Photon line	$\mathrm{i}D_F^{\mu\nu}$	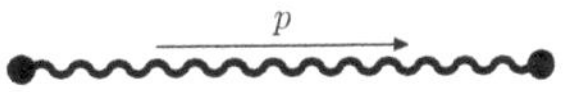
Fermion line	$\mathrm{i}S_F^{\mathrm{DC}}$	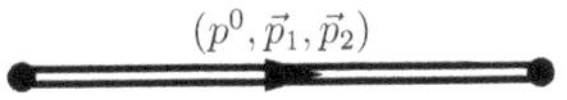
Fermion-photon vertex	$-\mathrm{i}\,e\,\gamma_\mu$	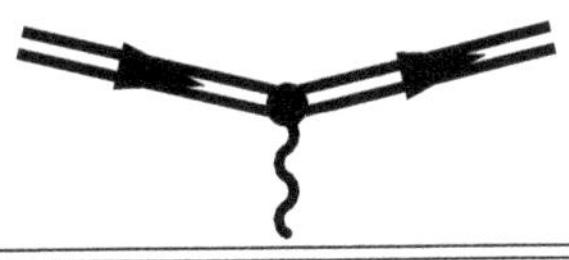

15.3 Binding Correction to the Lamb Shift

15.3.1 *Two-Coulomb-Vertex Scattering Amplitude*

We start from Eq. (15.15). After Fourier transformation, using the photon propagator in Feynman gauge according to $D_{\mu\nu}(k) = -g_{\mu\nu}/(k^2 + \mathrm{i}\epsilon)$ [see Eq. (9.121)], and

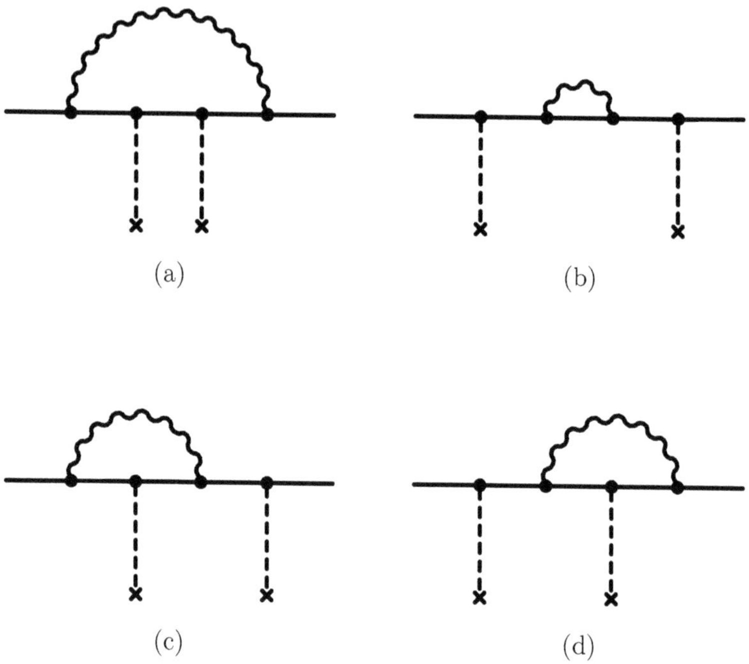

Fig. 15.2 The integrals for the two-Coulomb vertex, one-loop forward scattering amplitudes, to be discussed below, are actually two-loop integrals, because the three-momentum flowing into the diagrams has to flow out again. Coulomb interactions are denoted by dashed lines, with a space-like momentum exchange $(0, \vec{q})$. The photon in the loop carries all polarizations, and has a four-momentum k^μ.

taking into account Eq. (15.9), one finds

$$\Delta E_{\rm SE} = e^2 \int \frac{d^4 k}{(2\pi)^4 i} \frac{1}{k^2 + i\epsilon} \left\langle \overline{\psi} \left| \gamma^\mu \frac{1}{\not{p} - \not{k} - m + \gamma^0 V + i\epsilon} \gamma_\mu \right| \psi \right\rangle . \tag{15.17}$$

Here, we recall that V is the Coulomb potential. The $\alpha(Z\alpha)^5$ correction is described by the exchange of high-energy Coulomb photons, which is why we can expand the propagator in powers of the potential V, which amounts to an expansion in an increasing number of Coulomb vertices. Specifically, in Fourier space, one has $V = -4\pi Z\alpha/\vec{q}^2$ with $\vec{q}^2 \sim m^2$. Diagram (a) in Fig. 15.2 gives the energy shift

$$\Delta E_a = e^2 \int_{C_F} \frac{d^4 k}{(2\pi)^4 i} \frac{1}{k^2} \left\langle \overline{\psi} \left| \gamma^\mu \frac{1}{\not{p} - \not{k} - m} \gamma^0 V \frac{1}{\not{p} - \not{k} + \not{q} - m} \gamma^0 V \frac{1}{\not{p} - \not{k} - m} \gamma_\mu \right| \psi \right\rangle . \tag{15.18}$$

We here explicitly specify the Feynman contour C_F, which allows us to suppress the infinitesimal shifts $+i\epsilon$ in the propagators, making the notation more compact. Both the momenta of the Coulomb photons exchanged as well as the four-momenta of the virtual photon in the loop are hard, i.e., of the order of the electron mass. The momentum from the wave function entering the diagram is of the order of the bound-state momentum $Z\alpha m \ll m$ and can therefore be ignored to good approximation. This implies that we can approximate

$$p^\mu = (E, \vec{p}) \approx (m, \vec{0}) \equiv t^\mu \tag{15.19}$$

in all propagator denominators in Eq. (15.18). We can also approximate the relativistic wave function ψ by the value of the nonrelativistic wave function ϕ at the origin, $\phi(\vec{0}) = \phi(\vec{r} = \vec{0})$, multiplied by a fundamental bispinor u_s, and write

$$\psi(\vec{r}) \approx \phi(\vec{0})\, u_s, \qquad \phi(\vec{p}) \approx (2\pi)^3\, \delta^{(3)}(\vec{p})\, \phi(\vec{0}), \tag{15.20}$$

where ϕ denotes the nonrelativistic (Schrödinger–Coulomb) wave function (see Chap. 4). Here and in the following, we will implement the convention that $\phi(\vec{0}) = \phi(\vec{r} = \vec{0})$ denotes the value of the nonrelativistic wave function at the origin of coordinate space, not momentum space. The fundamental particle bispinors u_s with $s = \uparrow, \downarrow$ (spin up and spin down) are given as

$$u_\uparrow = (1, 0, 0, 0)^{\mathrm{T}}, \qquad u_\downarrow = (0, 1, 0, 0)^{\mathrm{T}}, \tag{15.21}$$

and they fulfill the following relation, relevant to the averaging over the electron spins,

$$\frac{1}{2} \sum_{s=\uparrow,\downarrow} u_s \otimes \overline{u}_s = \frac{\gamma^0 + 1}{4} = \frac{1}{2} \begin{pmatrix} \mathbb{1}_{2\times 2} & 0_{2\times 2} \\ 0_{2\times 2} & 0_{2\times 2} \end{pmatrix}. \tag{15.22}$$

For an S state, in the nonrelativistic approximation, the final result for the resulting energy shift must be independent of the spin state. Hence, we can write

$$\begin{aligned} \Delta E_a = {} & \frac{1}{2} \sum_{s=\uparrow,\downarrow} e^2 \int_{C_F} \frac{\mathrm{d}^4 k}{(2\pi)^4 \mathrm{i}} \int \frac{\mathrm{d}^3 q_1}{(2\pi)^3} \int \frac{\mathrm{d}^3 q_2}{(2\pi)^3} \int \frac{\mathrm{d}^3 p}{(2\pi)^3} \frac{1}{k^2} \\ & \times \phi^+(\vec{p} + \vec{q}_1 + \vec{q}_2)\, \overline{u}_s\, \gamma^\mu \frac{1}{\not{t} - \not{k} + \not{q}_1 + \not{q}_2 - m} \gamma^0 \left(-\frac{4\pi Z\alpha}{\vec{q}_1^{\,2}} \right) \\ & \times \frac{1}{\not{t} - \not{k} + \not{q}_2 - m} \gamma^0 \left(-\frac{4\pi Z\alpha}{\vec{q}_2^{\,2}} \right) \frac{1}{\not{t} - \not{k} - m} \gamma_\mu\, \phi(\vec{p})\, u_s\,. \end{aligned} \tag{15.23}$$

The Feynman slashes of the Coulomb exchange momenta are given as $\not{q}_1 = -\vec{\gamma} \cdot \vec{q}_1$, and $\not{q}_2 = -\vec{\gamma} \cdot \vec{q}_2$. There is no energy exchange in a Coulomb photon, and hence $(q_1)^0 = (q_2)^0 = 0$. One uses Eq. (15.20) and obtains

$$\begin{aligned} \Delta E_a = {} & e^2 \left|\phi(\vec{0})\right|^2 \int_{C_F} \frac{\mathrm{d}^4 k}{(2\pi)^4 \mathrm{i}} \int \frac{\mathrm{d}^3 q_1}{(2\pi)^3} \int \mathrm{d}^3 q_2 \int \mathrm{d}^3 p \frac{1}{k^2} \\ & \times \delta^{(3)}(\vec{p} + \vec{q}_1 + \vec{q}_2)\, \mathrm{Tr} \left[\gamma^\mu \frac{1}{\not{t} - \not{k} + \not{q}_1 + \not{q}_2 - m} \gamma^0 \left(-\frac{4\pi Z\alpha}{\vec{q}_1^{\,2}} \right) \right. \\ & \left. \times \frac{1}{\not{t} - \not{k} + \not{q}_2 - m} \gamma^0 \left(-\frac{4\pi Z\alpha}{\vec{q}_2^{\,2}} \right) \frac{1}{\not{t} - \not{k} - m} \gamma_\mu \frac{\gamma^0 + 1}{4} \right] \delta^{(3)}(\vec{p}), \end{aligned} \tag{15.24}$$

and carries out the $\int \mathrm{d}^3p$ and $\int \mathrm{d}^3q_2$ integrations,

$$\begin{aligned}\Delta E_a &= 64\,\pi^3\,\alpha\,(Z\alpha)^2\,\left|\phi(\vec{0})\right|^2 \int \frac{\mathrm{d}^3q}{(2\pi)^3} \int_{C_F} \frac{\mathrm{d}^4k}{(2\pi)^4\mathrm{i}}\,\frac{1}{\vec{q}^{\,4}}\,\frac{1}{k^2} \\ &\quad \times \mathrm{Tr}\left[\gamma^\mu\,\frac{1}{\not{t}-\not{k}-m}\,\gamma^0\,\frac{1}{\not{t}-\not{k}+\not{q}-m}\,\gamma^0\,\frac{1}{\not{t}-\not{k}-m}\,\gamma_\mu\,\frac{\gamma^0+1}{4}\right] \\ &= \frac{\alpha}{\pi}\,\frac{(Z\alpha)^5\,m}{n^3}\left(\frac{\mu}{m}\right)^3 \delta_{\ell 0}\,(4\pi K_a)\,, \end{aligned} \tag{15.25}$$

where we implicitly define the integral K_a. We have also used the relation $e^2 = 4\pi\alpha$. Based on the well-known value of the S-state wave function at the origin,

$$\left|\phi(\vec{0})\right|^2 = \frac{(Z\alpha\mu)^3}{\pi n^3}\,\delta_{\ell 0}\,, \tag{15.26}$$

we can write the dimensionless integral K_a as

$$\begin{aligned}K_a &= \int \frac{\mathrm{d}^3q}{(2\pi)^3} \int_{C_F} \frac{\mathrm{d}^4k}{\pi^2\mathrm{i}}\,\frac{1}{\vec{q}^{\,4}}\,\frac{m^2}{k^2} \\ &\quad \times \mathrm{Tr}\left[\gamma^\mu\,\frac{1}{\not{t}-\not{k}-m}\,\gamma^0\,\frac{1}{\not{t}-\not{k}+\not{q}-m}\,\gamma^0\,\frac{1}{\not{t}-\not{k}-m}\,\gamma_\mu\,\frac{\gamma^0+1}{4}\right], \end{aligned} \tag{15.27}$$

where we have used the trivial algebraic identity $64\pi^2/(2\pi)^4 = 4/\pi^2$. The reduced-mass dependence of the energy shift has been obtained consistently. In order to show that the integral K_a is dimensionless, we observe that the propagator denominators $\not{t}-\not{k}-m$ and $\not{t}-\not{k}+\not{q}-m$ carry powers of energy, respectively, which are canceled against the $\int \mathrm{d}^3q$ integral. The factor m^2/k^2 is dimensionless, as is the expression $\int \mathrm{d}^4k/\vec{q}^{\,4}$.

15.3.2 *Forward Scattering Amplitudes*

In Eq. (15.25), we have derived the energy shift ΔE_a, unrenormalized, for the diagram in Fig. 15.2(a). The total energy shift, due to the subdiagrams (a)–(d), is given as

$$\Delta E = \frac{\alpha}{\pi}\,\frac{(Z\alpha)^5\,m}{n^3}\left(\frac{\mu}{m}\right)^3 \delta_{\ell 0}\,(4\pi\mathcal{K})\,, \qquad \mathcal{K} = K_a + K_b + K_c + K_d\,, \tag{15.28}$$

where the integral K_a is given in Eq. (15.27). We compile the unrenormalized expressions for the dimensionless terms K_a and K_b, corresponding to the subdiagrams (a) and (b) in Fig. 15.2, respectively, employing both an infrared regulator mass λ as well as an ultraviolet regulator Λ, in the photon propagator,

$$K_a = \int \frac{\mathrm{d}^3q}{(2\pi)^3}\,\frac{1}{\vec{q}^{\,4}} \int_{C_F} \frac{\mathrm{d}^4k}{\pi^2\mathrm{i}} \left(\frac{1}{k^2-\lambda^2} - \frac{1}{k^2-\Lambda^2}\right) T_a(q,k)\,,$$

$$T_a(q,k) = m^2\,\mathrm{Tr}\left[\gamma^\mu \frac{1}{\not{t}-\not{k}-m}\gamma^0 \frac{1}{\not{t}-\not{k}+\not{q}-m}\gamma^0 \frac{1}{\not{t}-\not{k}-m}\gamma_\mu \frac{\gamma^0+1}{4}\right], \tag{15.29a}$$

$$K_b = \int \frac{\mathrm{d}^3q}{(2\pi)^3}\,\frac{1}{\vec{q}^{\,4}} \int_{C_F} \frac{\mathrm{d}^4k}{\pi^2\mathrm{i}} \left(\frac{1}{k^2-\lambda^2} - \frac{1}{k^2-\Lambda^2}\right) T_b(q,k)\,,$$

$$T_b(q,k) = m^2\,\mathrm{Tr}\left[\gamma^0 \frac{1}{\not{t}+\not{q}-m}\gamma^\mu \frac{1}{\not{t}-\not{k}+\not{q}-m}\gamma_\mu \frac{1}{\not{t}+\not{q}-m}\gamma^0 \frac{\gamma^0+1}{4}\right]. \tag{15.29b}$$

The expressions for subdiagrams (c) and (d) of Fig. 15.2 are obtained by rearranging the order of the Dirac γ matrices appropriately,

$$K_c = \int \frac{\mathrm{d}^3 q}{(2\pi)^3} \frac{1}{\vec{q}^{\,4}} \int_{C_F} \frac{\mathrm{d}^4 k}{\pi^2 \mathrm{i}} \left(\frac{1}{k^2 - \lambda^2} - \frac{1}{k^2 - \Lambda^2} \right) T_c(q,k) ,$$

$$T_c(q,k) = m^2 \,\mathrm{Tr}\left[\gamma^0 \frac{1}{\not{t} + \not{q} - m} \gamma^\mu \frac{1}{\not{t} - \not{k} + \not{q} - m} \gamma^0 \frac{1}{\not{t} - \not{k} - m} \gamma_\mu \frac{\gamma^0 + 1}{4} \right] , \tag{15.29c}$$

$$K_d = \int \frac{\mathrm{d}^3 q}{(2\pi)^3} \frac{1}{\vec{q}^{\,4}} \int_{C_F} \frac{\mathrm{d}^4 k}{\pi^2 \mathrm{i}} \left(\frac{1}{k^2 - \lambda^2} - \frac{1}{k^2 - \Lambda^2} \right) T_d(q,k) ,$$

$$T_d(q,k) = m^2 \,\mathrm{Tr}\left[\gamma^\mu \frac{1}{\not{t} - \not{k} - m} \gamma^0 \frac{1}{\not{t} - \not{k} + \not{q} - m} \gamma_\mu \frac{1}{\not{t} + \not{q} - m} \gamma^0 \frac{\gamma^0 + 1}{4} \right] . \tag{15.29d}$$

Here, we recall that $t^\mu = (m, 0, 0, 0)$ is a constant time-like vector [see Eq. (15.19)], k is the photon momentum four-vector $k^\mu = (\omega, \vec{k})$, while $q^\mu = (0, \vec{q})$ is the Coulomb exchange momentum. One derives the following identity,

$$(t - k + q)^2 - m^2 = -(\vec{k} - \vec{q}\,)^2 - X^2 , \qquad X^2 = 2m\omega - \omega^2 . \tag{15.30}$$

Furthermore, one can separate the Dirac traces into numerator and denominator structures,

$$T_a = \frac{N_a}{D_a} , \qquad T_b = \frac{N_b}{D_b} , \qquad T_c = \frac{N_c}{D_c} , \qquad T_d = \frac{N_d}{D_d} . \tag{15.31}$$

The mass dimension of all given terms T_a through T_d is -1. They can be written as the ratio of a numerator and a denominator, of mass dimensions five and six, respectively. For the diagram in Fig. 15.2(a), we have

$$\begin{aligned} N_a = & -2\vec{k}^2 m^3 + 8m^5 + 6\vec{k}^2 m^2 \omega - 12 m^4 \omega \\ & + 6m^3\omega^2 + 2m^2\omega^3 - 4m^3 \vec{k}\cdot\vec{q} - 4m^2\omega\,\vec{k}\cdot\vec{q} , \end{aligned} \tag{15.32}$$

$$D_a = (\vec{k}^2 + X^2)^2 \,[-(\vec{k} - \vec{q})^2 - X^2] . \tag{15.33}$$

The expressions for the diagram in Fig. 15.2(b) reads as follows,

$$N_b = 8m^5 + 2m^3\vec{q}^{\,2} + 8m^4\omega + 2m^2\vec{q}^{\,2}\omega - 8m^3\vec{k}\cdot\vec{q} , \tag{15.34}$$

$$D_b = \vec{q}^{\,4}\,[-(\vec{k} - \vec{q})^2 - X^2] . \tag{15.35}$$

For the diagram in Fig. 15.2(c), one has

$$\begin{aligned} N_c = & -4\vec{k}^2 m^3 + 8m^5 + 2m^3\vec{q}^{\,2} - 8m^4\omega \\ & - 2m^2\vec{q}^{\,2}\omega - 4m^3\omega^2 + 4m^2\omega\,\vec{k}\cdot\vec{q} , \end{aligned} \tag{15.36}$$

$$D_c = -\vec{q}^{\,2}\,(-\vec{k}^2 - X^2)\,[-(\vec{k} - \vec{q})^2 - X^2] . \tag{15.37}$$

The expressions for the diagram in Fig. 15.2(d) are the same as for the diagram in Fig. 15.2(c), and hence,

$$N_d = N_c , \qquad D_d = D_c . \tag{15.38}$$

What remains is to do the renormalization, and the integration over the photon loop momentum.

15.3.3 Dispersion Relation and Subtractions

The structure of the $\mathcal{K}$ term defined in Eq. (15.28) is as follows,

$$\mathcal{K} = \int \frac{d^3q}{(2\pi)^3} \frac{1}{\vec{q}^4} \int_{C_F} \frac{d^4k}{\pi^2 i} \left(\frac{1}{k^2 - \lambda^2} - \frac{1}{k^2 - \Lambda^2} \right) \mathcal{T}(q,k)$$
$$= \int \frac{d^3q}{(2\pi)^3} \frac{1}{\vec{q}^4} \mathcal{R}(\vec{q}^2) . \tag{15.39}$$

Here, $\mathcal{T} = T_a + T_b + T_c + T_d$ is the sum of the terms T_a through T_d, given in Eq. (15.31), including renormalizations. Furthermore, we have defined

$$\mathcal{R}(\vec{q}^2) = \int_{C_F} \frac{d^4k}{i\pi^2} \left(\frac{1}{k^2 - \lambda^2} - \frac{1}{k^2 - \Lambda^2} \right) \mathcal{T}(q,k) \tag{15.40}$$

as the expression obtained after the photon four-momentum integral. We recall that we define $q^\mu = (0, \vec{q})$ as the four-momentum pertaining to the Coulomb exchange. Therefore, the result for $\mathcal{R}(\vec{q}^2)$ only depends on the spatial components. Furthermore, for symmetry reasons, $\mathcal{R}(\vec{q}^2)$ only depends on the modulus of the momentum transfer. The idea now is that we consider what would happen in case $\vec{q}^2 = -Q^2$ becomes negative. This would correspond to a time-like (as opposed to space-like) fictitious Coulomb photon entering the diagram. In this case, the virtual photon and/or the virtual electron (if Q^2 is sufficiently large) in the diagrams in Fig. 15.2 could become on-shell, and the energy shift would acquire an imaginary part.

This consideration allows us to write a dispersion relation for $\mathcal{R}(\vec{q}^2)$. We use a counter-clockwise contour C, and encircle the branch cut for negative $\vec{q}^2$ in the positive sense. The contour extends from $\vec{q}^2 = -\infty$ to the origin, above the cut. The second part of the contour extends from $\vec{q}^2 = 0$ to $-\infty$, below the cut. We then complete the contour around the pole and obtain

$$\int_C d(Q^2) \frac{\mathcal{R}(Q^2)}{Q^2 - \vec{q}^2} = 2\pi i \, \mathcal{R}(\vec{q}^2) , \tag{15.41}$$

where the position of $\vec{q}^2$ in the complex plane in this formula is anywhere except on the negative real axis. But then, the contribution from the circle at infinite radius does not contribute, because the integrand falls off sufficiently fast. Then, with an infinitesimal imaginary part $i\epsilon$, one has (see Fig. 15.3)

$$\int_C d(Q^2) \frac{\mathcal{R}(Q^2)}{Q^2 - \vec{q}^2} = \int_{-\infty}^{0} d(Q^2) \frac{\mathcal{R}(Q^2 + i\epsilon)}{Q^2 - \vec{q}^2} - \int_{-\infty}^{0} d(Q^2) \frac{\mathcal{R}(Q^2 - i\epsilon)}{Q^2 - \vec{q}^2} . \tag{15.42}$$

We make the change of variable $Q^2 \to -Q^2$ and normalize the cut of the integrand as follows,

$$\mathcal{R}_C(Q^2) = \frac{1}{2\pi i} \left[\mathcal{R}(-Q^2 + i\epsilon) - \mathcal{R}(-Q^2 - i\epsilon) \right] , \qquad Q^2 > 0 . \tag{15.43}$$

Taking all information from Eqs. (15.41), (15.42) and (15.43) into account, one can write the relation

$$\mathcal{R}(\vec{q}^2) = -\int_0^\infty d(Q^2) \frac{\mathcal{R}_C(Q^2)}{\vec{q}^2 + Q^2} . \tag{15.44}$$

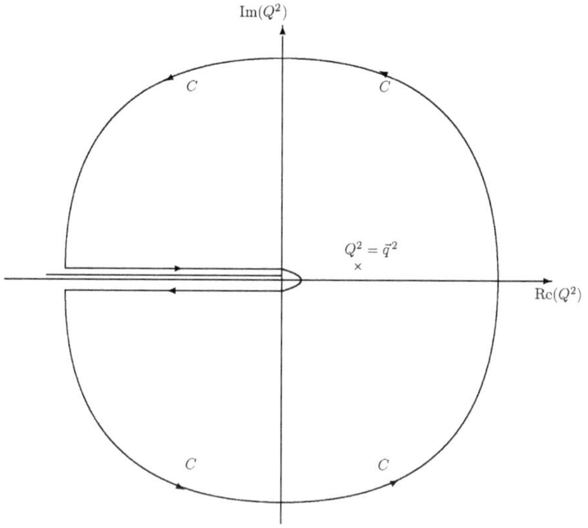

Fig. 15.3 The integration contour C used in Eqs. (15.41) and (15.42) encircles the pole at $Q^2 = q^2$ in the mathematically positive sense.

It comes as a surprise that, in order to calculate the energy shift according to Eq. (15.28), one needs to carry out additional subtractions. Namely, the formula (15.39) implies that the scaled dimensionless integral $\mathcal{K}$, being proportional to $\int \mathrm{d}^3q\,(1/\vec{q}^{\,4})\,\mathcal{R}(\vec{q})$, contains lower-order terms corresponding to double Coulomb exchange. We recall that the Coulomb potential, in momentum space, is proportional to $1/\vec{q}^{\,2}$. We have ignored the momentum of the incoming bound-state wave function, which otherwise acts as an infrared regulator in the calculation. Hence, the infrared divergences are naturally cut off at the bound-electron momentum scale, which is $Z\alpha\,m$. The integration of the infrared divergent terms with infrared cutoff would otherwise lead to contributions of proportional to $\int_{Z\alpha m}^{\infty} \mathrm{d}|\vec{q}|\,|\vec{q}|^{-2} \propto 1/(Z\alpha\,m)$, which is of lower order in the $(Z\alpha)$-expansion. Hence, we need to replace $\mathcal{K}$ by the subtracted integral

$$\mathcal{K}_{(1)} = \int \frac{\mathrm{d}^3q}{(2\pi)^3}\,\frac{1}{\vec{q}^{\,4}}\left[\mathcal{R}(\vec{q}^{\,2}) - \mathcal{R}(0)\right], \tag{15.45}$$

$$\mathcal{R}(\vec{q}^{\,2}) - \mathcal{R}(0) = \int_0^\infty \mathrm{d}(Q^2)\,\frac{\vec{q}^{\,2}\,\mathcal{R}_C(Q^2)}{Q^2(\vec{q}^{\,2} + Q^2)}\,. \tag{15.46}$$

The subtracted integral $\mathcal{K}_{(1)}$ evaluates to

$$\begin{aligned}
\mathcal{K}_{(1)} &= \int \frac{\mathrm{d}^3q}{(2\pi)^3} \int_0^\infty \mathrm{d}(Q^2)\,\frac{1}{\vec{q}^{\,2}(\vec{q}^{\,2} + Q^2)}\,\frac{\mathcal{R}_C(Q^2)}{Q^2} \\
&= \frac{4\pi}{8\pi^3}\int_0^\infty \mathrm{d}(Q^2)\int_0^\infty \mathrm{d}|\vec{q}|\,\frac{1}{\vec{q}^{\,2} + Q^2}\,\frac{\mathcal{R}_C(Q^2)}{Q^2} \\
&= \frac{4\pi}{8\pi^3}\int_0^\infty \mathrm{d}Q\,(2Q)\,\frac{\pi}{2Q}\,\frac{\mathcal{R}_C(Q^2)}{Q^2} = \frac{1}{2\pi}\int_0^\infty \mathrm{d}Q\,\frac{\mathcal{R}_C(Q^2)}{Q^2}\,.
\end{aligned} \tag{15.47}$$

There is another subtraction to be done in the Q-integral, which is also generated by a spurious infrared-divergent lower-order contribution. Infrared divergences proportional to $\int \mathrm{d}Q\, Q^2/Q^4 \to \int_{Z\alpha m}^{\infty} \mathrm{d}Q\, Q^{-2} \propto 1/(Z\alpha m)$ are cut off at the scale of the bound-state momentum. The infrared divergences describe lower-order terms in the $Z\alpha$-expansion. The end result is the double-subtracted integral

$$\mathcal{K}_{(2)} = \frac{1}{2\pi} \int_0^\infty \frac{\mathrm{d}Q}{Q^2} \left[\mathcal{R}_C(Q^2) - \mathcal{R}_C(0) \right]. \tag{15.48}$$

The energy shift (15.28) becomes

$$\Delta E = \frac{\alpha}{\pi} \frac{(Z\alpha)^5 m}{n^3} \left(\frac{\mu}{m} \right)^3 \delta_{\ell 0} \left(4\pi \mathcal{K}_{(2)} \right). \tag{15.49}$$

Our task now is to calculate $\mathcal{K}_{(2)}$, and that is not completely trivial.

15.3.4 *Example Integral*

Let us specialize Eq. (15.40) to diagram (c) of Fig. 15.2,

$$R_c(\vec{q}^{\,2}) = \int_{C_F} \frac{\mathrm{d}^4 k}{\mathrm{i}\pi^2} \left(\frac{1}{k^2 - \lambda^2} - \frac{1}{k^2 - \Lambda^2} \right) T_c(q,k)\,, \tag{15.50}$$

where T_c is given in Eqs. (15.31), (15.36) and (15.37). The diagram in Fig. 15.2(a) is ultraviolet finite. For the corresponding expressions for diagrams Fig. 15.2(b), Fig. 15.2(c), and Fig. 15.2(d), one has to subtract suitable integral representations of the mass counter term δm, and of the vertex renormalization Z_1.

After the renormalizations, one obtains ultraviolet convergent as well as infrared safe expressions, so that we can replace

$$\frac{1}{k^2 - \lambda^2} - \frac{1}{k^2 - \Lambda^2} \to \frac{1}{k^2} = \frac{1}{\omega^2 - \vec{k}^2}\,. \tag{15.51}$$

From diagram (c), one obtains a particularly instructive example integral whose cut, for negative $\vec{q}^{\,2} = -Q^2$, we would like to calculate for illustration. Let us define

$$\begin{aligned} J &= \frac{1}{\mathrm{i}\pi^2} \int_{C_F} \mathrm{d}^4 k \, \frac{1}{\omega^2 - \vec{k}^2} \, \frac{1}{X^2 + \vec{k}^2} \, \frac{1}{X^2 + (\vec{k} - \vec{q})^2} \\ &= \frac{1}{\mathrm{i}\pi^2} \int_{C_F} \mathrm{d}^4 k \, \frac{1}{k^2} \, \frac{1}{(t-k)^2 - m^2} \, \frac{1}{(t-k+q)^2 - m^2}\,, \end{aligned} \tag{15.52}$$

where we recall the definition of t from Eq. (15.19). We now use Feynman parameterization (see Sec. 10.3.5) in the form given in Eq. (10.159),

$$\frac{1}{ABC} = 2\int_0^1 \mathrm{d}x \int_0^x \mathrm{d}y \frac{1}{[A(1-x) + B(x-y) + Cy]^3} = 2\int_0^1 \mathrm{d}x \int_0^x \mathrm{d}y \frac{1}{[\mathcal{D}(x,y)]^3}\,, \tag{15.53}$$

where we implicitly define the denominator $\mathcal{D}(x,t)$. In our case, one has

$$A = k^2\,, \qquad B = (t-k)^2 - m^2\,, \tag{15.54}$$

$$C = (t+q-k)^2 - m^2 = (p-k)^2 - m^2\,, \qquad p \equiv t + q\,, \tag{15.55}$$

where we define p for our calculation. The denominator structure in Eq. (15.53) becomes, alternatively,

$$\begin{aligned}\mathcal{D} &= k^2(1-x) + [(k-t)^2 - m^2](x-y) + [(k-p)^2 - m^2]y \\ &= k^2(1-x+x-y+y) + (-2t\cdot k)(x-y) + (-2k\cdot p + p^2 - m^2]y \\ &= k^2 - 2k\cdot(t(x-y)+py) + (p^2-m^2).\end{aligned} \tag{15.56}$$

We now define

$$\widetilde{k} = k - (t(x-y)+py)\,, \tag{15.57}$$

to rewrite the denominator as

$$\begin{aligned}\mathcal{D} &= \widetilde{k}^2 - [t(x-y)+py]^2 + (p^2-m^2)y \\ &= \widetilde{k}^2 - m^2x^2 + (p^2-m^2)y - (p^2-m^2)y^2 \\ &= \widetilde{k}^2 - m^2x^2 + (m^2 - \vec{q}^{\,2} - m^2)y - (m^2 - \vec{q}^{\,2} - m^2)y^2 \\ &= \widetilde{k}^2 - \left[m^2x^2 + \vec{q}^{\,2}\, y\,(1-y)\right].\end{aligned} \tag{15.58}$$

We have used the relations

$$t^2 = m^2\,, \qquad t\cdot p = m^2\,, \qquad p^2 = m^2 - \vec{q}^{\,2}\,. \tag{15.59}$$

With the help of Eq. (10.138), one obtains the following integral for the total solid angle in Euclidean four-dimensional space,

$$\Omega_4 = \frac{2\pi^2}{\Gamma(2)} = 2\pi^2\,, \qquad \int_0^\infty \mathrm{d}k_E\, \frac{k_E^3}{(k_E^2+Y)^3} = \frac{1}{4Y}\,. \tag{15.60}$$

After a Wick rotation, $\int \mathrm{d}^4\widetilde{k} \to \mathrm{i}\int \mathrm{d}^4\widetilde{k}_E$ using the formula $\widetilde{k}^2 = -\widetilde{k}_E^2$, one obtains a convenient integral representation for J,

$$\begin{aligned}J &= \frac{2}{\mathrm{i}\pi^2}\int_0^1 \mathrm{d}x \int_0^x \mathrm{d}y \int \mathrm{d}^4\widetilde{k}\, \frac{1}{\left\{\widetilde{k}^2 - [m^2x^2 + \vec{q}^{\,2}\, y\,(1-y)]\right\}^3} \\ &= -\frac{2}{\pi^2}\int_0^1 \mathrm{d}x \int_0^x \mathrm{d}y \int \mathrm{d}^4\widetilde{k}_E\, \frac{1}{\left\{\widetilde{k}_E^2 + [m^2x^2 + \vec{q}^{\,2}\, y\,(1-y)]\right\}^3} \\ &= -\frac{2(2\pi^2)}{4\,\pi^2}\int_0^1 \mathrm{d}x \int_0^x \mathrm{d}y\, \frac{1}{m^2x^2 + \vec{q}^{\,2}\, y\,(1-y)} \\ &= -\int_0^1 \mathrm{d}x \int_0^1 \mathrm{d}s\, \frac{1}{m^2x + \vec{q}^{\,2}\, s\,(1-xs)}\,.\end{aligned} \tag{15.61}$$

In the last step, we have substituted $y = xs$, so that the integration interval becomes symmetric in x and s. It is clear that, for $\vec{q}^{\,2} = q^2 = -Q^2 < 0$, the above integral develops an imaginary part.

In the same spirit as given by Eqs. (15.41)–(15.43), we can write a dispersion relation

$$J_C(Q^2) = \frac{1}{2\pi\mathrm{i}}\left[J(-Q^2+\mathrm{i}\epsilon) - J(-Q^2-\mathrm{i}\epsilon)\right], \qquad Q^2 > 0\,, \tag{15.62a}$$

$$J(q^2) = -\int_0^\infty \mathrm{d}(Q^2)\, \frac{J_C(Q^2)}{q^2+Q^2}\,. \tag{15.62b}$$

Let us evaluate $J_C(Q^2)$ with care, with the help of the Sochocki–Plemelj prescription (4.356) (see Chap. 4),

$$
\begin{aligned}
J_C(Q^2) &= -\frac{1}{2\pi \mathrm{i}} \int_0^1 \mathrm{d}x \int_0^1 \mathrm{d}s \left(\frac{1}{m^2 x + (-Q^2 + \mathrm{i}\epsilon)\, s\,(1 - xs)} - (\mathrm{i}\epsilon \to -\mathrm{i}\epsilon) \right) \\
&= -\frac{1}{2\pi \mathrm{i}} \int_0^1 \mathrm{d}x \int_0^1 \mathrm{d}s \left(\frac{1}{m^2 x - Q^2 s\,(1 - xs) + \mathrm{i}\epsilon} - (\mathrm{i}\epsilon \to -\mathrm{i}\epsilon) \right) \\
&= \int_0^1 \mathrm{d}x \int_0^1 \mathrm{d}s\, \delta(m^2 x - Q^2 s\,(1 - xs)) \\
&= \int_0^1 \mathrm{d}s \int_0^1 \mathrm{d}x\, \delta(x(m^2 + Q^2 s^2) - sQ^2))\,. \qquad (15.63)
\end{aligned}
$$

We have used the fact that $s\,(1 - xs) \geq 0$ inside the integration interval. Now, let us see. In the integral over x, the Dirac-δ function peaks at

$$
x = x_0 = \frac{sQ^2}{m^2 + Q^2 s^2}\,, \qquad \delta(x(m^2 + Q^2 s^2) - sQ^2)) = \frac{1}{m^2 + Q^2 s^2}\, \delta(x - x_0)\,. \quad (15.64)
$$

The question now is whether or not the point x_0 still lies inside the integration interval $x \in (0,1)$. We thus need to consider the function

$$
J_C(Q^2) = \int_0^1 \mathrm{d}s \frac{1}{m^2 + Q^2 s^2} \Theta\left(1 - \frac{sQ^2}{m^2 + Q^2 s^2} \right). \qquad (15.65)
$$

Here, Θ is the step function. Let us consider the expression

$$
f(s,Q) = \frac{sQ^2}{m^2 + Q^2 s^2}\,, \qquad f(0,Q) = 0\,, \qquad f(1,Q) = \frac{Q^2}{m^2 + Q^2} \leq 1\,. \qquad (15.66)
$$

This means that, for $s = 0$ as well as $s = 1$, the Θ function in Eq. (15.65) is equal to unity. However, for sufficiently large Q, there is a region $s_1 < s < s_2$ where the expression $f(s,Q)$ is larger than unity. One finds that

$$
s = s_1 = \frac{1}{2} - \frac{1}{2}\sqrt{1 - \frac{4m^2}{Q^2}}\,, \qquad s = s_2 = \frac{1}{2} + \frac{1}{2}\sqrt{1 - \frac{4m^2}{Q^2}}\,. \qquad (15.67)
$$

Interestingly, the solutions exist only above the fermionic pair production threshold $Q > 2m$. This means that we must exclude, from the integration interval $s \in (0,1)$, the interval $s \in (s_1, s_2)$, but only if $Q > 2m$. So,

$$
J_C(Q^2) = \int_0^1 \mathrm{d}s \frac{1}{m^2 + Q^2 s^2} - \Theta(Q - 2m) \int_{s_1}^{s_2} \mathrm{d}s \frac{1}{m^2 + Q^2 s^2}\,. \qquad (15.68)
$$

The result can be expressed as follows,

$$
J_C(Q^2) = \frac{1}{mQ} \operatorname{arccot}\left(\frac{m}{Q} \right) - \Theta(Q - 2m) \frac{1}{mQ} \arccos\left(\frac{2m}{Q} \right). \qquad (15.69)
$$

The formula

$$
\arccos(y) = \frac{\mathrm{i}}{2} \ln\left(\frac{y - \mathrm{i}\sqrt{1 - y^2}}{y + \mathrm{i}\sqrt{1 - y^2}} \right) \qquad (15.70)
$$

is useful in the final simplifications.

All integrals pertaining to the renormalized integrand $\mathcal{R}(\vec{q}^2)$ can be evaluated using generalizations of the above result. Let us take, for example, the integral

$$\widetilde{J}(\vec{q}^2) = \int \frac{\mathrm{d}^4 k}{\pi^2 \mathrm{i}} \frac{1}{\omega^2 - k^2} \frac{1}{[\vec{k}^2 + X^2]^2} \frac{\omega \vec{k} \cdot \vec{q}}{(\vec{k} - \vec{q})^2 + X^2}, \tag{15.71}$$

which is obtained if one substitutes, for $\mathcal{T} = T_a + T_b + T_c + T_d$ in Eq. (15.40), terms from $T_a = N_a / D_a$, given in Eqs. (15.32) and (15.33). The cut of $\widetilde{J}$ is obtained as $\widetilde{J}_C(Q^2) = (2\pi\mathrm{i})^{-1} [\widetilde{J}(-Q^2 + \mathrm{i}\,\epsilon) - \widetilde{J}(-Q^2 - \mathrm{i}\,\epsilon)]$, where

$$\widetilde{J}_C(Q^2) = -\frac{Q^2}{4m} J_C(Q^2) + \frac{Q^2}{4m(m^2 + \vec{Q}^2)} + \Theta(Q - 2m) \frac{m}{Q^2} \sqrt{1 - \frac{4m^2}{Q^2}}. \tag{15.72a}$$

For other integrals, a partial fraction decompositions like

$$\frac{\omega}{(\omega^2 - \vec{k}^2)(\vec{k}^2 + X^2)} = \frac{1}{2m} \left(\frac{1}{\omega^2 - \vec{k}^2} + \frac{1}{\vec{k}^2 + X^2} \right) \tag{15.72b}$$

help in simplifying some of the integrals.

15.3.5 *Final Integration and A_{50} Coefficient*

One finally obtains the following results for the cut functions, defined in Eq. (15.44), for the diagrams in Fig. 15.2. The diagram is indicated as a superscript in the results

$$\begin{aligned} \mathcal{R}_C^{(a)}(Q^2) = & -m \frac{8m^6 + 18m^4 Q^2 + 12m^2 Q^4 + 3Q^6}{Q^2 (m^2 + Q^2)^2} \\ & + \frac{4m(-4m^4 - m^2 Q^2 + Q^4)\,\Theta(Q - 2m)}{Q^4 \sqrt{1 - \frac{4m^2}{Q^2}}} + m(8m^2 - Q^2)\, J_C(Q^2), \end{aligned} \tag{15.73}$$

$$\mathcal{R}_C^{(b)}(Q^2) = -\frac{m^3 (8m^4 + 10 m^2 Q^2 + Q^4)}{Q^2 (m^2 + Q^2)^2}, \tag{15.74}$$

$$\mathcal{R}_C^{(c)}(Q^2) = \frac{m^3 (4m^2 + 5Q^2)}{(m^2 + Q^2)^2} - \frac{4m\left(1 - \frac{4m^2}{Q^2}\right)\Theta(Q - 2m)}{(Q^2/m^2)\sqrt{1 - \frac{4m^2}{Q^2}}} - \frac{2m^3 (Q^2 - 4m^2)}{Q^2} J_C(Q^2), \tag{15.75}$$

$$\mathcal{R}_C^{(d)}(Q^2) = \mathcal{R}_C^{(c)}(Q^2). \tag{15.76}$$

The J_C function is given in Eq. (15.69). The function

$$\mathcal{R}_C = \mathcal{R}_C^{(a)} + \mathcal{R}_C^{(b)} + \mathcal{R}_C^{(c)} + \mathcal{R}_C^{(d)} \tag{15.77}$$

has mass dimension one and is given as

$$\begin{aligned} \mathcal{R}_C(Q^2) = & -m \frac{16m^4 + 20m^2 Q^2 + 3Q^4}{Q^2 (m^2 + Q^2)} + m \frac{4(Q^2 - 3m^2)(Q^2 + 4m^2)\Theta(Q - 2m)}{Q^4 \sqrt{1 - (4m^2)/Q^2}} \\ & + \frac{m(16m^4 + 4m^2 Q^2 - Q^4)}{Q^2} J_C(Q^2). \end{aligned} \tag{15.78}$$

It has the property

$$\mathcal{R}_C(Q^2) = -\frac{16}{3}m + \mathcal{O}(Q^2)\,, \tag{15.79}$$

which enters the formula (15.48). We recall from Eq. (15.48) that

$$\mathcal{K}_{(2)} = \frac{1}{2\pi}\int_0^\infty \frac{\mathrm{d}Q}{Q^2}\left[\mathcal{R}_C(Q^2) - \mathcal{R}_C(0)\right]. \tag{15.80}$$

The energy shift (15.49) can be written as

$$\Delta E = \frac{\alpha}{\pi}\frac{(Z\alpha)^5\, m}{n^3}\left(\frac{\mu}{m}\right)^3 A_{50}\,, \tag{15.81}$$

where the dimensionless coefficient A_{50} is

$$A_{50} = 4\pi\mathcal{K}_{(2)}\delta_{\ell 0} = 2\delta_{\ell 0}\int_0^\infty \frac{\mathrm{d}Q}{Q^2}\left[\mathcal{R}_C(Q^2) - \mathcal{R}_C(0)\right]. \tag{15.82}$$

The notation adopted for the coefficient A_{50} is explained in the following Sec. 15.4. The coefficient A_{50} can be expressed as a sum of a contribution $\mathcal{X}_1$ from the integration interval $0 < Q < 2m$, and a contribution $\mathcal{X}_2$ from the integration interval $2m < Q < \infty$,

$$\mathcal{X}_1 = \frac{5}{3} + \frac{9}{2}\arctan(2) + 2\mathrm{i}\,\chi_2(2\mathrm{i})\,, \tag{15.83}$$

$$\mathcal{X}_2 = -\frac{5}{3} + \frac{139\pi}{32} - \frac{9}{2}\arctan(2) - 2\pi\ln(2) - 2\mathrm{i}\,\chi_2(2\mathrm{i})\,. \tag{15.84}$$

Here, the Legendre χ function has been used,

$$\chi_\nu(z) = \frac{1}{2}\left[\mathrm{Li}_\nu(z) - \mathrm{Li}_\nu(-z)\right]. \tag{15.85}$$

In the end result, all the terms proportional to the Legendre χ cancel, and we have

$$A_{50} = (\mathcal{X}_1 + \mathcal{X}_2)\,\delta_{\ell 0} = 4\pi\,\delta_{\ell 0}\left(\frac{139}{128} - \frac{1}{2}\ln(2)\right) = \delta_{\ell 0}\,9.29112\ldots\,, \tag{15.86}$$

a result first derived in Ref. [340].

15.4 Higher-Order Terms

In the preceding discussion, we have obtained the self-energy binding correction to the Lamb shift of order $\alpha(Z\alpha)^5$ in Eq. (15.86), based on the analysis of the Feynman diagrams in Fig. 15.1, using dispersion relations. Technical difficulties mount as the power of $Z\alpha$ is being increased (this concerns the number of Coulomb vertices in the high-energy part, and the order of the multipole expansion in the low-energy part of the bound-state self-energy). The higher-order binding corrections in the order $\alpha(Z\alpha)^6 m$, for a number of excited states, were calculated in Refs. [163, 173, 337, 341–352].

We here concentrate on the $n = 1$ and $n = 2$ states, just to give a brief review of the results obtained over the last couple of decades. The (real part of the) energy shift ΔE_{SE} can be written as follows [540],

$$\Delta E_{\mathrm{SE}} = \frac{\alpha}{\pi} \frac{(Z\alpha)^4 m}{n^3} F(nL_j, Z\alpha), \tag{15.87}$$

where we assume the non-recoil limit. The semi-analytic expansion of the scaled self-energy function $F(nL_j, Z\alpha)$ about $Z\alpha = 0$ for a general atomic state with quantum numbers nL_j (in spectroscopic notation) is

$$\begin{aligned} F(nL_j, Z\alpha) = {} & A_{41}(nL_j) \ln[(Z\alpha)^{-2}] + A_{40}(nL_j) + (Z\alpha)\, A_{50}(nL_j) \\ & + (Z\alpha)^2 \left[A_{62}(nL_j) \ln^2[(Z\alpha)^{-2}] + A_{61}(nL_j) \ln[(Z\alpha)^{-2}] \right. \\ & \left. + G_{\mathrm{SE}}(nL_j, Z\alpha)\right] . \end{aligned} \tag{15.88}$$

Here, $G_{\mathrm{SE}}(nL_j, Z\alpha)$ is the so-called self-energy remainder function. In the limit of a vanishing nuclear charge, $Z \to 0$, one has $G_{\mathrm{SE}}(nL_j, Z\alpha) \to A_{60}(nL_j)$, which is another state-dependent coefficient. The A coefficients have two indices, the first of which denotes the power of $Z\alpha$ [including those powers implicitly contained in Eq. (15.87)], while the second index denotes the power of the logarithm $\ln[(Z\alpha)^{-2}]$. For P states, the coefficients A_{41}, A_{50} and A_{62} vanish, and we have

$$F(nP_j, Z\alpha) = A_{40}(nP_j) + (Z\alpha)^2 \left[A_{61}(nP_j) \ln[(Z\alpha)^{-2}] + G_{\mathrm{SE}}(nP_j, Z\alpha)\right] . \tag{15.89}$$

The higher-order binding corrections A_{62} and A_{61} were evaluated in [163, 341–349]. Together with results compiled in Chap. 4 and in Sec. 15.3, we have for the $1S$ ground state of hydrogen,

$$\begin{aligned} & A_{41}(1S_{1/2}) = \frac{4}{3}, \qquad A_{40}(1S_{1/2}) = \frac{10}{9} - \frac{4}{3} \ln k_0(1S), \\ & A_{50}(1S_{1/2}) = 4\pi \left[\frac{139}{128} - \frac{1}{2} \ln 2\right], \quad A_{62}(1S_{1/2}) = -1, \quad A_{61}(1S_{1/2}) = \frac{28}{3} \ln 2 - \frac{21}{20}. \end{aligned} \tag{15.90}$$

The numerical value of the Bethe logarithm $\ln k_0(1S)$ is given in Eq. (4.1) (to nine decimals) and in Eq. (4.337) to a very large numerical accuracy. For the metastable $2S$ (excited) state, we have

$$\begin{aligned} & A_{41}(2S_{1/2}) = \frac{4}{3}, \qquad A_{40}(2S_{1/2}) = \frac{10}{9} - \frac{4}{3} \ln k_0(2S), \\ & A_{50}(2S_{1/2}) = 4\pi \left[\frac{139}{128} - \frac{1}{2} \ln 2\right], \quad A_{62}(2S_{1/2}) = -1 \quad A_{61}(2S_{1/2}) = \frac{16}{3} \ln 2 + \frac{67}{30}. \end{aligned} \tag{15.91}$$

for the numerical value of $\ln k_0(2S)$, we refer to Eq. (4.337) and to Table 4.1. For the $2P_{1/2}$ and $2P_{3/2}$ states, we have the coefficients [173]

$$A_{40}(2P_{1/2}) = -\frac{1}{6} - \frac{4}{3} \ln k_0(2P), \qquad A_{61}(2P_{1/2}) = \frac{103}{108}, \tag{15.92}$$

$$A_{40}(2P_{3/2}) = \frac{1}{12} - \frac{4}{3} \ln k_0(2P), \qquad A_{61}(2P_{3/2}) = \frac{29}{90}. \tag{15.93}$$

Table 15.2 We present a table of A_{60} coefficients for higher excited atomic states with positive Dirac angular quantum number κ (i.e., $j = \ell - 1/2$). Data are taken from Ref. [352]. All decimal figures shown are significant.

A_{60} Coefficients for States with $\kappa > 0$

n	$A_{60}(n\mathrm{P}_{1/2})$	$A_{60}(n\mathrm{D}_{3/2})$	$A_{60}(n\mathrm{F}_{5/2})$	$A_{60}(n\mathrm{G}_{7/2})$
2	−0.998 904 402	–	–	–
3	−1.148 189 956	0.005 551 573	–	–
4	−1.195 688 142	0.005 585 985	0.002 326 988	–
5	−1.216 224 512	0.006 152 175	0.002 403 151	0.000 814 415
6	−1.226 702 391	0.006 749 745	0.002 531 636	0.000 827 468
7	−1.232 715 957	0.007 277 403	0.002 661 311	0.000 857 346

Table 15.3 In the analog of Table 15.1 for states with negative κ, we combine data from Ref. [352] with calculations reported in Ref. [312].

A_{60} Coefficients for States with $\kappa < 0$

n	$A_{60}(n\mathrm{S}_{1/2})$	$A_{60}(n\mathrm{P}_{3/2})$	$A_{60}(n\mathrm{D}_{5/2})$	$A_{60}(n\mathrm{F}_{7/2})$	$A_{60}(n\mathrm{G}_{9/2})$
1	−30.924 149 46	–	–	–	–
2	−31.840 465 09	−0.503 373 465	–	–	–
3	−31.702 501	−0.597 569 388	0.027 609 989	–	–
4	−31.561 922	−0.630 945 795	0.031 411 862	0.007 074 961	–
5	−31.455 393	−0.647 013 508	0.033 077 570	0.008 087 015	0.002 412 929
6	−31.375 130	−0.656 154 893	0.033 908 493	0.008 610 109	0.002 748 250
7	−31.313 224	−0.662 027 568	0.034 355 926	0.008 906 989	0.002 941 334

The evaluation of the A_{60} coefficient presented nearly insurmountable technical challenges in its evaluation. Ground-breaking work by PACHUCKI[2] (Ref. [163]) led the pathway to a resolution. The calculation of the A_{60} coefficient necessitates, among other things, the evaluation of so-called relativistic Bethe logarithms, which are generated by the relativistic corrections to the wave function. In order to derive the corresponding expressions, one consults Chap. 11 for the relativistic corrections to the Schrödinger–Coulomb Hamiltonian and applies the corrections to the wave function, and the Hamiltonian and reference-state energy in the propagator of the

[2]Krzysztof Pachucki (b. 1963).

nonrelativistic bound-state self-energy, as discussed in Sec. 4.6.2. From Refs. [267, 312], we have the numerically precise results

$$A_{60}(1S) = -30.924\,149\,46(1)\,, \qquad A_{60}(2S) = -31.840\,465\,09(1)\,. \tag{15.94}$$

From Ref. [351], we have the following results for P states,

$$A_{60}(2P_{1/2}) = -0.998\,904\,402(1)\,, \qquad A_{60}(2P_{3/2}) = -0.503\,373\,465(1)\,. \tag{15.95}$$

At nuclear charge $Z = 1$, we have the following results for $n = 1$ and $n = 2$ states, from the high-precision numerical study reported in Ref. [540]:

$$G_{\rm SE}(1S_{1/2}, Z = 1) = -30.290\;24(2)\,, \tag{15.96a}$$

$$G_{\rm SE}(2S_{1/2}, Z = 1) = -31.185\;15(9)\,, \tag{15.96b}$$

$$G_{\rm SE}(2P_{1/2}, Z = 1) = -0.973\;5(2)\,, \tag{15.96c}$$

$$G_{\rm SE}(2P_{3/2}, Z = 1) = -0.486\;5(2)\,. \tag{15.96d}$$

For higher excited S states, results have been indicated in Ref. [541], while for higher excited P states, one consults Ref. [542]. The calculation of the nonperturbative remainder function, in the expansion parameter $Z\alpha$, necessitates the use of advanced numerical techniques [543] and a suitable formulation of the numerical problem, in terms of virtual photon integration contours [257, 258].

15.5 Relativistic Recoil Correction

15.5.1 *Retardation and Salpeter Correction*

The physical essence of the relativistic recoil correction to atomic energy levels, which was calculated by Salpeter [534], is understood as follows. We had already stressed in our derivation of the Breit Hamiltonian in Sec. 12.3.3 [see the discussion following Eqs. (12.47)] that the poles of the photon propagator are being ignored in the derivation of the effective Hamiltonian. This leads, in the end, to a static Hamiltonian, beyond which we encounter self-energy and recoil effects. Expressed differently, the Fourier transform of a constant in frequency space has a Dirac-δ dependence in time, i.e., it is proportional to $\delta(t - t')$, i.e., to a non-retarded, instantaneous interaction. Retardation effects are induced by an expansion about the case of a constant in frequency space and correspond to the nontrivial ω-dependence of the propagator denominators.

In our discussions of the van der Waals and Casimir–Polder interactions and of the short-range and long-range forms of the atom-surface interactions (see Chap. 5), we took into account the fact that the atom needs to undergo a virtual transition to an excited state in order to be subject to any interaction at all; all results were proportional either to the static, or to the dynamic, polarizability. The necessity to consider retardation arose from the failure of the expansions of the "retardation exponentials" $\exp(-\omega R/c)$ and $\exp(-\omega z/c)$ in terms of their arguments, in the long-distance limits (see Sec. 5.4.1). In the case of photon exchange among two constituent particles of an atom, the situation is different. The "retardation exponential" is of the form $\exp(\mathrm{i}\vec{k}\cdot\vec{r})$, where $\vec{k}$ is the modulus of the wave vector of the exchanged photon, and $\vec{r}$ is an atomic coordinate. For a Coulomb exchange, we have $|\vec{k}| \sim Z\alpha\mu$ and $|\vec{r}| \sim 1/(Z\alpha\mu)$, and thus, the "retardation expansion" is unavailable for the higher-order terms beyond the Breit interaction. Note that the low-energy part of the self-energy correction discussed in Sec. 4.6 is mediated by even softer photons, with $|\vec{k}| \sim (Z\alpha)^2\mu$, whereas the Breit interaction and the dominant part of the Salpeter correction correspond to photons with $|\vec{k}| \sim Z\alpha\mu$.

The question then is, to which extent we can understand the Salpeter correction as the retardation correction to the Breit interaction. The answer is that the retardation expansion here comes from the fact that the exchange photon energy is large as compared to a typical virtual transition frequency of the atom. In the propagator denominators, we can thus expand

$$\frac{1}{H - E + \omega} \to \frac{1}{\omega} - \frac{H - E}{\omega^2} + \cdots, \tag{15.97}$$

where $\omega = |\vec{k}| \equiv k \sim Z\alpha\mu$ is the virtual photon frequency. We had already pointed out that within the retardation expansion, which is relevant to the middle-energy part of the recoil correction, the multipole expansion, i.e., the expansion of $exp(\mathrm{i}\vec{k}\cdot\vec{r})$ in powers of its argument, remains unavailable. However, instead, the expansion (15.97) leads to a tremendous simplification of the calculation.

We aim to calculate the energy shift of order $(Z\alpha)^5\,\mu^3/(m_1\,m_2)$, beyond the Breit interaction, as given by the diagrams in Fig. 15.4. This recoil correction is exclusively given by one- and two-photon exchange between the constituent particles of the system. We have to use a manifestly two-body QED Hamiltonian and analyze the exchange diagrams between the two particles. We denote by $e = -|e|$ the electron charge and consider one particle of charge $q_1 = e$, and assume there is another spin-1/2 particle (the nucleus) of charge $q_2 = -Ze$. For the low-energy part of the effect, we can use the total Hamiltonian of the two-body system as follows,

$$\begin{aligned} H = {}& \frac{(\vec{p}_1 - e\vec{A}(\vec{r}_1))^2}{2m_1} + \frac{(\vec{p}_2 + Ze\vec{A}(\vec{r}_2))^2}{2m_2} - \frac{Z\alpha}{r} \\ & - \frac{e}{2m_1}\vec{\sigma}_1\cdot\vec{B}(\vec{r}_1) + \frac{Ze}{2m_2}\vec{\sigma}_2\cdot\vec{B}(\vec{r}_2) + \sum_{\vec{k}\lambda} k\, \mathrm{a}^+_{\vec{k}\lambda}\, \mathrm{a}_{\vec{k}\lambda}\,. \end{aligned} \tag{15.98}$$

Here, the masses of the two particles are m_1 and m_2, and we define $r = |\vec{r}|$ where $\vec{r} = \vec{r}_1 - \vec{r}_2$. Also, $\vec{\sigma}_1$ and $\vec{\sigma}_2$ are the Pauli spin matrices for the two particles. The photon energy is $k = |\vec{k}|$. The unperturbed Hamiltonian is

$$H_0 = \frac{\vec{p}_1^{\,2}}{2m_1} + \frac{\vec{p}_2^{\,2}}{2m_2} - \frac{Z\alpha}{r} + \sum_{\vec{k}\lambda} k\, \mathrm{a}^+_{\vec{k}\lambda}\, \mathrm{a}_{\vec{k}\lambda}\,. \tag{15.99}$$

We now follow the discussion in Chap. 4 [see Eq. (4.1)] and in Secs. 13.2.2 and 13.2.3 and express H_0 in center-of-mass coordinates ($\vec{p}_1 = -\vec{p}_2$) and the sum of the Schrödinger and field Hamiltonians,

$$H_0 = \frac{\vec{p}^{\,2}}{2\mu} - \frac{Z\alpha}{r} + \sum_{\vec{k}\lambda} k\, \mathrm{a}^+_{\vec{k}\lambda}\, \mathrm{a}_{\vec{k}\lambda} = H_S + H_F\,. \tag{15.100}$$

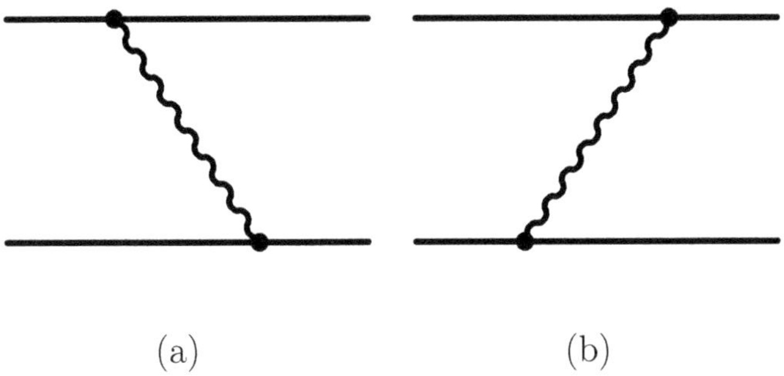

Fig. 15.4 The recoil correction to the Lamb shift involves a simple exchange of a virtual photons, but the two exchanges happen at manifestly different points in time; the two-body system is in a virtual excited state in between the exchanges, as displayed by the non-vertical photon lines. If the lines are made vertical and the contributions from the two diagrams are added, then the transverse part of the Breit interaction is recovered.

The reduced mass is $\mu = m_1 m_2/(m_1+m_2)$, and $\vec{p} = -\mathrm{i}\,\frac{\partial}{\partial\vec{r}}$ is the canonical momentum for the relative coordinate. Once a matrix element is formulated in terms of $\vec{r}$ only (no explicit appearance of $\vec{r}_1$ and $\vec{r}_2$), we can immediately replace $\vec{p}_1 \to \vec{p}$ and $\vec{p}_2 \to -\vec{p}$. The normalized eigenfunctions $\phi_n(\vec{r}) \equiv \psi_{n\ell m}(\vec{r})$ [see Eq. (4.60)], have the bound-state energies (n is a multi-index in our notation)

$$E_n = -\frac{(Z\alpha)^2\,\mu}{2\,n^2}. \tag{15.101}$$

The interaction Hamiltonian for a system of two spin-1/2 particles is given as follows,

$$\begin{aligned} H_I = & -\frac{e}{m_1}\,\vec{p}_1\cdot\vec{A}(\vec{r}_1) + \frac{Ze}{m_2}\,\vec{p}_2\cdot\vec{A}(\vec{r}_2) + \frac{e^2}{2m_1}\,\vec{A}(\vec{r}_1)^2 + \frac{(Ze)^2}{2m_2}\,\vec{A}(\vec{r}_2)^2 \\ & - \frac{e}{2m_1}\vec{\sigma}_1\cdot\vec{B}(\vec{r}_1) + \frac{Ze}{2m_2}\vec{\sigma}_2\cdot\vec{B}(\vec{r}_2)\,, \end{aligned} \tag{15.102}$$

where we courageously ignore the necessity to hermitize the interaction by anticommutators, $\vec{p}_1\cdot\vec{A}(\vec{r}_1) \to \frac{1}{2}\{\vec{p}_1, \vec{A}(\vec{r}_1)\}$, which would otherwise lead to a more clumsy notation. However, the sequence of the operators actually does not matter in the derivations to follow (up to the order of approximation relevant to our studies here). The details are left as an exercise to the reader.

15.5.2 *Low-Energy Part and Dipole Approximation*

The unperturbed state consists of the atom in the state $\phi_n(\vec{r})$ and zero photons in the radiation field. We know that the second-order perturbation with two couplings $\vec{p}_1\cdot\vec{A}(\vec{r}_1)$ and $\vec{p}_1\cdot\vec{A}(\vec{r}_1)$ affecting the same charged particle, yield self-energies already discussed in Sec. 4.6.2. The low-energy part is mediated by long-wavelength photons where we can still use the dipole approximation. In the extreme low-energy region, we can ignore the magnetic couplings from Eq. (15.102). The exchange contribution from the diagram in Fig. 15.4(a) is

$$\Delta E_a = \left\langle \left(-\frac{e}{m_1} \vec{p}_1 \cdot \vec{A}(\vec{r}_1) \right) \frac{1}{E_0 - H_0} \left(\frac{Ze}{m_2} \vec{p}_2 \cdot \vec{A}(\vec{r}_2) \right) \right\rangle . \tag{15.103}$$

We recall the vector field operator given in Eq. (4.309), within the dipole approximation,

$$\vec{A}(\vec{r}) \approx \vec{A}(\vec{r} = \vec{0}) = \sum_{\vec{k}\lambda} \frac{1}{\sqrt{2\omega_{\vec{k}} V}} \hat{\epsilon}_\lambda(\vec{k}) \left(\mathrm{a}_\lambda(\vec{k}) + \mathrm{a}^+_\lambda(\vec{k}) \right) . \tag{15.104}$$

In full analogy to Eqs. (4.289) and (4.311), one then obtains the low-energy part of the energy shift

$$\begin{aligned}
\Delta E_a &= -\frac{Ze^2}{m_1 m_2} \sum_m \sum_{\vec{k}\lambda} \frac{\epsilon^i_\lambda(\vec{k})\, \epsilon^j_\lambda(\vec{k})}{2\,k\,V} \frac{\left\langle \phi_n \left| p^i_1 \right| \phi_m \right\rangle \left\langle \phi_m \left| p^j_2 \right| \phi_n \right\rangle}{E_n - E_m - k} \\
&= -\frac{Ze^2}{m_1 m_2} \sum_m \int \frac{\mathrm{d}^3 k}{(2\pi)^3} \frac{1}{2\,k} \left(\delta^{ij} - \frac{k^i k^j}{\vec{k}^2} \right) \frac{\left\langle \phi_n \left| p^i_1 \right| \phi_m \right\rangle \left\langle \phi_m \left| p^j_2 \right| \phi_n \right\rangle}{E_n - E_m - k} \\
&= \frac{2}{3} \frac{Z\alpha}{\pi m_1 m_2} \int \mathrm{d}k\, k \left\langle \phi_n \left| p^i \frac{1}{E_n - H_S - k} p^i \right| \phi_n \right\rangle .
\end{aligned} \tag{15.105}$$

The second diagram leads to an equal energy shift $\Delta E_b = \Delta E_a$. We take the integration limits for the photon energy as zero and $Z\alpha\mu\eta$, where η is a dimensionless scale-separation parameter. So, the whole low-energy part is $\Delta E_L = \Delta E_a + \Delta E_b$ with

$$\begin{aligned}
\Delta E_L &= -\frac{4}{3} \frac{Z\alpha}{\pi m_1 m_2} \int_0^{Z\alpha\mu\eta} \mathrm{d}k\, k \left\langle \phi_n \left| p^i \frac{1}{H_S - E_n + k} p^i \right| \phi_n \right\rangle \\
&= -\frac{4}{3} \frac{Z\alpha}{\pi m_1 m_2} \int_0^{Z\alpha\mu\eta} \mathrm{d}k \left\langle \phi_n \left| p^i \frac{k + H_S - E_n - (H_S - E_n)}{H_S - E_n + k} p^i \right| \phi_n \right\rangle \\
&= \frac{4}{3} \frac{Z\alpha}{\pi m_1 m_2} \int_0^{Z\alpha\mu K} \mathrm{d}k \left\langle \phi_n \left| p^i \frac{(H_S - E_n)}{H_S - E_n + k} p^i \right| \phi_n \right\rangle + \mathrm{Ren} \\
&= \frac{4}{3} \frac{Z\alpha}{\pi m_1 m_2} \left\langle \phi_n \left| p^i \,(H_S - E_n) \ln\left(\frac{Z\alpha\mu\eta}{|H_S - E_n|} \right) p^i \right| \phi_n \right\rangle \\
&= \frac{8}{3} \frac{(Z\alpha)^5 \mu^3}{\pi n^3 m_1 m_2} \left[\ln\left(\frac{Z\alpha\mu\eta}{(Z\alpha)^2 \mu/2} \right) \frac{n^3}{4\mu^3 (Z\alpha)^4} \left\langle \phi_n \left| [p^i, [H_S - E_n, p^i]] \right| \phi_n \right\rangle \right. \\
&\quad \left. - \frac{n^3}{2\mu^3 (Z\alpha)^4} \left\langle \phi_n \left| p^i \,(H_S - E_S) \ln\left(\frac{|H_S - E_n|}{(Z\alpha)^2 \mu/2} \right) p^i \right| \phi_n \right\rangle \right]
\end{aligned} \tag{15.106}$$

where Ren are physically irrelevant renormalization terms as introduced in Sec. 4.6.2, and a principal-value prescription is understood for the k-integration. Now, we use the fact that a Dirac-δ potential is nonvanishing only for S states,

$$\frac{n^3}{4\mu^3 (Z\alpha)^4} \left\langle \phi_n \left| [p^i, [H_S - E_n, p^i]] \right| \phi_n \right\rangle = \delta_{\ell 0} . \tag{15.107}$$

We recall the definition of the Bethe logarithm from Eq. (4.317) and write the low-energy part as

$$\Delta E_L = \frac{8}{3} \frac{(Z\alpha)^5 \mu^3}{\pi n^3 m_1 m_2} \left[\ln\left(\frac{2\eta}{Z\alpha} \right) \delta_{\ell 0} - \ln k_0(n, \ell) \right] . \tag{15.108}$$

For systems with a heavy nucleus like hydrogen, the above correction is small, because m_2 (the proton mass) exceeds m_1 (the electron mass) by orders of magnitude. However, for $m_1 = m_2$ (positronium), the low-energy part of the one-photon exchange is numerically as important as the self-energy.

15.5.3 *Middle-Energy Part and Araki–Sucher Distribution*

In the one-photon exchange correction, the photon is emitted and absorbed at two different places, and the factors $e^{i\vec{k}\cdot\vec{r}_1}$ (for absorption) and $e^{-i\vec{k}\cdot\vec{r}_2}$ (for emission) therefore do not compensate each other as for the self-energy. The resulting exponential $e^{i\vec{k}\cdot(\vec{r}_1-\vec{r}_2)} = e^{i\vec{k}\cdot\vec{r}}$ serves as an ultraviolet cutoff for the photon energy integration and implies that we can eliminate the scale-separation parameter η, already by considering the middle-energy part ΔE_m, which is given by photons with $k \geq Z\alpha\mu\,\eta$. The magnetic term could be important but we will see it is not. Convergence for large photon energies is generated by a different mechanism as compared to the Lamb shift.

In the interaction Hamiltonian (15.102), we have included both the minimal $\vec{p}\cdot\vec{A}$ coupling as well as the $\vec{\sigma}\cdot\vec{B}$ coupling of the spin to the magnetic field of the exchange photon. Employing the same formalism as in the derivation of Eq. (15.105), but dropping the dipole approximation and adding the magnetic coupling, we obtain the expression

$$\begin{aligned}\Delta E_a' &= -\frac{Ze^2}{m_1 m_2}\int \frac{d^3k}{(2\pi)^3}\,\frac{1}{2\,k}\left(\delta^{ij} - \frac{k^i\,k^j}{\vec{k}^2}\right) \\ &\times \left\langle \phi_n \left| \left(p_1^i + \frac{i}{2}\,(\vec{\sigma}_1\times\vec{k})^i\right) e^{i\vec{k}\cdot\vec{r}_1}\,\frac{1}{E_n - H_S - k}\left(p_2^j - \frac{i}{2}\,(\vec{\sigma}_2\times\vec{k})^j\right) e^{-i\vec{k}\cdot\vec{r}_2} \right| \phi_n \right\rangle .\end{aligned} \tag{15.109}$$

We note the sign for the magnetic coupling of the photon emission operator, which is obtained from the expression

$$\left(p_2^j + \frac{1}{2}\,(\vec{\sigma}_2\times\vec{\nabla}_2)^j\right) e^{-i\vec{k}\cdot\vec{r}_2} = \left(p_2^j - \frac{i}{2}\,(\vec{\sigma}_2\times\vec{k})^j\right) e^{-i\vec{k}\cdot\vec{r}_2} . \tag{15.110}$$

It is left as an exercise to the reader to calculate the corresponding expression for virtual photon absorption, which also enters Eq. (15.110). The inclusion of the magnetic coupling is necessary because we are working in a regime where the exchange photon momentum is of order $p \sim k \sim Z\alpha\mu$, so that the expansion (15.97) is applicable,

$$\frac{1}{E_n - H_S - k} = -\frac{1}{k} + \frac{H_S - E_n}{k^2} + \mathcal{O}\left(\frac{1}{k^3}\right) . \tag{15.111}$$

The term $-1/k$ leads to the expression

$$\begin{aligned}\delta\widetilde{E}_a &= \frac{Ze^2}{m_1 m_2}\int \frac{d^3k}{(2\pi)^3}\,\frac{1}{2\,k^2}\left(\delta^{ij} - \frac{k^i\,k^j}{\vec{k}^2}\right) \\ &\times \left\langle \phi_n \left| \left(p_1^i + \frac{i}{2}\,(\vec{\sigma}_1\times\vec{k})^i\right) e^{i\vec{k}\cdot\vec{r}_1}\left(p_2^j - \frac{i}{2}\,(\vec{\sigma}_2\times\vec{k})^j\right) e^{-i\vec{k}\cdot\vec{r}_2} \right| \phi_n \right\rangle .\end{aligned} \tag{15.112}$$

This evaluates to

$$\delta\widetilde{E}_a = \frac{Z\alpha}{4m_1 m_2}\left\langle\phi_n\left|p_1^i\left(\frac{\delta^{ij}}{r}+\frac{r^i\,r^j}{r^3}\right)p_2^j\right|\phi_n\right\rangle - \frac{Z\alpha}{8m_1 m_2}\left(\frac{\vec{\sigma}_1\cdot\vec{\sigma}_2}{r^3} - 3\,\frac{(\vec{\sigma}_1\cdot\vec{r})\,(\vec{\sigma}_2\cdot\vec{r})}{r^5}\right)$$
$$+\frac{\pi Z\alpha}{3m_1 m_2}\,\vec{\sigma}_1\cdot\vec{\sigma}_2\,\delta^{(3)}(\vec{r})\,. \qquad (15.113)$$

The negative sign in front of the first term is obtained in the center-of-mass frame where the momentum operator assumes the form $\vec{p}\equiv\vec{p}_1=-\vec{p}_2$. The two remaining terms are easily identified as the spin-spin interaction, and the spin-spin contact interaction, which were already given for the helium system in Eqs. (13.33d) and (13.33e). As compared to the electron-electron interaction in helium, we here replace $\alpha\to -Z\alpha$. Furthermore, a factor 2 is supplied by the diagram in Fig. 15.4(b). We have thus explicitly verified that the first term in Eq. (15.111) exactly corresponds to a lower-order term from the Breit interaction, which has already been taken into account. In the middle-energy part, the next-to-leading term in Eq. (15.111) is of relative order $Z\alpha$ and contributes to the recoil correction.

We can thus express the middle-energy part generated by diagram (a) in Fig. 15.4 as

$$\Delta E_a = -\frac{Ze^2}{m_1\,m_2}\int_{Z\alpha\mu\eta}^{\infty}\frac{\mathrm{d}^3k}{(2\pi)^3}\,\frac{1}{2k^3}\left(\delta^{ij}-\frac{k^i\,k^j}{\vec{k}^2}\right)$$
$$\times\left\langle\phi_n\left|\left(p_1^i+\frac{\mathrm{i}}{2}\,(\vec{\sigma}_1\times\vec{k})^i\right)\mathrm{e}^{\mathrm{i}\,\vec{k}\cdot\vec{r}_1}\,(H_S-E_n)\left(p_2^j-\frac{\mathrm{i}}{2}\,(\vec{\sigma}_2\times\vec{k})^j\right)\mathrm{e}^{-\mathrm{i}\,\vec{k}\cdot\vec{r}_2}\right|\phi_n\right\rangle\,. \qquad (15.114)$$

We can now use the equation $(H_S-E_n)|\phi_n\rangle = 0$ and the commutator relation $A\,B\,A=\frac{1}{2}\,A^2\,B+\frac{1}{2}\,B\,A^2+\frac{1}{2}\,[A,[B,A]]$, and the fact that the reference state fulfills $(H_S-E_n)|\phi_n\rangle=0$, to write the matrix element in Eq. (15.114) in the form

$$M=\left\langle\left(p_1^i+\frac{\mathrm{i}}{2}\,(\vec{\sigma}_1\times\vec{k})^i\right)\mathrm{e}^{\mathrm{i}\,\vec{k}\cdot\vec{r}_1}\,(H_S-E_n)\left(p_2^j-\frac{\mathrm{i}}{2}\,(\vec{\sigma}_2\times\vec{k})^j\right)\mathrm{e}^{-\mathrm{i}\,\vec{k}\cdot\vec{r}_2}\right\rangle$$
$$=\frac{1}{2}\left\langle\left[\left(p_1^i+\frac{\mathrm{i}}{2}\,(\vec{\sigma}_1\times\vec{k})^i\right)\mathrm{e}^{\mathrm{i}\,\vec{k}\cdot\vec{r}_1},\left[H_S-E_n\,,\left(p_2^j-\frac{\mathrm{i}}{2}\,(\vec{\sigma}_2\times\vec{k})^j\right)\mathrm{e}^{-\mathrm{i}\,\vec{k}\cdot\vec{r}_2}\right]\right]\right\rangle\,. \qquad (15.115)$$

In the latter step, we suppress the reference state ϕ_n in the notation. Here, the photon wave vector $\vec{k}$ in the expression $\vec{\sigma}_i\times\vec{k}$ is of the order of $Z\alpha\mu$, so it not *a priori* permissible to neglect the magnetic terms. However, after the evaluation of the commutator $\left[(H_S-E_n)\,,\,\mathrm{i}\,(\vec{\sigma}_2\times\vec{k})^j\,\mathrm{e}^{-\mathrm{i}\,\vec{k}\cdot\vec{r}_2}\right]$, we are left with an expression that is only a function of $\vec{r}_2$, and therefore, the commutator of the spin-dependent interaction terms with the "left" expression in Eq. (15.115) vanishes. Using the formula $[A,BC]=[A,B]\,C+B\,[A,C]$ repeatedly, one finally obtains

$$M=\frac{1}{2}\left\langle\left[p_1^i\,\mathrm{e}^{\mathrm{i}\,\vec{k}\cdot\vec{r}_1},\left[H_S-E_n\,,\,p_2^j\,\mathrm{e}^{-\mathrm{i}\,\vec{k}\cdot\vec{r}_2}\right]\right]\right\rangle$$
$$=\frac{1}{2}\left\langle\left[p_1^i\,\mathrm{e}^{\mathrm{i}\,\vec{k}\cdot\vec{r}_1},\,\mathrm{e}^{-\mathrm{i}\,\vec{k}\cdot\vec{r}_2}\left[-\frac{Z\alpha}{r},\,p_2^j\right]\right]+\left[p_1^i\,\mathrm{e}^{\mathrm{i}\,\vec{k}\cdot\vec{r}_1},\,p_2^j\left[\frac{\vec{p}_2^{\,2}}{2m_2},\,\mathrm{e}^{-\mathrm{i}\,\vec{k}\cdot\vec{r}_2}\right]\right]\right\rangle$$
$$=\frac{1}{2}\left\langle\mathrm{e}^{\mathrm{i}\,\vec{k}\cdot\vec{r}_1}\left[p_1^i,\left[-\frac{Z\alpha}{r},\,p_2^j\right]\right]\mathrm{e}^{-\mathrm{i}\,\vec{k}\cdot\vec{r}_2}\right\rangle=\frac{1}{2}\left\langle\mathrm{e}^{\mathrm{i}\,\vec{k}\cdot\vec{r}}\left[p^i,\left[p^j,-\frac{Z\alpha}{r}\right]\right]\right\rangle\,. \qquad (15.116)$$

The resulting expression for $\Delta E_a'$ reads as

$$\Delta E_a' = \frac{Ze^2}{m_1 m_2} \int_{Z\alpha\mu\eta}^{\infty} \frac{d^3k}{(2\pi)^3} \frac{1}{2k^3} \left(\delta^{ij} - \frac{k^i k^j}{\vec{k}^2}\right) \left\{ \frac{1}{2} \left\langle e^{i\vec{k}\cdot\vec{r}} \left[p^i, \left[p^j, \frac{Z\alpha}{r}\right]\right]\right\rangle\right\} . \tag{15.117}$$

It is left as an exercise to the reader to verify that the diagram in Fig. 15.4(b) leads to an equal contribution, so that the middle-energy part reads as

$$\Delta E_M = \frac{Ze^2}{m_1 m_2} \int_{Z\alpha\mu\eta}^{\infty} \frac{d^3k}{(2\pi)^3} \frac{1}{2k^3} \left(\delta^{ij} - \frac{k^i k^j}{\vec{k}^2}\right) \left\langle \phi_n \left| e^{i\vec{k}\cdot\vec{r}} \left[p^i, \left[p^j, \frac{Z\alpha}{r}\right]\right]\right| \phi_n \right\rangle . \tag{15.118}$$

This expression has to be handled with extreme care, because

$$\left[p^i, \left[p^j, \frac{Z\alpha}{r}\right]\right] = \nabla^i \left(\nabla^j \left(-\frac{Z\alpha}{r}\right)\right) = \frac{Z\alpha}{r^3}\left(\delta^{ij} - \frac{3x^i x^j}{r^2}\right) + \frac{4\pi}{3}\delta^{ij} Z\alpha\, \delta^3(r) , \tag{15.119}$$

and the matrix element of $1/r^3$ is divergent for a reference S state at the origin. We thus face a rather peculiar situation and choose a different sequence for the d^3k and d^3r-integrations in Eq. (15.118), depending on the angular symmetry of the reference state. For S states, it is more convenient to do the evaluation of the matrix element first, as a function of $\vec{k}$, because the wave function is radially symmetric, and it is therefore possible to do the angular momentum algebra easily. For P states and higher angular momenta, by contrast, one does the k-integration first, which leads to a simple expression in coordinate space, yet, to a matrix element which is convergent only for states with $\ell \neq 0$.

Let us consider the $1S$ state first. We apply the *ansatz*

$$\begin{aligned} N^{ij} &= \left\langle 1S \left| e^{i\vec{k}\cdot\vec{r}} \left[p^i, \left[p^j, \frac{Z\alpha}{r}\right]\right]\right| 1S \right\rangle \\ &= \left\langle 1S \left| e^{i\vec{k}\cdot\vec{r}} \left[\frac{Z\alpha}{r^3}\left(\delta^{ij} - \frac{3x^i x^j}{r^2}\right) + \frac{4\pi}{3}\delta^{ij} Z\alpha\, \delta^{(3)}(\vec{r})\right]\right| 1S \right\rangle \\ &= A(k)\,\vec{k}^2\,\delta^{ij} + B(k)\,k^i\,k^j , \end{aligned} \tag{15.120}$$

where $A(k)$ and $B(k)$ can be determined based on suitable projection operations. We multiply both sides of Eq. (15.120) by δ^{ij} and obtain the relation

$$S = N^{ii} = \left\langle 1S \left| 4\pi Z\alpha\, \delta^{(3)}(\vec{r}) \right| 1S \right\rangle = 4\,(Z\alpha)^4 \mu^3 = (3\,A + B)\,\vec{k}^2 . \tag{15.121}$$

Second, we multiply both sides of Eq. (15.120) by $k^i k^j$ and obtain a second relation for A and B,

$$T = k^i k^j N^{ij} = \left\langle 1S \left| e^{i\vec{k}\cdot\vec{r}} \left(\frac{Z\alpha}{r^3}\left(\vec{k}^2 - \frac{3\,(\vec{k}\cdot\vec{r})^2}{r^2}\right) + \frac{4\pi}{3}\vec{k}^2 Z\alpha\, \delta^3(r)\right)\right| 1S \right\rangle = (A{+}B)\,\vec{k}^4 . \tag{15.122}$$

Even if the radial symmetry of the S state wave function leads to a mutual cancelation of the divergence at the origin in between the terms $\vec{k}^2 - 3(\vec{k}\cdot\vec{r})^2/r^2$, the matrix element T is still challenging to evaluate. We write it as

$$\begin{aligned} T &= \left\langle 1S \left| \mathrm{e}^{\mathrm{i}\vec{k}\cdot\vec{r}} \left(\frac{Z\alpha}{r^3} \left(\vec{k}^2 - \frac{3(\vec{k}\cdot\vec{r})^2}{r^2} \right) + \frac{4\pi}{3} \vec{k}^2 \, Z\alpha \, \delta^{(3)}(\vec{r}) \right) \right| 1S \right\rangle \\ &= \frac{4(Z\alpha)^4 \mu^3}{3} \vec{k}^2 + Z\alpha \vec{k}^2 \int_0^{2\pi} \mathrm{d}\phi \int_{-1}^{1} \mathrm{d}u \int_0^\infty \frac{\mathrm{d}r}{r} \left[\phi_{1S}(\vec{r})\right]^2 (1 - 3u^2) \mathrm{e}^{\mathrm{i}kru} \\ &= 4(Z\alpha)^4 \mu^3 \left(\vec{k}^2 + 4(Z\alpha)^2 \mu^2 \right) \left[1 - \frac{2Z\alpha\mu}{k} \operatorname{arccot}\left(\frac{2Z\alpha\mu}{k} \right) \right]. \end{aligned} \tag{15.123}$$

We can now solve Eqs. (15.121) and (15.122) for A and B as a function of S and T, with the result

$$\begin{aligned} A(k) &= \frac{4(Z\alpha)^5 \mu^4}{k^3} \left[\left(1 + \frac{4(Z\alpha\mu)^2}{\vec{k}^2} \right) \operatorname{arccot}\left(\frac{2Z\alpha\mu}{k} \right) - 2\frac{Z\alpha\mu}{k} \right], \\ B(k) &= \frac{4(Z\alpha)^4 \mu^3}{k^2} \left[-3\frac{Z\alpha\mu}{k} \left(1 + \frac{4(Z\alpha\mu)^2}{\vec{k}^2} \right) \operatorname{arccot}\left(\frac{2Z\alpha\mu}{k} \right) + 6\frac{(Z\alpha\mu)^2}{\vec{k}^2} + 1 \right]. \end{aligned} \tag{15.124}$$

After integration over k, the result for ΔE_M for the ground state is

$$\begin{aligned} \Delta E_M(1S) &= \frac{Ze^2}{m_1 m_2} \int_{Z\alpha\mu\eta}^\infty \frac{\mathrm{d}^3k}{(2\pi)^3} \frac{1}{2k^3} \left(\delta^{ij} - \frac{k^i k^j}{\vec{k}^2} \right) \left\langle \mathrm{e}^{\mathrm{i}\vec{k}\cdot\vec{r}} \left[p^i, \left[p^j, \frac{Z\alpha}{r} \right] \right] \right\rangle \\ &= \frac{Ze^2}{m_1 m_2} \int_{Z\alpha\mu\eta}^\infty \frac{\mathrm{d}^3k}{(2\pi)^3} \frac{1}{2k^3} \left(\delta^{ij} - \frac{k^i k^j}{\vec{k}^2} \right) \left(A(k)\, \vec{k}^2 \delta^{ij} + B(k)\, k^i k^j \right) \\ &= \frac{8}{3} \frac{(Z\alpha)^5 \mu^3}{\pi m_1 m_2} \left[-\ln\left(\frac{\eta}{2} \right) + \frac{4}{3} \right]. \end{aligned} \tag{15.125}$$

In the sum of the low-energy and middle-energy parts, the dependence on η cancels for the ground state,

$$\Delta E_{L+M}(1S) = \Delta E_L(1S) + \Delta E_M(1S) = \frac{8}{3} \frac{(Z\alpha)^5 \mu^3}{\pi m_1 m_2} \left[\ln\left(\frac{4}{Z\alpha} \right) + \frac{4}{3} - \ln k_0(1S) \right]. \tag{15.126}$$

For the weighted difference of S states, and for states with nonvanishing angular momenta, the d^3k integral is carried out first in Eq. (15.118). One has to calculate

$$J^{ij} = \int_{|\vec{k}| \geq Z\alpha\mu\eta} \frac{\mathrm{d}^3k}{(2\pi)^3} \frac{1}{2k^3} \left(\delta^{ij} - \frac{k^i k^j}{\vec{k}^2} \right) \exp(\mathrm{i}\vec{k}\cdot\vec{r}) = a(r)\, \delta^{ij} r^2 + b(r)\, r^i r^j. \tag{15.127}$$

Multiplication by δ^{ij} yields

$$J^{ij} \delta^{ij} = \int_{|\vec{k}| \geq Z\alpha\mu\eta} \frac{\mathrm{d}^3k}{(2\pi)^3} \frac{1}{2k^3} (3 - 1) \exp(\mathrm{i}\vec{k}\cdot\vec{r}) = (3a + b)\, r^2, \tag{15.128}$$

whereas a multiplication by $r^i r^j$ gives the result

$$J^{ij} r^i r^j = \frac{\mathrm{d}^3k}{(2\pi)^3} \frac{1}{2k^3} \left(r^2 - \frac{(\vec{k}\cdot\vec{r})^2}{k^2} \right) \exp(\mathrm{i}\vec{k}\cdot\vec{r}) = (a + b)\, r^4. \tag{15.129}$$

The two equations for two unknowns, $a(r)$ and $b(r)$, can now be solved as follows,

$$a(r) = \frac{5}{36\pi^2 r^2} - \frac{\gamma_E}{6\pi^2 r^2} - \frac{\ln(Z\alpha\mu\eta r)}{6\pi^2 r^2}, \qquad b(r) = \frac{1}{12\pi^2 r^2}. \tag{15.130}$$

Both results for $a(r)$ and $b(r)$ are given for small $Z\alpha\mu\eta$, which is sufficient for our purposes. The infrared divergence of $a(r)$ is cut off at $k = Z\alpha\mu\eta$, which leads to the logarithmic divergence in coordinate space for $r \to 0$. We now multiply the entire expression by the term

$$\left[p^i, \left[p^j, -\frac{Z\alpha}{r}\right]\right] = -Z\alpha \frac{3r^i r^j - \delta^{ij} r^2}{r^5} + \frac{4\pi}{3}\delta^{ij} Z\alpha\, \delta^{(3)}(\vec{r}), \tag{15.131}$$

where the delta term is irrelevant for non-S states. We finally have

$$J^{ij}\left[p^i, \left[p^j, -\frac{Z\alpha}{r}\right]\right] = -\frac{Z\alpha}{6\pi^2 r^3} + \frac{2Z\alpha\, \delta^{(3)}(\vec{r})}{3\pi}\left[1 - \gamma_E - \ln(Z\alpha\mu\eta r)\right]. \tag{15.132a}$$

For non-S states (they vanish at the origin), we can replace

$$J^{ij}\left[p^i, \left[p^j, -\frac{Z\alpha}{r}\right]\right] \to -\frac{Z\alpha}{6\pi^2 r^3}. \tag{15.132b}$$

For the $1/r^3$ term, for general hydrogenic states, one may write a regularizing distribution due to Araki[3] [537] and Sucher[4] [538]. It is canonically denoted as P, and is defined as follows,

$$\langle\phi|\, P\left(\frac{1}{r^3}\right) |\psi\rangle = \lim_{a\to 0} \int \mathrm{d}^3 r\, \phi^*(\vec{r}) \left[\frac{1}{r^3}\Theta(r-a) + 4\pi\delta^{(3)}(\vec{r})\, (\gamma_E + \ln(a))\right] \psi(\vec{r}). \tag{15.133}$$

Sandwiched between hydrogen eigenstates, this distribution has a defined expectation value for both S and non-S states. For S states, it has a finite expectation value, and its weighted difference for S states [see Eq. (15.135) below] describes the functional dependence of the $1/r^3$ matrix element. For non-S states, the expectation value of the Araki–Sucher distribution is just equal to the expectation value of the operator $1/(Z\alpha\mu r)^3$. The Araki–Sucher distribution has compensating terms to cut off the logarithmic divergence of the integral for small radial coordinate and is independent of a in the limit $a \to 0$. This can be verified if one assume that the rest of the integrand, namely, the product $\phi^*(\vec{r})\,\psi(\vec{r})$, approaches a constant in the limit $\vec{r} \to \vec{0}$.

As a function of the bound-state quantum numbers, we have

$$\begin{aligned}\left\langle P\left(\frac{1}{(Z\alpha\mu r)^3}\right)\right\rangle &\equiv \frac{\delta_{\ell 0}}{n^3}\left(\left\langle\!\!\left\langle nS\left|\frac{1}{(Z\alpha\mu r)^3}\right|nS\right\rangle\!\!\right\rangle - 4\ln 2\right) + (1-\delta_{\ell 0})\left\langle\frac{1}{(Z\alpha\mu r)^3}\right\rangle \\ &= \delta_{\ell 0}\frac{4}{n^3}\left[\ln\left(\frac{n}{2}\right) - \Psi(n) - \gamma + \frac{1}{2} - \frac{1}{2n}\right] + \frac{2}{n^3}\frac{1-\delta_{\ell 0}}{\ell(\ell+1)(2\ell+1)}.\end{aligned} \tag{15.134}$$

---

[3] Huzihiro Araki (b. 1932).

[4] Joseph Sucher (1930–2019).

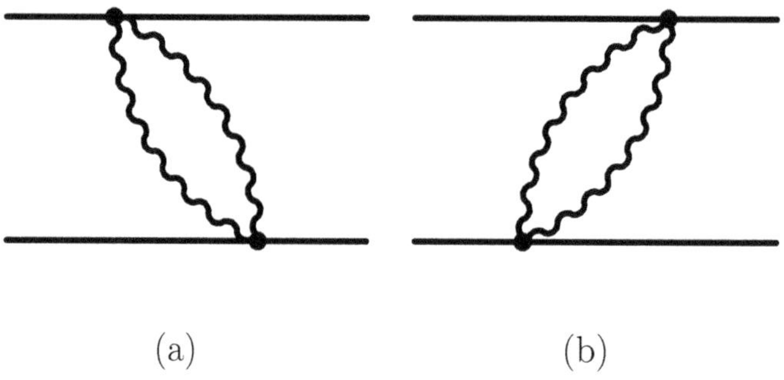

Fig. 15.5 The seagull contribution to the low-energy part of the recoil correction involves the exchange of four photons, two of which emerging from either fermion line. Diagrams (a) and (b) illustrate the different possible time orderings.

Here, the expectation value of the weighted difference of S states reads as

$$\left\langle\!\!\left\langle nS \left| \frac{1}{(Z\alpha\mu r)^3} \right| nS \right\rangle\!\!\right\rangle = n^3 \left\langle nS \left| \frac{1}{(Z\alpha\mu r)^3} \right| nS \right\rangle - \left\langle 1S \left| \frac{1}{(Z\alpha\mu r)^3} \right| 1S \right\rangle . \qquad (15.135)$$

The logarithmic term $-4\ln(2)$ in the S state contribution to the P distribution is chosen so that the non-S state expectation value has a particularly simple form. The complete result for the middle-energy part is

$$\Delta E_M = \frac{8}{3} \frac{(Z\alpha)^5 \mu^3}{\pi m_1 m_2} \left[\left(-\ln\eta + \frac{4}{3} \right) \delta_{\ell 0} - \frac{1}{4} \left\langle P\left(\frac{1}{(Z\alpha\mu r)^3} \right) \right\rangle \right] . \qquad (15.136)$$

Adding the low-energy part from Eq. (15.108) and the middle-energy part from Eq. (15.136), we have for $\Delta E_{L+M} = \Delta E_L + \Delta E_M$,

$$\Delta E_{L+M} = \frac{8}{3} \frac{(Z\alpha)^5 \mu^3}{\pi n^3 m_1 m_2} \left[\left(\ln\left(\frac{2}{Z\alpha} \right) + \frac{4}{3} \right) \delta_{\ell 0} - \frac{1}{4} \left\langle P\left(\frac{1}{(Z\alpha\mu r)^3} \right) \right\rangle - \ln k_0(n,\ell) \right] . \qquad (15.137)$$

The scale-separation parameter η has canceled.

15.5.4 *Seagull Part*

For the one-electron Lamb shift, the so-called seagull terms proportional to $\vec{A}^2$ in the interaction Hamiltonian (15.102) do not contribute to the energy shift because their contribution can be absorbed in a mass renormalization; the atom is not in an excited, virtual state in between the seagull photon emissions and absorptions. However, for a two-particle system, the seagull contribution (see Fig. 15.5) is non-trivial, because the emission and absorption processes involve two different particles. The seagull term involves two diagrams, each of which yield equal contributions, and so

$$\Delta E_S = 2 \times \frac{e^4}{4 m_1 m_2} \left\langle \phi_n \left| \vec{A}(\vec{r}_1)^2 \left(\frac{1}{E_0 - H_0} \right)' \vec{A}(\vec{r}_2)^2 \right| \phi_n \right\rangle . \qquad (15.138)$$

The atom undergoes an "instantaneous" emission and absorption process and therefore is not in an excited state. We can thus replace the expression $E_0 - H_0$ in Eq. (15.138) by $-q_1 - q_2$, where $q_1 = |\vec{q}_1|$ and $q_2 = |\vec{q}_2|$ are the momenta of the exchanged seagull photons. (H_0 contains photon field terms.)

Let us motivate the expression for this energy shift. From the form of the vector field operator, we know that each photon carries a factor $\sqrt{1/(2q)}$. As already stated, the propagator in the middle is just $1/(E_0 - H_0)' \to -1/(q_1 + q_2)$, from the field part of H_0, because the electron state has not changed, but the photon field undergoes a transition from a zero-photon state to a two-photon, virtual state. We now add a factor 2 because of the presence of two diagrams, and another factor 2 because we can, colloquially speaking, glue the photon lines emerging from the two vertices together in two different ways. The details are left as an exercise to the reader. The two transverse photon propagators and two factor $1/(2m_1)$ and $1/(2m_2)$ for the two seagull vertices lead to a momentum-space expression M, which reads

$$M = 2 \times 2 \times \frac{Ze^2}{2q_1} \frac{Ze^2}{2q_2} \frac{-1}{q_1 + q_2} \left(\delta^{ij} - \frac{q_1^i q_1^j}{q_1^2} \right) \left(\delta^{ij} - \frac{q_2^i q_2^j}{q_2^2} \right) \frac{1}{2m_1} \frac{1}{2m_2} . \tag{15.139}$$

This means that

$$\begin{aligned} E_S &= -\frac{Z^2 e^4}{4m_1 m_2} \int \frac{d^3 q_1}{(2\pi)^3} \int \frac{d^3 q_2}{(2\pi)^3} \int \frac{d^3 p}{(2\pi)^3} \frac{\phi^*(\vec{p} - \vec{q}_1)\, \phi(\vec{p} + \vec{q}_2)}{q_1\, q_2\, (q_1 + q_2)} \\ &\quad \times \left(\delta^{ij} - \frac{q_1^i q_1^j}{q_1^2} \right) \left(\delta^{ij} - \frac{q_2^i q_2^j}{q_2^2} \right) \\ &= -\frac{Z^2 e^4}{4m_1 m_2} \int \frac{d^3 q_1}{(2\pi)^3} \int \frac{d^3 q_2}{(2\pi)^3} \frac{\rho(\vec{q}_1 - \vec{q}_2)}{q_1\, q_2\, (q_1 + q_2)} \left(1 + \frac{(\vec{q}_1 \cdot \vec{q}_2)^2}{q_1^2 q_2^2} \right). \end{aligned} \tag{15.140}$$

Using the paradigm that convolution in momentum space is equivalent to multiplication in coordinate space, we recognize the Fourier transform of the charge density as

$$\rho(\vec{q}_1 + \vec{q}_2) = \int \frac{d^3 p}{(2\pi)^3} \phi^*(\vec{p} - \vec{q}_1)\, \phi(\vec{p} + \vec{q}_2) . \tag{15.141}$$

The UV divergence is created by the direction where $\vec{q}_1$ is parallel to $\vec{q}_2$, and so we do not need to regularize the second term. The first corresponds to the exchange of two photons, and so it has to be regularized according to

$$\frac{1}{q^2} \to \frac{1}{q^2} - \frac{1}{q^2 + \Lambda^2} = \frac{1}{q^2} \frac{\Lambda^2}{q^2 + \Lambda^2}, \qquad \Lambda = Z\alpha\mu\lambda . \tag{15.142}$$

The latter term would actually vanish were it not for the fact that it leads to a singular contribution at the origin in coordinate space which leads to a nonvanishing contribution for S states.

Just as for the middle-energy part, we need to consider the ground state, first. In that case, ρ is just the Fourier transform of the charge density of the ground state of the hydrogen atom and reads as follows,

$$\rho_{1S}(\vec{q}_1 + \vec{q}_2) = \frac{16(Z\alpha\mu)^4}{\left[(\vec{q}_1 + \vec{q}_2)^2 + 4(Z\alpha\mu)^2\right]^2} . \tag{15.143}$$

We now split the resulting expression into two parts,

$$E_S(1S) = -\frac{Z^2\,e^4}{4m_1 m_2}\int\frac{\mathrm{d}^3 q_1}{(2\pi)^3}\int\frac{\mathrm{d}^3 q_1}{(2\pi)^3}\,\frac{\rho(\vec{q}_1+\vec{q}_2)}{q_1\,q_2\,(q_1+q_2)}\left[2\frac{\Lambda^2}{q_1^2+\Lambda^2}\frac{\Lambda^2}{q_2^2+\Lambda^2}\right.$$
$$\left.+\frac{q_1^2 q_2^2-(\vec{q}_1\cdot\vec{q}_2)^2}{q_1^2\,q_2^2}\right] = A+B\,, \tag{15.144}$$

where A corresponds to the first term in brackets. An integration over the angles between $\vec{q}_1$ and $\vec{q}_2$ leads to the expression

$$A(1S) = -\frac{8(Z\alpha)^2\,(Z\alpha\mu)^4}{m_1 m_2 \pi^2}\int_0^\infty \mathrm{d}q_1\int_0^\infty \mathrm{d}q_2\,\frac{1}{q_1+q_2}$$
$$\times\left(\frac{1}{(q_1-q_2)^2+4(Z\alpha\mu)^2}\,\frac{\Lambda^2}{q_1^2+\Lambda^2}\,\frac{\Lambda^2}{q_2^2+\Lambda^2}-\frac{1}{(q_1+q_2)^2+4(Z\alpha\mu)^2}\right). \tag{15.145}$$

Only the first term requires a regularization. We now transform the integration variables as follows,

$$q_1 = \frac{q}{2}+k\,,\qquad q_2 = \frac{q}{2}-k\,, \tag{15.146a}$$

$$q = q_1+q_2\,,\qquad k = \tfrac{1}{2}(q_1-q_2)\,. \tag{15.146b}$$

$$J = \left|\det\begin{pmatrix}\frac{1}{2} & 1\\ \frac{1}{2} & -1\end{pmatrix}\right| = 1\,. \tag{15.146c}$$

The Jacobian of this transformation is just unity. The integration limits are $k \in (-q/2, q/2)$ and $q \in (0,\infty)$. In the terms involving the Λ parameters, when replacing q_1 and q_2 by q and k, we can replace $q_1 \to q/2$ and $q_2 \to q/2$ (and ignore k), because in a large subregion of the integration region, the difference $q_1 - q_2 = 2k$ has a negligible effect. In order to convince oneself, one expands in the terms proportional to k, and verifies that the subleading terms in that expansion fall off sufficiently fast with Λ to be ignored in the limit of large Λ. So,

$$A(nS) = -\frac{8(Z\alpha)^2(Z\alpha\mu)^4}{m_1 m_2\pi^2 n^3}\int_0^\infty \mathrm{d}q\int_{-q/2}^{q/2}\mathrm{d}k\frac{1}{q}\left(\frac{1}{4}\,\frac{\left(\frac{\Lambda^2}{q^2/4+\Lambda^2}\right)^2}{k^2+(Z\alpha\mu)^2}-\frac{1}{q^2+4(Z\alpha\mu)^2}\right)\delta_{\ell 0}$$
$$= -\frac{(Z\alpha)^5\,\mu^3}{m_1 m_2 \pi n^3}\,\{2\ln(\Lambda)-3\}\;\delta_{\ell 0}\mathcal{O}\left(\frac{1}{\Lambda}\right). \tag{15.147}$$

In writing Eq. (15.147), we anticipate that the contribution for the $1S$ state scales as $1/n^3$ for the excited states, except for a further term, which corresponds to the weighted difference of S states, and which can be evaluated separately. When adding this term to the above result and to the one below given in Eq. (15.148), we obtain the complete result for any state. For the remaining term B, we have the following expression,

$$B(1S) = -\frac{Z^2\,e^4}{4m_1 m_2}\int\frac{\mathrm{d}^3 q_1}{(2\pi)^3}\int\frac{\mathrm{d}^3 q_2}{(2\pi)^3}\,\frac{16(Z\alpha\mu)^4}{[(\vec{q}_1+\vec{q}_2)^2+4(Z\alpha\mu)^2]^2}\,\frac{\frac{(\vec{q}_1\cdot\vec{q}_2)^2}{q_1^2\,q_2^2}-1}{q_1\,q_2\,(q_1+q_2)}\,. \tag{15.148}$$

We use a parameterization of the $(\vec{q}_1, \vec{q}_2)$ space which can be useful in other contexts,

$$s = q_1 + q_2\,, \qquad t = q_1 - q_2\,, \qquad u = |\vec{q}_1 - \vec{q}_2|\,,$$
$$q_1 = \frac{s+t}{2}\,, \qquad q_2 = \frac{s-t}{2}\,, \qquad \frac{\mathrm{d}^3 q_1}{(2\pi)^3}\frac{\mathrm{d}^3 q_2}{(2\pi)^3} = 2\pi^2 (s^2 - t^2)\, u \,\mathrm{d}s\,\mathrm{d}t\,\mathrm{d}u. \qquad (15.149)$$

The integration limits are $0 \le t \le u \le s \le \infty$. One first integrates over t in the interval $(0, u)$, then over s in the interval (u, ∞), and finally one integrates u in the whole interval $(0, \infty)$. The latter variable takes the role of a homogeneity parameter. So,

$$\begin{aligned} B(nS) = & -\frac{(Z\alpha)^5 \mu^3}{4 m_1 m_2 n^3} \int_0^\infty \mathrm{d}u \int_u^\infty \mathrm{d}s \int_0^u \mathrm{d}t\, 2\pi^2 (s^2 - t^2)\, u \,\frac{1}{(u^2+4)^2}\,\frac{2}{s+t}\,\frac{2}{s-t}\,\frac{1}{s} \\ & \times \left[\left(\frac{2}{s+t}\right)^2 \left(\frac{2}{s-t}\right)^2 \left(u^2 - \left(\frac{s+t}{2}\right)^2 - \left(\frac{s-t}{2}\right)^2 \right)^2 \right] \delta_{\ell 0} \\ = & -\frac{(Z\alpha)^5 \mu^3}{m_1 m_2 \pi n^3}\,\frac{8}{3}\,\{\ln(2) - 1\}\,\delta_{\ell 0}\,. \end{aligned} \qquad (15.150)$$

For the weighted difference of S states and for any state with a nonvanishing angular momentum, we have $E_S \to C$ where C is proportional to the expectation value of the operator $1/r^3$,

$$C = -\frac{2(Z\alpha)^5 \mu^3}{m_1 m_2 \pi n^3} \left\{ \ln(n) - \Psi(n) - \gamma_{\mathrm{E}} - \frac{1}{2n} + \frac{1}{2} \right\} \delta_{\ell 0} - \frac{(Z\alpha)^5 \mu^3}{2 m_1 m_2 \pi} \left\langle \frac{1}{(Z\alpha\mu r)^3} \right\rangle (1 - \delta_{\ell 0}). \qquad (15.151)$$

The complete result for $E_S = A + B + C$ is

$$\begin{aligned} \Delta E_S = & -\frac{2(Z\alpha)^5 \mu^3}{m_1 m_2 \pi n^3} \left\{ \ln(n) - \Psi(n) - \gamma_{\mathrm{E}} - \frac{1}{2n} + \ln\lambda - \frac{17}{3} + \frac{4}{3}\ln(2) \right\} \delta_{\ell 0} \\ & - \frac{(Z\alpha)^5 \mu^3}{2 m_1 m_2 \pi} \left\langle \frac{1}{(Z\alpha\mu r)^3} \right\rangle (1 - \delta_{\ell 0})\,. \end{aligned} \qquad (15.152)$$

15.5.5 *High-Energy Part*

The contribution from the scattering amplitude involves the exchange of two photons (see Fig. 15.6). The starting expression for the energy shift can be derived based on the matching of the S-matrix element with the effective Hamiltonian (see Sec. 12.2) and on the averaging over the spin orientations of the S wave as outlined in Sec. 15.3.1. The result is

$$\begin{aligned} E_H = & -e^4 |\phi(\vec{0})|^2 \int \frac{\mathrm{d}^4 k}{(2\pi)^4 \mathrm{i}}\,\frac{1}{k^4}\, \mathrm{Tr}\left[\gamma^\mu \frac{1}{\not t_1 - \not k - m_1} \gamma^\nu \frac{\gamma^0 + 1}{4} \right] \\ & \times \left\{ \mathrm{Tr}\left[\gamma_\mu \frac{1}{\not t_2 + \not k - m_2} \gamma_\nu \frac{\gamma^0 + 1}{4} \right] + \mathrm{Tr}\left[\gamma_\nu \frac{1}{\not t_2 - \not k - m_2} \gamma_\mu \frac{\gamma^0 + 1}{4} \right] \right\}. \end{aligned} \qquad (15.153)$$

Here, $\not t_1 = \gamma^0 m_1$ and $\not t_2 = \gamma^0 m_2$ are time-like unit vectors parameterizing the two particles at rest. The loop momentum is on the order of the mass of the bound particles. One approximates the wave function as $\phi(\vec{r}) \approx \phi(\vec{0})$, which corresponds

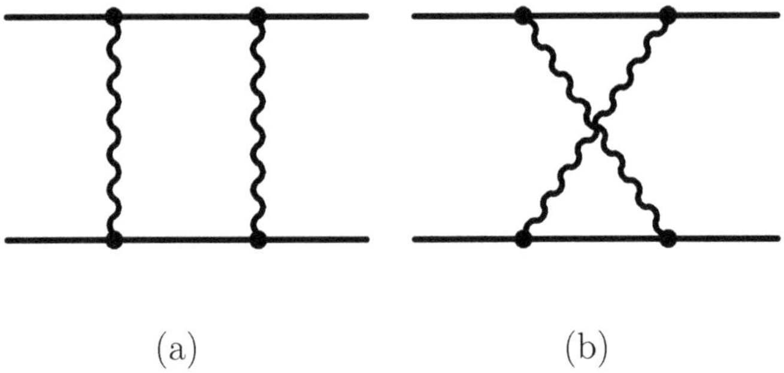

Fig. 15.6 The Feynman diagrams for the high-energy part of the recoil correction involve two-photon exchange. Diagrams (a) and (b) correspond to the two terms in the second line of Eq. (15.153).

to a Dirac-δ in momentum space. Essentially, one thereby ignores the incoming momentum of the wave function, which otherwise acts as an infrared regulator for the diagram. The approximation $\phi(\vec{r}) \approx \phi(\vec{0})$ thus entails the necessity to subtract from the scattering amplitude, certain spurious infrared divergent terms. Because only S state wave functions are nonvanishing at the origin, the entire high-energy part only contributes to S states.

In order to do the photon energy integration, one performs a Wick rotation in the integral

$$E_H = -\frac{128\pi(Z\alpha)^5 m_1 m_2 \mu^3}{n^3}\delta_{\ell 0}\int \frac{\mathrm{d}^4k}{(2\pi)^4\mathrm{i}}\,\frac{k^4+2\omega^4}{k^4} \times \frac{1}{k^2+2m_1\omega}\,\frac{1}{k^2-2m_1\omega}\,\frac{1}{k^2-2m_2\omega}\,\frac{1}{k^2+2m_2\omega}, \tag{15.154}$$

where $k^\mu = (\omega, \vec{k})$, with the result

$$E_H = -\frac{4\pi(Z\alpha)^5 m_1 m_2 \mu^3}{n^3}\delta_{\ell 0}\int \frac{\mathrm{d}^3k}{(2\pi)^3}\,\frac{1}{k^6\, m_1^2\, m_2^2} \times \left[k^3 + \frac{(k^4+6\,k^2\,m_1^2+8m_1^4)\,m_2^2}{\sqrt{k^2+m_1^2}\,(m_1^2-m_2^2)} + \frac{(k^4+6\,k^2\,m_2^2+8m_2^4)\,m_1^2}{\sqrt{k^2+m_2^2}\,(m_2^2-m_1^2)}\right], \tag{15.155}$$

where now $k = |\vec{k}|$. Just like in Sec. 15.5.3, various lower-order terms have to be subtracted. The leading subtraction term comes from the two-Coulomb-photon exchange in the on-shell approximation. We recall, from Chap. 5, that, for the exchange of hard photons, we can approximate the Schrödinger–Coulomb Green function as follows,

$$\left(\frac{1}{E-H_S}\right)' \approx -\frac{1}{H_S} \approx -\frac{2\mu}{\vec{k}^2}, \tag{15.156}$$

where the approximation is valid for hard momenta $\vec{k}$ on the order of the electron mass. We recall that the Coulomb potential is diagonal in coordinate space, but a function of the momentum transfer in momentum space,

$$\langle\vec{r}|V|\vec{r}'\rangle = V(\vec{r})\,\delta^{(3)}(\vec{r}-\vec{r}')\,, \qquad V(\vec{r}) = -\frac{Z\alpha}{|\vec{r}|}\,, \tag{15.157a}$$

$$\langle\vec{q}|V|\vec{q}'\rangle = V(\vec{q}-\vec{q}')\,, \qquad V(\vec{q}-\vec{q}') = -\frac{4\pi Z\alpha}{(\vec{q}-\vec{q}')^2}\,. \tag{15.157b}$$

The second-order energy perturbation for the exchange of two hard Coulomb photons can thus be written as

$$\left\langle V\left(\frac{1}{E-H}\right)' V\right\rangle = \int\frac{d^3k}{(2\pi)^3}\int\frac{d^3k'}{(2\pi)^3}\int\frac{d^3k''}{(2\pi)^3} \times \phi^+(\vec{k})\left(-\frac{4\pi Z\alpha}{(\vec{k}-\vec{k}')^2}\right)\left(-\frac{2\mu}{\vec{k}'^2}\right)\left(-\frac{4\pi Z\alpha}{(\vec{k}'-\vec{k}'')^2}\right)\phi(\vec{k}''). \tag{15.158}$$

Due to the hard nature of the Coulomb photons, we replace for the calculation of the subtraction term

$$-\frac{4\pi Z\alpha}{(\vec{k}-\vec{k}')^2} \to -\frac{4\pi Z\alpha}{\vec{k}'^2}\,, \qquad -\frac{4\pi Z\alpha}{(\vec{k}'-\vec{k}'')^2} \to -\frac{4\pi Z\alpha}{\vec{k}'^2}\,. \tag{15.159}$$

The integrations over d^3k and d^3k'' thus decouple. After Fourier transformation of the wave function into coordinate space (which results in its value at the origin), we obtain

$$\begin{aligned}\left\langle V\left(\frac{1}{E-H}\right)' V\right\rangle &\approx \left|\phi(\vec{r}=\vec{0})\right|^2 \int\frac{d^3k'}{(2\pi)^3}\left(-\frac{4\pi Z\alpha}{\vec{k}'^2}\right)\left(-\frac{2\mu}{\vec{k}'^2}\right)\left(-\frac{4\pi Z\alpha}{\vec{k}'^2}\right)\\ &= \int\frac{d^3k}{(2\pi)^3}\left(-\frac{32\pi(Z\alpha)^5\mu^4}{n^3}\frac{1}{\vec{k}^6}\right)\delta_{\ell 0}\,.\end{aligned} \tag{15.160}$$

We have used the fact that

$$\left|\phi(\vec{r}=\vec{0})\right|^2 = \frac{(Z\alpha)^3\mu^3}{\pi n^3}\,\delta_{\ell 0}\,. \tag{15.161}$$

Comparison of Eq. (15.160) with Eq. (15.155) reveals that the leading $1/\vec{k}^6$ asymptotics of the integrand in Eq. (15.155) indeed correspond to two-Coulomb-photon exchange, and need to be subtracted. A further subtraction term in Eq. (15.155) corresponds to the Breit Hamiltonian. The logarithmically divergent term has a factor $\Lambda^4/(\vec{k}^2+\Lambda^2)^2$ with $\Lambda = Z\alpha\mu\lambda$. Finally, one has

$$\Delta E_H = -\frac{(Z\alpha)^5\mu^3}{m_1 m_2 n^3}\delta_{\ell 0}\left\{\frac{7}{3} - 2\ln(Z\alpha\lambda) + \frac{2}{m_1^2-m_2^2}\left[m_1^2\ln\left(\frac{m_2}{\mu}\right) - m_2^2\ln\left(\frac{m_1}{\mu}\right)\right]\right\} \tag{15.162}$$

as the result for the high-energy part.

15.5.6 *Final Result*

The complete recoil correction of order $(Z\alpha)^5$, in first order in the mass ratio, is obtained as the sum of the low- and the middle-energy part, and of the seagull and high-energy terms. We have from Eqs. (15.108), (15.136), (15.152) and (15.162),

$$\Delta E = \Delta E_L + \Delta E_M + \Delta E_S + \Delta E_H$$
$$= \frac{(Z\alpha)^5 \mu^3}{\pi m_1 m_2 n^3} \left\{ \frac{2}{3} \delta_{\ell 0} \ln\left(\frac{1}{Z\alpha}\right) - \frac{8}{3} \ln k_0(n,\ell) - \frac{7}{6} n^3 \left\langle P\left(\frac{1}{(Z\alpha\mu r)^3}\right)\right\rangle \right.$$
$$\left. + \frac{62}{9} \delta_{\ell 0} - \frac{2\delta_{\ell 0}}{m_1^2 - m_2^2} \left[m_1^2 \ln\left(\frac{m_2}{\mu}\right) - m_2^2 \ln\left(\frac{m_1}{\mu}\right)\right]\right\}. \quad (15.163)$$

This result is independent of λ and η, as it should be, and the expectation value of the Araki–Sucher distribution P can be found in Eq. (15.134) for a general state.

15.6 Further Thoughts

Here are some suggestions for further thought.

(1) **Bound-Electron Self–Energy and Alternative Representations.** With the help of Eq. (9.121), $D_F^{\mu\nu}(k) = -g^{\mu\nu}/(k^2 + \mathrm{i}\epsilon)$, and a suitable identification of the binding potential V, show that Eq. (15.15) can alternatively be written as follows,

$$\Delta E_{\rm SE} = e^2 \int \frac{\mathrm{d}^4 k}{(2\pi)^4 \mathrm{i}} \frac{1}{k^2 + \mathrm{i}\epsilon} \left\langle \bar{\psi}_n \left| \gamma^\mu \frac{1}{\not p - \not k - m - \gamma^0 V} \gamma_\mu \right| \psi_n \right\rangle. \quad (15.164)$$

This is just the unrenormalized expression for the bound-state self-energy.

(2) **Spectral Representation and Classical Self–Energy.** In a spectral representation, we have

$$\mathcal{G}_F^{\rm DC}(E_n - \omega, \vec r, \vec r\,') = \sum_j \frac{\psi_j(\vec r)\, \psi_j^+(\vec r\,')}{E_j(1 - \mathrm{i}\delta) - E_n + \omega}. \quad (15.165)$$

Selecting the virtual state $j = n$ (equal to the reference state), and closing the contour so that the pole at zero photon energy is fully encircled, one obtains

$$\mathcal{G}_F^{\rm DC}(E_n - \omega, \vec r, \vec r\,') \to \frac{\psi_n(\vec r)\, \psi_n^+(\vec r)}{-\mathrm{i}E_n\delta + \omega} \to 2\pi \mathrm{i} \psi_n(\vec r)\, \psi_n^+(\vec r). \quad (15.166)$$

Of course, we assume that $E_n > 0$. The contribution of the reference-state pole to the self-energy becomes [see Eq. (15.15)]

$$\langle \psi^+ | H | \psi \rangle = -\mathrm{i}e^2 \int \frac{\mathrm{d}\omega}{2\pi} \int \frac{\mathrm{d}^3 k}{(2\pi)^3} D_{F\mu\nu}(\omega, \vec r - \vec r\,')$$
$$\times \left\langle \psi_n^+ \left| \alpha^\mu \mathrm{e}^{\mathrm{i}\vec k\cdot\vec r} \mathcal{G}_F(E_n - \omega, \vec r, \vec r\,')\, \alpha^\nu \mathrm{e}^{-\mathrm{i}\vec k\cdot\vec r\,'} \right| \psi_n \right\rangle$$
$$\to -\mathrm{i}e^2 \int \mathrm{d}^3 r \int \mathrm{d}^3 r' \int \frac{\mathrm{d}^3 k}{(2\pi)^3} D_{F00}(0, \vec k)$$
$$\times \psi_n^+(\vec r)\, \alpha^0 \mathrm{e}^{\mathrm{i}\vec k\cdot\vec r} \left\{ \frac{1}{2\pi} [2\pi \mathrm{i} \psi_n(\vec r)\, \psi_n^+(\vec r\,')] \right\} \alpha^0 \mathrm{e}^{-\mathrm{i}\vec k\cdot\vec r\,'} \psi_n(\vec r\,'). \quad (15.167)$$

Remembering that the photon propagator is given by [see Eq. (9.121)]

$$D_{F\mu\nu}(k) = -\frac{g_{\mu\nu}}{k^2 + \mathrm{i}\epsilon} = -\frac{g_{\mu\nu}}{\omega^2 - \vec{k}^2 + \mathrm{i}\epsilon}, \qquad D_{F00}(0,\vec{k}) = \frac{1}{\vec{k}^2} \tag{15.168}$$

one obtains

$$\begin{aligned}
\langle\psi^+|H|\psi\rangle &\to e^2 \int \mathrm{d}^3r \int \mathrm{d}^3r' \int \frac{\mathrm{d}^3k}{(2\pi)^3} \frac{\mathrm{e}^{\mathrm{i}\vec{k}\cdot(\vec{r}-\vec{r}\,')}}{\vec{k}^2} \psi_n^+(\vec{r})\,\psi_n(\vec{r})\,\psi_n^+(\vec{r}\,')\,\psi_n(\vec{r}\,') \\
&= \int \mathrm{d}^3r \int \mathrm{d}^3r' \,|\psi_n(\vec{r})|^2\, \frac{e^2}{4\pi|\vec{r}-\vec{r}\,'|}\, |\psi_n(\vec{r}\,')|^2 \\
&= \int \mathrm{d}^3r \int \mathrm{d}^3r' \,|\psi_n(\vec{r})|^2\, \frac{\alpha}{|\vec{r}-\vec{r}\,'|}\, |\psi_n(\vec{r}\,')|^2 \,.
\end{aligned} \tag{15.169}$$

The multiplying factor is α, not $Z\alpha$, because it is a self-interaction of the electron, not of the nucleus.

(3) **Renormalized Self–Energy.** How does the Ward identity (10.134) affect the renormalization of the Furry-picture relativistic bound-electron self-energy? It is possible that the renormalized version of the bound-state self-energy reads as follows,

$$\begin{aligned}
\Delta E_{\mathrm{SE}} = \mathrm{i}e^2 &\int \mathrm{d}^3r \int \mathrm{d}^3r' \int \frac{\mathrm{d}^4k}{(2\pi)^4} D_{F\mu\nu}(\omega,\vec{k}) \\
&\times \bar{\psi}_n(\vec{r})\,\gamma^\mu\,\mathrm{e}^{\mathrm{i}\vec{k}\cdot\vec{r}}\, S_F^{\mathrm{DC}}(E_n-\omega,\vec{r},\vec{r}\,')\,\gamma^\nu\,\mathrm{e}^{-\mathrm{i}\vec{k}\cdot\vec{r}\,'}\,\psi_n(\vec{r}) \\
&- \delta m \int \mathrm{d}^3r\,\bar{\psi}_n(\vec{r})\,\psi_n(\vec{r})\,.
\end{aligned} \tag{15.170}$$

Here, the mass counter term is the one defined in Eq. (10.261). Show that the renormalization of the bound-state self-energy derived in Sec. 15.2.2 does not require any additional renormalizations proportional to the Z_1 or Z_2 renormalization constants.

(4) **Nonrelativistic Self–Energy.** Show the equivalence of the relativistic self-energy (as discussed in Sec. 15.2.2) with the nonrelativistic version in Chap. 4, for bound states, in the nonrelativistic limit. Consider, in particular, Eqs. (6.100) and (6.99).

(5) **Two-Loop Binding Corrections.** Consider the generalization of the binding corrections of order $\alpha(Z\alpha)^5 m$, discussed in Sec. 15.3, to two-loop order, namely $\alpha^2(Z\alpha)^5 m$. The binding two-loop corrections have been evaluated by two independent groups, in Refs. [544–547] and in Refs. [548, 549]. It is instructive to study the role of the Fried–Yennie gauge in the treatment of infrared divergences. Gauge-invariant sets of diagrams can be treated separately, and the complete analytic form for the corrections of $\alpha(Z\alpha)^5 m$ is still unknown, as opposed to very accurate numerical results for the coefficients, which have been evaluated in Refs. [544–549]. An analytic evaluation of the entire correction of order $\alpha^2(Z\alpha)^5 m$ would be of interest.

(6) **Higher-Order Corrections.** In Sec. 15.4, we had discussed the evaluation of higher-order binding corrections of sixth order in $Z\alpha$ and beyond.

An informative review, with some explanatory remarks on particular higher-order terms of interest, is provided by EIDES,[5] GROTCH[6] and SHELYUTO[7] in Refs. [27, 550]. The leading two-loop triple logarithmic binding correction of order $\alpha^2(Z\alpha)^6 \ln^3[(Z\alpha)^{-2}]\, m$ (for S states) has been derived by KARSHENBOIM[8] in Ref. [551]. For two-loop logarithmic corrections of order $\alpha^2(Z\alpha)^6 \ln^n[(Z\alpha)^{-2}]\, m$, with $n = 0, 1, 2$, results can be found in Refs. [551–554]. A particular nonlogarithmic correction of order $\alpha^2(Z\alpha)^6\, m$, namely, the two-loop Bethe logarithm, constitutes the two-loop generalization of the effect discussed in Sec. 4.6.3. It has been calculated in Refs. [555–557]. Results for logarithmic energy corrections of orders $\alpha^2(Z\alpha)^7 \ln^2[(Z\alpha)^{-2}]\, m$, $\alpha^2(Z\alpha)^8 \ln^4[(Z\alpha)^{-2}]\, m$ and $\alpha^2(Z\alpha)^9 \ln^3[(Z\alpha)^{-2}]\, m$, have recently been derived in Refs. [554, 558, 559]. They compare favorably with numerical evaluations of the two-loop self-energy effect [560]. For a recent review on the general status of the higher-order corrections, we also mention Ref. [561]. Note that the two-loop slope of the Dirac form factor of the electron, as elucidated by BARBIERI,[9] MIGNACO[10] and REMIDDI,[11] is infrared finite [562, 563], in contrast to the one-loop slope discussed in Sec. 10.3. The three-loop slope of the Dirac form factor has been evaluated in Ref. [564], and the four-loop slope has recently been evaluated by LAPORTA[12] in Ref. [565]. The two-loop, three-loop, and four-loop slopes give rise to nonlogarithmic corrections of orders $\alpha^n(Z\alpha)^4\, m$, with $n = 2, 3, 4$, respectively. An independent confirmation of some of the recently obtained results, in particular, of the logarithmic corrections of orders $\alpha^2(Z\alpha)^7 \ln^2[(Z\alpha)^{-2}]\, m$, $\alpha^2(Z\alpha)^8 \ln^4[(Z\alpha)^{-2}]\, m$, and $\alpha^2(Z\alpha)^9 \ln^3[(Z\alpha)^{-2}]\, m$, with $n = 0, 1, 2$ (see Refs. [554, 558, 559]), could be of interest.

(7) **Hamiltonian and Hermiticity.** Reconsider the derivation of the relativistic recoil correction, and in particular, the Hamiltonian given in Eq. (15.102). In this expression, as already emphasized, we ignore the necessity to hermitize the interaction by anticommutators, $\vec{p}_1 \cdot \vec{A}(\vec{r}_1) \to \frac{1}{2}\{\vec{p}_1, \vec{A}(\vec{r}_1)\}$, which would otherwise lead to a more clumsy notation. Show that, in the derivation of the relativistic recoil correction, the hermitized version of the Hamiltonian leads to an equivalent result. When investigating the derivation, pay close attention to the transition from Eq. (15.145) to (15.147).

(8) **Vacuum Polarization and Dispersion Relation.** Be inspired by Sec. 15.3 and attempt to formulate a dispersion relation approach to the calculation of vacuum polarization, to supplement the treatment in Sec. 10.4.

[5] Mikhail Iosifovich Eides (b. 1947).
[6] Howard Grotch (b. 1940).
[7] Valery Aleksandrovich Shelyuto (b. 1955).
[8] Savely Grigorievich Karshenboim (b. 1960).
[9] Riccardo Barbieri (b. 1944).
[10] Juan Alberto Mignaco (1940–2001).
[11] Ettore Remiddi (b. 1941).
[12] Stefano Laporta (b. 1960).

Chapter 16

Bethe–Salpeter Equation

16.1 Overview

In earlier sections of this book, we have explored a variety of corrections to the basic description of a two-particle Coulombic bound state contained in the Dirac equation. We worked out higher-order corrections to one-electron energy levels in Chap. 11 through the use of the Foldy–Wouthuysen expansion of the Dirac–Coulomb interaction. Radiative corrections to the coupling of a single photon with one of the fermionic bound-state constituents were introduced through the use of form factors. The two-particle Breit Hamiltonian was described in Chap. 12. Additional two-body interactions were incorporated in Chap. 15, including the effect of radiative corrections to the coupling of two photons to one of the fermionic constituents. The latter aspect might be considered a two-photon form factor. We also discussed, in Chap. 15, recoil corrections coming from the exchange of a transverse photon (with retardation).

The Bethe–Salpeter approach allows us to analyze the energy levels of two-body bound systems to any desired order but presents us with an involved technical formalism which we would like to discuss here. The equation reminds us of the fact that the calculation of bound-state energy levels and scattering amplitudes is a rigorous and systematic procedure based directly on quantum field theory. In principle, the Bethe–Salpeter equation (BSEQ) can be used to obtain energies to any desired order of α, including all form factor, retardation, recoil, radiative-recoil, and every other type of correction. In practice, the procedures available to obtain these energies from the BSEQ are somewhat cumbersome, as we will see. In some sense, the BSEQ contains too many variables, being dependent on individual time variables for each constituent instead of simply an overall time. Also important to note is the wealth of scales: infrared ($E \to 0$), low-energy ($E \sim m\alpha^2$), middle-energy ($E \sim m\alpha$), high-energy ($E \sim m$), and ultraviolet ($E \to \infty$). All of these scales were present in earlier calculations and all are important for the evaluation of contributions using the BSEQ and must be dealt with. The challenge of separating contributions coming from different scales is a prime difficulty encountered in using the BSEQ for high-order calculations. Some early references to the BSEQ and related matters include [566–568], and reviews and an extensive list of references can be found in [569–571].

16.2 General Ideas

16.2.1 *Two-Body Bound States and Green Functions*

All methods for finding bound-state energies rely on identifying some quantity that depends on the energy and a procedure for extracting that energy value. So far in this book, we have obtained energy values as eigenvalues of some Hamiltonian operator which enters a Schrödinger equation or a generalization such as the Dirac equation, the Breit equation, or one or these supplemented with effective interactions obtained from scattering amplitudes. In the Bethe–Salpeter approach, the bound state energy can be extracted from information contained in the electron-proton to electron-proton (two-particle) Green function

$$G'(x_1, x_2; y_1, y_2) = \langle 0|\mathrm{T}\psi_1(x_1)\psi_2(x_2)\overline{\psi}_1(y_1)\overline{\psi}_2(y_2)|0\rangle. \tag{16.1}$$

(These are Heisenberg picture operators, so the Green function G' includes all radiative corrections but is not renormalized.) We are thinking of our bound state as hydrogen, but the procedure described here would apply as well to other bound states containing two spin-1/2 constituents, such as muonium, and with some modifications, also to positronium. We also assume that the fermions are elementary, so in this chapter we are not considering, say, the internal structure of the proton. The diagrammatic expansion of the 2-to-2 Green function is given in Fig. 16.1(a). Intuitively, if appropriately expressed in terms of the center of mass (CM) momentum, G' will have a pole when the CM energy equals that of a bound state, corresponding to coherent propagation of a two-body subsystem for an extended period of time. These bound-state poles are completely analogous to the poles found in the Coulomb Green function [see Eq. (4.124)], in the relativistic free-fermion propagator in Eq. (7.179), and in the Dirac–Coulomb propagator in Eq. (8.107). In order to fix ideas, let us note that the energy variable in the Schrödinger–Coulomb Green function of Eq. (4.124) is hidden in the denominator factor $(k + \ell + 1 - \nu)$. This Green function contains an energy pole because of the definition $\nu^2 = -Z^2\alpha^2\mu/(2E)$ found in Eq. (4.83).

Before displaying the bound-state poles in G' explicitly, it is useful to describe the transformation we use to go from coordinates for the individual particles to center-of-mass (CM) and relative space-time coordinates, as a generalization of the treatment in Sec. 13.2.2. These are

$$X = \xi_1 x_1 + \xi_2 x_2 \quad (\text{CM}), \qquad x = x_1 - x_2 \quad (\text{relative}), \tag{16.2a}$$

$$x_1 = X + \xi_2 x, \qquad x_2 = X - \xi_1 x. \tag{16.2b}$$

The transformation from individual particle coordinates x_1, x_2 to CM and relative variables X, x has unit Jacobian as long as $\xi_1 + \xi_2 = 1$. We will use

$$\xi_1 = \frac{m_1}{M}, \qquad \xi_2 = \frac{m_2}{M}, \qquad M = m_1 + m_2. \tag{16.3}$$

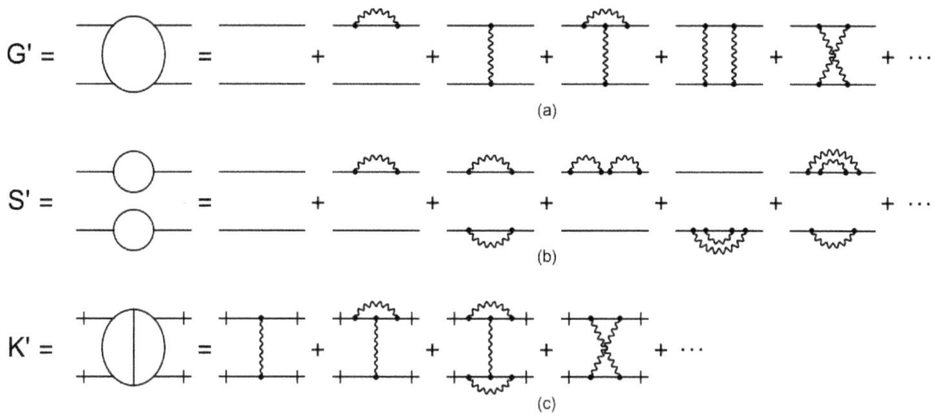

Fig. 16.1 Diagrammatic representation of the quantities that appear in the Bethe–Salpeter equation. Part (a) represents the full 2-to-2 Green function G'. The electron line is shown at the top and the proton at the bottom. Wiggly lines represent photons. Part (b) shows the product of a fully corrected electron propagator with a fully corrected proton propagator. Corrections on these lines are known as self-energy corrections. Part (c) shows the kernel of the BSEQ, symbolized by a circle with a vertical line through it. The kernel is two-particle irreducible, meaning that it cannot be cut into two disjoint pieces by a single cut on the electron line and a single cut on the proton line. It is also one-particle irreducible. The small vertical lines on the external legs indicate amputation — the kernel is defined with all external lines removed. (The amputated lines are present in the diagram purely for clarity.)

The corresponding CM and relative momenta are

$$P = p_1 + p_2 \quad (\text{CM}), \qquad p = \xi_2 p_1 - \xi_1 p_2 \quad (\text{relative}), \tag{16.4a}$$

$$p_1 = \xi_1 P + p, \qquad p_2 = \xi_2 P - p. \tag{16.4b}$$

The momentum transformation $p_1, p_2 \to P, p$ also has unit Jacobian, and we note that $x_1 \cdot p_1 + x_2 \cdot p_2 = X \cdot P + x \cdot p$, a relation that will prove useful when moving to momentum space.

The momentum-space version of the 2-to-2 Green function has the form

$$\begin{aligned}\overline{G}'(p_1, p_2; q_1, q_2) &= \int \mathrm{d}^4x_1\,\mathrm{d}^4x_2\,\mathrm{d}^4y_1\,\mathrm{d}^4y_2\,\mathrm{e}^{\mathrm{i}(p_1\cdot x_1 + p_2\cdot x_2 - q_1\cdot y_1 - q_2\cdot y_2)}\,G'(x_1, x_2; y_1, y_2)\\ &= \int \mathrm{d}^4X\,\mathrm{d}^4x\,\mathrm{d}^4Y\,\mathrm{d}^4y\,\mathrm{e}^{\mathrm{i}(P\cdot X + p\cdot x - Q\cdot Y - q\cdot y)}\tilde{G}'(X, x; Y, y), \end{aligned}\tag{16.5}$$

where $\tilde{G}'(X, x; Y, y) = G'(X + \xi_2 x, X - \xi_1 x; Y + \xi_2 y, Y - \xi_1 y)$. Translation invariance in coordinate space tells us that $\tilde{G}'(X, x; Y, y)$ is unaltered when X and Y are displaced by the same arbitrary amount. The immediate consequence is that the momentum space 2-to-2 Green function contains the energy-momentum conserving Dirac-δ function $\delta^{(4)}(P - Q)$ as a factor. In detail, we find that

$$\overline{G}'(p_1, p_2; q_1, q_2) = (2\pi)^4\,\delta^{(4)}(P - Q)\,G'(P; p, q) \tag{16.6}$$

where

$$G'(P; p, q) = \int \mathrm{d}^4X\,\mathrm{d}^4x\,\mathrm{d}^4y\,\mathrm{e}^{\mathrm{i}(P\cdot X + p\cdot x - q\cdot y)}G'(X, x; 0, y). \tag{16.7}$$

We are now prepared to display the bound-state poles in the momentum-space representation of G'. For times such that $x_1^0, x_2^0 > y_1^0, y_2^0$, we can insert a complete set of states into G', among which are the two-particle bound states $|\vec{P}, n, k\rangle$. These states of mass M_n and energy

$$\omega_n(\vec{P}) = (M_n^2 + \vec{P}^2)^{1/2} \tag{16.8}$$

are specified by n, their three-momentum $\vec{P}$, and some additional labels k describing possible degeneracy. (The mass M_n is the physical mass of the bound system including all corrections; n stands for the principal quantum number and all other quantum numbers that are necessary in order to determine the states that have that mass; and k summarizes the additional quantum numbers required to completely specify the state.) The contribution of the two-particle bound states to G' is

$$\begin{aligned} G'(x_1, x_2; y_1, y_2) = \Theta(X^0 - Y^0) \sum_{nk} \int \frac{\mathrm{d}^3 P}{(2\pi)^3\, 2\omega_n(\vec{P})} \\ \times \langle 0|\, \mathrm{T}\psi_1(x_1)\psi_2(x_2)\, |\vec{P}, n, k\rangle \langle \vec{P}, n, k|\, \mathrm{T}\overline{\psi}_1(y_1)\overline{\psi}_2(y_2)\, |0\rangle + \cdots, \end{aligned} \tag{16.9}$$

where the dots represent contributions from other types of states and contributions from other orderings of the time variables that don't give rise to particle-particle bound state poles. We note that the condition $x_1^0, x_2^0 > y_1^0, y_2^0$ actually implies a more stringent theta function, $\Theta(\min(x_1^0, x_2^0) - \max(y_1^0, y_2^0))$, rather than simply $\Theta(X^0 - Y^0)$. However, the use of the more exact theta function changes neither the positions nor the residues of the bound-state poles in G' (see also Sec. 16.5). The matrix elements that have appeared represent a new kind of form factor commonly known as the Bethe–Salpeter wave function:

$$\Psi_{nk}(\vec{P}; x_1, x_2) = \langle 0|\, \mathrm{T}\psi_1(x_1)\psi_2(x_2)\, |\vec{P}, n, k\rangle, \tag{16.10}$$

$$\overline{\Psi}_{nk}(\vec{P}; y_1, y_2) = \langle \vec{P}, n, k|\, \mathrm{T}\overline{\psi}_1(y_1)\overline{\psi}_2(y_2)\, |0\rangle. \tag{16.11}$$

These are curiously complicated wave functions as they each depend on two time variables. This complication can be reduced somewhat by use of CM and relative coordinates along with the explicit extraction of the dependence of the wave function on the CM coordinate. We make use of the field-theoretical four-momentum operator $\hat{P}$ to write

$$\begin{aligned} \Psi_{nk}(\vec{P}; x_1, x_2) &= \langle 0|\mathrm{T}\psi_1(X + \xi_2 x)\, \psi_2(X - \xi_1 x)|\vec{P}, n, k\rangle \\ &= \langle 0|\mathrm{T}\, \mathrm{e}^{\mathrm{i}\hat{P}\cdot X}\psi_1(\xi_2 x)\mathrm{e}^{-\mathrm{i}\hat{P}\cdot X}\, \mathrm{e}^{\mathrm{i}\hat{P}\cdot X}\psi_2(-\xi_1 x)\, \mathrm{e}^{-\mathrm{i}\hat{P}\cdot X}|\vec{P}, n, k\rangle \\ &= e^{-\mathrm{i}P_n\cdot X}\Psi_{nk}(\vec{P}, \xi_2 x, -\xi_1 x), \end{aligned} \tag{16.12}$$

where $P_n = (\omega_n(\vec{P}), \vec{P})$, and analogously

$$\overline{\Psi}_{nk}(\vec{P}; y_1, y_2) = \mathrm{e}^{\mathrm{i}P_n\cdot Y}\overline{\Psi}_{nk}(\vec{P}; \xi_2 y, -\xi_1 y). \tag{16.13}$$

So, the most important new feature compared with the usual nonrelativistic two-particle quantum mechanics is the dependence of the wave function on a relative time. We make use of the representation of Eq. (7.151) for the theta function (with $E = 0$ and $f(E) = 1$) to write (16.9) as

$$
\begin{aligned}
G'(x_1, x_2; y_1, y_2) &= \mathrm{i}\int \frac{\mathrm{d}P^0}{2\pi}\frac{\mathrm{e}^{-\mathrm{i}P^0(X^0-Y^0)}}{P^0+\mathrm{i}\epsilon}\sum_{nk}\int\frac{\mathrm{d}^3P}{(2\pi)^3 2\omega_n(\vec{P}\,)} \\
&\quad\times \Psi_{nk}(\vec{P},\xi_2 x, -\xi_1 x)\,\mathrm{e}^{-\mathrm{i}P_n\cdot X+\mathrm{i}P_n\cdot Y}\,\overline{\Psi}_{nk}(\vec{P},\xi_2 y,-\xi_1 y) + \cdots \\
&= \mathrm{i}\sum_{nk}\int \frac{\mathrm{d}^4P}{(2\pi)^4 2\omega_n(\vec{P}\,)}\frac{\mathrm{e}^{-\mathrm{i}P\cdot(X-Y)}}{P^0-\omega_n(\vec{P}\,)+\mathrm{i}\epsilon} \\
&\quad\times \Psi_{nk}(\vec{P},\xi_2 x,-\xi_1 x)\,\overline{\Psi}_{nk}(\vec{P},\xi_2 y,-\xi_1 y) + \cdots, \qquad (16.14)
\end{aligned}
$$

where $P_n = (\omega_n(\vec{P}), \vec{P}) = \big((M_n^2 + \vec{P}^2)^{1/2}, \vec{P}\big)$ is the physical (mass-shell) momentum and the replacement $P^0 \to P^0 - \omega_n(\vec{P}\,)$ was made in the P^0 integration variable in order to achieve the final form. The energy pole is most clearly revealed in momentum space. We combine Eqs. (16.7) and (16.14) to find

$$
G'(P;p,q) = \mathrm{i}\sum_{nk}\frac{\Psi_{nk}(\vec{P};p)\overline{\Psi}_{nk}(\vec{P};q)}{P^0-\omega_n(\vec{P}\,)+\mathrm{i}\epsilon} + \cdots, \qquad (16.15)
$$

where

$$
\Psi_{nk}(\vec{P},p) = \int \mathrm{d}^4x\,\mathrm{e}^{\mathrm{i}p\cdot x}\frac{\Psi_{nk}(\vec{P},\xi_2 x,-\xi_1 x)}{\sqrt{2\omega_n(\vec{P}\,)}}, \qquad (16.16)
$$

$$
\overline{\Psi}_{nk}(\vec{P},q) = \int \mathrm{d}^4y\,\mathrm{e}^{-\mathrm{i}q\cdot y}\frac{\overline{\Psi}_{nk}(\vec{P},\xi_2 y,-\xi_1 y)}{\sqrt{2\omega_n(\vec{P}\,)}}. \qquad (16.17)
$$

The bound state pole is at position $P^0 \to \omega_n(\vec{P}\,) = (M_n^2 + \vec{P}^{\,2})^{1/2}$, or, in the CM frame with $\vec{P} = 0$, it is at $P^0 \to M_n$. The extra terms represented by the dots do not contain a two-particle bound state pole. We have included the factor $(2\omega_n(\vec{P}\,))^{1/2}$ in the momentum space wave function for convenience later on.

We will use a graphical approach to derive the actual Bethe–Salpeter equation, as was done in the original Ref. [566]. The 2-to-2 Green function $G'(P;p,q)$ is shown in Fig. 16.2(a) with labels for the momenta of the external legs. The graphical expansion of G' was shown in Fig. 16.1(a) above. It contains all diagrams having one electron and one proton in the initial state, and one each in the final state, with photons (and vacuum polarization loops) attached in all possible ways. In our notation, the product of an arbitrary pair of 2-to-2 objects, say A and B, is obtained by integration over the internal relative momentum

$$
[AB]\,(P;p,q) = \int \frac{\mathrm{d}^4k}{(2\pi)^4}A(P;p,k)B(P;k,q)\,. \qquad (16.18)
$$

This is illustrated in Fig. 16.2(b). The inhomogeneous BSEQ is shown in the symbolic equation depicted in Fig. 16.2(c). The idea is that the individual diagrams in G' implicitly contained in the diagram on the left-hand side of the equation depicted in Fig. 16.2(c) are in one-to-one correspondence with the diagrams found in the sum

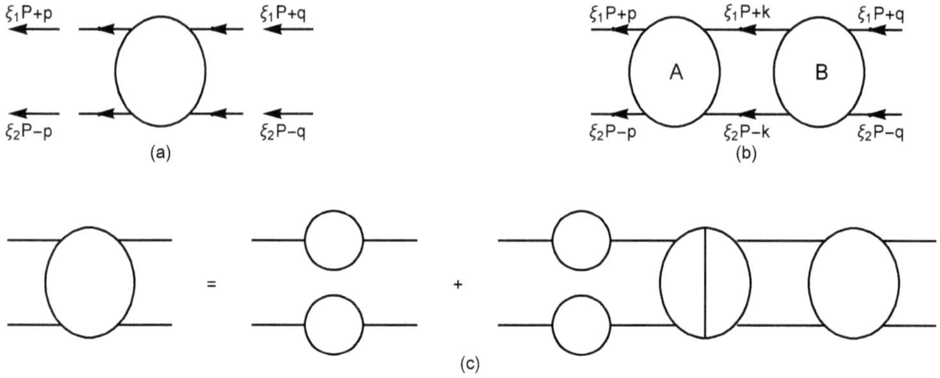

Fig. 16.2 Illustration of the diagrammatic language useful for discussion of the BSEQ. Panel (a) shows the 2-to-2 Green function with momentum labels on the external legs. A CM four-momentum P flows through the diagram, shared between the lines according to $\xi_1 P$ and $\xi_2 P$. The outgoing and incoming relative momenta are p and q. Part (b) represents the product of A and B where k is an integration variable representing the relative momentum of the intermediate state. Part (c) is the inhomogeneous BSEQ itself. The vertical line through the kernel is a reminder that it is two- (and one-)particle irreducible.

of the non-interacting 2-to-2 Green function S' of Fig. 16.1(b) and the product of S' times the two-particle irreducible kernel K' of Fig. 16.1(c) times G' itself (see also Sec. 16.5). In symbols, the BSEQ can be written as

$$G'(P;p,q) = S'(P;p,q) + \int \frac{d^4\ell}{(2\pi)^4} \frac{d^4k}{(2\pi)^4} S'(P;p,\ell)\, K'(P;\ell,k)\, G'(P;k,q)\,. \quad (16.19)$$

Because there is no possibility for momentum flow between the electron and proton lines in S', the relative momentum coming into the free propagator must equal that coming out, and we can write [see also Eq. (16.6)]

$$S'(P;p,q) = (2\pi)^4\, \delta^{(4)}(p-q)\, S'(P;p), \quad (16.20)$$

so that the BSEQ takes the slightly simpler form

$$G'(P;p,q) = S'(P;p,q) + S'(P;p) \int \frac{d^4k}{(2\pi)^4} K'(P;p,k)\, G'(P;k,q)\,. \quad (16.21)$$

As a shorthand, we can represent the BSEQ as

$$G' = S' + S'K'G', \quad (16.22)$$

with the integration over relative momentum implicit. We note that the graphical demonstration of the BSEQ works just as well in the reversed order, so as an alternative form we also have

$$G' = S' + G'K'S'. \quad (16.23)$$

Up until this point we have been dealing with unrenormalized quantities, but as discussed in Chap. 10, we can renormalize by *(i)* writing the equations in terms of the physical charge and mass, and *(ii)* multiplying by the appropriate electron and proton wave function renormalization constants Z_{2e} and Z_{2p} as defined in Eq. (10.111):

$$G' = Z_{2e}\, Z_{2p}\, G, \qquad S' = Z_{2e}\, Z_{2p}\, S, \qquad K' = (Z_{2e}\, Z_{2p})^{-1}\, K\,, \tag{16.24}$$

where the primed quantities are the original bare objects and their unprimed partners are the finite renormalized quantities. It is evident that Z_{2e} and Z_{2p} cancel out and the BSEQ

$$G = S + SKG = S + GKS \tag{16.25}$$

has the same form after renormalization that it had before, and we will continue to write it in the same way in terms of G, S, and K as we did in terms of G', S', and K'. No additional ultraviolet divergences are generated by the integrations implicit in the products contained in the BSEQ. (This is not the case when the BSEQ is used to describe a particle-antiparticle state such as positronium because the kernel contains a term representing the annihilation of the e^-– e^+ pair to a single photon [572, 573].) The graphical expressions for the quantities entering the BSEQ are best given in terms of the renormalized self-energy, vertex, and vacuum polarization functions. The graphical expansions are greatly compressed in terms of this more efficient language (see Fig. 16.3). As an explicit example, the non-interacting propagator S contains only one term and can be written as

$$S(P;p) = \left[\frac{\mathrm{i}}{\xi_1 \not{P} + \not{p} - m_1 - \Sigma_1(\xi_1 P + p) + \mathrm{i}\epsilon}\right]_1 \left[\frac{\mathrm{i}}{\xi_2 \not{P} - \not{p} - m_2 - \Sigma_2(\xi_2 P - p) + \mathrm{i}\epsilon}\right]_2 . \tag{16.26}$$

The $\Sigma_i(q)$ functions are the renormalized self-energies. The subscript labels on the square brackets are reminders that there are two fermion lines, each with its own Dirac indices, which must be distinguished. From now on we will deal exclusively with the renormalized theory.

A homogeneous equation for the Bethe–Salpeter wave functions is obtained as the residue of (16.25) at a bound state pole. From Eq. (16.15), we can write the pole term of G as follows,

$$G(P^0) \to \mathrm{i} \sum_k \frac{\Psi_{nk} \overline{\Psi}_{nk}}{P^0 - P^0_n}\,, \tag{16.27}$$

as P^0 approaches the bound-state energy $P^0_n = \omega_n(\vec{P}\,) = (M_n^2 + \vec{P}^{\,2})^{1/2}$, where M_n is the rest energy of the nth bound state. We now find that the residue of (16.22) [and of (16.25)] for P^0 near the bound-state pole position P^0_n is

$$\mathrm{i} \sum_k \Psi_{nk} \overline{\Psi}_{nk} = S(P^0_n) K(P^0_n)\, \mathrm{i} \sum_k \Psi_{nk} \overline{\Psi}_{nk}, \tag{16.28}$$

where the dependence of S and K on the CM energy P^0 is made explicit. The degenerate wave functions Ψ_{nk} for the various values of the degeneracy label k are assumed to be linearly independent, so we obtain the homogeneous BSEQ

$$\Psi_{nk} = S(P^0_n)\, K(P^0_n)\, \Psi_{nk}\,. \tag{16.29}$$

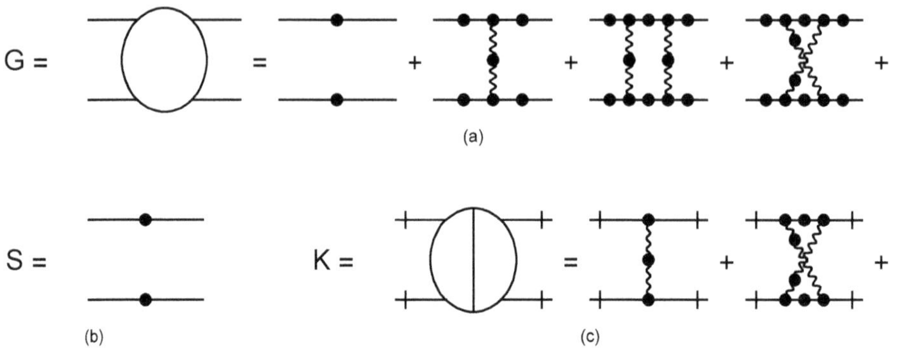

Fig. 16.3 Depiction of quantities entering the renormalized BSEQ. The 2-to-2 Green function is shown in (a), the non-interacting two particle Green function in (b), and the amputated two-particle-irreducible kernel in (c). The fermion and photon lines in this diagrams are adorned with filled dots to indicate that they are fully-corrected and renormalized propagators, while a dot on a fermion-photon vertex stands for a fully-corrected vertex function.

If we make the dependence on relative momentum explicit, the homogeneous BSEQ takes the form

$$\Psi_{nk}(\vec{P};p) = S(P_n;p) \int \frac{\mathrm{d}^4\ell}{(2\pi)^4} K(P_n;p,\ell)\, \Psi_{nk}(\vec{P};\ell)\,. \tag{16.30}$$

In order to derive a normalization condition for Bethe–Salpeter wave functions, and later on, a form of the BSEQ more convenient for calculations and a perturbative expansion for obtaining estimates of the bound state energies, it is necessary to develop some confidence with the manipulation of the BSEQ into a variety of equivalent forms. We can write formal solutions of the inhomogeneous BSEQ as given in Eqs. (16.22) and (16.23) as

$$G = (1 - SK)^{-1} S = S(1 - KS)^{-1}. \tag{16.31}$$

In our notation, the inverse of an operator A satisfies $AA^{-1} = 1$ if the function $A^{-1}(P;p,q)$ exists and satisfies

$$\int \frac{\mathrm{d}^4\ell}{(2\pi)^4} A(P;p,\ell)\, A^{-1}(P;\ell,q) = (2\pi)^4\, \delta^{(4)}(p-q)\,. \tag{16.32}$$

Using the general operator formula $(AB)^{-1} = B^{-1}A^{-1}$, we can express the 2-to-2 Green function as

$$G = (S^{-1} - K)^{-1}\,, \tag{16.33}$$

starting from either form of (16.31). Or, we can expand either form of (16.31) in a geometric series to find

$$G = S + S\,K\,S + S\,K\,S\,K\,S + \cdots. \tag{16.34}$$

Consideration of the diagrams shown in Fig. 16.1 will convince the reader that this expansion is correct. Finally, it is sometimes useful to define the truncated

Green function G_{T} by taking away the disconnected component and amputating the external legs:

$$G = S + S\,G_{\mathrm{T}}\,S\,. \tag{16.35}$$

Operator manipulations of the sort used above can be used to show that the truncated Green function satisfies BSEQs of its own:

$$G_{\mathrm{T}} = K + K\,S\,G_{\mathrm{T}} = K + G_{\mathrm{T}}\,S\,K\,, \tag{16.36}$$

with solutions

$$G_{\mathrm{T}} = (1 - KS)^{-1}K = K(1 - SK)^{-1}\,. \tag{16.37}$$

A geometric expansion, or a diagrammatic analysis, leads to the series expression

$$G_{\mathrm{T}} = K + K\,S\,K + K\,S\,K\,S\,K + \cdots\,. \tag{16.38}$$

We are now in a position to derive the orthonormality condition for Bethe–Salpeter wave functions. Ultimately, a normalization of the wave functions is implied by the fact that G satisfies an inhomogeneous equation given in Eq. (16.25). So, the normalization of G and also of its residues is fixed. We start by noting the expansion of G^{-1},

$$G^{-1}(P^0) = G^{-1}(P^0_n) + (P^0 - P^0_n)\,\frac{\mathrm{d}}{\mathrm{d}P^0}G^{-1}(P^0)\bigg|_{P^0_n} + \mathcal{O}(P^0 - P^0_n)^2\,, \tag{16.39}$$

where we have made the dependence of G^{-1} on the CM energy P^0 explicit. We consider the identity $G = GG^{-1}G$ and write out the residue at the bound-state pole P^0_n of each side, finding

$$\mathrm{i}\sum_k \Psi_{nk}\overline{\Psi}_{nk} = \mathrm{i}\sum_r \Psi_{nr}\overline{\Psi}_{nr}\,\frac{\mathrm{d}}{\mathrm{d}P^0}G^{-1}(P^0)\bigg|_{P^0_n}\,\mathrm{i}\sum_s \Psi_{ns}\overline{\Psi}_{ns}\,. \tag{16.40}$$

The homogeneous BSEQ was used to remove a potential second-order pole on the right-hand side. The linear independence of the states Ψ_{nk} for various values of k, and similarly of the states $\overline{\Psi}_{nk}$, allows us to write the orthonormality relation in the final form

$$\delta_{rs} = \mathrm{i}\overline{\Psi}_{nr}\,\frac{\mathrm{d}}{\mathrm{d}P^0}G^{-1}(P^0)\bigg|_{P^0_n}\,\Psi_{ns}\,, \tag{16.41}$$

where $G^{-1} = S^{-1} - K$.

To close this introductory section, we will take a look at the leading contribution to the BSEQ when the binding is due to a Coulomb potential. We will make a number of approximations and find in the end that the BSEQ is just an equation for a nonrelativistic Coulomb bound state embellished with relativistic, radiative, recoil, and retardation corrections. The main approximations are based on the facts that the binding energy in a nonrelativistic Coulombic bound state is small (of order $\mu\alpha^2$ where μ is the reduced mass), the typical momentum is of order $\mu\alpha$, and the radiative corrections are small of order α. We have explored these aspects in the preceding chapters. We commence our string of approximations by neglecting all self-energy corrections from the non-interacting electron-proton Green function $S(P;p)$. In this limit, $S(P;p)$ simply becomes the product of a free Dirac propagator for the electron and one for the proton,

$$S(P;p) \to S_0(P;p) = \left[\frac{\mathrm{i}}{\xi_1 \not{P} + \not{p} - m_1 + \mathrm{i}\epsilon}\right]_1 \left[\frac{\mathrm{i}}{\xi_2 \not{P} - \not{p} - m_2 + \mathrm{i}\epsilon}\right]_2 , \tag{16.42}$$

where the two particles, electron and proton, are denoted particles 1 and 2. It is important to maintain the distinction between the particles, as the first square bracket carries Dirac indices for particle 1 and the second carries Dirac indices for particle 2 (as denoted by subscripts on the square brackets). We also restrict the kernel to just one term describing the exchange of a single photon,

$$K(P;p,q) \to [-\mathrm{i}q_1\gamma_\mu]_1 \frac{-\mathrm{i}g_{\mu\nu}}{(p-q)^2 + \mathrm{i}\epsilon} [-\mathrm{i}q_2\gamma_\nu]_2 = \frac{-4\pi\mathrm{i}Z\alpha\, \gamma_1^\mu\, g_{\mu\nu}\, \gamma_2^\mu}{(p-q)^2 + \mathrm{i}\epsilon}, \tag{16.43}$$

where $q_1 q_2$ is the product of the charges of the two particles: $q_1 q_2 = -4\pi Z\alpha$. At this stage, we have made what is called the ladder approximation. The 2-to-2 Green function G has been approximated by a sum of terms containing 0, 1, 2, etc., photons connecting the electron and proton without crossing (like the rungs of a ladder). We make two additional approximations at this point. To begin with, we anticipate that the Bethe–Salpeter wave function has "large" and "small" components, just as the Dirac wave function does [see Eqs. (6.100), (6.101) and (8.11) in Sec. (8.2.2)]. When we expand the gamma matrix factor $\gamma_1^\mu g_{\mu\nu} \gamma_2^\nu = \gamma_1^0\gamma_2^0 - \vec{\gamma}_1\cdot\vec{\gamma}_2$, we see that the second term connects large to small components and will be small in the nonrelativistic limit. In fact, the spatial Dirac γ matrices typically make contributions on the order of the particle speed, which is $\mathcal{O}(\alpha)$ in a nonrelativistic Coulombic bound state. Our second approximation is the neglect of retardation. That is, we will work in the approximation that the potential is instantaneous. This approximation is equivalent to (via Fourier transformation) the neglect of the energy in the photon propagation factor (cf. the treatment of the Breit interaction in Chap. 12). Since the potential can only be instantaneous in one Lorentz frame (modulo spatial rotation), we choose to work in the CM frame where $\vec{P} = \vec{0}$, and $\not{P} = \gamma^0 P^0$. The homogeneous BSEQ, subject to all of these approximations, has the form [cf. Eq. (16.29)]

$$\Psi_{nk}(p) = \left[\frac{\mathrm{i}}{\xi_1 \not{P} + \not{p} - m_1 + \mathrm{i}\epsilon}\right]_1 \left[\frac{\mathrm{i}}{\xi_2 \not{P} - \not{p} - m_2 + \mathrm{i}\epsilon}\right]_2 \int \frac{\mathrm{d}^4 q}{(2\pi)^4} \frac{(-4\pi\mathrm{i}Z\alpha)\gamma_1^0\gamma_2^0}{-(\vec{p}-\vec{q}\,)^2} \Psi_{nk}(q), \tag{16.44}$$

where we have defined the CM wave function $\Psi_{nk}(p) \equiv \Psi_{nk}(\vec{P} = 0; p)$.

Quite surprisingly, the approximate BSEQ given in Eq. (16.44), while tremendously simplified compared to the full BSEQ, is still too complex to be solved directly. Some of the difficulty is related to the relative energy variable present in the wave function $\Psi_{nk}(p)$. The bound-state energy P^0 is independent of relative energy. The relative energy p^0 affects the residue of the pole in Eq. (16.27), not the position of the pole. The dependence on the relative energy is most conveniently eliminated by integrating it out and working in terms of a wave function dependent wholly on spatial variables,

$$\varphi_{nk}(\vec{p}\,) \equiv \int_{-\infty}^{\infty} \frac{\mathrm{d}p^0}{2\pi} \Psi_{nk}(p) \tag{16.45}$$

instead of with $\Psi_{nk}(p)$. Integration over the fermion propagation factor is facilitated by separating pole terms according to

$$\frac{\mathrm{i}}{\not{p} - m + \mathrm{i}\epsilon} = \frac{\mathrm{i}\,\Lambda_+(\vec{p}\,)\,\gamma^0}{p_0 - \omega(\vec{p}\,) + \mathrm{i}\epsilon} + \frac{\mathrm{i}\,\Lambda_-(\vec{p}\,)\,\gamma^0}{p_0 + \omega(\vec{p}\,) - \mathrm{i}\epsilon}, \tag{16.46}$$

where $\omega(\vec{p}\,) = (m^2 + \vec{p}^{\,2})^{1/2}$ and $\Lambda_\pm(\vec{p}\,)$ are the positive-energy and negative-energy projectors that we have already encountered in Eq. (14.18),

$$\Lambda_\pm(\vec{p}\,) \equiv \frac{1}{2\omega(\vec{p}\,)}\left[\omega(\vec{p}\,) \pm (m - \vec{\gamma}\cdot\vec{p}\,)\gamma^0\right] = \frac{1}{2\omega(\vec{p}\,)}\begin{pmatrix} \omega(\vec{p}\,) \pm m & \pm\vec{\sigma}\cdot\vec{p} \\ \pm\vec{\sigma}\cdot\vec{p} & \omega(\vec{p}\,) \mp m \end{pmatrix}. \tag{16.47}$$

The $\Lambda_\pm(\vec{p}\,)$ are projection operators satisfying

$$\Lambda_\pm(\vec{p}\,)\Lambda_\pm(\vec{p}\,) = \Lambda_\pm(\vec{p}\,), \qquad \Lambda_\pm(\vec{p}\,)\Lambda_\mp(\vec{p}\,) = 0, \qquad \Lambda_\pm(\vec{p}\,) + \Lambda_\mp(\vec{p}\,) = 1. \tag{16.48}$$

The relative energy integral is done by poles. It has four terms, two of which vanish upon integration (having poles on the same side of the axis), leaving

$$\begin{aligned}
\int_{-\infty}^{\infty} \frac{\mathrm{d}p^0}{2\pi} S_0(P;p) &= \int_{-\infty}^{\infty} \frac{\mathrm{d}p^0}{2\pi} \left[\frac{\mathrm{i}}{\xi_1\not{P} + \not{p} - m_1 + \mathrm{i}\epsilon}\right]_1 \left[\frac{\mathrm{i}}{\xi_2\not{P} - \not{p} - m_2 + \mathrm{i}\epsilon}\right]_2 \\
&= \int_{-\infty}^{\infty} \frac{\mathrm{d}p^0}{2\pi} \left[\frac{\mathrm{i}\,\Lambda_+(\vec{p}\,)\,\gamma^0}{\xi_1 P^0 + p^0 - \omega_1(\vec{p}\,) + \mathrm{i}\epsilon} + \frac{\mathrm{i}\,\Lambda_-(\vec{p}\,)\,\gamma^0}{\xi_1 P^0 + p^0 + \omega_1(\vec{p}\,) - \mathrm{i}\epsilon}\right]_1 \\
&\quad \times \left[\frac{\mathrm{i}\Lambda_+(-\vec{p}\,)\gamma^0}{\xi_2 P^0 - p^0 - \omega_2(\vec{p}\,) + \mathrm{i}\epsilon} + \frac{\mathrm{i}\Lambda_-(-\vec{p}\,)\gamma^0}{\xi_2 P^0 - p^0 + \omega_2(\vec{p}\,) - \mathrm{i}\epsilon}\right]_2 \\
&= \mathrm{i}\left\{\frac{\left[\Lambda_+(\vec{p}\,)\gamma^0\right]_1\left[\Lambda_+(-\vec{p}\,)\gamma^0\right]_2}{P^0 - \omega_1(\vec{p}\,) - \omega_2(\vec{p}\,)} - \frac{\left[\Lambda_-(\vec{p}\,)\gamma^0\right]_1\left[\Lambda_-(-\vec{p}\,)\gamma^0\right]_2}{P^0 + \omega_1(\vec{p}\,) + \omega_2(\vec{p}\,)}\right\},
\end{aligned} \tag{16.49}$$

where $\omega_i(\vec{p}\,) \equiv (m_i^2 + \vec{p}^{\,2})^{1/2}$. The approximate three-dimensional BSEQ (the integral of (16.44) over relative energy) takes the form

$$\varphi_{nk}(\vec{p}\,) = \left\{\frac{\Lambda_{1+}(\vec{p}\,)\Lambda_{2+}(-\vec{p}\,)}{P^0 - \omega_1(\vec{p}\,) - \omega_2(\vec{p}\,)} - \frac{\Lambda_{1-}(\vec{p}\,)\,\Lambda_{2-}(-\vec{p}\,)}{P^0 + \omega_1(\vec{p}\,) + \omega_2(\vec{p}\,)}\right\} \int \frac{\mathrm{d}^3q}{(2\pi)^3} \frac{-4\pi Z\alpha}{(\vec{p} - \vec{q}\,)^2} \varphi_{nk}(\vec{q}\,), \tag{16.50}$$

where $\Lambda_{\mathrm{i}\pm}(\vec{p}\,)$ is defined with $\omega_i(\vec{p}\,)$ and m_i instead of $\omega(\vec{p}\,)$ and m. This approximate BSEQ can be put into a form resembling a two-particle Dirac equation. We define the single-particle Dirac Hamiltonian

$$H_i(\vec{p}\,) \equiv \beta m_i + \vec{\alpha}\cdot\vec{p} = (m_i - \vec{\gamma}\cdot\vec{p}\,)\,\gamma^0 = \omega_i(\vec{p}\,)\,(\Lambda_{\mathrm{i}+}(\vec{p}\,) - \Lambda_{\mathrm{i}-}(\vec{p}\,)). \tag{16.51}$$

Under action of the operator $P^0 - H_1(\vec{p}\,) - H_2(-\vec{p}\,)$, the approximate BSEQ becomes

$$\begin{aligned}
&\left(P^0 - H_1(\vec{p}\,) - H_2(-\vec{p}\,)\right)\varphi_{nk}(\vec{p}\,) \\
&= \left(\Lambda_{1+}(\vec{p}\,)\Lambda_{2+}(-\vec{p}\,) - \Lambda_{1-}(\vec{p}\,)\Lambda_{2-}(-\vec{p}\,)\right) \int \frac{\mathrm{d}^3q}{(2\pi)^3} \frac{-4\pi Z\alpha}{(\vec{p} - \vec{q}\,)^2} \varphi_{nk}(\vec{q}\,).
\end{aligned} \tag{16.52}$$

The approximation to the BSEQ given in Eq. (16.52) is known as the "Salpeter equation" (see Ref. [534]). To wit, Salpeter actually derived a more general equation containing an arbitrary instantaneous kernel. The derivation for such a general kernel is no different from what we just did. The Salpeter equation is similar in form to the Breit equation [574], which can be written in a form much like (16.52), but missing the projection operators on the right-hand side.

Just a few additional approximations will bring us to a form amenable to actual solution. We start from (16.50), as (16.52) would require a two-particle Foldy–Wouthuysen expansion of the type discussed in Chaps. 11 and 12. We break $\varphi_{nk}(\vec{p})$ up into its large-large, large-small, etc., components using a seemingly trivial identity,

$$\begin{aligned}\varphi_{nk}(\vec{p}) &= \{\Lambda_{1+}(\vec{p}) + \Lambda_{1-}(\vec{p})\}\{\Lambda_{2+}(-\vec{p}) + \Lambda_{2-}(-\vec{p})\}\varphi_{nk}(\vec{p}) \\ &= \{\Lambda_{1+}(\vec{p})\Lambda_{2+}(-\vec{p}) + \Lambda_{1+}(\vec{p})\Lambda_{2-}(-\vec{p}) \\ &\quad + \Lambda_{1-}(\vec{p})\Lambda_{2+}(-\vec{p}) + \Lambda_{1-}(\vec{p})\Lambda_{2-}(-\vec{p})\}\varphi_{nk}(\vec{p}) \\ &= \varphi_{nk}^{++}(\vec{p}) + \varphi_{nk}^{+-}(\vec{p}) + \varphi_{nk}^{-+}(\vec{p}) + \varphi_{nk}^{--}(\vec{p}),\end{aligned} \tag{16.53}$$

where $\varphi_{nk}^{++}(\vec{p}) = \Lambda_{1+}(\vec{p})\Lambda_{2+}(-\vec{p})\varphi_{nk}(\vec{p})$, etc., and multiply both sides of (16.50) by $\Lambda_{1+}(\vec{p})\Lambda_{2+}(-\vec{p})$ to produce an equation for the large-large component of the wave function. We realize that $|\vec{p}|/m$ (and $|\vec{q}|/m$) are small in a nonrelativistic bound state, so $\Lambda_{\pm}(\vec{p}) \to \frac{1}{2}(1 \pm \gamma^0)$ and $P^0 - \omega_1(\vec{p}) - \omega_2(\vec{p})$ can be approximated as

$$P^0 - \omega_1(\vec{p}) - \omega_2(\vec{p}) \to (m_1 + m_2 + E) - \left(m_1 + \frac{\vec{p}^{\,2}}{2m_1} + \cdots\right) - \left(m_2 + \frac{\vec{p}^{\,2}}{2m_2} + \cdots\right) \approx E - \frac{\vec{p}^{\,2}}{2\mu}, \tag{16.54}$$

where $E = P^0 - m_1 - m_2$ is the binding energy and $\mu = (1/m_1 + 1/m_2)^{-1}$ is the reduced mass as before. In this limit, the equation satisfied by the large-large component is

$$\frac{\vec{p}^{\,2}}{2\mu}\varphi_{nk}^{++}(\vec{p}) + \int \frac{d^3q}{(2\pi)^3}\frac{(-4\pi Z\alpha)}{(\vec{p}-\vec{q})^2}\varphi_{nk}^{++}(\vec{q}) = E\,\varphi_{nk}^{++}(\vec{p}). \tag{16.55}$$

This is just the standard nonrelativistic Schrödinger–Coulomb equation (in momentum space). We note that the φ^{--} component describes states of two antiparticles, while the φ^{+-} and φ^{-+} components vanish in the nonrelativistic limit.

16.2.2 *Questions of Gauge and Transformations*

The BSEQ gives information about bound-state energies and scattering amplitudes, but our focus in this chapter is on energies. The energies appear as eigenvalues of the homogeneous equation (16.29) or as the positions of poles of the 2-to-2 Green function found by solving the inhomogeneous BSEQ, given in Eq. (16.25). Due to our specific focus on bound state energies, we can make use of a variety of options and transformations of the BSEQ that leave the energies unchanged. As one immediate example, the bound-state binding energy is independent of the Lorentz frame. The Lorentz invariance of QED implies that a bound state pole of G must have the covariant form

$$G(P;p,q) \to \frac{A}{P^2 - M_n^2} + B + \cdots , \tag{16.56}$$

for P near the four-momentum of a bound state, where M_n is an invariant (A and B are non-singular near this pole). The binding energy E_n of the bound state is defined through

$$M_n = m_1 + m_2 + E_n . \tag{16.57}$$

It constitutes an invariant. The equations we wrote down earlier for the Bethe–Salpeter wave function (16.29) and normalization condition (16.41) were specifically for the CM frame, but they could have been written for a general Lorentz frame. Other options and transformations that we will explore in this section are choice of gauge, the possibility of putting conditions on the relative time or energy variables, and the freedom associated with redefinitions of the non-interacting 2-to-2 Green function S.

Gauge invariance implies the independence of physical quantities from the choice of gauge (see Secs. 4.6.2 and 8.2.1). It is a fundamental feature of a QED. The basic interaction responsible for atomic structure is the Coulomb force. A physical photon such as one given off in an atomic transition propagates with two possible polarization states that are orthogonal to the direction of motion. These physical properties are captured most directly in the Coulomb gauge, which has the propagator [see Eq. (9.73)]

$$D_C^{00} = \frac{1}{\vec{k}^{\,2}} , \quad D_C^{0i} = D_C^{i0} = 0 , \quad D_C^{ij} = \frac{1}{k^2 + \mathrm{i}\epsilon} \left(\delta_{ij} - \frac{k^i k^j}{\vec{k}^{\,2}} \right) . \tag{16.58}$$

(We remember that Greek indices cover the range $0,1,2,3$, while Latin indices are spatial and cover the range $1,2,3$.) On the other hand, calculations that involve integration over loops containing virtual photons can be done much more simply and efficiently in a covariant gauge such as Feynman gauge, with the propagator [see Eq. (9.121)]

$$D_F^{\mu\nu} = -\frac{g^{\mu\nu}}{k^2 + \mathrm{i}\epsilon} . \tag{16.59}$$

Bethe and Salpeter used Coulomb gauge in their early work, and Salpeter [534] suggested that Feynman gauge was problematic in that unphysical corrections of anomalously low order in α appear that are only canceled when contributions from a number of graphs are considered together. A numerical exploration [575] using Feynman gauge confirms the appearance of energy corrections at order $\mathcal{O}(\mu\alpha^3 \ln(1/\alpha))$ as corrections to the Bohr energy levels $E_n^{\mathrm{Bohr}} = \mu\alpha^2/(2n^2)$. A systematic exploration [576] reveals contributions at the anomalously low orders $\mathcal{O}(\mu\alpha^3 \ln(1/\alpha))$ and $\mathcal{O}(\mu\alpha^3)$ when using Feynman gauge. Contributions at these orders are unwanted, as the first (relativistic) physical corrections in fact begin at $\mathcal{O}(\mu\alpha^4)$. In fact, these spurious contributions cancel, but only with the inclusion of an infinite number of additional terms in the kernel involving crossed photons. The problem can be seen to be associated with retardation present in the Feynman time-time propagator $D_F^{00} = 1/(\vec{k}^2 - k_0^2)$, as distinct from the instantaneous interaction mediated by the Coulomb gauge time-time propagator $D_C^{00} = 1/\vec{k}^{\,2}$. The moral is that Coulomb gauge should be used for the calculation of properties of nonrelativistic Coulombic bound states.

Insistence on the use of Coulomb gauge for bound-state calculations raises a number of issues associated with the fact that radiative corrections, and loop integrals in general, are much easier to evaluate in a covariant gauge. Fortunately, it is often possible to use a covariant gauge such as Feynman gauge for subsets of diagrams (under certain conditions), even when using Coulomb gauge for the calculation as a whole. As an immediate example, the vacuum polarization tensor [see Eq. (10.228)]

$$\Pi_{\mu\nu}(k) = \left(k^2 g_{\mu\nu} - k_\mu k_\nu\right) \Pi(k^2) \tag{16.60}$$

can be expressed in terms of a scalar function $\Pi(k^2)$ that is independent of gauge. The vacuum polarization tensor also satisfies the transversality condition $k^\mu \Pi_{\mu\nu}(k) = 0$, which can be useful in demonstrating a broader type of gauge independence because of the following connection between Coulomb gauge and Feynman gauge photon propagators (see Sec. 9.5.2)

$$D_C^{\mu\nu}(k) = D_F^{\mu\nu}(k) + \frac{1}{2k^2}\left(k^\mu f^\nu + f^\mu k^\nu\right), \quad f^0 = \frac{k^0}{\vec{k}^{\,2}}, \quad f^i = -\frac{k^i}{\vec{k}^{\,2}}. \tag{16.61}$$

In any subgraph where the terms involving f^μ can be shown to vanish (to some approximation), Feynman gauge can be used instead of Coulomb gauge for calculations. Many bound-state calculations involve vertex corrections. Since the fermions in a bound state are often nearly at rest, the on-shell version of the vertex function [see Eq. (10.96)]

$$\Gamma_\mu(p_2, p_1) = \gamma_\mu F_1(k^2) + \frac{\mathrm{i}}{2m}\sigma_{\mu\nu} k^\nu F_2(k^2) \tag{16.62}$$

can often be used. In this expression, $k = p_2 - p_1$, and the Dirac and Pauli form factors $F_1(k^2)$ and $F_2(k^2)$ are independent of gauge. The identification of gauge-invariant sets of diagrams that can be evaluated in Feynman gauge is a useful strategy for higher-order bound-state calculations.

The BSEQ is an intrinsically four-dimensional equation that, in the CM, depends on the total CM energy and either the relative energy and momentum variables (in momentum space) or the relative time and position variables (in coordinate space). The variables represented by relative energy or relative time are not essential for purposes of determining energy levels and greatly complicate the equations. Moreover, the bound-state energies are independent of the values of the relative coordinates. A number of ways have been developed to eliminate the relative energy (or time) variables and, so, to simplify the equations. One approach is to work with the wave function or 2-to-2 Green function in which the relative time variables have been set to zero. The corresponding momentum-space version of this condition is to work with the three-dimensional wave function of (16.45) instead of the full four-dimensional $\psi(p^0,\vec{p})$. A complete three-dimensional formalism and perturbation theory can be developed based on $t = 0$ quantities [534]. Some other works using $t = 0$ wave functions and Green functions include Refs. [2, 576–578].

Another line of approach is to put conditions on the relative energy, either setting the relative energy to zero or to a value that places one of the bound particles on its mass shell. A wide variety of bound-state formalisms have been constructed based on the general idea that the relative energy variable can be eliminated, and at the same time a simpler form for the non-interacting propagator can be used. Let us focus first on the second idea, which is that some simpler non-interacting 2-to-2 propagator S_0 can be used instead of S. One way to ensure that such a manipulation leaves the bound-state energies unchanged is to arrange matters so that the truncated Green function G_{T} is unchanged by the process. This is because G and G_{T} have bound-state poles at the same positions, as can be seen from (16.35). So, as long as the positions of the poles of G_{T} are unaffected, the energies are unchanged. If G_{T} stays constant, an alteration in S_0 must be accompanied by a compensating alteration in the kernel. Given the BSEQ of (16.36) and the new BSEQ for G_{T} in terms of S_0 and the new kernel $\overline{K}$

$$G_{\mathrm{T}} = \overline{K} + \overline{K}\, S_0\, G_{\mathrm{T}}, \tag{16.63}$$

we can eliminate G_{T} between the two using operator manipulations (represented here as simple algebra) to show that the new kernel must have the form

$$\overline{K} = \big(1 - K(S - S_0)\big)^{-1} K = K + K(S - S_0)K + \cdots. \tag{16.64}$$

The foregoing manipulation has often been used together with an astute choice of S_0 inspired by the physics of nonrelativistic bound states to produce a useful bound-state formalism.

Several bound-state formalisms have started with the observation that the full non-interacting Green function S that appears in the BSEQ contains far more information than is needed for a description of a nonrelativistic Coulombic bound state (at least to a first approximation). To start with, the radiative corrections found in the self-energy functions contained in S only affect bound-state energies at higher order. Considering now just the free 2-to-2 non-interacting propagator, the large-large (in the bispinor sense) component dominates for purposes of bound-state physics. We see from Eq. (16.49) that the dominant part of S is

$$S(P^0;p) \approx \frac{[\mathrm{i}\,\Lambda_+(\vec{p}\,)\,\gamma^0]_1}{\xi_1 P^0 + p^0 - \omega_1(\vec{p}\,) + \mathrm{i}\epsilon}\;\frac{[\mathrm{i}\,\Lambda_+(-\vec{p}\,)\,\gamma^0]_2}{\xi_2 P^0 - p^0 - \omega_2(\vec{p}\,) + \mathrm{i}\epsilon}\,. \tag{16.65}$$

With the choice $\xi_i = m_i/(m_1 + m_2)$ and under the condition that $|E| \ll |\vec{p}\,| \ll m_i$, we can see that the denominators are strongly peaked about $p^0 = 0$. The simplified non-interacting propagator

$$S_0(P^0;p) \equiv 2\pi\delta(p^0)\frac{-\mathrm{i}[\mathrm{i}\Lambda_+(\vec{p}\,)\gamma^0]_1[\mathrm{i}\Lambda_+(-\vec{p}\,)\gamma^0]_2}{P^0 - \omega_1(\vec{p}\,) - \omega_2(\vec{p}\,) + \mathrm{i}\epsilon} \tag{16.66}$$

expresses this small-p^0 behavior through a Dirac-δ function (as can be demonstrated by integrating $S(P^0;p)$ and $S_0(P^0;p)$ over the relative energy p_0). Use of S_0 instead of S requires a modification of the kernel, as seen in the preceding paragraph, but the modification just amounts to the addition of some (hopefully small) number of new contributions to the kernel. If we now set the relative energy equal to zero in (16.63), we find

$$G_{\mathrm{T}}\big|_{p^0=q^0=0} = \overline{K}\big|_{p^0=q^0=0} + (\overline{K}S_0G_{\mathrm{T}})\big|_{p^0=q^0=0}\,, \tag{16.67}$$

where all dependence on relative energy has been eliminated, and (16.63) becomes a purely three-dimensional equation for $G_{\mathrm{T}}(P^0;p^0=0,\vec{p},q^0=0,\vec{q}\,)$.

Many alternative formalisms to the one just described have been investigated. One group of alternatives has a S_0 similar to that of (16.66), with the same $2\pi\delta(p^0)$, an appropriate projection onto the large-large component, and a denominator that reduces to the nonrelativistic denominator $E-\vec{p}^{\,2}/(2\mu)$ in the nonrelativistic limit, but that differs in details. Another class of formalisms puts one of the particles onto its mass shell and uses the corresponding Dirac-δ function $2\pi\delta(\xi_1P^0+p^0-\omega_1(\vec{p}\,))$ in S_0. Many of these formalisms were developed to study nucleon-nucleon scattering, but they are applicable to bound-state QED as well. (There are a number of reviews and comparisons of the many options that have been investigated, for example in [336,579]. The conditions required of a formalism in order that it have the correct relativistic one-body limit (i.e., the Dirac equation) were discussed in [580].)

16.2.3 *Reference Kernels and Exact Solutions*

The key to performing higher-order calculations of properties of two-body bound states is to find an appropriate lowest order, or reference, problem with an exact solution, and to base a perturbative procedure on that foundation. Without an exactly soluble reference problem, even the lowest order wave function is uncertain, and it is difficult to be sure that all contributions to a physical property such as an energy level at a given perturbative order are accounted for. In this section, we will discuss strategies for obtaining exact solutions to a reference BSEQ and go through the details of one particular example. This will be followed in the next section by the development of a perturbation theory appropriate for extracting corrections to energy levels (and wave functions) from the BSEQ, and in the following section by the use of our sample exact solution in calculations.

The first step in finding a solvable reference problem is to choose a simplified non-interacting 2-to-2 propagator to use in place of

$$S(P;p) = \left[\frac{\mathrm{i}}{\xi_1\not{P}+\not{p}-m_1-\Sigma_1(\xi_1P+p)+\mathrm{i}\epsilon}\right]_1\left[\frac{\mathrm{i}}{\xi_2\not{P}-\not{p}-m_2-\Sigma_2(\xi_2P-p)+\mathrm{i}\epsilon}\right]_2, \tag{16.68}$$

where $\Sigma_i(p)$ is the self-energy function for the ith particle. We will call the simplified propagator S_0, and eliminate G between the BSEQ (16.25) and the modified BSEQ

$$G = S_0 + S_0\overline{K}G\,, \tag{16.69}$$

to arrive at the expression for the modified kernel $\overline{K}$:

$$\overline{K} = K + S_0^{-1} - S^{-1}. \tag{16.70}$$

The presence of the extra terms in $\overline{K}$ dependent on S_0 and S are required to compensate for the use of S_0 in (16.69). This manipulation only works if S_0^{-1} exists. An example of a simplified non-interacting propagator that does *not* have an inverse is Eq. (16.66) above. No inverse exists for the S_0 given in Eq. (16.66), because it contains projection operators, and a fundamental characteristic of projection operators is that they cannot be inverted. If a formalism utilizing a non-invertible S_0 is desired [such as the one given in Eq. (16.66)], one must work with the truncated Green function G_{T} instead of G. The BSEQ for G_{T} was given in (16.63), and the appropriate modified kernel (16.64) does not require S_0 to be invertible. It is not a disadvantage to work with G_{T} instead of G, as G_{T} has bound-state poles at the same positions as G. For the exact solution we choose to display here, however, we will work with G, not G_{T}, and use as our reference non-interacting propagator S_0 the free propagator

$$S_0(P;p) = \left[\frac{\mathrm{i}}{\xi_1 \not{P} + \not{p} - m_1 + \mathrm{i}\epsilon}\right]_1 \left[\frac{\mathrm{i}}{\xi_2 \not{P} - \not{p} - m_2 + \mathrm{i}\epsilon}\right]_2, \tag{16.71}$$

which is invertible. All self-energy corrections have been shifted into the kernel with minimal alteration of the propagator. The job of projection onto the relevant large-large subspace will be accomplished by our choice of lowest-order kernel.

We must now choose a lowest-order kernel K_0 so that the reference BSEQ

$$G_0 = S_0 + S_0\, K_0\, G_0 \tag{16.72}$$

can be solved exactly for G_0. Our solution will be based on the solution to the non-relativistic Schrödinger-Coulomb (NRSC) equation. K_0 must be carefully crafted so that, working together with S_0, Eq. (16.72) reduces to the nonrelativistic equation. Our plan is to use the reference BSEQ in the series form

$$\begin{aligned} G_0 &= S_0 + S_0\, K_0\, S_0 + S_0\, K_0\, S_0\, K_0\, S_0 + S_0\, K_0\, S_0\, K_0\, S_0\, K_0\, S_0 + \dots \\ &= S_0 + S_0 K_0 S_0 + S_0 \left[K_0\, S_0\, K_0 + K_0\, S_0\, K_0\, S_0\, K_0 + \dots\right] S_0 \\ &= S_0 + S_0 K_0 S_0 + S_0 H S_0, \end{aligned} \tag{16.73}$$

and sum the series

$$H \equiv K_0\, S_0\, K_0 + K_0\, S_0\, K_0\, S_0\, K_0 + \cdots. \tag{16.74}$$

This sum will be done by transforming it into the form of the corresponding sum for the nonrelativistic problem, which we must examine now. The NRSC Green function was discussed at length in Chap. 4. We review the results here in a slightly different notation useful for our present purposes. The NRSC Green function $g(E)$ satisfies the equation $(E - H_S)g(E) = 1$, which can be written either in coordinate or momentum space:

$$\left(E - \frac{\vec{p}^{\,2}}{2m} - V(|\vec{x}|)\right) g(E;\vec{x},\vec{y}\,) = \delta^{(3)}(\vec{x} - \vec{y}\,), \tag{16.75}$$

$$\left(E - \frac{\vec{p}^{\,2}}{2m}\right) g(E;\vec{p},\vec{q}\,) - \int \frac{\mathrm{d}^3\ell}{(2\pi)^3} V(\vec{p} - \vec{\ell}\,)\, g(E;\vec{\ell},\vec{q}\,) = (2\pi)^3 \delta^{(3)}(\vec{p} - \vec{q}\,), \tag{16.76}$$

with H_S given by Eq. (4.1) (except that we use a mass parameter m here instead of the reduced mass μ and will find that m is related to m_1 and m_2 in a non-trivial way), and $V(r) = -Z\alpha/r$ and $V(\vec{p}) = -4\pi Z\alpha/\vec{p}^{\,2}$. We will use the more flexible operator notation to write the solution to the inhomogeneous NRSC equation as

$$\begin{aligned} g(E) &= \left(E - \frac{\vec{p}^{\,2}}{2m} - V + \mathrm{i}\epsilon\right)^{-1} = (1 - s(E)V)^{-1}\, s(E) \\ &= s(E) + s(E)Vs(E) + s(E)Vs(E)Vs(E) + \cdots \\ &= s(E) + s(E)Vs(E) + s(E)h(E)s(E)\,, \end{aligned} \tag{16.77}$$

where

$$s(E;\vec{p},\vec{q}\,) = s(E;\vec{p}\,)(2\pi)^3\delta^{(3)}(\vec{p}-\vec{q}\,), \qquad s(E;\vec{p}\,) = \frac{1}{E - \vec{p}^{\,2}/(2m) + \mathrm{i}\epsilon}\,. \tag{16.78}$$

The free propagator is $s(E)$ and

$$h(E) \equiv Vs(E)V + Vs(E)Vs(E)V + \cdots \tag{16.79}$$

is a known function. (The exact form for $h(E)$ follows from the known form for $g(E)$.) The iϵ is needed to enforce the causal boundary conditions on $s(E)$ and $g(E)$. For future use, we write out the product $Vs(E)V$ in full:

$$[Vs(E)V]\,(\vec{p},\vec{q}\,) = \int \frac{\mathrm{d}^3\ell}{(2\pi)^3} V(\vec{p}-\vec{\ell}\,)s(E;\vec{\ell}\,)V(\vec{\ell}-\vec{q}\,). \tag{16.80}$$

The bound-state poles of the NRSC Green function can be displayed in

$$g(E;\vec{p},\vec{q}\,) \to \frac{\sum_k \psi_{nk}(\vec{p}\,)\psi^+_{nk}(\vec{q}\,)}{E - E_n}\,, \qquad E \to E_n\,, \tag{16.81}$$

where the index k labels the degeneracy.

We now return to our four-dimensional reference BSEQ shown in Eq. (16.72), with the goal of finding a reference kernel K_0 for which (16.72) can be solved exactly. As we will show, an appropriate reference kernel is

$$\begin{aligned} K_0(P^0;p,q) = {}& \left[\gamma^0\Lambda_+(\vec{p}\,)\frac{1}{2}(1+\gamma^0)\Lambda_+(\vec{q}\,)\right]_1 \left[\gamma^0\Lambda_+(-\vec{p}\,)\frac{1}{2}(1+\gamma^0)\Lambda_+(-\vec{q}\,)\right]_2 \\ & \times N(E;\vec{p}^{\,2})(-\mathrm{i})V(\vec{p}-\vec{q}\,)N(E;\vec{q}^{\,2})\,, \end{aligned} \tag{16.82}$$

where $V(\vec{p}) = -4\pi Z\alpha/\vec{p}^{\,2}$ is the usual (momentum-space) Coulomb potential and N is a function to be specified below. The point of the projection operators in K_0 and our coming choice of N is to make the sum H look like the corresponding sum for the NRSC Green function (16.79) discussed above. We start by evaluating the first term in H, which is $K_0\, S_0\, K_0 = [K_0\, S_0\, K_0]\,(E;p,q)$,

$$\begin{aligned} [K_0\, S_0\, K_0](E;p,q) &= \int \frac{d^4\ell}{(2\pi)^4} K_0(E;p,\ell)\, S_0(E;\ell)\, K_0(E;\ell,q) \\ &= \int \frac{d^4\ell}{(2\pi)^4} N(E;\vec{p}^{\,2})\,(-\mathrm{i})\,V(\vec{p}-\vec{\ell}\,)\,N^2(E;\vec{\ell}^{\,2})\,(-\mathrm{i})\,V(\vec{\ell}-\vec{q}\,)N(E;\vec{q}^{\,2}) \\ &\times \left[\gamma^0\Lambda_+(\vec{p}\,)\,\frac{1}{2}(1+\gamma^0)\,\Lambda_+(\vec{\ell}\,)\frac{\mathrm{i}}{\xi_1\not{P}+\not{\ell}-m_1+\mathrm{i}\epsilon}\gamma^0\Lambda_+(\vec{\ell}\,)\,\frac{1}{2}(1+\gamma^0)\Lambda_+(\vec{q}\,)\right]_1 \\ &\times \left[\gamma^0\Lambda_+(-\vec{p}\,)\,\frac{1}{2}(1+\gamma^0)\,\Lambda_+(-\vec{\ell}\,)\frac{\mathrm{i}}{\xi_2\not{P}-\not{\ell}-m_2+\mathrm{i}\epsilon}\gamma^0\,\Lambda_+(-\vec{\ell}\,)\,\frac{1}{2}(1+\gamma^0)\,\Lambda_+(-\vec{q}\,)\right]_2 . \end{aligned} \tag{16.83}$$

We use Eq. (16.49) to perform the integral over ℓ^0 (taking into account the imaginary infinitesimals),

$$\int \frac{d\ell^0}{2\pi} S_0(P;\ell) = \int \frac{d\ell^0}{2\pi} \left[\frac{\mathrm{i}}{\xi_1 \not{P} + \not{\ell} - m_1 + \mathrm{i}\epsilon}\right]_1 \left[\frac{\mathrm{i}}{\xi_2 \not{P} - \not{\ell} - m_2 + \mathrm{i}\epsilon}\right]_2$$

$$= \mathrm{i}\left\{\frac{\left[\Lambda_+(\vec{\ell})\gamma^0\right]_1 \left[\Lambda_+(-\vec{\ell})\gamma^0\right]_2}{P^0 - \omega_1(\vec{\ell}) - \omega_2(\vec{\ell}) + \mathrm{i}\epsilon} - \frac{\left[\Lambda_-(\vec{\ell})\gamma^0\right]_1 \left[\Lambda_-(-\vec{\ell})\gamma^0\right]_2}{P^0 + \omega_1(\vec{\ell}) + \omega_2(\vec{\ell}) - \mathrm{i}\epsilon}\right\}, \quad (16.84)$$

and note that the Λ_- projection operators are eliminated by the Λ_+ projections in K_0. A further simplification is achieved using the relation

$$\frac{1}{2}(1+\gamma^0)\Lambda_+(\vec{\ell})\frac{1}{2}(1+\gamma^0) = \left(\frac{\omega_i(\vec{\ell}) + m_i}{2\omega_i(\vec{\ell})}\right)\frac{1}{2}(1+\gamma^0). \quad (16.85)$$

We find that $K_0\, S_0\, K_0$ reduces to

$$\begin{aligned}
[K_0\, S_0\, K_0](E;p,q) = \int & \frac{\mathrm{d}^3\ell}{(2\pi)^3}(-\mathrm{i})V(\vec{p}-\vec{\ell}) \\
& \times \left\{\left(\frac{\omega_1(\vec{\ell}) + m_1}{2\omega_1(\vec{\ell})}\right) \frac{\mathrm{i}N^2(E;\vec{\ell}^{\,2})}{P^0 - \omega_1(\vec{\ell}) - \omega_2(\vec{\ell}) + \mathrm{i}\epsilon} \left(\frac{\omega_2(\vec{\ell}) + m_2}{2\omega_2(\vec{\ell})}\right)\right\} \\
& \times (-\mathrm{i})V(\vec{\ell}-\vec{q})N(E;\vec{p}^{\,2})\left[\gamma^0\Lambda_+(\vec{p})\frac{1}{2}(1+\gamma^0)\Lambda_+(\vec{q})\right]_1 \\
& \times \left[\gamma^0\Lambda_+(-\vec{p})\frac{1}{2}(1+\gamma^0)\Lambda_+(-\vec{q})\right]_2 N(E;\vec{q}^{\,2}).
\end{aligned} \quad (16.86)$$

An astute choice of $N(E;\vec{\ell})$ can turn the expression in the curly bracket of (16.86) into a nonrelativistic propagator. Following [581, 582], we use

$$\begin{aligned}
N^2(E;\vec{\ell}^{\,2}) \equiv & \left(\frac{2\omega_1(\vec{\ell})}{\omega_1(\vec{\ell}) + m_1}\right) \frac{2P^0((P^0)^2 - (m_1 - m_2)^2)}{(P^0 + \omega_1(\vec{\ell}) + \omega_2(\vec{\ell}))((P^0)^2 - (\omega_1(\vec{\ell}) - \omega_2(\vec{\ell}))^2)} \\
& \times \left(\frac{2\omega_2(\vec{\ell})}{\omega_2(\vec{\ell}) + m_2}\right).
\end{aligned} \quad (16.87)$$

After a little bit of algebra, one can confirm that

$$\left(\frac{\omega_1(\vec{\ell}) + m_1}{2\omega_1(\vec{\ell})}\right) \frac{N^2(E;\vec{\ell}^{\,2})}{P^0 - \omega_1(\vec{\ell}) - \omega_2(\vec{\ell}) + \mathrm{i}\epsilon} \left(\frac{\omega_2(\vec{\ell}) + m_2}{2\omega_2(\vec{\ell})}\right) = \left(\tilde{E} - \frac{\vec{\ell}^{\,2}}{2\tilde{m}} + \mathrm{i}\epsilon\right)^{-1} \equiv \tilde{s}(\tilde{E};\vec{\ell}), \quad (16.88)$$

which is the free nonrelativistic propagator, but with altered values for its nonrelativistic energy and its mass:

$$\tilde{E} = \frac{(P^0)^2 - M^2}{2P^0} = \frac{E(1 + E/(2M))}{1 + E/M}, \quad (16.89a)$$

$$\tilde{m} = \frac{(P^0)^2 - (m_1 - m_2)^2}{4P^0} = \frac{4m_1 m_2 + 2ME + E^2}{4(M+E)}, \quad (16.89b)$$

where we recall that $P^0 = M + E$ with E the nonrelativistic energy and $M = m_1 + m_2$. (Note that E and M here are to be distinguished from E_n and M_n, which are the nonrelativistic energy and total mass respectively of the state having quantum numbers n and k.) Exactly the same type of simplification occurs for $K_0S_0K_0S_0K_0$, $K_0S_0K_0S_0K_0S_0K_0$, etc., allowing us to perform the sum for H [see Eq. (16.74)]. We find that the infinite series for H contains within it the series for h, which is known. One has

$$\begin{aligned}
H(E;\vec p,\vec q\,) &= K_0S_0K_0 + K_0S_0K_0S_0K_0 + \cdots \\
&= \Bigg\{ \int \frac{d^3\ell}{(2\pi)^3}(-i)V(\vec p-\vec\ell\,)i\tilde s(\tilde E;\vec\ell\,)(-i)V(\vec\ell-\vec q\,) \\
&+ \int \frac{d^3\ell}{(2\pi)^3}\frac{d^3k}{(2\pi)^3}(-i)V(\vec p-\vec\ell\,)i\tilde s(\tilde E;\vec\ell\,)(-i)V(\vec\ell-\vec k\,)i\tilde s(\tilde E;\vec k\,)(-i)V(\vec k-\vec q\,)+\cdots\Bigg\} \\
&\times N(E;\vec p^{\,2})\left[\gamma^0\Lambda_+(\vec p\,)\frac{1}{2}(1+\gamma^0)\Lambda_+(\vec q\,)\right]_1\left[\gamma^0\Lambda_+(-\vec p\,)\frac{1}{2}(1+\gamma^0)\Lambda_+(-\vec q\,)\right]_2 N(E;\vec q^{\,2}) \\
&= -i\tilde h(\tilde E;\vec p,\vec q\,)N(E;\vec p^{\,2})\left[\gamma^0\Lambda_+(\vec p\,)\frac{1}{2}(1+\gamma^0)\Lambda_+(\vec q\,)\right]_1 \\
&\times\left[\gamma^0\Lambda_+(-\vec p\,)\frac{1}{2}(1+\gamma^0)\Lambda_+(-\vec q\,)\right]_2 N(E;\vec q^{\,2}) \,.
\end{aligned} \tag{16.90}$$

Here, $\tilde h(\tilde E)$ is obtained from $h(E)$ by the replacements $m \to \tilde m$, $E \to \tilde E$. We are in a position now to write down an exact formula for our reference 2-to-2 Green function $G_0(E)$ in terms of the known nonrelativistic Green function. The relations we use are

$$G_0 = S_0 + S_0\,K_0\,S_0 + S_0\,H\,S_0 \tag{16.91}$$

for the reference Green function, and

$$g = s + s\,V\,s + s\,h\,s\,, \qquad h = s^{-1}gs^{-1} - s^{-1} - V \tag{16.92}$$

for the nonrelativistic Green function. In all, we find that

$$\begin{aligned}
&G_0(E;p,q) = S_0(E;p,q) + S_0(E;p)\,K_0(E;p,q)\,S_0(E;q) \\
&- iN(E;\vec p^{\,2})\left\{\tilde s^{-1}(\tilde E,\vec p\,)\tilde g(\tilde E;\vec p,\vec q\,)\tilde s^{-1}(\tilde E;\vec q\,) - \tilde s^{-1}(E;\vec p,\vec q\,) - V(\vec p-\vec q\,)\right\} N(E;\vec q^{\,2}) \\
&\times S_0(E;p)\left[\gamma^0\Lambda_+(\vec p\,)\frac{1}{2}(1+\gamma^0)\Lambda_+(\vec q\,)\right]_1\left[\gamma^0\Lambda_+(-\vec p\,)\frac{1}{2}(1+\gamma^0)\Lambda_+(-\vec q\,)\right]_2 S_0(E;q)\,.
\end{aligned} \tag{16.93}$$

The only bound-state poles on the right-hand side of Eq. (16.93) are in the nonrelativistic Green function $\tilde g$, and they have the form given by (16.81). The bound state poles of $G_0(E)$ live in $\tilde g(\tilde E)$ and are located at the Bohr energies for mass $\tilde m$. That is, the poles are at values of E given by

$$\tilde E = \epsilon_n\Big|_{m\to\tilde m} = -\frac{\tilde m(Z\alpha)^2}{2n^2} \equiv \tilde m\beta_n\,, \qquad \beta_n \equiv -(Z\alpha)^2/(2n^2)\,, \tag{16.94}$$

and $\epsilon_n \equiv \beta_n m$ represents the Bohr energy. We use (16.89) to express $\tilde E$ and $\tilde m$ in terms of the physical binding energy E and masses, and find that the energy poles occur when $P^0 \to P^0_n$ (or, equivalently, $E \to E_n$) with

$$P^0_n = M\left(1 + \frac{2\mu\beta_n}{(1-\beta_n/2)M}\right)^{1/2} \tag{16.95}$$

and

$$E_n = P_n^0 - M = M\left(1 + \frac{2\mu\beta_n}{(1-\beta_n/2)M}\right)^{1/2} - M = \mu\beta_n + \frac{\mu\beta_n^2}{2}\left(1 - \frac{\mu}{M}\right) + \mathcal{O}(\alpha^6). \quad (16.96)$$

The pole term of $\tilde{g}(\tilde{E};\vec{p},\vec{q})$ is

$$\begin{aligned}\tilde{g}(\tilde{E};\vec{p},\vec{q}) &\to \sum_{\ell m}\frac{\psi_{n\ell m}(\vec{p})\psi^*_{n\ell m}(\vec{q})}{\tilde{E} - \epsilon_n|_{m\to\tilde{m}}}\\ &= \sum_{\ell m}\frac{\psi_{n\ell m}(\vec{p})\psi^*_{n\ell m}(\vec{q})}{\tilde{E} - \beta_n\tilde{m}} \to \sum_{\ell m}\frac{\psi_{n\ell m}(\vec{p})\psi^*_{n\ell m}(\vec{q})}{(1-\beta_n/2)(E-E_n)},\end{aligned} \quad (16.97)$$

as $E \to E_n$. We have used the definitions of $\tilde{E}$ and $\tilde{m}$ from Eq. (16.89), and the formula for E_n from Eq. (16.96). The corresponding pole term in $G_0(E;p,q)$

$$G_0(E;p,q) \to \mathrm{i}\sum_k \frac{\Psi^0_{nk}(p)\overline{\Psi}^0_{nk}(q)}{E - E_n} \quad (16.98)$$

allows us to define the reference wave function and its conjugate. We equate the pole terms in (16.98) and (16.93), using (16.97), to find

$$\begin{aligned}\sum_k \Psi^0_{nk}(p)\overline{\Psi}^0_{nk}(q) &= N(E_n;\vec{p}^{\,2})\tilde{s}^{-1}(\tilde{E}_n;\vec{p})\,\frac{\mathrm{i}^2\sum_{\ell m}\psi_{n\ell m}(\vec{p})\,\psi^*_{n\ell m}(\vec{q})}{1-\beta_n/2}\\ &\quad\times \tilde{s}^{-1}(\tilde{E}_n;\vec{q})N(E_n;\vec{q}^{\,2})\,S_0(P_n^0;p)\left[\gamma^0\Lambda_+(\vec{p})\frac{1}{2}(1+\gamma^0)\Lambda_+(\vec{q})\right]_1\\ &\quad\times\left[\gamma^0\Lambda_+(-\vec{p})\frac{1}{2}(1+\gamma^0)\Lambda_+(-\vec{q})\right]_2 S_0(P_n^0;q).\end{aligned} \quad (16.99)$$

In order to complete the factorization of the right-hand side into a sum of products, we note the Fierz-like identity

$$\frac{1}{2}(1+\gamma^0)_{aa'}\frac{1}{2}(1+\gamma^0)_{bb'} = \sum_{s_{1z},s_{2z}}\begin{pmatrix}\chi_{s_{1z}}\\0\end{pmatrix}_a\begin{pmatrix}\chi_{s_{2z}}\\0\end{pmatrix}_b\left(\chi^+_{s_{1z}},0\right)_{a'}\left(\chi^+_{s_{2z}},0\right)_{b'}, \quad (16.100)$$

where the sums are over the possible spin states of the fermions (+1/2 or −1/2) and χ_m is a fundamental two-component Pauli spinor defined in Eq. (6.42). We see that the degeneracy index k actually represents ℓ, m, s_{1z}, and s_{2z}, and that the reference wave functions are

$$\Psi^0_{nk}(p) = \eta\,\mathcal{S}(P_n^0;p)\psi_{n\ell m}(\vec{p})\,u_{1,s_{1z}}(\vec{p})\,u_{2,s_{2z}}(-\vec{p}), \quad (16.101a)$$

$$\overline{\Psi}^0_{nk}(p) = \eta\,\overline{u}_{1,s_{1z}}(\vec{p})\overline{u}_{2,s_{2z}}(-\vec{p})\psi^*_{n\ell m}(\vec{p})\,\mathcal{S}(P_n^0;p), \quad (16.101b)$$

where

$$\eta \equiv (1-\beta_n/2)^{-1/2}, \quad (16.102)$$

and

$$\begin{aligned}\mathcal{S}(P^0;p) &= \mathrm{i}\left(\frac{\omega_1(\vec{p})+m_1}{2\omega_1(\vec{p})}\right)^{1/2}\left(\frac{\omega_2(\vec{p})+m_2}{2\omega_2(\vec{p})}\right)^{1/2}\\ &\quad\times\frac{\left(\tilde{E} - \frac{\vec{p}^{\,2}}{2\tilde{m}}\right)N(E;\vec{p}^{\,2})}{(\xi_1P^0 + p^0 - \omega_1(\vec{p}) + \mathrm{i}\epsilon)\,(\xi_2P^0 - p^0 - \omega_2(\vec{p}) + \mathrm{i}\epsilon)},\end{aligned} \quad (16.103)$$

and $\psi_{n\ell m}$ are the usual nonrelativistic Schrödinger–Coulomb wave functions of Eq. (4.60). We use Dirac spinors defined by

$$u_{\mathrm{i},m}(\vec{p}) = \left(\frac{2\omega_i(\vec{p})}{\omega_i(\vec{p})+m_i}\right)^{1/2}\left[\Lambda_+(\vec{p})\begin{pmatrix}\chi_m\\0\end{pmatrix}\right]_i = \left(\frac{\omega_i(\vec{p})+m_i}{2\omega_i(\vec{p})}\right)^{1/2}\left[\begin{pmatrix}\chi_m\\ \dfrac{\vec{\sigma}\cdot\vec{p}\chi_m}{\omega_i(\vec{p})+m_i}\end{pmatrix}\right]_i \tag{16.104}$$

and having normalization

$$u^{+}_{\mathrm{i},s_z}(\vec{p})u_{\mathrm{i},s'_z}(\vec{p}) = \delta_{s_z,s'_z}. \tag{16.105}$$

The $\mathcal{S}$ function is strongly peaked near $p^0 = 0$ for nonrelativistic momentum and satisfies

$$\int\frac{\mathrm{d}p^0}{2\pi}\mathcal{S}(P^0;p) = \left(\frac{\omega_1(\vec{p})+m_1}{2\omega_1(\vec{p})}\right)^{-1/2}\left(\frac{\omega_2(\vec{p})+m_2}{2\omega_2(\vec{p})}\right)^{-1/2} N^{-1}(E;\vec{p}^{\,2}), \tag{16.106}$$

which is approximately one when E and the kinetic energies can be neglected compared to the rest masses. One can verify that the reference wave functions satisfy both the homogeneous reference BSEQ (16.29)

$$\Psi^0_{nk} = S_0(P^0_n)K_0(P^0_n)\Psi^0_{nk} \tag{16.107}$$

and the normalization condition (16.41),

$$\delta_{rs} = \mathrm{i}\overline{\Psi}^0_{nr}\frac{\mathrm{d}}{\mathrm{d}P^0}G_0^{-1}(P^0)\Big|_{P^0_n}\Psi^0_{ns} = \mathrm{i}\overline{\Psi}^0_{nr}\frac{\mathrm{d}}{\mathrm{d}P^0}\left(S_0^{-1}(P^0) - K_0(P^0)\right)\Big|_{P^0_n}\Psi^0_{ns}. \tag{16.108}$$

16.2.4 *Bound-State Perturbation Theory*

The purpose of bound-state perturbation theory is to find a systematic procedure for approximating the exact energy levels and wave functions (and the 2-to-2 Green function) based on the solution of a reference problem. The reference problem contains the dominant physics of the exact problem, but in a simplified form that allows the reference problem to be solved exactly. The difference between the exact and reference problems is in some sense *small*, and the perturbation theory we will construct will be in the form of a series in increasing powers of the small perturbation. For us, the exact problem is the BSEQ for the full renormalized 2-to-2 Green function G. This BSEQ has the form of (16.69),

$$G = S_0 + S_0\overline{K}G \tag{16.109}$$

with the free propagator S_0 shown in (16.71)) and a kernel $\overline{K}$ given by (16.70). The reference problem is a BSEQ of the same form, with the same free 2-to-2 free propagator S_0 but having an approximate kernel K_0,

$$G_0 = S_0 + S_0\,K_0\,G_0, \tag{16.110}$$

as in (16.72). We can multiply by S_0^{-1} on the left and G^{-1} or G_0^{-1} on the right to rewrite the two BSEQs (16.109) and (16.110) in the forms $S_0^{-1} = G^{-1} + \overline{K}$ and $S_0^{-1} =$

$G_0^{-1} + K_0$. We then eliminate the common factor S_0 to obtain $G^{-1} + \overline{K} = G_0^{-1} + K_0$, which can be solved for G,

$$G = G_0 + G_0\,\delta K\,G = G_0 + G_0\,\delta K\,G_0 + G_0\,\delta K\,G_0\,\delta K\,G_0 + \cdots, \tag{16.111}$$

where $\delta K \equiv \overline{K} - K_0$. The full Green function G has poles at the exact bound state energies with residues giving the exact wave functions, while the reference Green function G_0 has poles at the reference bound state energies with residues containing the reference wave functions. The task of perturbation theory is to extract from (16.111) explicit relations between the exact E, Ψ, $\overline{\Psi}$ and the reference E^0, Ψ^0, $\overline{\Psi}^0$.

There are several ways to develop the perturbative expansions for energy levels and wave functions coming from (16.111). We use the procedure outlined in [581]. There are two main differences between the perturbation theory that we will develop and the one used in ordinary nonrelativistic quantum mechanics. In quantum mechanics, Green functions, potentials, and wave functions depend on the coordinates of three-dimensional space (whether in coordinate or momentum representation), while our Green functions, kernels, and wave functions depend on the coordinates of four-dimensional space (the relative position or relative momentum variables). We point out that the dependence on relative time (or relative energy) is not a disadvantage. Rather, it is essential in order to include the effects of retardation. Of more immediate importance is the dependence of the perturbing kernel δK on the unknown CM energy. This is a new feature here that doesn't typically occur in ordinary quantum mechanics. Our procedure can account for an energy-dependent kernel in a natural way, but some new terms are introduced that are not present in the usual Rayleigh–Schrödinger perturbation theory of ordinary quantum mechanics.

Bound state energies appear as the positions of the poles of the 2-to-2 Green function, and we exploit this dependence to develop a perturbation theory. Near a bound-state pole E_n, G has the form

$$G(E) = \mathrm{i}\sum_k \frac{\Psi_{nk}\overline{\Psi}_{nk}}{E - E_n} + R(E), \tag{16.112}$$

where n indexes all the distinct energy levels, k labels the degeneracy, and the remainder function $R(E)$ is non-singular in the limit $E \to E_n$. Note that there will always be some degeneracy because (in the absence of coupling to external fields), the energies can never depend on the z component of the total angular momentum. The reference Green function G_0 has a similar expansion in terms of reference quantities

$$G_0(E) = \mathrm{i}\sum_k \frac{\Psi^0_{nk}\overline{\Psi}^0_{nk}}{E - E^0_n} + R^0(E), \tag{16.113}$$

as in (16.98). (When G_0 has more degeneracy than G, then the usual techniques of degenerate perturbation theory must be employed, such as using suitable states of G_0 in the first place or finding the suitable states by diagonalizing in the degenerate subspace. Our presentation here ignores this subtlety for now.) We assume that perturbative expansions exist for E_n and the wave functions Ψ_{nk}, $\overline{\Psi}_{nk}$ of the form

$$E_n = E_n^0 + E_n^1 + E_n^2 + \cdots, \tag{16.114a}$$

$$\Psi_{nk} = \Psi_{nk}^0 + \Psi_{nk}^1 + \Psi_{nk}^2 + \cdots, \tag{16.114b}$$

$$\overline{\Psi}_{nk} = \overline{\Psi}_{nk}^0 + \overline{\Psi}_{nk}^1 + \overline{\Psi}_{nk}^2 + \cdots, \tag{16.114c}$$

where the reference energies and wave functions were found in Sec. 16.2.3 above. The perturbative order is indicated by a superscript. In our calculations here, we will only carry terms of $\mathcal{O}(\delta K)^2$ and ignore any higher-order terms, although it is just a matter of patience to work out higher order contributions. For the left-hand side of (16.111), we have

$$\begin{aligned}
G(E) &= \frac{\mathrm{i}\Psi_{nk}\overline{\Psi}_{nk}}{E - E_n} = \frac{\mathrm{i}\left(\Psi_{nk}^0 + \Psi_{nk}^1 + \Psi_{nk}^2 + \cdots\right)\left(\overline{\Psi}_{nk}^0 + \overline{\Psi}_{nk}^1 + \overline{\Psi}_{nk}^2 + \cdots\right)}{(E - E_n^0)\left(1 - \frac{E_n^1 + E_n^2 + \cdots}{E - E_n^0}\right)} \\
&= \frac{\mathrm{i}\Psi_{nk}^0\overline{\Psi}_{nk}^0}{E - E_n^0} + \left\{\frac{E_n^1 \mathrm{i}\Psi_{nk}^0\overline{\Psi}_{nk}^0}{(E - E_n^0)^2} + \frac{\mathrm{i}(\Psi_{nk}^1\overline{\Psi}_{nk}^0 + \Psi_{nk}^0\overline{\Psi}_{nk}^1)}{E - E_n^0}\right\} \\
&\quad + \left\{\frac{(E_n^1)^2 \mathrm{i}\Psi_{nk}^0\overline{\Psi}_{nk}^0}{(E - E_n^0)^3} + \frac{E_n^2 \mathrm{i}\Psi_{nk}^0\overline{\Psi}_{nk}^0 + E_n^1 \mathrm{i}(\Psi_{nk}^1\overline{\Psi}_{nk}^0 + \Psi_{nk}^0\overline{\Psi}_{nk}^1)}{(E - E_n^0)^2}\right. \\
&\quad \left. + \frac{\mathrm{i}\Psi_{nk}^2\overline{\Psi}_{nk}^0 + \mathrm{i}\Psi_{nk}^1\overline{\Psi}_{nk}^1 + \mathrm{i}\Psi_{nk}^0\overline{\Psi}_{nk}^2}{(E - E_n^0)}\right\} + \mathcal{O}(\delta K)^3.
\end{aligned} \tag{16.115}$$

Summation over the degeneracy index k is implicit. (We have omitted the remainder term because it is non-singular in the limit $E \to E_n$ and doesn't contribute to the perturbative results for the energies or wave functions.) Our next step is to expand the right-hand side of (16.111). For this expansion we use

$$G_0(E) = \frac{\mathrm{i}\Psi_{nk}^0\overline{\Psi}_{nk}^0}{E - E_n^0} + \hat{G}_0(E) = \frac{\mathrm{i}\Psi_{nk}^0\overline{\Psi}_{nk}^0}{E - E_n^0} + \hat{G}_0 + (E - E_n^0)\,\hat{G}_0' + \mathcal{O}(E - E_n^0)^2, \tag{16.116}$$

where

$$\hat{G}_0(E) = G_0(E) - \frac{\mathrm{i}\Psi_{nk}^0\overline{\Psi}_{nk}^0}{E - E_n^0}, \qquad \hat{G}_0 = \hat{G}_0(E_n^0), \qquad \hat{G}_0' = \frac{d}{dE}\hat{G}_0\Big|_{E_n^0}, \tag{16.117}$$

and a similar expansion for δK:

$$\delta K(E) = \delta K_0 + (E - E_n^0)\,\frac{\mathrm{d}}{\mathrm{d}E}\delta K\Big|_{E_n^0} + \mathcal{O}(E - E_n^0)^2, \tag{16.118}$$

where $\delta K_0 \equiv \delta K(E_n^0)$. We find that

$$\begin{aligned}
G(E) &= G_0 + G_0\delta K G_0 + G_0\delta K G_0\delta K G_0 + \mathcal{O}(\delta K)^3 \\
&= \left(\frac{\mathrm{i}\Psi_{nk}^0\overline{\Psi}_{nk}^0}{E - E_n^0} + \hat{G}_0(E)\right) + \left(\frac{\mathrm{i}\Psi_{nk}^0\overline{\Psi}_{nk}^0}{E - E_n^0} + \hat{G}_0(E)\right)\delta K(E)\left(\frac{\mathrm{i}\Psi_{nk}^0\overline{\Psi}_{nk}^0}{E - E_n^0} + \hat{G}_0(E)\right) \\
&\quad + \left(\frac{\mathrm{i}\Psi_{nk}^0\overline{\Psi}_{nk}^0}{E - E_n^0} + \hat{G}_0(E)\right)\delta K(E)\left(\frac{\mathrm{i}\Psi_{nk}^0\overline{\Psi}_{nk}^0}{E - E_n^0} + \hat{G}_0(E)\right)\delta K(E) \\
&\quad \times \left(\frac{\mathrm{i}\Psi_{nk}^0\overline{\Psi}_{nk}^0}{E - E_n^0} + \hat{G}_0(E)\right) + \mathcal{O}(\delta K)^3,
\end{aligned} \tag{16.119}$$

which, upon expanding out $\hat{G}(E)$ and $\delta K(E)$, becomes

$$
\begin{aligned}
G(E) &= \left\{ \frac{\mathrm{i}\Psi^0_{nk}\overline{\Psi}^0_{nk}}{E - E^0_n} \right\} + \left\{ \frac{\mathrm{i}\Psi^0_{nk}\langle \mathrm{i}\delta K\rangle \overline{\Psi}^0_{nk}}{(E - E^0_n)^2} \right. \\
&\quad \left. + \frac{\hat{G}_0\delta K_0 \mathrm{i}\Psi^0_{nk}\overline{\Psi}^0_{nk} + \mathrm{i}\Psi^0_{nk}\overline{\Psi}^0_{nk}\delta K_0\hat{G}_0 + \mathrm{i}\Psi^0_{nk}\langle \mathrm{i}\delta K\rangle' \overline{\Psi}^0_{nk}}{(E - E^0_n)} \right\} \\
&\quad + \left\{ \frac{\mathrm{i}\Psi^0_{nk}\langle \mathrm{i}\delta K\rangle^2 \overline{\Psi}^0_{nk}}{(E - E^0_n)^3} + \frac{1}{(E - E^0_n)^2}\Big[\hat{G}_0\delta K_0 \mathrm{i}\Psi^0_{nk}\langle \mathrm{i}\delta K\rangle \overline{\Psi}^0_{nk} \right. \\
&\quad + \mathrm{i}\Psi^0_{nk}\langle \mathrm{i}\delta K\rangle \overline{\Psi}^0_{nk}\delta K_0\hat{G}_0 + \mathrm{i}\Psi^0_{nk}\langle \mathrm{i}\delta K\hat{G}_0\delta K\rangle \overline{\Psi}^0_{nk} + 2\mathrm{i}\Psi^0_{nk}\langle \mathrm{i}\delta K\rangle\langle \mathrm{i}\delta K\rangle' \overline{\Psi}^0_{nk} \Big] \\
&\quad + \frac{1}{E - E^0_n}\Big[\mathrm{i}\Psi^0_{nk}\overline{\Psi}^0_{nk}\big\{ \langle \mathrm{i}\delta K\rangle\langle \mathrm{i}\delta K\rangle'' + (\langle \mathrm{i}\delta K\rangle')^2 + \langle \mathrm{i}\delta K\hat{G}_0\delta K\rangle' \big\} \\
&\quad + \mathrm{i}\hat{G}_0\delta K_0\Psi^0_{nk}\overline{\Psi}^0_{nk}\delta K_0\hat{G}_0 + \mathrm{i}\hat{G}_0\delta K_0\Psi^0_{nk}\overline{\Psi}^0_{nk}\langle \mathrm{i}\delta K\rangle' + \mathrm{i}[\hat{G}_0\delta K]'\Psi^0_{nk}\overline{\Psi}^0_{nk}\langle \mathrm{i}\delta K\rangle \\
&\quad + \mathrm{i}\hat{G}_0\delta K_0\hat{G}_0\delta K_0\Psi^0_{nk}\overline{\Psi}^0_{nk} + \mathrm{i}\Psi^0_{nk}\overline{\Psi}^0_{nk}\delta K_0\hat{G}_0\langle \mathrm{i}\delta K\rangle' + \mathrm{i}\Psi^0_{nk}\overline{\Psi}^0_{nk}[\delta K\hat{G}_0]'\langle \mathrm{i}\delta K\rangle \\
&\quad \left. + \mathrm{i}\Psi^0_{nk}\overline{\Psi}^0_{nk}\delta K_0\hat{G}_0\delta K_0\hat{G}_0 \Big] \right\} + \mathcal{O}(\delta K)^3 + \mathcal{O}(E - E^0_n)^0,
\end{aligned} \tag{16.120}
$$

where

$$
\langle \mathrm{i}X\rangle \equiv \mathrm{i}\overline{\Psi}^0_{nk}\, X(E^0_n)\, \Psi^0_{nk}\,, \qquad \langle \mathrm{i}X\rangle' \equiv \mathrm{i}\overline{\Psi}^0_{nk}\, \left.\frac{\mathrm{d}X}{\mathrm{d}E}\right|_{E^0_n} \Psi^0_{nk}\,, \tag{16.121a}
$$

$$
\langle \mathrm{i}X\rangle'' \equiv \mathrm{i}\overline{\Psi}^0_{nk}\, \left.\frac{\mathrm{d}^2X}{\mathrm{d}E^2}\right|_{E^0_n} \Psi^0_{nk}\,, \qquad [Y]' \equiv \left.\frac{\mathrm{d}Y}{\mathrm{d}E}\right|_{E^0_n}. \tag{16.121b}
$$

Here, X could be δK or $\delta K\, \hat{G}_0\, \delta K$, which depend on energy and two relative momenta. The quantities $\langle \mathrm{i}X\rangle$, $\langle \mathrm{i}X\rangle'$, and $\langle \mathrm{i}X\rangle''$ are numbers, with no dependence on relative momentum. The quantity Y could be $\hat{G}_0\delta K$ or $\delta K\hat{G}_0$, which depend on energy and two relative momenta, while $[Y]'$ is a function only of the two relative momenta. Finally, we compare the two expansions (16.115) and (16.120) for G. These expansions are organized into powers of δK and $(E - E^0_n)$, and like powers in each of the two forms for G must match. The first correction to the energy is found by comparing the terms of $\mathcal{O}(\delta K \times (E - E^0_n)^{-2})$, with the result that $E^1_n = \langle \mathrm{i}\delta K\rangle$. The second correction comes from terms of $\mathcal{O}(\delta K^2 \times (E - E^0_n)^{-2})$, which leads to the relation $E^2_n = \langle \mathrm{i}\delta K\, \hat{G}_0\, \delta K\rangle + \langle \mathrm{i}\delta K\rangle\langle \mathrm{i}\delta K\rangle'$. Corrections to the wave function and conjugate wave function come from comparison of the $\mathcal{O}(\delta K \times (E - E^0_n)^{-1})$ and $\mathcal{O}(\delta K^2 \times (E - E^0_n)^{-1})$ terms. In all, we find that

$$
E_n = E^0_n + \langle \mathrm{i}\delta K\rangle + \langle \mathrm{i}\delta K\, \hat{G}_0\, \delta K\rangle + \langle \mathrm{i}\delta K\rangle\, \langle \mathrm{i}\delta K\rangle' + \mathcal{O}(\delta K)^3\,, \tag{16.122a}
$$

$$
\Psi_{nk} = \Psi^0_{nk} + \hat{G}_0\delta K\Psi^0_{nk} + \frac{1}{2}\langle \mathrm{i}\delta K\rangle'\, \Psi^0_{nk} + \mathcal{O}(\delta K)^2\,, \tag{16.122b}
$$

$$
\overline{\Psi}_{nk} = \overline{\Psi}^0_{nk} + \overline{\Psi}^0_{nk}\delta K\hat{G}_0 + \frac{1}{2}\langle \mathrm{i}\delta K\rangle'\, \overline{\Psi}^0_{nk} + \mathcal{O}(\delta K)^2. \tag{16.122c}
$$

We did not write out the $\mathcal{O}(\delta K)^2$ corrections to the wave functions, because we will not be using them. One should note that the corrected wave functions will not automatically be normalized. Terms with powers of $(E - E_n^0)$ other than -1 or -2 provide checks of the previous results but do not lead to new relations.

Because only the $1/(E - E_n^0)$ and $1/(E - E_n^0)^2$ terms in (16.120) are needed in order to work out perturbations to the energies and wave functions, it is possible to proceed in a more elegant manner using the residue theorem to extract from $G(E)$ only the required contributions [583–585]. We choose a small closed contour C_n in the complex energy plane that encloses (in a counter-clockwise way) the pole E_n of $G(E)$ and E_n^0 of $G_0(E)$ and no other pole of either. Application of the residue theorem to (16.112) gives

$$\mathrm{i}\Psi_{nk}\overline{\Psi}_{nk} = \oint_{C_n} \frac{\mathrm{d}E}{2\pi\mathrm{i}} G(E), \tag{16.123a}$$

$$E_n \mathrm{i}\Psi_{nk}\overline{\Psi}_{nk} = \oint_{C_n} \frac{\mathrm{d}E}{2\pi\mathrm{i}} E\, G(E) = E_n^0\, \mathrm{i}\Psi_{nk}\overline{\Psi}_{nk} + \oint_{C_n} \frac{\mathrm{d}E}{2\pi\mathrm{i}} (E - E_n^0)\, G(E). \tag{16.123b}$$

By way of illustration, we write out the results coming from (16.123) through $\mathcal{O}(\delta K)$,

$$\begin{aligned}\mathrm{i}\Psi_{nk}\overline{\Psi}_{nk} &= \mathrm{i}\Psi_{nk}^0\overline{\Psi}_{nk}^0 + \hat{G}_0\delta K_0 \mathrm{i}\Psi_{nk}^0\overline{\Psi}_{nk}^0 + \mathrm{i}\Psi_{nk}^0\overline{\Psi}_{nk}^0 \delta K_0 \hat{G}_0 \\ &\quad + \langle \mathrm{i}\delta K\rangle' \mathrm{i}\Psi_{nk}^0\overline{\Psi}_{nk}^0 + \mathcal{O}(\delta K)^2, \end{aligned} \tag{16.124a}$$

$$E_n \mathrm{i}\Psi_{nk}\overline{\Psi}_{nk} = E_n^0 \mathrm{i}\Psi_{nk}\overline{\Psi}_{nk} + \langle \mathrm{i}\delta K\rangle \mathrm{i}\Psi_{nk}^0\overline{\Psi}_{nk}^0 + \mathcal{O}(\delta K)^2, \tag{16.124b}$$

from which the results of (16.122a) follow [through $\mathcal{O}(\delta K)$]. It is an advantage, especially if higher orders of δK are required, to focus only on the terms in the expansion of (16.120) having orders $1/(E - E_n^0)$ and $1/(E - E_n^0)^2$.

16.2.5 *Energy Levels at $\mathcal{O}(Z\alpha)^4$ (BSEQ Approach)*

In this section, we obtain the corrections to the energy levels of (spin-1/2–spin-1/2) two-body Coulombic bound states at order α^4. These corrections are well known, and have been discussed already in Chaps. 11 and 12. Still, it is instructive to see them again from a different point of view, and coming from a formalism that can be extended in a systematic way to higher orders. These corrections arise from relativistic effects (relativistic kinetic energy, magnetic current-current interaction, and Darwin term) and interactions associated with spin (the spin-orbit interaction and a spin-spin tensor interaction). In the related case of a particle-antiparticle bound state, there is also a correction at this order arising from virtual annihilation to a single photon.

The energy levels of a Coulombic system are found by starting with the known solution of a reference problem. We use the solution shown in Sec. 16.2.3 and a perturbative expansion as in the preceding Sec. 16.2.4. The reference energy levels were given in Eq. (16.96), with the result

$$E_n^0 = M\left(1 + \frac{2\beta_n}{1-\beta_n/2}\frac{\mu}{M}\right)^{1/2} - M = \mu\beta_n + \frac{\mu\beta_n^2}{2}\left(1-\frac{\mu}{M}\right) + \mathcal{O}(\alpha^6), \tag{16.125}$$

where $\beta_n = -(Z\alpha)^2/(2n^2)$, while $\mu = m_1 m_2/M$ is the reduced mass and $M = m_1+m_2$ the total mass. The first of the $\mathcal{O}(\alpha^4)$ energy corrections comes from the $\mathcal{O}(\beta_n^2)$ term in the expansion of the reference energy.

Additional energy corrections at $\mathcal{O}(\alpha^4)$ come from first-order perturbation theory. In Eq. (16.122a), this is the correction containing only one factor of δK:

$$\Delta E \approx \langle \mathrm{i}\delta K\rangle . \tag{16.126}$$

Moreover, only two specific terms in δK are required. Let K_C and K_T represent the kernels for exchange of a single Coulomb photon and a single transverse photon. It turns out that the only terms in δK that contribute $\mathcal{O}(\alpha^4)$ energy corrections (for the particle-particle case) are $K_\mathrm{C} - K_0$, which is the difference between a Coulomb exchange and the reference kernel, and K_T. In the remainder of this section we will give the details of how the contributions of $K_\mathrm{C} - K_0$ and K_T can be reduced to an effective perturbing Hamiltonian that can be used with the NRSC wave functions to give the energy level corrections.

We begin with the transverse photon contribution $\Delta E_\mathrm{T} = \langle \mathrm{i}K_\mathrm{T}\rangle = \mathrm{i}\overline{\Psi}^0\, K_\mathrm{T}\, \Psi^0$. By making explicit the momentum dependence of wave functions and kernel and using the QED Feynman rules for K_T, we find the transverse energy contribution to be

$$\Delta E_\mathrm{T} = \mathrm{i}\int \frac{\mathrm{d}^4p}{(2\pi)^4}\frac{\mathrm{d}^4q}{(2\pi)^4}\overline{\Psi}^0(p)\left[-\mathrm{i}e\gamma^i\right]_1\left[-\mathrm{i}(-Ze)\gamma^j\right]_2 \frac{\mathrm{i}\delta^{\perp,ij}(\hat{k})}{k^2}\Psi^0(q), \tag{16.127}$$

where e is the electron charge, $-Ze$ is the charge of the positively charged particle, and $k = p - q$ is the photon momentum. The spatial part of the Coulomb gauge photon propagator given in Eq. (9.73) contains the transverse Dirac-δ function in momentum space [see also Eq. (2.54)],

$$\delta^{\perp,ij}(\vec{k}\,) = \delta^{ij} - \hat{k}^i\hat{k}^j = \delta^{ij} - \frac{k^i\,k^j}{\vec{k}^{\,2}}. \tag{16.128}$$

Inserting the form of the reference wave functions from (16.101), the ΔE_T energy correction becomes

$$\begin{aligned}\Delta E_\mathrm{T} = &-4\pi Z\alpha\,\eta^2\int\frac{\mathrm{d}^3p}{(2\pi)^3}\frac{\mathrm{d}^3q}{(2\pi)^3}\int\frac{\mathrm{d}p^0}{2\pi}\mathcal{S}(P_n^0;p)\int\frac{\mathrm{d}q^0}{2\pi}\mathcal{S}(P_n^0;q)\\ &\times\psi^*(\vec{p}\,)\,\overline{u}_1(\vec{p}\,)\gamma^i u_1(\vec{q}\,)\frac{\delta^{\perp,ij}(\vec{k}\,)}{k_0^2-\vec{k}^{\,2}}\,\overline{u}_2(-\vec{p}\,)\gamma^j u_2(-\vec{q}\,)\,\psi(\vec{q}\,),\end{aligned} \tag{16.129}$$

where $\eta = (1-\beta_n/2)^{-1/2}$ was defined on Eq. (16.102) and ψ, ψ^* are the nonrelativistic Schrödinger wave functions of Eq. (4.60). Subscripts were attached to the Dirac spinors to indicate which particle they belong to (1 for the electron, 2 for the positively charged one), but otherwise subscripts labeling the state being considered have been left implicit.

The correction ΔE_{T}, simple as it is, is still far to complex to be evaluated exactly. We shall have to make approximations and find in ΔE_{T} the $\mathcal{O}(\alpha^4)$ part. The approximations needed are based on considerations of the physical sizes and frequencies (or equivalently, momenta and energies) that characterize nonrelativistic Coulombic bound states. The physical size is set by the Bohr radius (as appropriate for an atom of reduced mass μ and positive charge number Z), which in natural units is $a_0 = 1/(Z\alpha\mu)$. The corresponding momentum scale is $Z\alpha\mu$, which characterizes the width of the wave functions in momentum space. The typical atomic energy scale $(Z\alpha)^2\mu$ is set by the Bohr energy levels. An examination of the positions of the p_0 poles in the reference wave functions [see Eqs. (16.101) and (16.103)], which are found at

$$p_0 = -\xi_1 P^0 + \omega_1(\vec{p}\,) = -m_1 - \xi_1 E + \sqrt{m_1^2 + \vec{p}^{\,2}} \approx -\xi_1 E + \vec{p}^{\,2}/(2m_1)\,, \tag{16.130a}$$

$$p_0 = \xi_2 P^0 - \omega_2(\vec{p}\,) = m_2 + \xi_2 E - \sqrt{m_2^2 + \vec{p}^{\,2}} \approx \xi_2 E - \vec{p}^{\,2}/(2m_2)\,, \tag{16.130b}$$

confirms the fact that the scale of the Bohr energy levels characterizes the relative energy dependence of the reference wave functions. In passing, we remark that, in deriving (16.130), we used the definition $\xi_i = m_i/M$.

We can now use our knowledge of the scales characterizing the problem to make a number of approximations. The first is to neglect $k_0^2 = (p_0 - q_0)^2$ relative to $\vec{k}^{\,2}$ in the transverse photon propagator. We have seen that k_0^2 is of $\mathcal{O}(\alpha^4)$, while $\vec{k}^{\,2}$ is of order $\mathcal{O}(\alpha^2)$ for typical energies and momenta of the system. So, to a first approximation, $k^2 = k_0^2 - \vec{k}^{\,2} \approx -\vec{k}^{\,2}$. With k_0 in the photon propagator out of the way, the relative energies p_0 and q_0 are decoupled, and we can use (16.106) to perform the energy integrals

$$\begin{aligned}
\int \frac{\mathrm{d}p_0}{2\pi} \mathcal{S}(P_n^0;p) &= \left(\frac{\omega_1(\vec{p}\,) + m_1}{2\omega_1(\vec{p}\,)}\right)^{-1/2} \left(\frac{\omega_2(\vec{p}\,) + m_2}{2\omega_2(\vec{p}\,)}\right)^{-1/2} N^{-1}(E_n;\vec{p}^{\,2}) \\
&= \left\{\frac{(P_n^0 + \omega_1(\vec{p}\,) + \omega_2(\vec{p}\,))((P_n^0)^2 - (\omega_1(\vec{p}\,) - \omega_2(\vec{p}\,))^2)}{2P_n^0\,((P_n^0)^2 - (m_1 - m_2)^2)}\right\}^{1/2} \\
&= 1 + \left(\frac{1}{m_1^2} + \frac{1}{m_2^2} - \frac{1}{m_1 m_2}\right)\frac{\vec{p}^{\,2}}{8} - \frac{\beta_n \mu}{4M} + \cdots \\
&= 1 + \mathcal{O}(\vec{p}^{\,2}) + \mathcal{O}(\alpha^2)\,,
\end{aligned} \tag{16.131}$$

where we have expanded the result of the integration for small α and small $\vec{p}/m_1$ and $\vec{p}/m_2$. The leading order contributions to the Dirac matrix elements are easy to work out:

$$\begin{aligned}
\overline{u}(\vec{p}\,)\gamma^i u(\vec{q}\,) &\approx \left(\chi^+\,, -\frac{\chi^+\vec{\sigma}\cdot\vec{p}}{2m}\right)\begin{pmatrix} 0 & \sigma^i \\ -\sigma^i & 0 \end{pmatrix}\begin{pmatrix} \chi \\ \dfrac{\vec{\sigma}\cdot\vec{q}\chi}{2m} \end{pmatrix} \\
&= \frac{1}{2m}\chi^+\left(\sigma^i\vec{\sigma}\cdot\vec{q} + \vec{\sigma}\cdot\vec{p}\sigma^i\right)\chi = \frac{1}{2m}\chi^+\left[(p+q)^i + \mathrm{i}\left(\vec{\sigma}\times\vec{k}\right)^i\right]\chi,
\end{aligned} \tag{16.132}$$

and $\overline{u}(-\vec{p})\gamma^i u(-\vec{q})$ is minus the same. The first term in (16.132), proportional to $\vec{p}+\vec{q}$, constitutes the *convection* term (see Secs. 11.4.2 and 12.5.3), as it has to do with the motion of a charged particle. The second term gives rise to a spin-dependent interaction. We neglect $\eta^2 = (1-\beta_n/2)^{-1}$, which is $1+\mathcal{O}(\alpha^2)$, and obtain

$$\Delta E_{\rm T} \approx -4\pi Z\alpha \int \frac{\mathrm{d}^3p}{(2\pi)^3}\frac{\mathrm{d}^3q}{(2\pi)^3} \times \psi^*(\vec{p})\chi_1^+\chi_2^+\left[\frac{(p+q)^i - \mathrm{i}\left(\vec{k}\times\vec{\sigma}\right)^i}{2m_1}\right]_1 \frac{\delta^{\perp,ij}(\vec{k})}{\vec{k}^{\,2}}\left[\frac{(p+q)^j - \mathrm{i}\left(\vec{k}\times\vec{\sigma}\right)^j}{2m_2}\right]_2 \chi_1\chi_2\psi(\vec{q}). \tag{16.133}$$

If we define an effective Hamiltonian through

$$\Delta E_{\rm A} \equiv \int \frac{\mathrm{d}^3p}{(2\pi)^3}\frac{\mathrm{d}^3q}{(2\pi)^3}\psi^*(\vec{p})\chi_1^+\chi_2^+ H_{\rm A}\chi_1\chi_2\psi(\vec{q}), \tag{16.134}$$

then the effective Hamiltonian for transverse photon exchange at $\mathcal{O}(\alpha^4)$ is

$$H_{\rm T} = \frac{\pi Z\alpha}{m_1m_2}\frac{1}{\vec{k}^{\,2}}\left\{\frac{-4}{\vec{k}^{\,2}}\left(\vec{p}^{\,2}\vec{q}^{\,2} - (\vec{p}\cdot\vec{q})^2\right) - 2\mathrm{i}(\vec{p}\times\vec{q})\cdot(\vec{\sigma}_1+\vec{\sigma}_2) + (\vec{k}\times\vec{\sigma}_1)\cdot(\vec{k}\times\vec{\sigma}_2)\right\}. \tag{16.135}$$

The first term in $H_{\rm T}$ comes completely from the coupling of convection currents and represents the basic magnetic interaction. The second term contributes part of the spin-orbit interaction, and the third term is a spin-spin tensor interaction. All contribute to the energy at exactly $\mathcal{O}(\alpha^4)$.

The Coulomb exchange contribution requires a little more effort to evaluate because $\langle \mathrm{i}K_{\rm C}\rangle$ and $\langle \mathrm{i}K_0\rangle$ both come in at $\mathcal{O}(\alpha^2)$ and only the difference contributes at the order $\mathcal{O}(\alpha^4)$ of interest. We start with the Coulomb exchange kernel

$$K_{\rm C} = \left[-\mathrm{i}e\gamma^0\right]_1 \frac{\mathrm{i}}{\vec{k}^{\,2}}\left[-\mathrm{i}(-Ze)\gamma^0\right]_2 = \frac{\mathrm{i}4\pi Z\alpha[\gamma^0]_1[\gamma^0]_2}{\vec{k}^{\,2}}, \tag{16.136}$$

with expectation value

$$\langle \mathrm{i}K_{\rm C}\rangle = \mathrm{i}\overline{\Psi}^0 K_{\rm C}\Psi^0 = -4\pi Z\alpha\eta^2\int \frac{\mathrm{d}^4p}{(2\pi)^4}\frac{\mathrm{d}^4q}{(2\pi)^4} \times \psi^*(\vec{p})\mathcal{S}(P_n^0;p)\frac{1}{\vec{k}^{\,2}}\mathcal{S}(P_n^0;q)\psi(\vec{q})\overline{u}_1(\vec{p})\gamma^0u_1(\vec{q})\overline{u}_2(-\vec{p})\gamma^0u_2(-\vec{q}). \tag{16.137}$$

There is no coupling between p^0 and q^0 since the Coulomb photon is instantaneous, and so the relative energy integrals can be done already. The result is given in (16.131). This time, we need to keep terms of $\mathcal{O}(\vec{p}^{\,2})$ and $\mathcal{O}(\alpha^2)$. The required Dirac matrix elements are

$$\begin{aligned}\overline{u}(\vec{p})\gamma^0u(\vec{q}) &\approx \left(\frac{\omega(\vec{p})+m}{2\omega(\vec{p})}\right)^{1/2}\left(\frac{\omega(\vec{q})+m}{2\omega(\vec{q})}\right)^{1/2}\left(\chi^+,\frac{\chi^+\vec{\sigma}\cdot\vec{p}}{\omega(\vec{p})+m}\right)\begin{pmatrix}\chi\\ \dfrac{\vec{\sigma}\cdot\vec{q}\chi}{\omega(\vec{q})+m}\end{pmatrix}\\ &= \left(\frac{\omega(\vec{p})+m}{2\omega(\vec{p})}\right)^{1/2}\left(\frac{\omega(\vec{q})+m}{2\omega(\vec{q})}\right)^{1/2}\chi^+\left[1+\frac{\vec{p}\cdot\vec{q}+\mathrm{i}(\vec{p}\times\vec{q})\cdot\vec{\sigma}}{(\omega(\vec{p})+m)(\omega(\vec{q})+m)}\right]\chi\\ &= \chi^+\left\{1-\frac{\vec{k}^{\,2}}{8m^2}+\frac{\mathrm{i}}{4m^2}(\vec{p}\times\vec{q})\cdot\vec{\sigma}\right\}\chi+\cdots,\end{aligned} \tag{16.138}$$

where the terms that were left out are of fourth order in the momenta. For $\langle \mathrm{i}K_\mathrm{C}\rangle$ we find

$$\begin{aligned}\langle \mathrm{i}K_\mathrm{C}\rangle = & -4\pi Z\alpha\,\eta^2 \int \frac{\mathrm{d}^3p}{(2\pi)^3}\frac{\mathrm{d}^3q}{(2\pi)^3} \\ & \times \psi^*(\vec{p}\,)\frac{1}{\vec{k}^{\,2}}\psi(\vec{q}\,)\left\{1+\left(\frac{1}{m_1^2}+\frac{1}{m_2^2}-\frac{1}{m_1m_2}\right)\frac{(\vec{p}^{\,2}+\vec{q}^{\,2})}{8}-\frac{\beta_n\mu}{2M}\right\} \\ & \times \chi_1^+\chi_2^+\left\{1-\left(\frac{1}{m_1^2}+\frac{1}{m_2^2}\right)\frac{\vec{k}^{\,2}}{8}+\frac{\mathrm{i}}{4}(\vec{p}\times\vec{q}\,)\cdot\left(\frac{\vec{\sigma}_1}{m_1^2}+\frac{\vec{\sigma}_2}{m_2^2}\right)\right\}\chi_1\chi_2+\cdots.\end{aligned} \tag{16.139}$$

The corresponding contribution to the effective Hamiltonian is

$$\begin{aligned}H_\mathrm{C} = & -\frac{4\pi Z\alpha}{\vec{k}^{\,2}}\eta^2\left\{1-\frac{\vec{k}^{\,2}}{8}\left(\frac{1}{m_1^2}+\frac{1}{m_2^2}\right)+\frac{(\vec{p}^{\,2}+\vec{q}^{\,2})}{8}\left(\frac{1}{m_1^2}+\frac{1}{m_2^2}-\frac{1}{m_1m_2}\right)\right. \\ & \left.-\frac{\beta_n\mu}{2M}+\frac{\mathrm{i}}{4}(\vec{p}\times\vec{q}\,)\cdot\left(\frac{\vec{\sigma}_1}{m_1^2}+\frac{\vec{\sigma}_2}{m_2^2}\right)\right\}.\end{aligned} \tag{16.140}$$

This term contributes initially at $\mathcal{O}(\alpha^2)$ with a remainder of $\mathcal{O}(\alpha^4)$. The leading order part will be canceled by the subtraction of the reference kernel.

Finally, we need to evaluate the expectation value of the reference kernel $\langle \mathrm{i}K_0\rangle$. One has

$$\begin{aligned}\langle \mathrm{i}K_0\rangle = & \,\mathrm{i}\overline{\Psi}^0K_0\Psi^0 = \mathrm{i}\eta^2\int\frac{d^4p}{(2\pi)^4}\frac{d^4q}{(2\pi)^4} \\ & \times\psi^*(\vec{p}\,)\mathcal{S}(P_n^0;p)N(E_n;\vec{p}^{\,2})(-\mathrm{i})V(\vec{p}-\vec{q}\,)N(E_n;\vec{q}^{\,2})\mathcal{S}(P_n^0;q)\psi(\vec{q}\,) \\ & \times\overline{u}_1(\vec{p}\,)\overline{u}_2(-\vec{p}\,)\left[\gamma^0\Lambda_+(\vec{p}\,)\frac{1+\gamma^0}{2}\Lambda_+(\vec{q}\,)\right]_1\left[\gamma^0\Lambda_+(-\vec{p}\,)\frac{1+\gamma^0}{2}\Lambda_+(-\vec{q}\,)\right]_2 u_1(\vec{q}\,)u_2(-\vec{q}\,).\end{aligned} \tag{16.141}$$

Using the explicit formulas for the projectors, one obtains

$$\begin{aligned}\langle \mathrm{i}K_0\rangle = & \,\mathrm{i}\eta^2\int\frac{d^4p}{(2\pi)^4}\frac{d^4q}{(2\pi)^4} \\ & \times\psi^*(\vec{p}\,)(-\mathrm{i})V(\vec{p}-\vec{q}\,)\psi(\vec{q}\,)\,\mathcal{S}(P_n^0;p)N(E_n;\vec{p}^{\,2})N(E_n;\vec{q}^{\,2})\mathcal{S}(P_n^0;q) \\ & \times\left(\frac{\omega_1(\vec{p}\,)+m_1}{2\omega_1(\vec{p}\,)}\right)^{1/2}\left(\frac{\omega_1(\vec{q}\,)+m_1}{2\omega_1(\vec{q}\,)}\right)^{1/2}\left(\frac{\omega_2(\vec{p}\,)+m_2}{2\omega_2(\vec{p}\,)}\right)^{1/2}\left(\frac{\omega_2(\vec{q}\,)+m_2}{2\omega_2(\vec{q}\,)}\right)^{1/2} \\ = & \,\eta^2\int\frac{\mathrm{d}^3p}{(2\pi)^3}\frac{\mathrm{d}^3q}{(2\pi)^3}\,\psi^*(\vec{p}\,)V(\vec{p}-\vec{q}\,)\psi(\vec{q}\,),\end{aligned} \tag{16.142}$$

where the last step was to perform the relative energy integrals using Eq. (16.106). The corresponding contribution to the effective Hamiltonian is

$$H_0 = \eta^2V(\vec{p}-\vec{q}\,) = -\frac{4\pi Z\alpha}{\vec{k}^{\,2}}\eta^2. \tag{16.143}$$

The complete effective Hamiltonian for $\mathcal{O}(\alpha^4)$ corrections is the sum of the transverse part H_T and the Coulomb part $H_\mathrm{C} - H_0$:

$$\begin{aligned} H_\mathrm{eff} = V(\vec{k}\,)\Bigg\{ & \frac{(\vec{p}^{\,2} + \vec{q}^{\,2})}{8}\left(\frac{1}{m_1^2} + \frac{1}{m_2^2} - \frac{1}{m_1 m_2}\right) - \frac{\beta_n \mu}{2M} \\ & - \frac{\vec{k}^{\,2}}{8}\left(\frac{1}{m_1^2} + \frac{1}{m_2^2}\right) + \frac{\left(\vec{p}^{\,2}\vec{q}^{\,2} - (\vec{p}\cdot\vec{q})^2\right)}{m_1 m_2 \vec{k}^{\,2}} \\ & + \frac{\mathrm{i}}{4}(\vec{p}\times\vec{q})\cdot\left(\frac{\vec{\sigma}_1}{m_1^2} + \frac{\vec{\sigma}_2}{m_2^2} + \frac{2(\vec{\sigma}_1 + \vec{\sigma}_2)}{m_1 m_2}\right) - \frac{(\vec{k}\times\vec{\sigma}_1)\cdot(\vec{k}\times\vec{\sigma}_2)}{4 m_1 m_2}\Bigg\} \end{aligned} \tag{16.144}$$

where $\vec{k} = \vec{p} - \vec{q}$. We were able to approximate η by one in the difference $H_C - H_0$ because the $\mathcal{O}(\alpha^2)$ contributions cancel. This $\mathcal{O}(\alpha^4)$ effective Hamiltonian agrees in all details with the one obtained in Ref. [582] using a closely related formalism (and, of course, with the results presented in Chap. 12). The first two terms can be seen to arise from the first relativistic correction to the particles' kinetic energy (as discussed in Chap. 17). The remaining contributions are the Darwin term, the magnetic interaction, the spin-orbit term, and the spin-spin tensor interaction (see also Chap. 12).

16.3 Relativistic Recoil Correction (BSEQ Approach)

16.3.1 *Exploring Integration Regions*

We now supplement the derivation of the relativistic recoil correction (Salpeter correction) outlined in Sec. 15.5 by a derivation based on the Bethe–Salpeter equation for QED. The recoil correction is relevant to a bound system such as hydrogen or muonium at order $(Z\alpha)^5$. The terms come from interactions involving particle exchange but not radiative corrections to one or the other particle's interaction (such as would be encoded in a form factor, in a first approximation). Energy corrections coming from a Dirac equation, which includes relativistic corrections, or the Breit equation commence at order $\mathcal{O}(Z\alpha)^4$ and go on to order $\mathcal{O}(Z\alpha)^6$ and higher even powers of $Z\alpha$ because they are ultimately obtained from expansions in the small quantity $v^2/c^2 \sim (Z\alpha)^2$. Terms at order $(Z\alpha)^5$ have their origin in quantum field theory and its description of the bound state as a two-body system.

Energy corrections of order $(Z\alpha)^5$ arise naturally from perturbing kernels involving the exchange of two photons, just as the one-photon exchange graphs of Sec. 16.2.5 gave corrections of order $(Z\alpha)^4$. The power counting that leads to this expectation includes one power of $Z\alpha$ for each exchanged photon and a factor of $(Z\alpha)^3$ coming from the wave function at the origin (or more generally, simply from the wave function). However, power counting for bound states is more subtle, and in fact all of the diagrams of Fig. 16.4 give contributions of order $(Z\alpha)^5$. The one-photon-exchange (1γE) diagrams contribute initially at order $(Z\alpha)^4$, as seen in the preceding section, but have higher-order contributions as well. Such behavior is characteristic of bound state graphs. Their effect commences at some initial order

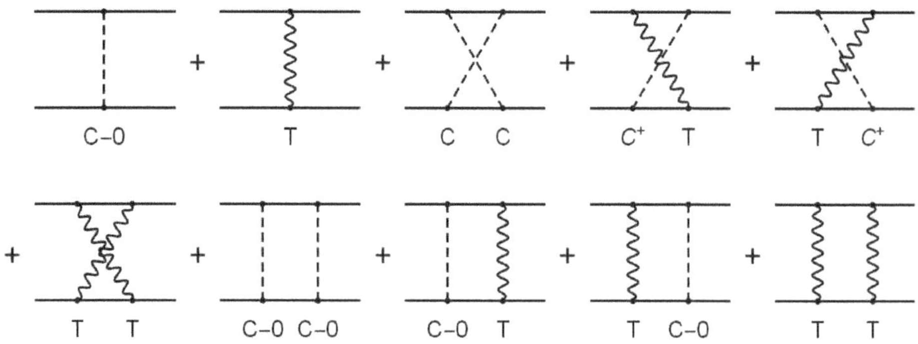

Fig. 16.4 The recoil graphs that contribute to bound state energies through order $(Z\alpha)^5$ are shown here. The type of exchange photon involved is indicated both graphically and with a label: C for Coulomb, 0 for reference, and T for transverse. These graphs all involve exchange photons only. Contributions with radiative corrections on one fermion line or the other as well as vacuum polarization effects are not considered here. The first six graphs shown are part of δK and appear in the expectation value $\langle i\delta K\rangle$. The final four graphs come from the second order perturbation theory term $\langle i\delta K\hat{G}_0\delta K\rangle$ with only the 0-potential part of $\hat{G}_0$: $\hat{G}_0 \to S_0$. The label C^+ indicates a sum of one, two, up to an infinite number of Coulomb photon exchanges, all in a ladder configuration with respect to each other and all crossed by the single transverse photon.

but they contribute at higher orders as well. More interesting are the diagrams involving C^+ where C^+ denotes a sum of terms involving exchange of one, two, etc., Coulomb photons in a ladder configuration. These diagrams all give contributions at order $(Z\alpha)^5$, no matter how many Coulomb photons are exchanged. The behavior here is similar to that discussed in Sec. 15.4 for the usual Lamb shift and will be fully explicated in the present context below.

As a first step in analyzing the order $(Z\alpha)^5$ recoil corrections, we note that the graphs in Fig. 16.4 with two exchange photons have an internal loop with a loop momentum that must be integrated over. That loop momentum can be either relativistic or nonrelativistic. (In this context, relativistic refers to values of energy and momentum comparable or greater than the smaller of the two masses.) In most cases, a nonrelativistic loop momentum leads to energy corrections of order $(Z\alpha)^6$ for these graphs. As a quick guide to this estimate, we note that a graph with two reference photons in a ladder configuration would give an energy contribution at order $(Z\alpha)^2$, the same as for a single reference photon exchange, because of the reference BSEQ: $\Psi^0 = S_0 K_0 \Psi^0 = S_0 K_0 S_0 K_0 \Psi^0 = \ldots$. For nonrelativistic momenta, the reference photon and the Coulomb photon are effectively the same and the difference $C-0$ (Coulomb minus reference photon) involves corrections proportional to the square of the momenta. Similarly, the transverse photon connects to a spatial Dirac γ matrix on each end, which leads to factors of the momenta at each end. The consequence is that the $(K_C - K_0)S_0(K_C - K_0)$, $(K_C - K_0)S_0K_{\mathrm{T}}$, and $K_{\mathrm{T}}S_0(K_C - K_0)$ graphs of Fig. 16.4 contribute at order $(Z\alpha)^6$ for nonrelativistic loop momenta. The reasoning is different for the crossed-Coulomb graph. If

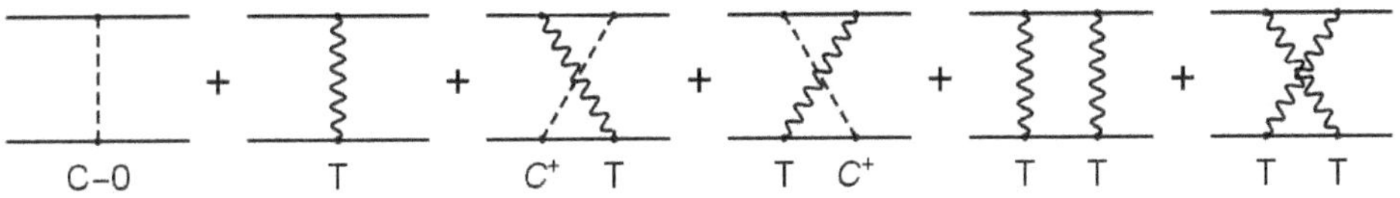

Fig. 16.5 We depict the recoil graphs that contribute to bound state energies through order $(Z\alpha)^5$ with a nonrelativistic loop momentum for the two-photon-exchange graphs.

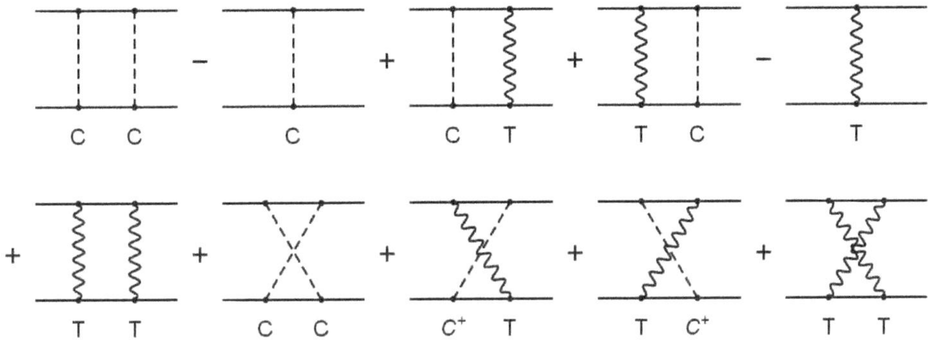

Fig. 16.6 We depict the recoil graphs that contribute to bound-state energies through order $(Z\alpha)^5$. These are the same graphs that were displayed in Fig. 16.4 but with the $C - 0$ terms expanded out and after use of the reference BSEQ.

we refer to the two poles in (16.46) as the particle and antiparticle poles, then in the integral over the loop energy for the crossed-Coulomb graph, the only nonvanishing contributions have one particle pole and one antiparticle pole, leading to an expression smaller in the numerator by a factor of $(Z\alpha)^2$ and less small in the denominator by $(Z\alpha)^2$ than the two-rung Coulomb ladder, i.e., a contribution of order $(Z\alpha)^6$. The graph with crossed Coulomb and transverse does contribute at order $(Z\alpha)^5$ in a way that we will discuss at length. Also, the two-transverse graphs contribute through the $[\Lambda_-]_1\,[\Lambda_-]_2$ part of S_0 because $\overline{u}(p_2)\gamma^i\Lambda_-(\vec{q}\,)\gamma^0\gamma^j u(p_1)$ is of order one for nonrelativistic particles, so the power counting gives a total contribution of order $(Z\alpha)^5$ in the same way as for relativistic loop momentum as will be discussed below. The complete set of diagrams with nonrelativistic loop momenta that contribute at order $(Z\alpha)^5$ is shown in Fig. 16.5.

When considering relativistic loop momentum, it is perhaps easier to put the order $(Z\alpha)^5$ graphs shown in Fig. 16.4 in the form of Fig. 16.6. The contributions here with relativistic loop momentum are shown in Fig. 16.7. The power counting is now easy. The fact that the loop momentum is relativistic means, by application of the uncertainty principle, that the space-time extent of the loop (roughly equal to $1/\mu$) is small compared to the size of the atom, which is set by the Bohr radius $1/(Z\alpha\mu)$. So, these contributions factorize into the product of a scattering amplitude calculated for external particles at rest [order $(Z\alpha)^2$] times the probability that the two particles making up the bound system will be in contact, the latter

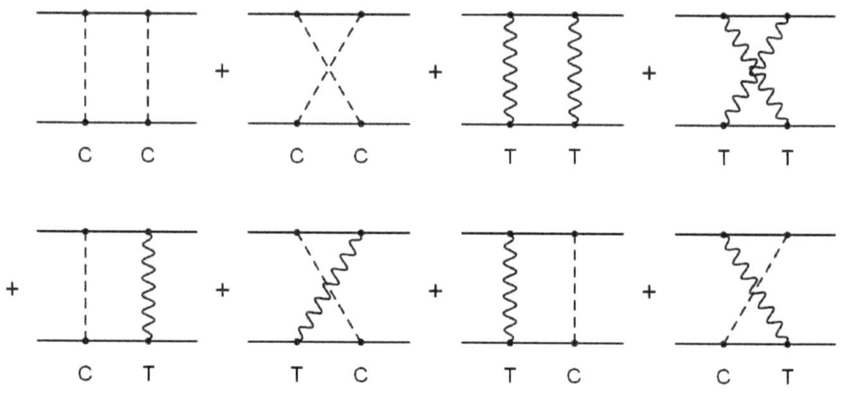

Fig. 16.7 The recoil graphs that contribute to bound state energies through order $(Z\alpha)^5$ with a relativistic loop momentum for the two-photon-exchange graphs.

being equal to $|\psi(\vec{r}=0)|^2 = \mathcal{O}(Z\alpha)^3$. This gives order $(Z\alpha)^5$ in all for the energy correction.

Our separation of the calculation into two regions (relativistic and nonrelativistic) brings about a complication: the relativistic and nonrelativistic contributions are separately divergent. The complete energy shift at order $(Z\alpha)^5$, and in fact each graph by itself, is completely finite and well defined, but the process of separation into regions engenders divergences due to the approximations that will be made appropriate for each region. The generation of divergences is inevitable when graphs are pulled apart into regions and simplified enough as to allow for analytic evaluation of the contribution at each order of $Z\alpha$. The alternative to this procedure of *divide-and-conquer* is to evaluate the graphs as a whole using numerical integration. The integrands are typically quite complex, multi-dimensional, and hard to integrate with good precision. So, analysis by regions is the usual approach.

We will use dimensional regularization (DR) to regularize the divergent integrals we encounter. We have encountered this method already in Secs. 10.3.6, 10.4.2 and 11.4.4. However, a more thorough discussion might still be useful. In general, among the various approaches to regularization than have been explored, DR does the least violence to the theory. It does not break gauge invariance and leaves the forms and algebraic properties of propagators and vertices unchanged, except for the need to interpret the dimensions of space-time (or energy-momentum space) as d-dimensional instead of four-dimensional. We imagine that d can take a general (possibly complex) value, that will tend to four at the end of the calculation after all divergences have canceled. In practice, we write $d = 4 - 2\varepsilon$ [see Eq. (10.172)] and let $\varepsilon \to 0$ at the end. It is also useful to define $D = d - 1 = 3 - 2\epsilon$ [see Eq. (10.141)] as the dimension of space when dealing, as here, with bound systems at rest in a specific (CM) frame. The regularized space-time dimension alters the space-time volume elements ($\mathrm{d}^d x$ and $\mathrm{d}^d p/(2\pi)^d$ instead of $\mathrm{d}^4 x$ and $\mathrm{d}^4 p/(2\pi)^4$), some algebraic properties (so that $g^{\mu}{}_{\mu} = d$, $\delta^{ii} = D$, etc.), but is mostly just hidden in the background until the time comes to actually do the integrals. One aspect of DR that we need to account for is the alteration of the units of some physical quantities in d dimensional space-time. While mass, energy, and spatial momentum all have units of mass, distances and times have units of inverse mass. The action is dimensionless in DR, the quantum fields for the fermions and photon have a dimension altered by a quantity of $\mathcal{O}(\varepsilon)$ from what they had in four dimensions. A consequence is that the quantum of charge is no longer dimensionless in d dimensions but must have units m^{ε} where m is a mass. The exact value chosen for m is arbitrary as it disappears in the limit $\varepsilon \to 0$, and we use this arbitrariness to our advantage. We define the charge quantum so that the product of the charges for the bound-state constituents is

$$q_1 q_2 = -Ze^2 = -4\pi Z\alpha\overline{\mu}^{2\varepsilon}\,, \qquad \overline{\mu}^2 = \mu^2 \frac{\mathrm{e}^{\gamma_E}}{4\pi}\,. \tag{16.145}$$

This definition is consistent with Eq. (10.173). We recognize γ_E as the Euler–Mascheroni constant $\gamma_E = 0.57721\,56649\ldots$. The reason for this definition, suggested by the modified minimal subtraction ("MS-bar", $\overline{\text{MS}}$), is that μ is arbitrary anyway so it doesn't really matter whether we include the extra factors of e^{γ_E} and 4π or not, but with the $\mathrm{e}^{\gamma_E}/(4\pi)$ factor the divergences arising from loop integrals take a much simpler form. Specifically, the expansion

$$\mathrm{e}^{\varepsilon\gamma_E}\Gamma(\varepsilon) = \frac{1}{\varepsilon} + \frac{1}{2}\zeta(2)\varepsilon - \frac{1}{3}\zeta(3)\varepsilon^2 + \frac{9}{40}\zeta^2(2)\varepsilon^3 + \mathcal{O}(\varepsilon^4) \tag{16.146}$$

is much simpler than the expansion of $\Gamma(\varepsilon)$ by itself. The factor of 4π acts to cancel part of the volume element $\mathrm{d}^d p/(2\pi)^d$ that remains after the momentum integral has been performed — as will be seen below. (The customary notation in field theory for the $\overline{\text{MS}}$ mass parameter is $\overline{\mu}$ instead of $\overline{\mu}$, but we reserve μ for the reduced mass as before.)

The technique we use to isolate the contributions of our bound state integrals coming from various parts of momentum space is called the *method of regions*. This approach is a powerful technique for the separation of scales and constitutes a generalization of the scale-separation parameter or "overlapping parameter" as described in Sec. 4.5.2. We describe the method of regions in some detail in the next section and apply it in the following sections.

16.3.2 *Method of Regions*

A central challenge of bound-state physics is the evaluation of integrals that contain important contributions from several different parts of the parameter space. The method of regions [586–590] is a tool that, in conjunction with DR, allows contributions from the various regions to be separated and evaluated. In this section we will describe the method of regions, work through a detailed example of its use, and discuss its application to bound-state physics.

The method of regions consists of the following steps: *(i)* Identify the relevant regions that contribute to the integral under consideration. *(ii)* In each region, expand the integrand in a Taylor series in terms of the quantities that are small in that region. *(iii)* In each region, integrate the expanded integral, term by term, over the whole integration domain of the integration variables. *(iv)* Add up the contributions from the various regions to regain the full integral. It is expected that the approximations appropriate for the various regions will lead to new divergences (ones not present in the original integral) that are automatically regulated by DR. New divergences generated in this way will cancel when the contributions of the different regions are combined. As is usual in DR, scaleless integrals are regulated to have the value zero. A particularly clear exposition of this fact, together with further explanatory remarks, is given in Chap. 8 of Ref. [201] [see the text following Eq. (8.52) of Ref. [201]].

We return to the example studied in Sec. 4.5.2,

$$F(\alpha) = \int_0^\infty \mathrm{d}\omega\, \mathrm{e}^{-\omega} \sqrt{\alpha^2 + \omega^2}\,, \tag{16.147}$$

where $\alpha \ll 1$ is a small parameter. We saw before how the function $F(\alpha)$ could be (approximately) evaluated by breaking the integration region into parts separated by an overlapping parameter ε where $\alpha \ll \varepsilon \ll 1$, and making appropriate approximations in each part. Now, we find the value of $F(\alpha)$ by use of the method of regions.

Our first step is to identity the relevant regions that contribute to $F(\alpha)$. We see that the integrand is a product of two terms, $\mathrm{e}^{-\omega}$ and $\sqrt{\alpha^2+\omega^2}$. The exponential clearly has a characteristic scale set by $\omega \sim 1$, while the square root has a characteristic scale set by $\omega \sim \alpha$. So for this example there are two regions to deal with, which we call the hard ($\omega \sim 1$) and soft ($\omega \sim \alpha$) regions.

Since we plan to use DR as a regularization, we need to dimensionally generalize our integral. Usually, for four-dimensional space-time, we generalize by the replacement $\mathrm{d}^4 p \to \mathrm{d}^d p = p^{d-1}\mathrm{d}p\,\mathrm{d}\Omega_p = p^{3-2\varepsilon}\mathrm{d}p\,\mathrm{d}\Omega_p$, where $\mathrm{d}\Omega_p$ represents the integral over the $d-1$ angles of p when using spherical coordinates. For our one-dimensional integral $F(\alpha)$, we dispense with angles and just work with the simplified dimensional generalization

$$F(\alpha, \varepsilon) \equiv \int_0^\infty \mathrm{d}\omega\, \omega^{-2\varepsilon} \mathrm{e}^{-\omega} \sqrt{\alpha^2 + \omega^2}. \tag{16.148}$$

Now, we apply the method of regions (and dimensional regularization). For the hard region, we can expand the square root using

$$\sqrt{\alpha^2 + \omega^2} = \omega\sqrt{1 + (\alpha/\omega)^2} = \omega \left\{1 + \frac{\alpha^2}{2\omega^2} - \frac{\alpha^4}{8\omega^4} + \mathcal{O}(\alpha^6)\right\}. \tag{16.149}$$

The required integrals can be done immediately using the integral representation of the gamma function:

$$\Gamma(n+1) = \int_0^\infty \mathrm{d}\omega\, \mathrm{e}^{-\omega}\, \omega^n\,. \tag{16.150}$$

For the hard part of $F(\alpha, \varepsilon)$, we find the expansion

$$\begin{aligned} F_H(\alpha, \varepsilon) &= \int_0^\infty d\omega\, e^{-\omega} \omega^{1-2\varepsilon} \left\{ 1 + \frac{\alpha^2}{2\omega^2} - \frac{\alpha^4}{8\omega^4} + \mathcal{O}(\alpha^6) \right\} \\ &= \Gamma(2 - 2\varepsilon) + \frac{\alpha^2}{2} \Gamma(-2\varepsilon) - \frac{\alpha^4}{8} \Gamma(-2 + 2\varepsilon) + \mathcal{O}(\alpha^6) \\ &= 1 + \frac{\alpha^2}{2} \left(-\frac{1}{2\varepsilon} - \gamma_E \right) - \frac{\alpha^4}{8} \left(-\frac{1}{4\varepsilon} + \frac{3}{4} - \frac{\gamma_E}{2} \right) + \mathcal{O}(\varepsilon) + \mathcal{O}(\alpha^6). \end{aligned} \quad (16.151)$$

For the soft region ω is small compared to one and we need to expand the exponential in a Taylor series. The dependence on α is displayed most clearly if we scale the ω variable using $\omega = \alpha x$. We find that

$$\begin{aligned} F_S(\alpha, \varepsilon) &= \int_0^\infty d\omega\, \omega^{-2\varepsilon} \left\{ 1 - \omega + \frac{\omega^2}{2} - \frac{\omega^3}{6} + \cdots \right\} (\alpha^2 + \omega^2)^{1/2} \\ &= \alpha^{2-2\varepsilon} \left\{ \chi(-2\varepsilon, 1/2) - \alpha\, \chi(1 - 2\varepsilon, 1/2) + \frac{\alpha^2}{2} \chi(2 - 2\varepsilon, 1/2) \right. \\ &\quad \left. - \frac{\alpha^3}{6} \chi(3 - 2\varepsilon, 1/2) + \ldots \right\} \\ &= \alpha^2 \left[\frac{1}{4\varepsilon} + \frac{1}{4} + \frac{1}{2} \ln\left(\frac{2}{\alpha}\right) \right] + \frac{\alpha^3}{3} + \alpha^4 \left[-\frac{1}{32\varepsilon} + \frac{1}{64} - \frac{1}{16} \ln\left(\frac{2}{\alpha}\right) \right] \\ &\quad - \frac{\alpha^5}{45} + \mathcal{O}(\varepsilon) + \mathcal{O}(\alpha^6), \end{aligned} \quad (16.152)$$

where we have used the integral formula

$$\chi(m, n) \equiv \int_0^\infty dx\, x^m (1 + x^2)^n = \frac{\Gamma\left(\frac{m+1}{2}\right) \Gamma\left(-\frac{m+1}{2} - n\right)}{2\Gamma(-n)}, \quad (16.153)$$

which converges when $m > -1$ and $m + 2n < -1$. We note in passing that the integral $\chi(m, n)$ fails to converge for $n = 1/2$ no matter what value we take for m. The integral $\chi(m, n)$ is defined instead by analytic continuation using Eq. (16.153), which is perfectly well defined when $n = 1/2$ except for occasional poles. We obtain the full $F(\alpha, \varepsilon)$ integral by adding the hard and soft parts:

$$\begin{aligned} F(\alpha, \varepsilon) &= F_H(\alpha, \varepsilon) + F_S(\alpha, \varepsilon) \\ &= 1 + \alpha^2 \left[\frac{1}{4} - \frac{\gamma_E}{2} + \frac{1}{2} \ln\left(\frac{2}{\alpha}\right) \right] + \frac{\alpha^3}{3} \\ &\quad + \alpha^4 \left[-\frac{5}{64} + \frac{\gamma_E}{16} - \frac{1}{16} \ln\left(\frac{2}{\alpha}\right) \right] - \frac{\alpha^5}{45} + \mathcal{O}(\varepsilon) + \mathcal{O}(\alpha^6). \end{aligned} \quad (16.154)$$

Since the original integral $F(\alpha)$ is defined as the limit of $F(\alpha, \varepsilon)$ as $\varepsilon \to 0$, we have completed our task, at least through terms of $\mathcal{O}(\alpha^5)$. Our expression (16.154) agrees with the earlier result (4.276) that was obtained using an overlapping parameter. It can also be confirmed by analytic integration of (16.147) followed by expansion in α, but the possibility of direct analytic integration of bound state integrals is seldom possible in practical calculations.

Table 16.1 Regions appearing in QED bound state calculations. Here m represents a typical mass — the reduced mass say — and α is the fine structure constant. (In this analysis we are not trying to distinguish between the possibly different scales set by m_1 and m_2 if they are distinct.)

region	notation	typical energy	typical momentum
hard	H	m	m
soft	S	$m\alpha$	$m\alpha$
potential	P	$m\alpha^2$	$m\alpha$
ultrasoft	US	$m\alpha^2$	$m\alpha^2$

The DR regions approach to the evaluation of multi-scale integrals has a number of attractive features compared to the use of overlapping parameters. One is the fact that each term that appears in each region after performing the Taylor expansion has a unique power of the small parameter (modulo possible logarithms). Also, integrals over the whole infinite integration range (even when supplemented by factors like $x^{-2\varepsilon}$) are often easier to evaluate and have more convenient analytic properties than the integrals over a partial range that would result from overlap parameters. For integrals resulting from quantum loops, the DR regions approach is even more advantageous, as it is awkward to decide just what parameter to have its range restricted by an overlapping parameter: the energy of one of the particles, or of another, or all; or perhaps not the energy at all but the three-momentum magnitude or the Euclidean four-momentum magnitude after Wick rotation. The advantages of DR regions are even more apparent when a single integral has more than just two relevant regions.

Experience has shown that there are four regions relevant to the physics of nonrelativistic Coulombic bound states. They are the *hard*, *soft*, *potential*, and *ultrasoft* regions with characteristic values of energy and spatial momenta as shown in Table 16.1. We take some time to understand something about the origin and nature of each of these regions. We look first at the wave function (16.101):

$$\Psi^0_{nk}(p) = \eta\, \mathcal{S}(P^0_n; p)\, \psi_{n\ell m}(\vec{p}\,)\, u_1(\vec{p}\,)\, u_2(-\vec{p}\,)\,, \tag{16.155}$$

and recall the definitions of η from Eq. (16.102) and $\mathcal{S}(P^0_n; p)$ from Eq. (16.103). The relevant part of this function for understanding the regions that are characteristic of a nonrelativistic bound state is

$$\begin{aligned}\Psi^0_{nk}(p) &\propto \frac{1}{(\xi_1 P^0_n + p^0 - \omega_1(\vec{p}\,) + \mathrm{i}\epsilon)(\xi_2 P^0_n - p^0 - \omega_2(\vec{p}\,) + \mathrm{i}\epsilon)}\psi_{n\ell m}(\vec{p}\,)\\ &\approx \frac{1}{\left(\xi_1 E_n + p^0 - \frac{\vec{p}^{\,2}}{2m_1} + \mathrm{i}\epsilon\right)\left(\xi_2 E_n - p^0 - \frac{\vec{p}^{\,2}}{2m_2} + \mathrm{i}\epsilon\right)}\psi_{n\ell m}(\vec{p}\,)\,.\end{aligned} \tag{16.156}$$

Here $\omega_i(\vec{p}) \equiv (m_i^2 + \vec{p}^{\,2})^{1/2}$, while E_n is the binding energy defined in Eq. (16.96). The expansion in the momenta implied in Eq. (16.156) is relevant for nonrelativistic motion of the constituents.

As described in Sec. 4.4.4, the dependence of the wave function on the magnitude of the spatial momentum has the typical behavior

$$\psi_{n\ell m}(\vec{p}) \propto \frac{1}{(\vec{p}^{\,2} + \gamma_n^2)^2}, \tag{16.157}$$

where $\gamma_n = \mu Z\alpha/n$. This specific form for the wave function is for the ground state, but all bound states involve powers of the factor $(\vec{p}^{\,2} + \gamma_n^2)$ in the denominator, setting the momentum scale $|\vec{p}\,| \sim \gamma_n \sim m\alpha$. The scale for the relative energy p^0 is determined by the positions of the poles in $\Psi^0(p)$. Given that both E_n and $\vec{p}^{\,2}/m$ have the scale γ_n^2/m, we see that the natural scale for the relative energy set by the wave function is $p^0 \sim \gamma_n^2/m \sim m\alpha^2$. So, the natural scale for bound-state processes, as set by the wave function, is of the *potential* kind according to the classification in Table 16.1. This is the region we assumed as the dominant region in the order-$(Z\alpha)^4$ calculations of Sec. 16.2.5. Other regions are engendered by the transverse photon propagator $1/(k^2 + \mathrm{i}\varepsilon) = [(k^0 - |\vec{k}\,| + \mathrm{i}\varepsilon)(k^0 + |\vec{k}\,| - \mathrm{i}\varepsilon)]^{-1}$ when retardation effects (those requiring the presence of k^0 in the propagator) are important. Both poles of the photon propagator set the photon energy magnitude equal to its spatial momentum magnitude. If that spatial momentum is $\sim m\alpha$, then the energy will be $\sim m\alpha$ as well, and the soft region is obtained. In a like manner, if the energy is $\sim m\alpha^2$, then the spatial momentum will be $\sim m\alpha^2$, giving the ultrasoft region. If we don't assume that the motion is nonrelativistic, then it is clear from the form of the fermion propagator $\mathrm{i}/(\not{p} - m + \mathrm{i}\varepsilon) \to 1/(p^2 - m^2 + \mathrm{i}\varepsilon)$ that relativistic values of p^0 and $\vec{p}$ $(p^0 \sim |\vec{p}\,| \sim m)$ can be important. This "hard" region also plays a role in the physics of QED bound states.

16.3.3 *Contribution from the Hard Region*

We are now ready to tackle the calculation of relativistic recoil corrections at $\mathcal{O}(\alpha^5)$. We already saw that we can separate these into ones involving a nonrelativistic loop momentum as shown in Fig. 16.5 and ones involving a relativistic loop momentum as in Fig. 16.7. The nonrelativistic ones can be further segregated into soft, potential, and ultrasoft contributions, as we will show. The relativistic ones represent hard region contributions. We will discuss the hard region contributions first, then move on to the nonrelativistic regions, and put our results together at the end.

The hard-region contribution refers to the part of the total energy shift coming from the region of momentum space where the photons and the two fermions connecting the photons are all hard. The diagrams contributing in this region are shown in Fig. 16.7. We will go through the calculation of these contributions pair by pair, starting with the first two diagrams Fig. 16.7. These are just the ones involving two Coulomb photons. In Fig. 16.8(a), we depict these two diagrams with momentum labels included. The associated wave functions on each side, $\overline{\Psi}^0(p)$ and

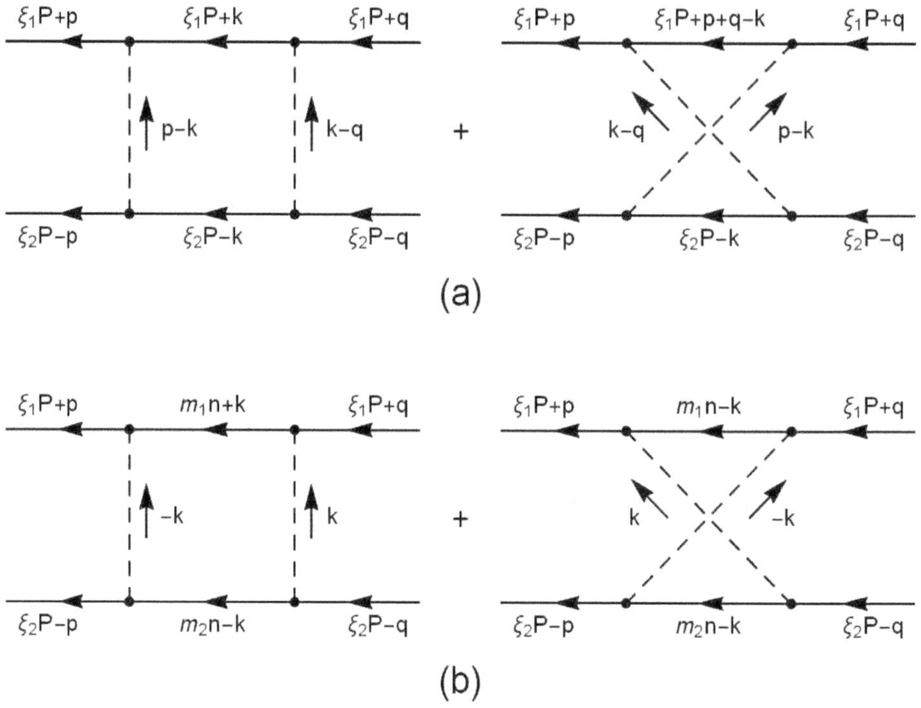

Fig. 16.8 These are the graphs — ladder plus crossed — involving two Coulomb photons that contribute to order $(Z\alpha)^5$ energy levels from the hard region. Part (a) shows the graphs with momentum labels included. Here $P = (m_1 + m_2 + E, \vec{0}\,)$ is the CM momentum, $\xi_i = m_i/(m_1 + m_2)$, and p, k, and q represent the relative momenta. Part (b) shows the same graphs but with hard region approximations indicated for the central loop integral, using $\xi_i P \pm k \to m_i n \pm k$ and $\xi_1 P + p + q - k \to m_1 n - k$ with $n \equiv (1, \vec{0}\,)$. Wave functions are implicitly present on both sides.

$\Psi^0(q)$, are assumed to be present but are not shown. The hard region approximations appropriate for the central loop involve the neglect of p and q compared to the loop momentum k as well as the neglect of the reference bound state energy E relative to $m_1 + m_2$ as shown in Fig. 16.8(b). The integrals over p, k, and q factorize in this approximation, and the central part is the two-particle scattering amplitude at rest due to the exchange of two Coulomb photons. For the wave function integrals, we use (16.131) and

$$\int \frac{\mathrm{d}^D p}{(2\pi)^D} \psi(\vec{p}\,) = \psi(\vec{r} = \vec{0}) = \psi(\vec{0})\,. \tag{16.158}$$

(A wave function at the origin $\psi(\vec{0})$ will be interpreted in the following as being taken at the origin of coordinate space, not momentum space.) We see that

$$\int \frac{\mathrm{d}^d p}{(2\pi)^d} \Psi^0(p) \approx \psi(\vec{0}) u_1(\vec{0}) u_2(\vec{0}), \tag{16.159a}$$

$$\int \frac{\mathrm{d}^d p}{(2\pi)^d} \overline{\Psi}^0(p) \approx \psi^*(\vec{0}) \overline{u}_1(\vec{0}) \overline{u}_2(\vec{0}), \tag{16.159b}$$

where the fundamental spinors $u_i(\vec{0})$ with $i = 1, 2$ are evaluated at the origin of momentum space. So, the contribution factorizes into the product of a free particle at-rest scattering amplitude times the probability of contact $|\psi(\vec{0})|^2$. Physically, the scattering event happens with relatively high energy and momentum, i.e., effectively instantaneously and at a point relative to the spatial size of the bound state. It remains to calculate the scattering-amplitude part of the diagram.

We apply the usual Feynman rules of QED to calculate the scattering amplitude, starting with the ladder (left side) diagram of Fig. 16.8(b):

$$\begin{aligned}
\Delta E_{\mathrm{CC;L}} &\approx \mathrm{i}|\psi(\vec{0})|^2 \int \frac{\mathrm{d}^d k}{(2\pi)^d} \overline{u}_1(\vec{0})(-\mathrm{i}e\gamma^0) \frac{\mathrm{i}}{m_1 \not{n} + \not{k} - m_1} (-\mathrm{i}e\gamma^0) u_1(\vec{0}) \\
&\quad \times \left(\frac{\mathrm{i}}{\vec{k}^{\,2}}\right)^2 \overline{u}_2(\vec{0})(\mathrm{i}Ze\gamma^0) \frac{\mathrm{i}}{m_2 \not{n} - \not{k} - m_2} (\mathrm{i}Ze\gamma^0) u_2(\vec{0}) \\
&= \mathrm{i}|\psi(\vec{0})|^2 \left(4\pi Z\alpha \overline{\mu}^{2\varepsilon}\right)^2 \int \frac{\mathrm{d}^d k}{(2\pi)^d} \left[(k^2 + 2\,k\cdot n\, m_1)(k^2 - 2\,k\cdot n\, m_2)(\vec{k}^{\,2})^2\right]^{-1} \\
&\quad \times \left(\overline{u}_1 \gamma^0 \left[m_1 \not{n} + \not{k} + m_1\right] \gamma^0 u_1\right) \left(\overline{u}_2 \gamma^0 \left[m_2 \not{n} - \not{k} + m_2\right] \gamma^0 u_2\right)
\end{aligned} \tag{16.160}$$

where $n^\mu = (1, \vec{0})$ is the timelike unit vector, and we have used the abbreviation here that $u = u(\vec{0}) = u(\vec{p} = 0)$. In terms of the fundamental Pauli spinors χ [see Eq. (6.42)], the Dirac spinor for a particle at rest is $u = (\chi, 0)^{\mathrm{T}}$. Since we are working out the spin-independent–that is, the spin averaged–energy shift in this section, we find for the Dirac expectation value

$$\overline{u} X u = \mathrm{Tr}[X u \overline{u}] \to \frac{1}{2} \sum_{\text{spin}} \mathrm{Tr}[X u \overline{u}] = \frac{1}{4} \mathrm{Tr}\left[X(\gamma^0 + 1)\right], \tag{16.161}$$

because the spin-averaged outer product can be evaluated as follows,

$$\frac{1}{2} \sum_{\text{spin}} u \otimes \overline{u} = \frac{1}{2} \begin{pmatrix} \sum_{\text{spin}} \chi \otimes \chi^+ & 0 \\ 0 & 0 \end{pmatrix} = \frac{1}{2} \begin{pmatrix} 1 & 0 \\ 0 & 0 \end{pmatrix} = \frac{1}{4}(\gamma^0 + 1), \tag{16.162}$$

where the Dirac matrices are represented in matrix form by giving the large-large, large-small, small-large, and small-small components (see also Chap. 15). We find after performing the spin averaging and the traces that

$$\Delta E_{\mathrm{CC;L}} \to \mathrm{i}|\psi(\vec{0})|^2 \left(4\pi Z\alpha \overline{\mu}^{2\varepsilon}\right)^2 \int \frac{\mathrm{d}^d k}{(2\pi)^d} \frac{(2m_1 + k^0)(2m_2 - k^0)}{(k^2 + 2\,k\cdot n\, m_1)(k^2 - 2\,k\cdot n\, m_2)(\vec{k}^{\,2})^2}. \tag{16.163}$$

We recall, for convenience, Eq. (16.145). The integral is most easily done using Feynman parameters, completing the square in the denominator, followed by a

Wick rotation and finally the actual integration as in the Appendix of Ref. [591] (see also Chaps. 11 and 15). Despite the fact that the integral looks like it would have an infrared divergence (because of the $(\vec{k}^{\,2})^2$ factor in the denominator), the dimensionally regularized result is finite in the limit $d \to 4$. Its value is

$$\Delta E_{\mathrm{CC;L}} = \frac{2}{3}(Z\alpha)^2|\psi(\vec{0})|^2\left(-\frac{2}{m_1^2} - \frac{1}{m_1 m_2} - \frac{2}{m_2^2}\right), \tag{16.164}$$

in agreement with (3.7) of Ref. [535], where a different form of regulation was used. (The potential divergence in (16.163) is a Coulomb singularity coming from the ladder configuration of Coulomb photons. Such a singularity is regulated to zero by DR [592].) The crossed-Coulomb calculation is done in the same way, leading to

$$\begin{aligned}\Delta E_{\mathrm{CC;X}} &= \mathrm{i}|\psi(\vec{0})|^2\left(4\pi Z\alpha\overline{\mu}^{2\varepsilon}\right)^2\int\frac{\mathrm{d}^d k}{(2\pi)^d}\frac{(2m_1 - k^0)(2m_2 - k^0)}{(k^2 - 2\,k{\cdot}n\,m_1)(k^2 - 2\,k{\cdot}n\,m_2)(\vec{k}^{\,2})^2}\\ &= \frac{2}{3}(Z\alpha)^2|\psi(\vec{0})|^2\left(\frac{2}{m_1^2} - \frac{1}{m_1 m_2} + \frac{2}{m_2^2}\right),\end{aligned} \tag{16.165}$$

in agreement with (3.9) of Ref. [535]. The total Coulomb contribution is the sum:

$$\Delta E_{\mathrm{CC}} = \Delta E_{\mathrm{CC;L}} + \Delta E_{\mathrm{CC;X}} = \frac{2}{3}(Z\alpha)^2|\psi(\vec{0})|^2\left(\frac{-2}{m_1 m_2}\right). \tag{16.166}$$

All graphs with one transverse photon and one Coulomb photon vanish in the hard region, as we now proceed to show. The same momentum assignments as in Fig. 16.8(b) are used, even though now not all photons are Coulomb ones. Consider the first graph on the bottom row of Fig. 16.7. The trace factor for the top fermion line is

$$\frac{1}{2}\sum_{\mathrm{spin}}\overline{u}_1\gamma^0(m_1\gamma^0 + \not{k} + m_1)\gamma^a u_1 = \frac{1}{4}\mathrm{Tr}\left[\gamma^0(m_1\gamma^0 + \not{k} + m_1)\gamma^a(\gamma^0 + 1)\right] = k^a, \tag{16.167}$$

where a is the spatial index for one end of the transverse photon propagator. Since that propagator carries the transverse delta factor $\delta^{\perp,ab}(\vec{k}\,) = \delta^{ab} - \hat{k}^a\hat{k}^b$ and $k^a\delta^{\perp,ab}(\vec{k}\,) = 0$, the contribution of this graph is zero. The remaining three hard-region graphs with one of each kind of photon vanish for the same reason.

We complete our hard-region calculation with the evaluation of the double transverse graphs. Again using the momentum assignments of Fig. 16.8(b), the transverse-transverse ladder graph gives

$$\begin{aligned}\Delta E_{\mathrm{TT;L}} &= \mathrm{i}\,|\psi(\vec{0})|^2\int\frac{\mathrm{d}^d k}{(2\pi)^d}\overline{u}_1(-\mathrm{i}e\gamma^a)\frac{\mathrm{i}}{m_1\not{n} + \not{k} - m_1}(-\mathrm{i}e\gamma^c)u_1\\ &\quad\times\frac{\mathrm{i}\delta^{\perp,ab}(\vec{k}\,)}{\vec{k}^{\,2}}\frac{\mathrm{i}\delta^{\perp,cd}(\vec{k}\,)}{\vec{k}^{\,2}}\overline{u}_2(\mathrm{i}Ze\gamma^b)\frac{\mathrm{i}}{m_2\not{n} - \not{k} - m_2}(\mathrm{i}Ze\gamma^d)u_2\\ &= \mathrm{i}\,|\psi(\vec{0})|^2\left(4\pi Z\alpha\overline{\mu}^{2\varepsilon}\right)^2\int\frac{\mathrm{d}^d k}{(2\pi)^d}\left[(k^2 + 2\,k{\cdot}n\,m_1)(k^2 - 2\,k{\cdot}n\,m_2)(\vec{k}^{\,2})^2\right]^{-1}\\ &\quad\times\delta^{\perp,ab}(\vec{k}\,)\delta^{\perp,cd}(\vec{k}\,)\left(\overline{u}_1\gamma^a\left[m_1\not{n} + \not{k} + m_1\right]\gamma^c u_1\right)\left(\overline{u}_2\gamma^b\left[m_2\not{n} - \not{k} + m_2\right]\gamma^d u_2\right).\end{aligned} \tag{16.168}$$

We perform the spin average as before and find that the traces simplify remarkably. The transverse-transverse ladder contribution becomes

$$\Delta E_{\mathrm{TT;L}} = \mathrm{i}|\psi(\vec{0})|^2 \left(4\pi Z\alpha\overline{\mu}^{2\varepsilon}\right)^2 \int \frac{\mathrm{d}^d k}{(2\pi)^d} \frac{(2-d)(k^0)^2}{(k^2+2\,k\cdot n\,m_1)(k^2-2\,k\cdot n\,m_2)(k^2)^2}. \tag{16.169}$$

The integral here is a standard covariant loop integral that can be performed with the usual methods, leading to the result

$$\Delta E_{\mathrm{TT;L}} = \frac{(Z\alpha)^2}{m_1 m_2}\overline{\mu}^{2\varepsilon}|\psi(\vec{0})|^2\left\{\frac{1}{2\varepsilon}+\frac{1}{2}-\frac{m_1\ln(m_2/\mu)+m_2\ln(m_1/\mu)}{m_1+m_2}+\mathcal{O}(\varepsilon)\right\} \tag{16.170}$$

after setting $d = 4-2\varepsilon$ and expanding about $\varepsilon = 0$. (Note that only one factor of $\mu^{2\varepsilon}$ was used in the expansion.) The result has an infrared divergence that is manifest as the pole at $\varepsilon = 0$. Because of the divergence, we cannot just set $d = 4$ in the wave function. Fortunately, all divergences will cancel before we need to know the value of the wave function in d dimensions, at which point we can go to the usual four-dimensional space-time and use the standard result for $\psi(\vec{0})$. An entirely analogous calculation for the transverse-transverse crossed contribution leads to

$$\begin{aligned}\Delta E_{\mathrm{TT;X}} &= \mathrm{i}|\psi(\vec{0})|^2 \left(4\pi Z\alpha\overline{\mu}^{2\varepsilon}\right)^2 \int \frac{\mathrm{d}^d k}{(2\pi)^d} \frac{(d-2)(k^0)^2}{(k^2+2\,k\cdot n\,m_1)(k^2+2\,k\cdot n\,m_2)(k^2)^2}\\ &= \frac{(Z\alpha)^2}{m_1 m_2}\overline{\mu}^{2\varepsilon}|\psi(\vec{0})|^2\left\{\frac{1}{2\varepsilon}+\frac{1}{2}-\frac{m_1\ln(m_2/\mu)-m_2\ln(m_1/\mu)}{m_1-m_2}+\mathcal{O}(\varepsilon)\right\}.\end{aligned} \tag{16.171}$$

The total transverse-transverse hard contribution is the sum of the ladder plus crossed terms:

$$\begin{aligned}\Delta E_{\mathrm{TT}} &= \Delta E_{\mathrm{TT;L}} + \Delta E_{\mathrm{TT;X}}\\ &= \frac{(Z\alpha)^2}{m_1 m_2}\overline{\mu}^{2\varepsilon}|\psi(\vec{0})|^2\left\{\frac{1}{\varepsilon}+1-2\frac{m_1^2\ln(m_2/\mu)-m_2^2\ln(m_1/\mu)}{m_1^2-m_2^2}+\mathcal{O}(\varepsilon)\right\}.\end{aligned} \tag{16.172}$$

We can now combine our Coulomb-Coulomb and transverse-transverse results to get the total contribution from the hard region. Adding the results from (16.166) and (16.172), we find

$$\begin{aligned}\Delta E_{\mathrm{hard}} &= \Delta E_{\mathrm{CC}} + \Delta E_{\mathrm{TT}}\\ &= \frac{(Z\alpha)^2}{m_1 m_2}\overline{\mu}^{2\varepsilon}|\psi(\vec{0})|^2\left\{\frac{1}{\varepsilon}-\frac{1}{3}-2\frac{m_1^2\ln(m_2/\mu)-m_2^2\ln(m_1/\mu)}{m_1^2-m_2^2}+\mathcal{O}(\varepsilon)\right\}.\end{aligned} \tag{16.173}$$

We anticipate that the infrared divergence exhibited here will cancel against an ultraviolet divergence in the nonrelativistic region to leave a finite total. Along with that cancellation of divergences, the μ factor in the logarithms will be replaced by one of the nonrelativistic scales to produce a logarithmic $\mathcal{O}((Z\alpha)^5\ln(Z\alpha))$ contribution to the overall recoil correction.

It is useful to point out that the central part of this calculation, that of the amplitude for scattering of two fermions at rest involving the exchange of two photons, could have been done in Feynman gauge instead of Coulomb gauge. The two-photon-exchange diagrams form a gauge-invariant set and any gauge can be used for the calculation. The Feynman gauge calculation is simpler as there is only one kind of photon (with propagator $-\mathrm{i}g^{\mu\nu}/k^2$) and one kind of vertex $(-\mathrm{i}e\gamma^\mu)$, so fewer terms appear and those that do are more compact and covariant. Of course, the result obtained using Feynman gauge is the same as the Coulomb gauge result of (16.173), demonstrating the gauge invariance of the hard region, separately.

16.3.4 *Soft Contribution from Two Transverse Photons*

We now turn to contributions coming from nonrelativistic regions, and start with the final two graphs of Fig. 16.5 involving the exchange of two transverse photons. We begin by looking at the top fermion line, having Dirac spinors at beginning and end, vertex factors, and the fermion propagator, which we assume carries relative momentum k. This fermion line makes a contribution

$$\begin{aligned} N_1 &\equiv \overline{u}_1(\vec{p}\,)\gamma^a \frac{1}{\xi_1 \not{P} + \not{k} - m_1 + \mathrm{i}\epsilon}\gamma^c u_1(\vec{q}\,) \\ &= \overline{u}_1(\vec{p}\,)\gamma^a \left\{ \frac{\Lambda_+(\vec{k}\,)\gamma^0}{\xi_1 P^0 + k^0 - \omega_1(\vec{k}\,) + \mathrm{i}\epsilon} + \frac{\Lambda_-(\vec{k}\,)\gamma^0}{\xi_1 P^0 + k^0 + \omega_1(\vec{k}\,) - \mathrm{i}\epsilon} \right\} \gamma^c u_1(\vec{q}\,) \\ &\approx \chi_1^+ \left\{ \frac{F^{ac}(\vec{p},\vec{k},\vec{q}\,)}{\xi_1 E + k^0 - \frac{\vec{k}^{\,2}}{2m_1} + \mathrm{i}\epsilon} + \frac{\sigma^a\sigma^c}{2m_1 + \xi_1 E + k^0 + \frac{\vec{k}^{\,2}}{2m_1} + \mathrm{i}\epsilon} \right\} \chi_1 \end{aligned} \tag{16.174}$$

where $F^{ac}(\vec{p},\vec{k},\vec{q}\,)$ is quadratic in the small momenta $\vec{p}$, $\vec{k}$, $\vec{q}$. The fermion propagator terms coming from Λ_+ give small contributions in any of the nonrelativistic regions [of order $\mathcal{O}(\alpha^2)$], and denominator factors are not able to completely reverse this tendency. Only the Λ_- terms contribute, and it turns out that these only contribute in the soft region of the loop momentum k. In this soft region, k^0 is $\mathcal{O}(\alpha)$ as are all the spatial momenta. The leading approximation to the two-transverse-photon graph in the ladder configuration is

$$\begin{aligned} \Delta E^{\rm soft}_{\rm TT;L} &= \mathrm{i}(4\pi Z\alpha\overline{\mu}^{2\varepsilon})^2 \int \frac{\mathrm{d}^D p}{(2\pi)^D}\frac{\mathrm{d}^D q}{(2\pi)^D}\psi^*(\vec{p}\,)\psi(\vec{q}\,)\frac{\chi_1^+\sigma^a\sigma^c\chi_1}{2m_1}\frac{\chi_2^+\sigma^b\sigma^d\chi_2}{2m_2} \\ &\quad\times \int \frac{\mathrm{d}^d k}{(2\pi)^d}\frac{\delta^{\perp,ab}(\vec{p}-\vec{k}\,)}{(k^0)^2 - (\vec{p}-\vec{k}\,)^2 + \mathrm{i}\epsilon}\frac{\delta^{\perp,cd}(\vec{k}-\vec{q}\,)}{(k^0)^2 - (\vec{k}-\vec{q}\,)^2 + \mathrm{i}\epsilon}. \end{aligned} \tag{16.175}$$

The spin average takes $\chi_1^+\sigma^a\sigma^c\chi_1 \to \delta^{ac}$, and similarly $\chi_2^+\sigma^b\sigma^d\chi_2 \to \delta^{bd}$. We make a shift in the integration variable $\vec{k} \to \vec{k} + \vec{p}$ and note that the crossed graph contribution at this order is identical to that of the ladder graph. In all, the spin-averaged recoil correction due to the soft region with two exchanged transverse photons is

$$\Delta E^{\rm soft}_{\rm TT} = \frac{(4\pi Z\alpha)^2}{2m_1 m_2}\overline{\mu}^{2\varepsilon}\int \frac{\mathrm{d}^D p}{(2\pi)^D}\frac{\mathrm{d}^D q}{(2\pi)^D}\psi^*(\vec{p}\,)\psi(\vec{q}\,)I(\vec{p},\vec{q}\,), \tag{16.176}$$

where $I(\vec{p},\vec{q}\,)$ is a divergent (in the ultraviolet) non-covariant integral that can be evaluated by the usual methods, giving

$$I(\vec{p},\vec{q}) = \mathrm{i}\,\overline{\mu}^{2\varepsilon} \int \frac{\mathrm{d}^d k}{(2\pi)^d} \frac{\delta^{\perp,ab}(\vec{k})}{k^2 + \mathrm{i}\epsilon} \frac{\delta^{\perp,ab}(\vec{k}+\vec{p}-\vec{q})}{(k^0)^2 - (\vec{k}+\vec{p}-\vec{q})^2 + \mathrm{i}\epsilon}$$
$$= \frac{1}{8\pi^2}\left\{-\frac{1}{\epsilon} + 2\ln\left(\frac{|\vec{p}-\vec{q}|}{\mu}\right) - \frac{8}{3}\ln 2 + \frac{5}{3}\right\}. \tag{16.177}$$

The resulting energy shift is

$$\Delta E^{\rm soft}_{\rm TT} = \frac{(Z\alpha)^2}{m_1 m_2}\overline{\mu}^{2\varepsilon}\left\{|\psi(\vec{0})|^2\left[-\frac{1}{\epsilon} - 2\ln\mu - \frac{8}{3}\ln 2 + \frac{5}{3}\right] + 2\big\langle \ln(|\vec{p}-\vec{q}|)\big\rangle_{\vec{p},\vec{q}}\right\}, \tag{16.178}$$

where $\langle \ln(|\vec{p}-\vec{q}|)\rangle_{\vec{p},\vec{q}}$ is the momentum space expectation value of $\ln(|\vec{p}-\vec{q}|)$, which equals the usual expectation value of the Fourier transform of $\ln k$:

$$\big\langle \ln(|\vec{p}-\vec{q}|)\big\rangle_{\vec{p},\vec{q}} = \int \frac{\mathrm{d}^D p}{(2\pi)^D}\frac{\mathrm{d}^D q}{(2\pi)^D}\psi^*(\vec{p})\ln(|\vec{p}-\vec{q}|)\psi(\vec{q})$$
$$= \int \mathrm{d}^D x\, \psi^*(\vec{x})\,\mathrm{FT}(\ln k)(\vec{x})\,\psi(\vec{x})$$
$$= \big\langle \mathrm{FT}(\ln k)\big\rangle \tag{16.179}$$

where $k = |\vec{k}|$. The result for this expectation value is given in the Appendix of Ref. [593]:

$$\big\langle \mathrm{FT}(\ln k)\big\rangle = \left(\frac{(\mu Z\alpha)^3}{\pi n^3}\right)\begin{cases} \ln\left(\dfrac{2\mu Z\alpha}{n}\right) + H_n + \dfrac{n-1}{2n} & (\ell = 0) \\ -\dfrac{1}{4\ell(\ell+1)(\ell+1/2)} & (\ell > 0) \end{cases}, \tag{16.180}$$

where $H_n = \sum_{j=1}^{n}\frac{1}{j}$ is the nth harmonic number. We note that the ultraviolet divergence arising from the two-transverse-exchange soft region just cancels the infrared divergence that was found in the hard region (16.173). The expectation value of the operator given in Eq. (16.180) is related to the Araki–Sucher distribution [see Eq. (15.133) and Refs. [537, 538].

16.3.5 *Soft Contribution from a Single Transverse Photon*

There are three contributions to recoil energies at order $(Z\alpha)^5$ involving the exchange of a single nonrelativistic transverse photon, as shown in Fig. 16.5. We consider first the contribution from the second graph in that figure containing an isolated transverse photon. This graph is shown in Fig. 16.9 including specifications for the fermion and photon momenta. If the left and right bound states surrounding that exchange photon in $\Delta E = \mathrm{i}\overline{\Psi}^0\delta K\Psi^0$ have relative momenta p and q, then the momentum flowing up the exchanged photon is $k \equiv p - q$. If k is soft, momentum conservation requires at least one of p or q be soft as well. The regions where the ordered set $\{p, k, q\}$ is {potential, soft, soft} or {soft, soft, potential} give the $\mathcal{O}(Z\alpha)^5$ contributions. Consider then the region {potential, soft, soft} in the sense

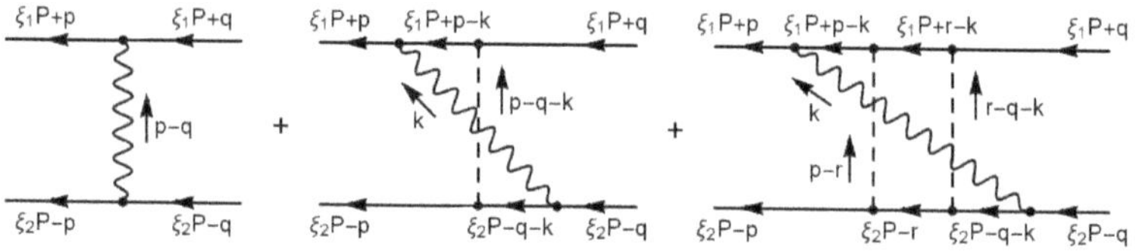

Fig. 16.9 Graphs containing a single transverse exchange photon crossing zero, one, or two Coulomb photons. Wave functions are implicitly present on both sides.

of Table 16.1. The single transverse exchange energy shift takes the approximate form

$$\begin{aligned}\Delta E_{\mathrm{T}}^{\mathrm{P,S,S}} &= \mathrm{i}\int \frac{\mathrm{d}^d p}{(2\pi)^d}\frac{\mathrm{d}^d q}{(2\pi)^d}\overline{\Psi}^0(p)\left[-\mathrm{i}e\gamma^a\right]_1 \frac{\mathrm{i}\delta^{\perp,ab}(\vec{k}\,)}{k^2+\mathrm{i}\epsilon}\left[\mathrm{i}Ze\gamma^b\right]_2 \Psi^0(q)\\ &\approx -4\pi Z\alpha\overline{\mu}^{2\varepsilon}\mathrm{i}\int \frac{\mathrm{d}^D p}{(2\pi)^D}\frac{\mathrm{d}^D q}{(2\pi)^D}\frac{dq_0}{2\pi}\psi^*(\vec{p}\,)\,\overline{u}_1(\vec{p}\,)\gamma^a u_1(\vec{q}\,)\,\overline{u}_2(-\vec{p}\,)\gamma^b u_2(-\vec{q}\,)\,\psi(\vec{q}\,)\\ &\quad\times \delta^{\perp,ab}(\vec{p}-\vec{q}\,)\left(E-\frac{\vec{q}^{\,2}}{2\mu}\right)\left[(q_0-|\vec{p}-\vec{q}\,|+\mathrm{i}\epsilon)(q_0+|\vec{p}-\vec{q}\,|-\mathrm{i}\epsilon)(q_0+\mathrm{i}\epsilon)(-q_0+\mathrm{i}\epsilon)\right]^{-1}.\end{aligned} \tag{16.181}$$

In obtaining this approximation, we performed the p^0 integral using

$$\int \frac{\mathrm{d}p^0}{2\pi}\Psi^0(p) \approx u_1(\vec{p}\,)u_2(-\vec{p}\,)\psi(\vec{p}\,) \tag{16.182}$$

or specifically its analog for $\overline{\Psi}^0$, which hold in the nonrelativistic limit. We also made the soft q^0 approximation that $q^0 \sim \alpha$ dominates terms of $\mathcal{O}(\alpha^2)$ such as p^0, E, and $\vec{q}^{\,2}$, and implemented a number of general nonrelativistic approximations based on the smallness of E, $\vec{p}^{\,2}/(2\mu)$, and $\vec{q}^{\,2}/(2\mu)$ compared to the reduced mass scale μ. Our next step is to perform the q^0 integration, which we do by poles. A subtle point is that the poles at $q^0 = 0$ are disallowed because they would take q^0 out of the soft region which would be inconsistent with the assumed softness of q. Furthermore, the Dirac expectation value takes the approximate form

$$\overline{u}_i(\vec{p}\,)\gamma^a u_i(\vec{q}\,) \approx \chi^+\frac{\sigma^a\vec{\sigma}\cdot\vec{q}+\vec{\sigma}\cdot\vec{p}\sigma^a}{2m_i}\chi \to \frac{p^a+q^a}{2m_i}, \tag{16.183}$$

where the last form was achieved by averaging over spin. We also note that $(p+q)^a = (2p-(p-q))^a \to 2p^a$ when next to a transverse delta. The energy shift takes the form

$$\Delta E_{\mathrm{T}}^{\mathrm{P,S,S}} = -\frac{4\pi Z\alpha\overline{\mu}^{2\varepsilon}}{m_1 m_2}\int \frac{\mathrm{d}^D p}{(2\pi)^D}\frac{\mathrm{d}^D q}{(2\pi)^D}\psi^*(\vec{p}\,)\frac{p^a p^b\delta^{\perp,ab}(\vec{p}-\vec{q}\,)}{2|\vec{p}-\vec{q}\,|^3}\left(E-\frac{\vec{q}^{\,2}}{2\mu}\right)\psi(\vec{q}\,). \tag{16.184}$$

It is difficult to continue this evaluation in a straightforward way because the $\vec{q}$ integration is divergent in the infrared (for $|\vec{q}| \sim |\vec{p}|$). In order to proceed, we would need to know the long-distance structure of the wave function in $D = 3-2\varepsilon$ dimensions. Since we do not know the D-dimensional wave function in that much detail, we will instead isolate the wave function from the infrared (IR) divergence by providing a buffer. Specifically, we use the nonrelativistic Schrödinger–Coulomb equation

$$\left(E-\frac{\vec{q}^{\,2}}{2\mu}\right)\psi(\vec{q}\,)=\int\frac{\mathrm{d}^D\ell}{(2\pi)^D}\left(\frac{-4\pi Z\alpha\mu^{2\varepsilon}}{(\vec{q}-\vec{\ell}\,)^2}\right)\psi(\vec{\ell}\,),\tag{16.185}$$

to insert a Coulomb photon between the soft transverse photon and the wave function, and then change integration variables according to $\vec{q}\to\vec{p}-\vec{k}$, $\vec{\ell}\to\vec{q}$ to find:

$$\Delta E_{\mathrm{T}}^{\mathrm{P,S,S}}=\frac{8\pi^2(Z\alpha)^2\overline{\mu}^{2\varepsilon}}{m_1m_2}\int\frac{\mathrm{d}^Dp}{(2\pi)^D}\frac{\mathrm{d}^Dq}{(2\pi)^D}$$
$$\times\,\psi^*(\vec{p}\,)\,p^ap^b\left(\overline{\mu}^{2\varepsilon}\int\frac{\mathrm{d}^Dk}{(2\pi)^D}\frac{\delta^{\perp,ab}(\vec{k}\,)}{|\vec{k}\,|^3(\vec{k}-\vec{p}+\vec{q}\,)^2}\right)\psi(\vec{q}\,).\tag{16.186}$$

The infrared divergent integral

$$I^{ab}(\vec{\ell}\,)\equiv\overline{\mu}^{2\varepsilon}\int\frac{\mathrm{d}^Dk}{(2\pi)^D}\frac{\delta^{\perp,ab}(\vec{k}\,)}{|\vec{k}\,|^3(\vec{k}-\vec{\ell}\,)^2}$$
$$=\overline{\mu}^{2\varepsilon}\int\frac{\mathrm{d}^Dk}{(2\pi)^D}\frac{\delta^{\perp,ab}(\vec{k}\,)}{|\vec{k}\,|^3}\left\{\frac{1}{\vec{k}^{\,2}+\vec{\ell}^{\,2}}+\frac{2\vec{k}\cdot\vec{\ell}}{(\vec{k}-\vec{\ell}\,)^2(\vec{k}^{\,2}+\vec{\ell}^{\,2})}\right\}\tag{16.187}$$

is not too difficult to evaluate in the latter form, since the first term in brackets is relatively simple (if divergent) and the second is finite (so the $D\to3$ limit may be safely taken). The result is

$$I^{ab}(\vec{\ell}\,)=\frac{1}{3\pi^2\vec{\ell}^{\,2}}\left\{\left(-\frac{1}{2\varepsilon}+\frac{1}{2}+\ln\left(\frac{2|\vec{\ell}\,|}{\mu}\right)\right)\delta^{ab}-\hat{\ell}^a\hat{\ell}^b\right\}+\mathcal{O}(\varepsilon).\tag{16.188}$$

(We are using $\vec{\ell}=\vec{p}-\vec{q}$.) If we include the contributions from both soft single-transverse-exchange regions: {potential, soft, soft} and {soft, soft, potential}, the total correction is

$$\Delta E_{\mathrm{T}}^{\mathrm{soft}}=\Delta E_{\mathrm{T}}^{\mathrm{P,S,S}}+\Delta E_{\mathrm{T}}^{\mathrm{S,S,P}}=\frac{8\pi^2(Z\alpha)^2\overline{\mu}^{2\varepsilon}}{m_1m_2}$$
$$\times\int\frac{\mathrm{d}^Dp}{(2\pi)^D}\frac{\mathrm{d}^Dq}{(2\pi)^D}\psi^*(\vec{p}\,)\left(p^ap^b+q^aq^b\right)I^{ab}(\vec{p}-\vec{q}\,)\,\psi(\vec{q}\,).\tag{16.189}$$

We will complete our examination of soft corrections by the evaluation of the crossed transverse-Coulomb graph — the middle one in Fig. 16.9. The relevant part of this graph is the region where the central loop (the photons and connecting fermions) are soft and both wave functions are potential. The full expression for the energy shift coming from this graph is

$$\Delta E_{\mathrm{TC}}=\mathrm{i}\int\frac{\mathrm{d}^dp}{(2\pi)^d}\frac{\mathrm{d}^dq}{(2\pi)^d}\frac{\mathrm{d}^dk}{(2\pi)^d}\overline{\Psi}^0(p)\left[(-\mathrm{i}e\gamma^a)\frac{\mathrm{i}}{\xi_1\not{P}+\not{p}-\not{k}-m_1+\mathrm{i}\epsilon}(-\mathrm{i}e\gamma^0)\right]_1$$
$$\times\frac{\mathrm{i}\delta^{\perp,ab}(\vec{k}\,)}{k^2+\mathrm{i}\epsilon}\frac{\mathrm{i}}{(\vec{p}-\vec{q}-\vec{k}\,)^2}\left[(\mathrm{i}Ze\gamma^0)\frac{\mathrm{i}}{\xi_2\not{P}-\not{q}-\not{k}-m_2+\mathrm{i}\epsilon}(\mathrm{i}Ze\gamma^b)\right]_2\Psi^0(q).\tag{16.190}$$

The nonrelativistic expansion of the electron propagator, for k soft and p potential, takes the form

$$\begin{aligned}\frac{\mathrm{i}}{\xi_1 \not{P} + \not{p} - \not{k} - m_1 + \mathrm{i}\epsilon} &= \frac{\mathrm{i}\Lambda_+(\vec{p}-\vec{k}\,)\gamma^0}{\xi_1 P^0 + p^0 - k^0 - \omega_1(\vec{p}-\vec{k}\,) + \mathrm{i}\epsilon} \\ &\quad + \frac{\mathrm{i}\Lambda_-(\vec{p}-\vec{k}\,)\gamma^0}{\xi_1 P^0 + p^0 - k^0 + \omega_1(\vec{p}-\vec{k}\,) - \mathrm{i}\epsilon} \\ &\approx \frac{\mathrm{i}\Lambda_+(\vec{p}-\vec{k}\,)\gamma^0}{-k^0 + \mathrm{i}\epsilon} + \frac{\mathrm{i}\Lambda_-(\vec{p}-\vec{k}\,)\gamma^0}{2m_1}, \end{aligned} \tag{16.191}$$

with a similar form for the positive fermion propagator. The first term in each dominates because $k^0/m_1 \sim \alpha$. All dependence on p^0 and q^0 is confined to the wave functions, so the relative energy p^0 and q^0 integrals can be done. We find the approximate soft region energy correction to be

$$\begin{aligned}\Delta E_{\mathrm{TC}}^{\mathrm{soft}} &\approx (4\pi Z\alpha\overline{\mu}^{2\varepsilon})^2 \mathrm{i} \int \frac{\mathrm{d}^D p}{(2\pi)^D} \frac{\mathrm{d}^D q}{(2\pi)^D} \frac{\mathrm{d}^d k}{(2\pi)^d} \psi^*(\vec{p}\,)\overline{u}_1(\vec{p}\,)\gamma^a \Lambda_+(\vec{p}-\vec{k}\,)\gamma^0\gamma^0 u_1(\vec{q}\,) \\ &\quad \times \frac{\delta^{\perp,ab}(\vec{k}\,)}{k^2 + \mathrm{i}\epsilon} \frac{1}{(\vec{p}-\vec{q}-\vec{k}\,)^2} \frac{1}{(-k^0 + \mathrm{i}\epsilon)^2} \overline{u}_2(-\vec{p}\,)\gamma^0 \Lambda_+(-\vec{q}-\vec{k}\,)\gamma^0\gamma^b u_2(-\vec{q}\,)\,\psi(\vec{q}\,). \end{aligned} \tag{16.192}$$

We perform the integration over k^0 using only the poles in the transverse photon propagator — use of the soft fermion poles would be inconsistent with the assumptions that the fermions are soft — and evaluate the spin-averaged Dirac expectation values:

$$\begin{aligned}\Delta E_{\mathrm{TC}}^{\mathrm{soft}} &\approx (4\pi Z\alpha\overline{\mu}^{2\varepsilon})^2 \mathrm{i} \int \frac{\mathrm{d}^D p}{(2\pi)^D} \frac{\mathrm{d}^D q}{(2\pi)^D} \frac{\mathrm{d}^D k}{(2\pi)^D} \\ &\quad \times \psi^*(\vec{p}\,) \frac{(2p-k)^a}{2m_1} \frac{-\mathrm{i}\delta^{\perp,ab}(\vec{k}\,)}{2|\vec{k}\,|^3(\vec{k}-\vec{p}+\vec{q}\,)^2} \frac{(-2q-k)^b}{2m_2} \psi(\vec{q}\,) \\ &= \frac{8\pi^2(Z\alpha)^2\overline{\mu}^{2\varepsilon}}{m_1 m_2} \int \frac{\mathrm{d}^D p}{(2\pi)^D} \frac{\mathrm{d}^D q}{(2\pi)^D} \psi^*(\vec{p}\,)\left(-p^a q^b\right) I^{ab}(\vec{p}-\vec{q}\,)\,\psi(\vec{q}\,). \end{aligned} \tag{16.193}$$

The result (16.193) must be doubled to account for the contribution of the graph like the middle one of Fig. 16.9, but reflected right-to-left.

The total spin-averaged recoil energy contribution coming from a soft exchanged transverse photon is the sum of (16.189) and twice (16.193). The spatial momentum factors combine according to $\left(p^a p^b - 2p^a q^b + q^a q^b\right) \to (p-q)^a (p-q)^b$ since $I^{ab}(\vec{p}-\vec{q}\,)$ of (16.188) is symmetric in a, b. For the total soft transverse energy contribution $(\mathrm{T} + \mathrm{TC} + \mathrm{CT})$ we find

$$\begin{aligned}\Delta E_{\mathrm{T,total}}^{\mathrm{soft}} &= \frac{8\pi^2(Z\alpha)^2\overline{\mu}^{2\varepsilon}}{m_1 m_2} \int \frac{\mathrm{d}^D p}{(2\pi)^D} \frac{\mathrm{d}^D q}{(2\pi)^D} \psi^*(\vec{p}\,)\,(p-q)^a (p-q)^b I^{ab}(\vec{p}-\vec{q}\,)\,\psi(\vec{q}\,) \\ &= \frac{8(Z\alpha)^2\overline{\mu}^{2\varepsilon}}{3m_1 m_2} \int \frac{\mathrm{d}^D p}{(2\pi)^D} \frac{\mathrm{d}^D q}{(2\pi)^D} \psi^*(\vec{p}\,)\left\{-\frac{1}{2\varepsilon} - \frac{1}{2} + \ln\left(\frac{2|\vec{p}-\vec{q}\,|}{\mu}\right)\right\}\psi(\vec{q}\,) \\ &= \frac{8(Z\alpha)^2}{3m_1 m_2}\overline{\mu}^{2\varepsilon}\left\{|\psi(\vec{0})|^2\left[-\frac{1}{2\varepsilon} - \frac{1}{2} + \ln\left(\frac{2}{\mu}\right)\right] + \left\langle \ln(|\vec{p}-\vec{q}\,|)\right\rangle_{\vec{p},\vec{q}}\right\}, \end{aligned} \tag{16.194}$$

where the expectation value of $\ln(|\vec{p}-\vec{q}\,|)$ was given in (16.179) and (16.180) above. We anticipate that the infrared divergence found in the soft region here will be canceled against an ultraviolet divergence from the ultrasoft region that remains to be worked out.

16.3.6 *Ultrasoft Contribution from a Single Transverse Photon*

The ultrasoft region makes a contribution at order $(Z\alpha)^5$ for all three graphs in Fig. 16.9. In fact, there are contributions at this order from graphs with any number of ladder Coulomb photons crossed by a single ultrasoft transverse photon. We will work out the effect of this infinite sum of terms using a procedure analogous to that used in the evaluation of the standard Lamb shift, and will find that the same Bethe logarithm makes its appearance here as it did there. We commence our work with an examination of the single transverse graph with no Coulomb photons and build up from there.

The contribution of the first term in Fig. 16.9 to the energy shift is

$$\Delta E_{\rm T}^{\rm US} = {\rm i}\int \frac{{\rm d}^d p}{(2\pi)^d}\frac{{\rm d}^d q}{(2\pi)^d}\overline{\Psi}^0(p)\left[-{\rm i}e\gamma^a\right]_1 \frac{{\rm i}\delta^{\perp,ab}(\vec{p}-\vec{q}\,)}{(p-q)^2+{\rm i}\epsilon}\left[{\rm i}Ze\gamma^b\right]_2\Psi^0(q)\,, \qquad (16.195)$$

where we have to remember that the transverse photon momentum $(p-q)$ is ultrasoft (and p and q are potential), since otherwise this expression is the same as what one would write down for the full graph with no regard to regions. The condition that $(p-q)$ be ultrasoft puts an effective constraint on the part of p–q space that contributes to the integral. The reference wave functions were given in (16.101) with $\mathcal{S}$ defined in (16.103). The nonrelativistic approximation to $\mathcal{S}$ is

$$\mathcal{S}(P;p) \approx \frac{{\rm i}s^{-1}(E;\vec{p}\,)}{\left(\xi_1 E + p^0 - \frac{\vec{p}^{\,2}}{2m_1} + {\rm i}\epsilon\right)\left(\xi_2 E - p^0 - \frac{\vec{p}^{\,2}}{2m_2} + {\rm i}\epsilon\right)} \qquad (16.196)$$

where $s(E;\vec{p}\,) = \left(E - \vec{p}^{\,2}/(2\mu) + {\rm i}\epsilon\right)^{-1}$ is the nonrelativistic propagator. The nonrelativistic approximation to the spin-averaged Dirac expectation values were worked out in (16.183). For the energy shift, we find

$$\begin{aligned}\Delta E_{\rm T}^{\rm US} &= -\frac{4\pi Z\alpha\overline{\mu}^{2\varepsilon}}{4m_1m_2}\int\frac{{\rm d}^D p}{(2\pi)^D}\frac{{\rm d}^D q}{(2\pi)^D}\\ &\quad\times\psi^*(\vec{p}\,)s^{-1}(E;\vec{p}\,)(p+q)^a\delta^{\perp,ab}(\vec{p}-\vec{q}\,)(p+q)^b s^{-1}(E;\vec{q}\,)\psi(\vec{q}\,)\\ &\quad\times\int\frac{{\rm d}p^0}{2\pi}\frac{dq^0}{2\pi}\Bigg[\left(\xi_1E+p^0-\frac{\vec{p}^{\,2}}{2m_1}+{\rm i}\epsilon_1\right)\left(\xi_2E-p^0-\frac{\vec{p}^{\,2}}{2m_2}+{\rm i}\epsilon_2\right)\\ &\quad\times\left(p^0-q^0-|\vec{p}-\vec{q}\,|+{\rm i}\epsilon_3\right)\left(p^0-q^0+|\vec{p}-\vec{q}\,|-{\rm i}\epsilon_3\right)\\ &\quad\times\left(\xi_1E+q^0-\frac{\vec{q}^{\,2}}{2m_1}+{\rm i}\epsilon_4\right)\left(\xi_2E-q^0-\frac{\vec{q}^{\,2}}{2m_2}+{\rm i}\epsilon_5\right)\Bigg]^{-1}.\end{aligned} \qquad (16.197)$$

We recall Eq. (16.145) and must now perform the integrals over p^0 and q^0. A new feature that appears here (and in more complicated relative energy integrals) is the need to choose an order for the positive infinitesimals. We must also choose which integral to do first and in which half plane to close the p^0 and q^0 contours. Any of these choices will do and will give the same result, but choices must be made. We choose to integrate over p^0 first and to close p^0 in the lower half plane (so, $p^0 \to -\xi_1 E + \vec{p}^{\,2}/(2m_1) - \mathrm{i}\epsilon_1$ and $p^0 \to q^0 + |\vec{p} - \vec{q}\,| - \mathrm{i}\epsilon_3$ are the two relevant pole positions). Then, with $\epsilon_1 < \epsilon_3$, we choose to close q_0 in the lower half plane as well (with relevant poles $q_0 \to -\xi_1 E + \vec{p}^{\,2}/(2m_1) + |\vec{p} - \vec{q}\,| - \mathrm{i}(\epsilon_1 + \epsilon_3)$ and $q^0 \to -\xi_1 E + \vec{q}^{\,2}/(2m_1) - \mathrm{i}\epsilon_4$). After some algebra, we find

$$\Delta E_{\mathrm{T}}^{\mathrm{US}} = -\frac{4\pi Z\alpha\overline{\mu}^{2\varepsilon}}{4m_1m_2}\int\frac{\mathrm{d}^Dp}{(2\pi)^D}\frac{\mathrm{d}^Dq}{(2\pi)^D}\psi^*(\vec{p}\,)(p+q)^a\delta^{\perp,ab}(\vec{k}\,)(p+q)^b\psi(\vec{q}\,)$$
$$\times\frac{-1}{2|\vec{k}\,|}\left\{\frac{1}{E-|\vec{k}\,|-\frac{\vec{p}^{\,2}}{2m_1}-\frac{\vec{q}^{\,2}}{2m_2}+\mathrm{i}\epsilon}+\frac{1}{E-|\vec{k}\,|-\frac{\vec{p}^{\,2}}{2m_2}-\frac{\vec{q}^{\,2}}{2m_1}+\mathrm{i}\epsilon}\right\}, \tag{16.198}$$

where $\vec{k} \equiv \vec{p} - \vec{q}$. The symmetry between m_1 and m_2 is apparent in this form. We are free to take $(p+q)^a = (2p-k)^a \to 2p^a$ and $(p+q)^b = (k+2q)^b \to 2q^b$ in the numerator because of the transverse delta factor and we note that the first denominator factor can be simplified using

$$E-|\vec{k}\,|-\frac{\vec{p}^{\,2}}{2m_1}-\frac{\vec{q}^{\,2}}{2m_2}+\mathrm{i}\epsilon=\left(E-|\vec{k}\,|-\frac{\vec{p}^{\,2}}{2\mu}+\mathrm{i}\epsilon\right)+\frac{\vec{p}^{\,2}-\vec{q}^{\,2}}{2m_2}=s^{-1}(E-|\vec{k}\,|;\vec{p}\,)+\frac{\vec{k}\cdot(\vec{p}+\vec{q}\,)}{2m_2}. \tag{16.199}$$

Now every term in $s^{-1}(E-|\vec{k}\,|;\vec{p}\,)$ is $\mathcal{O}(\alpha^2)$ while $\vec{k}\cdot(\vec{p}+\vec{q}\,)$ is $\mathcal{O}(\alpha^3)$ for $\vec{k}$ ultrasoft, so we can safely neglect the last term. A similar analysis gives the approximation $s^{-1}(E-|\vec{k}\,|;\vec{q}\,)$ for the second denominator in (16.198), so the ultrasoft contribution from transverse photon exchange becomes

$$\Delta E_{\mathrm{T}}^{\mathrm{US}} \approx \frac{4\pi Z\alpha\overline{\mu}^{2\varepsilon}}{2m_1m_2}\int\frac{\mathrm{d}^Dp}{(2\pi)^D}\frac{\mathrm{d}^Dq}{(2\pi)^D}\psi^*(\vec{p}\,)\frac{p^a\delta^{\perp,ab}(\vec{k}\,)q^b}{|\vec{k}\,|}\psi(\vec{q}\,)$$
$$\times\left\{s(E-|\vec{k}\,|;\vec{p}\,)+s(E-|\vec{k}\,|;\vec{q}\,)\right\}. \tag{16.200}$$

If we insert a factor of $1 = \int\frac{\mathrm{d}^Dk}{(2\pi)^D}\delta^{(D)}(\vec{p}-\vec{q}-\vec{k}\,)$ and recall the definition $s(E;\vec{p},\vec{q}\,) = s(E;\vec{p}\,)(2\pi)^D\delta^{(D)}(\vec{p}-\vec{q}\,)$, we can rewrite $\Delta E_{\mathrm{T}}^{\mathrm{US}}$ as

$$\Delta E_{\mathrm{T}}^{\mathrm{US}} \approx \frac{4\pi Z\alpha\overline{\mu}^{2\varepsilon}}{2m_1m_2}\int\frac{\mathrm{d}^Dp}{(2\pi)^D}\frac{\mathrm{d}^Dq}{(2\pi)^D}\frac{\mathrm{d}^Dk}{(2\pi)^D}\psi^*(\vec{p}\,)\frac{p^a\delta^{\perp,ab}(\vec{k}\,)q^b}{|\vec{k}\,|}\psi(\vec{q}\,)$$
$$\times\left\{s(E-|\vec{k}\,|;\vec{p},\vec{q}+\vec{k}\,)+s(E-|\vec{k}\,|;\vec{p}-\vec{k},\vec{q}\,)\right\}$$
$$\approx\frac{4\pi Z\alpha\overline{\mu}^{2\varepsilon}}{m_1m_2}\int\frac{\mathrm{d}^Dp}{(2\pi)^D}\frac{\mathrm{d}^Dq}{(2\pi)^D}\frac{\mathrm{d}^Dk}{(2\pi)^D}\psi^*(\vec{p}\,)\frac{p^a\delta^{\perp,ab}(\vec{k}\,)q^b}{|\vec{k}\,|}\psi(\vec{q}\,)s(E-|\vec{k}\,|;\vec{p},\vec{q}\,), \tag{16.201}$$

where the final approximation, using $|\vec{k}\,| \ll |\vec{p}\,|, |\vec{q}\,|$, was a consequence of the assumed ultrasoft nature of k. At this point it is best to pause in our evaluation of $\Delta E_{\mathrm{T}}^{\mathrm{US}}$ as we will want to combine it with additional ultrasoft contributions later on.

We move on to a consideration of the second graph in Fig. 16.9 where the transverse photon is ultrasoft while all other lines and wave functions are potential. The energy correction resulting from this graph, written without any approximations, was given in Eq. (16.190) above. When $\vec{k}$ was soft, we used Eq. (16.191) to approximate the electron propagator. Now that $\vec{k}$ is ultrasoft, the analogous approximation is

$$\frac{\mathrm{i}}{(\xi_1 \not{P} + \not{p} - \not{k}) - m_1 + \mathrm{i}\epsilon} \approx \frac{\mathrm{i}\Lambda_+(\vec{p}\,)\gamma^0}{\xi_1 E + p^0 - k^0 - \frac{\vec{p}^{\,2}}{2m_1} + \mathrm{i}\epsilon} + \frac{\mathrm{i}\Lambda_-(\vec{p}\,)\gamma^0}{2m_1}, \tag{16.202}$$

with a similar approximation for the other fermion propagator. Clearly the first terms give the dominant contribution. The energy shift from (16.190) with an ultrasoft transverse photon takes the form

$$\begin{aligned}\Delta E_{\mathrm{TC}}^{\mathrm{US}} &\approx 2(4\pi Z\alpha\overline{\mu}^{2\varepsilon})^2 \mathrm{i} \int \frac{\mathrm{d}^d p}{(2\pi)^d} \frac{\mathrm{d}^d q}{(2\pi)^d} \frac{\mathrm{d}^d k}{(2\pi)^d} \\ &\times \psi^*(\vec{p}\,)\mathcal{S}(P;p) \frac{\delta^{\perp,ab}(\vec{k}\,)}{(k^2+\mathrm{i}\epsilon)(\vec{p}-\vec{q}\,)^2} \mathcal{S}(P;q)\psi(\vec{q}\,) \\ &\times \overline{u}_1(\vec{p}\,)\gamma^a\Lambda_+(\vec{p}\,)\gamma^0\gamma^0 u_1(\vec{q}\,)\, \overline{u}_2(-\vec{p}\,)\gamma^0\Lambda_+(-\vec{q}\,)\gamma^0\gamma^b u_2(-\vec{q}\,) \\ &\times \left[\left(\xi_1 E + p^0 - k^0 - \frac{\vec{p}^{\,2}}{2m_1} + \mathrm{i}\epsilon\right)\left(\xi_2 E - q^0 - k^0 - \frac{\vec{q}^{\,2}}{2m_2} + \mathrm{i}\epsilon\right)\right]^{-1}.\end{aligned} \tag{16.203}$$

The two spin-averaged Dirac products have the approximate values p^a/m_1 and $-q^b/m_2$. We have anticipated the fact that the left-right reflected one-transverse one-Coulomb graph will give an equal contribution. We use (16.196) for the non-relativistic approximation to $\mathcal{S}$, and find

$$\begin{aligned}\Delta E_{\mathrm{TC}}^{\mathrm{US}} &\approx 2\frac{(4\pi Z\alpha\overline{\mu}^{2\varepsilon})^2}{m_1 m_2} \mathrm{i} \int \frac{\mathrm{d}^d p}{(2\pi)^d} \frac{\mathrm{d}^d q}{(2\pi)^d} \frac{\mathrm{d}^d k}{(2\pi)^d} \\ &\times \psi^*(\vec{p}\,)\, s^{-1}(E;\vec{p}\,) \frac{p^a \delta^{\perp,ab}(\vec{k}\,) q^b}{(\vec{p}-\vec{q}\,)^2} s^{-1}(E;\vec{q}\,)\,\psi(\vec{q}\,) \\ \times &\left[\left(\xi_1 E + p^0 - \frac{\vec{p}^{\,2}}{2m_1} + \mathrm{i}\epsilon_1\right)\left(\xi_2 E - p^0 - \frac{\vec{p}^{\,2}}{2m_2} + \mathrm{i}\epsilon_2\right)\left(k^0 - |\vec{k}\,| + \mathrm{i}\epsilon_3\right)\left(k^0 + |\vec{k}\,| - \mathrm{i}\epsilon_3\right)\right. \\ &\times \left(\xi_1 E + p^0 - k^0 - \frac{\vec{p}^{\,2}}{2m_1} + \mathrm{i}\epsilon_4\right)\left(\xi_2 E - q^0 - k^0 - \frac{\vec{q}^{\,2}}{2m_2} + \mathrm{i}\epsilon_5\right) \\ &\times \left.\left(\xi_1 E + q^0 - \frac{\vec{q}^{\,2}}{2m_1} + \mathrm{i}\epsilon_6\right)\left(\xi_2 E - q^0 - \frac{\vec{q}^{\,2}}{2m_2} + \mathrm{i}\epsilon_7\right)\right]^{-1},\end{aligned} \tag{16.204}$$

where again we have employed a separate infinitesimal imaginary part for each distinct propagator. We perform the relative energy integrals (with any convenient choice of order, half-planes to close in, and relative sizes of infinitesimals) to find

$$\begin{aligned}\Delta E_{\rm TC}^{\rm US} &\approx 2\frac{(4\pi Z\alpha\overline{\mu}^{2\varepsilon})^2}{m_1 m_2}\mathrm{i}\int\frac{\mathrm{d}^D p}{(2\pi)^D}\frac{\mathrm{d}^D q}{(2\pi)^D}\frac{\mathrm{d}^D k}{(2\pi)^D}\\ &\times\psi^*(\vec{p})s^{-1}(E;\vec{p})\frac{p^a\delta^{\perp,ab}(\vec{k})q^b}{(\vec{p}-\vec{q})^2}s^{-1}(E;\vec{q})\psi(\vec{q})\\ &\times(-\mathrm{i})^3\Big[s^{-1}(E;\vec{p})\,(2|\vec{k}|)\,s^{-1}(E-|\vec{k}|;\vec{p})s^{-1}(E-|\vec{k}|;\vec{q})s^{-1}(E;\vec{q})\Big]^{-1}\\ &= \frac{4\pi Z\alpha\overline{\mu}^{2\varepsilon}}{m_1 m_2}\int\frac{\mathrm{d}^D p}{(2\pi)^D}\frac{\mathrm{d}^D q}{(2\pi)^D}\frac{\mathrm{d}^D k}{(2\pi)^D}\psi^*(\vec{p})\frac{p^a\delta^{\perp,ab}(\vec{k})q^b}{|\vec{k}|}\psi(\vec{q})\\ &\times s(E-|\vec{k}|;\vec{p})V(\vec{p}-\vec{q})s(E-|\vec{k}|;\vec{q}).\end{aligned}\tag{16.205}$$

Again, we pause to consider additional ultrasoft graphs that contribute at order $(Z\alpha)^5$ before continuing the evaluation of this contribution.

The third graph of Fig. 16.9 shows a pair of ladder Coulomb photons crossed by a transverse photon. When the transverse photon is ultrasoft, we will find that this graphs contributes at the same order $(Z\alpha)^5$ level as the other graphs in Fig. 16.9. Instead of evaluating this diagram from scratch, we focus on the changes between the one- and two-Coulomb graphs. We see that the two-Coulomb contribution can be obtained from the one-Coulomb contribution (16.190) by the replacement

$$\begin{aligned}\frac{(-\mathrm{i}e\gamma^0)_1(\mathrm{i}Ze\gamma^0)_2}{(\vec{p}-\vec{q}-\vec{k})^2} &\to \int\frac{\mathrm{d}^d r}{(2\pi)^d}\frac{1}{(\vec{p}-\vec{r})^2}\left[(-\mathrm{i}e\gamma^0)\frac{\mathrm{i}}{\xi_1\not{P}+\not{r}-\not{k}-m_1+\mathrm{i}\epsilon}(-\mathrm{i}e\gamma^0)\right]_1\\ &\times\left[(\mathrm{i}Ze\gamma^0)\frac{\mathrm{i}}{\xi_2\not{P}-\not{r}-m_2+\mathrm{i}\epsilon}(\mathrm{i}Ze\gamma^0)\right]_2\frac{\mathrm{i}}{(\vec{r}-\vec{q}-\vec{k})^2},\end{aligned}\tag{16.206}$$

or, with nonrelativistic approximations,

$$\begin{aligned}\frac{1}{(\vec{p}-\vec{q}-\vec{k})^2} &\to 4\pi Z\alpha\overline{\mu}^{2\varepsilon}(-\mathrm{i})\int\frac{\mathrm{d}^d r}{(2\pi)^d}\frac{1}{(\vec{p}-\vec{r})^2}\frac{1}{\xi_1 E+r_0-k^0-\frac{(\vec{r}-\vec{k})^2}{2m_1}+\mathrm{i}\epsilon}\\ &\times\frac{1}{\xi_2 E-r_0-\frac{\vec{r}^{\,2}}{2m_2}+\mathrm{i}\epsilon}\frac{1}{(\vec{r}-\vec{q}-\vec{k})^2}\\ &= -4\pi Z\alpha\overline{\mu}^{2\varepsilon}\int\frac{\mathrm{d}^D r}{(2\pi)^D}\frac{1}{(\vec{p}-\vec{r})^2}\frac{1}{E-k^0-\frac{(\vec{r}-\vec{k})^2}{2m_1}-\frac{\vec{r}^{\,2}}{2m_2}+\mathrm{i}\epsilon}\frac{1}{(\vec{r}-\vec{q}-\vec{k})^2}.\end{aligned}\tag{16.207}$$

For ultrasoft k, the replacement takes the simple form,

$$\frac{1}{(\vec{p}-\vec{q})^2}\to -4\pi Z\alpha\overline{\mu}^{2\varepsilon}\int\frac{\mathrm{d}^D r}{(2\pi)^D}\frac{1}{(\vec{p}-\vec{r})^2}s(E-k^0;\vec{r})\frac{1}{(\vec{r}-\vec{q})^2},\tag{16.208}$$

or

$$V(\vec{p}-\vec{q})\to\int\frac{\mathrm{d}^D r}{(2\pi)^D}V(\vec{p}-\vec{r})s(E-k^0;\vec{r})V(\vec{r}-\vec{q}) = [VsV](E-k^0;\vec{p},\vec{q})\tag{16.209}$$

in the compact notation where the internal convolution integrals are implicit. The resulting energy shift is thus

$$\Delta E^{\mathrm{US}}_{\mathrm{TCC}} \approx \frac{4\pi Z\alpha\overline{\mu}^{2\varepsilon}}{m_1 m_2}\int \frac{\mathrm{d}^D p}{(2\pi)^D}\frac{\mathrm{d}^D q}{(2\pi)^D}\frac{\mathrm{d}^D k}{(2\pi)^D}\psi^*(\vec{p})\frac{p^a\delta^{\perp,ab}(\vec{k})q^b}{|\vec{k}|}\psi(\vec{q})$$
$$\times [sVsVs](E-|\vec{k}|;\vec{p},\vec{q}). \tag{16.210}$$

It is now clear that graphs with additional ladder Coulomb photons can be handled in the same way, each time giving extra factors of V times s to be included in the string seen in (16.210).

The total contribution of ultrasoft transverse photon exchange at order $(Z\alpha)^5$ is the sum of (16.201), (16.205), (16.210), plus all diagrams with additional ladder Coulomb photons. The infinite series of terms can be summed using (16.77) to give

$$\Delta E^{\mathrm{US}}_{\mathrm{T(total)}} \approx \frac{4\pi Z\alpha\overline{\mu}^{2\varepsilon}}{m_1 m_2}\int \frac{\mathrm{d}^D p}{(2\pi)^D}\frac{\mathrm{d}^D q}{(2\pi)^D}\frac{\mathrm{d}^D k}{(2\pi)^D}\psi^*(\vec{p})\frac{p^a\delta^{\perp,ab}(\vec{k})q^b}{|\vec{k}|}\psi(\vec{q})$$
$$\times\left[s+sVs+sVsVs+sVsVsVs+\cdots\right](E-|\vec{k}|;\vec{p},\vec{q})$$
$$= \frac{4\pi Z\alpha\overline{\mu}^{2\varepsilon}}{m_1 m_2}\int \frac{\mathrm{d}^D p}{(2\pi)^D}\frac{\mathrm{d}^D q}{(2\pi)^D}\frac{\mathrm{d}^D k}{(2\pi)^D}\psi^*(\vec{p})\frac{p^a\delta^{\perp,ab}(\vec{k})q^b}{|\vec{k}|}\psi(\vec{q})g(E-|\vec{k}|;\vec{p},\vec{q}), \tag{16.211}$$

where $g(E,\vec{p},\vec{q})$ is the NRSC Green function of (16.75) and (16.81). The reader has no doubt noticed the extremely close relationship between this result and the related expression involving the NRSC Green function and an ultrasoft transverse photon that appeared in our study of the Lamb shift (see Chap. 4 for the low-energy part of the self-energy). As for the Lamb shift work, we use the sum-over-states form (16.81) for the Green function. It is useful now to also include an index (which we denote as n), in order to specify what state we are finding the energy shift of, and we will convert to Dirac bra/ket notation when convenient. In addition, we must point out that, although the early work in this chapter, including the definition of the NRSC Green function shown in (16.75)–(16.81) was written down in three dimensional space, we now realize that the dimensionally continued version of that Green function is required.

For the ultrasoft energy shift of the nth state, we resort to a derivation akin to the one employed in Secs. 4.6.2 and 15.5.2, but this time, starting from momentum space. We have

$$\Delta E^{\mathrm{US}}_{\mathrm{T(total)}} \approx \frac{4\pi Z\alpha\overline{\mu}^{2\varepsilon}}{m_1 m_2}\int \frac{\mathrm{d}^D p}{(2\pi)^D}\frac{\mathrm{d}^D q}{(2\pi)^D}\frac{\mathrm{d}^D k}{(2\pi)^D}\frac{\delta^{\perp,ab}(\vec{k})}{|\vec{k}|}$$
$$\times\sum_{n'}\frac{\psi_n^*(\vec{p})p^a\psi_{n'}(\vec{p})\,\psi_{n'}^*(\vec{q})q^b\psi_n(\vec{q})}{E_n-|\vec{k}|-E_{n'}+\mathrm{i}\epsilon}$$
$$= \frac{4\pi Z\alpha\overline{\mu}^{2\varepsilon}}{m_1 m_2}\left(\frac{D-1}{D}\right)\frac{\Omega_D}{(2\pi)^D}\sum_{n'}\int d|\vec{k}|\,|\vec{k}|^{D-2}\frac{\langle n|p^a|n'\rangle\langle n'|p^a|n\rangle}{(E_n-E_{n'}+\mathrm{i}\epsilon)-|\vec{k}|}, \tag{16.212}$$

where we appeal to Eq. (10.138) for the definition of the solid angle in D dimensions. One finally arrives at the following result, where we use the shorthand notation $X(D) = \Gamma(D-1)\Gamma(2-D)(D-1)/D$ for an intermediate step,

$$\begin{aligned}\Delta E^{\rm US}_{\rm T(total)} &= -\frac{4\pi Z\alpha\overline{\mu}^{2\varepsilon}}{m_1 m_2} X(D)\frac{\Omega_D}{(2\pi)^D}\sum_{n'}\langle n|p^a|n'\rangle\langle n'|p^a|n\rangle\left[-(E_n - E_{n'} + \mathrm{i}\varepsilon)\right]^{D-2} \\ &= -\frac{4\pi Z\alpha}{m_1 m_2}\sum_{n'}\langle n|p^a|n'\rangle\langle n'|p^a|n\rangle\frac{(E_n - E_{n'})}{6\pi^2} \\ &\quad\times\left\{\frac{1}{\varepsilon} - 2\ln\left(\frac{-2(E_n - E_{n'} + \mathrm{i}\varepsilon)}{\mu}\right) + \frac{5}{3} + \mathcal{O}(\varepsilon)\right\}. \end{aligned} \tag{16.213}$$

In the reductions shown above, we integrated the transverse delta function over $D-1$ angles using

$$\int \mathrm{d}\Omega_D \delta^{\perp,ab}(\vec{k}\,) = \int \mathrm{d}\Omega_D\left(\delta^{ab} - \hat{k}^a\hat{k}^b\right) = \Omega_D\left(\delta^{ab} - \frac{\delta^{ab}}{D}\right) = \Omega_D\left(\frac{D-1}{D}\right)\delta^{ab}. \tag{16.214}$$

The integral over $|\vec{k}\,|$ was done using the general formula

$$\int_0^\infty \mathrm{d}x\frac{x^a}{(m + x^b)^c} = m^{\frac{a+1}{b}-c}\frac{\Gamma\left(\frac{a+1}{b}\right)\Gamma\left(c - \frac{a+1}{b}\right)}{b\,\Gamma(c)}. \tag{16.215}$$

The logarithm obtained by performing the small-ε expansion has an argument that can be negative (when $E_n > E_{n'}$), leading to an imaginary part for the logarithm in that case. We define the logarithm in the usual way with a cut on the negative real axis, so that

$$\ln\left(-\frac{E_n - E_{n'}}{\mu} - \mathrm{i}\epsilon\right) = \ln\left(\frac{|E_n - E_{n'}|}{\mu}\right) - \mathrm{i}\pi\Theta(E_n > E_{n'}), \tag{16.216}$$

where Θ is the Heaviside step function. The imaginary part gives the correction to the decay rate from state n (cf. Sec. 3.2.3)

$$\Gamma = -2\,\mathrm{Im}(\Delta E) = \frac{8Z\alpha}{3m_1 m_2}\sum_{\substack{n' \\ (E_{n'}<E_n)}}\langle n|p^a|n'\rangle\langle n'|p^a|n\rangle(E_n - E_{n'}), \tag{16.217}$$

while the real part gives the actual energy correction

$$\begin{aligned}\Delta E^{\rm US}_{\rm T(total)} &= \frac{4Z\alpha}{3\pi m_1 m_2}\sum_{n'}\langle n|p^a|n'\rangle\langle n'|p^a|n\rangle(E_n - E_{n'})\left\{-\frac{1}{2\varepsilon} + \ln\left(\frac{2|E_n - E_{n'}|}{\mu}\right) - \frac{5}{6}\right\} \\ &= \frac{4Z\alpha}{3\pi m_1 m_2}\left\langle n\left|p^a(E_n - H)\left\{-\frac{1}{2\varepsilon} + \ln\left(\frac{2|E_n - H|}{\mu}\right) - \frac{5}{6}\right\}p^a\right|n\right\rangle. \end{aligned} \tag{16.218}$$

As in our earlier discussion of the Lamb shift in Sec. 4.6.3, we use the reduction formula

$$\begin{aligned}\langle n|p^a(E_n - H)p^a|n\rangle &= -\frac{1}{2}\langle n|\left[p^a, \left[p^a, E_n - H\right]\right]|n\rangle \\ &= -\frac{1}{2}\langle n|\nabla^2 V(\vec{x}\,)|n\rangle = -2\pi Z\alpha\overline{\mu}^{2\varepsilon}|\psi(\vec{0})|^2. \end{aligned} \tag{16.219}$$

We recall Eq. (10.150) for the definition of the D-dimensional Coulomb potential, which satisfies Eq. (10.146):

$$\vec{\nabla}^2 V = 4\pi Z\alpha \overline{\mu}^{2\varepsilon} \delta^{(D)}(\vec{r}\,) \tag{16.220}$$

in D dimensions (see also Ref. [313] and Chap. 17). The (finite) Bethe log definition (4.317) leads us to the result

$$\langle n | p^a (E_n - H) \ln\left(\frac{2|E_n - H|}{\mu(Z\alpha)^2}\right) p^a | n\rangle = -\frac{2\mu^3 (Z\alpha)^4}{n^3} \ln k_0(n,\ell)\,. \tag{16.221}$$

Finally, we write the ultrasoft correction in the form

$$\Delta E^{\rm US}_{\rm T(total)} = \frac{8(Z\alpha)^2}{3m_1 m_2} \overline{\mu}^{2\varepsilon} |\psi(\vec{0})|^2 \left\{\frac{1}{2\varepsilon} + \frac{5}{6} + \ln\left(\frac{\overline{\mu}}{\mu(Z\alpha)^2}\right)\right\} - \frac{8\mu^3(Z\alpha)^5}{3\pi m_1 m_2 n^3} \ln k_0(n,\ell), \tag{16.222}$$

keeping in mind the distinction between the reduced mass μ and the mass parameter of dimensional regularization $\overline{\mu}$.

16.3.7 *Sum of the Contributions*

We have now completed the calculation of all recoil corrections to the spin-averaged energy levels at order $(Z\alpha)^5$. It just remains to sum up the results. We note first that all divergences cancel between pairs of terms. The two-transverse-exchange divergences cancel between the infrared divergence in the hard region (16.172) and the ultraviolet divergence in the soft region (16.178). The total of all two-transverse-photon exchange contributions is

$$\begin{aligned}\Delta E_{\rm TT} &= \frac{\mu^3(Z\alpha)^5}{\pi m_1 m_2 n^3}\left\{-\frac{8}{3}\ln 2 + \frac{8}{3} - 2\frac{m_1^2 \ln(m_2/\mu) - m_2^2 \ln(m_1/\mu)}{m_1^2 - m_2^2}\right\}\delta_{\ell=0} \\ &\quad + \frac{2(Z\alpha)^2}{m_1 m_2}\left\langle \ln\left(\frac{|\vec{p}-\vec{q}\,|}{\mu}\right)\right\rangle_{\vec{p},\vec{q}} \\ &= \frac{\mu^3(Z\alpha)^5}{\pi m_1 m_2 n^3}\left\{\left[2\ln\left(\frac{2Z\alpha}{n}\right) + 2H_n + \frac{n-1}{n} - \frac{8}{3}\ln 2 + \frac{8}{3}\right.\right. \\ &\quad \left.\left. - 2\frac{m_1^2 \ln(m_2/\mu) - m_2^2 \ln(m_1/\mu)}{m_1^2 - m_2^2}\right]\delta_{\ell=0} - \frac{\delta_{\ell\neq 0}}{2\ell(\ell+1)(\ell+1/2)}\right\}.\end{aligned} \tag{16.223}$$

The one-transverse-exchange divergences cancel between the infrared divergence in the soft region (16.194) and the ultraviolet divergence in the ultrasoft region (16.222). The total of these two one-transverse-exchange contributions is

$$\begin{aligned}\Delta E_{\rm T} &= \frac{8\mu^3(Z\alpha)^5}{3\pi m_1 m_2 n^3}\left\{\ln\left(\frac{2}{(Z\alpha)^2}\right) + \frac{1}{3}\right\}\delta_{\ell=0} + \frac{8(Z\alpha)^2}{3m_1 m_2}\left\langle \ln\left(\frac{|\vec{p}-\vec{q}\,|}{\mu}\right)\right\rangle_{\vec{p},\vec{q}} \\ &\quad - \frac{8\mu^3(Z\alpha)^5}{3\pi m_1 m_2 n^3}\ln k_0(n,\ell) \\ &= \frac{8\mu^3(Z\alpha)^5}{3\pi m_1 m_2 n^3}\left\{\left[\ln\left(\frac{4}{Z\alpha n}\right) + H_n + \frac{n-1}{2n} + \frac{1}{3}\right]\delta_{\ell=0}\right. \\ &\quad \left. - \ln k_0(n,\ell) - \frac{\delta_{\ell\neq 0}}{4\ell(\ell+1)(\ell+1/2)}\right\}.\end{aligned} \tag{16.224}$$

Finally, the two-Coulomb exchange contribution from the hard region has the finite value found in (16.166):

$$\Delta E_{\rm CC} = \frac{\mu^3 (Z\alpha)^5}{\pi m_1 m_2 n^3} \left\{ -\frac{4}{3} \right\}. \tag{16.225}$$

The total spin-averaged recoil correction at order $(Z\alpha)^5$ is the sum of these three:

$$\begin{aligned} \Delta E_{\rm recoil}^{\rm spin\text{-}average} &= \Delta E_{\rm TT} + \Delta E_{\rm T} + \Delta E_{\rm CC} \\ &= \frac{\mu^3 (Z\alpha)^5}{\pi m_1 m_2 n^3} \left\{ \left[-\frac{2}{3} \ln(Z\alpha) + \frac{14}{3} \left(\ln\left(\frac{2}{n}\right) + H_n - \frac{1}{2n} \right) + \frac{41}{9} \right.\right. \\ &\left.\left. - 2 \frac{m_1^2 \ln(m_2/\mu) - m_2^2 \ln(m_1/\mu)}{m_1^2 - m_2^2} \right] \delta_{\ell=0} - \frac{8}{3} \ln k_0(n,\ell) - \frac{7 \delta_{\ell \neq 0}}{6\ell(\ell+1)(\ell+1/2)} \right\}. \end{aligned} \tag{16.226}$$

This result verifies Eq. (15.163) on the basis of an independent derivation, based on the Bethe–Salpeter equation.

16.4 Hyperfine Structure of S States (BSEQ Approach)

As a second illustration of the Bethe–Salpeter equation, we calculate the leading order $(Z\alpha)^5$ correction to the hfs (hyperfine structure, or hyperfine splitting) of S states. Specifically, we will be considering the difference between the spin energy levels ($S = 1$ minus $S = 0$) for states with no angular momentum. A look at the spin-averaged order $(Z\alpha)^5$ energy shift (16.226) suggests that the leading correction to the hfs at this order will include a logarithmic enhancement $\ln(m_2/m_1)$. We consider a logarithmically enhanced correction to the hfs mediated by two-photon exchange (recoil) as depicted in Fig. 16.10. This correction is actually suppressed for systems with a very large mass ratio of nucleus to orbiting particle, such as atomic hydrogen (because the large mass occurs in the denominator), but is extremely relevant for positronium and muonic hydrogen. (In atomic hydrogen, the numerically dominant radiative correction for S states is mediated by a radiative correction to the vertex, as summarized in Eq. (12.137).) Let us proceed to evaluate the nuclear-spin dependent part of the relativistic recoil correction. The logarithmic terms involving the masses in (16.226) have their origin in the hard (high-energy) region, and indeed that is the only region that can produce such enhancement. The reason is that logarithmic terms such as $\ln(m_2/m_1)$ come from the part of the hard region of integration spanning from m_1 to m_2, as in the relation $\int_{m_1}^{m_2} {\rm d}x/x = \ln(m_2/m_1)$. The leading $(Z\alpha)^5$ recoil correction to the hfs also comes from the hard region.

As an introduction to the hfs calculation, we first obtain the leading order $(Z\alpha)^4$ Fermi result for the hfs splitting. In contrast to Sec. 12.5.3, we assume both involved particles to be Dirac particles with g factors equal to $g \approx 2$ (ignoring radiative corrections). This leading effect is contained in the effective Hamiltonian of (16.144). A glance at $H_{\rm eff}$ shows that the spin dependence is confined to the last two terms. The angular dependence of these two terms is given by $(\vec{p} \times \vec{q})/\vec{k}^{\,2}$ and $k_i k_j/\vec{k}^{\,2}$

where $\vec{k} = \vec{p} - \vec{q}$. Upon taking spherical averages of these terms (as appropriate for S states), we find that the first vanishes while the second is proportional to δ^{ij}. Consequently, it is only the second that contributes for S states. In more detail, the matrix element $\langle(\vec{p}\times\vec{q}\,)/\vec{k}^{\,2}\rangle_{\vec{p},\vec{q}}$ is given by

$$\begin{aligned}\left\langle\frac{\vec{p}\times\vec{q}}{\vec{k}^{\,2}}\right\rangle_{\vec{p},\vec{q}} &= \int\frac{d^3p}{(2\pi)^3}\frac{d^3q}{(2\pi)^3}\psi^+(\vec{p}\,)\frac{\vec{p}\times\vec{q}}{\vec{k}^{\,2}}\psi(\vec{q}\,)\\ &= \int\frac{d^3p}{(2\pi)^3}\frac{d^3q}{(2\pi)^3}d^3y\,d^3x\,\psi^+(\vec{y}\,)e^{-i\vec{p}\cdot\vec{y}}\frac{(\vec{p}-\vec{q}\,)\times(-i\vec{\nabla}_x)}{\vec{k}^{\,2}}e^{i\vec{q}\cdot\vec{x}}\psi(\vec{x}\,)\\ &= \int d^3y\,d^3x\,\psi^+(\vec{y}\,)\int\frac{d^3k}{(2\pi)^3}e^{-i\vec{k}\cdot\vec{x}}\frac{\vec{k}\times(-i\vec{\nabla}_x)}{\vec{k}^{\,2}}\int\frac{d^3p}{(2\pi)^3}e^{i\vec{p}\cdot(\vec{x}-\vec{y}\,)}\psi(\vec{x}\,)\\ &= \int d^3y\,d^3x\,\psi^+(\vec{y}\,)\frac{-i\vec{x}\times(-i\vec{\nabla}_x)}{4\pi r^3}\delta^{(3)}(\vec{x}-\vec{y}\,)\psi(\vec{x}\,) = \frac{-i}{4\pi}\int d^3x\,\psi^+(\vec{x}\,)\frac{\vec{L}}{r^3}\psi(\vec{x}\,),\end{aligned} \tag{16.227}$$

where $r = |\vec{x}\,|$ and $\vec{L} = \vec{x}\times(-i\vec{\nabla}_x)$ is the orbital angular momentum. This term does not contribute for states with $\ell = 0$. On the other hand, the $k^ik^j/\vec{k}^{\,2}$ term, which arises from the Fermi spin-spin interaction, does contribute for S states. In fact, we find this contribution to be

$$\begin{aligned}\Delta E^{\text{Fermi}} &= \left\langle -V(\vec{k}\,)\frac{\vec{k}\times\vec{\sigma}_1\cdot\vec{k}\times\vec{\sigma}_2}{4m_1m_2}\right\rangle^{\text{hfs}}_{\vec{p},\vec{q}}\\ &= \frac{\pi Z\alpha}{m_1m_2}\left\langle\frac{\vec{k}^{\,2}\vec{\sigma}_1\cdot\vec{\sigma}_2 - \vec{k}\cdot\vec{\sigma}_1\vec{k}\cdot\vec{\sigma}_2}{\vec{k}^{\,2}}\right\rangle^{\text{hfs}}_{\vec{p},\vec{q}}\\ &= \frac{\pi Z\alpha}{m_1m_2}\left\langle\frac{2}{3}\vec{\sigma}_1\cdot\vec{\sigma}_2 + \frac{\vec{k}^{\,2}\delta^{ij} - 3k^ik^j}{3\vec{k}^{\,2}}\sigma_1^i\sigma_2^j\right\rangle^{\text{hfs}}_{\vec{p},\vec{q}}.\end{aligned} \tag{16.228}$$

Now the second term above has vanishing angular average, so gives a vanishing contribution in S states. It is an operator characterized by angular momentum $\ell = 2$. We can get the spin expectation value $\langle\vec{\sigma}_1\cdot\vec{\sigma}_2\rangle = 2\langle(\vec{S}_1 + \vec{S}_2)^2 - \vec{S}_1^{\,2} - \vec{S}_2^{\,2}\rangle = 2\{2\delta_{S=1} - 3/4 - 3/4\} = 4\delta_{S=1} - 3$, since $\vec{S}^{\,i} = \frac{1}{2}\vec{\sigma}^{\,i}$. The hyperfine difference for the spin expectation $\langle\vec{\sigma}_1\cdot\vec{\sigma}_2\rangle$ is thus 4, and the Fermi term is

$$\Delta E^{\text{Fermi}} = \frac{\pi Z\alpha}{m_1m_2}\left\langle\frac{2}{3}\vec{\sigma}_1\cdot\vec{\sigma}_2\right\rangle^{\text{hfs}}_{\vec{p},\vec{q}} = \frac{\pi Z\alpha}{m_1m_2}\frac{2}{3}\left(4|\psi(\vec{0})|^2\right) = \frac{8\mu^3(Z\alpha)^4}{3m_1m_2n^3}\delta_{\ell,0}. \tag{16.229}$$

Consistency with Eq. (12.137) can be shown for a proton nucleus, since in that case we must replace one of the $Z\alpha$ factors as $Z\alpha \to \alpha\,g_N/2$ [see also Eq. (12.123)]. We move on to the evaluation of the leading order $(Z\alpha)^5$ recoil correction to the S state hfs. As discussed above this contribution is logarithmic and comes from the hard region exclusively, so our task reduces to finding the free particle hfs difference multiplied by the probability of spatial contact between the two particles. In order to deal with spins, it is convenient to consider the positive particle as an antiparticle as illustrated in Fig. 16.10. (This treatment of spins is also appropriate for particle-antiparticle bound systems such as positronium.) With this choice, a spinor for one

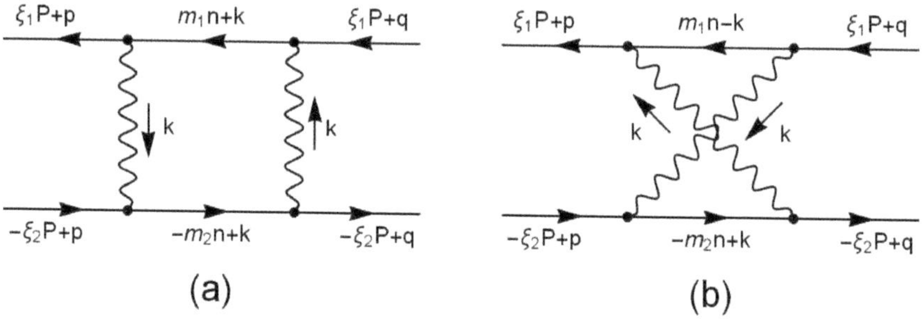

Fig. 16.10 Two-photon exchange contributions to hard scattering of a particle (mass m_1, on top) and an antiparticle (mass m_2, bottom). The ladder and crossed graphs are shown as parts (a) and (b). Momentum labels are consistent with the direction of the fermion line. For the antiparticle, this implies that $+\xi_2 P - q$ is the incoming momentum on the right and $+\xi_2 P - p$ is the outgoing momentum on the left. For all momenta involved in the hard loop, the appropriate approximations have been made: p and q have been dropped relative to k and $\xi_i P$ has been approximated as $m_i n$ where $n = (1, \vec{0})$.

of the fermions occur paired with a barred spinor for the other fermion and can be conveniently combined.

In order to describe the spin structure, we start from the QED particle spin states and the corresponding antiparticle states, which are

$$u_+ = \begin{pmatrix} 1 \\ 0 \\ 0 \\ 0 \end{pmatrix}, \quad u_- = \begin{pmatrix} 0 \\ 1 \\ 0 \\ 0 \end{pmatrix}, \quad v_+ = \begin{pmatrix} 0 \\ 0 \\ 0 \\ 1 \end{pmatrix}, \quad v_- = \begin{pmatrix} 0 \\ 0 \\ -1 \\ 0 \end{pmatrix}. \tag{16.230}$$

The particle spinors here (the u's) are the usual spin up and down spinors for particles essentially at rest (we note that only the internal loop is hard, the external particles are part of an atom and so are nonrelativistic) while the corresponding antiparticle spinors (the v's) were obtained from the u's by multiplication by the charge conjugation matrix [see Sec. 2-4-2 of Ref. [2] and Eq. (7.68)]:

$$C = \begin{pmatrix} 0 & -\mathrm{i}\sigma^2 \\ -\mathrm{i}\sigma^2 & 0 \end{pmatrix} = \begin{pmatrix} 0 & 0 & 0 & -1 \\ 0 & 0 & 1 & 0 \\ 0 & -1 & 0 & 0 \\ 1 & 0 & 0 & 0 \end{pmatrix}. \tag{16.231}$$

The barred spinors ($\overline{u} = u^+\gamma^0$, $\overline{v} = v^+\gamma^0$) are

$$\overline{u}_+ = \begin{pmatrix} 1 \; 0 \; 0 \; 0 \end{pmatrix}, \; \overline{u}_- = \begin{pmatrix} 0 \; 1 \; 0 \; 0 \end{pmatrix}, \; \overline{v}_+ = \begin{pmatrix} 0 \; 0 \; 0 \; -1 \end{pmatrix}, \; \overline{v}_- = \begin{pmatrix} 0 \; 0 \; 1 \; 0 \end{pmatrix}. \tag{16.232}$$

Two-particle states $(u\overline{v})^{SM}$ of total spin $S = 0$ and $S = 1$ and z component of total spin M are formed using Clebsch–Gordan coefficients for combining two constituents of spin-1/2:

$$(u\overline{v})^{0,0} = \frac{1}{\sqrt{2}}\left(u_+\overline{v}_- - u_-\overline{v}_+\right) = \begin{pmatrix} 0 & \chi^{00} \\ 0 & 0 \end{pmatrix}, \tag{16.233a}$$

$$(u\overline{v})^{1,1} = u_+\overline{v}_+ = \begin{pmatrix} 0 & \chi^{11} \\ 0 & 0 \end{pmatrix}, \tag{16.233b}$$

$$(u\overline{v})^{1,0} = \frac{1}{\sqrt{2}}\left(u_+\overline{v}_- + u_-\overline{v}_+\right) = \begin{pmatrix} 0 & \chi^{10} \\ 0 & 0 \end{pmatrix}, \tag{16.233c}$$

$$(u\overline{v})^{1,-1} = u_-\overline{v}_- = \begin{pmatrix} 0 & \chi^{1-1} \\ 0 & 0 \end{pmatrix}. \tag{16.233d}$$

These states have the form of tensor products: particle operators act on them from the left while antiparticle operators act from the right. These Dirac two-particle states contain within them the Pauli two-particle tensor-product states [594],

$$\chi^{00} = \frac{1}{\sqrt{2}}\begin{pmatrix} 1 & 0 \\ 0 & 1 \end{pmatrix}, \ \chi^{11} = \begin{pmatrix} 0 & -1 \\ 0 & 0 \end{pmatrix}, \ \chi^{10} = \frac{1}{\sqrt{2}}\begin{pmatrix} 1 & 0 \\ 0 & -1 \end{pmatrix}, \ \chi^{1-1} = \begin{pmatrix} 0 & 0 \\ 1 & 0 \end{pmatrix}, \tag{16.234}$$

or equivalently

$$\chi^{00} = \frac{1}{\sqrt{2}}\mathbb{1}_{2\times 2}, \qquad \chi^{1M} = \frac{1}{\sqrt{2}}\vec{\sigma}\cdot\vec{e}_M, \tag{16.235}$$

where we refer to the spherical basis vectors in Eq. (6.22). For clarity, we remember that, component-wise,

$$\vec{e}_\pm = \mp\frac{1}{\sqrt{2}}(1, \pm\mathrm{i}, 0), \qquad \vec{e}_0 = (0, 0, 1). \tag{16.236}$$

The unit vectors $\hat{e}_M$ form an orthonormal basis, i.e., we have $\vec{e}^{\,*}_M \cdot \vec{e}_{M'} = \delta_{MM'}$, and $\sum_M (\vec{e}_M)^i (\vec{e}^{\,j}_M)^* = \delta^{ij}$. We will use (16.233) to organize our calculation of the spin combinations appropriate to $S = 0$ and $S = 1$.

The calculation of the leading order $(Z\alpha)^5$ recoil effect proceeds much as the hard region part of the spin-averaged Salpeter correction from before. The main difference is in the treatment of the positive particle (as an antiparticle) and the need to distinguish among the different spin states instead of just summing over them. Starting with the transverse-transverse ladder kernel of Fig. 16.10(a), we find a spin-dependent energy contribution

$$\begin{aligned}
\Delta E_{\mathrm{TT;L}} &= \mathrm{i}|\psi(\vec{0})|^2(-1)\int \frac{\mathrm{d}^d k}{(2\pi)^d}\overline{u}_{1f}(-\mathrm{i}e\gamma^a)\frac{\mathrm{i}}{m_1 \not{n} + \not{k} - m_1}(-\mathrm{i}e\gamma^c)u_{1i} \\
&\quad\times \frac{\mathrm{i}\delta^{\perp,ab}(\vec{k}\,)}{k^2}\frac{\mathrm{i}\delta^{\perp,cd}(\vec{k}\,)}{k^2}\overline{v}_{2i}(\mathrm{i}Ze\gamma^d)\frac{\mathrm{i}}{-m_2\not{n} + \not{k} - m_2}(\mathrm{i}Ze\gamma^b)v_{2f} \\
&= -\mathrm{i}|\psi(\vec{0})|^2\left(4\pi Z\alpha\overline{\mu}^{2\varepsilon}\right)^2\int \frac{\mathrm{d}^d k}{(2\pi)^d}N_L \\
&\quad\times\left[(k^2 + 2k\cdot n\, m_1)(k^2 - 2k\cdot n\, m_2)(k^2)^2(\vec{k}^{\,2})^2\right]^{-1},
\end{aligned} \tag{16.237}$$

where $n = (1,0,0,0)$ is the time-like unit vector and

$$\begin{aligned} N_L &= (\vec{k}^{\,2})^2 \delta^{\perp,ab}(\vec{k}\,)\,\delta^{\perp,cd}(\vec{k}\,)\,\overline{u}_{1f}\gamma^a(m_1\not{n} + \not{k} + m_1)\gamma^c u_{1i} \\ &\quad \times \overline{v}_{2i}\gamma^d(-m_2\not{n} + \not{k} + m_2)\gamma^b v_{2f} \\ &= (\vec{k}^{\,2}\delta^{ab} - k^a k^b)(\vec{k}^{\,2}\delta^{cd} - k^c k^d) \\ &\quad \times \mathrm{Tr}\Big[\gamma^a(m_1\not{n} + \not{k} + m_1)\gamma^c\big(u\overline{v}\big)^{SM}\gamma^d(-m_2\not{n} + \not{k} + m_2)\gamma^b\big(v\overline{u}\big)^{SM}\Big]. \end{aligned} \tag{16.238}$$

We note the presence of a fermionic factor (-1) in Eq (16.237). We use Eqs. (16.233) and (16.234) above to write the two-particle spin states $(u\overline{v})^{SM}$ and $(v\overline{u})^{SM}$ as

$$\big(u\overline{v}\big)^{SM} = \begin{pmatrix} 0 & \chi^{SM} \\ 0 & 0 \end{pmatrix} = \frac{1}{2\sqrt{2}}(\not{n} + 1)\,\Xi(S,M)\,, \tag{16.239}$$

where

$$\Xi(S,M) = \gamma^5\,, \qquad \Xi(1,M) = -\not{e}_M \tag{16.240}$$

and

$$\big(v\overline{u}\big)^{SM} = \gamma^0\left[\big(u\overline{v}\big)^{SM}\right]^+\gamma^0 = -\frac{1}{2\sqrt{2}}\,[\Xi(S,M)]^+\,(\not{n}+1), \tag{16.241}$$

where $e_M^\mu = (0, \vec{e}_M)$ is a spatial unit vector, with $n \cdot e_M = 0$, and $e^*_M \cdot e_{M'} = -\vec{e}^{\,*}_M \cdot \vec{e}_{M'} = -\delta_{M,M'}$. Furthermore, we recall that γ_5 is a matrix that anticommutes with the other Dirac-γ matrices [see Eqs. (7.14) and (7.15)]

$$\gamma_5 \equiv \gamma^5 \equiv \begin{pmatrix} 0 & \mathbb{1}_{2\times2} \\ \mathbb{1}_{2\times2} & 0 \end{pmatrix}, \quad \gamma^0 = \begin{pmatrix} \mathbb{1}_{2\times2} & 0 \\ 0 & -\mathbb{1}_{2\times2} \end{pmatrix}, \quad \vec{\gamma} = \begin{pmatrix} 0 & \vec{\sigma} \\ -\vec{\sigma} & 0 \end{pmatrix}. \tag{16.242}$$

We will use this definition of γ_5 even in d dimensions. [Other definitions of γ_5 are more useful for other purposes, particularly when dealing with chiral symmetries (see also Sec. 10.7).] For the transverse-transverse crossed kernel of Fig. 16.10(b) we find

$$\begin{aligned} \Delta E_{\mathrm{TT;X}} &= -\,\mathrm{i}|\psi(\vec{0})|^2\left(4\pi Z\alpha\overline{\mu}^{2\varepsilon}\right)^2 \int \frac{\mathrm{d}^d k}{(2\pi)^d} \\ &\quad \times N_X\Big[(k^2 - 2k\cdot n m_1)(k^2 - 2k\cdot n m_2)(k^2)^2(\vec{k}^{\,2})^2\Big]^{-1} \end{aligned} \tag{16.243}$$

where we recall Eq. (16.145) and define

$$\begin{aligned} N_X &= (\vec{k}^{\,2}\delta^{ab} - k^a k^b)(\vec{k}^{\,2}\delta^{cd} - k^c k^d) \\ &\quad \times \mathrm{Tr}\Big[\gamma^a(m_1\not{n} - \not{k} + m_1)\gamma^c\big(u\overline{v}\big)^{SM}\gamma^b(-m_2\not{n} + \not{k} + m_2)\gamma^d\big(v\overline{u}\big)^{SM}\Big]. \end{aligned} \tag{16.244}$$

In each case, we calculated the traces separately for $S = 1$ and $S = 0$ and subtracted to get the hfs contribution. These trace results were then integrated over k using Feynman parameters. The sum of ladder plus crossed contributions is finite ($1/\varepsilon$ terms in the ladder and crossed terms cancel), giving the final result for the transverse-transverse hfs:

$$\Delta E^{\mathrm{hfs}}_{\mathrm{TT}} = \frac{8}{3}|\psi(\vec{0})|^2 (Z\alpha)^2 \frac{\ln(m_1/m_2)}{m_1^2 - m_2^2}\,. \tag{16.245}$$

The kernels involving dissimilar photon propagators, one transverse and one Coulomb, also lead to a finite hfs difference:

$$\Delta E^{\rm hfs}_{\rm TC+CT} = -\frac{32}{3}|\psi(\vec{0})|^2 (Z\alpha)^2 \frac{\ln(m_1/m_2)}{m_1^2 - m_2^2}. \qquad (16.246)$$

Finally, the Coulomb-Coulomb term is independent of spin, so its contribution to the hfs difference vanishes:

$$\Delta E^{\rm hfs}_{\rm CC} = 0. \qquad (16.247)$$

The complete hfs result is the sum of these: TT+TC+CT+CC:

$$\Delta E^{\rm hfs} = -8|\psi(\vec{0})|^2 (Z\alpha)^2 \frac{\ln(m_1/m_2)}{m_1^2 - m_2^2} = -\frac{8(Z\alpha)^5 \mu^3}{\pi n^3} \frac{\ln(m_1/m_2)}{m_1^2 - m_2^2} \delta_{\ell,0}. \qquad (16.248)$$

Since we assume that the mass ratio m_2/m_1 is large, we have found the dominant relativistic recoil correction to the hfs difference at order $(Z\alpha)^5$.

The calculation just completed was essentially a scattering calculation with free incoming and outgoing particles (in fact, the incoming and outgoing particles are taken to be at rest) so gauge invariance applies and we could as well have used Feynman gauge. In fact, Feynman gauge is simpler. There is no need to break up the calculation into transverse and Coulomb parts, and the integrals are covariant. As an illustration, we will use Feynman gauge to obtain the $S = 1$ and $S = 0$ hard region energy contributions individually. In analogy to (16.237) above, the Feynman gauge ladder contribution is

$$\begin{aligned}\Delta E_{\rm Feyn;L} &= {\rm i}|\psi(\vec{0})|^2(-1)\int \frac{{\rm d}^d k}{(2\pi)^d}\overline{u}_{1f}(-{\rm i}e\gamma^\mu)\frac{\rm i}{m_1\not{n} + \not{k} - m_1}(-{\rm i}e\gamma^\nu)u_{1i} \\ &\quad\times\left(\frac{-{\rm i}}{k^2}\right)\left(\frac{-{\rm i}}{k^2}\right)\overline{v}_{2i}({\rm i}Ze\gamma_\nu)\frac{\rm i}{-m_2\not{n} + \not{k} - m_2}({\rm i}Ze\gamma_\mu)v_{2f} \\ &= -{\rm i}|\psi(\vec{0})|^2\left(4\pi Z\alpha\overline{\mu}^{2\varepsilon}\right)^2\int\frac{{\rm d}^d k}{(2\pi)^d}N_L\Big[(k^2 + 2k\cdot nm_1)(k^2 - 2k\cdot nm_2)(k^2)^2\Big]^{-1}\end{aligned} \qquad (16.249)$$

where

$$N_L = {\rm Tr}\Big[\gamma^\mu(m_1\not{n} + \not{k} + m_1)\gamma^\nu\big(u\overline{v}\big)^{SM}\gamma_\nu(-m_2\not{n} + \not{k} + m_2)\gamma_\mu\big(v\overline{u}\big)^{SM}\Big]. \qquad (16.250)$$

We work through the usual steps of replacing $(u\overline{v})$ and $(v\overline{u})$ by Eqs. (16.239) and (16.241), bringing the γ_5 factors together and removing them, performing the traces (in d dimensions), and finally doing the Feynman integrals. We then go through the process again with the crossed graphs. In fact, the crossed expressions are very similar to the ladder ones, with only the changes that the second set of μ, ν indices in the traces are swapped and $m_1 n + k \to m_1 n - k$ in the electron propagator. We add the ladder and crossed terms to find results for the $S = 0$ and $S = 1$ energy shifts:

$$\Delta E^{S=0} = \frac{|\psi(\vec{0})|^2 (Z\alpha)^2 \overline{\mu}^{2\varepsilon}}{m_1 m_2} \times \left\{ \frac{1}{\varepsilon} - \frac{1}{3} - 2\frac{m_1(m_1+3m_2)\ln(m_2/\mu) - m_2(3m_1+m_2)\ln(m_1/\mu)}{m_1^2 - m_2^2} + \mathcal{O}(\varepsilon) \right\}, \tag{16.251a}$$

$$\Delta E^{S=1} = \frac{|\psi(\vec{0})|^2 (Z\alpha)^2 \overline{\mu}^{2\varepsilon}}{m_1 m_2} \left\{ \frac{1}{\varepsilon} - \frac{1}{3} - 2\frac{m_1 \ln(m_2/\mu) + m_2 \ln(m_1/\mu)}{m_1 + m_2} + \mathcal{O}(\varepsilon) \right\}. \tag{16.251b}$$

As we have seen before, the (infrared) divergences here will be canceled by ultraviolet divergences from other corrections to the energies. We are now in a position to obtain again the hfs correction:

$$\Delta E^{\rm hfs} = \Delta E^{S=1} - \Delta E^{S=0} = -8\,|\psi(\vec{0})|^2\,(Z\alpha)^2 \frac{\ln(m_1/m_2)}{m_1^2 - m_2^2}, \tag{16.252}$$

in accord with the Coulomb gauge result (16.248). As a quick consistency check, we also calculate the spin average energy shift. The spin average in d dimensions is

$$\begin{aligned} \Delta E^{\rm avg} &= \frac{\Delta E^{S=0} + (d-1)\Delta E^{S=1}}{d} \\ &= \frac{|\psi(\vec{0})|^2 (Z\alpha)^2 \overline{\mu}^{2\varepsilon}}{m_1 m_2} \left\{ \frac{1}{\varepsilon} - \frac{1}{3} - 2\frac{m_1^2 \ln(m_2/\mu) - m_2^2 \ln(m_1/\mu)}{m_1^2 - m_2^2} + \mathcal{O}(\varepsilon) \right\}, \end{aligned} \tag{16.253}$$

in exact accord with the spin-averaged hard region result obtained in Eq. (16.173). (We note that in order to calculate the d-dimensional spin average, we must account for the $d-1$ spin sub-states having $S = 1$ since one way to label these states is by the vector $\hat{\epsilon}_M$, which has $d-1$ independent components.)

16.5 Further Thoughts

(1) **Green Function and Bound-State Poles.** Reconsider Eq. (16.9), in the form

$$G'(x_1, x_2; y_1, y_2) = \Theta(X^0 - Y^0) \sum_{n,k} \int \frac{\mathrm{d}^3 P}{(2\pi)^3\, 2\omega_n(\vec{P})} \times \langle 0|\, \mathrm{T}\psi_1(x_1)\psi_2(x_2)\, |\vec{P}, n, k\rangle \langle \vec{P}, n, k|\, \mathrm{T}\overline{\psi}_1(y_1)\overline{\psi}_2(y_2)\, |0\rangle + \cdots. \tag{16.254}$$

In the discussion following Eq. (16.9), we pointed out that the condition $x_1^0, x_2^0 > y_1^0, y_2^0$ actually implies a more stringent theta function, in the form $\Theta(\min(\mathrm{x}_1^0, \mathrm{x}_2^0) - \max(\mathrm{y}_1^0, \mathrm{y}_2^0))$, rather than simply $\Theta(X^0 - Y^0)$. Show that use of the more exact theta function changes neither the positions nor the residues of the bound state poles in G'. *Hint:* Use

$$\min(x_1^0, x_2^0) = X^0 - \xi(x^0)|x^0|, \tag{16.255}$$
$$\max(y_1^0, y_2^0) = Y^0 + [1 - \xi(y^0)]\,|y^0|, \tag{16.256}$$
$$\xi(x^0) = \begin{cases} \xi_1 & x^0 > 0 \\ \xi_2 & x^0 < 0 \end{cases}. \tag{16.257}$$

The theta function will be $\Theta\big(X^0 - Y^0 - \xi(x^0)|x^0| - [1 - \xi(y^0)]\,|y^0|\big)$ instead of simply $\Theta(X^0 - Y^0)$. Show that the final form of (16.14) contains an extra factor of $\exp\big[\mathrm{i}\big(P^0 - \omega_n(\vec{P})\big)\big(\xi(x^0)|x^0| + (1 - \xi(y^0))|y^0|\big)\big]$, which changes neither the position of the pole at $P^0 = \omega_n(\vec{P})$ nor its residue.

(2) **Inhomogeneous Bethe–Salpeter Equation.** Consider the inhomogeneous BSEQ as shown in the symbolic equation depicted in Fig. 16.2(c). Show that the individual diagrams in G' implicitly contained in the diagram on the left-hand side of the equation given in Fig. 16.2(c) are in one-to-one correspondence with the sum of two terms. The first of these is the non-interacting 2-to-2 Green function S' of Fig. 16.1(b), and the second is the product of S' times the two-particle irreducible kernel K' of Fig. 16.2(c) times G' itself.

(3) **Alternative Form of the Bethe–Salpeter Equation.** Show the equivalence of Eqs. (16.22) and (16.23).

(4) **Series Representation of the Two-Particle Green Function.** Show Eq. (16.34).

(5) **Schrödinger–Coulomb Hamiltonian.** Show that Eq. (16.55) is the momentum-space transform of the eigenvalue equation corresponding to the Schrödinger–Coulomb Hamiltonian given in Eq. (4.1).

(6) **Hyperfine Structure for Equal Masses.** Calculate the correction (16.248) in the limit of positronium ($m_1 = m_2$) and evaluate it numerically for positronium.

(7) **Two-Time Green Function Method.** Compare the matching of scattering amplitudes, outlined in Sec. 12.2.1 and elsewhere in Chap. 12 and Chap. 15 in this treatise with the use of contour integrals for the derivation of expressions pertaining to energy shifts [see Eq. (16.123)]. Then, consult the two-time Green function method introduced by SHABAEV[1] in Ref. [595] for additional interesting interconnections.

[1] Vladimir Moiseevich Shabaev (b. 1959).

Chapter 17

NRQED: An Effective Field Theory for Atomic Physics

17.1 Overview

In this chapter, we develop a general approach to the physics of nonrelativistic Coulombic bound states based on an effective nonrelativistic quantum field theory. Specifically, we describe and illustrate the use of Non-Relativistic Quantum Electrodynamics (NRQED). The use of effective theories to describe low energy physics has a long history (for an overview, see Ref. [596]). Effective field theory has been used in atomic physics, at least implicitly, since the earliest days. The explicit construction of an effective nonrelativistic theory was initiated by CASWELL[1] and LEPAGE[2] in Ref. [597] with the introduction of NRQED, which has become a dominant framework for modern analytic and semi-analytic calculations of higher-order corrections to bound-state properties. Some early uses of NRQED and developments of the approach are given in Refs. [45, 329, 598–603]. NRQED and related work have been reviewed in [588, 598, 604].

There are two general approaches to the development of effective low-energy quantum field theories. In the *top down* approach, one assumes the knowledge of some more general field theory valid at higher energies and integrates out the high energy field modes, leaving an effective low-energy theory (see also Chap. 18). In principle, using the top down approach, one is able to derive all details of the low-energy theory from the more general theory. In practice, such derivations can be challenging. In the *bottom up* approach, knowledge of an overarching theory is not assumed. Instead, one simply writes down a Lagrangian for the effective field theory (EFT) consistent with general principles such as Hermiticity, translational and rotational symmetry, gauge invariance, and parity and time reversal invariance. The various terms in the EFT contain some undetermined coefficients, the WILSON[3] *coefficients*, that are determined phenomenologically. Often, an intermediate strategy is employed. The overarching theory is assumed to be known. In our case it is QED, or more generally the Standard Model. The EFT is created as in the

[1] William Edward Caswell (1947–2001).
[2] Gerard Peter Lepage (b. 1952).
[3] Kenneth Geddes Wilson (1936–2013).

bottom up approach, as the most general quantum field theory subject to the desired conditions and symmetries. The various terms of the EFT carry coefficients that are determined by the requirement that the EFT and the overarching theory agree when used to calculate some set of processes for which both are valid. The process of determining the coefficients thus is independent of the phenomenological aspects. In this context, the Wilson coefficients are often termed *matching coefficients*, since their values are found through a matching procedure.

Low-energy effective theories such as NRQED have two characteristic and related attributes. First, they are not renormalizable. Renormalizability is an extremely restrictive condition that effective theories do not satisfy. As one simple example of an unrenormalizable term in the NRQED Lagrangian one could mention the operator representing the *Darwin* or *zitterbewegung* term,

$$\mathcal{L}_{\mathrm{D}} = -\frac{q}{8m^2} c_{\mathrm{D}} \psi^+ \left(\vec{\nabla}^2 A^0\right) \psi \,. \tag{17.1}$$

For the case $q = e$ (electron field), where ψ represents the electron-field operator and m the electron mass, this operator contributes part of the Lamb shift (see Chap. 11). It contains a dimensionless constant c_{D} that is only determined through the matching operation. The presence of the inverse mass squared in the multiplicative factor makes clear that the operator dimension of $\psi^+ (\vec{\nabla}^2 A^0) \psi$ is too high for this term to be renormalizable. Specifically, the mass dimension of $\psi^+ (\vec{\nabla}^2 A^0) \psi$ is six, consistent with a mass dimension of ψ of 3/2, a mass dimension of the $\vec{\nabla}^2$ operator of 2, and a mass dimension of qA^0 (potential energy) of one. A trivial calculation shows that the mass dimension of the operator given in Eq. (17.1) is $2 \times \frac{3}{2} + 2 + 1 = 6$. Because the mass dimension of $\mathcal{L}_D$ must be 4, the coefficient multiplying the operators must have dimension -2, consistent with the explicit factor of $1/m^2$. This is but one of an infinite number of such operators having higher and higher dimension, and more and more inverse masses in their coefficients. They involve a corresponding infinitude of matching constants that are only known by virtue of the matching procedure. As a general remark, higher-dimension operators give rise to higher-order high-energy divergences in the amplitudes they are involved in, leading to a theory that is not renormalizable. The second and related point is that these terms, with increasingly negative powers of mass in their coefficients, necessarily make smaller and smaller numerical contributions to whatever quantity is being calculated. This is because every explicit mass in the denominator must be balanced by a factor of the characteristic energy or momentum in the numerator. In a nonrelativistic effective theory of Coulombic bound states, the characteristic energies and momenta are small compared to the constituent masses, being typically of order α or α^2 times a characteristic mass. So, even though there are infinitely many operators in the effective Lagrangian, only a well-defined and finite number of them contribute at any given power of the fine-structure constant. Under these conditions, lack of renormalizability is not a disadvantage.

In this chapter, we show how the matching coefficients can be obtained, and use our results to develop a convenient description of nonrelativistic Coulombic two-fermion bound states. We aim to demonstrate that NRQED contains the physics of these bound systems in an extremely natural way. We also develop the Bethe–Salpeter equation (BSEQ) of NRQED. As is to be expected, the NRQED BSEQ is identical to the nonrelativistic Schrödinger equation (in leading order). It also provides a framework for including relativistic and radiative corrections. Perturbation theory in NRQED is more convenient than in QED for a number of reasons. There are more NRQED operators, but those operators are simpler to work with and are more directly related to identifiable physical effects. It is easier using NRQED to estimate the order that each term in perturbation theory initially contributes, and much easier to find the higher-order contributions that are present in any given term of the series. These points are illustrated through examples. Specifically, in the following, we use NRQED to derive the effective Hamiltonian that could be used to find all energy levels at $\mathcal{O}(\alpha^4)$. The spin-averaged $\mathcal{O}(Z\alpha)^5$ recoil correction, and the leading $\mathcal{O}(Z\alpha)^5$ hyperfine corrections for atoms consisting of an electron and an elementary positive constituent are also derived.

17.2 Basics of NRQED

17.2.1 *NRQED Lagrangian and Feynman Rules*

The NRQED Lagrangian is constructed from basic fields representing the nonrelativistic fermions and the photons. The quantum field for the ith spin-1/2 fermion is a (nonrelativistic, two-component) Pauli-spinor valued field $\psi_i(x)$ having mass m_i and charge q_i (see Chap. 6). There are no antiparticle degrees of freedom contained in $\psi_i(x)$ (in contrast to the Dirac formalism described in Chaps. 7 and 8). So, if we want to describe both an electron and a positron, then we need to do this using separate and independent fields for each of them. Typically, we include just two fermion fields. One of them describes a positively charged particle, the other, a negatively charged particle. The terms in the NRQED Lagrangian have various inverse powers of the fermion masses. As discussed earlier, terms with higher negative powers of the masses tend to give smaller effects, so we can organize the terms entering $\mathcal{L}$ according to the power of mass. We will focus on recoil corrections of $\mathcal{O}(Z\alpha)^5$. For these contributions, it is sufficient (with one exception) to include terms of mass order $1/m^2$ (but not higher). We describe the construction of the NRQED Lagrangian following the discussion of Ref. [605] and then use that Lagrangian to derive the corresponding NRQED Feynman rules.

The principle of gauge invariance greatly restricts the number of possible terms that can enter the effective Lagrangian $\mathcal{L}$. We assume that $\mathcal{L}$ is invariant under the local gauge transformation [see Eq. (8.2) and the surrounding discussion in Chap. 8]

$$\psi(x) \to \mathrm{e}^{\mathrm{i}q\Lambda(x)}\psi(x)\,, \tag{17.2}$$

while the Hermitian adjoint transforms as follows,

$$\psi^{+}(x) \to \psi'^{+}(x) = \psi^{+}(x)\mathrm{e}^{-\mathrm{i}q\Lambda(x)}\,. \tag{17.3}$$

Table 17.1 We consider the behavior of the building blocks of the NRQED Lagrangian under the parity and time reversal symmetries. The time variable and the scalar potential are scalars, while the gradient operator and the vector potential are vectors. The magnetic field is an axial vector. For time reversal, one must remember that the time reversal operator is represented as an anti-unitary operator involving complex conjugation and that the vector potential is produced by moving charges. The behavior of $\vec{\sigma}$ is determined by noting that $\psi^+\vec{\sigma}\psi$ transforms as an angular momentum (odd under time reversal, even under parity).

	$\mathrm{i}\mathcal{D}_t$	$\mathrm{i}\vec{\mathcal{D}}$	A^0	$\vec{A}$	$\vec{E}$	$\vec{B}$	$\vec{\sigma}$
P	+	−	+	−	−	+	+
T	+	−	+	−	+	−	−

This implies the presence of a gauge field $A^\mu(x)$ with the corresponding transformation

$$A_\mu(x) \to A'_\mu(x) = A_\mu(x) - \partial_\mu \Lambda(x) . \tag{17.4}$$

In place of the usual derivative $\partial_\mu = \partial/\partial x^\mu$, $\mathcal{L}$ will be constructed using the covariant derivative $\mathcal{D}_\mu = \partial_\mu + \mathrm{i}\, q\, A_\mu$ defined in Eq. (8.3). We here denote the charge as q rather than e, inspired by the development of the NRQED formalism, which originated as an effective field theory, describing the low-energy (bound-state) limit of quantum electrodynamics. In the literature on NRQED, the charge of the particle is usually denoted by the general symbol q rather than the more special electron charge e, and we follow this convention in the current discussion. The gauge field can enter the Lagrangian through the Faraday tensor

$$F_{\mu\nu} = \left(-\frac{\mathrm{i}}{q}\right) [\mathcal{D}_\mu, \mathcal{D}_\nu] = \partial_\mu A_\nu - \partial_\nu A_\mu , \tag{17.5}$$

or other gauge invariant forms. Specifically, the electric field $\vec{E}$, and the magnetic induction field $\vec{B}$,

$$\vec{E} = -\vec{\nabla} A^0 - \frac{\partial \vec{A}}{\partial t} , \qquad \vec{B} = \vec{\nabla} \times \vec{A} , \tag{17.6}$$

are gauge invariant and enter the NRQED Lagrangian.

We also require our NRQED Lagrangian to respect the symmetries of rotational invariance, parity, and time reversal, which it inherits from QED. Rotational symmetry is built in if we construct the NRQED Lagrangian from scalar and vector quantities in a rotationally invariant way. The scalars involved will be A^0, $\mathrm{i}\mathcal{D}_t$, and $\psi_i^+\psi_i$, while the relevant vectors are $\vec{A}$, $\vec{E}$, $\vec{B}$, $\vec{\mathcal{D}} = \vec{\nabla} - iq\vec{A}$, and $\psi_i^+\vec{\sigma}\psi_i$. The behavior of the various building blocks under action of the discrete symmetries is shown in Table 17.1.

Upon taking the various symmetries into account, we find that the Lagrangian for NRQED adequate for energy levels through $\mathcal{O}(Z\alpha)^5$ can be written as

$$\mathcal{L} = \sum_i \psi_i^+ \left\{ \mathrm{i}\mathcal{D}_t + \frac{\vec{\mathcal{D}}^2}{2m_i} + \frac{\vec{\mathcal{D}}^4}{8m_i^3} + c_{\mathrm{F}i}\frac{q_i}{2m_i}\vec{\sigma}_i \cdot \vec{B} + c_{\mathrm{D}i}\frac{q_i}{8m_i^2}\left(\vec{\mathcal{D}}\cdot\vec{E} - \vec{E}\cdot\vec{\mathcal{D}}\right) \right.$$
$$\left. + c_{\mathrm{S}i}\frac{\mathrm{i}q_i}{8m_i^2}\vec{\sigma}_i \cdot \left(\vec{\mathcal{D}}\times\vec{E} - \vec{E}\times\vec{\mathcal{D}}\right) + \cdots \right\} \psi_i - \frac{1}{4}F_{\mu\nu}F^{\mu\nu} + \cdots$$
$$+ \psi_{1a}^+\psi_{2c}^+ \left\{ \frac{d_{\mathrm{S}}}{m_1 m_2}\delta_{ab}\,\delta_{cd} + \frac{d_{\mathrm{V}}}{m_1 m_2}\vec{\sigma}_{ab}\cdot\vec{\sigma}_{cd} \right\} \psi_{1b}\psi_{2d} + \cdots, \tag{17.7}$$

where i is an index used to distinguish the electron ($i = 1$) from the positive particle (proton or positive muon, $i = 2$). Of course, the covariant derivative in Eq. (17.7) acts on the argument x of $\psi_i(x)$, where $\psi_i(x)$ is a second-quantized (Schrödinger–Pauli) field, and $\vec{E} = \vec{E}(x)$ and $\vec{B} = \vec{B}(x)$ are second-quantized field operators. The index i enumerates the particle *species* that enter the Lagrangian. (In contrast to our treatment of bound systems in Chap. 12, the fermions that make up the atoms here are "second quantized", *i.e.* they are quantized fields subject to creation and annihilation.) The parameters q_i and m_i denote the charge and mass of the ith particle; and a, b, c, d are spinor indices for the Pauli spinors and the Pauli sigma matrices, defined in Eq. (4.343). For example, the term given in Eq. (17.1) is obtained from the full Lagrangian (17.7) according to

$$c_{\mathrm{D}i}\frac{q_i}{8m_i^2}\,\psi_i^+\left(\vec{\mathcal{D}}\cdot\vec{E} - \vec{E}\cdot\vec{\mathcal{D}}\right)\psi_i \to -c_{\mathrm{D}i}\frac{q_i}{8m_i^2}\,\psi_i^+(\vec{\nabla}^2 A^0)\,\psi_i\,, \tag{17.8}$$

upon picking the term $\vec{E} \to -\vec{\nabla}A^0$ from Eq. (17.6). At this order, we only need to include terms up to quadratic in denominator masses except for the single $1/m^3$ term coming from the nonrelativistic expansion of the kinetic energy,

$$\sqrt{m^2 + \vec{\pi}^{\,2}} - m = \frac{\vec{\pi}^{\,2}}{2m} - \frac{\vec{\pi}^{\,4}}{8m^3} + \cdots, \tag{17.9}$$

with $\vec{\pi} = -\mathrm{i}\vec{\mathcal{D}}$, which contributes at $\mathcal{O}(Z\alpha)^4$. Values for the Fermi c_{F}, Darwin c_{D}, spin-orbit c_{S}, and scalar d_{S} and vector d_{V} Wilson coefficients will be found by matching, i.e., by requiring agreement between QED and NRQED for a set of low-energy processes that can be readily computed from both theories.

The Feynman rules for NRQED are obtained from the Lagrangian in the usual way. There are propagators for the fermions and photons (Coulomb and transverse) found by isolating the terms in $\mathcal{L}$ quadratic in the field operators and working in momentum space. The rules for interactions are found by considering the interaction terms in $\mathrm{i}\mathcal{L}$ one by one and imagining the presence of factors $\mathrm{e}^{\mathrm{i}\vec{p}_1\cdot\vec{x}}$ for fields carrying momentum $\vec{p}_1$ in and $\mathrm{e}^{-\mathrm{i}\vec{p}_2\cdot\vec{x}}$ for outgoing $\vec{p}_2$. Derivatives $\vec{\nabla}$ are replaced by $\mathrm{i}\vec{p}_1$ for incoming and $-\mathrm{i}\vec{p}_2$ for outgoing momenta. The basic Coulomb interaction is contained in $\mathrm{i}\mathcal{D}_t$. Temporarily dropping the index i labeling fermion type, the fermion-fermion-photon term coming from $\mathrm{i}\mathcal{D}_t$ is $\mathrm{i}\mathcal{L} \to \mathrm{i}\psi^+\mathrm{i}(\mathrm{i}qA_0)\psi$, leading to the Feynman rule $-\mathrm{i}q$ for the Coulomb interaction (the field A^0 and the fermion operators are of course not part of the Feynman rule). The coupling to a transverse photon is contained in the $\vec{\mathcal{D}}^2$ term. Specifically,

$$\mathrm{i}\mathcal{L} \to \mathrm{i}\psi^+ \frac{1}{2m}\left(\vec{\nabla} - \mathrm{i}q\vec{A}\right)^2 \psi \to \frac{q}{2m}\psi^+\left(\vec{\nabla}\cdot\vec{A} + \vec{A}\cdot\vec{\nabla}\right)\psi. \tag{17.10}$$

The corresponding rule for the interaction of a fermion with a spatial photon of index j is $\frac{q}{2m}\left[\mathrm{i}(k+p_1)^j + \mathrm{i}(p_1)^j\right] = \frac{\mathrm{i}q}{2m}(p_2+p_1)^j$. This coupling can be called *convective* because of its dependence on the fermion motion (see Secs. 11.4.2 and 12.5.3). Another example, containing some new features, is the Fermi coupling (proportional to c_{F}) to a photon carrying spatial index k. The interaction Lagrangian gives $\mathrm{i}\mathcal{L} \to \mathrm{i}c_{\mathrm{F}}\frac{q}{2m}\psi^+\sigma^i\epsilon^{ijk}\nabla^j A^k\psi$ where the $\nabla^j \equiv \partial_j$ derivative acts only on A^k. The corresponding Feynman rule is $\mathrm{i}c_{\mathrm{F}}\frac{q}{2m}\sigma^i\epsilon^{ijk}(\mathrm{i}k^j) = -c_{\mathrm{F}}\frac{q}{2m}\sigma^i\epsilon^{ijk}k^j$, where the incoming photon momentum is $\vec{k} = \vec{p}_2 - \vec{p}_1$. One subtlety concerns the presence of the antisymmetric symbol ϵ^{ijk} in the Feynman rules. The problem is that the definition of ϵ^{ijk} is tied to the three dimensional nature of space. When we generalize the number of spatial dimensions to D with $D \neq 3$, ϵ^{ijk} no longer makes sense. One solution to this difficulty is to replace the product $\epsilon^{ijk}\sigma^k$ by σ^{ij} where

$$\sigma^{ij} \equiv -\frac{\mathrm{i}}{2}\left[\sigma^i, \sigma^j\right]. \tag{17.11}$$

We note that a set of sigma matrices satisfying $\{\sigma^i, \sigma^j\} = 2\delta^{ij}$ exists even in D dimensions (for any positive integer D), and that σ^{ij} as defined generates a representation of the rotation group in D dimensions [606]. In three dimensions we find that $\sigma^{ij} = \epsilon^{ijk}\sigma^k$ and $\sigma^i = \frac{1}{2}\epsilon^{ijk}\sigma^{jk}$, so the replacement works when $D = 3$ and provides a convenient generalization to $D \neq 3$. The Feynman rules that we will need for our calculations are displayed in Fig. 17.1.

17.2.2 *Matching Coefficients*

The NRQED Lagrangian comes with a number of coefficients: c_{S}, c_{D}, c_{F}, etc. Since NRQED is an effective theory for an exactly known more general theory, QED, these coefficients are calculable. We will use the matching technique to work out their values. The idea of matching is to find some appropriate low-energy processes that can in principle be calculated using both QED and NRQED since QED is valid for all energy scales while NRQED is valid for the low energy processes under consideration. The NRQED results will involve the coefficients while the QED results are fully determined, so the values of the various coefficients can be found. These coefficients are often referred to as *matching coefficients* because of this procedure for finding their values.

In order to find all corrections at $\mathcal{O}(\alpha^5)$ we would need values for the spin-orbit, Darwin, and Fermi matching coefficients c_{S}, c_{D}, and c_{F} through order α, the vacuum polarization coefficient c_{VP} through order α, and the four-fermion *contact terms* d_{S} and d_{V} through $\mathcal{O}(\alpha^2)$. The explicit processes that we will focus on in this chapter are the recoil corrections at $\mathcal{O}(Z\alpha)^5$, for which only d_{S} and d_{V} are

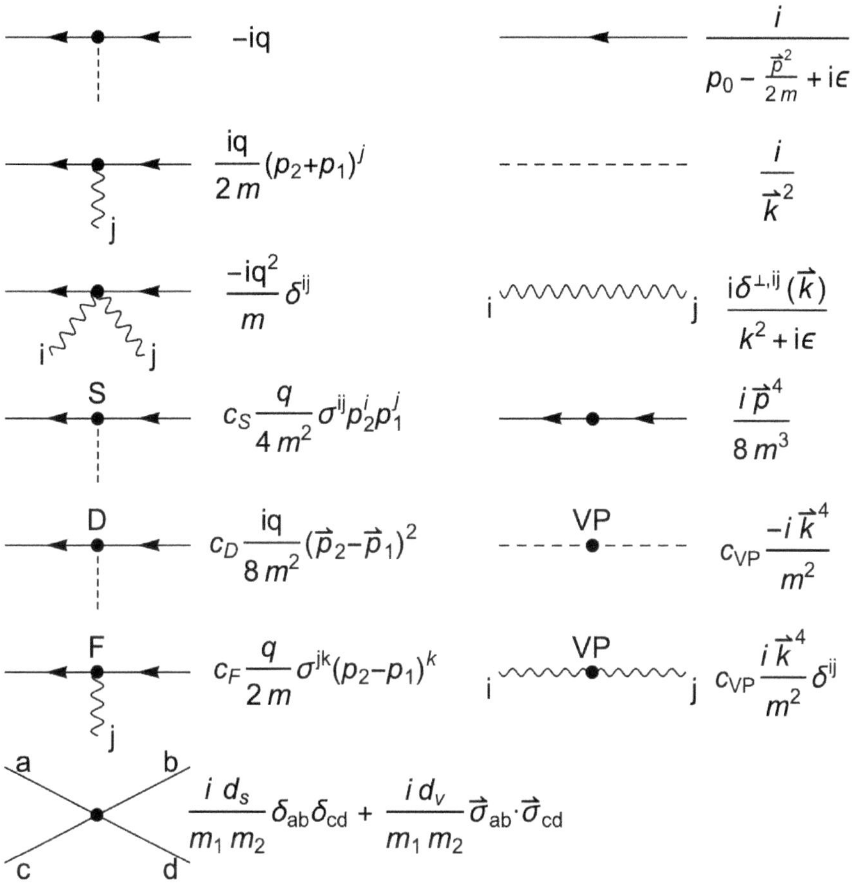

Fig. 17.1 The Feynman rules of NRQED are given here. Solid lines represent fermions, dashed lines stand for Coulomb photons, and wiggly lines for transverse photons. The NRQED vertices are represented by dots. Incoming and outgoing momenta on the fermion line (with mass m) are represented by $\vec{p}_1$ and $\vec{p}_2$ respectively, and $\vec{k} = \vec{p}_2 - \vec{p}_1$ is the incoming momentum of the photon. Indices $i, j, k, \ldots$ are spatial indices while $a, b, c, \ldots$ are spin indices. In the four-fermion contact term, m_1 and m_2 represent the masses of the two particles involved. Additional Feynman rules exist for higher order terms in the effective Lagrangian.

required, but in order to illustrate the procedure we will describe the evaluation of the other coefficients as well.

The coefficients for terms involving a single fermion (incoming plus outgoing) interacting with a single photon can be found by considering the scattering of an on-shell fermion from an external electromagnetic source at linear order in the electromagnetic field. We considering only low-energy incoming and outgoing fermions so that NRQED is available for performing the calculations. The one-photon matching condition is illustrated in Fig. 17.2(a), where the large solid dots represent all radiative corrections. The fermion momentum labels in this diagram

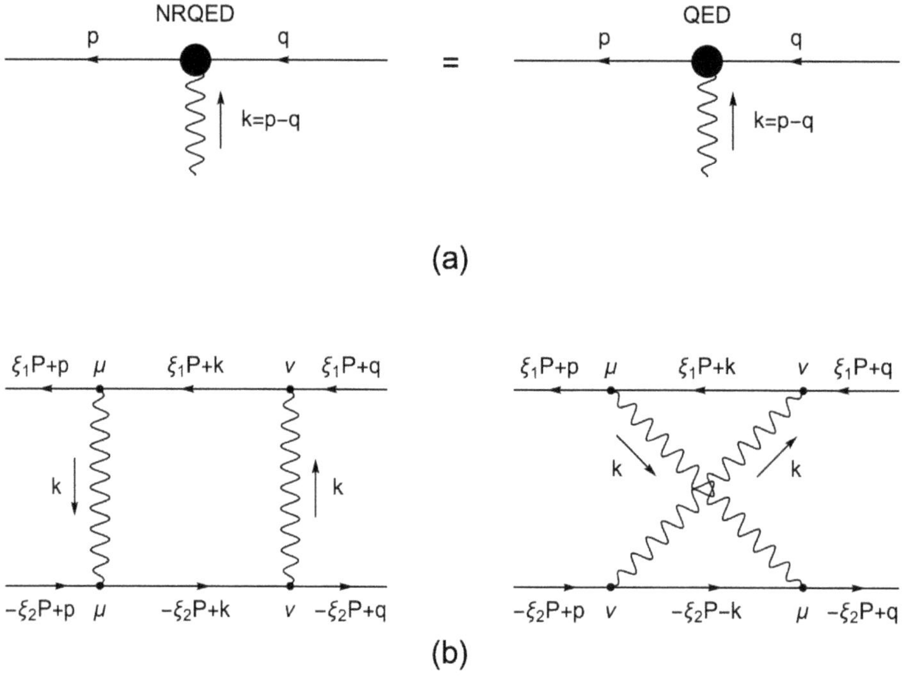

Fig. 17.2 Diagrams entering into the matching calculation for NRQED. Part (a) shows the matching condition for interactions involving a single fermion interacting with a photon. As long as all momenta are nonrelativistic, both NRQED and QED describe this interaction equally well. On the NRQED side, this amplitude involves the matching coefficients c_F, c_D, c_S, etc., while on the QED side the full effect of fermion-photon interactions is contained in the Dirac and Pauli form factors. Part (b) shows QED diagrams contributing to the interaction between two fermions. The electron on the top is shown as a fermion, while the positive particle (positron, proton, positive muon, etc.) is shown as an antiparticle on the bottom. The incoming and outgoing relative momenta (p, q) are negligible compared to the exchange momentum k. In the extreme nonrelativistic limit with at-rest or nearly-at-rest fermions, this interaction can be equally well described by QED or by NRQED through the four-fermion contact term involving d_S and d_V.

are interpreted as follows. For NRQED, p^0 is the nonrelativistic kinetic energy $p^0 = \vec{p}^{\,2}/(2m) + \mathcal{O}\left(\vec{p}^{\,4}/m^3\right)$. For QED, $p^0 = \sqrt{\vec{p}^{\,2} + m^2}$ is the total energy. In either case we consider only the nonrelativistic limit $|\vec{p}\,|/m \ll 1$. The same considerations apply to the four-vector q. Radiative corrections do not contribute in the NRQED vertex function [600] making the NRQED side of the matching equality easy to evaluate. A remark is in order. When expanding about $\vec{p} = 0$, $p^0 = 0$, which is appropriate for matching calculations in NRQED, the integrals in the vertex function all vanish in dimensional regularization since there is no nonrelativistic scale. This consideration becomes obvious when one takes into account that the evaluation of NRQED loop integrals proceeds in the momentum range $k \in (0, \epsilon)$ with $\epsilon \ll m$, much in the spirit of the evaluation of the low-energy part of the bound-electron self-energy outlined in Chap. 4. Hence, it is always possible to expand in the ratio

k/m, which renders a number of important integrals scaleless and hence zero in dimensional regularization.

Working directly from the Feynman rules given in Fig. 17.1, we find for the 0 and j components of the NRQED scattering amplitude shown in Fig. 17.2(a) the results ($\vec{k} = \vec{p} - \vec{q}$)

$$\mathcal{M}^0_{\rm NRQED} = -q\,\eta_f^+ \left\{1 - \frac{\vec{k}^{\,2}}{8m^2}c_{\rm D} + \frac{i\sigma^{\ell m}p^\ell q^m}{4m^2}c_{\rm S} + \cdots\right\}\eta_i, \tag{17.12}$$

$$\mathcal{M}^j_{\rm NRQED} = -q\,\eta_f^+ \left\{-\frac{(p^j + q^j)}{2m} + \frac{i\sigma^{jk}k^k}{2m}c_{\rm F} + \cdots\right\}\eta_i, \tag{17.13}$$

where η_f and η_i are Pauli spinors for the spin states of the outgoing and incoming particles. For the result given in Eq. (17.12), we sum the contributions from the first, fourth and fifth Feynman rule in the first column of Fig. 17.1, while for Eq. (17.13) we include contributions from the second and sixth rule in that first column. The dots represent the contributions of higher-order operators giving terms of $\mathcal{O}(1/m^4)$ in $\mathcal{M}^0$ and $\mathcal{O}(1/m^3)$ in $\mathcal{M}^j$. For the QED side, single-photon scattering is a standard process whose result can be expressed in terms of form factors, and these form factors are known to high precision. The QED scattering amplitude is

$$\mathcal{M}^\mu_{\rm QED} = -q\,\overline{u}(p)\left\{\gamma^\mu F_1(k^2) + \frac{i\sigma^{\mu\nu}k_\nu}{2m}F_2(k^2)\right\}u(q)\,. \tag{17.14}$$

The fermion momenta are on-shell and the photon momentum is $k = p - q$. We use the nonrelativistic spinor normalization

$$u^+(p)\,u(p) = 1, \qquad \overline{u}(p)\,u(p) = m/E\,, \tag{17.15}$$

so that

$$u(p) = N\begin{pmatrix}\eta \\ \dfrac{\vec{\sigma}\cdot\vec{p}}{E+m}\eta\end{pmatrix}, \qquad N = \sqrt{\frac{E+m}{2E}}\,, \tag{17.16a}$$

$$\overline{u}(p) = N\left(\eta^+, -\eta^+\frac{\vec{\sigma}\cdot\vec{p}}{E+m}\right). \tag{17.16b}$$

Note that the spinors used here are manifestly different from the helicity basis spinors encountered in Sec. 7.3.1. For the *nonrelativistic* particles being considered here, it is appropriate to use solutions that reduce to the fundamental Pauli spinors χ [see Eq. (6.42)], and are normalized to one particle per unit volume. Because the fermions are nonrelativistic, we will expand everything in the small quantities $\vec{p}^{\,2}/m^2$ and $\vec{q}^{\,2}/m^2$. The expansion for the spinors is

$$u(p) = N\begin{pmatrix}\eta \\ \dfrac{\vec{\sigma}\cdot\vec{p}}{E+m}\eta\end{pmatrix} \approx \begin{pmatrix}\left[1 - \dfrac{p^2}{8m^2} + \dfrac{11p^4}{128m^4} + \mathcal{O}\left(\dfrac{p^6}{m^6}\right)\right]\eta \\ \left[1 - \dfrac{3p^2}{8m^2} + \mathcal{O}\left(\dfrac{p^4}{m^4}\right)\right]\dfrac{\vec{\sigma}\cdot\vec{p}}{2m}\eta\end{pmatrix}, \tag{17.17}$$

with an analogous expansion for $\overline{u}(p)$. The Dirac and Pauli form factors $F_1(k^2)$, $F_2(k^2)$ have the expansions

$$F_j(k^2) = F_j(0) + k^2 F_j'(0) + \frac{1}{2}(k^2)^2 F_j''(0) + \cdots, \tag{17.18}$$

$$\equiv \overline{F}_j + \frac{k^2}{m^2}\overline{F}_j' + \frac{1}{2}\left(\frac{k^2}{m^2}\right)^2 \overline{F}_j'' + \cdots, \tag{17.19}$$

$$= \overline{F}_j - \frac{\vec{k}^{\,2}}{m^2}\overline{F}_j' + \frac{(k^0)^2}{m^2}\overline{F}_j' + \frac{\vec{k}^{\,4}}{2m^4}\overline{F}_j'' + \cdots, \tag{17.20}$$

with an obvious notation for the quantities with bars. Here, we have

$$k^0 \approx \frac{\vec{p}^{\,2}}{2m} - \frac{\vec{q}^{\,2}}{2m} \tag{17.21}$$

for on-shell nonrelativistic incoming and outgoing fermions. Results for the coefficients in these expansions are known (see below). The nonrelativistic expansions of the QED amplitudes give

$$\begin{aligned} \mathcal{M}^0_{\text{QED}} &= -q\,\eta_f^+ \left\{1 - \frac{\vec{k}^{\,2}}{8m^2} + \frac{\mathrm{i}\sigma^{ab}p^a q^b}{4m^2} + \cdots\right\}\eta_i \times \left\{\overline{F}_1 - \frac{\vec{k}^{\,2}}{m^2}\overline{F}_1' + \cdots\right\} \\ &\quad - q\eta_f^+ \left\{\frac{\mathrm{i}\sigma^{ab}p^a q^b}{2m^2} - \frac{\vec{k}^{\,2}}{4m^2} + \cdots\right\}\eta_i \times \left\{\overline{F}_2 + \cdots\right\}, \end{aligned} \tag{17.22}$$

$$\begin{aligned} \mathcal{M}^j_{\text{QED}} &= -q\,\eta_f^+ \left\{-\frac{p^j + q^j}{2m} + \frac{\mathrm{i}\sigma^{ja}k^a}{2m} + \cdots\right\}\eta_i \times \left\{\overline{F}_1 + \cdots\right\} \\ &\quad - q\eta_f^+ \left\{\frac{\mathrm{i}\sigma^{ja}k^a}{2m} + \cdots\right\}\eta_i \times \left\{\overline{F}_2 + \cdots\right\}. \end{aligned} \tag{17.23}$$

A comparison of the NRQED and QED results for the scattering amplitudes gives values for the NRQED coefficients in terms of the form factors:

$$1 = \overline{F}_1, \tag{17.24a}$$

$$c_{\text{D}} = \overline{F}_1 + 2\overline{F}_2 + 8\overline{F}_1', \tag{17.24b}$$

$$c_{\text{S}} = \overline{F}_1 + 2\overline{F}_2, \tag{17.24c}$$

$$c_{\text{F}} = \overline{F}_1 + \overline{F}_2. \tag{17.24d}$$

The form factors themselves are power series in α, for which the first several terms are known. We worked out their values through one-loop order in Sec. 10.3.6 and found the results [see Eq. (10.190)]

$$\overline{F}_1 = 1, \tag{17.25a}$$

$$\overline{F}_1' = \frac{\alpha}{\pi}\left\{-\frac{1}{6\epsilon} - \frac{1}{6}\ln\left(\frac{\mu^2}{m^2}\right) - \frac{1}{8}\right\} + \mathcal{O}(\alpha^2), \tag{17.25b}$$

$$\overline{F}_2 = \frac{\alpha}{2\pi} + \mathcal{O}(\alpha^2). \tag{17.25c}$$

For the matching coefficients, we obtain

$$c_{\mathrm{D}} = 1 - \frac{4\alpha}{3\pi}\left\{\frac{1}{\epsilon} + \ln\left(\frac{\mu^2}{m^2}\right)\right\} + \mathcal{O}(\alpha^2), \tag{17.26a}$$

$$c_{\mathrm{S}} = 1 + \frac{\alpha}{\pi} + \mathcal{O}(\alpha^2), \tag{17.26b}$$

$$c_{\mathrm{F}} = 1 + \frac{\alpha}{2\pi} + \mathcal{O}(\alpha^2). \tag{17.26c}$$

In a similar way, the vacuum polarization coefficient c_{VP} can be obtained. We compare the NRQED and QED results for the static potential between two charges at small values of $\vec{k}^{\,2}/m^2$, as corrected by vacuum polarization. From NRQED we find

$$\begin{aligned} V^{\mathrm{NRQED}} &\to \mathrm{i}(-\mathrm{i}q_1)\left\{\frac{\mathrm{i}}{\vec{k}^{\,2}} + \frac{\mathrm{i}}{\vec{k}^{\,2}}\left(c_{\mathrm{VP}}\frac{-\mathrm{i}\vec{k}^{\,4}}{m^2}\right)\frac{\mathrm{i}}{\vec{k}^{\,2}} + \cdots\right\}(-\mathrm{i}q_2) \\ &= \frac{q_1 q_2}{\vec{k}^{\,2}}\left\{1 + c_{\mathrm{VP}}\frac{\vec{k}^{\,2}}{m^2} + \cdots\right\}, \end{aligned} \tag{17.27}$$

while from standard QED (see Sec. 10.4.3), we have

$$V^{\mathrm{QED}} \to \frac{q_1 q_2}{\vec{k}^{\,2}}\left\{1 + \frac{\alpha}{15\pi}\frac{\vec{k}^{\,2}}{m^2} + \cdots\right\}. \tag{17.28}$$

Comparison of the NRQED and QED forms yields the value

$$c_{\mathrm{VP}} = \frac{\alpha}{15\pi}. \tag{17.29}$$

The two-photon-exchange contact term requires more work. The relevant diagrams are shown in Fig. 17.2(b). The exchanged photons and fermions in the inner loop are high energy, with momenta of the order of the fermion masses. So, as far as atomic distance scales are concerned, these interactions are point-like. The NRQED representation of these hard corrections is contained in the four-fermion contact term, with amplitude

$$\mathrm{i}\mathcal{M}^{\mathrm{NRQED}} = \mathrm{i}\eta^{+}_{1fa}\,\eta^{+}_{2fc}\left\{\frac{d_{\mathrm{S}}}{m_1 m_2}\delta_{ab}\,\delta_{cd} + \frac{d_{\mathrm{V}}}{m_1 m_2}\vec{\sigma}_{ab}\cdot\vec{\sigma}_{cd}\right\}\eta_{1ib}\,\eta_{2id}\,. \tag{17.30}$$

Here η_{1ib}, for example, is the bth component of the Pauli spinor describing the incoming spin state of fermion 1. We find it convenient to calculate the amplitudes directly for spin-0 and spin-1 combinations of the two spin-1/2 fermion spin states. (Since this is a point-like interaction, S states are the only ones that will contribute, so the total angular momentum is just due to the spins of the two fermionic constituents.) The relativistic (QED) two-particle spin states for a particle-antiparticle system were found in Sec. 16.4. Here, we need the nonrelativistic (NRQED) two-particle spin states (no antiparticle involved). Initially, we work in three spatial dimensions, where the total spin operator is the usual $\vec{S}_{\mathrm{tot}} = \vec{S}_1 + \vec{S}_2$ with $\vec{S}_i = \frac{1}{2}\vec{\sigma}_i$ being the spin operator for the ith particle. Let ξ by the desired two-particle spin state, and take the action of a general operator $A_1 B_2$ on ξ to be

$$\left[A^{(1)}B^{(2)}\xi\right]_{ab} = A_{aa'}\,B_{bb'}\,\xi_{a'b'} = \left[A\xi B^{\mathrm{T}}\right]_{ab}. \tag{17.31}$$

We here introduce the ξ matrices as auxiliary devices in the calculation of the two-particle matrix elements, in a distant analogy to Casimir's trick for the calculation of Dirac traces. The introduction of the ξ matrices, together with the calculational prescription (17.31), eliminates the need for a separate consideration of the projection operators onto the two-particle states. The explicit forms for the two-particle ξ states of spin $S = 0$ and $S = 1$ are

$$\xi^{00} = \frac{1}{\sqrt{2}}\begin{pmatrix} 0 & 1 \\ -1 & 0 \end{pmatrix}, \quad \xi^{11} = \begin{pmatrix} 1 & 0 \\ 0 & 0 \end{pmatrix}, \quad \xi^{10} = \frac{1}{\sqrt{2}}\begin{pmatrix} 0 & 1 \\ 1 & 0 \end{pmatrix}, \quad \xi^{1-1} = \begin{pmatrix} 0 & 0 \\ 0 & 1 \end{pmatrix}. \tag{17.32}$$

One can verify that the spin eigenvalues of these states have been identified properly by making some quick calculations:

$$S^z_{\rm tot}\xi^{SM} = (S^z_1 + S^z_2)\,\xi^{SM} = \frac{1}{2}\sigma^3\xi^{SM} + \xi^{SM}\frac{1}{2}(\sigma^3)^{\rm T} = M\xi^{SM}, \tag{17.33}$$

$$\begin{aligned}(\vec{S}_{\rm tot})^2\xi^{SM} &= \left(\vec{S}_1 + \vec{S}_2\right)^2\xi^{SM} = \left(\frac{3}{4}\right)\xi^{SM} + \frac{1}{2}\sigma^i\,\xi^{SM}(\sigma^i)^{\rm T} + \xi^{SM}\left(\frac{3}{4}\right) \\ &= S(S+1)\,\xi^{SM}.\end{aligned} \tag{17.34}$$

The multiplet structure can be verified by showing

$$\begin{aligned}S^{\pm}_{\rm tot}\xi^{SM} &= \frac{1}{2}\left(\sigma^x_{\rm tot} \pm {\rm i}\sigma^y_{\rm tot}\right)\xi^{SM} = \frac{1}{2}\left(\sigma^x \pm {\rm i}\sigma^y\right)\xi^{SM} + \xi^{SM}\frac{1}{2}\left(\sigma^x \pm {\rm i}\sigma^y\right)^{\rm T} \\ &= \sqrt{(S \mp M)(S \pm M + 1)}\,\xi^{S(M\pm 1)}.\end{aligned} \tag{17.35}$$

These two-particle spin states are orthonormal and complete:

$${\rm Tr}\left[\left(\xi^{SM}\right)^+\xi^{S'M'}\right] = \delta_{SS'}\delta_{MM'}, \qquad \sum_{SM}\xi^{SM}_{ab}\left(\xi^{SM}_{cd}\right)^* = \delta_{ac}\delta_{bd}. \tag{17.36}$$

A more convenient representation of the two-particle spin states can be obtained by action of the 2×2 "charge conjugation" matrix $c = -{\rm i}\sigma_2$ on the indices of the second (positively charged) particle. We define this new representation as

$$\chi^{SM} = \xi^{SM}c. \tag{17.37}$$

The χ^{SM} matrices are just the ones found in Eq. (16.234) for the relativistic two-particle spin states consisting of particle and antiparticle:

$$\chi^{00} = \frac{1}{\sqrt{2}}\mathbb{1}_{2\times 2}, \qquad \chi^{1M} = \frac{1}{\sqrt{2}}\vec{\sigma}\cdot\vec{e}_M, \tag{17.38}$$

where the polarization vectors in the spherical basis are $\vec{e}_M = \frac{\mp 1}{\sqrt{2}}(1, \pm{\rm i}, 0)$, and $\vec{e}_0 = (0, 0, 1)$. The expectation value of a spin operator $A_1 B_2$ takes the convenient form

$$\xi^+ A_1 B_2\,\xi = {\rm Tr}\left[\xi^+ A\xi B^{\rm T}\right] = {\rm Tr}\left[\chi^+ A\chi c^{-1}B^{\rm T}c\right] = {\rm Tr}\left[\chi^+ A\chi\widetilde{B}\right], \tag{17.39}$$

where $\widetilde{B}$ is obtained from B by reversing the order of all σ matrices in B and changing the sign of each. The connection between $\widetilde{B}$ and B follows from the properties of the charge conjugation matrix $c^+ = c^{-1} = c^{\rm T} = -c$ and the relationship

$$c^{-1}(\sigma^i)^{\rm T}c = -\sigma^i. \tag{17.40}$$

At this stage we promote our result to D dimensions and take (17.39) to give the definition of an NRQED two-particle spin expectation value for general D.

We are now in a position to exploit the matching conditions to obtain coefficients d_S and d_V. We start with the NRQED calculation. For the scalar expectation value we find

$$\left(\xi_{ac}^{SM}\right)^* \delta_{ab}\delta_{cd}\,\xi_{bd}^{SM} = \mathrm{Tr}\left[\left(\xi^{SM}\right)^+ \xi^{SM}\right] = 1, \tag{17.41}$$

while the vector expectation value is

$$\xi_{ac}^{*SM}(\sigma^k)_{ab}(\sigma^k)_{cd}\,\xi_{bd}^{SM} = \mathrm{tr}\left[\left(\chi^{SM}\right)^+ \sigma^k\chi^{SM}(-\sigma^k)\right] = \begin{cases} -D & (S=0) \\ D-2 & (S=1) \end{cases}. \tag{17.42}$$

We have used (17.39), $\widetilde{\sigma}^k = -\sigma^k$ and the D-dimensional sigma matrix relations $\{\sigma^a, \sigma^b\} = 2\delta^{ab}$, and $\sigma^k\sigma^k = D$, as well as $\sigma^k\sigma^n\sigma^k = -(D-2)\sigma^n$ with an understood D-dimensional summation over k. The D-dimensional trace formulas are $\mathrm{tr}[\mathbb{1}_{2\times2}] = 2$, $\mathrm{tr}[\sigma^a] = 0$, and $\mathrm{tr}[\sigma^a\sigma^b] = 2\delta^{ab}$. For the complete NRQED expectation value, we find

$$\begin{aligned}\mathcal{M}_{\mathrm{NRQED}}^{SM} &= \left(\xi_{ac}^{SM}\right)^* \left\{\frac{d_S}{m_1m_2}\delta_{ab}\,\delta_{cd} + \frac{d_V}{m_1m_2}\vec{\sigma}_{ab}\cdot\vec{\sigma}_{cd}\right\}\xi_{bd}^{SM} \\ &= \frac{d_S}{m_1m_2} + \frac{d_V}{m_1m_2}\begin{cases} -D & \text{if } S=0 \\ D-2 & \text{if } S=1 \end{cases},\end{aligned} \tag{17.43}$$

which implies

$$d_S = \frac{m_1m_2}{2(D-1)}\left\{(D-2)\mathcal{M}_{\mathrm{NRQED}}^{S=0} + D\mathcal{M}_{\mathrm{NRQED}}^{S=1}\right\}, \tag{17.44a}$$

$$d_V = \frac{m_1m_2}{2(D-1)}\left\{-\mathcal{M}_{\mathrm{NRQED}}^{S=0} + \mathcal{M}_{\mathrm{NRQED}}^{S=1}\right\}. \tag{17.44b}$$

On the QED side, the graphs of Fig. 17.2(b) give the scattering amplitude

$$\begin{aligned}\mathrm{i}\mathcal{M}_{\mathrm{QED}} = (-1)\int\frac{\mathrm{d}^dk}{(2\pi)^d}\,\overline{u}_{1f}\,(-\mathrm{i}q_1\gamma^\mu)\,\frac{\mathrm{i}}{m_1\not{n}+\not{k}-m_1}\,(-\mathrm{i}q_1\gamma^\nu)\,u_{1i}\left(\frac{-\mathrm{i}}{k^2}\right)^2 \\ \times\left\{\overline{v}_{2\mathrm{i}}\,(-\mathrm{i}q_2\gamma_\nu)\,\frac{\mathrm{i}}{-m_2\not{n}+\not{k}-m_2}\,(-\mathrm{i}q_2\gamma_\mu)\,v_{2f}\right. \\ \left.+\,\overline{v}_{2\mathrm{i}}\,(-\mathrm{i}q_2\gamma_\mu)\,\frac{\mathrm{i}}{-m_2\not{n}-\not{k}-m_2}\,(-\mathrm{i}q_2\gamma_\nu)\,v_{2f}\right\},\end{aligned} \tag{17.45}$$

which contains a Fermi antisymmetry factor (-1). The imaginary infinitesimals in propagator denominators required to enforce Feynman boundary conditions are implicit. The amplitude can be written as

$$\begin{aligned}\mathcal{M}_{\mathrm{QED}} = \mathrm{i}(q_1q_2)^2\int\frac{\mathrm{d}^dk}{(2\pi)^d}\left[\left(k^2+2k\cdot n\,m_1\right)\left(k^2\right)^2\right]^{-1}\mathrm{Tr}\Bigg[\gamma^\mu\left(m_1\not{n}+\not{k}+m_1\right)\gamma^\nu \\ \times\left(u_{1\mathrm{i}}\overline{v}_{2\mathrm{i}}\right)\left\{\frac{\gamma_\nu\left(-m_2\not{n}+\not{k}+m_2\right)\gamma_\mu}{k^2-2k\cdot n\,m_2} + \frac{\gamma_\mu\left(-m_2\not{n}-\not{k}+m_2\right)\gamma_\nu}{k^2+2k\cdot n\,m_2}\right\}\left(v_{2f}\overline{u}_{1f}\right)\Bigg].\end{aligned} \tag{17.46}$$

In this case, we have $q_1 q_2 = +4\pi Z\alpha\overline{\mu}^{2\epsilon}$ [see Eq. (16.145)]. We have converted the product of two spinor bilinears into a single trace. Also, we make use of (16.233) written as

$$(u_1\overline{v}_2)^{SM} = \begin{pmatrix} 0 & \chi^{SM} \\ 0 & 0 \end{pmatrix} = \frac{1}{\sqrt{2}}\begin{Bmatrix} \gamma_5 \\ -\not{e}_M \end{Bmatrix}\frac{1}{2}\left(-\not{n}+1\right) = \frac{1}{\sqrt{2}}\,\frac{1}{2}\left(\not{n}+1\right)\begin{Bmatrix} \gamma_5 \\ -\not{e}_M \end{Bmatrix} \qquad (17.47)$$

for $S = 0$ (above) and $S = 1$ (below), where $\gamma_5 = \begin{pmatrix} 0 & \mathbb{1}_{2\times 2} \\ \mathbb{1}_{2\times 2} & 0 \end{pmatrix}$. The specific definition of γ_5 used here satisfies $\gamma_5^2 = 1$ and $\{\gamma_5, \gamma^\mu\} = 0$. Other definitions of γ_5 are more useful for theories involving chiral symmetries, but for our purposes, the present definition is convenient (see also Sec. 10.7). One also has $(u_1\overline{v}_2)^+ = \gamma^0 v_2 \overline{u}_1 \gamma^0$, so that

$$(v_2\overline{u}_1)^{SM} = \begin{pmatrix} 0 & 0 \\ -\chi^{+SM} & 0 \end{pmatrix} = \frac{1}{\sqrt{2}}\begin{Bmatrix} -\gamma_5 \\ -\not{e}_M^* \end{Bmatrix}\frac{1}{2}\left(\not{n}+1\right) = \frac{1}{\sqrt{2}}\,\frac{1}{2}\left(-\not{n}+1\right)\begin{Bmatrix} -\gamma_5 \\ -\not{e}_M^* \end{Bmatrix}. \qquad (17.48)$$

We use the fact that e_M is a spatial unit vector: $e_M^* \cdot e_M = -1$ (with no sum on M), and $e_M^* \cdot n = n \cdot e_M = 0$. The two diagrams of Fig. 17.2(b) will be evaluated in turn. The first (ladder) graph contributes

$$\mathcal{M}_{\mathrm{L;QED}} = \mathrm{i}(4\pi Z\alpha\overline{\mu}^{2\epsilon})^2 \int \frac{\mathrm{d}^d k}{(2\pi)^d} N_{\mathrm{L}} \left[(k^2 + 2\,k\cdot n\, m_1)(k^2 - 2\,k\cdot n\, m_2)(k^2)^2\right]^{-1}, \qquad (17.49)$$

where

$$\begin{aligned} N_{\mathrm{L}} &= \frac{1}{8}\mathrm{Tr}\left[\gamma^\mu(m_1\not{n} + \not{k} + m_1)\gamma^\nu(\not{n}+1)\begin{Bmatrix} -\gamma_\nu(m_2\not{n} - \not{k} + m_2)\gamma_\mu \\ \not{e}_M\gamma_\nu(-m_2\not{n} + \not{k} + m_2)\gamma_\mu\not{e}_M^* \end{Bmatrix}(\not{n}+1)\right] \\ &= \begin{Bmatrix} 2(d-1)k^2 + (d-2)^2(k\cdot n)^2 + 2(m_1 - m_2)k\cdot n - 4m_1 m_2 \\ (d-3)[2(e_m^*\cdot k)(k\cdot e_m) + k^2] + (d-4)^2(k\cdot n)^2 + 2(m_1 - m_2)k\cdot n - 4m_1 m_2 \end{Bmatrix}, \end{aligned} \qquad (17.50)$$

valid for $\begin{Bmatrix} S = 0 \\ S = 1 \end{Bmatrix}$. Similarly, the crossed contribution is

$$\mathcal{M}_{\mathrm{X;QED}} = \mathrm{i}(4\pi Z\alpha\overline{\mu}^{2\epsilon})^2 \int \frac{\mathrm{d}^d k}{(2\pi)^d} N_{\mathrm{X}} \left[(k^2 + 2k\cdot n m_1)(k^2 + 2k\cdot n m_2)(k^2)^2\right]^{-1} \qquad (17.51)$$

with

$$\begin{aligned} N_{\mathrm{X}} &= \frac{1}{8}\mathrm{Tr}\left[\gamma^\mu(m_1\not{n} + \not{k} + m_1)\gamma^\nu(\not{n}+1)\begin{Bmatrix} -\gamma_\mu(m_2\not{n} + \not{k} + m_2)\gamma_\nu \\ \not{e}_M\gamma_\mu(-m_2\not{n} - \not{k} + m_2)\gamma_\nu\not{e}_M^* \end{Bmatrix}(\not{n}+1)\right] \\ &= \begin{Bmatrix} 2(d-3)k^2 + (d-2)(d-4)(k\cdot n)^2 - 2(m_1 + m_2)k\cdot n - 4m_1 m_2 \\ 2(d-5)k^2 + R(d,k,n) + 4(d-3)(e_M^*\cdot k)(k\cdot e_M) - 2(m_1 + m_2)k\cdot n - 4m_1 m_2 \end{Bmatrix} \end{aligned} \qquad (17.52)$$

where $R(d,k,n) = (d^2 - 10d + 20)(k \cdot n)^2$. The d-dimensional momentum space integrals can be done using Feynman parameters, to give

$$\mathcal{M}_{\rm QED} = \frac{(Z\alpha)^2\overline{\mu}^{2\epsilon}}{m_1 m_2} \times \begin{Bmatrix} -\dfrac{1}{\epsilon}+\dfrac{1}{3}+\dfrac{m_1^2\ln(m_2^2/\mu^2)-m_2^2\ln(m_1^2/\mu^2)-3m_1m_2\ln(m_1^2/m_2^2)}{m_1^2-m_2^2}+\mathcal{O}(\epsilon) \\ -\dfrac{1}{\epsilon}+\dfrac{1}{3}+\dfrac{m_1\ln(m_2^2/\mu^2)+m_2\ln(m_1^2/\mu^2)}{m_1+m_2}+\mathcal{O}(\epsilon) \end{Bmatrix}, \tag{17.53}$$

for $\begin{Bmatrix} S=0 \\ S=1 \end{Bmatrix}$ as the total of the ladder plus crossed contributions. These results are consistent with the hard region results of the previous chapter, in Eq. (16.251). We note that the energy correction is related to the expectation value of the amplitude by $\Delta E = -|\psi(0)|^2\mathcal{M}$. Finally, we can obtain $d_{\rm S}$ and $d_{\rm V}$ from the QED amplitudes using (17.44). We find that [see Eq. (16.145) for the relation of μ and $\overline{\mu}$]

$$d_{\rm S} = (Z\alpha)^2\overline{\mu}^{2\epsilon}\left\{-\frac{1}{\epsilon}+\frac{1}{3}+\frac{m_1^2\ln\left(\dfrac{m_2^2}{\mu^2}\right)-m_2^2\ln\left(\dfrac{m_1^2}{\mu^2}\right)}{m_1^2-m_2^2}\right\},$$

$$d_{\rm V} = (Z\alpha)^2\overline{\mu}^{2\epsilon}\frac{m_1m_2\ln\left(\dfrac{m_1^2}{m_2^2}\right)}{m_1^2-m_2^2}. \tag{17.54}$$

Our results for $d_{\rm S}$ and $d_{\rm V}$ are consistent with those of Ref. [592]. The difference in the sign of $d_{\rm S}$ is due to a difference in how the positively charged particle fields are defined.

17.2.3 *Bethe–Salpeter Equation of NRQED*

As in standard QED, the Bethe–Salpeter equation provides a convenient means for working out the energy levels of two-body bound systems in the context of NRQED. In Chap. 16, we have discussed, for the ordinary Bethe–Salpeter equation, the 2-to-2 Green functions, the two-body irreducible kernel, the inhomogeneous BSEQ, the presence of bound-state poles, the homogeneous BSEQ, the normalization conditions, a solvable reference problem, and a perturbation theory built upon that solution. These are present in analogous forms for NRQED. Here, we describe the NRQED approach.

The 2-to-2 Green function of NRQED is defined as for QED:

$$G(x_1,x_2;y_1,y_2) = \langle 0|\mathrm{T}\psi_1(x_1)\psi_2(x_2)\overline{\psi}_1(y_1)\overline{\psi}_2(y_2)|0\rangle, \tag{17.55}$$

except that there is no need to renormalize G as hard-region effects do not exist in NRQED by construction, and the rest energies have been extracted from the fields:

$$\widetilde{\psi}_i(x) = e^{-im_it}\psi_i(x). \tag{17.56}$$

Here $\widetilde{\psi}_i(x)$ is a Pauli fermion field with the standard time dependence, while $\psi_i(x)$ is a field with the dominant time dependence factored out. The practical consequence of this redefinition is that the Feynman rules for the NRQED propagator of a fermion, say fermion 1, carrying energy $\xi_1 P_0 + p_0 = m_1 + \xi_1 E + p_0$, and spatial momentum $\vec{p}$ (in the CM), is $\mathrm{i}/(\xi_1 E + p_0 - \vec{p}^{\,2}/(2m_1) + \mathrm{i}\epsilon)$. The transformation to CM and relative coordinates goes through as before, as does the proof that the 2-to-2 Green function contains poles at the positions of the bound-state energies. The bound state pole in the 2-to-2 Green function (in the CM) takes a form analogous to Eq. (16.15),

$$G(E;p,q) = \mathrm{i}\sum_{n,k} \frac{\Psi_{nk}(p)\overline{\Psi}_{nk}(q)}{E - E_n + \mathrm{i}\epsilon} + \cdots. \tag{17.57}$$

The inhomogeneous BSEQ of NRQED is

$$G = S' + S'K'G. \tag{17.58}$$

The graphical definitions of G, S', and K' are shown in Fig. 17.3. As was the case for QED, each of G, S', K' has an infinite number of terms. As in the previous chapter, we choose to rewrite the BSEQ in terms of a simpler non-interacting 2-to-2 reference propagator S, where for S we use the uncorrected product

$$S(E;p) = \mathrm{i}S_1(E;p) \times \mathrm{i}S_2(E;-p) = \frac{\mathrm{i}}{\xi_1 E + p_0 - \frac{\vec{p}^{\,2}}{2m_1} + \mathrm{i}\epsilon} \times \frac{\mathrm{i}}{\xi_2 E - p_0 - \frac{\vec{p}^{\,2}}{2m_2} + \mathrm{i}\epsilon}. \tag{17.59}$$

This reference propagator has the crucial property that

$$\int \frac{\mathrm{d}p_0}{2\pi}\, S(E;p) = \int \frac{\mathrm{d}p_0}{2\pi} \frac{\mathrm{i}}{\xi_1 E + p_0 - \frac{\vec{p}^{\,2}}{2m_1} + \mathrm{i}\epsilon} \times \frac{\mathrm{i}}{\xi_2 E - p_0 - \frac{\vec{p}^{\,2}}{2m_2} + \mathrm{i}\epsilon} = \mathrm{i}\, s(E;\vec{p}\,), \tag{17.60}$$

where $s(E;\vec{p}\,)$ is the nonrelativistic propagator

$$s(E;\vec{p}\,) = \frac{1}{E - \frac{\vec{p}^{\,2}}{2\mu} + \mathrm{i}\epsilon}. \tag{17.61}$$

So, the integral over the relative energy of the reference noninteracting 2-to-2 propagator is just equal to the imaginary unit times the standard free nonrelativistic Green's function for an effective particle with reduced mass μ. We simplify the BSEQ by replacing the fully-corrected propagator S' by the reference propagator S so that all self-energy corrections are moved to the modified kernel K,

$$G = S + S\,K\,G\,, \tag{17.62}$$

where now

$$K = K' + S^{-1} - (S')^{-1}. \tag{17.63}$$

The reference propagator S and modified kernel K' are depicted in parts (d) and (e) of Fig. 17.3.

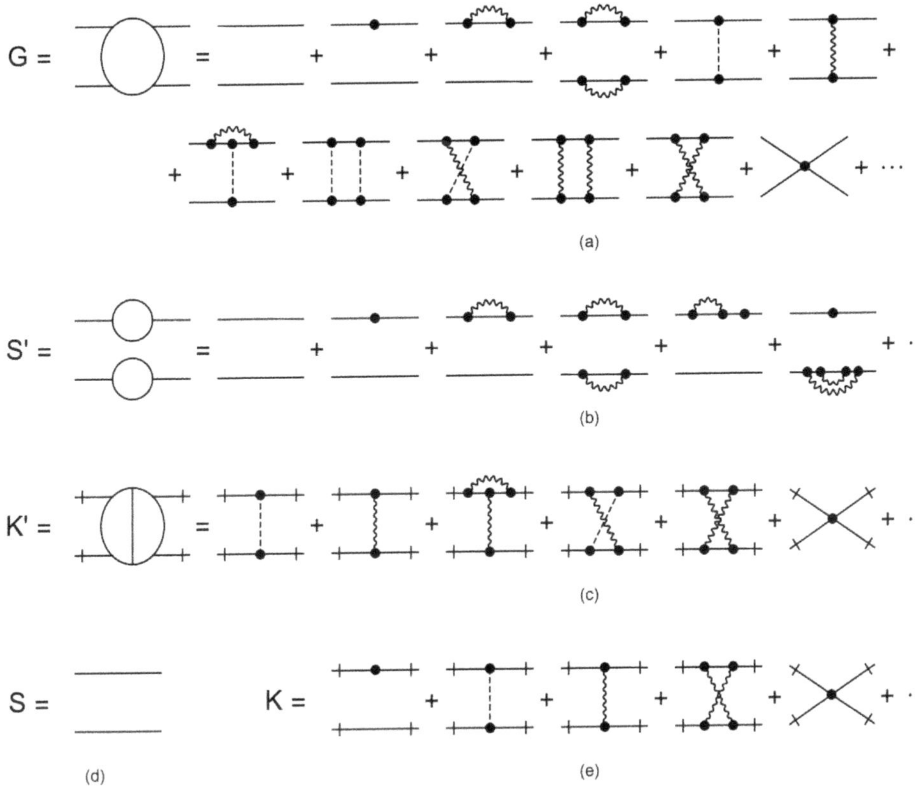

Fig. 17.3 Green functions and kernel entering into the NRQED Bethe–Salpeter equation. Part (a) represents the full 2-to-2 Green function G of NRQED. The electron line is shown at the top and the proton at the bottom. Dashed and wiggly lines represent Coulomb and transverse photons, respectively. The various vertices of NRQED are represented collectively by the dots. Part (b) shows S', the product of a fully corrected electron propagator with a fully corrected proton propagator. Part (c) shows K', the two-particle irreducible kernel of the NRQED BSEQ. As before, the external fermion lines are shown with vertical hashes to indicate that they are amputated. These lines are shown only for clarity. Part (d) shows the free 2-to-2 propagator S, and part (e) shows the modified kernel K for use in a BSEQ with propagator S. Only a suggestive subset of diagrams is shown.

Our next goal is to find an approximation to the NRQED BSEQ that can be solved exactly for use as the basis of a perturbative expansion. The NRQED reference equation is analogous to the one obtained in full QED in the previous chapter, but significantly simpler. As a reference kernel we can make the simplest possible choice, namely, an unmodified Coulomb interaction between the two charged particles. That is, we choose the reference kernel as follows,

$$K_0(E;p,q) = (-\mathrm{i}q_1)(-\mathrm{i}q_2)\frac{\mathrm{i}}{(\vec{p}-\vec{q}\,)^2} = -\mathrm{i}V(\vec{p}-\vec{q}\,)\,. \tag{17.64}$$

Here, $V(\vec{k}\,) = -4\pi Z\alpha\overline{\mu}^{2\epsilon}/\vec{k}^{\,2}$ is the usual momentum-space Coulomb potential (in

D dimensions) using Eq. (16.145). As we will show, this reference BSEQ

$$G_0 = S + SK_0G_0 \tag{17.65}$$

can be solved, at least for $D = 3$. As in the previous chapter, we build a solution for the reference 2-to-2 propagator using the series

$$\begin{aligned} G_0 &= (1 - SK_0)^{-1} S = S + SK_0S + SK_0SK_0S + \cdots \\ &= S + SK_0S + S(K_0SK_0 + K_0SK_0SK_0 + \cdots) S = S + SK_0S + SHS, \end{aligned} \tag{17.66}$$

where $H = K_0SK_0 + K_0SK_0SK_0 + \cdots$. The terms in H implicitly involve integrals over the full d-dimensional space-time, but because of (17.60) and the fact that the reference potential is energy-independent, they can all be reduced to purely spatial integrals of $D = d - 1$ dimensions. The first term of H is

$$\begin{aligned} [K_0\, S\, K_0](E; p, q) &= \int \frac{\mathrm{d}^d k}{(2\pi)^d} K_0(E; p, k) S(E; k) K_0(E; k, q) \\ &= \int \frac{\mathrm{d}^D k}{(2\pi)^D} (-\mathrm{i}) V(\vec{p} - \vec{k}\,) s(E; \vec{k}\,) V(\vec{k} - \vec{q}\,) = -\mathrm{i}[V\, s\, V](E; \vec{p}, \vec{q}\,). \end{aligned} \tag{17.67}$$

Likewise, the second term of H reduces as

$$[K_0SK_0SK_0]\,(E; p, q) = -\mathrm{i}\,[VsVsV]\,(E; \vec{p}, \vec{q}\,), \tag{17.68}$$

and so on. Comparison with (16.79) and (16.77) shows that $H(E)$ is essentially the corresponding sum that enters the nonrelativistic Schrödinger-Coulomb Green function. One has $H(E) = -\mathrm{i}\,h(E)$ where

$$g(E) = s(E) + s(E)Vs(E) + s(E)h(E)s(E) \tag{17.69}$$

is the Schrödinger equation of standard quantum mechanics. Given the known solution for $g(E)$, we also have the solution to our NRQED reference problem:

$$G_0 = S + SK_0S + S(-\mathrm{i}h)S. \tag{17.70}$$

The bound-state poles of g have the form

$$g(E; \vec{p}, \vec{q}\,) \to \frac{\sum \psi_n(\vec{p}\,)\, \psi_n^+(\vec{q}\,)}{E - E_n^0}, \tag{17.71}$$

where the sum is over the degenerate quantum numbers labeling orbital angular momentum and spin. There are no poles in s or sVs, so Eq. (17.69) implies that the poles of h are

$$h(E; \vec{p}, \vec{q}\,) \to \sum_n \frac{s(E_n^0; \vec{p}\,)^{-1} \psi_n(\vec{p}\,) \psi_n^+(\vec{q}\,) s(E_n^0; \vec{q}\,)^{-1}}{E - E_n^0}. \tag{17.72}$$

These same bound-state poles are present in the reference Green's function G_0 as seen by (17.70), so that

$$\begin{aligned} G_0(E; p, q) &\to -\mathrm{i} \sum_n \frac{S(E_n^0; p)\, s(E_n^0; \vec{p}\,)^{-1}\, \psi_n(\vec{p}\,)\, \psi_n^+(\vec{q}\,)\, s(E_n^0; \vec{q}\,)^{-1}\, S(E_n^0; q)}{E - E_n^0} \\ &\equiv \mathrm{i} \sum_n \frac{\Psi_n^0(p) \overline{\Psi}_n^0(q)}{E - E_n^0}. \end{aligned} \tag{17.73}$$

This allows us to reduce the problem of the calculation of the two-particle wave function to the calculation of a reference wave function $\psi_n(\vec{p}\,)$, which is a function of the relative momentum $\vec{p}$ only,

$$\Psi^0_n(p) = -\mathrm{i}\, S(E^0_n;p)\, s(E^0_n;\vec{p}\,)^{-1}\, \psi_n(\vec{p}\,), \tag{17.74a}$$

$$\overline{\Psi}^0_n(p) = -\mathrm{i}\, \psi^+_n(\vec{p}\,)\, s(E^0_n;\vec{p}\,)^{-1}\, S(E^0_n;p)\,. \tag{17.74b}$$

Here, $\Psi^0_n(p)$ and $\overline{\Psi}^0_n(p)$ correspond to the nonrelativistic limit of the wave functions which had been calculated in Eq. (16.101) for the relativistic Bethe–Salpeter formalism. Note, however, that $\overline{\Psi}^0_n(p)$ in Eq. (17.74) does not contain a γ^0 matrix [cf. Eq. (7.34)]. The γ^0 matrix enters the relativistic (four-component) bispinor formalism, but is absent in the nonrelativistic (two-component) spinor formalism that is used in NRQED.

The reference wave functions contain the nonrelativistic wave functions in a straightforward (multiplicative) way, and satisfy the integral identities

$$\begin{aligned}\int \frac{\mathrm{d}p_0}{2\pi} \Psi^0_n(p) &= -\mathrm{i} \int \frac{\mathrm{d}p_0}{2\pi} S(E^0_n;p) s(E^0_n;\vec{p}\,)^{-1} \psi_n(\vec{p}\,) \\ &= -\mathrm{i}\left[\mathrm{i} s(E^0_n;\vec{p}\,)\right] s(E^0_n;\vec{p}\,)^{-1} \psi_n(\vec{p}\,) = \psi_n(\vec{p}\,)\,,\end{aligned} \tag{17.75}$$

and

$$\int \frac{\mathrm{d}p_0}{2\pi}\, \overline{\Psi}^0_n(p) = \psi^+_n(\vec{p}\,)\,. \tag{17.76}$$

The reader should recognize how much more naturally the nonrelativistic Coulomb problem fits into the formalism of the quantum field theory NRQED compared to the contortions required to adapt the Coulomb problem to the relativistic theory of QED. Additional aspects of this simplicity are illustrated below.

The bound-state equation satisfied by the reference wave functions can be obtained from the pole part of

$$1 = G_0^{-1} G_0, \tag{17.77}$$

which gives the reference homogeneous Bethe–Salpeter equations:

$$G_0^{-1}(E^0_n)\, \Psi^0_n = \overline{\Psi}^0_n\, G_0^{-1}(E^0_n) = 0\,. \tag{17.78}$$

The reference Green's function has the value $G_0 = (1 - S K_0)^{-1}\, S$, so that

$$G_0^{-1} = S^{-1}\,(1 - S K_0) = S^{-1} - K_0. \tag{17.79}$$

The reference Bethe–Salpeter equation $(S^{-1} - K_0)\Psi^0_n = 0$, multiplied by S, takes the form

$$\Psi^0_n(p) = S(E^0_n;p) \int \frac{\mathrm{d}^d q}{(2\pi)^d} K_0\left(E^0_n;p,q\right) \Psi^0_n(q)\,. \tag{17.80}$$

We use the fact that the reference kernel is energy independent, so we can perform the q^0 integral on the wave function with Eq. (17.75), and integrate both sides over relative energy p^0, to find

$$\psi_n(\vec{p}\,) = \mathrm{i} s(E^0_n;\vec{p}\,) \int \frac{\mathrm{d}^D q}{(2\pi)^D} (-\mathrm{i})\, V(\vec{p} - \vec{q}\,)\, \psi_n(\vec{q}\,)\,, \tag{17.81}$$

or

$$\left(E_n^0 - \frac{\vec{p}^{\,2}}{2\mu}\right)\psi_n(\vec{p}) = s(E_n^0;\vec{p})^{-1}\psi_n(\vec{p}) = \int \frac{\mathrm{d}^D q}{(2\pi)^D} V(\vec{p}-\vec{q})\psi_n(\vec{q}). \tag{17.82}$$

So, the reference BSEQ in NRQED reduces exactly to the usual Schrödinger-Coulomb equation.

Finally, we consider the normalization condition for the reference wave functions. It can be obtained by expanding the identity

$$G_0 = G_0\, G_0^{-1}\, G_0 \tag{17.83}$$

about the bound-state pole. Near the pole, one has

$$\mathrm{i}\frac{\Psi_n^0\overline{\Psi}_n^0}{E-E_n^0} + \cdots = \left(\mathrm{i}\frac{\Psi_n^0\overline{\Psi}_n^0}{E-E_n^0} + \cdots\right) \times \left(G_0^{-1}(E_n^0) + (E-E_n^0)\left.\frac{\mathrm{d}G_0^{-1}}{\mathrm{d}E}\right|_{E=E_n^0} + \cdots\right)\left(\mathrm{i}\frac{\Psi_n^0\overline{\Psi}_n^0}{E-E_n^0} + \cdots\right). \tag{17.84}$$

The term in the expansion with a single pole gives the normalization condition:

$$1 = \mathrm{i}\overline{\Psi}_n^0 \left.\frac{\mathrm{d}G_0^{-1}}{dE}\right|_{E=E_n^0} \Psi_n^0 = \mathrm{i}\int \frac{\mathrm{d}^d p}{(2\pi)^d}\frac{\mathrm{d}^d q}{(2\pi)^d}\overline{\Psi}_n^0(p) \left.\frac{\mathrm{d}G_0^{-1}(E;p,q)}{\mathrm{d}E}\right|_{E=E_n^0} \Psi_n^0(q). \tag{17.85}$$

Since $G_0^{-1} = S^{-1} - K_0$ and K_0 is independent of E, only the S^{-1} term needs to be kept in the energy derivative. The S^{-1} term satisfies

$$S^{-1}(E;p,q) = S(E;p)^{-1}(2\pi)^d\delta^{(d)}(p-q)\,, \tag{17.86}$$

$$S^{-1}(E;p) = \left[\mathrm{i}S_1(E;p)\,\mathrm{i}S_2(E;-p)\right]^{-1} = -\left(\xi_1 E + p^0 - \frac{\vec{p}^{\,2}}{2m_1} + \mathrm{i}\epsilon\right)\left(\xi_2 E - p^0 - \frac{\vec{p}^{\,2}}{2m_2} + \mathrm{i}\epsilon\right). \tag{17.87}$$

So, one has

$$\left.\frac{\mathrm{d}G_0^{-1}(E;p,q)}{\mathrm{d}E}\right|_{E=E_n^0} = -(2\pi)^d\,\delta^{(d)}(p-q)\left(\xi_1 S_2(E_n^0;-p)^{-1} + S_1(E_n^0;p)^{-1}\xi_2\right), \tag{17.88}$$

and the normalization condition becomes

$$1 = \mathrm{i}\int \frac{\mathrm{d}^d p}{(2\pi)^d}\frac{\mathrm{d}^d q}{(2\pi)^d}\overline{\Psi}_n^0(p)\left.\frac{\mathrm{d}G_0^{-1}(E;p,q)}{\mathrm{d}E}\right|_{E=E_n^0}\Psi_n^0(q) = -\mathrm{i}\int \frac{\mathrm{d}^d p}{(2\pi)^d}\overline{\Psi}_n^0(p)\left(\xi_1 S_2(E_n^0;-p)^{-1} + S_1(E_n^0;p)^{-1}\xi_2\right)\Psi_n^0(p)\,. \tag{17.89}$$

With the help of Eq. (17.74), we can write this as

$$\begin{aligned} 1 &= \mathrm{i}\int \frac{\mathrm{d}^d p}{(2\pi)^d}\psi_n^+(\vec{p})s(E_n^0;\vec{p})^{-1}\left(\xi_1 S_1(E_n^0;p)^2 S_2(E_n^0;-p) + \xi_2 S_1(E_n^0;p)S_2(E_n^0;p)^2\right) \\ &\quad\times s(E_n^0;\vec{p})^{-1}\psi_n(\vec{p}) \\ &= \mathrm{i}\int \frac{\mathrm{d}^D p}{(2\pi)^D}\psi_n^+(\vec{p})s(E_n^0;\vec{p})^{-1}\left(-\mathrm{i}\xi_1 s(E_n^0;\vec{p})^2 - \mathrm{i}\xi_2 s(E_n^0;\vec{p})^2\right)s(E_n^0;\vec{p})^{-1}\psi_n(\vec{p}) \\ &= \int \frac{\mathrm{d}^D p}{(2\pi)^D}\psi_n^+(\vec{p})\psi_n(\vec{p}), \end{aligned} \tag{17.90}$$

which is the usual normalization. This agreement provides a check of the procedure. In the reduction shown above we have used the integral formula

$$\int \frac{\mathrm{d}p_0}{2\pi}[S_1(E;p)]^2\, S_2(E,-p) = \int \frac{\mathrm{d}p_0}{2\pi} S_1(E;p)\,[S_2(E;-p)]^2 = -\mathrm{i}\, s(E;\vec{p})^2. \quad (17.91)$$

17.3 Applications of NRQED

17.3.1 *Energy Levels at $\mathcal{O}(Z\alpha)^4$ (NRQED Approach)*

In this section, we consider energy level corrections to two-fermion bound states such as hydrogen or muonium using NRQED. In Chap. 16, we obtained the same result using a QED-based formalism in Eq. (16.144), but seeing the NRQED derivation allows for a direct comparison of the two approaches. All diagrams that contribute to the energy levels of nonrelativistic two-fermion Coulombic bound states through $\mathcal{O}(\alpha^5)$ are shown in Fig. 17.4. Only a subset of these enter at $\mathcal{O}(\alpha^4)$. These are, specifically, the diagrams in Figs. 17.4(a)–(f). (There is a contribution at $\mathcal{O}(\alpha^4)$ to the energy levels of fermion-antifermion bound states such as positronium from the contact term Fig. 17.4(f) accounting for virtual annihilation into a single photon, but we will not consider this contribution here.)

We aim to discuss the $\mathcal{O}(\alpha^4)$ energy contributions one by one, first noting that the fermion momenta are all from the *potential* region (in the sense of Table 16.1), as enforced by the wave function. (Graphs with virtual fermions that are not part of the wave function do not contribute at this order.) Also, the exchanged photons are exclusively potential at this order. We will explore the contributions of non-potential fermions and photons when we move on to $\mathcal{O}(\alpha^5)$.

The relativistic kinetic energy correction of Fig. 17.4(a) is the only one that requires a bit of special handling as the corresponding kernel is produced by the $S^{-1} - (S')^{-1}$ term of (17.63) instead of coming directly from an interaction diagram in K'. In order to sort this out, we consider the corrected single-particle propagator $S_i(E;p)$,

$$S_i(E;p) = \frac{1}{\xi_i E + p_0 - \frac{\vec{p}^{\,2}}{2m_i} - \Sigma_i(E;p) + \mathrm{i}\epsilon}, \quad (17.92)$$

where $\Sigma_i(E;p) = -\frac{\vec{p}^{\,4}}{8m_i^3} + \cdots$ is the NRQED self-energy of particle i. The fully corrected non-interacting 2-to-2 propagator has the form

$$S'(E;p,q) = \frac{\mathrm{i}^2\,(2\pi)^4\delta^{(4)}(p-q)}{\left[\xi_1 E + p_0 - \frac{\vec{p}^{\,2}}{2m_1} - \Sigma_1(E;p) + \mathrm{i}\epsilon\right]\left[\xi_2 E - p_0 - \frac{\vec{p}^{\,2}}{2m_2} - \Sigma_2(E;-p) + \mathrm{i}\epsilon\right]}, \quad (17.93)$$

so that the contribution to K from the $\mathcal{O}(\vec{p}^{\,4})$ part (fourth-order correction to the

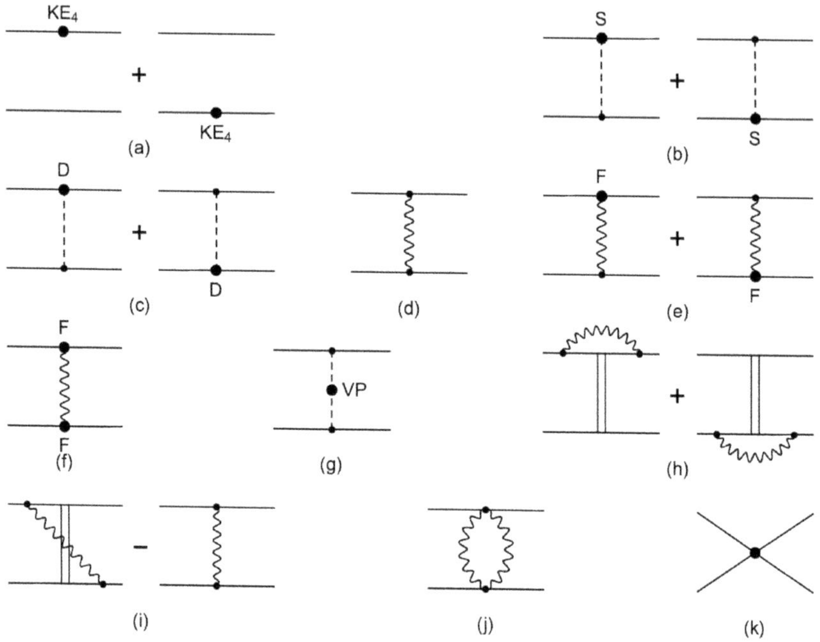

Fig. 17.4 Diagrams contributing to NRQED energy levels at orders α^4 [diagrams (a)–(f)] and α^5 [diagrams (b)–(k)]. (Diagram k also contributes at $\mathcal{O}(\alpha^4)$ for particle-antiparticle bound states.) The various contributions are (a) the first relativistic correction to the kinetic energy, (b) spin-orbit, (c) Darwin, (d) magnetic interaction due to transverse photon exchange, (e) a spin-orbit contribution from the Fermi interaction, (f) Fermi-Fermi tensor interaction, (g) vacuum polarization, (h) transverse photon spanning zero, one, two, etc. Coulomb photons (which are represented collectively as a vertical double line), (i) transverse photon exchange crossing one, two, etc. Coulomb photons, (j) seagull interaction, and (k) four-fermion contact interaction.

kinetic energy, appropriately named K_4) is

$$[S^{-1} - (S')^{-1}](E;p,q) \to K_{K_4} \equiv \left\{ \frac{\vec{p}^{\,4}}{8m_1^3} S_2(E;-p)^{-1} + S_1(E;p)^{-1} \frac{\vec{p}^{\,4}}{8m_2^3} \right\} (2\pi)^4 \, \delta^{(4)}(p-q) \,. \tag{17.94}$$

For the corresponding energy shifts, we use

$$\Delta E = \langle \mathrm{i}\delta K \rangle = \mathrm{i} \int \frac{\mathrm{d}^4 p}{(2\pi)^4} \frac{\mathrm{d}^4 q}{(2\pi)^4} \, \overline{\Psi}^0(p) \, \delta K(E;p,q) \, \Psi^0(q) \,, \tag{17.95}$$

with wave functions given by Eq. (17.74) as

$$\Psi^0(p) = \mathrm{i} S_1(E^0;p) S_2(E^0;-p) s(E^0;\vec{p})^{-1} \psi(\vec{p}) \,, \tag{17.96a}$$

$$\overline{\Psi}^0(p) = \mathrm{i} \psi^+(\vec{p}) s(E^0;\vec{p})^{-1} S_1(E^0;p) S_2(E^0;-p) \,, \tag{17.96b}$$

to find

$$\begin{aligned}\Delta E_{K_4} &= \mathrm{i}\int \frac{\mathrm{d}^4p}{(2\pi)^4}\frac{\mathrm{d}^4q}{(2\pi)^4}\overline{\Psi}^0(p)K_{K_4}(E^0;p,q)\Psi^0(q) \\ &= \mathrm{i}\int \frac{\mathrm{d}^4p}{(2\pi)^4}\mathrm{i}\psi^+(\vec{p}\,)s(E^0;\vec{p}\,)^{-1}\left\{\frac{\vec{p}^{\,4}}{8m_1^3}S_1(E^0;p)^2S_2(E^0;-p)\right. \\ &\quad \left. + S_1(E^0;p)S_2(E^0;-p)^2\frac{\vec{p}^{\,4}}{8m_2^3}\right\}\mathrm{i}s(E^0;\vec{p}\,)^{-1}\psi(\vec{p}\,) \\ &= \int \frac{\mathrm{d}^3p}{(2\pi)^3}\psi^+(\vec{p}\,)\left\{-\frac{\vec{p}^{\,4}}{8m_1^3} - \frac{\vec{p}^{\,4}}{8m_2^3}\right\}\psi(\vec{p}\,). \end{aligned} \tag{17.97}$$

The relative energy integrals were done by way of (17.91). It follows that the effective Hamiltonian coming from the relativistic kinetic energy correction is

$$H_{\text{eff}}(K_4) = \left(-\frac{\vec{p}^{\,4}}{8m_1^3} - \frac{\vec{p}^{\,4}}{8m_2^3}\right)(2\pi)^3\,\delta^{(3)}(\vec{p}-\vec{q}\,). \tag{17.98}$$

The remaining contributions to the effective Hamiltonian arise more directly from the corresponding diagrams of Fig. 17.4. The kernels involved are all independent of relative energy, so the energy integrals in $\langle \mathrm{i}\delta K\rangle$ can be done trivially using Eqs. (17.75) and (17.76). In general, for kernel contributions independent of relative energy, we find that

$$\begin{aligned}\langle \mathrm{i}\delta K\rangle &= \mathrm{i}\int \frac{\mathrm{d}^4p}{(2\pi)^4}\frac{\mathrm{d}^4q}{(2\pi)^4}\overline{\Psi}^0(p)\,\delta K(\vec{p},\vec{q}\,)\,\Psi^0(q) \\ &= \mathrm{i}\int \frac{\mathrm{d}^3p}{(2\pi)^3}\frac{\mathrm{d}^3q}{(2\pi)^3}\psi^+(\vec{p}\,)\,\delta K(\vec{p},\vec{q}\,)\,\psi(\vec{q}\,), \end{aligned} \tag{17.99}$$

and we identify

$$\delta H_{\text{eff}} = \mathrm{i}\delta K \tag{17.100}$$

as the corresponding contribution to the effective Hamiltonian. The kernel for the first spin-orbit correction, from the first part of Fig. 17.4(b), is

$$K(\mathrm{SO}) = \frac{q_1}{4m_1^2}(\vec{p}\times\vec{q}\,)\cdot\vec{\sigma}_1\frac{\mathrm{i}}{(\vec{p}-\vec{q}\,)^2}(-\mathrm{i}q_2) = -\pi Z\alpha\frac{\vec{p}\times\vec{q}}{(\vec{p}-\vec{q}\,)^2}\cdot\frac{\vec{\sigma}_1}{m_1^2}, \tag{17.101}$$

and the contribution from the second part of Fig. 17.4(b) is analogous. The total spin-orbit effective Hamiltonian is

$$H_{\text{eff}}(\mathrm{SO}) = -\mathrm{i}\pi Z\alpha\frac{\vec{p}\times\vec{q}}{(\vec{p}-\vec{q}\,)^2}\cdot\left(\frac{\vec{\sigma}_1}{m_1^2} + \frac{\vec{\sigma}_2}{m_2^2}\right). \tag{17.102}$$

We note that where there are no divergences, we can work with $D = 3$ and take $\sigma^{\ell m}p^\ell q^m \to \vec{\sigma}\cdot(\vec{p}\times\vec{q}\,)$, and to the order of interest we can take $c_\mathrm{S} = c_\mathrm{D} = c_\mathrm{F} = 1$.

The Darwin correction, from Fig. 17.4(c), has the form

$$\begin{aligned}H_{\text{eff}}(\mathrm{D}) &= \mathrm{i}\times\left\{\frac{\mathrm{i}q_1}{8m_1^2}(\vec{p}-\vec{q}\,)^2\frac{\mathrm{i}}{(\vec{p}-\vec{q}\,)^2}(-\mathrm{i}q_2) + (-\mathrm{i}q_1)\frac{\mathrm{i}}{(\vec{p}-\vec{q}\,)^2}\frac{iq_2}{8m_2^2}(-\vec{p}+\vec{q}\,)^2\right\} \\ &= \frac{\pi Z\alpha}{2}\left(\frac{1}{m_1^2} + \frac{1}{m_2^2}\right). \end{aligned} \tag{17.103}$$

The kernel for the magnetic interaction of Fig. 17.4(d) is

$$K(\mathrm{M}) = \frac{\mathrm{i}q_1}{2m_1}(p+q)^i \frac{\mathrm{i}\delta^{\perp,ij}(\vec{p}-\vec{q}\,)}{(p-q)^2+\mathrm{i}\epsilon} \frac{\mathrm{i}q_2}{2m_2}(-p-q)^j\,. \tag{17.104}$$

We can use the transversality of $\delta^{\perp,ij}(\vec{p}-\vec{q}\,)$ and the identity $\vec{k} = \vec{p}-\vec{q}$ to simplify the convection factors as follows, $(p+q)^i \to (2p-k)^i \to 2p^i$, and $(-p-q)^j \to (-k-2q)^j \to -2q^j$. The transverse photon propagator contains the four-vector photon momentum squared as opposed to the three-momentum squared of the Coulomb photon. This photon denominator gives rise to soft and ultrasoft contributions in addition to the usual potential-region contribution, but the potential region still dominates and is the only one that enters at $\mathcal{O}(\alpha^4)$. The same neglect of *retardation* holds for the remaining $\mathcal{O}(\alpha^4)$ contributions. Consequently, at $\mathcal{O}(\alpha^4)$, we replace $(p-q)^2$ by its spatial contribution $-(\vec{p}-\vec{q}\,)^2$. The magnetic effective Hamiltonian takes the form

$$H_{\mathrm{eff}}(\mathrm{M}) = -\frac{4\pi Z\alpha}{m_1 m_2} \frac{\vec{p}^{\,2}\vec{q}^{\,2} - (\vec{p}\cdot\vec{q}\,)^2}{(\vec{p}-\vec{q}\,)^4}. \tag{17.105}$$

The Fermi spin-orbit contribution of Fig. 17.4(e) involves a convection interaction with one fermion combined with a Fermi (spin-dependent) interaction on the other and exchange of a transverse photon. For the corresponding kernel, we find

$$\begin{aligned} K(\mathrm{FSO}) &= \frac{q_1}{2m_1}\sigma_1^{ik}(p-q)^k \frac{\mathrm{i}\delta^{\perp,ij}(\vec{p}-\vec{q}\,)}{-(\vec{p}-\vec{q}\,)^2} \frac{\mathrm{i}q_2}{2m_2}(-p-q)^j \\ &+ \frac{\mathrm{i}q_1}{2m_1}(p+q)^i \frac{\mathrm{i}\delta^{\perp,ij}(\vec{p}-\vec{q}\,)}{-(\vec{p}-\vec{q}\,)^2} \frac{q_2}{2m_2}\sigma_2^{jk}(-p+q)^k\,. \end{aligned} \tag{17.106}$$

The transverse delta function reduces to the usual Kronecker delta because of the antisymmetry of σ^{ij}. Because this contribution is finite we can use the three-dimensional reduction $\sigma^{ij} = -\frac{\mathrm{i}}{2}\left[\sigma^i, \sigma^j\right] \to \epsilon^{ijk}\sigma^k$. The corresponding effective Hamiltonian is

$$H_{\mathrm{eff}}(\mathrm{FSO}) = -2\mathrm{i}\pi Z\alpha \frac{\vec{p}\times\vec{q}}{(\vec{p}-\vec{q}\,)^2} \cdot \frac{(\vec{\sigma}_1+\vec{\sigma}_2)}{m_1 m_2}. \tag{17.107}$$

Finally, the Fermi-Fermi, or "tensor", contribution of Fig. 17.4(f) is

$$H_{\mathrm{eff}}(\mathrm{FF}) = \frac{\pi Z\alpha}{m_1 m_2} \frac{[(\vec{p}-\vec{q}\,)\times\vec{\sigma}_1]\cdot[(\vec{p}-\vec{q}\,)\times\vec{\sigma}_2]}{(\vec{p}-\vec{q}\,)^2}. \tag{17.108}$$

We have now obtained all terms that contribute to energy levels at $\mathcal{O}(\alpha^4)$ (subject to the exception for particle-antiparticle systems that have an additional contact contribution at this order). The complete (momentum space) effective Hamiltonian at $\mathcal{O}(\alpha^4)$ is

$$\begin{aligned} H_{\mathrm{eff}} &= \left(-\frac{\vec{p}^{\,4}}{8m_1^3} - \frac{\vec{p}^{\,4}}{8m_2^3}\right)(2\pi)^3\delta^{(3)}(\vec{k}\,) + \frac{\pi Z\alpha}{2}\left(\frac{1}{m_1^2}+\frac{1}{m_2^2}\right) \\ &- \frac{4\pi Z\alpha}{m_1 m_2}\frac{\vec{p}^{\,2}\vec{q}^{\,2}-(\vec{p}\cdot\vec{q}\,)^2}{\vec{k}^{\,4}} - \mathrm{i}\pi Z\alpha\frac{\vec{p}\times\vec{q}}{\vec{k}^{\,2}}\cdot\left(\frac{\vec{\sigma}_1}{m_1^2}+\frac{\vec{\sigma}_2}{m_2^2}+\frac{2(\vec{\sigma}_1+\vec{\sigma}_2)}{m_1 m_2}\right) \\ &+ \frac{\pi Z\alpha}{m_1 m_2}\frac{(\vec{k}\times\vec{\sigma}_1)\cdot(\vec{k}\times\vec{\sigma}_2)}{\vec{k}^{\,2}}. \end{aligned} \tag{17.109}$$

All terms that enter here can be immediately identified as identical to the corresponding terms in the QED version of the effective Hamiltonian (16.144) except for the first term in (17.109), coming from the relativistic kinetic energy, which is not obvious. We note that on the QED side, the correction to the Bohr energy level comes from the reference energy (16.125) as well from the QED effective Hamiltonian. The comparison must include the extra contribution $\frac{\mu\beta_n^2}{2}\left(1-\frac{\mu}{M}\right)$ coming from the reference energy where $\beta_n = -(Z\alpha)^2/(2n^2)$ as defined in Eq. (16.94). In other words, agreement between the QED and NRQED versions of the $\mathcal{O}(\alpha^4)$ effective Hamiltonian requires

$$\begin{aligned}\frac{\mu\beta_n^2}{2}\left(1-\frac{\mu}{M}\right)&+\int\frac{\mathrm{d}^3p}{(2\pi)^3}\frac{\mathrm{d}^3q}{(2\pi)^3}\psi^+(\vec{p})V(\vec{k})\\ &\times\left\{\frac{(\vec{p}^{\,2}+\vec{q}^{\,2})}{8}\left(\frac{1}{m_1^2}+\frac{1}{m_2^2}-\frac{1}{m_1m_2}\right)-\frac{\beta_n\mu}{2M}\right\}\psi(\vec{q})\\ &=\int\frac{\mathrm{d}^3p}{(2\pi)^3}\psi^+(\vec{p})\left(-\frac{\vec{p}^{\,4}}{8m_1^3}-\frac{\vec{p}^{\,4}}{8m_2^3}\right)\psi(\vec{p}).\end{aligned} \tag{17.110}$$

This condition can be verified using $E_n^0 = \mu\beta_n$, $\langle V\rangle = 2E_n^0$, and $\langle\vec{p}^{\,2}\rangle = -2\mu E_n^0$,

$$\frac{1}{m_1^2}+\frac{1}{m_2^2}-\frac{1}{m_1m_2}=\frac{1}{\mu^2}-\frac{3}{m_1m_2},\qquad \frac{1}{m_1^3}+\frac{1}{m_2^3}=\frac{1}{\mu^3}-\frac{3}{\mu^2M}, \tag{17.111}$$

and the momentum-space Schrödinger equation

$$\vec{p}^{\,2}\psi(\vec{p}) = 2\mu\left\{E_n^0\psi(\vec{p})-\int\frac{\mathrm{d}^3q}{(2\pi)^3}V(\vec{p}-\vec{q})\psi(\vec{q})\right\}, \tag{17.112}$$

applied to one factor of $\vec{p}^{\,2}$ in each term on the right-hand side. As in Chap. 16, the actual energy levels at $\mathcal{O}(Z\alpha)^4$ can be found by evaluating some standard Coulomb expectation values.

17.3.2 *Relativistic Recoil Correction (NRQED Approach)*

Here we will obtain the spin-averaged relativistic recoil corrections to energy levels at $\mathcal{O}(Z\alpha)^5$ using NRQED. In the end, we will compare the QED (BSEQ) and NRQED calculations. The spin averaging will be performed quite differently in NRQED compared to the QED method of Chap. 16. Before, in our QED calculation, we averaged the spins of the two particles separately. Here, we average the combined spins using

$$X^{\mathrm{avg}} \equiv \frac{X^{00}+\sum_M X^{1M}}{D+1} \tag{17.113}$$

where X^{SM} is the expectation value of an operator X evaluated in a state of spin S and z-component M. We note that there are D (the number of spatial dimensions) possible values for the z-component M in a D-dimensional space when $S = 1$. The recoil corrections of interest here all involve a single inverse power of m_2. We

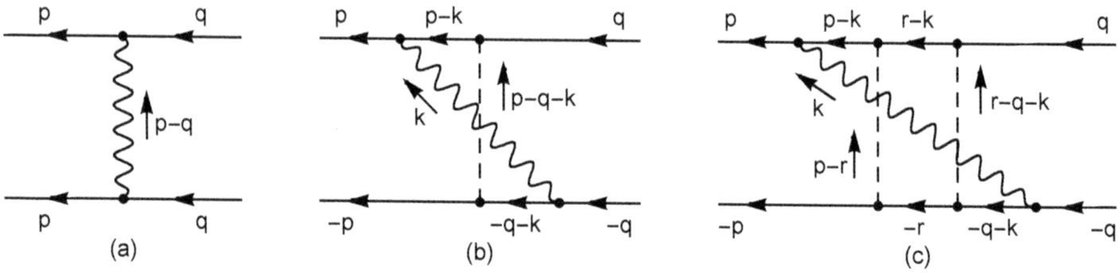

Fig. 17.5 NRQED kernels containing a single transverse exchange photon crossing zero, one, or two Coulomb photons, showing momentum assignments. The particle i fermion line implicitly caries an energy $\xi_i E$ in addition to the relative momentum shown.

consider the ratio m_1/m_2 to be small and ignore terms containing higher powers of m_2 in the denominator.

The kernels that contribute to the spin-averaged recoil energies at $\mathcal{O}(Z\alpha)^5$ are shown in Figs. 17.4(d), (i), (j), and (k). The relativistic kinetic energy from the $\vec{p}^{\,4}/m^3$ term is pure $\mathcal{O}(Z\alpha)^4$, while the next relativistic correction involves the operator $\vec{p}^{\,6}/m^5$, which is pure $\mathcal{O}(Z\alpha)^6$. None of these contribute at $\mathcal{O}(Z\alpha)^5$ nor do they have the right power of m_2 for our current interest. The spin-orbit and Darwin kernels of Figs. 17.4(b) and (c) have α^5 contributions, because the matching coefficients $c_{\rm S}$ and $c_{\rm D}$ contain terms which are $\mathcal{O}(\alpha/\pi)$, but the powers of m_2 are either $(m_2)^0 = 1$ (when the spin-orbit or Darwin vertex affects particle 1) or $(m_2)^{-2}$ (when the spin-orbit or Darwin vertex affects particle 2). In neither case do we find the desired $1/m_2$ dependence. The kernels of Figs. 17.4(e) and (f) do give corrections of $\mathcal{O}(Z\alpha)^5$ with $1/m_2$ mass dependence, but their spin average vanishes. For, say, the second graph of Fig. 17.4(e), the spin average involves the trace $\mathrm{Tr}\left[\left(\chi^{SM}\right)^+ \chi^{SM}\sigma^{ij}\right]$, which vanishes because $\left(\chi^{00}\right)^+ \chi^{00} = 1/2$, $\sum_M \chi^{+1M}\chi^{1M} = D/2$, and $\mathrm{Tr}[\sigma^{ij}] = 0$. Similar considerations show that all spin averages for Figs. 17.4(e) and (f) vanish. Neither the vacuum polarization kernel of Fig. 17.4(g) nor either of the Lamb shift terms of Fig. 17.4(h) have the right m_2 dependence to be of interest here. We now focus our attention on the kernels of Figs. 17.4(d), (i), (j), and (k) that do contribute spin-averaged recoil corrections at $\mathcal{O}(Z\alpha)^5$.

We begin our calculation with the magnetic interaction of Fig. 17.4(d), shown with momentum assignments in Fig. 17.5(a). Application of the NRQED Feynman rules leads to the energy correction

$$\Delta E_{\rm T} = \mathrm{i}\int \frac{\mathrm{d}^d p}{(2\pi)^d}\frac{\mathrm{d}^d q}{(2\pi)^d}\overline{\Psi}^0(p)\frac{\mathrm{i}q_1}{2m_1}(p+q)^i\frac{\mathrm{i}\delta^{\perp,ij}(\vec{p}-\vec{q}\,)}{(p-q)^2+\mathrm{i}\epsilon}\frac{\mathrm{i}q_2}{2m_2}(-p-q)^j\Psi^0(q). \tag{17.114}$$

When all the momenta are potential, this contribution is $\mathcal{O}(Z\alpha)^4$ as seen in the previous section. Higher order corrections from this region are $\mathcal{O}(Z\alpha)^6$. For our present calculation of $(Z\alpha)^5$ recoil corrections, the relevant regions for $\{p, p-q, q\}$ are {P,S,S}, {S,S,P}, and {P,US,P} (using the notation introduced in Table 16.1).

We start with the soft correction from region {P,S,S} and make the appropriate soft corrections to $\Psi^0(q)$ and k^2:

$$\Delta E_{\mathrm{T}}^{\mathrm{P,S,S}} = \frac{\mathrm{i}4\pi Z\alpha\overline{\mu}^{2\epsilon}}{4m_1m_2}\int\frac{\mathrm{d}^dp}{(2\pi)^d}\frac{\mathrm{d}^dq}{(2\pi)^d} \times\overline{\Psi}^0(p)\frac{(2p^i)\,\delta^{\perp,ij}(\vec{p}-\vec{q}\,)\,(2p^j)}{q_0^2-(\vec{p}-\vec{q}\,)^2+\mathrm{i}\epsilon}\frac{s^{-1}(E^0;\vec{q}\,)}{(q_0+\mathrm{i}\epsilon)(-q_0+\mathrm{i}\epsilon)}\psi(\vec{q}\,), \tag{17.115}$$

where we have used $(p+q)^i = (2p-(p-q))^i \to (2p)^i$ to simplify the convection factors. We use (17.76) to perform the p^0 integral and recall that the $q^0 = 0$ poles are disallowed in the q^0 integral. After performing the relative energy integrals, we find the contribution

$$\Delta E_{\mathrm{T}}^{\mathrm{P,S,S}} = -\frac{4\pi Z\alpha\overline{\mu}^{2\epsilon}}{m_1m_2}\int\frac{\mathrm{d}^Dp}{(2\pi)^D}\frac{\mathrm{d}^Dq}{(2\pi)^D}\psi^+(\vec{p}\,)\frac{p^ip^j\,\delta^{\perp,ij}(\vec{p}-\vec{q}\,)}{2|\vec{p}-\vec{q}\,|^3}\left(E^0-\frac{\vec{q}^{\,2}}{2\mu}\right)\psi(\vec{q}\,), \tag{17.116}$$

which is identical to the corresponding contribution (16.184) found in the previous chapter using the BSEQ of QED after a number of approximations. The manipulations described following Eq. (16.184) go through just as before, and we find the total result ({P,S,S}+{S,S,P}) from single transverse photon exchange to be given by (16.189) as before,

$$\Delta E_{\mathrm{T}}^{\mathrm{soft}} = \frac{8\pi^2(Z\alpha)^2\overline{\mu}^{2\epsilon}}{m_1m_2}\int\frac{\mathrm{d}^Dp}{(2\pi)^D}\frac{\mathrm{d}^Dq}{(2\pi)^D}\psi^+(\vec{p}\,)\left(p^ip^j+q^iq^j\right)I^{ij}(\vec{p}-\vec{q}\,)\,\psi(\vec{q}\,), \tag{17.117}$$

where the integral $I^{ab}(\vec{k}\,)$ was given in Eq. (16.188). For the ultrasoft contribution from the {P,US,P} region, we have directly that Eq. (17.114) takes the form of Eq. (16.197). We perform the relative energy integrals and insert an extra integration with corresponding Dirac-δ function (contained within $s(E-|\vec{k}\,|;\vec{p},\vec{q}\,)$), as in the previous chapter to find [see Eq. (16.201)],

$$\Delta E_{\mathrm{T}}^{\mathrm{US}} = \frac{4\pi Z\alpha\overline{\mu}^{2\epsilon}}{m_1m_2}\int\frac{\mathrm{d}^Dp}{(2\pi)^D}\frac{\mathrm{d}^Dq}{(2\pi)^D}\frac{\mathrm{d}^Dk}{(2\pi)^D}\psi^+(\vec{p}\,)\frac{p^i\delta^{\perp,ij}(\vec{k}\,)q^j}{|\vec{k}\,|}\psi(\vec{q}\,)s(E-|\vec{k}\,|;\vec{p},\vec{q}\,). \tag{17.118}$$

This completes, for now, our reduction of the magnetic interaction of Fig. 17.4(d).

We move on to consider the kernels represented by Fig. 17.4(i) containing one, two, etc., ladder Coulomb photons crossed by a single transverse photon. All of these kernels contribute at the order of interest when the crossing transverse photon is ultrasoft. This gives rise to a term that is closely related to the usual Lamb shift. There is also a contribution from the first of these kernels, involving a single Coulomb photon, when the inner loop is soft. We begin our analysis with this soft contribution. The relevant kernel is represented by the diagram shown in Fig. 17.5(b). Working with the NRQED Feynman rules, we write down the corresponding energy contribution

$$\Delta E_{\rm TC}^{\rm soft} = {\rm i}\int \frac{{\rm d}^d p}{(2\pi)^d}\frac{{\rm d}^d k}{(2\pi)^d}\frac{{\rm d}^d q}{(2\pi)^d}\overline{\Psi}^0(q)\frac{{\rm i}q_1}{2m_1}(2p-k)^i\frac{\rm i}{\xi_1 E + p^0 - k^0 - \frac{(\vec p-\vec k)^2}{2m_1} + {\rm i}\epsilon}$$
$$\times(-{\rm i}q_1)\frac{{\rm i}\delta^{\perp,ij}(\vec k)}{k^2+{\rm i}\epsilon}\frac{\rm i}{(\vec p-\vec q-\vec k)^2}(-{\rm i}q_2)$$
$$\times\frac{\rm i}{\xi_2 E - q^0 - k^0 - \frac{(-\vec q-\vec k)^2}{2m_2} + {\rm i}\epsilon}\frac{{\rm i}q_2}{2m_2}(-2q-k)^j\Psi^0(q). \qquad (17.119)$$

We make the appropriate "soft-k" approximations and perform the p_0 and q_0 relative energy integrals using (17.75) and (17.76). The k^0 integral is done using poles, remembering that the $k^0 = 0$ poles are disallowed. We find

$$\Delta E_{\rm TC}^{\rm soft} = \frac{-{\rm i}(4\pi Z\alpha\overline{\mu}^{2\epsilon})^2}{m_1 m_2}\int\frac{{\rm d}^D p}{(2\pi)^D}\frac{{\rm d}^d k}{(2\pi)^d}\frac{{\rm d}^D q}{(2\pi)^D}$$
$$\times\psi^+(\vec p)\frac{1}{(-k^0+{\rm i}\epsilon)(-k^0+{\rm i}\epsilon)}\frac{p^i q^j\delta^{\perp,ij}(\vec k)}{(k^0)^2-\vec k^2+{\rm i}\epsilon}\frac{1}{(\vec p-\vec q-\vec k)^2}\psi(\vec q)$$
$$= -\frac{(4\pi Z\alpha\overline{\mu}^{2\epsilon})^2}{m_1 m_2}\int\frac{{\rm d}^D p}{(2\pi)^D}\frac{{\rm d}^D k}{(2\pi)^D}\frac{{\rm d}^D q}{(2\pi)^D}\psi^+(\vec p)\frac{p^i q^j\delta^{\perp,ij}(\vec k)}{2|\vec k|^3(\vec p-\vec q-\vec k)^2}\psi(\vec q)$$
$$= \frac{8\pi^2(Z\alpha)^2\overline{\mu}^{2\epsilon}}{m_1 m_2}\int\frac{{\rm d}^D p}{(2\pi)^D}\frac{{\rm d}^D q}{(2\pi)^D}\psi^+(\vec p)\left(-p^i q^j\right)I^{ij}(\vec p-\vec q)\,\psi(\vec q), \qquad (17.120)$$

which is identical to (16.193) of the previous chapter. As before, this contribution must be doubled to account for the additional kernels like those of Fig. 17.4(i), but with the transverse photon slanting down to the left. The soft single transverse contribution of (17.117) combined with the crossed transverse-Coulomb contribution of (17.120) here gives a total soft single-transverse contribution of

$$\Delta E_{\rm T(total)}^{\rm soft} = \frac{8(Z\alpha)^2}{3m_1 m_2}\overline{\mu}^{2\epsilon}\left\{|\psi(0)|^2\left[-\frac{1}{2\epsilon}-\frac{1}{2}+\ln\left(\frac{2}{\mu}\right)\right]+\left\langle\ln(|\vec p-\vec q|)\right\rangle_{\vec p,\vec q}\right\}, \qquad (17.121)$$

analogous to the derivation of Eq. (16.194).

Continuing our evaluation of the kernels represented by Fig. 17.4(i), we consider the case where the transverse photon is ultrasoft. We will start with the contribution involving the exchange of a single Coulomb photon crossed by the ultrasoft transverse photon. Using the momentum assignments of diagram in Fig. 17.5(b) and the approximations appropriate to an ultrasoft k, we find the contribution

$$\Delta E_{\rm TC}^{\rm US} = {\rm i}\int\frac{{\rm d}^d p}{(2\pi)^d}\frac{{\rm d}^d k}{(2\pi)^d}\frac{{\rm d}^d q}{(2\pi)^d}\overline{\Psi}^0(p)\frac{{\rm i}q_1}{2m_1}(2p)^i\frac{\rm i}{\xi_1 E + p^0 - k^0 - \frac{\vec p^2}{2m_1}+{\rm i}\epsilon}(-{\rm i}q_1)$$
$$\times\frac{{\rm i}\delta^{\perp,ij}(\vec k)}{k^2+{\rm i}\epsilon}\frac{\rm i}{(\vec p-\vec q)^2}(-{\rm i}q_2)\frac{\rm i}{\xi_2 E - q^0 - k^0 - \frac{\vec q^2}{2m_2}+{\rm i}\epsilon}\frac{{\rm i}q_2}{2m_2}(-2q^j)\,\Psi^0(q).$$
$$(17.122)$$

This integral, once doubled to account for the reflected kernels, is precisely the integral displayed in (16.204). The result after integration over relative energies is the same as was shown in (16.205),

$$\Delta E^{\rm US}_{\rm TC} = \frac{4\pi Z\alpha\overline{\mu}^{2\epsilon}}{m_1 m_2}\int\frac{{\rm d}^D p}{(2\pi)^D}\frac{{\rm d}^D q}{(2\pi)^D}\frac{{\rm d}^D k}{(2\pi)^D}\psi^+(\vec{p}\,)\frac{p^i\,\delta^{\perp,ij}(\vec{k}\,)\,q^j}{|\vec{k}\,|}\psi(\vec{q}\,)$$
$$\times\, s(E-|\vec{k}\,|;\vec{p}\,)V(\vec{p}-\vec{q}\,)s(E-|\vec{k}\,|;\vec{q}\,)\,. \qquad (17.123)$$

For the two-Coulomb case of Fig. 17.5(c), we follow the same procedure as in the previous chapter, not re-calculating the whole graph but rather just analyzing the changes between the one-Coulomb graph of Fig. 17.5(b) and the two-Coulomb graph of Fig. 17.5(c). The only difference is that the Coulomb propagator in Fig. 17.5(b) is replaced by a loop integration containing two fermions and two Coulomb photons (remembering that k is ultrasoft):

$$\frac{1}{(\vec{p}-\vec{q}-\vec{k}\,)^2}\to\int\frac{{\rm d}^d r}{(2\pi)^d}\frac{1}{(\vec{p}-\vec{r}\,)^2}\frac{\rm i}{\xi_1 E + r^0 - k^0 - \frac{\vec{r}^{\,2}}{2m_1}+{\rm i}\epsilon}(-{\rm i}q_1)$$
$$\times\frac{\rm i}{(\vec{r}-\vec{q}\,)^2}\frac{\rm i}{\xi_2 E - r^0 - \frac{\vec{r}^{\,2}}{2m_2}+{\rm i}\epsilon}(-{\rm i}q_2)$$
$$= -4\pi Z\alpha\overline{\mu}^{2\epsilon}\int\frac{{\rm d}^D r}{(2\pi)^D}\frac{1}{(\vec{p}-\vec{r}\,)^2}s(E-k^0;\vec{r}\,)\frac{1}{(\vec{r}-\vec{q}\,)^2}, \qquad (17.124)$$

or, equivalently,

$$V(\vec{p}-\vec{q}\,)\to[V\,s\,V]\,(E-k^0;\vec{p},\vec{q}\,). \qquad (17.125)$$

When we combine results for the kernels with no Coulomb photons (17.118), one Coulomb photon (17.123), and two and more Coulomb photons using (17.125) and its generalization to the many-Coulomb case, we find

$$\Delta E^{\rm US}_{\rm T(total)} = \frac{4\pi Z\alpha\overline{\mu}^{2\epsilon}}{m_1 m_2}\int\frac{{\rm d}^D p}{(2\pi)^D}\frac{{\rm d}^D q}{(2\pi)^D}\frac{{\rm d}^D k}{(2\pi)^D}\psi^+(\vec{p}\,)\frac{p^i\delta^{\perp,ij}(\vec{k}\,)q^j}{|\vec{k}\,|}\psi(\vec{q}\,)$$
$$\times\,[s+sVs+sVsVs+sVsVsVs+\cdots]\,(E-|\vec{k}\,|;\vec{p},\vec{q}\,)$$
$$= \frac{4\pi Z\alpha\overline{\mu}^{2\epsilon}}{m_1 m_2}\int\frac{{\rm d}^D p}{(2\pi)^D}\frac{{\rm d}^D q}{(2\pi)^D}\frac{{\rm d}^D k}{(2\pi)^D}\psi^+(\vec{p}\,)\frac{p^i\delta^{\perp,ij}(\vec{k}\,)q^j}{|\vec{k}\,|}\psi(\vec{q}\,)g(E-|\vec{k}\,|;\vec{p},\vec{q}\,).$$
$$(17.126)$$

The analysis now proceeds just as in the previous chapter, leading to the same result as (16.222):

$$\Delta E^{\rm US}_{\rm T(total)} = \frac{8(Z\alpha)^2}{3m_1 m_2}\overline{\mu}^{2\epsilon}|\psi(0)|^2\left\{\frac{1}{2\epsilon}+\frac{5}{6}+\ln\left(\frac{\mu}{\mu(Z\alpha)^2}\right)\right\} - \frac{8\mu^3(Z\alpha)^5}{3\pi m_1 m_2 n^3}\ln k_0(n,\ell).$$
$$(17.127)$$

At this point, we recall the difference between the reduced mass μ and the renormalization scales μ and $\overline{\mu}$, according to the discussion surrounding Eq. (16.145). We have now completed the evaluation of all spin-averaged recoil corrections of $\mathcal{O}(Z\alpha)^5$

involving the exchange of a single transverse photon. There are two classes of corrections: the soft-region terms of (17.121) and the ultrasoft region terms of (17.127). The infrared divergence of the soft term exactly cancels the ultraviolet divergence of the ultrasoft term, leading to a finite total. The sum of these two contributions is

$$\Delta E_{\mathrm{T}} = \frac{8(Z\alpha)^2}{3m_1m_2}\left\{|\psi(0)|^2\left[\ln\left(\frac{2}{\mu(Z\alpha)^2}\right)+\frac{1}{3}-\ln k_0(n,\ell)\right]+\big\langle\ln(|\vec{p}-\vec{q}\,|)\big\rangle\right\}, \quad (17.128)$$

where $k^0(n,\ell)$ is the Bethe log discussed at length in Chap. 4, and the expectation value $\langle\ln(|\vec{p}-\vec{q}\,|)\rangle$ is given in (16.180). The total single-transverse-photon contribution of Figs. 17.4(d) and (i) reduces to

$$\begin{aligned}\Delta E_{\mathrm{T}} = \frac{8\mu^3(Z\alpha)^5}{3\pi m_1m_2n^3}\Bigg\{&\left[\ln\left(\frac{4}{Z\alpha n}\right)+H_n-\frac{1}{2n}+\frac{5}{6}\right]\delta_{\ell=0}\\ &-\ln k_0(n,\ell)-\frac{\delta_{\ell\neq 0}}{4\ell(\ell+1)(\ell+1/2)}\Bigg\}. \end{aligned} \quad (17.129)$$

We recall from Eq. (16.180) that $H_n = \sum_{j=1}^{n}\frac{1}{j}$ is the nth harmonic number.

Graphs involving the exchange of two transverse photons also contribute to the spin-averaged recoil corrections at $\mathcal{O}(Z\alpha)^5$. These are contained in the seagull term Fig. 17.4(j) and the contact term Fig. 17.4(k). The contact term includes the effect of the exchange of two photons of all varieties, transverse and Coulomb, but only in the hard region. Our calculation of the matching coefficients for the contact term was done in Feynman gauge, but the values of these coefficients are independent of gauge, so the calculation could have been done in Coulomb gauge if desired, with the same result.

The seagull graph of Fig. 17.4(j) makes the following contribution:

$$\begin{aligned}\Delta E_{\text{seagull}} = \mathrm{i}\left(\frac{1}{2}\right)\int &\frac{\mathrm{d}^dp}{(2\pi)^d}\frac{\mathrm{d}^dk}{(2\pi)^d}\frac{\mathrm{d}^dq}{(2\pi)^d}\\ &\times\overline{\Psi}^0(p)\left(\frac{-\mathrm{i}q_1^2}{m_1}\right)\left(\frac{-\mathrm{i}q_2^2}{m_2}\right)\frac{\mathrm{i}\delta^{\perp,ij}(\vec{k}\,)}{k^2+\mathrm{i}\epsilon}\frac{\mathrm{i}\delta^{\perp,ij}(\vec{k}+\vec{p}-\vec{q}\,)}{(k+p-q)^2+\mathrm{i}\epsilon}\Psi^0(q)\,. \end{aligned} \quad (17.130)$$

The initial factor of 1/2 is a symmetry factor needed because the two vertices in the seagull diagram are connected by identical propagators. This term is clearly a recoil correction because of the explicit factor of m_2 in the denominator. The leading order is $(Z\alpha)^5$ when the two wave functions are potential but the inner loop is soft. For this region, we must approximate $(k+p-q)^2 \to (k^0)^2-(\vec{k}+\vec{p}-\vec{q}\,)^2$, so the relative energy integrals factorize. We use Eqs. (17.75) and (17.76) to do the p^0 and q^0 integrals, arriving at

$$\Delta E_{\text{seagull}} = \frac{(4\pi Z\alpha)^2}{2m_1m_2}\overline{\mu}^{2\epsilon}\int\frac{\mathrm{d}^Dp}{(2\pi)^D}\frac{\mathrm{d}^Dq}{(2\pi)^D}\psi^+(\vec{p}\,)\,I(\vec{p},\vec{q}\,)\,\psi(\vec{q}\,)\,, \quad (17.131)$$

where $I(\vec{p},\vec{q}\,)$ contains the k integration and was defined in Eq. (16.177). We see that the NRQED seagull graph contains exactly the same information, and has the same value (16.178), as the soft two-transverse-photon contribution discussed in Sec. 16.3.4. For the seagull contribution, we find

$$\Delta E_{\text{seagull}} = \frac{(Z\alpha)^2}{m_1 m_2}\bar{\mu}^{2\epsilon}\left\{|\psi(0)|^2\left[-\frac{1}{\epsilon} - 2\ln\mu - \frac{8}{3}\ln 2 + \frac{5}{3}\right] + 2\big\langle \ln(|\vec{p}-\vec{q}|)\big\rangle_{\vec{p},\vec{q}}\right\}. \tag{17.132}$$

The contact term shown in Fig. 17.4(k) makes an energy contribution

$$\Delta E_{\text{contact}} = \mathrm{i}\int \frac{\mathrm{d}^d p}{(2\pi)^d}\frac{\mathrm{d}^d q}{(2\pi)^d}\overline{\Psi}^0(p)\left\{\frac{\mathrm{i}d_{\mathrm{S}}}{m_1 m_2} + \frac{\mathrm{i}d_{\mathrm{V}}}{m_1 m_2}\vec{\sigma}_1\cdot\vec{\sigma}_2\right\}\Psi^0(q). \tag{17.133}$$

The reference wave functions contain the spin state factors ξ^{SM} shown in Eq. (17.32), so the expectation value for $\Delta E_{\text{contact}}$ reduces to the spin expectation value of the contact interaction (17.43), proportional to $\mathcal{M}^{SM}_{\text{NRQED}}$, times the wave function at the spatial origin,

$$\Delta E_{\text{contact}} = -|\psi(0)|^2\mathcal{M}^{SM}_{\text{NRQED}} = -\frac{|\psi(0)|^2}{m_1 m_2}\left\{d_{\mathrm{S}} + d_{\mathrm{V}}\left[-D\delta_{S=0} + (D-2)\delta_{S=1}\right]\right\}. \tag{17.134}$$

Values for d_{S} and d_{V} were given in Eq. (17.54). It follows that the spin-averaged energy shift due to the contact interaction is

$$\begin{aligned}\Delta E^{\text{avg}}_{\text{contact}} &= \frac{1}{D+1}\left\{\Delta E^{S=0}_{\text{contact}} + D\Delta E^{S=1}_{\text{contact}}\right\}\\ &= -\frac{|\psi(0)|^2}{m_1 m_2}\left\{d_{\mathrm{S}} + \frac{D(D-3)}{D+1}d_{\mathrm{V}}\right\}\\ &= \frac{(Z\alpha)^2}{m_1 m_2}\bar{\mu}^{2\varepsilon}|\psi(0)|^2\left\{\frac{1}{\epsilon} - \frac{1}{3} - 2\frac{m_1^2\ln(m_2/\mu) - m_2^2\ln(m_1/\mu)}{m_1^2 - m_2^2} + \mathcal{O}(\epsilon)\right\}.\end{aligned} \tag{17.135}$$

(We note that d_{V} doesn't contribute.) A glance back at Eq. (16.173) shows that this result is identical to the QED hard contribution computed using Coulomb gauge in the previous chapter.

We have now completed the calculation of all contributions to the spin-averaged recoil correction at $\mathcal{O}(Z\alpha)^5$ coming from the exchange of two soft transverse photons (seagull term) plus the exchange of two hard photons of any sort (contact term). We note that the ultraviolet $-1/\epsilon$ from the soft region cancels the infrared $1/\epsilon$ from the hard region, leaving a finite result. We use (16.180) for the logarithmic expectation value. The sum of the two-photon-exchange terms is

$$\begin{aligned}\Delta E_{\gamma\gamma} = \frac{\mu^3(Z\alpha)^5}{\pi m_1 m_2 n^3}\Bigg\{\Bigg[&2\ln\left(\frac{Z\alpha}{n}\right) - \frac{2}{3}\ln 2 + 2H_n - \frac{1}{n} + \frac{7}{3}\\ &-2\frac{m_1^2\ln(m_2/\mu) - m_2^2\ln(m_1/\mu)}{m_1^2 - m_2^2}\Bigg]\delta_{\ell=0} - \frac{\delta_{\ell\neq 0}}{2\ell(\ell+1)(\ell+1/2)}\Bigg\}.\end{aligned} \tag{17.136}$$

The grand total for the spin-averaged recoil correction at $\mathcal{O}(Z\alpha)^5$ is the sum of the one-transverse exchange term ΔE_{T} of (17.129) and the two-photon exchange term $\Delta E_{\gamma\gamma}$ of (17.136),

$$\Delta E^{\text{spin-average}}_{\text{recoil}} = \frac{\mu^3(Z\alpha)^5}{\pi m_1 m_2 n^3}\left\{\left[\frac{2}{3}\ln\left(\frac{1}{Z\alpha}\right) + \frac{14}{3}\left(\ln\left(\frac{2}{n}\right) + H_n - \frac{1}{2n}\right) + \frac{41}{9}\right.\right.$$
$$\left.\left. - 2\frac{m_1^2\ln(m_2/\mu) - m_2^2\ln(m_1/\mu)}{m_1^2 - m_2^2}\right]\delta_{\ell=0} - \frac{8}{3}\ln k_0(n,\ell) - \frac{7\delta_{\ell\neq 0}}{6\ell(\ell+1)(\ell+1/2)}\right\}.$$

This result is in complete agreement with (16.226) found using QED.

17.3.3 *Hyperfine Structure of S States (NRQED Approach)*

The leading $\mathcal{O}(Z\alpha)^5$ corrections to the hfs come exclusively from the exchange of two hard-region photons that are represented in NRQED by the contact term. This leading effect includes a factor of $\ln(m_2/m_1)$, which is large when $m_2 \gg m_1$. Such logarithms arise from integration over a range spanning m_1 to m_2, all of which is in the hard region. The contact-term energy correction has already been obtained in (17.134). The hfs correction is

$$\Delta E^{\text{hfs}}_{\text{contact}} = -\frac{|\psi(0)|^2}{m_1 m_2}2(D-1)d_{\text{V}} \to -8|\psi(0)|^2(Z\alpha)^2\frac{\ln(m_1/m_2)}{m_1^2 - m_2^2}, \tag{17.137}$$

in agreement with the result (16.248) obtained using QED and Coulomb gauge in the previous chapter and with known results, found, e.g., in Sec. 10.1.1 of Ref. [27].

17.4 Discussion

In Chaps. 16 and 17, we have displayed two versions of a bound-state formalism, both based on quantum field theory and both available for the calculation of relativistic, recoil, and other field-theory corrections to the properties of two-body Coulombic bound states. In each case, we have used the formalism to calculate an effective Hamiltonian that could be used to find the energy levels at $\mathcal{O}(Z\alpha)^4$, the spin-averaged recoil corrections at $\mathcal{O}(Z\alpha)^5$, and the leading $\mathcal{O}(Z\alpha)^5$ correction to the hyperfine structure of S states. These results were worked out in detail to illustrate the types of considerations needed to derive results for these corrections. We have two main observations to make about the QED and NRQED versions of the Bethe–Salpeter formalism.

First, we notice that the two formalisms have many common features. The two approaches are instances of the same basic development and are based on the same solvable nonrelativistic Coulomb problem. The perturbative expansions have the same structure. The reference wave functions consist of a spin state, the same nonrelativistic hydrogen-like wave function, and features that accentuate the two-body nature of the bound states, viz., $[(\xi_1 E + p^0 - \vec{p}^{\,2}/(2m_1) + \mathrm{i}\epsilon)\,(\xi_2 E - p^0 - \vec{p}^{\,2}/(2m_2) + \mathrm{i}\epsilon)]^{-1}$ for NRQED and $[(\xi_1 P^0 + p^0 - \omega_1(\vec{p}) + \mathrm{i}\epsilon)(\xi_2 P^0 - p^0 - \omega_2(\vec{p}) + \mathrm{i}\epsilon)]^{-1}$ for QED. The calculations of energy contributions boil down to the analogous manipulations eventually because the underlying physics is the same in both cases; yet, the NRQED approach allows us to formulate the results in a more systematic fashion, one of the advantages being that results obtained for high-energy contributions to one process are summarized in an effective operator and can be used for the calculation of other processes without having to go through the calculations again.

Second, and strikingly, we notice the huge disparity in the complexity of the formalisms. In both cases we are required to employ the nonrelativistic Schrödinger–Coulomb problem as the heart of a two-body quantum field theory formalism, but the fact that all relativistic effects have been stripped off from NRQED to be included as perturbations makes the fit for NRQED direct and natural, as opposed to the exceedingly unnatural nature of the fit for QED for the study of nonrelativistic bound systems (where we are allowed to expand in the coupling parameter $Z\alpha$). To begin with, we note the difference in the fundamental variables used to construct the theories: two-component Pauli spinor fields for NRQED versus four-component Dirac spinor fields for QED. The NRQED fermions are simply second-quantized versions of the nonrelativistic wave functions, while the QED fermions contain antiparticle components that are only remotely relevant for the physics of nonrelativistic bound states. The core reason for NRQED's relative simplicity is that the hard, i.e., relativistic, aspects of particle interactions are isolated from the low-energy binding aspects by the encapsulation of the hard effects in matching coefficients and the corresponding effective interactions (spin-orbit, Darwin, Fermi, contact, etc.). The actual calculation of the matching coefficients can be challenging, but it is no more challenging than absolutely necessary, given that the matching coefficients can be obtained from standard gauge-independent scattering calculations involving free external particles. The NRQED Lagrangian contains many more terms than that for QED, but those terms represent physical effects that are in a form that leads to relatively straightforward evaluation instead of being hidden inside of a much more complicated QED bound state expectation value that must be teased apart before actual evaluation. The relative simplicity of the NRQED calculations has been illustrated by the calculations of spin-averaged recoil corrections and hyperfine contributions given above. Finally, there is an aspect that did not come up in the calculations shown: each region of each diagram (for instance, the {P,US,P} region of the single transverse exchange of Fig. 17.5(a)) has not only a leading contribution that was calculated here, but higher order contributions as well. It is relatively straightforward to extract these higher-order contributions from NRQED, but extremely difficult in a QED-based formalism. For all these reasons, NRQED in one form or another has been the basis of most modern calculations of higher-order corrections to the properties of two-particle Coulombic bound states.

17.5 Further Thoughts

Here are some suggestions for further thought.

(1) **Spin and Rotations.** One particular entry in Table 17.1 concerns the symmetries of the spin current under parity (+) and time reversal (−). It is a good exercise to derive the behavior of the spin current under parity and under rotations, within full QED, noting the transformation of the Dirac wave function under parity, $\psi^{\mathcal{P}}(x) = P\,\psi(x_P)$, with $P = \gamma^0$ (see Sec. 7.2.3). Another good exercise is to derive the spinor representation of the parity transformation, which is γ^0, to apply it to the wave function, and to rederive this result again. To this end, show that $\vec{\gamma} \to -\vec{\gamma}$ under parity, while $\gamma^0 \to \gamma^0$ under parity. Consider Sec. 7.2.2 and especially Eq. (7.30). Derive the Lorentz generator for a full rotation about the z axis, which has $\omega_{12} = \omega_3 = \theta = 2\pi$. Show that the spinor representation of the rotation by 360° about the z axis is

$$S = \exp\left[-\frac{\mathrm{i}}{4}\omega_{\mu\nu}\sigma^{\mu\nu}\right] = \exp\left[-i\pi\Sigma^3\right]. \tag{17.138}$$

In view of

$$\exp\left[-\mathrm{i}\pi\begin{pmatrix}\sigma^3 & 0\\ 0 & \sigma^3\end{pmatrix}\right] = -\begin{pmatrix}\mathbb{1}_{2\times2} & 0\\ 0 & \mathbb{1}_{2\times2}\end{pmatrix} = -\mathbb{1}_{4\times4}\,, \tag{17.139}$$

justify why spinors change sign upon a rotation by a full 360°. Now, consider the effect of a parity transformation in the space of spin matrices, according to the formalism introduced in Sec. 7.2.3, and show that the spin does not change sign.

(2) **Darwin and Spin-Orbit Feynman Rules.** The Feynman rule for the Fermi vertex was derived in the text. Work out the Feynman rules for the Darwin and spin-orbit vertices in an analogous way, starting from the NRQED Lagrangian.

(3) **Determination of Matching Coefficients.** Consider the scattering of a low-energy electron in an external electromagnetic field. By comparing the QED and NRQED expressions for the scattering amplitude, express the Fermi, Darwin, and spin-orbit matching coefficients c_F, c_D, and c_S in terms of the Dirac and Pauli form factors F_1 and F_2. Verify the results in Eq. (17.24).

(4) **NRQED and Higher-Order Corrections.** Consider self-energy corrections to the energy levels of a single-electron atom like hydrogen. Study the terms in the NRQED Lagrangian given in Eq. (17.7), and investigate how these will affect the corrections to the low-energy part of the self-energy, discussed in the velocity gauge in Sec. 4.6.2. Consider Refs. [163, 170, 173, 315], and verify the consistency of the results obtained in the cited references with those that would otherwise be derived from the Lagrangian in Eq. (17.7), and from the formalism introduced in Chap. 11. Also, compare to the results reported in Ref. [326].

(5) **Araki–Sucher Distribution.** Consider Eq. (17.121) and clarify the role of the matrix element of $\ln(|\vec{p}-\vec{q}\,|)$ in comparison to the Araki–Sucher distribution given in Eq. (15.133).

Chapter 18

Fermionic Determinants and Effective Lagrangians

18.1 Overview

The title of Ref. [607] reads as "Effective Interactions are Effective Interactions". With this wisdom, which follows the general principle that underlines trivial statements like "a soccer ball is spherical" (in German: *"Der Ball ist rund!"*), we start the discussion of effective Lagrangians. The rationale behind the statement is as follows: Once you have a Lagrangian, use it in any way imaginable! Put it to the test! Plug into it all kinds of fields that you can imagine, provided the field configuration is within the physical regime covered by the effective Lagrangian, i.e., in the regime for which the Lagrangian is valid. The general principle which underlies the calculation of all "effective" Lagrangians is as follows: In general, quantum processes happen at different length scales. In general, one strives to "integrate out" the processes that happen on short length scales (high-energy scales) and to formulate an effective interaction which describes the low-energy degrees of freedom, i.e., the processes at low energy scales. The quantum perturbations due to the high-energy degrees of freedom are summarized in an "effective Lagrangian" which describes the modification of the long-distance physics, due to the presence of the short-range interactions. We had already followed this principle in Chap. 17, in the derivation of the effective NRQED Lagrangian, after the integration of the short-distance physics.

The secret then is that the fields appearing in the effective Lagrangian can be interpreted as classical fields, or quantum fields (field operators), or sums of both. This is because in deriving the effective interaction, one actually ignores the quantum fluctuations of the "low-energy" degrees of freedom and only considers the fluctuations of the "high-energy" degrees of freedom.

Within field theory, the effective Lagrangian can generally be calculated in the following steps [130, 608]. *(i)* One writes the complete Lagrangian of the system, including the free Lagrangian for the light, and the free Lagrangian for the heavy degrees of freedom, and the interaction terms. *(ii)* The path integral representing the generating function is typically dominated by the zero-field configuration for both light and heavy degrees of freedom. *(iii)* One identifies the field for which one would like to calculate the effective interaction as a classical field. *(iv)* The path integral needs to be taken over the configurations of the "heavy" field, while the "light" degrees of freedom are kept as classical "sources". This path integral can be calculated by a systematic expansion about the saddle point (in general, the Gaussian saddle point where the fields vanish), or in some cases, an all-order calculation is possible. Taking the logarithm of the path integral over the generating functional, one then obtains the effective Lagrangian for the "light" degrees of freedom.

Here, we shall attempt to illustrate this program on the basis of the Heisenberg–Euler effective Lagrangian, where the "light" degrees of freedom are the (massless) photons, and the "heavy" degrees of freedom are the electrons and positrons. We shall do so in a rather *ad hoc* approach, without recourse to the path integral formalism. In Sec. 18.2, we discuss the derivation of the Heisenberg–Euler effective Lagrangian of quantum electrodynamics [29, 87, 88, 609], as well as convenient representations of the expansions in both perturbative as well as nonperturbative regimes, and in Sec. 18.3, we present applications of the formulas, in the context of a modified speed of light in strong background fields, as well as in the context of bound-state theory (long-range asymptotic tail of the Wichmann–Kroll potential). Other applications include simplified calculation of various cross sections using the effective Lagrangian, which can be obtained by simply using the negative of the Lagrangian as the Hamiltonian density, in order to calculate the S matrix of a scattering process, with essentially arbitrary photon fields in the initial and final states. These latter applications are left as an exercise for the reader.

18.2 Derivation of the Heisenberg–Euler Effective Lagrangian

18.2.1 *Heisenberg–Euler Lagrangian for Electric Fields*

In our route toward the derivation of the effective action [87, 88, 610, 611], we first take one step back and recall the general rationale behind the derivation of the photon emission statistics for a given classical current distribution, as discussed in Sec. 9.6.2. If we wish to calculate the photon emission, then we first need to know the current $J^\mu(x)$, and then we calculate the creation of the quanta of the electromagnetic field. The S matrix must be analyzed for the case where J^μ is classical, but A_μ is quantized. In the evaluation of the QED effective action, the situation will be opposite: The classical electromagnetic field configuration $A^\mu(x)$ is given, and we calculate the influence of the quantized fermionic fields onto the equations of motion of the photon field, i.e., the modification of the Maxwell equations. In the former case, we assume the classical currents to be strong, and we generate a few quanta of the radiation field. In the latter case, we inherently assume the electromagnetic fields to be strong, and calculate the influence of the virtual quanta of the fermion field onto the classical electromagnetic fields.

For our purposes, we write the QED interaction Lagrangian, introduced in

Sec. 10.1, as follows,

$$\mathcal{L}_I(x) = -e\,J^\mu_{\rm in}(x)A^{\rm cl}_\mu(x) = -e\,\overline{\psi}_{\rm in}(x)\gamma^\mu\psi_{\rm in}(x)\,A^{\rm cl}_\mu(x) = -e\,\overline{\psi}_{\rm in}(x)\not{A}^{\rm cl}(x)\,\psi_{\rm in}(x)\,. \tag{18.1}$$

Here, $J^\mu_{\rm in}(x) = e\overline{\psi}_{\rm in}(x)\gamma^\mu\psi_{\rm in}(x)$ is the four-vector current, while $\psi_{\rm in}(x)$ is the fermionic field operator. The in fields ensure that we match the free fermion fields with the non-interacting field operators in the infinite past, where the vector potential $A^{\rm cl}_\mu(x)$ is supposed to be adiabatically switched off. The quantity $\not{A}^{\rm cl}(x) = \gamma^\mu A^{\rm cl}_\mu(x)$ is the Feynman slash of the classical vector potential. The interaction Lagrangian is equal to the negative of the interaction Hamiltonian, $\mathcal{L}_I(x) = -\mathcal{H}_I(x)$. All field operators are formulated in the interaction picture. Following the discussion in Sec. 9.6.2, we can write down the time-ordered expression for the S matrix,

$$S[\,A^{\rm cl}\,] = T\exp\left[-{\rm i}\,e\int {\rm d}^4x\,\overline{\psi}_{\rm in}(x)\not{A}^{\rm cl}(x)\,\psi_{\rm in}(x)\right]. \tag{18.2}$$

The amplitude for emitting no pair (vacuum-persistence amplitude) is given by

$$\begin{aligned}\langle 0\,{\rm in}\,|S|\,0\,{\rm in}\rangle &= \sum_{n=0}^{\infty}\frac{(-{\rm i}e)^n}{n!}\int {\rm d}^4x_1\cdots\int {\rm d}^4x_n\\ &\quad\times\left\langle 0\left|T\left[\overline{\psi}_{\rm in}(x_1)\not{A}^{\rm cl}(x_1)\psi_{\rm in}(x_1)\cdots\overline{\psi}_{\rm in}(x_n)\not{A}^{\rm cl}(x_n)\psi_{\rm in}(x_n)\right]\right|0\right\rangle.\end{aligned} \tag{18.3}$$

From here on, we leave out the classical (cl) and interaction picture, in-field (in) labels. In order to proceed further, we have a careful look at the contribution to the vacuum persistence amplitude $\langle 0\,{\rm in}\,|S|\,0\,{\rm in}\rangle$ from the term with definite n in Eq. (18.3). We have interactions at points $(x_1,\dots,x_n)$. For the Wick theorem (see Sec. 9.6.1) to lead to a nonvanishing contribution, we need an operator $\overline{\psi}(x_\ell)$ to be contracted with an operator $\psi(x_k)$. In addition, the field operator $\psi(x_k)$ gets multiplied by some $\not{A}(x_k)$. The summation over spinor indices yields an expression $\sum_\alpha A_\mu(x_k)\gamma^\mu_{\alpha_k\alpha}\psi_\alpha(x_k)$. In the fully quantized framework of the Wick theorem with the fermionic field operators, the contraction has to proceed via some $\overline{\psi}(x_\ell)$, to yield some expressions like $-{\rm i}e\left\langle 0\left|T\left[\not{A}(x_k)\psi(x_k)\overline{\psi}(x_\ell)\right]\right|0\right\rangle$.

We now define a matrix whose form is inspired by the necessity to contract the expression $\sum_\alpha A_\mu(x_k)\gamma^\mu_{\alpha_k\alpha}\psi_\alpha(x_k)$, taken at spinor index α_k and at the space-time coordinate x_k, with a field operator, taken at spinor index α_ℓ and at space-time coordinate x_ℓ,

$$C(\alpha_k,x_k,\alpha_\ell,x_\ell) = -{\rm i}e\sum_\alpha\left\langle 0\left|T\left[A_\mu(x_k)\,\gamma^\mu_{\alpha_k\alpha}\,\psi_\alpha(x_k)\overline{\psi}_{\alpha_\ell}(x_\ell)\right]\right|0\right\rangle. \tag{18.4}$$

Now, applying the fermionic Wick theorem, $S_0[A]$ can be written

$$\begin{aligned}S_0[A] &= \sum_{n=0}^{\infty}\frac{1}{n!}\int {\rm d}^4x_1\cdots\int {\rm d}^4x_n\\ &\quad\times\sum_P(-1)^n\,\varepsilon_P\sum_{\alpha_1,\dots,\alpha_n}C(\alpha_1,x_1,\alpha_{P_1},x_{P_1})\cdots C(\alpha_n,x_n,\alpha_{P_n},x_{P_n}),\end{aligned} \tag{18.5}$$

where P is a permutation of the indices $1, \dots, n$, and ε_P is the sign of the permutation. One might wonder about the origin of the factor $(-1)^n$. The $-ie$ from the expansion of the exponential in Eq. (18.2) is already absorbed in the prefactor of the matrix C. So, one might assume that there should be no factor $(-1)^n$, and also, the origin of the factor ε_P is unclear. Both factors are of combinatorial origin. The reason for its occurrence lies in the sign of the permutations ε_P, which depends on n in quite a subtle fashion. Roughly speaking, one may illustrate the occurrence by considering the prefactor of a "standard term" $C(\alpha_1, x_1, \alpha_2, x_2)\, C(\alpha_2, x_2, \alpha_3, x_3) \cdots C(\alpha_n, x_n, \alpha_1, x_1)$, which has a direction of indices $1 \to 2 \to 3 \to \cdots \to n \to 1$, and corresponds to a "standard" cyclic perturbation where all numbers advance one place, while the last one flips back to unity; on other words, one has $P_1 = 2$, $P_2 = 3$, $P_3 = 4, \dots, P_n = 1$. The permutation P connected with the "standard term" fulfills $\varepsilon_P = (-1)^{n+1}$, i.e., it is odd if n is even, and vice versa. When calculating the prefactor of the "standard term" $C(\alpha_1, x_1, \alpha_2, x_2)\, C(\alpha_2, x_2, \alpha_3, x_3) \cdots C(\alpha_n, x_n, \alpha_1, x_1)$, we have to take into that according to the Wick theorem, this term acquires an additional negative prefactor -1, because it describes a closed fermion loop, or in other words, the "last" fermion operator has to be commuted with an odd number $(2n+1)$ of other fermion operators, and this always generates an additional factor (-1). The "standard term" must obtain a prefactor (-1), however, $\varepsilon_P = (-1)^{n+1}$ for the "standard permutation". This works provided we supply an addition prefactor of $(-1)^n$ in the order n. For permutations other than the "standard one", the additional minus signs generated by the Wick theorem compensate against the additional minus signs due to the permutations, and the prefactor $(-1)^n$ is retained.

It is very instructive to illustrate the prefactor on the basis of an example calculation, say, for $n = 2$ and $n = 3$, by way of example. We shall investigate the contraction leading to a "standard term" which is given by a "cyclic contraction" of the fermionic field operators, with the with the "first" field operator becoming the "last" after the permutation. For $n = 2$, this means that

$$
\left\langle 0 \left| T\left[\overline{\psi}(x_1) \not{A}(x_1) \psi(x_1) \overline{\psi}(x_2) \not{A}(x_2) \psi(x_2) \right] \right| 0 \right\rangle
$$
$$
= (-1)^2 \, \varepsilon_{P_2} \, \left\langle 0 \left| T\left[\hat{A}(x_1) \psi(x_1) \overline{\psi}(x_2) \right] \right| 0 \right\rangle \left\langle 0 \left| T\left[\hat{A}(x_2) \psi(x_2) \overline{\psi}(x_1) \right] \right| 0 \right\rangle , \quad (18.6)
$$

because $(-1)^2 \, \varepsilon_{P_2} = -1$ where P_2 is the "standard permutation" for $n = 2$,

$$
P_2 = \begin{pmatrix} 1 & 2 \\ 2 & 1 \end{pmatrix} , \qquad \varepsilon_{P_2} = -1 . \quad (18.7)
$$

We reemphasize that the prefactor in Eq. (18.6) can otherwise be understood as a "standard prefactor (-1) resulting from a closed fermion loop", which, in our example case, is written as the product of $(-1)^2$ and $\varepsilon_{P_2} = -1$. For $n = 3$, the prefactor is obtained in the form $(-1)^n \varepsilon_{P_n}$ with $n = 3$, i.e.

$$\begin{aligned}\Big\langle 0\Big| T\Big[\overline{\psi}(x_1)\hat{A}(x_1)\psi(x_1)\overline{\psi}(x_2)\hat{A}(x_2)\psi(x_2)\overline{\psi}(x_3)\hat{A}(x_3)\psi(x_3)\Big]\Big| 0\Big\rangle \\ = (-1)^3 \varepsilon_{P_3} \left\langle 0\left|T\left[\hat{A}(x_1)\psi(x_1)\overline{\psi}(x_2)\right]\right|0\right\rangle \\ \times \left\langle 0\left|T\left[\hat{A}(x_2)\psi(x_2)\overline{\psi}(x_3)\right]\right|0\right\rangle \left\langle 0\left|T\left[\hat{A}(x_3)\psi(x_3)\overline{\psi}(x_1)\right]\right|0\right\rangle .\end{aligned} \tag{18.8}$$

Here, we have $(-1)^3 \varepsilon_{P_3} = -1$ where the "standard permutation" is

$$P_3 = \begin{pmatrix} 1\,2\,3 \\ 2\,3\,1 \end{pmatrix}, \qquad \varepsilon_{P_3} = +1 . \tag{18.9}$$

For general n, we define the "standard permutation" as

$$P_n = \begin{pmatrix} 1\ 2\ 3\ \cdots\ n \\ 2\ 3\ \cdots\ n\ 1 \end{pmatrix}, \qquad \varepsilon_{P_n} = (-1)^{n+1} . \tag{18.10}$$

The prefactors given in Eqs. (18.6) and (18.8) are manifestations of the general prefactor $(-1)^n \varepsilon_P$ in Eq. (18.5). They allow us to identify $S_0[A]$ with a determinant. Indeed, for any matrix, one has the identity

$$\begin{aligned}\det(1-\Gamma) &= \exp\left[\mathrm{tr}\ln(1-\Gamma)\right] = 1 - \sum_a \Gamma_{aa} \\ &+ \frac{(-1)^2}{2}\sum_{ab}(\Gamma_{aa}\Gamma_{bb} - \Gamma_{ab}\Gamma_{ba}) + \dots \\ &+ \frac{(-1)^n}{n!}\sum_{\alpha_1,\dots,\alpha_n}\sum_P \varepsilon_P \Gamma_{\alpha_1\alpha_{P_1}}\cdots\Gamma_{\alpha_n\alpha_{P_n}} + \dots .\end{aligned} \tag{18.11}$$

So, if we associate the integrals over the $\mathrm{d}x_i$ in Eq. (18.5) with sums over a discretized space $\int \mathrm{d}x_k \to \Delta x \sum_\ell x_{k,\ell}$, in the limit of an infinite number of integration points (Riemannian definition of an integral), then we have

$$S_0[A] = \det(1-\Gamma) , \tag{18.12}$$

where Γ is a matrix with two spinor and two space-time coordinate indices,

$$\Gamma_{\alpha_k x_k \alpha_\ell x_\ell} = \langle \alpha_k, x_k \,|\Gamma|\, \alpha_\ell, x_\ell\rangle = C(\alpha_k, x_k, \alpha_\ell, x_\ell) . \tag{18.13}$$

We have four spinor indices for every coordinate index. Now we can even write Γ as a matrix, bearing in mind that a matrix acting on an infinite-dimensional Hilbert space in fact is an operator. So, we need to find the operator representation of Γ, bearing in mind that the determinant of a matrix acting in an infinite-dimensional space is a Fredholm determinant. The Hilbert space is spanned by the fundamental eigenkets $|\alpha, x\rangle$, which represent the spinor indices and the coordinates. The operators X_μ and P_μ, which are denoted here by capital letters, act on the eigenkets as follows,

$$X_\mu|\alpha, x\rangle = x_\mu|\alpha, x\rangle , \qquad \langle \alpha, x\,|P_\mu|\,\varphi\rangle = \mathrm{i}\frac{\partial}{\partial x^\mu}\langle \alpha, x|\varphi\rangle , \tag{18.14}$$

where $|\varphi\rangle$ is a state in the Hilbert space. The fundamental commutator is $[X^\mu, P^\nu] = -\mathrm{i}\, g^{\mu\nu}$. We normalize the eigenkets for the coordinate and the momentum space in such a way that we can directly associate the integrations with summations if we replace the Kronecker deltas with Dirac delta functions. This is accomplished as follows,

$$\langle \alpha, x | \alpha', x' \rangle = \delta_{\alpha\alpha'}\, \delta^{(4)}(x - x')\,, \qquad \langle \alpha, p | \alpha', p' \rangle = \delta_{\alpha\alpha'}\, \delta^{(4)}(p - p')\,,$$
$$\langle \alpha, x | \alpha', p' \rangle = \frac{\delta_{\alpha\alpha'}\, \mathrm{e}^{\mathrm{i}\, p'\cdot x}}{(2\pi)^2}\,. \tag{18.15}$$

The transformation from coordinate to momentum space (corresponding to a definition of the wave functions) is given by the last of the above equations. The orthogonality relations in Eq. (18.15) hold for an integration measure obtained from the limiting process of replacing sums by integrals, without any additional factors of $1/(2\pi)^4$, in either coordinate or momentum space,

$$\langle \alpha, x | \alpha', x' \rangle = \sum_{\alpha''} \int \mathrm{d}^4 p'' \, \langle \alpha, x | \alpha'', p'' \rangle \langle \alpha'', p'' | \alpha', x' \rangle = \delta_{\alpha\alpha'}\, \delta^{(4)}(x - x')\,, \tag{18.16a}$$
$$\langle \alpha, p | \alpha', p' \rangle = \sum_{\alpha''} \int \mathrm{d}^4 x'' \, \langle \alpha, p | \alpha'', x'' \rangle \langle \alpha'', x'' | \alpha', p' \rangle = \delta_{\alpha\alpha'}\, \delta^{(4)}(p - p')\,. \tag{18.16b}$$

With these considerations, we can now easily write down the expression for the vacuum persistence amplitude, having identified integrations as the limiting process of summations over the coordinates,

$$S_0[A] = \det\left[\mathbb{1} - e \not{A}(X) \frac{1}{\not{P} - m + \mathrm{i}\epsilon}\right] = \det\left[\left(\not{P} - e\not{A}(X) - m + \mathrm{i}\epsilon\right) \frac{1}{\not{P} - m + \mathrm{i}\epsilon}\right]$$
$$= \exp\left[\mathrm{Tr} \ln\left(\left(\not{P} - e\not{A}(X) - m + \mathrm{i}\epsilon\right) \frac{1}{\not{P} - m + \mathrm{i}\epsilon}\right)\right]. \tag{18.17}$$

Here, the classical field $A(X)$ is a function of the space-time coordinate X, which constitutes an operator.

It could appear that the expression for $S_0[A]$ from Eq. (18.17) is not gauge invariant. Indeed, under a gauge transformation, one transforms $A_\mu(X) \to A_\mu(X) - \partial_\mu \Lambda(X)$ (see Sec. 8.2.1). The operator $\not{P} - e\not{A}(X) - m + \mathrm{i}\epsilon$ transforms as

$$\not{P} - e\not{A}(X) - m + \mathrm{i}\epsilon \to \mathrm{e}^{\mathrm{i}e\,\Lambda(X)} \left(\not{P} - e\not{A}(X) - m + \mathrm{i}\epsilon\right) \mathrm{e}^{-\mathrm{i}e\,\Lambda(X)}\,. \tag{18.18}$$

A gauge transformation, however, is not complete unless one also transforms the wave functions, in the sense that

$$\langle \alpha, x | \alpha', p' \rangle = \frac{1}{(2\pi)^2}\, \delta_{\alpha\alpha'}\, \mathrm{e}^{-\mathrm{i}\, p'\cdot x} \to \frac{1}{(2\pi)^2}\, \delta_{\alpha\alpha'}\, \mathrm{e}^{-\mathrm{i}\, p'\cdot x}\, \mathrm{e}^{\mathrm{i}e\,\Lambda(x)}\,. \tag{18.19}$$

The eigenkets of the position operator thus acquire a phase under the gauge transformation,

$$|\alpha, x\rangle \to |\alpha, x\rangle\, \mathrm{e}^{-\mathrm{i}e\,\Lambda(x)}\,. \tag{18.20}$$

However, one can argue that the operators in (18.17) do not know anything about the gauge transformation of the wave functions, and therefore, it remains to show that Eq. (18.17) is gauge invariant *without* the transformation of the wave functions. In the sense of a matrix transformation, we observe that the transformation $|x\rangle \to \mathrm{e}^{-\mathrm{i}e\,\Lambda(x)}\,|x\rangle$ is unitary, because it leaves the scalar products of the fundamental eigenkets invariant,

$$\left\langle \alpha, x \left| \mathrm{e}^{\mathrm{i}e\,(\Lambda(x')-\Lambda(x))} \right| \alpha', x' \right\rangle = \mathrm{e}^{\mathrm{i}e\,(\Lambda(x')-\Lambda(x))}\, \langle \alpha, x | \alpha', x' \rangle = \langle \alpha, x | \alpha', x' \rangle \,. \qquad (18.21)$$

Any unitary transformation leaves the determinant of a matrix invariant, and the same applies to the Fredholm determinant (determinant over an infinite-dimensional Hilbert space) from Eq. (18.17). We conclude that the expression (18.17) is gauge invariant without any additional gauge transformation of the wave functions or eigenkets. In order to better understand the role of the unitary character of the gauge transformation, we write the logarithm of the effective action as follows,

$$\ln S_0[A] = \ln\det\left(\not{P} - e\not{A}(X) - m + \mathrm{i}\epsilon\right) - \ln\det\left(\not{P} - m + \mathrm{i}\epsilon\right) . \qquad (18.22)$$

The gauge transformation then acts only on the first term, and it is unitary in the Hilbert space, which leaves the determinant unaltered.

In order to evaluate $S_0[A]$, we are facing the problem of getting rid of the spinor structures, which make it impossible to separate the numerator $\not{P} - e\not{A}(X) - m + \mathrm{i}\epsilon$ from the denominator $\not{P} - m + \mathrm{i}\epsilon$. We should at least try to change the denominator into a scalar quantity, so that it becomes possible to separate numerator and denominator by an integral representation, which in turns makes it possible to evaluate the traces explicitly, e.g., in coordinate space. We are doing this with a trick, which entails the charge conjugation operation. It turns out to be advantageous to use charge conjugation invariance of the vacuum persistence amplitude (the vacuum should be as likely to return to its state no matter what the sign of the interacting charges is), and to add the charge conjugated vacuum persistence amplitude to its original expression. The physical reason is that the vacuum persistence amplitude cannot depend on the sign of the charge of the electron, or in other words, in a fictitious world in which the electron charge changes sign, just as many electron-positron pairs would be produced as with the physical electron charge. We recall the defining property of the charge conjugation (7.49), which reads as $C\gamma^\mu C^{-1} = -(\gamma^\mu)^{\mathrm{T}}$. So, the charge conjugation invariance implies that

$$\begin{aligned} \ln S_0[A] &= \mathrm{Tr}\ln\left[\left(\not{P} - e\not{A} - m + \mathrm{i}\epsilon\right)\frac{1}{\not{P} - m + \mathrm{i}\epsilon}\right] \\ &= \mathrm{Tr}\ln\left[C\left(\not{P} - e\not{A} - m + \mathrm{i}\epsilon\right)C^{-1}\,\frac{1}{C(\not{P} - m + \mathrm{i}\epsilon)C^{-1}}\right] = \ln S_0[A]^{C} . \quad (18.23) \end{aligned}$$

Let us transform the charge-conjugated expression,

$$
\begin{aligned}
\ln S_0[A]^C &= \operatorname{Tr}\ln\left[C\left(\not{P} - e\not{A} - m + \mathrm{i}\epsilon\right)C^{-1}\,\frac{1}{C(\not{P} - m + \mathrm{i}\epsilon)C^{-1}}\right] \\
&= \operatorname{Tr}\ln\left[\left(-\not{P}^{\mathrm{T}} + e\not{A}^{\mathrm{T}} - m + \mathrm{i}\epsilon\right)\,\frac{1}{\left(-\not{P}^{\mathrm{T}} - m + \mathrm{i}\epsilon\right)}\right] \\
&= \operatorname{Tr}\ln\left[\left(\frac{1}{(\not{P} + m - \mathrm{i}\epsilon)}\left(\not{P} - e\not{A} + m - \mathrm{i}\epsilon\right)\right)^{\mathrm{T}}\right] \\
&= \operatorname{Tr}\ln\left[\frac{1}{(\not{P} + m - \mathrm{i}\epsilon)}\left(\not{P} - e\not{A} + m - \mathrm{i}\epsilon\right)\right]. \qquad (18.24)
\end{aligned}
$$

Here, we have used the fact that the matrix exponential fulfills the condition $\exp(\mathcal{A}^{\mathrm{T}}) = [\exp(\mathcal{A})]^{\mathrm{T}}$, and that the trace of a matrix is equal to the trace of its transpose. The matrix logarithm is the inverse operation of the matrix exponential. Thus, the logarithm of the transpose of a matrix is equal to the transpose of the logarithm of the same matrix. We now add the vacuum persistence amplitude and its charge conjugate,

$$
\begin{aligned}
\ln S_0[A] &= \frac{1}{2}\left(\ln S_0[A] + \ln S_0[A]^C\right) \\
&= \frac{1}{2}\operatorname{Tr}\ln\left[\left(\not{P} - e\not{A} - m + \mathrm{i}\epsilon\right)\frac{1}{\not{P} - m + \mathrm{i}\epsilon}\,\frac{1}{\not{P} + m - \mathrm{i}\epsilon}\left(\not{P} - e\not{A} + m - \mathrm{i}\epsilon\right)\right] \\
&= \frac{1}{2}\operatorname{Tr}\ln\left[\left(\not{P} - e\not{A} - m + \mathrm{i}\epsilon\right)\,\frac{1}{P^2 - m^2 + \mathrm{i}\epsilon}\,\left(\not{P} - e\not{A} + m - \mathrm{i}\epsilon\right)\right]. \qquad (18.25)
\end{aligned}
$$

There might be a question regarding the transformation. Namely, in general, the matrix exponential $\exp(\mathcal{A})\exp(\mathcal{B})$ is only equal to $\exp(\mathcal{A} + \mathcal{B})$ if A and B commute. Hence, we cannot claim that $\ln(\mathcal{A}\mathcal{B})$ is equal to $\ln(\mathcal{A}) + \ln(\mathcal{B})$. Setting $\mathcal{A} = S_0[A]$ and $\mathcal{B} = S_0[A]^C$, we might encounter a problem, in the transformations leading to Eq. (18.25). However, the product of Fredholm determinants fulfills $\det(\mathcal{A}\mathcal{B}) = \det(\mathcal{A})\det(\mathcal{B})$. Taking the logarithm and using the fact that $\ln(\det\mathcal{A}) = \operatorname{Tr}\ln(\mathcal{A})$, one can then show Eq. (18.25). Otherwise, one has to be extremely careful with traces in infinite-dimensional vector space. E.g., the relation $\operatorname{Tr}(\mathcal{A}\mathcal{B}) = \{\operatorname{Tr}(\mathcal{B}^+\mathcal{A}^+)\}^*$ holds, but we cannot simply assert that $\operatorname{Tr}(\mathcal{A}\mathcal{B}) = \operatorname{Tr}(\mathcal{B}\mathcal{A})$. Consider, e.g., in one-dimensional quantum mechanics, with a coordinate q and a momentum operator p, the trace $\operatorname{Tr}(qp - pq) = \operatorname{Tr}(\mathrm{i}\,\mathbb{1})$, where $\mathbb{1}$ is the unit operator.

We now insert a γ^0 matrix, use the identities $\ln(\det\mathcal{A}) = \operatorname{Tr}\ln(\mathcal{A})$, as well as $\det(\mathcal{A}^+) = (\det\mathcal{A})^*$ and $\det(\mathcal{A}\mathcal{B}) = \det(\mathcal{A})\det(\mathcal{B})$, and $\gamma^0\gamma^\mu\gamma^0 = (\gamma^\mu)^+$, repeatedly. The net result is to transform $\ln S_0[A]$ into the form

$$
\begin{aligned}
\ln S_0[A] &= \frac{1}{2}\,\mathrm{Tr}\ln\left[\gamma^0\left(\not{P}-e\not{A}-m+\mathrm{i}\epsilon\right)\gamma^0\frac{1}{P^2-m^2+\mathrm{i}\epsilon}\gamma^0\left(\not{P}-e\not{A}+m-\mathrm{i}\epsilon\right)\gamma^0\right] \\
&= \frac{1}{2}\ln\det\left(\left(\not{P}-e\not{A}-m-\mathrm{i}\epsilon\right)^{+}\left(\frac{1}{P^2-m^2-\mathrm{i}\epsilon}\right)^{+}\left(\not{P}-e\not{A}+m+\mathrm{i}\epsilon\right)^{+}\right) \\
&= \frac{1}{2}\ln\det\left(\left(\not{P}-e\not{A}+m+\mathrm{i}\epsilon\right)\left(\not{P}-e\not{A}-m-\mathrm{i}\epsilon\right)\frac{1}{P^2-m^2-\mathrm{i}\epsilon}\right)^{*} \\
&= \frac{1}{2}\,\mathrm{Tr}\ln\left[\left(\left(\not{P}-e\not{A}\right)^2-m^2+\mathrm{i}\epsilon\right)\frac{1}{P^2-m^2+\mathrm{i}\epsilon}\right]. \qquad (18.26)
\end{aligned}
$$

We have not completely eliminated the spinor structure, but in view of the identity

$$
\left(\not{P}-e\not{A}\right)^2 = (P-e\,A)^2 + \frac{e}{2}\,\sigma_{\mu\nu}\,F^{\mu\nu}\,, \qquad F^{\mu\nu} = \partial^\mu F^\nu - \partial^\nu F^\mu\,, \qquad (18.27)
$$

it has been significantly simplified. Here, the spin matrices are given as $\sigma_{\mu\nu} = \frac{\mathrm{i}}{2}\,[\gamma_\mu,\gamma_\nu]$, according to Eq. (7.30).

Up to now, we have only considered the vacuum persistence amplitude $S_0[A]$, which is a functional of the electromagnetic field configuration $A = A(X)$, where $X = (X^\mu)$ is an operator, but the stated aim of the current derivation is to calculate the effective action of the electromagnetic field. Let us go back to the definition of the vacuum persistence amplitude as the time-ordered product,

$$
S_0[A] = \mathrm{T}\left\langle 0\left|\exp\left[\mathrm{i}\int \mathrm{d}^4x\,\mathcal{L}_I(x)\right]\right|0\right\rangle. \qquad (18.28)
$$

The time evolution operator is calculated in the interaction picture. We assume that the electromagnetic field is adequately described as a classical field, whereas the quantum fluctuations of the fermionic field are described by the fermionic determinant. The vacuum persistence amplitude finds a natural interpretation as an exponentiated integral over the effective action of the classical electromagnetic field which generates the interaction,

$$
S_0[A] = \exp\left[\mathrm{i}\int \mathrm{d}^4x\,\mathcal{L}_{\mathrm{eff}}(x)\right]. \qquad (18.29)
$$

In other words, having kept the field configuration as a variable entity in our derivation, we are now in the position to interpret the result of our calculation as an effective Lagrangian for the remaining (classical) electromagnetic fields.

We can now use the integral representation of the logarithm,

$$
\ln\left(\frac{a}{b}\right) = -\int_0^\infty \frac{\mathrm{d}s}{s}\left[\mathrm{e}^{\mathrm{i}s(a+\mathrm{i}\epsilon)} - \mathrm{e}^{\mathrm{i}s(b+\mathrm{i}\epsilon)}\right], \qquad (18.30)
$$

together with the integral (18.30) and the matching (18.29), to write the effective action as

$$
\begin{aligned}
\int \mathrm{d}^4x\,\mathcal{L}_{\mathrm{eff}}(x) &= \frac{\mathrm{i}}{2}\int \mathrm{d}^4x\int_0^\infty \frac{\mathrm{d}s}{s}\,\mathrm{e}^{-\mathrm{i}s(m^2-\mathrm{i}\epsilon)}\,\mathrm{tr}\left(\left\langle x\left|\mathrm{e}^{\mathrm{i}s\,(\not{P}-e\not{A})^2}\right|x\right\rangle - \left\langle x\left|\mathrm{e}^{\mathrm{i}s\,P^2}\right|x\right\rangle\right) \\
&= \frac{\mathrm{i}}{2}\int \mathrm{d}^4x\int_0^\infty \frac{\mathrm{d}s}{s}\,\mathrm{e}^{-\mathrm{i}s(m^2-\mathrm{i}\epsilon)}\left[M(A)-M(0)\right]. \qquad (18.31)
\end{aligned}
$$

Here, tr is now reduced to the trace over the (4×4)-spinor space, but not anymore to the trace over the coordinate space; we have replaced $\mathrm{Tr}\to \mathrm{tr}\int \mathrm{d}^4x$. For slowly varying fields, the integral over d^4x can be dropped on both sides of Eq. (18.31). The central object of the next considerations therefore is the matrix element

$$M(A) = \mathrm{tr}\left\langle x \left| \mathrm{e}^{\mathrm{i}s\,(\not{P}-e\not{A})^2} \right| x \right\rangle = \mathrm{tr}\left\langle x \left| \exp\left(\mathrm{i}s\,(P-e\,A)^2 + \frac{e}{2}\sigma_{\mu\nu}\,F^{\mu\nu} \right) \right| x \right\rangle, \quad (18.32)$$

which leaves only the expression containing $\sigma_{\mu\nu}\,F^{\mu\nu}$ within spinor algebra. We evaluate it for a spatially constant electric field.

$$A^3(x) = -E\,t\,, \qquad F^{03} = -F^{30} = -E\,, \quad (18.33)$$

while all other components of the field-strength tensor vanish, and the electric field vector points in the positive z-direction, $\vec{E} = (0,0,E)$. Furthermore,

$$\sigma_{03} = \frac{\mathrm{i}}{2}\left[\gamma_0,\gamma_3\right] = -\mathrm{i}\gamma^0\,\gamma^3 = -\mathrm{i}\alpha^3\,, \quad (18.34)$$

where $\alpha^i = \gamma^0\,\gamma^i$ [see Eqs. (7.14) and (7.18)]. The spinor trace thus leads to

$$\begin{aligned}
\mathrm{tr}\left[\exp\left(-s\,\frac{e}{2}\,\sigma_{\mu\nu}\,F^{\mu\nu}\right)\right] &= \mathrm{tr}\left[\exp\left(-s\,e\,\alpha^3\,E\right)\right] \\
&= \mathrm{tr}\left[\left(\mathbb{1}_{4\times4} - s\,e\,\alpha^3\,E + \frac{\mathbb{1}_{4\times4}}{2!}\,(seE)^2 - \frac{1}{3!}(seE)^3\alpha^3 + \cdots\right)\right] \\
&= 4\left(1 + \frac{1}{2!}(seE)^2 + \cdots\right) = 4\sum_{n=0}^{\infty}\frac{(seE)^{2n}}{(2n)!} = 4\,\cosh(seE)\,. \quad (18.35)
\end{aligned}$$

The result $4\cosh(seE)$ commutes with $(P-eA)^2$, and therefore

$$M(A) = 4\,\cosh(seE)\,\left\langle x \left| \exp\left(\mathrm{i}s\,(P-e\,A)^2\right) \right| x \right\rangle. \quad (18.36)$$

We can write the scalar quantity $(P-eA)^2$ as

$$\begin{aligned}
(P-eA)^2 &= (P^0)^2 - (P^1)^2 - (P^2)^2 - (P^3+eEt)^2 \\
&= \mathrm{e}^{-\mathrm{i}P^0P^3/(eE)}\left[(P^0)^2 - (P^1)^2 - (P^2)^2 - (eEt)^2\right]\mathrm{e}^{\mathrm{i}P^0P^3/(eE)}\,, \quad (18.37)
\end{aligned}$$

where we have used the notation $P = (P^0, P^x \equiv P^1, P^y \equiv P^2, P^z \equiv P^3)$. The identity (18.37) follows from the Campbell–Baker–Hausdorff formula (9.149), which we use in the form

$$\exp(\mathcal{A})\,\mathcal{B}\,\exp(-\mathcal{A}) = \mathcal{B} + [\mathcal{A},\mathcal{B}] + \frac{1}{2!}\,[\mathcal{A},[\mathcal{A},\mathcal{B}]] + \cdots, \quad (18.38)$$

so that, with a as $a = P^3/(eE)$,

$$\begin{aligned}
\mathrm{e}^{-\mathrm{i}aP^0}\,t^2\mathrm{e}^{\mathrm{i}aP^0} &= \mathrm{e}^{a\partial_t}\,t^2\mathrm{e}^{-a\partial_t} \\
&= t^2 + [a\,\partial_t, t^2] + \frac{1}{2}\,[a\partial_t,[a\partial_t,t^2]] = t^2 + 2\,a\,t + a^2 = (t+a)^2\,. \quad (18.39)
\end{aligned}$$

Now, $U = \exp[-\mathrm{i}P^0P^3/(eE)]$ mediates a unitary transformation of Hilbert space of momentum eigenstates just like the previously discussed unitarity transformation of the coordinate-space eigenkets. Setting $p_1 = (\omega_1, \vec{p}_1)$, $p_2 = (\omega_2, \vec{p}_2)$ and $\vec{p}_\perp = (p^x, p^y, 0)$, one can insert a basis of momentum eigenkets as follows,

$$\begin{aligned}\frac{M(A)}{4\cosh(seE)} &= \left\langle x \left| \exp\left\{ \mathrm{i}s\left[(P^0)^2 - (P^1)^2 - (P^2)^2 - (P^3 + eEt)^2 \right] \right] \right| x \right\rangle \\ &= \left\langle x \left| \exp\left[\mathrm{i}s\, \mathrm{e}^{-\mathrm{i}P^0 P^3/(eE)} \left[(P^0)^2 - (P^1)^2 - (P^2)^2 - (eEt)^2 \right] \mathrm{e}^{\mathrm{i}P^0 P^3/(eE)} \right] \right| x \right\rangle \\ &= \int \mathrm{d}^4 p_1 \int \mathrm{d}^4 p_2 \, \langle x | p_1 \rangle \left\langle p_1 \left| \mathrm{e}^{\mathrm{i}sU[(P^0)^2 - (P^1)^2 - (P^2)^2 - (eEt)^2]U^{-1}} \right| p_2 \right\rangle \langle p_2 | x \rangle .\end{aligned} \tag{18.40}$$

We have carefully distinguished momentum operators (written in capital letters) from momentum eigenvalues (written in lowercase letters) and used the fact that

$$\langle x | p_1 \rangle \, \langle p_2 | x \rangle = \frac{1}{(2\pi)^4} \exp[\mathrm{i}(p_2 - p_1) \cdot x] , \tag{18.41}$$

which is compatible with Eq. (18.15).

Then, with $U = \exp[-\mathrm{i}P^0 P^3/(eE)]$, the vector structure in the matrix element in the integrand of Eq. (18.40) can be disentangled as follows,

$$\begin{aligned}\mathcal{M} &= \left\langle p_1 \left| \mathrm{e}^{\mathrm{i}sU[(P^0)^2 - (P^1)^2 - (P^2)^2 - (eEt)^2]U^{-1}} \right| p_2 \right\rangle \\ &= \left\langle p_1 \left| \exp\left[\mathrm{i}\, s \, \exp[-\mathrm{i}P^0 p_2^z/(eE)] \left[(P^0)^2 - (p_2^x)^2 - (p_2^y)^2 - (eEt)^2 \right] \right. \right. \\ &\quad \left. \left. \times \exp[\mathrm{i}P^0 p_2^z/(eE)] \right] \right| p_2 \right\rangle \\ &= \mathrm{e}^{-\mathrm{i}s\vec{p}_\perp^{\,2}} \, \delta^{(3)}(\vec{p}_1 - \vec{p}_2) \left\langle \omega_1 \left| \exp\left[\mathrm{i}\, s \, \exp[-\mathrm{i}P^0 p_2^z/(eE)] \left[(P^0)^2 - (eEt)^2 \right] \right. \right. \\ &\quad \left. \left. \times \exp[\mathrm{i}P^0 p_2^z/(eE)] \right] \right| \omega_2 \right\rangle .\end{aligned} \tag{18.42}$$

Here, $\vec{p}_\perp = p_2^x \, \hat{\mathrm{e}}_x + p_2^y \, \hat{\mathrm{e}}_y$. The Dirac-$\delta$ function $\delta^{(3)}(\vec{p}_1 - \vec{p}_2)$ ensures that both momenta are equal and can be written as $\vec{p}$ after one of the integrations has been performed. Furthermore, we have identified the matrix element as being evaluated between eigenstates of the four-momentum $|p_i\rangle = |\omega_i, \vec{p}_i\rangle$, with $\mathrm{i} = 1, 2$. Then, $\langle \omega_1 | \omega_2 \rangle = \delta(\omega_1 - \omega_2)$ and $\langle \vec{p}_1 | \vec{p}_2 \rangle = \delta^{(3)}(\vec{p}_1 - \vec{p}_2)$.

After all the algebra corresponding to the momentum operators has been performed, we can finally use the representation $P^0 = -\mathrm{i}\partial/\partial t$ and write $\mathcal{M}$ as follows,

$$\begin{aligned}\mathcal{M} &= \mathrm{e}^{-\mathrm{i}s\vec{p}_\perp^{\,2}} \, \delta^{(3)}(\vec{p}_1 - \vec{p}_2) \left\langle \omega_1 \left| \exp\left[\mathrm{i}\, s \, \exp\left[\frac{p_2^z}{eE} \frac{\partial}{\partial t} \right] \left[(P^0)^2 - (eEt)^2 \right] \right. \right. \\ &\quad \left. \left. \times \exp\left[-\frac{p_2^z}{eE} \frac{\partial}{\partial t} \right] \right] \right| \omega_2 \right\rangle \\ &= \mathrm{e}^{-\mathrm{i}s\vec{p}_\perp^{\,2}} \, \delta^{(3)}(\vec{p}_1 - \vec{p}_2) \left\langle \omega_1 \left| \exp\left[\mathrm{i}\, s \left[(P^0)^2 - (p_2^z + eEt)^2 \right] \right] \right| \omega_2 \right\rangle .\end{aligned} \tag{18.43}$$

Inserting this result into Eq. (18.40), and rewriting the remaining integration variable in momentum space as $\vec{p} \equiv \vec{p}_1$, one obtains

$$\begin{aligned}\frac{M(A)}{4\cosh(seE)} &= \frac{1}{(2\pi)^4}\int \mathrm{d}^3p \int \mathrm{d}\omega_1 \int \mathrm{d}\omega_2 \exp\left[\mathrm{i}(\omega_2-\omega_1)\,t\right] \\ &\quad\times \mathrm{e}^{-\mathrm{i}s\vec{p}_\perp^2}\left\langle \omega_1 \left| \mathrm{e}^{\mathrm{i}s[(P^0)^2-(p^z+eEt)^2]} \right| \omega_2 \right\rangle , \\ &= \frac{1}{(2\pi)^4}\int \mathrm{d}^3p \int \mathrm{d}\omega_1 \int \mathrm{d}\omega_2 \exp\left[\mathrm{i}(\omega_2-\omega_1)\left(t-\frac{p^z}{eE}\right)\right] \\ &\quad\times \mathrm{e}^{-\mathrm{i}s\vec{p}_\perp^2}\left\langle \omega_1 \left| \mathrm{e}^{\mathrm{i}s[(P^0)^2-(eEt)^2]} \right| \omega_2 \right\rangle , \end{aligned} \tag{18.44}$$

where the time variable has been shifted by the replacement $t \to t - p^z/(eE)$, which is permissible because we assume a constant electric field (in time). The d^3p integral can easily be carried out, splitting the integration into a transverse integral over $\mathrm{d}^2p_\perp = \mathrm{d}p^x\,\mathrm{d}p^y$ and an integral over $\mathrm{d}p^z$.

$$\begin{aligned}\int \frac{\mathrm{d}^3p}{(2\pi)^4}\exp\left[\mathrm{i}(\omega_1-\omega_2)\frac{p^z}{eE}\right]\mathrm{e}^{-\mathrm{i}s\vec{p}_\perp^2} &= \int \frac{\mathrm{d}^2p_\perp}{(2\pi)^3}\int \frac{\mathrm{d}p^3}{2\pi}\exp\left[\mathrm{i}(\omega_1-\omega_2)\frac{p^3}{eE}\right]\mathrm{e}^{-\mathrm{i}s\vec{p}_\perp^2} \\ = \int_0^\infty \frac{\mathrm{d}p_\perp\,p_\perp}{(2\pi)^2}\,eE\,\delta(\omega_1-\omega_2)\,\mathrm{e}^{-\mathrm{i}sp_\perp^2} &= \frac{eE}{8\pi^2\,\mathrm{i}\,s}\,\delta(\omega_1-\omega_2)\,. \end{aligned} \tag{18.45}$$

In view of $(P^0)^2 = -\partial_t^2$, we have

$$M(A) = \cosh(seE)\,\frac{eE}{2\pi^2\,\mathrm{i}\,s}\int \mathrm{d}\omega \left\langle \omega \left| \mathrm{e}^{\mathrm{i}s[-\partial_t^2-(eEt)^2]} \right| \omega \right\rangle . \tag{18.46}$$

We can now identify

$$\int \mathrm{d}\omega \left\langle \omega \left| \mathrm{e}^{\mathrm{i}s[-\partial_t^2-(eEt)^2]} \right| \omega \right\rangle = \mathrm{Tr}\left(\mathrm{e}^{\mathrm{i}sH_{\mathrm{osc}}}\right) = \int \mathrm{d}t \left\langle t \left| \mathrm{e}^{\mathrm{i}s[-\partial_t^2-(eEt)^2]} \right| t \right\rangle , \tag{18.47}$$

simply as the trace of a quantum-mechanical operator, which can be taken in the frequency or the time representation. The fact that there are no additional prefactors to consider, can be traced to our normalization of the four-momentum eigenstates given in Eqs. (18.15) and (18.41).

The operator H_{osc} is that of a harmonic oscillator whose "coordinate variable" is the time

$$H_{\mathrm{osc}} = -\frac{1}{2m_0}\frac{\partial^2}{\partial\xi^2} + \frac{1}{2}\,m_0\,\omega_0^2\xi^2\,, \qquad \xi \equiv t\,, \tag{18.48}$$

and which has an imaginary eigenfrequency, $m_0 = 1/2$ and $\omega_0 = 2\,\mathrm{i}\,eE$. Now, the trace can be taken either in the frequency, or the time representation, or in the representation of energy eigenfunctions, which as we know display a well-known discrete spectrum of eigenfrequencies $(n+1/2)\omega_0$. The trace therefore is equal to

$$\mathrm{Tr}\left(\mathrm{e}^{\mathrm{i}sH_{\mathrm{osc}}}\right) = \sum_{n=0}^{\infty}\mathrm{e}^{\mathrm{i}s(n+1/2)\omega_0} = \sum_{n=0}^{\infty}\mathrm{e}^{-2seE(n+1/2)} = \frac{1}{2\sinh(seE)}\,. \tag{18.49}$$

Finally, we obtain a simple and compact result for $M(A)$, as defined in Eq. (18.32),

$$M(A) = \coth(seE)\,\frac{eE}{4\pi^2\,\mathrm{i}\,s}\,, \qquad M(0) = \frac{1}{4\pi^2\,\mathrm{i}\,s^2}\,, \tag{18.50}$$

where the latter result is obtained in the limit $E \to 0$. Inserting this compact result for $M(A)$ into Eq. (18.31), one obtains

$$\int \mathrm{d}^4x \, \mathcal{L}_{\rm eff}(x) = \frac{\mathrm{i}}{2} \int \mathrm{d}^4x \int_0^\infty \frac{\mathrm{d}s}{s} \, \mathrm{e}^{-\mathrm{i}s(m^2-\mathrm{i}\epsilon)} \, [M(A) - M(0)]$$
$$= \int \mathrm{d}^4x \, \frac{1}{8\pi^2} \int_0^\infty \frac{\mathrm{d}s}{s^2} \, \mathrm{e}^{-\mathrm{i}s(m^2-\mathrm{i}\epsilon)} \left(eE \coth(seE) - \frac{1}{s} \right). \qquad (18.51)$$

We can now drop the d^4x integration and obtain

$$\mathcal{L}_{\rm eff} = \frac{1}{8\pi^2} \int_0^\infty \frac{\mathrm{d}s}{s^2} \, \mathrm{e}^{-\mathrm{i}s(m^2-\mathrm{i}\epsilon)} \left(eE \coth(seE) - \frac{1}{s} \right). \qquad (18.52)$$

The leading term in the expansion of the hyperbolic cotangent in the integrand for $s \to 0$ leads to a $1/s^3$ divergence; this divergence is exactly compensated by the term $M(0)$.

However, there is a second divergence, in view of the expansion

$$e\,E \coth(se\,E) = \frac{1}{s} + \frac{1}{3} s(e\,E)^2 + \mathcal{O}(s^2), \qquad (18.53)$$

for small proper time s. We recall that the unperturbed Maxwell Lagrangian reads as $\mathcal{L}^{[0]} = \frac{1}{2}\left(\vec{E}^2 - \vec{B}^2\right)$, which for $\vec{B} = \vec{0}$ reduces to the expression $\vec{E}^2/2$. The logarithmic divergence generated by the $1/s$ term, for small proper time s, multiplies the square of the electric field $\vec{E}^2$. According to the renormalization program discussed in Sec. 10.6, it necessitates a Z_3 renormalization.

Let us study this in more detail. The Maxwell Lagrangian $\mathcal{L}^{[0]}$, for an electric field, plus the effective Lagrangian (18.52), produces the following term, proportional to $\vec{E}^2$,

$$\mathcal{L}^{[0]} + \mathcal{L}_{\rm eff} \sim \left(\frac{1}{2} + \frac{e^2}{8\pi^2} \int_\epsilon^\infty \frac{\mathrm{d}s}{3s} \, \mathrm{e}^{-\mathrm{i}s(m^2-\mathrm{i}\epsilon)} \right) \vec{E}^2 \,. \qquad (18.54)$$

Here, ϵ is a small proper time cutoff. The renormalization program dictates that a counter term has to be added to the original Maxwell Lagrangian, in terms of a Z_3 renormalization, so that the complete theory, formulated in terms of the renormalized, finite, physical fields, produces finite Green functions and in our case, a finite effective Lagrangian. This is fulfilled if we set, on the one-loop level,

$$(Z_3 - 1)^{[1]} = -\frac{e^2}{8\pi^2} \int_\epsilon^\infty \frac{\mathrm{d}s}{3s} \, \mathrm{e}^{-\mathrm{i}s(m^2-\mathrm{i}\epsilon)} \,. \qquad (18.55)$$

The counter term is perturbatively small, of order $e^2 \propto \alpha$, but logarithmically divergent for small ϵ. The unrenormalized and renormalized electric fields are related by

$$\vec{E}_0 = \sqrt{Z_3}\vec{E} \approx \left(1 - \frac{1}{2} \frac{e^2}{8\pi^2} \int_\epsilon^\infty \frac{\mathrm{d}s}{3s} \, \mathrm{e}^{-\mathrm{i}s(m^2-\mathrm{i}\epsilon)} \right) \vec{E} \,. \qquad (18.56)$$

One may now interpret $\vec{E}$ in Eq. (18.52) as the unrenormalized, bare electric field $\vec{E}_0$, use Eq. (18.56), and then write the result in terms of the physical electric field strength. If this is done, to one-loop order, i.e., cutting off the expansion at order

$e^2 \propto \alpha$, then it has the same effect as a simple subtraction of the logarithmically divergent term under the integral sign in Eq. (18.52). The result of this operation is

$$\mathcal{L}_{\text{eff}} = \frac{1}{8\pi^2} \int_0^\infty \frac{\mathrm{d}s}{s^2} \, \mathrm{e}^{-\mathrm{i}s(m^2 - \mathrm{i}\epsilon)} \left(eE \coth(seE) - \frac{e^2 E^2 s}{3} - \frac{1}{s} \right). \tag{18.57}$$

The renormalization counter term given in Eq. (18.55) restores the Maxwell Lagrangian in the low-energy (weak-field) limit without additional, finite renormalizations; the Lagrangian $\mathcal{L}^{[0]} + \mathcal{L}_{\text{eff}}$, with $\mathcal{L}_{\text{eff}}$ given in Eq. (18.57), has the leading terms $\frac{1}{2} E^2 + \mathcal{O}(E^4)$, with the prefactor of the quadratic term being exactly $\frac{1}{2}$. This corresponds to the on-mass-shell renormalization; the renormalization-group invariance of the QED effective action is analyzed in greater detail in Sec. 19.4.1.

The QED interaction Lagrangian does not contain derivative terms; therefore, the QED interaction Hamiltonian is equal to the negative of the QED interaction Lagrangian. In Sec. 3.3.1, we have learned that a decay width can be described as a complex resonance energy, according to the identification $E = \mathrm{Re}\, E - \frac{\mathrm{i}}{2} \Gamma$, where Γ is the width. The QED effective action as written in Eq. (18.57) has a manifestly nonvanishing imaginary part, which can thus be interpreted according to

$$\mathcal{L}_{\text{eff}} = \mathrm{Re}\left[\mathcal{L}_{\text{eff}}\right] + \frac{\mathrm{i}}{2}\, w \,, \tag{18.58}$$

where $w = w(x)$ is the pair-production amplitude per space-time volume, i.e., the probability of producing an electron-positron pair in a very intense, static, electric field. The vacuum persistence probability, which is the absolute square of the vacuum persistence amplitude, then decays according to a time evolution proportional to $\exp[-\int_V \mathrm{d}^3 r\, w(t, \vec{r})]$, where we have written $w(x) = w(t, \vec{r})$, and V is the reference volume. We can alternatively write the imaginary part of $\mathcal{L}_{\text{eff}}$ as the real part of the following expression,

$$w = -\frac{1}{4\pi^2} \int_0^\infty \frac{\mathrm{d}s}{s^2} \, \mathrm{Re}\left[\mathrm{i}\,\mathrm{e}^{-\mathrm{i}s(m^2 - \mathrm{i}\epsilon)}\right] \left(eE \coth(seE) - \frac{e^2 E^2 s}{3} - \frac{1}{s} \right). \tag{18.59}$$

Both factors in the integrand change sign under the transformation $s \to -s$, in the limit of $\epsilon \to 0$. Therefore, the integrand is even in s, and we may write

$$w = -\frac{1}{8\pi^2} \,\mathrm{Re} \int_{-\infty}^\infty \frac{\mathrm{d}s}{s^2} \, \mathrm{i}\,\mathrm{e}^{-\mathrm{i}sm^2 - \epsilon|s|} \left(eE \coth(seE) - \frac{e^2 E^2 s}{3} - \frac{1}{s} \right), \tag{18.60}$$

where of course, $|s|$ in the infinitesimal suppression factor in the exponential denotes the modulus of s. From now on, we assume $eE > 0$. Because of the structure of the argument of the exponential, the integration contour for s must be closed in the lower complex half plane; the singularities are at $s = -\mathrm{i}\, s'$, with $s' = n\pi/(eE)$, with positive integer $n > 0$. The singularity at $n = 0$ is canceled by the counterterm $-1/s$. Near the singularities, the hyperbolic cotangent can be expanded as follows,

$$\coth(seE) = \frac{1}{seE + \mathrm{i}n\pi} + \frac{1}{3}\,(seE + \mathrm{i}n\pi) + \mathcal{O}\left[(seE + \mathrm{i}n\pi)^2\right], \tag{18.61}$$

so that

$$\operatorname*{Res}_{s=-\frac{\mathrm{i}n\pi}{eE}} \coth(seE) = \frac{1}{eE} \,. \tag{18.62}$$

Collecting, carefully, all the prefactors, one finally obtains the expression

$$\begin{aligned} w &= -\frac{1}{8\pi^2} \sum_{n=1}^{\infty} \mathrm{i}\,(-2\pi\mathrm{i}) \left(-\mathrm{i}\frac{eE}{n\pi}\right)^2 \exp\left(-m^2 \frac{n\pi}{eE}\right) \\ &= \sum_{n=1}^{\infty} \frac{\alpha\, E^2}{n^2\pi^2} \exp\left(-\frac{n\pi m^2}{eE}\right), \end{aligned} \tag{18.63}$$

where we have used $e^2 = 4\pi\alpha$. This is the nonperturbative (in the QED coupling e) result for the pair production probability in a vacuum dressed by an intensive electric field of magnitude E. It is known as the SCHWINGER[1] mechanism [89].

18.2.2 *Heisenberg–Euler Lagrangian for General Fields*

The result (18.57) needs to be generalized to cases where both the electric as well as the magnetic fields are nonvanishing [87, 88]. It is intuitively clear that the general result must be expressible in terms of Lorentz invariants of the fields. From Chap. 8 of Ref. [81], we recall the field-strength tensor

$$F^{\mu\nu} = \begin{pmatrix} 0 & -E^x & -E^y & -E^z \\ E^x & 0 & -cB^z & cB^y \\ E^y & cB^z & 0 & -cB^x \\ E^z & -cB^y & cB^x & 0 \end{pmatrix}, \tag{18.64}$$

and its dual

$$\widetilde{F}^{\mu\nu} = \frac{1}{2}\,\epsilon^{\mu\nu\rho\delta}\, F_{\rho\delta} \,. \tag{18.65}$$

It reads, in components,

$$\widetilde{F}^{\mu\nu} = \begin{pmatrix} 0 & -cB^x & -cB^y & -cB^z \\ cB^x & 0 & E^z & -E^y \\ cB^y & -E^z & 0 & E^x \\ cB^z & E^y & -E^x & 0 \end{pmatrix}. \tag{18.66}$$

The general result for the QED effective action can be expressed in term of the eigenvalues ("secular invariants") of the field strength tensor. As a (4×4) matrix, $F^{\mu\nu}$ has four eigenvalues, expressible as $\pm a$, $\pm b$. They are expressed as follows,

$$a = \sqrt{\sqrt{\mathcal{F}^2 + \mathcal{G}^2} + \mathcal{F}}\,, \qquad b = \sqrt{\sqrt{\mathcal{F}^2 + \mathcal{G}^2} - \mathcal{F}}\,, \tag{18.67}$$

in terms of the Lorentz invariants $\mathcal{F}$ and $\mathcal{G}$,

$$\mathcal{F} = \frac{1}{4}\, F_{\mu\nu}\, F^{\mu\nu} = \frac{1}{2}\left(\vec{B}^{\,2} - \vec{E}^{\,2}\right) = \frac{1}{2}\left(a^2 - b^2\right), \tag{18.68a}$$

$$\mathcal{G} = \frac{1}{4}\, F_{\mu\nu}\, \widetilde{F}^{\mu\nu} = -\vec{E} \cdot \vec{B} = \pm ab \,. \tag{18.68b}$$

[1] Julian Seymour Schwinger (1918–1994).

For $\vec{E} = \vec{0}$, we have $b = 0$ and $a = |\vec{B}|$, whereas for $\vec{B} = \vec{0}$ we have $a = 0$ and $b = |\vec{E}|$. More generally, we have $b = |\vec{E}|$, and $a = |\vec{B}|$, if and only if $\vec{B}$ is (anti-)parallel to $\vec{E}$.

For $\mathcal{G} > 0$, it is possible to choose a Lorentz frame with $\vec{E} \parallel -\vec{B}$. For $\mathcal{G} < 0$, it is possible to choose a Lorentz frame with $\vec{E} \parallel \vec{B}$. For a plane-wave, monochromatic, traveling laser wave, we have $\mathcal{G} = 0$, and therefore $b = 0$. As we shall soon see, this implies that a plane, monochromatic laser wave cannot produce electron-positron pairs. In any case, because a and b are positive definite, we have $a\,b = |\vec{E}\cdot\vec{B}| > 0$, which clarifies the sign ambiguity in (18.68b). The Maxwell Lagrangian is given by

$$\mathcal{L}^{[0]} = -\mathcal{F} = -\frac{1}{4} F_{\mu\nu} F^{\mu\nu} = \frac{1}{2}\left(\vec{E}^2 - \vec{B}^2\right) = \frac{1}{2}\left(b^2 - a^2\right). \tag{18.69}$$

The QED effective action $\mathcal{L}_{\rm eff}$ is conveniently expressed as follows,

$$\begin{aligned}\mathcal{L}_{\rm eff} &= \frac{e^2}{8\pi^2}\int_0^\infty \frac{\mathrm{d}s}{s}\,\mathrm{e}^{-\mathrm{i}s\,(m^2-\mathrm{i}\epsilon)}\left[ab\coth(eas)\cot(ebs) - \frac{a^2-b^2}{3} - \frac{1}{(e\,s)^2}\right] \qquad (18.70)\\ &= -\frac{e^2}{8\pi^2}\lim_{\epsilon,\eta\to 0^+}\int_\eta^{\mathrm{i}\infty+\eta}\frac{\mathrm{d}s}{s}\,\mathrm{e}^{-(m^2-\mathrm{i}\epsilon)s}\left[ab\coth(eas)\cot(ebs) - \frac{a^2-b^2}{3} - \frac{1}{(es)^2}\right],\end{aligned}$$

where in the second step, the rotation of the contour gives rise to an overall minus sign in front of the integral. An expansion in powers of α gives rise to the following leading terms,

$$\mathcal{L} = \mathcal{L}^{[0]} + \mathcal{L}^{[1]}, \qquad \mathcal{L}^{[1]} = \frac{2\,\alpha^2}{45m^4}\left(\vec{E}^2 - \vec{B}^2\right)^2 + \frac{14\,\alpha^2}{45m^4}\left(\vec{E}\cdot\vec{B}\right)^2, \tag{18.71}$$

where $\mathcal{L}^{[0]}$ has been given in Eq. (18.69). We should note that an expansion of Eq. (18.70) in powers of α gives rise to a divergent series [612]. This is an expansion in interactions with external fields, which can be visualized by Feynman diagrams. Borel summability of the divergent series, in the sense of Eq. (3.4), has been studied and shown, both within the ordinary [612] as well as within the distributional sense for the case of an electric field [126,613]. The divergent character of the perturbative expansions does not limit their predictive power [614].

For a number of applications, it is useful to express the nonperturbative expression (18.70) in terms of special functions [615,616]. In particular, the real and imaginary parts can be expressed as

$$\mathrm{Re}\,\mathcal{L}_{\rm eff} = -\frac{e^2}{4\pi^3}ab\sum_{n=1}^{\infty}\left[a_n + d_n\right], \tag{18.72a}$$

$$a_n = \frac{\coth\left(\frac{n\pi b}{a}\right)}{n}\left[\mathrm{Ci}\left(\frac{n\pi m^2}{ea}\right)\cos\left(\frac{n\pi m^2}{ea}\right) + \mathrm{si}\left(\frac{n\pi m^2}{ea}\right)\sin\left(\frac{n\pi m^2}{ea}\right)\right],$$

$$d_n = -\frac{\coth\left(\frac{n\pi a}{b}\right)}{2n}\left[\mathrm{e}^{n\pi m^2/(eb)}\,\mathrm{Ei}\left(-\frac{n\pi m^2}{eb}\right) + \mathrm{e}^{-n\pi m^2/(eb)}\,\mathrm{Ei}\left(\frac{n\pi m^2}{eb}\right)\right],$$

$$\mathrm{Im}\,\mathcal{L}_{\rm eff} = \frac{e^2\,ab}{8\pi^2}\sum_{n=1}^{\infty}\frac{1}{n}\coth\left(\frac{n\pi a}{b}\right)\exp\left(-\frac{n\pi m^2}{eb}\right). \tag{18.72b}$$

On the occasion, we recall that m in these formulas denotes the electron mass, not an extraneous summation index. The "asymmetry" in these formulas regarding cosine and sine integrals is not arbitrary; the generally accepted definitions for these integrals [159] are indeed asymmetric:

$$\mathrm{Ci}(z) = -\int_z^\infty \mathrm{d}t\,\frac{\cos(t)}{t} \qquad \mathrm{si}(z) = -\int_z^\infty \mathrm{d}t\,\frac{\sin(t)}{t} = \mathrm{Si}(z) - \frac{\pi}{2}\,, \tag{18.73}$$

whereas $\mathrm{Si}(z) = \int_0^z dt\, t^{-1}\sin(t)$ and we assume that $z > 0$. The exponential integral is given as a principal-value expression,

$$\mathrm{Ei}(u) = -(\mathrm{P.V.})\int_{-u}^\infty \frac{\mathrm{e}^{-t}}{t}\,\mathrm{d}t\,, \qquad u \in \mathbb{R}\,. \tag{18.74}$$

Under an appropriate deformation of the contour, a unified representation for both the real and the imaginary parts is obtained [617],

$$\mathcal{L}_{\mathrm{eff}} = \lim_{\epsilon\to 0^+} -\frac{e^2}{4\pi^3}ab\sum_{n=1}^\infty (b_n + c_n)\,, \tag{18.75a}$$

$$b_n = -\frac{\coth\left(\frac{n\pi b}{a}\right)}{2n}\left[\exp\left(-\mathrm{i}\frac{n\pi m^2}{ea}\right)\Gamma\left(0,-\mathrm{i}\frac{n\pi m^2}{ea}\right)\right.$$
$$\left.+\exp\left(\mathrm{i}\frac{n\pi m^2}{ea}\right)\Gamma\left(0,\mathrm{i}\frac{n\pi m^2}{ea}\right)\right], \tag{18.75b}$$

$$c_n = \frac{\coth\left(\frac{n\pi a}{b}\right)}{2n}\left[\exp\left(\frac{n\pi m^2}{eb}\right)\Gamma\left(0,\frac{n\pi m^2}{eb}\right)\right.$$
$$\left.+\exp\left(-\frac{n\pi m^2}{eb}\right)\Gamma\left(0,-\frac{n\pi m^2}{eb}+\mathrm{i}\epsilon\right)\right]. \tag{18.75c}$$

Here, we make extensive use of the incomplete Gamma function defined as [see Eq. (6.5.3) of [159]]

$$\Gamma(a,z) = \int_z^\infty \mathrm{d}t\,\mathrm{e}^{-t}\,t^{a-1}\,. \tag{18.76}$$

For $a = 0$, the quantity $\Gamma(0,z)$ as a function of z has a branch cut along the negative real z axis, and we assume that $\lim_{\epsilon\to 0^+}\mathrm{Im}\,\Gamma(0,-x+\mathrm{i}\,\epsilon) = -\pi$, for $x > 0$. The effective action has branch cuts along the positive and negative b-axis as well as along the positive and negative imaginary a-axis. For more details, the reader is referred to [618]. The incomplete Gamma function $\Gamma(0,z)$ is defined in the entire complex plane with a cut along the negative real axis [159, 618].

18.3 Application of the Heisenberg–Euler Lagrangian

18.3.1 *Modification of the Speed of Light in Background Fields*

Now that we have evaluated the QED effective action, and especially the phenomenologically important perturbative terms from Eq. (18.71), the question is how to employ the concept in experimentally relevant calculations. From Ref. [81] and other textbooks on classical electrodynamics, we know that the homogeneous

Maxwell equations are an inherent property of the field strength tensor and thus do not change if we take into account the quantum corrections,

$$\vec{\nabla} \times \vec{E} + \frac{\partial}{\partial t}\vec{B} = 0\,, \qquad \vec{\nabla} \cdot \vec{B} = 0\,. \tag{18.77}$$

In curved space-times, the homogeneous Maxwell equations can be traced to so-called Bianchi identities fulfilled by the field-strength tensor [see Eq. (8.262) of Ref. [81]]. However, the inhomogeneous Maxwell equations, which couple the sources to the fields, are derived from an action principle. This is true in flat as well as curved space-times [see Eq. (8.258) of Ref. [81]]. The inhomogeneous Maxwell equations thus receive a quantum correction.

For convenience, we recall the Maxwell Lagrangian, the one-loop correction, and the modified Lagrangian which couples to the sources, as follows,

$$\mathcal{L} = \mathcal{L}^{[0]} + \mathcal{L}^{[1]} - J^\mu A_\mu\,, \qquad \mathcal{L}^{[0]} = \tfrac{1}{2}\left(\vec{E}^2 - \vec{B}^2\right), \tag{18.78a}$$

$$\mathcal{L}^{[1]} = \frac{2\,\alpha^2}{45 m^4}\left(\vec{E}^2 - \vec{B}^2\right)^2 + \frac{14\,\alpha^2}{45 m^4}\left(\vec{E}\cdot\vec{B}\right)^2, \tag{18.78b}$$

$$F^{\mu\nu} = \frac{\partial \mathcal{L}^{[0]}}{\partial(\partial_\mu A_\nu)}\,, \qquad \mathcal{F}^{\mu\nu} = \frac{\partial\left(\mathcal{L}^{[0]} + \mathcal{L}^{(1)}\right)}{\partial(\partial_\mu A_\nu)}\,. \tag{18.78c}$$

The variational equations become

$$\frac{\partial}{\partial x^\mu}\mathcal{F}^{\mu\nu} = \frac{\partial}{\partial x^\mu}\frac{\partial\left(\mathcal{L}^{[0]} + \mathcal{L}^{[1]}\right)}{\partial(\partial_\mu A_\nu)} = J^\nu\,. \tag{18.79}$$

The tensor $\mathcal{F}^{\mu\nu}$ is composed of the dielectric displacement field $\vec{D}$ (in place of the electric field $\vec{E}$) and of the magnetic field $\vec{H}$ (in place of the magnetic induction field $\vec{B}$). These are the effective fields in the presence of the background field. The field-strength tensor $F_{\mu\nu} = \partial_\mu A_\nu - \partial_\nu A_\mu$ contains the $\vec{E}$ and $\vec{B}$ fields, and we obtain the dielectric displacement $\vec{D}$ and the magnetic field strength $\vec{H}$ by differentiating the effective Lagrangian. In Cartesian components, we have, for spatial indices $i, j = 1, 2, 3$,

$$D^i = \frac{\partial}{\partial E^i}\left(\mathcal{L}^{[0]} + \mathcal{L}^{[1]}\right) = \varepsilon_r^{ij}\,E^j\,, \tag{18.80a}$$

$$H^i = -\frac{\partial}{\partial B^i}\left(\mathcal{L}^{[0]} + \mathcal{L}^{[1]}\right) = \left(\mu_r^{-1}\right)^{ij}\,B^j\,. \tag{18.80b}$$

We recall that $\epsilon = \mu_0 = 1$ in our units. The radiatively corrected Maxwell equations assume the form

$$\vec{\nabla} \cdot \vec{D} = \rho\,, \qquad \vec{\nabla} \times \vec{H} - \frac{\partial}{\partial t}\vec{D} = \vec{J}\,. \tag{18.81}$$

We have formulated the dielectric displacement $D^i = \varepsilon_r^{ij}\,E^j$ and the magnetic field $H^i = (\mu_r^{-1})^{ij}\,B^j$ in terms of the electric field $\vec{E}$ and the magnetic induction $\vec{B}$, and the vacuum-induced relative permittivity tensor ε_r^{ij} and vacuum-induced permeability tensor $(\mu_r^{-1})^{ij}$. The Cartesian components of the (relative) permittivity and permeability tensors are given as follows,

$$\varepsilon_r^{ij} = \delta^{ij} + \frac{4\alpha^2}{45\,m^4}\left[2(\vec{E}^2 - \vec{B}^2)\,\delta^{ij} + 7\,B^iB^j\right], \tag{18.82a}$$

$$\left(\mu_r^{-1}\right)^{ij} = \delta^{ij} + \frac{4\alpha^2}{45\,m^4}\left[2(\vec{E}^2 - \vec{B}^2)\,\delta^{ij} - 7\,E^iE^j\right], \tag{18.82b}$$

$$\mu_r^{ij} = \delta^{ij} - \frac{4\alpha^2}{45\,m^4}\left[2(\vec{E}^2 - \vec{B}^2)\,\delta^{ij} - 7\,E^iE^j\right]. \tag{18.82c}$$

In the last step, we have calculated the matrix inverse to the leading order in α^2. Terms of order α^4 and higher are neglected in Eq. (18.82).

It is very instructive to derive the effective wave equations for an electromagnetic wave propagating under the presence of a large magnetic background field. To this end, we replace, in the field equations,

$$\vec{E} \to \delta\vec{E}\,, \qquad \vec{D} \to \delta\vec{D}\,, \qquad \vec{B} \to \vec{B} + \delta\vec{B}\,, \qquad \vec{H} \to \vec{H} + \delta\vec{H}\,, \tag{18.83}$$

where the fields denoted by capital letters are the background fields, and the fields denoted by the lowercase letters denote a small perturbation. From Eqs. (18.80a) and (18.82a), we obtain the dielectric displacement,

$$\delta D^i = \left(1 - \frac{8}{45}\frac{\alpha^2\,\vec{B}^2}{m^4}\right)\delta E^i + \frac{28}{45}\frac{\alpha^2 B^i\,B^j}{m^4}\,\delta E^j\,, \tag{18.84}$$

whereas Eqs. (18.80b) and (18.82b) imply that there is an additional term when we expand to first order in the $\delta\vec{H}$,

$$\begin{aligned} H^i + \delta H^i &= \left(\delta^{ij} - \frac{8}{45}\frac{\alpha^2(\vec{B}+\delta\vec{B})^2}{m^4}\,\delta^{ij}\right)\left(B^j + \delta B^j\right) \\ &\approx \left(1 - \frac{8}{45}\frac{\alpha^2\,\vec{B}^2}{m^4}\right)B^i + \left\{\left(1 - \frac{8}{45}\frac{\alpha^2\,\vec{B}^2}{m^4}\right)\delta B^i - \frac{16}{45}\frac{\alpha^2 B^i\,B^j}{m^4}\,\delta B^j\right\}. \end{aligned} \tag{18.85}$$

The term in curly brackets in Eq. (18.85) is identified as δH^i. We thus have the following relative, effective permittivity and permeability tensors (which we denote as ε^{ij} and μ^{ij}). These are relevant to the perturbative fields,

$$\delta D^i = \varepsilon^{ij}\,\delta E^j\,, \qquad \delta H^i = \left(\mu^{-1}\right)^{ij}\,\delta B^j\,, \tag{18.86a}$$

$$\varepsilon^{ij} = \delta^{ij}\left(1 - \frac{8}{45}\frac{\alpha^2\,\vec{B}^2}{m^4}\right) + \frac{28}{45}\frac{\alpha^2\,B^iB^j}{m^4}\,, \tag{18.86b}$$

$$\left(\mu^{-1}\right)^{ij} = \delta^{ij}\left(1 - \frac{8}{45}\frac{\alpha^2\,\vec{B}^2}{m^4}\right) - \frac{16}{45}\frac{\alpha^2\,B^iB^j}{m^4}\,, \tag{18.86c}$$

$$\mu^{ij} = \delta^{ij}\left(1 + \frac{8}{45}\frac{\alpha^2\,\vec{B}^2}{m^4}\right) + \frac{16}{45}\frac{\alpha^2\,B^iB^j}{m^4}\,. \tag{18.86d}$$

Let us assume that $\delta\vec{E}$, $\delta\vec{D}$, $\delta\vec{B}$ and $\delta\vec{H}$ describe a propagating wave with wave vector $\vec{k}$ and angular frequency ω, in a strong homogeneous background field $\vec{B}$. In a source-free region of space, the homogeneous and inhomogeneous Maxwell equations for a wave, propagating with a wave vector $\vec{k}$, then imply that

$$\vec{k} \times \delta\vec{E} - \omega\,\delta\vec{B} = \vec{0}\,, \qquad \vec{k}\cdot\delta\vec{B} = 0\,, \qquad \vec{k}\cdot\delta\vec{D} = 0\,, \qquad \vec{k}\times\delta\vec{H} + \omega\,\delta\vec{D} = \vec{0}\,. \tag{18.87}$$

From this point onwards, we can treat the wave propagation problem just like in wave optics. The background field $\vec{B}$ and the propagation vector $\vec{k}$ span a reference plane. The $\delta\vec{E}$ field, which defines the polarization vector of the electric field (unit vector in the direction of $\delta\vec{E}$), can either lie in this plane or it can be perpendicular. If $\delta\vec{E}$ is perpendicular to the reference plane, then we speak of the transverse ($\perp$) mode. In case $\delta\vec{H}$ is perpendicular to the reference plane, we have a longitudinal ($\|$) mode. The strong background magnetic field defines a susceptibility tensor, and we would like to find a solution to the effective wave equations as listed in Eq. (18.87). For the transverse mode (where the electric field is perpendicular to the background magnetic field), this solution reads as follows,

$$\delta\vec{E} = \delta\vec{E}_\perp = \mathcal{E}_0\,\hat{k}\times\hat{B}\,, \qquad \delta\vec{H} = \mathcal{E}_0\,\chi\left(n_\perp\,(\hat{k}\cdot\hat{B})\,\hat{k} - n_\perp^{-1}\,\hat{B}\right), \tag{18.88a}$$

$$n_\perp = 1 + \frac{8}{45}\frac{\alpha^2\,\vec{B}^2}{m^4}\sin^2\theta\,, \qquad \theta = \angle(\vec{B},\vec{k})\,, \qquad \chi = 1 - \frac{8}{45}\frac{\alpha^2\,\vec{B}^2}{m^4}\,. \tag{18.88b}$$

Here, $\mathcal{E}_0$ is a field amplitude, which is multiplied by the vector product of the unit vectors $\hat{k}\times\hat{B}$. As usual, $\hat{k} = \vec{k}/|\vec{k}|$ denotes the unit vector. Let us verify that indeed, $n_\perp$ has the physical meaning of a refractive index. To this end, we use the equation $\vec{k}\times\delta\vec{H} = -\omega\,\delta\vec{D}$ and observe that, by virtue of the definition of the perpendicular mode,

$$\delta D^i = \varepsilon^{ij}\,\delta E^j = \mathcal{E}_0\,\varepsilon^{ij}\,(\hat{k}\times\hat{B})^j = \mathcal{E}_0\,\chi\,\delta^{ij}\,(\hat{k}\times\hat{B})^j = \mathcal{E}_0\,\chi\,(\hat{k}\times\hat{B})^i\,. \tag{18.89}$$

We can now write the vector product $\vec{k}\times\delta\vec{H}$ either using Eq. (18.88a), or using Eq. (18.87), and compare,

$$\vec{k}\times\delta\vec{H} = -\mathcal{E}_0\,k\,n_\perp^{-1}\,\chi\,(\hat{k}\times\hat{B}) = -\omega\,\delta\vec{D} = -\mathcal{E}_0\,\omega\,\chi\,(\hat{k}\times\hat{B})\,. \tag{18.90}$$

So, indeed,

$$-\,k\,n_\perp^{-1}\,\chi\,(\hat{k}\times\hat{B}) = -\omega\,\chi\,(\hat{k}\times\hat{B})\,. \tag{18.91}$$

We therefore have, for the phase velocity,

$$\frac{\omega}{k} = n_\perp^{-1} = c_\perp\,, \tag{18.92}$$

where $c_\perp$ is the speed of light in the prepared medium, along the specific polarization axis (we recall that in our unit system, we have $c = 1$ where c is the speed of light in vacuum). For the parallel mode ($\|$ mode, with the magnetic field of the traveling wave being perpendicular to the background magnetic field), the solution reads as follows,

$$\delta\vec{H} = \delta\vec{H}_\perp = \mathcal{E}_0\,\hat{k}\times\hat{B}\,, \qquad \delta\vec{E} = \frac{\mathcal{E}_0}{\chi}\left(-n_\|\,(\hat{k}\cdot\hat{B})\,\hat{k} + \frac{1}{n_\|}\,\hat{B}\right), \tag{18.93a}$$

$$n_\| = 1 + \frac{14}{45}\frac{\alpha^2\,\vec{B}^2}{m^4}\sin^2\theta\,, \qquad \theta = \angle(\vec{B},\vec{k})\,, \qquad \chi = 1 - \frac{8}{45}\frac{\alpha^2\,\vec{B}^2}{m^4}\,. \tag{18.93b}$$

In the presence of the strong background field, the vacuum becomes birefringent ($n_\perp \neq n_\|$), and the two different refractive indices are $n_\perp$ and $n_\|$, with the results given in Eqs. (18.88b) and (18.93b). The speed of light along the ordinary axis (the "fast" axis) of the vacuum is $c/n_\perp$. Along the extraordinary ("slow") axis, the speed of light is $c/n_\|$. Very strong magnetic fields can be found in so-called neutron stars, and some current high-precision experiments currently attempt to detect the vacuum birefringence effects caused by the excitation of virtual electron-positron pairs, as described by the effective Lagrangian [619, 620].

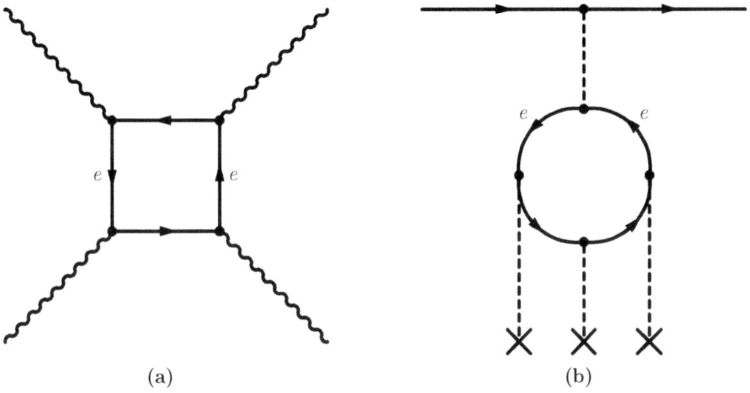

Fig. 18.1 The left Feynman diagram (a) is a light-by-light scattering diagram which gives rise to the perturbative terms of fourth order in the electromagnetic fields, as listed in Eq. (18.71). The diagram (b) on the right is the Wichmann–Kroll correction to the Coulomb interaction, which is described by a quantum correction to the Coulomb photon, due to three (or more) interactions of the electron-positron pair in the loop, with the external Coulomb field.

18.3.2 *Effective Lagrangian and Wichmann–Kroll Potential*

In Sec. 10.4, we have analyzed the one-loop correction to the photon propagator, due to a vacuum-polarization loop, as well as the concomitant correction to the Coulomb potential (Uehling potential), and we had mentioned that there exist further corrections to the Coulomb interaction, which are described by diagrams with more Coulomb vertices; these are known as the Wichmann–Kroll corrections [90,91]. The result for the fourth-order terms in the Heisenberg–Euler effective Lagrangian given in Eq. (18.71) corresponds to the light-by-light scattering Feynman diagram of Fig. 18.1. We should thus clarify the connection of the Heisenberg–Euler Lagrangian to the Wichmann–Kroll potential, which in leading order also entails the same diagram, but in a different kinematic region. This is very important, because in the spirit of effective Lagrangians, we should be able to use the Heisenberg–Euler Lagrangian in order to minimize the effective action for the quantum-corrected Coulomb field, obtaining at least an asymptotic form of the Wichmann–Kroll correction.

The Coulomb potential Φ generates a static electric field $\vec{E} = -\vec{\nabla}\Phi$. We should thus start from the effective action for a pure electric field $\vec{E}$, generated by a radially symmetric potential Φ, so that the Lagrangian (18.71) with the source term (18.78) reads as

$$\mathcal{L} = \frac{1}{2}\vec{E}^2 + \frac{2\alpha^2}{45m^4}\vec{E}^4 - \rho\,\Phi = \frac{1}{2}\left(\vec{\nabla}\Phi\right)^2 + \frac{2\alpha^2}{45m^4}\left(\vec{\nabla}\Phi\right)^4 - \rho\,\Phi\,. \tag{18.94}$$

The variational principle demands that the Coulomb potential, including the quantum correction, should minimize the action, which is the space-time integral of the Lagrangian. Because the charge configuration is static, the action S is obtained as the integral of $\mathcal{L}$ over all space,

$$\begin{aligned} S[\Phi] &= \int \mathrm{d}^3r \left(\frac{1}{2}\left(\vec{\nabla}\Phi\right)^2 + \frac{2\alpha^2}{45m^4}\left(\vec{\nabla}\Phi\right)^4 - \rho\,\Phi\right) \\ &= 4\pi\int_0^\infty \mathrm{d}r\, r^2 \left(\frac{1}{2}\left(\partial_r\Phi\right)^2 + \frac{2\alpha^2}{45m^4}\left(\partial_r\Phi\right)^4 - \rho\,\Phi\right), \end{aligned} \tag{18.95}$$

where we assume that $\Phi(\vec{r}) = \Phi(r)$ is radially symmetric,

$$\partial_r\Phi = \frac{\partial}{\partial r}\Phi(r) = \left|\vec{\nabla}\Phi(\vec{r})\right|\,. \tag{18.96}$$

We now calculate the variation

$$S[\Phi + \delta\Phi] - S[\Phi] = \int_0^\infty \mathrm{d}r\, r^2\,\delta\Phi\left[\left(-\partial_r^2 - \frac{2}{r}\,\partial_r\right)\Phi - \frac{8\alpha^2}{45m^4}\left(-\partial_r - \frac{2}{r}\right)\left(\partial_r\Phi\right)^3 - \rho\right]. \tag{18.97}$$

Here, we have used the following identity which is valid for arbitrary, radially symmetric functions $A(r)$ and $B(r)$,

$$\int_0^\infty \mathrm{d}r\, r^2\,\left(\partial_r A(r)\right)\,B(r) = \int_0^\infty \mathrm{d}r\, r^2\, A(r)\left(-\partial_r - \frac{2}{r}\right)B(r)\,. \tag{18.98}$$

The variational equation thus reads as

$$\left(-\partial_r^2 - \frac{2}{r}\,\partial_r\right)\Phi + \frac{8\alpha^2}{45m^4}\left(-\partial_r - \frac{2}{r}\right)\left(\partial_r\Phi\right)^3 = \rho\,. \tag{18.99}$$

For a radially symmetric function $A(\vec{r}) = A(r)$, we have $\vec{\nabla}^2 A(\vec{r}) = \left[\partial_r^2 + (2/r)\,\partial_r\right]A(r)$, and thus, the variational equation can be reformulated as follows,

$$\vec{\nabla}^2\Phi + \frac{8\alpha^2}{45m^4}\left(\partial_r + \frac{2}{r}\right)\left(\partial_r\Phi\right)^3 = -\rho\,. \tag{18.100}$$

We now assume that $\Phi = \Phi_C + \Xi$ is given by the sum of the Coulomb potential Φ_C and a quantum correction Ξ. The charge density of the nucleus and the Coulomb potential are given by

$$\rho(\vec{r}) = Z\,|e|\,\delta^{(3)}(\vec{r})\,, \qquad \Phi_C(\vec{r}) = \frac{Z\,|e|}{4\pi\,r}\,, \qquad \vec{\nabla}^2\,\Phi_C(\vec{r}) = -\rho(\vec{r})\,. \tag{18.101}$$

We note that in view of the nonlinearity of Eq. (18.100), the absolute normalization of the potential matters in the current calculation; it would thus be a mistake to replace $\Phi \to e\,\Phi = -Z\alpha/r$, where e is the electron charge. Now, we enter into Eq. (18.100) with an *ansatz*

$$\Phi = \Phi_C + \Xi\,, \qquad \left(\partial_r^2 + \frac{2}{r}\,\partial_r\right)\Xi + \frac{8\,\alpha^2}{45 m^4}\left(\partial_r + \frac{2}{r}\right)\left(\partial_r \Phi_C\right)^3 = 0\,. \tag{18.102}$$

Here, we have inserted the first-order solution Φ_C into the nonlinear corrective quantum term, whereas Ξ is used in the leading, nonlinear term. We have also used the fact that $(\partial_r + 2/r)\,(1/r) = 0$. In order to justify this approximation, we recall that no further variation is necessary; we simply have to solve Eq. (18.100) with a suitable *ansatz* $\Phi = \Phi_C + \Xi$, where Ξ is the correction to be determined. It is straightforward to observe that Eq. (18.102) is solved by a potential proportional to r^{-5}, and we finally obtain

$$\Phi = \Phi_C + \Xi = \frac{Z\,|e|}{4\pi\,r}\left(1 - \frac{2}{225}\,\frac{\alpha}{\pi}\,\frac{(Z\alpha)^2}{(m\,r)^4}\right), \tag{18.103a}$$

$$V = e\,\Phi = -\frac{Z\alpha}{r} + \frac{2}{225}\,\frac{\alpha}{\pi}\,\frac{(Z\alpha)^3}{m^4\,r^5}\,. \tag{18.103b}$$

The latter term in the above expression is equal to the long-distance tail of the Wichmann–Kroll potential [90, 91]. We note that a tail of the form $1/r^5$ for the Wichmann–Kroll potential is different in its functional dependence from the Uehling potential, which decays exponentially at large distances [379]. The term of order $(\alpha/\pi)\,(Z\alpha)^3/(m^4\,r^5)$ generates an energy shift proportional to $\alpha\,(Z\alpha)^8\,m$. We should note that this term does not constitute the "leading" term in the Wichmann–Kroll correction for S states in hydrogen and other, simple atomic systems. The leading Wichmann–Kroll term is otherwise known to be proportional to a Dirac-δ function and generates an energy shift of the order $\alpha\,(Z\alpha)^6\,m$.

Our variational calculation shows, at the same time, the usefulness of the effective Lagrangian for bound-state calculations, but also, its limitations: Namely, near the atomic nucleus, the Coulomb potential and its derivative, the Coulomb field, vary considerably on the length scale of an electron Compton wavelength, which exceeds the operational parameters for the Heisenberg–Euler Lagrangian. So, it is impossible to use the QED effective action in order to obtain the Dirac–δ term of the Wichmann–Kroll potential [90, 91], due to the very starting point used in the calculation of the effective Lagrangian. In bound-state calculations, the usefulness of effective field-theory methods typically is limited. They are useful in order to obtain the asymptotic limits of corrections, but they can seldomly be used in order to obtain complete results. At some risk to oversimplification, we can say that bound-state theory is a "Rallye Monte Carlo", which puts many more refined, and more ramified demands on theory than the plain "Formula One" course of loop correction in the field theory of free quantum states. The latter simply entail "high-profile specifications" without the need to deal with the "narrow curves" of bound-state theory, encountered near the singularities of the Coulomb potential, or the "narrow curves" due to the electron-electron repulsive force. Still, our treatment serves to convince ourselves that the non-exponential decay of the Wichmann–Kroll potential at long distances is fully consistent with the principles of field theory. We reemphasize that this asymptotic behavior is in sharp contrast to the exponential asymptotic behavior of the (one-loop) Uehling potential for large distances [379].

18.4 Further Thoughts

Here are some suggestions for further thought.

(1) **Matrix Exponential and Logarithm.** Show that the matrix exponential and matrix logarithm have the properties

$$\exp(C\,\mathcal{A}\,C^{-1}) = C\,\exp(\mathcal{A})\,C^{-1}\,, \qquad \ln(C\,\mathcal{A}\,C^{-1}) = C\,\ln(\mathcal{A})\,C^{-1}\,. \tag{18.104}$$

On account of these identities, ramify the transformation (18.26).

(2) **Effective Action and Divergent Series.** Calculate the first few terms, or even, the general terms (accompanying the nth power of α) in the asymptotic expansion of the effective Lagrangian (18.57) in powers of the electric field, and convince yourself of the divergent character of the expansion. Try to find a graphical representation for the terms in the expansion, conceivably in terms of Feynman diagrams.

(3) **Effective Action and Electric Field.** Find the limit as $a \to 0$, $b = |\vec{E}|$, of the general expressions of the real part, and the imaginary part, of the effective action as given in Eqs. (18.72a) and (18.72b). Write the result as a function of the field strength $E = |\vec{E}|$. For the real part, convince yourself of the consistency of your result with the original expression given in the integral representation (18.57), conceivably comparing to a numerical evaluation of the integral given in Eq. (18.57). For the imaginary part, you should obtain a result consistent with Eq. (18.63).

(4) **Effective Action and Vacuum Birefringence.** Derive both Eqs. (18.88) and (18.93) corresponding to vacuum birefringence in a strong magnetic field.

(5) **Effective Action and Wichmann–Kroll Potential.** Attempt to derive Eq. (18.102) by examining the Euler–Lagrange equations corresponding to the Lagrangian (18.95).

(6) **Applications of the Light-by-Light Scattering Tensor.** Study a few applications (and historical notes on the possible observation of photon splitting) recorded in Ref. [621]. In particular, investigate whether or not, in some applications related to photon production in heavy-ion collisions, virtual light-by-light scattering processes can dominate over tree-level amplitudes [622, 623].

Chapter 19

Renormalization-Group Equations

19.1 Orientation and Motivation

19.1.1 *Renormalization on Different Scales*

Quantum electrodynamics is a quantum field theory, and one of the paradigmatic developments in this area, continued over the last couple of decades, concerns the so-called renormalization group which establishes connections between the behavior of quantum electrodynamics at varying distance, momentum, or energy scales. As a final part of this treatise, an overview is to be given regarding the applications of related concepts to precision physics. Indeed, we shall attempt, below, to give an illustrative overview of the main ideas that have been pursued over the last couple of decades, in the field of renormalization-group transformations in the context of gauge theories. This discussion will be illustrated by a few concrete calculations in the following.

The renormalization program implies that in the canonical approach, we start from the original Lagrangian and do perturbation theory. The perturbative corrections yield further terms which need to be taken into account. Thus, we identify the original Lagrangian as the bare Lagrangian. The multiplicative renormalization implies that the original bare Lagrangian, written in terms of the bare quantities, can be rewritten as the same Lagrangian, but in terms of the renormalized quantities. If $\mathcal{L}$ stands for the mapping of the parameters of the theory to the Lagrangian, then, for a scalar theory with a bare field ϕ_0 (renormalized field ϕ), and a bare coupling g_0 (renormalized coupling g), we would have the schematic relationship

$$\mathcal{L}_0 = \mathcal{L}(\phi_0, g_0, \dots), \quad \mathcal{L}_R = \mathcal{L}(\phi, g, \dots) + \mathcal{L}_{\mathrm{CT}}(\phi, g, \dots). \tag{19.1}$$

Here, $\mathcal{L}_0$ is the bare Lagrangian, $\mathcal{L}_R$ is the renormalized Lagrangian, and the counter terms $\mathcal{L}_{\mathrm{CT}}$ are obtained from the loop corrections, i.e., they represents those contributions which need to be reabsorbed into the physical parameters of the theory. The multiplicative nature of the renormalization is manifest in the following relations,

$$\phi_0 = Z_\phi \, \phi, \qquad g_0 = Z_g \, g, \tag{19.2}$$

which relate the bare parameters of the theory to the physical, renormalized ones.

A different interpretation of the renormalization program is as follows. We start from the Lagrangian of the theory, and, evaluating the loop corrections, we see that we cannot maintain the desired interpretation of the physical parameters in the Lagrangian as the physical, renormalized fields. However, we would like to keep this interpretation. We can thus simply replace the parameters in the original Lagrangian of the theory by the physical parameters, provided we renormalize them accordingly. Here, Eq. (19.1) is recalled as

$$\mathcal{L}_R = \mathcal{L}(Z_\phi \phi, Z_g g, \dots) = \mathcal{L}(\phi, g, \dots) + \mathcal{L}_{\rm CT}(\phi, g, \dots)\,. \tag{19.3}$$

The counter terms shift the parameters back to the original, physical ones by the additional perturbations created. The new terms which need to be added to the Lagrangian are proportional to the ones that were there before, and therefore, after the renormalization, we end up with the same Lagrangian that we had before, but under the replacement $(\phi_0 \to Z_\phi \phi, g_0 \to Z_g g, \dots)$. The identification of the physical parameters is fixed by a set of renormalization conditions. In contrast to our discussion in Chap. 10, we here leave out the subscript R for the renormalized quantities.

The question now is whether some additional insight can be drawn from the process of renormalization. The most obvious route to additional insight, originally taken by Gell-Mann and Low [624], and later explained in greater detail by Akhiezer and Berestestkii [4], starts from the following consideration. We have already realized that the electron charge is "screened" by a dynamically induced "virtual charge" of opposite sign, i.e., the bare electron charge, visible at large momenta or very close approach, is larger than the electron charge seen from afar. The increase of the electron charge with the momentum scale is encoded in the vacuum polarization. In Sec. 10.4, we had evaluated the modified Coulomb potential $V_C(k^2)$ with $k^2 = -\vec{k}^2$ as follows (e_0 is the bare charge),

$$\begin{aligned} V_C(k^2) &= \frac{e^2}{k^2(Z_3 - \Pi(k^2))} = \frac{e^2}{k^2(1 + \Pi(0) - \Pi(k^2))} = \frac{e^2}{k^2(1 - \Pi_{\rm R}(k^2))}\,, \\ e^2 &= Z_3\, e_0^2\,, \qquad Z_3 = (1 + \Pi(0))\,. \end{aligned} \tag{19.4}$$

Here, Z_3 is a renormalization constant (see Chap. 10) which depends on the cutoff scale Λ; for $\Lambda \to \infty$, formally, we have $Z_3 \to 0$. However, we know that physically, the cutoff scale Λ cannot be sent to infinity. The renormalized vacuum polarization function $\overline{\Pi}(k^2) = \Pi(k^2) - \Pi(0)$ only diverges logarithmically for $k^2 \to \infty$; we can thus write down the "effective" interaction at the scale k^2 as

$$e_{\rm eff}^2 = \frac{e_{\rm phys}^2}{1 - \Pi(0)}\,, \tag{19.5}$$

where $e_{\rm phys}^2$ is the physical electron charge, renormalized so that it represents the elementary charge that governs the cross section of low-energy Thomson scattering (we have previously denoted this charge as $e = e_{\rm phys}$). Yet, at the same time, we know that Λ only acts as a cutoff parameter for the logarithmic divergence of the vacuum polarization function $\omega(k^2)$, i.e., the leading logarithmic divergence of $\omega(0)$ as a function of Λ, and of $\overline{\omega}(k^2)$ as a function of k^2 are not independent. This consideration suggests that, at least in the asymptotic regime $k^2 \gg m^2$, the dependence of Z_3 on Λ and of $e^2_{\rm eff}$ on k^2 must be interrelated. Furthermore, for $k^2 \gg m^2$, one might conjecture that the "running" of the charge with the momentum scale must be independent of m^2; the concrete numerical value of m^2 cannot be important in the asymptotic, high-energy regime. Indeed, as we shall see in the following, it is possible to express the "running" of the charge of high energy with the momentum scale exclusively as a function of the effective charge at a reference scale. Let us formulate this statement in terms of a "running" of the QED coupling $\alpha = e^2/(4\pi)$. The QED β function expresses the fact that the "running" of the coupling with the scale is a function of the coupling at the reference momentum scale itself,

$$k^2 \frac{\partial}{\partial k^2}\alpha(k^2) = \beta\left(\alpha(k^2)\right) . \tag{19.6}$$

Because the behavior of the β function is determined by the dependence of the Z_3 renormalization constant on the cutoff parameter, the renormalization-group (RG) analysis assigns "additional physical significance" to the process of renormalization, beyond the simple statement that the "infinities are reabsorbed" into the physical parameters of the theory observed at a specific scale. The renormalization group is called a "group" because the β function actually allows us to evolve the coupling, say, from the scale k_1^2 to the scale k_2^2 to the scale k_3^2, by integrating Eq. (19.6) with respect to k^2, starting from a reference scale k_1^2, where the value of $\alpha(k_1^2)$ needs to be "matched" against observations (e.g., QED at the Z pole).

For QED, the theory is canonically matched against the physical theory at long distances and on the mass shell of the electron (on-mass-shell renormalization). This means that we have measured the electron charge at long distances (fine-structure constant), and the free electron mass, and then, we construct QED theory and match the theory against the physically observed parameters on mass shell. However, we could also have measured the electron charge at a momentum scale equal to the weak interaction scale (Z scale) near $k^2 \approx m_Z^2 \approx 91\,\mathrm{GeV}^2$, and the electron mass at a different scale, and matched the QED Lagrangian against the different set of observed parameters at the Z scale. This would be a no less fundamental matching. In dimensional regularization, one naturally introduces a momentum scale μ into the theory, as a parameter into the counterterms. The counterterms therefore depend on the renormalization scale μ, and one can describe the running of the coupling etc. by a change in the renormalization scale μ. Irrespective of where we carry out the matching, the perturbative evolution of the coupling with the scale must still be described, at least approximately, by the same multiplicative factors that we originally invoked in order to describe the evolution of the electron charge with momentum, starting from zero momentum. We shall see that this is universally the case in the limit where the distance scale is much shorter than the electron Compton wavelength, i.e., in a regime where we can ignore the electron mass compared to the momentum transfer. QED in that sense is a set of possible theories, with the physical QED theory picked out by nature through the somewhat arbitrary selection of the value of the fine-structure constant α by a mood of nature. The actual value of α is not fixed by any more fundamental consideration. Again, the running of the coupling due to the vacuum polarization therefore is not more or less fundamental than a conceivable change of α, that is, the fine-structure constant in the limit of zero momentum transfer.

As one approaches the fundamental, UV, scale by carrying out the renormalization at ever shorter distance scales, one approaches the microscopic distance of the bare theory, and the renormalized parameters become more and more equal to the bare parameters. In QED, since the coupling is ever growing with the scale, the bare coupling is supposed to be very large, or even infinite if the renormalized coupling has a finite, nonzero value (which it has, since the fine-structure constant has a nonvanishing value $\alpha \neq 0$). There might have been hope that the running of the charge is a fundamental process that eventually leads us to a potential understanding of the unrenormalized, bare charge at very small distance scales. However, that is not the case: rather, we are taught by the running of the coupling, that the evolution of the charge at smaller distance needs to be described by a more fundamental theory. In that sense, QED theory is the effective theory, in the infrared, of a more fundamental theory that acts on very short distance scales (the Planck scale!?!), and whose details are hidden from us. We cannot hope to learn anything about the more fundamental interactions that govern the Planck scale, from the infrared limit of the theory, which is QED.

These consideration eventually lead to an additional interpretation of a quantum field theory itself: Namely, a field theory actually has the value of the coupling as a free parameter. Through the β function, a field theory predicts how the effective coupling evolves with the momentum scale at which processes are observed; it does not fix the value of the coupling at a reference scale; this latter observation must still be made independently, and necessitates an experiment. For QED, we are in the lucky position that the couplings remain finite in the infrared, so that we can match the electron charge with the observed electron charge for large-distance interactions. For other gauge theories, like quantum chromodynamics (QCD), the effective couplings actually diverge in the infrared (at least within a perturbative formalism), and it is thus impossible to fix the QCD coupling constant by observing, say, the interaction of two quarks in the limit in which they are far apart (this limit otherwise serves as a valid ground for fixing the value of the QED coupling). The question then is whether one can generalize the scale transformation argument, originally presented above, to a more general setting.

19.1.2 *Renormalization-Group Equations*

The formulation of RG equations may proceed along three different lines. *(i)* The original approach of Gell-Mann and Low was to formulate a specific function that characterizes the RG evolution of a physical quantity of interest (in their case, the running of the QED coupling, which is formulated in terms of the logarithmic derivative of the coupling with respect to the momentum scale), and to show that it is independent of any other physical parameters in a specified, asymptotic limit (in their case, independent of the electron mass for very short distance scales). The RG equation for the quantity of interest then follows immediately. *(ii)* The approach of Callan and Symanzik is based on the observation that differentiations of generating vertex functions and other "inverted" Green functions with respect to mass scales generate mass insertions. Paradigmatically, we can formulate this observation as

$$\frac{\partial}{\partial m_0^2}\Gamma^{(n)}\bigg|_{g_0=\text{const.}} = \Gamma^{(n,1)}\,, \tag{19.7}$$

where $\Gamma^{(n)}$ is an n-vertex function, and $\Gamma^{(n,1)}$ is a vertex function with a mass insertion. The mass insertion, for QED as well as for other theories (N-vector model), necessitates the introduction of yet another renormalization constant, called Z_0 or $Z_{\psi\psi}$ in the case of QED. The application of the operator $m_R\,\partial/\partial m_R$,

$$\Gamma_0^{(n)} = Z^{-1}\Gamma_R^{(n)}\,, \qquad m_R\,\frac{\partial}{\partial m_R}\Gamma_0^{(n)} = m_R\,\frac{\partial m_0^2}{\partial m_R}\,\frac{\partial}{\partial m_0^2}\Gamma_0^{(n)} = \left(m_R\,\frac{\partial m_0^2}{\partial m_R}\right)\Gamma_0^{(n,1)}\,, \tag{19.8}$$

then leads to equations of the Callan–Symanzik type, where upon renormalization, both sides of the equation are finally expressed in terms of renormalized quantities. The paradigm is that one compares theories at different length scales, while keeping the bare coupling constant and while varying the bare mass of the theory in such a way that a meaningful equation relating vertex functions is obtained. In appropriate asymptotic limits, one may then discard the right-hand side on dimensional grounds. Typically, the mass insertion implies that for very large momenta, the function $\Gamma_0^{(n,1)}$ falls off more rapidly than the function $\Gamma_0^{(n)}$, which in turn implies that one can ignore the inhomogeneous term in the Callan–Symanzik equation in an appropriate asymptotic limit, which in some cases enables us to recover the approach *(i)* from approach *(ii)*. Again, the Callan–Symanzik equations compares two theories valid at different renormalization scales while keeping the bare theory invariant (up to a variation of the mass). The approach *(iii)* of Zinn-Justin [130] is based on the introduction of a cutoff Λ, and upon the comparison of bare (!) theories at the cutoff scale. The Zinn-Justin equation expresses the following fact: Near the scale Λ of the bare theory, it is possible to vary the cutoff scale Λ, the value of the coupling g_0, and the normalization of the field as given by the anomalous dimension η, while keeping the physics at the phenomenological scale of the theory, i.e., in the infrared, invariant. This is expressed in Eq. (13.13) of Ref. [130]. One then expresses the fact that the bare theory, at scale Λ/λ, can be identified with the physical, renormalized theory at scale Λ/λ, provided one applies a scale-dependent renormalization $Z = Z(\lambda)$ to the "bare" theories which are being compared to each other. The "bare" theory at scale Λ/λ is then being identified with the renormalized theory at that scale, modulo the renormalizations. The latter consideration allows us to identify the β functions of the theory, initially formulated in terms of logarithmic derivatives with respect to the cutoff Λ, in terms of derivatives with respect to the scale λ, of the renormalized coupling, thus restoring the intuitive picture of the β function (description of the RG evolution of the coupling with the scale).

As a last remark, we briefly comment on so-called functional RG equations which express the evolution of the effective Hamiltonian of a theory with the cutoff scale. Specifically, the Polchinski [625] and Wegner–Houghton [626] equations are based on the general idea that if one integrates out the fluctuations of the field ϕ_1 in the path integral, one obtains an equation for the field ϕ_2. In QED, this approach, where ϕ_1 plays the role of the fermion field, leads to an effective interaction for the electromagnetic field (see Chap. 18). Now, Polchinski's realization is based on the observation that this approach can be applied to the field ϕ_1 itself if one splits the field fluctuations in the following way: Namely, one first isolates those fluctuations that happen at momentum scales $k \geq \Lambda$ (field ϕ_2) and integrates them out. As a function of Λ, one then obtains an equation for the evolution of the effective potential $\mathcal{V} = \mathcal{V}_\Lambda$ as a function of the scale Λ (obtaining an equation for the field ϕ_1). One can then solve this equation using "trial functions" which represent nothing but an *ansatz* for the "running effective potential" as a function of the cutoff scale Λ. In this way, one can even study the evolution of the effective potential with distance (or momentum) scales for theories which are not renormalizable in the perturbative sense, because the *ansatz* that one chooses for the effective potential may contain, e.g., periodic terms in the fields like $\cos\phi$ which are not renormalizable perturbatively. In a typical practical application, the functional RG equations are excellent tools for the study of the phase structure of perturbatively nonrenormalizable, field-theoretical models. One interesting question to ask is: Does one of the couplings (expansion coefficients for the trial function for the effective potential) reverse its asymptotic behavior in the infrared, as a function of the value chosen for this coupling in the ultraviolet?

The Wetterich equation [130] is obtained from the Polchinski equation by a Legendre transformation. It expresses the running of the effective action Γ_k with the scale k in terms of the derivative of a regulator function $\partial_k R_k$, and of the complete generating functional for the two-point function $\Gamma_k^{(2)}$. A summation (trace) over the quantum fluctuations beyond the regulator scale and a supertrace over the bosonic and fermionic degrees of freedom of the model is implied.

Of course, in the perturbative limit, the functional equations reproduce the findings of the perturbative expansions, and it is hard to obtain more accurate predictions for, e.g., the scale-dependent wave function renormalization constant from a nonperturbative equation as compared to the perturbative approach, because the Polchinski equation nevertheless introduces an ambiguity into the calculation (exact shape of the smooth cutoff function). The Wegner–Houghton equation is difficult to solve while taking into account all the renormalizations of the couplings and fields during the evolution. In principle, the full knowledge of the effective potential at a given distance scale, obtained by a solution of the Polchinski or Wegner–Houghton equations, would allow us to obtain the RG evolution of all correlation functions. However, this program has not yet been completed to greater accuracy than by a perturbative approach for any nontrivial field theory of practical interest.

We proceed as follows. In Sec. 19.2, we first take a step back and discuss scale transformations and their effect on solutions of the Schrödinger–Coulomb and Dirac–Coulomb problems. Indeed, the scaling equations will be very similar in form to the Callan–Symanzik RG equations discussed further down. We are then in the position to treat actual renormalization-group transformations in Sec. 19.3, starting first from a demonstration of the "existence" of the QED β function, and then, continuing to the calculation of a few terms in both the Gell-Mann–Low approach, and in the Callan–Symanzik approach. The connection of the Gell-Mann–Low and Callan–Symanzik approaches amounts to a change in the renormalization scheme; this statement will be cast into mathematical formulas. Finally, in Sec. 19.4, we discuss a number of applications of the developed formalism, first, in the context of the RG-invariance of the QED effective action, then, in terms of RG-optimized perturbation theory, before concluding with a discussion of conceivable connections of the physical value of the fine-structure constant with the RG analysis of QED.

19.2 Scale Transformations

19.2.1 *From Poor Man's Scaling to Nontrivial Scaling*

Scaling transformations are not renormalization-group transformations; however, there is a connection. By a scaling transformation, one can understand how a physical system develops when one of the parameters undergoes a scale change, which may result in a different hierarchy of terms: namely, on different scales, "small", "perturbative" terms may suddenly become dominant. E.g., for a quartic anharmonic oscillator, the harmonic term x^2 in the potential is dominant for small x, whereas the "perturbative" term $g\,x^4$ is dominant for large x, or, large coupling constant $g \gg 1$. In the context of bound-state problems, one can derive scaling laws for the solutions of the Schrödinger–Coulomb, and Dirac–Coulomb problems, which describe the evolution of bound-state solutions as a function of the nuclear charge number Z and the electron mass m. These relations will be used in the following, in order to derive compact formulas for a number of matrix elements, evaluated on relativistic bound states. Renormalization-group transformations are scale transformations which connect the physical parameters of a field theory on different length or momentum scales, due to the dependence of the renormalization constants on the reference scale at which the parameters are matched against observations; i.e., renormalization-group transformations constitute a special, "distinguished" class of scale transformations.

Let us start with a brief digression on "poor man's scaling", which is a scale transformation relevant to the analysis of quantum anharmonic oscillators. From the theory of anharmonic oscillators generalized to the anharmonic potentials, we are familiar with so-called poor man's scaling which mediates the transition from the weak-coupling perturbative regime, where the main contribution to the energy is from the unperturbed Hamiltonian, to the strong-coupling regime, where the energy is dominated by the anharmonic term.

To this end, we consider the Hamiltonian operator of the one-dimensional quartic anharmonic oscillator,

$$H = -\frac{1}{2}\frac{\partial^2}{\partial x^2} + \frac{1}{2}x^2 + g\,x^4 = -\frac{1}{2}\frac{\partial^2}{\partial x^2} + \frac{1}{2}x^2 + \mathcal{O}(g)\,, \tag{19.9}$$

where x is the coordinate. For $g \to 0$, the eigenvalues approximate those of the unperturbed harmonic oscillator Hamiltonian $-\frac{1}{2}\,\partial_x^2 + \frac{1}{2}\,x^2$, and are thus approximate by the formula $E_n \sim n + 1/2 + \mathcal{O}(g)$, where n is a nonnegative integer. However, for large coupling $g \to \infty$, the functional form of the eigenvalues is totally different, the "perturbative" term $g\,x^4$ becomes dominant, and a scaling transformation must be used in order to describe the large-coupling asymptotics. We scale the coordinate in Eq. (19.9) as $x \to g^{-1/6}\,x$, to obtain a Hamiltonian whose eigenvalues are still the same as those of the original Hamiltonian (19.9), but the functional form is very different,

$$H = g^{1/3}\left(-\frac{1}{2}\frac{\partial^2}{\partial x^2} + x^4 + \frac{x^2}{2\,g^{2/3}}\right) = g^{1/3}\left(-\frac{1}{2}\frac{\partial^2}{\partial x^2} + x^4 + \mathcal{O}(g^{-2/3})\right). \tag{19.10}$$

Here, the last term in round brackets acts as a perturbation that vanishes for $g \to \infty$. The energy levels of H, for $g \to \infty$, are well approximated by the formula $E_n \sim g^{1/3}\,\epsilon_n$, where ϵ_n is the nth level of the Hamiltonian operator $-\frac{1}{2}\,\partial_x^2 + x^4$. The scale transformation $x \to g^{-1/6}\,x$, which reveals the large-coupling asymptotics of the energy eigenvalues, is otherwise known as poor man's scaling.

19.2.2 *Scale Transformation and Schrödinger Hamiltonian*

Using the idea of "poor man's scaling", we aim to derive a set of identities, based on scale transformations of the Schrödinger–Coulomb, and Dirac–Coulomb Hamiltonians, to gain some additional insight into the physics and the scaling properties of the physical system. The systems under investigation are more complex than a simple anharmonic oscillator. We indicate the scaling relations for the Schrödinger–Coulomb, and the Dirac–Coulomb case; they have not yet appeared in the literature up to this point, to the best of our knowledge. Let us start from the Schrödinger–Coulomb Hamiltonian with an eigenvector $|\phi\rangle$,

$$H_S = \frac{\vec{p}^{\,2}}{2\mu} - \frac{Z\alpha}{r}\,, \qquad H_S|\phi\rangle = E_S|\phi\rangle\,, \tag{19.11}$$

where μ is the reduced mass of the system. We can factor out the reduced mass as follows,

$$H_S = (Z\alpha)^2 \mu \left(-\frac{1}{2} \vec{\nabla}_\rho^2 - \frac{1}{\rho} \right), \qquad \vec{\rho} = Z\alpha\mu\vec{r}\,. \tag{19.12}$$

The first-order perturbations of the eigenstate wave function under an infinitesimal variation of the mass μ and the nuclear charge number Z read as follows,

$$\mu \frac{\partial}{\partial\mu}|\phi\rangle = \left(\frac{1}{E_S - H_S}\right)' \left(-\frac{\vec{p}^{\,2}}{2\mu}\right)|\phi\rangle\,, \tag{19.13a}$$

$$Z \frac{\partial}{\partial Z}|\phi\rangle = \left(\frac{1}{E_S - H_S}\right)' \left(-\frac{Z\alpha}{r}\right)|\phi\rangle\,. \tag{19.13b}$$

Here, the reduced Green function [see Eq. (5.7)] is denoted with a prime.

While the identities are intuitively obvious, the derivation is somewhat nontrivial in the details. Let us write the nonrelativistic kinetic term as $H_0 = \vec{p}^{\,2}/(2\mu)$ and consider the Schrödinger Hamiltonian $H_S = \vec{p}^{\,2}/(2\mu) + V = H_0 + V$. For the Coulomb potential $V = -Z\alpha/r$, we have $Z\partial V/\partial Z = V$. The Schrödinger equation can be written as

$$(E_S - H_0)\,|\phi\rangle = V|\phi\rangle\,. \tag{19.14}$$

Let us differentiate this equation with respect to Z, taking into account that the free Hamiltonian H_0 is independent of Z. A careful consideration of both sides of the equation leads to the identity,

$$Z\frac{\partial E_S}{\partial Z}|\phi\rangle + (E_S - H_0)\,Z\frac{\partial}{\partial Z}|\phi\rangle = Z\frac{\partial V}{\partial Z}|\phi\rangle + V\,Z\frac{\partial}{\partial Z}|\phi\rangle = V|\phi\rangle + V\,Z\frac{\partial}{\partial Z}|\phi\rangle. \tag{19.15}$$

This can be written as

$$Z\frac{\partial E_S}{\partial Z}|\phi\rangle + (E_S - H_0 - V)\,Z\frac{\partial}{\partial Z}|\phi\rangle = V|\phi\rangle\,. \tag{19.16}$$

We project this equation onto the Hilbert space orthogonal to the eigenstate $|\phi\rangle$. To this end, we first observe that $(Z\,\partial E_S/\partial Z)|\phi\rangle$ is proportional to $|\phi\rangle$, while $Z\frac{\partial}{\partial Z}|\phi\rangle$ is orthogonal to $|\phi\rangle$. The latter observation can be made by recognizing that $Z\frac{\partial}{\partial Z}\langle\phi|\phi\rangle = 0$, due to conservation of the norm. We can now obtain a complete result for $Z\frac{\partial}{\partial Z}|\phi\rangle$ by simply inverting the operator $(E_S - H_0 - V)$ in the space orthogonal to $|\phi\rangle$, i.e., by applying the reduced Green function to the right-hand side. Thus,

$$Z\frac{\partial}{\partial Z}|\phi\rangle = \left(\frac{1}{E_S - H_0 - V}\right)' V|\phi\rangle = \left(\frac{1}{E_S - H_S}\right)' V|\phi\rangle\,. \tag{19.17}$$

This proves Eq. (19.13b). Analogous relations can be used in order to show Eq. (19.13a). Adding the two perturbations given in Eqs. (19.13a) and (19.13b), one obtains

$$\left(\frac{1}{E_S - H_S}\right)' \left(-\frac{\vec{p}^{\,2}}{2\mu} + \frac{Z\alpha}{r}\right)|\phi\rangle = -E_S\left(\frac{1}{E_S - H_S}\right)'|\phi\rangle = 0\,, \tag{19.18}$$

because the reference state is excluded from the sum over virtual states, by virtue of the definition of the reduced Green function. We thus obtain the scaling equation

$$\left(\mu\,\frac{\partial}{\partial\mu} - Z\,\frac{\partial}{\partial Z}\right)|\phi\rangle = 0\,. \tag{19.19}$$

For the ground state of hydrogen, the wave function in coordinate space is given as

$$\phi(\vec{r}) = \frac{(Z\alpha\mu)^{3/2}}{\sqrt{\pi}}\,\exp(-Z\alpha\mu r)\,. \tag{19.20}$$

The quantities Z and μ enter the formula for $\phi(\vec{r})$ only via the combination $Z\alpha\mu$, it is evident that indeed, the scaling equation (19.19) is fulfilled.

19.2.3 *Scale Transformation of the Dirac Hamiltonian*

Let us now turn our attention to the Dirac Hamiltonian with an eigenstate $|\psi\rangle$,

$$H_D = \vec{\alpha}\cdot\vec{p} + \beta\,m - \frac{Z\alpha}{r}\,, \qquad H_D|\psi\rangle = E_D|\psi\rangle\,, \tag{19.21}$$

where the Dirac energy is denoted as E_D. The Dirac equation is a one-particle equation, and we thus work with the electron mass m, not with the reduced mass μ. The first-order perturbations under a change of the mass and the charge number are given as follows,

$$m\,\frac{\partial}{\partial m}|\psi\rangle = \left(\frac{1}{E_D - H_D}\right)'\beta\,m|\psi\rangle\,, \tag{19.22a}$$

$$Z\,\frac{\partial}{\partial Z}|\psi\rangle = \left(\frac{1}{E_D - H_D}\right)'\left(-\frac{Z\alpha}{r}\right)|\psi\rangle\,. \tag{19.22b}$$

From the eigenvector property we have, immediately,

$$\left(\frac{1}{E_D - H_D}\right)'\left(\vec{\alpha}\cdot\vec{p} + \beta\,m - \frac{Z\alpha}{r}\right)|\psi\rangle = 0\,, \tag{19.23}$$

if $|\psi\rangle$ is an eigenvector, $H_D|\psi\rangle = E_D|\psi\rangle$. Combining Eqs. (19.22) and (19.23), and (8.95), we obtain

$$\begin{aligned}\left(m\,\frac{\partial}{\partial m} + Z\,\frac{\partial}{\partial Z}\right)|\psi\rangle &= \left(\frac{1}{E_D - H_D}\right)'(-\vec{\alpha}\cdot\vec{p})\,|\psi\rangle \\ &= \left(\frac{1}{E_D - H_D}\right)'(-\mathrm{i}\,[H_D - E_D, \vec{r}\,]\cdot\vec{p})\,|\psi\rangle \\ &= \vec{r}\cdot\frac{\partial}{\partial\vec{r}}\,|\psi\rangle - |\psi\rangle\,\mathrm{i}\,\langle\psi^+|\vec{r}\cdot\vec{p}|\psi\rangle + \mathrm{i}\left(\frac{1}{E_D - H_D}\right)'\vec{r}\,[H_D - E_D, \vec{p}\,]\;|\psi\rangle.\end{aligned} \tag{19.24}$$

We have used the identity $[1/(E_D - H_D)]'\,(E_D - H_D) = 1 - |\psi\rangle\,\langle\psi^+|$ (for absolute clarity, the Hermitian adjoint ψ^+ rather than the Dirac adjoint $\psi^+\,\gamma^0$ is being indicated. The matrix element $\mathrm{i}\,\langle\psi|\vec{r}\cdot\vec{p}|\psi\rangle = \langle\psi|\vec{r}\cdot\frac{\partial}{\partial\vec{r}}|\psi\rangle$ can be evaluated for a general eigenvector $|\psi\rangle$, as follows,

$$\left\langle\psi^+\left|\vec{r}\cdot\frac{\partial}{\partial\vec{r}}\right|\psi\right\rangle = \left\langle\psi^+\left|\left[\vec{r}, \frac{\partial}{\partial\vec{r}}\right]\right|\psi\right\rangle + \left\langle\psi^+\left|\frac{\partial}{\partial\vec{r}}\cdot\vec{r}\right|\psi\right\rangle = -3 - \left\langle\psi^+\left|\vec{r}\cdot\frac{\partial}{\partial\vec{r}}\right|\psi\right\rangle, \tag{19.25}$$

where we have used an integration by parts. This gives immediately

$$\mathrm{i}\,\langle\psi^+|\vec{r}\cdot\vec{p}|\psi\rangle = -\frac{3}{2}\,. \tag{19.26}$$

The commutator on the right-hand side of Eq. (19.24) is easily evaluated,

$$\mathrm{i}\,\vec{r}\cdot[H_D - E, \vec{p}] = \mathrm{i}\,\vec{r}\cdot\left[-\frac{Z\alpha}{r}, \vec{p}\right] = \vec{r}\cdot\frac{\partial}{\partial\vec{r}}\frac{Z\alpha}{r} = -\frac{Z\alpha}{r}\,. \tag{19.27}$$

We thus have the relation,

$$\left(m\frac{\partial}{\partial m} + Z\frac{\partial}{\partial Z}\right)|\psi\rangle = \vec{r}\cdot\frac{\partial}{\partial\vec{r}}|\psi\rangle + \frac{3}{2}|\psi\rangle + \left(\frac{1}{E_D - H_D}\right)'\left(-\frac{Z\alpha}{r}\right)|\psi\rangle\,. \tag{19.28}$$

Finally, using Eq. (19.22b) in reverse, we obtain the result

$$\left(m\frac{\partial}{\partial m} - \vec{r}\cdot\frac{\partial}{\partial\vec{r}} - \frac{3}{2}\right)|\psi\rangle = 0\,. \tag{19.29}$$

This equation constitutes the analogue of the relation (19.19) within the fully relativistic formalism.

19.3 Renormalization Group Transformations

19.3.1 *Functional Equations in the Asymptotic Regime*

From Eq. (10.111), using the Ward–Takahashi identity $Z_1 = Z_2$, we infer that the physical charge e of the electron and the bare charge e_0 are related by the renormalization constant Z_3, as follows,

$$e_\mathrm{R} = \mathcal{Z}^{1/2}\,e_0\,, \qquad \mathcal{Z} = Z_3 = \mathcal{Z}(\lambda) \qquad e_\mathrm{R} = e(\lambda) = e_\lambda\,. \tag{19.30}$$

Let us now compare finite, renormalized quantities at three different scales, denoted here by the symbols 1, 2 and 3 in a somewhat lax notation,

$$e^2_{\lambda_3} = \frac{\mathcal{Z}(\lambda_3)}{\mathcal{Z}(\lambda_2)}\,e^2_{\lambda_2} = \frac{\mathcal{Z}(\lambda_3)}{\mathcal{Z}(\lambda_2)}\,\frac{Z(\lambda_2)}{Z(\lambda_1)}\,e^2_{\lambda_1} = \underbrace{F(\lambda_3^2,\lambda_2^2)\,F(\lambda_2^2,\lambda_1^2)}_{=F(\lambda_3^2,\lambda_1^2)}\,e^2_{\lambda_1}\,, \tag{19.31}$$

where the $F(\lambda_a^2,\lambda_b^2) = Z_3(\lambda_a)/Z_3(\lambda_b)$ denotes the ratio of renormalization constants.

Let us assume that we have matched the parameter e^2_λ at the momentum scale λ against the observed, physical electron charge square at the same momentum scale. Then, the F function can be written in terms of all of its arguments as follows,

$$F = F(k^2,\lambda^2) = F\left(e^2_\lambda, \frac{k^2}{\lambda^2}, \frac{m^2}{\lambda^2}\right)\,, \tag{19.32}$$

i.e., it describes the evolution of the charge from the scale λ^2 to the scale k^2. We observe that since F is itself dimensionless, it can only depend on dimensionless arguments, and furthermore, we explicitly indicate the value of e^2_λ as a third argument, because this value is determined by observation and represents "experimental" input. Up to a prefactor $-g_{\mu\nu}$, the renormalized photon propagator at the momentum scale k^2 has the structure (see Chap. 9)

$$G^{\gamma}(k^2) = \frac{1}{k^2} F\left(e_\lambda^2, \frac{k^2}{\lambda^2}, \frac{m^2}{\lambda^2}\right) = \frac{F(k^2)}{k^2}, \tag{19.33}$$

where the superscript γ denotes the photon. The function $F(k^2)$ describes the running of the charge but it is not the running charge itself; once the reference scale λ and the effective charge square e_λ^2 at the reference scale are fixed, the theory becomes predictive. After the fixing of the reference scale, F is a function of only one variable, namely, k^2. In order to obtain the actual value of the running charge at a specific momentum scale, one must multiply $F(k^2)$ by the value of the running charge at a reference momentum scale, to obtain the expression $e_k^2\, G^{\gamma}(k^2)$. Now, for $k^2 = \lambda^2$, we have fixed the physical charge square to be e_λ^2, and so

$$F(k^2 = \lambda^2) = 1, \qquad e_\lambda^2\, G^{\gamma}(k^2 = \lambda^2) = \frac{e_\lambda^2\, F(\lambda^2)}{\lambda^2} = \frac{e_\lambda^2}{\lambda^2}, \tag{19.34}$$

and the photon propagator at the scale $k^2 = \lambda^2$ has the desired form. When the three arguments of F are fully written out, the condition $F(k^2 = \lambda^2) = 1$ implies that

$$F\left(e_\lambda^2, 1, \frac{m^2}{\lambda^2}\right) = 1. \tag{19.35}$$

Before we now explore the properties of the renormalization group function F in the asymptotic regime, we illustrate its properties by a number of further example calculations. First, we observe the we can obtain the invariant charge square e_k^2 at the scale k both using scale transformations $\lambda \to k$ as well as $\lambda' \to k$, a fact which is expressed as follows,

$$e_k^2 = e_\lambda^2\, F\left(e_\lambda^2, \frac{k^2}{\lambda^2}, \frac{m^2}{\lambda^2}\right) = e_{\lambda'}^2\, F\left(e_{\lambda'}^2, \frac{k^2}{\lambda'^2}, \frac{m^2}{\lambda'^2}\right). \tag{19.36}$$

The effective charge e_k^2 thus defined is called the invariant charge: We can either start the RG evolution at scale λ and transform to scale k via the action of the F function, or we can start from the scale λ' and go over to scale k by the action of an F function of different arguments. In any case, since e_k^2 is dimensionless, it must be expressed as a function of dimensionless arguments. Equation (19.36) also is consistent with Eq. (19.31), in the form $e_{\lambda_3}^2 = F(\lambda_3^2, \lambda_1^2)\, e_{\lambda_1}^2$, upon the identification $1 \to \lambda, \lambda'$ and $3 \to k^2$. The group structure, evident from Eq. (19.31),

$$F(\lambda_3^2, \lambda_2^2)\, F(\lambda_2^2, \lambda_1^2) = F(\lambda_3^2, \lambda_1^2), \tag{19.37}$$

is expressed as follows in terms of the F functions of three dimensionless arguments,

$$F\left(e_{\lambda'}^2, \frac{k^2}{\lambda'^2}, \frac{m^2}{\lambda'^2}\right) F\left(e_\lambda^2, \frac{\lambda'^2}{\lambda^2}, \frac{m^2}{\lambda^2}\right) = F\left(e_\lambda^2, \frac{k^2}{\lambda^2}, \frac{m^2}{\lambda^2}\right). \tag{19.38}$$

This relation exhibits the group structure of the RG transformation in a particularly clear way. We start from e_λ^2 at scale λ and transform to scale λ' and then to scale k^2 (left-hand side) or directly from scale λ to the scale k^2 (right-hand side).

One more alternative point of view is as follows. From the invariance of the bare charge e_0 under the RG transformation,

$$e_0^2 = [Z_3(\lambda')]^{-1}\, e_{\lambda'}^2\,, \qquad e_0^2 = [Z_3(\lambda)]^{-1}\, e_\lambda^2\,, \tag{19.39}$$

we have

$$\frac{e_k^2}{[Z_3(\lambda')]^{-1}\, e_{\lambda'}^2} = \frac{e_k^2}{[Z_3(\lambda)]^{-1}\, e_\lambda^2}\,,$$

$$\Rightarrow \quad Z_3(\lambda') F\left(e_{\lambda'}^2, \frac{k^2}{\lambda'^2}, \frac{m^2}{\lambda'^2}\right) = Z_3(\lambda)\, F\left(e_\lambda^2, \frac{k^2}{\lambda^2}, \frac{m^2}{\lambda^2}\right). \tag{19.40}$$

This is the another form of Eq. (19.36). The function F describes the evolution of the renormalization constant Z_3, and therefore, of the electron charge square, with the scale, and therefore we also have the relation

$$F\left(e_\lambda^2, \frac{k^2}{\lambda^2}, \frac{m^2}{\lambda^2}\right) = \frac{Z_3(\lambda')}{Z_3(\lambda)} F\left(e_{\lambda'}^2, \frac{k^2}{\lambda'^2}, \frac{m^2}{\lambda'^2}\right) = F\left(e_\lambda^2, \frac{\lambda'^2}{\lambda^2}, \frac{m^2}{\lambda^2}\right) F\left(e_{\lambda'}^2, \frac{k^2}{\lambda'^2}, \frac{m^2}{\lambda'^2}\right) \tag{19.41}$$

since $Z_3(\lambda')/Z_3(\lambda) = F(\lambda'^2, \lambda^2) = F(e_\lambda^2, \frac{\lambda'^2}{\lambda^2}, \frac{m^2}{\lambda^2})$ according to Eq. (19.32), which illustrates once more the group structure given in Eq. (19.38).

We now want to describe the flow of the effective QED coupling with the momentum scale, starting from the invariant charge described by Eq. (19.36) and the group structure (19.38) of the RG transformations. Interchanging $\lambda \leftrightarrow \lambda'$, the group structure (19.38) can be written as follows,

$$F\left(e_\lambda^2, \frac{k^2}{\lambda^2}, \frac{m^2}{\lambda^2}\right) F\left(e_{\lambda'}^2, \frac{\lambda^2}{\lambda'^2}, \frac{m^2}{\lambda'^2}\right) = F\left(e_{\lambda'}^2, \frac{k^2}{\lambda'^2}, \frac{m^2}{\lambda'^2}\right). \tag{19.42}$$

With

$$x = \frac{k^2}{\lambda'^2}\,, \qquad y = \frac{m^2}{\lambda'^2}\,, \qquad t = \frac{\lambda^2}{\lambda'^2}\,, \tag{19.43}$$

we can write Eq. (19.42) as follows, after a multiplication of both sides by a factor $e_{\lambda'}^2$,

$$F\left(e_\lambda^2, \frac{x}{t}, \frac{y}{t}\right) e_{\lambda'}^2\, F\left(e_{\lambda'}^2, t, y\right) = e_{\lambda'}^2\, F\left(e_{\lambda'}^2, x, y\right). \tag{19.44}$$

We can eliminate the quantity e_λ^2 in the argument of the first function, using the invariant charge formula (19.36),

$$e_\lambda^2 = e_{\lambda'}^2\, F\left(e_{\lambda'}^2, \frac{\lambda^2}{\lambda'^2}, \frac{m^2}{\lambda'^2}\right) = e_{\lambda'}^2\, F\left(e_{\lambda'}^2, t, y\right). \tag{19.45}$$

So, Eq. (19.44) is rewritten as follows,

$$e_{\lambda'}^2\, F\left(e_{\lambda'}^2, t, y\right) F\left(e_{\lambda'}^2\, F\left(e_{\lambda'}^2, t, y\right), \frac{x}{t}, \frac{y}{t}\right) = e_{\lambda'}^2\, F\left(e_{\lambda'}^2, x, y\right). \tag{19.46}$$

Now, the parameter $e^2 \equiv e^2_{\lambda'}$ is fixed by an observation, and therefore

$$e^2\,F\left(e^2,t,y\right)\;F\left(e^2\,F\left(e^2,t,y\right),\frac{x}{t},\frac{y}{t}\right)=e^2\,F\left(e^2,x,y\right). \tag{19.47}$$

We now consider the asymptotic regime $y \to 0$, which in view of $y = m^2/\lambda'^2$ is either the massless limit or the limit $\lambda' \to \infty$, i.e., the limit of large momenta. Endowing the functions in the asymptotic regime with a superscript "as", we can drop the last argument $y \to 0$ in the formulas, and we also replace the first argument by $\alpha = e^2/(4\pi)$,

$$F^{\rm as}\left(\alpha,x\right)\equiv F\left(e^2,x,0\right), \tag{19.48a}$$

$$G^{\rm as}(\alpha,x)\equiv\alpha F^{\rm as}\left(\alpha,x\right)=\alpha F^{\rm as}\left(\alpha,t\right)\;F^{\rm as}\left(\alpha\,F^{\rm as}\left(\alpha,t\right),\frac{x}{t}\right). \tag{19.48b}$$

Equation (19.48b) is equivalent to Eq. (19.47) in the asymptotic region, and we define the effective coupling $G^{\rm as}$ in the asymptotic region as indicated. Then, a recursive application of the definition (19.48b) leads to the result

$$G^{\rm as}\left(G^{\rm as}\left(\alpha,t\right),\frac{x}{t}\right)=G^{\rm as}\left(\alpha\,F^{\rm as}\left(\alpha,t\right),\frac{x}{t}\right)=\alpha\,F^{\rm as}\left(\alpha,\,t\right)\;F^{\rm as}\left(\alpha\,F^{\rm as}\left(\alpha,t\right),\frac{x}{t}\right), \tag{19.49}$$

and the latter expression is easily identified as the right-hand side of Eq. (19.48b), so that

$$G^{\rm as}(\alpha,x)=G^{\rm as}\left(G^{\rm as}(\alpha,t),\frac{x}{t}\right),\qquad \alpha_x=G^{\rm as}\left(\alpha_t,\frac{x}{t}\right), \tag{19.50}$$

where in the latter step we have identified $\alpha_x = G^{\rm as}(\alpha,x)$ as the effective coupling at the scale x. The solution to the RG equation (19.50) needs to be found, now. Indeed, Eq. (19.50) is solved by any function of the form

$$G^{\rm as}(\alpha,x)=f\left(x\,f^{-1}(\alpha)\right), \tag{19.51}$$

but this is not obvious.

Let us first apply f^{-1} to both sides of the defining Eq. (19.51). We get

$$f^{-1}\left(G^{\rm as}(\alpha,x)\right)=x\,f^{-1}(\alpha). \tag{19.52}$$

We now go back to the original RG equation (19.50) and also apply f^{-1},

$$f^{-1}\left(G^{\rm as}(\alpha,x)\right)=f^{-1}\left(G^{\rm as}\left(G^{\rm as}(\alpha,t),\;\frac{x}{t}\right)\right). \tag{19.53}$$

Using Eq. (19.52) for both sides of Eq. (19.53) (twice, recursively, to the right-hand side) yields the relation

$$x\,f^{-1}\left(\alpha\right)=\frac{x}{t}\,f^{-1}\left(G^{\rm as}(\alpha,t)\right)=\frac{x}{t}\,t\,f^{-1}\left(\alpha\right)=x\,f^{-1}\left(\alpha\right), \tag{19.54}$$

which verifies the solution of the RG equation (19.50). We now look at the derivative of the asymptotic effective coupling $G^{\rm as}$ with the scale x,

$$\frac{\partial}{\partial x}G^{\rm as}(\alpha,x)=\frac{\partial}{\partial x}f\left(x\,f^{-1}\left(\alpha\right)\right)=f^{-1}\left(\alpha\right)\,f'\left(x\,f^{-1}\left(\alpha\right)\right). \tag{19.55}$$

If we supply an additional prefactor x, then the logarithmic derivative of the coupling can be expressed as a function of a single variable, namely, the coupling as given by Eq. (19.51),

$$x\,\frac{\partial}{\partial x} G^{\rm as}(\alpha,x) = \left(x\,f^{-1}(\alpha)\right)\,f'\left(x\,f^{-1}(\alpha)\right) = \beta\left(f\left(x\,f^{-1}(\alpha)\right)\right) = \beta\left(G^{\rm as}(\alpha,x)\right)\,. \tag{19.56}$$

Evidently, this expression is a function, named the β function, of a single variable, namely, $x\,f^{-1}(\alpha)$, or alternatively $f\left(x\,f^{-1}(\alpha)\right)$, which is $\alpha_x = G^{\rm as}(\alpha,x)$. We have derived the result that the logarithmic derivative of the effective coupling with the (square of the) momentum scale, for large momenta, is described by a function of the value of the effective coupling at that scale (and nothing else),

$$\alpha_x = G^{\rm as}(\alpha,x)\,, \qquad x\,\frac{\partial}{\partial x}\alpha_x = x\,\frac{\partial}{\partial x}G^{\rm as}(\alpha,x) = \beta\left(G^{\rm as}(\alpha,x)\right) = \beta\left(\alpha_x\right)\,. \tag{19.57}$$

This is one central result of the renormalization group when applied to gauge theories. The result (19.56) is the basis for the definition of the β function of QED, which describes the running of the effective charge with the momentum scale.

We now assume that β is positive, based on the one-loop value from Eq. (19.60) below, and verify consistency with the different possible kinematic regions (time-like momentum transfer $k^2 > 0$ and space-like exchange momenta $k^2 = -\vec{q}^2 < 0$. Indeed, we shall assume that the β function is positive for both (time-like as well as space-like) physically relevant domains of the asymptotic, effective coupling $G^{\rm as}(\alpha,x)$. We first recall the definition of the x variable from Eq. (19.43), i.e., $x = k^2/\lambda'^2$. Here, k^2 is the "final" momentum of the RG evolution step. The most relevant physical domain for k^2 is the deep Euclidean region $k^2 = -\vec{q}^2$, with $\vec{q}^2 \gg m^2$, i.e., the region of a "strong Coulomb potential" with a "hard Coulomb photon exchange". In this kinematic region, pair production is impossible. For negative x, the value of the effective coupling grows with the modulus of x, i.e., it grows with x as x becomes more negative, and both x as well as $\partial_x\alpha_x$ are negative. This implies that the product $x\,\partial_x\alpha_x$ is positive for both relevant kinematic regions. Let us now fix $\lambda'^2 > 0$ as a fixed reference scale for the momentum square. Then, we can identify the logarithmic differential operator $x\,\partial/\partial x$ with the operator $k^2\partial/\partial(k^2)$.

Let us also convince ourselves that the same β function can be used in order to describe the evolution of the coupling in the manifestly Euclidean region, setting the reference scale $\lambda'^2 = m^2$. Then, $x = k^2/m^2 = -\vec{q}^{\,2}/m^2$, and the RG evolution can be expressed in terms of $\xi = \vec{q}^{\,2}/m^2$ as an independent variable,

$$\begin{gathered} x \to k^2 = -\vec{q}^{\,2} = -\xi\,, \qquad \alpha_x = \alpha_{-\xi} \equiv \alpha(\xi)\,, \\ \xi\frac{\partial}{\partial\xi}\alpha(\xi) = \xi\frac{\partial}{\partial\xi}\alpha_{-\xi} = x\,\frac{\partial}{\partial x}\alpha_x = \beta\left(\alpha_{-\xi}\right) = \beta\left(\alpha(\xi)\right)\,, \end{gathered} \tag{19.58}$$

so that the RG evolution with the reference momentum scale $\vec{q}^2$ is indeed given by the β function,

$$\vec{q}^{\,2}\,\frac{\partial}{\partial\vec{q}^{\,2}}\alpha(\vec{q}^{\,2}) = \beta\left(\alpha(\vec{q}^{\,2})\right)\,, \qquad \alpha(\vec{q}^{\,2}) = \alpha\left(\frac{\vec{q}^{\,2}}{m^2}\right)\,. \tag{19.59}$$

This implies that we can alternatively interpret the RG evolution operator $x\,\partial_x$ in the equation $x\,\partial_x\alpha = \beta(\alpha_x)$ as a differential operator in $x = k^2/m^2$ or in $x = \vec{q}^2/m^2$, provided we use an appropriate definition of α_x, namely, the value of α_x at $x = k^2/m^2$ or $x = -\vec{q}^2/m^2$.

In the above considerations, we have implicitly assumed that the value of α from which we start the RG transformation at zero momentum transfer is the physical value of the fine-structure constant. This is known as the MOM (momentum) renormalization scheme. (We are somewhat using the RG method in a region *out of its initial specification*, i.e., in the non-asymptotic region. For more remarks on this issue, we refer to the text following Eq. (19.74).) The result for the β function has been given to three loops by Baker and Johnson in Ref. [627], and the four-loop result has been indicated by Gorishny, Kataev, Larin and Surguladze in Ref. [628],

$$\beta(\alpha_x) = \beta_{\rm MOM}(\alpha_x) = \frac{\alpha_x^2}{3\pi} + \frac{\alpha_x^3}{4\pi^2} + \frac{\alpha_x^4}{\pi^3}\left(\frac{1}{3}\,\zeta(3) - \frac{101}{288}\right)$$
$$+ \frac{\alpha_x^5}{\pi^4}\left(-\frac{5}{3}\,\zeta(5) + \frac{1}{3}\,\zeta(3) + \frac{93}{128}\right) + \mathcal{O}(\alpha_x^6)\,. \qquad (19.60)$$

Note that our definition of the β function differs by an overall factor 4π from the one given in Eq. (2.5) of Ref. [628]. The RG evolution equation can be solved as follows,

$$x\frac{\partial}{\partial x}\alpha_x = \beta(\alpha_x)\,, \qquad \int_{G^{\rm as}(\alpha,x_1)}^{G^{\rm as}(\alpha,x_2)} \frac{{\rm d}z}{\beta(z)} = \ln\left(\frac{x_2}{x_1}\right), \qquad (19.61)$$

where $\alpha_x = G^{\rm as}(\alpha, x)$. Here, we come to an understanding of the often used jargon of "matching and running": namely, when the coupling $G^{\rm as}(\alpha, x_1)$ is "matched" against the observed coupling strength at that scale, then the evolution of the coupling to $G^{\rm as}(\alpha, x_2)$ at the scale x_2 is given by the QED beta function. We should stress, here, that the RG flow is formulated only for the beta function of the coupling; the flow of the mass is being ignored. The role of this approximation will be clarified below.

Let us indicate a method of solving the RG evolution, perturbatively. With the β function begin held constant, we have approximately

$$\ln\left(\frac{x_2}{x_1}\right) = \int_{G^{\rm as}(\alpha,x_1)}^{G^{\rm as}(\alpha,x_2)} \frac{{\rm d}z}{\beta(z)} \approx \frac{G^{\rm as}(\alpha,x_2) - G^{\rm as}(\alpha,x_1)}{\beta(G^{\rm as}(\alpha,x_1))}\,, \qquad (19.62)$$

if x_2 and x_1 are not too far apart. Then, we have

$$G^{\rm as}(\alpha,x_2) \approx G^{\rm as}(\alpha,x_1) + \ln\left(\frac{x_2}{x_1}\right)\beta(G^{\rm as}(\alpha,x_1))\,. \qquad (19.63)$$

A better approximation for the function in the integrand of Eq. (19.62) is

$$\frac{1}{\beta(z)} = \frac{1}{\beta(z_0)} - \frac{\beta'(z_0)}{[\beta(z_0)]^2}\,(z - z_0) + \cdots, \qquad z_0 = G^{\rm as}(\alpha,x_1)\,. \qquad (19.64)$$

Here, terms of order $(z - z_0)^2$ and higher have been ignored. Integrating the RG evolution to second order in $z - z_0$, we find that

$$\begin{aligned}G^{\rm as}(\alpha, x_2) - G^{\rm as}(\alpha, x_1) &= \frac{1}{1!} \ln\left(\frac{x_2}{x_1}\right) \beta(G^{\rm as}(\alpha, x_1)) \\ &+ \frac{1}{2!}\left[\ln\left(\frac{x_2}{x_1}\right)\right]^2 \beta(G^{\rm as}(\alpha, x_1))\, \beta'(G^{\rm as}(\alpha, x_1)) + \cdots .\end{aligned} \tag{19.65}$$

This result easily generalizes to all orders,

$$\begin{aligned}G^{\rm as}(\alpha, x_2) - G^{\rm as}(\alpha, x_1) &= \frac{1}{1!} \ln\left(\frac{x_2}{x_1}\right) \left\{\beta(z) \frac{\partial}{\partial z} z\right\}\bigg|_{z=G^{\rm as}(\alpha,x_1)} \\ &+ \frac{1}{2!}\left[\ln\left(\frac{x_2}{x_1}\right)\right]^2 \left\{\beta(z) \frac{\partial}{\partial z} \beta(z) \frac{\partial}{\partial z} z\right\}\bigg|_{z=G^{\rm as}(\alpha,x_1)} + \cdots \\ &= \sum_{n=1}^{\infty} \frac{1}{n!}\left[\ln\left(\frac{x_2}{x_1}\right)\right]^n \left\{\left[\beta(z) \frac{\partial}{\partial z}\right]^n z\right\}\bigg|_{z=G^{\rm as}(\alpha,x_1)} .\end{aligned} \tag{19.66}$$

This equation defines $G^{\rm as}(\alpha, x_2)$ as a power series in the coupling $G^{\rm as}(\alpha, x_1) = \alpha_{x_1}$.

19.3.2 *Renormalization Group of Callan and Symanzik*

In contrast to the approach of Gell-Mann and Low, and Callan–Symanzik equation also implies a variation of the mass parameter of the physical theory. The Gell-Mann low β function can only be applied in the asymptotic region of large momenta, and it describes only the RG flow of the coupling. If we wish to use the QED β function in the non-asymptotic region, we also have to include the running of the mass parameter. This is done in the approach of Callan and Symanzik [130], which is based on a variation of the renormalized and/or bare masses of the bare theory, while keeping the bare coupling and cutoff scale Λ constant.

According to the principle of renormalization, the effective charge at the scale $x = -\vec{q}^2/m^2$ can be written in terms of a renormalization constant Z_3, which in turn can be expressed in terms of the renormalized value α_x and the ratio Λ^2/m^2, i.e., in terms of the renormalized quantities,

$$\alpha_x = Z_3\left(\frac{\Lambda^2}{m^2}, \alpha_x\right) \alpha_0 . \tag{19.67}$$

As $\Lambda \to \infty$, the bare coupling which is the effective, high-energy coupling, grows indefinitely, and to keep α_x finite, we have to demand that $Z_3 \to 0$ for $\Lambda \to \infty$. In order to fix ideas, we observe that this implies that α_x actually increases when we increase the mass parameter m^2 in the denominator of the Z_3 renormalization constant; this otherwise corresponds to a decrease of Λ and a corresponding increase in Z_3 and α_x.

Callan and Symanzik observed that the logarithmic derivative with respect to the mass of the effective charge at scale x can be expressed in terms of α_x alone, at least in the limit $\Lambda \to \infty$. The derivative is to be taken at constant cutoff Λ and constant bare coupling α_0, but varying renormalized mass m (and, as a consequence, varying bare mass m_0),

$$
\begin{aligned}
m^2 \frac{\partial}{\partial m^2} \alpha_x \bigg|_{\alpha_0,\Lambda=\text{const.}} &= m^2 \frac{\partial}{\partial m^2} \left(Z_3\left(\frac{\Lambda^2}{m^2}, \alpha_x\right) \alpha_0 \right) \bigg|_{\alpha_0,\Lambda=\text{const.}} \\
&= \alpha_0\, m^2 \frac{\partial}{\partial m^2} \left(Z_3\left(\frac{\Lambda^2}{m^2}, \alpha_x\right) \right) \bigg|_{\alpha_0,\Lambda=\text{const.}} \\
&= \frac{\alpha_x}{Z_3\left(\frac{\Lambda^2}{m^2}, \alpha_x\right)}\, m^2 \frac{\partial}{\partial m^2} \left(Z_3\left(\frac{\Lambda^2}{m^2}, \alpha_x\right) \right) \bigg|_{\alpha_0,\Lambda=\text{const.}} \\
&= \alpha_x\, m^2 \frac{\partial}{\partial m^2} \ln \left(Z_3\left(\frac{\Lambda^2}{m^2}, \alpha_x\right) \right) \bigg|_{\alpha_0,\Lambda=\text{const.}} \\
&= -\, \alpha_x\, \Lambda^2 \frac{\partial}{\partial \Lambda^2} \ln \left(Z_3\left(\frac{\Lambda^2}{m^2}, \alpha_x\right) \right) \bigg|_{\alpha_0,m=\text{const.}} \overset{\Lambda\to\infty}{=} \widetilde{\beta}(\alpha_x)\,.
\end{aligned}
\tag{19.68}
$$

The relation of $\widetilde{\beta}$ to the β function defined in Sec. 19.3.1 will be clarified below. As already noted, α_x shrinks with growing Λ, which explains why $\widetilde{\beta}(\alpha_x)$ still is a positive quantity if x describes a space-like momentum transfers.

The derivation of the Callan–Symanzik equation starts from the basic relation fulfilled by the effective charge, which we write as follows,

$$
G(\alpha, q^2, m^2) = G\left(Z_3\left(\frac{\Lambda^2}{m^2}, \alpha\right) \alpha_0, q^2, m^2 \right) = \lim_{\Lambda\to\infty} G(\alpha_0, q^2, m_0^2, \Lambda^2)\,. \tag{19.69}
$$

This equation might raise some questions, because the right-hand side appears to be independent of Λ, while the expression in the middle appears to be manifestly Λ-dependent. The rationale is that, once the cutoff Λ and the physical coupling α are fixed, the renormalization constant Z_3 which relates the bare and the renormalized coupling is also determined, by the relation

$$
\alpha = Z_3\left(\frac{\Lambda^2}{m^2}, \alpha\right) \alpha_0\,, \tag{19.70}
$$

where α_0 is the bare coupling, i.e., the coupling at scale Λ. The same relation must hold in the limit $\Lambda \to \infty$, and this justifies the functional form of Eq. (19.69).

We now vary the mass while keeping the bare coupling and the scale Λ constant. There is an additional ingredient to the formulation of the Callan–Symanzik equation which we have to discuss. Without derivation, we here indicate that the differentiation of the inverse $1/G$ of the effective charge, as given by vacuum polarization diagrams, generates mass insertions in the Feynman diagrams, often denoted by a cross in the diagrammatic language. We can define an inhomogeneity function

$$
Z_0^{-1}\, \Delta\left(\alpha, \frac{q^2}{m^2}\right) = \lim_{\Lambda\to\infty} m_0 \frac{\partial}{\partial m_0} G^{-1}(\alpha_0, q^2, m_0^2, \Lambda^2)\,, \tag{19.71}
$$

where again we take account of the running of the mass from its bare value m_0 to its renormalized value m. The right-hand side can only depend on the parameters that the left-hand side depends upon. Note that the mass insertion does not correspond to any available term in the original QED Lagrangian. A term that generates a mass insertion can be formulated, if desired, by an infinitesimal addition to the QED Lagrangian (or generating functional), but it necessitates the introduction of a further renormalization constant in addition to the vertex renormalization Z_1, the wave function renormalization Z_2, and the charge renormalization Z_3. The new, additional renormalization constant is conveniently referred to as Z_0. We now vary the mass parameter while keeping the bare coupling and the scale Λ constant. A change in the mass scale of the left-hand side of the coupling given in Eq. (19.69) results in the relation

$$\begin{aligned} m\frac{\partial}{\partial m}G^{-1}\bigg|_{\alpha_0,\Lambda=\text{const.}} &= m\frac{\partial}{\partial m}G^{-1}\left(Z_3\left(\frac{\Lambda^2}{m^2},\alpha\right)\alpha_0, q^2, m^2\right) \\ &= \alpha_0\left(2m^2\frac{\partial}{\partial m^2}Z_3\left(\frac{\Lambda^2}{m^2},\alpha\right)\right)\bigg|_{\alpha_0,\Lambda}\frac{\partial}{\partial\alpha}G^{-1}\left(\alpha,q^2,m^2\right) \\ &\quad + m\frac{\partial}{\partial m}G^{-1}\left(\alpha,q^2,m^2\right) \\ &= \frac{2\,\alpha\,m^2}{Z_3\left(\frac{\Lambda^2}{m^2},\alpha\right)}\frac{\partial}{\partial m^2}Z_3\left(\frac{\Lambda^2}{m^2},\alpha\right)\bigg|_{\alpha_0,\Lambda}\frac{\partial}{\partial\alpha}G^{-1}\left(\alpha,q^2,m^2\right) \\ &\quad + m\frac{\partial}{\partial m}G^{-1}\left(\alpha,q^2,m^2\right) \\ &= \left(2\,\widetilde{\beta}(\alpha)\frac{\partial}{\partial\alpha} + m\frac{\partial}{\partial m}\right)G^{-1}\left(\alpha,q^2,m^2\right). \end{aligned} \tag{19.72}$$

Our definition of the $\widetilde{\beta}(\alpha)$ function,

$$\widetilde{\beta}(\alpha) = \frac{\alpha\,m^2}{Z_3\left(\frac{\Lambda^2}{m^2},\alpha\right)}\frac{\partial}{\partial m^2}Z_3\left(\frac{\Lambda^2}{m^2},\alpha\right)\bigg|_{\alpha_0,\Lambda}, \tag{19.73}$$

differs by a factor two from the definition in Eq. (13.25) of Ref. [2]. As we shall see later, our definition ensures that the first two terms of the β function are renormalization-scheme independent. For the left-hand side of Eq. (19.72), we can rewrite the equation in terms of differentiations with respect to the bare mass, and, using Eq. (19.71),

$$\begin{aligned} &\lim_{\Lambda\to\infty} m\frac{\partial}{\partial m}G^{-1}(\alpha_0,q^2,m_0^2,\Lambda^2)\bigg|_{\alpha_0,\Lambda} \\ &= \lim_{\Lambda\to\infty}\frac{m}{m_0}\frac{\partial m_0}{\partial m}\underbrace{\left(m_0\frac{\partial}{\partial m_0}G^{-1}\left(\alpha_0,q^2,m_0^2,\Lambda^2\right)\bigg|_{\alpha_0,\Lambda}\right)}_{=Z_0^{-1}\,\Delta(\alpha,q^2/m^2)} \\ &= \underbrace{\lim_{\Lambda\to\infty}\left(\frac{m}{m_0}\frac{\partial m_0}{\partial m}Z_0^{-1}\right)}_{=1+\delta(\alpha)}\Delta\left(\alpha,\frac{q^2}{m^2}\right) = (1+\delta(\alpha))\;\Delta\left(\alpha,\frac{q^2}{m^2}\right), \end{aligned} \tag{19.74}$$

where we have used Eq. (19.71). The Callan–Symanzik renormalization-group equation thus reads as

$$\left(2\,\widetilde{\beta}(\alpha)\frac{\partial}{\partial\alpha}+m\,\frac{\partial}{\partial m}\right)\,G^{-1}\left(\alpha,\frac{q^2}{m^2}\right)=(1+\delta(\alpha))\;\;\Delta\left(\alpha,\frac{q^2}{m^2}\right),\tag{19.75}$$

where the right-hand side describes higher-order mass insertions. The validity is not restricted to the asymptotic region; the price to pay is an additional mass insertion (nonvanishing right-hand side), and the necessity to calculate the additional renormalization constant Z_0 for the field due to the mass insertion. Meaningful results can be inferred from Eq. (19.75) in the asymptotic regime of $|q^2|\gg m^2$, where the right-hand side of Eq. (19.75) is suppressed by at least one power of q^2/m^2 as compared to the terms on the left-hand side.

19.3.3 *Connection of Callan–Symanzik and Gell-Mann–Low*

In the asymptotic regime of large momenta $q^2\to\pm\infty$, the inhomogeneous term in the Callan–Symanzik RG equation (19.75) is suppressed, and the coupling assumes its asymptotic behavior, $G\to G^{\rm as}$. Then, the Callan–Symanzik equation reduces to the following statement,

$$\left(2\,\widetilde{\beta}(\alpha)\frac{\partial}{\partial\alpha}+m\,\frac{\partial}{\partial m}\right)\,(G^{\rm as})^{-1}\left(\alpha,\frac{q^2}{m^2}\right)=0\,.\tag{19.76}$$

As before, the physically most interesting limit concerns the deep Euclidean regime, $q^2=-\vec{q}^{\,2}\to-\infty$. In view of the chain rule, $\frac{\partial}{\partial x}[1/f(x)]=-f'(x)/[f(x)]^2$, this implies that

$$\left(m\,\frac{\partial}{\partial m}+2\,\widetilde{\beta}(\alpha)\frac{\partial}{\partial\alpha}\right)\,G^{\rm as}\left(\alpha,\frac{q^2}{m^2}\right)=0\,.\tag{19.77}$$

Using the identity

$$m\,\frac{\partial}{\partial m}\left(\frac{q^2}{m^2}\right)=-2\frac{q^2}{m^2}\,,\tag{19.78}$$

this can be reformulated as

$$\left(-2\left(\frac{q^2}{m^2}\right)\frac{\partial}{\partial(q^2/m^2)}+2\,\widetilde{\beta}(\alpha)\frac{\partial}{\partial\alpha}\right)\,G^{\rm as}\left(\alpha,\frac{q^2}{m^2}\right)=0\,.\tag{19.79}$$

Defining $x=q^2/m^2$, this relation translates into the RG equation for the flow of the QED coupling in the asymptotic regime,

$$x\,\frac{\partial}{\partial x}\,G^{\rm as}\,(\alpha,x)=\widetilde{\beta}(\alpha)\frac{\partial}{\partial\alpha}\,G^{\rm as}\,(\alpha,x)\,.\tag{19.80}$$

Now, this needs to be compared to the Gell-Mann–Low RG equation (19.57),

$$x\frac{\partial}{\partial x}G^{\rm as}(\alpha,x)=\beta(G^{\rm as}(\alpha,x))\,.\tag{19.81}$$

Using $\alpha_x=G^{\rm as}\,(\alpha,x)$, Eqs. (19.80) and (19.81) imply the relation

$$\beta(\alpha_x)=\widetilde{\beta}(\alpha)\,\frac{\partial\alpha_x}{\partial\alpha}\,.\tag{19.82}$$

Intuitively, one would assume that since both Eqs. (19.80) and (19.81) describe the RG flow of the coupling, the connection between the equations should be found in a Jacobian factor, and this is indeed the case,

$$\beta(\alpha_x) = x\,\frac{\partial \alpha_x}{\partial x} = \left(x\,\frac{\partial \alpha}{\partial x}\right)\left(\frac{\partial \alpha_x}{\partial \alpha}\right) = \widetilde{\beta}(\alpha)\,\frac{\partial \alpha_x}{\partial \alpha}\,. \tag{19.83}$$

A change in the definition of the coupling amounts to a change in the renormalization scheme. Up to the Jacobian which describes the transition among different renormalization schemes, we see that the Callan–Symanzik formalism and the Gell-Mann Low formalism are actually equivalent in the regime of large momentum transfers. Conversely, the Callan–Symanzik equation clarifies how to augment the RG flow in the non-asymptotic regime, taking into account the running of the mass and the inhomogeneous terms from the Callan–Symanzik equation, generated by the mass insertions.

We briefly indicate the available results for the β function in various renormalization schemes, without going into any further detail. In the on-shell scheme, the three-loop result for the QED beta function has been given by de Rafael and Rosner in Ref. [629], and the result has been generalized to four loops by Gorishny, Kataev, Larin and Surguladze in Refs. [628, 630],

$$\begin{aligned}\widetilde{\beta}_{\rm OS}(\alpha) &= \frac{\alpha^2}{3\pi} + \frac{\alpha^3}{4\pi^2} - \frac{121}{288}\,\frac{\alpha^4}{\pi^3} \\ &\quad + \frac{\alpha^5}{\pi^4}\left(\frac{5561}{10368} - \frac{23}{18}\,\zeta(3) + \frac{4}{3}\,\ln 2\,\zeta(2) - \frac{7}{16}\,\zeta(3)\right) + \mathcal{O}(\alpha^6)\,. \end{aligned} \tag{19.84}$$

It is perhaps useful to clarify that the authors of Ref. [628] refer to the β function as the "Gell-Mann–Low function", in a slight departure from the usual naming conventions. In the minimal subtraction (MS) scheme, we have from Ref. [628] the four-loop result

$$\beta_{\rm MS}(\alpha) = \frac{\alpha^2}{3\pi} + \frac{\alpha^3}{4\pi^2} - \frac{31}{288}\,\frac{\alpha^4}{\pi^3} + \frac{\alpha^5}{\pi^4}\left(-\frac{13}{36}\,\zeta(3) - \frac{2785}{31104}\right) + \mathcal{O}(\alpha^6)\,. \tag{19.85}$$

Another result of particular interest concerns so-called "quenched" QED. In the case of "perturbative quenching", the related Feynman diagrams with internal photon vacuum polarization diagrams are not considered and therefore the coupling constant of QED is renormalized only by the subset of vacuum polarization subgraphs with one external fermion loop and $0 \le n \le \infty$ internal photon lines. It can be shown, that within this approximation the QED vacuum polarization function and the related contribution to the QED β-function do not depend on the choice of the renormalization scheme [631]. A partial five-loop result for the β function $\beta_{\rm Q}(\alpha)$ subgraphs in quenched QED has been given in Refs. [632–634],

$$\beta_{\rm Q}(\alpha) = \frac{\alpha^2}{3\,\pi} + \frac{\alpha^3}{4\,\pi^2} - \frac{1}{32}\,\frac{\alpha^4}{\pi^3} - \frac{23}{128}\,\frac{\alpha^5}{\pi^4} + \frac{\alpha^6}{\pi^5}\left(\frac{4157}{6144} + \frac{1}{8}\,\zeta(3)\right) + \mathcal{O}(\alpha^7)\,. \tag{19.86}$$

The analytic structure of the 5-loop result of Ref. [11] differs from the previously known terms; it contains no $\zeta(3)$-term in the 4-loop term, but $\zeta(3)$ does appear in the 5-loop coefficient.

Surprisingly, an inspection of Eqs. (19.60), (19.84), (19.85) and (19.86) shows that the terms of order α^2 and α^3 in the otherwise scheme-dependent results for the QED β function, in fact, are the same. This is not accidental. We had previously noted the Jacobian character of the scheme dependence in Eq. (19.83), which is manifest in the relations

$$\beta(\alpha_x) = \widetilde{\beta}(\alpha)\,\frac{\partial \alpha_x}{\partial \alpha}\,, \qquad \beta(\alpha_x) = x\,\frac{\partial \alpha_x}{\partial x}\,, \qquad \widetilde{\beta}(\alpha) = x\,\frac{\partial \alpha}{\partial x}\,. \tag{19.87}$$

We can set, by perturbative expansion,

$$\alpha_x = \alpha\,\left(1 + a_1\,\alpha + a_2\,\alpha^2 + \mathcal{O}(\alpha^3)\right)\,, \quad \frac{\partial \alpha_x}{\partial \alpha} = 1 + 2\,a_1\,\alpha + 3\,a_2\,\alpha^2 + \mathcal{O}(\alpha^3)\,. \tag{19.88}$$

The Taylor expansion can of course be inverted to read

$$\alpha = \alpha_x\,\left(1 - a_1\,\alpha_x + (2a_1^2 - a_2)\,\alpha_x^2 + \mathcal{O}(\alpha_x^3)\right)\,. \tag{19.89}$$

The $\widetilde{\beta}$ function in the transformed scheme is given as

$$\widetilde{\beta}(\alpha) = b_2\,\alpha^2 + b_3\,\alpha^3 + b_4\,\alpha^4 + \mathcal{O}(\alpha^5)\,, \tag{19.90}$$

where b_2 is the one-loop, and b_3 is the two-loop coefficient. The scheme independence of the first two perturbative coefficients of the QED β function can be verified as follows,

$$\begin{aligned}
\beta(\alpha_x) = \widetilde{\beta}(\alpha)\,\frac{\partial \alpha_x}{\partial \alpha} &= \left(b_2\,\alpha^2 + b_3\,\alpha^3 + \mathcal{O}(\alpha^4)\right)\left(1 + 2\,a_1\,\alpha + 3\,a_2\,\alpha^2 + \mathcal{O}(\alpha^3)\right) \\
&= b_2\,\alpha^2 + (b_3 + 2\,b_2\,a_1)\,\alpha^3 + \mathcal{O}(\alpha^4) \\
&= b_2\,(\alpha_x - a_1\,\alpha_x^2)^2 + (b_3 + 2\,b_2\,a_1)\,\alpha_x^3 + \mathcal{O}(\alpha_x^4) \\
&= b_2\,\alpha_x^2 + b_3\,\alpha_x^3 + \mathcal{O}(\alpha_x^4)\,.
\end{aligned} \tag{19.91}$$

This shows that the first two coefficients of the QED β function are scheme-independent, which is what we wanted to show. The next higher-order term is

$$\begin{aligned}
\beta(\alpha_x) &= b_2\,\alpha_x^2 + b_3\,\alpha_x^3 + b_4\,\alpha_x^4 + \left(-a_1^2\,b_2 + a_2\,b_2 - a_1\,b_3\right)\,\alpha_x^4 + \mathcal{O}(\alpha_x^5) \\
&= \widetilde{\beta}(\alpha_x) + \left(-a_1^2\,b_2 + a_2\,b_2 - a_1\,b_3\right)\,\alpha_x^4 + \mathcal{O}(\alpha_x^5)\,.
\end{aligned} \tag{19.92}$$

In order to fix ideas, we should mention that an obvious requirement on the β function is that the mapping $\alpha \to \alpha_x$ must be infinitely differentiable, and it must be strictly monotonous so that no artificial zeros or zeros of the derivatives of the β functions are being introduced.

19.4 RG Equations and Applications

19.4.1 *Effective Action and Renormalization Group*

The QED effective action is a Lagrangian, written in terms of the electromagnetic fields, which describes a quantum-corrected Maxwell Lagrangian. In the strong-field limit, the mass of the electron becomes irrelevant, and one may thus ask the question to which extent the effective charge regime can be related to the analytic form of the

QED effective Lagrangian, at least in the asymptotic regime. Indeed, connections between the strong-field and short-distance limits of quantum electrodynamics, and the effective Lagrangian, have been studied in Refs. [635, 636]. The first aspect to notice is that the product

$$e\,A^\mu = \left(Z_3^{1/2}\,e_0\right)\,\frac{A_0^\mu}{Z_3^{1/2}} = e_0\,A_0^\mu \tag{19.93}$$

actually is invariant under RG transformations. We can intuitively understand that, in the asymptotic region, the Maxwell Lagrangian must contain the effective charge $\alpha(t)$ of the renormalization group "time" t, via the renormalization of the field and the coupling $e^2 = 4\,\pi\,\alpha$,

$$\mathcal{L}_0 = -\frac{1}{4}\,F_0^{\mu\nu}\,F_{0,\mu\nu} \to -\frac{1}{4Z_3}\,F^{\mu\nu}\,F_{\mu\nu} = -\frac{1}{4}\,\frac{\alpha}{\alpha(t)}\,F^{\mu\nu}\,F_{\mu\nu}\,. \tag{19.94}$$

This structure is rigorously shown in Ref. [637] (see also the references therein). The parameter t has to be defined to be consistent with the strong-field asymptotic behavior of the QED effective Lagrangian,

$$t = \frac{1}{4}\,\ln\left(\frac{\alpha\,|F^2|}{\mu^4}\right) \tag{19.95}$$

where $F^2 = F^{\mu\nu}\,F_{\mu\nu} = 2(\vec{B}^2 - \vec{E}^2)$. Because t is proportional to the logarithm of the renormalization scale μ, and in view of the identity $\mathrm{d}\ln(\mu) = \mu^{-1}\mathrm{d}\mu$, the β function can be expressed as

$$2\beta(\alpha(t)) = \frac{\mathrm{d}}{\mathrm{d}t}\alpha(t)\,, \qquad t = \int_\alpha^{\alpha(t)} \frac{\mathrm{d}\alpha'}{2\beta(\alpha')}\,. \tag{19.96}$$

The factor two is a consequence of our definition (19.59). We recall, from Eq. (19.60), the first two perturbative terms of the β function as $\beta(\alpha) = b_2\,\alpha^2 + b_3\,\alpha^3 + \cdots$, and obtain from the RG evolution,

$$\frac{1}{\alpha(t)} = \frac{1}{\alpha} - 2\,\beta_2\,t - 2\,\alpha\,\beta_3\,t + \mathcal{O}(t^2)\,, \tag{19.97}$$

so that the structure of the QED effective Lagrangian (including the leading Maxwell term) reads as follows,

$$\mathcal{L} = -\frac{1}{4}\,F^2 + \frac{\alpha\,\beta_2}{8}\,F^2\,\ln\left(\frac{\alpha\,|F^2|}{\mu^4}\right) + \frac{\alpha^2\,\beta_3}{8}\,F^2\,\ln\left(\frac{\alpha\,|F^2|}{\mu^4}\right) + \mathcal{O}(\alpha^3)\,. \tag{19.98}$$

For QED, the asymptotic behavior of the one-loop, and two-loop Lagrangian is known from the calculations of Ritus [635, 636]. For a constant, strong, magnetic background field, we have

$$\mathcal{L}^{(1)} = \frac{\alpha\,\vec{B}^2}{12\pi}\,\ln\left(\frac{\alpha\,\vec{B}^2}{m^4}\right)\,, \qquad \mathcal{L}^{(2)} = \frac{\alpha^2\,\vec{B}^2}{16\pi^2}\,\ln\left(\frac{\alpha\,\vec{B}^2}{m^4}\right)\,. \tag{19.99}$$

Using the relation $F^2 = 2\,\vec{B}^2$, which is valid for a purely magnetic background field, we can read off the coefficients

$$\beta_2 = \frac{1}{3\pi}\,, \qquad \beta_3 = \frac{1}{4\pi^2}\,, \tag{19.100}$$

in agreement with the result indicated in Eq. (19.60). (We recall that the first two coefficients of the QED β function actually are scheme-independent.) We here identify the electron mass with the renormalization scale μ.

One might thus think that the QED effective action would be an ideal vehicle for studying the RG evolution of the effective charge, possibly even superior to the QED β function. However, that is not the case. First, we recall that the RG evolution discussed above relies on the asymptotic regime of strong fields, which is equivalent to the approximation made previously in deriving the Gell-Mann–Low β function. Second, and this point is less obvious, the RG evolution of α from the QED effective action, and from the Gell-Mann–Low β function is not quite as universally matched as one would otherwise expect. In Ref. [637], one considers self-dual (sd) background fields with $F_{\mu\nu} = \mathcal{F}_{\mu\nu} = \widetilde{\mathcal{F}}_{\mu\nu}$, where

$$\mathcal{F}_{\mu\nu} = \widetilde{\mathcal{F}}_{\mu\nu} = \frac{1}{2}\,\epsilon_{\mu\nu\rho\delta}\,\mathcal{F}^{\rho\delta}\,, \qquad \mathcal{F}^2 \equiv \mathcal{F}^{\mu\nu}\,\mathcal{F}_{\mu\nu}\,. \tag{19.101}$$

These fields are still constant, and strong, but self-dual, so that the dual field strength tensor equals the field strength tensor. In a self-dual background field [see Eqs. (2.30) and (2.31) of Ref. [637]], the strong-field asymptotics do not match the coefficients of the β functions at all,

$$\mathcal{L}^{(1)}_{\rm sd} = -\frac{\alpha\,\mathcal{F}^2}{48\pi}\,\ln\left(\frac{\alpha\,|\mathcal{F}^2|}{m^4}\right), \qquad \mathcal{L}^{(2)}_{\rm sd} = -\frac{\alpha^2\,\mathcal{F}^2}{16\pi^2}\,\ln\left(\frac{\alpha\,|\mathcal{F}^2|}{m^4}\right). \tag{19.102}$$

The explanation for the apparent mismatch of the strong-field behavior of the QED effective Lagrangian in a self-dual background, given in Ref. [637], is subtle and depends on the loop order. At one loop, one can argue that in a self-dual background, as opposed to a constant magnetic field, so-called zero modes contribute to the effective action. These zero modes are virtual states of the Dirac field with a specific helicity, and they contribute to the fermionic determinant, but only in the self-dual background field, which has both electric and magnetic field components. The zero modes come from the states of a massless Dirac field, which always is in a specific helicity state.

For the two-loop, self-dual term in Eq. (19.102), a different explanation is offered in Ref. [637]. Namely, it is argued that the RG equation fulfilled by the Lagrangian in the self-dual background should read as follows,

$$\left(\mu\,\frac{\partial}{\partial\mu} + \beta(\alpha)\,\frac{\partial}{\partial\alpha} - \gamma_m\,m\,\frac{\partial}{\partial m}\right)\mathcal{L}(\alpha, m, \mu, e\,\mathcal{F}_{\mu\nu}) = 0\,, \tag{19.103}$$

in contrast to the magnetic-field case, where the mass does not run,

$$\left(\mu\,\frac{\partial}{\partial\mu} + \beta_m(\alpha)\,\frac{\partial}{\partial\alpha}\right)\mathcal{L}(\alpha, m, \mu, e\vec{B}) = 0\,. \tag{19.104}$$

In both Eqs. (19.103) and (19.104), the quantities $e\,\mathcal{F}_{\mu\nu}$ viz. $e\vec{B}$ act as spectators under the RG evolution. Furthermore, μ is the renormalization scale, which also enters the asymptotic form of the one-loop Lagrangian,

$$\mathcal{L}^{(1)}_{\rm sd} = -\frac{\mathcal{F}}{4} - \frac{\alpha\,\mathcal{F}^2}{48\pi}\,\ln\left(\frac{\alpha\,|\mathcal{F}^2|}{m^4}\right) + \frac{\alpha\,\mathcal{F}^2}{12\pi}\,\ln\left(\frac{m^2}{\mu^2}\right). \tag{19.105}$$

One can now read off the anomalous mass dimension as follows,

$$\gamma_m = -\frac{\mu}{m}\frac{\partial m}{\partial \mu} = -\frac{3}{2}\frac{\alpha}{\pi} + \cdots. \tag{19.106}$$

This implies that at two-loop order, one can then trade the anomalous dimension term in Eq. (19.103) for a mass-dependent β function,

$$\begin{aligned}\beta_m(\alpha) &= \frac{2}{3}\frac{\alpha^2}{\pi} + \frac{1}{2}\frac{\alpha^3}{\pi^2} + \gamma_m\frac{\alpha^2}{\pi} + \mathcal{O}(\alpha^4)\\ &= \frac{2}{3}\frac{\alpha^2}{\pi} + \frac{1}{2}\frac{\alpha^3}{\pi^2} - \frac{3}{2}\frac{\alpha^3}{\pi^2} + \mathcal{O}(\alpha^4) = \frac{2}{3}\frac{\alpha^2}{\pi} - \frac{\alpha^3}{\pi^2} + \mathcal{O}(\alpha^4)\end{aligned} \tag{19.107}$$

which perfectly matches the two-loop coefficient of the strong-field limit of the self-dual spinor case given in Eq. (19.102). The correspondence between the strong-field limit and the β function in the self-dual spinor case at two-loop only applies to the β function β_m of an implicitly electron-mass-dependent regularization scheme. The mass-dependent scheme is indicated for the self-dual spinor case because of the presence of the zero modes which inhibit a direct massless limit. Via the running one-loop electron mass in the Dirac propagator insertion of the two-loop effective action, the anomalous mass dimension creeps into the renormalization-group analysis of the QED effective action, in the case of a self-dual background field. While the matching of the QED β function with the strong-field limit of the QED effective action does not universally hold for all types of background fields, one can find a very straightforward connection for the case of a magnetic background, and a rather subtle connection for the self-dual background field case.

19.4.2 *Brodsky–Lepage–Mackenzie Scale Setting*

We start from the two-loop result for the anomalous magnetic moment of the electron, which has been obtained in Refs. [54, 639–641]. Combining the one-loop vertex calculation of the electron g factor [2] with the two-loop result, we have

$$a_e = \frac{g_e - 2}{2} = \frac{\alpha}{2\pi}\left(1 + C_1\frac{\alpha}{\pi} + \mathcal{O}(\alpha^2)\right), \tag{19.108}$$

where the two-loop coefficient C_1 has the numerical value

$$C_1 = \frac{197}{72} + \zeta(2) - 6\zeta(2)\ln(2) + \frac{3}{2}\zeta(3) = -0.657. \tag{19.109}$$

The muon anomalous magnetic moment is different,

$$a_\mu = \frac{g_\mu - 2}{2} = \frac{\alpha}{2\pi}\left(1 + D_1\frac{\alpha}{\pi} + \mathcal{O}(\alpha^2)\right), \tag{19.110}$$

where

$$D_1 \approx C_1 + \frac{1}{3}\left[\ln\left(\frac{m_\mu^2}{m_e^2}\right) - \frac{25}{18}\right]. \tag{19.111}$$

This result holds if we assume that the two-loop difference of the muon anomalous magnetic moment and the electron anomalous magnetic moment is mostly due to

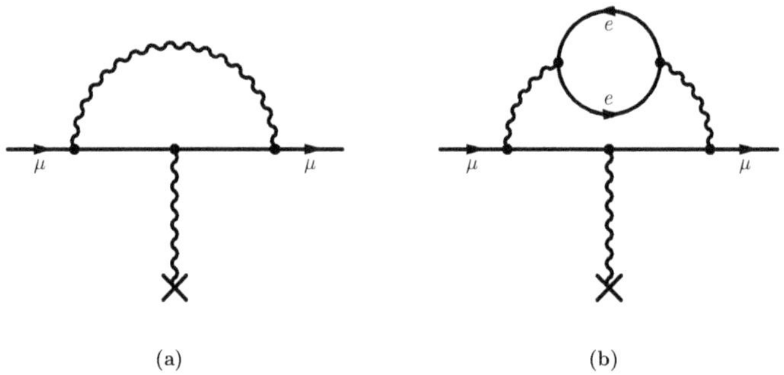

Fig. 19.1 Insertion of the vacuum-polarization loop [Fig. (b)] into the vertex correction [Fig. (a)] relevant to the muon $g_\mu - 2$. This effect can be described, to good accuracy, by an optimally fixed momentum scale in the asymptotic form of the one-loop vacuum polarization. The interaction with the external magnetic field is denoted by the external photon line which ends in a cross.

(electron) vacuum polarization loops. We thus ignore a numerically small term, which is obtained when the electron loop is replaced by a muon loop. From the RG analysis of the running charge, we know that the formula

$$\alpha(Q^2) = \frac{\alpha}{1 - \frac{\alpha}{\pi}\left(\frac{1}{3}\ln\left(\frac{Q^2}{m_e^2}\right) - \frac{5}{3}\right) + \cdots} \tag{19.112}$$

describes the running coupling constant at the scale $Q^2 = -\vec{q}^2$, in the deep Euclidean region. The paradigm of scale setting now is as follows: Let us consider the difference of the muon anomalous magnetic moment and the electron anomalous magnetic moment. The question is whether we can find an "optimum" scale Q_*^2 so that the difference in the two-loop terms in the muon versus electron anomalous magnetic moments can be obtained by inserting the running charge at the optimum scale Q_*^2, into the one-loop term of the electron anomalous magnetic moment, i.e., the matching criterion for Q_*^2 is as follows,

$$a_\mu - a_e = \frac{\alpha(Q_*^2)}{2\pi} - \frac{\alpha}{2\pi} + \mathcal{O}(\alpha^3) = \frac{\alpha^2}{2\pi^2}(D_1 - C_1) + \mathcal{O}(\alpha^3)\,. \tag{19.113}$$

The notion is that the running charge $\alpha(Q_*^2)$ describes the two-loop vacuum-polarization insertions indirectly, by a change in the effective coupling used for the calculation of the one-loop effect, with a scale Q_*^2 matched so that upon reexpansion of $\alpha(Q_*^2)$, the two-loop vacuum polarization diagram in Fig. 19.1 is described. This is why the remainder term in Eq. (19.113) is assumed to be of order α^3, not α^2.

Expanding the running charge (19.112) to first order in α, the criterion then is that

$$\alpha(Q_*^2) - \alpha(Q^2) = \frac{\alpha}{\pi}\left(\frac{1}{3}\ln\left(\frac{Q_*^2}{m_e^2}\right) - \frac{5}{3}\right) = \frac{\alpha}{\pi}\frac{1}{3}\left[\ln\left(\frac{m_\mu^2}{m_e^2}\right) - \frac{25}{18}\right], \tag{19.114}$$

which is solved by

$$Q_*^2 = m_\mu^2 \exp\left(\frac{5}{6}\right) \gg m_e^2\,. \tag{19.115}$$

The hope then is that, inserting the running charge at the optimum scale Q_*^2 into the two-loop result for the anomalous magnetic moment of the electron,

$$a_\mu - a_e \approx \frac{1}{2\pi}\left[\alpha(Q_*^2) - \alpha\right] + \frac{1}{2\pi^2} C_1 \left[\alpha^2(Q_*^2) - \alpha^2\right] + \cdots \qquad (19.116)$$

one obtains an approximation to the three-loop effects. In an ideal scenario, one might thus assume that the terms denoted by the ellipsis $(\dots)$ in Eq. (19.116) are of order $\mathcal{O}(\alpha^4)$, not $\mathcal{O}(\alpha^3)$.

The perturbation series for difference of the anomalous magnetic moments of the muon and electron is given as

$$a_\mu^{\rm QED} - a_e^{\rm QED} = 1.09\left(\frac{\alpha}{\pi}\right)^2 + 22.87\left(\frac{\alpha}{\pi}\right)^3 + 127.00\left(\frac{\alpha}{\pi}\right)^4 + \mathcal{C}_\mu\left(\frac{\alpha}{\pi}\right)^5 + \mathcal{O}(\alpha^6), \qquad (19.117)$$

with explicit numerical result for the coefficients through four-loop order. It has been verified by Kataev and Starshenko that the higher-order terms are well represented by inserting the running charge into lower-order diagram [34, 642]. The coefficients in the series (19.117) grow drastically in numerical magnitude with an increase of the loop order. This is a manifestation of the general divergence of perturbation theory in QED [132], which is especially relevant to the current situation because the muon anomalous magnetic moment is obtained from an optimum momentum scale which is much larger than the electron mass scale [see Eq. (19.115)], and therefore the effect of higher-order vacuum-polarization insertions is considerably enhanced.

RG improvement is relevant to the tenth-order (in e) graphs, which are of fifth order in α, and contribute to the next higher-order analytic term not included in Eq. (19.117). In Ref. [643] a group led by KINOSHITA[1] has obtained an estimate of $\mathcal{C}_\mu \approx 570(14)$ for the term of order $(\alpha/\pi)^5$ to be included into Eq. (19.117). Karshenboim [638] has shown that the previous analysis was somewhat incomplete; he analyses the graph in Fig. 19.2(c) and obtains an additional contribution of $\mathcal{C}_\mu \approx 176(35)$ for the coefficient. It is worth noting that not all graphs which contribute in fifth order in α can be analyzed by RG-improved perturbation theory. The topologically new graphs in Figs. 19.2(e) and (f), whose contribution has been estimated in Ref. [638], give additional contributions of $\mathcal{C}_\mu \approx 85 \pm 85$ and $\mathcal{C}_\mu \approx \pm 40$, respectively. The total result obtained by adding all of the mentioned contributions reads as

$$\mathcal{C}_\mu = 930(170)\,, \qquad (19.118)$$

which is an impressive magnitude for a perturbative coefficient; with a twinkle in the eye and with some humor, we can state that the tenth-order contribution to the anomalous magnetic moment of the muon currently holds the "world record" for the largest (in numerical magnitude) perturbative coefficient obtained for any process in field theory, when the perturbation series is expressed in terms of a "reasonable" expansion parameter, namely, in our case, the QED coupling α. Very recently, a

[1]Toichiro Kinoshita (b. 1925).

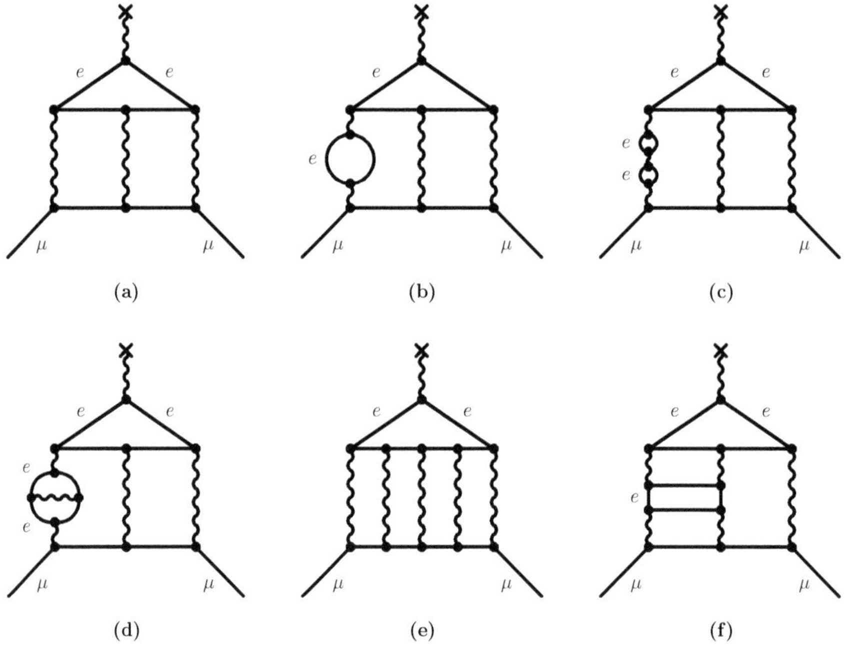

Fig. 19.2 Representative diagrams illustrating the calculation of five-loop contributions (order α^5) to the anomalous magnetic moment of the muon. Diagram (a) is a three-loop "skeleton" diagram. The number of internal lines is $I = 9$, the number of vertices is $V = 7$, and the loop order is $I - V + 1 = 3$. The four-loop diagram (b) results from a one-loop vacuum-polarization insertion into diagram (a), which can be estimated based on the running charge, inserted into diagram (a). Diagrams (c) and (d) can be estimated based on the two-loop running of the effective charge, inserted into diagram (a). Finally, diagrams (e) and (f) cannot be estimated based on the renormalization-group equations for the running charge. Expressed in terms of the coefficient C_μ [see Eq. (19.117)], their contributions have been estimated (see Ref. [638]) as 185 ± 15 and ± 40, respectively. The electron lines and muon lines are denoted by e and μ, respectively.

complete calculation (without estimation) of the coefficient has been presented in Ref. [644] with the result

$$\mathcal{C}_\mu = 753.29(1.04)\,, \tag{19.119}$$

for the QED contribution to $\mathcal{C}_\mu$ (contributions beyond QED are numerically small). This result is in fair numerical agreement with the estimate (19.118).

The BLM scale setting is a form of an RG improved perturbation theory. We notice that the expression (19.112) for the running coupling has the form of a so-called [1/1]-Padé approximant (for an explanation of the notation, see Ref. [124]), where the running coupling is replaced by a rational function of a different expansion parameter. Furthermore, one knows that if the running were perfect, we could represent e.g. the anomalous magnetic moment of the muon as $\alpha(Q^*)/(2\pi)$ where Q^* is just the value of α at the perfectly matched scale Q^*. Of course, in their original work [92], Brodsky, Lepage and MacKenzie explicitly state that the matching only makes sense for the vacuum-polarization diagram insertions. They do not include topologically new graphs which can contribute in each higher order, such as those shown in Fig. 19.2. Yet, we could be a little bold. Using the effective coupling, we effectively replace the perturbation series by a Padé approximant to itself, which upon reexpansion yields back the perturbation series whose convergence was to be improved. This is known an accuracy-through-order relation [124]. As an open problem, we mention here the conceivable replacement of the rational function fixing the effective charge (19.112), by a Padé approximant or even a Weniger transformation [645], which might improve the scale setting procedure slightly. We also notice that the continuous readjustment of the optimum scale for the RG improvement, order-by-order in perturbation theory, is reminiscent of the celebrated order-dependent mappings originally introduced by Seznec and Zinn-Justin [646].

19.5 Further Thoughts

Here are some suggestions for further thought.

(1) **Predictive Limits.** In view of the considerations reported in Sec. 3.4.2, extrapolate some numerical coefficients for the anomalous magnetic moment of the muon and try to determine the predictive limits of quantum electrodynamics. Also, consult Ref. [130]. Note that estimates contain the factor $\Gamma(k/2)$ in the kth loop order in QED as opposed to $\Gamma(k)$ (see Sec. 3.4.2).

(2) **Beta Function at Large Coupling.** Consult Refs. [647,648] for quite recent work on the large-order behavior of the coefficients of the QED β function. It is claimed in Ref. [648] that the perturbative expansion of the β function has the asymptotic form

$$\beta(\alpha) \sim C \sum_K A^{-K}\,\Gamma(N/2+6)\,\alpha^K\,, \qquad A = 4.886\,, \tag{19.120}$$

in accordance with Eq. (3.105). Based on this conjecture, and on the numerically observed trend in the large-coupling behavior of the summations (using order-dependent mappings [130]) of the first few partial sum of the QED β function, it has been conjectured [647] that the entire β function has the asymptotic behavior

$$\beta(\alpha) \sim \beta_\infty\,\alpha^\lambda\,, \qquad \beta_\infty = 1.0 \pm 0.3\,, \qquad \lambda = 1.0 \pm 0.1\,. \tag{19.121}$$

The exact equality $\beta_\infty = \lambda = 1$ has been conjectured in Ref. [648]. A confirmation of this result by independent methods would otherwise imply a linear growth of the effective QED coupling at large momenta and be consistent with what is expected for Grand Unification.

(3) **Toward Grand Unification.** Study the RG evolution of QED, and the electroweak sector of the Standard Mode, and of quantum chromodynamics (QCD). Why do we conjecture the confluence of all coupling parameters in nature at very high energy scales, and how is this related to Grand Unification?

(4) **Novel Interactions.** Explore weak and other beyond-Standard-Model interaction effects in atomic systems. According to Eq. (6.100), the atomic wave function can be represented as follows,

$$\psi = \begin{pmatrix} \phi \\ \dfrac{\vec{\sigma} \cdot \vec{p}}{2m} \phi \end{pmatrix}, \qquad \psi^+ = \begin{pmatrix} \phi^+ \\ \phi^+ \dfrac{\vec{\sigma} \cdot \vec{p}}{2m} \end{pmatrix}, \qquad \overline{\psi} = \psi^+ \gamma^0 . \tag{19.122}$$

Here, m is the bound particle mass (for us, the electron) and ϕ is the Schrödinger–Pauli wave function, while ψ is the relativistic Dirac wave function. The representation is valid for essentially nonrelativistic systems, with first relativistic corrections in the lower components.

Consider a hypothetical novel particle X_1 with mass M_{X_1} which has an axial coupling to the electron and a vector coupling to the quarks in the nucleus. Convince yourself that one can obtain a parity-violating (PV) interaction Hamiltonian by combining the $\mu = 0$ axial component of the electron current with the $\mu = 0$ vector component of the quark current, as follows,

$$H^{(1)}_{\rm PV} = \overline{\psi} \gamma^0 g^A_e \gamma^5 \psi \frac{1}{{M_{X_1}}^2} \overline{\psi}_{q_i} \gamma^0 g^V_{q_{ii}} \psi_{q_i} , \tag{19.123}$$

with suitable identifications of the parameters.

The γ^5 matrix in atomic theory is almost exclusively used in the Dirac representation, and is given in Eq. (7.15). Derive, in the nonrelativistic limit, the relevant effective Hamiltonian,

$$H^{(1)}_{\rm PV} = \frac{g^A_e g^V_{q_{ii}}}{M^2_X} \left\langle \overline{\psi}_{q_i} \gamma^0 \psi_{q_i} \right\rangle \left(\frac{\vec{\sigma} \cdot \vec{p}}{m} \delta^{(3)}(\vec{r}) + \delta^{(3)}(\vec{r}) \frac{\vec{\sigma} \cdot \vec{p}}{m} \right) \tag{19.124}$$

to be evaluated on the atomic wave function. Convince yourself that the corresponding term for the weak interaction is

$$H^{(1)}_{\rm PV} = \frac{G_F}{4\sqrt{2}} \frac{1}{m} Q^{(1)}_W(Z,N) \left(\frac{\vec{\sigma} \cdot \vec{p}}{m} \delta^{(3)}(\vec{r}) + \delta^{(3)}(\vec{r}) \frac{\vec{\sigma} \cdot \vec{p}}{m} \right) . \tag{19.125}$$

Here, $Q^{(1)}_W(Z,N)$ is the weak charge of the nucleus. Compare with Refs. [649, 650].

One can also obtain a parity-violating (PV) interaction Hamiltonian by combining the $\mu = 1,2,3$ vector component of the electron current, with the $\mu = 1,2,3$ axial vector component of the quark current, as follows,

$$H^{(2)}_{\rm PV} = \overline{\psi} \gamma^0 g^V_{l_{aa}} \vec{\gamma} \psi \frac{1}{{M_{X_1}}^2} \overline{\psi}_{q_i} \gamma^0 g^A_{q_{ii}} \vec{\gamma} \gamma^5 \psi_{q_i} , \tag{19.126}$$

again, with a suitable identifications of the parameters. Convince yourself that one can assume that

$$\left\langle \overline{\psi}_{q_i} \gamma^0 g^A_{q_{ii}} \vec{\gamma} \gamma^5 \psi_{q_i} \right\rangle = \sum_q G^{(q)}_A(Z,N) \, \delta^{(3)}(\vec{r}) \left\langle I, I'_3 \left| \vec{I} \right| I, I_3 \right\rangle , \tag{19.127}$$

where $\vec{I}$ is the nuclear spin vector. In view of $\overline{\psi}\gamma^0 = \psi^+$, this leads to

$$H_{\rm PV}^{(2)} = \frac{g_e^V \sum_q G_A^{(q)}(Z,N)}{M_{X_1}^2} \left(\frac{\vec{\sigma}\cdot\vec{p}}{m} \vec{\sigma}\cdot\vec{I}\,\delta^{(3)}(\vec{r}) + \delta^{(3)}(\vec{r})\,\frac{\vec{\sigma}\cdot\vec{p}}{m}\vec{\sigma}\cdot\vec{I} \right), \tag{19.128}$$

to be evaluated on the atomic wave function.

In fact, the corresponding term for the weak interaction is

$$H_{\rm PV}^{(2)} = \frac{G_F}{4\sqrt{2}}\frac{1}{m} Q_W^{(2)}(Z,N) \left(\frac{\vec{\sigma}\cdot\vec{p}}{m}\,\delta^{(3)}(\vec{r}) + \delta^{(3)}(\vec{r})\,\frac{\vec{\sigma}\cdot\vec{p}}{m} \right), \tag{19.129}$$

where $Q_W^{(2)}(Z,N)$ is an anapole moment of the nucleus. Be inspired by Ref. [651].

(5) **Higher-Order Corrections.** Generalizing results reported in Refs. [56–59] and Ref. [55], evaluate the electron anomalous magnetic moment to such high loop orders that you are able to check about the asymptotic large-order limit, where the series should divergence factorially. This task might last a lifetime!

Bibliography

[1] R. P. Feynman, *QED: The Strange Theory of Light and Matter.* Princeton University Press, Princeton, NJ (2006).

[2] C. Itzykson and J.-B. Zuber, *Quantum Field Theory.* McGraw-Hill, New York (1980).

[3] V. B. Berestetskii, E. M. Lifshitz and L. P. Pitaevskii, *Quantum Electrodynamics, Volume 4 of the Course on Theoretical Physics*, 2nd edn. Pergamon Press, Oxford, UK (1982).

[4] A. I. Akhiezer and V. B. Berestetskii, *Quantum Electrodynamics.* Nauka, Moscow (1969).

[5] I. Białynicki-Birula and Z. Białynicka-Birula, *Quantum Electrodynamics.* Pergamon Press, Oxford (1975).

[6] H. A. Bethe and E. E. Salpeter, *Quantum Mechanics of One- and Two-Electron Atoms.* Springer, Berlin (1957).

[7] W. Heitler, *Quantum Theory of Radiation.* Oxford University Press, New York (1950).

[8] H. A. Bethe and R. Jackiw, *Intermediate Quantum Mechanics.* Perseus, New York (1986).

[9] J. Schwinger, *Particles, Sources and Fields.* Addison-Wesley, Reading, MA (1970).

[10] S. S. Schweber, *An Introduction to Relativistic Quantum Field Theory.* Harper & Row, New York, NY (1961).

[11] R. P. Feynman, *The Theory of Fundamental Processes.* Perseus, Cambridge, Massachusetts (1998).

[12] W. Greiner and J. Reinhardt, *Quantum Electrodynamics.* Springer, New York (1992).

[13] J. M. Jauch and F. Rohrlich, *The Theory of Photons and Electrons*, 2nd edn. Springer, Heidelberg (1980).

[14] A. R. Edmonds, *Angular Momentum in Quantum Mechanics.* Princeton University Press, Princeton, New Jersey (1957).

[15] D. M. Brinks and G. R. Satchler, *Angular Momentum.* Oxford University Press, Oxford (1994).

[16] D. A. Varshalovich, A. N. Moskalev and V. K. Khersonskii, *Quantum Theory of Angular Momentum.* World Scientific, Singapore (1988).

[17] M. E. Rose, *Relativistic Electron Theory.* J. Wiley & Sons, New York, NY (1961).

[18] E. Merzbacher, *Quantum Mechanics.* J. Wiley & Sons, New York (1961).

[19] B. H. Bransden and C. J. Joachain, *Physics of Atoms and Molecules.* Benjamin Cummings, New York (2000).
[20] M. Weissbluth, *Atoms and Molecules.* Academic Press, New York (1978).
[21] W. R. Johnson, *Atomic Structure Theory.* Springer, Heidelberg (2010).
[22] I. P. Grant, *Relativistic Quantum Theory of Atoms and Molecules.* Springer, Heidelberg (2007).
[23] I. Lindgren and J. Morrison, *Atomic Many-Body Theory.* Springer, Heidelberg (1986).
[24] L. N. Labzowsky, G. L. Klimchirskaya and Y. Y. Dmitriev, *Relativistic Effects in the Spectra of Atomic Systems.* Taylor and Francis, London (1993).
[25] J. J. Sakurai, *Modern Quantum Mechanics.* Addison-Wesley, Reading, MA (1994).
[26] J. J. Sakurai, *Advanced Quantum Mechanics.* Addison-Wesley, Reading, MA (1967).
[27] M. I. Eides, H. Grotch and V. A. Shelyuto, *Theory of Light Hydrogenic Bound States (Springer Tracts in Modern Physics 222).* Springer, Berlin, Heidelberg, New York (2007).
[28] P. W. Milonni, *The Quantum Vacuum.* Academic Press, San Diego (1994).
[29] W. Dittrich and H. Gies, *Probing the Quantum Vacuum (Springer Tracts in Modern Physics 166).* Springer, Berlin, Heidelberg, New York (2000).
[30] P. J. Mohr, B. N. Taylor and D. B. Newell, CODATA recommended values of the fundamental physical constants: 2010, *Rev. Mod. Phys.* **84**, pp. 1527–1605 (2012).
[31] See the URL http://physics.nist.gov/cuu/constants (2021).
[32] L. Essen, R. W. Donaldson, M. J. Bangham and E. G. Hope, Frequency of the Hydrogen Maser, *Nature (London)* **229**, pp. 110–111 (1971).
[33] L. Essen, R. W. Donaldson, E. G. Hope and M. J. Bangham, Hydrogen Maser Work at the National Physical Laboratory, *Metrologia* **9**, pp. 128–137 (1973).
[34] S. G. Karshenboim, Precision physics of simple atoms: QED tests, nuclear structure and fundamental constants, *Phys. Rep.* **422**, pp. 1–63 (2005).
[35] N. Kolachevsky, M. Fischer, S. G. Karshenboim and T. W. Hänsch, High-Precision Optical Measurement of the $2S$ Hyperfine Interval in Atomic Hydrogen, *Phys. Rev. Lett.* **92**, p. 033003 (2004).
[36] S. G. Karshenboim and V. G. Ivanov, Hyperfine structure of the ground state and first excited states in light hydrogen-like atoms and high-precision tests of QED, *Eur. Phys. J. D* **19**, pp. 13–23 (2002).
[37] U. D. Jentschura and V. A. Yerokhin, Quantum electrodynamic corrections to the hyperfine structure of excited S states, *Phys. Rev. A* **73**, p. 062503 (2006).
[38] W. Liu, M. G. Boshier, S. Dhawan, O. van Dyck, P. Egan, X. Fei, M. Grosse Perdekamp, V. W. Hughes, M. Janousch, K. Jungmann, D. Kawall, F. G. Mariam, C. Pillai, R. Prigl, G. zu Putlitz, I. Reinhard, W. Schwarz, P. A. Thompson and K. A. Woodle, High Precision Measurements of the Ground State Hyperfine Structure Interval of Muonium and of the Muon Magnetic Moment, *Phys. Rev. Lett.* **82**, pp. 711–714 (1999).
[39] V. A. Yerokhin and U. D. Jentschura, Electron Self-Energy in the Presence of a Magnetic Field: Hyperfine Splitting and g Factor, *Phys. Rev. Lett.* **100**, p. 163001 (2008).
[40] V. A. Yerokhin and U. D. Jentschura, Self-energy correction to the hyperfine splitting and the electron g factor in hydrogenlike ions, *Phys. Rev. A* **81**, p. 012502 (2010).

[41] A. P. Mills and G. H. Bearman, New Measurement of the Positronium Hyperfine Interval, *Phys. Rev. Lett.* **34**, pp. 246–250 (1975).

[42] A. P. Mills, Line-shape effects in the measurement of the positronium hyperfine interval, *Phys. Rev. A* **27**, pp. 262–267 (1983).

[43] M. W. Ritter, P. O. Egan, V. W. Hughes and K. A. Woodle, Precision determination of the hyperfine-structure interval in the ground state of positronium. v, *Phys. Rev. A* **30**, pp. 1331–1338 (1984).

[44] K. Pachucki, Effective Hamiltonian approach to the bound state: Positronium hyperfine structure, *Phys. Rev. A* **56**, pp. 297–304 (1997).

[45] A. Czarnecki, K. Melnikov and A. Yelkhovsky, Positronium Hyperfine Splitting: Analytical Value at $O(m\alpha^6)$, *Phys. Rev. Lett.* **82**, pp. 311–314 (1999).

[46] R. J. Hill, New Value of m_μ/m_e from Muonium Hyperfine Splitting, *Phys. Rev. Lett.* **86**, pp. 3280–3283 (2001).

[47] K. Melnikov and A. S. Yelkhovsky, $\mathcal{O}(\alpha^3 \ln \alpha)$ Corrections to Muonium and Positronium Hyperfine Splitting, *Phys. Rev. Lett.* **86**, pp. 1498–1501 (2001).

[48] A. Ishida, T. Namba, S. Asai, T. Kobayashi, H. Saito, M. Yoshida, K. Tanakac and A. Yamamoto, New precision measurement of hyperfine splitting of positronium, *Phys. Lett. B* **734**, pp. 338–344 (2014).

[49] R. S. van Dyck, Jr., P. B. Schwinberg and H. G. Dehmelt, New High-Precision Comparison of Electron and Positron g Factors, *Phys. Rev. Lett.* **59**, pp. 26–29 (1987).

[50] B. Odom, D. Hanneke, B. D'Urso and G. Gabrielse, New Measurement of the Electron Magnetic Moment Using a One–Electron Quantum Cyclotron, *Phys. Rev. Lett.* **97**, p. 030801 (2006), [Erratum Phys. Rev. Lett. **99**, 039902 (2007)].

[51] G. Gabrielse, D. Hanneke, T. Kinoshita, M. Nio and B. Odom, New Determination of the Fine Structure Constant from the Electron g Value and QED, *Phys. Rev. Lett.* **97**, p. 030802 (2006), [Erratum Phys. Rev. Lett. **99**, 039902 (2007)].

[52] D. Hanneke, S. Fogwell and G. Gabrielse, Measurement of the Electron Magnetic Moment and the Fine Structure Constant, *Phys. Rev. Lett.* **100**, p. 120801 (2008).

[53] J. Schwinger, On Quantum-Electrodynamics and the Magnetic Moment of the Electron, *Phys. Rev.* **73**, pp. 416–417 (1948).

[54] C. M. Sommerfield, Magnetic Dipole Moment of the Electron, *Phys. Rev.* **107**, pp. 328–329 (1957).

[55] S. Laporta, High-precision calculation of the 4-loop contribution to the electron $g-2$ in QED, *Phys. Lett. B* **772**, pp. 232–238 (2017).

[56] T. Aoyama, M. Hayakawa, T. Kinoshita and M. Nio, Tenth-Order QED Contribution to the Electron $g-2$ and an Improved Value of the Fine-Structure Constant, *Phys. Rev. Lett.* **109**, p. 111807 (2012).

[57] T. Aoyama, M. Hayakawa, T. Kinoshita and M. Nio, Tenth-order electron anomalous magnetic moment: Contribution of diagrams without closed lepton loops, *Phys. Rev. D* **91**, p. 033006 (2015), [Erratum Phys. Rev. D **96**, 019901 (2017)].

[58] T. Aoyama, T. Kinoshita and M. Nio, Revised and improved value of the QED tenth-order electron anomalous magnetic moment, *Phys. Rev. D* **97**, p. 036001 (2018).

[59] P. A. Baikov, A. Maier and P. Marquard, The QED vacuum polarization function at four loops and the anomalous magnetic moment at five loops, *Nucl. Phys. B* **877**, pp. 647–661 (2013).

[60] P. J. Mohr, D. B. Newell and B. N. Taylor, CODATA recommended values of the fundamental physical constants: 2014, *Rev. Mod. Phys.* **88**, p. 035009 (2016).

[61] E. Tiesinga, P. J. Mohr, D. B. Newell and B. N. Taylor, CODATA recommended values of the fundamental physical constants: 2018, (2021).

[62] M. Niering, R. Holzwarth, J. Reichert, P. Pokasov, T. Udem, M. Weitz, T. W. Hänsch, P. Lemonde, G. Santarelli, M. Abgrall, P. Laurent, C. Salomon and A. Clairon, Measurement of the Hydrogen 1S–2S Transition Frequency by Phase Coherent Comparison with a Microwave Cesium Fountain Clock, *Phys. Rev. Lett.* **84**, pp. 5496–5499 (2000).

[63] M. Fischer, N. Kolachevsky, M. Zimmermann, R. Holzwarth, T. Udem, T. W. Hänsch, M. Abgrall, J. Grünert, I. Maksimovic, S. Bize, H. Marion, F. Pereira Dos Santos, P. Lemonde, G. Santarelli, P. Laurent, A. Clairon, C. Salomon, M. Haas, U. D. Jentschura and C. H. Keitel, New Limits on the Drift of Fundamental Constants from Laboratory Measurements, *Phys. Rev. Lett.* **92**, p. 230802 (2004).

[64] C. G. Parthey, A. Matveev, J. Alnis, B. Bernhardt, A. Beyer, R. Holzwarth, A. Maistrou, R. Pohl, K. Predehl, T. Udem, T. Wilken, N. Kolachevsky, M. Abgrall, D. Rovera, C. Salomon, P. Laurent and T. W. Hänsch, Improved Measurement of the Hydrogen 1S–2S Transition Frequency, *Phys. Rev. Lett.* **107**, p. 203001 (2011).

[65] O. Arnoult, F. Nez, L. Julien and F. Biraben, Optical frequency measurement of the 1S-3S two-photon transition in hydrogen, *Eur. Phys. J. D* **60**, pp. 243–256 (2010).

[66] D. C. Yost, A. Matveev, A. Grinin, E. Peters, L. Maisenbacher, A. Beyer, R. Pohl, N. Kolachevsky, K. Khabarova, T. W. Hänsch and T. Udem, Spectroscopy of the hydrogen 1S–3S transition with chirped laser pulses, *Phys. Rev. A* **93**, p. 042509 (2016).

[67] For an interactive database of hydrogen and deuterium transition frequencies, see the URL http://physics.nist.gov/hdel (2018).

[68] U. D. Jentschura, E.-O. Le Bigot, J. Evers, P. J. Mohr and C. H. Keitel, Relativistic and radiative energy shifts for Rydberg states, *J. Phys. B* **38**, pp. S97–S105 (2005).

[69] J. S. Rigden, *Hydrogen: The Essential Element.* Harvard University Press, Harvard, MA (2002).

[70] J. D. Franson, Apparent correction to the speed of light in a gravitational potential, *New J. Phys.* **16**, p. 065008 (2014).

[71] U. D. Jentschura, Gravitational correction to vacuum polarization, *Phys. Rev. A* **91**, p. 022112 (2015).

[72] B. Bertotti, L. Iess and P. Tortora, A test of general relativity using radio links with the Cassini spacecraft, *Nature (London)* **425**, pp. 374–376 (2003).

[73] U. D. Jentschura, Gravitational effects in g factor measurements and high-precision spectroscopy: Limits of Einstein's equivalence principle, *Phys. Rev. A* **98**, p. 032508 (2018).

[74] J. D. Bekenstein, Fine-structure constant: Is it really a constant? *Phys. Rev. D* **25**, pp. 1527–1539 (1982).

[75] U. D. Jentschura and I. Nándori, Attempts at a determination of the fine-structure from first principles: A brief historical overview, *Eur. Phys. J. H* **39**, pp. 591–613 (2014).

[76] P. A. M. Dirac, Classical theory of radiating electrons, *Proc. Roy. Soc. London, Ser. A* **167**, pp. 148–169 (1938).

[77] S. Bize, S. A. Diddams, U. Tanaka, C. E. Tanner, W. H. Oskay, R. E. Drullinger, T. E. Parker, T. P. Heavner, S. R. Jefferts, L. Hollberg, W. M. Itano and J. C. Bergquist, Testing the Stability of Fundamental Constants with the $^{199}Hg^{+}$ Single-Ion Optical Clock, *Phys. Rev. Lett.* **90**, p. 150802 (2003).

[78] E. Peik, B. Lipphardt, H. Schnatz, T. Schneider, C. Tamm and S. G. Karshenboim, Limit on the Present Temporal Variation of the Fine Structure Constant, *Phys. Rev. Lett.* **93**, p. 170801 (2004).

[79] R. M. Godun, P. B. R. Nisbet-Jones, J. M. Jones, S. A. King, L. A. M. Johnson, H. S. Margolis, K. Szymaniec, S. N. Lea, K. Bongs and P. Gill, Frequency Ratio of Two Optical Clock Transitions in $^{171}Yb^{+}$ and Constraints on the Time Variation of Fundamental Constants, *Phys. Rev. Lett.* **113**, p. 210801 (2014).

[80] N. Huntemann, B. Lipphardt, C. Tamm, V. Gerginov, S. Weyers and E. Peik, Improved Limit on a Temporal Variation of m_p/m_e from Comparisons of Yb^{+} and Cs Atomic Clocks, *Phys. Rev. Lett.* **113**, p. 210802 (2014).

[81] U. D. Jentschura, *Advanced Classical Electrodynamics: Green Functions, Regularizations, Multipole Decompositions.* World Scientific, Singapore (2017).

[82] P. A. M. Dirac, The Quantum Theory of the Electron, *Proc. Roy. Soc. London, Ser. A* **117**, pp. 610–624 (1928a).

[83] P. A. M. Dirac, The Quantum Theory of the Electron. Part II, *Proc. Roy. Soc. London, Ser. A* **118**, pp. 351–361 (1928b).

[84] S. Goudsmit, Über den Grundterm des Neonspektrums, *Z. Phys.* **32**, pp. 111–112 (1925).

[85] G. E. Uhlenbeck and S. Goudsmit, Ersetzung der Hypothese vom unmechanischen Zwang durch eine Forderung bezüglich des inneren Verhaltens jedes einzelnen Elektrons, *Naturwissenschaften* **13**, pp. 953–954 (1925).

[86] R. G. Milner, *A Short History of Spin*, Contribution to the XVth International Workshop on Polarized Sources, Targets, and Polarimetry, September 9–13, 2013, Charlottesville, Virginia, USA, e-print arXiv:1311.5016 (2013).

[87] W. Heisenberg and H. Euler, Folgerungen aus der Diracschen Theorie des Positrons, *Z. Phys.* **98**, pp. 714–732 (1936).

[88] W. Heisenberg and H. Euler, An English translation of Ref. [87] was published as arXiv:physics/0605038v1 [physics.hist-ph] (translation by W. Korolevski and H. Kleinert, Freie Universität Berlin) (2006).

[89] J. Schwinger, On Gauge Invariance and Vacuum Polarization, *Phys. Rev.* **82**, pp. 664–679 (1951).

[90] E. H. Wichmann and N. M. Kroll, Vacuum Polarization in a Strong Coulomb Field, *Phys. Rev.* **96**, pp. 232–234 (1954).

[91] E. H. Wichmann and N. M. Kroll, Vacuum Polarization in a Strong Coulomb Field, *Phys. Rev.* **101**, pp. 843–859 (1956).

[92] S. J. Brodsky, G. P. Lepage and P. B. Mackenzie, On the elimination of scale ambiguities in perturbative quantum chromodynamics, *Phys. Rev. D* **28**, pp. 228–235 (1983).

[93] P. J. Mohr, B. N. Taylor and D. B. Newell, CODATA recommended values of the fundamental physical constants: 2006, *Rev. Mod. Phys.* **80**, pp. 633–730 (2008).

[94] H. Müller, S.-W. Chiow, Q. Long, C. Vo and S. Chu, A new photon recoil measurement: towards a determination of the fine-structure constant, *Appl. Phys. B* **84**, pp. 633–642 (2006).

[95] R. Bouchendira, P. Cladé, S. Guellati-Khélifa, F. Nez and F. Biraben, New Determination of the Fine Structure Constant and Test of the Quantum Electrodynamics, *Phys. Rev. Lett.* **106**, p. 080801 (2011).

[96] R. H. Parker, C. Yu, W. Zhong, B. Estey and H. Müller, Measurement of the fine-structure constant as a test of the Standard Model, *Science* **360**, pp. 191–195 (2011).

[97] U. D. Jentschura, S. Kotochigova, E.-O. Le Bigot, P. J. Mohr and B. N. Taylor, Precise Calculation of Transition Frequencies of Hydrogen and Deuterium Based on a Least-Squares Analysis, *Phys. Rev. Lett.* **95**, p. 163003 (2005).

[98] S. Sturm, F. Köhler, J. Zatorski, A. Wagner, Z. Harman, G. Werth, W. Quint, C. H. Keitel and K. Blaum, High-precision measurement of the atomic mass of the electron, *Nature (London)* **506**, pp. 467–470 (2014).

[99] W. Nolting, *Grundkurs Theoretische Physik 1: Klassische Mechanik*. Springer, Berlin, Heidelberg (2013).

[100] H. H. Rosenbrock, A variational principle for quantum mechanics, *Phys. Lett. A* **100**, pp. 343–346 (1986).

[101] A. S. Davydov, *Quantum Mechanics*. Pergamon Press, Oxford (1968).

[102] I. Mills, On the units of radian and cycle for the quantity plane angle, *Metrologia* **53**, pp. 991–997 (2016).

[103] B. P. Leonard, Comment on 'On the units radian and cycle for the quantity plane angle', *Metrologia* **53**, pp. 1281–1285 (2016).

[104] E. Buckingham, On Physically Similar Systems; Illustrations of the Use of Dimensional Equations, *Phys. Rev.* **4**, pp. 345–376 (1914).

[105] I. M. Mills, P. J. Mohr, T. J. Quinn, B. N. Taylor and E. R. Williams, Redefinition of the kilogram: a decision whose time has come, *Metrologia* **42**, pp. 71–80 (2005).

[106] I. M. Mills, P. J. Mohr, T. J. Quinn, B. N. Taylor and E. R. Williams, Redefinition of the kilogram, ampere, kelvin and mole: a proposed approach to implementing CIPM recommendation I (CI-2005), *Metrologia* **43**, pp. 227–246 (2006).

[107] P. J. Mohr, Defining units in the quantum based SI, *Metrologia* **45**, pp. 129–133 (2008).

[108] I. M. Mills, P. J. Mohr, T. J. Quinn, B. N. Taylor and E. R. Williams, Adapting the International System of Units to the twenty-first century, *Proc. Roy. Soc. London, Ser. A* **369**, pp. 3907–3924 (2011).

[109] S. Knotts, P. J. Mohr and W. D. Phillips, An Introduction to the New SI, *Physics Teacher* **55**, pp. 16–21 (2017).

[110] S. Schlamminger, D. Haddad, F. Seifert, L. S. Chao, D. B. Newell, R. Liu, R. L. Steiner and J. R. Pratt, Determination of the Planck constant using a watt balance with a superconducting magnet system at the National Institute of Standards and Technology, *Metrologia* **51**, pp. S15–S24 (2014).

[111] T. D. Lee, *Particle Physics and Introduction to Field Theory*. Harwood Publishers, Newark, NJ (1981).

[112] U. D. Jentschura and C. H. Keitel, Radiative Corrections in Laser–Dressed Atoms: formalism and applications, *Ann. Phys. (N.Y.)* **310**, pp. 1–55 (2004).

[113] M. O. Scully and M. S. Zubairy, *Quantum Optics*. Cambridge University Press, Cambridge, UK (1997).

[114] U. D. Jentschura, J. Evers, M. Haas and C. H. Keitel, Lamb Shift of Laser-Dressed Atomic States, *Phys. Rev. Lett.* **91**, p. 253601 (2003).

[115] C. Cohen-Tannoudji, J. Dupont-Roc and G. Grynberg, *Atom–Photon Interactions: Basic Processes and Applications*. J. Wiley & Sons, New York (1992).

[116] P. J. Mohr, G. Plunien and G. Soff, QED corrections in heavy atoms, *Phys. Rep.* **293**, pp. 227–369 (1998).

[117] M. Haas, U. D. Jentschura, C. H. Keitel, N. Kolachevsky, M. Herrmann, P. Fendel, M. Fischer, Th. Udem, R. Holzwarth, T. W. Hänsch, M. O. Scully and G. S. Agarwal, Two-photon excitation dynamics in bound two-body Coulomb systems including AC Stark shift and ionization, *Phys. Rev. A* **73**, p. 052501 (2006).

[118] J. Sucher, S–Matrix Formalism for Level–Shift Calculations, *Phys. Rev.* **107**, pp. 1448–1449 (1957).

[119] B. M. Smirnov and M. I. Chibisov, The breaking up of atomic particles by an electric field and by electron collisions, *Zh. Éksp. Teor. Fiz.* **49**, pp. 841–851 (1965), [JETP **22**, 585 (1966)].

[120] T. Yamabe, A. Tachibana and H. J. Silverstone, Theory of the ionization of the hydrogen atom by an external electrostatic field, *Phys. Rev. A* **16**, pp. 877–890 (1977).

[121] R. J. Damburg and V. V. Kolosov, An asymptotic approach to the Stark effect for the hydrogen atom, *J. Phys. B* **11**, p. 1921 (1978).

[122] H. J. Silverstone, B. G. Adams, J. Cizek and P. Otto, Stark Effect in Hydrogen: Dispersion Relation, Asymptotic Formulas, and Calculation of the Ionization Rate via High-Order Perturbation Theory, *Phys. Rev. Lett.* **43**, pp. 1498–1501 (1979).

[123] U. D. Jentschura, Resummation of the divergent perturbation series for a hydrogen atom in an electric field, *Phys. Rev. A* **64**, p. 013403 (2001).

[124] E. Caliceti, M. Meyer-Hermann, P. Ribeca, A. Surzhykov and U. D. Jentschura, From Useful Algorithms for Slowly Convergent Series to Physical Predictions Based on Divergent Perturbative Expansions, *Phys. Rep.* **446**, pp. 1–96 (2007).

[125] V. Franceschini, V. Grecchi and H. J. Silverstone, Complex energies from real perturbation series for the LoSurdo-Stark effect in hydrogen by Borel-Padé approximants, *Phys. Rev. A* **32**, pp. 1338–1340 (1985).

[126] U. D. Jentschura, Resummation of nonalternating divergent perturbative expansions, *Phys. Rev. D* **62**, p. 076001 (2000).

[127] R. J. Damburg and V. V. Kolosov, A hydrogen atom in a uniform electric field, *J. Phys. B* **9**, pp. 3149–3158 (1976).

[128] R. J. Damburg and V. V. Kolosov, A hydrogen atom in a uniform electric field. III, *J. Phys. B* **12**, pp. 2637–2643 (1979).

[129] R. Guida and J. Zinn-Justin, Critical exponents of the N-vector model, *J. Phys. A* **31**, pp. 8103–8121 (1998).

[130] J. Zinn-Justin, *Quantum Field Theory and Critical Phenomena*, 4th edn. Oxford University Press, Oxford (2002).

[131] J. Zinn-Justin and U. D. Jentschura, Order-dependent mappings: Strong coupling behaviour from weak coupling expansions in non-Hermitian theories, *J. Math. Phys.* **51**, p. 072106 (2010).

[132] F. J. Dyson, Divergence of Perturbation Theory in Quantum Electrodynamics, *Phys. Rev.* **85**, pp. 631–632 (1952).

[133] G. A. Baker, B. G. Nickel and D. I. Meiron, Critical indices from perturbation analysis of the Callan–Symanzik equation, *Phys. Rev. B* **17**, pp. 1365–1374 (1978).

[134] M. Reed and B. Simon, *Methods of Modern Mathematical Physics, Volume I: Functional Analysis*. Academic Press, San Diego, California (1980).
[135] M. Reed and B. Simon, *Methods of Modern Mathematical Physics, Volume II: Fourier Analysis, Self-Adjointness*. Academic Press, San Diego, California (1978a).
[136] M. Reed and B. Simon, *Methods of Modern Mathematical Physics, Volume III: Scattering Theory*. Academic Press, San Diego, California (1978b).
[137] M. Reed and B. Simon, *Methods of Modern Mathematical Physics, Volume IV: Analysis of Operators*. Academic Press, San Diego, California (1978c).
[138] C. M. Bender and T. T. Wu, Large-Order Behavior of Perturbation Theory, *Phys. Rev. Lett.* **27**, pp. 461–465 (1971).
[139] U. D. Jentschura, A. Surzhykov and J. Zinn-Justin, Unified Treatment of Even and Odd Anharmonic Oscillators of Arbitrary Degree, *Phys. Rev. Lett.* **102**, p. 011601 (2009).
[140] U. D. Jentschura, A. Surzhykov and J. Zinn-Justin, Multi-instantons and exact results III: Unification of even and odd anharmonic oscillators, *Ann. Phys. (N.Y.)* **325**, pp. 1135–1172 (2010).
[141] J. Zinn-Justin and U. D. Jentschura, Multi-instantons and exact results I: Conjectures, WKB expansions, and instanton interactions, *Ann. Phys. (N.Y.)* **313**, pp. 197–267 (2004a).
[142] J. Zinn-Justin and U. D. Jentschura, Multi-instantons and exact results II: Specific cases, higher-order effects, and numerical calculations, *Ann. Phys. (N.Y.)* **313**, pp. 269–325 (2004b).
[143] J. Zinn-Justin, Perturbation Series at Large Orders in Quantum Mechanics and Field Theories: Application to the Problem of Resummation, *Phys. Rep.* **70**, pp. 109–167 (1981).
[144] An overview of recent activity in the field was given at the conference on "Resurgence and Transseries in Quantum, Gauge and String Theories" held at CERN, see also http://ep-news-cern.ch/CERN_TH_Institute_Resurgence (2014).
[145] G. V. Dunne and M. Ünsal, Resurgence and dynamics of $O(N)$ and Grassmannian sigma models, *J. High Energy Phys.* **1509**, p. 199 (2015).
[146] C. Itzykson, G. Parisi and J.-B. Zuber, Asymptotic estimates in quantum electrodynamics, *Phys. Rev. D* **16**, pp. 996–1013 (1977).
[147] R. Balian, C. Itzykson, J.-B. Zuber and G. Parisi, Asymptotic estimates in quantum electrodynamics. II, *Phys. Rev. D* **17**, pp. 1041–1052 (1978).
[148] E. B. Bogomolny and V. A. Fateyev, The Dyson instability and asymptotics of the perturbation series, *Phys. Lett. B* **76**, pp. 210–212 (1978).
[149] E. B. Bogomolny and Y. A. Kubyshin, Asymptotic Estimates for Graphs With a Fixed Number of Loops in Quantum Electrodynamics. The Choice of the Form of the Steepest Descent Solutions, *Yad. Fiz.* **34**, pp. 1535–1546 (1981), [Sov. J. Nucl. Phys. **34**, 853–858 (1981)].
[150] E. B. Bogomolny and Y. A. Kubyshin, Asymptotic Estimates for Graphs With a Fixed Number of Fermionic Loops in Quantum Electrodynamics. The Extremal Configurations with the Symmetry Group $O(2) \times O(3)$, *Yad. Fiz.* **35**, pp. 202–212 (1982), [Sov. J. Nucl. Phys. **35**, 114–119 (1982)].
[151] S. Wolfram, *The Mathematica Book*, 4th edn. Cambridge University Press, Cambridge, UK (1999).

[152] L. I. Schiff, *Quantum Mechanics.* McGraw-Hill, New York (1955).

[153] E. T. Whittaker and G. N. Watson, *A Course of Modern Analysis.* Cambridge University Press, Cambridge, UK (1944).

[154] H. Bateman, *Higher Transcendental Functions*, Vol. 1. McGraw-Hill, New York (1953).

[155] F. W. J. Olver, *Asymptotics and Special Functions.* Academic Press, New York, NY (1974).

[156] H. Bateman, *Higher Transcendental Functions*, Vol. 2. McGraw-Hill, New York (1953a).

[157] H. Bateman, *Higher Transcendental Functions*, Vol. 3. McGraw-Hill, New York (1953b).

[158] R. A. Swainson and G. W. F. Drake, A unified treatment of the non-relativistic and relativistic hydrogen atom I: the wavefunctions, *J. Phys. A* **24**, pp. 79–94 (1991).

[159] M. Abramowitz and I. A. Stegun, *Handbook of Mathematical Functions*, 10th edn. National Bureau of Standards, Washington, D. C. (1972).

[160] L. D. Landau and E. M. Lifshitz, *Quantum Mechanics, Volume 3 of the Course on Theoretical Physics.* Pergamon Press, Oxford, UK (1958).

[161] N. Shafer and R. Bersohn, Photoionization cross section of hydrogenic atoms: A generating function solution, *Phys. Rev. A* **42**, pp. 1313–1317 (1990), [Erratum Phys. Rev. A **44**, 7855–7856 (1991)].

[162] M. Gavrila and A. Costescu, Retardation in the Elastic Scattering of Photons by Atomic Hydrogen, *Phys. Rev. A* **2**, pp. 1752–1758 (1970), [Erratum Phys. Rev. A **4**, 1688 (1971)].

[163] K. Pachucki, Higher-Order Binding Corrections to the Lamb Shift, *Ann. Phys. (N.Y.)* **226**, pp. 1–87 (1993).

[164] I. S. Gradshteyn and I. M. Ryzhik, *Tables of Series, Integrals, and Products.* Academic Press, San Diego (1994).

[165] A. P. Prudnikov, Y. A. Brychkov and O. I. Marichev, *Integrals and Series, Volume I: Elementary Functions.* Taylor and Francis, Amsterdam (1998a).

[166] A. P. Prudnikov, Y. A. Brychkov and O. I. Marichev, *Integrals and Series, Volume II: Special Functions.* Taylor and Francis, London (1998b).

[167] A. P. Prudnikov, Y. A. Brychkov and O. I. Marichev, *Integrals and Series, Volume III: More Special Functions.* Taylor and Francis, London (2002).

[168] A. P. Prudnikov, Y. A. Brychkov and O. I. Marichev, *Integrals and Series, Volume IV: Direct Laplace Transforms.* Taylor and Francis, London (1998).

[169] A. P. Prudnikov, Y. A. Brychkov and O. I. Marichev, *Integrals and Series, Volume V: Inverse Laplace Transforms.* Taylor and Francis, London (2002).

[170] U. D. Jentschura, *Master Thesis: The Lamb Shift in Hydrogenlike Systems, [in German: Theorie der Lamb–Verschiebung in wasserstoffartigen Systemen],.* University of Munich, 1996 (unpublished (see e-print hep-ph/0305065)).

[171] V. Yakhontov, Relativistic Linear Response Wave Functions and Dynamic Scattering Tensor for the $ns_{1/2}$-States in Hydrogen–Like Atoms, *Phys. Rev. Lett.* **91**, p. 093001 (2003).

[172] U. D. Jentschura and M. Haas, Reexamining blackbody shifts for hydrogenlike ions, *Phys. Rev. A* **78**, p. 042504 (2008).

[173] U. Jentschura and K. Pachucki, Higher-order binding corrections to the Lamb shift of $2P$ states, *Phys. Rev. A* **54**, pp. 1853–1861 (1996).
[174] J. Schwinger, Coulomb Green's Function, *J. Math. Phys.* **5**, pp. 1606–1608 (1964).
[175] H. A. Bethe, The Electromagnetic Shift of Energy Levels, *Phys. Rev.* **72**, pp. 339–341 (1947).
[176] E. A. Power and S. Zienau, Coulomb Gauge in Non-Relativistic Quantum Electro–Dynamics and the Shape of Spectral Lines, *Phil. Trans. R. Soc. Lond. A* **251**, pp. 427–454 (1959).
[177] U. D. Jentschura, Non-resonant two-photon transitions in length and velocity gauges, *Phys. Rev. A* **94**, p. 022117 (2016).
[178] W. E. Lamb, Fine Structure of the Hydrogen Atom. III, *Phys. Rev.* **85**, pp. 259–276 (1952).
[179] D. H. Kobe, Question of Gauge: Nonresonant Two-Photon Absorption, *Phys. Rev. Lett.* **40**, pp. 538–540 (1978).
[180] H. R. Reiss, *Limitations of gauge invariance*, e-print arXiv:1302.1212 [quant-ph] (2013).
[181] R. Barbieri and J. Sucher, General Theory of Radiative Corrections to Atomic Decay Rates, *Nucl. Phys. B* **134**, pp. 155–168 (1978).
[182] U. D. Jentschura, Non-uniform convergence of two-photon decay rates for excited atomic states, *J. Phys. A* **40**, pp. F223–F227 (2007).
[183] U. D. Jentschura, Two-photon decays reexamined: cascade contributions and gauge invariance, *J. Phys. A* **41**, p. 155307 (2008).
[184] S. Salomonson and P. Öster, Solution of the pair equation using a finite discrete spectrum, *Phys. Rev. A* **40**, pp. 5559–5567 (1989).
[185] B. J. Wundt, *Quantum electrodynamics and fundamental constants*, Ph.D. thesis, Missouri University of Science and Technology, Rolla (2011 (unpublished)).
[186] U. D. Jentschura and P. J. Mohr, Calculation of hydrogenic Bethe logarithms for Rydberg states, *Phys. Rev. A* **72**, p. 012110 (2005).
[187] The full set of values is available at arXiv.org/quant-ph/0504002 (2005).
[188] W. Gordon, Zur Berechnung der Matrizen beim Wasserstoffatom, *Ann. Phys. (Leipzig)* **394**, pp. 1031–1056 (1929).
[189] W. J. Karzas and R. Latter, Electron Radiative Transitions in a Coulomb Field, *Astrophysical J. Suppl.* **6**, pp. 167–212 (1961).
[190] M. Stobbe, Zur Quantenmechanik photoelektrischer Prozesse, *Ann. Phys. (Leipzig)* **7**, pp. 661–715 (1930).
[191] H. A. Bethe, L. M. Brown and J. R. Stehn, Numerical Value of the Lamb Shift, *Phys. Rev.* **77**, pp. 370–374 (1950).
[192] S. P. Goldman and G. W. F. Drake, High accuracy for atomic calculations involving logarithmic sums, *Phys. Rev. A* **61**, p. 052513 (2000).
[193] S. V. Aksenov, M. A. Savageau, U. D. Jentschura, J. Becher, G. Soff and P. J. Mohr, Application of the combined nonlinear-condensation transformation to problems in statistical analysis and theoretical physics, *Comput. Phys. Commun.* **150**, pp. 1–20 (2003).
[194] H. A. Kramers, *Die Grundlagen der Quantentheorie/Quantentheorie des Elektrons und der Strahlung*. Akademische Verlagsgesellschaft, Leipzig (1938), [English translation: *Quantum Mechanics* (North Holland, Amsterdam, 1958))].

[195] S. Pasternack, On the mean value of r^s for Keplerian systems, *Proc. Natl. Acad. Sci. USA* **23**, pp. 91–94 (1937), [Erratum Proc. Natl. Acad. Sci. USA **23**, 250 (1937)].

[196] U. D. Jentschura and A. Surzhykov, Relativistic calculation of the two-photon decay rate of highly-excited ionic states, *Phys. Rev. A* **77**, p. 042507 (2008).

[197] U. D. Jentschura, Separation of transitions with two quantum jumps from cascades, *Phys. Rev. A* **81**, p. 012112 (2010).

[198] A. Lapierre, U. D. Jentschura, J. R. Crespo López-Urrutia, J. Braun, G. Brenner, H. Bruhns, D. Fischer, A. J. González Martinez, Z. Harman, W. R. Johnson, C. H. Keitel, V. Mironov, C. J. Osborne, G. Sikler, R. Soria Orts, V. Shabaev, H. Tawara, I. I. Tupitsyn, J. Ullrich and A. Volotka, Relativistic Electron Correlation, Quantum Electrodynamics and the Lifetime of the $1s^2 2s^2 2p\, ^2P^{\circ}_{3/2}$ Level in Boronlike Argon, *Phys. Rev. Lett.* **95**, p. 183001 (2005).

[199] K. Pachucki, A. Czarnecki, U. D. Jentschura and V. A. Yerokhin, Complete two-loop correction to the bound-electron g factor, *Phys. Rev. A* **72**, p. 022108 (2005).

[200] J. L. Friar, Nuclear Finite-Size Effects in Light Muonic Atoms, *Ann. Phys. (N.Y.)* **122**, pp. 151–196 (1979).

[201] H. Kleinert and V. Schulte-Frohlinde, *Critical Properties of ϕ^4-Theories*. World Scientific, Singapore (2007).

[202] J. L. Friar and I. Sick, Zemach moments for hydrogen and deuterium, *Phys. Lett. B* **579**, pp. 285–289 (2004).

[203] R. Pohl, A. Antognini, F. Nez, F. D. Amaro, F. Biraben, J. M. R. Cardoso, D. S. Covita, A. Dax, S. Dhawan, L. M. P. Fernandes, A. Giesen, T. Graf, T. W. Hänsch, P. Indelicato, L. Julien, C.-Y. Kao, P. Knowles, E.-O. Le Bigot, Y.-W. Liu, J. A. M. Lopes, L. Ludhova, C. M. B. Monteiro, F. Mulhauser, T. Nebel, P. Rabinowitz, J. M. F. dos Santos, L. A. Schaller, K. Schuhmann, C. Schwob, D. Taqqu, J. F. C. A. Veloso and F. Kottmann, The size of the proton, *Nature (London)* **466**, pp. 213–216 (2010).

[204] A. Antognini, F. Nez, K. Schuhmann, F. D. Amaro, F. Biraben, J. M. R. Cardoso, D. S. Covita, A. Dax, S. Dhawan, M. Diepold, L. M. P. Fernandes, A. Giesen, A. L. Gouvea, T. Graf, T. W. Hänsch, P. Indelicato, L. Julien, C.-Y. Kao, P. Knowles, F. Kottmann, E.-O. Le Bigot, Y.-W. Liu, J. A. M. Lopes, L. Ludhova, C. M. B. Monteiro, F. Mulhauser, T. Nebel, P. Rabinowitz, J. M. F. dos Santos, L. A. Schaller, C. Schwob, D. Taqqu, J. F. C. A. Veloso, J. Vogelsang and R. Pohl, Proton Structure from the Measurement of 2S-2P Transition Frequencies of Muonic Hydrogen, *Science* **339**, pp. 417–420 (2013).

[205] U. D. Jentschura, Proton Radius, Darwin-Foldy Term and Radiative Corrections, *Eur. Phys. J. D* **61**, pp. 7–14 (2011).

[206] J. C. Bernauer, P. Achenbach, C. Ayerbe Gayoso, R. Böhm, D. Bosnar, L. Debenjak, M. O. Distler, L. Doria, A. Esser, H. Fonvieille, J. M. Friedrich, J. Friedrich, M. Gómez Rodríguez de la Paz, M. Makek, H. Merkel, D. G. Middleton, U. Müller, L. Nungesser, J. Pochodzalla, M. Potokar, S. Sánchez Majos, B. S. Schlimme, S. Sirca, T. Walcher and M. Weinriefer, High–Precision Determination of the Electric and Magnetic Form Factors of the Proton, *Phys. Rev. Lett.* **105**, p. 242001 (2010).

[207] M. O. Distler, J. C. Bernauer and T. Walcher, The RMS charge radius of the proton and Zemach moments, *Phys. Lett. B* **696**, pp. 343–347 (2011).

[208] U. D. Jentschura and V. Debierre, Long-range tails in van der Waals interactions of excited-state and ground-state atoms, *Phys. Rev. A* **95**, p. 042506 (2017).

[209] U. D. Jentschura, C. M. Adhikari and V. Debierre, Virtual Resonant Emission and Oscillatory Long–Range Tails in van der Waals Interactions of Excited States: QED Treatment and Applications, *Phys. Rev. Lett.* **118**, p. 123001 (2017).

[210] H. Safari and M. R. Karimpour, Body-Assisted van der Waals Interaction between Excited Atoms, *Phys. Rev. Lett.* **114**, p. 013201 (2015).

[211] M. Donaire, R. Guérout and A. Lambrecht, Quasiresonant van der Waals Interaction between Nonidentical Atoms, *Phys. Rev. Lett.* **115**, p. 033201 (2015).

[212] C. M. Adhikari, V. Debierre, A. Matveev, N. Kolachevsky and U. D. Jentschura, Long-range interactions of hydrogen atoms in excited states. I. $2S$–$1S$ interactions and Dirac–δ perturbations, *Phys. Rev. A* **95**, p. 022703 (2017).

[213] H. B. G. Casimir and D. Polder, The Influence of Retardation on the London-van-der-Waals Forces, *Phys. Rev.* **73**, pp. 360–372 (1948).

[214] E. M. Lifshitz, The Theory of Molecular Attractive Forces between Solids, *Zh. Éksp. Teor. Fiz.* **29**, pp. 94–110 (1955), [Sov. Phys. JETP **2**, 73–83 (1956)].

[215] X.-P. Jiang, F. Toigo and M. W. Cole, The dispersion force of physical absorption: I. Local theory, *Surface Science* **145**, pp. 281–293 (1984a).

[216] X.-P. Jiang, F. Toigo and M. W. Cole, The dispersion force of physical absorption: II. Nonlocal theory, *Surface Science* **148**, pp. 21–36 (1984b).

[217] J. M. Hutson, P. W. Fowler and E. Zaremba, Quadrupolar contributions to the atom-surface van der Waals interaction, *Surface Science* **175**, pp. L775–L781 (1986).

[218] L. Spruch and Y. Tikochinsky, Elementary approximate derivations of some retarded Casimir interactions involving one or two dielectric walls, *Phys. Rev. A* **48**, pp. 4213–4222 (1993).

[219] Z.-C. Yan, J. F. Babb, A. Dalgarno and G. W. F. Drake, Variational calculations of dispersion coefficients for interactions among H, He, and Li atoms, *Phys. Rev. A* **54**, pp. 2824–2833 (1996).

[220] Z.-C. Yan, A. Dalgarno and J. F. Babb, Long-range interactions of lithium atoms, *Phys. Rev. A* **55**, pp. 2882–2887 (1997).

[221] A. Salam and T. Thirunamachandran, A new generalization of the Casimir-Polder potential to higher electric multipole polarizabilities, *J. Chem. Phys.* **104**, pp. 5094–5099 (1996).

[222] M. Marinescu and L. You, Casimir-Polder long-range interaction potentials between alkali-metal atoms, *Phys. Rev. A* **59**, pp. 1936–1954 (1999).

[223] A. Salam, Comment on "Casimir-Polder long-range interaction potentials between alkali-metal atoms", *Phys. Rev. A* **62**, p. 026701 (2000).

[224] R. Marani, L. Cognet, V. Savalli, N. Westbrook, C. I. Westbrook and A. Aspect, Using atomic interference to probe atom-surface interactions, *Phys. Rev. A* **61**, p. 053402 (2000).

[225] H. B. Chan, V. A. Aksyuk, R. N. Kleiman, D. J. Bishop and F. Capasso, Quantum Mechanical Actuation of Microelectromechanical Systems by the Casimir Force, *Science* **291**, pp. 1941–1944 (2001a).

[226] H. B. Chan, V. A. Aksyuk, R. N. Kleiman, D. J. Bishop and F. Capasso, Nonlinear Micromechanical Casimir Oscillator, *Phys. Rev. Lett.* **87**, p. 211801 (2001b).

[227] J. N. Munday and F. Capasso, Precision measurement of the Casimir-Lifshitz force in a fluid, *Phys. Rev. A* **75**, p. R060102 (2007).

[228] F. Capasso, J. N. Munday, D. Iannuzzi and H. B. Chan, Casimir Forces and Quantum Electrodynamical Torques: Physics and Nanomechanics, *IEEE J. Sel. Top. Quantum Electron.* **13**, pp. 400–414 (2007).

[229] G. Barton, Frequency shifts near an interface: inadequacy of two-level atomic models, *J. Phys. B* **7**, pp. 2134–2142 (1974).

[230] U. D. Jentschura, Long-range atom-wall interactions and mixing terms: Metastable hydrogen, *Phys. Rev. A* **91**, p. 010502(R) (2015).

[231] R. C. Hilborn, Einstein coefficients, cross sections, f values, dipole moments, and all that, *Am. J. Phys.* **50**, pp. 982–986 (1982).

[232] M. Goeppert-Mayer, On Closed Shells in Nuclei. II", *Phys. Rev.* **75**, pp. 1269–1270 (1949).

[233] U. D. Jentschura, Self–Energy Correction to the Two-Photon Decay Width in Hydrogenlike Atoms, *Phys. Rev. A* **69**, p. 052118 (2004).

[234] U. D. Jentschura, Virtual Resonant States in Two-Photon Decay Processes: Lower-Order Terms, Subtractions, and Physical Interpretations, *Phys. Rev. A* **79**, p. 022510 (2009).

[235] L. N. Labzowsky, A. V. Shonin and D. A. Solovyev, QED calculation of E1M1 and E1E2 transition probabilities in one-electron ions with arbitrary nuclear charge, *J. Phys. B* **38**, pp. 265–278 (2005).

[236] L. Labzowsky, D. Solovyev, G. Plunien and G. Soff, Two-photon E1M1 and E1E2 transitions between 2p and 1s levels in hydrogen, *Eur. Phys. J. D* **37**, pp. 335–343 (2006).

[237] Y. Itin and S. Reches, Decomposition of third-order constitutive tensors, *e-print arXiv:2009.10752*, published in Mathematics and Mechanics of Solids, article https://doi.org/10.1177/10812865211016530 (2021).

[238] G. Łach, M. DeKieviet and U. D. Jentschura, Multipole Effects in atom-surface interactions: A theoretical study with an application to He–α-quartz, *Phys. Rev. A* **81**, p. 052507 (2010).

[239] E. D. Palik, *Handbook of Optical Constants of Solids.* Academic Press, San Diego (1985).

[240] C. M. Adhikari, V. Debierre and U. D. Jentschura, Long-range interactions of hydrogen atoms in excited states. III. nS–$1S$ interactions for $n \geq 3$, *Phys. Rev. A* **96**, p. 032702 (2017).

[241] W. L. Wiese and J. R. Fuhr, Accurate Atomic Transition Probabilities for Hydrogen, Helium, and Lithium, *J. Phys. Chem. Ref. Data* **38**, pp. 565–720 (2009).

[242] I. I. Sobel'man, *Introduction to the Theory of Atomic Spectra.* Pergamon Press, Oxford (1972).

[243] I. I. Sobel'man, *Atomic Spectra and Radiative Transitions.* Springer, New York (1979).

[244] E. U. Condon and G. H. Shortley, *The Theory of Atomic Spectra.* Cambridge University Press, Cambridge, England (1935).

[245] L. D. Landau and E. M. Lifshitz, *Quantenmechanik (Band III der Lehrbuchreihe über Theoretische Physik von L. D. Landau und E. M. Lifshitz).* Akademie-Verlag, Berlin (1979).

[246] H. Goldstein, *Klassische Mechanik.* Akademische Verlagsgesellschaft, Frankfurt am Main (1963).

[247] B. de Beauvoir, C. Schwob, O. Acef, L. Jozefowski, L. Hilico, F. Nez, L. Julien, A. Clairon and F. Biraben, Metrology of the hydrogen and deuterium atoms: Determination of the Rydberg constant and Lamb shifts, *Eur. Phys. J. D* **12**, pp. 61–93 (2000).

[248] H. van Hees, *Introduction to Relativistic Quantum Field Theory*, Lecture Scriptum, available at http://itp.uni-frankfurt.de/~hees/publ/lect.pdf (2003).

[249] M. Srednicki, *Quantum Field Theory.* Cambridge University Press, Cambridge (2007).

[250] H. Kleinert, *Particles and Quantum Fields.* World Scientific, Singapore (2016).

[251] U. D. Jentschura, Antimatter Gravity: Second Quantization and Lagrangian Formalism, *Physics* **2**, pp. 397–411 (2020).

[252] U. D. Jentschura and B. J. Wundt, Localizability of tachyonic particles and neutrinoless double beta decay, *Eur. Phys. J. C* **72**, p. 1894 (2012).

[253] U. D. Jentschura and B. J. Wundt, From Generalized Dirac Equations to a Candidate for Dark Energy, *ISRN High Energy Physics* **2013**, p. 374612 (2013).

[254] K. Pachucki, U. D. Jentschura and M. Pfützner, Radiative orbital electron capture by the atomic nucleus, *Phys. Rev. C* **75**, p. 055502 (2007).

[255] U. D. Jentschura and J. H. Noble, Nonrelativistic limit of the Dirac–Schwarzschild Hamiltonian: Gravitational *Zitterbewegung* and gravitational spin-orbit coupling, *Phys. Rev. A* **88**, p. 022121 (2013).

[256] M. E. Peskin and D. V. Schroeder, *An Introduction to Quantum Field Theory.* Perseus, Cambridge, Massachusetts (1995).

[257] P. J. Mohr, Self-Energy Radiative Corrections in Hydrogen-Like Systems, *Ann. Phys. (N.Y.)* **88**, pp. 26–51 (1974a).

[258] P. J. Mohr, Numerical Evaluation of the $1S_{1/2}$ Radiative Level Shift, *Ann. Phys. (N.Y.)* **88**, pp. 52–87 (1974b).

[259] U. D. Jentschura, I. Nándori and R. Ehrlich, Calculation of the decay rate of tachyonic neutrinos against charged-lepton-pair and neutrino-pair Cerenkov radiation, *J. Phys. G* **44**, p. 105201 (2017).

[260] N. H. March, The Virial Theorem for Dirac's Equation, *Phys. Rev.* **92**, pp. 481–482 (1953).

[261] H. Hellmann, *Einführung in die Quantenchemie.* Franz Deuticke, Leipzig (1937).

[262] R. P. Feynman, Forces in molecules, *Phys. Rev.* **56**, p. 340 (1939).

[263] S. P. Goldman and G. W. F. Drake, Relativistic sum rules and integral properties of the Dirac equation, *Phys. Rev. A* **25**, pp. 2877–2881 (1982).

[264] V. M. Shabaev, Generalizations of the virial relations for the Dirac equation in a central field and their application to the Coulomb field, *J. Phys. B* **24**, pp. 4479–4488 (1991).

[265] V. M. Shabaev, Virial relations for the dirac equation and their applications to calculations of hydrogen-like atoms, in S. G. Karshenboim and V. B. Smirnov (eds.), *Precision Physics of Simple Atomic Systems — Lecture Notes in Physics Vol. 627.* Springer, Berlin, pp. 97–113 (2003).

[266] G. S. Adkins, Dirac–Coulomb energy levels and expectation values, *Am. J. Phys.* **76**, pp. 579–584 (2008).

[267] U. D. Jentschura, P. J. Mohr and G. Soff, Calculation of the Electron Self-Energy for Low Nuclear Charge, *Phys. Rev. Lett.* **82**, pp. 53–56 (1999).

[268] V. A. Yerokhin, P. Indelicato and V. M. Shabaev, Two-Loop Self-Energy Correction in High-Z Hydrogenlike Ions, *Phys. Rev. Lett.* **91**, p. 073001 (2003).

[269] V. A. Yerokhin, P. Indelicato and V. M. Shabaev, Two-loop self-energy correction in a strong Coulomb nuclear field, *Zh. Éksp. Teor. Fiz.* **128**, pp. 322–336 (2005), [JETP Lett. **101**, 280 (2005)].

[270] V. A. Yerokhin and A. V. Maiorova, Calculations of QED Effects with the Dirac Green Function, *Symmetry* **12**, p. 800 (2020).

[271] D. J. Hylton, The reduced Dirac Green function for the Coulomb potential, *J. Math. Phys.* **25**, pp. 1125–1132 (1984).

[272] E. Lötstedt and U. D. Jentschura, Recursive algorithm for generalized Bessel functions: Numerical access to Dirac–Volkov solutions, *Phys. Rev. E* **79**, p. 026707 (2009).

[273] H. R. Reiss and J. H. Eberly, Green's Function in Intense–Field Electrodynamics, *Phys. Rev.* **151**, pp. 1058–1066 (1966).

[274] V. I. Ritus, Vacuum polarization correction to elastic electron and muon scattering in an intense field and pair electro- and muoproduction, *Nucl. Phys. B* **44**, pp. 236–252 (1972).

[275] H. Mitter, quantum electrodynamics in laser fields, *Acta Phys. Austriaca Suppl.* **14**, pp. 397–468 (1975).

[276] S. Schnez, E. Lötstedt, U. D. Jentschura and C. H. Keitel, Laser-assisted bremsstrahlung for circular and linear polarization, *Phys. Rev. A* **75**, p. 053412 (2007).

[277] C. W. Misner, K. S. Thorne and J. A. Wheeler, *Gravitation*. W. H. Freeman, New York (1973).

[278] D. R. Brill and J. A. Wheeler, Interaction of Neutrinos and Gravitational Fields, *Rev. Mod. Phys.* **29**, pp. 465–479 (1957).

[279] D. G. Boulware, Spin-1/2 quantum field theory in Schwarzschild space, *Phys. Rev. D* **12**, pp. 350–367 (1975).

[280] M. Soffel, B. Müller and W. Greiner, Particles in a stationary spherically symmetric gravitational field, *J. Phys. A* **10**, pp. 551–560 (1977).

[281] U. D. Jentschura, Gravitationally Coupled Dirac Equation for Antimatter, *Phys. Rev. A* **87**, p. 032101 (2013), [Erratum Phys. Rev. A **87**, 069903(E) (2013)].

[282] M. Bojowald, *Canonical Gravity and Applications*. Cambridge University Press, Cambridge (2011).

[283] A. K. Gorbatsievich, *Quantum mechanics in general relativity. Basic principles and elementary applications*. Nauka i Technika, Minsk (1985).

[284] J. Yepez, *Einstein's vierbein field theory of curved space*, e-print arXiv:1106.2037 [gr-qc] (2011).

[285] A. Založnik and N. S. Mankoč Borštnik, *Kaluza–Klein theory*, advanced seminar 4 at the University of Ljubljana, in the physics department. Available from the URL http://reprints.gravitywaves.com/Kaluza-Klein/Zaloznik-2012-Hamber_KaluzaKleinTheory_(1992).

[286] U. D. Jentschura and J. H. Noble, Foldy–Wouthuysen transformation, scalar potentials and gravity, *J. Phys. A* **47**, p. 045402 (2014).

[287] Y. N. Obukhov, Spin, Gravity and Inertia, *Phys. Rev. Lett.* **86**, pp. 192–195 (2001).

[288] E. Borel, *Lectures on Divergent Series*, Translation LA-6140-TR. Los Alamos Scientific Laboratory, Los Alamos (1975).

[289] V. M. Red'kov, Ricci Coefficients in Covariant Dirac Equation, Symmetry Aspects and Newman–Penrose Approach, *Nonlinear Phenomena in Complex Systems* **7**, pp. 250–262 (2011).

[290] A. S. Eddington, *The Mathematical Theory of Relativity.* Cambridge University Press, Cambridge, England (1924).

[291] M. D. Pollock, On the Dirac equation in curved space-time, *Acta Phys. Pol. B* **41**, pp. 1827–1845 (2010).

[292] C. Amole *et al.* [ALPHA Collaboration], Description and first application of a new technique to measure the gravitational mass of antihydrogen, *Nat. Commun.* **4**, p. 1785 (2013).

[293] R. P. Martínez-y Romero, Relativistic hydrogen atom revisited, *Am. J. Phys.* **68**, pp. 1050–1055 (2000).

[294] P. J. Mohr, Solutions of the Maxwell equations and photon wave functions, *Ann. Phys. (N.Y.)* **325**, pp. 607–663 (2010).

[295] P. Langacker, *The Standard Model and Beyond.* CRC Press, Boca Raton, FL (2017).

[296] E. L. Shirley, M. Furst and U. Arp, Invited Article: Refined analysis of synchrotron radiation for NIST's SURF III facility, *Rev. Sci. Instr.* **89**, p. 041301 (2018).

[297] W. G. Bickley, L. J. Comrie, J. C. P. Miller, D. H. Sadler and A. J. Thompson, *Bessel functions, Part II, functions of positive integer order*, vol. X of *Mathematical tables*, Cambridge University Press, Cambridge (1960).

[298] W. Gautschi, Computational Aspects of Three-Term Recurrence Relations, *SIAM Rev.* **9**, pp. 24–82 (1967).

[299] H. I. Arcos, V. C. de Andrade and J. G. Pereira, Torsion and Gravitation: A New View, *Int. J. Mod. Phys. D* **13**, pp. 807–818 (2004).

[300] M. Dafermos, Spherically symmetric spacetimes with a trapped surface, *Class. Quantum Grav.* **22**, pp. 2221–2232 (2005).

[301] V. Fock and D. Iwanenko, Über eine mögliche geometrische Deutung der relativistischen Quantentheorie, *Z. Phys.* **56**, pp. 798–802 (1929).

[302] V. Fock, Geometrisierung der Diracschen Theorie des Elektrons, *Z. Phys.* **57**, pp. 261–277 (1929).

[303] V. Fock and D. Ivanenko, Géométrie quantique linéaire et déplacement parallèle, *C. R. Acad. Sci. Paris* **188**, pp. 1470–1472 (1929).

[304] O. S. Ivanitskaya, N. V. Mitskiévic and Y. S. Vladimirov, Reference Frames and Gravitational Effects in the General Theory of Relativity, in J. Kovalevsky and V. A. Brumberg (eds.), *Proceedings of the 114th Symposium of the International Astronomical Union held in Leningrad, USSR, May 1985.* Kluwer, Dordrecht, pp. 177–186 (1985).

[305] B. J. Wundt and U. D. Jentschura, Sources, potentials and fields in Lorenz and Coulomb gauge: Cancellation of instantaneous interactions for moving point charges, *Ann. Phys. (N.Y.)* **327**, pp. 1217–1230 (2012).

[306] U. D. Jentschura, K. Hencken and V. G. Serbo, Revisiting unitarity corrections for electromagnetic processes in collisions of relativistic nuclei, *Eur. Phys. J. C* **58**, pp. 281–289 (2008).

[307] F. Olness and R. Scalise, Regularization, renormalization, and dimensional analysis:

Dimensional regularization meets freshman E&M, *Am. J. Phys.* **76**, pp. 306–312 (2011).

[308] B. J. Wundt, C. T. Munger and U. D. Jentschura, Quantum Dynamics in Atomic–Fountain Experiments for Measuring the Electric Dipole Moment of the Electron with Improved Sensitivity, *Phys. Rev. X* **2**, p. 041009 (2012).

[309] N. Dombey and A. D. Kennedy, A calculation of the electron anapole moment, *Phys. Lett. B* **91**, pp. 428–430 (1980).

[310] M. I. Eides and V. A. Shelyuto, One-loop electron vertex in Yennie gauge, *Eur. Phys. J. C* **21**, pp. 489–494 (2001).

[311] G. S. Adkins, R. N. Fell and J. Sapirstein, Two-loop renormalization of Feynman gauge QED, *Phys. Rev. D* **63**, p. 125009 (2001).

[312] U. D. Jentschura, A. Czarnecki and K. Pachucki, Nonrelativistic QED approach to the Lamb shift, *Phys. Rev. A* **72**, p. 062102 (2005).

[313] G. S. Adkins, The hydrogen atom in $D = 3 - 2\epsilon$ dimensions, *Phys. Lett. A* **382**, pp. 1545–1549 (2018).

[314] U. D. Jentschura, G. Soff, V. G. Ivanov and S. G. Karshenboim, Bound $\mu^+\mu^-$ system, *Phys. Rev. A* **56**, pp. 4483–4495 (1997).

[315] A. Czarnecki, U. D. Jentschura and K. Pachucki, Calculation of the One- and Two-Loop Lamb Shift for Arbitrary Excited Hydrogenic States, *Phys. Rev. Lett.* **95**, p. 180404 (2005), [Erratum Phys. Rev. Lett. **95**, 199903 (2005)].

[316] J. D. Bjorken and S. D. Drell, *Relativistic Quantum Mechanics*. McGraw-Hill, New York (1964).

[317] L. L. Foldy and S. A. Wouthuysen, On the Dirac Theory of Spin 1/2 Particles and Its Non-Relativistic Limit, *Phys. Rev.* **78**, pp. 29–36 (1950).

[318] E. Eriksen, Foldy–Wouthuysen Transformation. Exact Solution with Generalization to the Two-Particle Problem, *Phys. Rev.* **111**, pp. 1011–1016 (1958).

[319] E. Eriksen and M. Kolsrud, Canonical Transformations of Dirac's Equation to Even Forms, *Supplemento al Nuovo Cimento* **18**, p. 1 (1958).

[320] N. Nicolaevici, Comment on "Spin, Gravity, and Inertia", *Phys. Rev. Lett.* **89**, p. 068902 (2002).

[321] Y. N. Obukhov, Reply to the Comment on "Spin, Gravity, and Inertia", *Phys. Rev. Lett.* **89**, p. 068903 (2002).

[322] K. Pachucki, Higher-order effective Hamiltonian for light atomic systems, *Phys. Rev. A* **71**, p. 012503 (2005).

[323] V. Patkos, V. A. Yerokhin and K. Pachucki, Higher-order recoil corrections for triplet states of the helium atom, *Phys. Rev. A* **94**, p. 052508 (2016).

[324] J. Zatorski and K. Pachucki, Electrodynamics of finite-size particles with arbitrary spin, *Phys. Rev. A* **82**, p. 052520 (2010).

[325] M. Haidar, Z.-X. Zhong, V. I. Korobov and J.-P. Karr, Nonrelativistic QED approach to the fine- and hyperfine-structure corrections of order $m\alpha^6$ and $m\alpha^6(m/M)$: Application to the hydrogen atom, *Phys. Rev. A* **101**, p. 022501 (2020).

[326] R. J. Hill, G. Lee, G. Paz and M. P. Solon, NRQED Lagrangian at order $1/M^4$, *Phys. Rev. D* **87**, p. 053017 (2013).

[327] F. Gesztesy, B. Thaller and H. Grosse, Efficient Method for Calculating Relativistic Corrections for Spin-1/2 Particles, *Phys. Rev. Lett.* **50**, pp. 625–628 (1983).

[328] J. E. Drummond, The anharmonic oscillator: perturbation series for cubic and quartic energy distortion, *J. Phys. A* **14**, pp. 1651–1661 (1981).

[329] A. Czarnecki, K. Melnikov and A. S. Yelkhovsky, Positronium S-state spectrum: Analytic results at $O(m\alpha^6)$, *Phys. Rev. A* **59**, pp. 4316–4330 (1999).

[330] J. M. Brown, R. J. Buenker, A. Carrington, C. Di Lauro, R. N. Dixon, R. W. Field, J. T. Hougen, W. Hüttner, K. Kuchitsu, M. Mehring, A. J. Merer, T. A. Miller, M. Quack, D. A. Ramsay, L. Veseth and R. N. Zare, Remarks on the signs of g factors in atomic and molecular Zeeman spectroscopy, *Mol. Phys.* **98**, pp. 1597–1601 (2000).

[331] U. D. Jentschura, Fifth Force and Hyperfine Splitting in Bound Systems, *Phys. Rev. A* **101**, p. 062503 (2020).

[332] R. P. Feynman, Space-Time Approach to Quantum Electrodynamics, *Phys. Rev.* **76**, pp. 769–789 (1949).

[333] R. P. Feynman, Relativistic Cut-Off for Quantum Electrodynamics, *Phys. Rev.* **74**, p. 1430 (1948).

[334] N. M. Kroll and W. E. Lamb, On the Self–Energy of a Bound Electron, *Phys. Rev.* **75**, p. 388 (1949).

[335] J. B. French and V. F. Weisskopf, The Electromagnetic Shift of Energy Levels, *Phys. Rev.* **75**, p. 1240 (1949).

[336] J. Sapirstein and D. R. Yennie, Theory of Hydrogenic Bound States, in T. Kinoshita (ed.), *Quantum Electrodynamics, Advanced Series on Directions in High Energy Physics, Vol. 7*, Vol. 7. World Scientific, Singapore, pp. 560–672 (1990).

[337] B. J. Wundt and U. D. Jentschura, Reparameterization invariance of NRQED self-energy corrections and improved theory for excited D states in hydrogenlike systems, *Phys. Lett. B* **659**, pp. 571–575 (2008).

[338] M. Baranger, Relativistic Corrections to the Lamb Shift, *Phys. Rev.* **84**, pp. 866–867 (1951), [Erratum Phys. Rev. **85**, 1064 (1952)].

[339] R. Karplus, A. Klein and J. Schwinger, Electrodynamics Displacement of Atomic Energy Levels. II. Lamb Shift, *Phys. Rev.* **86**, pp. 288–301 (1952).

[340] M. Baranger, H. A. Bethe and R. P. Feynman, Relativistic Correction to the Lamb Shift, *Phys. Rev.* **92**, p. 482 (1953).

[341] H. M. Fried and D. R. Yennie, New Techniques in the Lamb Shift Calculation, *Phys. Rev.* **112**, pp. 1391–1404 (1958).

[342] H. M. Fried and D. R. Yennie, Higher Order Terms in the Lamb Shift Calculation, *Phys. Rev. Lett.* **4**, p. 583 (1960).

[343] A. J. Layzer, New Theoretical Value for the Lamb Shift, *Phys. Rev. Lett.* **4**, p. 580 (1960).

[344] A. J. Layzer, Free Propagator Expansion in the Evaluation of the Lamb Shift. I, *J. Math. Phys.* **2**, pp. 292–307 (1961a).

[345] A. J. Layzer, Free Propagator Expansion in the Evaluation of the Lamb Shift. II, *J. Math. Phys.* **2**, pp. 308–323 (1961b).

[346] G. W. Erickson and D. R. Yennie, Radiative Level Shifts, I. Formulation and Lowest Order Lamb Shift, *Ann. Phys. (N.Y.)* **35**, pp. 271–313 (1965a).

[347] G. W. Erickson and D. R. Yennie, Radiative Level Shifts II. Higher Order Contributions to the Lamb Shift, *Ann. Phys. (N.Y.)* **35**, pp. 447–510 (1965b).

[348] G. W. Erickson, Improved Lamb–Shift Calculation for All Values of Z, *Phys. Rev. Lett.* **27**, pp. 780–783 (1971).

[349] J. Sapirstein, Higher-Order Binding Corrections to the Lamb Shift, *Phys. Rev. Lett.* **47**, pp. 1723–1725 (1981).

[350] U. D. Jentschura, G. Soff and P. J. Mohr, Lamb shift of 3P and 4P states and the determination of α, *Phys. Rev. A* **56**, pp. 1739–1755 (1997).

[351] U. D. Jentschura, E.-O. Le Bigot, P. J. Mohr, P. Indelicato and G. Soff, Asymptotic Properties of Self-Energy Coefficients, *Phys. Rev. Lett.* **90**, p. 163001 (2003).

[352] U. D. Jentschura, Techniques in Analytic Lamb Shift Calculations, *Mod. Phys. Lett. A* **20**, pp. 2261–2276 (2005).

[353] A. D. Fokker, The Geodesic Precession: A Consequence of Einstein's Theory of Gravitation, *Kon. Akad. Weten. Amsterdam. Proc.* **23**, pp. 729–738 (1920).

[354] I. B. Khriplovich, Spinning Relativistic Particles in External Fields, in C. Lämmerzahl, C. W. F. Everitt and F. W. Hehl (eds.), *Gyros, Clocks, Interferometers...: Testing Relativistic Gravity in Space (Lecture Notes in Physics 562)*. Springer, Heidelberg, pp. 109–128 (2001).

[355] I. B. Khriplovich, Spinning Relativistic Particles in External Fields, *Acta Phys. Pol. B Proc. Suppl.* **1**, pp. 197–214 (2008).

[356] C. W. F. Everitt, D. B. DeBra, B. W. Parkinson, J. P. Turneaure, J. W. Conklin, M. I. Heifetz, G. M. Keiser, A. S. Silbergleit, T. Holmes, J. Kolodziejczak, M. Al-Meshari, J. C. Mester, B. Muhlfelder, V. G. Solomonik, K. Stahl, P. W. Worden, W. Bencze, S. Buchman, B. Clarke, A. Al-Jadaan, H. Al-Jibreen, J. Li, J. A. Lipa, J. M. Lockhart, B. Al-Suwaidan, M. Taber and S. Wang, Gravity Probe B: Final Results of a Space Experiment to Test General Relativity, *Phys. Rev. Lett.* **106**, p. 221101 (2011).

[357] U. D. Jentschura, Equivalence principle for antiparticles and its limitations, *Int. J. Mod. Phys. A* **34**, p. 1950180 (2019).

[358] M. Charlton, S. Eriksson and G. M. Shore, *Antihydrogen and Fundamental Physics (Spring Briefs in Fundamental Physics)*. Springer Nature, Cham, Switzerland (2020).

[359] K. Pachucki, Long-wavelength quantum electrodynamics, *Phys. Rev. A* **69**, p. 052502 (2004).

[360] U. D. Jentschura and V. A. Yerokhin, QED corrections of order $\alpha(Z\alpha)^2 E_F$ to the hyperfine splitting of $P_{1/2}$ and $P_{3/2}$ states in hydrogenlike ions, *Phys. Rev. A* **81**, p. 012503 (2010).

[361] U. D. Jentschura, V. Debierre, C. M. Adhikari, A. Matveev and N. Kolachevsky, Long-range interactions of excited hydrogen atoms. II. Hyperfine-resolved $2S$–$2S$ system, *Phys. Rev. A* **95**, p. 022704 (2017).

[362] K. Pachucki and S. G. Karshenboim, Nuclear-spin-dependent recoil correction to the Lamb shift, *J. Phys. B* **28**, pp. L221–L224 (1995).

[363] A. Huber, Th. Udem, B. Gross, J. Reichert, M. Kourogi, K. Pachucki, M. Weitz and T. W. Hänsch, Hydrogen-Deuterium $1S$–$2S$ isotope shift and the structure of the deuteron, *Phys. Rev. Lett.* **80**, pp. 468–471 (1998).

[364] J. L. Friar, J. Martorell and D. W. L. Sprung, Nuclear sizes and the isotope shift, *Phys. Rev. A* **56**, pp. 4579–4586 (1997).

[365] K. Pachucki, Electrodynamics of a compound system with relativistic corrections, *Phys. Rev. A* **76**, p. 022106 (2007).

[366] K. Pachucki, Nuclear mass correction to the magnetic interaction of atomic systems, *Phys. Rev. A* **78**, p. 012504 (2008).

[367] W. A. Barker and F. N. Glover, Reduction of relativistic two-particle wave equations to approximate forms. iii, *Phys. Rev.* **99**, pp. 317–324 (1955).

[368] M. Phillips, The Effect of Nuclear Motion on Atomic Magnetic Moments, *Phys. Rev.* **76**, pp. 1803–1804 (1949).

[369] R. A. Hegstrom, Nuclear-Mass and Anomalous-Moment Corrections to the Hamiltonian Constant External Magnetic Field, *Phys. Rev. A* **7**, pp. 451–456 (1973).

[370] K. Pachucki, Finite nuclear mass corrections to electric and magnetic interactions in diatomic molecules, *Phys. Rev. A* **81**, p. 032505 (2010).

[371] U. D. Jentschura, M. Puchalski and P. J. Mohr, Thermal Correction to the Molar Polarizability of a Boltzmann Gas, *Phys. Rev. A* **84**, p. 064102 (2011).

[372] U. Jentschura and E. Lippert, NMR-spektroskopische Untersuchungen über die Eigenassoziation von Karbonsäuren in inerten Lösungsmitteln I. Berechnung thermodynamischer Größen für Essigsäure, *Ber. Bunsenges. Phys. Chem.* **75**, pp. 556–564 (1971a).

[373] U. Jentschura and E. Lippert, NMR-spektroskopische Untersuchungen über die Eigenassoziation von Karbonsäuren in inerten Lösungsmitteln II. Thermodynamische Größen der Propionsäure, *Ber. Bunsenges. Phys. Chem.* **75**, pp. 782–787 (1971b).

[374] D. Ziessow, U. Jentschura and E. Lippert, ^{17}O–NMR–Untersuchung der Wasserstoffbrückenbindung, *Ber. Bunsenges. Phys. Chem.* **75**, pp. 901–904 (1971).

[375] A. Rudzinski, M. Puchalski and K. Pachucki, Relativistic, QED, and nuclear mass effects in the magnetic shielding of 3He, *Phys. Rev. Lett.* **107**, p. 043004 (2011).

[376] G. A. Miller, Defining the proton radius: A unified treatment, *Phys. Rev. C* **99**, p. 035202 (2019).

[377] K. Pachucki, Theory of the Lamb shift in muonic hydrogen, *Phys. Rev. A* **53**, pp. 2092–2100 (1996).

[378] R. Pohl, F. Nez, F. D. Amaro, F. Biraben, J. M. R. Cardoso, D. S. Covita, A. Dax, S. Dhawan, M. Diepold, A. Giesen, A. L. Gouvea, T. Graf, T. W. Hänsch, P. Indelicato, L. Julien, P. Knowles, F. Kottmann, E.-O. Le Bigot, Y.-W. Liu, J. A. M. Lopes, L. Ludhova, C. M. B. Monteiro, F. Mulhauser, T. Nebel, P. Rabinowitz, J. M. F. dos Santos, L. A. Schaller, K. Schuhmann, C. Schwob, D. Taqqu, J. F. C. A. Veloso and A. Antognini, Laser spectroscopy of muonic deuterium, *Science* **353**, pp. 669–673 (2016).

[379] U. D. Jentschura, Lamb Shift in Muonic Hydrogen. —I. Verification and Update of Theoretical Predictions, *Ann. Phys. (N.Y.)* **326**, pp. 500–515 (2011a).

[380] U. D. Jentschura, Lamb Shift in Muonic Hydrogen. —II. Analysis of the Discrepancy of Theory and Experiment, *Ann. Phys. (N.Y.)* **326**, pp. 516–533 (2011b).

[381] U. D. Jentschura, Relativistic Reduced-Mass and Recoil Corrections to Vacuum Polarization in Muonic Hydrogen, Muonic Deuterium and Muonic Helium Ions, *Phys. Rev. A* **84**, p. 012505 (2011c).

[382] R. A. Hegstrom, Relativistic treatment of the shielding of the electron and proton magnetic dipole moments in atomic hydrogen, *Phys. Rev.* **184**, pp. 17–22 (1969), [Erratum Phys. Rev. A **1**, 536–537 (1970)].

[383] H. Grotch and R. A. Hegstrom, Hydrogenic Atoms in a Magnetic Field, *Phys. Rev. A* **4**, pp. 59–69 (1971).

[384] R. A. Hegstrom, Magnetic moment of atomic lithium, *Phys. Rev. A* **11**, pp. 421–426 (1975).

[385] R. A. Hegstrom, *g* factors and related magnetic properties of molecules. Formulation of theory and calculations for H_2^+, HD^+, and D_2^+, *Phys. Rev. A* **19**, pp. 17–30 (1979).
[386] V. A. Yerokhin, K. Pachucki, Z. Harman and C. H. Keitel, QED Theory of the Nuclear Magnetic Shielding in Hydrogenlike Ions, *Phys. Rev. Lett.* **107**, p. 043004 (2011).
[387] V. A. Yerokhin, K. Pachucki, Z. Harman and C. H. Keitel, QED calculation of the nuclear magnetic shielding for hydrogenlike ions, *Phys. Rev. A* **85**, p. 022512 (2012).
[388] K. Pachucki, Simple derivation of helium Lamb shift, *J. Phys. B* **31**, pp. 5123–5133 (1998).
[389] C. Schwartz, Fine Structure of Helium, *Phys. Rev.* **134**, pp. A1181–A1187 (1964).
[390] G. W. F. Drake and S. P. Goldman, Bethe logarithms for Ps^-, H^-, and heliumlike atoms, *Can. J. Phys.* **77**, pp. 835–845 (1999).
[391] R. van Rooij, J. S. Borbely, J. Simonet, M. D. Hoogerland, K. S. E. Eikema, R. A. Rozendaal and W. Vassen, Frequency Metrology in Quantum Degenerate Helium: Direct Measurement of the $2^3S_1 \rightarrow 2^1S_0$ Transition, *Science* **333**, pp. 196–198 (2011).
[392] E. A. Hylleraas, Über den Grundzustand des Heliumatoms, *Z. Phys.* **48**, p. 469 (1928).
[393] E. A. Hylleraas, Neue Berechnung der Energie des Heliums in Grundzustande, sowie des tiefsten Terms von Ortho–Helium, *Z. Phys.* **54**, pp. 347–336 (1929).
[394] E. A. Hylleraas, Über den Grundterm der Zweielektronenprobleme von H^-, He, Li^+, Be^{++} usw. *Z. Phys.* **65**, pp. 209–225 (1930).
[395] E. A. Hylleraas, The Schrödinger Two-Electron Atomic Problem, *Adv. Quant. Chem.* **1**, pp. 1–33 (1964).
[396] V. I. Korobov and S. V. Korobov, Bethe logarithm for the 1^1S and 1^1S states of helium, *Phys. Rev. A* **59**, p. 3394 (1999).
[397] V. I. Korobov, Coulomb three-body bound-state problem: Variational calculations of nonrelativistic energies, *Phys. Rev. A* **61**, p. 064503 (2000).
[398] V. I. Korobov, Nonrelativistic ionization energy for the helium ground state, *Phys. Rev. A* **66**, p. 024501 (2002).
[399] G. W. F. Drake, High Precision Variational Calculations for the $1s^2\ ^1S$ State of H^- and the $1s^2\ ^1S$, $1s2s\ ^1S$ and $1s2s\ ^3S$ States of Helium, *Nucl. Instrum. Methods Phys. Res. B* **31**, pp. 7–13 (1988).
[400] G. W. F. Drake, New variational techniques and high precision eigenvalues for helium, in W. R. Johnson, P. J. Mohr and J. Sucher (eds.), *AIP Conf. Proc. 189*. A. I. P., Woodbury, NY, pp. 146–164 (1989).
[401] C. Schwartz, Experiment and Theory in Computations of the He Atom Ground State, *Int. J. Mod. Phys. E* **15**, pp. 877–888 (2002).
[402] C. Schwartz, *Further Computations of the He Atom Ground State*, e-print math-ph/0605018 (2006).
[403] G. W. F. Drake and Z. C. Yan, Energies and relativistic corrections for the Rydberg states of helium: Variational results and asymptotic analysis, *Phys. Rev. A* **46**, pp. 2378–2409 (1992).
[404] G. W. F. Drake, High Precision Calculations for Helium, in G. W. F. Drake (ed.), *Springer Handbook of Atomic, Molecular and Optical Physics*. Springer, New York, pp. 199–220 (2006).

[405] G. W. F. Drake, Angular integrals and radial recurrence relatiosn for two-electron matrix elements in Hylleraas coordinates, *Phys. Rev. A* **18**, pp. 820–826 (1978).
[406] K. Pachucki, Improved Result for Helium 2^3S_1 Ionization Energy, *Phys. Rev. Lett.* **84**, pp. 4561–4564 (2000).
[407] D. H. Bailey, *A portable high performance multiprecision package*, NASA Ames Tech. Rep. RNR-90-022 (1990).
[408] D. H. Bailey, Multiprecision translation and execution of Fortran programs, *ACM Trans. Math. Soft.* **19**, pp. 288–319 (1993).
[409] D. H. Bailey, A Fortran-90 based multiprecision system, *ACM Trans. Math. Soft.* **21**, pp. 379–387 (1995).
[410] J. S. Sims and S. A. Hagstrom, High-Precision Hy–CI Variational Calculations for the Ground State of Neutral Helium and Helium-Like Ions, *Int. J. Quantum Chem.* **90**, pp. 1600–1609 (2002).
[411] G. W. F. Drake, High precision calculations and qed effects for two- and three-electron atoms, *Phys. Scr. T* **46**, pp. 116–124 (1993).
[412] G. W. F. Drake and W. C. Martin, Ionization energies and quantum electrodynamic effects in the lower $1sns$ and $1snp$ levels of neutral helium (^{4}He I), *Can. J. Phys.* **76**, pp. 679–698 (1998).
[413] D. C. Morton, Q. Wu and G. W. F. Drake, Energy levels for stable isotopes of atomic helium (^{4}He I and ^{3}He I, *Can. J. Phys.* **84**, pp. 83–105 (2006).
[414] V. A. Yerokhin, V. Patkóš, M. Puchalski and K. Pachucki, QED calculation of ionization energies of $1s\,nd$ states in helium, *Phys. Rev. A* **102**, p. 012807 (2020).
[415] V. A. Yerokhin, V. Patkóš and K. Pachucki, Relativistic corrections to the Bethe logarithm for the 2^3S and 2^3P states of He, *Phys. Rev. A* **98**, p. 032503 (2018), erratum Phys. Rev. A **103**, 029901(E) (2021)].
[416] V. A. Yerokhin, V. Patkóš and K. Pachucki, Nonradiative $\alpha^7 m$ QED effects in the Lamb shift of helium triplet states, *Phys. Rev. A* **101**, p. 062516 (2020), erratum Phys. Rev. A **103**, 029902(E) (2021)].
[417] V. Patkóš, V. A. Yerokhin and K. Pachucki, Radiative $\alpha^7\, m$ QED contribution to the helium Lamb shift, *Phys. Rev. A* **103**, p. 012803 (2021a).
[418] V. Patkóš, V. A. Yerokhin and K. Pachucki, Complete $\alpha^7\, m$ Lamb shift for helium triplet states, *Phys. Rev. A* **103**, p. 042809 (2021b).
[419] T. Kato, On the Eigenfunctions of Many-Particle Systems in Quantum Mechanics, *Commun. Pure Appl. Math.* **10**, pp. 151–177 (1957).
[420] J. D. Baker, D. E. Freund, R. N. Hill and J. D. Morgan, Radius of convergence and analytic behavior of the $1/Z$ expansion, *Phys. Rev. A* **41**, pp. 1247–1273 (1990).
[421] R. J. Drachman, A new global operator for two-particle delta functions, *J. Phys. B* **14**, pp. 2733–2738 (1981).
[422] Z.-C. Yan, W. Nörtershäuser and G. W. F. Drake, High Precision Theory for Li and Be^+: QED Shifts and Isotope Shifts, *Phys. Rev. Lett.* **100**, p. 243002 (2008).
[423] C. Lhuillier, J. P. Faroux and N. Billy, High-precision measurement of the Zeeman effect for the triplet P states in helium, *J. Phys. (Paris)* **37**, pp. 335–354 (1976).
[424] S. A. Lewis and V. W. Hughes, Higher-order relativistic contributions to the Zeeman effect in helium and heliumlike ions, *Phys. Rev. A* **11**, pp. 383–384 (1975).
[425] J. M. Anthony and K. J. Sebastian, Relativistic corrections to the Zeeman effect in heliumlike ions, *Phys. Rev. A* **48**, pp. 3792–3810 (1993).

[426] Z. C. Yan and G. W. F. Drake, High-precision calculations of the Zeeman effect in the 2^3P_J, 2^1P_1, $^3S_1$ and 3^3P_J states of helium, *Phys. Rev. A* **50**, pp. R1980–R1983 (1994).
[427] M. Puchalski and U. D. Jentschura, Quantum Electrodynamic Corrections to the $\boldsymbol{g}$ Factor of Helium $\boldsymbol{P}$ States, *Phys. Rev. A* **86**, p. 022508 (2012).
[428] V. M. Shabaev, I. I. Tupitsyn and V. A. Yerokhin, QEDMOD: Fortran program for calculating the model Lamb-shift operator, *Comput. Phys. Commun.* **189**, pp. 175–181 (2015).
[429] C. Froese Fischer, *The Hartree-Fock method for atoms.* Wiley–VCH, New York (1977).
[430] C. Froese Fischer, T. Brage and P. Jönsson, *Computational Atomic Structure: An MCHF Approach.* CRC Press, Boca Raton (1997).
[431] I. Lindgren, *Relativistic Many-Body Theory.* Springer, Heidelberg (2011).
[432] A. Hibbert, Charlotte Froese Fischer—Her Work and Her Impact, *Atoms* **7**, p. 107 (2019).
[433] A. Derevianko, Reconciliation of the Measurement of Parity Nonconservation in Cs with the Standard Model, *Phys. Rev. Lett.* **85**, pp. 1618–1621 (2000).
[434] A. Derevianko, Correlated many-body treatment of the Breit interaction with application to cesium atomic properties and parity violation, *Phys. Rev. A* **65**, p. 012106 (2001).
[435] D. R. Hartree, The Wave Mechanics of an Atom with a Non-Coulomb Central Field. Part I. Theory and Methods, *Math. Proc. Camb. Phil. Soc.* **24**, pp. 89–110 (1928a).
[436] D. R. Hartree, The Wave Mechanics of an Atom with a Non-Coulomb Central Field. Part II. Some Results and Discussion, *Math. Proc. Camb. Phil. Soc.* **24**, pp. 111–132 (1928b).
[437] V. Fock, Näherungsmethode zur Lösung des quantenmechanischen Mehrkörperproblems, *Z. Phys.* **61**, pp. 126–148 (1930).
[438] R. D. Cowan, *The Theory of Atomic Structure and Spectra.* University of California Press, Berkeley (1981).
[439] T. Gomez, T. Nagayama, C. Fontes, D. Kilcrease, S. Hansen, M. Montgomery and D. Winget, Matrix Methods for Solving Hartree-Fock Equations in Atomic Structure Calculations and Line Broadening, *Atoms* **6**, p. 22 (2018).
[440] C. C. J. Roothaan, New Developments in Molecular Orbital Theory, *Rev. Mod. Phys.* **23**, pp. 69–89 (1951).
[441] J. P. Desclaux, A. J. Freeman and J. V. Mallow, Unrestricted Dirac–Fock Theory: Relativistic Determination of Core Polarization Hyperfine Interactions, *J. Magn. Magn. Mater.* **5**, pp. 265–276 (1977).
[442] M. S. Safronova and W. R. Johnson, All-Order Methods for Relativistic Atomic Structure Calculations, *Adv. At. Mol. Opt. Phys.* **55**, pp. 191–233 (2008).
[443] M. S. Safronova, A. Derevianko and W. R. Johnson, Relativistic many-body calculations of energy levels, hyperfine constants, and transition rates for sodiumlike ions, $Z = 11$–16, *Phys. Rev. A* **58**, pp. 1016–1028 (1998).
[444] M. S. Safronova, W. R. Johnson and A. Derevianko, Relativistic many-body calculations of energy levels, hyperfine constants, electric-dipole matrix elements, and static polarizabilities for alkali-metal atoms, *Phys. Rev. A* **60**, pp. 4476–4487 (1999).
[445] C. D. Anderson, The Positive Electron, *Phys. Rev.* **43**, pp. 491–494 (1933).

[446] G. E. Brown and D. G. Ravenhall, On the interaction of two electrons, *Proc. Roy. Soc. London, Ser. A* **208**, pp. 552–559 (1951).
[447] I. P. Grant, Relativistic calculation of atomic structures, *Adv. Phys.* **19**, pp. 747–811 (1970).
[448] I. P. Grant, General program to calculate angular momentum coefficients in relativistic atomic structure, *Comput. Phys. Commun.* **5**, pp. 263–282 (1973).
[449] K. G. Dyall, I. P. Grant, C. T. Johnson, F. A. Parpia and E. P. Plummer, GRASP: A General-purpose Relativistic Atomic Structure Program, *Comput. Phys. Commun.* **55**, pp. 425–456 (1989).
[450] J. S. Greenberg and W. Greiner, Search for the sparking of the vacuum, *Phys. Today* **35**, pp. 24–32 (1982).
[451] R. Jauregui, C. F. Bunge and E. Ley-Koo, Upper bounds to the eigenvalues of the no-pair Hamiltonian, *Phys. Rev. A* **55**, pp. 1781–1784 (1997).
[452] J. Maruani, The Dirac Electron: From Quantum Chemistry to Holistic Cosmology, *J. Chin. Chem. Soc.* **63**, pp. 33–48 (2016).
[453] P. Jönsson, F. A. Parpia and C. Froese Fischer, HFS92: A program for relativistic atomic hyperfine structure calculations, *Comput. Phys. Commun.* **96**, pp. 301–310 (1996).
[454] F. A. Parpia, C. Froese Fischer and I. P. Grant, GRASP92: A package for large-scale relativistic atomic structure calculations, *Comput. Phys. Commun.* **94**, pp. 249–271 (1996).
[455] C. Froese Fischer, G. Gaigalas and Y. Ralchenko, Some corrections to GRASP92, *Comput. Phys. Commun.* **175**, pp. 738–744 (2006).
[456] P. Jönsson, X. He, C. Froese Fischer and I. P. Grant, The GRASP2K relativistic atomic structure package, *Comput. Phys. Commun.* **177**, pp. 597–622 (2007).
[457] J. P. Desclaux, A multiconfiguration relativistic Dirac–Fock program, *Comput. Phys. Commun.* **9**, pp. 31–45 (1975).
[458] P. Indelicato and J. P. Desclaux, Multiconfiguration Dirac–Fock calculations of transition energies with QED corrections in three-electron ions, *Phys. Rev. A* **42**, pp. 5139–5149 (1990).
[459] P. Indelicato, Projection operators in multiconfiguration Dirac–Fock calculations: Application to the ground state of heliumlike ions, *Phys. Rev. A* **51**, pp. 1132–1145 (1995).
[460] V. A. Yerokhin and V. M. Shabaev, private communication (2006).
[461] W. F. Perger and M. Idrees, Relativistic calculation of specific mass shifts for Ar^+, Ni, Kr^+, and Ce^+ using a multi-configuration Dirac–Fock approach, *Comput. Phys. Commun.* **85**, pp. 389–397 (1995).
[462] M. I. Bhatti, M. Bucardo and W. F. Perger, MCDF calculations of the specific mass shift in helium-like ions, *J. Phys. B* **34**, pp. 223–232 (2001).
[463] W. H. King, *Isotope Shifts in Atomic Spectra*. Springer, New York, NY (1984).
[464] I. I. Tupitsyn, V. M. Shabaev, J. R. Crespo López-Urrutia, I. Draganić, R. Soria Orts and J. Ullrich, Relativistic calculations of isotope shifts in highly charged ions, *Phys. Rev. A* **68**, p. 022511 (2003).
[465] F. A. Parpia, M. Tong and C. F. Fischer, Relativistic calculations of nuclear motional effects in many-electron atoms, *Phys. Rev. A* **46**, pp. 3717–3724 (1992).

[466] A. Derevianko, B. Ravaine and W. R. Johnson, Relaxation effect and radiative corrections in many-electron atoms, *Phys. Rev. A* **69**, p. 054502 (2004).

[467] https://www.tcd.ie/Physics/people/Cormac.McGuinness/Cowan/ (2021).

[468] J.-F. Wyart, On the Interpretation of Complex Atomic Spectra by Means of the Parametric Racah-Slater Method and Cowan Codes, *Can. J. Phys.* **89**, pp. 451–456 (2011).

[469] A. Kramida, Cowan Code: 50 Years of Growing Impact on Atomic Physics, *Atoms* **7**, p. 64 (2019).

[470] A. Kramida, Y. Ralchenko and J. Reader, NIST ASD Team. NIST Atomic Spectra Database (version 5.8), [Online]. Available: https://physics.nist.gov/asd [Fri Sep 10 2021]. National Institute of Standards and Technology, Gaithersburg, MD. DOI: https://doi.org/10.18434/T4W30F (2020).

[471] J. E. Hansen, P. H. M. Uylings and A. J. J. Raassen, Parametric Fitting with Orthogonal Operators, *Phys. Scr.* **37**, pp. 664–672 (1988).

[472] A. J. J. Raassen and P. H. M. Uylings, The Use of Complete Sets of Orthogonal Operators in Spectroscopic Studies, *Phys. Scr. T* **65**, pp. 84–87 (1996).

[473] V. I. Azarov and R. R. Gayasov, The Third Spectrum of Rhenium (Re III): Analysis of the $(5d^5+5d^46s)-(5d^46p+5d^36s6p)$ Transition Array, *At. Data Nucl. Data Tables* **121-122**, pp. 306–344 (2018).

[474] V. I. Azarov, W.-U. L. Tchang-Brillet and R. R. Gayasov, Analysis of the Spectrum of the $(5d^6+5d^56s)-(5d^56p+5d^46s6p)$ Transitions of Two Times Ionized Osmium (Os III), *At. Data Nucl. Data Tables* **121-122**, pp. 345–377 (2018).

[475] J.-F. Wyart, A. J. J. Raassen, P. H. M. Uylings and Y. N. Joshi, Spectra of High-Z Ions of Stellar Interest. A Theoretical Study of $(d+s)^8$ Mixed Configurations in $5d$- and $4d$- Elements, *Phys. Scr. T* **47**, pp. 59–64 (1993).

[476] C. Froese Fischer, A Multiconfiguration Hartree–Fock program, *Comput. Phys. Commun.* **1**, pp. 151–166 (1969).

[477] C. Froese Fischer, A general multi-configuration Hartree–Fock program, *Comput. Phys. Commun.* **14**, pp. 145–153 (1978).

[478] C. Froese Fischer, The MCHF atomic-structure package, *Comput. Phys. Commun.* **64**, pp. 369–398 (1991).

[479] C. Froese Fischer, G. Tachiev, G. Gaigalas and M. R. Godefroid, An MCHF Atomic-Structure Package for Large-Scale Calculations, *Comput. Phys. Commun.* **176**, pp. 559–579 (2007).

[480] https://github.com/tcarette/ATSP2K_plus (2021).

[481] A. Hibbert, CIV3—A General Program to Calculate Configuration Interaction Wave Functions and Electric-Dipole Oscillator Strengths, *Comput. Phys. Commun.* **9**, pp. 141–172 (1975).

[482] A. Gaupp, P. Kuske and H. J. Andrä, Accurate lifetime measurements of the lowest $^2P_{1/2}$ states in neutral lithium and sodium, *Phys. Rev. A* **26**, pp. 3351–3359 (1982).

[483] T. Brage, C. Froese Fischer and P. Jönsson, Effects of core-valence and core-core correlation on the line strength of the resonance lines in Li I and Na I, *Phys. Rev. A* **49**, pp. 2181–2184 (1994).

[484] U. Volz, M. Majerus, H. Liebel, A. Schmitt and H. Schmoranzer, Precision Lifetime Measurements on Na I $3p^2P_{1/2}$ and $3p^2P_{3/2}$ and by Beam–Gas–Laser Spectroscopy, *Phys. Rev. Lett.* **76**, pp. 2862–2865 (1996).

[485] C. W. Oates, K. R. Vogel and J. L. Hall, High Precision Linewidth Measurement of Laser-Cooled Atoms: Resolution of the Na $3p^2P_{3/2}$ Lifetime Discrepancy, *Phys. Rev. Lett.* **76**, pp. 2866–2869 (1996).
[486] K. M. Jones, P. S. Julienne, P. D. Lett, W. D. Phillips, E. Tiesinga and C. J. Williams, Measurement of the atomic Na($3P$) lifetime and of retardation in the interaction between two atoms bound in a molecule, *Europhys. Lett.* **35**, pp. 85–90 (1996).
[487] W. Li, A. M. Amarsi, A. Papoulia, J. Ekman and P. Jönsson, Extended theoretical transition data in C I – IV, *Mon. Not. Roy. Astron. Soc.* **502**, pp. 3780–3799 (2021a).
[488] W. Li, P. Jönsson and C. Froese Fischer, private communication (2021b).
[489] G. Racah, Theory of Complex Spectra. I. *Phys. Rev.* **61**, pp. 186–197 (1942a).
[490] G. Racah, Theory of Complex Spectra. II. *Phys. Rev.* **62**, pp. 438–462 (1942b).
[491] G. Racah, Theory of Complex Spectra. III. *Phys. Rev.* **63**, pp. 367–382 (1943).
[492] I. P. Grant, Relativistic atomic structure: past, present and future, *J. Phys. B* **43**, p. 074033 (2010).
[493] P. Jönsson, G. Gaigalas, J. Bieroń, C. Froese Fischer and I. P. Grant, New Version: GRASP2K Relativistic Atomic Structure Package, *Comput. Phys. Commun.* **184**, pp. 2197–2203 (2013).
[494] I. P. Grant, Gauge invariance and relativistic radiative transitions, *J. Phys. B* **7**, pp. 1458–1475 (1974).
[495] I. P. Grant, A program to calculate angular momentum coefficients in relativistic atomic structure - revised version, *Comput. Phys. Commun.* **11**, pp. 397–405 (1976).
[496] I. P. Grant and N. C. Pyper, Breit interaction in multi-configuration relativistic atomic calculations, *J. Phys. B* **9**, pp. 761–774 (1976).
[497] N. C. Pyper, I. P. Grant and N. Beatham, A new program for calculating matrix elements of one-particle operators in jj-coupling, *Comput. Phys. Commun.* **15**, pp. 387–400 (1978).
[498] I. P. Grant, B. J. McKenzie, P. H. Norrington, D. F. Mayers and N. C. Pyper, An atomic multiconfigurational Dirac–Fock package, *Comput. Phys. Commun.* **21**, pp. 207–231 (1980).
[499] C. Froese Fischer, M. Godefroid, T. Brage, P. Jönsson and G. Gaigalas, Advanced multiconfiguration methods for complex atoms: I. Energies and wave functions, *J. Phys. B* **49**, p. 182004 (2016).
[500] O. Gorceix and P. Indelicato, Effect of the complete Breit interaction on two-electron ion energy levels, *Phys. Rev. A* **37**, pp. 1087–1094 (1988).
[501] E. Lindroth and A. M. Mårtensson-Pendrill, Further analysis of the complete Breit interaction, *Phys. Rev. A* **39**, pp. 3794–3802 (1989).
[502] P. Indelicato, Effects of the Breit interaction on the 1s binding energy of superheavy elements, *J. Phys. B* **19**, pp. 1719–1730 (1986).
[503] D. Rudolph, U. Forsberg, P. Golubev, L. G. Sarmiento, A. Yakushev, L.-L. Andersson, A. Di Nitto, C. E. Düllmann, J. M. Gates, K. E. Gregorich, C. J. Gross, F. P. Heßberger, R.-D. Herzberg, J. Khuyagbaatar, J. V. Kratz, K. Rykaczewski, M. Schädel, S. Åberg, D. Ackermann, M. Block, H. Brand, B. G. Carlsson, D. Cox, X. Derkx, K. Eberhardt, J. Even, C. Fahlander, J. Gerl, E. Jäger, B. Kindler, J. Krier, I. Kojouharov, N. Kurz, B. Lommel, A. Mistry, C. Mokry, H. Nitsche, J. P. Omtvedt, P. Papadakis, I. Ragnarsson, J. Runke, H. Schaffner, B. Schausten,

P. Thörle-Pospiech, T. Torres, T. Traut, N. Trautmann, A. Türler, A. Ward, D. E. Ward and N. Wiehl, Spectroscopy of Element 115 Decay Chains, *Phys. Rev. Lett.* **111**, p. 112502 (2013).
[504] F. G. Werner and J. A. Wheeler, Superheavy Nuclei, *Phys. Rev.* **109**, pp. 126–144 (1958).
[505] P. Indelicato, J. Bieroń and P. Jönsson, Are MCDF calculations 101% correct in the super-heavy elements range? *Theor. Chem. Acta* **129**, pp. 495–505 (2011).
[506] I. Angeli, A consistent set of nuclear rms charge radii: properties of the radius surface $R(N, Z)$, *At. Data Nucl. Data Tables* **87**, pp. 185–206 (2004).
[507] M. F. Gu, The Flexible Atomic Code, *Can. J. Phys.* **86**, pp. 675–689 (2008).
[508] M. J. Vilkas, K. Koc and Y. Ishikawa, Relativistic Multireference Møller-Plesset Perturbation Theory Based on Multiconfigurational Dirac–Fock Reference Functions, *Chem. Phys. Lett.* **296**, pp. 68–76 (1998).
[509] S. Fritzsche, A fresh computational approach to atomic structures, processes and cascades, *Comput. Phys. Commun.* **240**, pp. 1–14 (2019).
[510] https://github.com/OpenJAC/JAC.jl (2021).
[511] J. Bezanson, A. Edelman, S. Karpinski and V. B. Shah, Julia: A Fresh Approach to Numerical Computing, *SIAM Rev.* **59**, pp. 65–98 (2017).
[512] J. C. Slater, A Simplification of the Hartree–Fock Method, *Phys. Rev.* **81**, pp. 385–390 (1951).
[513] W. Kohn, Nobel Lecture: Electronic structure of matter-wave functions and density functionals, *Rev. Mod. Phys.* **71**, pp. 1253–1266 (1935).
[514] J. A. Pople, Nobel Lecture: Quantum chemical models, *Rev. Mod. Phys.* **71**, pp. 1267–1274 (1999).
[515] R. K. Nesbet, Kinetic energy in density-functional theory, *Phys. Rev. A* **58**, pp. R12–R15 (1998).
[516] P. Hohenberg and W. Kohn, Inhomogeneous Electron Gas, *Phys. Rev.* **136**, pp. B864–B871 (1964).
[517] W. Kohn and L. J. Sham, Self-Consistent Equations Including Exchange and Correlation Effects, *Phys. Rev.* **140**, pp. A1133–A1138 (1965).
[518] L. H. Thomas, The calculation of atomic fields, *Math. Proc. Camb. Phil. Soc.* **23**, pp. 542–548 (1927).
[519] E. Fermi, Eine statistische Methode zur Bestimmung einiger Eigenschaften des Atoms und ihre Anwendung auf die Theorie des periodischen Systems der Elemente, *Z. Phys.* **48**, pp. 73–79 (1927a).
[520] E. Fermi, Un Metodo Statistico per la Determinazione di alcune Priopriet à dell'Atomo, *Atti R. Accad. Naz. Lincei* **6**, pp. 602–607 (1927b).
[521] C. F. von Weizsäcker, Zur Theorie der Kernmassen, *Z. Phys.* **96**, pp. 431–458 (1935).
[522] W. J. Hehre, W. A. Lathan, R. Ditchfield, M. D. Newton and J. A. Pople, Gaussian 70 (Quantum Chemistry Program Exchange, Program No. 237) (1970).
[523] M. J. Frisch, G. W. Trucks, H. B. Schlegel, G. E. Scuseria, M. A. Robb, J. R. Cheeseman, G. Scalmani, V. Barone, G. A. Petersson, H. Nakatsuji, X. Li, M. Caricato, A. V. Marenich, J. Bloino, B. G. Janesko, R. Gomperts, B. Mennucci, H. P. Hratchian, J. V. Ortiz, A. F. Izmaylov, J. L. Sonnenberg, D. Williams-Young, F. Ding, F. Lipparini, F. Egidi, J. Goings, B. Peng, A. Petrone, T. Henderson, D. Ranasinghe, V. G. Zakrzewski, J. Gao, N. Rega, G. Zheng, W. Liang,

M. Hada, M. Ehara, K. Toyota, R. Fukuda, J. Hasegawa, M. Ishida, T. Nakajima, Y. Honda, O. Kitao, H. Nakai, T. Vreven, K. Throssell, J. A. Montgomery, Jr., J. E. Peralta, F. Ogliaro, M. J. Bearpark, J. J. Heyd, E. N. Brothers, K. N. Kudin, V. N. Staroverov, T. A. Keith, R. Kobayashi, J. Normand, K. Raghavachari, A. P. Rendell, J. C. Burant, S. S. Iyengar, J. Tomasi, M. Cossi, J. M. Millam, M. Klene, C. Adamo, R. Cammi, J. W. Ochterski, R. L. Martin, K. Morokuma, O. Farkas, J. B. Foresman and D. J. Fox, Gaussian 16 Revision C.01, Gaussian Inc. Wallingford CT, available at https://gaussian.com (2016).

[524] DIRAC, a relativistic ab initio electronic structure program, Release DIRAC21, written by R. Bast, A. S. P. Gomes, T. Saue, L. Visscher, and H. J. Aa. Jensen, with contributions from I. A. Aucar, V. Bakken, K. G. Dyall, S. Dubillard, U. Ekström, E. Eliav, T. Enevoldsen, E. Faßhauer, T. Fleig, O. Fossgaard, L. Halbert, E. D. Hedegård, T. Helgaker, B. Helmich–Paris, J. Henriksson, M. Iliaš, Ch. R. Jacob, S. Knecht, S. Komorovský, O. Kullie, J. K. Lærdahl, C. V. Larsen, Y. S. Lee, N. H. List, H. S. Nataraj, M. K. Nayak, P. Norman, G. Olejniczak, J. Olsen, J. M. H. Olsen, A. Papadopoulos, Y. C. Park, J. K. Pedersen, M. Pernpointner, J. V. Pototschnig, R. di Remigio, M. Repisky, K. Ruud, P. Sałek, B. Schimmelpfennig, B. Senjean, A. Shee, J. Sikkema, A. Sunaga, A. J. Thorvaldsen, J. Thyssen, J. van Stralen, M. L. Vidal, S. Villaume, O. Visser, T. Winther, and S. Yamamoto (available at http://dx.doi.org/10.5281/zenodo.4836496, see also http://www.diracprogram.org) (2021).

[525] T. Saue, R. Bast, A. S. P. Gomes, H. J. A. Jensen, L. Visscher, I. A. Aucar, R. Di Remigio, K. G. Dyall, E. Eliav, E. Faßhauer, T. Fleig, L. Halbert, E. D. Hedegård, B. Helmich-Paris, M. Iliaš, C. R. Jacob, S. Knecht, J. K. Laerdahl, M. L. Vidal, M. K. Nayak, M. Olejniczak, J. M. H. Olsen, M. Pernpointner, B. Senjean, A. S. A. Sunaga and J. N. P. van Stralen, The DIRAC code for relativistic molecular calculations, *J. Chem. Phys.* **152**, p. 204104 (2020).

[526] A. S. P. Gomes, K. G. Dyall and L. Visscher, Relativistic double-zeta, triple-zeta, and quadruple-zeta basis sets for the lanthanides La–Lu, *Theor. Chem. Acc.* **127**, pp. 369–381 (2010).

[527] P. S. Bagus, E. S. Ilton, R. L. Martin, H. J. A. Jensen and S. Knecht, Spin-orbit coupling in actinide cations, *Chem. Phys. Lett.* **546**, pp. 58–62 (2012).

[528] https://www.msg.chem.iastate.edu/gamess (2021).

[529] M. W. Schmidt, K. K. Baldridge, J. A. Boatz, S. T. Elbert, M. S. Gordon, J. H. Jensen, S. Koseki, N. Matsunaga, K. A. Nguyen, S. Su, T. L. Windus, M. Dupuis and J. A. Montgomery, General Atomic and Molecular Electronic Structure System, *J. Comput. Chem.* **14**, pp. 1347–1363 (1993).

[530] M. S. Gordon and M. W. Schmidt, Advances in electronic structure theory: GAMESS a decade later, in C. E. Dykstra, G. Frenking, K. S. Kim and G. E. Scuseria (eds.), *Theory and applications of computational chemistry: the first forty years.* Elsevier, Amsterdam, pp. 1167–1189 (2005).

[531] http://ambermd.org (2021).

[532] D. A. Case, T. E. Cheatham, T. Darden, H. Gohlke, R. Luo, K. M. Merz, A. Onufriev, C. Simmerling, B. Wang and R. J. Woods, The Amber biomolecular simulation programs, *J. Comput. Chem.* **26**, pp. 1668–1688 (2005).

[533] R. Soloman-Ferrer, D. A. Case and R. C. Walker, An overview of the Amber

biomolecular simulation package, *WIREs Comput. Mol. Sci.* **3**, pp. 198–210 (2013).
[534] E. E. Salpeter, Mass Corrections to the Fine Structure of Hydrogen-Like Atoms, *Phys. Rev.* **87**, pp. 328–343 (1952).
[535] T. Fulton and P. C. Martin, Two-Body system in Quantum Electrodynamics. Energy Levels of Positronium, *Phys. Rev.* **95**, pp. 811–822 (1954).
[536] G. W. Erickson, Energy Levels of One-Electron Atoms, *J. Phys. Chem. Ref. Data* **6**, pp. 831–869 (1977).
[537] H. Araki, Quantum–Electrodynamical Corrections to Energy–Levels of Helium, *Prog. Theor. Phys. (Kyoto)* **17**, pp. 619–642 (1957).
[538] J. Sucher, Energy Levels of the Two-Electron Atom to Order α^3 ry; Ionization Energy of Helium, *Phys. Rev.* **109**, pp. 1010–1011 (1958).
[539] W. H. Furry, On Bound States and Scattering in Positron Theory, *Phys. Rev.* **81**, pp. 115–124 (1951).
[540] U. D. Jentschura, P. J. Mohr and G. Soff, Electron self energy for the K and L shells at low nuclear charge, *Phys. Rev. A* **63**, p. 042512 (2001).
[541] U. D. Jentschura and P. J. Mohr, Electron Self Energy for Higher Excited S Levels, *Phys. Rev. A* **69**, p. 064103 (2004).
[542] U. D. Jentschura and P. J. Mohr, Self-energy values for P states in hydrogen and low-Z hydrogenlike ions, *Phys. Rev. A* **72**, p. 014103 (2005).
[543] U. D. Jentschura, P. J. Mohr, G. Soff and E. J. Weniger, Convergence acceleration via combined nonlinear-condensation transformations, *Comput. Phys. Commun.* **116**, pp. 28–54 (1999).
[544] M. I. Eides, H. Grotch and D. A. Owen, Coulomb line vacuum polarization corrections to Lamb shift of order $\alpha^2 (Z\alpha)^5 m$, *Phys. Lett. B* **294**, pp. 115–119 (1992).
[545] M. I. Eides, H. Grotch and P. Pebler, $\alpha^2(Z\alpha)^5 m$ contribution to the hydrogen Lamb shift from virtual light-by-light scattering, *Phys. Rev. A* **50**, pp. 144–170 (1994).
[546] M. I. Eides and V. A. Shelyuto, Corrections of order $\alpha^2(Z\alpha)^5$ to the hyperfine splitting and the Lamb shift, *Phys. Rev. A* **52**, pp. 954–961 (1995).
[547] M. I. Eides, H. Grotch and V. A. Shelyuto, Contribution of order $\alpha^2 (Z\alpha)^5$ to the Lamb shift, *Phys. Rev. A* **55**, pp. 2447–2449 (1997).
[548] K. Pachucki, Contributions to the binding, two-loop correction to the Lamb shift, *Phys. Rev. A* **48**, pp. 2609–2614 (1993).
[549] K. Pachucki, Complete Two-Loop Binding Correction to the Lamb Shift, *Phys. Rev. Lett.* **72**, pp. 3154–3157 (1994).
[550] M. I. Eides, H. Grotch and V. A. Shelyuto, Theory of light hydrogenlike atoms, *Phys. Rep.* **342**, pp. 63–261 (2001).
[551] S. G. Karshenboim, Two-loop logarithmic corrections in the hydrogen Lamb shift, *J. Phys. B* **29**, pp. L29–L31 (1996).
[552] K. Pachucki, Logarithmic two-loop corrections to the Lamb shift in hydrogen, *Phys. Rev. A* **63**, p. 042503 (2001).
[553] U. D. Jentschura and I. Nándori, Double-Logarithmic Two-Loop Self-Energy Corrections to the Lamb Shift, *Phys. Rev. A* **66**, p. 022114 (2002).
[554] S. G. Karshenboim, A. Ozawa, V. A. Shelyuto, R. Szafron and V. G. Ivanov, The Lamb shift of the 1s state in hydrogen: Two-loop and three-loop contributions, *Phys. Lett. B* **795**, pp. 432–437 (2019).
[555] K. Pachucki and U. D. Jentschura, Two-Loop Bethe Logarithm Correction in Hydrogenlike Atoms, *Phys. Rev. Lett.* **91**, p. 113005 (2003).

[556] U. D. Jentschura, Two-Loop Bethe logarithms for higher excited S levels, *Phys. Rev. A* **70**, p. 052108 (2004).

[557] U. D. Jentschura, Two-Loop Bethe Logarithms for non-S Levels, *Phys. Rev. A* **74**, p. 062517 (2006).

[558] S. G. Karshenboim and V. G. Ivanov, Higher-order logarithmic contributions to the Lamb shift in hydrogen, deuterium, and He^+, *Phys. Rev. A* **98**, p. 022522 (2018).

[559] S. G. Karshenboim, A. Ozawa and V. G. Ivanov, Higher-order logarithmic corrections and the two-loop self-energy of a $1s$ electron in hydrogen, *Phys. Rev. A* **100**, p. 032515 (2019).

[560] V. A. Yerokhin, Two-loop self-energy in the Lamb shift of the ground and excited states of hydrogenlike ions, *Phys. Rev. A* **97**, p. 052209 (2018).

[561] V. A. Yerokhin, V. Patkóš and K. Pachucki, Theory of the Lamb Shift in Hydrogen and Light Hydrogen-Like Ions, *Ann. Phys. (Berlin)* **531**, p. 1800324 (2019).

[562] R. Barbieri, J. A. Mignaco and E. Remiddi, Electron Form Factors up to Fourth Order. -I, *Nuovo Cimento* **11**, pp. 824–864 (1972a).

[563] R. Barbieri, J. A. Mignaco and E. Remiddi, Electron Form Factors up to Fourth Order. -II, *Nuovo Cim. A* **11**, pp. 865–916 (1972b).

[564] K. Melnikov and T. van Ritbergen, Three-Loop Slope of the Dirac Form Factor and the $1S$ Lamb Shift in Hydrogen, *Phys. Rev. Lett.* **84**, pp. 1673–1676 (2000).

[565] S. Laporta, High-precision calculation of the 4-loop QED contribution to the slope of the Dirac form factor, *Phys. Lett. B* **800**, p. 135137 (2020).

[566] E. E. Salpeter and H. A. Bethe, A Relativistic Equation for Bound-State Problems, *Phys. Rev.* **84**, pp. 1232–1242 (1951).

[567] M. Gell-Mann and F. Low, Bound States in Quantum Field Theory, *Phys. Rev.* **84**, pp. 350–354 (1951).

[568] D. Lurié, A. J. Macfarlane and Y. Takahashi, Normalization of Bethe–Salpeter Wave Functions, *Phys. Rev.* **140**, pp. B1091–B1099 (1965).

[569] N. Nakanishi, A General Survey of the Theory of the Bethe–Salpeter Equation, *Prog. Theor. Phys. Suppl.* **43**, pp. 1–81 (1969).

[570] S. J. Brodsky, Quantum electrodynamics and the theory of the hydrogenic atom, in M. Chrétien and E. Lipworth (eds.), *Atomic Physics and Astrophysics, Lectures given at the Summer Institute in Theoretical Physics, 1969, at Brandeis University, Vol. 1.* Gordon and Breach, New York, NY, pp. 92–169 (1971).

[571] M.-T. Noda, N. Nakanishi and N. Setô, Bibliography of the Bethe–Salpeter equation, *Prog. Theor. Phys. Suppl.* **95**, pp. 78–111 (1988).

[572] R. Barbieri and E. Remiddi, Solving the Bethe–Salpeter equation for positronium, *Nucl. Phys. B* **141**, pp. 413–422 (1978).

[573] G. S. Adkins, Application of the bound state formalism to positronium, in W. R. Johnson, P. J. Mohr and J. Sucher (eds.), *AIP Conf. Proc. 189.* A. I. P., Woodbury, NY, pp. 65–92 (1989).

[574] G. Breit, The effect of retardation on the interaction of two electrons, *Phys. Rev.* **34**, pp. 553–573 (1929).

[575] J. L. Gammel and M. T. Menzel, Bethe–Salpeter Equation: Numerical Experience with a Hydrogenlike Atom, *Phys. Rev. A* **7**, pp. 858–866 (1973).

[576] S. Love, A Study of Gauge Properties of the Bethe–Salpeter Equation for Two-Fermion Electromagnetic Bound State Systems, *Ann. Phys. (N.Y.)* **113**, pp. 153–176 (1978).

[577] R. Karplus and A. Klein, Electrodynamic Displacement of Atomic Energy Levels. III. The Hyperfine Structure of Positronium, *Phys. Rev.* **87**, pp. 848–858 (1952).
[578] A. A. Logunov and A. N. Tavkhelidze, Quasi-Optical Approach in Quantum Field Theory, *Nuovo Cimento* **29**, pp. 380–399 (1963).
[579] R. M. Woloshyn and A. D. Jackson, Comparison of three-dimensional relativistic scattering equations, *Nucl. Phys. B* **64**, pp. 269–288 (1973).
[580] E. D. Cooper and B. K. Jennings, Obtaining the one-body limit from the relativistic two-body equation, *Nucl. Phys. B* **500**, pp. 553–562 (1989).
[581] G. T. Bodwin and D. R. Yennie, Hyperfine splitting in positronium and muonium, *Phys. Rep.* **43**, pp. 267–303 (1978).
[582] W. E. Caswell and G. P. Lepage, Reduction of the Bethe–Salpeter equation to an equivalent Schrödinger equation, with applications, *Phys. Rev. A* **18**, pp. 810–819 (1978).
[583] T. Kato, On the convergence of the perturbative method. I. *Prog. Theor. Phys. (Kyoto)* **4**, pp. 514–523 (1949).
[584] A. Messiah, *Quantum Mechanics II.* North-Holland, Amsterdam (1962).
[585] G. P. Lepage, Analytic bound-state solutions in a relativistic two-body formalism with applications in muonium and positronium, *Phys. Rev. A* **16**, pp. 863–876 (1977).
[586] M. Beneke and V. A. Smirnov, Asymptotic expansion of Feynman integrals near threshold, *Nucl. Phys. B* **522**, pp. 321–344 (1998).
[587] V. A. Smirnov, Problems of the strategy of regions, *Phys. Lett. B* **465**, pp. 226–234 (1999).
[588] A. H. Hoang, Heavy quarkonium dynamics, in M. Shifman (ed.), *At the Frontier of Particle Physics/Handbook of QCD, Vol. 4.* World Scientific, Singapore, pp. 2216–2331 (2002).
[589] V. A. Smirnov, *Evaluating Feynman Integrals (Springer Tracts in Modern Physics 211).* Springer, Berlin, Heidelberg, New York (2005).
[590] V. A. Smirnov, *Analytic Tools for Feynman Integrals (Springer Tracts in Modern Physics 250).* Springer, Heidelberg (2012).
[591] G. S. Adkins, One-loop renormalization of Coulomb-gauge QED, *Phys. Rev. D* **27**, pp. 1814–1820 (1983).
[592] A. Pineda and J. Soto, Matching at one loop for the four-quark operators in NRQCD, *Phys. Rev. D* **58**, p. 114011 (1998).
[593] S. Titard and F. J. Ynduráin, Rigorous QCD evaluation of spectrum and ground state properties of heavy $q\bar{q}$ systems, with a precision determination of m_b, $M(\eta_b)$, *Phys. Rev. D* **49**, pp. 6007–6025 (1994).
[594] G. S. Adkins and R. N. Fell, Bound-state formalism for positronium, *Phys. Rev. A* **60**, pp. 4461–4475 (1999).
[595] V. M. Shabaev, Two-time Green's function method in quantum electrodynamics of high-Z few-electron atoms, *Phys. Rep.* **356**, pp. 119–228 (2002).
[596] A. A. Petrov and A. E. Blechman, *Effective Field Theories.* World Scientific, Singapore (2016).
[597] W. E. Caswell and G. P. Lepage, Effective Lagrangians for Bound State Problems in QED, QCD, and Other Field Theories, *Phys. Lett. B* **167**, pp. 437–442 (1986).
[598] G. P. Lepage, Atomic Physics in QED and QCD, in D. J. Wineland, C. E. Wieman

and S. J. Smith (eds.), *AIP Conf. Proc. 323*. A. I. P., Woodbury, NY, pp. 18–29 (1994).

[599] T. Kinoshita and M. Nio, Radiative corrections to the muonium hyperfine structure: The $\alpha^2 (Z\alpha)$ correction, *Phys. Rev. D* **53**, pp. 4909–4929 (1996).

[600] A. V. Manohar, Heavy quark effective theory and nonrelativistic QCD Lagrangian to order α_s/m^3, *Phys. Rev. D* **56**, pp. 230–237 (1997).

[601] A. H. Hoang, P. Labelle and S. M. Zebarjad, Single Photon Annihilation Contributions to the Positronium Hyperfine Splitting to Order $m_e \alpha^6$, *Phys. Rev. Lett.* **79**, pp. 3387–3390 (1997).

[602] P. Labelle, Effective field theories for QED bound states: Extending nonrelativistic QED to study retardation effects, *Phys. Rev. D* **58**, p. 093013 (1998).

[603] A. Pineda and J. Soto, Effective Field Theory for Ultrasoft Momenta in NRQCD and NRQED, *Nucl. Phys. B (Proc. Suppl.)* **64**, pp. 428–432 (1998).

[604] N. Brambilla, A. Pineda, J. Soto and A. Vairo, Effective-field theories for heavy quarkonium, *Rev. Mod. Phys.* **77**, pp. 1423–1496 (2005).

[605] G. Paz, An introduction to NRQED, *Mod. Phys. Lett. A* **30**, p. 1550128 (2015).

[606] A. Zee, *Group Theory in a Nutshell for Physicists*. Princeton University Press, Princeton, New Jersey (2016).

[607] B. R. Holstein, Effective Interactions are Effective Interactions, e-print hep-ph/0010033, didactic lectures on effective field theory given at Hadron2000, Caraguatatuba, Brazil (2000).

[608] R. P. Feynman and A. R. Hibbs, *Quantum Mechanics and Path Integrals*. McGraw-Hill, New York (1965).

[609] W. Dittrich and M. Reuter, *Effective Lagrangians in Quantum Electrodynamics – Lecture Notes in Physics Vol. 220*. Springer, Berlin, Heidelberg, New York (1985).

[610] F. Sauter, Über das Verhalten eines Elektrons im homogenen elektrischen Feld nach der relativistischen Theorie Diracs, *Z. Phys.* **69**, pp. 742–764 (1931).

[611] F. Sauter, Zum "Kleinschen Paradoxon", *Z. Phys.* **73**, pp. 547–552 (1932).

[612] V. I. Ogievetsky, A possible interpretation of the perturbation theory series in QFT, *Proc. Acad. Sci. USSR* **109**, pp. 919–922 (1956), [in Russian].

[613] E. Caliceti, Distributional Borel summability for vacuum polarization by an external field, *J. Math. Phys.* **44**, pp. 2026–2036 (2003).

[614] U. D. Jentschura, J. Becher, E. J. Weniger and G. Soff, Resummation of QED Perturbation Series by Sequence Transformations and the Prediction of Perturbative Coefficients, *Phys. Rev. Lett.* **85**, pp. 2446–2449 (2000).

[615] S. R. Valluri, D. Lamm and W. J. Mielniczuk, Comments on the evaluation of the effective potential in quantum electrodynamics, *Phys. Rev. D* **25**, pp. 2729–2735 (1982).

[616] S. R. Valluri, D. R. Lamm and W. J. Mielniczuk, Applications of the representation of the Heisenberg–Euler Lagrangian by means of special functions, *Can. J. Phys.* **71**, pp. 389–397 (1993).

[617] S. R. Valluri, U. D. Jentschura and D. R. Lamm, The study of the Heisenberg–Euler Lagrangian and some of its applications, in *AIP Conf. Proc. 687*. A. I. P., Woodbury, NY, pp. 203–214 (2003).

[618] U. D. Jentschura, H. Gies, S. R. Valluri, D. R. Lamm and E. J. Weniger, QED effective action revisited, *Can. J. Phys.* **80**, pp. 267–284 (2002).

[619] A. Cadène, P. Berceau, M. Fouché, R. Battesti and C. Rizzo, Vacuum magnetic linear birefringence using pulsed fields: status of the BMV experiment, *Eur. Phys. J. D* **68**, p. 16 (2014).

[620] F. Della Valle, A. Ejlli, U. Gastaldi, G. Messineo, E. Milotti, R. Pengo, G. Ruoso and G. Zavattini, The PVLAS experiment: measuring vacuum magnetic birefringence and dichroism with a birefringent Fabry–Perot cavity, *Eur. Phys. J. C* **76**, p. 24 (2016).

[621] A. B. Arbuzov, V. V. Bytev, E. A. Kuraev, E. Tomasi-Gustafsson and Y. M. Bystritskiy, Some QED Processes: Light-by-Light and Moeller Scattering, *Phys. Part. Nucl.* **42**, pp. 79–92 (2011).

[622] I. F. Ginzburg, U. D. Jentschura and V. G. Serbo, Dominance of virtual Delbrück scattering for the photon emission by nuclei in relativistic electron-nucleus and nucleus-nucleus collisions, *Phys. Lett. B* **658**, pp. 125–129 (2008a).

[623] I. F. Ginzburg, U. D. Jentschura and V. G. Serbo, Large contribution of virtual Delbrück scattering to the emission of photons by relativistic nuclei in nucleus-nucleus and electron-nucleus collisions, *Eur. Phys. J. C* **54**, p. 267 (2008b).

[624] M. Gell-Mann and F. E. Low, Quantum Electrodynamics at Small Distances, *Phys. Rev.* **95**, pp. 1300–1312 (1954).

[625] J. Polchinski, Renormalization and effective Lagrangians, *Nucl. Phys. B* **231**, pp. 269–295 (1984).

[626] F. J. Wegner and A. Houghton, Renormalization Group Equation for Critical Phenomena, *Phys. Rev. A* **8**, pp. 401–412 (1973).

[627] M. Baker and K. Johnson, Quantum Electrodynamics at Small Distances, *Phys. Rev.* **183**, pp. 1292–1299 (1969).

[628] S. G. Gorishny, A. L. Kataev, S. A. Larin and L. R. Surguladze, The analytic four-loop corrections to the QED β-function in the MS scheme and to the QED ψ-function. Total reevaluation, *Phys. Lett. B* **256**, pp. 81–86 (1991).

[629] E. de Rafael and J. L. Rosner, Short-Distance Behavior of Quantum Electrodynamics and the Callan–Symanzik Equation for the Photon Propagator, *Ann. Phys. (N.Y.)* **82**, pp. 369–406 (1974).

[630] S. G. Gorishny, A. L. Kataev and S. A. Larin, The three-loop QED photon vacuum polarization function the MS-scheme and the four-loop QED β-function in the on-shell scheme, *Phys. Lett. B* **273**, pp. 141–144 (1991), [Erratum Phys. Lett. B **275** (1992) 512].

[631] D. J. Broadhurst, A. L. Kataev and O. V. Tarasov, Analytical on-shell QED results: 3-loop vacuum polarization, 4-loop β-function and the muon anomaly, *Phys. Lett. B* **298**, pp. 445–452 (1993).

[632] P. A. Baikov, K. G. Chetyrkin and J. H. Kühn, *Massless propagators: applications in QCD and QED*, Proceedings of 8th International Symposium on Radiative Corrections (RADCOR 2007): Applications of Quantum Field Theory to Phenomenology, 1–6 October 2007, Florence, Italy; Proceedings of Science (RAD COR 2007) 023 (2007).

[633] P. A. Baikov, K. G. Chetyrkin, A. V. Smirnov, V. A. Smirnov and M. Steinhauser, Quark and Gluon Form Factors to Three Loops, *Phys. Rev. Lett.* **102**, p. 212002 (2009).

[634] P. A. Baikov, K. G. Chetyrkin and J. H. Kühn, Adler Function, Bjorken Sum Rule,

and the Crewther Relation to Order α_s^4 in a General Gauge Theory, *Phys. Rev. Lett.* **104**, p. 132004 (2010).
[635] V. I. Ritus, Lagrangian of an intensive electromagnetic field and quantum electrodynamics at small distances, *JETP* **42**, pp. 774–782 (1975), [Zh. Eksp. Teor. Fiz. **69**, 1517–1535 (1975)].
[636] V. I. Ritus, Connection between strong-field quantum electrodynamics with short-distance quantum electrodynamics, *JETP* **46**, pp. 423–430 (1977), [Zh. Eksp. Teor. Fiz. **73**, 807–821 (1977)].
[637] G. V. Dunne, H. Gies and C. Schubert, Zero modes, beta functions and IR/UV interplay in higher-loop QED, *J. High Energy Phys.* **11**, p. 032 (2002).
[638] S. G. Karshenboim, Tenth-order contributions to the muon anomalous magnetic moment, *Phys. At. Nucl.* **56**, pp. 857–858 (1993), [Yad. Fiz. **56**, 252–254 (1993)].
[639] R. Karplus and N. M. Kroll, Fourth-Order Correction in Quantum Electrodynamics and the Magnetic Moment of the Electron, *Phys. Rev.* **77**, p. 536 (1950).
[640] C. M. Sommerfield, The Magnetic Moment of the Electron, *Ann. Phys. (N.Y.)* **5**, pp. 26–57 (1958).
[641] G. S. Adkins, Calculation of the electron magnetic moment in Fried–Yennie-gauge QED, *Phys. Rev. D* **39**, pp. 3798–3801 (1989).
[642] A. L. Kataev and V. V. Starshenko, Renormalization-group-inspired approaches and estimates of the tenth-order corrections to the muon anomaly in QED, *Phys. Rev. D* **52**, pp. 402–409 (1995).
[643] T. Kinoshita, B. Nizic and Y. Okamoto, Eighth-order QED contribution to the anomalous magnetic moment of the muon, *Phys. Rev. D* **41**, pp. 593–610 (1990).
[644] T. Aoyama, M. Hayakawa, T. Kinoshita and M. Nio, Complete Tenth-Order QED Contribution to the Muon $g-2$, *Phys. Rev. Lett.* **109**, p. 111808 (2012).
[645] E. J. Weniger, Nonlinear sequence transformations for the acceleration of convergence and the summation of divergent series, *Comput. Phys. Rep.* **10**, pp. 189–371 (1989).
[646] R. Seznec and J. Zinn-Justin, Summation of divergent series by order dependent mappings: Application to the anharmonic oscillator and critical exponents in field theory, *J. Math. Phys.* **20**, p. 1398 (1979).
[647] I. M. Suslov, Gell-Mann–Low Function in QED, *Pis'ma v ZhETF* **74**, pp. 211–215 (2001), [JETP Lett. **74**, 191–195 (2001)].
[648] I. M. Suslov, Exact Asymptotic Form for the β Function in Quantum Electrodynamics, *Zh. Éksp. Teor. Fiz.* **135**, pp. 1129–1133 (2009), [JETP **108**, 980–984 (2009)].
[649] M. A. Bouchiat and C. C. Bouchiat, Weak Neutral Currents in Atomic Physics, *Phys. Lett. B* **48**, pp. 111–114 (1974).
[650] J. Guéna, M. Lintz and M.-A. Bouchiat, Proposal for high-precision atomic-parity-violation measurements by amplification of the asymmetry by stimulated emission in a transverse electric and magnetic field pump–probe experiment, *J. Opt. Soc. Am. B* **22**, pp. 21–28 (2005).
[651] G. W. Botz, D. Bruß and O. Nachtmann, Polarization Rotation Effects due to Parity Violation in Atoms, *Ann. Phys. (N.Y.)* **240**, pp. 107–202 (1994).

Index

www.ingramcontent.com/pod-product-compliance
Ingram Content Group UK Ltd.
Pitfield, Milton Keynes, MK11 3LW, UK
UKHW051044150926
8624IPUK00006B/388

9 789811 252259